Human Anatomy and Physiology

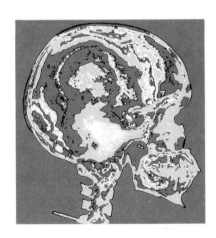

Human Anatomy

and Physiology

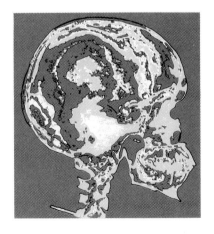

John W. Hole, Jr.
Rio Hondo College

Second Edition

ωcb
Wm. C. Brown Company Publishers
Dubuque, Iowa

Human Anatomy and Physiology

Cover photo

The photo on the cover of this text has been reproduced from an ordinary X ray through a photographic technique known as density slicing by contrast (or equidensity). A series of high, medium, and low contrast negatives were made from the X ray to isolate the various gradations of grays. These negatives were then arbitrarily assigned a color, reregistered, and a color photograph made. The curious protrusion at the bottom of the photo is actually a necklace worn by the model.

Copyright © 1978, 1981 by Wm. C. Brown Company Publishers

Library of Congress Catalog Card Number: 80–65515
ISBN 0–697–04597–8

Printed in the United States of America

2-04597-03

Eighth Printing, 1983

Book Team

Editor
John Stout

Designers
David A. Corona / Marla A. Schafer

Production Editors
Mary Ellen Landwehr / Mary M. Monner

Visual Research Editor
Mary M. Heller

Permissions Editor
Mavis M. Oeth

Senior Medical Illustrator
Diane L. Nelson, M.H.P.E.

Associate Medical Illustrators
Dany Richards Fields
Joanne Goldin
Kathy Jung
June Hill Pedigo

For Velma and Wesley Hole,
my mom and dad

Contents

Expanded Contents

Unit **2**

Support and Movement   149

Unit **3**

Integration and Coordination 265

Contents

Contents

Unit 5 The Human Life Cycle 697

Plates

The number following the plate description refers to the text page preceding the location of the plate.

Charts

Reviewers

The following persons provided valuable suggestions and comments:

Edward M. Barnett
Kellogg Community College

William Bednar
C. S. Mott Community College

Colin E. Campbell
Pima Community College

Jessie Dolson
Delta College

Hester Fassel
Iowa State University

Yola Forbes
Iowa State University

John L. Frehn
Illinois State University

Cecilia Valle Gonzales
St. Philip's College

Terry E. Greathouse
Cuyahoga Community College

Joe Harber
San Antonio College

John P. Harley
Eastern Kentucky University

Eugene S. Horowitz
Queensborough Community College

Anne Lesak
Moraine Valley Community College

Robert E. Nabors
Tarrant County Junior College

Richard Northrup
Delta College

Joseph R. Powell
Florida Junior College

Margaret Howarth Przyogda
Middlesex County College

Ed Reschke
Muskegon Community College

Ethel Sloane
University of Wisconsin, Milwaukee

Jane McNamara Bieber
Virginia Commonwealth University

Anne Denner
Indiana State University, Evansville

Joyce M. Dungan
University of Evansville

Jerry O. Erkert
Santa Fe Community College

James Ezell
J. Sargeant Reynolds Community College

William Garretson
Valencia Community College

Mary Etta Hight
Marshall University

Edward C. Hurlbut
Mesa College

Kenneth L. Jones
Mt. San Antonio College

Thomas S. Kaufman
Montgomery College

Jack Kildebeck
Bakersfield College

Donald S. Kisiel
Suffolk County Community College

Mary Linda Lundgren
Community College of Denver

Margaret May
Virginia Commonwealth University

Mary Lou Mulvihill
William Rainey Harper College

Patricia M. O'Mahoney
University of Southern Maine

Harry S. Reasor
Miami-Dade Community College

Jo Ann Robertson
Western Illinois University

Curtis Robinson
Milwaukee Area Technical College

Maggie Sample
Valencia Community College

Elise Schoenfeld
University of Albuquerque

Louis Squitieri
Bronx Community College

G. Arthur Stephens
Arapahoe Community College

Michael J. Timmons
Moraine Valley Community College

Kent M. Van De Graaff
Brigham Young University

Preface

The second edition of **Human Anatomy and Physiology** continues to provide accurate, current information about the structure and function of the human body in an interesting and readable manner. It is especially designed for students pursuing careers in allied health fields, who have minimal backgrounds in physical and biological sciences.

Organization

The text is organized in units, each containing several chapters. These chapters are arranged traditionally beginning with a discussion of the physical basis of life and proceeding through levels of increasing complexity.

Unit 1
is concerned with the structure and function of cells and tissues; it introduces membranes as organs and the integumentary system as an organ system.

Unit 2
deals with the skeletal and muscular systems that support and protect body parts and make movements possible.

Unit 3
concerns the nervous and endocrine systems that integrate and coordinate body functions.

Unit 4
discusses the digestive, respiratory, circulatory, lymphatic, and urinary systems. These systems obtain nutrients and oxygen from the external environment; transport these substances internally; utilize them as energy sources, structural materials, and essential components in metabolic reactions; and excrete the resulting wastes.

Unit 5
describes the male and female reproductive systems, their functions in producing offspring, and the growth and development of this offspring. The final chapter concerns genetics and explains the determination of individual traits and their passage from parents to offspring.

Biological Themes

In addition to being organized according to levels of increasing complexity the narrative emphasizes the following: the complementary nature of structure and function, homeostasis and homeostatic regulating mechanisms, the interaction between humans and their environments, metabolic processes, responses to stress, pathological disorders, and phases in the human life cycle.

Special Features of the Second Edition

Included in this edition is information of particular interest to allied health students. Such topics as the following are discussed: radioactive isotopes, inflammation and wound healing, muscles of the perineum, spinal cord injuries, general stress syndrome, vomiting reflex, desirable weights, partial pressures of blood gases, cardiac arrhythmias, central venous pressure, immune responses, fluid compartments, electrolyte balance, birth control, fetal development, and medical genetics.

Major sections in nearly every chapter have been rewritten to provide more complete, accurate, and current information with a greater emphasis on physiological mechanisms. These sections include homeostasis, structure of cell membranes, osmotic pressure, metabolic pathways, body temperature regulation, structure of bone tissue, muscle contraction mechanism, cell membrane potential, nerve impulse conduction, autonomic nervous system, hormonal actions, liver functions, cardiac cycle, and renal counter-current mechanism.

Topics new to the second edition include exchange reactions, intercellular junctions, facilitated diffusion, beta oxidation, summation of twitches, multiple motor unit summation, neuropeptides, action potential, crossed-extensor reflex, extrapyramidal tracts, biofeedback, visual pathways, high and low density lipoproteins, autoimmune diseases, juxtaglomerular apparatus, and limited transport capacity.

In addition, over forty new boxed asides are included to help students bridge classroom experiences with professional practice. These boxes provide brief discussions of such topics as marrow transplantation, total hip replacement, drug effects on neurotransmitters, drug effects on the autonomic nervous system, reduction of stress effects, pernicious anemia, lactase deficiency, nutrition and aging, allergens and asthma, leukemia, lipoproteins and atherosclerosis, disseminated intervascular clotting, bundle-branch block, angina pectoris, pulmonary edema, tissue matching, metabolic alkalosis, prostatic cancer, infertility, twinning, fetal alcohol syndrome, sickle-cell anemia, and Tay-Sachs disease.

Human Anatomy and Physiology, second edition, provides separate chapters dealing with nutrition and metabolism, the lymphatic system, human growth and development, and human genetics. For students requiring additional information, brief discussions of cellular respiration, hydrogen ion concentrations, and action potential are included in the Appendixes. To help students locate color plates and charts of special interest within the text, the preliminary section contains a complete listing of these. (See pp. xix, xx).

Pedagogical Devices

The text includes an unusually large number of pedagogical devices intended to involve students in the learning process, to stimulate their interests in the subject matter, and to help them relate their classroom knowledge to their future clinical experiences. For an annotated listing of these devices, see Aids to the Reader, which follows this Preface.

Color illustrations have been added to chapters 8 and 17, which explain the skeletal and cardiovascular systems. These will help the student grasp skeletal distinctions and understand cardiovascular structures and functions. Several color plates appear in addition to the figures in chapters 8 and 17. These plates are listed for ready reference in the preliminary matter (see p. xix), and they are also referred to within the chapter narrative so that students can easily locate them at appropriate times.

Readability

Readability is an important asset of this text. The writing style is intentionally informal and easy to read. Technical vocabulary has been minimized, and summary paragraphs and review questions occur frequently within the narrative. Numerous illustrations, summary charts, and flow diagrams are carefully positioned near the discussions they complement.

Other features provided to increase readability and to aid student understanding include these:

Unit introductions
giving overviews of major groups of chapters.

Bold and italic type
identifying words and ideas of particular importance.

Glossary of terms
with phonetic pronunciations, supplying easy access to word definitions.

Aids to the reader
a section preceding chapter 1, which describes ways the reader can employ the various pedagogical devices to best advantage.

Supplementary Materials

Supplementary materials designed to help the instructor plan class work and presentations and to aid students in their learning activities are also available. They include:

Instructor's Resource Manual and Test Item File
by John W. Hole, Jr., which contains chapter overviews, instructional techniques, suggested schedules, discussions of chapter elements, lists of related films, and directories of suppliers of audiovisual and laboratory materials. It also contains approximately fifty items for each chapter of the text, designed to evaluate student understanding.

A Student Study Guide to Accompany Human Anatomy and Physiology

by Nancy Corbett, Thomas Jefferson University, Philadelphia, which contains chapter overviews, chapter objectives, focus questions, mastery tests, study activities, and answer keys corresponding to the chapters of the text.

Transparencies

a set of 50 acetate transparencies designed to complement classroom lectures or to be used for short quizzes.

Anatomy and Physiology Laboratory Textbook

by Harold Benson and Stanley Gunstream, Pasadena City College (Pasadena, Cal.), in complete and short editions.

Aids to the Reader

This textbook includes a variety of aids to the reader that should make your study of human anatomy and physiology more effective and enjoyable. These aids are included to help you master the basic concepts of human anatomy and physiology that are needed before progressing to more difficult material.

Unit Introductions

Each unit opens with a brief description of the general content of the unit, and a list of chapters included within the unit. (See p. 3 for an example.) This introduction provides an overview of chapters that make up a unit and tells how the unit relates to the other aspects of human anatomy and physiology.

Chapter Introductions

Each chapter introduction previews the chapter's contents and relates that chapter to the others within the unit. (See p. 5.)

After reading an introduction, browse through the chapter, paying particular attention to topic headings and illustrations, so that you get a feeling for the kinds of ideas that are included in the chapter.

Chapter Outlines

The chapter outline includes all the major topic headings and subheadings within the body of the chapter. (See p. 6.) It provides an overview of the chapter's contents and helps you locate sections dealing with particular topics.

Chapter Objectives

Before you begin to study a chapter, carefully read the chapter objectives. (See p. 6.) These indicate what you should be able to do after mastering the information within the narrative. The review activities at the end of each chapter (see p. 16) are phrased like detailed objectives, and it is helpful to read them also before beginning your study. Both sets of objectives are guides that indicate important sections of the narrative.

Key Terms

The list of terms and their phonetic pronunciations, given at the beginning of each chapter, helps build your science vocabulary. The words included in these lists are used within the chapter and are likely to be found in subsequent chapters as well. (See p. 7.) There is an explanation of phonetic pronunciation on page G-1 of the Glossary.

Aids to Understanding Words

Aids to understanding words, at the beginning of each chapter, also helps build your vocabulary. This section includes a list of word roots, stems, prefixes, and suffixes that help you discover word meanings. Each root and an example word, which uses that root, are defined. (See p. 7.)

Knowing the roots from these lists will help you discover and remember scientific word meanings.

Review Questions within the Narrative

Review questions occur at the ends of major sections within each chapter. (See p. 8.) When you reach such questions, try to answer them. If you succeed, then you probably understand the previous discussion and are ready to proceed. If you have difficulty, reread that section before proceeding.

Illustrations and Charts

Numerous illustrations and charts occur in each chapter and are placed near their related textual discussion. They are designed to help you visualize structures and processes, to clarify complex ideas, to summarize sections of the narrative, or to present pertinent data.

As was mentioned, it is a good idea to skim through the chapter before you begin reading it, paying particular attention to these figures. Then, as you read for detail, carefully look at each figure and use it to add to your understanding.

Sometimes the figure legends contain questions that will help you apply your knowledge to the object or process the figure illustrates. The ability to apply information to new situations is of prime importance. These questions concerning the illustrations will provide practice in this skill.

Boxed Information

Short paragraphs set off in boxes of colored type also occur throughout each chapter. (See p. 10.) These asides often contain information that will help you apply the ideas presented in the narrative to clinical situations. Some of the boxes contain information about changes that occur in the body's structure and function as a person passes through the various phases of the human life cycle. These will help you understand how certain body conditions change as a person grows older.

Clinical Terms

At the ends of certain chapters are lists of related terms and phonetic pronunciations that are sometimes used in clinical situations. (See p. 35.) Although these lists and the word definitions are often brief, they will be a useful addition to your understanding of medical terminology.

Chapter Summaries

A summary in outline form occurs at the end of each chapter to help you review the major ideas presented in the narrative. (See p. 15.) Scan this section a few days after you have read the chapter. If you find portions that seem unfamiliar, reread the related sections of the narrative.

Application of Knowledge

These questions at the end of each chapter (see p. 37) help you gain experience in applying information to a few clinical situations. Discuss your answers with other students or with an instructor.

Review Activities

The review activities at the end of each chapter (see p. 16) will check your understanding of the major ideas presented in the narrative. After studying the chapter, read the review activities; if you can perform the tasks suggested, you have accomplished the goals of the chapter. If not, reread the sections of the narrative that need clarification.

Suggestions for Additional Reading

The suggestions for additional reading at the end of each chapter (see p. 16) will help you locate library materials that extend your understanding of topics discussed within the chapter. If a particular idea interests you, check the list of readings for items related to it.

Appendixes, Glossary, and Index

The Appendixes, following chapter 23, contain a variety of useful information. They include the following:
    1. Brief explanations of cellular respiration, hydrogen ion concentration, and action potential to supplement the discussions of these topics within the chapter narratives. (See pp. A–1, A–5, A–7.)

2. Lists of various units of measurements and their equivalents together with a description of how to convert one unit into another. (See p. A–9.)

3. Lists of clinical laboratory tests commonly performed on human blood and urine. These lists include test names, normal adult values, and brief descriptions of their clinical significance. (See p. A–11.)

The Glossary defines the more important textual terms and provides their phonetic pronunciations. (See p. G–1.) It also contains an explanation of phonetic pronunciation on page G–1.

The Index is complete and comprehensive.

Acknowledgments

I want to express my thanks to the many users of the first edition of *Human Anatomy and Physiology* who provided helpful suggestions for improving the textbook. I also gratefully acknowledge the contributions of those who reviewed all or portions of the manuscript for the second edition and supplied detailed criticisms and valuable ideas for changes. These reviewers are listed on a separate page in the introductory matter of this text. Special appreciation is extended to Professor Edwin Reschke for the many excellent photomicrographs he made available.

I also want to thank the many members of the Wm. C. Brown Company staff who worked so tirelessly in the various phases of planning and producing the book. I am especially grateful to John Stout, editor; Mary Ellen Landwehr and Mary Monner, production editors; Marla Schafer, designer; and Mary Heller, visual research manager. They handled the innumerable details involved with the production of the book and were available to provide assistance and encouragement whenever it was needed.

It is a special pleasure to acknowledge the contributions of Diane Nelson, medical illustration coordinator, for her excellent drawings, which add so much to the usefulness and attractiveness of the book.

Finally, I want to express my heartfelt appreciation to Mary Schambach, a former student who read the entire manuscript, to Karen Lynn, my daughter who helped with the typing, and to my wife Shirley who was cheerfully patient while I was busy at my desk.

John W Hole Jr

Human Anatomy and Physiology

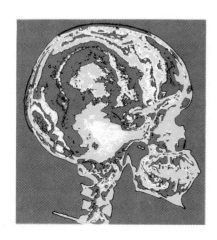

ncipit liber primus huius ope
ris Et est Anathomia huis scia
e Doctrina b

Unit 1
Levels of Organization

The chapters of unit 1 introduce the study of human anatomy and physiology. They are concerned with the ways the human is similar to other living organisms and the ways the structure and function of the human body are dependent on interactions between its parts—chemical substances, cells, tissues, organs, and organ systems.

This unit includes

An Introduction to Human Anatomy and Physiology

1 Human anatomy and physiology is the study of the structures and functions of human body parts.

Since a human is a particular kind of living organism (*Homo sapiens*), it has a variety of traits that are shared by other organisms. For example, a human, like all organisms, carries on life processes and thus demonstrates the characteristics of life. It also has needs that must be met if it is to survive, and its life depends upon *homeostasis*, or the maintenance of a relatively stable internal environment.

A discussion of traits that humans have in common with other organisms can provide a beginning for the study of anatomy and physiology, and this is the purpose of chapter 1.

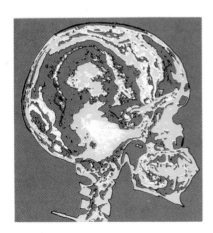

Chapter Outline

Chapter Objectives

After you have studied this chapter, you should be able to

1. Define *anatomy* and *physiology* and explain how they are related.
2. List and describe the major characteristics of life.
3. List and describe the major needs of organisms.
4. Define *homeostasis* and explain its importance to survival.
5. Describe a homeostatic mechanism.
6. Explain what is meant by levels of complexity.
7. Complete the review activities at the end of this chapter.

absorption (ab-sorp'shun)

anatomy (ah-nat'o-me)

assimilation (ah-sim''ĭ-la'shun)

atom (at'om)

cell (sel)

circulation (ser-ku-la'shun)

digestion (di-jes'chun)

excretion (ek-skre'shun)

growth (grōth)

homeostasis (ho''me-ō-sta'sis)

metabolism (mĕ-tab'o-lizm)

molecule (mol'ĕ-kūl)

negative feedback (neg'ah-tiv fēd'bak)

organ (or'gan)

organelle (or''gan-el')

organism (or'gah-nizm)

physiology (fiz''e-ol'o-je)

reproduction (re''pro-duk'shun)

respiration (res''pĭ-ra'shun)

system (sis'tem)

tissue (tish'u)

ana-, apart: *ana*tomy—the study of structure, which often involves cutting or removing body parts.

cret-, to separate: ex*cret*ion—the separation or removal of wastes from the body.

homeo-, the same: *homeo*stasis—the maintenance of a stable internal environment.

-logy, the study of: physio*logy*—the study of body functions.

meta-, a change: *meta*bolism—the chemical changes that occur within the body.

-stasis, standing still: homeo*stasis*—the maintenance of a stable internal environment.

-tomy, cutting: ana*tomy*—the study of structure, which often involves cutting or removing body parts.

The accent marks used in the pronunciation guides are derived from a simplified system of phonetics standard in medical usage. The single accent (') denotes the major stress. Emphasis is placed on the most heavily pronounced syllable in the word. The double accent ('') indicates secondary stress. A syllable marked with a double accent receives less emphasis than the syllable that carries the main stress, but more emphasis than neighboring unstressed syllables.

The study of the human body has a long and interesting history. Its beginnings stretch back to our earliest ancestors, for they must have been curious about their body parts and functions as we are today. At first their interests most likely concerned injuries and illnesses, because healthy bodies demand little attention from their owners. Even primitive people suffered from occasional aches and pains, injured themselves, bled, broke bones, and developed diseases. When they were sick or felt pain, these people probably sought relief by visiting priests or medicine men. The treatment they received, however, must have been less than satisfactory, because primitive doctors relied heavily on superstitions and notions about magic and religion to help their patients. Still, these early medical workers began to discover useful ways of examining and treating the sick. They also began learning how to use certain herbs and potions to affect body functions.

In ancient times it was generally believed that natural processes were controlled by spirits and supernatural forces that humans could not understand. About 2500 years ago, attitudes began to change, and the belief that natural processes were caused by forces that humans could understand grew in popularity.

This new idea stimulated people to look more closely at the world around them. They began to ask questions and seek answers. In this way the stage was set for the development of modern science. As techniques were developed for making accurate observations and performing careful experiments, knowledge of the human body expanded rapidly. Figure 1.1 shows an illustration from a classic Renaissance work on human anatomy.

---

1. *What factors interfered with the systematic study of the human body in ancient times?*
2. *What concept stimulated people to begin studying the natural world?*
3. *How did this concept spark the beginning of modern science?*

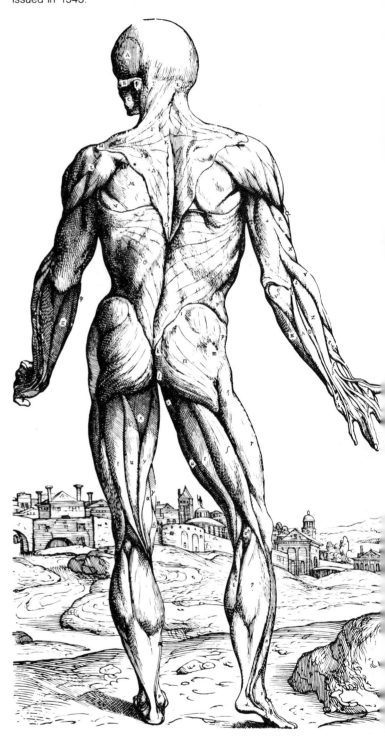

Fig. 1.1 The study of the human body has a long history, as indicated by this illustration from the second book of *De Humani Corporis Fabrica* by Andreas Vesalius, issued in 1543.

Anatomy and Physiology

Anatomy is the branch of science that deals with the structure of body parts—their forms and arrangements. Anatomists observe body parts grossly and microscopically and describe them as accurately and in as much detail as possible. **Physiology**, on the other hand, is concerned with the functions of body parts—what they do and how they do it. Physiologists are interested in finding out how such parts carry on life processes. In addition to using the same observational techniques as the anatomists, physiologists are likely to conduct experiments and make use of complex laboratory equipment.

It is difficult to separate the topics of anatomy and physiology because the structures of body parts are so closely associated with their functions. These parts are arranged to form a well-organized unit—the human organism—and each part plays a role in the operation of the unit as a whole. This role, which is the part's function, depends upon the way the part is constructed—that is, the way its subparts are organized. The arrangement of parts in the human hand with its long, jointed fingers is related to the function of grasping objects. The blood vessels of the heart are designed to transport blood to and from its hollow chambers; the heart's powerful muscular walls are structured to contract and cause the blood to move out of the chambers; the valves associated with the vessels and chambers ensure that the blood will move in the correct direction. The shape of the mouth is related to the function of receiving food; the teeth are designed to break solids into smaller pieces; the muscular tongue and cheeks are constructed to help mix food particles with saliva and prepare it for swallowing. (See fig. 1.2.)

While reading about the human body, keep the connection between form and function of body parts in mind. Although the relationship between the anatomy and the physiology of a particular part is not always obvious, such a relationship is sure to exist.

---

1. Why are physiologists more likely to perform laboratory experiments than anatomists?
2. Why is it difficult to separate the topics of anatomy and physiology?
3. List several additional examples to illustrate the idea that the structure of a body part is closely related to its function.

Fig. 1.2 The structures of body parts are closely related to their functions. (a) The hand is adapted for grasping; (b) the heart for pumping blood; and (c) the mouth for receiving food.

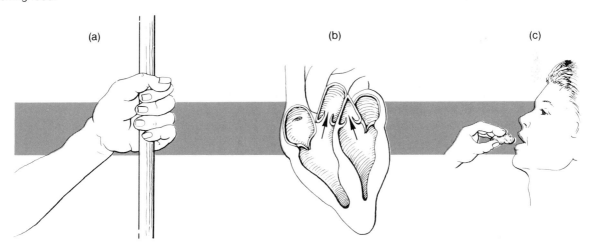

(a)   (b)   (c)

The Characteristics of Life

Before beginning a detailed study of anatomy and physiology, it is helpful to consider some of the traits humans share with other living things, or *organisms*. Such *characteristics of life* include the following:

1. **Movement.** Movement usually refers to a change in an organism's position or to its traveling from one place to another. However, the term also applies to the motion of parts, such as the heart, inside an organism's body.

2. **Responsiveness.** Responsiveness (or irritability) is the ability of an organism to sense changes taking place inside or outside its body and to react to these changes. Pulling away from a hot stove demonstrates responsiveness; drinking water to quench thirst is a response to loss of water in the tissues.

3. **Growth.** Growth is an increase in body size, usually without any important change in shape. It occurs whenever an organism produces new body materials faster than the old ones are worn out or used up.

4. **Reproduction.** Reproduction is the process of making a new individual, as when parents produce an offspring. It also indicates the process by which microscopic cells of the body produce others like themselves, as when parts are repaired or replaced following an injury.

5. **Respiration.** Respiration is the process of obtaining oxygen through the functions of body parts, using oxygen in releasing energy from food substances, and transporting carbon dioxide wastes from body parts to the outside.

6. **Digestion.** Digestion is the process by which various food substances are chemically changed into simpler forms that can be absorbed and used by body parts.

7. **Absorption.** Absorption is the passage of the digestive products through the membranes that line the digestive organs and into the body fluids.

8. **Circulation.** Circulation is the movement of substances from place to place within the body by means of the body fluids.

9. **Assimilation.** Assimilation is changing absorbed substances into forms that are chemically different from those that entered the body fluids.

10. **Excretion.** Excretion is removal of wastes that are produced by body parts as a result of their activities.

Each of these characteristics of life—in fact, everything an organism does—depends upon chemical changes that occur within body parts. Taken together, these chemical reactions are called **metabolism**.

The characteristics shared by all living organisms are summarized in chart 1.1.

Chart 1.1 Characteristics of life

Process	Examples
Movement	Change in position of the body or body part; motion of an internal organ.
Responsiveness	Reaction to a change taking place inside or outside the body.
Growth	Increase in body size.
Reproduction	Production of new organisms.
Respiration	Obtaining oxygen, using oxygen in releasing energy from foods, and transporting carbon dioxide from cells to the outside of the body.
Digestion	Breakdown of food substances into simpler forms.
Absorption	Passage of digestive products through membranes and into body fluids.
Circulation	Movement of substances from place to place in body fluids.
Assimilation	Changing of absorbed substances into chemically different forms.
Excretion	Removal of wastes produced by metabolic reactions.

Vital signs are among the more common observations made by physicians and nurses working with patients. These observations include body temperature measurements, blood pressure readings, and rates and types of pulse and breathing movements. There is a close relationship between these signs and the characteristics of life, since vital signs are the results of *metabolic activities*. In fact, death is recognized by the absence of such signs. More specifically, a person who has died will lack spontaneous muscular movements (including those of breathing muscles), responses to stimuli (even the most painful that can be ethically applied), reflexes (such as the knee-jerk reflex and pupillary reflexes of the eye), and brain waves (demonstrated by a flat encephalogram, which reflects a lack of metabolic activity in the brain).

1. What are the characteristics of life?
2. How are the characteristics of life related to metabolism?

The Maintenance of Life

With the exception of an organism's reproductive structures, which function to ensure that its particular form of life will continue into the future, the structures and functions of all body parts are directed toward achieving one goal—the maintenance of the life of the organism.

Needs of Organisms

Life is fragile, and it depends upon the presence of certain factors for its existence. These factors include the following:

1. **Water.** Water is the most abundant chemical substance within the body, and it is necessary for a variety of metabolic processes. In the case of humans, it is taken in as liquid and as part of foods and is produced as a by-product of certain metabolic reactions. Water is lost largely through the skin and lungs by evaporation and through the kidneys as urine is formed and excreted.

2. **Food.** Food is a general term for substances that provide the body with necessary chemicals in addition to water. Some of these chemicals are used as energy sources, others as raw materials for building new living matter, and still others help regulate vital chemical reactions.

3. **Oxygen.** Oxygen is a gas that makes up about one-fifth of ordinary air. It is used in releasing energy from food substances. The energy, in turn, drives the metabolic processes in all body parts.

4. **Heat.** Heat is a form of energy. The reading obtained from the measurement of heat intensity is called the *temperature*. Heat is a by-product of metabolic reactions, and the rate at which such reactions occur is partly governed by the amount of heat present. Generally, the higher the temperature, the more rapidly the reactions take place.

5. **Pressure.** Pressure is a force that creates a pressing or compressing action. For example, the force acting on the outside of the body due to the weight of air is called *atmospheric pressure*. This pressure plays an important part in breathing.

Although organisms need water, food, oxygen, heat, and pressure, the presence of these factors alone is not enough to ensure their survival. The quantities and the qualities of such factors are important, too. For example, the amount of water entering and leaving an organism must be regulated, as must the concentration of oxygen reaching body parts. Similarly, survival depends on the quality of food available; it must supply the correct chemicals and must be available in adequate amounts.

Homeostasis

As an organism moves from place to place or as the climate changes, vital factors in its external environment change, too. However, if the organism is to survive, the conditions within its body must remain relatively stable. In other words, body parts function efficiently only when their supplies of water, food, and oxygen, and the conditions of heat and pressure remain within certain limits. Biologists use the term **homeostasis** to describe the maintenance of such a *stable internal environment.*

To better understand this idea, imagine a room equipped with a furnace and an air conditioner. Suppose the room temperature is to remain near 20°C (68°F), so the thermostat is adjusted to a set point of 20°C. Since a thermostat is sensitive to temperature changes, once it is adjusted it will signal the furnace to start whenever the room temperature drops below the set point. If the temperature rises above the set point, the thermostat will cause the furnace to stop and the air conditioner to start. As a result, a relatively constant temperature will be maintained in the room. (See fig. 1.3.)

A *homeostatic mechanism* in the human body achieves similar results in regulating body temperature. The thermostat is a temperature-sensitive region of the brain. In healthy persons the set point of this region is at or near 37°C (98.6°F).

If a person is exposed to a cold environment and the body temperature begins to drop, the brain's thermostat can sense this change and trigger heat-generating and heat-conserving activities. For example, small groups of muscles may be stimulated to contract involuntarily, an action called shivering. Such muscular contractions produce heat as a by-product, which helps to warm the body. At the same time, blood vessels in the skin may be signaled to constrict so that less warm blood reaches the skin, and heat that might otherwise be lost is held in deeper tissues.

If a person is becoming overheated, the brain may trigger a series of changes that leads to increased loss of body heat. It may stimulate the sweat glands of the skin to secrete watery perspiration; as the water evaporates from the surface, some heat is carried away and the skin is cooled. Also, blood vessels in the skin may be caused to dilate so blood from deeper tissues reaches the surface in greater volume and loses some of its heat to the outside.

Fig. 1.3 A thermostat that can signal an air conditioner and a furnace to turn on or off helps to maintain a constant room temperature. This system represents a homeostatic mechanism.

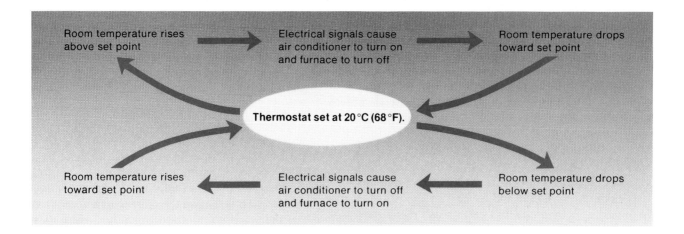

Another homeostatic mechanism functions to regulate blood pressure in vessels leading away from the heart. In this instance, pressure-sensitive parts in the walls of these vessels are stimulated if the blood pressure increases above normal. When this happens, the sensors signal the brain, and it signals the heart, causing the heart chambers to contract more slowly and with less force. Because of decreased heart action, less blood enters the blood vessels, and the pressure inside the vessels decreases. If the blood pressure is dropping below normal, the brain signals the heart to contract more rapidly and with greater force so that the pressure in the vessels increases.

In these examples, homeostasis is maintained by a self-regulating control mechanism that can sense changes away from the normal set point and can cause reactions that tend to return conditions to normal. Since the changes away from normal stimulate changes to occur in the opposite direction, the responses are called *negative*, and the homeostatic control mechanism is said to act by a process of *negative feedback*, which is discussed in chapter 12. (See fig. 1.4.)

Details about these and other homeostatic mechanisms are presented in later chapters. The maintenance of homeostasis is of primary importance to survival, and it is not surprising that metabolic activities are largely directed toward maintaining stable internal conditions.

1. What is the relationship between the use of oxygen and the production of body heat?
2. Why is homeostasis so important to survival?
3. Describe two homeostatic mechanisms.

Fig. 1.4 A negative feedback mechanism helps control blood pressure.

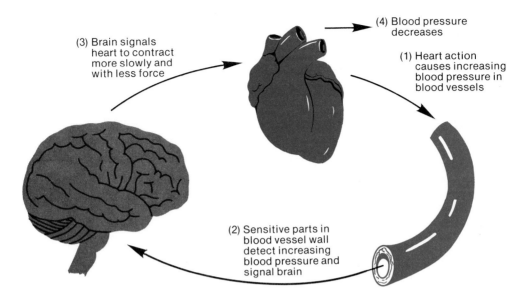

(3) Brain signals heart to contract more slowly and with less force

(4) Blood pressure decreases

(1) Heart action causes increasing blood pressure in blood vessels

(2) Sensitive parts in blood vessel wall detect increasing blood pressure and signal brain

Levels of Complexity

Early investigators focused their attention on the larger body structures, for they were limited in their abilities to observe small parts. Studies of small parts had to wait for the invention of magnifying lenses and microscopes, which came into use about 400 years ago. Once these tools were available, it was soon discovered that larger body structures were made up of smaller parts, which, in turn, were composed of still smaller parts.

Today scientists recognize that all materials, including those that comprise the human body, are composed of chemical substances. These substances are made up of tiny, invisible particles called **atoms**, which are commonly bonded together to form larger particles called **molecules**; small molecules may be combined in complex ways to form giant molecules.

Within the human organism the basic unit of structure and function is a microscopic part called a **cell**. Although individual cells vary in size, shape, and specialized functions, all have certain traits in common. For instance, all cells contain tiny parts called **organelles** that carry on specific activities. These organelles are composed of aggregates of giant molecules, including those of such substances as proteins, carbohydrates, and lipids.

Cells are organized into layers or masses that have common functions. Such a group of cells constitutes a **tissue**. Groups of different tissues form **organs**—complex structures with special functions—while groups of organs are arranged into **organ systems**. Organ systems make up an **organism**.

The human body is, thus, composed of parts that vary in *level of complexity*, as is illustrated in figure 1.5. Atoms are less complex than molecules; molecules are less complex than organelles; organelles are less complex than cells; cells are less complex than tissues; tissues are less complex than organs; organs are less complex than organ systems; and organ systems are less complex than the whole organism.

1. *How does the human body illustrate levels of complexity?*
2. *Define* organism.

Fig. 1.5 A human body is composed of parts within parts, which vary in complexity.

LEVELS OF COMPLEXITY

LEVEL

EXAMPLES

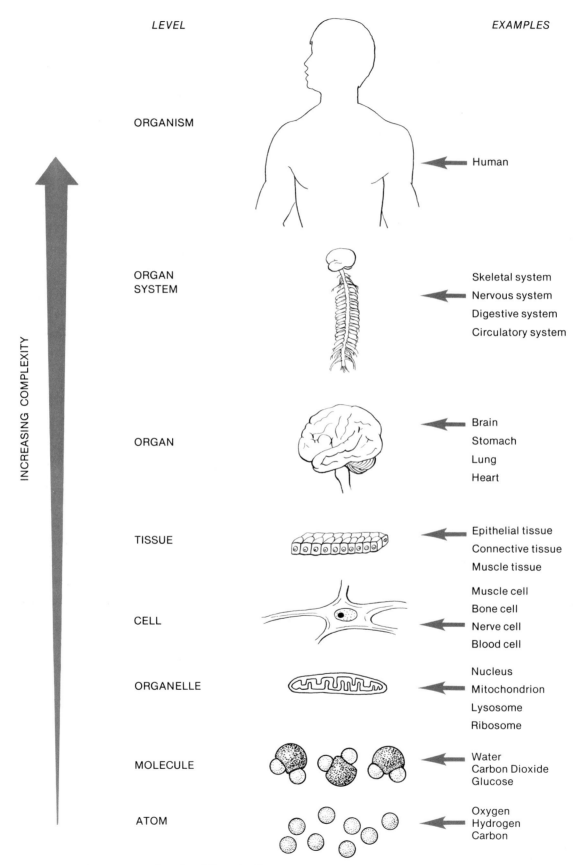

ORGANISM

Human

ORGAN SYSTEM

Skeletal system
Nervous system
Digestive system
Circulatory system

ORGAN

Brain
Stomach
Lung
Heart

TISSUE

Epithelial tissue
Connective tissue
Muscle tissue

CELL

Muscle cell
Bone cell
Nerve cell
Blood cell

ORGANELLE

Nucleus
Mitochondrion
Lysosome
Ribosome

MOLECULE

Water
Carbon Dioxide
Glucose

ATOM

Oxygen
Hydrogen
Carbon

INCREASING COMPLEXITY

Some of the Medical and Applied Sciences

cardiology (kar″de-ol′o-je)—branch of medical science dealing with the heart and heart diseases.

dermatology (der″mah-tol′o-je)—study of skin and its diseases.

endocrinology (en″do-kri-nol′o-je)—study of hormone-secreting glands and the diseases involving them.

epidemiology (ep″ĭ-de″me-ol′o-je)—study of infectious diseases, their distribution and control.

gastroenterology (gas″tro-en″ter-ol′o-je)—study of the stomach and intestines and diseases involving these organs.

geriatrics (jer″e-at′riks)—branch of medicine dealing with aged persons and their medical problems.

gerontology (jer″on-tol′o-je)—study of the process of aging and the problems of elderly persons.

gynecology (gi″nĕ-kol′o-je)—study of the female reproductive system and its diseases.

hematology (hem″ah-tol′o-je)—study of blood and blood diseases.

immunology (im″u-nol′o-je)—study of body resistance to disease.

neurology (nu-rol′o-je)—study of the nervous system in health and disease.

obstetrics (ob-stet′riks)—branch of medicine dealing with pregnancy and childbirth.

oncology (ong-kol′o-je)—study of tumors.

ophthalmology (of″thal-mol′o-je)—study of the eye and eye diseases.

orthopedics (or″tho-pe′diks)—branch of medicine dealing with the muscular and skeletal systems and problems of these systems.

otolaryngology (o″to-lar″in-gol′o-je)—study of the ear, throat, larynx, and diseases of these parts.

pathology (pah-thol′o-je)—study of body changes produced by diseases.

pediatrics (pe″de-at′riks)—branch of medicine dealing with children and their diseases.

pharmacology (fahr″mah-kol′o-je)—study of drugs and their uses in the treatment of diseases.

psychiatry (si-ki′ah-tre)—branch of medicine dealing with the mind and its disorders.

radiology (ra″de-ol′o-je)—study of X rays and radioactive substances and their uses in the diagnosis and treatment of diseases.

toxicology (tok″sĭ-kol′o-je)—study of poisonous substances and their effects upon body parts.

urology (u-rol′o-je)—branch of medicine dealing with the kidneys and urinary system and their diseases.

Chapter Summary

Introduction
Understanding of natural processes grew slowly until the development of modern science.

Anatomy and Physiology
1. Anatomy deals with the structures of body parts.
2. Physiology deals with the functions of these parts.
3. The structure and functions of a part are closely related.

Characteristics of Life
Traits shared by all living things are called characteristics of life.
1. Movement—changes in body position or motion of internal parts.
2. Responsiveness—sensing and reacting to internal or external changes.
3. Growth—increase in size.
4. Reproduction—production of offspring.
5. Respiration—obtaining oxygen, using oxygen to release energy from foods, and transporting carbon dioxide from body parts to the outside.
6. Digestion—changing food substances into forms that can be absorbed and used by cells.
7. Absorption—passage of digestive products into body fluids.
8. Circulation—movement of substances in body fluids.
9. Assimilation—changing of substances into chemically different forms.
10. Excretion—removal of metabolic wastes.

Taken together these activities constitute metabolism.

Maintenance of Life
The structures and functions of body parts are directed toward maintaining the life of the organism.
1. Needs of organisms
   a. Water is needed for a variety of metabolic processes.
   b. Food is needed to supply energy, to provide raw materials for building new living matter, and to supply chemicals necessary in vital reactions.
   c. Oxygen is used in releasing energy from food materials.
   d. Heat is needed to promote chemical reactions.
   e. Pressure is needed for breathing.

2. Homeostasis
   a. If an organism is to survive, the conditions within its body must remain stable.
   b. Internal conditions are regulated by homeostatic mechanisms such as those that control body temperature and blood pressure.
   c. Homeostatic mechanisms act by negative feedback.
   d. Metabolic activities are largely directed toward maintaining homeostasis.

Levels of Complexity

The body is composed of parts that vary in complexity.
1. Material substances are composed of atoms.
2. Atoms bond together to form molecules.
3. Organelles contain aggregates of giant molecules.
4. Cells, which are composed of organelles, are the fundamental units of structure and function within the body.
5. Cells are organized into layers or masses called tissues.
6. Tissues are organized into organs.
7. Organs are arranged into organ systems.
8. Organ systems constitute the organism.

Review Activities

1. Distinguish between anatomy and physiology.
2. Explain the relationship between the form and function of human body parts.
3. List and describe ten characteristics of life.
4. Define *metabolism*.
5. List and describe five needs of the human organism.
6. Distinguish between heat and temperature.
7. Define *atmospheric pressure*.
8. Explain how body temperature and blood pressure are controlled.
9. Explain why homeostatic mechanisms are said to use negative feedback.
10. Explain what is meant by levels of complexity.
11. List the levels of complexity within the human.

Suggestions for Additional Reading

Enger, E. D., et. al. 1979. *Concepts in biology*. Dubuque: Wm. C. Brown.

Langley, L. L. 1965. *Homeostasis*. New York: Reinhold Publishing.

Lockhart, R. D. 1966. *Living anatomy*. 5th ed. London: Faber and Faber.

Majno, G. 1975. *The healing hand: man and wound in the ancient world*. Cambridge, Mass.: Harvard Univ. Press.

Singer, C. 1957. *A short history of anatomy and physiology*. New York: Dover Publications.

Singer, C. A., and Underwood, E. A. 1962. *A short history of medicine*. New York: Oxford Univ. Press.

Szent-Györgyi, A. 1972. *The living state*. New York: Academic Press.

The Chemical Basis of Life

2 As chapter 1 explained, an organism is composed of parts that vary in complexity. These include organ systems, organs, tissues, cells, and cellular organelles.

Organelles are in turn composed of chemical substances made up of atoms and groups of atoms called molecules. Similarly, atoms are composed of still smaller units called protons, neutrons, and electrons.

A study of living material at its lower levels of complexity must be concerned with the chemical substances that form the structural basis of all matter and that interact in the metabolic processes carried on within organisms.

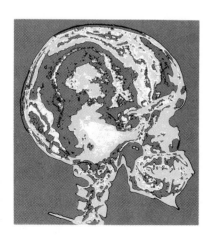

Chapter Outline

Chapter Objectives

After you have studied this chapter, you should
be able to

1. Explain how the study of living material is
   dependent on the study of chemistry.
2. Describe the relationships between matter,
   atoms, and molecules.
3. Discuss how atomic structure is related to
   the ways in which atoms interact.
4. Explain how molecular and structural formulas
   are used to symbolize the composition of
   compounds.
5. Describe what is meant by a chemical
   reaction.
6. Discuss the concept of pH.
7. List the major groups of inorganic substances
   that are common in cells.
8. Describe the general roles played in cells
   by various types of organic compounds.
9. Complete the review activities at the end
   of this chapter.

Key Terms

atom (at′om)
carbohydrate (kar″bo-hi′drāt)
decomposition (de″kom-po-zish′un)
electrolyte (e-lek′tro-līt)
formula (fōr′mu-lah)
inorganic (in″or-gan′ik)
ion (i′on)
isotope (i′so-tōp)
lipid (lip′id)
molecule (mol′ĕ-kūl)
nucleic acid (nu-kle′ik)
organic (or-gan′ik)
protein (pro′te-in)
synthesis (sin′thĕ-sis)

Aids to Understanding Words

bio-, life: *bio*chemistry—branch of science dealing with the chemistry of life forms.

di-, two: *di*saccharide—a compound whose molecules are composed of two saccharide units bonded together.

glyc-, sweet: *glyc*ogen—a complex carbohydrate composed of sugar molecules bonded together.

iso-, equal: *iso*tope—atom that has the same atomic number as another atom but a different atomic weight.

lip-, fat: *lip*ids—group of organic compounds that includes fats.

-lyt, dissolvable: electro*lyte*—substance that dissolves in water and releases ions.

mono-, one: *mono*saccharide—a compound whose molecules consist of a single saccharide unit.

nucle-, kernel: *nucle*us—central core of an atom.

poly-, many: *poly*unsaturated—molecule that has many double bonds between its carbon atoms.

sacchar-, sugar: mono*saccharide*—sugar molecule composed of a single saccharide unit.

syn-, together: *syn*thesis—process by which substances are bonded together to form new substances.

-valent, having power: electro*valent* bond—chemical bond created by the attraction between two ions with opposite electrical charges.

Chemistry is the branch of science dealing with the composition of substances and the changes that take place in their composition. Although it is possible to study anatomy without much reference to chemistry, knowledge of chemistry is essential for understanding physiology, because body functions involve chemical changes that occur within cells.

As interest in the chemistry of living organisms grew, and knowledge in this area expanded, a new subdivision of science called biological chemistry or *biochemistry* emerged.

Biochemical studies have been important not only in helping to explain physiological processes, but also in making possible the development of many new drugs and methods for treating diseases.

---

1. Why is a knowledge of chemistry essential to an understanding of physiology?
2. What is biochemistry?

Structure of Matter

Matter is anything that has weight and takes up space. This includes all the solids, liquids, and gases in our surroundings as well as in our bodies.

Elements and Atoms

Studies of matter have revealed that all things are composed of basic substances called **elements.** At present, 106 such elements are known, although most naturally occurring matter is made up of only about 90 of them. Among these elements are such common materials as iron, copper, silver, gold, aluminum, carbon, hydrogen, and oxygen. Although some elements exist in a pure form, they occur more frequently in mixtures or in chemical combinations of 2 or more elements known as **compounds.**

About 20 elements are needed by living things, and of these oxygen, carbon, hydrogen, and nitrogen make up more than 95% of the human body. A list of the more abundant elements in the body is shown in chart 2.1. Notice that each element is represented by a one- or two-letter symbol.

Elements are composed of tiny, invisible particles called **atoms.** Although the atoms that make up each element are very similar to each other, they differ from the atoms that compose other elements. Atoms vary in size, weight, and the ways they interact with each other. Some, for instance, are capable of combining with atoms like themselves or with other kinds of atoms, while others lack this ability.

Chart 2.1 Most abundant elements in the human body

Major Elements	Symbol	Approximate Percentage of the Human Body
Oxygen	O	65.0
Carbon	C	18.5
Hydrogen	H	9.5
Nitrogen	N	3.2
Calcium	Ca	1.5
Phosphorus	P	1.0
Potassium	K	0.4
Sulfur	S	0.3
Chlorine	Cl	0.2
Sodium	Na	0.2
Magnesium	Mg	0.1

Trace Elements	Symbol	
Cobalt	Co	
Copper	Cu	
Fluorine	F	
Iodine	I	Less than 0.1
Iron	Fe	
Manganese	Mn	
Zinc	Zn	

Fig. 2.1 An atom of lithium includes 3 electrons in motion around a nucleus, which contains 3 protons and 4 neutrons.

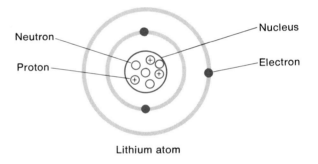

Lithium atom

Atomic Structure

An atom consists of a central core called the **nucleus** and one or more **electrons** that are in constant motion around the nucleus. The nucleus itself contains some relatively large particles called **protons** and **neutrons,** whose weights are about equal. (See fig. 2.1.)

Electrons, which are extremely small and have almost no weight, carry a single, negative electrical charge (e^-), while protons each carry a single, positive electrical charge (p^+). Neutrons are uncharged and thus are electrically neutral (n^0).

Chart 2.2 Atomic structure of elements 1 through 12

Element	Symbol	Atomic Number	Atomic Weight	Protons	Neutrons	Electrons in Shells		
						First	Second	Third
Hydrogen	H	1	1	1	0	1		
Helium	He	2	4	2	2	2	(inert)	
Lithium	Li	3	7	3	4	2	1	
Beryllium	Be	4	9	4	5	2	2	
Boron	B	5	11	5	6	2	3	
Carbon	C	6	12	6	6	2	4	
Nitrogen	N	7	14	7	7	2	5	
Oxygen	O	8	16	8	8	2	6	
Fluorine	F	9	19	9	10	2	7	
Neon	Ne	10	20	10	10	2	8	(inert)
Sodium	Na	11	23	11	12	2	8	1
Magnesium	Mg	12	24	12	12	2	8	2

Chart 2.3 Isotopes of oxygen

Isotope	Protons	Neutrons	Atomic Number	Atomic Weight	Type of Isotope
Oxygen-14	8	6	8	14	Radioactive
Oxygen-15	8	7	8	15	Radioactive
Oxygen-16*	8	8	8	16	Stable
Oxygen-17	8	9	8	17	Stable
Oxygen-18	8	10	8	18	Stable
Oxygen-19	8	11	8	19	Radioactive

*Most abundant isotope.

Since the atomic nucleus contains the protons, this part of an atom is always positively charged. However, as the number of electrons outside the nucleus is equal to the number of protons, a complete atom is uncharged.

The atoms of different elements contain different numbers of protons. The number of protons in the atoms of a particular element is called the *atomic number* of that element. Hydrogen, for example, whose atoms contain one proton, has the atomic number 1; and carbon, whose atoms have six protons, has the atomic number 6.

The weight of an atom of an element is due primarily to the protons and neutrons in its nucleus, since the electrons have so little weight. For this reason, an atom of carbon with six protons and six neutrons weighs about 12 times as much as an atom of hydrogen, which has only one proton and no neutrons.

The number of protons plus the number of neutrons in each atom is approximately equal to the *atomic weight* of an element. Thus, the atomic weight of hydrogen is 1, and the atomic weight of carbon is 12. (See chart 2.2.)

Isotopes

All the atoms of a particular element have the same atomic number, because they have the same number of protons and electrons. However, the atoms of an element may vary in the number of neutrons in their nuclei and thus vary in atomic weight. Chart 2.3 shows that all oxygen atoms have 8 protons in their nuclei. Some, however, have 8 neutrons (atomic weight 16), others have 9 neutrons (atomic weight 17), and still others have 10 neutrons (atomic weight 18). Atoms that have the same atomic numbers, but different atomic weights, are called **isotopes** of an element. Since a sample of an element is likely to include more than one isotope, the atomic weight of the element is the average weight of the isotopes present.

The ways atoms interact with one another are due largely to the number of electrons they possess. Since the number of electrons in an atom is equal to its number of protons, all the isotopes of a particular element have the same number of electrons and react chemically in the same manner. Therefore, any of the isotopes of oxygen might play a role in the metabolic reactions of an organism.

Fig. 2.2 Scintillation counters such as this are used to detect the presence of radioactive isotopes and to measure the amount of radiation they are emitting.

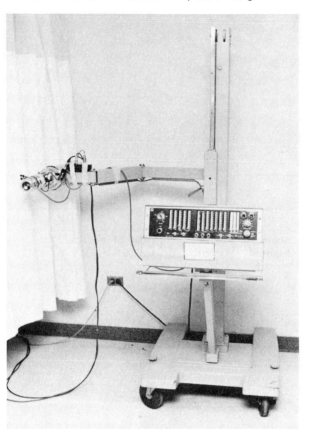

Fig. 2.3 Radiation from radioactive cobalt may be used to treat patients with cancerous conditions.

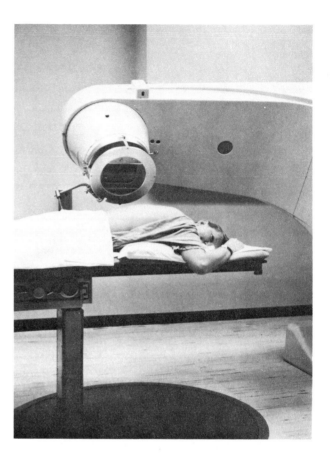

Although some of the isotopes of an element may be stable, others may have unstable atomic nuclei that tend to decompose, releasing energy or pieces of themselves. Such unstable isotopes are called *radioactive*, and the energy or atomic fragments they give off are called *atomic radiations*. Examples of elements that have radioactive isotopes include oxygen, iodine, iron, phosphorus, and cobalt.

Since it is possible to detect the presence of atomic radiation by using special equipment, such as Geiger-Müller counters and scintillation counters, radioactive substances are useful in studying life processes. A radioactive isotope, for example, can be introduced into an organism and then traced with detectors as it enters into metabolic activities. Since the human thyroid gland makes special use of iodine in its metabolic processes, radioactive iodine has been used to study its functions. (See fig. 2.2.)

Atomic radiation can also cause changes in the structures of various chemical substances and in this way can alter vital processes within cells. This is the reason radioactive elements such as cobalt are sometimes used to treat cancers. The radiation coming from cobalt can affect the chemicals within the cancerous cells and cause their deaths. Since cancer cells are more susceptible to such damage than normal cells, the cancer cells are destroyed more rapidly than the normal ones. (See fig. 2.3.)

1. What is the relationship between matter and elements?
2. What elements are most common in the human body?
3. How are electrons, protons, and neutrons positioned within an atom?
4. How are radioactive isotopes used to study life processes?

Fig. 2.4 The single electron of a hydrogen atom is located in its first shell; the 2 electrons of a helium atom fill its first shell; and the 3 electrons of a lithium atom are arranged with 2 in the first shell and 1 in the second.

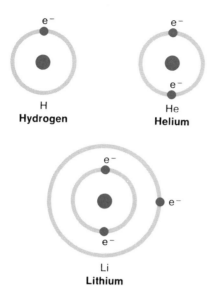

H
Hydrogen

He
Helium

Li
Lithium

Fig. 2.5 A diagram of a sodium atom.

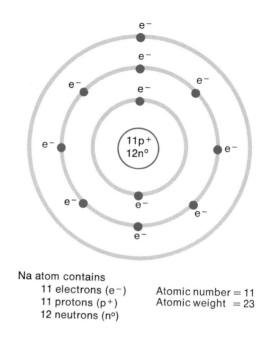

Na atom contains
11 electrons (e^-)
11 protons (p^+)
12 neutrons (n^o)

Atomic number = 11
Atomic weight = 23

Bonding of Atoms

When atoms combine with other atoms, their electrons are responsible for the interaction. When atoms combine they either gain or lose electrons, or share electrons with other atoms.

The electrons of an atom are arranged in one or more *shells* around the nucleus. The maximum number of electrons that each inner shell can hold is as follows:

First shell (closest to the nucleus)	2
Second shell	8
Third shell	8

The third shell can hold 8 electrons for elements up to atomic number 20. More complex atoms may have as many as 18 electrons in the third shell.

For convenience, we can represent the arrangements of electrons within the shells of atoms by using simplified diagrams such as those in figure 2.4. Notice that the single electron of a hydrogen atom is located in the first shell; the 2 electrons of a helium atom fill its first shell; the 3 electrons of a lithium atom are arranged with 2 in the first shell and 1 in the second shell.

Atoms such as helium, whose outermost electron shells are filled, have stable configurations and are chemically inactive (inert). Atoms with incompletely filled outer shells, such as those of hydrogen or lithium, tend to gain, lose, or share electrons in ways that empty or fill their outer shells. In this way they achieve stable configurations.

An atom of sodium, for example, has 11 electrons, arranged as shown in figure 2.5: 2 in the first shell, 8 in the second shell, and 1 in the third shell. This atom tends to lose the single electron in its outer shell, which leaves the second shell filled and the configuration stable.

A chlorine atom has 17 electrons arranged with 2 in the first shell, 8 in the second shell, and 7 in the third shell. An atom of this type will tend to accept a single electron, thus filling its outer shell and achieving a stable form.

Since each sodium atom tends to lose a single electron and each chlorine atom tends to accept a single electron, sodium and chlorine atoms will react together. During this reaction, a sodium atom loses an electron and is left with 11 protons (11+) in its nucleus and only 10 electrons (10−). As a result, the atom develops a net electrical charge of 1+ and is symbolized Na^+. At the same time, a chlorine atom gains an electron, which leaves it with 17 protons (17+) in its nucleus and 18 electrons (18−). Thus it develops a net electrical charge of 1−. Such an atom is symbolized Cl^-.

The Chemical Basis of Life

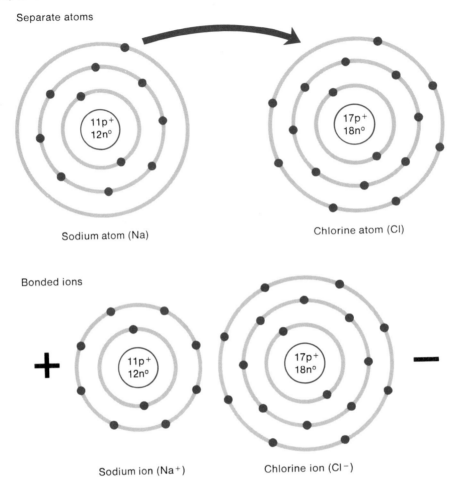

Fig. 2.6 If a sodium atom loses an electron to a chlorine atom, the sodium atom becomes a sodium ion and the chlorine atom becomes a chlorine ion. These oppositely charged particles are attracted to one another and become bonded by an electrovalent bond.

Separate atoms

Sodium atom (Na)

Chlorine atom (Cl)

Bonded ions

Sodium ion (Na^+)

Chlorine ion (Cl^-)

Atoms that have become electrically charged by gaining or losing electrons are called **ions**, and two ions with opposite charges are electrically attracted to one another. When this happens, a chemical bond called an *electrovalent bond* (ionic bond) is created between them, as shown in figure 2.6. When sodium ions (Na^+) and chlorine ions (Cl^-) become bonded in this manner, the compound sodium chloride (NaCl, common salt) is formed.

Similarly, a hydrogen atom may lose its single electron and become a hydrogen ion (H^+). Such an ion can bond with a chlorine ion (Cl^-) to form hydrogen chloride (HCl, hydrochloric acid).

Atoms can also bond together by sharing electrons rather than by exchanging them. A hydrogen atom, for example, has 1 electron in its first shell, but needs 2 to achieve a stable configuration. It may fill this shell by combining with another hydrogen atom in such a way that the 2 atoms share a pair of electrons. As figure 2.7 shows, the 2 electrons then encircle the nuclei of both atoms so that each achieves a stable form. In this case, the atoms are held together by a *covalent bond*.

Carbon atoms, with 2 electrons in their first shells and 4 electrons in their second shells, always form covalent bonds when they unite with other atoms. In fact, carbon atoms may bond to other carbon atoms in such a way that 2 atoms share 1, 2, or

Fig. 2.7 A hydrogen molecule is formed when two hydrogen atoms share a pair of electrons and become bonded by a covalent bond.

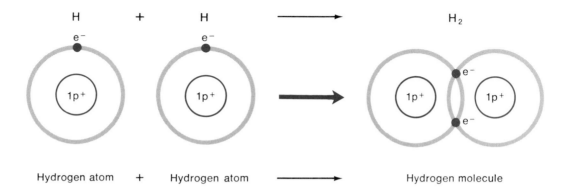

Hydrogen atom   +   Hydrogen atom   ⟶   Hydrogen molecule

Fig. 2.8 Hydrogen molecules (H_2) combine with oxygen molecules (O_2) to form water molecules (H_2O).

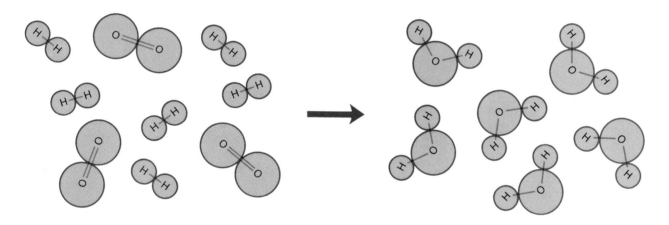

3 pairs of electrons. If 1 pair of electrons is shared, the resulting bond is called a *single covalent bond*; if 2 pairs of electrons are shared, the bond is called a *double covalent bond*; and if 3 pairs of electrons are shared, the bond is called a *triple covalent bond*.

Molecules and Compounds

When 2 or more atoms bond together, they form a new kind of particle called a **molecule**. When atoms of the same element combine, they produce molecules of that element. The gases of hydrogen (H_2), oxygen (O_2), and nitrogen (N_2) contain such molecules.

When atoms of different elements combine, molecules of substances called **compounds** form. Two atoms of hydrogen, for example, can combine with one atom of oxygen to produce a molecule of the compound water (H_2O), as shown in figure 2.8. Sugar, table salt, natural gas, alcohol, and most drugs are examples of compounds.

A molecule of a compound always contains definite kinds and numbers of atoms. A molecule of water (H_2O), for instance, always contains 2 hydrogen atoms and 1 oxygen atom. If 2 hydrogen atoms combine with 2 oxygen atoms, the compound formed is not water, but hydrogen peroxide (H_2O_2).

Chart 2.4 contains a summary of some particles of matter.

Chart 2.4 Some particles of matter

Name	Characteristic
Atom	Smallest particle of an element that has the properties of that element.
Electron (e^-)	Extremely small particle with almost no weight; carries a negative electrical charge and is in constant motion around an atomic nucleus.
Proton (p^+)	Relatively large atomic particle; carries a positive electrical charge and is found within the nucleus of an atom.
Neutron (n^0)	Particle with about the same weight as a proton; uncharged and thus electrically neutral; found within the nucleus of an atom.
Ion	Atom that is electrically charged because it has gained or lost 1 or more electrons.
Molecule	Particle formed by the chemical union of 2 or more atoms.

1. *What is an ion?*
2. *Describe two ways in which atoms may combine with other atoms.*
3. *Distinguish between a molecule and a compound.*

Formulas. The numbers and kinds of atoms in a molecule can be represented by a *molecular formula.* Such a formula consists of the symbols of the elements in the molecule together with numbers to indicate how many atoms of each element are present. For example, the molecular formula for water is H_2O, which means there are 2 atoms of hydrogen and 1 atom of oxygen in each molecule. The molecular formula for the sugar called glucose is $C_6H_{12}O_6$, which means there are 6 atoms of carbon, 12 atoms of hydrogen, and 6 atoms of oxygen in a molecule.

Usually the atoms of each element will form a specific number of bonds—hydrogen atoms form single bonds, oxygen atoms form 2 bonds, nitrogen atoms form 3 bonds, and carbon atoms form 4 bonds. The bonding capacity of these atoms can be represented by using symbols and lines as follows:

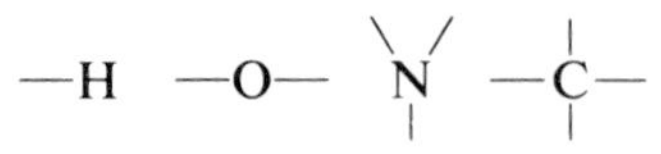

$$-H \quad -O- \quad \overset{\diagup}{\underset{|}{N}} \quad -\overset{|}{\underset{|}{C}}-$$

These representations can be used to show how atoms are bonded and arranged in various molecules. Illustrations of this type are called *structural formulas.* (See fig. 2.9 and color plates 1 and 2.)

Fig. 2.9 Structural formulas of molecules of hydrogen (H_2), oxygen (O_2), water (H_2O), and carbon dioxide (CO_2).

Chemical Reactions

When atoms or molecules react together, bonds between atoms are formed or broken. As a result, new combinations of atoms are created, and a chemical reaction has occurred.

Two kinds of chemical reactions are **synthesis** and **decomposition**. In a *synthetic* reaction, 2 or more atoms bond together to form a molecule, as when atoms of hydrogen and oxygen bond together to form molecules of water. Such a reaction is symbolized in this way:

$$A + B \rightarrow AB$$

In a *decomposition* reaction the bonds within a molecule break so that simpler molecules, atoms, or ions form. Thus, molecules of water can decompose to yield hydrogen and oxygen. Decomposition is symbolized as follows:

$$AB \rightarrow A + B$$

Synthetic reactions are particularly important in the growth of body parts and the repair of worn or damaged tissues, which involve the buildup of larger molecules from smaller ones. When food substances are digested or energy is released through respiration, the processes that occur are largely decomposition reactions.

A third type of chemical reaction is an *exchange reaction,* in which parts of 2 different kinds of molecules trade positions with one another. The reaction is symbolized as follows:

$$AB + CD \rightarrow AD + CB$$

An exchange reaction occurs when an acid reacts with a base to produce water and a salt.

Acids, Bases, and Salts

Some compounds release ions (ionize) when they dissolve in water or react with water molecules. Sodium chloride (NaCl), for example, releases sodium ions (Na^+) and chlorine ions (Cl^-) when it dissolves, as is represented by the following:

$$NaCl \rightarrow Na^+ + Cl^-$$

(See fig. 2.10.)

Fig. 2.10 When crystals of table salt (NaCl) dissolve in water, ions of sodium (Na$^+$) and ions of chlorine (Cl$^-$) are released.

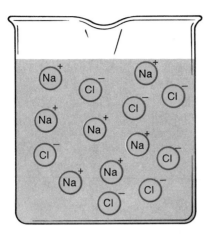

Chart 2.5 Types of electrolytes

	Characteristic	Examples
Acid	Ionizes to release hydrogen ions (H+).	Carbonic acid, hydrochloric acid, acetic acid, phosphoric acid
Base	Ionizes to release ions that combine with hydrogen ions.	Sodium hydroxide, potassium hydroxide, magnesium hydroxide, aluminum hydroxide
Salt	Substance formed by the reaction between an acid and a base.	Sodium chloride, aluminum chloride, magnesium sulfate

Since the resulting solution contains electrically charged particles (ions), it will conduct an electric current. Substances that ionize in water are, therefore, called **electrolytes**. Electrolytes that release hydrogen ions (H$^+$) in water are **acids**. For example, the compound, carbonic acid, ionizes in water to release hydrogen ions and bicarbonate ions (HCO$_3^-$).[1]

$$H_2CO_3 \rightarrow H^+ + HCO_3^-$$

Electrolytes that release ions that combine with hydrogen ions are called **bases**. The compound, sodium hydroxide, for example, ionizes in water to release hydroxyl ions (OH$^-$).

$$NaOH \rightarrow Na^+ + OH^-$$

The hydroxyl ions, in turn, combine with hydrogen ions to form water; thus, sodium hydroxide is a base.

1. Some ions, such as HCO$_3^-$ ions, contain two or more atoms. However, such a group behaves like a single atom and usually remains unchanged in a chemical reaction.

Acids that release hydrogen ions and bases that release hydroxyl ions can react to form water and electrolytes called **salts**. For example, hydrochloric acid and sodium hydroxide react to form water and sodium chloride. This reaction is symbolized as follows:

$$HCl + NaOH \rightarrow H_2O + NaCl$$

Chart 2.5 summarizes the 3 types of electrolytes—acids, bases, and salts.

Acid and Base Concentrations

The chemical reactions involved with life processes are often affected by the presence of hydrogen and hydroxyl ions; therefore the concentrations of these ions in body fluids is important. Such concentrations are measured in units of **pH**.

The pH scale includes values from 0 to 14. A solution with a pH of 7.0, the midpoint of the scale, contains equal numbers of hydrogen and hydroxyl ions and is said to be *neutral*. (See fig. 2.11.) A solution that contains more hydrogen than hydroxyl

Fig. 2.11 As the concentration of hydrogen ions (H$^+$) increases, a solution becomes more acidic, and the pH value decreases. As the concentration of hydroxyl ions (OH$^-$) increases, a solution becomes more basic, and the pH value increases.

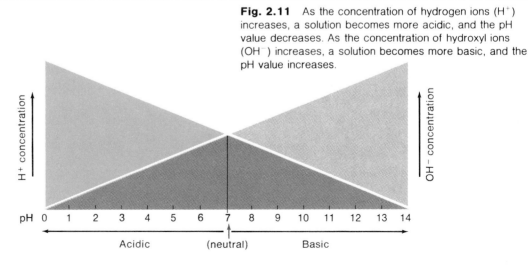

Fig. 2.12 Approximate pH values of some common substances.

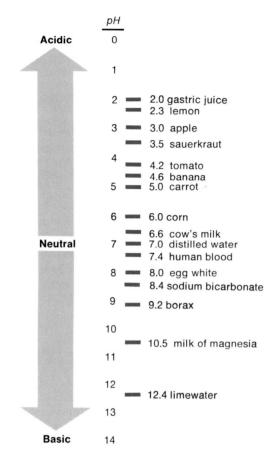

ions has a pH less than 7.0 and is *acidic*, while one with fewer hydrogen than hydroxyl ions has a pH above 7.0 and is *basic* (alkaline). The diagram in figure 2.11 shows the pH range. Figure 2.12 indicates the pH value of some common substances.

There is a tenfold difference in the hydrogen ion concentration between each whole number on the pH scale. That is, a solution of pH 4.0 contains 0.0001 grams of hydrogen ions per liter, and a solution of pH 3.0 contains 0.001 grams of hydrogen ions per liter. See the Appendix for a more detailed explanation of pH.

1. What is meant by a formula?
2. Describe three kinds of chemical reactions.
3. Compare the characteristics of an acid with those of a base.
4. What is meant by pH?

Chemical Constituents of Cells

The cellular chemicals, which enter into reactions or are produced by them, can be divided into two large groups. Except for a few simple molecules, those that contain carbon atoms are called **organic**, and those that lack carbon are called **inorganic**.

Generally, inorganic substances will dissolve in water or react with water to produce ions; thus they are electrolytes. Some organic compounds will dissolve in water also, but they are more likely to dissolve in organic liquids like ether or alcohol. Those that will dissolve in water usually do not release ions, and are, therefore, called *nonelectrolytes*.

Inorganic Compounds

Among the inorganic compounds common in cells are water, oxygen, carbon dioxide, and certain salts.

Water. Water (H_2O) is the most abundant compound in living material and is responsible for about two-thirds of the weight of an adult human. It is the major ingredient of blood and other body fluids, including those within cells.

When substances dissolve in water, relatively large pieces of the material break into smaller ones, and eventually molecular-sized particles or ions result. These tiny particles are much more likely to react with one another than were the original large pieces. Consequently, most metabolic reactions occur in water.

Water also plays an important role in the transportation of chemicals within the body. The aqueous portion of blood, for example, carries many vital substances, such as oxygen, sugars, salts, and vitamins, from the organs of digestion and respiration to the body cells. Water also carries waste materials, such as carbon dioxide and urea, from these cells to the kidneys, which remove them from the blood and excrete them to the outside of the body.

In addition, water can absorb and transport heat. For instance, heat released from muscle cells during exercise can be carried by the blood from deeper parts to the surface. At the same time, water released by skin cells in the form of perspiration can absorb heat from underlying parts and carry it away by evaporation.

Chart 2.6 Inorganic substances common in cells

Substance	Symbol or Formula	Function
Water	H_2O	Major ingredient of body fluids; medium in which most biochemical reactions occur; transports various chemical substances; helps regulate body temperature.
Oxygen	O_2	Used in the release of energy from glucose molecules during respiration.
Carbon dioxide	CO_2	Waste product that results from the respiration of glucose molecules; reacts with water to form carbonic acid.
Ions of inorganic salts	Na^+, Cl^-, K^+, Ca^{+2}, Mg^{+2}, PO_4^{-3}, CO_3^{-2}, HCO_3^-, SO_4^{-2}	Play important roles in various metabolic processes including those involved in the maintenance of water balance, blood clotting, bone development, cell formation, energy transfer within cells, and muscle and nerve actions.

Oxygen. Molecules of oxygen (O_2) enter the body through the respiratory organs and are transported to body cells by the blood. The oxygen is used by cell organelles in the process of releasing energy from glucose and other molecules, and the energy that is released is used to drive the cell's metabolic activities. A continuing supply of oxygen is necessary for cell survival, and ultimately for the survival of the organism.

Carbon Dioxide. Carbon dioxide (CO_2) is one of the small, carbon-containing compounds of the inorganic group. It is produced as a waste product when energy is released from glucose and other molecules. As it moves from the cells into the surrounding body fluids and blood, most of the carbon dioxide reacts with water to form a weak acid (carbonic acid, H_2CO_3). This acid ionizes, releasing hydrogen ions (H^+) and bicarbonate ions (HCO_3^-), which the blood carries to the respiratory organs. Here the chemical reactions reverse, and carbon dioxide gas is given off into the air in the lungs.

Inorganic Salts. Various kinds of inorganic salts are common in body parts and fluids. They are the sources of many necessary ions, including ions of sodium (Na^+), chlorine (Cl^-), potassium (K^+), calcium (Ca^{+2}), magnesium (Mg^{+2}), phosphate (PO_4^{-3}), carbonate (CO_3^{-2}), bicarbonate (HCO_3^-), and sulfate (SO_4^{-2}). These ions play important roles in metabolic processes including those involved in maintenance of proper water concentrations in body fluids, blood clotting, bone development, cell formation, energy transfer within cells, and muscle and nerve functions.

These electrolytes must not only be present, but they must be present in the proper concentrations, both inside and outside cells, if body functions are to occur effectively. Such a condition is called **electrolyte balance**. In certain diseases, this balance tends to be lost, and modern medical treatment places considerable emphasis on restoring it.

Chart 2.6 summarizes the function of some of the inorganic compounds that commonly occur in cells.

---

1. *What is the difference between an organic and an inorganic molecule?*
2. *What is the difference between an electrolyte and a nonelectrolyte?*
3. *Define* electrolyte balance.

Organic Compounds

Important groups of organic compounds found in cells include carbohydrates, lipids, proteins, and nucleic acids.

Carbohydrates. These compounds provide much of the energy cells need. They also supply cells with materials to build certain cell structures and are often stored as reserve energy supplies.

Carbohydrate molecules contain atoms of carbon, hydrogen, and oxygen. These molecules usually have twice as many hydrogen as oxygen atoms, the same ratio of hydrogen to oxygen that occurs in water molecules (H_2O). This ratio is easy to see in the molecular formulas of glucose ($C_6H_{12}O_6$) and sucrose or table sugar ($C_{12}H_{22}O_{11}$).

The carbon atoms of carbohydrate molecules are bonded in chains. The number of atoms in these chains varies with the kind of carbohydrate. Some have 3 carbon atoms in a chain, while others have 4, 5, 6, or more. The carbohydrates with shorter chains are called **sugars**. Those with 6 carbon atoms are called *simple sugars* or *monosaccharides* and are the building blocks of more complex carbohydrate molecules. The simple sugars include glucose, fructose, and galactose. Figure 2.13 and color plate 3 illustrate the molecular structure of glucose.

In complex carbohydrates, the simple sugars are bonded to form molecules of varying lengths, as shown in figure 2.14. Some, like sucrose, maltose, and lactose, are *double sugars* or *disaccharides*, whose molecules each contain two building blocks. Others are made up of many simple sugar units bonded together to form *polysaccharides*, such as plant starch. Starch molecules consist of highly branched chains, each containing from 24 to 30 glucose units. Since the number of glucose units within a starch molecule may vary, starch is symbolized by the formula $(C_6H_{10}O_5)_x$ with *x* equal to the number of glucose units in the molecule.

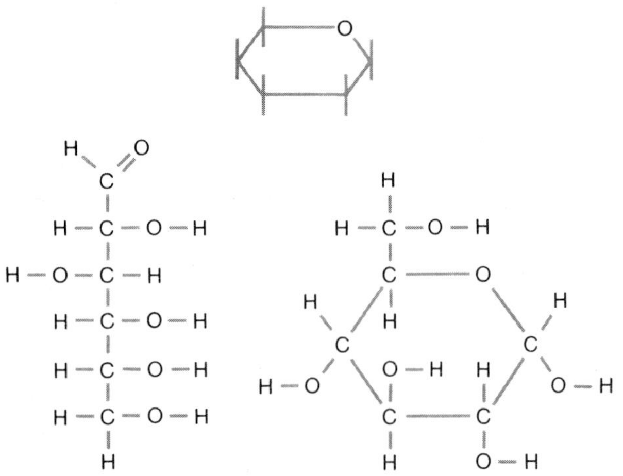

Animals, including humans, synthesize a substance similar to starch called *glycogen*. Its molecules also consist of branched chains of sugar units, but each chain contains only about a dozen glucose units.

Lipids. A second important group of organic compounds is the **lipids**. These include a number of substances, such as phospholipids and cholesterol, that have vital functions in cells, but the most common lipids are *fats*. Fats are used to build cell parts and to supply energy for cell activities. In fact, fat molecules can supply more energy, gram for gram, than carbohydrate molecules.

Like carbohydrates, fat molecules are composed of carbon, hydrogen, and oxygen atoms. They contain, however, a much smaller proportion of oxygen than do carbohydrates. This fact can be illustrated by the formula for the fat, tristearin, $C_{57}H_{110}O_6$, which obviously has many times more hydrogen atoms than oxygen atoms. Carbohydrates usually have only twice as many hydrogen as oxygen atoms.

The building blocks of fat molecules are **fatty acids** and **glycerol** (glycerin). These smaller molecules are bonded so that each glycerol molecule is combined with 3 fatty acid molecules. The result is a single fat molecule, as shown in figure 2.15. (Also see color plate 4.)

Although the glycerol portion of every fat molecule is the same, there are many kinds of fatty acids and, therefore, many kinds of fats. Fatty acid molecules differ in the lengths of their carbon atom chains and in the way the carbon atoms are bonded together. In some cases, the carbon atoms are all joined by single carbon-carbon bonds. A fat that contains this type of fatty acid is *saturated*; that is, each carbon atom is bonded to as many hydrogen atoms as possible and is thus saturated with hydrogen atoms. Other fatty acids have one or more double bonds between carbon atoms, and the fats that contain them are

Fig. 2.13 *Left*, some glucose molecules ($C_6H_{12}O_6$) have a straight chain of carbon atoms; *right*, more commonly, glucose molecules form a ring structure, which can be symbolized with this shape:

Fig. 2.14 (*a*) A simple sugar molecule, such as glucose, consists of one 6-carbon atom building block; (*b*) a double sugar molecule, such as sucrose (table sugar), consists of two building blocks; (*c*) a complex carbohydrate molecule, such as starch, consists of many building blocks.

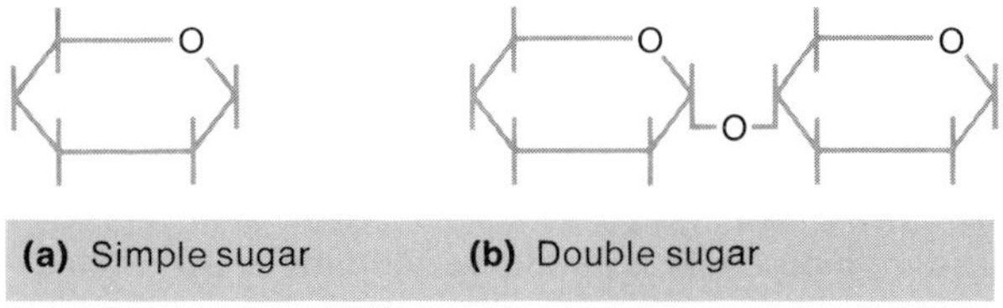

(a) Simple sugar   (b) Double sugar

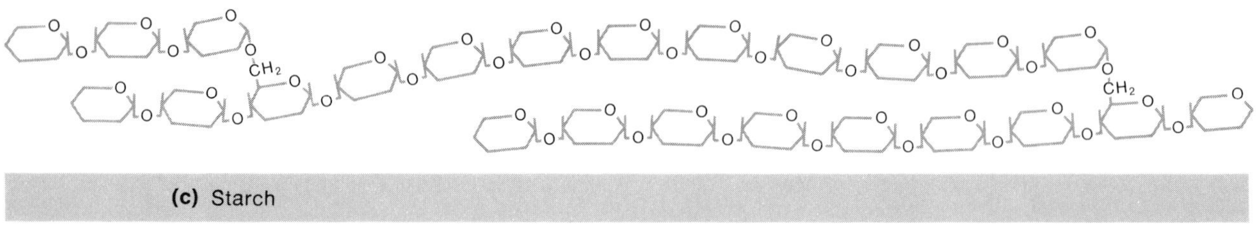

(c) Starch

Fig. 2.15 A fat molecule consists of a glycerol portion and three fatty acid portions.

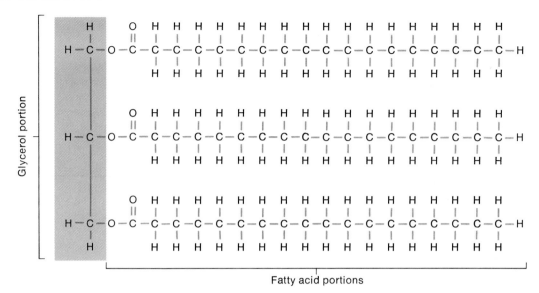

Glycerol portion

Fatty acid portions

unsaturated. Fat molecules with many double-bonded carbon atoms are *polyunsaturated.* (See fig. 2.16.)

Thus each kind of fat molecule has particular kinds of fatty acids bonded to its glycerol portion, and each has special properties. Saturated fats like those in lard, for example, tend to be solids at room temperature, while unsaturated fats like those in salad oils tend to be liquids.

Proteins. A third important group of organic compounds found in cells is **proteins**. They serve as structural materials, are often used as energy sources, and play vital roles in all metabolic processes by acting as **enzymes**. Enzymes are proteins that have special properties and can speed up specific chemical reactions without being changed or used up themselves.

Like carbohydrates and lipids, proteins contain atoms of carbon, hydrogen, and oxygen. In addition, they always contain nitrogen atoms, and sometimes sulfur atoms as well. The building blocks of proteins are smaller molecules called **amino acids**. (See fig. 2.17 and color plate 5.) About 20 different amino acids occur commonly in proteins. They are bonded together in chains, varying in length from less than 100 to more than 50,000 amino acids. The hemoglobin molecules, which are responsible for the red color of blood, each contain a protein portion that includes 574 amino acid units arranged in four chains.

Each type of protein contains specific numbers and kinds of amino acids arranged in particular sequences. Consequently different kinds of protein molecules have different shapes. (See fig. 2.18 and color plate 6.) The molecules of some (fibrous proteins) are long and threadlike. They have special functions, such as binding body parts together or forming the fine threads of a blood clot. Other protein molecules (globular proteins) are coiled or folded into compact masses. Enzyme molecules are usually of this type.

The special shapes of protein molecules can be altered by exposure to excessive amounts of heat, radiation, electricity, or various chemicals. When this occurs, the molecules lose their unique shapes, become disorganized (as shown in fig. 2.19), and are said to be *denatured.* At the same time, they lose their unique properties and behave differently. For example, the protein in egg white is denatured when it is heated. This treatment causes the egg white to change from a liquid to a solid—a change that cannot be reversed. Similarly, if cellular proteins are denatured, they are permanently changed and become nonfunctional. Such actions may threaten the life of the cells.

Fig. 2.16 What are the major differences between (a) a molecule of saturated fat and (b) a molecule of unsaturated fat?

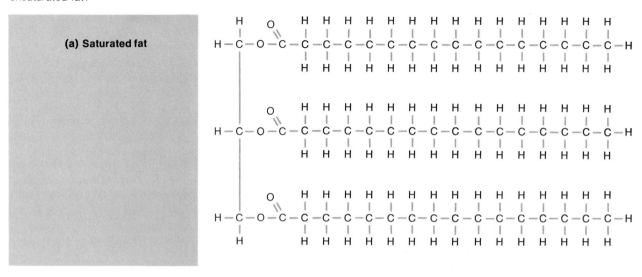

(a) Saturated fat

Fig. 2.17 Some representative amino acids and their structural formulas. Each amino acid molecule has a particular shape due to the arrangement of its parts.

Amino acid	Structural formulas	Amino acid	Structural formulas
Alanine		Phenylalanine	
Valine		Tyrosine	
Cysteine		Histidine	

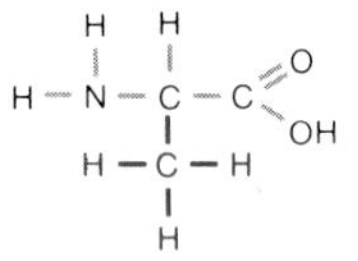

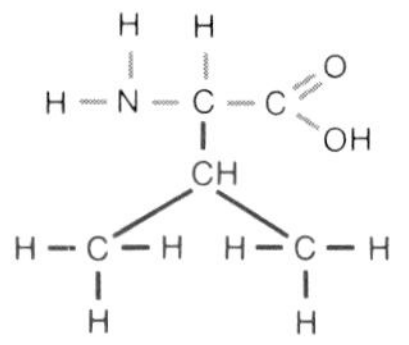

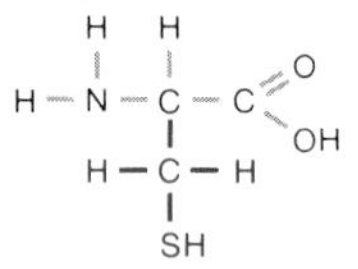

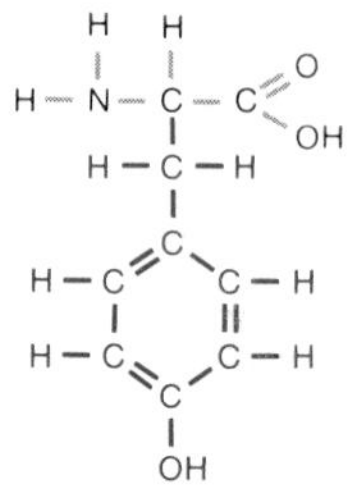

Levels of Organization

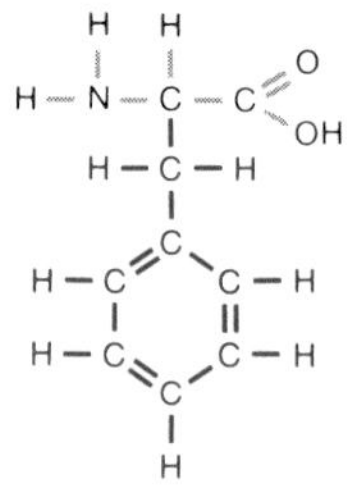

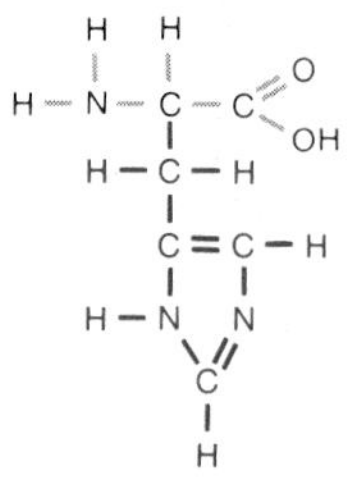

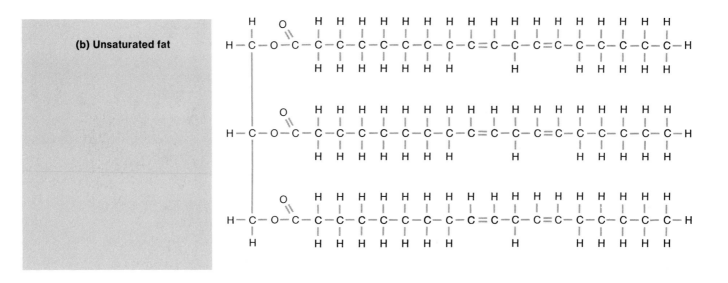
(b) Unsaturated fat

Fig. 2.18 Each different shape in this chain represents a different amino acid molecule. The whole chain represents a portion of a protein molecule.

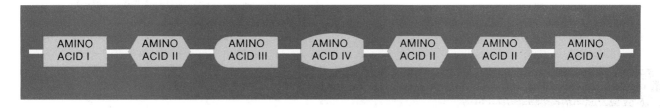

Fig. 2.19 What factors may cause the bonds of (a) a protein molecule to break so that (b) it loses its unique shape?

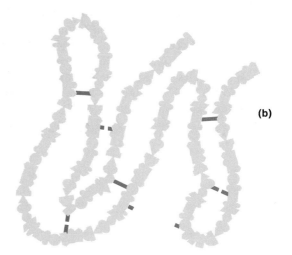

(a)

(b)

Fig. 2.20 A nucleotide consists of a 5-carbon sugar (S), a phosphate group (P), and an organic base (B).

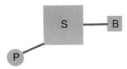

Fig. 2.21 A nucleic acid molecule consists of nucleotides bonded together in a chain.

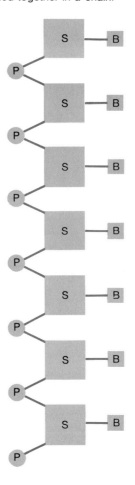

Nucleic Acids. Of all the compounds present in cells, the most fundamental are the **nucleic acids**. These organic substances control cell activities in very special ways.

Nucleic acid molecules are generally very large and complex. They contain atoms of carbon, hydrogen, oxygen, nitrogen, and phosphorus, which are bonded into building blocks called **nucleotides**. Each nucleotide consists of a 5-carbon *sugar* (ribose or deoxyribose), a *phosphate group*, and one of several *organic bases* (fig. 2.20). A nucleic acid molecule consists of many nucleotides bonded together in a chain, as shown in figure 2.21 and color plate 7.

There are two major types of nucleic acids. One type, composed of molecules whose nucleotides contain ribose sugar, is called **RNA** (ribonucleic acid). The nucleotides of the second type contain deoxyribose sugar; these are called **DNA** (deoxyribonucleic acid). Figure 2.22 compares the structure of these two 5-carbon sugars.

DNA molecules store information in a kind of molecular code. This information tells cell parts how to construct specific protein molecules, and these proteins, acting as enzymes, are responsible for all of the metabolic reactions that occur in cells.

DNA molecules also have a unique ability to duplicate (or replicate) themselves. They duplicate whenever a cell reproduces, and each newly formed cell receives an exact copy of its parent's DNA molecules. These molecules contain the instructions for carrying on vital life processes.

Chapter 4 discusses storage of this information in nucleic acid molecules, its use in the manufacture of protein molecules, and the function of these proteins in metabolic reactions.

Chart 2.7 summarizes the 4 groups of organic compounds.

Fig. 2.22 How are the molecules of ribose and deoxyribose different? How are they similar to a molecule of glucose?

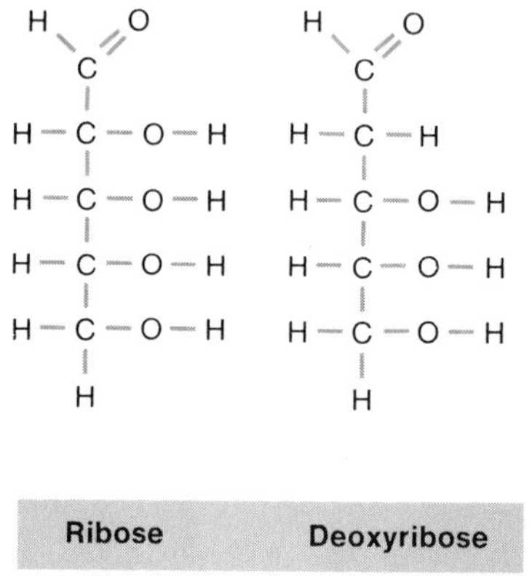

Ribose	**Deoxyribose**

Chart 2.7 Organic compounds in cells

Compound	Elements Present	Building Blocks	Functions	Examples
Carbohydrates	C,H,O	Simple sugar	Energy, cell structure	Glucose, starch
Fats	C,H,O	Glycerol, fatty acids	Energy, cell structure	Tristearin, tripalmitin
Proteins	C,H,O,N (often S)	Amino acids	Cell structure, enzymes, energy	Albumins, hemoglobin
Nucleic acids	C,H,O,N,P	Nucleotides	Stores information, controls cell activities	RNA, DNA

1. *Compare the chemical composition of carbohydrates, lipids, proteins, and nucleic acids.*
2. *How does an enzyme affect a chemical reaction?*
3. *What is likely to happen to a protein molecule that is exposed to excessive heat or radiation?*
4. *What is the function of DNA?*

Clinical Terms Related to the Use of Radioactive Isotopes

alpha particle (al′fah par′tĭ-kl)—positively charged particle, consisting of two protons and two neutrons, given off by certain radioactive isotopes.

beta particle (ba′tah par′tĭ-kl)—negatively charged particle (electron) given off by certain radioactive isotopes; sometimes used to treat diseased tissues on or near the surface of organs.

cobalt 60 (ko′bawlt)— radioactive isotope of the element, cobalt, used as a source of gamma radiation in the treatment of cancer.

cobalt unit (ko′bawlt u′nit)—therapy apparatus that contains cobalt 60.

curie (ku′re)—a unit of radioactivity used to measure the amount of radiation emitted from a radioactive isotope.

dosimetry (do-sim′ĕ-tre)—the process of measuring radiation exposure or dose.

gamma ray (gam′ah ra)—a form of radiation, similar to X ray, that is given off by certain radioactive isotopes; the most penetrating form of radiation produced by isotopes.

half-life (haf′ lif)—the time required for a radioactive isotope to lose one-half of its radioactivity.

iodine 131 (i′o-din)—a radioactive isotope of the element, iodine, sometimes used in testing or treating the thyroid gland.

ionizing radiation (i′on-ĭ-zing ra-de-a′shun)—any form of radiation capable of causing atoms to become electrically charged ions.

iron 59 (i′ern)—a radioactive isotope of the element, iron, sometimes used to test bone marrow.

irradiation (i-ra′′de-a′shun)—the process of exposing something to radiation.

nuclear medicine (nu′kle-ar med′ĭ-sin)—the branch of medicine concerned with the use of radiation.

potassium 42 (po-tas′e-um)—a radioactive isotope of the element potassium sometimes used to study brain tumors.

rad (rad)—a unit used in measuring the amount of radiation exposure; *r*adiation *a*bsorbed *d*ose.

radiation sickness (ra′′de-a′shun sik′nes)—a disease usually resulting from the effects of radiation on the digestive organs.

scan (skan)—a recording of the radiation emitted by radioactive isotopes within certain tissues.

Chapter Summary

Introduction
Chemistry deals with the composition of substances and changes in their composition.
Biochemistry is the chemistry of living things.

Structure of Matter
1. Elements and atoms
   a. Naturally occurring matter is composed of about 90 elements, which are composed of atoms.
   b. Atoms of different elements vary in size, weight, and ways of interacting.

2. Atomic structure
   a. An atom consists of electrons surrounding an atomic nucleus, which contains protons and neutrons.
   b. Electrons are negatively charged, protons positively charged, and neutrons uncharged.
   c. The atomic number of an element is equal to the number of protons in each atom; the atomic weight is equal to the number of protons plus the number of neutrons in each atom.
3. Isotopes
   a. Isotopes are atoms with the same atomic number but different atomic weights (due to differing numbers of neutrons).
   b. All the isotopes of an element react chemically in the same manner.
   c. Some isotopes are radioactive and release radiation that can create chemical changes and cause the death of cells.
4. Bonding of atoms
   a. Electrons move in shells around atomic nuclei.
   b. Atoms with filled outer shells are inactive, while atoms with incompletely filled outer shells tend to gain, lose, or share electrons and thus achieve a stable configuration.
   c. Atoms that lose electrons become positively charged; atoms that gain electrons become negatively charged.
   d. Ions with opposite charges are attracted to one another and become bonded by electrovalent bonds; atoms that share electrons become bonded by covalent bonds.
5. Molecules and compounds
   a. When 2 or more atoms of the same element become bonded, a molecule of that element is formed; when atoms of different kinds become bonded, a molecule of a compound is formed.
   b. Molecules contain definite kinds and numbers of atoms.
6. Formulas
   a. The numbers and kinds of atoms in a molecule can be represented by a molecular formula.
   b. The arrangement of atoms within a molecule can be represented by a structural formula.
7. Chemical reactions
   a. When a chemical reaction occurs, bonds between atoms are broken or created.
   b. Three kinds of chemical reaction are synthesis, in which larger molecules are formed from smaller particles; decomposition, in which smaller particles are formed from larger molecules; and exchange reactions, in which the parts of 2 molecules trade positions.
8. Acids, bases, and salts
   a. Compounds that ionize when they dissolve in water are electrolytes.
   b. Electrolytes that release hydrogen ions are acids, and those that release hydroxyl or other ions that react with hydrogen ions are bases.
   c. Acids and bases that release hydroxyl ions react together to form water and electrolytes called salts.
9. Acid and base concentrations
   a. The concentration of hydrogen (H^+) and hydroxyl (OH^-) ions in a solution can be represented by pH.
   b. A solution with equal numbers of H^+ and OH^- is neutral, with a pH of 7.0; a solution with more H^+ than OH^- is acidic (pH less than 7.0); a solution with fewer H^+ than OH^- is basic (pH more than 7.0).

Chemical Constituents of Cells

Molecules containing carbon atoms are organic and are usually nonelectrolytes; those lacking carbon atoms are inorganic and are usually electrolytes.
1. Inorganic compounds
   a. Water is the most abundant compound in cells and serves as a substance in which chemical reactions occur; it also transports chemicals and heat and helps to release excess body heat.
   b. Oxygen is used in releasing energy from glucose and other molecules.
   c. Carbon dioxide is produced when energy is released from glucose and other molecules.
   d. Inorganic salts provide ions needed in metabolic processes and must be balanced for effective functioning.
2. Organic compounds
   a. Carbohydrates provide much of the energy needed by cells; their basic building blocks are simple sugar molecules.
   b. Lipids, such as fats, phospholipids, and cholesterol, supply energy also; their basic building blocks are molecules of glycerol and fatty acids.
   c. Proteins serve as structural materials, energy sources, and enzymes.
      (1) Enzymes initiate or speed chemical reactions without being changed themselves.
      (2) The building blocks of proteins are amino acids.
      (3) Different kinds of proteins vary in the numbers and kinds of amino acids they contain, and in the sequences in which these amino acids are arranged.
      (4) Protein molecules can be denatured by exposure to excessive heat, radiation, electricity, or various chemicals.
   d. Nucleic acids are the most fundamental chemicals in cells, since they control cell activities.
      (1) The 2 major kinds are RNA and DNA.
      (2) They are composed of building blocks called nucleotides.

(3) DNA molecules store information that tells a cell how to construct protein molecules, including enzymes.

(4) DNA molecules are duplicated and passed from parent to offspring cells during cell reproduction.

Application of Knowledge

1. Explain how various carbohydrate molecules, such as sugars and starch, differ in complexity.

2. Using the criteria given on pages 30–31 to distinguish saturated from unsaturated fats, list all of the sources of saturated and unsaturated fats you have eaten during the past 24 hours.

3. Read the labels on the packages of several foods that are advertised as being high in polyunsaturates. What substances seem to be the sources of the polyunsaturates in these foods?

Review Activities

1. Distinguish between chemistry and biochemistry.
2. Define *matter*.
3. Explain the relationship between elements and atoms.
4. List the 4 most abundant elements in the human body.
5. Describe the major parts of an atom.
6. Distinguish between protons and neutrons.
7. Explain why a complete atom is electrically neutral.
8. Define *isotope*.
9. Explain what is meant by atomic radiation.
10. Explain how electrons are arranged within an atom.
11. Distinguish between an electrovalent bond and a covalent bond.
12. Distinguish between a single covalent bond and a double covalent bond.
13. Explain the relationship between molecules and compounds.
14. Distinguish between a molecular formula and a structural formula.
15. Describe three major types of chemical reactions.
16. Define *acid, base, salt* and *electrolyte*.
17. Explain what is meant by pH, and describe the pH scale.
18. Distinguish between organic and inorganic substances.
19. Describe the roles played by water and by oxygen in the human body.
20. Explain how carbon dioxide is transported within the body.
21. List several ions needed by cells, and describe their general functions.
22. Define *electrolyte balance*.
23. Describe the general characteristics of carbohydrates.
24. Distinguish between simple and complex carbohydrates.
25. Describe the general characteristics of lipids.
26. Distinguish between saturated and unsaturated fats.
27. Describe the general characteristics of proteins.
28. Define *enzyme*.
29. Explain how protein molecules may become denatured.
30. Describe the general characteristics of nucleic acids.
31. Explain the general functions of nucleic acids.

Suggestions for Additional Reading

Baker, J. J. W., and Allen, G. E. 1974. *Matter, energy, and life.* 3rd ed. Palo Alto: Addison-Wesley.

Frieden, E. July 1972. The chemical elements of life. *Scientific American.*

Koshland, D. E. Oct. 1973. Protein shape and biological control. *Scientific American.*

Lehninger, A. 1970. *Biochemistry.* New York: Worth.

Mahler, H. R., and Cordes, E. H. 1971. *Biological chemistry.* 2nd ed. New York: Harper and Row.

Maugh, T. H. 1973. Trace elements, a growing appreciation of effects on man. *Science* 181:253.

Orten, J. M., and Neuhaus, O. W. 1970. *Biochemistry.* 8th ed. St. Louis: C. V. Mosby.

Roller, A. 1974. *Discovering the basis of life: an introduction to molecular biology.* New York: McGraw-Hill.

Sackheim, G. I. 1974. *Introduction to chemistry for biology students.* Chicago: Educational Methods.

Wood, W. B., et al. 1974. *Biochemistry.* Menlo Park, Calif.: W. A. Benjamin.

The Cell

3 Matter is composed of atoms and molecules, and the interactions among these particles constitute the metabolic reactions of organisms. The cell, however, constitutes the fundamental unit of life.

   The human body is composed entirely of cells, the products of cells, and various fluids. These cells represent the basic structural units of the body in that they are the building blocks from which all larger parts are formed. They are also the functional units, since whatever a body part can do is the result of activities within its cells.

Cells account for the shape, organization, and construction of the body and for carrying on its life processes. In addition, they can reproduce, and thus provide the new cells needed for growth and development and for the replacement of worn and injured tissues.

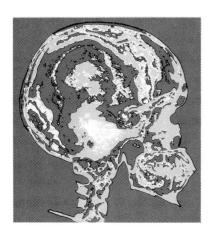

Chapter Outline

Chapter Objectives

After you have studied this chapter, you should be able to

1. Describe the general characteristics of a cell.
2. List the major organelles of a cell.
3. Describe the structure and function of each cytoplasmic organelle.
4. Explain how substances move through cell membranes.
5. Describe the life cycle of a cell and explain how a cell reproduces.
6. Explain how the surface-volume relationship is related to the control of cell reproduction.
7. List the major characteristics of cancer.
8. Complete the review activities at the end of this chapter.

active transport   (ak'tiv trans'port)

centrosome   (sen'tro-sōm)

chromosome   (kro'mo-sōm)

cytoplasm   (si'to-plazm)

differentiation   (dif''er-en''she-a'shun)

diffusion   (dĭ-fu'zhun)

endoplasmic reticulum
    (en'do-plaz'mik rĕ-tik'u-lum)

equilibrium   (e''kwĭ-lib're-um)

filtration   (fil-tra'shun)

Golgi apparatus   (gol'je ap''ah-ra'tus)

lysosome   (li'so-sōm)

mitochondrion   (mi''to-kon'dre-un); plural,
    mitochondria   (mi''to-kon'dre-ah)

mitosis   (mi-to'sis)

nucleolus   (nu-kle'o-lus)

nucleus   (nu'kle-us)

osmosis   (oz-mo'sis)

permeable   (per'me-ah-bl)

phagocytosis   (fag''o-si-to'sis)

pinocytosis   (pi''no-si-to'sis)

ribosome   (ri'bo-sōm)

cyt-, cell: *cyt*oplasm—fluid that occupies the space between the cell membrane and the nuclear membrane.

endo-, within: *endo*plasmic reticulum—network of membranes within the cytoplasm.

hyper-, above: *hyper*tonic—a solution that has a greater concentration of dissolved particles than another solution.

hypo-, below: *hypo*tonic—a solution that has a lesser concentration of dissolved particles than another solution.

inter-, between: *inter*phase—stage that occurs between mitotic divisions of a cell.

iso-, equal: *iso*tonic—a solution that has a concentration of dissolved particles equal to that of another solution.

lys-, to break up: *lys*osome—organelle that contains enzymes capable of breaking up molecules of protein and carbohydrate.

mit-, thread: *mit*osis—process during which threadlike chromosomes appear within a cell.

phag-, to eat: *phag*ocytosis—process by which a cell takes in solid particles.

plas-, to form: hyper*plas*ia—formation of excessive numbers of cells.

pino-, to drink: *pino*cytosis—process by which a cell takes in tiny droplets of liquid.

-som, body: ribo*som*e—tiny, spherical organelle.

The trillions of cells that make up a human body have much in common, yet those in different tissues vary in a number of ways. For example, they vary considerably in size.

Cell sizes are measured in units called *microns*. A micron equals 1/1000th of a millimeter and is symbolized by the Greek letter mu (μ). Measured in microns, a human egg cell is about 140 μ in diameter and is just barely visible to an unaided eye. (See fig. 3.1.) This is large when compared to a red blood cell, which is about 8 μ in diameter, or the most common white blood cells, which vary from 10–12 μ in diameter. On the other hand, smooth muscle cells can be between 20–500 μ long.

Human cells also vary in shape, and typically their shapes are closely related to their functions (figure 3.2). For instance, nerve cells often have long, threadlike extensions that transmit nerve impulses from one part of the body to another. The epithelial cells that line the inside of the mouth serve to shield underlying cells. These protective cells are thin, flattened, and tightly packed, somewhat like the tiles of a floor. Muscle cells, which function to pull parts closer together, are slender and rodlike, with their ends attached to the parts they move.

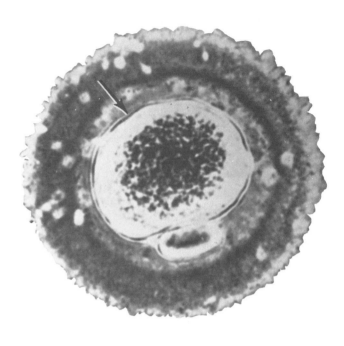

Fig. 3.1 A human egg cell is a relatively large cell, visible to the unaided eye as a tiny speck.

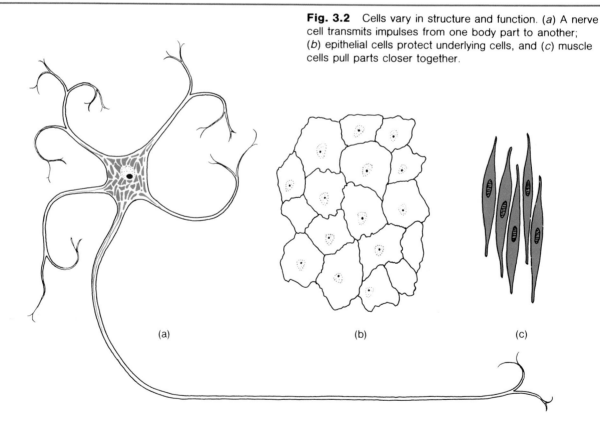

Fig. 3.2 Cells vary in structure and function. (*a*) A nerve cell transmits impulses from one body part to another; (*b*) epithelial cells protect underlying cells, and (*c*) muscle cells pull parts closer together.

(a) (b) (c)

A Composite Cell

Since cells vary so greatly in size, shape, and function, it is not possible to describe a typical cell. For purposes of discussion, however, it is convenient to imagine that one exists. Such a composite cell would contain parts observed in many kinds of cells, even though some of these cells lack parts included in the imagined structure. Thus, the cell shown in figure 3.3 and described in the following sections is not real. Instead, it is a composite cell—one that includes most known cell structures.

Commonly a cell consists of two major parts, one within the other and each surrounded by a thin membrane. The inner portion is called the *cell nucleus,* and it is enclosed by a *nuclear membrane.* A mass of fluid called *cytoplasm* surrounds the nucleus and is encircled by a cytoplasmic or *cell membrane.*

Within the cytoplasm are other membranes that separate it into small subdivisions. These include networks of membranes and membranes that mark off tiny, distinct parts called *cytoplasmic organelles.* These organelles perform specific functions necessary for cell survival.

---

1. What is meant by the statement that the cell is the basic structural and functional unit of the body?
2. Give two examples to illustrate the idea that the shape of a cell is related to its function.
3. Name the two major parts of a cell.

Cell Membrane

The **cell membrane** is the outermost limit of the cell, but it is more than a simple envelope surrounding the cellular contents; it is an actively functioning part of the living material. The membrane is extremely thin—visible only with the aid of an electron microscope—but is flexible and somewhat elastic. It typically has a complex surface with many outpouchings and infoldings that provide extra surface area. The membrane quickly seals minute breaks, but if it is extensively damaged, the cell contents escape, and the cell dies.

In addition to maintaining the wholeness of the cell, the membrane serves as a "gateway" through which chemicals enter and leave. This gate acts in a special manner; it allows some substances to pass and excludes others. When a membrane functions in this way, it is *selectively permeable.* A *permeable* membrane, on the other hand, is one that allows all materials to pass through.

The mechanism by which the membrane accomplishes its selective function is not well understood. It is known, however, that the mechanism involves the chemical nature of the membrane—which is largely protein and lipid (usually phospholipid and cholesterol) with a small amount of carbohydrate.

The way molecules are positioned within the cell membrane is unknown. Evidence indicates that the structural qualities of the membrane are due primarily to the molecules of phospholipids, which are arranged in two layers (bilayer). (See fig. 3.4.)

Although there appear to be only a few types of lipid molecules, there are many kinds of proteins in the membrane. Some of these protein molecules are quite large and seem to extend across the lipid layer; other kinds are located on the outer surface of the lipid layer and project outward from it; still other kinds reside on the cytoplasmic side of the membrane. (See fig. 3.5.)

The phospholipid molecules and the protein molecules embedded among them can move sideways in any direction, so that instead of being a rigid structure, the membrane is flexible and acts like a thin film of liquid—a liquid of phospholipid molecules with protein molecules floating in it.

The carbohydrates of the membrane all seem to be associated with the outer surface, where some of them are combined with lipids (glycolipids) and others are combined with proteins (glycoproteins).

Since different kinds of molecules are located on the inner and outer surfaces of the membrane, it is not surprising that the two surfaces have functional differences. For instance, various molecules on the outer surface function as receptor sites that can combine with specific chemicals, such as hormones, when they occur in the cell's surroundings. Other outer surface molecules serve as markers by which a cell can be recognized by neighboring cells. Many of the proteins of the inner surface function as enzymes that speed chemical reactions, such as those that help vital substances through the membrane.

Because the membrane is largely phospholipid, molecules that are soluble in lipids can easily pass through it. On the other hand, molecules of substances like water, which do not dissolve in lipids, cannot penetrate the phospholipid layers. Water and other small molecules, however, can pass through the large protein molecules that span the thickness of the membrane and serve as passageways or "pores."

Fig. 3.3 A composite cell and its major organelles.

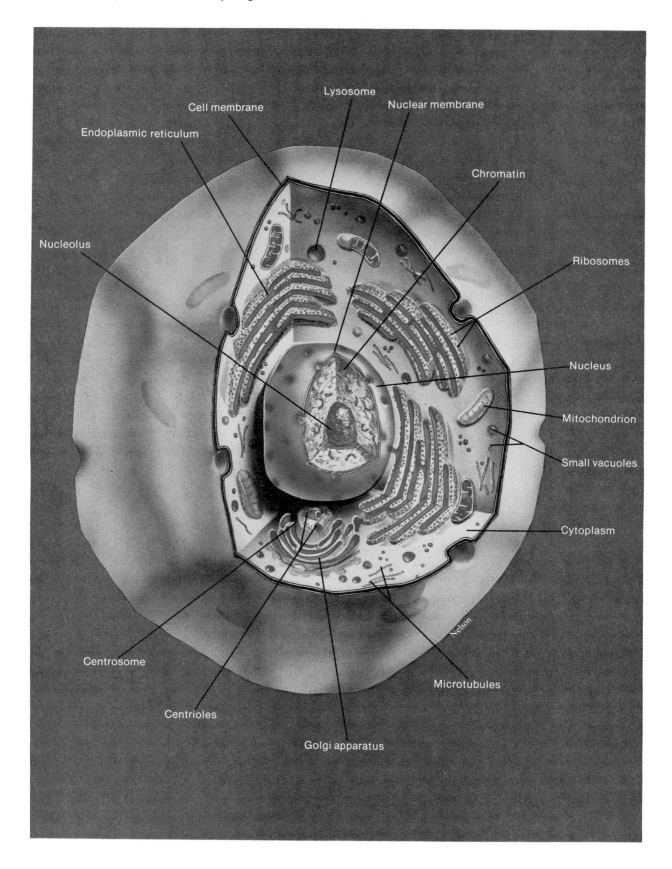

Lysosome

Cell membrane

Nuclear membrane

Endoplasmic reticulum

Chromatin

Nucleolus

Ribosomes

Nucleus

Mitochondrion

Small vacuoles

Cytoplasm

Centrosome

Microtubules

Centrioles

Golgi apparatus

Fig. 3.4 A cell membrane appears layered in an image produced by an electron microscope.

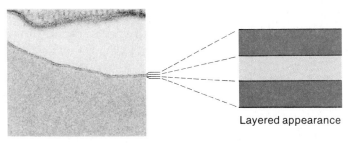

Layered appearance

Electron microscope image

Many cells, such as blood cells, are not in direct contact with their neighbors because fluid-filled space (extracellular space) separates them. In other tissues, the cells are tightly packed, and the membranes of these cells are commonly connected by *intercellular junctions*. One type of specialized junction, a *desmosome,* serves to "spot weld" adjacent skin cells so they form a structural unit. The membranes of certain other cells, such as those in heart muscle, are interconnected by *gap junctions* in the form of tubular channels. These channels link the cytoplasm of adjacent cells and allow nutrients and other molecules to be exchanged between them. (See fig. 3.6.)

Cytoplasm

When viewed through an ordinary laboratory microscope, **cytoplasm** usually appears as a clear, thick liquid with specks scattered throughout. An electron microscope (fig. 3.7), which produces much greater magnification and resolution, reveals cytoplasm to be highly structured and filled with networks of membranes and other organelles.

The activities of a cell occur largely in its cytoplasm. It is there that food molecules are received, processed, and used. In other words, cytoplasm is a site of metabolic reactions, in which the following **cytoplasmic organelles** play specific roles.

1. **Endoplasmic reticulum.** The endoplasmic reticulum is a complex network of interconnected membranes. These membranes form flattened sacs, elongated canals, and fluid-filled vesicles. The network is connected to the cell membrane, the nuclear membrane, and certain cytoplasmic organelles. Thus, it is widely distributed through the cytoplasm. It seems to function as a tubular communication system through which molecules can be transported from one cell part to another.

Fig. 3.5 The cell membrane is composed primarily of phospholipids with proteins scattered throughout the lipid layer and associated with the surfaces.

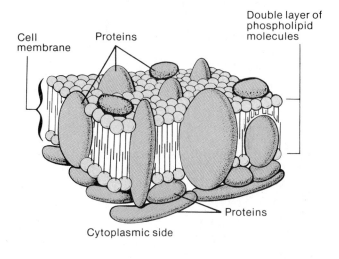

Cell membrane

Proteins

Double layer of phospholipid molecules

Proteins

Cytoplasmic side

Fig. 3.6 Some cells are joined by intercellular junctions such as desmosomes, which serve as "spot welds," or gap junctions, which allow movement between the cytoplasm of the cells.

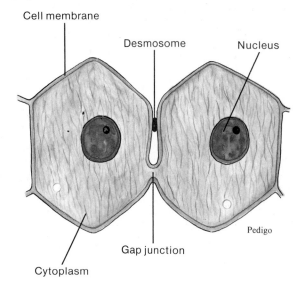

Cell membrane

Desmosome

Nucleus

Pedigo

Gap junction

Cytoplasm

The endoplasmic reticulum also functions in the synthesis of chemicals. Often a large proportion of its outer membranous surface has numerous tiny, spherical organelles called *ribosomes* attached to it, and for this reason is termed *rough* endoplasmic reticulum (fig. 3.8). The remaining surface, which lacks ribosomes, is called *smooth* endoplasmic reticulum. Ribosomes function in the synthesis of proteins, while smooth endoplasmic reticulum contains enzymes involved in manufacturing lipid molecules.

2. **Ribosomes.** Although many ribosomes are attached to membranes of the endoplasmic reticulum, others occur as free particles scattered throughout the cytoplasm. In both cases, these ribosomes are composed of protein and RNA molecules, and they function in the synthesis of protein molecules. Some of the resulting proteins are used for building cell structure, while others act as enzymes, and still others are transported to the cell membrane and secreted outside the cell.

3. **Golgi apparatus.** The Golgi apparatus is usually centrally located near the nucleus. It is composed of a group of flattened, membranous sacs whose membranes are continuous with those of the endoplasmic reticulum (fig. 3.9).

The Golgi apparatus is thought to be involved in the processing of proteins synthesized by the endoplasmic reticulum and in the "packaging" of these proteins for secretion. In some cells, for example, the Golgi apparatus seems to manufacture carbohydrate molecules that are combined with protein molecules. The resulting *glycoproteins* are then packaged in bits of the membrane from the Golgi apparatus (fig. 3.10). Then the package may move to the cell membrane where its contents are released to the outside. The Golgi apparatus may also store such packaged molecules.

Fig. 3.8 (*a*) An electron micrograph of endoplasmic reticulum magnified about 100,000 times; (*b*) rough endoplasmic reticulum has ribosomes attached to its surface, while (*c*) smooth endoplasmic reticulum lacks ribosomes.

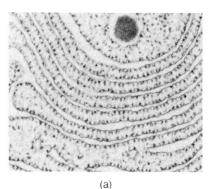

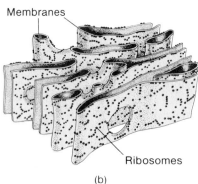

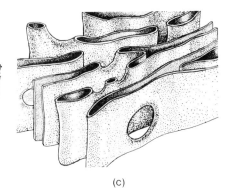

Membranes

Ribosomes

(a)

(b)

(c)

Fig. 3.9 (*a*) An electron micrograph of a Golgi apparatus magnified about 40,000 times; (*b*) the Golgi apparatus consists of membranous sacs that are continuous with the endoplasmic reticulum.

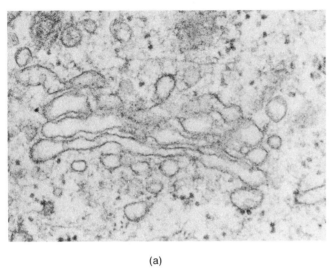

(a)

(b)

Fig. 3.10 Portions of the Golgi apparatus function in packaging glycoproteins, which are then moved to the cell membrane and released as a secretion.

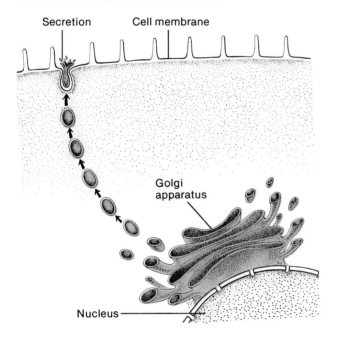

1. *What is selective permeability?*
2. *Describe the structure of a cell membrane.*
3. *What are the functions of the endoplasmic reticulum?*
4. *Describe the way in which the Golgi apparatus functions.*

4. **Mitochondria.** Mitochondria are relatively large, fluid-filled sacs (fig. 3.11). They vary in size, can change shape, and often move about in the cytoplasm.

The membrane surrounding a mitochondrion has an outer and an inner layer. The inner layer is folded to form partitions called *cristae* within the saclike structure. Small, stalked particles that contain enzymes are connected to the cristae. These enzymes control some of the chemical reactions by which energy is released from glucose and other molecules. The mitochondria also function in transforming this energy into a form that is usable by cell parts. For this reason they are sometimes called the "powerhouses" of cells. More details concerning the energy-releasing function of mitochondria can be found in chapter 4.

Fig. 3.11 (a) An electron micrograph of a mitochondrion magnified about 40,000 times. (b) What is the function of the cristae that form partitions within this saclike organelle?

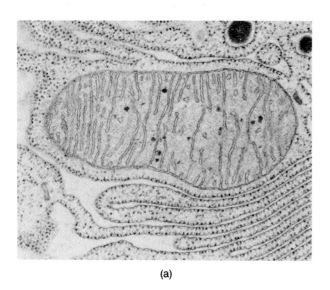

(a)

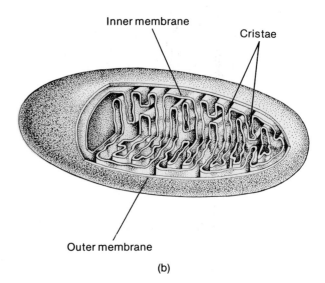

Inner membrane

Cristae

Outer membrane

(b)

5. **Lysosomes.** Lysosomes are sometimes difficult to identify because their shapes vary so greatly. However, they commonly appear as tiny, membranous sacs. Inside these sacs are powerful enzymes that are capable of breaking down molecules of protein, carbohydrates, and nucleic acids. Such enzymes function to digest various particles that enter cells. Certain white blood cells, for example, can engulf foreign substances, such as bacteria. Lysosomes then fuse with the engulfed substances, and lysosomal enzymes digest the bacteria. Consequently, white blood cells aid in preventing bacterial infections.

Lysosomes also function in the destruction of worn cell parts. In fact, sometimes they are responsible for destroying entire, injured cells that have been engulfed by scavenger cells. How the lysosomal membrane is able to withstand being digested itself is not understood.

6. **Peroxisomes.** Peroxisomes are membranous vesicles that resemble lysosomes in size and shape. They occur in most cells and contain certain enzymes, including a group called *peroxidases.* The action of the most abundant peroxidase, catalase, is discussed in chapter 4.

7. **Centrosome.** A centrosome (central body) is located in the cytoplasm near the Golgi apparatus and the nucleus. It is nonmembranous, and consists of two rodlike structures called *centrioles,* which in turn contain tiny tubelike parts, or microtubules (fig. 3.12). The centrioles lie at right angles to each other and function in cell reproduction. During this process, the centrioles move away from one another and take positions on either side of the nucleus. There they aid in the distribution of the chromosomes, which carry genetic information, to the newly forming daughter cells.

Centrioles also function in initiating the formation of tiny, moving, hairlike projections called *cilia* that extend outward from the surfaces of some cell membranes and in the formation of the taillike *flagella* of sperm cells.

8. **Vacuoles.** Vacuoles are membranous sacs, varying greatly in size. They may be formed by an action of the cell membrane in which a portion of the membrane folds inward and pinches off. As a result, a small, bubblelike vacuole, containing some liquid or solid material that was outside the cell a moment before, appears in the cytoplasm.

9. **Fibrils** and **microtubules.** Two types of thin, threadlike processes likely to be found in the cytoplasm are fibrils and microtubules. The fibrils are tiny

Plate 1 Space-filling model of a water molecule, H₂O. The white parts represent hydrogen atoms, and the red parts represent oxygen.

Plate 2 Space-filling model of a carbon dioxide molecule, CO₂. The black part represents a carbon atom.

Plate 3 (a) Space-filling model of a glucose molecule; (b) reverse side.

Plate 4 Space-filling model of a fatty acid molecule.

Plate 5 Space-filling model of the amino acid, glycine. The blue part represents a nitrogen atom.

1

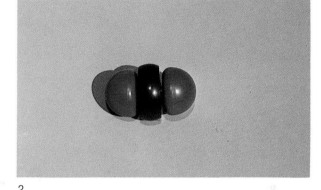

2

3(a)

(b)

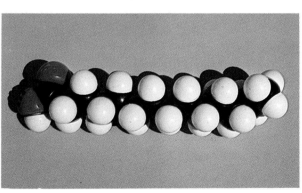

4

5

Plate 6 Space-filling model of a molecule of collagen, a protein composed of many amino acids.

Plate 7 Space-filling model of a portion of a DNA molecule.

6

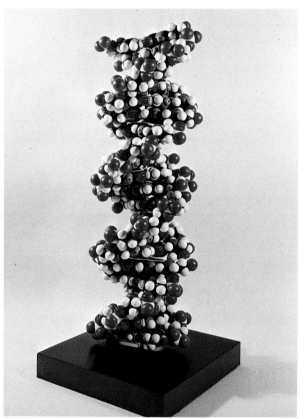

7

Fig. 3.12 (a) An electron micrograph of a centriole magnified about 120,000 times. The small circular parts that comprise this structure are microtubules. (b) A microtubule is composed of globular proteins.

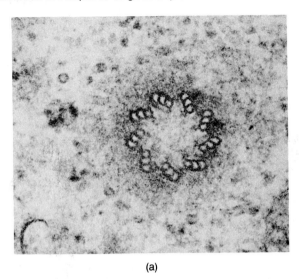

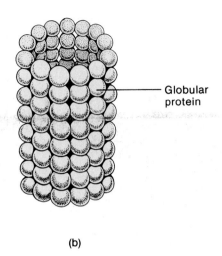

Globular protein

(a)

(b)

rods, and the microtubules are tubular structures composed of globular proteins. Both seem to function in providing support and stability to the cytoplasm and thus help to maintain the shape of the cell. In addition, fibrils, which are most highly developed in muscle cells (*myofibrils*), function in the contraction mechanism of these cells. Microtubules appear in cells during cell reproduction and aid in the distribution of chromosomes to the newly forming cells. They may also function to direct the flow of cytoplasm in cells that move about and may be involved in transporting secretory products to the cellular surfaces. (See fig. 3.13.)

In addition to these functional organelles, cytoplasm may also contain masses of lifeless chemical substances called *inclusions*. Most commonly, inclusions consist of cellular products such as stored nutrients, pigments, and substances that are being prepared for secretion.

1. Why are mitochondria sometimes called "powerhouses" of cells?
2. How do lysosomes within certain white blood cells function?
3. Name two functions of fibrils and microtubules.
4. Distinguish between organelles and inclusions.

Fig. 3.13 Microtubules, such as those shown in this electron micrograph, aid in the distribution of chromosomes during cellular reproduction.

Fig. 3.14 (a) An electron micrograph of a cell nucleus magnified about 20,000 times. It contains a nucleolus and masses of chromatin. (b) The nuclear membrane is porous and allows substances to pass between the nucleus and the cytoplasm.

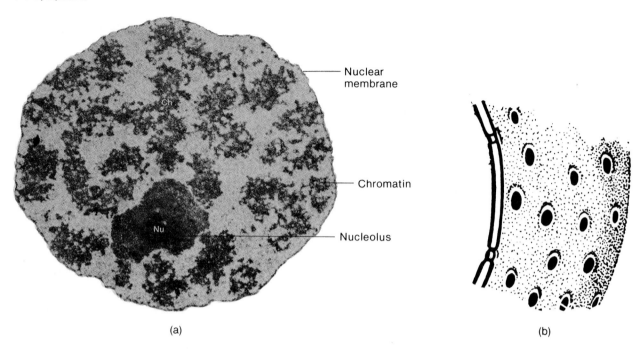

Nuclear membrane

Chromatin

Nucleolus

(a)

(b)

Cell Nucleus

A **cell nucleus** (fig. 3.14) is usually located near the center of the cytoplasm. It is a relatively large, spherical structure and functions to direct the activities of the cell.

The nucleus is enclosed in a **nuclear membrane,** which is porous and allows substances to move between the nucleus and the cytoplasm. The nucleus contains a fluid (nucleoplasm) in which other structures float. These structures include the following:

1. **Nucleolus.** A nucleolus, or little nucleus, is a small, dense body composed largely of RNA and protein. It has no surrounding membrane and is thought to function in the production of ribosomes. Once the ribosomes are formed, they seem to migrate through the pores in the nuclear membrane and enter the cytoplasm. The nucleus of a cell that synthesizes large amounts of protein may contain several nucleoli.

2. **Chromatin.** Chromatin consists of loosely coiled fibers that are present in the nuclear fluid. When the cell begins to undergo the reproductive process, these fibers become more tightly coiled and are transformed into tiny, rodlike *chromosomes.* Chromatin fibers are composed of protein and DNA molecules, which in turn are organized into tiny, beadlike

particles called *nucleosomes.* The DNA molecules contain the information that directs the cell in carrying out its life processes.

Chart 3.1 summarizes the structures and functions of the cellular organelles.

Aging of cells. Although the ways cells are influenced by aging are poorly understood, studies of aged tissues indicate that certain organelles may be altered with age. For example, within the nucleus, the chromatin and chromosomes may show changes such as clumping, shrinking, or fragmenting, and the number and size of the nucleoli may increase. In the cytoplasm, the mitochondria may undergo changes in shape and number, and the Golgi apparatus may become fragmented. Lipid inclusions tend to accumulate in the cytoplasm, while those containing glycogen disappear.

1. *How are the nuclear contents separated from the cytoplasm of a cell?*
2. *What is the function of the nucleolus?*
3. *What is chromatin?*

Chart 3.1 Structure and function of cellular organelles

Organelle	Structure	Function
Cell membrane	Membrane composed of protein and lipid molecules.	Maintains wholeness of cell and controls passage of materials in and out of cell.
Endoplasmic reticulum	Network of interconnected membrane forming sacs and canals.	Transports materials within the cell, provides attachment for ribosomes, and synthesizes lipids.
Ribosomes	Particles composed of protein and RNA molecules.	Synthesize proteins.
Golgi apparatus	Group of flattened, membranous sacs.	Synthesizes carbohydrates and packages protein molecules for secretion.
Mitochondria	Membranous sacs with inner partitions.	Release energy from food molecules and transform energy into usable form.
Lysosomes	Membranous sacs.	Digest molecules of foreign substances and of worn and damaged cells.
Peroxisomes	Membranous vesicles.	Contain certain enzymes.
Centrosome	Nonmembranous structure composed of two rodlike centrioles.	Helps distribute chromosomes to daughter cells during mitosis and initiates formation of cilia.
Vacuoles	Membranous sacs.	Contain various substances that recently entered the cell.
Fibrils and microtubules	Thin rods and tubules.	Provide support to cytoplasm and help move objects within the cytoplasm.
Nuclear membrane	Porous membrane that separates nuclear contents from cytoplasm.	Maintains wholeness of the nucleus and controls passage of materials between nucleus and cytoplasm.
Nucleolus	Dense, nonmembranous body composed of protein and RNA molecules.	Forms ribosomes.
Chromatin	Fibers composed of protein and DNA molecules.	Contains cellular information for carrying on life processes.

Movements through Cell Membranes

The cell membrane provides a surface through which substances enter and leave the cell. More specifically, oxygen and food molecules enter a cell through this membrane, while carbon dioxide and other wastes leave through it. These movements involve physical (or nonliving) processes such as diffusion, osmosis, and filtration, and physiological (or living) mechanisms such as active transport, phagocytosis, and pinocytosis.

Diffusion

Diffusion is the process by which molecules or ions become scattered or are spread from regions where they are in higher concentrations toward regions where they are in lower concentrations. As a rule, this phenomenon involves the movement of molecules or ions in gases or liquids.

Actually, molecules in gases and molecules and ions in body fluids are constantly moving at high speeds. Each of these particles travels in a separate path along a straight line until it collides and bounces off some other particle. Then it moves in another direction, only to collide again and change direction once more. Such motion is haphazard, but it accounts for the mixing of molecules that commonly occurs when different kinds of substances are put together.

For example, when some sugar is put into a glass of water, as illustrated in figure 3.15, the sugar will seem to remain at the bottom for a while. Then it slowly disappears into solution. As this happens, the moving water and sugar molecules are colliding haphazardly with one another, and in time the sugar and water molecules will be evenly mixed. This mixing occurs by diffusion—the sugar molecules spread from where they are in higher concentration toward the regions where they are less concentrated. Eventually the sugar becomes uniformly distributed in the water. This condition is called *equilibrium*.

To better understand how diffusion accounts for the movement of various molecules through a cell membrane, imagine a container of water that is separated into two compartments by a permeable membrane (fig. 3.16). This membrane has numerous pores that are large enough for water and sugar molecules to pass through. Sugar molecules are placed in one

Fig. 3.15 (a, b, c) If a sugar cube is placed in water, it slowly disappears. As this happens, sugar and water molecules collide haphazardly with one another, and the sugar molecules diffuse from regions where they are more concentrated toward regions where they are less concentrated. (d) Eventually, the sugar molecules are evenly distributed throughout the water, and a state of equilibrium is achieved.

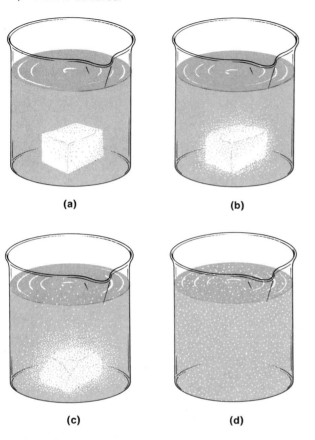

(a)   (b)

(c)   (d)

compartment (A) but not in the other (B). Although the sugar molecules will move in all directions, more will diffuse from compartment A (where they are in greater concentration) through the pores in the membrane and into compartment B (where they are in lesser concentration) than will move in the other direction. At the same time, water molecules will tend to diffuse from compartment B (where they are in greater concentration) through the pores into compartment A (where they are in lesser concentration). Eventually, equilibrium will be achieved when there are equal numbers of water and of sugar molecules in each compartment.

Similarly, oxygen molecules diffuse through cell membranes and enter cells if these molecules are more highly concentrated on the outside than on the inside. Carbon dioxide molecules, too, diffuse through cell membranes and leave cells if they are more concentrated on the inside than on the outside. It is by diffusion that oxygen and carbon dioxide molecules are exchanged between the air and the blood in the lungs and between the blood and the cells of various tissues.

Since the cell membrane is composed primarily of lipids, molecules that are soluble in lipids, such as oxygen, carbon dioxide, and fatty acids, can diffuse readily through it. Substances that are not soluble in lipids, such as glucose, cannot pass through in this manner, but must be carried through by special carrier molecules. In this process, called *facilitated diffusion,* a glucose molecule must combine with a carrier molecule at the surface of the membrane. The resulting compound is soluble in lipid and can diffuse through the membrane. On the other side, the glucose is released from the carrier, and the carrier can return

Fig. 3.16 (1) The container is separated into two compartments by a permeable membrane. Compartment A contains water and sugar molecules, while compartment B contains only water molecules. (2) As a result of molecular motions over time, sugar molecules will tend to diffuse from compartment A (where they are in higher concentration) to compartment B. Water molecules will tend to diffuse from compartment B (where they are in higher concentration) to compartment A. (3) Eventually, equilibrium will be achieved.

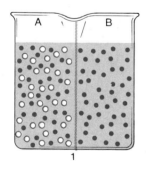

1

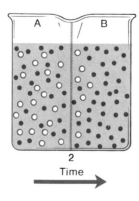

2

Time

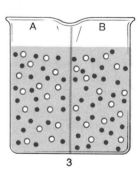

3

to the other side of the membrane to pick up another glucose molecule. (See fig. 3.17.)

Facilitated diffusion is similar to simple diffusion in that it can only cause movement of molecules from regions of higher concentration toward regions of lower concentration. The rate at which facilitated diffusion can occur, however, is limited by the number of carrier molecules in the cell membrane.

Diffusion can be used to separate smaller molecules from larger ones in a liquid. This process, called *dialysis,* is employed in the artificial kidney (fig. 3.18). When this device is operating, blood is passed through a long, coiled tubing composed of porous cellophane. The size of the pores allows smaller molecules in blood, such as urea, to pass out through the cellophane; while larger molecules, like those of blood proteins, remain inside. The tubing is submerged in a tank of wash solution, which contains varying concentrations of chemicals. The solution is low, for instance, in concentrations of substances that should leave the blood and higher in concentrations of those that should remain in the blood.

Since it is desirable for an artificial kidney to remove blood urea, the wash solution must have a lower concentration of urea than the blood does; it is also desirable to maintain the blood glucose concentration, so the concentration of glucose in the wash solution must be kept at least equal to that of the blood. Thus, by altering the concentrations of molecules in the wash solution, it is possible to control those that will diffuse out of the blood and those that will remain in it.

Osmosis

Osmosis is a special kind of diffusion. It occurs whenever *water* molecules diffuse from a region of higher concentration to a region of lower concentration through a selectively permeable membrane, such as a cell membrane.

Ordinarily, the concentrations of water molecules are equal on either side of a cell membrane. Sometimes, however, the water on one side has more chemicals dissolved in it than the water on the other side. For example, if there is a greater concentration of glucose in the water outside a cell, there must be a lesser concentration of water there, because the glucose molecules occupy space that would otherwise contain water molecules. Under such conditions, water molecules diffuse from inside the cell (where they are in higher concentration) toward the outside (where they are in lesser concentration).

Fig. 3.17 Some substances are moved through the cell membrane by facilitated diffusion in which carrier molecules transport the substance from a region of higher concentration to one of lower concentration.

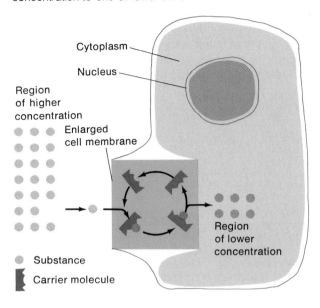

Fig. 3.18 An artificial kidney uses dialysis to separate smaller molecules of waste substances like urea from larger molecules in blood.

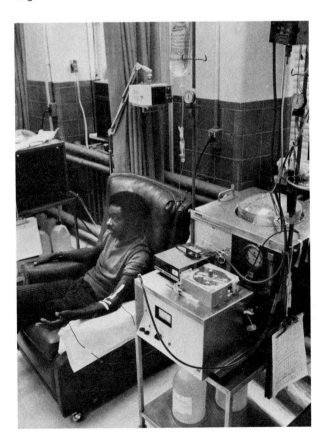

The Cell   53

This process, shown in figure 3.19, is similar to the one in figure 3.16, illustrating the diffusion of water and glucose molecules through a permeable membrane. In the case of osmosis, however, the membrane is selectively permeable—water molecules pass through readily, but glucose molecules do not.

Note in figure 3.19 that, as osmosis occurs, the volume of water on side *A* increases. This increase in volume would be resisted if pressure were applied to the surface of the liquid on side *A*. The amount of pressure needed to stop osmosis in such a case is called *osmotic pressure.* Thus, the osmotic pressure of a solution is a potential pressure and is due to the presence of nondiffusable solute particles in that solution.

When osmosis occurs, water tends to move toward the region of greater osmotic pressure. Since the amount of osmotic pressure depends upon the difference in concentration of dissolved particles on opposite sides of the membrane, the greater the difference, the greater the tendency for water to move toward the region of higher solute concentration.

If some cells were put into a water solution that had a greater concentration of dissolved particles (higher osmotic pressure) than did the cells, the cells would begin to shrink. In time equilibrium might be achieved, and the shrinking would cease. Solutions of this type, in which more water leaves a cell than enters it because the concentration of dissolved particles is greater outside the cell, are called **hypertonic.**

Conversely, if there is a greater concentration of dissolved particles inside a cell than in the water around it, water will diffuse into the cell. As this happens, water accumulates within the cell, and it begins to swell. Although cell membranes are somewhat elastic, they may swell so much that they burst. Solutions in which more water enters a cell than leaves it because of a lesser concentration of dissolved particles outside the cell are called **hypotonic.**

A solution that contains the same concentration of dissolved particles as a cell is said to be **isotonic** to that cell. Figure 3.20 illustrates these three types of solutions.

Fig. 3.19 (1) The container is separated into two compartments by a selectively permeable membrane. At first, compartment *A* contains water and sugar molecules, while compartment *B* contains only water molecules. As a result of molecular motions, water molecules will tend to diffuse by osmosis from compartment *B* into compartment *A*. The sugar molecules will remain in compartment *A* because they are too large to pass through the pores of the membrane. (2) Since more water is entering compartment *A* than is leaving it, water will accumulate in this compartment, and the level of the liquid will rise.

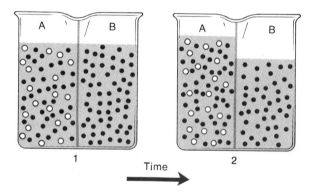

Time

It is important to control the concentration of solute in solutions that are infused into body tissues or blood. Otherwise, osmosis may cause cells to swell or shrink, and they may be damaged. For instance, if red blood cells are exposed to distilled water (which is hypotonic to them), water will diffuse into the cells and they will burst (hemolyze). On the other hand, if red blood cells are exposed to 0.9% NaCl solution (normal saline), the cells will remain unchanged because this solution is isotonic to human cells.

Fig. 3.20 (*a*) If a red blood cell is placed in a hypertonic solution, more water leaves the cell than enters it, and it shrinks; (*b*) in a hypotonic solution, more water enters than leaves the cell, and it swells and may burst; (*c*) in an isotonic solution, water enters and leaves the cell in equal amounts, and its size remains unchanged.

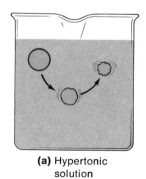

(a) Hypertonic solution

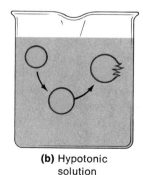

(b) Hypotonic solution

(c) Isotonic solution

Filtration

The passage of substances through membranes by diffusion or osmosis is due to movements of the molecules of those substances. In other instances, molecules are forced through membranes by hydrostatic pressure that is greater on one side of the membrane than on the other. This process is called **filtration**.

Filtration is commonly used to separate solids from a liquid. One method is to pour a mixture onto some filter paper in a funnel (fig. 3.21). The paper serves as a porous membrane through which the liquid can pass, leaving the solids behind. The force of gravity pulls the liquid through the paper membrane.

Similarly, in the body, tissue fluid is formed when water and dissolved substances are forced out through the membranes of blood capillaries, while larger particles such as blood protein molecules are left inside (fig. 3.22). The force for this movement comes from blood pressure, which is greater within the vessel than outside. Filtration also takes place in the kidneys, when water and dissolved wastes are forced out of blood vessels and into kidney tubules. This is the first step in the formation of urine, and once again the force for filtration is supplied by blood pressure.

1. What kinds of substances move most readily through a cell membrane by diffusion?
2. Explain the differences between diffusion, facilitated diffusion, and osmosis.
3. How is osmotic pressure related to hypertonic, isotonic, and hypotonic solutions?
4. Explain how filtration occurs within the body.

Active Transport

When molecules or ions pass through cell membranes by diffusion or osmosis, their net movement is from regions of higher concentration to regions of lower concentration. Sometimes, however, the net movement of particles passing through membranes is in the opposite direction; that is, they move from a region of lower concentration to one of higher concentration.

It is known, for example, that sodium ions can diffuse through cell membranes. Yet the concentration of these ions typically remains many times greater on the outside of cells (in the extracellular fluid) than on the inside (in the intracellular fluid).

Fig. 3.21 In this example of filtration, gravity provides the force that pulls water through filter paper, while the tiny openings in the paper hold the solids behind.

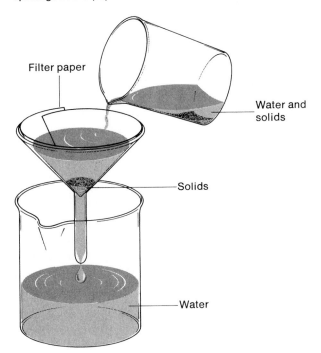

Filter paper

Water and solids

Solids

Water

Fig. 3.22 In this example of filtration, smaller molecules are forced through the membranes of a capillary by blood pressure, while larger molecules remain behind.

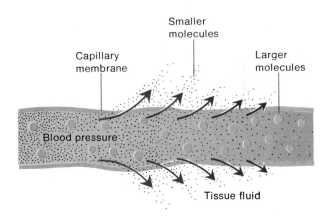

Smaller molecules

Larger molecules

Capillary membrane

Blood pressure

Tissue fluid

Fig. 3.23 What process is illustrated in this diagram? What do the items labeled *C* represent?

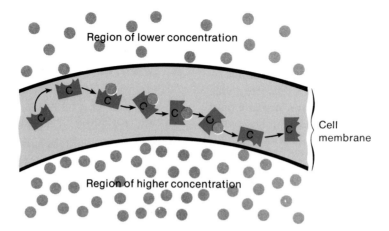

Region of lower concentration

Cell membrane

Region of higher concentration

Furthermore, sodium ions are continually moved from the regions of lower concentration (inside) to regions of higher concentration (outside). Movement of this type is called **active transport.** It depends on life processes within cells and requires energy. In fact, it is estimated that up to 40% of a cell's energy supply may be used for active transport of particles through its membranes.

Although the details are poorly understood, the mechanism of active transport is similar to facilitated diffusion in that it involves specific carrier molecules, probably protein molecules found within cell membranes. As figure 3.23 shows, these carrier molecules combine at the membrane's surface with the particles being transported. Then they move through the membrane with their "passengers" attached. Once on the other side, the transported molecules are released, and the carriers return for another load.

Particles that are carried across cell membranes by active transport include various sugars and amino acids as well as a variety of ions such as those of sodium, potassium, calcium, and hydrogen. Movements of this type are important to cell survival and are involved with the maintenance of homeostasis.

Pinocytosis

The word **pinocytosis,** which means *cell drinking,* refers to the process by which cells take in tiny droplets of liquid from their surroundings (fig. 3.24). When this happens, a small portion of a cell membrane becomes indented. The open end of the tubelike part thus formed soon seals off, and the result is a small vacuole that becomes detached from the surface and moves into the cytoplasm.

For a time, the vacuolar membrane, which was part of the cell membrane, separates its contents from the rest of the cell; but eventually the membrane breaks down and the liquid inside becomes part of the cytoplasm. In this way, a cell is able to take in water and dissolved particles that otherwise might be unable to pass through the cell membrane because of large molecular size.

Phagocytosis

Phagocytosis (*cell eating*) describes a process that is essentially the same as pinocytosis. In phagocytosis, however, the material taken into the cell is solid rather than liquid.

Certain kinds of cells, including some white blood cells, are called *phagocytes* because they can take in tiny solid particles such as bacterial cells. When a phagocyte first encounters such a particle, the object becomes attached to its cell membrane. Portions of the phagocytic cell project outward, surround the object, and slowly draw it inside. The part of the cell membrane surrounding the solid detaches from the surface, and a vacuole containing the particle is formed (fig. 3.25).

Commonly, a lysosome then combines with such a vacuole, and its digestive enzymes cause the enclosed solid to be decomposed (fig. 3.26). In this way cells can destroy bacteria that might otherwise cause infections, dispose of foreign objects like dust particles, or remove damaged cells or cell parts that are no longer functional.

Fig. 3.24 A cell may take in a tiny droplet of fluid from its surroundings by pinocytosis.

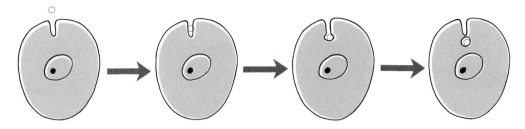

Fig. 3.25 A cell may take in a solid particle from its surroundings by phagocytosis.

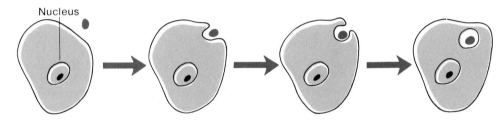

Fig. 3.26 When a lysosome combines with a vacuole containing a phagocytized particle, its digestive enzymes destroy the particle.

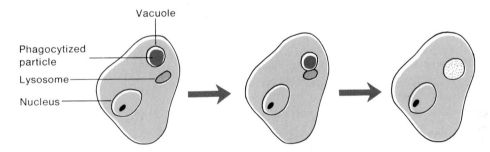

Both pinocytosis and phagocytosis depend upon life processes and require the use of cellular energy. Although pinocytosis is thought to provide only a minor route for substances entering cells, phagocytosis is an important line of defense against invasion by disease-causing microorganisms.

Chart 3.2 summarizes the various types of movement through cell membranes.

1. What type of mechanism is responsible for maintaining unequal concentrations of electrolytes on opposite sides of a cell membrane?
2. What factors are necessary for the action of this mechanism?
3. What is the difference between pinocytosis and phagocytosis?

Chart 3.2 Movements through cell membranes

	Characteristics	Source of Energy	Example
Physical Processes			
Diffusion	Molecules or ions move from regions of higher concentration toward regions of lower concentration.	Molecular motion	Exchange of gases in lungs.
Facilitated diffusion	Molecules are moved through a membrane from a region of higher concentration to one of lower concentration by carrier molecules.	Molecular motion	Movement of glucose through a cell membrane.
Osmosis	Water molecules move from regions of higher concentration to regions of lower concentration through a selectively permeable membrane.	Molecular motion	Water enters a cell placed in distilled water.
Filtration	Molecules are forced from regions of higher pressure to regions of lower pressure.	Blood pressure	Molecules leave blood capillaries.
Physiological Processes			
Active transport	Molecules or ions are carried through membranes by other molecules.	Cellular energy	Movement of various ions and amino acids through membranes.
Pinocytosis	Membrane acts to engulf minute droplets of liquid from surroundings.	Cellular energy	Membrane forms vacuole containing liquid and dissolved particles.
Phagocytosis	Membrane acts to engulf solid particles from surroundings.	Cellular energy	White blood cell membrane engulfs bacterial cell.

Life Cycle of a Cell

The series of changes that a cell undergoes from the time it is formed until it reproduces is called its *life cycle* (fig. 3.27). Superficially, this cycle seems rather simple—a newly formed cell grows for a time and then splits in half to form two daughter cells. Yet the details of the cycle are quite complex. Growth, for example, requires that the young cell obtain nutrients from its surroundings, utilize them in the manufacture of new living materials, and synthesize many vital compounds.

Mitosis

Cell reproduction involves the dividing of a cell into two portions and includes two separate processes: (1) division of the nuclear parts, which is called **mitosis** (karyokinesis); and (2) division of the cytoplasm (cytokinesis).

Division of the nuclear parts by mitosis is, of necessity, very precise, since the nucleus contains information, in the form of DNA molecules, that "tells" cell parts how to function. Each daughter cell must have a copy of this information in order to survive.

Fig. 3.27 Life cycle of a cell.

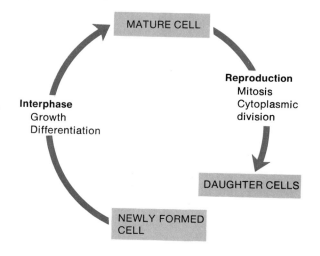

Thus, the DNA molecules of the parent cell must be duplicated, and the duplicate sets must be distributed equally to the daughter cells. Once this has been accomplished, the cytoplasm and its parts can be divided.

Although mitosis is often described in terms of stages, the process is really a continuous one without marked changes between one step and the next. (See fig. 3.28.) The idea of stages is useful, however, to indicate the sequence in which major events occur.

Fig. 3.28 Mitosis is a continuous process during which the nuclear parts of a cell are divided into two equal portions. After reading about mitosis, identify the phases of the process in this diagram.

These stages include the following:

1. **Prophase.** One of the first indications that a cell is going to reproduce is the appearance of *chromosomes*. These structures form from chromatin in the nucleus as chromatin threads condense into tightly coiled, rodlike parts. The resulting chromosomes contain DNA and protein molecules. Sometime earlier (during interphase, which is described on page 61) the DNA molecules became duplicated, and each chromosome is consequently composed of two identical portions (chromatids), which temporarily remain fastened together.

The centrioles of the centrosome move apart and travel to opposite sides of the cytoplasm. Soon the nuclear membrane and the nucleolus disappear. Microtubules (spindle fibers) form in the cytoplasm, and they become associated with the centrioles and chromosomes. (See fig. 3.29.)

2. **Metaphase.** The chromosomes become lined up in an orderly fashion about midway between the centrioles, apparently as a result of microtubule activity. Microtubules also are attached to the chromosome parts so that a microtubule accompanying one centriole is attached to one part of a chromosome, while a microtubule accompanying the other centriole is attached to the other part. Soon the identical chromosome parts, which have been fastened together, become separated as individual chromosomes. (See fig. 3.30.)

3. **Anaphase.** The separated chromosomes now move in opposite directions, and once again, the movement seems to result from microtubule activity. In this case, the microtubules appear to shorten and pull their attached chromosomes toward the centrioles at opposite sides of the cell. (See fig. 3.31.)

4. **Telophase.** This final stage of mitosis is said to begin when the chromosomes complete their migration toward the centrioles. It is much like prophase, but in reverse. As the chromosomes approach the centrioles, they begin to elongate and change from rodlike into threadlike structures. A nuclear membrane forms around each chromosome set, and nucleoli appear within the newly formed nuclei. Finally, the microtubules disappear and the centrioles become duplicated. (See fig. 3.32.) Telophase is accompanied by cytoplasmic division in which the cytoplasm is split into two portions.

Chart 3.3 summarizes the events that take place during these phases in mitosis.

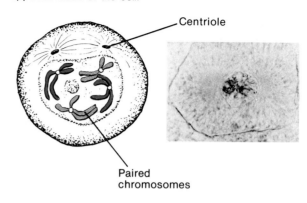

Fig. 3.29 In prophase, paired chromosomes form from chromatin in the nucleus, and the centrioles move to opposite sides of the cell.

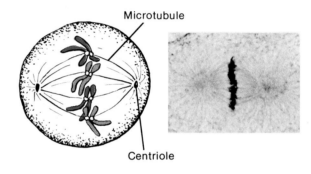

Fig. 3.30 In metaphase, the chromosomes become lined up midway between the centrioles, and the duplicated parts of the chromosomes are separated.

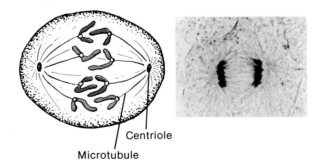

Fig. 3.31 In anaphase, microtubules that have become attached to the chromosome parts pull these parts toward the centrioles.

Fig. 3.32 In telophase, the chromosomes elongate and become chromatin threads, and the cytoplasm begins to divide.

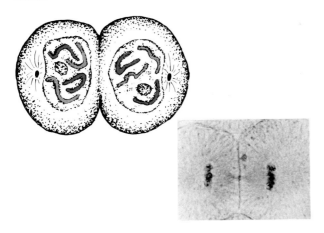

Fig. 3.33 Following mitosis, the cytoplasm of the cells is divided into two portions. How are the resulting cells similar to the parent cell? How are they different?

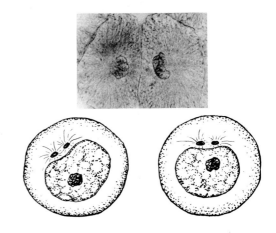

Chart 3.3	Major events in mitosis
Stage	**Major Events**
Prophase	Chromatin forms chromosomes, centrioles move to opposite sides of cytoplasm, nuclear membrane and nucleolus disappear, microtubules appear and become associated with centrioles and duplicate parts of chromosomes.
Metaphase	Chromosomes become arranged midway between the centrioles, microtubules become attached to duplicate parts of chromosomes, duplicate parts of chromosomes become separated.
Anaphase	Microtubules shorten and pull individual chromosomes toward centrioles.
Telophase	Chromosomes elongate and form chromatin threads, nuclear membranes appear around each chromosome set, nucleoli appear, microtubules disappear.

Cytoplasmic Division

The mechanism responsible for cytoplasmic division (cytokinesis) is not well understood, but the cell membrane begins to constrict during telophase and then pinches inward between the daughter nuclei. As a result, about half the cytoplasmic organelles are distributed to each new cell.

Although the newly formed cells may differ slightly in size and number of cytoplasmic parts, they have identical chromosomes and thus contain identical DNA information. Except for size, they are copies of the parent cell. (See fig. 3.33.)

Interphase

Once formed, the daughter cells usually begin growing. At the same time, various cell parts become duplicated. In the nucleus, the chromosomes may be doubling (at least in cells that will soon reproduce); and in the cytoplasm, new ribosomes, lysosomes, mitochondria, and membranes are appearing. This stage in the life cycle is called **interphase**, and it lasts until the cell begins to undergo mitosis.

Many kinds of cells within the body are constantly growing and reproducing. Such activity is responsible for the growth and development of an embryo into a child, and a child into an adult. It is also responsible for the replacement of cells that are worn out or lost due to injury or disease.

Cell Differentiation

Since all body cells are formed by mitosis and contain the same DNA information, it might be expected that they look and act alike; obviously, they do not.

A human begins life as a single cell—a fertilized egg cell. This cell reproduces to form 2 daughter cells; they in turn divide into 4 cells, the 4 become 8, and so forth. Then, sometime during development, the cells begin to *specialize*. That is, they develop special structures or begin to function in different ways. Some become skin cells, others become bone cells, and others become nerve cells. (See fig. 3.34.)

Fig. 3.34 During development, the numerous body cells are produced from a single fertilized egg cell by mitosis. As these cells undergo differentiation, they become different kinds of cells that carry on specialized functions.

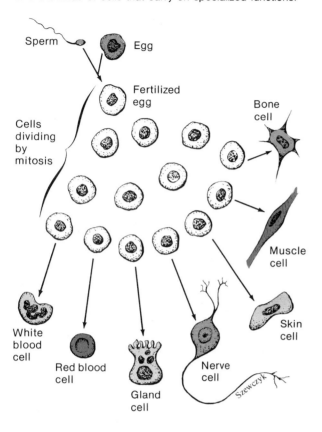

Sperm
Egg
Fertilized egg
Cells dividing by mitosis
Bone cell
Muscle cell
Skin cell
White blood cell
Red blood cell
Gland cell
Nerve cell

The process by which cells develop different characteristics in structure and function is called **differentiation.** The mechanism responsible for this phenomenon is not well understood, but it may involve the repression of some of the DNA information. Thus, the DNA information needed for general cell activities may be "switched on" in both nerve and bone cells. The information related to specific bone cell functions, however, may be repressed or "switched off" in the nerve cells. Similarly, the information related to specific nerve cell functions would be repressed in bone cells.

Although the mechanism of differentiation is obscure, the results are obvious. Cells of many kinds are produced, and each kind carries on specialized functions: skin cells protect underlying tissues, red blood cells carry oxygen, nerve cells transmit impulses. Each type of cell somehow helps the others, and aids in the survival of the organism.

1. Why is it important that the division of nuclear materials during mitosis be so precise?
2. Describe the events that occur during mitosis.
3. Name the process by which some cells become muscle cells and others become nerve cells.
4. How does DNA seem to control this process?

Control of Cell Reproduction

The existence of a control mechanism for cell reproduction is evident in the fact that some cells reproduce continually, some occasionally, and some not at all.

Skin cells, blood-forming cells, and the cells that line the intestine, for example, reproduce continually throughout life. Those that compose some organs, such as the liver, seem to reproduce until a particular number of cells are present, and then they cease reproducing. Interestingly, if the number of liver cells is reduced by injury or surgery, the remaining cells are somehow stimulated to reproduce again. Still other cells, such as nerve cells, apparently lose their ability to reproduce as they become differentiated; therefore, damage to nerve cells is likely to result in permanent loss of nerve function.

How the reproductive capacities of cells are controlled is not well understood, but the mechanism may involve the release of *growth-inhibiting substances.* Such a substance might slow or stop the growth and reproduction of particular cells when their numbers reach a certain level.

Surface-Volume Relationships. Another factor involved with the control of cell reproduction is the relationship between a cell's membrane surface area and its volume. The quantity of nutrients needed to maintain a cell is directly related to the volume of its living material. The quantity of nutrients that can enter the cell is related directly to the surface area of its membrane. As a cell grows, however, its membrane surface area increases proportionately less than its volume. Consequently, the surface area will eventually become inadequate for the needs of the living material inside.

A cell can solve this growth problem by dividing. The resulting daughter cells are smaller than the parent cell and thus have a more favorable surface-volume relationship (fig. 3.35).

Fig. 3.35 When a cell divides, the amount of surface exposed to the surroundings is increased.

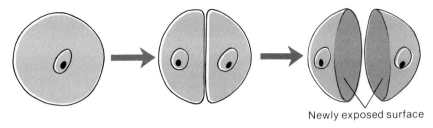

Newly exposed surface

Fig. 3.36 When normal cells (a) become cancerous, (b) they give rise to large cell populations and develop abnormal structural characteristics. Note how the cancerous cells have become disorganized.

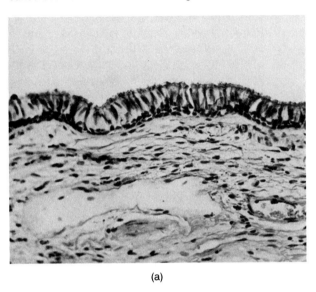

(a)

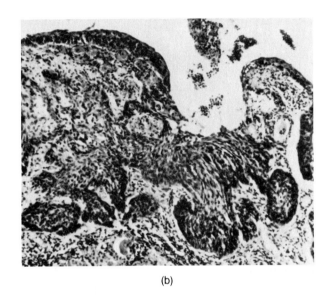

(b)

Cancer

Cancers are thought to be a group of closely related diseases, and as such all cancers have certain characteristics in common. These include the following:

1. **Hyperplasia.** Hyperplasia is the uncontrolled reproduction of cells. Although the rate of reproduction among cancer cells is usually unchanged, they are not responsive to normal controls on cell numbers. As a result, cancer cells eventually give rise to large cell populations. (See fig. 3.36.)

2. **Anaplasia.** The word anaplasia is used to describe the appearance of abnormalities in cellular structure. Typically, cancer cells resemble undifferentiated or primitive cells. That is, they fail to develop the specialized structure of the kind of cell they represent. Also, they fail to function in expected ways.

Certain white blood cells, for example, normally function in resisting bacterial infections. When cells of this type become cancerous, they do not function effectively, and the cancer victim becomes more subject to infectious diseases. Cancer cells also are likely to form jumbled masses rather than becoming arranged in orderly groups like normal cells.

3. **Metastasis.** Metastasis has a tendency to spread. Normal cells are usually cohesive; that is, they stick together in groups of similar kinds. Cancer cells often become detached from their cellular mass and may then be carried away from their place of origin and may establish new cancerous growths in other parts of the body. Metastasis is the characteristic most closely associated with the word *malignant*, which suggests a power to threaten life, and is often used to distinguish a cancerous growth from a noncancerous (benign) one. (See fig. 3.36.)

The Cell

63

The cause or causes of cancers remain largely unknown. There is, however, increasing evidence that the cause or causes are complex. Among the factors that may be involved are exposure to various chemicals or to harmful radiation, virus infections, changes in DNA structure, or deficiencies in the body's ability to resist disease. Actually, two or more of these factors may be needed to cause a cancer, or perhaps different kinds of cancers are caused by different factors or combinations of factors.

When untreated, cancer cells eventually accumulate in large numbers. They may damage normal cells by effectively competing with them for nutrients. At the same time, cancer cells may invade vital organs and interfere with their functions, obstruct important passageways, or penetrate blood vessels and cause internal bleeding.

1. *What factors seem to control the rate at which cells reproduce?*
2. *In what ways are cancer cells different from normal cells?*
3. *What are the expected consequences of an untreated cancer?*

Chapter Summary

Introduction
Cells vary in size and shape.
The shapes of cells are closely related to their functions.

A Composite Cell
1. A cell includes a nucleus, cytoplasm, and a cell membrane.
2. Cytoplasmic organelles perform specific vital functions.
3. Cell membrane
   a. Forms the outermost limit of the living material.
   b. Acts as a selectively permeable passageway that controls the movements of substances between the cell and its surroundings.
   c. Composed of protein, phospholipid, and carbohydrate molecules that have special functions.
   d. Inner and outer surfaces have different structure and functions.
   e. Some substances pass through the lipid portion; others pass through protein "pores."
   f. Some cells are connected by means of membrane junctions.

4. Cytoplasm
   a. Endoplasmic reticulum forms a network of membranes that provides a tubular communication system in the cytoplasm and an attachment for ribosomes; also functions in the synthesis of lipids.
   b. Ribosomes are particles of protein and RNA that function in protein synthesis.
   c. Golgi apparatus is composed of a group of flattened, membranous sacs that manufacture carbohydrates and package glycoproteins for secretion.
   d. Mitochondria are membranous sacs that contain enzymes involved with releasing energy from food molecules and transforming energy into a usable form.
   e. Lysosomes are membranous sacs containing digestive enzymes that can destroy foreign substances and worn or damaged cell parts.
   f. Peroxisomes are membranous vesicles that contain certain enzymes.
   g. The centrosome is a nonmembranous structure that aids in the distribution of chromosomes during cell reproduction and initiates cilia formation.
   h. Vacuoles are membranous sacs that contain substances that recently entered the cell.
   i. Fibrils and microtubules are threadlike processes that provide support and stability to cytoplasm, act in muscle contraction, and aid in the reproduction of cells.
   j. Cytoplasm may also contain nonliving cellular products called inclusions.
5. Cell nucleus
   a. Enclosed in a nuclear membrane that controls the movement of substances between the nucleus and cytoplasm.
   b. A nucleolus is a dense body of protein and RNA that functions in the production of ribosomes.
   c. Chromatin is composed of loosely coiled fibers of protein and DNA that is organized into nucleosomes and becomes chromosomes during cell reproduction.

Movements through Cell Membranes
Movement may involve physical or physiological processes.
1. Diffusion
   a. A scattering of molecules or ions from regions of higher toward regions of lower concentration.
   b. Responsible for exchanges of oxygen and carbon dioxide within the body.
   c. Some substances move through membranes by facilitated diffusion that involves the use of carrier molecules.
   d. A form of diffusion called dialysis is used to separate wastes from blood in an artificial kidney.

2. Osmosis
   a. A form of diffusion in which water molecules diffuse from regions of higher to lower concentration through a selectively permeable membrane.
   b. Osmotic pressure is due to the number of particles dissolved in a solution.
   c. Cells lose water when placed in hypertonic solutions and gain water when placed in hypotonic solutions.
   d. Solutions that contain the same concentration of dissolved particles as a cell are called isotonic.
3. Filtration
   a. Involves movement of molecules from regions of higher toward regions of lower hydrostatic pressure.
   b. Blood pressure causes filtration through capillary membranes.
4. Active transport
   a. Responsible for movement of molecules or ions from regions of lower to regions of higher concentration.
   b. Requires cellular energy and involves action of carrier molecules in the cell membrane.
5. Pinocytosis is a process by which the cell membrane engulfs tiny droplets of liquid.
6. Phagocytosis is a process by which the cell membrane engulfs solid particles.

Life Cycle of a Cell
1. Mitosis
   a. Division and distribution of nuclear parts to daughter cells during cell reproduction.
   b. Stages include prophase, metaphase, anaphase, and telophase.
2. Cytoplasmic division is a process by which cytoplasm is divided into two portions following mitosis.
3. Interphase
   a. Stage in life cycle when cell grows and forms new organelles.
   b. Terminates when cell begins to undergo mitosis.
4. Cell differentiation involves the development of specialized structures and functions.

Control of Cell Reproduction
1. Surface-volume relationships
   a. As a cell grows, its surface area increases to a lesser degree than its volume, and eventually the area becomes inadequate for the needs of the living material within the cell.
   b. When a cell divides, the daughter cells have more favorable surface-volume relationships.
2. Cancer
   a. Involves a change in the normal control on cell reproduction.

   b. Characteristics include hyperplasia, anaplasia, and metastasis.
   c. Causative factors may include chemicals, radiation, viruses, changes in DNA molecules, and deficiencies in the body's resistance to disease.

Application of Knowledge

1. Which process—diffusion, osmosis, or filtration—is most closely related to each of the following situations?
   a. A person is given an injection of isotonic glucose solution.
   b. A person with extremely low blood pressure stops producing urine.
   c. The concentration of urea in the wash solution of an artificial kidney is decreased.

2. What characteristic of cell membranes may account for the observation that fat-soluble substances like chloroform and ether cause rapid effects upon cells?

3. A person who has been exposed to excessive amounts of X ray may develop a decrease in white blood cell number and an increase in susceptibility to infections. In what way are these effects related?

Review Activities

1. Use specific examples to illustrate how cells vary in size.

2. Describe how the shapes of nerve and muscle cells are related to their functions.

3. Name the major portions of a cell and describe their relationship to one another.

4. Discuss the structure and functions of a cell membrane.

5. Distinguish between organelles and inclusions.

6. Define selectively permeable.

7. Describe the structures and functions of each of the following:
   a. endoplasmic reticulum
   b. ribosome
   c. Golgi apparatus
   d. mitochondrion
   e. lysosome
   f. centrosome
   g. vacuole
   h. fibril and microtubule

8. Define inclusion.

9. Describe the structure of the nucleus and the functions of its parts.

10. Distinguish between diffusion and facilitated diffusion.

11. Explain how diffusion aids in the exchange of gases within the body.

12. Describe how dialysis is used to separate wastes from blood in an artificial kidney.

13. Define *osmosis*.

14. Define *osmotic pressure*.

15. Distinguish between solutions that are hypertonic, hypotonic, and isotonic.

16. Define *filtration* and explain how it is involved with the movement of substances through capillary membranes.

17. Explain why active transport is called a physiological process while diffusion is a physical process.

18. Explain the function of carrier molecules in active transport.

19. Distinguish between pinocytosis and phagocytosis.

20. Describe the life cycle of a cell.

21. Name the two processes included in cell reproduction.

22. Describe the major events of mitosis.

23. Define *interphase*.

24. Define *differentiation* and explain how it may involve the repression of DNA information.

25. Explain how surface-volume relationships may function in the control of cell reproduction.

26. Describe the general characteristics of cancers.

27. Name some of the factors that may cause cancers.

Suggestions for Additional Reading

Avers, C. J. 1976. *Cell biology*. New York: D. Van Nostrand Co.

Berns, M. W. 1977. *Cells*. New York: Holt, Rinehart and Winston.

Bingham, C. A. July 1978. The cell cycle and cancer chemotherapy. *American Journal of Nursing*.

Braun, A. C. 1975. *The biology of cancer*. Reading, Mass.: Addison-Wesley.

Capaldi, R. A. March 1974. A dynamic model of cell membranes. *Scientific American*.

Corbett, T. H. 1977. *Cancer and chemicals*. Chicago: Nelson-Hall.

Danielli, J. F. August 1973. The bilayer hypothesis of membrane structure. *Hosp. Prac.*

DeWitt, W., and Brown, E. R. 1977. *Biology of the cell*. Philadelphia: W.B. Saunders.

Dustin, P. August 1980. Microtubules. *Scientific American*.

Fox, C. F. February 1972. Structure of cell membranes. *Scientific American*.

Lenz, T. L. 1971. *Cell fine structure*. Philadelphia: W. B. Saunders.

Lodish, H. F., and Rothman, J. E. January 1979. The assembly of cell membranes. *Scientific American*.

Luria, S. E. December 1975. Colicins and the energetics of cell membranes. *Scientific American*.

McElroy, W. D. 1976. *Modern cell biology*, 2nd ed. Englewood Cliffs, N.J.: Prentice-Hall.

Mazia, D. January 1974. The cell cycle. *Scientific American*.

Nicolson, G. L. March 1979. Cancer metastasis. *Scientific American*.

Novikoff, A. B., and Holtzman, E. 1970. *Cells and organelles*. New York: Holt, Rinehart and Winston.

Rothman, J. E., and Lenard, J. 1977. Membrane asymmetry. *Science*, 195:747–748.

Satir, B. October 1975. The final steps in secretion. *Scientific American*.

Sloboda, R. D. 1980. The role of microtubules in cell structure and cell division. *American Scientist* 68:290.

Staehelin, L. A., and Hull, B. E. May 1978. Junctions between living cells. *Scientific American*.

Swanson, C. P., and Webster, P. 1977. *The cell*, 4th ed. Englewood Cliffs, N.J.: Prentice-Hall.

Cellular Metabolism

4 Cellular metabolism includes all the chemical reactions that occur within the cells of an organism. These reactions, for the most part, involve the utilization of food substances such as carbohydrates, lipids, and proteins, and they are of two major types. In one type of reaction, molecules of nutrients are converted into simpler forms by changes that are accompanied by the release of energy. In the other type, the nutrient molecules are used in constructive processes that produce cell structural materials or molecules that store energy.

In either case, metabolic reactions are controlled by organic catalysts called *enzymes*, which are synthesized in cells. The production of these enzymes is, in turn, controlled by genetic information held within molecules of DNA, which instructs a cell how to synthesize specific kinds of enzyme molecules.

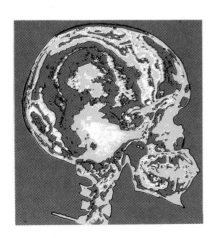

Chapter Outline

Chapter Objectives

After you have studied this chapter, you should be able to

1. Define *anabolic* and *catabolic metabolism.*
2. Explain how enzymes control metabolic processes.
3. Explain how chemical energy is released by respiratory processes.
4. Describe how energy is made available for cellular activities.
5. Describe the general metabolic pathways of carbohydrates, lipids, and proteins.
6. Explain how genetic information is stored within nucleic acid molecules and how this information is used in the control of cellular processes.
7. Explain how genetic information can be altered and how such a change may affect a cell or an organism.
8. Complete the review activities at the end of this chapter.

aerobic respiration　(a-er-ō′bik res″pĭ-ra′shun)

anabolic metabolism　(an″ah-bol′ik mĕ-tab′o-lizm)

anaerobic respiration
　　(an″a-er-ō′bik res″pĭ-ra′shun)

catabolic metabolism
　　(kat″ah-bol′ik mĕ-tab′o-lizm)

coenzyme　(ko-en′zīm)

deamination　(de-am″ĭ-na′shun)

dehydration synthesis
　　(de″hi-dra′shun sin′thĕ-sis)

DNA

energy　(en′er-je)

enzyme　(en′zīm)

gene　(jēn)

genetic code　(jĕ-net′ik kōd)

hydrolysis　(hi-drol′ĭ-sis)

mutation　(mu-ta′shun)

oxidation　(ok″sĭ-da′shun)

RNA

substrate　(sub′strāt)

aer-, air:　*aer*obic respiration—respiratory process that requires oxygen.

an-, without:　*an*aerobic respiration—respiratory process that proceeds without oxygen.

ana-, up:　*ana*bolic metabolism—cellular processes in which smaller molecules are used to build larger ones.

cata-, down:　*cata*bolic metabolism—cellular processes in which larger molecules are broken down into smaller ones.

co-, with:　*co*enzyme—substance that unites with a protein and completes the structure of an enzyme molecule.

de-, undoing:　*de*amination—process by which nitrogen-containing portions of amino acid molecules are removed.

-genic, giving rise to:　muta*genic* factor—factor that gives rise to a mutation.

mut-, change:　*mut*ation—change in the genetic information of a cell.

-strat, spread out:　sub*strat*e—substance on which an enzyme acts.

sub-, under:　*sub*strate—substance on which an enzyme acts.

zym-, causing to ferment:　en*zym*e—protein that initiates or speeds up a chemical reaction without being changed itself.

Although a living cell may appear to be idle, it is actually very active, because the numerous metabolic processes needed to maintain its life are occurring all the time.

Metabolic Processes

Metabolic processes, which include all of the chemical reactions that take place in a cell, can be divided into two major types. One type, called *anabolic*, involves the buildup of larger molecules from smaller ones and utilizes energy. The other, called *catabolic*, involves the breakdown of larger molecules into smaller ones and releases energy.

Anabolic Metabolism

Anabolic metabolism includes all the constructive processes used to manufacture substances needed for cellular growth and repair. For example, simple sugar molecules (monosaccharides) are joined in liver and muscle cells to form larger carbohydrate molecules by an anabolic process called *dehydration synthesis.*

In dehydration synthesis of a carbohydrate, an −OH (hydroxyl group) from a monosaccharide molecule and an −H (hydrogen atom) from another are removed. These particles react to produce a water molecule, and the monosaccharides become bonded by a shared oxygen atom, as shown in figure 4.1.

Similarly, glycerol and fatty acid molecules are bonded by dehydration synthesis in fat tissue cells to form fat molecules. In this case, 3 hydrogen atoms are removed from a glycerol molecule, and an −OH group is removed from each of 3 fatty acid molecules, as shown in figure 4.2. The result is 3 water molecules and a single fat molecule, whose glycerol and fatty acid portions are bonded by shared oxygen atoms.

When 2 amino acid molecules unite, an −OH is removed from one and a hydrogen atom from the other. A water molecule is formed, and the amino acid molecules are joined by a bond between a carbon atom and a nitrogen atom (fig. 4.3). This type of bond, which is called a *peptide bond*, holds the amino acids together. Two amino acids bonded together form a *dipeptide,* while many bonded in a chain form a *polypeptide.* Generally, a polypeptide consisting of 100 or more amino acid molecules is called a *protein.*

Catabolic Metabolism

Physiological processes in which larger molecules are broken down into smaller ones constitute **catabolic metabolism.** An example of such a process is *hydrolysis,* which can bring about the decomposition of carbohydrates, lipids, and proteins.

When a molecule of one of these substances is hydrolyzed, a water molecule is used, and the original molecule is split into two simpler parts. The hydrolysis of the disaccharide sucrose, for instance, results in molecules of 2 monosaccharides, glucose and fructose.

$$C_{12}H_{22}O_{11} + H_2O \rightarrow C_6H_{12}O_6 + C_6H_{12}O_6$$
(sucrose)   (water)   (glucose)   (fructose)

In this case, the bond between the simple sugars within the sucrose molecule is broken, and the water molecule supplies a hydrogen atom to one sugar molecule and a hydroxyl group to the other. Thus, hydrolysis is the reverse of dehydration synthesis.

$$\text{Disaccharide} + \text{Water} \underset{\substack{\text{Dehydration} \\ \text{synthesis}}}{\overset{\text{Hydrolysis}}{\rightleftharpoons}} \text{Monosaccharide} + \text{Monosaccharide}$$

Hydrolysis, more commonly called *digestion,* occurs in various regions of the digestive tract and is responsible for the breakdown of carbohydrates into monosaccharides, fats into glycerol and fatty acids, and proteins into amino acids (fig. 4.4).

Fig. 4.1 What type of reaction is represented by this diagram?

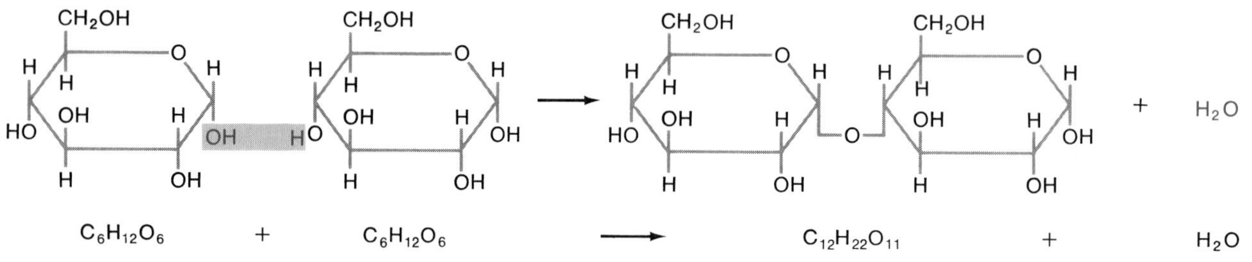

Fig. 4.2 A glycerol molecule and three fatty acid molecules may become bonded by dehydration synthesis to form a fat molecule.

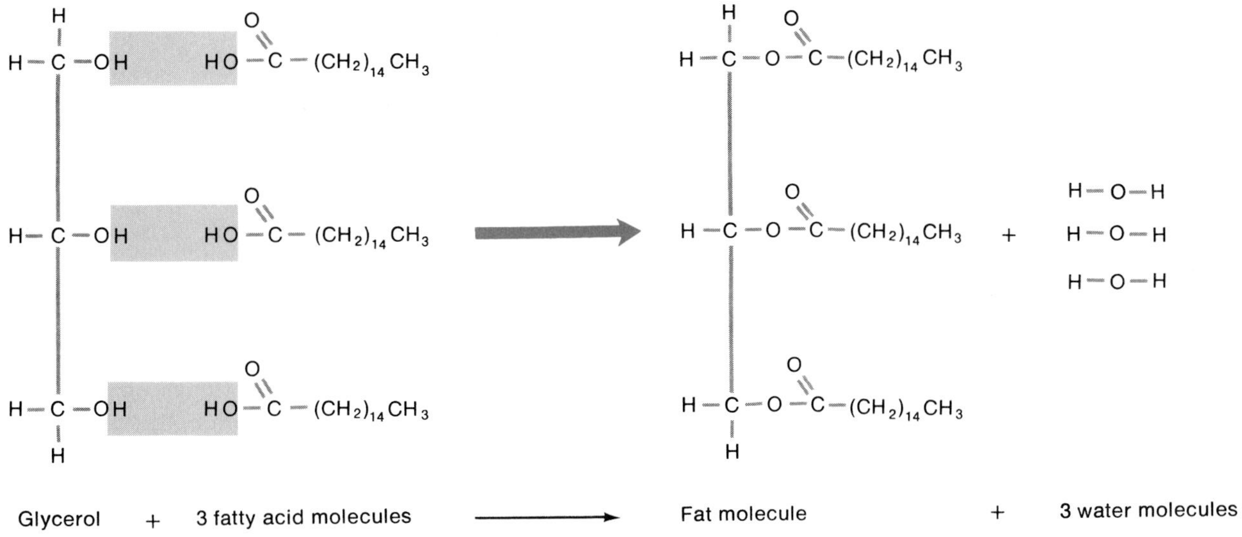

Glycerol + 3 fatty acid molecules ⟶ Fat molecule + 3 water molecules

Fig. 4.3 When two amino acid molecules are united by dehydration synthesis, a peptide bond is formed between a carbon atom and a nitrogen atom.

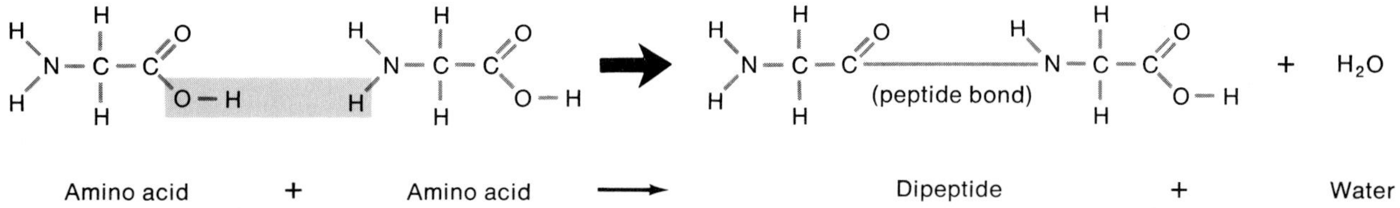

Amino acid + Amino acid ⟶ Dipeptide + Water

Fig. 4.4 Hydrolysis causes the decomposition of (a) carbohydrates into monosaccharides; (b) fats into glycerol and fatty acids; and (c) proteins into amino acids.

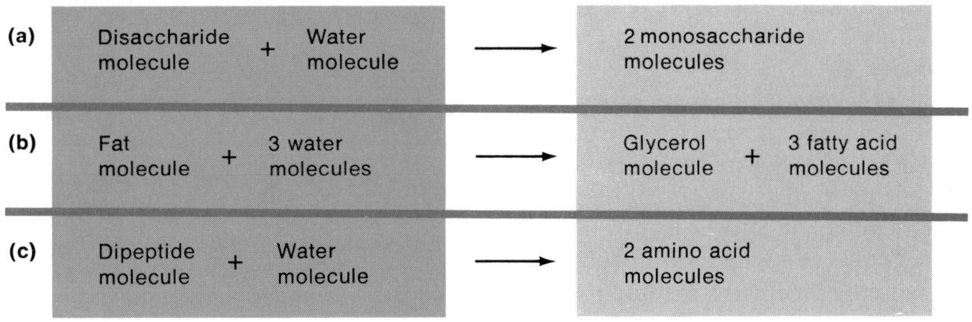

(a) Disaccharide molecule + Water molecule ⟶ 2 monosaccharide molecules

(b) Fat molecule + 3 water molecules ⟶ Glycerol molecule + 3 fatty acid molecules

(c) Dipeptide molecule + Water molecule ⟶ 2 amino acid molecules

Both catabolic and anabolic metabolism are carried on continually in cells. However, these activities are carefully balanced so that the breakdown or energy-releasing reactions occur at rates that are adjusted to the needs of the building-up or energy-utilizing reactions.

---

1. Name a substance formed by the anabolic metabolism of monosaccharides. Of amino acids. Of glycerol and fatty acids.
2. What general functions does anabolic metabolism serve?
3. Distinguish between dehydration synthesis and hydrolysis.

Control of Metabolic Reactions

Although different kinds of cells may conduct specialized metabolic processes, all cells perform certain basic reactions. These reactions include the buildup and breakdown of carbohydrates, lipids, proteins, and nucleic acids.

Like other chemical reactions, metabolic reactions generally require a certain amount of energy before they will occur. (This is why heat is commonly used to increase the rates of chemical reactions in laboratories.) The temperature conditions that exist in cells, however, are usually too mild to promote the reactions needed to support life; but this is not a problem because cells contain enzymes.

Enzymes and Their Actions

Enzymes are proteins that promote chemical reactions within cells by somehow lowering the amount of energy (activation energy) needed to initiate these reactions. Thus, in the presence of enzymes, metabolic reactions are speeded up. Enzymes are needed in very small quantities because as they function, they are not changed or used up, and can, therefore, function again and again.

The reaction promoted by a particular enzyme is very specific. Each enzyme acts only on a particular substance, which is called its **substrate.** For example, the substrate of one enzyme called *catalase* (a peroxidase in the peroxisomes of cells) is hydrogen peroxide, a harmful substance sometimes produced in cells. This enzyme's only function is to cause the decomposition of hydrogen peroxide into water and oxygen. In this way it helps to prevent an accumulation of hydrogen peroxide that might otherwise damage cells.

You may have noticed the action of the enzyme catalase if you have ever used hydrogen peroxide to cleanse a wound. Injured cells contain catalase, and when hydrogen peroxide contacts them, bubbles of oxygen are released. This is useful since the resulting foam aids in removing debris from inaccessible parts of the wound.

Cellular metabolism includes hundreds of different chemical reactions, and each of these reactions is controlled by a specific kind of enzyme. Thus, there must be hundreds of different kinds of enzymes present in every cell. How each enzyme "recognizes" its specific substrate is not well understood, but this ability seems to involve the shapes of molecules. That is, each enzyme's polypeptide chain is thought to be twisted and coiled into a unique three-dimensional form, that somehow fits the special shape of a substrate molecule. In short, an enzyme is thought to fit its substrate much as a key fits a particular lock.

It is also believed that during an enzyme-controlled reaction particular regions of the enzyme molecule called *active sites* temporarily combine with portions of the substrate, creating a substrate-enzyme complex. At the same time, the interaction between the molecules seems to distort or strain chemical bonds within the substrate, which increases the likelihood of a change in the substrate molecule. When the substrate is changed, the product of the reaction appears, and the enzyme is released unaltered. (See fig. 4.5.)

This activity can be summarized as follows:

$$\text{Substrate molecule} + \text{Enzyme molecule} \rightarrow \text{Substrate-enzyme complex} \rightarrow \text{Product (changed substrate)} + \text{Enzyme molecule}$$

The speed of an enzyme-controlled reaction is related to the number of enzyme and substrate molecules present in the cell. However, the efficiency of different kinds of enzymes varies greatly. For example, some seem to be able to process only a few substrate molecules per second while others can process thousands or even millions of substrate molecules in a second.

Some of the catabolic enzymes that cause the digestion (hydrolysis) of food molecules are listed in chart 4.1 with their substrates and products. Note that most of the enzyme names end in -ase.

Fig. 4.5 (a) The shape of an enzyme molecule (b) fits the shape of the substrate molecule. (c) When the substrate molecule becomes temporarily combined with the enzyme a chemical reaction occurs. The result is (d) product molecules and (e) an unaltered enzyme.

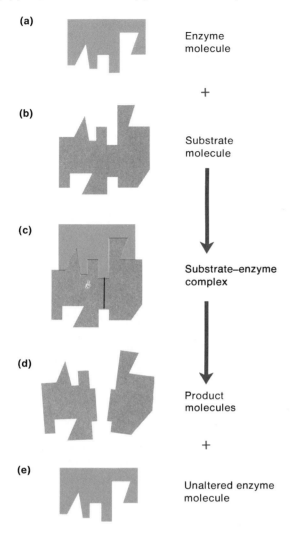

(a) Enzyme molecule

+

(b) Substrate molecule

(c) Substrate–enzyme complex

(d) Product molecules

+

(e) Unaltered enzyme molecule

Coenzymes

Often the protein portion of an enzyme is inactive until it is combined with an additional substance. This part is needed to complete the active form of the enzyme molecule or bind the enzyme to its substrate. Such a substance is called a **coenzyme.** A coenzyme may be an ion of an element such as copper, iron, or zinc, or it may be a relatively small organic molecule. Most vitamins are coenzymes of the latter type.

Vitamins are substances that usually cannot be synthesized in sufficient quantities by human cells and therefore must be obtained in the diet. Since most vitamins function as coenzymes and can, like enzymes, be used again and again, they are needed by cells in very small quantities.

Chart 4.1   Some digestive enzymes

Enzyme	Substrate	Product
Amylase	Starch	Disaccharides
Pepsin	Proteins	Peptides
Lipase	Fats	Fatty acids and glycerol
Nuclease	Nucleic acids	Nucleotides
Pepidase	Peptides	Amino acids
Maltase	Disaccharide (maltose)	Monosaccharides
Lactase	Disaccharide (lactose)	Monosaccharides
Sucrase	Disaccharide (sucrose)	Monosaccharides
Nucleotidase	Nucleotides	Sugars (ribose and deoxyribose), phosphates, and organic bases

Factors That Alter Enzymes

Except for their coenzyme portions, enzymes are proteins, and like other proteins, they can be denatured by exposure to excessive heat, radiation, electricity, or certain chemicals. For example, many enzymes become inactive at 45°C, and nearly all of them are denatured at 55°C. Chemicals that cause enzymes to be denatured are considered poisons. They may cause the death of cells by affecting enzymes responsible for vital metabolic reactions. Cyanides, for instance, interfere with respiratory enzymes and damage cells by halting their energy-releasing processes.

The antibiotic drug penicillin acts by interfering with enzymes that function in the production of cell walls. These walls surround and protect normal bacterial cells. In the presence of penicillin, bacterial cell walls are not properly produced, and the affected bacteria are less able to survive. Thus, penicillin may be used to control the growth of bacteria that might otherwise cause an infection.

1. What is an enzyme?
2. How does an enzyme recognize its substrate?
3. What is the role of a coenzyme?
4. What factors are likely to alter the rates of metabolic reactions?

Energy for Metabolic Reactions

Energy is the capacity to produce changes in matter or the ability to move something—to do work. Therefore, energy is recognized by what it can do; it is utilized whenever changes take place. Common forms of energy include heat, light, sound, electrical energy, mechanical energy, and chemical energy.

Although energy cannot be created or destroyed, it can be converted from one form to another. Sunlight is converted to heat when it is absorbed by skin; an ordinary incandescent light bulb changes electrical energy to heat and light; and an automobile engine converts the chemical energy in gasoline to heat and mechanical energy.

Release of Chemical Energy

Most metabolic processes use chemical energy, which is held in the bonds between the atoms of molecules and is released when these bonds are broken. For example, the chemical energy of many substances can be released by burning. Such a reaction must be started, and this is usually accomplished by applying heat to activate the burning process. As the substance burns, molecular bonds are broken, and energy escapes as heat and light.

Similarly, glucose molecules are "burned" in cells during respiration, although the process is more correctly called **oxidation.** The energy released by the oxidation of glucose is used to promote cellular metabolism. There are, however, some important differences between the oxidation of substances inside cells and the burning of substances outside them.

Burning usually requires a relatively large amount of energy to activate the process, and most of the energy released escapes as heat or light. In cells, the oxidation process is initiated by enzymes that reduce the *activation energy* needed. Also, by transferring energy to special energy-carrying molecules, cells are able to capture and store about half of the energy released. The rest escapes as heat, which helps maintain body temperature.

The process by which energy is released from glucose molecules and transferred to other molecules is quite complex. It involves a number of chemical reactions that must occur in a particular sequence, each one controlled by a separate kind of enzyme. Some of these enzymes are in the cell's cytoplasm, while others are in the mitochondria.

---

1. What is energy?
2. How does cellular oxidation differ from burning?

Fig. 4.6 Anaerobic respiration occurs in the cytoplasm in the absence of oxygen, and aerobic respiration occurs in the mitochondria in the presence of oxygen. Both processes cause the release of energy that is captured by ATP molecules.

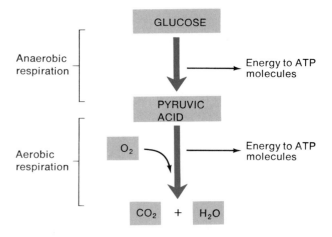

Anaerobic Respiration

When a 6-carbon glucose molecule is decomposed in the process of cellular respiration, enzymes act to break it into two 3-carbon pyruvic acid molecules. This phase of the process (glycolysis) occurs in the cytoplasm, and because it takes place in the absence of oxygen, it is called **anaerobic respiration.**

Although some energy is needed to activate the anaerobic phase of respiration, more energy is released than is used. A portion of the excess is transferred to molecules of an energy-carrying substance called **ATP** (adenosine triphosphate). (See fig. 4.6.)

Aerobic Respiration

Following the anaerobic phase of respiration, oxygen must be available for further molecular breakdown to occur. For this reason, the second phase is called **aerobic respiration.** It takes place within the mitochondria, and as a result, considerably more energy is transferred to ATP molecules.

When the decomposition of a glucose molecule is complete, carbon dioxide molecules and hydrogen atoms remain. The carbon dioxide diffuses out of the cell as a waste, and the hydrogen atoms combine with oxygen to form water molecules. Thus, the final products of glucose oxidation are carbon dioxide, water, and energy. (See fig. 4.6.)

Fig. 4.7 An ATP molecule consists of an adenine portion, a ribose portion, and three phosphates.

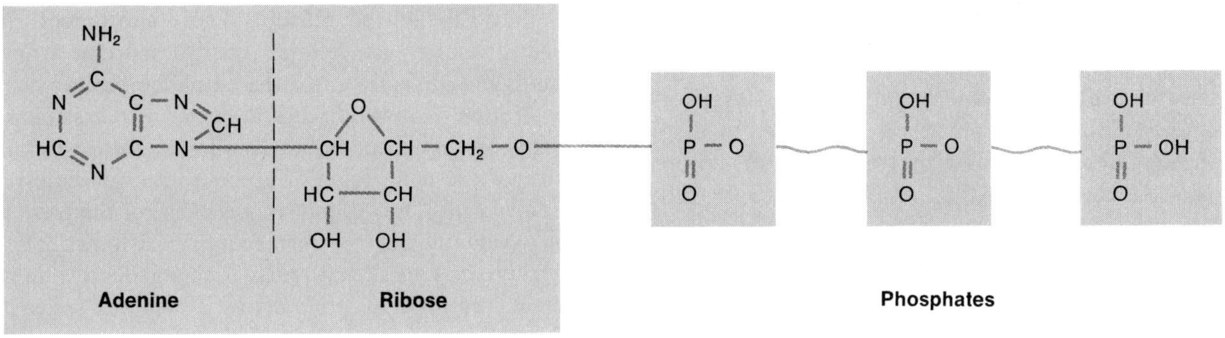

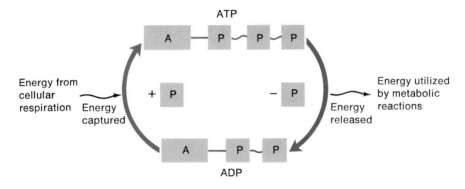

Fig. 4.8 What is the significance of this cyclic process?

ATP Molecules

For each glucose molecule that is decomposed, 38 molecules of ATP are produced. Two of these are the result of anaerobic respiration, while the rest are formed during the aerobic phase.

Each ATP molecule consists of three main parts—an adenine portion, a ribose portion, and 3 phosphates in a chain. (Figure 4.7). As energy is released during cellular respiration, some of it is stored in the bond of the end phosphate. When energy is needed for some metabolic process, the terminal phosphate bond of an ATP molecule is broken, and the energy stored in this bond is released. Such energy is used for a variety of functions including muscle contractions, active transport mechanisms, and synthesizing compounds.

An ATP molecule that has lost its terminal phosphate is called an **ADP** (adenosine diphosphate) molecule. An ADP molecule can be converted back into an ATP by capturing some energy and a phosphate. Thus, as figure 4.8 shows, ATP and ADP molecules shuttle back and forth between the energy-releasing reactions of cellular respiration and the energy-utilizing reactions of the cell.

Although ATP is not the only kind of energy-carrying molecule within a cell, it is the primary one, and without a source of ATP most cells die quickly.

---

1. What is meant by anaerobic respiration? Aerobic respiration?
2. What happens to the energy released by these processes?
3. What are the by-products of these reactions?
4. What is the function of ATP molecules?

Metabolic Pathways

The anabolic and catabolic reactions that occur in cells usually involve a number of different steps that must occur in a particular sequence. For example, the anaerobic phase of cellular respiration, by which glucose is converted to pyruvic acid, involves 10 separate reactions. Since each reaction is controlled by a specific kind of enzyme, these enzymes must act in proper order. Such precision of activity suggests that the enzymes are arranged in precise ways. It is believed that those responsible for aerobic respiration are located in tiny stalked particles on the membranes within the mitochondria, and that they are positioned in the exact sequence of the reactions they control.

Such a sequence of enzyme-controlled reactions that leads to the production of particular products is called a **metabolic pathway.** (See fig. 4.9.) Typically these pathways are interconnected so that molecules of a certain kind of substance may enter more than one pathway as the substance is metabolized. For example, carbohydrate molecules from foods may enter catabolic pathways and be used to supply energy, or they may enter anabolic pathways and be stored as glycogen or fat. (See fig. 4.10.)

Carbohydrate Pathways

The average human diet consists largely of carbohydrates, which are changed by digestion to monosaccharides such as glucose. These substances are used primarily as cellular energy sources, which means they usually enter the catabolic pathways of respiration. As was discussed earlier, the first phase of this process is anaerobic. It is called *glycolysis* and involves the conversion of glucose into molecules of *pyruvic acid.* During the second phase of the process the pyruvic acid is transformed into a substance called *acetyl coenzyme A.* It, in turn, is changed into a number of intermediate products by a complex series of chemical reactions, shown in figure 4.11, known as the **citric acid cycle** (Kreb's cycle). As these changes occur, energy is released and some of it is transferred to molecules of ATP, while the rest is lost as heat. The end products of the oxidation process are molecules of carbon dioxide and water. The citric acid cycle is also discussed in chapter 15. (See the Appendix for a more detailed description of cellular respiration.) Any excess glucose may enter anabolic pathways and be converted into storage forms such as glycogen.

Fig. 4.9 A metabolic pathway consists of a series of enzyme-controlled reactions leading to a product.

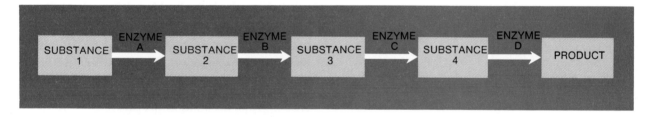

Fig. 4.10 Carbohydrates from foods are changed into monosaccharides by hydrolysis. The resulting molecules may enter catabolic pathways and be used as energy sources, or they may enter anabolic pathways and be converted to glycogen or fat.

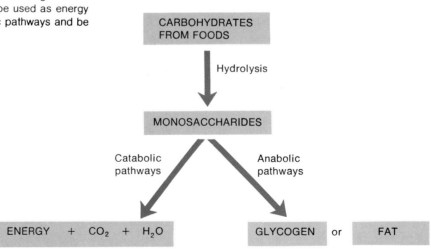

Fig. 4.11 Aerobic respiration, the second phase of cellular respiration, includes a complex series of chemical reactions called the citric acid cycle. What are the major end products of aerobic respiration?

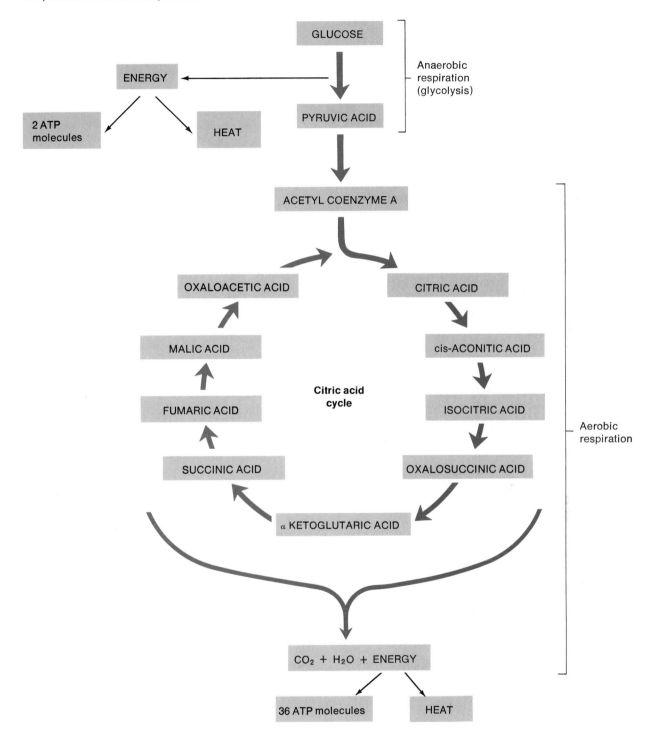

Although cells generally can produce glycogen, the liver and muscle cells store the greatest amounts. Following a meal, when the blood glucose level is relatively high, liver cells obtain glucose from the blood and convert it to glycogen. Between meals, when the blood glucose level is lower, the reaction is reversed, and glucose is released into the blood. This mechanism ensures that various tissues will continue to have an adequate supply of glucose molecules to support their respiratory processes.

Glucose can also be converted into fat molecules, which are later deposited in fat tissues. This happens when a person takes in more carbohydrates than can be stored as glycogen or are needed for normal activities. The body has an almost unlimited capacity to perform this type of anabolic metabolism, so an excessive intake of nutrients can result in overweight (obesity).

Lipid Pathways

Although foods may contain lipids in the form of phospholipids or cholesterol, the most common dietary lipids are *triglycerides*. As was explained in chapter 2, the molecules of these lipids consist of a glycerol portion and 3 fatty acids.

The metabolism of lipids is controlled mainly by the liver, which can remove them from the circulating blood and alter their molecular structures. For example, the liver can shorten or lengthen the carbon chains of fatty acid molecules or introduce double bonds into these chains, thus, converting fatty acids from one form to another.

Lipids provide for a variety of physiological functions, however, they are used mainly to supply energy. Fats are a particularly concentrated form of energy because gram for gram, they contain more than twice as much chemical energy as carbohydrates or proteins.

Before energy can be released from a triglyceride molecule it must undergo hydrolysis. As shown in figure 4.12, the resulting fatty acid portions can then be converted into molecules of acetyl coenzyme A by a series of reactions called *beta oxidation*. In these reactions, fatty acid molecules are broken down by mitochondria into segments containing 2 carbon atoms each. Some of these segments are converted into acetyl coenzyme A molecules; others are converted into compounds called *ketone bodies*, which later can be changed to acetyl coenzyme A and oxidized by means of the citric acid cycle. The glycerol portions of the triglyceride molecules can also enter metabolic pathways leading to the citric acid cycle, or they can be used to synthesize glucose.

Another possibility for glycerol and fatty acid molecules resulting from the hydrolysis of fats is to be changed back into fat molecules by anabolic processes and stored in fat tissue. Additional fat molecules can be synthesized from excess molecules of glucose or amino acids.

When ketone bodies are formed faster than they can be decomposed, some of them are eliminated through the lungs and kidneys. Consequently, the breath and urine may develop a fruity odor due to the presence of a ketone called *acetone*. This sometimes happens when a person fasts or diets to lose weight. Persons suffering from diabetes mellitus are also likely to metabolize excessive amounts of fats, and they too may have acetone in the breath and urine. At the same time, they may develop a serious imbalance in pH called *acidosis* due to an accumulation of acidic ketone bodies.

Protein Pathways

When dietary proteins are digested, the resulting amino acids are absorbed into the blood and transported to various body cells. Many of these amino acids are bonded by cellular enzymes to form protein molecules again, which function in a variety of ways. Some of the proteins are incorporated into cell parts, others serve as enzymes, and still others are used to supply energy.

If protein molecules are to be used as energy sources, they must first be broken down into amino acids. Then the amino acids undergo *deamination*, a process that occurs in the liver and involves removing the nitrogen-containing portions ($-NH_2$ groups) from the amino acids. These $-NH_2$ groups are later converted into a waste substance called *urea*, which is excreted in the urine.

Depending upon the amino acids involved, the remaining deaminated portions of the amino acid molecules are decomposed by one of several pathways—all of which lead to the formation of acetyl coenzyme A. As before, the acetyl coenzyme A can enter the citric acid cycle, and as energy is released some of it is stored in molecules of ATP. (See fig. 4.13.) If energy is not needed immediately, the deaminated portions of the amino acids may be changed by still other metabolic pathways into glucose or fat molecules.

Fig. 4.12 Fats from foods are digested into glycerol and fatty acids. These molecules may enter catabolic pathways and be used as energy sources.

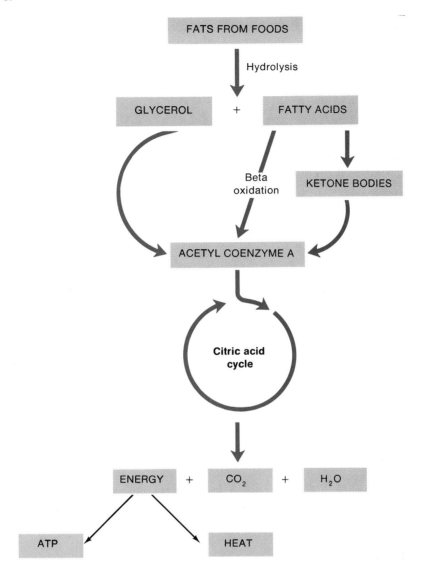

Glucose can be changed back into amino acids if certain nitrogen-containing molecules are available. However, about 8 necessary amino acids cannot be synthesized in human cells by such processes and so must be provided in the diet. For this reason, they are called *essential amino acids*. These are discussed in chapter 14.

1. *What is meant by a metabolic pathway?*
2. *How are carbohydrates used within cells?*
3. *What must happen to fat molecules before they can be used as energy sources?*
4. *How do cells use proteins?*

Fig. 4.13 Proteins from foods are digested into amino acids, but before these molecules can be used as energy sources they must be deaminated.

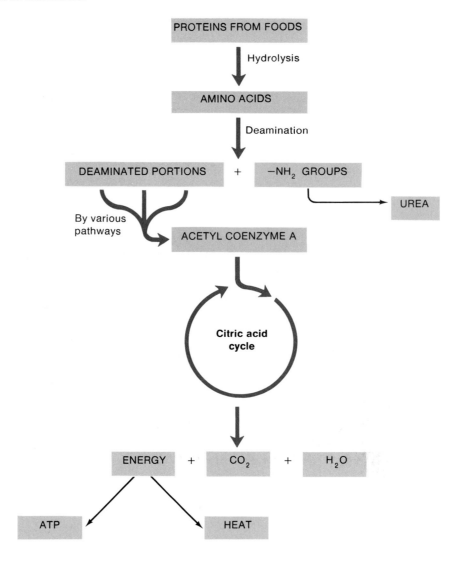

Nucleic Acids and Protein Synthesis

Since enzymes control the metabolic processes that enable cells to survive, the cells must possess information for producing these enzymes. Such information is held in DNA molecules in the form of a genetic code. This code "instructs" cells how to synthesize specific protein molecules.

Genetic Information

Children resemble their parents because of inherited traits; but what is actually passed from parents to a child is *genetic information*. This information is received in the form of DNA molecules from the parents' sex cells, and as the offspring develops it is passed from cell to cell by mitosis. Genetic information "tells" the cells of the developing body how to construct specific protein molecules, which in turn function as structural materials, enzymes, or other vital substances.

The portion of a DNA molecule that contains the genetic information for making one kind of protein is called a **gene.** Thus, inherited traits are determined by the genes contained in the sex cells that fused to form the first cell of the body. These genes instructed the cells to synthesize the particular enzymes needed to control metabolic pathways.

Fig. 4.14 Each nucleotide of a nucleic acid consists of a 5-carbon sugar (S), a phosphate group (P), and an organic base (B).

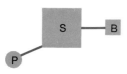

Fig. 4.15 The nucleotides of a DNA strand are bonded to form a sugar-phosphate backbone. The organic bases of the nucleotides extend from this backbone and are weakly bonded to the bases of the second strand.

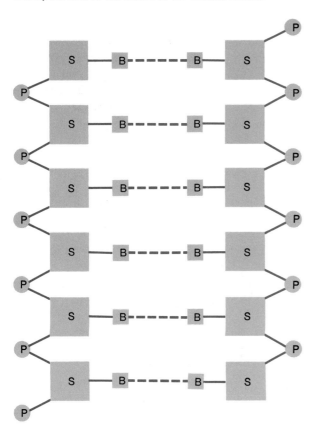

Fig. 4.16 The ladder of a double-stranded DNA molecule is twisted into the form of a double helix.

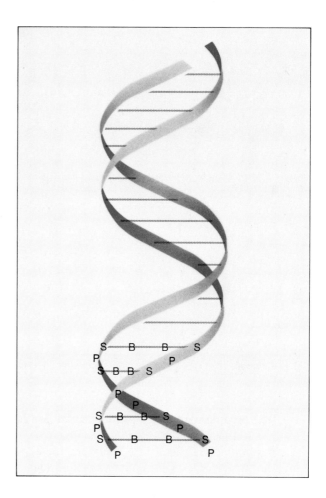

DNA Molecules

Molecules of DNA (deoxyribonucleic acid) are relatively large and are able to duplicate themselves. Each molecule consists of 2 strands twisted into a double spiral or *helix*. As mentioned earlier, the building blocks of nucleic acids are nucleotides, each of which contains a 5-carbon sugar (ribose or deoxyribose), a phosphate group, and one of several organic bases (fig. 4.14).

The nucleotides of a **DNA** strand are bonded together so that the sugar and phosphate portions form a long chain, or "backbone" (fig. 4.15). The organic bases stick out from this backbone and are bonded weakly to those of the second strand. The resulting structure is something like a ladder in which the uprights represent the sugar and phosphate backbones of the 2 strands, and the crossbars represent the organic bases. In addition, the molecular ladder is twisted to form a *double helix* (fig. 4.16 and color plate 7).

Fig. 4.17 Each nucleotide of a DNA molecule contains one of four organic bases: adenine, thymine, cytosine, or guanine.

Fig. 4.18 A strand of DNA consists of a chain of nucleotides arranged in a particular sequence. Within the chain there are four kinds of nucleotides: adenine nucleotide (A); thymine nucleotide (T); cytosine nucleotide (C); and guanine nucleotide (G).

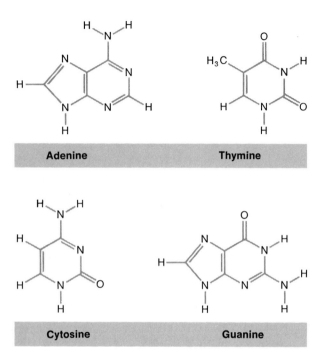

Adenine

Thymine

Cytosine

Guanine

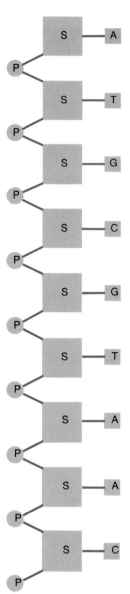

The organic base of a DNA nucleotide can be one of four kinds: *adenine* (A), *thymine* (T), *cytosine* (C), or *guanine* (G) (fig. 4.17). Therefore, there are only four kinds of DNA nucleotides: adenine nucleotide, thymine nucleotide, cytosine nucleotide, and guanine nucleotide.

A strand of DNA consists of nucleotides arranged in a particular sequence (fig. 4.18). Moreover, the nucleotides of one strand are paired in a special way with those of the other strand. Only certain organic bases have the molecular shapes needed to fit together so that their nucleotides can bond with one another. Specifically, an adenine will bond only to a thymine, and a cytosine will bond only to a guanine. As a consequence of such base pairing, a DNA strand possessing the base sequence T, C, A, G would have to be bonded to a second strand with the complementary base sequence A, G, T, C (fig. 4.19). It is the particular sequence of base pairs that encodes the genetic information held in a DNA molecule.

Fig. 4.19 The nucleotides of a double-stranded molecule of DNA are paired so that an adenine nucleotide of one strand is bonded to a thymine nucleotide of the other strand, and a guanine nucleotide of one is bonded to a cytosine nucleotide of the other. The dotted lines represent weak bonds between the paired bases of these nucleotides.

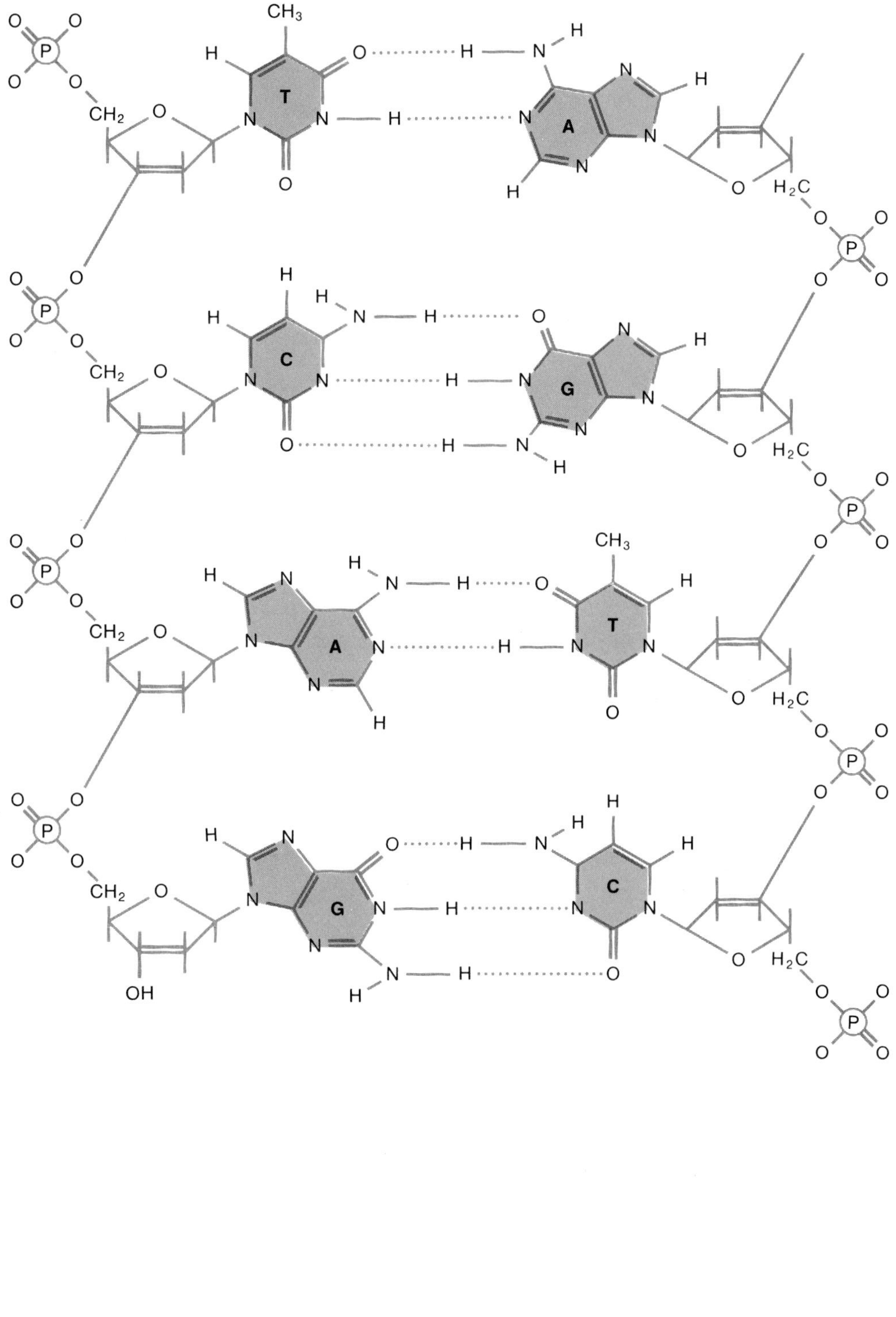

The Genetic Code

Genetic information is information for synthesizing proteins, and since proteins consist of 20 different amino acids bonded in particular sequences, the genetic information must tell how to position the amino acids correctly in a polypeptide chain.

It is believed that each of the 20 different amino acids is represented in a DNA molecule by a particular sequence of 3-nucleotide groups. That is, the sequence C, G, A in a DNA strand represents one kind of amino acid; the sequence G, C, A represents another kind, and T, T, A still another kind. (See chart 4.2.)

Thus, the sequence in which the nucleotide groups are arranged within a DNA molecule can denote the arrangement of amino acids within a protein molecule. This method of storing information used for the synthesis of particular protein molecules is termed the **genetic code.**

Although DNA molecules are located in the chromatin within a cell's nucleus, protein synthesis occurs in the cytoplasm. Therefore the genetic information must somehow be transferred from the nucleus into the cytoplasm. This transfer of information is the function of certain RNA molecules.

RNA Molecules

RNA (ribonucleic acid) molecules differ from DNA molecules in several ways. For example, RNA molecules are usually single-stranded, and their nucleotides contain ribose rather than deoxyribose sugar. Like DNA, RNA nucleotides each contain one of four organic bases, but while adenine, cytosine, and guanine nucleotides occur in both DNA and RNA, thymine nucleotides are found only in DNA. In its place, RNA molecules contain *uracil* (U) nucleotides. (See fig. 4.20.)

One step in the transfer of information from the nucleus to the cytoplasm involves the synthesis of a type of RNA called **messenger RNA** (mRNA). As it is produced, a double-stranded section of a DNA molecule seems to unwind and pull apart. At the same time, the weak bonds between the base pairs of this section are broken. A molecule of messenger RNA is then formed of nucleotides that are complementary to those arranged along the exposed strand of DNA.

Chart 4.2 Genetic code for certain amino acids

Amino Acid	DNA Code	Messenger RNA Code
Alanine	CGA	GCU
Arginine	GCA	CGU
Asparagine	TTA	AAU
Glutamine	GTT	CAA
Histidine	GTA	CAU
Phenylalanine	AAA	UUU

Fig. 4.20 Each nucleotide of an RNA molecule contains one of four organic bases: adenine, uracil, cytosine, or guanine. How is the group of nucleotides similar to those found in DNA molecules? How is it different?

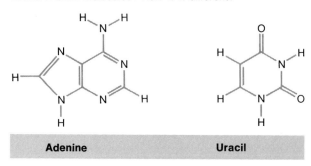

Adenine   Uracil

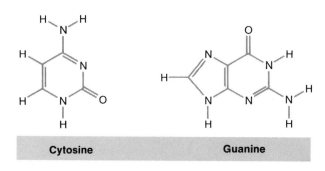

Cytosine   Guanine

For example, if the sequence of DNA bases is A, T, G, C, G, T, A, A, C, then the complementary bases in the developing RNA molecule would be U, A, C, G, C, A, U, U, G, as shown in figure 4.21. In this way, an RNA molecule is synthesized that contains the information for arranging the amino acids of a protein molecule in the sequence dictated by the DNA "master information."

Fig. 4.21 When an RNA molecule is synthesized beside a strand of DNA, complementary nucleotides bond as in a double-stranded molecule of DNA with one exception: RNA contains uracil nucleotides (U) in place of thymine nucleotides (T).

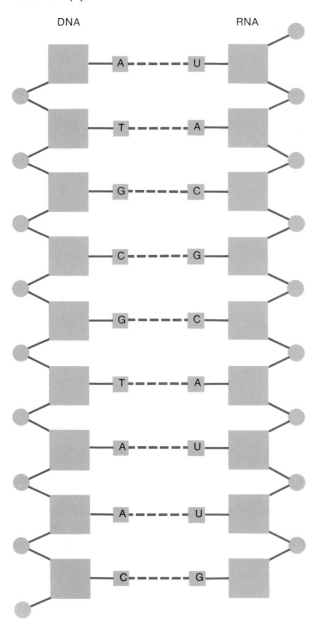

DNA   RNA

A - - - - - U
T - - - - - A
G - - - - - C
C - - - - - G
G - - - - - C
T - - - - - A
A - - - - - U
A - - - - - U
C - - - - - G

Once they are formed, messenger RNA molecules, each of which consists of hundreds or even thousands of nucleotides, can move out of the nucleus through tiny pores in the nuclear membrane and enter the cytoplasm (fig. 4.22). There they become associated with groups of ribosomes and act as patterns or templates for the synthesis of protein molecules.

Fig. 4.22 After copying a section of DNA information, a messenger RNA molecule (mRNA) moves out of the nucleus and enters the cytoplasm. There it becomes associated with a ribosome.

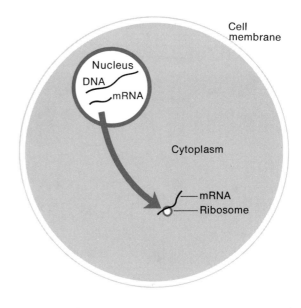

Cell membrane
Nucleus
DNA
mRNA
Cytoplasm
mRNA
Ribosome

Protein Synthesis. Before a protein molecule can be synthesized, the correct amino acids must be present in the cytoplasm to serve as building blocks. Furthermore, these amino acids must be positioned in the proper locations along a strand of messenger RNA. The positioning of amino acid molecules is the function of a second kind of RNA called **transfer RNA** (tRNA).

Since 20 different kinds of amino acids are involved in protein synthesis, there must be at least 20 different kinds of transfer RNA molecules to serve as guides. However, before a transfer RNA molecule can pick up its particular kind of amino acid, the amino acid must be activated. This is a function of ATP, which provides the energy needed to form a bond between the amino acid and its transfer RNA molecule.

Each type of transfer RNA, which is a relatively small molecule, contains 3 nucleotides in a particular sequence. These nucleotides can only bond to the complementary set of 3 nucleotides of a messenger RNA molecule. In this way, the transfer RNA carries its amino acid to a correct position on a messenger RNA strand. These actions occur on the surface of ribosomes.

Fig. 4.23 Molecules of transfer RNA bring specific amino acids and place them in the sequence determined by the nucleotides in the messenger RNA molecule. These amino acids bond and form the polypeptide chain of a protein molecule.

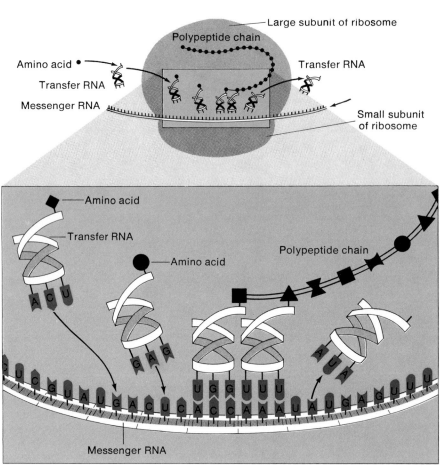

Each ribosome is a tiny particle of 2 unequal-sized subunits composed of *ribosomal RNA* (rRNA) and protein. By its smaller subunit a ribosome binds to a molecule of messenger RNA near a set of 3 nucleotides. This action allows a transfer RNA molecule with the complementary set of 3 nucleotides to recognize its correct location on the messenger RNA. The transfer RNA molecule brings the amino acid it carries into position and becomes temporarily joined to the ribosome. Then the transfer RNA molecule releases its amino acid and returns to the cytoplasm. (See fig. 4.23.)

This process is repeated again and again as the messenger RNA moves through the ribosome, and the amino acids, which are released by the transfer RNA molecules, are added one at a time to a developing protein molecule. The enzymes necessary for the bonding of the amino acids are found in the larger subunit of the ribosome. This subunit also seems to hold the growing chain of amino acids as it develops.

Actually, a molecule of messenger RNA is usually passing through several ribosomes at the same time. Thus, several protein molecules, each in a different stage of development, may be present at any given moment.

When a protein is completed it is released and becomes a separate functional molecule. The transfer RNA molecules can pick up other amino acids from the cytoplasm, and the messenger RNA molecule can function again and again.

1. *What is the function of DNA?*
2. *How is information carried from the nucleus to the cytoplasm.*
3. *How are protein molecules synthesized?*

Changes in Genetic Information

When a cell reproduces, each daughter cell needs a copy of the parent cell's genetic information so that it will be able to synthesize the proteins necessary for cell structure and metabolism. DNA molecules can duplicate themselves, and this duplicating occurs during the interphase of the cell's life cycle. As a result of mitosis, the duplicate sets of DNA-containing chromosomes are evenly divided between the forming daughter cells.

The trillions of cells in an adult body are the result of such mitotic divisions, and in almost all cases these divisions are extremely precise. However, occasionally a mistake is made, and the genetic information is somehow altered. Such a change in genetic information is called a **mutation.**

Nature of Mutations

Mutations can originate in a number of ways. Sometimes they seem to arise spontaneously. At other times they result from exposure to factors such as radiation or various chemicals. In any event, the consequence is the same—genetic information is changed—and if the cell in which the mutation occurs survives, the mutation is passed to future generations of cells.

One kind of mutation involves the breaking of a chromosome. When this happens, a chromosome part may be lost or become attached to another chromosome. Another type of mutation appears when there is a change in the sequence of base pairs within a DNA molecule. Perhaps a base pair is left out during the process of chromosome duplication. In either instance, the instructions encoded in the original molecule are changed, and it is likely that the cell carrying the mutation will be unable to synthesize some protein. Figure 4.24 illustrates such a change.

Effects of Mutations

The effects of mutations vary greatly. At one extreme there is little or no effect. Perhaps the mutant cell is unable to manufacture an enzyme that is relatively unimportant, and the cell continues to function effectively. At the other extreme, mutations cause cell death. In such a case, the cell may be unable to produce an enzyme needed for energy release, so it cannot survive. The mutations of most concern are those between these extremes. They result in some decrease in cell efficiency, and although the cell may live, it has difficulty functioning normally.

Fig. 4.24 (*a*) The DNA code for the amino acid histidine is GTA. (*b*) If something happens to change the guanine in this section of the molecule to thymine, the DNA code is changed to that for the amino acid, asparagine. Such a change would be a mutation.

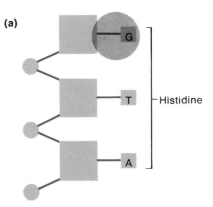

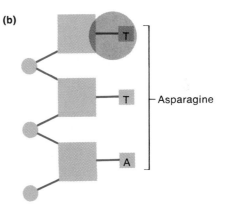

It also makes a difference whether a mutation occurs in a body cell of an adult or in a cell that is part of a developing embryo. In an adult, the mutant cell might not be noticed because there are many normally functioning cells around it. In the embryo, however, the affected cell might be the ancestor of great numbers of cells that are forming the body of a child. In fact, all the cells of the child's body could be defective if the mutation were present in the fertilized egg.

Well over one hundred human diseases are known to be related to defects in the genetic information of cells. The majority of these involve an inability to produce specific enzymes. As an example, the disease called phenylketonuria (PKU) is caused by a mutation, passed from parents to child, that results in a defective enzyme. The normal enzyme functions in the metabolic breakdown of the amino acid phenylalanine. When the defective enzyme is present, phenylalanine is only partially metabolized, and a toxic substance accumulates. This substance interferes with the normal activity of the child's nervous system. If the condition is untreated, it usually results in severe mental retardation. Treatment for PKU involves restricting the intake of foods containing phenylalanine.

Mutagenic Factors

Although the rate at which spontaneous mutations occur varies greatly from gene to gene, certain external factors increase the mutation rates. These are called *mutagenic factors,* and they include radiation from radioactive substances, X rays, ultraviolet light, and a variety of chemicals.

Cancers sometimes develop in various tissues following exposure to mutagenic factors. While the causes of cancers are still largely unknown, some investigators believe that changes in the genetic information of cells are involved, and it is known that mutagenic factors can produce such changes.

It is difficult to avoid contact with mutagenic factors in everyday living. Skin is exposed to ultraviolet light whenever we are in sunlight; X rays and radioactive isotopes are often used in the diagnosis and treatment of diseases; mutagenic chemicals occur in such commonly used substances as tobacco and various petroleum products. Yet with the knowledge that these factors are hazardous to cells of present and future generations, the only wise course is to limit exposure to them whenever possible.

Some investigators believe that cellular aging is related to changes in intracellular molecules that may occur from random damage to the structure of DNA molecules and to various molecules involved with the synthesis of proteins. As a result, highly specialized, nonreproducing cells like those of the brain may accumulate numerous errors in the amino acid sequences of the proteins they manufacture. This could lead to a reduction in the efficiency of metabolic processes and have a marked effect upon cellular behavior.

1. What is a mutation?
2. What factors cause mutations?
3. Why is a mutation in a cell of an embryo likely to be more significant than one in a cell of an adult?
4. How can the number of mutations be reduced?

Chapter Summary

Introduction
A cell continuously carries on metabolic processes.

Metabolic Processes
1. Anabolic metabolism
   a. Consists of constructive processes (such as dehydration synthesis) in which smaller molecules are used to build larger ones.
   b. In dehydration synthesis, hydrogen atoms and hydroxyl groups are removed, water is formed, and smaller molecules become bonded by shared atoms.
   c. Carbohydrates are synthesized from monosaccharides, fats are synthesized from glycerol and fatty acids, and proteins are synthesized from amino acids.
2. Catabolic metabolism
   a. Consists of decomposition processes (such as hydrolysis) in which larger molecules are broken down into smaller ones.
   b. In hydrolysis, a water molecule is used to supply a hydrogen atom to one portion of a molecule and a hydroxyl group to a second portion; the bond between these two portions is broken.
   c. Carbohydrates are decomposed into monosaccharides, fats are decomposed into glycerol and fatty acids, and proteins are decomposed into amino acids.

Control of Metabolic Reactions

Metabolic reactions are controlled by enzymes.

1. Enzymes and their actions
   a. Enzymes are proteins that initiate or speed reactions without being changed themselves.
   b. An enzyme acts upon a specific substrate.
   c. The shape of an enzyme molecule seems to fit the shape of its substrate molecule.
   d. When an enzyme combines with its substrate, the substrate is changed, resulting in a product, while the enzyme is unaltered.
   e. The rate of enzyme-controlled reactions depends upon the number of enzyme and substrate molecules present and the efficiency of the enzyme.
2. Coenzymes
   a. Coenzymes are necessary parts of some enzyme molecules.
   b. A coenzyme may be an ion or a small organic molecule.
   c. Vitamins, which act as coenzymes, usually cannot be synthesized by human cells and must be obtained from foods.
3. Factors that alter enzymes
   a. Enzymes are proteins and can be denatured.
   b. Factors that may denature enzymes include excessive heat, radiation, electricity, and certain chemical substances.

Energy for Metabolic Reactions

Energy is a capacity to produce change or to do work.

Common forms of energy include heat, light, sound, electrical energy, mechanical energy, and chemical energy.

Energy can be converted from one form into another, though it cannot be created or destroyed.

1. Release of chemical energy
   a. Most metabolic processes utilize chemical energy that is released when molecular bonds are broken.
   b. The energy released from glucose during respiration is used to promote cellular metabolism.
   c. Cellular respiration is controlled by enzymes in the cytoplasm and mitochondria.
2. Anaerobic respiration
   a. The first phase of glucose decomposition occurs in the cytoplasm and is anaerobic.
   b. Some of the energy released is transferred to molecules of ATP.
3. Aerobic respiration
   a. The second phase of glucose decomposition occurs within the mitochondria and is aerobic.
   b. Considerably more energy is transferred to ATP molecules during this phase than during the anaerobic phase.
   c. The final products of glucose decomposition are carbon dioxide, water, and energy.

4. ATP molecules
   a. Thirty-eight molecules of ATP are produced for each glucose molecule that is decomposed.
   b. Energy is stored in the bond of the terminal phosphate of each ATP molecule.
   c. When energy is needed by a cellular process, the terminal phosphate bond of an ATP molecule is broken and the stored energy released.
   d. An ATP molecule that loses its terminal phosphate becomes an ADP molecule.
   e. An ADP can be converted to an ATP by capturing some energy and a phosphate.

Metabolic Pathways

Metabolic processes usually involve a number of steps that must occur in the correct sequence.

A sequence of enzyme-controlled reactions is called a metabolic pathway.

Typically metabolic pathways are interconnected.

1. Carbohydrate pathways
   a. Carbohydrates may enter catabolic pathways and be used as energy sources.
   b. When present in excess, carbohydrates may enter anabolic pathways and be converted to glycogen or fat.
2. Lipid pathways
   a. Most dietary fats are triglycerides.
   b. Before fats can be used as an energy source they must be converted into glycerol and fatty acids.
   c. Fatty acids can be changed to acetyl coenzyme A and ketone bodies which in turn can be oxidized by the citric acid cycle.
   d. Fats can be synthesized from glycerol and fatty acids and from excess glucose or amino acids.
3. Protein pathways
   a. Proteins are used as building materials for cellular parts, as enzymes, and as energy sources.
   b. Before proteins can be used as energy sources, they must be decomposed into amino acids, and the amino acids must be deaminated.
   c. The deaminated portions of amino acids can be broken down into carbon dioxide and water, or can be converted into glucose or fat.
   d. About 8 essential amino acids cannot be synthesized by human cells and must be obtained in foods.

Nucleic Acids and Protein Synthesis

DNA molecules contain information that instructs a cell how to synthesize the correct enzymes.

1. Genetic information
   a. Inherited traits result from DNA information that is passed from parents to child.
   b. A gene is a portion of a DNA molecule that contains the genetic information for making one kind of protein.
2. DNA molecules
   a. A DNA molecule consists of two strands of nucleotides twisted into a double helix.
   b. The nucleotides of a DNA strand are arranged in a particular sequence and are paired with those of the second strand.
3. The genetic code
   a. The sequence of nucleotides in a DNA molecule represents the sequence of amino acids in a protein molecule.
   b. Genetic information is transferred from the nucleus to the cytoplasm by RNA molecules.
4. RNA molecules
   a. RNA molecules are usually a single strand, contain ribose instead of deoxyribose, and contain uracil nucleotides in place of thymine nucleotides.
   b. Messenger RNA molecules, which are synthesized in the nucleus, contain a nucleotide sequence that is complementary to that of an exposed strand of DNA.
   c. Messenger RNA molecules move into the cytoplasm, become associated with ribosomes, and act as patterns for the synthesis of protein molecules.
5. Protein synthesis
   a. Molecules of transfer RNA serve to position amino acids along a strand of messenger RNA, and these amino acids become bonded to form a protein molecule.
   b. A ribosome binds to a messenger RNA molecule and allows a transfer RNA molecule to recognize its correct position on the messenger RNA.
   c. The ribosome contains enzymes needed for the synthesis of the developing protein and holds the protein until it is completed.

Changes in Genetic Information

During mitosis, duplicate sets of chromosomes, which contain genetic information, are evenly divided between the forming daughter cells.

Information within a DNA molecule is occasionally altered or mutated.

1. Nature of mutations
   a. Once genetic information has been changed, the mutation may be passed to future generations of cells.
   b. Mutations may involve the breaking of a chromosome or a change in the sequence of bases within a DNA molecule.
2. Effects of mutations
   a. The mutations that cause a decrease in cell efficiency are of most concern.
   b. A mutation that occurs in a cell of a developing embryo may affect a great number of cells in the forming body.
3. Mutagenic factors
   a. Factors that increase the mutation rate are called mutagenic.
   b. Mutagenic factors, which include various forms of radiation and a variety of chemicals, may stimulate cancerous growths.

Application of Knowledge

1. Since enzymes are proteins, they may be denatured. Relate this to the fact that a very high fever accompanying an illness may be life-threatening.

2. Some reducing diets drastically limit the dieter's intake of carbohydrates but allow liberal use of fat and protein foods. What changes would such a diet cause in the cellular metabolism of the dieter? What changes could be noted in the urine of such a person?

Review Activities

1. Distinguish between anabolic and catabolic metabolism.

2. Distinguish between dehydration synthesis and hydrolysis.

3. Explain what is meant by a peptide bond.

4. Define *enzyme*.

5. Describe how an enzyme is thought to interact with its substrate.

6. List three factors related to the rates of enzyme-controlled reactions.

7. Define *coenzyme*.

8. Explain why humans need vitamins in their diets.

9. Explain how an enzyme may be denatured.

10. Define *energy*.

11. Explain how oxidation of molecules inside cells differs from burning of substances.

12. Distinguish between anaerobic and aerobic respiration.

13. Explain the importance of ATP to cellular processes.

14. Describe the relationship between ATP and ADP.

15. Explain what is meant by a metabolic pathway.

16. Describe what happens to carbohydrates that enter catabolic pathways.

17. Explain how fats may serve as energy sources.

18. Define *deamination* and explain its importance.

19. Explain why some amino acids are essential.

20. Explain what is meant by genetic information.

21. Describe the relationship between a DNA molecule and a gene.

22. Describe the general structure of a DNA molecule.

23. Explain how genetic information is stored in a DNA molecule.

24. Distinguish between messenger RNA and transfer RNA.

25. Explain the function of a ribosome in protein synthesis.

26. Define *mutation* and explain how mutations may originate.

27. Explain how a mutation may affect a cell or an organism.

28. Explain what is meant by a mutagenic factor and list several such factors.

Suggestions for Additional Reading

Baker, J. J., and Allen, G. E. 1974. *Matter, energy and life*. Palo Alto: Addison-Wesley.

Becker, W. M. 1977. *Energy and the living cell*. Philadelphia: Lippincott.

Bender, M. L., and Brubacher, L. J. 1973. *Catalysis and enzyme action*. New York: McGraw-Hill.

Brown, D. D. August 1973. The isolation of genes. *Scientific American*.

Carter, L. C. 1973. *Guide to cellular energetics*. San Francisco: W. H. Freeman.

Eyre, D. R. 1980. Collagen: Molecular diversity in the body's protein scaffold. *Science* 207:1315.

Goldsby, R. A. 1977. *Cells and energy*, 3rd ed. New York: Macmillan.

Koshland, D. E. October 1973. Protein shape and biological control. *Scientific American*.

Lehnigher, A. 1971. *Bioenergetics*, 2nd ed. Reading, Mass.: Benjamin.

Lieber, C. S. March 1976. The metabolism of alcohol. *Scientific American*.

Maniatis, T., and Ptashne, M. January 1976. A DNA operator-repressor system. *Scientific American*.

Mirsky, A. E. June 1968. The discovery of DNA. *Scientific American*.

Roller, A. 1974. *Discovering the basis of life*. New York: McGraw-Hill.

Smith, A. E. 1976. *Protein biosynthesis*. New York: Wiley.

Watson, J. D. 1968. *The double helix*. New York: Antheneum Publishers.

Tissues

5 Cells are the basic units of structure and function within the human organism, and they are organized into groups and layers called *tissues*.

Each type of tissue is composed of similar cells that are specialized to carry on particular functions. For example, epithelial tissues form protective coverings and function in secretion and absorption. Connective tissues provide support for softer body parts and bind structures together. Muscle tissue is responsible for producing body movements, and nerve tissue is specialized to conduct impulses that help to control and coordinate body activities.

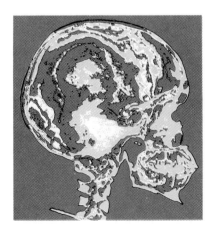

Chapter Outline

Chapter Objectives

After you have studied this chapter, you should be able to

1. Describe the general characteristics and functions of epithelial tissue.
2. Name the major types of epithelium and identify an organ in which each is found.
3. Explain how glands are classified.
4. Describe the general characteristics of connective tissue.
5. List the major types of connective tissues that occur within the body.
6. Describe the major functions of each type of connective tissue.
7. Distinguish between the three types of muscle tissue.
8. Describe the general characteristics and functions of nerve tissue.
9. Complete the review activities at the end of this chapter.

adipose tissue   (ad'ĭ-pōs tish'u)

cartilage   (kar'tĭ-lij)

connective tissue   (kŏ-nek'tiv tish'u)

epithelial tissue   (ep''ĭ-the'le-al tish'u)

fibroblast   (fi'bro-blast)

fibrous tissue   (fi'brus tish'u)

macrophage   (mak'ro-fāj)

muscle tissue   (mus'el tish'u)

nerve tissue   (nerv tish'u)

neuroglia   (nu-rog'le-ah)

neuron   (nu'ron)

osteocyte   (os''te-o-sīt'')

reticuloendothelial tissue
   (rĕ-tik''u-lo-en''do-the'le-al tish'u)

vascular tissue   (vas'ku-lar tish'u)

adip-, fat: *adip*ose tissue—tissue that stores fat.

-cyt, cell: osteo*cyt*e—a bone cell.

epi-, upon: *epi*thelial tissue—tissue that covers all body surfaces.

-glia, glue: neuro*glia*—cells that bind nerve tissue together.

inter-, between: *inter*calated disc—band located between the ends of adjacent cardiac muscle cells.

macro-, large: *macro*phage—a large phagocytic cell.

neuro-, nerve: *neuro*n—a nerve cell.

osseo-, bone: *osseo*us tissue—bone tissue.

phago-, to eat: *phago*cyte—a cell that engulfs and destroys foreign particles.

pseudo-, false: *pseudo*stratified epithelium—tissue whose cells appear to be arranged in layers, but are not.

squam-, scale: *squam*ous epithelium—tissue whose cells appear flattened or scalelike.

stratum, layer: *strat*ified epithelium—tissue whose cells occur in layers.

stria-, groove: *stria*ted muscle—tissue whose cells are characterized by alternating light and dark cross-markings.

In all complex organisms, cells are organized into layers or groups called **tissues**. Although the cells of different tissues vary in size, shape, arrangement, and function, those within a tissue are quite similar.

Usually tissue cells are separated by nonliving, intercellular materials, which they secrete. These intercellular materials vary in composition from one tissue to another and may take the form of solids, semisolids, or liquids. For example, bone tissue cells are separated by a solid intercellular substance, while blood tissue cells are separated by a liquid.

The tissues of the human body include four major types: *epithelial tissues,* which cover or line all body surfaces; *connective tissues,* which bind and support body parts; *muscle tissues,* which cause parts to move, and *nerve tissue,* which is sensitive to changes and transmits impulses from one area to another. These tissues are organized into organs that have specialized functions.

This chapter is concerned primarily with various types of epithelial and connective tissues. Other kinds of tissues will be discussed in later chapters.

---

1. What is meant by a tissue?
2. List the four major types of tissues.

Epithelial Tissues

General Characteristics. **Epithelial tissues** are widespread throughout the body. They cover all body surfaces—inside and out—forming the outer layer of the skin and the inner lining of body cavities, covering the organs inside these cavities and lining the chambers of hollow organs. They are also the major tissues of glands.

Since epithelium covers parts and lines hollow organs, it always has a free surface—one that is exposed to the outside or to an open space internally. The underside of the tissue is anchored to connective tissue by a thin, nonliving layer called the *basement membrane.*

Epithelial tissues lack blood vessels. For this reason the skin sometimes does not bleed when its surface is cut superficially. However, epithelial cells are nourished by substances that diffuse from underlying connective tissues, which are well supplied with blood vessels.

Although the cells of some tissues have limited abilities to reproduce, those of epithelium reproduce readily. Injuries to epithelium are likely to heal rapidly as new cells replace lost or damaged ones. Skin cells and the cells that line the stomach and intestines, for example, are continually being damaged and replaced.

Epithelial cells are tightly packed, and there is little intercellular material between them. Often they are attached by desmosomes. (See chapter 3.) Consequently, these cells can provide effective protective barriers, such as the skin and the lining of the mouth. Other epithelial functions include secretion, absorption, excretion, and sensory reception.

Epithelial tissues are classified according to the specialized shapes, arrangements, and functions of their cells. For example, epithelial tissues that are composed of single layers of cells are called *simple,* those with many layers of cells are said to be *stratified,* those with thin, flattened cells are called *squamous,* those with cubelike cells are called *cuboidal,* and those with elongated cells are called *columnar.*

Simple Squamous Epithelium

As its name implies, **simple squamous epithelium** consists of a single layer of thin, flattened cells. These cells fit tightly together, somewhat like floor tiles, and their nuclei are usually broad and thin. (See fig. 5.1.)

As a rule, substances pass rather easily through this type of tissue, and it often occurs where diffusion and filtration are taking place. For instance, simple squamous epithelium lines the air sacs of the lungs where oxygen and carbon dioxide gases are exchanged. It also forms the walls of capillaries, lines the insides of blood and lymph vessels, and covers the membranes that line body cavities.

Simple Cuboidal Epithelium

Simple cuboidal epithelium consists of a single layer of cube-shaped cells. These cells usually have centrally located spherical nuclei. (See fig. 5.2.)

This tissue covers the ovaries, lines the kidney tubules, and lines the ducts of various glands, such as the salivary glands, thyroid gland, pancreas, and liver. In the kidneys, it functions in secretion and absorption, while in the glands it is concerned with the secretion of the glandular products.

Fig. 5.1 Simple squamous epithelium consists of a single layer of tightly packed flattened cells.

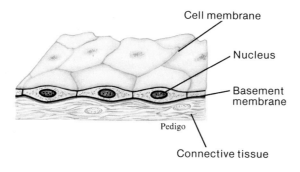

Cell membrane

Nucleus

Basement membrane

Pedigo

Connective tissue

Fig. 5.2 Simple cuboidal epithelium consists of a single layer of tightly packed cube-shaped cells.

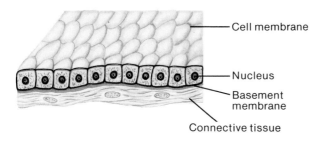

Cell membrane

Nucleus

Basement membrane

Connective tissue

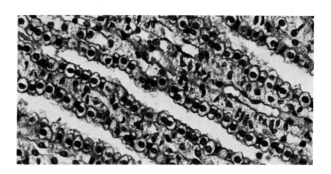

Fig. 5.3 Simple columnar epithelium consists of a single layer of elongated cells. What is the function of the goblet cells found among these columnar cells?

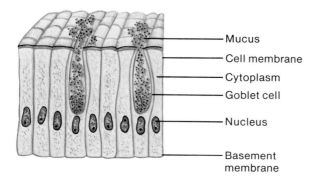

Mucus

Cell membrane

Cytoplasm

Goblet cell

Nucleus

Basement membrane

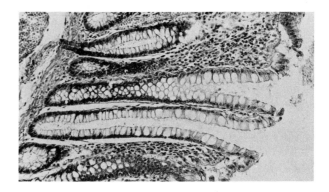

Simple Columnar Epithelium

The cells of **simple columnar epithelium** are elongated; that is, they are longer than they are wide. This tissue is composed of a single layer of cells whose nuclei are usually located at about the same level near the basement membrane.

Simple columnar epithelium occurs in the linings of the uterus and various organs of the digestive tract, including the stomach and intestines. Since its cells are elongated, the tissue is relatively thick, providing protection for underlying tissues. It also functions in the secretion of digestive fluids and in the absorption of nutrient molecules resulting from the digestion of foods.

Typically, there are specialized, flask-shaped glandular cells scattered along the columnar cells of this tissue. These cells, called *goblet cells,* secrete a thick, protective fluid called *mucus* onto the free surface of the tissue. (See fig. 5.3 and color plate 8.)

Fig. 5.4 Pseudostratified columnar epithelium appears stratified because the nuclei of various cells are located at different levels. This tissue contains goblet cells, and the columnar cells are ciliated.

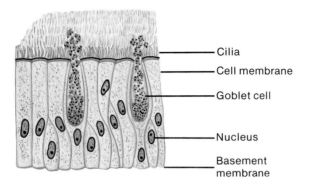

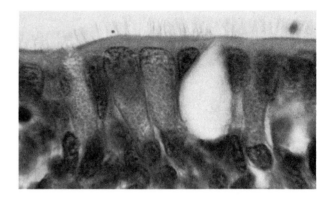

— Cilia

— Cell membrane

— Goblet cell

— Nucleus

— Basement membrane

Fig. 5.5 Stratified squamous epithelium consists of many layers of cells. The cells in the deeper layers, where reproduction occurs, are nearly cuboidal; however, as the newer cells grow, older ones become flattened as they are pushed toward the surface.

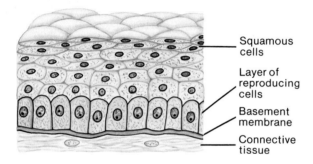

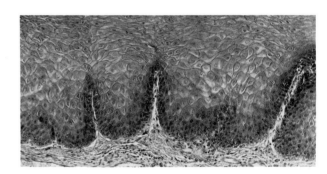

Squamous cells

Layer of reproducing cells

Basement membrane

Connective tissue

Pseudostratified Columnar Epithelium

The cells of **pseudostratified columnar epithelium** appear to be stratified or layered, but they are not. The layered effect occurs because nuclei are located at two or more levels within the cells of the tissue. However, the cells, which vary in shape, all reach the basement membrane, even though some of them may not contact the free surface.

These cells commonly possess microscopic hairlike projections called *cilia*. The cilia extend from the free surfaces of the cells, moving constantly. Goblet cells are also scattered throughout this tissue, and the mucus they secrete is swept along by the activity of the cilia. (See fig. 5.4.)

Pseudostratified columnar epithelium is found lining the passages of the respiratory system and in various tubes of the reproductive systems. In the respiratory passages, the mucus-covered linings are sticky and tend to trap particles of dust and microorganisms that enter with the air. The cilia of the columnar cells move the mucus and its captured particles upward and out of the airways. In the reproductive tubes, the cilia aid in moving sex cells from one region to another.

Stratified Squamous Epithelium

Stratified squamous epithelium consists of many layers of cells, making this tissue relatively thick. Only the cells near the free surface, however, are likely to be flattened. Those in the deeper layers, where cellular reproduction occurs, are usually cuboidal or columnar. As the newer cells grow, older ones are pushed further and further outward, and they tend to become flattened. (See fig. 5.5 and color plate 9.)

In the case of the skin, the older outermost layers of cells (epidermis) accumulate a protein called *keratin*. As this happens, the cells become hardened and die. This action produces a covering of dry, tough, protective material that prevents the escape of water from underlying tissues and the entrance of various microorganisms.

Chart 5.1   Epithelial tissues

Type	Function	Location
Simple squamous epithelium	Filtration, diffusion, osmosis	Air sacs of lungs, walls of capillaries, linings of blood and lymph vessels
Simple cuboidal epithelium	Secretion, absorption	Surface of ovaries, linings of kidney tubules, and linings of ducts of various glands
Simple columnar epithelium	Protection, secretion, absorption	Linings of uterus and tubes of the digestive tract
Pseudostratified columnar epithelium	Protection, secretion, movement of mucus and cells	Linings of respiratory passages and various tubes of the reproductive systems
Stratified squamous epithelium	Protection	Outer layers of skin, linings of mouth cavity, throat, vagina, and anal canal

Stratified squamous epithelium also lines the mouth cavity, throat, vagina, and anal canal. In these parts, the tissue is not keratinized; it remains moist, and the cells on the free surfaces remain alive.

Chart 5.1 summarizes the characteristics of the different types of epithelial tissue.

---

1. List the general characteristics of epithelial tissue.
2. Explain how epithelial tissues are classified.
3. Describe the general functions of each type of epithelium.

Glandular Epithelium

Glandular epithelium is composed of cells that are specialized to manufacture and secrete various substances into ducts or into body fluids. Such glandular cells are usually found within columnar or cuboidal epithelium, and one or more of them constitutes a *gland*. Those glands that secrete their products into ducts that open onto some internal or external surface are called **exocrine glands**. Those that secrete into tissue fluid or blood are called **endocrine glands**.

Although a gland may be a single cell (unicellular gland), most glands are composed of many cells (multicellular gland). These more complex glands can be classified according to the arrangement of their epithelial cells. For example, exocrine glands that consist of simple epithelial-lined tubes opening to the surface are called *tubular* glands, while those that are composed of one or more saclike dilations connected to the surface by narrowed secretory ducts are called *alveolar* glands (fig. 5.6).

Fig. 5.6 (*a*) A single-celled gland is represented by a mucus-secreting goblet cell; (*b*) tubular glands are multicellular structures that consist of simple epithelial-lined tubes; (*c*) alveolar glands are composed of saclike dilations connected to a secretory duct.

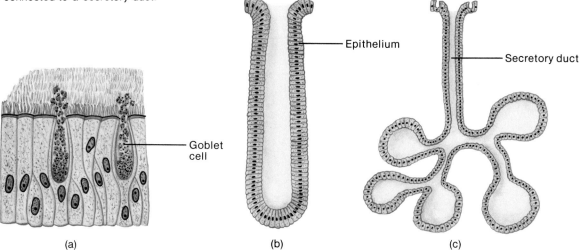

(a) (b) (c)

Fig. 5.7 (*a*) Cells of merocrine glands secrete without a loss of cytoplasm; (*b*) those of apocrine glands lose portions of their cells; and (*c*) holocrine glands release whole cells filled with secretory products.

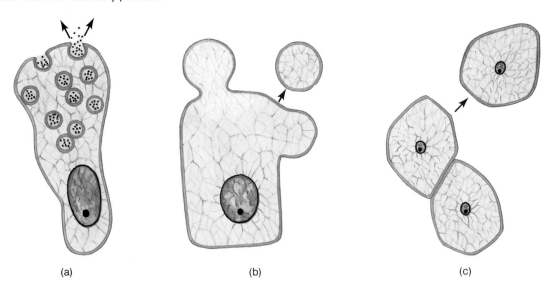

(a)　　　　(b)　　　　(c)

Chart 5.2　Types of glandular secretions

Type	Description of Secretion	Example
Merocrine glands	A fluid cellular product that is released through the cell membrane	Salivary glands, pancreatic glands, certain sweat glands of the skin
Apocrine glands	Cellular product and portions of the free ends of glandular cells that are pinched off during secretion	Mammary glands, certain sweat glands of the skin
Holocrine glands	Entire cells that are laden with secretory products	Sebaceous glands of the skin

Glandular secretions are classified according to whether they are a cellular product or portions of the glandular cells. Glands that release fluid cellular products through cell membranes without the loss of cytoplasm are called *merocrine* glands. Those that release entire cells filled with secretory products are called *holocrine* glands, while those of an intermediate type that lose small portions of their glandular cell bodies during secretion are called *apocrine* glands. (See fig. 5.7.) Chart 5.2 summarizes these glands and their secretions.

Most secretory cells are merocrine, and they can be further subdivided as either *serous cells* or *mucous cells*. The secretion of serous cells is typically watery and has a high concentration of enzymes. Such cells are common in the glands of the digestive tract. Mucous cells secrete the thicker fluid called *mucus*. This substance is rich in the glycoprotein mucin and is secreted abundantly from the inner linings of the digestive and respiratory tubes.

1. *Distinguish between exocrine and endocrine glands.*
2. *Explain how glands are classified according to their structures; according to their secretions.*

Connective Tissues

General Characteristics. **Connective tissues** occur in all parts of the body. They bind structures together, provide support and protection, serve as frameworks, fill spaces, store fat, and produce blood cells. Connective tissue cells are usually further apart than epithelial cells, and they have an abundance of intercellular material between them. This material consists of fibers and a thick, gellike fluid called the *matrix.*

Connective tissue cells are able to reproduce. In most cases, they have good blood supplies and are well nourished. Although some connective tissues, such as bone and cartilage, are quite rigid, loose connective tissue, adipose connective tissue, and fibrous connective tissue are more flexible.

Major Cell Types. Connective tissues contain a variety of cell types. Some of them are called resident cells because they are usually present in relatively stable numbers. These include fibroblasts, macrophages, and mast cells. Another group known as wandering cells temporarily appear in the tissues, usually in response to an injury or infection. The wandering cells include several types of white blood cells.

The fibroblast is the most common kind of cell in connective tissue. It is relatively large and usually star-shaped. Fibroblasts function to produce white and yellow fibers of protein in the intercellular materials of connective tissues.

The white fibers are composed of the protein *collagen,* which is the major structural protein of the body. (It is also the main ingredient of leather and gelatin.) Collagenous fibers occur in bundles and are strong, flexible, and only slightly elastic.

Yellow fibers are composed of the protein *elastin,* which has less strength than collagen but is highly elastic. These fibers can be stretched and will return to their original length. They commonly branch to form dense networks and are abundant in tissues that must normally stretch.

When skin is exposed to sunlight excessively, its connective tissue fibers tend to lose their elasticity, and the skin becomes increasingly stiff. In time it may sag and wrinkle. On the other hand, the skin of a healthy, well-nourished person usually shows minimal change with age, provided it remains covered.

Fig. 5.8 Macrophages are common in connective tissues where they function as scavenger cells.

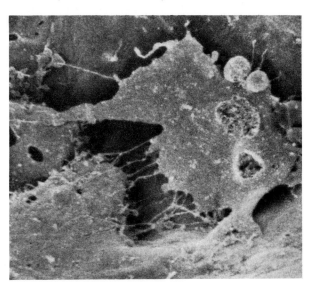

Macrophages (histiocytes) are almost as numerous as fibroblasts in some connective tissues. They are usually attached to fibers and are specialized to carry on phagocytosis—the process by which solid particles are engulfed and destroyed. Since they function as scavenger cells that can clear foreign particles from tissues, macrophages represent an important defense against infectious bacteria. They also play a role in immunity, which is discussed in chapter 18. (See fig. 5.8.)

Mast cells are relatively large cells that are widely distributed in connective tissues, and are usually located near blood vessels. Their function is not well understood, however, they are known to contain *heparin,* a compound that prevents blood clotting. They also contain *histamine,* a substance that promotes some of the reactions associated with allergies such as asthma and hayfever. (See chapter 18 for more information about allergies.)

Loose Connective Tissue

Loose connective tissue forms delicate, thin membranes throughout the body. The cells of this tissue, which are mainly fibroblasts, are located some distance apart and are separated by a jellylike intercellular matrix that contains many white and yellow fibers (fig. 5.9 and color plate 10).

This tissue binds the skin to the underlying organs and fills spaces between muscles. It lies beneath most layers of epithelium, where its numerous blood vessels provide nourishment for the epithelial cells.

Sometimes tissue fluids accumulate in loose connective tissues and cause them to swell. Such a condition, called *edema,* occurs for a variety of reasons, including damage to blood vessels and changes in the osmotic pressure of the blood.

Adipose Tissue

Adipose tissue, commonly called fat, is really a specialized form of loose connective tissue. Certain cells within loose connective tissue (adipocytes) store fat in droplets within their cytoplasm. At first these cells resemble fibroblasts, but as they accumulate fat they become swollen, and their nuclei are pushed to one side. (See fig. 5.10.) When they occur in such large numbers that other cell types are crowded out, they form the tissue called adipose tissue.

Adipose tissue is found beneath the skin and in the spaces between the muscles. It also occurs around the kidneys, behind the eyeballs, in certain abdominal membranes, on the surface of the heart, and around various joints.

Adipose tissue serves as a protective cushion for joints and some organs, such as the kidneys. It also functions as a heat insulator beneath the skin and stores energy in fat molecules.

Fig. 5.9 Loose connective tissue contains numerous fibroblasts that produce white and yellow fibers.

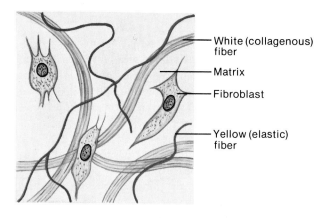

- White (collagenous) fiber
- Matrix
- Fibroblast
- Yellow (elastic) fiber

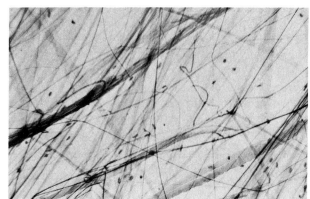

Fig. 5.10 Adipose tissue cells contain large fat droplets, causing the nuclei to be pushed close to the cell membranes. How is this tissue related to loose connective tissue?

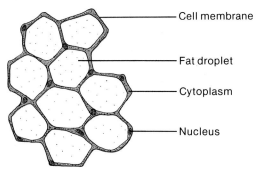

- Cell membrane
- Fat droplet
- Cytoplasm
- Nucleus

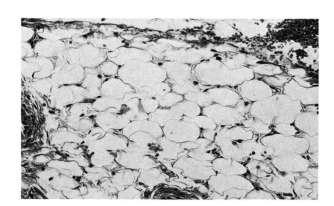

The amount of adipose tissue present in the body is usually related to a person's diet, for excess food substances are likely to be converted to fat and stored in adipose cells. During a period of fasting, adipose cells may lose their fat droplets, shrink in size, and become more like fibroblasts again.

Some investigators believe that overnutrition during infancy increases the total number of adipose cells, and that this increase may cause susceptibility to obesity later in life.

Fibrous Connective Tissue

Fibrous connective tissue contains many, closely packed, thick, collagenous fibers and a fine network of elastic fibers. It has relatively few cells, almost all of which are fibroblasts (fig. 5.11).

Since white fibers are very strong, this type of tissue can withstand pulling forces, and it often functions to bind body parts together. **Tendons,** which connect muscles to bones, and **ligaments,** which connect bones to bones at joints, are composed of fibrous connective tissue. This type of tissue also occurs in the protective white layer of the eyeball and in the deeper portions of the skin.

The blood supply to fibrous connective tissue is relatively poor, so tissue repair occurs slowly. This is why a sprain, which involves damage to the tissues surrounding a joint, may take some time to heal.

1. What are the general characteristics of connective tissue?
2. What is the primary function of fibroblasts?
3. What are the characteristics of collagen and elastin?
4. How is loose connective tissue related to adipose tissue?

Cartilage

Cartilage is one of the rigid connective tissues. It supports parts, provides frameworks and attachments, and protects underlying tissues.

The intercellular material of cartilage is abundant and is composed largely of fibers embedded in a solid matrix. The matrix contains a gellike substance rich in a protein-polysaccharide complex (chondromucoprotein). The cartilage cells (chondrocytes) occupy small chambers called *lacunae,* which are completely surrounded by matrix.

Although cartilage lacks a direct blood supply, there are blood vessels in the connective tissues that surround it. The cartilage cells obtain nutrients from these vessels by diffusion through the gellike matrix. This lack of a direct blood supply apparently is related to the slow rate of cellular reproduction and repair that is characteristic of cartilage.

Fig. 5.11 Fibrous connective tissue consists largely of tightly packed collagenous fibers.

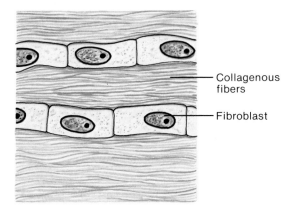

Collagenous fibers

Fibroblast

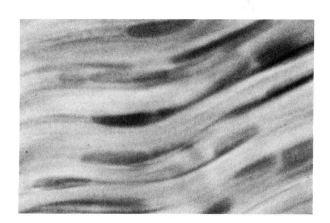

Fig. 5.12 Hyaline cartilage consists of cells located in lacunae, and an intercellular material containing very fine, white fibers.

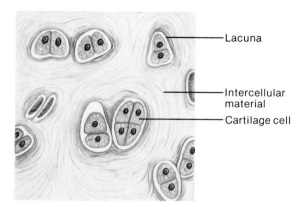

Lacuna

Intercellular material

Cartilage cell

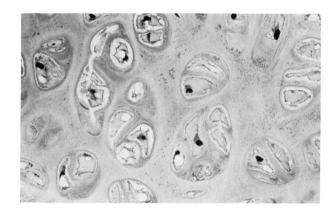

Fig. 5.13 Elastic cartilage contains fine, white fibers and many yellow elastic fibers in its intercellular material.

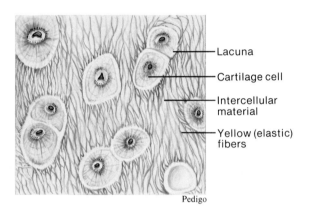

Lacuna

Cartilage cell

Intercellular material

Yellow (elastic) fibers

Pedigo

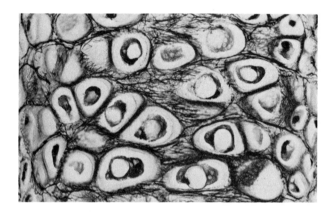

There are three kinds of cartilage, and each contains a different type of intercellular material. Hyaline cartilage has very fine white fibers in its matrix; elastic cartilage contains a dense network of yellow fibers, while fibrocartilage contains many large white fibers.

Hyaline cartilage (fig. 5.12), the most common type, looks somewhat like milk glass. It occurs on the ends of bones in many joints, in the soft part of the nose, and in the supporting rings of the respiratory passages. In an embryo, many of the skeletal parts are formed at first of hyaline cartilage, which is later replaced by bone.

Elastic cartilage (fig. 5.13), which is more flexible than hyaline cartilage because it contains many yellow fibers, provides the framework for the external ears and also occurs in parts of the larynx.

Fibrocartilage (fig. 5.14), a very tough tissue, often serves as a shock absorber for structures that are subjected to pressure. For example, fibrocartilage forms pads (intervertebral discs) between the individual parts of the backbone. It also forms protective cushions in the knees and between the bones of the pelvic girdle.

Bone

Bone (osseous tissue) is the most rigid of the connective tissues. Its hardness is due largely to the presence of mineral salts, such as calcium phosphate and calcium carbonate, in the intercellular matrix. This material also contains a considerable amount of collagen.

Bone provides an internal support for body structures. It protects vital parts in the cranial and thoracic cavities, and serves as an attachment for muscles. It also functions in forming blood cells and in storing various inorganic salts.

Fig. 5.14 Fibrocartilage contains many large, white fibers in the intercellular material.

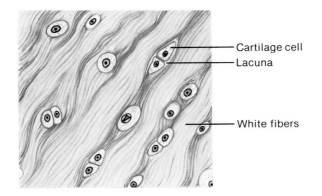

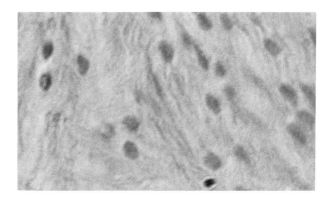

Cartilage cell
Lacuna

White fibers

Fig. 5.15 Bone matrix is deposited in concentric layers around haversian canals. The bone cells are located in lacunae and are interconnected by means of canaliculi.

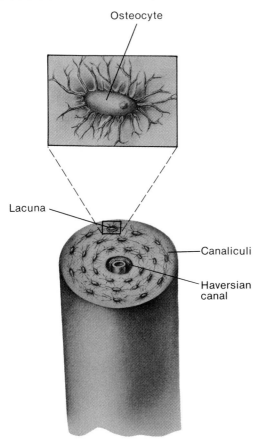

Osteocyte

Lacuna

Canaliculi

Haversian canal

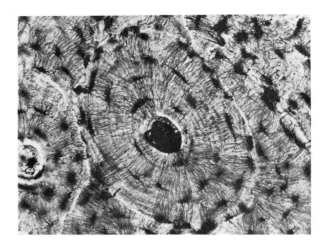

Bone matrix is deposited in thin layers called *lamellae,* which most commonly are arranged in concentric patterns around tiny longitudinal tubes called *haversian canals.* Bone cells called *osteocytes* are located in lacunae, rather evenly spaced between the lamellae. Consequently, they too are arranged in patterns of concentric circles. (See fig. 5.15 and color plate 11.)

Each haversian canal contains a blood vessel, so that every bone cell is fairly close to a nutrient supply. In addition, the bone cells have numerous cytoplasmic processes that extend outward and pass through minute tubes (*canaliculi*) in the matrix. These cellular processes meet those of nearby cells or connect to haversian canals, and as a result, materials can move rapidly between blood vessels and bone cells. Thus, in spite of its inert appearance, bone is a very active tissue that heals much more rapidly than cartilage.

Chart 5.3 lists the characteristics of the major types of connective tissue.

Other Connective Tissues

Other important connective tissues include blood (vascular tissue) and reticuloendothelial tissue.

Blood is composed of cells that are suspended in fluid intercellular matrix called *blood plasma*. These cells include red blood cells, white blood cells, and some cellular fragments called platelets. Most of the cells are formed by special tissues (hematopoietic tissues) found in the hollow parts of certain bones. (See fig. 5.16 and color plate 12.)

Chart 5.3 Connective tissues

Type	Function	Location
Loose connective tissue	Binds organs together, holds tissue fluids	Beneath the skin, between muscles, beneath most epithelial layers
Adipose tissue	Protection, insulation, and storage of fat	Beneath the skin, around the kidneys, behind the eyeballs, on the surface of the heart
Fibrous connective tissue	Binds organs together	Tendons, ligaments, skin
Hyaline cartilage	Support, protection, provides framework	Ends of bones, nose, rings in walls of respiratory passages
Elastic cartilage	Support, protection, provides framework	Framework of external ear and part of the larynx
Fibrocartilage	Support, protection	Between bony parts of backbone, pelvic girdle, and knee
Bone	Support, protection, provides framework	Bones of skeleton

Fig. 5.16 Blood tissue consists of an intercellular fluid in which red blood cells, white blood cells, and blood platelets are suspended.

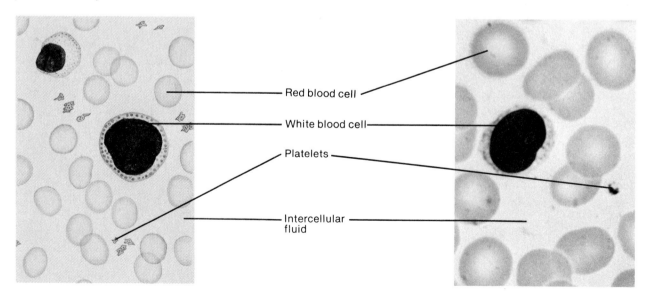

Red blood cell

White blood cell

Platelets

Intercellular fluid

Of the blood cells, only the red cells function entirely within the blood vessels. The white cells migrate through capillary walls and enter connective tissues where they carry on their major activities. In this way, blood transports cells from the hollow parts of bones to other connective tissues where they usually reside until they die.

Reticuloendothelial tissue is composed of a variety of specialized cells that are widely scattered throughout the body. As a group, these cells are phagocytic, that is, they function to ingest and destroy foreign particles, such as microorganisms, that may invade the body. Thus, they are particularly important in defending the body against infection.

The reticuloendothelial cells include types found in the blood, lungs, brain, bone marrow, and lymph glands. The most common ones, however, are *macrophages*. Typically a macrophage remains in a fixed position until it "senses" a foreign particle. Then, the phagocyte becomes motile, moves toward the invader, and may engulf and destroy it. Once the invader has been destroyed, the macrophage becomes fixed again.

Reticuloendothelial tissue is discussed in more detail in chapter 18.

---

1. *Describe the general characteristics of cartilage.*
2. *Explain why injured bone heals more rapidly than cartilage.*
3. *Why is blood considered a connective tissue?*
4. *What do the cells of reticuloendothelial tissue have in common?*

Muscle Tissues

General Characteristics. **Muscle tissues** are contractile—their elongated cells or *muscle fibers* can change shape by becoming shorter and thicker. As they contract, the fibers pull at their attached ends and cause body parts to move. The three types of muscle tissue are skeletal muscle, smooth muscle, and cardiac muscle.

Skeletal Muscle Tissue

Skeletal muscle (fig. 5.17 and color plate 13) is found in muscles that usually are attached to bones and can be controlled by conscious effort. For this reason, it is said to be under *voluntary* control. The cells are long and threadlike, with alternating light and dark cross-markings called *striations*. Each cell, or muscle fiber, has many nuclei, located just beneath its cell membrane. When the cell is stimulated, it contracts and then relaxes. Usually a muscle fiber is stimulated to contract by an action of a nerve fiber. If the connecting nerve fiber is damaged, it may not be possible to control the muscle fiber's action, and the muscle fiber is paralyzed.

Skeletal muscles are responsible for moving the head, trunk, and limbs. They produce the movements people use in communicating with each other—facial expressions, writing, talking, and singing—and they are involved in such actions as chewing, swallowing, and breathing.

Fig. 5.17 Skeletal muscle tissue is composed of striated muscle fibers that contain many nuclei.

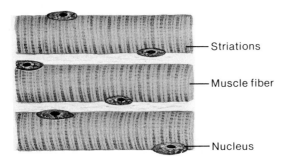

- Striations
- Muscle fiber
- Nucleus

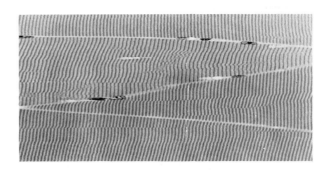

Fig. 5.18 Smooth muscle tissue is formed of spindle-shaped cells, each containing a single nucleus.

- Nucleus
- Cytoplasm
- Cell membrane

Fig. 5.19 How is this cardiac muscle tissue similar to skeletal muscle? How is it similar to smooth muscle?

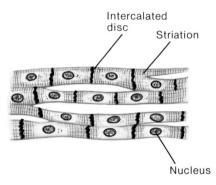

Intercalated disc
Striation
Nucleus

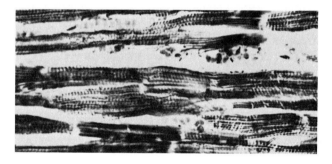

Smooth Muscle Tissue

Smooth muscle (fig. 5.18) is called smooth because it lacks striations. This tissue is found in the walls of hollow internal organs such as the stomach, intestines, urinary bladder, uterus, and blood vessels. Unlike skeletal muscle, smooth muscle usually cannot be stimulated to contract by conscious efforts. Its contractions result from *involuntary* nerve or gland activity. Smooth muscle cells are shorter than those of skeletal muscle, and they each have a single, centrally located nucleus. Smooth muscles are responsible for the movements that force food through the digestive tube, constrict blood vessels, and empty the urinary bladder.

Cardiac Muscle Tissue

Cardiac muscle tissue (fig. 5.19 and color plate 14) occurs only in the heart. Its cells, which are striated, are joined end to end. The resulting fibers are branched and interconnected in complex networks. Each cell has a single nucleus. At its end, where it touches another cell, there is a specialized intercellular junction called an *intercalated disc,* which occurs only in cardiac tissue.

Cardiac muscle, like smooth muscle, is controlled *involuntarily* and, in fact, can continue to function without being stimulated by nerve impulses. This tissue makes up the bulk of the heart and is responsible for pumping blood through the heart chambers and into the blood vessels.

1. *List the general characteristics of muscle tissue.*
2. *Distinguish between skeletal, smooth, and cardiac muscle tissue.*

Fig. 5.20 Nerve tissue includes (*a*) neurons, which function to transmit impulses, and (*b*) neuroglial cells, which support and bind the tissue together.

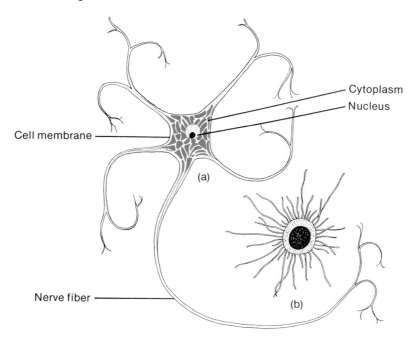

Cytoplasm
Nucleus
Cell membrane
(a)
Nerve fiber
(b)

Nerve Tissue

Nerve tissue is found in the brain, spinal cord, and associated nerves. The basic cells of this tissue are called nerve cells, or *neurons,* and of all body cells they seem to be the most highly specialized. Neurons are sensitive to certain types of changes in their surroundings, and they respond by transmitting nerve impulses along cytoplasmic extensions to other neurons or to muscles or glands.

As a result of the extremely complex patterns by which neurons are connected with each other and with various body parts, they are able to coordinate and regulate many body functions.

In addition to neurons, nerve tissue contains **neuroglial cells** (fig. 5.20 and color plate 15). These cells function to support and bind the nerve tissue together, carry on phagocytosis, and aid in supplying nutrients to neurons by connecting them to blood vessels.

Chart 5.4 summarizes the general characteristics of muscle and nerve tissues.

Chart 5.4   Muscle and nerve tissues

Type	Function	Location
Skeletal muscle tissue	Voluntary movements of skeletal parts	Muscles attached to bones
Smooth muscle tissue	Involuntary movements of internal organs	Walls of hollow internal organs
Cardiac muscle tissue	Heart movements	Heart muscle
Nerve tissue	Sensitivity and conduction of nerve impulses	Brain, spinal cord, and nerves

1. *Describe the general characteristics of nerve tissue.*
2. *Distinguish between neurons and neuroglial cells.*

Chapter Summary

Introduction
Cells of complex organisms are arranged in tissues.
Cells are separated by intercellular materials whose composition varies from solids to liquids.
The four major types of human tissues are epithelial tissue, connective tissue, muscle tissue, and nerve tissue.

Epithelial Tissues
1. General characteristics
   a. Covers all free body surfaces and comprises various glands.
   b. Anchored to connective tissue by basement membrane, lacks blood vessels, contains little intercellular material, replaced continuously.
   c. Functions in protection, secretion, absorption, excretion, and sensory reception.
2. Simple squamous epithelium
   a. Consists of a single layer of thin, flattened cells through which substances pass rather easily.
   b. Functions in exchanges of gases in the lungs; lines blood vessels, lymph vessels, and various membranes within the thorax and abdomen.
3. Simple cuboidal epithelium
   a. Consists of a single layer of cube-shaped cells.
   b. Carries on secretion and absorption in the kidneys and various glands.
4. Simple columnar epithelium
   a. Composed of elongated cells whose nuclei are located near the basement membrane.
   b. Lines the uterus and digestive tract where it functions in protection, secretion, and absorption.
   c. Contains goblet cells that secrete mucus.
5. Pseudostratified columnar epithelium
   a. Appears stratified because the nuclei are located at two or more levels within the cells.
   b. May have cilia that function to move mucus or cells over the surface of the tissue.
   c. Lines various tubes of the respiratory and reproductive systems.
6. Stratified squamous epithelium
   a. Composed of many cell layers.
   b. Protects underlying cells from harmful environmental effects.
   c. Covers the skin and lines the mouth, throat, vagina, and anal canal.
7. Glandular epithelium
   a. Composed of cells that are specialized to secrete substances.
   b. One or more cells can constitute a gland.
      (1) Exocrine glands secrete into ducts.
      (2) Endocrine glands secrete into tissue fluid or blood.
   c. Glands are classified according to the arrangement of their cells.
      (1) Tubular glands consist of simple epithelial-lined tubes.
      (2) Alveolar glands consist of saclike dilations connected to the surface by narrowed ducts.
   d. Glandular secretions are classified according to the contents of the secretions.
      (1) Merocrine glands secrete watery fluids without loss of cytoplasm. Most secretory cells, including serous cells and mucous cells, are merocrine.
         (a) Serous cells secrete watery fluid with a high enzyme content.
         (b) Mucous cells secrete mucus.
      (2) Apocrine glands lose portions of their cells during secretion.
      (3) Holocrine glands release cells filled with secretory products.

Connective Tissues
1. General characteristics
   a. Connects, supports, protects, provides frameworks, fills spaces, stores fat, and produces blood cells.
   b. Consists of cells, usually some distance apart, and considerable intercellular material composed of fibers and a gellike matrix.
2. Major cell types
   a. Fibroblasts produce collagenous and elastic fibers.
   b. Macrophages function as phagocytes.
   c. Mast cells contain heparin and histamine.
3. Loose connective tissue
   a. Forms thin membranes between organs and binds them together.
   b. Found beneath the skin and between muscles.
   c. Contains tissue fluids in intercellular spaces.
4. Adipose tissue
   a. A specialized form of loose connective tissue that stores fat, provides a protective cushion, and functions as a heat insulator.
   b. Found beneath the skin, in certain abdominal membranes, and around the kidneys, heart, and various joints.
5. Fibrous connective tissue
   a. Composed largely of strong, collagenous fibers that bind parts together.
   b. Found in tendons, ligaments, eyes, and skin.
6. Cartilage
   a. Provides support and framework for various parts.
   b. Intercellular material is largely composed of fibers and a gellike matrix.
   c. Lacks a direct blood supply and is slow to heal following an injury.
   d. Major types are hyaline cartilage, elastic cartilage, and fibrocartilage.
   e. Occurs at ends of various bones, in the ear, in the larynx, and in pads between bones of the backbone, pelvic girdle, and knees.

7. Bone
   a. Intercellular matrix contains mineral salts and collagen.
   b. Cells arranged in concentric circles around haversian canals and interconnected by canaliculi.
   c. An active tissue that heals rapidly following an injury.
8. Other connective tissues
   a. Blood
      (1) Composed of cells suspended in fluid.
      (2) Formed by special tissue in the hollow parts of bones.
      (3) Transports white blood cells to connective tissues.
   b. Reticuloendothelial tissue
      (1) Composed of a variety of phagocytic cells widely distributed in body organs.
      (2) Defends the body against invasion by microorganisms.

Muscle Tissues
1. General characteristics
   a. Contractile tissue that moves parts attached to it.
   b. Three types are skeletal, smooth, and cardiac muscle tissues.
2. Skeletal muscle tissue
   a. Attached to bones and controlled by conscious effort.
   b. Cells or muscle fibers are long and threadlike with alternating light and dark cross-markings.
   c. Muscle fibers contract when stimulated by nerve action and relax immediately.
3. Smooth muscle tissue
   a. Found in walls of hollow internal organs.
   b. Usually controlled by involuntary activity.
4. Cardiac muscle tissue
   a. Found only in the heart.
   b. Cells joined by intercalated discs and arranged in branched, interconnecting networks.
   c. Controlled by involuntary activity.

Nerve Tissue
1. Found in brain, spinal cord, and nerves.
2. Neurons
   a. Sensitive to changes and respond by transmitting nerve impulses to other neurons or to other body parts.
   b. Function in coordinating and regulating body activities.
3. Neuroglial cells
   a. Bind and support nerve tissue.
   b. Carry on phagocytosis.
   c. Connect neurons to blood vessels.

Application of Knowledge

1. The sweeping action of the ciliated epithelium that lines the respiratory passages is inhibited by excessive exposure to tobacco smoke. What special problems is this likely to create for a smoker?

2. Joints, such as the elbow, shoulder, and knee, contain considerable amounts of cartilage and fibrous connective tissue. How is this related to the fact that joint injuries are often very slow to heal?

3. There is a group of disorders called collagenous diseases that are characterized by deterioration of connective tissues. Why would you expect such diseases to produce widely varying symptoms?

Review Activities

1. Define *tissue*.

2. Name the four major types of tissue found in the human body, and list the general functions of each.

3. Describe the general characteristics of epithelial tissue.

4. Distinguish between a simple epithelium and a stratified epithelium.

5. Explain how the structure of simple squamous epithelium is related to its function.

6. Name an organ in which each of the following tissues is found, and give the function of the tissue in each case:
   a. Simple squamous epithelium
   b. Simple cuboidal epithelium
   c. Simple columnar epithelium
   d. Pseudostratified columnar epithelium
   e. Stratified squamous epithelium

7. Define *gland*.

8. Distinguish between an exocrine gland and an endocrine gland.

9. Explain how glands are classified according to the arrangement of their cells.

10. Explain how glands are classified according to the nature of their cellular secretions.

11. Distinguish between a serous cell and a mucous cell.

12. Describe the general characteristics of connective tissue.

13. Describe three major types of connective tissue cells.

14. Distinguish between collagen and elastin.

15. Explain the relationship between loose connective tissue and adipose tissue.

16. Explain how the quantity of adipose tissue in the body is related to diet.

17. Explain why injured fibrous connective tissue and cartilage are usually slow to heal.

18. Name the major types of cartilage, and describe their differences and similarities.

19. Describe how bone cells are arranged in bone tissue.

20. Describe vascular tissue.

21. Describe the composition of blood.

22. Define *reticuloendothelial tissue*.

23. Describe the general characteristics of muscle tissue.

24. Distinguish between skeletal, smooth, and cardiac muscle tissues.

25. Describe the general characteristics of nerve tissue.

26. Distinguish between neurons and neuroglial cells.

Suggestions for Additional Reading

Bevelander, G., and Ramaley, J. A. 1974. *Essentials of histology.* 7th ed. St. Louis: C. V. Mosby Co.

Bloom, W. B., and Fawcett, D. W. 1975. *A textbook of histology.* 10th ed. Philadelphia: W. B. Saunders.

DiFiore, M. S. M. 1974. *An atlas of human histology.* 4th ed. Philadelphia: Lea and Febiger.

Follis, B. D., and Ashworth, R. D. 1970. *Textbook of human histology.* Boston: Little, Brown and Co.

Leeson, T. S., and Leeson, C. R. 1970. *Histology.* 2nd ed. Philadelphia: W. B. Saunders.

Reith, E. J., and Ross, M. H. 1970. *Atlas of descriptive histology.* 2nd ed. New York: Harper and Row.

Weiss, L., and Greep, R. O. 1977. *Histology.* 4th ed. New York: McGraw-Hill.

Plate 8 Photomicrograph of simple columnar epithelium from the lining of the small intestine (duodenum). Note the goblet cells.

Plate 9 Photomicrograph of stratified squamous epithelium from the lining of the esophagus.

Plate 10 Photomicrograph of loose connective tissue. What components of this tissue can you identify?

Plate 11 What structures can you identify in this photomicrograph of a section of bone tissue?

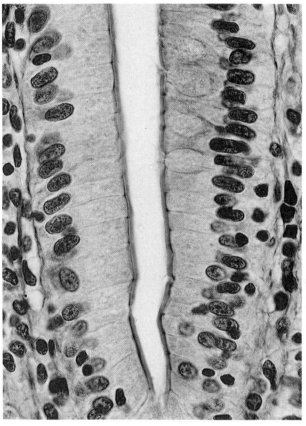

8

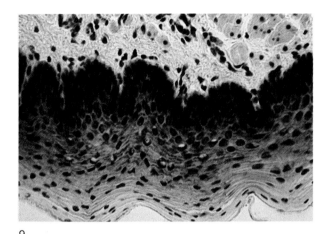

9

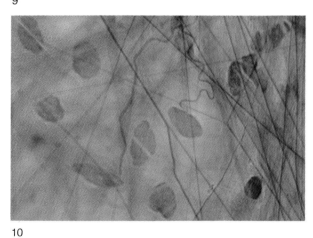

10

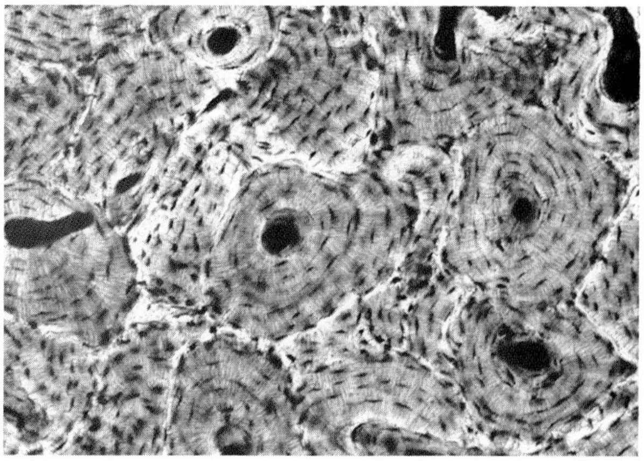

11

Plate 12 Photomicrograph of human blood cells. Note the two white blood cells with stained nuclei, surrounded by red blood cells.

Plate 13 Photomicrograph of skeletal muscle tissue. Note the striations of the individual muscle fibers.

Plate 14 Photomicrograph of cardiac muscle tissue. The dark lines are intercalated disks.

Plate 15 The dark cell in this photomicrograph is a neuron from the cerebellum of the brain. Note the numerous nerve fibers associated with this cell.

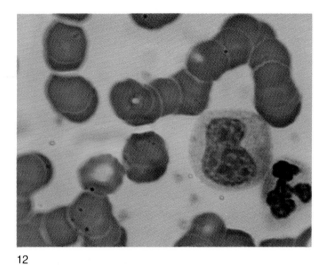

12

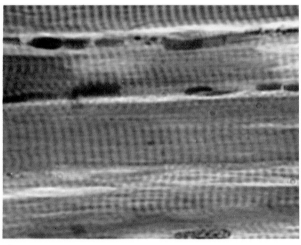

13

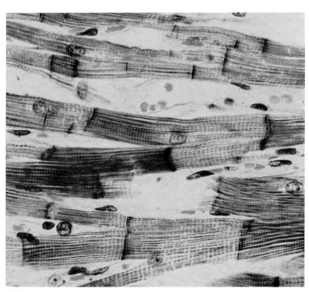

14

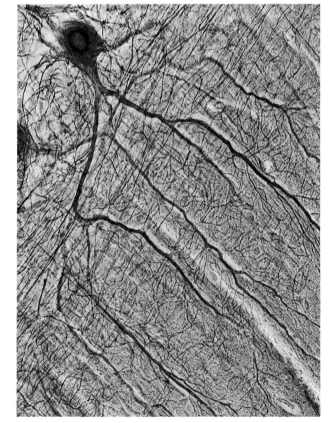

15

The Skin and the Integumentary System

6 The previous chapters dealt with the lower levels of complexity within the human organism—tissues, cells, cellular organelles, and the chemical substances that form these parts.

Chapter 6 explains how tissues are grouped to form organs and how related organs comprise organ systems. In this explanation the skin provides an example of an organ. It acts together with hair follicles, sebaceous glands, and sweat glands in providing a variety of vital functions. As a group, these organs constitute the *integumentary system*.

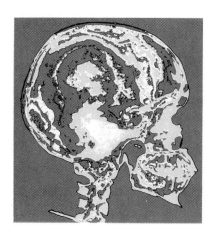

Chapter Outline

Chapter Objectives

After you have studied this chapter, you should be able to

1. Describe the four major types of membranes.
2. Describe the structure of the various layers of the skin.
3. List the general functions of each of these layers.
4. Describe the accessory organs associated with the skin.
5. Explain the functions of each accessory organ.
6. Explain how the skin functions in regulating body temperature.
7. Summarize the factors that determine skin color.
8. Discuss how the tissues of the skin respond to injuries.
9. Complete the review activities at the end of this chapter.

conduction   (kon-duk'shun)

convection   (kon-vek'shun)

cutaneous membrane (ku-ta'ne-us mem'brān)

dermis   (der'mis)

epidermis   (ep''ĭ-der'mis)

evaporation   (e-vap''o-ra'shun)

hair follicle   (hār fol'ĭ-kl)

inflammation   (in''flah-ma'shun)

integumentary   (in-teg-u-men'tar-e)

keratinization   (ker''ah-tin''ĭ-za'shun)

melanin   (mel'ah-nin)

mucous membrane   (mu'kus mem'brān)

radiation   (ra''de-a'shun)

sebaceous gland   (se-ba'shus gland)

serous membrane   (se'rus mem'brān)

subcutaneous   (sub''ku-ta'ne-us)

sweat gland   (swet gland)

alb-, white: *alb*ino—condition characterized by a lack of pigment.

cubit-, to lie down: de*cubit*us ulcer—condition that develops as a result of lying in one position for a prolonged time.

cut-, skin: sub*cut*aneous—beneath the skin.

derm-, skin: *derm*is—inner layer of the skin.

epi, upon: *epi*dermis—outer layer of the skin.

follic-, small bag: hair *follic*le—tubelike depression in which a hair develops.

inflamm-, set on fire: *inflamm*ation—condition characterized by reddened, warm, swollen tissues.

kerat-, horn: *kerat*in—protein produced as epidermal cells die and harden.

melan-, black: *melan*in—dark pigment produced by certain epidermal cells.

por-, channel: *por*e—opening by which a sweat gland communicates to the surface.

seb-, grease: *seb*aceous gland—gland that secretes an oily substance.

Two or more kinds of tissues grouped together and performing specialized functions constitute an organ. Thus, the sheetlike structures called *membranes* that cover body surfaces and line body cavities are organs.

Types of Membranes

There are four major types of membranes: *serous, mucous, cutaneous,* and *synovial membranes.* Usually these structures are relatively thin. Serous, mucous, and cutaneous membranes are composed of epithelial tissue and some underlying connective tissue; synovial membranes are composed entirely of connective tissue.

Serous membranes line the body cavities that lack openings to the outside. They form the inner linings of the thorax and abdomen, and they cover the organs within these cavities. Such a membrane consists of a layer of simple squamous epithelium (mesothelium) covering a thin layer of loose connective tissue. It secretes a watery *serous fluid,* which helps to lubricate the surfaces of the membrane.

Mucous membranes line the cavities and tubes that open to the outside of the body. These include the oral and nasal cavities and the tubes of the digestive, respiratory, urinary, and reproductive systems. Mucous membrane consists of epithelium overlying a layer of loose connective tissue; however, the type of epithelium varies with the location of the membrane. For example, stratified squamous epithelium lines the oral cavity, pseudostratified columnar epithelium lines part of the nasal cavity, and simple columnar epithelium lines the small intestine. Mucous membrane secretes the thick fluid called *mucus.*

The **cutaneous membrane** is an organ of the integumentary system and is more commonly called *skin.* It is described in detail in the next section of this chapter.

Synovial membranes form the inner linings of joint cavities between the ends of bones at freely movable joints. These membranes usually include fibrous connective tissue overlying loose connective tissue and adipose tissue. They secrete a thick, colorless *synovial fluid* into the joint cavity. This fluid lubricates the ends of the bones within the joint.

The Skin and Its Tissues

The skin is the largest and probably the most versatile organ of the body. It functions as a protective covering, aids in the regulation of body temperature, houses sensory receptors, and synthesizes various chemical substances. Also, persons are recognized and their ages are often judged by their skin characteristics.

The skin is composed of two distinct layers of tissues. The outer layer, called the **epidermis**, contains stratified squamous epithelial cells. The inner layer, or **dermis**, is thicker than the epidermis, and it includes a variety of tissues, such as epithelial tissue, fibrous connective tissue, smooth muscle tissue, nerve tissue, and blood. (See fig. 6.1.)

Beneath the dermis are masses of loose connective and adipose tissues that bind the skin to underlying organs. This is called the **subcutaneous layer**.

Subcutaneous injections are administered into the layer beneath the skin. *Intradermal injections*, on the other hand, are injected between layers of tissues within the skin. Subcutaneous injections and intramuscular injections, which are administered into muscles, are sometimes called hypodermic injections.

---

1. Name the four types of membranes and explain how they differ.
2. List the general functions of the skin.
3. Describe the outer layer of the skin; the inner layer.
4. Describe the tissue that underlies the skin.

The Epidermis

The deepest cells of the epidermis, which are next to the dermis, form a layer called the *stratum germinativum.* These cells are nourished by dermal blood vessels and are capable of reproducing. As they divide, older epidermal cells are slowly pushed away from the dermis toward the surface of the skin. The further the cells travel, the poorer their nutrient supply becomes, and in time they die.

Meanwhile, the membranes of the cells become thickened and develop numerous intercellular junctions (desmosomes) that fasten them to adjacent cells. At the same time the cells undergo a hardening process called **keratinization**, during which the cytoplasm develops strands of tough, fibrous, waterproof protein called *keratin.* As a result, many layers of tightly packed dead cells accumulate in the outer portion of the epidermis. This outermost layer is the *stratum corneum,* and the dead cells that compose it are easily rubbed away.

Other layers of epidermal cells occur between the stratum germinativum and the stratum corneum. As shown in figure 6.2, they include the *stratum lucidum,* the *stratum granulosum,* and the *stratum*

Fig. 6.1 A section of skin. Why is the skin considered an organ?

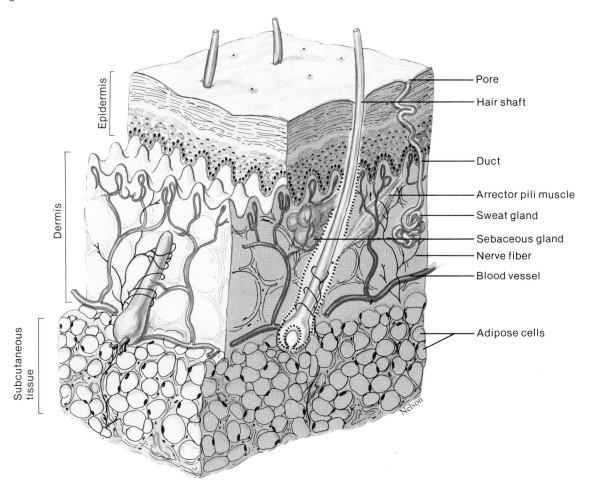

Epidermis

Dermis

Subcutaneous tissue

Pore

Hair shaft

Duct

Arrector pili muscle

Sweat gland

Sebaceous gland

Nerve fiber

Blood vessel

Adipose cells

Nelson

Fig. 6.2 The various layers of the epidermis are characterized by changes that occur in cells as they are pushed toward the surface of the skin.

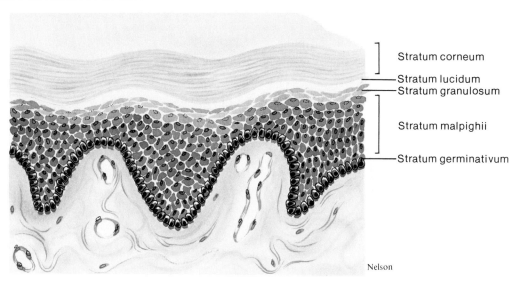

Stratum corneum

Stratum lucidum

Stratum granulosum

Stratum malpighii

Stratum germinativum

Nelson

Chart 6.1   Layers of the epidermis

Layer	Location	Characteristics
Stratum corneum (horny layer)	Outermost layer	Many layers of keratinized, dead epithelial cells that are flattened and nonnucleated.
Stratum lucidum	Beneath the stratum corneum	A thin, translucent layer found only in the skin of the palms and soles.
Stratum granulosum	Beneath the stratum lucidum	One or more layers of granular cells that contain shrunken fibers of keratin and shriveled nuclei.
Stratum malpighii	Beneath the stratum granulosum	Many layers of cells with centrally located, large oval nuclei and developing fibers of keratin.
Stratum germinativum (basal cell layer)	Deepest layer	A single row of columnar cells that undergo mitosis; this layer also includes pigment-producing melanocytes.

malpighii. The cells of these layers are characterized by changes they undergo as they are pushed toward the surface of the skin. Chart 6.1 describes their characteristics.

In healthy skin, the production of epidermal cells is closely balanced with the loss of stratum corneum, so that skin seldom wears away completely. In fact, the rate of cellular reproduction tends to increase in regions where the skin is being rubbed or pressed regularly. This response causes the growth of calluses on the palms and soles as well as the development of corns on the toes when poorly fitting shoes excessively rub the skin.

The epidermis has important protective functions. It shields the moist underlying tissues against water loss, mechanical injury, and the effects of harmful chemicals. When it is unbroken, the epidermis prevents the entrance of many disease-causing microorganisms. **Melanin**, a pigment that occurs in deeper layers of the epidermis and is produced by specialized cells known as *melanocytes*, absorbs light energy (fig. 6.3). In this way, it helps to protect deeper cells from the damaging effects of ultraviolet rays of sunlight.

Although melanocytes are usually located in the deepest portion of the epidermis and are the only cells that can produce melanin, the pigment also occurs in nearby epidermal cells. This happens because melanocytes have long cellular extensions that pass upward between neighboring epidermal cells, and they can transfer granules of melanin into these other cells by a process sometimes called *cytocrine secretion*. Consequently, nearby epidermal cells may contain more melanin than the melanocytes.

Fig. 6.3   The melanin produced by melanocytes helps to protect deeper cells against the effects of ultraviolet light.

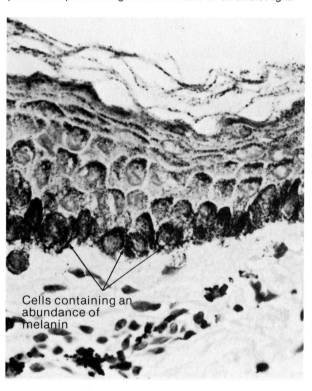

Cells containing an abundance of melanin

The protective stratum corneum of an infant's skin is relatively thin. Consequently, infant skin is more easily injured and more easily penetrated by chemicals than that of an adult.

1. Explain how the epidermis is formed.
2. What factors help prevent the loss of body fluids through the skin?
3. What is the function of melanin?

The Dermis

The dermis binds the epidermis to the underlying tissues. It is composed largely of fibrous connective tissue, which includes tough white (collagenous) fibers and yellow (elastin) fibers surrounded by a gellike matrix. Networks of these fibers give the dermis its strength and elasticity. Since the fibers run in definite directions over the body surface, they are responsible for producing lines of tension in the skin (fig. 6.4).

Lines of tension produce different patterns in different regions of the skin, and their presence is of special interest to surgeons. Skin that is cut parallel to the lines tends to gape less than skin cut across the lines. Thus, a sutured parallel incision is likely to heal more rapidly and result in less scar tissue than a transverse incision

Blood vessels in the dermis supply nutrients to all skin cells, including those of the epidermis. Interference with the flow of blood is likely to result in the death of cells. For example, when a person lies in one position for a prolonged time, the weight of the body pressing against the bed interferes with the skin's blood supply. As a consequence, cells may die and *decubitus ulcers* (bedsores) may appear.

Decubitus ulcers usually occur in the skin overlying bony projections, such as on the hip, heel, elbow, or shoulder. These sores can often be prevented by changing the position of the body frequently or by massaging the skin to stimulate blood flow in regions associated with bony prominences. Likewise, a diet rich in proteins and other necessary nutrients and an adequate intake of fluids to maintain blood volume help in preventing this condition.

Dermal blood vessels also aid in regulating body temperature. This function is explained in a later section of this chapter.

Fig. 6.4 Dermal fibers are arranged in patterns that produce lines of tension in the skin.

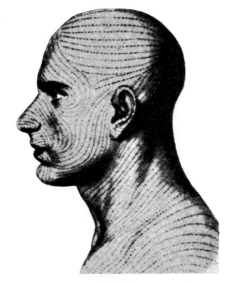
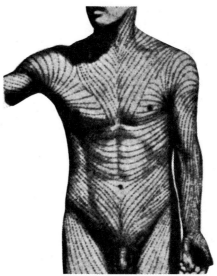

Nerve Tissue. There are numerous nerve fibers scattered throughout the dermis. Some of them (motor fibers) carry impulses to dermal muscles and glands, causing these structures to react. Others (sensory fibers) carry impulses away from specialized sensory receptors located within the dermis (fig. 6.5).

One set of dermal receptors (pacinian corpuscles) is stimulated by heavy pressure, while another set (Meissner's corpuscles) is sensitive to light touch. Still other receptors are stimulated by temperature changes or by factors that can damage tissues. Sensory receptors are further discussed in chapter 11.

Vitamin D Production

Skin cells also can play an important role in the production of vitamin D, which is necessary for the normal development of bones and teeth. This vitamin can be formed from a substance (dehydrocholesterol) that is synthesized by skin cells when they are exposed to sun or ultraviolet light. Since natural foods are generally poor sources of vitamin D, many people depend on their skin to provide it. On the other hand, vitamin D is commonly added to foods such as homogenized and evaporated milk.

A deficiency of vitamin D may lead to a disease called *rickets* in which the bones of the legs, ribs, sternum, and vertebrae may be deformed. This condition can be largely prevented by providing special sources of vitamin D in the diet—especially for children.

---

1. *What kinds of tissues make up the dermis?*
2. *How do dermal blood vessels help to regulate body temperature?*
3. *What role does the skin play in the production of vitamin D?*

Fig. 6.5 Various types of sensory receptors, such as touch receptors, are located in the dermis.

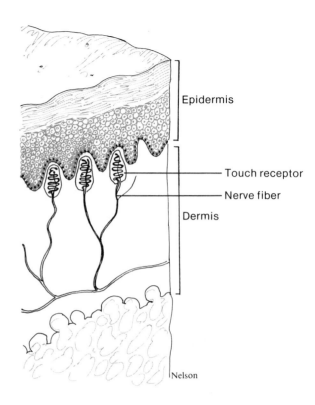

Epidermis

Touch receptor

Nerve fiber

Dermis

Nelson

Appendages and Glands of the Skin

Hair Follicles

Hair is present on all skin surfaces except the palms, soles, nipples, and glans penis, however it is not always well developed. For example, it is very fine on the forehead and the underside of the arm.

Each hair develops from a group of epidermal cells at the base of a tubelike depression called a **hair follicle**. This follicle extends from the surface into the dermis and may pass into the subcutaneous layer. The epidermal cells at its base receive nourishment from nearby dermal blood vessels, and as these cells divide and grow, older cells are pushed toward the surface. The cells that move upward and away from their nutrient supply become keratinized and die. Their remains constitute the shaft of a developing hair. In other words, a hair (fig. 6.6) is composed of dead epidermal cells.

Fig. 6.6 A hair grows from the base of a hair follicle when epidermal cells undergo cell division and older cells move upward and become keratinized.

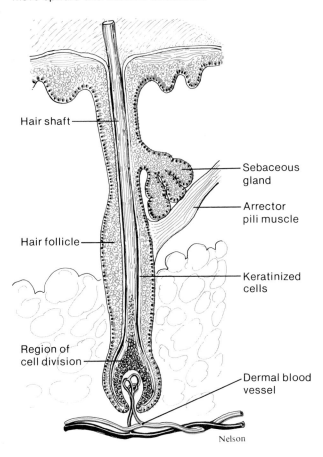

Nelson

Fig. 6.7 Acne results from overactive sebaceous glands, whose secretions tend to cause the openings of hair follicles to become dilated and plugged.

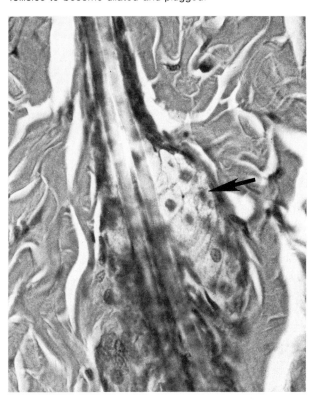

Usually a hair grows for a time and then undergoes a resting period during which it remains anchored in its follicle. Later a new hair begins to grow from the base of the follicle, and the old hair is pushed outward and drops off. Sometimes, however, the hairs of the scalp are not replaced, and the result is baldness (alopecia). This condition is commonly due to heredity and is most likely to occur in males.

Hair color is determined by the type and amount of pigment produced by the cells at the base of the hair follicles. For example, dark hair contains an abundance of melanin, while blond hair contains an intermediate quantity; white hair lacks this pigment. Bright red hair contains an iron pigment that does not occur in hair of any other color. A mixture of pigmented and unpigmented hair usually appears gray.

A bundle of smooth muscle cells, forming the *arrector pili muscle*, is attached to each hair follicle. This muscle is positioned so that the hair within the follicle stands on end when the muscle contracts. If a person is emotionally upset or very cold, nerve impulses may stimulate the arrector pili muscles to contract, causing gooseflesh or goose bumps.

Sebaceous Glands

Sebaceous glands are holocrine glands that are usually associated with hair follicles. They produce an oily secretion called *sebum*, which is a mixture of fatty material and cellular debris. Sebum is secreted into hair follicles and helps to keep the hair and skin soft, pliable, and waterproof.

Sebaceous glands are also responsible for *acne* (acne vulgaris), a common adolescent disorder. In this condition, the sebaceous glands in certain body regions become overactive in response to hormonal changes, and their excessive secretions cause the openings of hair follicles to become dilated and plugged (fig. 6.7). The result is blackheads; if pus-forming bacteria are present, pimples or boils may develop in the follicles.

Nails

Nails are protective coverings on the ends of fingers and toes. Each nail is produced by layers of specialized epithelial cells that are continuous with the epithelium of the skin. The whitish, half-moonshaped region (lunula) at the base of a nail is its most active growing portion. The cells formed in this region undergo keratinization and become part of the nail (fig. 6.8). The keratin they contain, however, is harder than that formed in the epidermis and hair follicles.

Sweat Glands

Sweat glands occur in nearly all regions of the skin, but are most numerous in the palms and soles. Each gland consists of a tiny tube that originates as a ball-shaped coil in the dermis or subcutaneous layer. The coiled portion of the gland is closed at the end and is lined with sweat-secreting epithelial cells.

Some sweat glands, the *apocrine glands* (fig. 6.9), respond to emotional stress. Apocrine secretions typically have odors, and the glands are considered to be scent glands. They are responsible for some skin regions becoming moist when a person is emotionally upset or uneasy. These glands are most numerous in the armpits and groin and are usually connected with hair follicles. They also become active when a person is sexually stimulated.

Other sweat glands, the *eccrine glands* (fig. 6.9), respond primarily to elevated body temperatures due to environmental heat or physical exercise. These glands are common on the forehead, neck, and back, where they produce profuse sweating on hot days and during physical exertion.

The fluid secreted by these glands is carried upward by a tubular part that opens at the surface as a pore. Although it is mostly water, this fluid contains small quantities of salts and certain wastes, such as urea and uric acid. Thus, the secretion of sweat is, to a limited degree, an excretory function.

With advancing age, there is a reduction in sweat gland activity, and in the very old, sweat glands may be replaced by fibrous tissues. Similarly, there is a decrease in sebaceous gland activity with age, so that the skin of elderly persons tends to be dry and lack oils.

Fig. 6.8 A nail is produced by epithelial cells that reproduce and undergo keratinization in the lunula region of the nail.

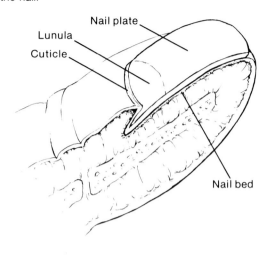

Fig. 6.9 How do the functions of eccrine sweat glands and apocrine sweat glands differ?

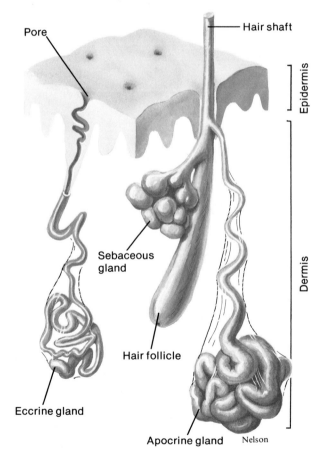

1. Explain how a hair forms.
2. What causes gooseflesh?
3. What is the function of the sebaceous glands?
4. How does the composition of a fingernail differ from that of a hair?
5. Describe the functions of the sweat glands.

Regulation of Body Temperature

Normally, the body temperature remains close to 37°C (98.6°F), and the maintenance of a constant temperature requires that the amount of heat lost by the body be balanced by the amount it produces. The skin plays a key role in the homeostatic mechanism associated with the regulation of body temperature.

Heat Production and Loss

Heat is a by-product of cellular respiration, so the more active cells are also the greater heat producers. These include muscle cells and the cells of certain glands such as the liver.

When heat production is excessive, nerve impulses stimulate skin structures to react in ways that promote heat loss. For example, during physical exercise, active muscles release heat, and the blood carries away the excess. Parts of the nervous system sense the increasing blood temperature and signal muscles in the walls of dermal blood vessels to relax. As the blood vessels dilate, more heat-carrying blood enters the dermis, and some of the heat is likely to escape to the outside of the body.

The primary means of body heat loss is **radiation**, by which heat in the form of *infrared heat rays*, escapes from warmer surfaces to cooler surroundings. These rays radiate in all directions, much like those from the bulb of a heat lamp.

Lesser amounts of heat are lost by conduction and convection. In **conduction** heat moves from the body directly into the molecules of cooler objects in contact with its surface. For example, heat is lost by conduction into the seat of a chair when a person sits down. The heat loss continues as long as the chair is cooler than the body surface in contact with it.

Heat is also lost by conduction to the air molecules that contact the body. As air becomes heated it moves away from the body, carrying heat with it, and is replaced by cooler air moving inward. This type of continuous circulation of air over a warm surface is called **convection**.

Still another means of body heat loss is **evaporation**. When the body temperature is rising above normal, the nervous system stimulates eccrine sweat glands to become active and to release fluid onto the surface of the skin. As this fluid evaporates (changes from a liquid to a gas), it carries heat away from the surface. The cooling effect of evaporation can be experienced by rubbing some water on the skin and waiting for the water to disappear. Cooling by evaporation is also used when a feverish patient receives a sponge bath to help lower the body temperature.

If heat is lost excessively, as may occur in a very cold environment, parts of the nervous system trigger different responses in the skin structures. For example, muscles in the walls of dermal blood vessels are stimulated to contract; this decreases the flow of heat-carrying blood through the skin and helps to reduce heat loss by radiation, conduction, and convection. At the same time, the sweat glands remain inactive, decreasing heat loss by evaporation. If the body temperature continues to drop, the nervous system may stimulate muscle fibers in the skeletal muscles throughout the body to contract slightly. This action requires an increase in the rate of cellular respiration and produces heat as a by-product. If this response does not raise the body temperature to normal, small groups of muscles may be stimulated to contract rhythmically with still greater force, and the person begins to shiver, generating more heat.

In addition to the skin, the circulatory and respiratory systems function in body temperature regulation. For example, if the body temperature is rising above normal, the heart is stimulated to beat faster. This causes more blood to be moved from the deeper, heat-generating tissues into the skin. At the same time, the breathing rate is increased, and more heat is lost from the lungs as an increased volume of air is moved in and out.

Figure 6.10 summarizes some of the actions involved in the body temperature regulating mechanism.

Fig. 6.10 Some actions involved in body temperature regulation.

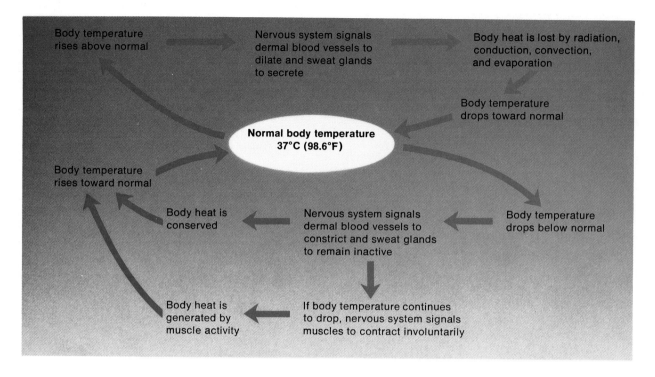

Problems in Temperature Regulation

Unfortunately, the body's temperature-regulating mechanism does not always operate satisfactorily, and the consequences may be unpleasant and dangerous. For example, since air can hold only a limited amount of water vapor, on a hot, humid day the air may become saturated with it. At such times, the sweat glands may be activated, but the sweat will not evaporate. The skin becomes wet, but the person remains hot and uncomfortable. In addition, if the air temperature is high, heat is lost less effectively by radiation. In fact, if the air temperature is higher than the body temperature, the person may gain heat from the surroundings, and the body temperature may rise even higher.

Whenever sweat is produced profusely, there is danger that an *electrolyte imbalance* may develop. This can occur because sweat contains certain essential electrolytes, and if they are lost in excess, a person may experience an unpleasant condition called heat exhaustion. Its symptoms include dizziness, headache, muscle cramps, and nausea; they are treated by providing needed salts to restore electrolyte balance.

1. How does the body lose heat?
2. What actions help the body to conserve heat?
3. What is the primary cause of heat exhaustion?

Skin Color

Skin color, like hair color, is due largely to the presence or absence of pigment produced by epidermal melanocytes. The amount of pigment synthesized by these cells is influenced by a variety of factors.

Genetic Factors

Each person inherits genes for the production of melanin. Those whose genes cause a relatively large amount of pigment to be produced have dark skins, while those whose genes cause less pigment to be formed have lighter complexions. Still others inherit mutant genes, and their cells are unable to manufacture melanin. As a consequence, their skins remain nonpigmented, and they are termed *albinos* (see fig. 6.11).

Fig. 6.11 An albino lacks the ability to produce melanin because of the presence of a mutant gene.

Environmental Factors

Environmental factors such as sunlight, ultraviolet light from sunlamps, or X rays can affect skin color by causing melanin to darken and by stimulating epidermal melanocytes to produce more pigment. This is why sunbathing results in skin tanning. Unless the exposure to sunlight is continued, however, the tan is eventually lost, as pigmented epidermal cells become keratinized and are worn away.

Blood in the dermal vessels causes additional skin color. For example, when the blood is well oxygenated, the blood pigment *hemoglobin* is bright red, making the skins of light-complexioned persons and albinos appear pinkish. On the other hand, when the blood oxygen concentration is low, hemoglobin is dark red and the skin appears bluish—a condition called *cyanosis*.

The state of the blood vessels also has an effect on skin color. If the vessels are dilated, more blood enters the dermis, and the skin becomes redder than usual. This may happen when a person is overheated, embarrassed, or under the influence of alcohol. Conversely, conditions that produce blood vessel constriction cause blanching of the skin. Thus, if the body temperature is dropping abnormally or if a person is frightened, the skin may appear pale.

The presence of a yellow-orange plant pigment called *carotene*, which is especially common in yellow vegetables, may cause the skin to have a yellowish color, since it accumulates in the adipose tissue of the subcutaneous layer.

Various diseases may affect skin color. In certain forms of liver disease, bile pigments may occur in the skin and cause it to appear yellowish. A person with anemia may have pale skin due to a decrease in the concentrations of hemoglobin in blood cells.

Some middle-aged and older people have a chronic skin disorder called *acne rosacea*. This condition is characterized by redness of the nose, cheeks, chin, and forehead that results from persistently dilated dermal blood vessels. It is sometimes accompanied by eruptions similar to those associated with adolescent acne. Although acne rosacea seems to be inherited, the problem may be aggravated by certain foods and beverages, including alcohol.

1. What factors influence skin color?
2. Which of these factors are genetic? Which are environmental?

Responses to Injuries

In functioning as a protective covering for deeper tissues, skin occupies an exposed position and is frequently broken or damaged by sunlight, heat, or corrosive chemicals.

Inflammation is a tissue response to injury or stress in which blood vessels in the affected area become dilated. At the same time, the permeability of these vessels increases, and there is a tendency for excessive amounts of fluids to leave the blood vessels and enter the damaged tissues. Thus, when skin is injured it may become reddened, swollen, warm, and painful to touch. On the other hand, the dilated blood vessels provide more nutrients and oxygen to the tissues, which may aid the healing process.

Wound Healing

The process by which damaged skin heals depends on the extent of the injury. If a break in the skin is shallow, epithelial cells along its margin are stimulated to reproduce more rapidly than usual, and the newly formed cells simply fill the gap.

If the injury extends into the dermis or subcutaneous layer, other tissues become involved, and the repair process is more complex. Blood vessels are broken, and the blood that escapes forms a clot in the wound. Tissue fluids seep into the area and become dried. The blood clot and the dried fluids form a scab, which covers and protects the damaged tissues. Before long, fibroblasts from connective tissue at the wound margins migrate into the region and begin forming new fibers. These fibers tend to bind the edges of the wound together. The closer the edges of the wound, the sooner the gap can be filled; this is one reason for suturing or otherwise closing a large break in the skin.

As the healing process continues, blood vessels send out new branches that grow into the area beneath the scab. Phagocytic cells remove dead cells and other debris. Eventually the damaged tissues are replaced, and the scab sloughs off. If the wound was extensive, the newly formed connective tissue may appear on the surface as a scar.

In large, open wounds, the healing process may be accompanied by the formation of *granulations* that develop in the exposed tissues as small, rounded masses. Each of these granulations consists of a new branch of a blood vessel and a cluster of connective tissue cells that are being nourished by the vessel. Figure 6.12 shows the stages in the healing of a wound.

As with other types of injuries, the skin's response to a burn depends on the amount of damage it sustains. If the skin is only lightly burned, as in sunburn, it may simply become warm and reddened as dermal blood vessels become dilated. If the damage is more severe, tissue fluids may accumulate in spaces between the dermal and epidermal layers, so that the skin becomes blistered.

If large areas of epidermis are destroyed by burning, the replacement process may involve the epithelial cells that line the hair follicles. These epithelial cells may be able to grow out of the follicles that survive in the deeper skin layers, and spread over the surface of the exposed dermis. If, however, the follicles are completely destroyed, the only way the epidermis can be replaced is through growth from the margins of the injured tissues. In such extreme cases, a thin layer of skin may be removed from another part of a person's body and transplanted in the injured area. This procedure is known as a *skin graft*.

Burns can be classified according to the degree of thermal damage sustained by the skin.

First-degree burns result in dilation of dermal blood vessels and redness of the skin (erythema), followed by shedding of the surface layers (desquamation).

Second-degree burns are characterized by the appearance of blisters due to fluid escaping from dermal capillaries that accumulate beneath the outer layers of epidermal cells. Recovery from such burns is usually complete, and scar tissue does not form.

Third-degree burns destroy the full thickness of the skin and may also damage the subcutaneous tissue. As a result, ulcerating wounds appear, and healing involves the formation of scar tissue.

---

1. How does the process of inflammation promote healing?
2. Describe how the skin heals following a shallow wound?
3. Describe how the skin heals following a burn that destroys a large area of epidermis.

Fig. 6.12 (*a*) If normal skin is (*b*) injured deeply, (*c*) blood escapes from dermal blood vessels, and (*d*) a blood clot soon forms. (*e*) The blood clot and dried tissue fluid form a scab that protects the damaged region. (*f*) Later, blood vessels send out branches, and fibroblasts migrate into the area. (*g*) The fibroblasts produce new connective tissue fibers, and when the skin is largely repaired, the scab sloughs off.

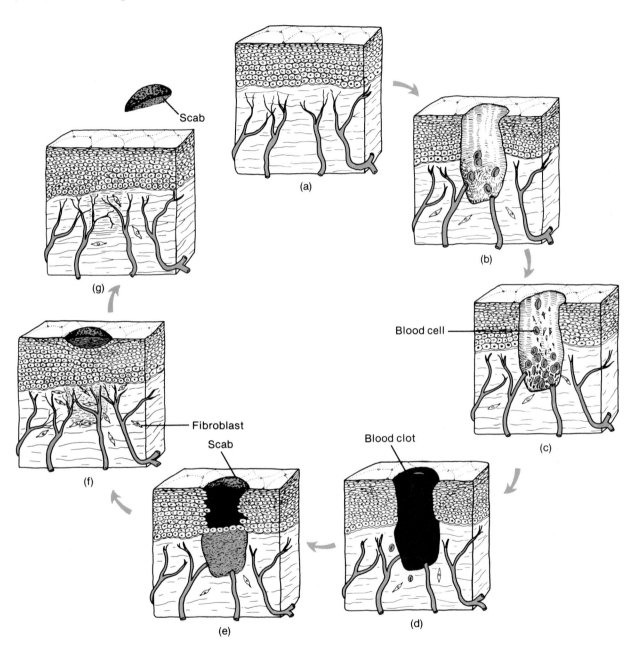

Common Skin Disorders

acne (ak'ne)—a disease of the sebaceous glands accompanied by blackheads and pimples.

alopecia (al″o-pe'she-ah)—loss of hair.

athlete's foot (ath'-lētz foot) (tinea pedis)—a fungus infection usually involving the skin of the toes and soles.

birthmark (berth' mark)—vascular tumor involving the skin and subcutaneous tissues.

boil (boil) (furuncle)—a bacterial infection involving a hair follicle and/or sebaceous glands.

carbuncle (kar' bung-kl)—a bacterial infection similar to a boil, spreading into the subcutaneous tissues.

cyst (sist)—a liquid-filled sac or capsule.

dermatitis (der″mah-ti'tis)—an inflammation of the skin.

eczema (ek'zĕ-mah)—a noncontagious skin rash often accompanied by itching, blistering, and scaling.

erythema (er″ĭ-the'mah)—reddening of the skin, due to dilation of dermal blood vessels, in response to injury or inflammation.

impetigo (im″pĕ-ti'go)—a contagious disease of bacterial origin characterized by pustules that rupture and become covered with loosely held crusts.

keloids (ke'loid)—fibrous tumors usually initiated by an injury.

mole (mōl) (nevi)—skin tumor that usually is pigmented; colors range from yellow to brown or black.

pediculosis (pĕ-dik″u-lo'sis)—a disease produced by infestation of lice.

pruritus (proo-ri'tus)—itching of the skin without eruptions.

psoriasis (so-ri'ah-sis)—a chronic skin disease characterized by red patches covered with silvery scales.

pustule (pus'tūl)—elevated, pus-filled area.

scabies (ska'bēz)—a disease resulting from an infestation of mites.

seborrhea (seb″o-re'ah)—a disease characterized by hyperactivity of the sebaceous glands and accompanied by greasy skin and dandruff.

shingles (shing'gelz) (herpes zoster)—disease caused by a viral infection of certain spinal nerves, characterized by severe localized pain and blisters in the skin areas supplied by the affected nerves.

ulcer (ul'ser)—an open sore.

wart (wort)—a flesh-colored raised area caused by a virus infection.

Chapter Summary

Introduction

The integumentary system comprises the skin, together with its appendages and glands.

Organs are composed of two or more kinds of tissues.

Types of Membranes

1. Serous membranes
   a. Organs that line body cavities that lack openings to the outside.
   b. They are composed of epithelium and loose connective tissue.
   c. They secrete watery fluid that lubricates membrane surfaces.
2. Mucous membranes
   a. Organs that line cavities and tubes that open to the outside.
   b. They are composed of various kinds of epithelium and loose connective tissue.
   c. They secrete mucus.
3. The cutaneous membrane is the skin.
4. Synovial membranes
   a. Organs that line joint cavities.
   b. They are composed of fibrous connective tissue overlying loose connective tissue and adipose tissue.
   c. They secrete synovial fluid that lubricates the ends of bones at joints.

The Skin and Its Tissues

Skin functions as a protective covering, aids in regulating body temperature and blood pressure, houses sensory receptors, and synthesizes various chemicals.

It is composed of an epidermis, a dermis, and a subcutaneous layer beneath the dermis.

1. The epidermis
   a. A layer composed of striated squamous epithelium.
   b. The deepest layer, called stratum germinativum, contains cells undergoing mitosis.
   c. Epidermal cells undergo keratinization as they are pushed toward the surface.
   d. The outermost layer, stratum corneum, is composed of dead epidermal cells.
   e. The production of epidermal cells is balanced with the rate at which they are lost.
   f. Epidermis functions to protect underlying tissues against water loss, mechanical injury, and the effects of harmful chemicals.
   g. Melanin protects underlying cells from the effects of ultraviolet light.
   h. Melanocytes transfer melanin to nearby epidermal cells.

2. The dermis
   a. A layer composed largely of fibrous connective tissue that binds the epidermis to underlying tissues.
   b. Patterns of fibers produce lines of tension in the skin.
   c. Dermal blood vessels supply nutrients to all skin cells, and interference with blood flow may result in death of these cells.
   d. Nerve tissue is scattered through the dermis.
      (1) Some dermal nerve fibers carry impulses to muscles and glands of the skin.
      (2) Other dermal nerve fibers are associated with various sensory receptors in the skin.
3. Vitamin D production
   a. Vitamin D is needed for normal development of bones and teeth.
   b. Vitamin D can be synthesized from a substance produced in skin cells when they are exposed to ultraviolet light.

Appendages and Glands of the Skin

1. Hair follicles
   a. Hair occurs in nearly all skin.
   b. Each hair develops from epidermal cells at the base of a tubelike hair follicle.
   c. As newly formed cells develop and grow, older cells are pushed toward the surface; during this process they die and undergo keratinization.
   d. A hair usually grows for a while, undergoes a resting period, and is replaced by a new hair.
   e. Hair color is determined by the type and amount of pigment in its cells.
   f. A bundle of smooth muscle cells is attached to each hair follicle.
2. Sebaceous glands
   a. Usually associated with hair follicles.
   b. They secrete sebum, which helps keep skin and hair soft and waterproof.
   c. Overactive sebaceous glands cause adolescent acne.
3. Nails
   a. Protective covers on the ends of fingers and toes.
   b. They are produced by epidermal cells that undergo keratinization.
4. Sweat glands
   a. Located in nearly all regions of the skin.
   b. Each gland consists of a coiled tube.
   c. Apocrine glands respond to emotional stress, while eccrine glands respond to elevated body temperature.
   d. Sweat is primarily water, but also contains salts and waste products.

Regulation of Body Temperature

Normal body temperature is about 37°C (98.6°F).
1. Heat production and loss
   a. Heat is a by-product of cellular respiration.
   b. When the body temperature rises above normal, dermal blood vessels dilate and sweat glands secrete sweat.
   c. Heat is lost to the outside by radiation, conduction, convection, and evaporation.
   d. If the body temperature drops below normal, dermal blood vessels constrict and sweat glands become inactive.
   e. During excessive heat loss, skeletal muscles are stimulated to contract involuntarily; this increases cellular respiration and produces additional heat.
2. Problems in temperature regulation
   a. When the air is saturated with water, sweat may fail to evaporate and the body temperature may remain elevated.
   b. If sweat is produced excessively, an electrolyte imbalance may develop.

Skin Color

Skin color is due largely to the presence of pigment in the epidermis.
1. Genetic factors
   a. Each person inherits genes for melanin production.
   b. Dark skin is due to genes that cause large amounts of melanin to be produced; lighter skin is due to genes that cause lesser amounts of melanin to form.
   c. Mutant genes may cause a lack of melanin in the skin.
2. Environmental factors
   a. Environmental factors that influence skin color include sunlight, ultraviolet light, and X ray, all of which darken melanin and stimulate melanin production.
   b. Blood in dermal vessels may cause the skin of light-complexioned persons to appear pinkish or bluish.
   c. Carotene in the subcutaneous layer may cause the skin to appear yellowish.
   d. Various diseases may affect skin color.

Responses to Injuries

Skin is often damaged by exposure to sunlight, heat, or corrosive chemicals.
1. Inflammation
   a. Inflammation is a response to injury or stress in which blood vessels dilate and become more permeable.
   b. Fluids seep into damaged tissues, which become reddened, warm, swollen, and painful to touch.

c. The inflammation reaction may aid healing by providing additional nutrients and oxygen to the affected tissues.

2. Wound healing
   a. A simple break in the skin may be repaired by cells growing inward from the wound margin.
   b. Repair of a deep wound involves blood clotting, scab formation, growth of new blood vessels, and development of new connective tissue.
   c. Phagocytic cells remove dead cells and debris from the area.

3. Burns
   a. Minor burns cause dermal blood vessels to dilate and tissues to become reddened.
   b. In more severe burns, tissue fluid enters the layers of the skin and blisters form.
   c. Epidermal cells from hair follicles may grow out over the surface of exposed dermis and produce a new epidermal covering.

Application of Knowledge

1. What special problems would result from a loss of 50% of a person's functional skin surface? How might this person's environment be modified to compensate partially for such a loss?

2. Read the nutritional information labels on a variety of food packages, and prepare a list of those that represent particularly good sources of vitamin D.

3. A premature infant typically lacks subcutaneous adipose tissue. Also, the surface area of an infant's small body is relatively large compared to its volume. How do you think these factors influence such an infant's ability to regulate its body temperature?

Review Activities

1. Define *integumentary system*.

2. Distinguish between serous and mucous membranes.

3. Explain the function of synovial membranes.

4. Distinguish between the epidermis and the dermis.

5. Describe the subcutaneous layer and its function.

6. Explain what happens to epidermal cells as they undergo keratinization.

7. List the functions of the epidermis.

8. Describe the function of melanocytes.

9. Define *lines of tension*.

10. Explain the primary causes of decubitus ulcers.

11. Review the functions of dermal nerve tissue.

12. Describe the importance of vitamin D and explain how it is produced in the skin.

13. Distinguish between a hair and a hair follicle.

14. Review how hair color is determined.

15. Explain the function of sebaceous glands.

16. Describe how nails are formed.

17. Distinguish between apocrine and eccrine glands.

18. Explain how body heat is lost by radiation.

19. Distinguish between conduction and convection.

20. Explain how sweat glands function in regulating body temperature.

21. Describe the body's responses to decreasing body temperature.

22. Review how the humidity of the air may interfere with body temperature regulation.

23. Explain how excessive sweating may promote the development of an electrolyte imbalance.

24. List the factors that influence skin color.

25. Define *inflammation* and tell how it may aid the healing process.

26. Describe the steps in deep wound healing.

27. Explain how the hair follicle cells may help replace epidermis that has been destroyed by burning.

Suggestions for Additional Reading

Bickers, D. R., and Kappas, A. May 1974. Metabolic and pharmacologic properties of the skin. *Hosp. Prac.*

Elden, H. R., ed. 1971. *Biophysical properties of the skin.* New York: John Wiley & Sons.

Hardy, J. D., et al. 1970. *Physiological and behavioral temperature regulation.* Springfield: Charles C. Thomas.

Heller, H. C., et al. August 1978. The thermostat of vertebrate animals. *Scientific American.*

Loomis, W. F. December 1970. Rickets. *Scientific American.*

Marples, M. J. January 1969. Life on the human skin. *Scientific American.*

Moncrief, J. A. Burns. 1973. *New Eng. Jour. Med.* 228:444.

Montagna, W. February 1965. The skin. *Scientific American.*

Nicoll, P. A., et al. 1972. The physiology of the skin. *Ann. Rev. Physio.* 34:177.

Ross, R. June 1969. Wound healing. *Scientific American.*

Rushmer, R. L., et al. 1966. The skin. *Science* 154:343.

Tregear, R. T. 1966. *Physical functions of skin.* New York: Oxford Univ. Press.

Wurtman, R. J. July 1975. The effects of light on the human body. *Scientific American.*

Body Organization

7 The human body is a complex structure composed of many parts with specialized functions. In order to communicate with each other more effectively, investigators over the ages have developed a set of terms to name body features—cavities, membranes, organs, and organ systems. Another set of terms has been devised to describe the positions of body parts with respect to each other, imaginary planes passing through these structures, and various body regions.

The purpose of this chapter is to introduce a number of these terms and to consider the human body as a whole before beginning a detailed study of its individual parts.

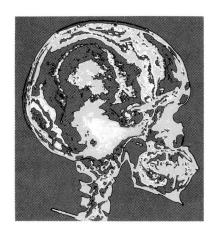

Chapter Outline

Chapter Objectives

After you have studied this chapter, you should be able to

1. Describe the location of the major body cavities.
2. List the organs located in each of the body cavities.
3. Name the membranes associated with the thoracic and abdominopelvic cavities.
4. Name the major organs systems of the body and list the organs associated with each.
5. Describe the general functions of each organ system.
6. Properly use the terms that describe relative positions, body sections, and body regions.
7. Complete the review activities at the end of this chapter.

abdominal (ab-do′mĭn-al)

appendicular (ap″en-dik′u-lar)

axial (ak′se-al)

cranial (kra′ne-al)

parietal (pah-ri′ĕ-tal)

pelvic (pel′vik)

pericardial (per″ĭ-kar′de-al)

peritoneal (per″ĭ-to-ne′al)

pleural (ploo′ral)

spinal (spi′nal)

thoracic (tho-ras′ik)

visceral (vis′er-al)

append-, to hang something: *append*icular—pertaining to the arms and legs.

cardi-, heart: peri*cardi*um—membrane that surrounds the heart.

cran-, helmet: *cran*ial—pertaining to the portion of the skull that surrounds the brain.

dors-, back: *dors*al—a position toward the back of the body.

hormon-, to excite: *hormon*e—chemical that excites action in a specific tissue.

mediastin-, middle: *mediastin*um—membranous septum located between the lungs.

nas-, nose: *nas*al—pertaining to the nose.

orb-, circle: *orb*ital—pertaining to the portion of skull that encircles the eye.

pariet-, wall: *pariet*al membrane—membrane that lines the wall of a cavity.

pelv-, basin: *pelv*ic cavity—basin-shaped cavity enclosed by the pelvic bones.

peri-, around: *peri*cardial membrane—membrane that surrounds the heart.

pleur-, rib: *pleur*al membrane—membrane that encloses the lungs within the rib cage.

thorac-, chest: *thorac*ic—pertaining to the chest.

ventr-, belly: *ventr*al—a position toward the front of the body.

The human organism can be divided into an **axial portion,** which includes the head, neck, and trunk, and an **appendicular portion,** which includes the arms and legs.

Body Cavities and Membranes

Body Cavities

Within the axial portion of the body there are two major cavities—a **dorsal cavity** and a larger **ventral cavity.** These cavities are filled with organs, which as a group are called *visceral organs.* The dorsal cavity can be subdivided into two parts—the **cranial cavity,** which houses the brain; and the **spinal cavity,** which contains the spinal cord and is surrounded by the sections of the backbone (vertebrae). The ventral cavity consists of a **thoracic cavity** and an **abdominopelvic cavity.** These major body cavities are shown in figure 7.1.

The thoracic cavity is separated from the lower abdominopelvic cavity by a broad, thin muscle called the **diaphragm.** When it is at rest, this muscle curves upward into the thorax like a dome. When it contracts during inhalation, it presses down on the abdominal organs. The wall of the thoracic cavity is composed of skin, skeletal muscles, and various bones. The viscera within it include the lungs and a membranous partition or septum called the *mediastinum.* It separates the thorax into two compartments that contain the right and left lungs. The rest of the thoracic viscera—heart, esophagus, trachea, and thymus gland—are located within the mediastinum.

The abdominopelvic cavity, which includes an upper abdominal portion and a lower pelvic portion, extends from the diaphragm to the floor of the pelvis. Its wall consists primarily of skin, skeletal muscles, and bones. The visceral organs of the abdomen include the stomach, liver, spleen, pancreas, gall bladder, kidneys, and most of the small and large intestines.

Fig. 7.1 Major body cavities.

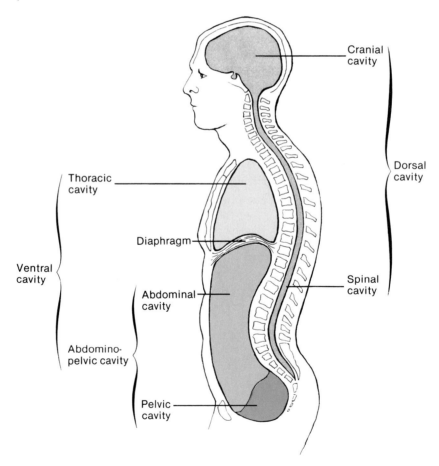

The **pelvic cavity** is the portion of the abdominopelvic cavity enclosed by the pelvic bones. It contains the terminal end of the large intestine, the rectum, the urinary bladder, and the internal reproductive organs.

In addition to the relatively large dorsal and ventral body cavities, there are several smaller ones within the head. They include the following (see fig. 7.2 and color plate 16):

1. *Oral cavity*, containing the teeth and tongue.

2. *Nasal cavity*, located within the nose and divided into right and left portions by a nasal septum. Several air-filled *sinuses* are connected to the nasal cavity.

3. *Orbital cavities*, containing the eyes and associated skeletal muscles and nerves.

4. *Middle ear cavities*, containing the middle ear bones.

Fig. 7.2 The smaller cavities within the head include the oral, nasal, orbital, and middle ear cavities, and several sinuses connected to the nasal cavity.

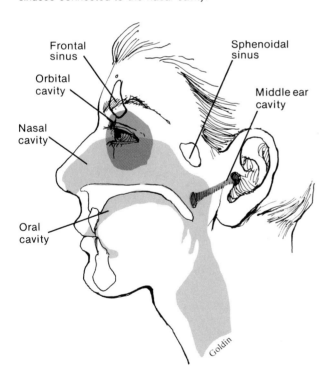

Thoracic and Abdominopelvic Membranes

The walls of the right and left thoracic compartments, which contain the lungs, are lined with a serous membrane called the *parietal pleura*. The lungs themselves are covered by a similar membrane called the *visceral pleura*.

The parietal and visceral **pleural membranes** are separated by a thin film of watery serous fluid that they secrete, and although there is normally no actual space between these membranes, a possible or potential space called the *pleural cavity* exists between them.

The heart, which is located in the broadest portion of the mediastinum, is surrounded by serous membranes called **pericardial membranes.** A thin *visceral pericardium* (epicardium) covers the heart's surface and is separated from the much thicker, fibrous *parietal pericardium* by a small amount of serous fluid. The potential space between these membranes is called the *pericardial cavity*. Figure 7.3 shows the membranes associated with the heart and lungs. (See color plate 17.)

In the abdominopelvic cavity, the serous membranes are called **peritoneal membranes.** A *parietal peritoneum* lines the wall, and a *visceral peritoneum* covers each organ in the abdominal cavity. The potential space between these membranes is called the *peritoneal cavity*. (See fig. 7.4 and color plates 16 and 18.)

---

1. What is meant by visceral organ?
2. What organs occupy the dorsal cavity? The ventral cavity?
3. Name the cavities of the head.
4. Describe the location of the membranes associated with the thoracic and abdominopelvic cavities.

Fig. 7.3 A transverse section through the thorax reveals the serous membranes associated with the heart and lungs.

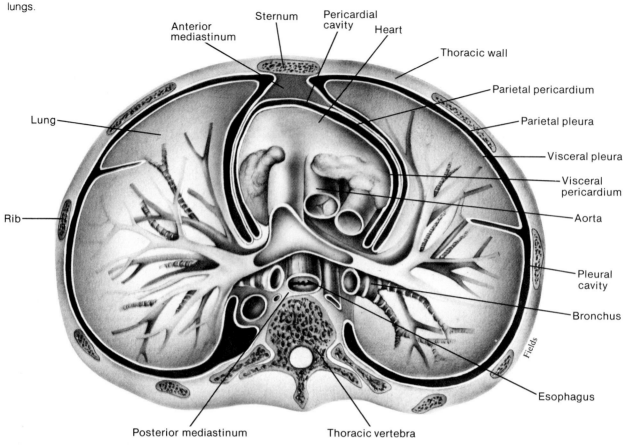

Anterior mediastinum

Sternum

Pericardial cavity

Heart

Thoracic wall

Parietal pericardium

Parietal pleura

Visceral pleura

Visceral pericardium

Aorta

Pleural cavity

Bronchus

Esophagus

Lung

Rib

Fields

Posterior mediastinum

Thoracic vertebra

Fig. 7.4 A transverse section through the abdomen at the level of the stomach. How would you distinguish between the parietal and the visceral peritoneum?

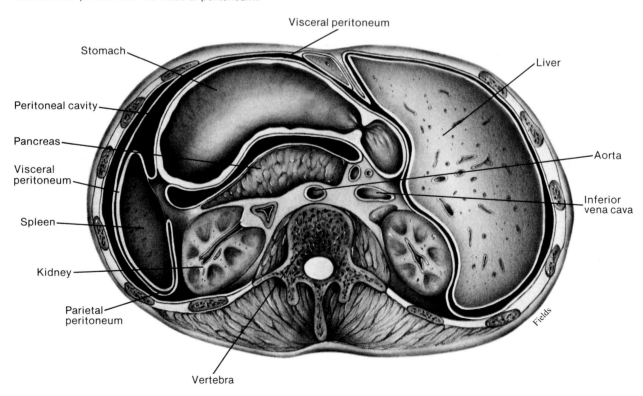

Visceral peritoneum

Stomach

Liver

Peritoneal cavity

Pancreas

Aorta

Visceral peritoneum

Inferior vena cava

Spleen

Kidney

Parietal peritoneum

Fields

Vertebra

Fig. 7.5 The integumentary system forms the body covering.

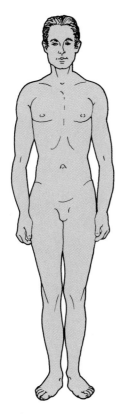

Integumentary
system

Organ Systems

As described in chapters 1 through 6, the human body is composed of parts within parts—atoms within molecules, molecules within organelles, organelles within cells, cells within tissues, and tissues within organs. The chapters that follow discuss organ systems and the way these units make up the human organism.

Body Covering

The organs of the **integumentary system** (discussed in chapter 6) include the skin and accessory organs such as the hair, nails, sweat glands, and sebaceous glands. These parts protect underlying tissues, help regulate the body temperature, house a variety of sensory receptors, and synthesize certain cellular products. (See fig. 7.5.)

Support and Movement

The organs of the skeletal and muscular systems function to support and move body parts. (See fig. 7.6.)

The **skeletal system** consists of the bones and the ligaments and cartilages that bind the bones together. These parts provide frameworks and protective shields for softer tissues, serve as attachments for muscles, and act together with muscles when body parts move. Tissues within bones also produce blood cells and store inorganic salts.

The muscles, are the organs of the **muscular system.** By contracting and pulling their ends closer together, they furnish the forces that cause body movements. They also function in maintaining posture and are largely responsible for the production of body heat.

Integration and Coordination

In order to function as a unit, body parts must be integrated and coordinated. Their activities must be controlled and adjusted, from time to time, so that homeostasis is maintained. The nervous and endocrine systems function to maintain homeostasis by means of electrochemical signals called *impulses,* which are carried on nerve fibers, and by the secretion of chemical substances called *hormones,* which are transported in body fluids. (See fig. 7.7.)

The **nervous system** consists of the brain, spinal cord, and associated nerves. Some of the nerve cells within the nervous system detect changes that occur inside and outside the body. Other nerve cells receive the impulses transmitted from these sensory units and interpret and act on the information received. Still other nerve cells carry impulses from the brain or spinal cord to muscles or glands and stimulate these parts to contract or to secrete various products.

The **endocrine system** includes all the glands that secrete **hormones.** The hormones, in turn, travel away from the gland in some body fluid such as blood. Usually a particular hormone only affects a particular group of cells, which is called its target tissue. The effect of a hormone is to alter the function of the target tissue in some way and thus aid in controlling metabolism.

The organs of the endocrine system include the pituitary, thyroid, parathyroid, and adrenal glands, as well as the pancreas, ovaries, testes, pineal gland, and thymus gland.

Fig. 7.6 Organ systems associated with support and movement.

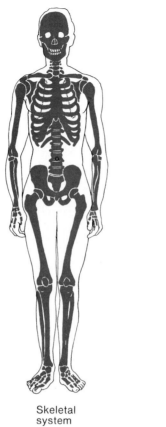

Skeletal
system

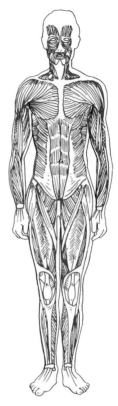

Muscular
system

Fig. 7.7 Organ systems associated with integration and coordination.

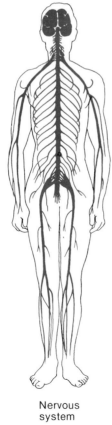

Nervous
system

Endocrine
system

Levels of Organization

Processing and Transporting

The organs of several systems are involved with processing and transporting nutrients, oxygen, and various wastes. The organs of the **digestive system** receive foods from the outside and convert their molecules into forms that can pass through cell membranes. These organs also produce hormones amd transport undigested materials to the outside of the body.

The digestive system includes the mouth, teeth, salivary glands, pharynx, esophagus, stomach, liver, gallbladder, pancreas, small intestine, and large intestine.

The organs of the **respiratory system** provide for the intake and output of air and for the exchange of gases between the blood and the air. More specifically, oxygen passes from air in the lungs into the blood, and carbon dioxide leaves the blood and enters the air. The nasal cavity, pharynx, larynx, trachea, bronchi, and lungs are parts of this system.

The **circulatory system** includes the heart, arteries, veins, capillaries, and blood. The heart functions as a muscular pump that forces blood through the blood vessels. The blood serves as a fluid for transporting oxygen, nutrients, hormones, and wastes. It carries oxygen from the lungs and nutrients from the digestive organs to all body cells, where these substances are used in metabolic processes. It also transports hormones from various endocrine glands to their target tissues and wastes from body cells to the excretory organs, where they are removed from the blood and released to the outside.

The **lymphatic system** is sometimes considered to be part of the circulatory system. It is composed of the lymphatic vessels, lymph fluid, lymph nodes, thymus gland, and spleen. This system transports fluid from the tissues back to the bloodstream. It also aids in defending the body against infections caused by microorganisms and in the transport of certain lipids away from the digestive organs.

The **urinary system** consists of the kidneys, ureters, urinary bladder, and urethra. The kidneys remove various wastes from the blood, and assist in maintaining fluid and electrolyte balance. The product of these activities is urine. Other portions of the urinary system are concerned with storing urine and transporting it to the outside of the body.

Sometimes the urinary system is called the *excretory system*. However, excretion or waste removal is also a function of the respiratory, digestive, and integumentary systems.

The systems associated with transporting and processing are shown in figure 7.8.

Reproduction

Reproduction is the process of producing offspring. Cells reproduce when they undergo mitosis and give rise to daughter cells. The **reproductive system** of an organism, however, is concerned with the production of new organisms, each of which is composed of innumerable cells.

The male reproductive system includes the scrotum, testes, epididymides, vasa deferentia, seminal vesicles, prostate gland, bulbourethral glands, penis, and urethra. These parts are concerned with the production and maintenance of the male sex cells or sperm. They also function to transfer these cells into the female reproductive tract.

The female reproductive system consists of the ovaries, fallopian tubes, uterus, vagina, and vulva. These parts produce and maintain the female sex cells or eggs, receive the male cells, and transport these cells. The female parts also provide for the support and development of embryos and function in the birth process.

Figure 7.9 illustrates female and male reproductive systems.

---

1. Name each of the major organ systems and list the organs of each system.
2. Describe the general functions of each organ system.

Descriptive Terms

In describing body parts and the relationships between them, it is convenient to use terms with precise meanings. Some of these terms concern the relative positions of parts, others relate to imaginary planes along which cuts may be made, and still others are used to describe various regions of the body.

When such terms are used, it is assumed that the body is in the **anatomical position,** that is it is standing erect, the face forward, the arms are at the sides, and the palms are forward.

Fig. 7.8 Organ systems associated with processing and transporting.

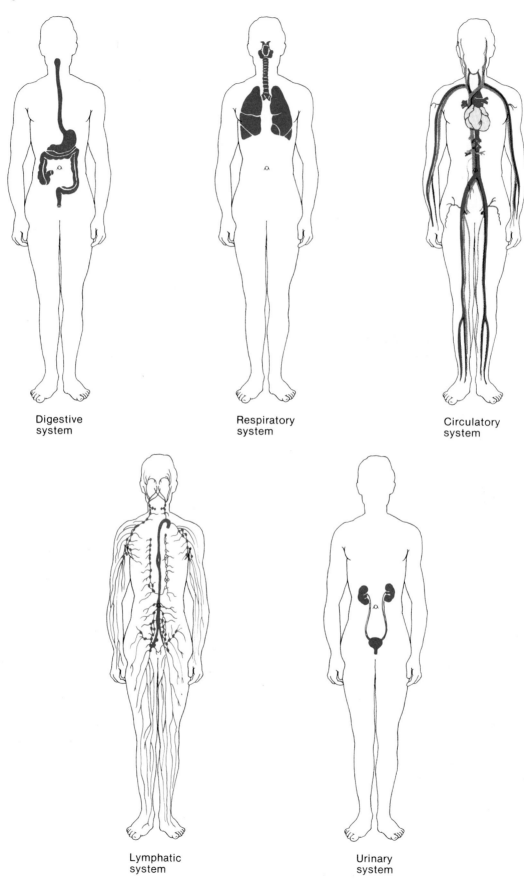

Digestive
system

Respiratory
system

Circulatory
system

Lymphatic
system

Urinary
system

Levels of Organization

Fig. 7.9 Organ systems associated with reproduction.

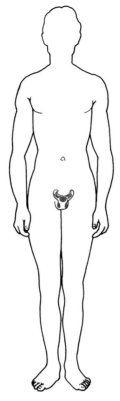
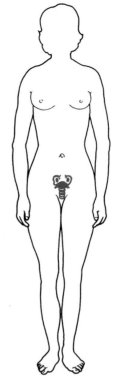

Male reproductive
system

Female reproductive
system

Relative Position

Terms of relative position are used to describe the location of one part with respect to another. They include the following:

1. **Superior** means a part is above, or closer to the head, than another part. (The thoracic cavity is superior to the abdominopelvic cavity.)

2. **Inferior** means situated below something, or toward the feet. (The neck is inferior to the head.)

3. **Anterior** (or *ventral*) means toward the front. (The eyes are anterior to the brain.)

4. **Posterior** (or *dorsal*) is the opposite of anterior; it means toward the back. (The pharynx is posterior to the oral cavity.)

5. **Medial** relates to an imaginary midline dividing the body into equal right and left halves. A part is medial if it is closer to this line than another part. (The nose is medial to the eyes.)

6. **Lateral** means toward the side with respect to the imaginary midline. (The ears are lateral to the eyes.)

7. **Proximal** is used to describe a part that is closer to a point of attachment or to the trunk of the body than something else. (The elbow is proximal to the wrist.)

8. **Distal** is the opposite of proximal. It means something is further from the point of attachment or the trunk than something else. (The fingers are distal to the wrist.)

9. **Superficial** means situated near the surface. (The epidermis is the superficial layer of the skin.) **Peripheral** also means outward or near the surface. It is used to describe the location of certain blood vessels and nerves. (Nerves that branch from the brain and spinal cord are peripheral nerves.)

10. **Deep** is used to describe more internal parts. (The dermis is the deep layer of the skin.)

Figure 7.10 illustrates some of these terms.

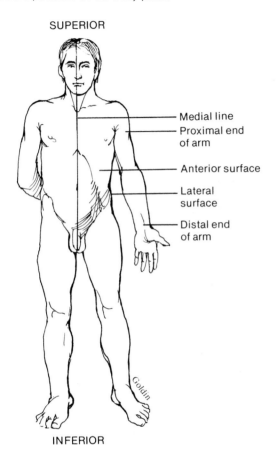

SUPERIOR

— Medial line
— Proximal end of arm
— Anterior surface
— Lateral surface
— Distal end of arm

INFERIOR

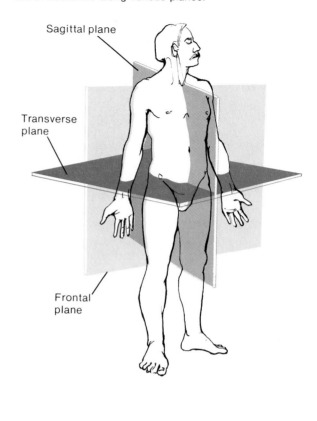

Sagittal plane

Transverse plane

Frontal plane

Fig. 7.12 Tubular parts, such as blood vessels, may be cut in (*a*) cross section (*b*) oblique section, or (*c*) longitudinal section.

Body Sections

To observe the relative locations and arrangements of internal parts, it is necessary to cut or section the body along various planes. (See fig. 7.11.) These terms are used to describe such planes and sections:

1. **Sagittal.** This refers to a lengthwise cut that divides the body into right and left portions. If the section passes along the midline and divides the body into equal halves, it is called *mid-sagittal*.

2. **Transverse.** A transverse cut is one that divides the body into superior and inferior portions.

3. **Frontal** (or *coronal*). A frontal section divides the body into anterior and posterior portions.

Sometimes individual organs such as a blood vessel or the spinal cord are sectioned. In this case, a cut across the structure is called a *cross section*, a lengthwise cut is called a *longitudinal section*, and an angular cut is called an *oblique section* (fig. 7.12).

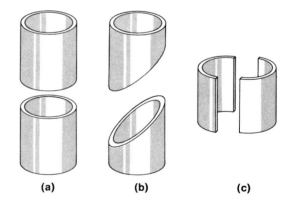

(a)   (b)   (c)

Fig. 7.13 The abdominal area is subdivided into nine regions. How do the names of these regions describe their locations?

ABDOMINAL REGIONS

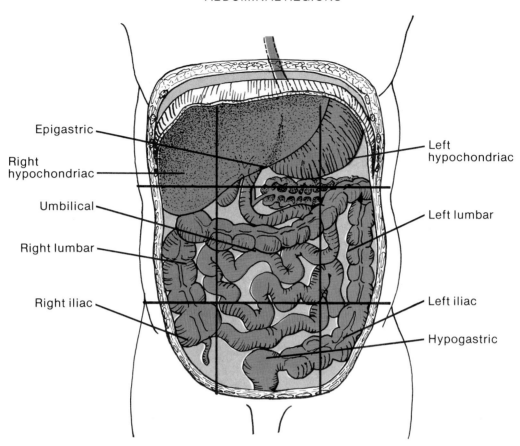

Epigastric

Right hypochondriac

Umbilical

Right lumbar

Right iliac

Left hypochondriac

Left lumbar

Left iliac

Hypogastric

Body Regions

A number of terms designate body regions or refer to specific parts. The abdomen, for example, is subdivided as shown in figure 7.13. The upper middle portion is called the **epigastric region,** the central portion is the **umbilical region,** and the lower middle portion is the **hypogastric region.** On either side of the epigastric region is a **hypochondriac region;** on each side of the umbilical region is a **lumbar region;** on each side of the hypogastric region is an **iliac region.**

Some other terms commonly used to indicate body regions include the following. (See figure 7.14.)

antebrachium (an″te-bra′ke-um)—pertaining to the forearm.

antecubital (an″te-ku′bĭ-tal)—pertaining to the space in front of the elbow.

axillary (ak′sĭ-ler″e)—pertaining to the armpit.

brachial (bra′ke-al)—pertaining to the arm.

buccal (buk′al)—pertaining to the mouth or inner surfaces of the cheeks.

celiac (se′le-ak)—pertaining to the abdomen.

cephalic (sĕ-fal′ik)—pertaining to the head.

cervical (ser′vĭ-kal)—pertaining to the neck.

costal (kos′tal)—pertaining to the ribs.

cubital (ku′bĭ-tal)—pertaining to the forearm.

cutaneous (ku-ta′ne-us)—pertaining to the skin.

dorsum (dor′sum)—pertaining to the back.

frontal (frun′tal)—pertaining to the forehead.

gluteal (gloo′te-al)—pertaining to the buttocks.

groin (groin)—pertaining to the depressed area of the abdominal wall near the thigh.

inguinal (ing′gwĭ-nal)—pertaining to the groin.

lumbar (lum′bar)—pertaining to the loins, the region of the lower back between the ribs and the pelvis.

mammary (mam′er-e)—pertaining to the breasts.

occipital (ok-sip′ĭ-tal)—pertaining to the back of the head.

ophthalmic (of-thal′mik)—pertaining to the eyes.

palmar (pahl′mar)—pertaining to the palm.

pectoral (pek′tor-al)—pertaining to the chest or breast.

pelvic (pel′vik)—pertaining to the pelvis.

perineal (per″ĭ-ne′al)—pertaining to the perineum, the region between the anus and the external reproductive organs.

plantar (plan′tar)—pertaining to the sole.

popliteal (pop″lĭ-te′al)—pertaining to the area behind the knee.

Fig. 7.14 Some terms used to describe various body regions.

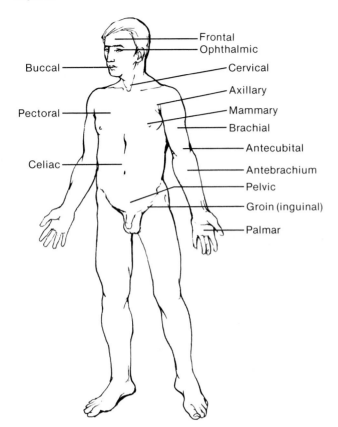

Frontal
Ophthalmic
Buccal
Cervical
Axillary
Mammary
Pectoral
Brachial
Antecubital
Antebrachium
Celiac
Pelvic
Groin (inguinal)
Palmar

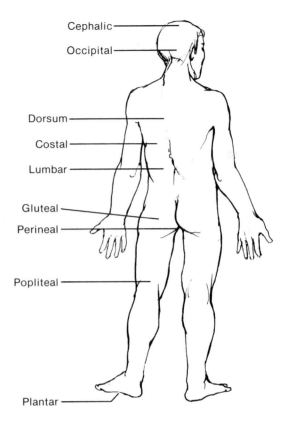

Cephalic
Occipital
Dorsum
Costal
Lumbar
Gluteal
Perineal
Popliteal
Plantar

---

1. *Describe the anatomical position.*
2. *Using the appropriate terms, describe the relative positions of several body parts.*
3. *Describe three types of sections.*
4. *Describe the location of the six regions of the abdomen.*

Chapter Summary

Introduction
Terms have been devised to describe body features, positions of parts, imaginary planes, and body regions.

Body Cavities and Membranes
1. Body cavities
   a. The axial portion of the body contains the dorsal and ventral cavities.
      (1) Dorsal cavity includes the cranial and spinal cavities.
      (2) Ventral cavity includes the thoracic and abdominopelvic cavities, which are separated by the diaphragm.
   b. Body cavities are filled with visceral organs.
   c. Other body cavities include the oral, nasal, orbital, and middle ear cavities.
2. Thoracic and abdominopelvic membranes
   a. Thoracic membranes
      (1) Pleural membranes line the thoracic cavity and cover the lungs.
      (2) Pericardial membranes surround the heart and cover its surface.
      (3) The pleural and pericardial cavities are potential spaces between these membranes.
   b. Abdominopelvic membranes
      (1) Peritoneal membranes line the abdominopelvic cavity and cover the organs inside.
      (2) The peritoneal cavity is a potential space between these membranes.

Organ Systems
1. Body covering
   a. The integumentary system provides the body covering.
   b. Includes the skin, hair, nails, sweat glands, and sebaceous glands.

c. Functions to protect underlying tissues, regulate body temperature, house sensory receptors, and synthesize various cellular products.
2. Support and movement
   a. Skeletal system
      (1) Composed of bones and parts that bind bones together.
      (2) Provides framework, protective shields, and attachments for muscles; produces blood cells and stores inorganic salts.
   b. Muscular system
      (1) Includes the muscles of the body.
      (2) Responsible for body movements, maintenance of posture, and production of body heat.
3. Integration and coordination
   a. Nervous system
      (1) Consists of the brain, spinal cord, and nerves.
      (2) Functions to receive impulses from sensory parts, interpret these impulses, and act on them by causing muscles or glands to respond.
   b. Endocrine system
      (1) Consists of glands that secrete hormones.
      (2) Hormones help regulate metabolism by stimulating target tissues.
      (3) Includes the pituitary, thyroid, parathyroid, and adrenal glands; the pancreas, ovaries, testes, pineal gland, and thymus gland.
4. Processing and transporting
   a. Digestive system
      (1) Receives foods and converts their molecules into forms that can pass through cell membranes.
      (2) Certain portions also produce hormones.
      (3) Digestive tube transports undigested materials to the outside of the body.
      (4) Includes the mouth, teeth, salivary glands, pharynx, esophagus, stomach, liver, gallbladder, pancreas, small intestine, and large intestine.
   b. Respiratory system
      (1) Provides for the intake and output of air and for the exchange of gases between the blood and the air.
      (2) Includes the nasal cavity, pharynx, larynx, trachea, bronchi, and lungs.
   c. Circulatory system
      (1) Includes the heart, which pumps blood, and the blood vessels, which carry blood to and from body parts.
      (2) Blood transports oxygen, nutrients, hormones, and wastes.
   d. Lymphatic system
      (1) Composed of lymphatic vessels, lymph nodes, thymus, and spleen.
      (2) Transport lymph from tissues to bloodstream, aids in defending the body against infections and in the transport of lipids.
   e. Urinary system
      (1) Includes the kidneys, ureters, urinary bladder, and urethra.
      (2) Filters wastes from the blood and helps maintain fluid and electrolyte balance.
5. Reproduction
   a. Concerned with the production of new organisms.
   b. Male reproductive system includes the scrotum, testes, epididymides, vasa deferentia, seminal vesicles, prostate gland, bulbourethral glands, penis, and urethra, which produce, maintain, and transport male sex cells.
   c. Female reproductive system includes the ovaries, fallopian tubes, uterus, vagina, and vulva, which produce, maintain, and transport female sex cells.

Descriptive Terms
1. Relative position
   Terms are used to describe the location of one part with respect to another part.
2. Body sections
   Planes along which the body may be cut to observe the relative locations and arrangements of internal parts.
3. Body regions
   Areas designated by special terms.

Application of Knowledge

1. When a woman is pregnant, what changes would occur in the relative sizes of the abdominal and pelvic portions of the abdominopelvic cavity?

2. Suppose two individuals are afflicted with benign (noncancerous) tumors that produce symptoms because they occupy space and crowd adjacent organs. If one of these persons has the tumor in the ventral cavity and the other has it in the dorsal cavity, which would be likely to develop symptoms first? Why?

3. When a patient's liver or spleen is being examined by palpation (using the examiner's hands and fingers), why is the patient usually instructed to take a deep breath?

Review Activities

Part A

1. Distinguish between the axial and the appendicular portions of the body.

2. Distinguish between the dorsal and ventral body cavities and name the smaller cavities that occur within each.

3. Explain what is meant by a visceral organ?

4. Describe the mediastinum.

5. Describe the location of the oral, nasal, orbital, and middle ear cavities.

6. Distinguish between a parietal and a visceral membrane.

7. Name the major organ systems and describe the general functions of each.

8. List the organs that comprise each organ system.

Part B

1. Name the body cavity in which each of the following organs is located:
   a. Stomach
   b. Heart
   c. Brain
   d. Liver
   e. Trachea
   f. Rectum
   g. Spinal cord
   h. Esophagus
   i. Spleen
   j. Urinary bladder

2. Write complete sentences using each of the following terms correctly:
   a. Superior
   b. Inferior
   c. Anterior
   d. Posterior
   e. Medial
   f. Lateral
   g. Proximal
   h. Distal
   i. Superficial
   j. Peripheral
   k. Deep

3. Prepare a sketch of a human body and use lines to indicate each of the following sections:
   a. Midsagittal
   b. Transverse
   c. Frontal

4. Prepare a sketch of the abdominal area and indicate the location of each of the following regions:
   a. Epigastric
   b. Umbilical
   c. Hypogastric
   d. Hypochondriac
   e. Lumbar
   f. Iliac

5. Name the region or body part to which each of the following terms pertains:
   a. Antebrachium
   b. Antecubital
   c. Axillary
   d. Brachial
   e. Buccal
   f. Celiac
   g. Cephalic
   h. Cervical
   i. Costal
   j. Cubital
   k. Cutaneous
   l. Dorsum
   m. Frontal
   n. Gluteal
   o. Groin
   p. Inguinal
   q. Lumbar
   r. Mammary
   s. Occipital
   t. Ophthalmic
   u. Palmar
   v. Pectoral
   w. Pelvic
   x. Perineal
   y. Plantar
   z. Popliteal

Suggestions for Additional Reading

Basmajian, J. V. 1975. *Grant's method of anatomy*, 9th ed. Baltimore: Williams and Wilkins.

Clemente, C. D. 1975. *Anatomy: a regional atlas of the human body*. Philadelphia: Lea and Febiger.

Goss, C. M., ed. 1973. *Gray's anatomy of the human body*, 29th American ed. Philadelphia: Lea and Febiger.

Hamilton, W. J., ed. 1976. *Textbook of human anatomy*, 2nd ed. St. Louis: C. V. Mosby.

Hollinshead, W. H. 1974. *Textbook of anatomy*, 3rd ed. New York: Harper and Row.

Leeson, T. S., and Leeson, C. R. 1972. *Human structure*. Philadelphia: Saunders.

McMinn, R. M. H., and Hutchings, R. T. 1977. *Color atlas of human anatomy*. Chicago: Year Book Medical Publishers.

Woodburne, R. T. 1973. *Essentials of human anatomy*, 5th ed. New York: Oxford Univ. Press.

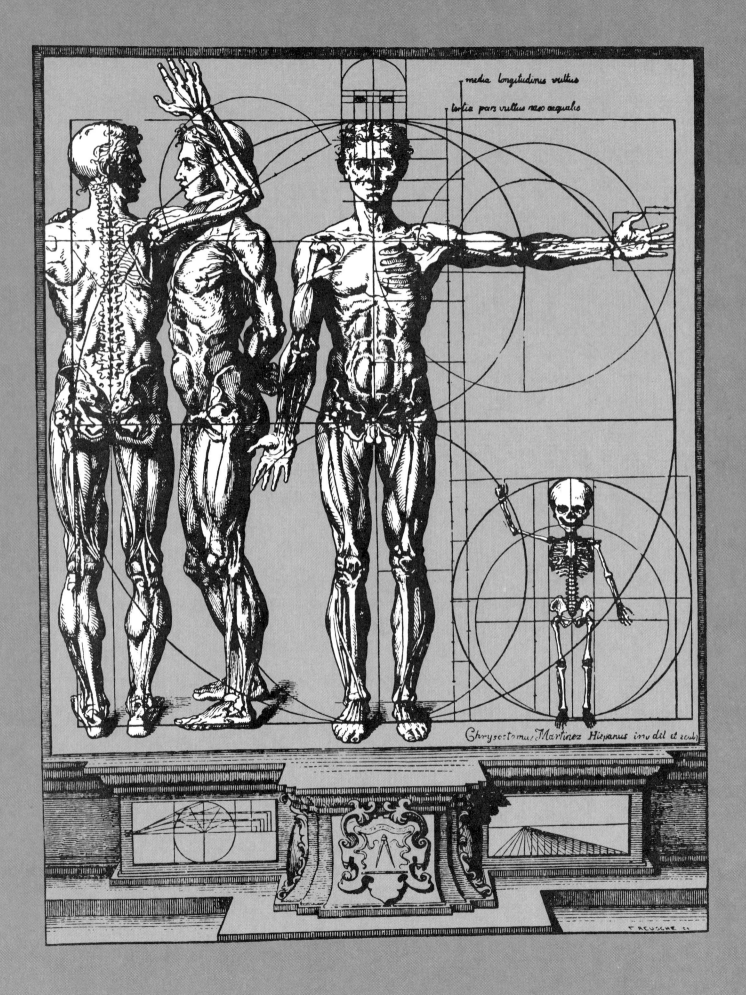

media longitudinis vultus

tertia pars vultus naso aequalis

Chrysostomus Martinez Hispanus inv del et sculp

Unit 2
Support and Movement

The chapters of unit 2 deal with the structures and functions of the skeletal and muscular systems. They describe how the organs of the skeletal system support and protect other body parts and how they function with organs of the muscular system to enable body parts to move. They also describe how skeletal structures participate in the formation of blood and in the storage of inorganic salts, and how the muscular tissues act to produce body heat and to move body fluids.

This unit includes

Chapter 8: The Skeletal System
Chapter 9: The Muscular System

The Skeletal System

8 The bones of the skeleton are the organs of the *skeletal system*. These organs, which are composed of several kinds of tissues, are classified according to their shapes and the ways they develop.

Since bones are rigid structures, they provide support and protection for softer tissues, and they act together with skeletal muscles to make body movements possible. They also manufacture blood cells and store inorganic salts.

The shapes of individual bones are closely related to their functions. Projections provide places for the attachments of muscles, tendons, and ligaments; openings serve as passageways for blood vessels and nerves, and the ends of bones are modified to articulate with other bones at joints.

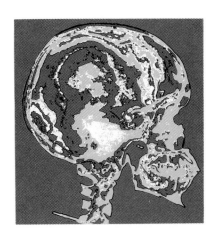

Chapter Objectives

After you have studied this chapter, you should be able to

1. Classify bones according to their shapes and name an example from each group.
2. Describe the general structure of a bone and list the functions of its parts.
3. Distinguish between membranous and cartilaginous bones and explain how such bones grow and develop.
4. Describe the effects of sunlight, nutrition, hormonal secretions, and exercise on bone development.
5. Name the general types of fractures and explain how a fracture is repaired.
6. Discuss the major functions of bones.
7. Distinguish between the axial and appendicular skeletons and name the major parts of each.
8. Locate and identify the bones and the major features of the bones that comprise the skull, vertebral column, thoracic cage, pectoral girdle, upper limb, pelvic girdle, and lower limb.
9. List three types of joints, describe their characteristics, and name an example of each.
10. List six types of freely movable joints and describe the actions of each.
11. Complete the review activities at the end of this chapter.

articular cartilage   (ar-tik′u-lar kar′tĭ-lij)

cartilaginous bone   (kar″tĭ-laj′ĭ-nus bōn)

compact bone   (kom′pakt bōn)

diaphysis   (di-af′ĭ-sis)

epiphyseal disk   (ep″ĭ-fiz′e-al disk)

epiphysis   (e-pif′ĭ-sis)

fracture   (frak′tūr)

haversian system   (ha-ver′shan sis′tem)

hematopoiesis   (hem″ah-to-poi-e′sis)

lever   (lev′er)

marrow   (mar′o)

medullary cavity   (med′u-lār″e kav′ĭ-te)

membranous bone   (mem′brah-nus bōn)

osteoblast   (os′te-o-blast)

osteoclast   (os′te-o-klast)

osteocyte   (os′te-o-sīt)

periosteum   (per″e-os′te-um)

spongy bone   (spun′je bōn)

acetabul-, vinegar cup: *acetabul*um—depression of the os coxa that articulates with the head of the femur.

ax-, an axis: *ax*ial skeleton—upright portion of the skeleton that supports the head, neck, and trunk.

carp-, wrist: *carp*als—bones of the wrist.

-clast, broken: osteo*clast*—cell that breaks down bone tissue.

condyl-, knob: *condyl*e—a rounded, bony process.

corac-, beaklike: *corac*oid process—beaklike process of the scapula.

cribr-, sievelike: *cribr*iform plate—portion of the ethmoid bone with many small openings.

crist-, ridge: *crist*a galli—a bony ridge that projects upward into the cranial cavity.

fov-, pit: *fov*ea capitis—a pit in the head of a femur.

gladi-, sword: *gladi*olus—middle portion of the bladelike sternum.

glen-, joint socket: *glen*oid cavity—a depression in the scapula that articulates with the head of the humerus.

hema-, blood: *hema*toma—a blood clot.

meat-, passage: auditory *meat*us—canal of the temporal bone that leads inward to parts of the ear.

odont-, tooth: *odont*oid process—a toothlike process of the second cervical vertebra.

os-, bone: *os* coxa—a hip bone.

poie-, making: hemato*poie*sis—process by which blood cells are formed.

An individual bone of the skeletal system is composed of a variety of tissues, including bone tissue, cartilage, fibrous connective tissue, blood, and nerve tissue. Because there is so much nonliving material present in the matrix of the bone tissue, the whole organ may appear to be inert. Bone, however, is composed of very active tissues.

Bone Structure

Although bones vary greatly in size and shape (fig. 8.1), they have much in common both macroscopically (as seen with the unaided eye) and microscopically (as seen through a microscope).

Classification of Bones

Bones can be classified according to their shapes as long, short, flat, and irregular. (See fig. 8.2.)

1. **Long bones** have long longitudinal axes and expanded ends, such as those of arm and leg bones.

2. **Short bones** are similar in shape to long bones, but they are smaller and have less prominent ends. The bones of the wrists and ankles, for example, are of this type.

3. **Flat bones** are platelike structures with broad surfaces, such as the ribs, the scapulae, and the bones of the skull that form a protective wall around the brain.

4. **Irregular bones** have a variety of shapes and are usually connected to several other bones. Irregular bones include the vertebrae that comprise the backbone and many of the facial bones.

In addition to these four groups of bones, some authorities recognize a fifth group called the *round* or *sesamoid* bones. The members of this group are usually small, and they often occur in tendons adjacent to joints. The kneecap (patella) is an example of a very large sesamoid bone.

Parts of a Long Bone

In describing the macroscopic structure of bone, a long bone will be used as an example. (See fig. 8.3.) At each end of such a bone there is an expanded portion called an **epiphysis**, which articulates or forms a joint with another bone. On its outer face the epiphysis is coated with a layer of hyaline cartilage called **articular cartilage**. The shaft of the bone, which is located between the epiphyses, is called the **diaphysis**.

Except for the articular cartilage on its ends, the bone is completely enclosed by a tough, vascular covering of fibrous tissue called the **periosteum**. This membrane is firmly attached to the bone, and its fi-

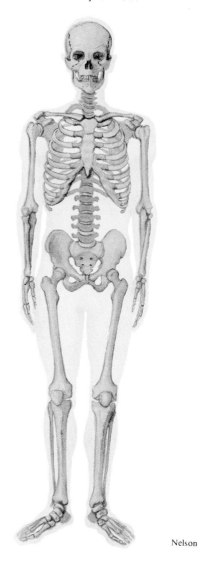

Fig. 8.1 The individual bones are the organs of the skeletal system. Name those you know.

Nelson

bers are continuous with various ligaments and tendons that are connected to it. The periosteum also functions in the formation and repair of bone tissue.

As a rule, such a bone is not symmetrical, but has a shape closely related to its functions. Bony projections called *processes*, for example, provide sites for the attachment of ligaments and tendons; grooves and openings serve as passageways for blood vessels and nerves; a depression of one bone might articulate with a process of another.

While observing a longitudinal section of the bone, it becomes apparent that the structure of the tissue varies from place to place. The wall of the diaphysis, for example, is composed of tightly packed tissue called *compact bone*. This type of bone appears solid, strong, and resistant to bending.

Fig. 8.3 The major parts of a long bone.

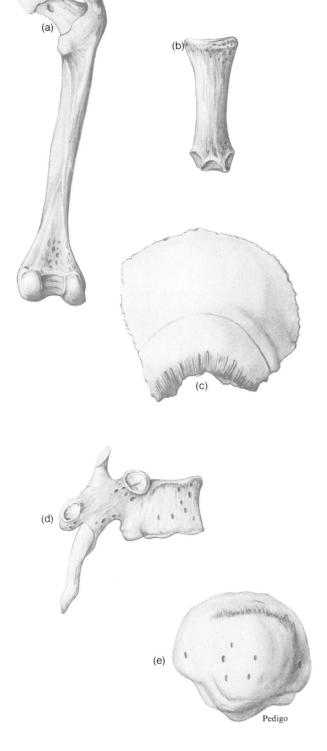

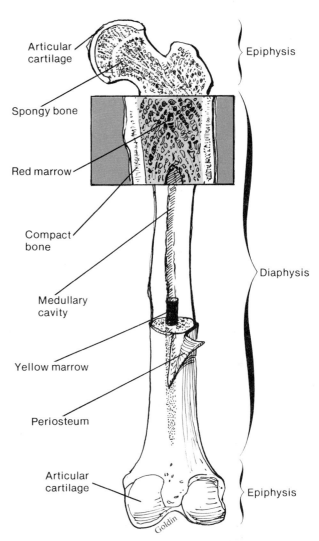

The epiphyses are composed largely of *spongy* (cancellous) *bone* with thin layers of compact bone on their surfaces. Spongy bone consists of numerous branching bony plates called *trabeculae*. Irregular interconnecting spaces occur between the trabeculae, and help reduce the weight of the bone. Spongy bone provides strength where it is needed because trabeculae are most highly developed in the regions of the epiphyses that are subjected to the greatest forces of compression.

Both compact and spongy tissues are usually present in bones. Short, flat, and irregular bones typically consist of a mass of spongy bone that is either covered by a layer of compact bone or sandwiched between plates of compact bone.

The compact bone in the diaphysis of a long bone forms a rigid tube with a hollow chamber called

The Skeletal System 155

the *medullary cavity*. This cavity is continuous with the spaces of the spongy bone. All of these areas are lined with a thin, cellular layer called *endosteum* and are filled with a specialized type of connective tissue called *marrow*. The marrow in the medullary cavity of an adult bone is usually of a type called *yellow marrow*, which functions as fat-storage tissue. The marrow in the spaces of spongy bone is likely to be *red marrow*, which functions to produce various types of blood cells.

Microscopic Structure

As was discussed in chapter 5, bone cells (osteocytes) are located in minute, bony chambers called lacunae which are arranged in concentric circles around haversian canals. The intercellular material of bone tissue is largely collagen, which gives it strength and resilience, and inorganic salts, which make it hard and resistant to crushing.

In compact bone, the osteocytes and layers of intercellular material clustered concentrically around a haversian canal form a cylinder-shaped unit called a *haversian system* (osteon). Many of these units cemented together form the substance of compact bone.

Each haversian canal contains one or two small blood vessels (usually capillaries) surrounded by some loose connective tissue. Blood in these vessels nourishes the cells associated with the haversian canal.

Haversian canals travel longitudinally through bone tissue. They are interconnected by transverse channels called *Volkmann's canals*. These canals contain larger blood vessels by which the vessels in the haversian canals communicate with the surface of the bone and the medullary cavity. (See fig. 8.4.)

Although spongy bone is also composed of osteocytes and intercellular material, the trabeculae are usually very thin, and therefore are not arranged around haversian canals. Instead, the bone cells are nourished by diffusion of substances into canaliculi that lead from the bone cells to the surface of the spongy bones.

1. Explain how bones are classified.
2. How do compact and spongy bone differ?
3. List five major parts of a long bone.
4. Distinguish between red and yellow bone marrow.
5. Explain how bone cells are nourished.

Fig. 8.4 Compact bone is composed of haversian systems cemented together.

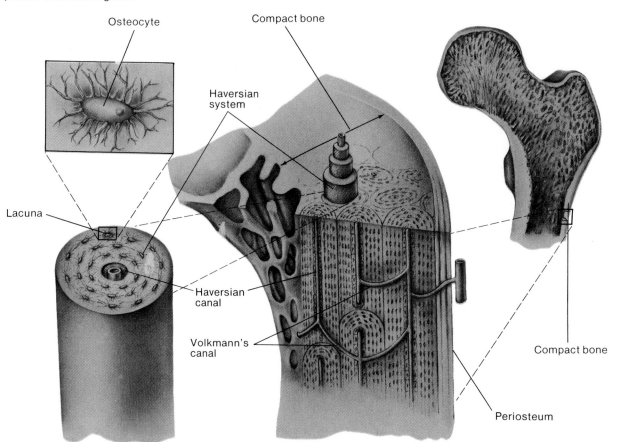

Bone Growth and Development

Although parts of the skeletal system begin to form during the first few weeks of life, bony structures continue to grow and develop into adulthood. These bones form in one of two ways. Some appear first as layers of membranous connective tissues, and are called membranous bones. Others begin as masses of cartilage that are later replaced by bone tissue. They are called cartilaginous bones.

Membranous Bones

The **membranous bones** include certain flat bones of the skull. During their development, sheetlike masses of connective tissues appear in the sites of the future bones. (See fig. 8.5.) Then clusters of bone-forming cells called **osteoblasts** become active within these membranes, and deposit bony intercellular materials around themselves. As a result, spongy bone tissue is produced in all directions within the membranes.

Fig. 8.5 The tissues of this miscarried fetus have been cleared and the developing bones have been selectively stained. Certain of the flat bones of the skull are membranous bones, while the others are cartilaginous.

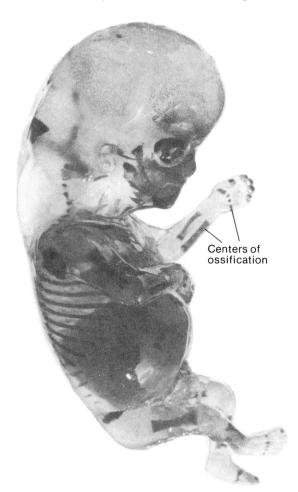

Centers of ossification

Eventually the cells of the membranous tissues that persist outside the developing bone become the periosteum. At the same time osteoblasts on the inside of the periosteum form a layer of compact bone over the surface of the newly built spongy bone. When the osteoblasts are surrounded by matrix and isolated in lacunae, they are called **osteocytes**.

Cartilaginous Bones

Most of the bones of the skeleton are **cartilaginous bones**. They develop from masses of hyaline cartilage with shapes similar to the future bony structures. These cartilaginous models grow rapidly for a while, and then begin to undergo extensive changes. In a long bone, for example, the changes begin in the center of the diaphysis where the cartilage slowly breaks down and disappears. About the same time, a periosteum forms from connective tissues on the outside of the developing diaphysis. Blood vessels and osteoblasts from the periosteum invade the disintegrating cartilage, and spongy bone is formed in its place. This region of bone formation is called the *primary ossification center*, and bone tissue develops from it toward the ends of the cartilaginous structure.

Meanwhile, osteoblasts from the periosteum deposit a thin layer of compact bone around the primary ossification center. The epiphyses of the developing bone remain cartilaginous and continue to grow. Later, *secondary ossification centers* appear in the epiphyses, and spongy bone forms in all directions from them. As bone is deposited in the diaphysis and in the epiphysis, a band of cartilage, called the **epiphyseal disk**, is left between the two ossification centers.

The cartilaginous cells of the epiphyseal disk are arranged in four layers, as shown in figure 8.6 and color plate 19. The first layer, closest to the end of the epiphysis, is composed of resting cells. Although these cells are not actively participating in the growing process, this layer does anchor the epiphyseal disk to the bony tissue of the epiphysis.

The second layer contains rows of numerous young cells that are undergoing mitosis. As new daughter cells are produced, and as intercellular material is formed around them, the cartilaginous disk thickens.

The rows of older cells, which are left behind when new cells appear, form the third layer. These cells enlarge and cause the epiphyseal disk to thicken still more. Consequently, the length of the entire bone increases. At the same time, calcium salts accumulate in the intercellular substance of this layer, and the oldest of the enlarged cells begin to die.

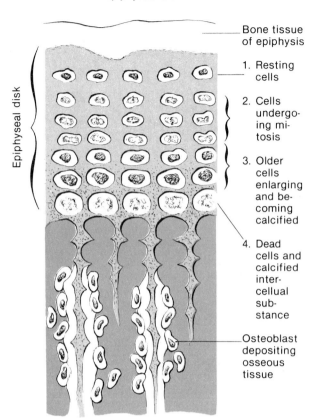

Fig. 8.6 The cartilaginous cells of an epiphyseal disk are arranged in four layers. How does this diagram relate to the function of the epiphyseal disk?

Bone tissue of epiphysis

1. Resting cells

2. Cells undergoing mitosis

3. Older cells enlarging and becoming calcified

4. Dead cells and calcified intercellual substance

Osteoblast depositing osseous tissue

Epiphyseal disk

The fourth layer of the epiphyseal disk is quite thin and is composed largely of dead cells and calcified intercellular substance. The intercellular substance nearest the diaphysis is broken down, apparently by the action of giant cells called **osteoclasts**. Soon, bone-building osteoblasts invade this region and deposit bone tissue in place of the calcified cartilage.

A long bone will grow in length while the cartilaginous cells of the epiphyseal disk are active. However, once the ossification centers of the diaphysis and epiphysis come together and the epiphyseal disk becomes ossified, growth in length is no longer possible in that end of the bone.

While a bone is growing in length, the outside is being surrounded by compact bone and some of the spongy bone within the diaphysis begins to disappear. This breakdown of tissue is the result of resorption by osteoclasts, and it leaves a space in the diaphysis that becomes the medullary cavity and later fills with marrow. The bone in the central regions of the epiphyses remains spongy, and the hyaline cartilage on the ends of the epiphyses persists throughout life as articular cartilage. Figure 8.7 illustrates the stages of bone development. Chart 8.1 lists the ages at which various bones become ossified.

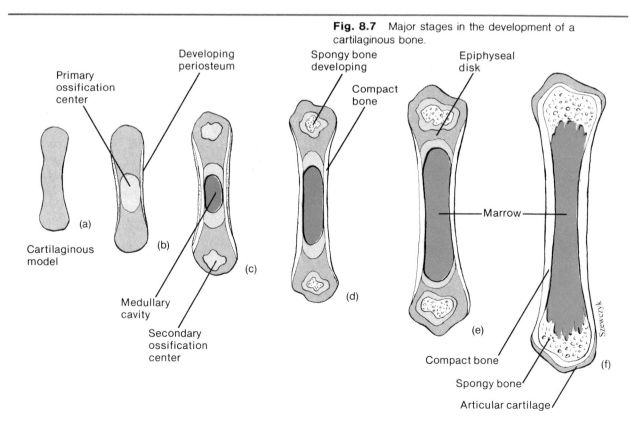

Fig. 8.7 Major stages in the development of a cartilaginous bone.

Developing periosteum

Primary ossification center

Spongy bone developing

Epiphyseal disk

Compact bone

(a)

Cartilaginous model

(b)

(c)

Medullary cavity

Secondary ossification center

(d)

Marrow

(e)

Compact bone

Spongy bone

Articular cartilage

(f)

Szewczyk

Chart 8.1 Ossification timetable

Age	Occurrence
Third month of embryonic development	Ossification in long bones beginning
Fourth month	Most primary ossification centers have appeared in the diaphyses of bones.
Birth to 5 years	Secondary ossification centers appear in the epiphyses.
5 to 12 years in females, or 5 to 14 years in males	Ossification is spreading rapidly from the ossification centers and various bones are becoming ossified.
17 to 20 years	Bones of the upper limbs and scapulae become completely ossified.
18 to 23 years	Bones of the lower limbs and os coxae become completely ossified.
23 to 25 years	Bones of the sternum, clavicles, and vertebrae become completely ossified.
By 25 years	Nearly all bones are completely ossified.

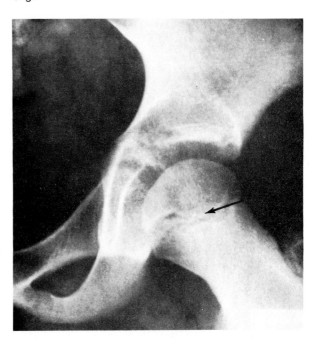

Fig. 8.8 The presence of an epiphyseal disk in a child's bone is an indication that the bone is still growing in length.

It is possible to determine whether a child's long bones are still growing by examining an X-ray photograph to see if the epiphyseal disks are present. (See fig. 8.8.) If a disk is damaged before it becomes ossified, growth of the long bone may cease prematurely, or if growth continues, it may be uneven. For this reason, injuries to the epiphyses of a young person's bones are of special interest. On the other hand, an epiphysis is sometimes altered surgically in order to equalize growth of bones that are developing at very different rates.

Factors Affecting Bone Growth and Development

Bone development is influenced by a number of factors, including nutrition, exposure to sunlight, hormonal secretions, and physical exercise. For example, exposure of skin to the ultraviolet portion of sunlight is favorable to bone development, because the skin can produce vitamin D when it is exposed to such radiation. Vitamin D is necessary for the proper absorption of calcium in the small intestine. In the absence of this vitamin, calcium is poorly absorbed, the inorganic salt portion of bone tissue is deficient in calcium, and the bones are likely to be deformed.

Although vitamin D is relatively uncommon in natural foods, it is readily available in milk and milk products fortified with vitamin D.

Vitamins A and C also are needed for normal bone growth and development. A dietary deficiency of vitamin A may result in retardation of bone development, while an excess of this vitamin may cause painful swellings associated with the long bones. Dietary deficiencies of vitamin C also may inhibit bone development. In this case, osteoblasts produce less collagen in the intercellular material of the bone tissue, and the resulting bones are abnormally slender and fragile.

Hormones that affect bone growth and development include those secreted by the pituitary gland, thyroid gland, parathyroid glands, adrenal glands, and ovaries or testes. The pituitary gland, for instance, secretes a chemical called **growth hormone**, which stimulates cellular activity in the epiphyseal disks. In the absence of this hormone, the long bones of the limbs fail to develop normally, and the child may become a pituitary dwarf. Such a person is very short, but has normal body proportions. If excessive amounts of growth hormone are released before the epiphyseal disks are ossified, the affected person may become a giant. (See fig. 8.9.)

Fig. 8.9 Failure of the pituitary gland to secrete growth hormone results in a pituitary dwarf; excessive secretion of this hormone may result in a pituitary giant.

A deficiency of thyroid hormone delays growth and development of bones, and this can also result in a type of dwarfism in children.

Physical exercise influences bone structure by causing stress on bone tissue. For example, when skeletal muscles contract, they pull at their attachments on bones, and the resulting stress stimulates the bone tissue to thicken and strengthen (hypertrophy). Conversely, with lack of exercise the same bone tissue undergoes a wasting process and tends to become thinner and weaker (atrophy).

Aging is often accompanied by a loss of bone tissue. For example, in the condition called *osteoporosis*, which is promoted by malnutrition, lack of exercise, and various hormonal imbalances, the spaces and canals within the bones become abnormally enlarged and filled with fibrous and fatty tissues. Consequently, the bones are more easily fractured than before, and they may break spontaneously because they are no longer able to sustain the weight of the body. An elderly person with osteoporosis may suffer a spontaneous fracture of a hip or the collapse of a vertebra in the lower back.

1. Describe the development of a membranous bone.
2. Explain how cartilaginous bones develop.
3. What nutritional factors influence bone development?
4. How does exercise affect bone structure?

Fractures

Although a **fracture** may involve injury to cartilaginous structures, it is usually defined as a break in a bone. A fracture can be classified according to its cause and the nature of the break sustained. For example, a break due to injury is a *traumatic* fracture, while one resulting from disease is a *spontaneous* or *pathologic* fracture.

If a broken bone is exposed to the outside by an opening in the skin, the injury is termed a *compound fracture*. Such a fracture is accompanied by the added danger of infection, since microorganisms almost surely enter through the broken skin. On the other hand, if the break is protected by uninjured skin, it is called a *simple fracture*. Figure 8.10 shows several types of fractures.

Repair of a Fracture. Whenever a bone is broken, blood vessels within the bone and its periosteum are ruptured, and the periosteum is likely to be torn. Blood escaping from the broken vessels spreads through the damaged area and soon forms a blood clot, or *hematoma*. As vessels in surrounding tissues dilate, these tissues become swollen and inflamed.

Plate 16 CAT scans of (*a*) the head and (*b*) the abdomen. What organs can you identify in these CAT scans?

Plate 17 Cross section of the thorax.

Plate 18 Cross section of the abdomen.

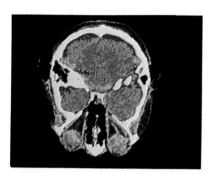

16(a)

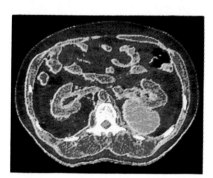

16 (b)

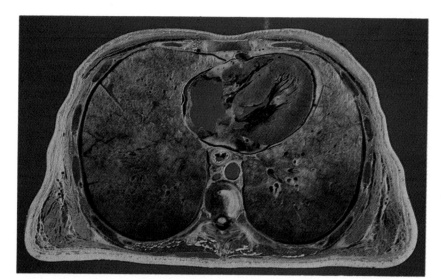

17

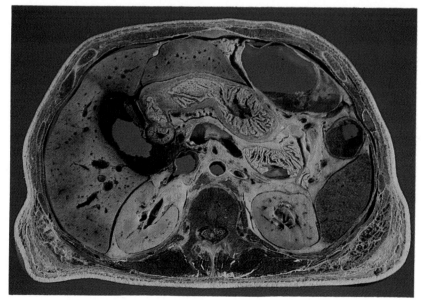

18

Plate 19 Photomicrograph of a longitudinal section of
the epiphyseal disk of a developing bone.

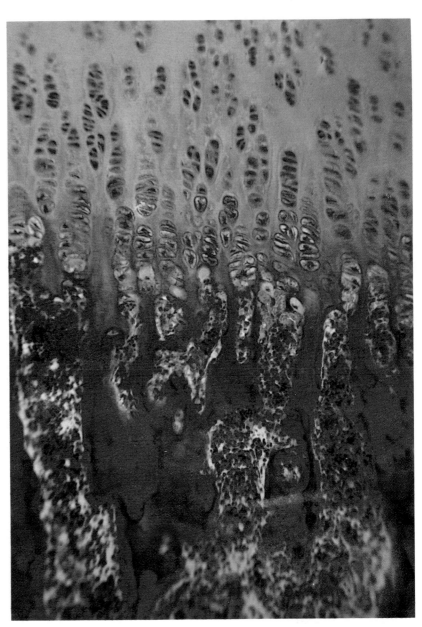

19

Fig. 8.10 Various types of traumatic fractures.

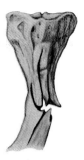

A *greenstick* fracture is incomplete, and the break occurs on the convex surface of the bend in the bone

A *fissured* fracture involves an incomplete longitudinal break

A *comminuted* fracture is complete and results in several bony fragments

A *transverse* fracture is complete, and the break occurs at a right angle to the axis of the bone

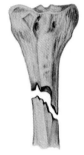

An *oblique* fracture occurs at an angle other than a right angle to the axis of the bone

Pedigo

A *spiral* fracture is caused by twisting a bone excessively

Eventually the hematoma is invaded by blood vessels and osteoblasts originating from the periosteum, and granulation tissue begins to develop. The osteoblasts multiply rapidly in the regions close to the developing blood vessels, building spongy bone nearby. In regions further from a blood supply, fibroblasts produce masses of fibrocartilage.

Meanwhile, phagocytic cells begin to remove the blood clot as well as any dead or damaged cells in the affected area. Osteoclasts also appear and resorb bone fragments, thus aiding in "cleaning up" debris.

In time a large amount of fibrocartilage fills the gap between the ends of the broken bone, and this mass is termed a cartilaginous *callus*. The callus is later replaced by bone tissue in much the same way as the hyaline cartilage of a developing bone is replaced. That is, the cartilaginous callus is broken down, the area is invaded by blood vessels and osteoblasts, and the space is filled with a bony callus.

Usually more bone is produced at the site of a healing fracture than is needed to replace the damaged tissues. However, osteoclasts are able to remove the excess, and the final result of the repair process is a bone shaped very much like the original one. Figure 8.11 shows the steps in the healing of a fracture.

The rate at which a fracture is repaired depends on several factors. For instance, if the ends of the broken bone are close together, healing is more rapid than if they are far apart. This is the reason for setting fractured bones and for using casts or metal pins to keep the broken ends together. Also, some bones naturally heal more rapidly than others. The long bones of the arms, for example, may heal in half the time required by the leg bones. Furthermore, as age increases, so does the time required for healing. Figure 8.12 shows X rays of a healing fracture.

Fig. 8.11 Major steps in the repair of a fracture.

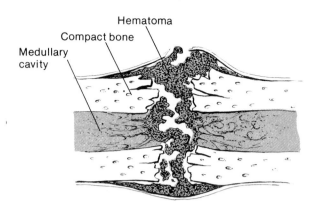

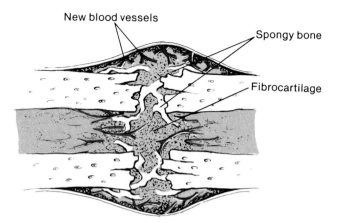

(a) Blood escapes from ruptured blood vessels and forms a hematoma

(b) Spongy bone forms in regions close to developing blood vessels, and fibrocartilage forms in more distant regions

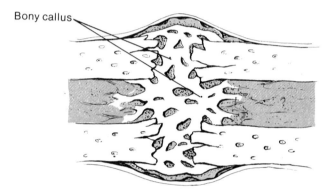

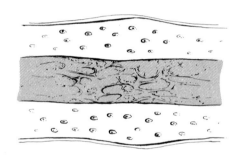

(c) Fibrocartilage is replaced by a bony callus

(d) Osteoclasts remove excess bony tissue, making new bone structure much like the original

Fig. 8.12 X ray of a healing fracture.

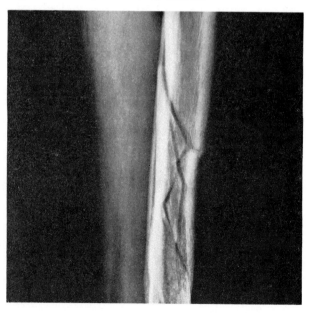

When a child's bone is fractured, growth may be stimulated at the epiphyseal disks of that bone as well as at the site of the fracture. Consequently, a fractured limb may be an inch or so longer than its companion after healing. Although such a difference may eventually even out, it is sometimes necessary to "staple" the epiphysis of an affected bone until growth in the other limb compensates for the difference in lengths.

1. Explain how fractures can be classified according to causes; according to the nature of the break.
2. List the major steps in the repair of a fracture.

Fig. 8.13 Three types of levers. (a) A first-class lever is used in a pair of scissors; (b) a second-class lever is used in a wheelbarrow; (c) a third-class lever is used in a pair of forceps.

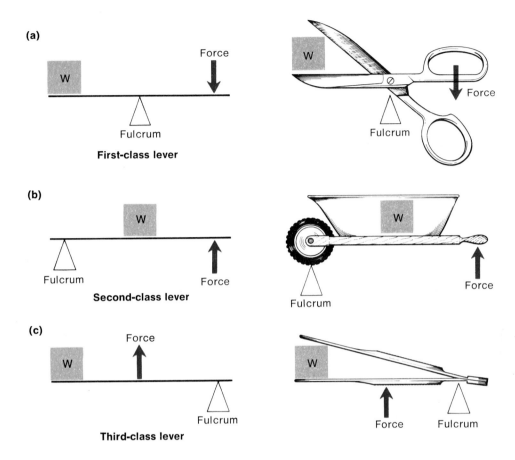

(a)

Force

W

Fulcrum

First-class lever

Fulcrum

Force

(b)

W

Fulcrum

Force

Second-class lever

Fulcrum

Force

(c)

Force

W

Fulcrum

Third-class lever

W

Force

Fulcrum

Functions of Bones

Skeletal parts provide shape and support for body structures. They also act as levers that aid body movements, produce blood cells, and store various inorganic salts.

Support and Protection

Bones give shape to structures such as the head, face, thorax, and limbs. They also provide support and protection. For example, the bones of the feet and legs, pelvis and backbone support the weight of the body. The bones of the skull protect the eyes, ears, and brain. Those of the rib cage and shoulder girdle protect the heart and lungs, while the bones of the pelvic girdle protect the lower abdominal and internal reproductive organs.

Lever Actions

Whenever limbs or other body parts are moved, bones and muscles function together as simple mechanical devices called *levers*. Such a lever has four basic components: (a) a rigid bar or rod; (b) a pivot or fulcrum on which the bar turns; (c) an object or weight that is moved; (d) a force that supplies energy for the movement of the bar.

A playground seesaw is a lever. The board of the seesaw serves as a rigid bar that rocks on a pivot near its center. The person on one end of the board represents the weight that is moved, while the person at the opposite end supplies the force needed for moving the board and its rider.

There are three kinds of levers, and they differ in the arrangements of their parts, as shown in figure 8.13. A first-class lever is one whose parts are arranged like those of the seesaw. Its pivot is located between the weight and the force, making the sequence of parts weight-pivot-force. Other examples

Fig. 8.14 (a) When the arm is bent at the elbow, a third-class lever is employed; (b) when the arm is straightened at the elbow, a first-class lever is used.

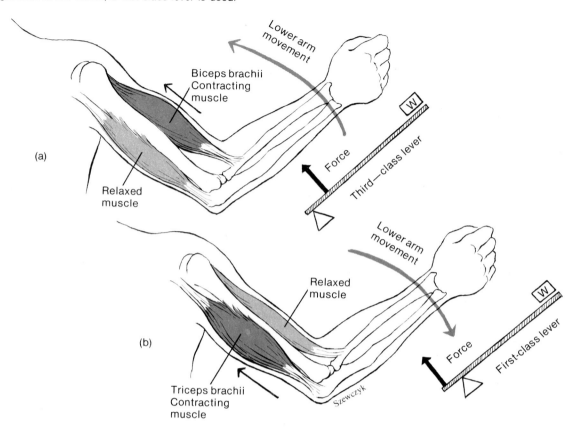

(a)

Lower arm movement

Biceps brachii Contracting muscle

Relaxed muscle

Force

W

Third—class lever

(b)

Lower arm movement

Relaxed muscle

Force

W

First-class lever

Triceps brachii Contracting muscle

Szewczyk

of first-class levers are scissors and hemostats (used to clamp blood vessels closed).

The parts of a second-class lever are arranged in the sequence pivot-weight-force, as in a wheelbarrow.

The parts of a third-class lever are arranged in the sequence pivot-force-weight. This type of lever is illustrated when eyebrow tweezers or forceps are used to grasp an object.

The actions of bending and straightening the arm at the elbow involve bones and muscles functioning together as levers, as illustrated in figure 8.14. When the arm is bent, the lower arm bones represent the rigid bar, the elbow joint is the pivot, the hand is the weight that is moved, and the force is supplied by muscles on the anterior side of the upper arm. One of these muscles, the *biceps brachii,* is attached by a tendon to a process on the *radius* bone in the lower arm, a short distance below the elbow. Since the parts of this lever are arranged in the sequence pivot-force-weight, it is an example of a third-class lever.

When the arm is straightened at the elbow, the lower arm bones again serve as a rigid bar, and the elbow joint as the pivot. However, this time the force is supplied by the *triceps brachii,* a muscle located on

the posterior side of the upper arm. A tendon of this muscle is attached to a process of the *ulna* bone at the point of the elbow. Thus, the parts of the lever are arranged weight-pivot-force, and it is an example of a first-class lever.

Although many lever arrangements occur throughout the skeletal-muscular systems, they are not always easy to identify. Nevertheless, these levers provide advantages in movements. The parts of some levers, such as those that function in moving the limbs, are arranged in ways that produce rapid motions, while others, such as those that move the head, aid in maintaining posture with minimal effort.

Blood Cell Formation

The process of blood cell formation is called **hematopoiesis**. Very early in life it occurs in a structure called a *yolk sac*, which lies outside the body of a human embryo. Later in development, blood cells are manufactured in the liver and spleen, and still later they are formed in the marrow within bones.

Marrow is a soft, netlike mass of connective tissue found in the medullary cavities of long bones, in the irregular spaces of spongy bone, and in the larger haversian canals of bone tissue.

Fig. 8.15 The parathyroid and thyroid glands function to control the level of blood calcium.

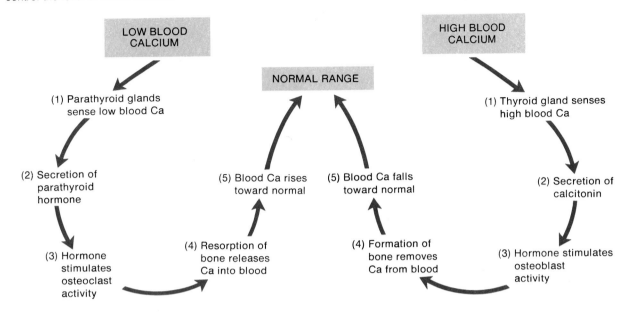

There are two kinds of marrow—red and yellow. *Red marrow* functions in the formation of red blood cells (erythrocytes), certain white blood cells (leukocytes), and blood platelets (thrombocytes). It is red because of the red, oxygen-carrying pigment called **hemoglobin**, which is contained within the red blood cells it produces.

Red marrow occupies the cavities of most bones in a newborn child, but with age more and more of it is replaced by yellowish, fat-storage cells that are inactive in blood cell production.

In an adult, red marrow is found primarily in the spongy parts of the bones of the skull, ribs, sternum, clavicles, vertebrae, and pelvis. If the blood cell supply is deficient, yellow marrow may change back into red marrow and become active in blood cell production. Blood cell formation is described in more detail in chapter 16.

Red marrow may be damaged or destroyed by excessive exposure to radiation, adverse drug reactions, or the presence of cancerous tissues. The treatment of this condition sometimes involves a bone marrow transplant.

In this procedure, normal red marrow cells are removed from the spongy bone of a donor by means of a hollow needle and syringe. These cells are then injected into the recipient's blood with the hope that they will lodge in the bone spaces normally inhabited by red marrow and will, in time, replace the damaged tissue.

Storage of Inorganic Salts

The intercellular matrix of bone tissue contains large quantities of calcium salts. These are mostly in the form of tiny crystals of a type of calcium phosphate called *hydroxyapatite*, which has the formula $3Ca_3(PO_4)_2 \cdot Ca(OH)_2$.

Calcium is needed for a number of metabolic processes, including blood clot formation, nerve impulse conduction, and muscle cell contraction. When a low blood calcium condition exists, the *parathyroid glands* respond by releasing *parathyroid hormone*. This hormone stimulates osteoclasts to break down bone tissue, and as a result, calcium salts are released into the blood. On the other hand, if the blood calcium level is excessively high, the *thyroid gland* responds by releasing a hormone called *calcitonin*. Its effect is opposite that of parathyroid hormone; it inhibits osteoclast activity and stimulates osteoblasts to form bone tissue. As a result, the excessive calcium is stored in bone matrix. (See fig. 8.15.)

In addition to storing calcium and phosphorus, bone tissue stores lesser amounts of magnesium and sodium. Bones also tend to accumulate certain metallic elements such as lead and radium, which are not normally present in the body but are sometimes ingested accidentally.

During childhood, lead poisoning sometimes results from the ingestion of paint chips that contain lead pigments. Although modern interior paints do not contain lead, the woodwork and painted plaster of houses constructed prior to 1940 may be covered with many layers of lead-pigmented paints. Such paints may peel off and be eaten by preschool children.

Strontium is another element that may be concentrated in bone tissue. A radioactive isotope of this element, strontium-90, is a by-product of nuclear reactions, such as those that occur in atomic explosions and nuclear power plants. If strontium-90 is released into the environment, it may be taken in by plants and animals because it is chemically similar to calcium and can be used metabolically by organisms in the same ways they use calcium. Humans may ingest strontium-90 by drinking milk from cows that fed upon plants exposed to nuclear fallout. When this happens, some of the strontium-90 accumulates in the human bones, and nearby cells are subjected to its radiation. Since radiation from strontium-90 can cause mutations, such exposure may result in the development of abnormal cells associated with bone cancers or leukemias.

1. Name three major functions of bones.
2. Explain how body parts form a first-class lever; a third-class lever.
3. Explain how the level of blood calcium is controlled.
4. List seven inorganic elements that may be stored in bone tissue.

Organization of the Skeleton

Number of Bones

Although the number of bones in a human skeleton is often reported to be 206, the number actually varies from person to person. Some people may lack certain bones, while others have extra ones. For example, the flat bones of the skull usually grow together and become tightly joined along irregular lines called **sutures**. Occasionally extra bones called *wormian* or *sutural bones* develop in these sutures (fig. 8.16). Also extra sesamoid bones may develop in tendons, where they apparently function to reduce friction in places where tendons pass over bony prominences. (See chart 8.2.)

Fig. 8.16 Wormian bones are extra bones that sometimes develop in sutures between the flat bones of the skull.

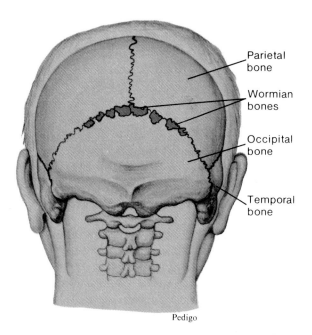

Parietal bone

Wormian bones

Occipital bone

Temporal bone

Pedigo

Chart 8.2 Bones of the adult skeleton

1. Axial Skeleton

   a. Skull   22 bones

        8 cranial bones, which include
           frontal   1
           parietal   2
           occipital   1
           temporal   2
           sphenoid   1
           ethmoid   1

        13 facial bones, which include
           maxilla   2
           palatine   2
           zygomatic   2
           lacrimal   2
           nasal   2
           vomer   1
           inferior nasal concha   2

        1 mandible

   b. Middle ear bones   6 bones
        malleus   2
        incus   2
        stapes   2

   c. Hyoid   1 bone
        hyoid bone   1

   d. Vertebral column   26 bones
        cervical vertebra   7
        thoracic vertebra   12
        lumbar vertebra   5
        sacrum   1
        coccyx   1

   e. Thoracic cage   25 bones
        rib   24
        sternum   1

2. Appendicular Skeleton

   a. Pectoral girdle   4 bones
        scapula   2
        clavicle   2

   b. Upper limbs   60 bones
        humerus   2
        radius   2
        ulna   2
        carpal   16
        metacarpal   10
        phalanx   28

   c. Pelvic girdle   2 bones
        os coxa   2

   d. Lower limbs   60 bones
        femur   2
        tibia   2
        fibula   2
        patella   2
        tarsal   14
        metatarsal   10
        phalanx   28

                              Total   206 bones

Divisions of the Skeleton

For purposes of study, it is convenient to divide the skeleton into two major portions—an axial skeleton and an appendicular skeleton. (See fig. 8.17.)

The **axial skeleton** consists of the bony and cartilaginous parts that support and protect the organs of the head, neck, and trunk. These parts include the following:

1. **Skull**. The skull is composed of the brain case, or *cranium*, and the facial bones.

2. **Hyoid bone**. The hyoid bone (fig. 8.18) is located in the neck between the lower jaw and the larynx. It does not articulate with any other bones, but is fixed in position by muscles and ligaments. The hyoid bone supports the tongue and serves as an attachment for certain muscles that help to move the tongue and function in swallowing.

3. **Vertebral column**. The vertebral column, or backbone, consists of many vertebrae separated by cartilaginous *intervertebral disks*. This column forms the central axis of the skeleton. At its distal end, several vertebrae are fused to form the **sacrum**, which is part of the pelvis. A small, rudimentary tailbone called the **coccyx** is attached to the end of the sacrum.

4. **Thoracic cage**. The thoracic cage protects the organs of the thorax and the upper abdomen. It is composed of twelve pairs of ribs that articulate posteriorly with thoracic vertebrae. It also includes the **sternum** (breastbone), to which most of the ribs are attached anteriorly.

The **appendicular skeleton** consists of the bones of the limbs and those that anchor the limbs to the axial skeleton. It includes the following:

1. **Pectoral girdle**. The pectoral girdle is formed by a **scapula** (shoulder blade) and a **clavicle** (collarbone) on both sides of the body. The pectoral girdle connects the bones of the arms to the axial skeleton and aids in arm movements.

2. **Upper limbs** (arms). Each upper limb consists of a **humerus**, or upper arm bone, and two lower arm bones—a **radius** and an **ulna**. These three bones articulate with each other at the elbow joint. At the distal end of the radius and ulna, there are eight **carpals**, or wrist bones. The bones of the palm are called **metacarpals**, and the finger bones are **phalanges**.

3. **Pelvic girdle**. The pelvic girdle is formed by two **os coxae** (hip bones), which are attached to each other anteriorly and to the sacrum posteriorly. They connect the bones of the legs to the axial skeleton and, with the sacrum and coccyx, form the **pelvis**, which protects the lower abdominal and internal reproductive organs.

Fig. 8.17 The skeleton can be divided into axial and appendicular portions.

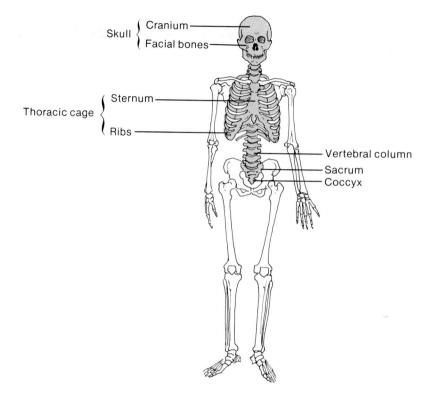

AXIAL SKELETON

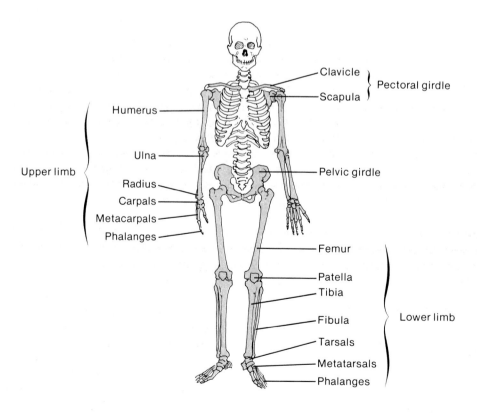

APPENDICULAR SKELETON

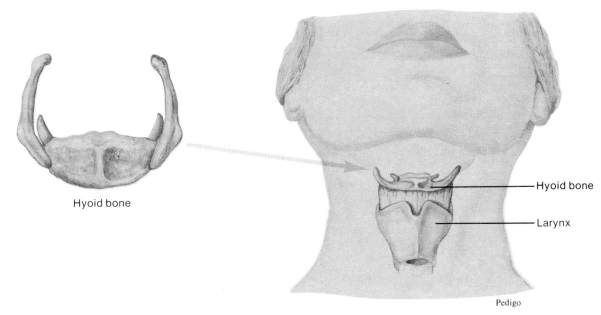

Hyoid bone

Hyoid bone

Larynx

Pedigo

4. **Lower limbs** (legs). Each lower limb consists of a **femur** (thigh bone) and two lower leg bones—a large **tibia** (shinbone) and a slender **fibula** (calf bone). These three bones articulate with each other at the knee joint, where the **patella**, or kneecap, covers the anterior surface. At the distal ends of the tibia and fibula, there are seven **tarsals**, or ankle bones. The bones of the foot are called **metatarsals**, while those of the toes (like the fingers) are **phalanges**.

1. *Distinguish between the axial and the appendicular skeletons.*
2. *Name the bones that each includes.*

Terms Used to Describe Skeletal Structures

The following terms are commonly used to describe features of skeletal structures. Each definition includes an example and the number of a figure illustrating the example.

condyle (kon′dil)—a rounded process that usually articulates with another bone (occipital condyle of the occipital bone) (fig. 8.22).

crest (krest)—a narrow, ridgelike projection (iliac crest of the ilium) (fig. 8.55).

epicondyle (ep″ĭ-kon′dil)—a projection situated above a condyle (medial epicondyle of the humerus) (fig. 8.50).

facet (fas′et)—a small, nearly flat surface (costal facet of a thoracic vertebra) (fig. 8.38).

fontanel (fon″tah-nel′)—a soft spot in the skull where membranes cover the space between bones (anterior fontanel between the frontal and parietal bones) (fig. 8.33).

foramen (fo-ra′men)—an opening through a bone that usually serves as a passageway for blood vessels, nerves, or ligaments (foramen magnum of the occipital bone) (fig. 8.22).

fossa (fos′ah)—a relatively deep pit or depression (olecranon fossa of the humerus) (fig. 8.50).

fovea (fo′ve-ah)—a tiny pit or depression (fovea capitis of the femur) (fig. 8.60).

head (hed)—an enlargement on the end of a bone that usually articulates with another bone (head of the humerus) (fig. 8.50).

meatus (me-a′tus)—a tubelike passageway within a bone (auditory meatus of the ear) (fig. 8.21).

process (pros′es)—a prominent projection on a bone (mastoid process of the temporal bone) (fig. 8.21).

sinus (si′nus)—a cavity or hollow space within a bone (frontal sinus of the frontal bone) (fig. 8.26).

suture (soo′cher)—a line of union between bones (lambdoid suture between the occipital and parietal bones) (fig. 8.21).

trochanter (tro-kan′ter)—a relatively large process (greater trochanter of the femur) (fig. 8.60).

trochlea (trok′le-ah)—a pulley-shaped part (trochlea of the humerus) (fig. 8.50).

tubercle (tu′ber-kl)—a small, knoblike process (tubercle of a rib) (fig. 8.43).

tuberosity (tu″bĕ-ros′ĭ-te)—a knoblike process usually larger than a tubercle (radial tuberosity of the radius) (fig. 8.51).

Fig. 8.19 Anterior view of the skull.

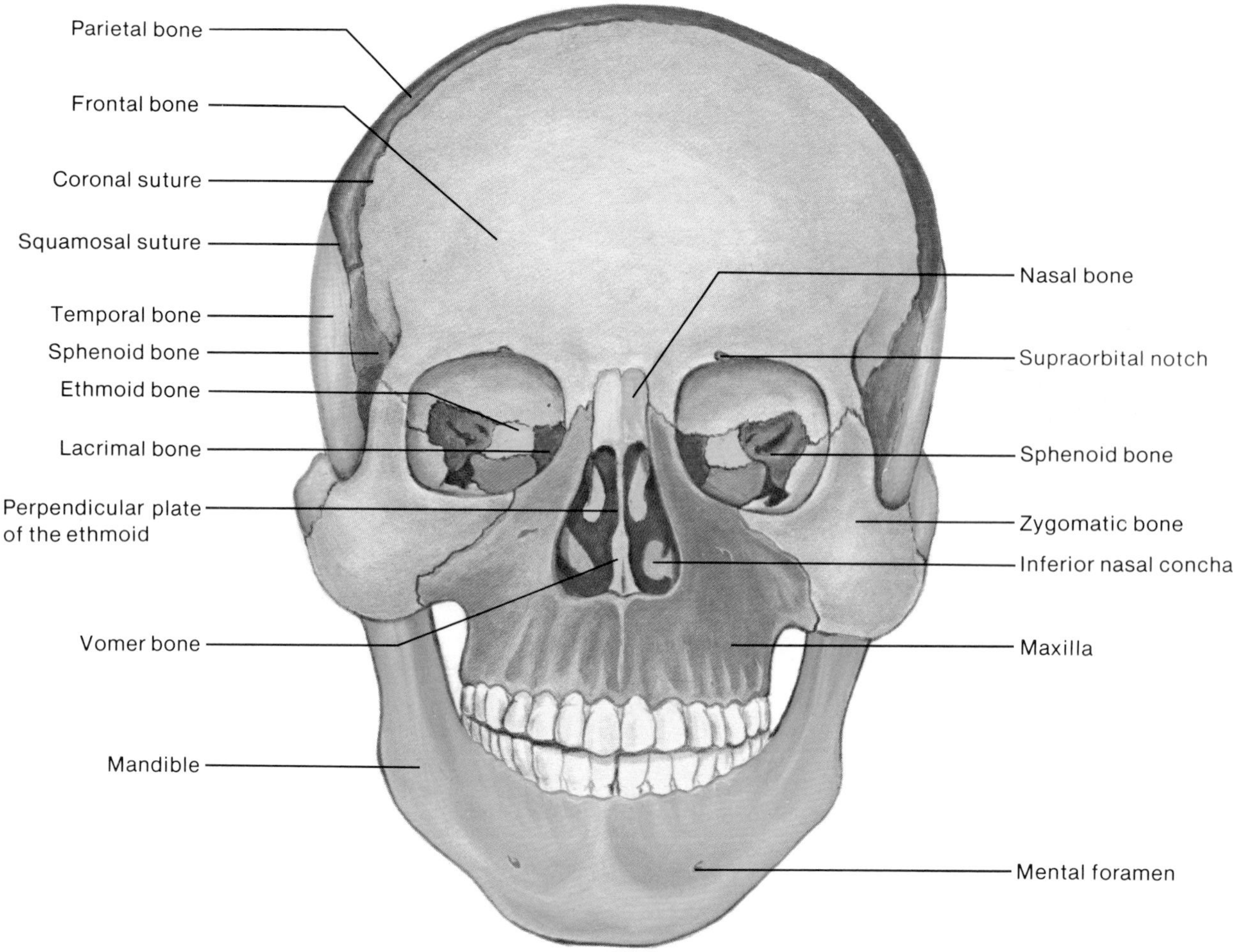

Parietal bone

Frontal bone

Coronal suture

Squamosal suture

Temporal bone

Sphenoid bone

Ethmoid bone

Lacrimal bone

Perpendicular plate of the ethmoid

Vomer bone

Mandible

Nasal bone

Supraorbital notch

Sphenoid bone

Zygomatic bone

Inferior nasal concha

Maxilla

Mental foramen

The Skull

A human skull usually consists of twenty-two bones that, except for the lower jaw, are firmly interlocked along sutures. Eight of these immovable bones make up the cranium, and thirteen form the facial skeleton. The **mandible**, or lower jawbone, is a movable bone held to the cranium by ligaments. (See figs. 8.19, 8.20, 8.21, 8.22.)

The Cranium

The **cranium** encloses and protects the brain, and its surface provides attachments for various muscles that make chewing and head movements possible. Some of the cranial bones contain air-filled cavities called *sinuses*, which are lined with mucous membranes and are connected by passageways to the nasal cavity.

Sinuses reduce the weight of the skull and increase the intensity of the voice by serving as resonant sound chambers.

The eight bones of the cranium (chart 8.3) are as follows:

1. **Frontal bone**. The frontal bone forms the anterior portion of the skull above the eyes and includes the forehead, the roof of the nasal cavity, and the roofs of the orbits of the eyes. On the upper margin of each orbit, the frontal bone is marked by a *supraorbital foramen* (or *supraorbital notch* in some skulls), through which blood vessels and nerves pass to the tissues of the forehead. Within the frontal bone are two *frontal sinuses*, one above each eye near the midline. Although it is a single bone in adults, the frontal bone develops in two parts. These halves grow together and are usually completely fused by the fifth or sixth year of age.

Fig. 8.20 The orbit of the eye includes both cranial and facial bones.

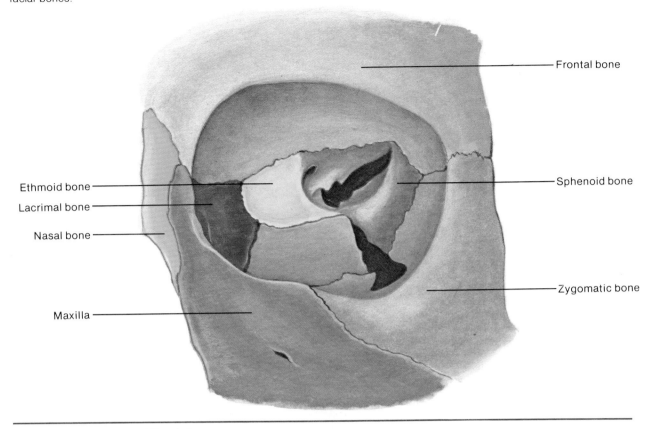

Frontal bone

Ethmoid bone

Lacrimal bone

Nasal bone

Sphenoid bone

Zygomatic bone

Maxilla

Fig. 8.21 Lateral view of the skull.

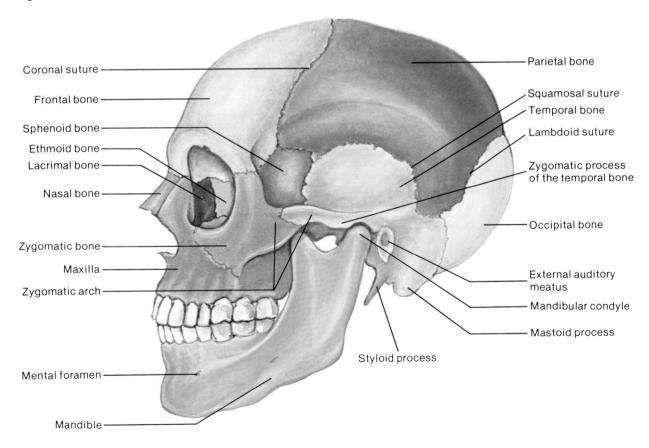

Coronal suture

Frontal bone

Sphenoid bone

Ethmoid bone

Lacrimal bone

Nasal bone

Zygomatic bone

Maxilla

Zygomatic arch

Mental foramen

Mandible

Parietal bone

Squamosal suture

Temporal bone

Lambdoid suture

Zygomatic process of the temporal bone

Occipital bone

External auditory meatus

Mandibular condyle

Mastoid process

Styloid process

The Skeletal System   171

Fig. 8.22 Inferior view of the skull.

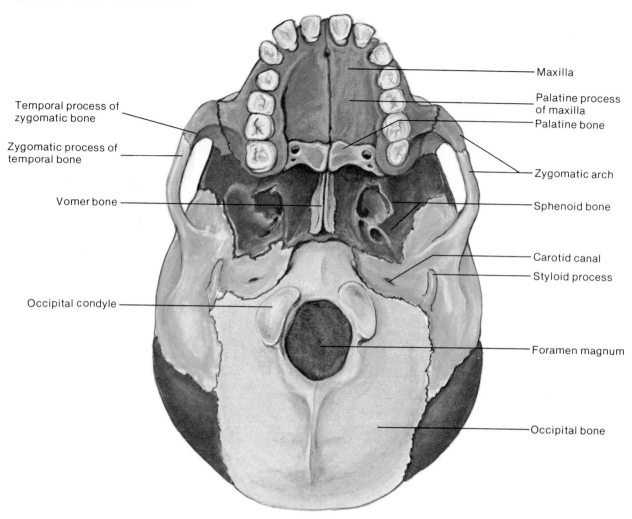

Temporal process of zygomatic bone

Zygomatic process of temporal bone

Vomer bone

Occipital condyle

Maxilla

Palatine process of maxilla

Palatine bone

Zygomatic arch

Sphenoid bone

Carotid canal

Styloid process

Foramen magnum

Occipital bone

Chart 8.3 Cranial bones

Name and Number	Description	Special Features
Frontal 1	Forms forehead, roof of nasal cavity, and roofs of orbits	Supraorbital foramen, frontal sinuses
Parietal 2	Form side walls and roof of cranium	
Occipital 1	Forms back of skull and base of cranium	Foramen magnum, occipital condyles
Temporal 2	Form side walls and floor of cranium	External auditory meatus, mandibular fossa, mastoid process, styloid process, zygomatic process
Sphenoid 1	Forms parts of base of cranium, sides of skull, and floors and sides of orbits	Sella turcica, sphenoidal sinuses
Ethmoid 1	Forms parts of roof and walls of nasal cavity, floor of cranium, and walls of orbits	Cribriform plates, perpendicular plate, superior and middle nasal conchae, ethmoidal sinuses, crista galli

Fig. 8.23 Lateral surface of the left temporal bone. What sensory structures are located within this bone?

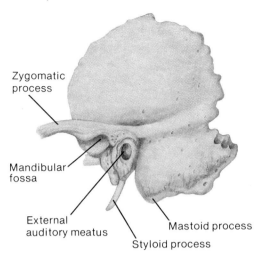

Zygomatic process

Mandibular fossa

External auditory meatus

Styloid process

Mastoid process

Fig. 8.24 Posterior surface of the sphenoid bone as viewed from above.

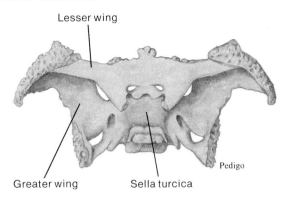

Lesser wing

Greater wing

Sella turcica

Pedigo

2. **Parietal bones**. One *parietal bone* is located on each side of the skull just posterior to the frontal bone. It is shaped like a curved plate and has four borders. Together, the parietal bones form the bulging sides and roof of the cranium. They are fused in the midline along the *sagittal suture*, and they meet the frontal bone along the *coronal suture*.

3. **Occipital bone**. The occipital bone joins the parietal bones along the *lambdoid suture*. It forms the back of the skull and the base of the cranium. There is a large opening on its lower surface called the *foramen magnum*, through which nerve fibers from the brain pass and enter the vertebral canal to become part of the spinal cord. A rounded process called an *occipital condyle* is located on each side of the foramen magnum. These condyles articulate with the first vertebra of the vertebral column.

4. **Temporal bones**. A temporal bone (fig. 8.23) on each side of the skull joins the parietal bone along a *squamosal suture*. The temporal bones form parts of the sides and the base of the cranium. Located near the inferior margin of each one is an opening to a canal, the *external auditory meatus*, which leads inward to parts of the ear. The temporal bones also house the internal ear structures and have depressions, the *mandibular fossae* (glenoid fossae), that articulate with processes of the mandible. Below each external auditory meatus, there are two projections—a rounded *mastoid process* and a long, pointed *styloid process*. The mastoid process provides an attachment

for certain muscles of the neck, while the styloid process serves as an anchorage for muscles associated with the tongue and pharynx. An opening near the mastoid process, the *carotid canal*, transmits a branch of the carotid artery, while an opening between the temporal and occipital bones, the *jugular foramen*, accommodates the jugular vein. (See fig. 8.22.)

The mastoid process is of clinical interest because it may become infected. The tissues in this region of the temporal bone contain a number of interconnected air cells, lined with mucous membranes, that communicate with the middle ear. These spaces sometimes become inflamed when microorganisms spread from an infected middle ear (*otitis media*). The resulting mastoid infection, called *mastoiditis*, is of particular concern because the membranes that surround the brain are close by and may also become infected.

A *zygomatic process* projects anteriorly from the temporal bone in the region of the external auditory meatus. It joins the *zygomatic bone* and helps form the prominence of the cheek.

5. **Sphenoid bone**. The sphenoid bone (fig. 8.24) is wedged between several other bones in the anterior portion of the cranium. It consists of a central part

and two winglike structures that extend laterally toward each side of the skull. This bone helps form the base of the cranium, the sides of the skull, and the floors and sides of the orbits. Along the midline within the cranial cavity, a portion of the sphenoid bone rises up and forms a saddle-shaped mass called *sella turcica* (Turk's saddle). The depression of this saddle is occupied by the pituitary gland, which hangs from the base of the brain by a stalk.

The sphenoid bone contains two *sphenoidal sinuses*, which lie side by side and are separated by a bony septum that projects downward into the nasal cavity.

6. **Ethmoid bone**. The ethmoid bone (fig. 8.25) is located in front of the sphenoid bone. It consists of two masses, one on each side of the nasal cavity, which are joined horizontally by thin *cribriform plates*. These plates form part of the roof of the nasal cavity, and nerves associated with the sense of smell pass through tiny openings in them. Portions of the ethmoid bone also form sections of the cranial floor, orbital walls, and nasal cavity walls. A *perpendicular plate* projects downward in the midline from the cribriform plates to form the bulk of the nasal septum.

Delicate scroll-shaped plates called *superior* and *middle nasal conchae* project inward from the lateral portions of the ethmoid bone toward the perpendicular plate. These bones, which are also called *turbinate bones*, support mucous membranes that line the nasal cavity. The mucous membranes, in turn, begin the processes of moistening, warming, and filtering air as it enters the respiratory tract. The lateral portions of the ethmoid bone contain many small air spaces, the *ethmoidal sinuses*. Various structures in the nasal cavity are shown in fig. 8.26.

Projecting upward into the cranial cavity between the cribriform plates is a triangular process of the ethmoid bone called the *crista galli* (cock's comb). This process serves as an attachment for membranes that enclose the brain. A view of the cranial cavity is shown in figure 8.27.

Fig. 8.25 (*a*) Ethmoid bone viewed from above and (*b*) from behind.

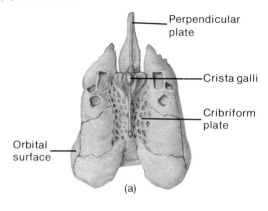

Perpendicular plate

Crista galli

Cribriform plate

Orbital surface

(a)

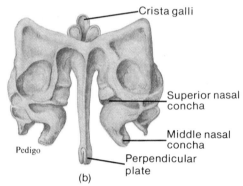

Crista galli

Superior nasal concha

Middle nasal concha

Pedigo

Perpendicular plate

(b)

Fig. 8.26 Lateral wall of the nasal cavity.

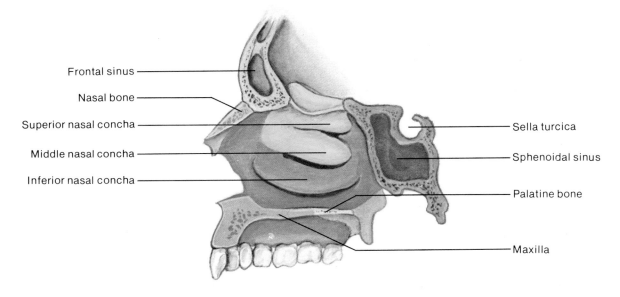

Frontal sinus

Nasal bone

Superior nasal concha

Middle nasal concha

Inferior nasal concha

Sella turcica

Sphenoidal sinus

Palatine bone

Maxilla

Fig. 8.27 Floor of the cranial cavity viewed from above.

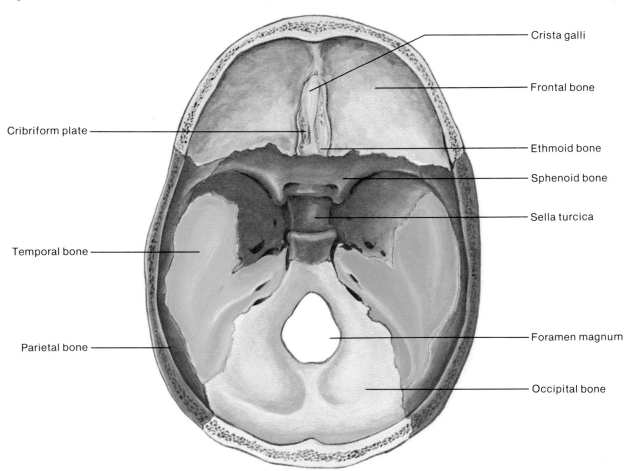

Crista galli

Frontal bone

Cribriform plate

Ethmoid bone

Sphenoid bone

Sella turcica

Temporal bone

Parietal bone

Foramen magnum

Occipital bone

The **facial skeleton** consists of thirteen immovable bones and a movable lower jawbone. In addition to forming the basic shape of the face, these bones provide attachments for various muscles that move the jaw and control facial expressions.

The bones of the facial skeleton are as follows:

1. **Maxillary bones.** The maxillary bones form the upper jaw; together, they are the keystone of the face, for all the other immovable facial bones articulate with them.

Portions of the maxillae comprise the anterior roof of the mouth (*hard palate*), the floors of the orbits, and the sides and floor of the nasal cavity. They also contain the sockets of the upper teeth. Inside the maxillae, lateral to the nasal cavity, are *maxillary sinuses* (antrum of Highmore). These spaces are the largest of the sinuses, and they extend from the floor of the orbits to the roots of the upper teeth. Figure 8.28 shows the maxillary and other sinuses. See chart 8.4 for a summary of these sinuses.

During development, portions of the maxillae called *palatine processes* grow together and fuse along the midline to form the anterior section of the hard palate.

Chart 8.4 Sinuses of the cranial and facial bones

Sinuses	Number	Location
Frontal sinuses	2	Frontal bone above each eye and near the midline
Sphenoidal sinuses	2	Sphenoid bone above the posterior portion of the nasal cavity
Ethmoidal sinuses	2 groups of small air cells	Ethmoid bone on either side of the upper portion of the nasal cavity
Maxillary sinuses	2	Maxillary bones lateral to the nasal cavity and extending from the floor of the orbits to the roots of the upper teeth

Sometimes the fusion of the palatine processes of the maxillae is incomplete at the time of birth; the result is called a *cleft palate*. Infants with this deformity may have trouble sucking because of the opening that remains between the oral and nasal cavities.

Fig. 8.28 Locations of the sinuses.

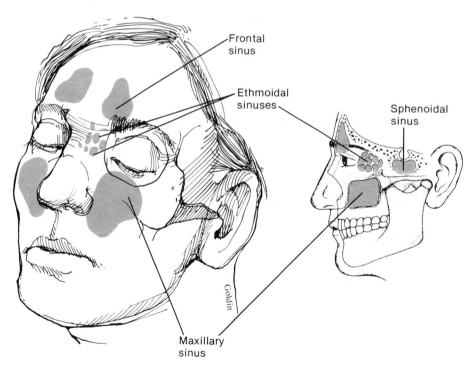

Frontal sinus

Ethmoidal sinuses

Sphenoidal sinus

Maxillary sinus

Goldin

2. **Palatine bones**. The palatine bones (fig. 8.29) are located behind the maxillae. Each bone is roughly L-shaped. The horizontal portions serve as both the posterior section of the hard palate and the floor of the nasal cavity. The perpendicular portions help form the lateral walls of the nasal cavity.

3. **Zygomatic bones**. The zygomatic bones (malar bones) are responsible for the prominences of the cheeks below and to the sides of the eyes. These bones also help form the lateral walls and the floors of the orbits. Each bone has a *temporal process*, which extends posteriorly to join the zygomatic process of a temporal bone. Together these processes form a *zygomatic arch*. (See figs. 8.21 and 8.22.)

4. **Lacrimal bones**. A lacrimal bone is a thin, scalelike structure located in the medial wall of each orbit between the ethmoid bone and the maxilla. A groove in its anterior portion leads from the orbit to the nasal cavity, providing a pathway for a tube that carries tears from the eye to the nasal cavity.

5. **Nasal bones**. The nasal bones are long, thin, and nearly rectangular. They lie side by side and are fused at the midline, where they form the bridge of the nose. These bones serve as attachments for the cartilaginous tissues that are largely responsible for the shape of the nose.

6. **Vomer bone**. The thin, flat vomer bone is located along the midline within the nasal cavity. Posteriorly it joins the perpendicular plate of the ethmoid bone, and together they form the nasal septum (fig. 8.30).

Fig. 8.29 The horizontal portions of the palatine bones form the posterior section of the hard palate, and the perpendicular portions help form the lateral walls of the nasal cavity.

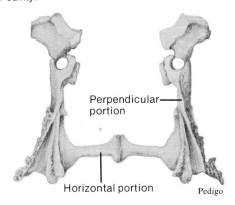

Perpendicular portion

Horizontal portion

Pedigo

Fig. 8.30 Sagittal section of the skull.

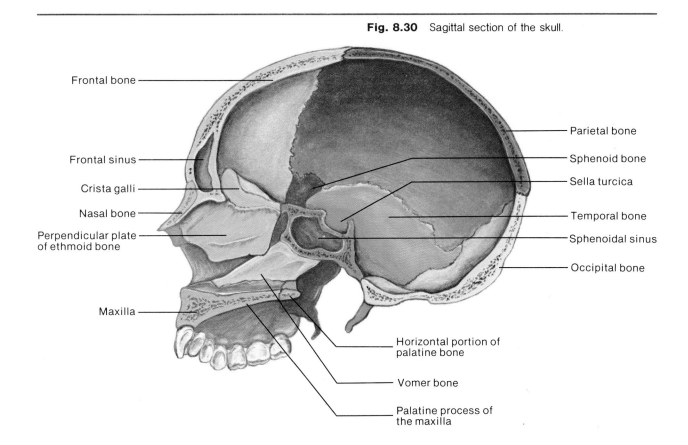

Frontal bone

Frontal sinus

Crista galli

Nasal bone

Perpendicular plate of ethmoid bone

Maxilla

Parietal bone

Sphenoid bone

Sella turcica

Temporal bone

Sphenoidal sinus

Occipital bone

Horizontal portion of palatine bone

Vomer bone

Palatine process of the maxilla

Fig. 8.31 Lateral view of the mandible.

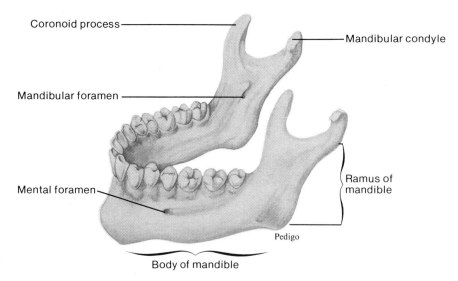

Coronoid process

Mandibular condyle

Mandibular foramen

Mental foramen

Ramus of mandible

Pedigo

Body of mandible

7. **Inferior nasal conchae.** The inferior nasal conchae are fragile, scroll-shaped bones attached to the lateral walls of the nasal cavity. They are the largest of the conchae and are positioned below the conchae of the ethmoid bone (figs. 8.19 and 8.26). Like the superior and middle conchae, the inferior conchae provide support for mucous membranes within the nasal cavity.

8. **Mandible.** The mandible, or lower jawbone, consists of a horizontal, horseshoe-shaped body with a flat *ramus* projecting upward at each end. The rami, in turn, are divided into two processes—a posterior *mandibular condyle* and an anterior *coronoid process* (fig. 8.31). The mandibular condyles articulate with the mandibular fossae of the temporal bones, while the coronoid processes serve as attachments for muscles used in chewing. Other large chewing muscles are inserted on the lateral surfaces of the rami.

The mandible develops in two halves that are connected by fibrous tissues before birth. During the second year they usually fuse along the midline. The body of the mandible, which bears the sockets for the lower teeth, lengthens as a child grows and thus makes room for the larger adult teeth.

On the medial side of the mandible, near the center of each ramus, is a *mandibular foramen*. This opening admits blood vessels and a nerve, which supply the roots of the lower teeth. Dentists commonly inject anesthetic into the tissues near this foramen to temporarily block nerve impulse conduction and cause the teeth on that side of the jaw to become insensitive. Branches of these vessels and nerve emerge from the mandible through the *mental foramen*, which opens on the outside near the point of the jaw. They supply the tissues of the chin and lower lip.

Chart 8.5 contains a descriptive summary of the fourteen facial bones. Various features of these bones can be seen in the X rays in figure 8.32.

Teeth are sometimes lost because of inadequate care or gum disease. Without the lower teeth, the mandibular border that bears the sockets tends to be resorbed. Consequently, the chin of a toothless person appears more pronounced and can be raised much closer to the nose than before.

Chart 8.5 Facial bones

Name and Number	Description	Special Features
Maxilla 2	Form upper jaw, anterior roof of mouth, floors of orbits, and sides and floor of nasal cavity	Sockets of upper teeth, maxillary sinuses, palatine process
Palatine 2	Form posterior roof of mouth, and floor and lateral walls of nasal cavity	
Zygomatic 2	Form prominences of cheeks, and lateral walls and floors of orbits	Temporal process
Lacrimal 2	Form part of medial walls of orbits	Groove that leads from orbit to nasal cavity
Nasal 2	Form bridge of nose	
Vomer 1	Forms part of nasal septum	
Inferior nasal concha 2	Extend into nasal cavity from its lateral walls	
Mandible 1	Forms lower jaw	Body, ramus, mandibular condyle, coronoid process, mandibular foramen, mental foramen

Fig. 8.32 (*a*) X ray of the skull from the front and (*b*) from the side. What features of the cranium and facial skeleton do you recognize?

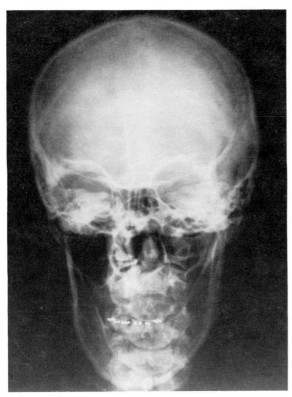

(a)

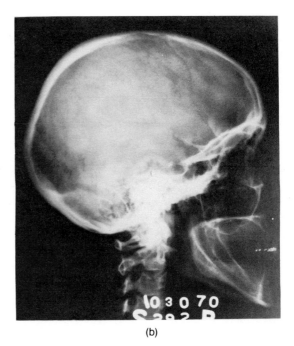

(b)

Fig. 8.33 (a) Lateral view and (b) superior view of the infantile skull.

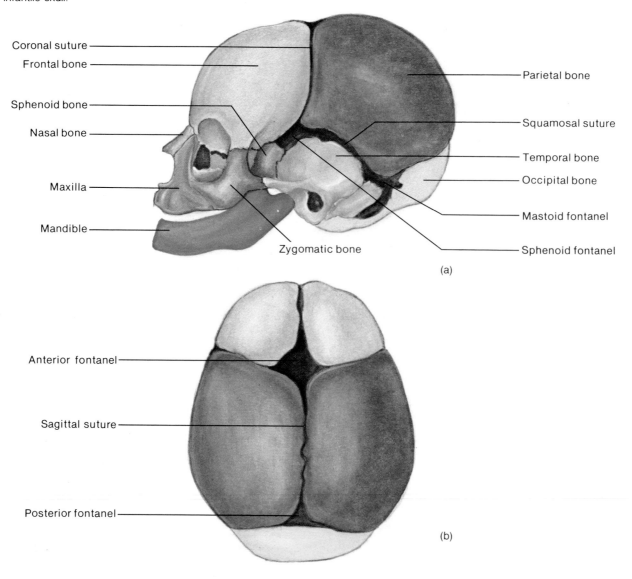

Coronal suture
Frontal bone
Sphenoid bone
Nasal bone
Maxilla
Mandible
Zygomatic bone

Parietal bone
Squamosal suture
Temporal bone
Occipital bone
Mastoid fontanel
Sphenoid fontanel

(a)

Anterior fontanel
Sagittal suture
Posterior fontanel

(b)

The Infantile Skull

At birth the skull is incompletely developed, and the cranial bones are separated by fibrous membranes. These membranous areas are called **fontanels** or, more commonly, soft spots. They permit some movement between the bones, so that the skull is partially compressible and can change shape slightly. This action enables an infant's skull to pass more easily through the birth canal. Eventually, the fontanels close as the cranial bones grow together. The posterior fontanel usually closes within a few months after birth, but the anterior one may not close until the middle or end of the second year.

Other characteristics of an infantile skull (shown in fig. 8.33) include a relatively small face with a prominent forehead and large orbits. The jaw and nasal cavity are small, the sinuses are incompletely formed, and the frontal bone is in two parts. The skull bones are thin, but they are also somewhat flexible and thus are less easily fractured than adult bones.

1. Locate and name each of the bones of the cranium.
2. Locate and name each of the bones of the face.
3. Explain how an adult skull differs from that of an infant.

The Vertebral Column

The **vertebral column** extends from the skull to the pelvis and forms the vertical axis of the skeleton. It is composed of many bony parts called **vertebrae**. These are separated by masses of fibrocartilage called *intervertebral disks* and are connected to one another by ligaments. The vertebral column supports the head and the trunk of the body, yet is flexible enough to permit movements, such as bending forward, backward, or to the side, and turning or rotating on the central axis. It also protects the spinal cord, which passes through a *vertebral canal* formed by openings in the vertebrae.

In an infant there are thirty-three separate bones in the vertebral column. Five of these bones eventually fuse to form the sacrum, and four others join to become the coccyx. As a result, an adult vertebral column has twenty-six parts.

Normally the vertebral column has four curvatures that give it a degree of resiliency. The names of the curves correspond to the regions in which they occur, as shown in figure 8.34. The *thoracic* and *pelvic curvatures* are concave anteriorly and are called primary curves. The *cervical curvature* in the neck and the *lumbar curvature* in the lower back are convex anteriorly and are secondary curves.

A Typical Vertebra.   Although the vertebrae in different regions of the vertebral column have special characteristics, they also have features in common. Thus, a typical vertebra (fig. 8.35) has a drum-shaped *body* (centrum) that forms a thick, anterior portion of the bone. A longitudinal row of these vertebral bodies supports the weight of the head and trunk. The intervertebral disks, which separate adjacent vertebrae, are fastened to the roughened upper and lower surfaces of the bodies. These disks cushion and soften the forces created by such movements as walking and jumping, which might otherwise fracture vertebrae or jar the brain.

Projecting posteriorly from each vertebral body are two short stalks called *pedicles*. They form the sides of the *vertebral foramen*. Two plates called *laminae* arise from the pedicles and fuse in the back to become a *spinous process*. The pedicles, laminae, and spinous process together complete a bony *vertebral arch* around the vertebral foramen, through which the spinal cord passes.

If the laminae of the vertebrae fail to unite during embryonic development, a condition called *spina bifida* results, and the contents of the vertebral canal may protrude outward. This problem occurs most frequently in the lumbosacral region.

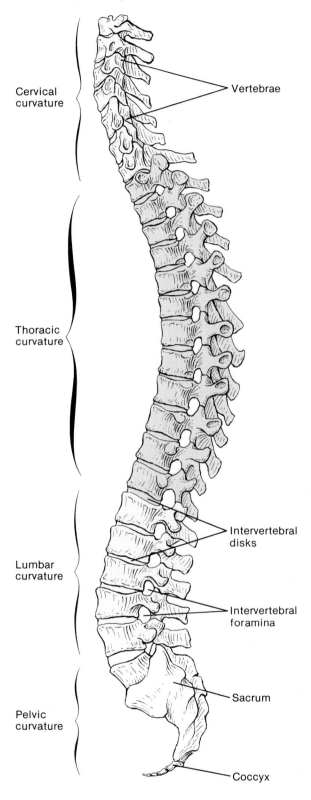

Fig. 8.34 The curved vertebral column consists of many vertebrae separated by intervertebral disks.

Cervical curvature

Vertebrae

Thoracic curvature

Lumbar curvature

Intervertebral disks

Intervertebral foramina

Pelvic curvature

Sacrum

Coccyx

Between the pedicles and laminae of a typical vertebra is a *transverse process* that projects laterally and toward the back. Various ligaments and muscles are attached to the dorsal spinous process and the transverse processes. Projecting upward and downward from each vertebral arch, there are *superior* and *inferior articulating processes*. These processes bear cartilage-covered facets by which each vertebra is joined to the one above and the one below it.

On the lower surfaces of the vertebral pedicles are notches that align to create openings, called *intervertebral foramina*. These openings provide passageways for spinal nerves that pass between adjacent vertebrae and connect to the spinal cord. (See fig. 8.36.)

Cervical Vertebrae

Seven **cervical vertebrae** comprise the bony axis of the neck. Although these are the smallest of the vertebrae, their bone tissues are denser than those in any other region of the vertebral column.

The transverse processes of the cervical vertebrae are distinctive because they have *transverse foramina*, which serve as passageways for arteries leading to the brain. Also, the spinous processes of the second through the fifth cervical vertebrae are uniquely forked (bifid). These processes provide attachments for various muscles.

Two of the cervical vertebrae, shown in figure 8.37, are of special interest. The first vertebra, or **atlas**, supports and balances the head. It has practically no body or spine, and appears as a bony ring with two transverse processes. On its upper surface, the atlas has two kidney-shaped facets that articulate with the occipital condyles of the skull.

The second cervical vertebra, or **axis**, bears a toothlike *odontoid process* on its body. This process projects upward and lies in the ring of the atlas. As the head is turned from side to side, the atlas pivots around the odontoid process.

Thoracic Vertebrae

The twelve **thoracic vertebrae** are larger than those in the cervical region. They have long, pointed spinous processes that slope downward, and facets on the sides of their bodies that articulate with ribs.

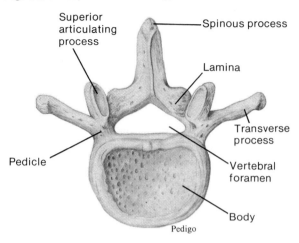

Fig. 8.35 Superior view of a typical vertebra.

Superior articulating process — Spinous process — Lamina — Transverse process — Vertebral foramen — Body — Pedicle

Pedigo

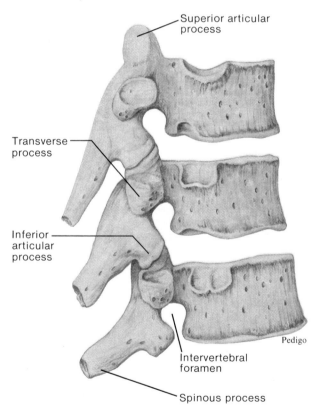

Fig. 8.36 Adjacent vertebrae are joined at facets on their articular processes.

Superior articular process — Transverse process — Inferior articular process — Intervertebral foramen — Spinous process

Pedigo

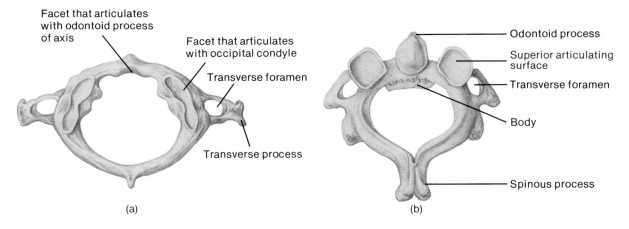

Fig. 8.37 How do the structures of the (*a*) atlas and (*b*) axis function together to allow movement of the head?

Facet that articulates with odontoid process of axis

Facet that articulates with occipital condyle

Transverse foramen

Transverse process

(a)

Odontoid process

Superior articulating surface

Transverse foramen

Body

Spinous process

(b)

Lumbar Vertebrae

There are five **lumbar vertebrae** in the small of the back (loins). Since these bones must support more weight than the vertebrae above them, it is not surprising that the lumbars have larger and stronger bodies. The transverse processes of these vertebrae project backward at relatively sharp angles, while their short, thick spinous processes are directed nearly horizontally.

Figure 8.38 compares the structures of the lumbar, thoracic, and cervical vertebrae.

The Sacrum

The **sacrum** is a triangular structure, composed of five fused vertebrae, that forms the base of the vertebral column. The spinous processes of these fused bones are represented by a ridge of tubercles. To the sides of the tubercles are rows of openings, the *dorsal sacral foramina*, through which nerves and blood vessels pass. (See fig. 8.39.)

The sacrum is wedged between the os coxae of the pelvis and is united to them by fibrocartilage at the *sacroiliac joints*. The weight of the body is transmitted to the legs through the pelvic girdle at these joints.

The sacrum forms the posterior wall of the pelvic cavity. The upper anterior margin of the sacrum, which represents the body of the first sacral vertebra, is called the *sacral promontory*. This projection can be felt during a vaginal examination and is used as a guide in determining the size of the pelvis. This measurement is helpful, because an infant must pass through a woman's pelvic cavity during birth.

The vertebral foramina of the sacral vertebrae form the *sacral canal*, which continues through the sacrum to an opening of variable size at the tip called the *sacral hiatus*. This foramen exists because the laminae of the last sacral vertebra are not fused.

Although the sacral hiatus is normally covered by fibrous tissue, an anesthetic is sometimes injected through it into the sacral canal in order to reduce pain during childbirth. This procedure is called *caudal anesthesia* or *caudal block*.

Fig. 8.38 Superior views of (*a*) a cervical vertebra; (*b*) a thoracic vertebra; and (*c*) a lumbar vertebra.

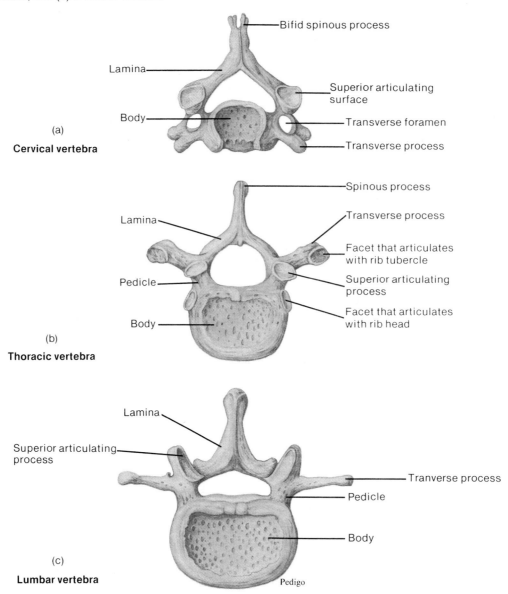

(a)
Cervical vertebra

Bifid spinous process
Lamina
Superior articulating surface
Body
Transverse foramen
Transverse process

(b)
Thoracic vertebra

Spinous process
Lamina
Transverse process
Facet that articulates with rib tubercle
Pedicle
Superior articulating process
Facet that articulates with rib head
Body

(c)
Lumbar vertebra

Lamina
Superior articulating process
Tranverse process
Pedicle
Body
Pedigo

The Coccyx

The **coccyx**, or tailbone, is the lowest part of the vertebral column and is composed of four fused vertebrae. It is attached by ligaments to the margins of the sacral hiatus. When a person is sitting, pressure is exerted upon the coccyx, and it moves forward acting somewhat like a shock absorber. Sitting down with great force sometimes causes the coccyx to be fractured or dislocated.

Chart 8.6 summarizes the bones of the vertebral column.

Conditions of Clinical Interest

A common vertebral problem involves changes in the intervertebral disks. Each disk is composed of a tough, outer layer of fibrocartilage (anulus fibrosus) and an elastic central mass (nucleus pulposus). As a person ages, these disks tend to undergo degenerative changes in which the central masses lose their firmness and the outer layers become thinner and weaker and develop cracks. Extra pressure, as when a person falls or lifts a heavy object, can break the outer layers of the disks and allow the central masses to squeeze out. Such a rupture may cause pressure on the spinal cord or on spinal nerves that branch from it. This

Fig. 8.39 Posterior view of the sacrum and coccyx.

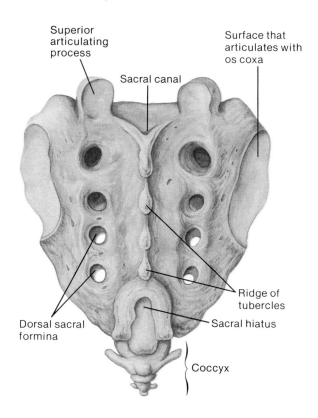

Superior articulating process

Sacral canal

Surface that articulates with os coxa

Dorsal sacral formina

Ridge of tubercles

Sacral hiatus

Coccyx

Chart 8.6	Bones of the vertebral column	
Bones	**Number**	**Special Features**
Cervical vertebrae	7	Transverse foramina, facets of atlas articulate with occipital condyles of skull, odontoid process of axis articulates with atlas, spinous processes of second through fifth vertebrae are bifid
Thoracic vertebrae	12	Pointed spinous processes that slope downward, facets that articulate with ribs
Lumbar vertebrae	5	Large bodies; transverse processes that project backward at sharp angles; short, thick spinous processes directed nearly horizontally
Sacrum	5 vertebrae fused into 1 bone	Dorsal sacral foramina, sacral promontory, sacral canal, sacral hiatus
Coccyx	4 vertebrae fused into 1 bone	

condition—a slipped or ruptured disk—may cause back pain and numbness or loss of muscular function in the parts innervated by the affected spinal nerves.

A ruptured disk may eventually be resorbed or repaired by healing processes, but if there is persistent and intolerable discomfort, a damaged disk may be removed surgically.

Sometimes problems develop in the curvatures of the vertebral column because of poor posture, injury, or disease. (See fig. 8.40). For example, if an exaggerated thoracic curvature appears, the person has rounded shoulders and a hunchback. This condition, called *kyphosis*, occasionally develops in adolescents who are active in strenuous athletic activities. Unless the problem is corrected before bone growth is complete, the vertebral column may be permanently deformed.

Fig. 8.40 (a) Kyphosis is characterized by an exaggerated thoracic curvature; (b) scoliosis involves an abnormal lateral curvature; (c) lordosis results from an accentuated lumbar curvature.

(a)   (b)   (c)

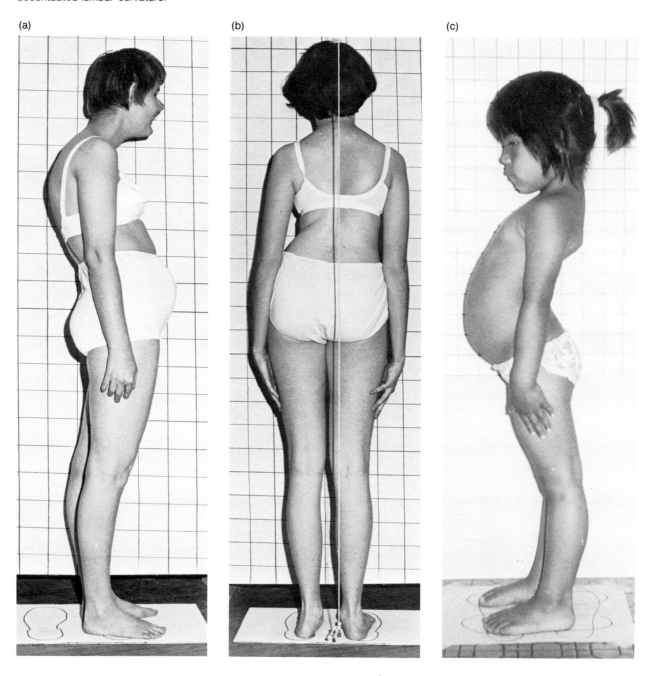

Sometimes the vertebral column develops an abnormal lateral curvature, so that one hip or shoulder is lower than the other. At the same time, the thoracic and abdominal organs may be displaced or compressed. This condition is called *scoliosis*, and although it most commonly develops in adolescent females without known cause, it also may accompany such diseases as poliomyelitis, rickets, or tuberculosis.

If the vertebral column develops an accentuated lumbar curvature, the deformity is called *lordosis* or swayback.

In addition to problems involving the intervertebral disks and curvatures, the vertebral column is subject to degenerative diseases, fracturing, dislocations of its bony parts, and to stretching and tearing of the ligaments that hold the column together.

Fig. 8.41 The thoracic cage includes the thoracic vertebrae, the sternum, the ribs, and the costal cartilages that attach the ribs to the sternum.

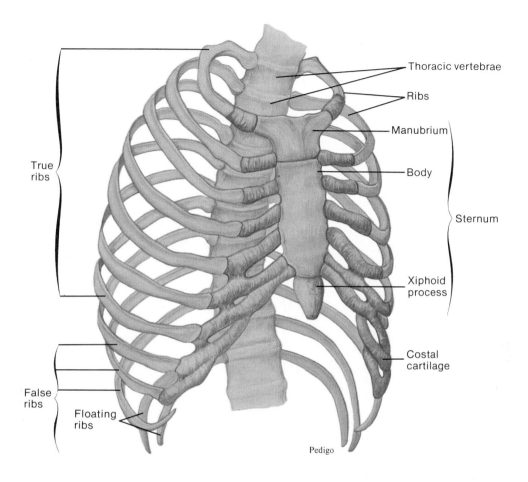

Thoracic vertebrae

Ribs

Manubrium

Body

Sternum

Xiphoid process

Costal cartilage

True ribs

False ribs

Floating ribs

Pedigo

As a person ages, the intervertebral disks tend to become smaller and more rigid, and the vertebral bodies are likely to be fractured by compression. Consequently, the height of an elderly person may be decreased, and the thoracic curvature of the vertebral column may be accentuated.

---

1. Describe the structure of the vertebral column.
2. Explain the difference between the vertebral column of an adult and that of an infant.
3. Describe a typical vertebra.
4. How do the structures of a cervical, a thoracic, and a lumbar vertebra differ?

The Thoracic Cage

The **thoracic cage** includes the ribs, the thoracic vertebrae, the sternum (breastbone), and the costal cartilages by which the ribs are attached to the sternum. These parts support the shoulder girdle and arms, protect the visceral organs in the thoracic and upper abdominal cavities, and play a role in breathing. (See figs. 8.41 and 8.42.)

The Ribs

Regardless of sex, each person usually has twelve pairs of ribs—one pair attached to each of the twelve thoracic vertebrae. Occasionally, however, there may be extra ribs associated with the cervical or lumbar vertebrae.

The first seven rib pairs, the *true ribs*, join the sternum directly by their costal cartilages. The remaining five pairs are called *false ribs* because their cartilages do not reach the sternum directly. Instead, the cartilages of the upper three false ribs join the cartilages of the ribs next above, while the last two rib pairs have no cartilaginous attachments to the sternum. These last two pairs are sometimes called *floating ribs*.

A typical rib (fig. 8.43) has a long, slender shaft that curves around the chest and slopes downward. On the posterior end is an enlarged head by which the rib articulates with a facet on the body of its own vertebra and usually with the body of the next higher vertebra. Also near the head are tubercles that articulate with the transverse process of the vertebra.

The costal cartilages are composed of hyaline cartilage. They are attached to the anterior ends of the ribs and continue in line with them toward the sternum.

The Sternum

The **sternum**, or breastbone, is located along the midline in the anterior portion of the thoracic cage. It is a flat, elongated bone that develops in three parts—an upper *manubrium*, a middle *body* (gladiolus), and a lower *xiphoid process* that projects downward (fig. 8.41). The sides of the manubrium and the body are notched where they articulate with costal cartilages. The manubrium also articulates with the clavicles by facets on its superior border.

The red marrow within the spongy bone of the sternum functions in blood-cell formation into adulthood. Since the sternum has a thin covering of compact bone and is easy to reach, samples of its blood-cell-forming tissue may be removed for use in diagnosing diseases. This procedure, a *sternal puncture*, involves suctioning some marrow through a hollow needle.

---

1. What bones make up the thoracic cage?
2. Describe a typical rib.
3. What are the differences between true, false, and floating ribs?

Fig. 8.42 X ray of the thoracic cage viewed from the front.

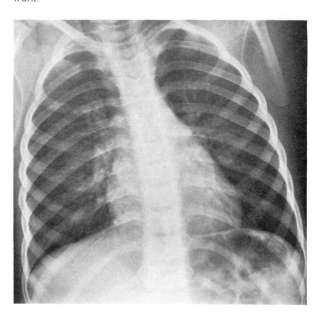

Fig. 8.43 A typical rib.

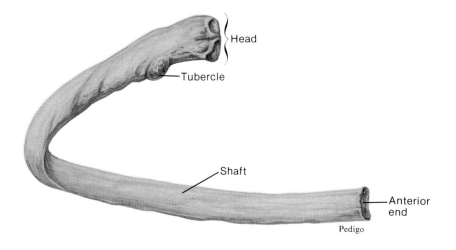

Head

Tubercle

Shaft

Anterior end

Pedigo

Support and Movement

The Pectoral Girdle

The **pectoral girdle**, or shoulder girdle, is composed of four parts—two clavicles, or collarbones, and two scapulae, or shoulder blades. Although the word *girdle* suggests a ring-shaped structure, the pectoral girdle is an incomplete ring. It is open in the back between the scapulae, and its bones are separated in the front by the sternum. However, the pectoral girdle supports the arms and serves as an attachment for several muscles that move the arms. (See figs. 8.44 and 8.45.)

The Clavicles

The **clavicles** are slender, rodlike bones with elongated S-shapes (fig. 8.46). They are located at the base of the neck and run horizontally between the sternum and shoulders. The medial (or sternal) ends of the clavicles articulate with the manubrium, while the lateral (or acromial) ends join processes of the scapulae.

The clavicles act as braces for the freely movable scapulae, and thus they help to hold the shoulders in place. They also provide attachments for muscles of the arms, chest, and back. As a result of

Fig. 8.44 The pectoral girdle, to which the arms are attached, consists of a clavicle and a scapula on either side.

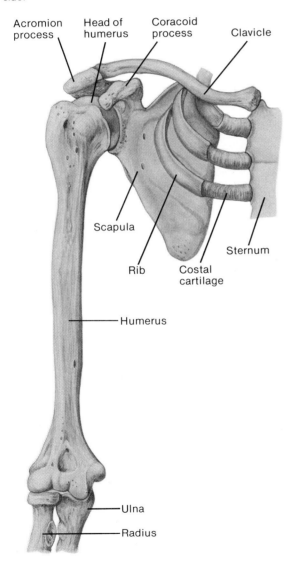

Fig. 8.45 X ray of the right shoulder region viewed from the front.

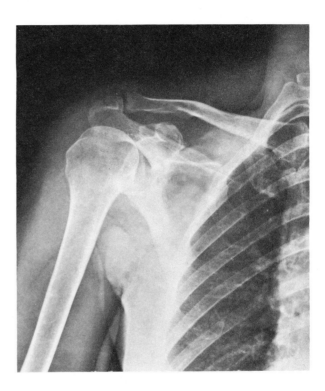

Fig. 8.46 The right clavicle viewed from above.

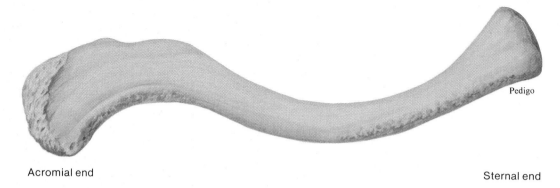

Acromial end

Sternal end

its elongated double curve the clavicle is structurally weak. If it is compressed from the end due to abnormal pressure on the shoulder, it is likely to fracture.

The Scapulae

The **scapulae** are broad, somewhat triangular bones located on either side of the upper back. They have flat bodies with concave anterior surfaces. The posterior surface of each scapula is divided into unequal portions by a *spine*. This spine leads to a *head*, which bears two processes—an *acromion process* that forms the tip of the shoulder and a *coracoid process* that curves forward and downward below the clavicle (fig. 8.47). The acromion process articulates with a clavicle and provides attachments for muscles of the arm and chest. The coracoid process also provides attachments for arm and chest muscles.

On the head of the scapula, between the processes, is a depression called the *glenoid cavity*. It articulates with the head of the upper arm bone (humerus).

---

1. *What bones form the pectoral girdle?*
2. *What is the function of the pectoral girdle?*

The Upper Limb

The bones of the upper limb form the framework of the arm, wrist, palm, and fingers. They also provide attachments for muscles and function in levers for moving limb parts. These bones include a humerus, a radius, an ulna, and several carpals, metacarpals, and phalanges. (See figs. 8.48 and 8.49.)

Fig. 8.47 (*a*) Posterior surface of the scapula; (*b*) medial view showing the glenoid cavity that articulates with the head of the humerus.

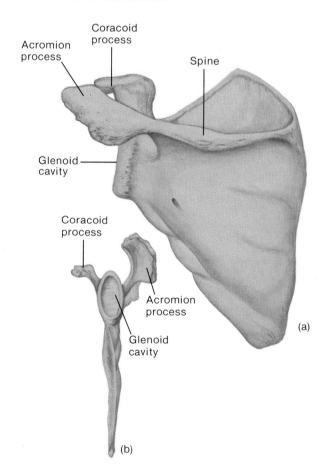

Fig. 8.48 Frontal view of the left arm in the anatomical position.

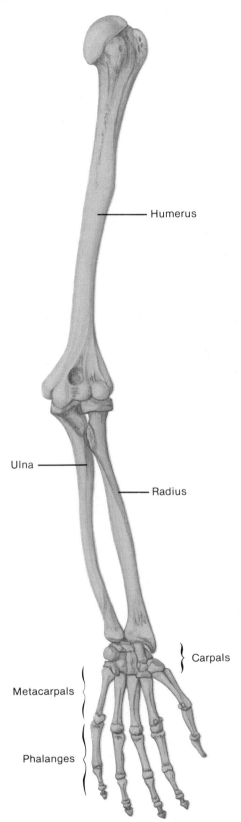

Humerus

Ulna

Radius

Carpals

Metacarpals

Phalanges

Fig. 8.49 X ray of the left elbow and lower arm viewed from the front.

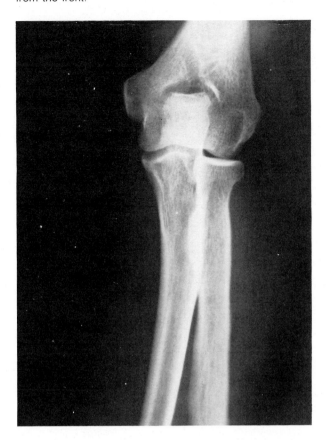

The Humerus

The **humerus** (fig. 8.50) is a heavy bone that extends from the scapula to the elbow. At its upper end, it has a smooth, rounded head that fits into the glenoid cavity of the scapula. Just below the head, there are two processes—a *greater tubercle* on the lateral side and a *lesser tubercle* anteriorly. These tubercles provide attachments for muscles that move the arm at the shoulder. Between them is a narrow furrow, the *intertubercular groove*, through which a tendon passes from a muscle in the upper arm (biceps brachii) to the shoulder.

Just below the head and the tubercles of the humerus is a region called the *surgical neck*, so named because fractures commonly occur there. Near the middle of the bony shaft on the lateral side, there is a rough V-shaped area called the *deltoid tuberosity*. It provides an attachment for the muscle (deltoid) that raises the arm horizontally to the side.

At the lower end of the humerus, there are two smooth condyles—a knoblike *capitulum* on the lateral side and a pulley-shaped *trochlea* on the medial side. The capitulum articulates with the radius at the elbow, while the trochlea joins the ulna.

Fig. 8.50 (*a*) Posterior surface and (*b*) anterior surface of the left humerus.

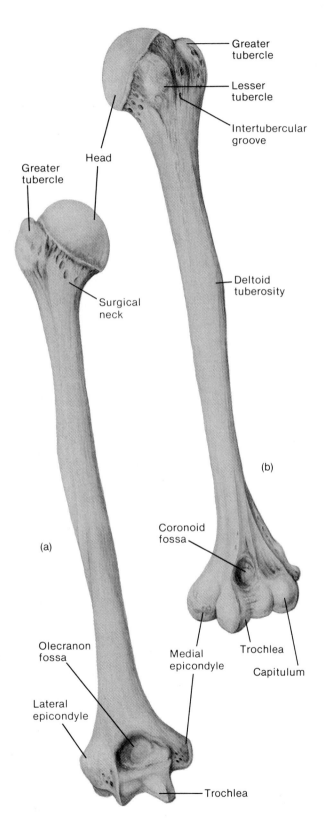

Greater tubercle

Lesser tubercle

Intertubercular groove

Head

Greater tubercle

Surgical neck

Deltoid tuberosity

(b)

(a)

Coronoid fossa

Olecranon fossa

Medial epicondyle

Trochlea

Capitulum

Lateral epicondyle

Trochlea

Fig. 8.51 The head of the radius articulates with the radial notch of the ulna, and the head of the ulna articulates with the ulnar notch of the radius.

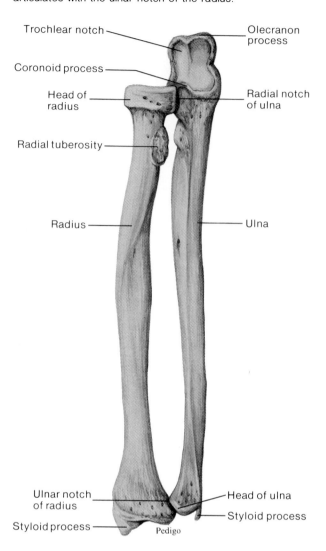

Trochlear notch

Olecranon process

Coronoid process

Head of radius

Radial notch of ulna

Radial tuberosity

Radius

Ulna

Ulnar notch of radius

Head of ulna

Styloid process

Styloid process

Pedigo

Above the condyles on either side are *epicondyles*, which provide attachments for muscles and ligaments of the elbow. Between the epicondyles anteriorly there is a depression, the *coronoid fossa*, that receives a process of the ulna (coronoid process) when the elbow is bent. Another depression on the posterior surface, the *olecranon fossa*, receives an ulnar process (olecranon process) when the arm is straightened at the elbow.

The Radius

The **radius**, located on the thumb side of the lower arm, is somewhat shorter than its companion, the ulna. (Fig. 8.51 shows these bones.) The radius extends from the elbow to the wrist and crosses over the ulna when the hand is turned so that the palm faces backward.

Fig. 8.52 The left hand viewed from the back.

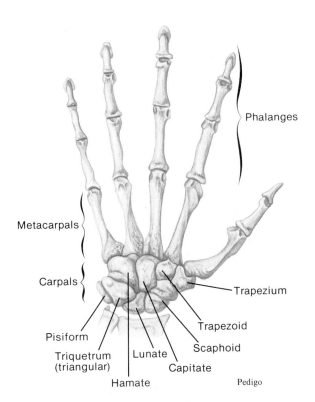

Phalanges

Metacarpals

Carpals

Pisiform

Triquetrum
(triangular)

Hamate

Lunate

Capitate

Scaphoid

Trapezoid

Trapezium

Pedigo

Fig. 8.53 X ray of the left hand. Note the small sesamoid bone associated with the joint at the base of the thumb.

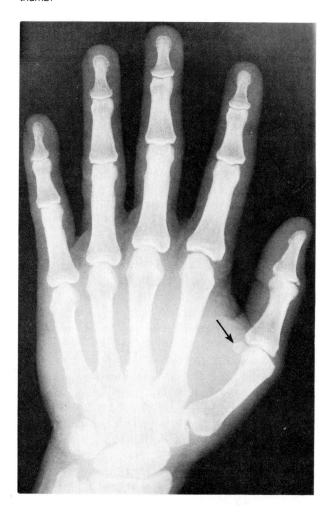

A thick, disklike head at the upper end of the radius articulates with the capitulum of the humerus and a notch of the ulna (radial notch). This arrangement allows the radius to rotate freely.

On the radial shaft, just below the head, is a process called the *radial tuberosity*. It serves as an attachment for a muscle (biceps brachii) that functions to bend the arm at the elbow. At the lower end of the radius, a lateral *styloid process* provides attachments for ligaments of the wrist.

The Ulna

The **ulna** is longer than the radius and overlaps the end of the humerus posteriorly. At its upper end, the ulna has a wrenchlike opening, the *trochlear notch* (semilunar notch), that articulates with the trochlea of the humerus. There is a process on either side of this notch. The one above it, the *olecranon process*, provides an attachment for the muscle (triceps brachii) that straightens the arm at the elbow. During this movement, the olecranon process of the ulna fits into the olecranon fossa of the humerus. Similarly, the coronoid process, just below the trochlear notch, fits into the coronoid fossa of the humerus when the elbow is bent.

At the lower end of the ulna, its knoblike head articulates with a notch of the radius (ulnar notch) laterally and with a disk of fibrocartilage inferiorly. (See fig. 8.51.) This disk, in turn, joins a wrist bone (triquetrum). A medial *styloid process* at the distal end of the ulna provides attachments for ligaments of the wrist.

The Hand

The hand is composed of a wrist, a palm, and five fingers. (See figs. 8.52 and 8.53.) The skeleton of the wrist consists of eight small **carpal bones** that are firmly bonded in two rows of four bones each. The resulting compact mass is called a *carpus*. The carpus is rounded on its proximal surface, where it articulates with the radius and with the fibrocartilaginous

Chart 8.7 Bones of the pectoral girdle and upper limbs

Name and Number	Location	Special Features
Clavicle 2	Base of neck, between sternum and scapula	Sternal end, acromial end
Scapula 2	Upper back, forming part of the shoulder	Body, spine, head, acromion process, coracoid process
Humerus 2	Upper arm, between scapula and elbow	Head, greater tubercle, lesser tubercle, intertubercular groove, surgical neck, deltoid tuberosity, capitulum, trochlea, medial epicondyle, lateral epicondyle, coronoid fossa, olecranon fossa
Radius 2	Lateral side of lower arm, between elbow and wrist	Head, radial tuberosity, styloid process
Ulna 2	Medial side of lower arm, between elbow and wrist	Trochlear notch, olecranon process, head, styloid process
Carpal 16	Wrist	Arranged in two rows of four bones each
Metacarpal 10	Palm	One in line with each finger
Phalanx 28	Finger	Three in each finger, two in each thumb

disk on the ulnar side. The carpus is concave anteriorly, forming a canal through which tendons and nerves extend to the palm. Its distal surface articulates with the metacarpal bones.

Five **metacarpal bones**, one in line with each finger, form the framework of the palm. These bones are cylindrical, with rounded distal ends that make the knuckles on a clenched fist. The metacarpals articulate proximally with the carpals and distally with the phalanges. The metacarpal on the lateral side is the most freely movable; it permits the thumb to oppose the fingers when something is grasped in the hand.

The **phalanges** are the bony elements of the fingers. There are three in each finger—a proximal, a middle, and a distal phalanx—and two in the thumb. Thus, there are fourteen finger bones in each hand.

Chart 8.7 summarizes the bones of the pectoral girdle and upper limbs.

---

1. *Locate and name each of the bones of the upper limb.*
2. *Explain how these bones articulate with one another.*

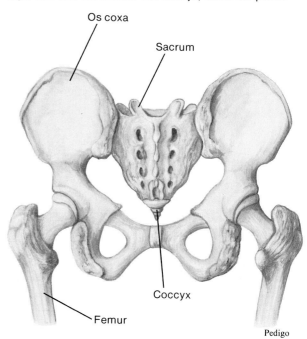

Fig. 8.54 Posterior view of the pelvic girdle, which consists of two os coxae; provides an attachment for the legs; and with the sacrum and coccyx, forms the pelvis.

Os coxa

Sacrum

Coccyx

Femur

Pedigo

The Pelvic Girdle

The **pelvic girdle** consists of the two os coxae, or hip bones, that articulate with each other anteriorly and with the sacrum posteriorly. (See fig. 8.54.) With the sacrum and coccyx, the pelvic girdle forms the ring-like pelvis. The pelvis provides a stable support for the trunk of the body and attachments for the legs.

The weight of the body is transmitted through the pelvis to the legs and then onto the ground. The pelvis also protects the urinary bladder, the distal end of the large intestine, and the internal reproductive organs.

Fig. 8.55 Lateral surface of the right os coxa.

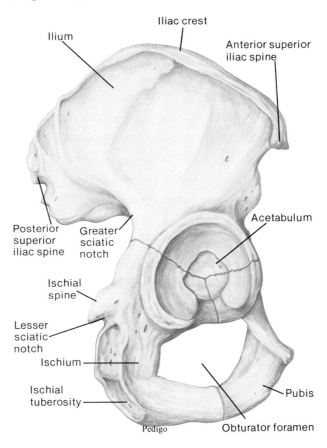

Iliac crest

Ilium

Anterior superior
iliac spine

Posterior
superior
iliac spine

Greater
sciatic
notch

Acetabulum

Ischial
spine

Lesser
sciatic
notch

Ischium

Ischial
tuberosity

Pedigo

Pubis

Obturator foramen

The Os Coxae

Each **os coxa** develops from three parts—an ilium, an ischium, and a pubis. These parts fuse in the region of a cup-shaped cavity called the *acetabulum*. This depression is on the lateral surface of the hip bone, and it receives the rounded head of the femur (thigh bone). (See fig. 8.55.)

The **ilium**, which is the largest and uppermost portion of the os coxa, flares outward to form the prominence of the hip. The margin of this prominence is called the *iliac crest*.

Posteriorly, the ilium joins the sacrum at the *sacroiliac joint*. A projection of the ilium, the *anterior superior iliac spine*, can be felt lateral to the groin. This spine provides attachments for ligaments and muscles and is an important surgical landmark.

The **ischium**, which forms the lowest portion of the os coxa, is L-shaped with its angle, the *ischial tuberosity*, pointing posteriorly and downward. This tuberosity has a rough surface that provides attach-

ments for ligaments and leg muscles. It also supports the weight of the body when a person is sitting. Above the ischial tuberosity, near the junction of the ilium and ischium, is a sharp projection called the *ischial spine*. This spine, which can be felt during a vaginal examination, is used as a guide for determining the size of the pelvis. The distance between the ischial spines represents the shortest diameter of the pelvic outlet.

The **pubis** constitutes the anterior portion of the os coxa. The two pubic bones come together in the midline to create a joint called the *symphysis pubis*. The angle formed by these bones below the symphysis is the *pubic arch*.

A portion of each pubis passes posteriorly and downward to join an ischium. Between the bodies of these bones on either side there is a large opening, the *obturator foramen*, which is the largest foramen in the skeleton.

Fig. 8.56 The female pelvis is wider in all diameters and roomier than that of the male.

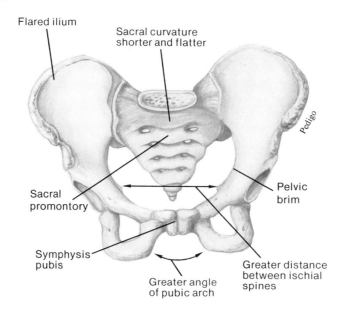

Flared ilium

Sacral curvature shorter and flatter

Pedigo

Sacral promontory

Pelvic brim

Symphysis pubis

Greater distance between ischial spines

Greater angle of pubic arch

The True and False Pelves

If a line is drawn along each side of the pelvis from the sacral promontory downward and anteriorly to the upper margin of the symphysis pubis, it marks the *pelvic brim*. This margin, which can be traced in figure 8.56, separates the lower, or true, pelvis from the upper, or false, pelvis.

The *false pelvis* is bounded posteriorly by the lumbar vertebrae, laterally by the flared parts of the iliac bones, and anteriorly by the abdominal wall. Although the false pelvis helps to support the abdominal organs, it has no particular importance in childbirth.

The *true pelvis* is bounded posteriorly by the sacrum and coccyx, and laterally and anteriorly by the lower ilium, ischium, and pubis bones. This portion of the pelvis surrounds a short, canallike cavity that has an upper inlet and a lower outlet. This cavity is of special interest because an infant passes through it during childbirth.

Sexual Differences in Pelves

The primary differences between adult male and female pelves are related to the function of the female pelvis as a birth canal. Usually the female iliac bones are more flared than those of the male, and consequently the female hips are broader. The angle of the female pubic arch is greater; there is more distance between the ischial spines and the ischial tuberosities; and the sacral curvature is shorter and flatter. Thus, the female pelvic cavity is usually wider in all diameters and is roomier than that of the male. Also, the bones of the female pelvis are lighter, more delicate, and show less evidence of muscle attachments. (See fig. 8.57.) Chart 8.8 summarizes the sexual differences of the skeleton.

---

1. Locate and name each of the bones of the pelvis.
2. Explain what is meant by the false pelvis and the true pelvis.
3. How can a male and female pelvis be distinguished?

Fig. 8.57 (*a*) X ray of a female pelvis and (*b*) X ray of a male pelvis viewed from the front.

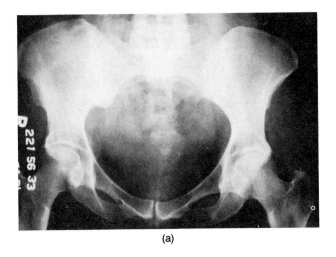

(a)

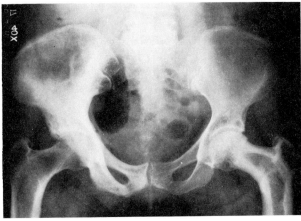

(b)

The Lower Limb

The bones of the lower limb form the frameworks of the leg, ankle, instep, and toes. They include a femur, a tibia, a fibula, and several tarsals, metatarsals, and phalanges. (See figs. 8.58 and 8.59.)

The Femur

The **femur**, or thigh bone, is the longest bone in the body and extends from the hip to the knee. A large, rounded head at its upper end projects medially into the acetabulum of the os coxa. On the head, a pit called the *fovea capitis* marks the attachment of a ligament that holds the femur into the acetabulum. Just below the head, there is a constriction, or neck, and two large processes—an upper, lateral *greater trochanter* and a lower, medial *lesser trochanter*. These processes provide attachments for muscles of the legs and buttocks. On the posterior surface in the middle third of the shaft, there is a longitudinal crest called the *linea aspera*. This rough strip serves as an attachment for several muscles. (See fig. 8.60.)

At the lower end of the femur, two rounded processes, the *lateral* and *medial condyles*, articulate with the tibia of the lower leg. A **patella** (kneecap) also articulates with the femur on its distal anterior surface.

The patella is a flat sesamoid bone located in a tendon that passes anteriorly over the knee. (See fig. 8.58.) Because of its position, the patella controls the angle at which this tendon continues toward the tibia, and so it functions in lever actions associated with lower leg movements.

Chart 8.8 Some sexual differences of the skeleton

Part	Sexual Differences
Skull	Female skull is relatively smaller and lighter, and its muscular attachments are less conspicuous. The female forehead is longer vertically, the facial area is rounder, the jaw is smaller, and the mastoid process is less prominent than that of a male.
Pelvis	Female pelvic bones are lighter, thinner, and have less obvious muscular attachments. The obturator foramina and the acetabula are smaller, and farther apart than those of a male.
Pelvic cavity	Female pelvic cavity is wider in all diameters and is shorter, roomier, and less funnel-shaped. The distances between the ischial spines and between the ischial tuberosities are greater than in the male.
Sacrum	Female sacrum is relatively wider, the sacral promontory projects forward to a lesser degree, and the sacral curvature is bent more sharply posteriorly than in a male.
Coccyx	Female coccyx is more movable than that of a male.

The Tibia

The **tibia**, or shinbone, is the larger of the two lower leg bones and is located on the medial side. Its upper end is expanded into a *medial* and a *lateral condyle*, which have concave surfaces and articulate with the condyles of the femur. Below the condyles, on the anterior surface, is a process called the *tibial tuberosity*, which provides an attachment for the *patellar ligament*—a continuation of the patella-bearing tendon. A prominent *anterior crest* extends downward

Fig. 8.58 The left lower limb viewed from the front.

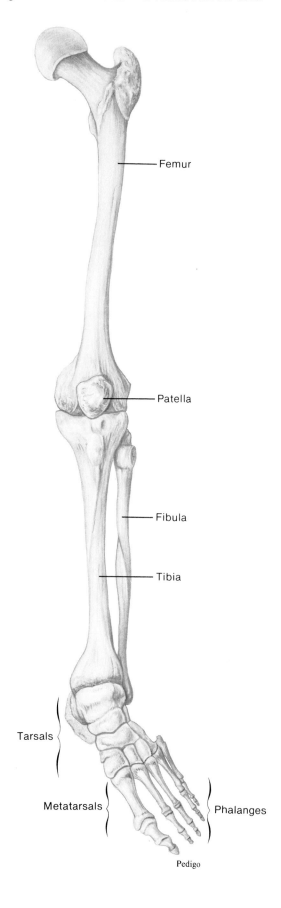

- Femur
- Patella
- Fibula
- Tibia
- Tarsals
- Metatarsals
- Phalanges

Pedigo

Fig. 8.59 X ray of the left knee showing the ends of the femur, tibia, and fibula.

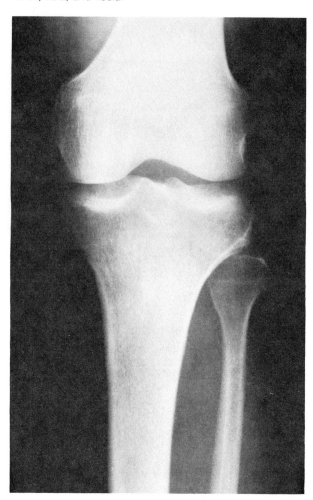

from the tuberosity and serves as an attachment for connective tissues in the lower leg.

At its lower end, the tibia expands to form a prominence on the inner ankle called the *medial malleolus*, which serves as an attachment for ligaments. On its lateral side is a depression that articulates with the fibula. The inferior surface of its distal end articulates with a large bone (the talus) in the foot.

The Fibula

The **fibula** is a long, slender bone located on the lateral side of the tibia. Its ends are slightly enlarged into an upper head and a lower *lateral malleolus*. The head articulates with the tibia just below the lateral condyle; however, it does not enter into the knee joint and does not bear any body weight. The lateral malleolus articulates with the ankle and forms a prominence on the lateral side.

Figure 8.61 shows the tibia and the fibula.

Fig. 8.60 (*a*) Anterior surface and (*b*) posterior surface of the left femur.

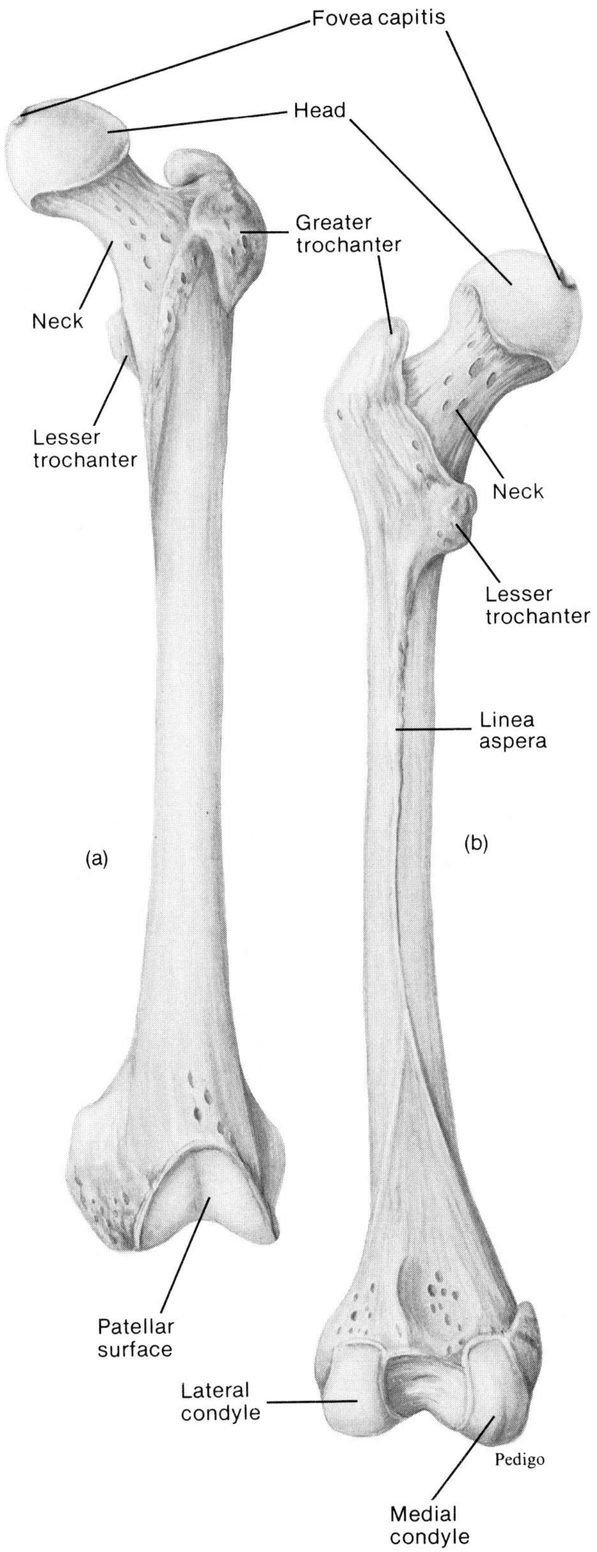

Fovea capitis

Head

Greater trochanter

Neck

Lesser trochanter

Neck

Lesser trochanter

Linea aspera

(a)

(b)

Patellar surface

Lateral condyle

Pedigo

Medial condyle

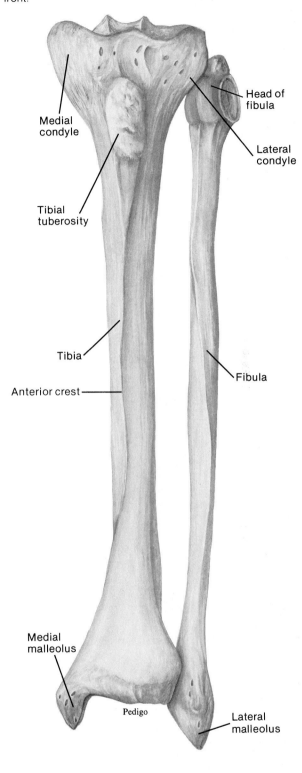

Fig. 8.61 Bones of the left lower leg viewed from the front.

Medial condyle

Head of fibula

Lateral condyle

Tibial tuberosity

Tibia

Fibula

Anterior crest

Medial malleolus

Pedigo

Lateral malleolus

Fig. 8.62 The talus moves freely where it articulates with the tibia and fibula.

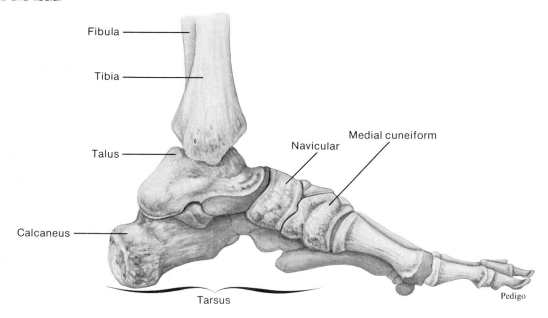

Fibula

Tibia

Talus

Navicular

Medial cuneiform

Calcaneus

Tarsus

Pedigo

Fig. 8.63 X ray of the left foot viewed from the medial side.

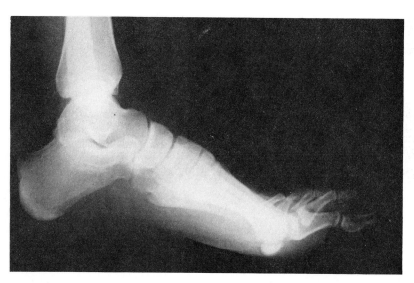

The Foot

The foot consists of an ankle, an instep, and five toes. The ankle is composed of seven **tarsal bones**, forming a group called the *tarsus*. These bones are arranged so that one of them, the **talus**, can move freely where it joins the tibia and fibula. The remaining tarsal bones are bound firmly together, forming a mass on which the talus rests. Figures 8.62 and 8.63 show views of the foot.

The largest of the ankle bones, the **calcaneus** or heel bone, is located below the talus where it projects backward to form the base of the heel. The calcaneus helps support the weight of the body and provides an attachment for muscles that move the foot.

The instep consists of five, elongated **metatarsal bones** that articulate with the tarsus. (See figs. 8.64 and 8.65.) The heads at the distal ends of these bones form the ball of the foot. The tarsals and metatarsals are arranged and bound by ligaments to form the arches of the foot. A longitudinal arch extends from

Fig. 8.64 The left foot viewed from above.

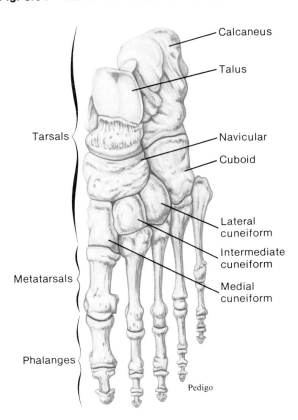

Calcaneus

Talus

Navicular

Cuboid

Tarsals

Lateral cuneiform

Intermediate cuneiform

Medial cuneiform

Metatarsals

Phalanges

Pedigo

Fig. 8.65 X ray of the left foot viewed from above.

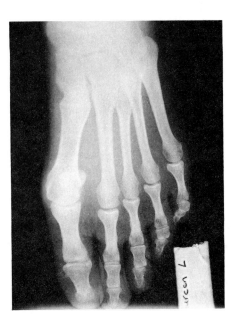

Fig. 8.66 The arches of the feet provide stable, springy supports for the body.

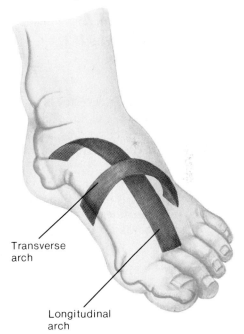

Transverse arch

Longitudinal arch

the heel to the toe, and a transverse arch stretches across the foot (fig. 8.66). These arches provide a stable, springy base for the body. Sometimes, however, the tissues that bind the metatarsals become weakened, producing fallen arches or flat feet.

The **phalanges** of the toes are similar to those of the fingers. They are in line with the metatarsals and articulate with them. There are three phalanges in each toe, except the great toe which has only two.

Chart 8.9 summarizes the bones of the pelvic girdle and lower limbs.

1. Locate and name each of the bones of the lower limb.
2. Explain how these bones articulate with one another.
3. Describe how the foot is adapted to support the body.

Chart 8.9 Bones of the pelvic girdle and lower limbs

Name and Number	Location	Special Features
Os coxa 2	Hip, articulating with each other anteriorly and with the sacrum posteriorly	Ilium, iliac crest, anterior superior iliac spine, ischium, ischial tuberosity, ischial spine, obturator foramen, acetabulum, pubis
Femur 2	Upper leg, between the hip and knee	Head, fovea capitis, neck, greater trochanter, lesser trochanter, linea aspera, lateral condyle, medial condyle
Patella 2	Anterior surface of knee	
Tibia 2	Medial side of lower leg, between knee and ankle	Medial condyle, lateral condyle, tibial tuberosity, anterior crest, medial malleolus
Fibula 2	Lateral side of lower leg, between knee and ankle	Head, lateral malleolus
Tarsal 14	Ankle	Freely movable talus that articulates with lower leg bones and six other bones bound firmly together
Metatarsal 10	Instep	One in line with each toe, arranged and bound by ligaments to form arches
Phalanx 28	Toe	Three in each toe, two in the great toe

The Joints

Joints (articulations) are junctions between bones. Although they vary considerably in structure, they can be classified according to the amount of movement they make possible. On this basis, three general groups can be identified—immovable joints, slightly movable joints, and freely movable joints.

Immovable Joints

Immovable joints (synarthroses) occur between bones that come into close contact with one another. The bones at such joints are separated by a thin layer of fibrous tissue or cartilage, as in the case of a *suture* between a pair of flat bones of the cranium (fig. 8.67). The periosteums on the outer and inner surfaces of these bones bridge the gaps between them and form the primary bond at the suture. No active movement takes place at an immovable joint.

Slightly Movable Joints

The bones of *slightly movable joints* (amphiarthroses) are connected by disks of fibrocartilage or by ligaments. The vertebrae of the vertebral column, for instance, are separated by joints of this type (fig. 8.68). The articulating surfaces of the vertebrae are covered by layers of hyaline cartilage and are separated from adjacent bodies by intervertebral disks of

Fig. 8.67 (*a*) The joints between the bones of the cranium are immovable and are called sutures; (*b*) the bones at a suture are separated by a thin layer of connective tissue.

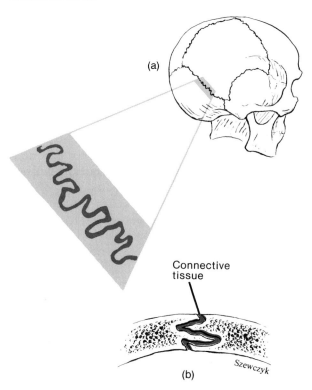

Connective tissue

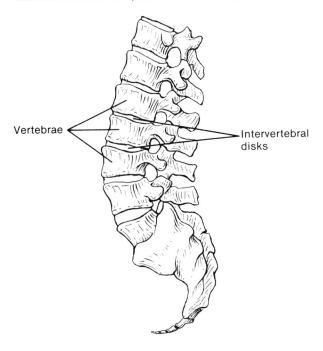

Vertebrae

Intervertebral disks

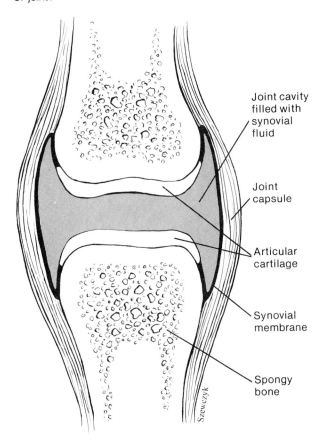

Joint cavity filled with synovial fluid

Joint capsule

Articular cartilage

Synovial membrane

Spongy bone

fibrocartilage. Such joints allow a limited amount of movement, as when the back is bent forward, to the side, or is twisted. Other examples of slightly movable joints include the symphysis pubis, sacroiliac joint, and the joint in the lower leg between the distal ends of the tibia and fibula. In the leg, the bones are held together by ligaments.

The symphysis pubis and the sacroiliac joint become more mobile during pregnancy. Consequently, the diameter of the pelvic outlet is increased, and an infant may pass through the pelvic canal more easily. This relaxation of the pelvic joints is thought to result from hormonal changes associated with pregnancy.

Freely Movable Joints

Most joints within the skeletal system are *freely movable* (diarthroses) and have more complex structures than immovable or slightly movable joints.

The ends of the bones at a freely movable joint are covered with hyaline cartilage (articular cartilage) and are held together by a surrounding, tubelike capsule of fibrous tissue. This *joint capsule* is composed of an outer layer of ligaments and an inner lining of synovial membrane, which secretes synovial fluid that acts as a joint lubricant. For this reason, freely movable joints are often called *synovial joints*. (See fig. 8.69.)

Some freely movable joints have flattened, shock-absorbing pads of fibrocartilage between the articulating surfaces of the bones. The knee joint, for example, contains pads called *semilunar cartilages*. Such joints may also have closed, fluid-filled sacs called **bursae** associated with them. Bursae are lined with synovial membrane, which may be continuous with the synovial membranes of nearby joint cavities.

Fig. 8.70 (*a*) Some freely movable joints, such as that of the shoulder, have bursae associated with them; (*b*) the articulating ends of the bones at the knee are separated by synovial fluid and semilunar cartilages.

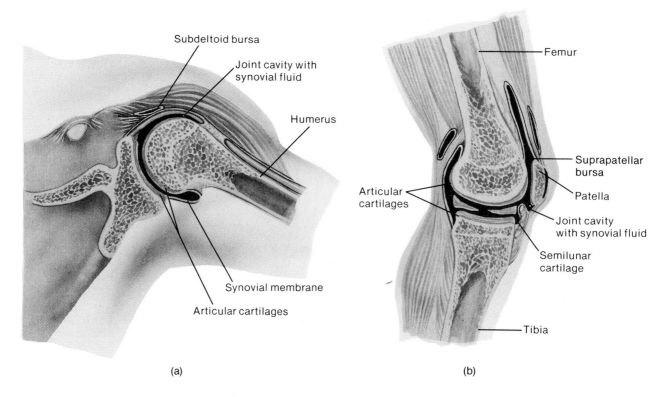

(a)

(b)

Bursae are commonly located between the skin and underlying bony prominences, as in the case of the patella of the knee or the olecranon process of the elbow. They aid in the movement of tendons that pass over these bony parts or over other tendons. (See fig. 8.70.)

The articulating bones of freely movable joints have a variety of shapes, making a number of different movements possible. These joints, shown in figure 8.71, can be classified as follows:

1. **Ball and socket joints.** A ball and socket joint consists of a bone with a ball-shaped head that articulates with a cup-shaped socket of another bone. Such a joint allows for a wider range of motion than does any other kind. Movements in all planes, as well as rotational movement around a central axis, are possible. The hip and shoulder contain joints of this type.

2. **Condyloid joints.** In a condyloid joint, an oval-shaped condyle of one bone fits into an elliptical cavity of another bone, as in the case of the joints between the metacarpals and phalanges. This type of joint allows a variety of movements in different planes; rotational movements, however, are not possible.

3. **Gliding joints.** The articulating surfaces of gliding joints are nearly flat or only slightly curved. Such joints are found between some wrist bones, some ankle bones, and between the articular processes of adjacent vertebrae. They allow sliding and twisting movements.

4. **Hinge joints.** In a hinge joint, the convex surface of one bone fits into the concave surface of another, as in the case of the elbow, the knee, and the joints of the phalanges. This type of joint allows movement in one plane only, like the motion of a single-hinged door.

5. **Pivot joints.** In a pivot joint, a cylindrical surface of one bone rotates within a ring formed of bone and fibrous tissue. The movement at such a joint is limited to rotation about a central axis. The joint between the proximal ends of the radius and the ulna is of this type.

6. **Saddle joints.** A saddle joint is formed between bones whose articulating surfaces have both concave and convex regions. The surfaces of one bone fit the complementary surfaces of the other. This arrangement allows a wide variety of movements, as in the case of the joint between a carpal (trapezium) and the metacarpal of the thumb.

Fig. 8.71 Types of freely movable joints: (a) ball and socket joint of the hip; (b) condyloid joint between metacarpals and phalanges of the hand; (c) gliding joint between tarsals of the ankle; (d) hinge joint of the elbow; (e) pivot joint between proximal ends of the radius and ulna; (f) saddle joint between trapezium of the wrist and metacarpal of the thumb.

Os coxa

(a) Ball and socket joint

Femur

Tarsals

(c) Gliding joint

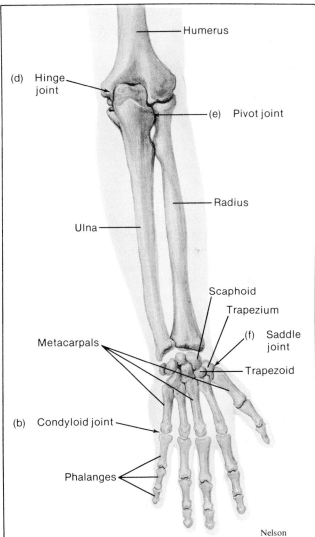

Humerus

(d) Hinge joint

(e) Pivot joint

Radius

Ulna

Scaphoid

Trapezium

(f) Saddle joint

Trapezoid

Metacarpals

(b) Condyloid joint

Phalanges

Nelson

The Skeletal System   205

Chart 8.10 Types of joints

Type	Description	Possible Movements	Example
Immovable	Articulating bones in close contact and separated by a thin layer of fibrous tissue or cartilage	No active movement	Suture between bones of the cranium
Slightly movable	Articulating bones separated by disks of fibrocartilage	Limited movements as when back is bent or twisted	Joints between vertebrae, symphysis pubis, sacroiliac joint
Freely movable	Articulating bones surrounded by joint capsule of ligaments and synovial membranes; ends of articulating bones covered by hyaline cartilage and separated by synovial fluid		
1. Ball and socket	Ball-shaped head of one bone articulates with cup-shaped socket of another	Movements in all planes and rotation	Shoulder, hip
2. Condyloid	Oval-shaped condyle of one bone articulates with elliptical cavity of another	Variety of movements in different planes, but no rotation	Joints between metacarpals and phalanges
3. Gliding	Articulating surfaces are nearly flat or slightly curved	Sliding or twisting	Joints between various bones of wrist and ankle
4. Hinge	Convex surface of one bone articulates with concave surface of another	Up and down motion in one plane	Elbow, knee, joints of phalanges
5. Pivot	Cylindrical surface of one bone articulates with ring of bone and fibrous tissue	Rotation	Joint between proximal ends of radius and ulna
6. Saddle	Articulating surfaces have both concave and convex regions; surface of one bone fits complementary surface of another	Variety of movements	Joint between carpal and metacarpal of thumb

Consult chart 8.10 for a summary of the various types of joints.

Disorders of Joints

Joints are subjected to considerable stress because they function in a variety of body movements and are used almost continuously. Dislocations and sprains most commonly occur during strenuous physical activity. Joints also may be affected by inflammation and a number of degenerative diseases, including arthritis.

Dislocations. A dislocation involves the displacement of bones at a joint, usually as a result of a fall or some other abnormal body movement. The joints of the shoulders, knees, elbows, fingers, and mandible are common sites for such injuries.

Sprains. Sprains are the result of stretching or tearing the ligaments at a joint, without dislocating the bones. They are usually caused by forcefully wrenching or twisting a joint.

Bursitis. Bursitis is an inflammation of a bursa and may be due to excessive use of a joint or to stress on the bursa. Tennis elbow is a form of bursitis that involves the bursa near the olecranon process, while housemaid's knee is bursitis involving the patellar bursa.

Arthritis. Arthritis is a condition that causes inflamed, swollen, and painful joints. Although there are several different types of arthritis, the most common forms are *rheumatoid arthritis*, and *osteoarthritis*.

In rheumatoid arthritis, which is the most painful and crippling of the arthritic diseases, the synovial membrane of a freely movable joint becomes inflamed and grows thicker. This change is usually followed by damage to the articular cartilages on the ends of the bones and an invasion of the joint by fibrous tissues. These fibrous tissues increasingly interfere with joint movements, and in time these tissues may become ossified so that the articulating bones are fused together. The cause of rheumatoid arthritis is unknown.

Osteoarthritis is a degenerative disease that occurs as a result of aging and affects a large percentage of persons over 60 years of age. In this condition, the articular cartilages soften and disintegrate gradually so that the articular surfaces become roughened. Consequently, the joints are sore and less movement is possible. Osteoarthritis is most likely to affect joints that have received the greatest use over the years, such as those in the knees and the lower regions of the vertebral column.

To help remedy joint injury or disease, it is sometimes desirable to replace the articulating parts with artificial (prosthetic) devices. The hip joint, for example, can be totally replaced. In this procedure, the acetabulum is replaced by a low-friction polyethylene socket cemented into the os coxa, and the head of the femur is replaced by a metallic, ball-shaped part. A damaged knee joint can be replaced in a similar manner.

1. Describe the characteristics of the three major types of joints.
2. List six different types of freely movable joints.
3. Explain how joints may change as a result of stress or aging.

Some Clinical Terms Related to the Skeletal System

achondroplasia (a-kon″dro-pla′ze-ah)—an inherited condition in which the formation of cartilaginous bone is retarded. The result is a type of dwarfism.

acromegaly (ak″ro-meg′ah-le)—a condition due to an overproduction of growth hormone in adults and characterized by abnormal enlargement of facial features, hands, and feet.

ankylosis (ang″kĭ-lo′sis)—abnormal stiffness of a joint, often due to damage of joint membranes from chronic rheumatoid arthritis.

arthralgia (ar-thral′je-ah)—pain in a joint.

arthrocentesis (ar″thro-sen-te′sis)—puncture and removal of fluid from a joint cavity.

epiphysiolysis (ep″ĭ-fiz″e-ol′ĭ-sis)—a separation or loosening of the epiphysis from the diaphysis of a bone.

gout (gowt)—a metabolic disease in which excessive uric acid in the blood may be deposited in the joints, causing them to become inflamed, swollen, and painful.

hemarthrosis (hem″ar-thro′sis)—blood in a joint cavity.

laminectomy (lam″ĭ-nek′to-me)—surgical removal of the posterior arch of a vertebra, usually to relieve the symptoms of a ruptured intervertebral disk.

lumbago (lum-ba′go)—a dull ache in the lumbar region of the back.

orthopedics (or″tho-pe′diks)—the science of prevention, diagnosis, and treatment of diseases and abnormalities involving the skeletal and muscular systems.

ostalgia (os-tal′je-ah)—pain in a bone.

ostectomy (os-tek′to-me)—surgical removal of a bone.

osteitis (os″te-i′tis)—inflammation of bone tissue.

osteochondritis (os″te-o-kon-dri′tis)—inflammation of bone and cartilage tissues.

osteogenesis (os″te-o-jen′ĕ-sis)—the development of bone.

osteogenesis imperfecta (os″te-o-jen′ĕ-sis im-per-fek′ta)—a congenital condition characterized by the development of deformed and abnormally brittle bones.

osteoma (os″te-o′mah)—a tumor composed of bone tissue.

osteomalacia (os″te-o-mah-la′she-ah)—a softening of adult bone due to a disorder in calcium and phosphorus metabolism, usually caused by a deficiency of vitamin D.

osteomyelitis (os″te-o-mi″ĕ-li′tis)—inflammation of bone caused by the action of bacteria or fungi.

osteonecrosis (os″te-o-ne-kro′sis)—death of bone tissue. This condition occurs most commonly in the head of the femur in elderly persons and may be due to obstructions in arteries that supply the bone.

osteopathology (os″te-o-pah-thol′o-je)—the study of bone diseases.

rheumatism (roo′mah-tizm)—an inflammation of the fibrous tissue surrounding a joint. Although an affected joint feels tender and stiff, the articulating surfaces usually remain healthy and recovery can be complete.

Chapter Summary

Introduction

Individual bones are the organs of the skeleton system. Bones are composed of very active tissues.

Bone Structure

1. Classification of bones
   Bones are grouped according to their shapes—long, short, flat, or irregular.
2. Parts of a long bone
   a. Epiphyses are covered with articular cartilage and articulate with other bones.
   b. The shaft of a bone is called diaphysis.
   c. Except for the articular cartilage, a bone is covered by a periosteum.
   d. Compact bone provides strength and resistance to bending.
   e. Spongy bone provides strength where needed and reduces the weight of bone.
   f. The diaphysis contain a medullary cavity filled with marrow.

3. Microscopic structure
    a. Compact bone contains haversian systems bonded together.
    b. Haversian canals contain blood vessels that nourish the cells of haversian systems.
    c. Volkmann's canals connect haversian canals transversely and communicate with the bone's surface and the medullary cavity.
    d. Cells of spongy bone are nourished by diffusion from the surface of the bony plates.

Bone Growth and Development

1. Membranous bones
    a. Include certain flat bones of the skull.
    b. Develop from layers of connective tissues.
    c. Bone tissue is formed by osteoblasts within the membranous layers.
    d. Membranous tissues give rise to an inner and outer periosteum.
2. Cartilaginous bones
    a. Include most of the bones of the skeletal system.
    b. Develop first as hyaline cartilage that is later replaced by bone tissue.
    c. Primary ossification center appears in the diaphysis, while secondary ossification centers appear in the epiphyses.
    d. An epiphyseal disk remains between the primary and secondary ossification centers.
    e. An epiphyseal disk consists of layers of cells: resting cells, young reproducing cells, older enlarging cells, and dying cells.
    f. Cartilaginous tissue is broken down by the action of osteoclasts.
    g. A long bone continues to grow in length until the epiphyseal disks become ossified.
3. Factors affecting bone growth and development
    a. Exposure to sunlight favors bone growth.
    b. Deficiencies of vitamin A, C, or D result in abnormal development.
    c. Insufficient secretion of pituitary growth hormone may result in dwarfism; excessive secretion may result in giantism.
    d. Deficiency of thyroid hormone delays bone growth.
    e. Exercise causes bones to thicken and strengthen.
4. Fractures
    a. A fracture may be traumatic or spontaneous, compound or simple.
    b. Repair process includes the following steps:
        (1) Bleeding and formation of a hematoma.
        (2) Invasion of hematoma by blood vessels and osteoblasts, which produce spongy bone, and fibroblasts, which produce fibrocartilage.
        (3) Removal of blood clot and damaged tissues by phagocytosis.
        (4) Appearance of cartilaginous callus.
        (5) Replacement of cartilaginous callus by bony callus.
        (6) Removal of excess bone by osteoclasts.

Functions of Bones

1. Support and protection
    a. Skeletal parts provide shape and form for body structures.
    b. They support and protect softer, underlying tissues.
2. Lever actions
    a. Bones and muscles function together as levers.
    b. A lever consists of a rod, pivot (fulcrum), weight that is moved, and a force that supplies energy.
    c. Parts of a first-class lever are arranged weight-pivot-force; parts of a second-class lever are arranged pivot-weight-force; parts of a third-class lever are arranged pivot-force-weight.
3. Blood cell formation
    a. At different ages, hematopoiesis occurs in the yolk sac, liver and spleen, and red bone marrow.
    b. Red marrow functions in the production of red blood cells, white blood cells, and blood platelets.
    c. If the blood cell supply is deficient, yellow marrow may become red marrow and function in hematopoiesis.
4. Storage of inorganic salts
    a. The intercellular material of bone tissue contains large quantities of calcium in the form of hydroxyapatite.
    b. A hormone from the parathyroid glands stimulates osteoclasts to resorb bone, thus releasing salts.
    c. Calcitonin promotes the storage of calcium in bones.
    d. Bone tissues may accumulate lead, radium, or strontium.
        (1) Strontium-90 is radioactive but acts metabolically like calcium.
        (2) Its radiation may cause cell mutation.

Organization of the Skeleton

1. Number of bones
    a. Usually there are 206 bones in the human skeleton, but the number may vary.
    b. Extra bones in sutures are called wormian bones.
2. Divisions of the skeleton
    a. The skeleton can be divided into axial and appendicular portions.
    b. The axial skeleton consists of the skull, hyoid bone, vertebral column, and thoracic cage.
    c. The appendicular skeleton consists of the pectoral girdle, upper limbs, pelvic girdle, and lower limbs.

The Skull

Consists of 22 bones, which include 8 cranial bones, 13 facial bones, and 1 mandible.

1. The cranium
   a. The cranium encloses and protects the brain and provides attachments for muscles.
   b. Some cranial bones contain air-filled sinuses that help to reduce the weight of the skull.
   c. Cranial bones include frontal bone, parietal bones, occipital bone, temporal bones, sphenoid bone, and ethmoid bone.
2. The facial skeleton
   a. Facial bones provide the basic shape of the face and attachments for muscles.
   b. Facial bones include maxillary bones, palatine bones, zygomatic bones, lacrimal bones, nasal bones, vomer bone, inferior nasal conchae, and mandible.
3. The infantile skull
   a. Incompletely developed bones are separated by fontanels that enable the infantile skull to change shape slightly during birth.
   b. Proportions of the infantile skull are different from those of adult skull, and its bones are less easily fractured.

The Vertebral Column

Extends from the skull to the pelvis.

Composed of vertebrae separated by intervertebral disks.

Protects the spinal cord.

Consists of 33 bones in an infant and 26 bones in an adult.

Has 4 curvatures—cervical, thoracic, lumbar, and pelvic.

1. A typical vertebra
   a. Consists of a body, pedicles, laminae, spinous process, transverse processes, and superior and inferior articulating processes.
   b. Notches on the lower surfaces of the pedicles provide intervertebral foramina through which spinal nerves pass.
2. Cervical vertebrae
   a. Comprise the bones of the neck.
   b. Transverse processes bear transverse foramina.
   c. Atlas (first vertebra) supports and balances the head.
   d. Odontoid process of the axis (second vertebra) provides a pivot for the atlas when the head is turned from side to side.
3. Thoracic vertebrae
   a. Thoracic vertebrae are larger than cervical vertebrae.
   b. Long spinous processes slope downward, and facets on the sides of bodies articulate with the ribs.
4. Lumbar vertebrae
   a. Vertebral bodies are large and strong.
   b. Transverse processes project back at sharp angles, and spinous processes are directed horizontally.

5. The sacrum
   a. A triangular structure that bears rows of dorsal sacral foramina.
   b. United with the os coxae at the sacroiliac joints.
   c. Sacral promontory provides a guide for determining the size of the pelvis.
6. The coccyx
   a. Lowest part of the vertebral column.
   b. Acts as a shock absorber when a person sits.
7. Conditions of clinical interest
   a. Intervertebral disks change with age and may be ruptured when a person falls or lifts heavy objects.
   b. A ruptured disk may affect the spinal cord or spinal nerves and cause pain, numbness, or loss of muscular functions.
   c. Abnormal curvatures of the vertebral column include kyphosis, scoliosis, or lordosis.

The Thoracic Cage

Includes the ribs, thoracic vertebrae, sternum, and costal cartilages.

Supports the shoulder girdle and arms, protects visceral organs, and functions in breathing.

1. The ribs
   a. Twelve pairs of ribs are attached to the thoracic vertebrae.
   b. Costal cartilages of true ribs join the sternum directly; those of the false ribs join indirectly or lack cartilaginous attachments, as do the floating ribs.
   c. A typical rib bears a shaft, head, and tubercles that articulate with the vertebrae.
2. The sternum
   a. Consists of a manubrium, body, and xiphoid process.
   b. Articulates with costal cartilages and clavicles.

The Pectoral Girdle

Composed of two clavicles and two scapulae.

An incomplete ring that supports the arms and provides attachments for muscles that move the arms.

1. The clavicles
   a. Rodlike bones that run horizontally between the sternum and shoulder.
   b. Function to hold the shoulders in place and provide attachments for muscles.
2. The scapulae
   a. Broad, triangular bones with bodies, spines, heads, acromion processes, coracoid processes, and glenoid cavities.
   b. Articulate with the humerus of each arm, and provide attachments for muscles of the arms and chest.

The Upper Limb

Bones provide frameworks for arms, wrists, palms, and fingers.

Provides attachments for muscles and functions in levers that move the limb and its parts.

1. The humerus
   a. Extends from the scapula to the elbow.
   b. Bears a head, greater tubercle, lesser tubercle, intertubercular groove, surgical neck, deltoid tuberosity, capitulum, trochlea, epicondyles, coronoid fossa, and olecranon fossa.
2. The radius
   a. Located on the thumb side of the lower arm between the elbow and wrist.
   b. Bears a head, radial tuberosity, and styloid process.
3. The ulna
   a. Longer than the radius, overlaps the humerus posteriorly.
   b. Bears a trochlear notch, olecranon process, head, and styloid process.
   c. Articulates with the radius laterally and with a disk of fibrocartilage inferiorly.
4. The hand
   a. Composed of a wrist, palm, and 5 fingers.
   b. Includes 8 carpals that form a carpus, 5 metacarpals, and 14 phalanges.

The Pelvic Girdle

Consists of two os coxae that articulate with each other anteriorly and with the sacrum posteriorly.

The sacrum, coccyx, and pelvic girdle form the pelvis.

Provides support for body weight and attachments for muscles, and protects visceral organs.

1. The os coxae
   Each os coxa consists of an ilium, ischium, and pubis, which are fused in the region of the acetabulum.
   a. The ilium
      (1) Largest portion of the os coxa, joins the sacrum at the sacroiliac joint.
      (2) Bears an iliac crest and anterior superior iliac spine.
   b. The ischium
      (1) Lowest portion of the os coxa.
      (2) Bears an ischial tuberosity and ischial spine.
   c. The pubis
      (1) Anterior portion of the os coxa.
      (2) Pubis bones are fused anteriorly at the symphysis pubis.
2. The true and false pelves
   a. True pelvis is below the pelvic brim; false pelvis is above it.
   b. True pelvis functions as a birth canal; false pelvis helps to support abdominal organs.
3. Sex differences in pelves
   a. Differences between male and female pelves are related to the function of the female pelvis as a birth canal.
   b. Female pelvis is more flared; pubic arch is broader; distance between the ischial spines and the ischial tuberosities is greater; and sacral curvature is shorter.

The Lower Limb

Bones provide frameworks for the leg, ankle, instep, and toes.

1. The femur
   a. Extends from the knee to the hip.
   b. Bears a head, fovea capitis, neck, greater trochanter, lesser trochanter, linea aspera, lateral condyle, and medial condyle.
   c. Patella articulates with its anterior surface.
2. The tibia
   a. Located on the medial side of the lower leg.
   b. Bears medial and lateral condyles, tibial tuberosity, anterior crest, and medial malleolus.
   c. Articulates with the talus of the ankle.
3. The fibula
   a. Located on the lateral side of the tibia.
   b. Bears a head and lateral malleolus that articulates with the ankle.
4. The foot
   a. Consists of an ankle, instep, and 5 toes.
   b. Includes 7 tarsals that form the tarsus, 5 metatarsals, and 14 phalanges.

The Joints

Can be classified on the basis of the amount of movement they make possible.

1. Immovable joints
   a. Bones in close contact, separated by a thin layer of fibrous tissue or cartilage, as in a suture.
   b. No active movements are possible.
2. Slightly movable joints
   a. Bones connected by disks of fibrocartilage or by ligaments, as the vertebrae.
   b. Allows a limited amount of movement.
3. Freely movable joints
   a. Bones of the joint are covered with hyaline cartilage and held together by a fibrous capsule.
   b. Joint capsule consists of an outer layer of ligaments and an inner lining of synovial membrane that secretes lubricating synovial fluid.
   c. Some, such as the knee, contain pads of fibrocartilage that absorb shocks.
   d. Bursae are often located between the skin and underlying bony prominences, where they aid in the movement of tendons.
   e. Freely movable joints include several types: ball and socket, condyloid, gliding, hinge, pivot, and saddle.

4. Disorders of joints
   a. Joints are subjected to stress and may be injured by strenuous activity, inflammation, and degenerative diseases.
   b. Common joint disorders include dislocations, sprains, bursitis, and arthritis.

Application of Knowledge

1. What steps do you think should be taken to reduce the chances of persons accumulating abnormal metallic elements such as lead, radium, and strontium in their bones?
2. Why do you think incomplete, longitudinal fractures of bone shafts (greenstick fractures) are more common in children than in adults?

Review Activities

Part A

1. List four groups of bones, based upon their shapes, and name an example from each group.
2. Sketch a typical long bone and label its epiphyses, diaphysis, medullary cavity, periosteum, and articular cartilages.
3. Distinguish between spongy and compact bone.
4. Explain how haversian canals and Volkmann's canals are related.
5. Explain how the development of membranous bone differs from that of cartilaginous bone.
6. Distinguish between osteoblasts and osteoclasts.
7. Explain the function of an epiphyseal disk.
8. Describe the effects of vitamin deficiencies on bone development.
9. Explain the causes of pituitary dwarfism and giantism.
10. Explain the effect of exercise on bone structure.
11. Distinguish between traumatic and spontaneous fractures, and between simple and compound fractures.
12. List the major steps in the repair of a fracture.
13. Provide several examples to illustrate how bones support and protect body parts.
14. Describe a lever, and explain how its parts may be arranged to form first-, second-, and third-class levers.
15. Describe the functions of red and yellow bone marrow.
16. Explain the mechanism that regulates the level of blood calcium.

17. Explain how strontium-90 may be incorporated into bone tissues, and discuss the effect it may have on such tissues.
18. Distinguish between the axial and appendicular skeletons.
19. List the bones that form the pectoral and the pelvic girdles.
20. Explain the importance of fontanels.
21. Describe a typical vertebra.
22. Explain the differences between cervical, thoracic, and lumbar vertebrae.
23. Describe the locations of the sacroiliac joint, the sacral promontory, and the sacral hiatus.
24. Distinguish between kyphosis, scoliosis, and lordosis.
25. Distinguish between the true and false pelves.
26. Explain the differences between male and female pelves.
27. Describe an immovable joint, a slightly movable joint, and a freely movable joint.
28. Distinguish between a dislocation and a sprain.
29. Distinguish between rheumatoid arthritis and osteoarthritis.

Part B
Match the parts listed in column I with the bones listed in column II.

	I		II
1.	Coronoid process	A.	Ethmoid bone
2.	Cribriform plate	B.	Frontal bone
3.	Foramen magnum	C.	Mandible
4.	Mastoid process	D.	Maxillary bone
5.	Palatine process	E.	Occipital bone
6.	Sella turcica	F.	Temporal bone
7.	Supraorbital foramen	G.	Sphenoid bone
8.	Temporal process	H.	Zygomatic bone

	I		II
9.	Acromion process	I.	Femur
10.	Deltoid tuberosity	J.	Fibula
11.	Greater trochanter	K.	Humerus
12.	Lateral malleolus	L.	Radius
13.	Medial malleolus	M.	Scapula
14.	Olecranon process	N.	Sternum
15.	Radial tuberosity	O.	Tibia
16.	Xiphoid process	P.	Ulna

Suggestions for Additional Reading

Aufranc, O. E., and Turner, R. H. October 1971. Total replacement of the arthritic hip. *Hosp. Pract.*

Bourne, G. W., ed. 1972. *The biochemistry and physiology of bone.* 2nd ed. New York: Academic Press.

Chisolm, J. J. February 1971. Lead poisoning. *Scientific American.*

Evans, F. G., ed. 1966. *Studies in the anatomy and function of bones and joints.* New York: Springer-Verlag.

Hall, B. K. 1970. Cellular differentiation in skeletal tissue. *Biol. Rev.* 45:455.

Harris, W. H., and Heaney, R. P. 1970. *Skeletal renewal and metabolic bone disease.* Boston: Little, Brown, and Co.

Loomis, W. F. Rickets. December 1970. *Scientific American.*

Sonstegard, D. A., et. al. January 1978. The surgical replacement of the human knee joint. *Scientific American.*

Trueta, J. 1968. *Studies in the development and decay of the human frame.* Philadelphia: W. B. Saunders.

Vaughan, J. M. 1975. *The physiology of bone.* 2nd ed. New York: Oxford Univ. Press.

The Muscular System

9 Muscles, the organs of the *muscular system,* account for nearly half of the body weight. They consist largely of muscle cells, which are specialized to undergo muscular contractions during which the chemical energy of nutrients is converted into mechanical energy or movement.

When muscle cells contract, they pull on the parts that they are attached to. This action usually causes some movement, as when the joints of the legs are flexed and extended during walking. At other times, muscular contractions resist motion, as when they help to hold body parts in postural positions. Muscles are also responsible for the movement of body fluids such as blood and urine, and they function in heat production, which aids in maintaining body temperature.

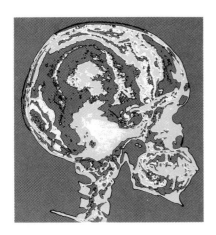

Chapter Outline

Chapter Objectives

After you have studied this chapter, you should be able to

1. Describe the structure of a skeletal muscle.
2. Name the major parts of a skeletal muscle fiber and describe the function of each part.
3. Explain the major events that occur during muscle fiber contraction.
4. Explain how energy is supplied to the muscle fiber contraction mechanism, how oxygen debt develops, and how a muscle may become fatigued.
5. Distinguish between a twitch and a sustained contraction, and explain how various types of muscular contractions are used to produce body movements and maintain posture.
6. Distinguish between the structure and function of a multiunit smooth muscle and a visceral smooth muscle.
7. Compare the fiber contraction mechanisms of skeletal, smooth, and cardiac muscles.
8. Explain how the locations of skeletal muscles are related to the movements they produce and how muscles interact in producing such movements.
9. Identify and describe the location of the major skeletal muscles of each body region and describe the action of each muscle.
10. Complete the review activities at the end of this chapter.

actin (ak′tin)

antagonist (an-tag′o-nist)

depolarization (de-po″lar-i-za′shun)

fascia (fash′e-ah)

insertion (in-ser′shun)

motor neuron (mo′tor nu′ron)

motor unit (mo′tor u′nit)

myofibril (mi″o-fi′bril)

myogram (mi′o-gram)

myosin (mi′o-sin)

neurotransmitter (nu″ro-trans′mit-er)

origin (or′ĭ-jin)

oxygen debt (ok′sĭ-jen det)

polarization (po″lar-i-za′shun)

prime mover (prīm moov′er)

sarcomere (sar′ko-mēr)

synergist (sin′er-jist)

threshold stimulus (thresh′old stim′u-lus)

calat-, something inserted: inter*calat*ed disk—membranous band that separates adjacent cardiac muscle cells.

erg-, work: syn*erg*ist—muscle that works together with a prime mover to produce a movement.

fasc-, a bundle: *fasc*iculus—a bundle of muscle fibers.

-gram, something written: myo*gram*—recording of a muscular contraction.

hyper-, over, more: muscular *hyper*trophy—enlargement of muscle fibers.

inter-, between: *inter*calated disk—membranous band that separates adjacent cardiac muscle cells.

iso-, equal: *iso*tonic contraction—contraction during which the tension in a muscle remains unchanged.

laten-, hidden: *laten*t period—period between the time a stimulus is applied and the beginning of a muscle contraction.

myo-, muscle: *myo*fibril—contractile fiber of a muscle cell.

reticul-, a net: sarcoplasmic *reticul*um—network of membranous channels within a muscle fiber.

syn-, together: *syn*ergist—muscle that works together with a prime mover to produce a movement.

tetan-, stiff: *tetan*ic contraction—sustained muscular contraction.

-tonic, stretched: iso*tonic* contraction—contraction during which the tension in a muscle remains unchanged.

-troph, well fed: muscular hyper*trophy*—enlargement of muscle fibers.

voluntar-, of one's free will: *voluntar*y muscle—muscle that can be controlled by conscious effort.

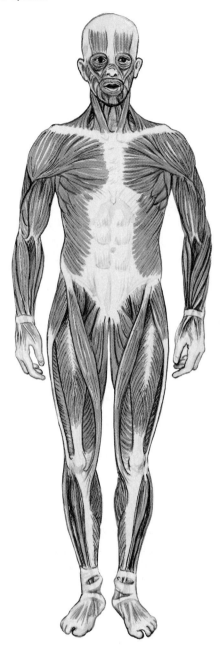

Although there are three types of muscle tissue within the body—skeletal muscle, smooth muscle, and cardiac muscle—this chapter is primarily concerned with skeletal muscle. Smooth muscle and cardiac muscle are briefly discussed and presented again in later chapters. Smooth muscle occurs in the walls of hollow visceral organs, so it is described in the chapters on the digestive, respiratory, and circulatory systems; cardiac muscle, which is found only in the heart, is discussed in the chapter on the cardiovascular system.

Structure of a Skeletal Muscle

A skeletal muscle, an organ of the muscular system, is composed of several kinds of tissue. These include skeletal muscle tissue, nerve tissue, blood, and various connective tissues (fig. 9.1).

Connective Tissue Coverings

An individual skeletal muscle is separated from adjacent muscles and held into position by layers of fibrous connective tissue called **fascia.** This fascia is part of a network of connective tissue that extends throughout the skeletal muscle system and its attachments with other parts.

Sometimes the connective tissues surrounding a muscle project beyond the end of the muscle fibers to become part of a cordlike **tendon.** Fibers in the tendon intertwine with those in the periosteum of the bone, thus attaching the muscle to bone. In other cases, the connective tissues associated with a muscle form broad fibrous sheets called **aponeuroses,** which may be attached to the coverings of adjacent muscles. (See fig. 9.29.)

The layer of connective tissue that closely surrounds a skeletal muscle is called the *epimysium.* Other layers of connective tissue called the *perimysium* extend inward from the epimysium and separate the muscle tissue into small compartments. These compartments contain bundles of skeletal muscle fibers called *fascicles.* Each muscle fiber within a fascicle is surrounded by a layer of connective tissue in the form of a thin, delicate covering called *endomysium.* (See fig. 9.2.)

Thus, all parts of a skeletal muscle are invested in layers of connective tissue. This arrangement allows the parts to have some independent movement. Also, numerous blood vessels and nerves pass through these layers.

Skeletal Muscle Fibers

A skeletal muscle fiber, which represents a single cell, is the contractile unit of a muscle. Each fiber is a thin elongated cylinder with rounded ends, and it may extend the full length of the muscle. Just beneath its *sarcolemma* (cell membrane), the cytoplasm, or *sarcoplasm,* of the fiber contains many small, oval nuclei and mitochondria. Also within the sarcoplasm are numerous threadlike **myofibrils** that lie parallel to one another. (See fig. 9.3.)

The myofibrils play a fundamental role in the muscle contraction mechanism. They contain two kinds of protein filaments—thick ones composed of the protein **myosin** and thin ones composed of the protein **actin.** The arrangement of these filaments in

Fig. 9.2 (*a*) A skeletal muscle is composed of a variety of tissues including various layers of connective tissue. (*b*) Fascia covers the surface of the muscle, epimysium lies beneath the fascia, and perimysium extends into the structure of the muscle and separates muscle cells into bundles called fasciculi. (*c*) Individual muscle fibers within a fasciculus are separated by endomysium.

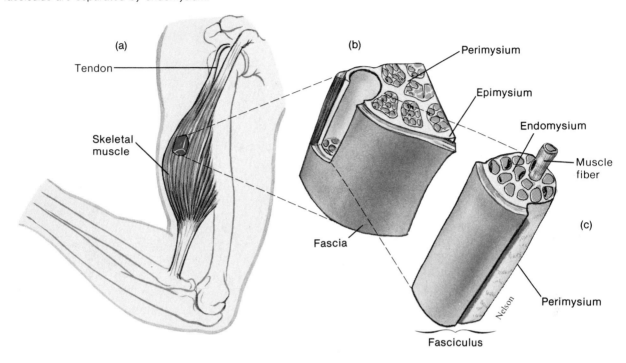

Fig. 9.3 A skeletal muscle fiber may extend the full length of a muscle.

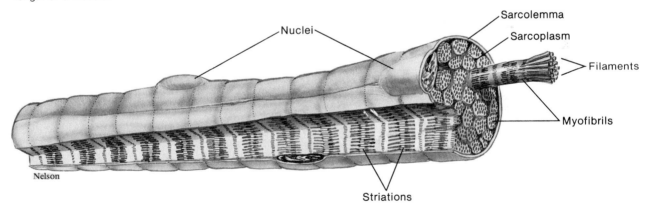

Fig. 9.4 (*a*) A skeletal muscle fiber contains numerous myofibrils, each consisting of (*b*) units called sarcomeres. (*c*) The characteristic striations of a sarcomere are due to the arrangement of actin and myosin filaments.

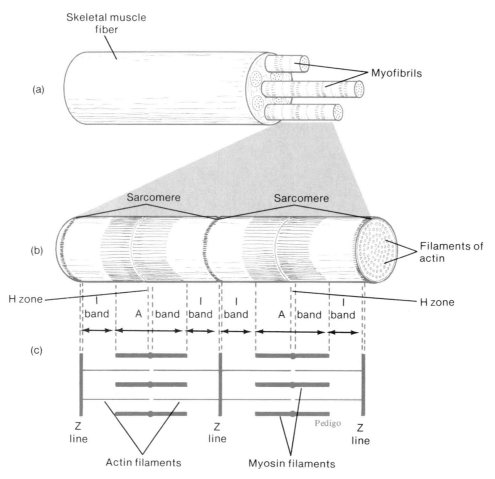

Fig. 9.5 Try to identify the bands of the striations in this electron micrograph of myofibrils.

repeated units called **sarcomeres** produces the characteristic alternating light and dark striations of muscle fiber. (See figs. 9.4 and 9.5.)

Myosin filaments are located primarily within the dark portions, or *A bands,* of the sarcomeres, while actin filaments occur primarily in the light areas, or *I bands*. The actin filaments, however, also extend into the A bands, and when the muscle fiber contracts, the actin filaments slide further into these bands.

The actin filaments seem to be attached to Z lines at the ends of the sarcomeres. These Z lines extend across the muscle fiber so that the sarcomeres of adjacent myofibrils lie side by side. The consequent regular arrangement of the sarcomeres and the A and I bands within the sarcomeres causes the muscle fiber to appear striated under high magnification.

Fig. 9.6 Within the sarcoplasm of a skeletal muscle fiber are a network of sarcoplasmic reticulum and a system of transverse tubules.

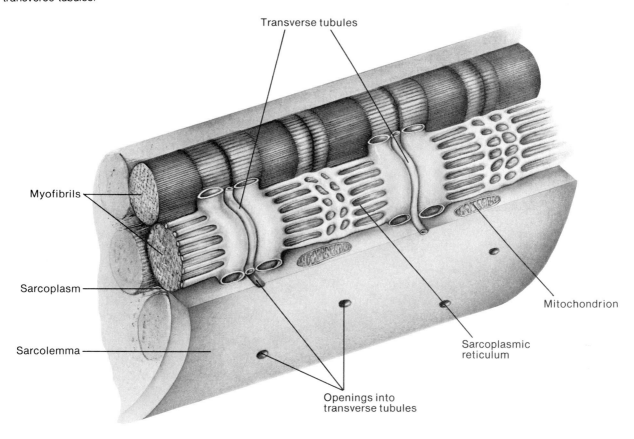

Transverse tubules

Myofibrils

Sarcoplasm

Sarcolemma

Mitochondrion

Sarcoplasmic reticulum

Openings into transverse tubules

If a muscle fiber is stretched beyond its normal limit, the ends of the actin filaments within the A bands are pulled apart. As a result, light H zones appear in the centers of the A bands.

Within the cytoplasm of a muscle fiber is a network of membranous channels that surrounds each myofibril and runs parallel to it. This is the **sarcoplasmic reticulum,** and it is similar to the endoplasmic reticulum of other cells. *Transverse tubules* extend inward, as invaginations, from the fiber's membrane. These tubules contain extracellular fluid and have closed ends that terminate within the I bands of the sarcomeres. Each transverse tubule contacts specialized enlarged portions (cisternae) of the sarcoplasmic reticulum near the region where the actin and myosin filaments overlap. These parts function in activating the muscle contraction mechanism when the fiber is stimulated. (See fig. 9.6.)

1. *Describe the general structure of a skeletal muscle.*
2. *Explain why skeletal muscle fibers appear striated under high magnification.*
3. *Explain the relationship between sarcoplasmic reticulum and transverse tubules.*

The Neuromuscular Junction

Each skeletal muscle fiber is connected to a fiber from a nerve cell. Such a nerve fiber is a branch of a **motor neuron** that extends outward from the brain or spinal cord. Usually a skeletal muscle fiber will contract only when it is stimulated by an action of a motor neuron.

Fig. 9.7 A neuromuscular junction includes the end of a motor neuron and the motor end plate of a muscle fiber. The neurotransmitter is stored in vesicles at the end of the nerve fiber.

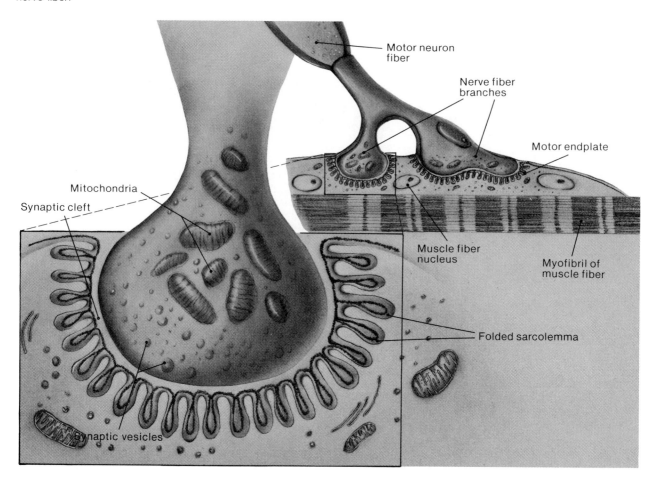

The site where the nerve fiber and muscle fiber meet is called a **neuromuscular junction** (myoneural junction). At this junction the muscle fiber membrane is specialized to form a **motor end plate.** In this region, muscle fiber nuclei and mitochondria are abundant and the sarcolemma is highly folded. (See fig. 9.7.)

The end of the motor nerve fiber is branched, and the ends of these branches project into recesses (synaptic clefts) of the muscle fiber membrane. The cytoplasm at the ends of the nerve fibers is rich in mitochondria and contains many tiny vesicles (synaptic vesicles) that store chemicals called **neurotransmitters.**

When a nerve impulse traveling from the brain or spinal cord reaches the end of a motor nerve fiber, some of the vesicles release neurotransmitter into the gap between the nerve and the motor end plate. This action stimulates the muscle fiber to contract.

Motor Units

Although a muscle fiber usually has a single motor end plate, the nerve fibers of motor neurons are highly branched, and one motor fiber may connect to many muscle fibers. Furthermore, when the motor nerve fiber transmits an impulse, all the muscle fibers it is connected to are stimulated to contract simultaneously. Together, a motor neuron and the muscle fibers that it controls constitute a **motor unit.** (See fig. 9.8.)

The number of muscle fibers in a motor unit varies considerably. The fewer muscle fibers in the motor units, however, the finer the movements that can be produced in a particular muscle. For example, the motor units of the muscles that move the eyes, which may contain fewer than ten muscle fibers per unit, can produce very slight movements. Conversely, the motor units of the large muscles in the back may include a hundred or more muscle fibers. When these units are stimulated, the movements produced are coarse in comparison with those of the eye muscles.

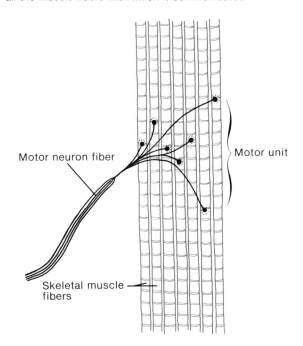

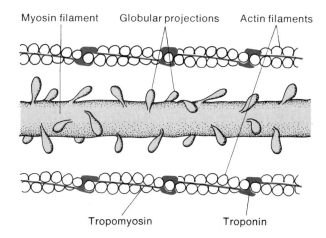

Skeletal Muscle Contraction

A muscle fiber contraction is a complex action involving a number of cell parts and chemical substances. The final result is a sliding movement within the myofibrils in which the filaments of actin and myosin merge. When this happens, the muscle fiber is shortened, and it pulls on its attachments.

Role of Actin and Myosin

A myosin molecule is composed of two protein strands twisted around one another to form an elongated helix. Globular protein parts project from the myosin. These globular parts can react with actin filaments and form cross-bridges with them. This reaction between the actin and myosin filaments generates the force involved with shortening the myofibrils during a contraction.

An actin molecule also consists of double strands of protein twisted into a helix, and it has ADP molecules attached to its surface. These ADP molecules serve as active sites for the formation of cross-bridges with the globular parts of myosin molecules.

Two other proteins, **tropomyosin** and **troponin,** are also associated with actin. The tropomyosin molecules occupy the longitudinal grooves of the actin helix, and each has a troponin molecule attached to its surface. (See fig. 9.9.)

When a muscle fiber is at rest, these tropomyosin-troponin complexes somehow inhibit the active sites on the actin molecules and thus, prevent the formation of cross-bridges. If *calcium ions* are present, however, the ions bind to the troponin, and this modifies the position of the tropomyosin. As the tropomyosin molecules move, the active sites on the actin filaments are exposed, and cross-bridges can form between the actin and myosin filaments.

It is not known how the formation of cross-bridges results in shortening of myofibrils. One theory (ratchet theory) suggests that the head of a myosin cross-bridge can attach to an actin active site and bend slightly, pulling the actin filament with it. Then the head may release, straighten itself, and combine with another active site further down the actin filament. Presumably this cycle can be repeated again and again as the actin filament is moved toward the center of the sarcomere. (See fig. 9.10.)

Stimulus for Contraction

A skeletal muscle fiber normally does not contract until it is stimulated by a motor nerve fiber releasing neurotransmitter at the motor end plate. In skeletal muscle, the neurotransmitter is a compound called **acetylcholine,** which is synthesized in the cytoplasm of the motor neuron and stored in vesicles. When a nerve impulse reaches the end of the nerve fiber, many of these vesicles discharge their contents into the gap between the nerve fiber and the motor end plate. (See fig. 9.7.)

The acetylcholine diffuses rapidly across the gap and causes a change in the muscle fiber membrane. As a result, an impulse, very much like a nerve impulse (described in chapter 10), passes in all directions over the surface of the muscle fiber membrane and through the transverse tubules, deep into the fiber.

Fig. 9.10 (*a*) In the presence of calcium ions, cross-bridges form between filaments of actin and myosin; (*b, c*) as a result, the actin filaments are pulled toward the center of the A band, causing the myofibril to contract.

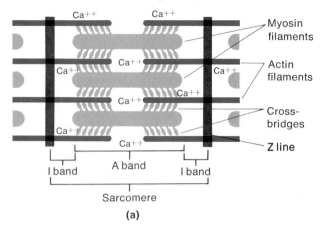

(a)

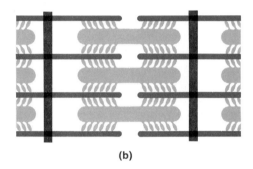

(b)

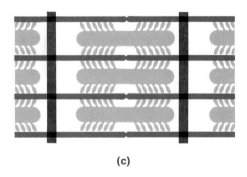

(c)

In response to this signal, the enlarged portions of the sarcoplasmic reticulum in contact with the transverse tubules release calcium ions into the cytoplasm of the muscle fiber. When calcium ions are present in relatively high concentrations, cross-bridges form between the actin and myosin filaments, and a contraction occurs.

The contraction continues while the calcium ions are present, but they are quickly moved back into the sarcoplasmic reticulum by active transport. Consequently, within a few thousandths of a second following contraction, a skeletal muscle fiber relaxes.

Meanwhile, the acetylcholine that stimulated the muscle fiber in the first place is rapidly decomposed by the action of an enzyme called **cholinesterase,** which is present in the neuromuscular junction.

Chart 9.1 Major events of muscle contraction and relaxation

Muscle Fiber Contraction

1. Stimulation occurs when acetylcholine is released from the end of a motor neuron.
2. Acetylcholine diffuses across gap at neuromuscular junction.
3. Muscle fiber membrane is changed and an impulse travels deep into the fiber through the transverse tubules.
4. Calcium ions are released from sarcoplasmic reticulum and bind to troponin molecules.
5. Tropomyosin molecules move and expose specific sites on actin filaments.
6. Cross-bridges form between actin and myosin filaments.
7. Actin filaments slide inward along myosin filaments.
8. Muscle fiber shortens as contraction occurs.

Muscle Fiber Relaxation

1. Cholinesterase causes acetylcholine to decompose, and muscle fiber membrane is no longer stimulated.
2. Calcium ions are actively transported into the sarcoplasmic reticulum.
3. Cross-bridges between actin and myosin filaments are broken.
4. Actin and myosin filaments slide apart.
5. Muscle fiber lengthens as it relaxes and its resting state is reestablished.
6. Troponin and tropomyosin molecules inhibit the interaction between actin and myosin filaments.

This action prevents a single nerve impulse from causing a continued stimulation of the muscle fiber.

Chart 9.1 summarizes the major events leading to muscle contraction and relaxation.

In the disease called *myasthenia gravis,* the motor nerve fibers seem unable to secrete enough acetylcholine to stimulate muscle fibers in the usual manner. The victim suffers from muscular weakness and may have difficulty contracting the muscles associated with eye movements, facial expressions, chewing, speaking, and breathing. This condition is sometimes treated with a drug (neostigmine) that inactivates cholinesterase. When the cholinesterase is inactivated, acetylcholine tends to accumulate at the neuromuscular junctions with successive nerve impulses, and enough may accumulate to stimulate muscle fiber contraction.

1. Describe a neuromuscular junction.
2. List four proteins associated with myofibrils and explain their relationships.
3. Explain how the filaments of a myofibril interact during muscle contraction.
4. Explain how a motor nerve impulse can trigger a muscle contraction.

Fig. 9.11 Energy released by cellular respiration may be used to promote the synthesis of ATP or to synthesize creatine phosphate. Later, energy from creatine phosphate may be used to promote ATP synthesis.

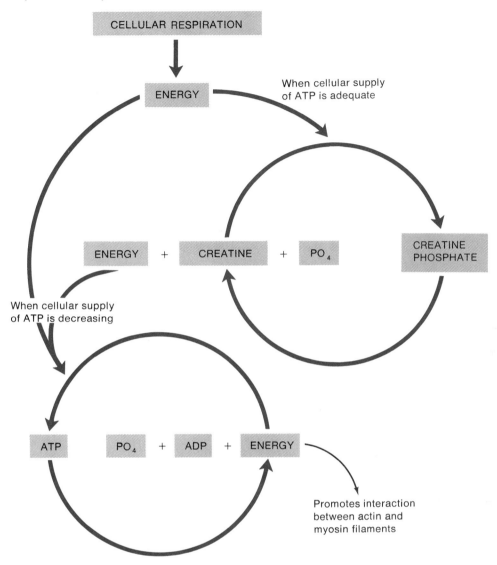

Energy Sources for Contraction

The energy used in muscle fiber contraction comes from ATP molecules, which are supplied by numerous mitochondria positioned close to the myofibrils. The globular portions of the myosin filaments contain an enzyme called **ATPase** that causes ATP to decompose into ADP and phosphate and, at the same time, to release some energy. This energy makes possible the reaction between the actin and myosin filaments.

Although ATP is the direct supplier of energy for the contraction mechanism, it is not the only energy-carrying molecule within the muscle cells. A substance called **creatine phosphate,** which also contains high-energy phosphate bonds, is actually four

to six times more abundant than ATP. Creatine phosphate, however, cannot directly supply energy for a cell's energy-utilizing reactions. Instead, it acts to store energy released from mitochondria whenever sufficient amounts of ATP are already present.

At times when ATP is being decomposed, energy from creatine phosphate can be transferred to ADP molecules, which are quickly converted back into ATP. The amount of ATP and creatine phosphate present in a skeletal muscle, however, is usually not sufficient to support maximal muscle activity for more than a few seconds. Consequently, the muscle fibers in an active muscle soon become dependent upon cellular respiration as a source of energy for synthesizing ATP. (See fig. 9.11.)

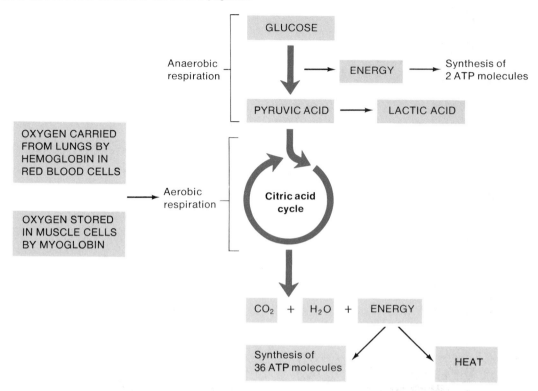

Oxygen Supply and Cellular Respiration

As is described in chapter 4, the early phase of cellular respiration occurs in the cytoplasm and is *anaerobic,* taking place in the absence of oxygen. This phase involves a partial breakdown of energy-supplying molecules such as glucose and results in a gain of only a few ATP molecules. The complete breakdown of glucose occurs in the mitochondria and is *aerobic.* This process, which includes the complex series of reactions of the *citric acid cycle,* produces a relatively large number of ATP molecules.

The oxygen needed to support aerobic respiration is carried from the lungs to body cells by the blood. It is transported within the red blood cells loosely bonded to molecules of hemoglobin, the pigment responsible for the color of blood. In the regions of the body cells where the oxygen concentration is relatively low, oxygen is released from the hemoglobin and becomes available for cellular respiration.

Another pigment, **myoglobin,** is synthesized in the muscle cells and is responsible for the reddish brown color of skeletal muscle tissue. It has properties similar to hemoglobin in that it can combine loosely with oxygen. In fact, myoglobin has a greater attraction for oxygen than does hemoglobin, and it apparently functions to store oxygen in muscle tissue, at least temporarily. This ability to store oxygen seems to reduce a muscle's need for a continuous blood supply during muscular contraction, which may be accompanied by a decreased blood flow. (See fig. 9.12.)

Oxygen Debt

When a person is resting or is moderately active, the ability of the respiratory and circulatory systems to supply oxygen to the skeletal muscles is usually adequate to support aerobic respiration. When skeletal muscles are used strenuously for even a minute or two, however, the respiratory and circulatory systems usually cannot supply oxygen efficiently enough to meet the needs of aerobic respiration, and the muscle fibers must depend more and more on the anaerobic phase to obtain energy.

In anaerobic respiration, glucose molecules are changed to *pyruvic acid.* The pyruvic acid is then converted to *lactic acid,* which diffuses out of the muscle fibers and is carried to the liver by the blood.

Liver cells change the lactic acid into *glucose,* which can be stored as glycogen for future use or be utilized immediately. The conversion of lactic acid into glucose also requires energy from ATP. During strenuous exercise, the available oxygen is used primarily to synthesize the ATP needed for the muscle fiber contraction rather than to make ATP for changing lactic acid into glucose. Consequently, as lactic

Fig. 9.13 The lactic acid that accumulates in muscles as a result of anaerobic respiration can be converted back into glucose by liver cells.

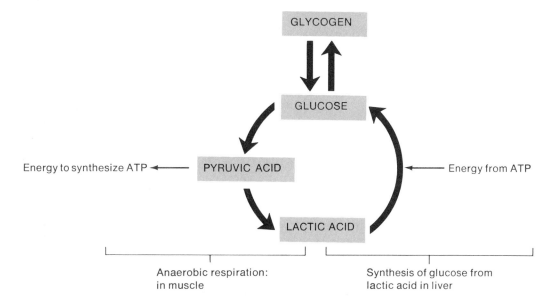

acid accumulates, a person develops an **oxygen debt** that must be paid off at a later time. The amount of this oxygen debt is equal to the amount of oxygen needed by the liver cells to convert the accumulated lactic acid into glucose plus the amount needed by muscle cells to resynthesize ATP and creatine phosphate to their original concentrations. (See fig. 9.13.)

Usually a person feels out of breath and continues breathing heavily for a while following vigorous exercise. During this time the oxygen debt is paid off, and then normal breathing can be resumed.

Muscle Fatigue

If a muscle is exercised strenuously for a prolonged period, it may lose its ability to contract and is said to be *fatigued.* This condition may result from an interruption in the muscle's blood supply or from an exhaustion of the supply of acetylcholine in its motor nerve fibers. Muscle fatigue, however, is usually due to lactic acid that has accumulated in the muscle as a result of anaerobic respiration. The lactic acid causes factors, such as pH, to change so that muscle fibers are no longer responsive to stimulation.

Occasionally a muscle becomes fatigued and cramps at the same time. A cramp is a painful condition in which the muscle contracts spasmodically, but does not relax completely. This condition seems to be due to a lack of ATP, which is needed to move calcium ions back into the sarcoplasmic reticulum and break the connections between actin and myosin filaments before relaxation in muscle fibers can occur.

Tolerance to the effects of lactic acid varies, so that some persons experience muscular fatigue more quickly than others. Resistance to this effect also varies among athletes; as a group, however, they can exercise and produce less lactic acid than nonathletes. In part, this is because the strenuous exercise of physical training stimulates new capillaries to grow within the muscles. Thus, more oxygen and nutrients can be supplied to the muscle fibers.

A few hours after death, the skeletal muscles undergo a partial contraction that causes the joints to become fixed. This condition, *rigor mortis,* may continue for 72 hours or more. It seems to result from a lack of ATP in the muscle fibers, which prevents relaxation. Thus, the actin and myosin filaments of the muscle fibers remain bonded together until the muscles begin to decompose.

Heat Production

Heat is produced as a by-product of cellular respiration, so it is generated by all active cells. Since muscle tissue represents such a large proportion of the total body mass, it is a particularly important source of heat.

Actually only about 25% of the energy released by cellular respiration is available for use in metabolic processes. The rest is lost as heat. Thus, whenever muscles are active, large amounts of heat are released.

This heat is transported to other tissues by the blood and helps to maintain body temperature. If heat is present in excess, mechanisms that promote heat loss (described in chapter 6) are activated, and homeostasis is maintained.

---

1. *What substances provide energy for muscle contraction?*
2. *What is the relationship between lactic acid and oxygen debt?*
3. *Why are athletes less likely to experience muscle fatigue than nonathletes?*

Muscular Responses

One way to observe muscle contraction is to remove a single muscle fiber from a skeletal muscle and connect it to a device that senses and records changes in the fiber's length. In such experiments, the muscle fiber is usually stimulated with an electrical stimulator that is capable of producing stimuli of varying strength and frequency.

Threshold Stimulus. By exposing an isolated muscle fiber to a series of stimuli of increasing strength, it can be shown that the fiber remains unresponsive until a certain strength of stimulation is applied. This minimal strength needed to elicit a contraction is called the **threshold stimulus.**

All-or-None Response. When a muscle fiber is exposed to a stimulus of threshold strength (or above), it responds to its fullest extent. Increasing the strength of the stimulus does not affect the degree to which the fiber contracts. In other words, there are no partial contractions of a muscle fiber—if it contracts at all, it will contract completely. This phenomenon is called the **all-or-none response.**

Since the muscle fibers within a muscle are organized into motor units, and each motor unit is controlled by a single motor neuron, all the muscle fibers in a motor unit are stimulated at the same time. Consequently, a motor unit also responds in an all-or-none way. A whole muscle, however, does not behave like this, because it is composed of many motor units that respond to different thresholds of stimulation. Thus, when a given stimulus is applied to a muscle, some motor units may respond while others do not.

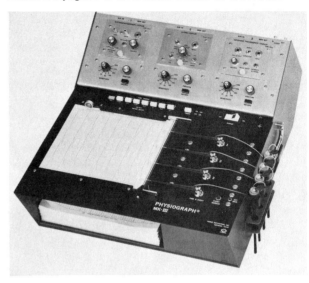

Fig. 9.14 An apparatus, such as this, can be used to record a myogram of an isolated muscle as it contracts.

Twitch Contraction

To demonstrate how a whole muscle responds to stimulation, a skeletal muscle can be removed from a frog or other laboratory animal and mounted in an apparatus similar to the one shown in figure 9.14. The muscle is stimulated electrically, and when it contracts, it pulls on a lever. The lever's movement is recorded, and the resulting pattern is called a **myogram.**

If a muscle is exposed to a single stimulus of sufficient strength to activate some of its motor units, the muscle will contract and then relax. This action—a single contraction that lasts only a fraction of a second—is called a **twitch.** A twitch produces a myogram like that in figure 9.15. It is apparent from this record that the muscle response did not begin immediately following the stimulation. There is a lag between the time that the stimulus was applied and the time that the muscle responded. This time lag is called the **latent period.** In a frog muscle, the latent period lasts for about 0.01 second, and it is even shorter in a human muscle.

The latent period is followed by a *period of contraction* during which the muscle pulls at its attachments, and a *period of relaxation* during which it returns to its former length. (See fig. 9.15.)

If a muscle is exposed to two stimuli (of threshold strength or above) in quick succession, it may respond with a twitch to the first stimulus but not to the second. This is because it takes an instant following a contraction for the muscle fibers to reestablish

Fig. 9.15 A myogram of a single muscle twitch.

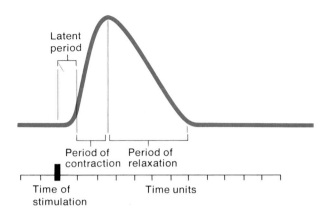

Fig. 9.16 Myograms of (*a*) a series of twitches; (*b*) a summation of twitches; and (*c*) a tetanic contraction.

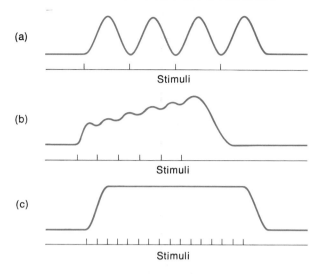

necessary electrolyte concentrations. Thus, there is a very brief moment following stimulation during which a muscle remains unresponsive. This time is called the **refractory period.**

Sustained Contractions

A muscle exposed to a series of threshold stimuli may undergo a series of twitches. However if the frequency of stimulation is increased, a point is reached when the muscle is unable to complete its relaxation period before the next stimulus in the series arrives. Then, the contractions begin to combine and the muscle contraction is sustained. This combination of twitches is called *summation of twitches* or *wave summation.*

At the same time that twitches are combining, the strength of the contractions may be increasing. This is due to the presence of motor units within the muscle that vary in the number and size of their muscle fibers. The smaller motor units, which have finer fibers, are most easily stimulated and tend to respond earlier in the series of stimuli. The larger motor units, which contain thicker fibers, respond later and produce more forceful contractions. The production of such a sustained contraction of increasing strength is called *multiple motor unit summation.* When the resulting forceful, sustained contraction lacks even partial relaxation, it is termed a **tetanic contraction** (tetany). (See fig. 9.16.)

Although twitches may occur occasionally in human skeletal muscles, as when an eye blinks, such contractions are of limited use. More commonly contractions are sustained, and they are smooth rather than irregular or jerky because a mechanism operates within the spinal cord to stimulate contractions in different sets of motor units at different moments. Thus, while some motor units are contracting, others are relaxing.

Tetanic contractions occur frequently in skeletal muscles during everyday activities. In many cases the condition occurs only in a portion of a muscle. When a person lifts a weight or walks, for example, sustained contractions are maintained in arm or leg muscles for varying lengths of time. These contractions are responses to a rapid series of stimuli transmitted from the brain and spinal cord on motor neuron fibers.

Actually, even when a muscle appears to be at rest, a certain amount of sustained contraction is occurring in its fibers. This is called **muscle tone** (tonus), and it is a response to nerve impulses originating in the spinal cord from moment to moment. These impulses travel to small numbers of muscle fibers and stimulate them into action. The result is a continuous state of partial contraction.

Muscle tone is particularly important in maintaining posture. Tautness in the muscles of the neck, trunk, and legs enable a person to stand or sit. If these muscles suddenly become relaxed, as happens when a person loses consciousness, the body will collapse. Although muscle tone is maintained in health, it is lost if motor nerve fibers are cut or if diseases interfere with the conduction of nerve impulses.

When skeletal muscles are contracted very forcefully they may generate up to 50 pounds of pull for each square inch of muscle cross section. Consequently, large muscles such as those in the thigh can pull with several hundred pounds of force. Occasionally this force is so great that the tendons of muscles are torn away from their attachments on bones.

Fig. 9.17 (*a*) Isotonic contractions occur when a muscle contracts and shortens; (*b*) isometric contractions occur when a muscle contracts but does not shorten.

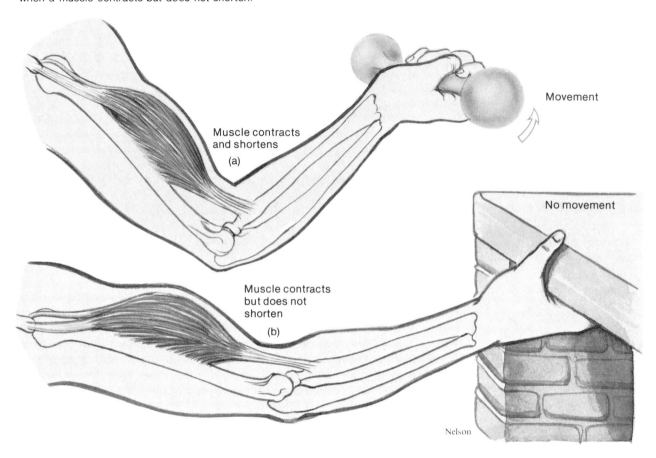

Muscle contracts and shortens
(a)

Movement

Muscle contracts but does not shorten
(b)

No movement

Nelson

Isometric and Isotonic Contractions

Sometimes a skeletal muscle contracts but the parts to which it is attached do not move. This happens, for instance, when a person pushes against the wall of a building. Tension within the muscles increases, but the wall does not move, and the muscles remain the same length. Contractions of this type are called **isometric.** Isometric contractions occur continuously in postural muscles that function to stabilize skeletal parts and hold the body upright.

At other times, muscles shorten when they contract. For example, if a person lifts an object, the tautness in the muscles remains unchanged, their attached ends pull closer together, and the object is moved. This type of contraction is termed **isotonic.** Figure 9.17 illustrates these two types of contraction.

Skeletal muscles can contract either isometrically or isotonically, and most body actions involve both types of contraction. In walking, for instance, certain leg muscles contract isometrically and keep the limb stiff as it touches the ground, while other muscles contract isotonically, causing the leg to bend and lift upward.

Isometric exercises are usually avoided with patients suffering from heart disease or those suspected of having heart disease, because such exercises tend to cause an excessive rise in arterial blood pressure.

Use and Disuse of Skeletal Muscles

Skeletal muscles are very responsive to use and disuse. Muscles that are forcefully exercised either isometrically or isotonically tend to enlarge. This phenomenon, which is termed **muscular hypertrophy,** involves an increase in size of individual muscle fibers rather than an increase in the number of fibers. The mitochondria within these fibers are stimulated to reproduce, the sarcoplasmic reticula become more extensive, and new filaments of actin and myosin are produced.

As a result of hypertrophy a muscle becomes capable of more forceful contractions because the strength of a contraction is directly related to the diameter of the muscle fibers.

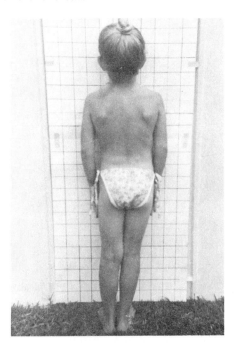

Physical training involving exercises such as running usually produces relatively slight increases in the sizes and strengths of the muscles used. However, training that requires near maximal contractions—such as weight lifting—induces muscular hypertrophy.

A muscle that is not used or is only used to produce weak contractions tends to decrease in size and strength. There is a reduction in both the number of capillaries and the number of mitochondria in the muscle fibers. The networks of sarcoplasmic reticula become less extensive, and the size of the actin and myosin filaments decreases. This condition is called **atrophy**, and it commonly occurs when limbs are immobilized by casts or when accidents or diseases interfere with motor nerve impulses. A muscle that cannot be exercised may decrease to less than one-half its usual size within a few months (fig. 9.18.)

Muscles whose motor neurons are severed undergo atrophy relatively rapidly. The muscle fibers not only decrease in size, but also become fragmented and degenerated, and are replaced by connective tissues.

Muscular strength generally increases with age until a peak is reached in the midtwenties. Thereafter there is a gradual loss of strength. The reasons for this loss are not well understood, but they seem to involve changes in dietary habits (including a decreased intake of potassium which is common in elderly persons), a decline in the secretion of certain sex hormones, and a decrease in respiratory and circulatory efficiency.

1. *Define threshold stimulus.*
2. *What is meant by an all-or-none response?*
3. *Distinguish between a twitch and a sustained contraction.*
4. *Explain the difference between isometric and isotonic contractions.*
5. *What effect is repeated exercise that involves maximal contractions likely to have on the size of muscle fibers?*

Smooth Muscles

Although the contractile mechanisms of smooth and cardiac muscles are essentially the same as that of skeletal muscles, the cells of these tissues have important structural and functional differences.

Smooth Muscle Fibers

As is discussed in chapter 5, smooth muscle cells are shorter than those of skeletal muscle, and they have single, centrally located nuclei. These cells are elongated with tapering ends. Smooth muscle cells contain filaments of *actin* and *myosin* in myofibrils that extend the lengths of the cells. However, these filaments are very thin and more randomly arranged than in skeletal muscle. Consequently, the cells lack striations. They also apparently lack transverse tubules, and their sarcoplasmic reticula are not well developed.

There are two major types of smooth muscles. In one type, called **multiunit smooth muscle,** the muscle fibers are less well organized and occur as separate fibers rather than in sheets. Smooth muscle of this type is found in the irises of the eyes and in the walls of blood vessels. Typically it contracts only after stimulation by motor nerve impulses. (See fig. 9.19.)

Fig. 9.19 Multiunit smooth muscles control the size of the pupil of the eye. (a) When one set of muscles within the iris contracts, the pupil is constricted; (b) when another set of muscles contracts, the pupil is dilated.

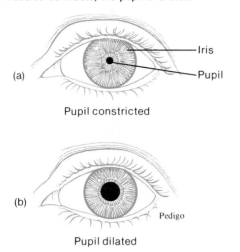

Pupil constricted

Pupil dilated

Fig. 9.20 Visceral smooth muscles within the walls of tubular digestive organs are responsible for the peristaltic waves that help move the contents of the tube along its length.

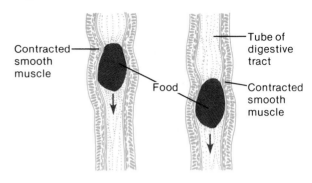

The second type of smooth muscle is called **visceral.** It is composed of sheets of spindle-shaped cells that are in close contact with one another and are positioned so the thick portion of each cell is next to the thin parts of adjacent cells. Visceral smooth muscle is the more common type and is found in the walls of hollow visceral organs such as the stomach, intestines, urinary bladder, and uterus. Usually there are two thicknesses of smooth muscle in the walls of these organs. The fibers of the outer coats are directed longitudinally, while those of the inner coats are arranged circularly. These muscular layers are responsible for the changes in size and shape that occur in visceral organs as they carry on their special functions.

The fibers of visceral smooth muscles are capable of stimulating each other. Consequently, when one fiber is stimulated, the impulse moving over its surface may excite adjacent fibers, which in turn stimulate still others. Visceral smooth muscles also display *rhythmicity*—a pattern of repeated contractions. This phenomenon is due to the presence of self-exciting fibers from which spontaneous impulses travel periodically into the surrounding muscle tissue.

These two features of visceral smooth muscle—transmission of impulses from cell to cell and rhythmicity—are largely responsible for the wavelike motion called **peristalsis** that occurs in various tubular organs (fig. 9.20). Peristalsis involves alternate contractions and relaxations of the longitudinal and circular muscles. These movements help force the contents of a tube along its length. In the intestines, for example, peristaltic waves move masses of food substances through the tube and mix them with digestive fluids. Similar activity in the ureters moves urine from the kidneys to the urinary bladder.

Smooth Muscle Contraction

Smooth muscle contraction resembles skeletal muscle contraction in a number of ways. Both involve reactions of actin and myosin, both are triggered by membrane impulses and the release of calcium ions, and both use energy from ATP molecules. There are, however, significant differences in the actions of these muscles. Smooth muscle is slower to contract and to relax. On the other hand, it can maintain a forceful contraction for a longer time with a given amount of ATP than can skeletal muscle. Unlike skeletal muscle, smooth muscle fibers can change length without changing tautness; because of this, smooth muscles in the stomach and intestinal walls can stretch as these organs become filled, while the pressure inside remains unchanged.

1. Describe the two major types of smooth muscle.
2. How does smooth muscle produce peristalsis?
3. How does smooth muscle contraction differ from that of skeletal muscle?

Cardiac Muscle

Cardiac muscle occurs only in the heart. It is composed of striated cells that are joined end to end forming fibers. These fibers are interconnected in branching, three-dimensional networks. Each cell contains a single nucleus and numerous filaments of actin and myosin. It also has a well-developed sarcoplasmic reticulum, a system of transverse tubules, and many mitochondria.

The opposing ends of cardiac muscle cells are separated by cross-bands called *intercalated disks*. These bands are the result of elaborate junctions of the membranes at the cell's boundary. They help to hold adjacent cells together and transmit the force of contraction from cell to cell. The intercalated disks also seem to have very low resistance to the passage of impulses, so that cardiac muscle fibers transmit impulses relatively rapidly. (See fig. 9.21.)

When one portion of the cardiac muscle network is stimulated, the impulse passes to the other fibers of the net, and the whole structure contracts as a unit; that is, the network responds to stimulation in an all-or-none manner. Cardiac muscle is also self-exciting and rhythmic. Consequently, a pattern of contraction and relaxation is repeated again and again and is responsible for the rhythmic contractions of the heart.

Once it is stimulated, the cardiac muscle remains contracted for a longer time than does skeletal muscle. Also unlike skeletal muscle, cardiac muscle remains refractory until a contraction is completed, so that sustained or tetanic contractions do not occur in the heart muscle.

Chart 9.2 summarizes the characteristics of the three types of muscles.

1. How does cardiac muscle differ from smooth muscle?
2. How does it differ from skeletal muscle?
3. In what ways are these three tissues similar?

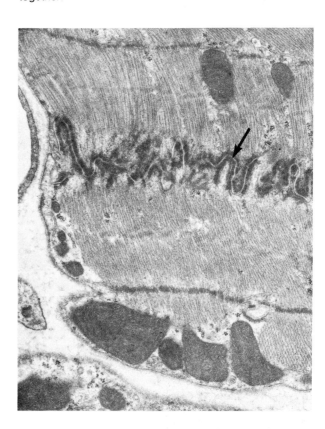

Fig. 9.21 The intercalated disks of cardiac muscle, shown in this electron micrograph, hold adjacent cells together.

Chart 9.2 Types of muscle tissue

	Skeletal	Smooth	Cardiac
Major location	Skeletal muscles	Walls of hollow visceral organs	Wall of heart
Major function	Movement of bones at joints, maintenance of posture	Movement of visceral organs, peristalsis	Pumping action of heart
Cellular characteristics			
Striations	Present	Absent	Present
Nucleus	Multinuclei	Single nucleus	Single nucleus
Special features	Transverse tubule system well developed	Lacks transverse tubules	Transverse tubule system well developed, adjacent cells separated by intercalated disks
Mode of control	Voluntary	Involuntary	Involuntary
Contraction characteristics	Contracts and relaxes relatively rapidly	Contracts and relaxes relatively slowly, self-exciting, rhythmic	Network of fibers contracts as a unit, self-exciting, rhythmic, remains refractory until contraction is completed

Body Movements

Skeletal muscles are responsible for a great variety of body movements. The action of each muscle depends largely upon the kind of joint it is associated with and the way it is attached on either side of a joint.

Origin and Insertion

Usually one end of a skeletal muscle is fastened to a relatively immovable or fixed part, and the other end is connected to a movable part on the other side of a joint. The immovable end is called the **origin** of the muscle, and the movable one is its **insertion.** When a muscle contracts, its insertion is pulled toward its origin. (See fig. 9.22.)

Some muscles have more than one origin or insertion. The *biceps brachii* in the upper arm, for example, has two origins. This is reflected in its name, *biceps,* which means *two heads.* As figure 9.23 shows, one head is attached to the coracoid process of the scapula, while the other one arises from a tubercle above the glenoid cavity of the scapula. The muscle extends along the front surface of the humerus and is inserted by means of a tendon on the radial tuberosity of the radius. When the biceps brachii contracts, its insertion is pulled toward its origin, and the arm bends at the elbow.

Sometimes the end of a muscle changes its function in different body movements; that is, the origin may become the insertion or vice versa. For instance, a large muscle in the chest, called *pectoralis major,* connects the humerus to bones of the thorax. If the arm moves across the chest, the end of the muscle attached to the humerus acts as the insertion. When a person hangs from a bar to do chinning exercises, however, the arm becomes fixed and the body is pulled up. In this movement, the end attached to the humerus acts as the origin.

Similarly, the *latissimus dorsi* muscle in the back, shown in figure 9.22 has its origin on the pelvic girdle and vertebral column and its insertion on the proximal end of the humerus. Usually it functions to pull the arm toward the back as in rowing a boat or swimming. During chinning exercises, however, the humerus becomes the origin of the muscle and the end attached to the pelvis and vertebral column moves.

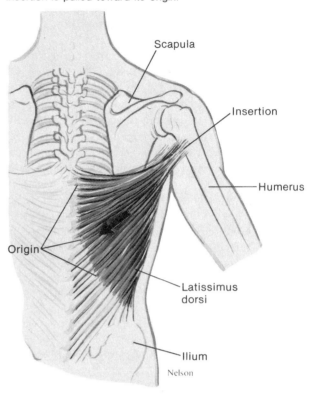

Fig. 9.22 When the fibers of a muscle contract, its insertion is pulled toward its origin.

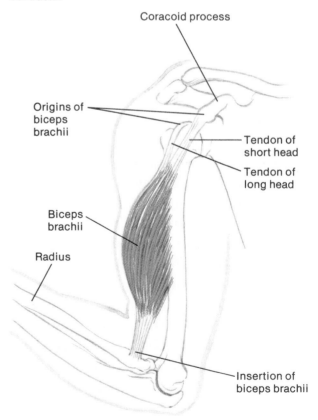

Fig. 9.23 The biceps brachii has two heads that originate on the scapula. This muscle is inserted on the radius by means of a tendon. What movement results as it contracts?

Interaction of Skeletal Muscles

Skeletal muscles almost always function in groups rather than singly. Consequently, when a particular body movement occurs, a person must do more than command a single muscle to contract; instead that person wills the movement to occur, and the appropriate group of muscles responds to the decision.

By carefully observing body movements, it is possible to determine the particular roles of various muscles. For instance, when the arm is lifted horizontally away from the side, a *deltoid* muscle is responsible for most of the movement, and so is said to be the **prime mover.** However, while the prime mover is acting, certain nearby muscles also are contracting. They help to hold the shoulder steady and in this way make the action of the prime mover more effective. Muscles that assist the prime mover are called **synergists.**

Still other muscles act as **antagonists** to prime movers. These muscles are capable of resisting a prime mover's action and are responsible for movement in the opposite direction—the antagonist of the prime mover that raises the arm can lower the arm, or the antagonist of the prime mover that bends the arm can straighten it. If both a prime mover and its antagonist contract simultaneously, the part they act upon remains rigid. Consequently, smooth body movements depend upon the antagonists' giving way to the actions of the prime movers whenever the prime movers are contracting.

---

1. *Distinguish between the origin and insertion of a muscle.*
2. *Define* prime mover.
3. *What is the function of a synergist? An antagonist?*

Types of Movements

Muscles acting at freely movable joints produce body movements in different directions and in different planes. The following terms are used to describe such movements. (See figs. 9.24, 9.25, and 9.26.)

flexion (flek′shun)—bending parts at a joint so that the angle between them is decreased and the parts come closer together (bending the leg at the knee).

extension (ek-sten′shun)—straightening parts at a joint so that the angle between them is increased and the parts move further apart (straightening the leg at the knee).

hyperextension (hi″per-ek-sten′shun)—excessive extension of the parts at a joint, beyond the anatomical position (bending the head back beyond the upright position).

abduction (ab-duk′shun)—moving a part away from the midline (lifting the arm horizontally to form a right angle with the side of the body).

adduction (ah-duk′shun)—moving a part toward the midline (returning the arm from the horizontal position to the side of the body).

rotation (ro-ta′shun)—moving a part around an axis (twisting the head from side to side).

circumduction (ser″kum-duk′shun)—moving a part so that its end follows a circular path (moving the finger in a circular motion without moving the hand).

supination (soo″pĭ-na′shun)—turning the hand so the palm is upward.

pronation (pro-na′shun)—turning the hand so the palm is downward.

eversion (e-ver′zhun)—turning the foot so the sole is outward.

inversion (in-ver′zhun)—turning the foot so the sole is inward.

protraction (pro-trak′shun)—moving a part forward (thrusting the chin forward).

retraction (re-trak′shun)—moving a part backward (pulling the chin backward).

elevation (el″ĕ-va′shun)—raising a part (shrugging the shoulders).

depression (de-presh′un)—lowering a part (drooping the shoulders).

Fig. 9.24 Muscle movements illustrating flexion, extension, hyperextension, abduction, and adduction.

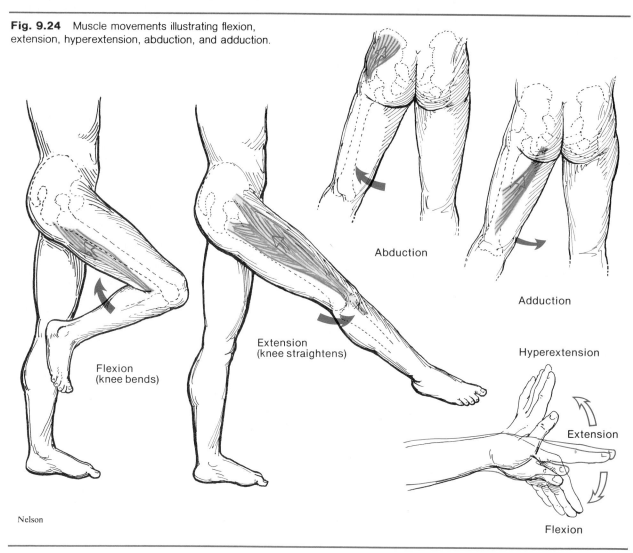

Abduction

Adduction

Flexion
(knee bends)

Extension
(knee straightens)

Hyperextension

Extension

Flexion

Nelson

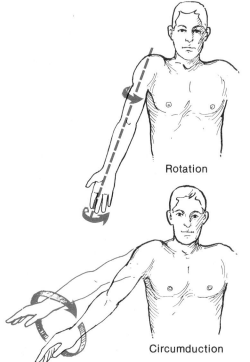

Rotation

Circumduction

Fig. 9.25 Muscle movements illustrating rotation and circumduction, supination and pronation.

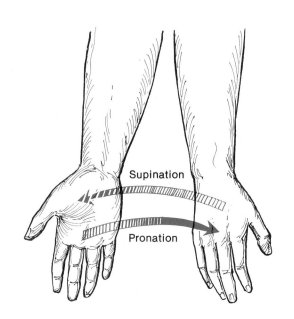

Supination

Pronation

Support and Movement

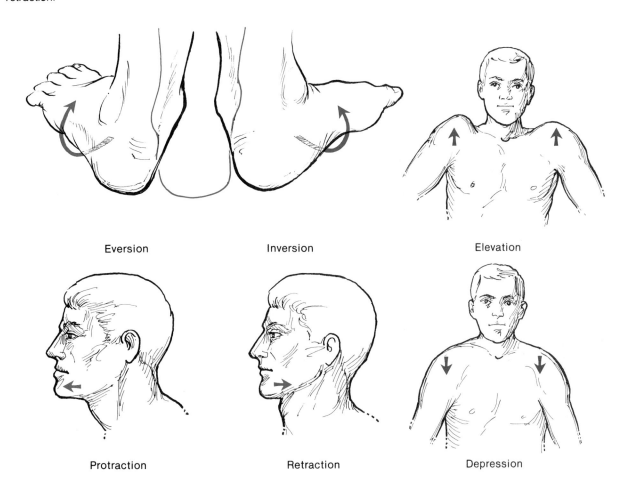

Eversion   Inversion   Elevation

Protraction   Retraction   Depression

Major Skeletal Muscles

The following section describes the locations, actions, and attachments of some of the major skeletal muscles of the body. (Figures 9.27 and 9.28 show the locations of superficial skeletal muscles.) Notice that the names of these muscles often describe them in some way. A name may indicate a muscle's size, shape, location, action, number of attachments, or the direction of its fibers, as in the following examples:

pectoralis major—a muscle of large size (*major*) located in the pectoral region or chest.

deltoid—shaped like a delta or triangle.

extensor digitorum—acts to extend the digits (fingers or toes).

biceps brachii—a muscle with two heads (*biceps*), or points of origin, and located in the brachium or arm.

sternocleidomastoid—attached to the sternum, clavicle, and mastoid process.

external oblique—located near the outside, with fibers that run obliquely or in a slanting direction.

Fig. 9.27 Anterior view of superficial skeletal muscles.

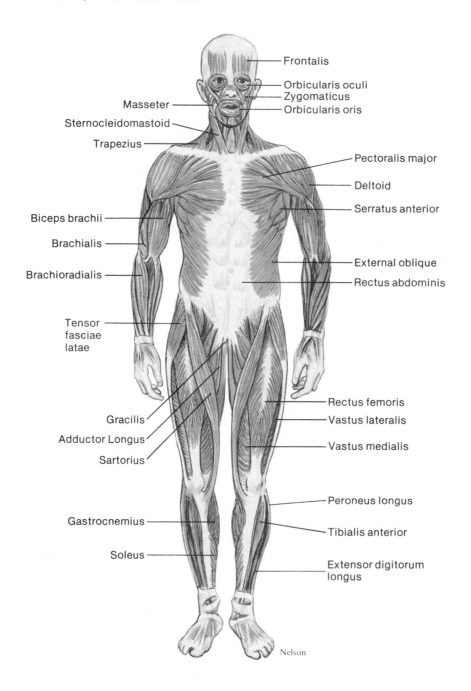

Frontalis

Orbicularis oculi
Zygomaticus
Orbicularis oris

Masseter

Sternocleidomastoid

Trapezius

Pectoralis major

Deltoid

Serratus anterior

Biceps brachii

Brachialis

Brachioradialis

External oblique

Rectus abdominis

Tensor
fasciae
latae

Rectus femoris

Vastus lateralis

Gracilis

Adductor Longus

Vastus medialis

Sartorius

Peroneus longus

Gastrocnemius

Tibialis anterior

Soleus

Extensor digitorum
longus

Nelson

Fig. 9.28 Posterior view of superficial skeletal muscles.

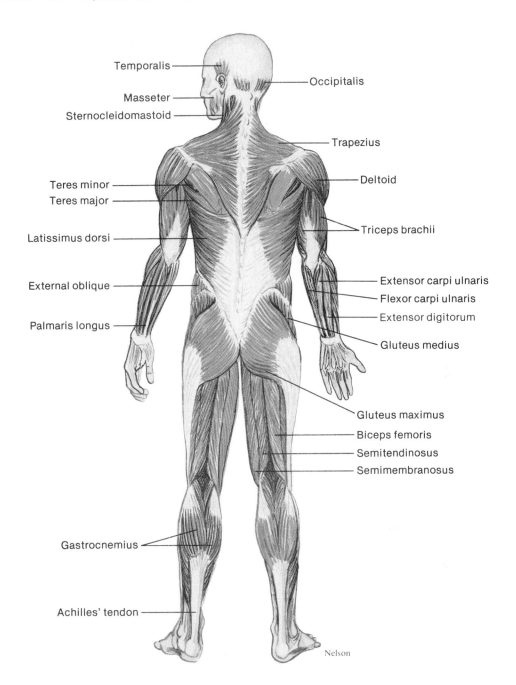

Temporalis

Masseter

Sternocleidomastoid

Occipitalis

Trapezius

Teres minor

Teres major

Deltoid

Latissimus dorsi

Triceps brachii

External oblique

Extensor carpi ulnaris

Palmaris longus

Flexor carpi ulnaris

Extensor digitorum

Gluteus medius

Gluteus maximus

Biceps femoris

Semitendinosus

Semimembranosus

Gastrocnemius

Achilles' tendon

Nelson

Muscles of Facial Expression

A number of small muscles that lie beneath the skin of the face and scalp enable us to communicate feelings through facial expression. Many of these muscles are located around the eyes and mouth, and they are responsible for such expressions as surprise, sadness, anger, fear, disgust, and pain. As a group, the muscles of facial expression connect the bones of the skull to connective tissue in various regions of the overlying skin. They include the following. (See fig. 9.29 and chart 9.3.)

Epicranius
Orbicularis oculi
Orbicularis oris
Buccinator
Zygomaticus
Platysma

The **epicranius** covers the upper part of the cranium and consists of two muscular parts—the *frontalis,* which lies over the frontal bone, and the *occipitalis,* which lies over the occipital bone. These parts are united by a broad, tendinous membrane called the *epicranial aponeurosis,* which covers the cranium like a cap. Contraction of the epicranius raises the eyebrows and causes the skin of the forehead to wrinkle horizontally, as when a person expresses surprise. Headaches often result from sustained contraction of this muscle.

The **orbicularis oculi** is a ringlike band of muscle, called a *sphincter muscle,* that surrounds the eye. It lies in the subcutaneous tissue of the eyelid and causes the eye to close or blink. At the same time it compresses the nearby tear or *lacrimal gland,* aiding the flow of tears over the surface of the eye. Contraction of the orbicularis oculi also causes the appearance of folds or crow's feet that radiate laterally from the corner of the eye.

Chart 9.3   Muscles of facial expression

Muscle	Origin	Insertion	Action
Epicranius	Occipital bone	Skin and muscles around eye	Raises eyebrow
Orbicularis oculi	Maxillary and frontal bones	Skin around eye	Closes eye
Orbicularis oris	Muscles near the mouth	Skin of central lip	Closes lips, protrudes lips
Buccinator	Outer surfaces of maxilla and mandible	Orbicularis oris	Compresses cheeks inward
Zygomaticus	Zygomatic bone	Orbicularis oris	Raises corner of mouth
Platysma	Fascia in upper chest	Lower border of mandible	Draws angle of mouth downward

Fig. 9.29   Muscles of expression and mastication.

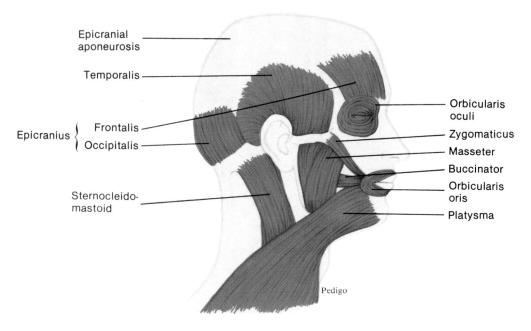

The **orbicularis oris** is a sphincter muscle that encircles the mouth. It lies between the skin and the mucous membranes of the lips, extending upward to the nose and downward to the region between the lower lip and chin. The orbicularis oris is sometimes called the kissing muscle because it causes the lips to close and pucker.

The **buccinator** is located in the wall of the cheek. Its fibers are directed forward from the bones of the jaws to the angle of the mouth, and when they contract the cheek is compressed inward. This action helps to hold food in contact with the teeth when a person is chewing. The buccinator also aids in blowing air out of the mouth, and for this reason, it is sometimes called the trumpeter muscle.

The **zygomaticus** extends from the zygomatic arch downward to the corner of the mouth. When it contracts, the corner of the mouth is drawn up, as in smiling or laughing.

The **platysma** is a thin, sheetlike muscle whose fibers extend from the chest upward over the neck to the face. It functions to pull the angle of the mouth downward, as in pouting. The platysma also helps to lower the mandible.

Muscles of Mastication

Chewing movements are produced by four pairs of muscles that are attached to the mandible. Two of these pairs act to close the lower jaw, as in biting, while the others cause side-to-side grinding motions as well as the movements of opening, closing, and protruding the jaw. The muscles of mastication include the following. (See figs. 9.29, 9.30, and chart 9.4.)

> *Masseter*
> *Temporalis*
> *Medial pterygoid*
> *Lateral pterygoid*

The **masseter** is a thick, flattened muscle that can be felt just in front of the ear when the teeth are clenched. Its fibers extend downward from the zygomatic arch to the mandible. The masseter functions primarily to raise the jaw, but also can control the rate at which the jaw falls open in response to gravity.

The **temporalis** is a fan-shaped muscle located on the side of the skull above and in front of the ear. Its fibers, which also raise the jaw, pass downward beneath the zygomatic arch to the mandible.

The **medial pterygoid** extends back and downward from the sphenoid, palatine, and maxilla bones to the ramus of the mandible. It functions to close the jaw and to pull it sideways in grinding movements.

The fibers of the **lateral pterygoid** are directed forward from the region just below the mandibular condyle to the sphenoid bone. This muscle aids in opening the mouth and can pull the jaw forward, making it protrude. (See fig. 9.30 and chart 9.4.)

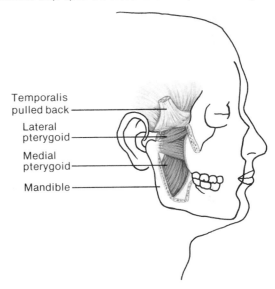

Fig. 9.30 The medial pterygoid muscles help close the jaw and can pull it to one side. The lateral pterygoid muscles help open the mouth and can protrude the jaw.

Temporalis pulled back

Lateral pterygoid

Medial pterygoid

Mandible

Chart 9.4 Muscles of mastication

Muscle	Origin	Insertion	Action
Masseter	Lower border of zygomatic arch	Lateral surface of mandible	Raises jaw
Temporalis	Temporal bone	Coronoid process and anterior ramus of mandible	Raises jaw
Medial pterygoid	Sphenoid, palatine, and maxilla bones	Medial surface of mandible	Closes jaw, pulls jaw sideways
Lateral pterygoid	Sphenoid bone	Anterior surface of mandibular condyle	Pulls jaw forward

Chart 9.5   Muscles that move the head

Muscle	Origin	Insertion	Action
Sternocleido-mastoid	Anterior surface of sternum and upper surface of clavicle	Mastoid process of temporal bone	Pulls head to one side, pulls head toward chest, or raises sternum
Splenius capitis	Spinous processes of lower cervical and upper thoracic vertebrae	Mastoid process of temporal bone	Rotates head, bends head to one side, or brings head into upright position
Semispinalis capitis	Processes of lower cervical and upper thoracic vertebrae	Occipital bone	Extends head, bends head to one side, or rotates head
Longissimus capitis	Processes of lower cervical and upper thoracic vertebrae	Mastoid process of temporal bone	Extends head, bends head to one side, or rotates head

Chart 9.6   Muscles that move the pectoral girdle

Muscle	Origin	Insertion	Action
Trapezius	Occipital bone and spines of cervical and thoracic vertebrae	Clavicle, spine, and acromion process of scapula	Raises scapula, pulls scapula medially, or pulls scapula and shoulder downward
Rhomboideus major	Spines of upper thoracic vertebrae	Medial border of scapula	Raises and adducts scapula
Levator scapulae	Transverse processes of cervical vertebrae	Medial margin of scapula	Elevates scapula
Serratus anterior	Outer surfaces of upper ribs	Medial border of scapula	Pulls scapula anteriorly and downward
Pectoralis minor	Sternal ends of upper ribs	Coracoid process of scapula	Pulls scapula forward and downward or raises ribs

Muscles That Move the Head

Head movements result from the actions of paired muscles in the neck and upper back. These muscles are responsible for flexing, extending, and rotating the head. They include the following. (See figs. 9.31, 9.33, and chart 9.5.)

Sternocleidomastoid

Splenius capitis

Semispinalis capitis

Longissimus capitis

The **sternocleidomastoid** is a long muscle in the side of the neck that extends upward from the thorax to the base of the skull behind the ear. When the sternocleidomastoid on one side contracts, the face is turned to the opposite side. When both muscles contract, the head is bent toward the chest. If the head is fixed in position by other muscles, the sternocleidomastoids can raise the sternum—an action that aids forceful inhalation.

The **splenius capitis** is a broad, straplike muscle located in the back of the neck. It connects the base of the skull to the vertebrae in the neck and upper thorax. A splenius capitus acting singly causes the head to rotate and bend toward one side. Acting together, these muscles bring the head into an upright position.

The **semispinalis capitis** is a broad, sheetlike muscle extending upward from the vertebrae in the neck and thorax to the occipital bone. It functions to extend the head, bend it to one side, or rotate it.

The **longissimus capitis** is a narrow band of muscle that ascends from the vertebrae of the neck and thorax to the temporal bone. It also functions to extend the head, bend it to one side, or rotate it.

Muscles That Move the Pectoral Girdle

The muscles that move the pectoral girdle are closely associated with those that move the upper arm. A number of these chest and shoulder muscles connect the scapula to nearby bones and act to move the scapula upward, downward, forward, and backward. They include the following. (See figs. 9.32, 9.33, and chart 9.6.)

Trapezius

Rhomboideus major

Levator scapulae

Serratus anterior

Pectoralis minor

Fig. 9.31 Deep muscles in the back of the neck help to move the head. (The splenius capitis is removed on the left to show the underlying muscles.)

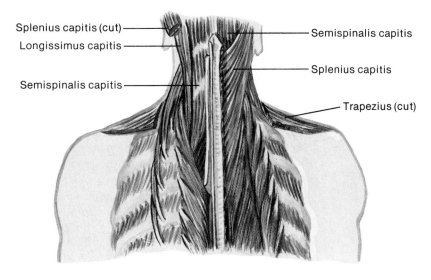

Splenius capitis (cut)

Longissimus capitis

Semispinalis capitis

Semispinalis capitis

Splenius capitis

Trapezius (cut)

Fig. 9.32 Muscles of the posterior shoulder. (The trapezius is removed on the right to show underlying muscles.)

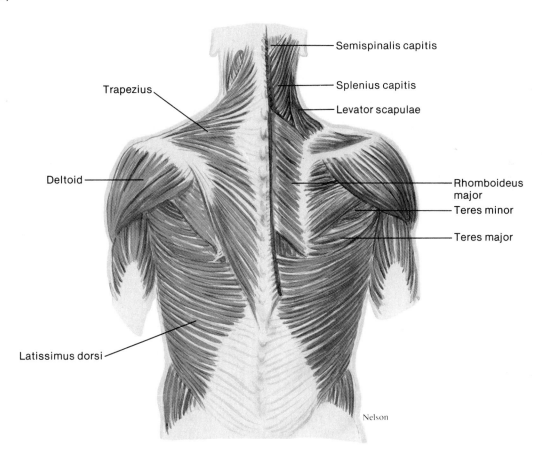

Semispinalis capitis

Trapezius

Splenius capitis

Levator scapulae

Deltoid

Rhomboideus major

Teres minor

Teres major

Latissimus dorsi

Nelson

The **trapezius** is a large, triangular muscle in the upper back that extends horizontally from the base of the skull and vertebral column to the shoulder. Its fibers are arranged in three groups—upper, middle, and lower. The upper fibers raise the scapula and shoulder, as when the shoulders are shrugged to express a feeling of indifference. The middle fibers pull the scapula toward the vertebral column, and the lower fibers draw the scapula and shoulder downward. When the shoulder is fixed in position by other muscles, the trapezius can pull the head backward or to one side.

The **rhomboideus major** connects the upper thoracic vertebrae to the scapula. It acts to raise the scapula and to adduct it.

The **levator scapulae** is a straplike muscle that runs almost vertically through the neck, connecting the cervical vertebrae to the scapula. It functions to elevate the scapula.

The **serratus anterior** is a broad, curved muscle located on the side of the chest. It arises as fleshy slips on the upper ribs and extends along the medial wall of the axilla to the scapula. It functions to pull the scapula downward and anteriorly and is used to thrust the shoulder forward, as when pushing something.

The **pectoralis minor** is a thin, flat muscle that lies beneath the larger pectoralis major. It extends laterally and upward from the ribs to the scapula and serves to pull the scapula forward and downward. When other muscles fix the scapula in position, the pectoralis minor can raise the ribs and thus aid forceful inhalation.

Fig. 9.33 Muscles of the anterior chest. (The left pectoralis major is removed to show the pectoralis minor.)

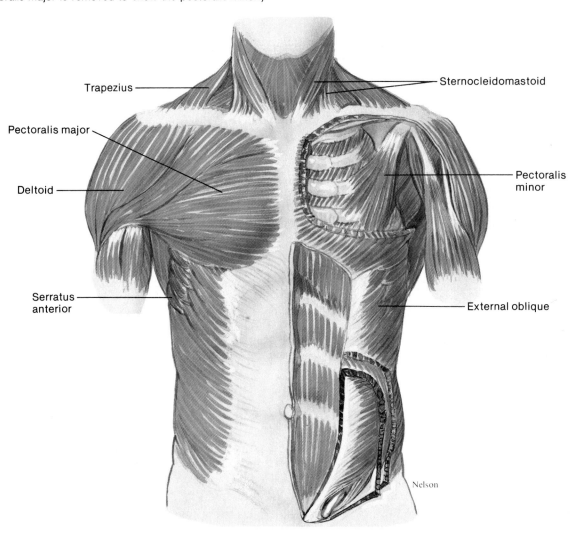

Trapezius

Pectoralis major

Deltoid

Serratus anterior

Sternocleidomastoid

Pectoralis minor

External oblique

Nelson

Muscles That Move the Upper Arm

The upper arm is one of the more freely movable parts of the body. Its many movements are made possible by muscles that connect the humerus to various regions of the pectoral girdle, ribs, and vertebral column. (See figs. 9.32, 9.33, 9.34, and 9.35.) These muscles can be grouped according to their primary actions—flexion, extension, abduction, and rotation—as follows. (See chart 9.7.)

Flexors
  Coracobrachialis
  Pectoralis major
Extensors
  Teres major
  Latissimus dorsi
Abductors
  Supraspinatus
  Deltoid
Rotators
  Subscapularis
  Infraspinatus
  Teres minor

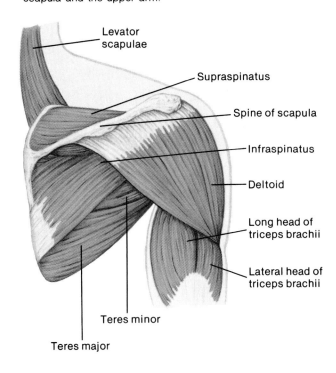

Fig. 9.34 Muscles of the posterior surface of the scapula and the upper arm.

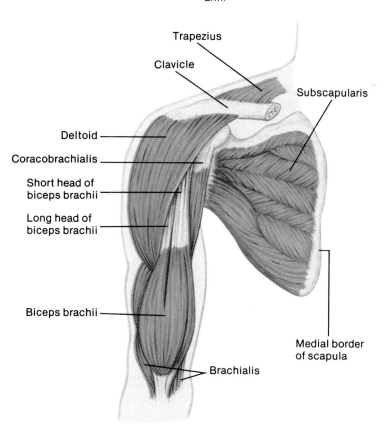

Fig. 9.35 Muscles of the anterior shoulder and the upper arm.

Chart 9.7 Muscles that move the upper arm

Muscle	Origin	Insertion	Action
Coracobrachialis	Coracoid process of scapula	Shaft of humerus	Flexes and adducts the upper arm
Pectoralis major	Clavicle, sternum, and costal cartilages of upper ribs	Intertubercular groove of humerus	Pulls arm forward and across chest, rotates humerus, or adducts arm
Teres major	Lateral border of scapula	Intertubercular groove of humerus	Extends humerus, or adducts and rotates arm medially
Latissimus dorsi	Spines of sacral, lumbar, and lower thoracic vertebrae, iliac crest, and lower ribs	Intertubercular groove of humerus	Extends, adducts, and rotates humerus medially, or pulls the shoulder downward and back
Supraspinatus	Posterior surface of scapula	Greater tubercle of humerus	Abducts the upper arm
Deltoid	Acromion process, spine of the scapula, and the clavicle	Deltoid tuberosity of humerus	Abducts upper arm, extends humerus, or flexes humerus
Subscapularis	Anterior surface of scapula	Lesser tubercle of humerus	Rotates arm medially
Infraspinatus	Posterior surface of scapula	Greater tubercle of humerus	Rotates arm laterally
Teres minor	Lateral border of scapula	Greater tubercle of humerus	Rotates arm laterally

Flexors. The **coracobrachialis** extends from the scapula to the middle of the humerus along its medial surface. It acts to flex and adduct the upper arm.

The **pectoralis major** is a thick, fan-shaped muscle located in the upper chest. Its fibers extend from the center of the thorax through the armpit to the humerus. This muscle functions primarily to pull the upper arm forward and across the chest. It can also rotate the humerus medially and adduct the arm from a raised position.

Extensors. The **teres major** connects the scapula to the humerus. It extends the humerus and can also adduct and rotate the upper arm medially.

The **latissimus dorsi** is a wide, triangular muscle that curves upward from the lower back, around the side, and to the armpit. It can extend, adduct, and rotate the humerus medially. It also acts to pull the shoulder downward and back. This muscle is used to pull the arm back in swimming, climbing, and rowing movements.

Abductors. The **supraspinatus** is located in the depression above the spine of the scapula on its posterior surface. It connects the scapula to the greater tubercle of the humerus and acts to abduct the upper arm.

The **deltoid** is a thick, triangular muscle that covers the shoulder joint. It connects the clavicle and scapula to the lateral side of the humerus and acts to abduct the upper arm. The deltoid's posterior fibers can extend the humerus, and its anterior fibers can flex the humerus.

Rotators. The **subscapularis** is a large, triangular muscle that covers the anterior surface of the scapula. It connects the scapula to the humerus and functions to rotate the upper arm medially.

The **infraspinatus** occupies the depression below the spine of the scapula on its posterior surface. The fibers of this muscle attach the scapula to the humerus and act to rotate the upper arm laterally.

The **teres minor** is a small muscle connecting the scapula to the humerus. It acts to rotate the upper arm laterally.

Muscles That Move the Forearm

Most forearm movements are produced by muscles that connect the radius or ulna to the humerus or pectoral girdle. A group of muscles located along the anterior surface of the humerus act to flex the elbow, while a single posterior muscle serves to extend this joint. Other muscles cause movements at the radioulnar joint and are responsible for rotating the forearm.

The muscles that move the forearm include the following. (See figs. 9.34, 9.35, 9.36, and chart 9.8.)

Flexors
Biceps brachii
Brachialis
Brachioradialis
Extensor
Triceps brachii
Rotators
Supinator
Pronator teres
Pronator quadratus

Chart 9.8   Muscles that move the forearm

Muscle	Origin	Insertion	Action
Biceps brachii	Coracoid process and tubercle above glenoid cavity of scapula	Radial tuberosity of radius	Flexes arm at elbow and rotates the hand laterally
Brachialis	Anterior shaft of humerus	Coronoid process of ulna	Flexes arm at elbow
Brachio-radialis	Distal lateral end of humerus	Lateral surface of radius above styloid process	Flexes arm at elbow
Triceps brachii	Tubercle below glenoid cavity and lateral and medial surfaces of humerus	Olecranon process of ulna	Extends arm at elbow
Supinator	Lateral epicondyle of humerus and crest of ulna	Lateral surface of radius	Rotates forearm laterally
Pronator teres	Medial epicondyle of humerus	Lateral surface of radius	Rotates arm medially
Pronator quadratus	Anterior distal end of ulna	Anterior distal end of radius	Rotates arm medially

Flexors. The **biceps brachii** is a fleshy muscle that forms a long, rounded mass on the anterior side of the upper arm. It connects the scapula to the radius and functions to flex the arm at the elbow and to rotate the hand laterally (supination), as when a person turns a doorknob or screwdriver.

The **brachialis** is a large muscle beneath the biceps brachii. It connects the shaft of the humerus to the ulna and is the strongest flexor of the elbow.

The **brachioradialis** connects the humerus to the radius and aids in flexing the elbow.

Extensor. The **triceps brachii** has three heads and is the only muscle on the back of the upper arm. It connects the humerus and scapula to the ulna and is the primary extensor of the elbow.

Rotators. The **supinator** is a short muscle whose fibers run from the ulna and the lateral end of the humerus to the radius. It assists the biceps brachii in rotating the forearm laterally (supination).

The **pronator teres** is a short muscle connecting the medial end of the humerus to the radius. It functions to rotate the arm medially, as when the hand is turned so the palm is facing downward (pronation).

The **pronator quadratus** runs from the distal end of the ulna to the distal end of the radius. It assists the pronator teres in rotating the arm medially.

Muscles That Move the Wrist, Hand, and Fingers

Many muscles are responsible for wrist, hand, and finger movements. They originate from the distal end of the humerus and from the radius and ulna. The two major groups of these muscles are flexors on the anterior side of the forearm and extensors on the posterior side. These muscles include the following. (See figs. 9.36, 9.37, and chart 9.9.)

> Flexors
>> *Flexor carpi radialis*
>> *Flexor carpi ulnaris*
>> *Palmaris longus*
>> *Flexor digitorum profundus*
>
> Extensors
>> *Extensor carpi radialis longus*
>> *Extensor carpi radialis brevis*
>> *Extensor carpi ulnaris*
>> *Extensor digitorum*

Flexors. The **flexor carpi radialis** is a fleshy muscle that runs medially on the anterior side of the forearm. It extends from the distal end of the humerus into the hand, where it is attached to metacarpal bones. The flexor carpi radialis functions to flex and abduct the wrist.

The **flexor carpi ulnaris** is located along the medial border of the forearm. It connects the distal end of the humerus and the proximal end of the ulna to carpal and metacarpal bones. It serves to flex and adduct the wrist.

The **palmaris longus** is a slender muscle located on the medial side of the forearm between the flexor carpi radialis and the flexor carpi ulnaris. It connects the distal end of the humerus to fascia of the palm and functions to flex the wrist.

The **flexor digitorum profundus** is a large muscle that connects the ulna to the distal phalanges. It acts to flex the distal joints of the fingers, as when a fist is made.

Fig. 9.36 Muscles of the anterior forearm.

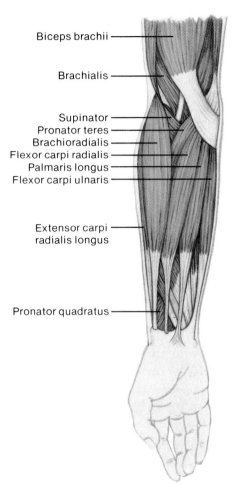

Biceps brachii

Brachialis

Supinator
Pronator teres
Brachioradialis
Flexor carpi radialis
Palmaris longus
Flexor carpi ulnaris

Extensor carpi
radialis longus

Pronator quadratus

Fig. 9.37 Muscles of the posterior forearm.

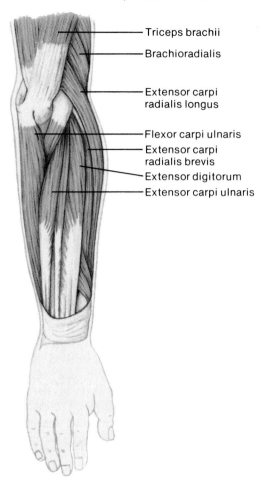

Triceps brachii

Brachioradialis

Extensor carpi
radialis longus

Flexor carpi ulnaris
Extensor carpi
radialis brevis
Extensor digitorum
Extensor carpi ulnaris

Chart 9.9 Muscles that move the wrist, hand, and fingers

Muscle	Origin	Insertion	Action
Flexor carpi radialis	Medial epicondyle of humerus	Base of second and third metacarpals	Flexes and abducts wrist
Flexor carpi ulnaris	Medial epicondyle of humerus and olecranon process	Carpal and metacarpal bones	Flexes and adducts wrist
Palmaris longus	Medial epicondyle of humerus	Fascia of palm	Flexes wrist
Flexor digitorum profundus	Anterior surface of ulna	Bases of distal phalanges in fingers two through five	Flexes distal joints of fingers
Extensor carpi radialis longus	Distal end of humerus	Base of second metacarpal	Extends wrist and abducts hand
Extensor carpi radialis brevis	Lateral epicondyle of humerus	Base of second and third metacarpals	Extends wrist and abducts hand
Extensor carpi ulnaris	Lateral epicondyle of humerus	Base of fifth metacarpal	Extends and adducts wrist
Extensor digitorum	Lateral epicondyle of humerus	Posterior surface of phalanges in fingers two through five	Extends fingers

Fig. 9.38 Muscles of the abdominal wall.

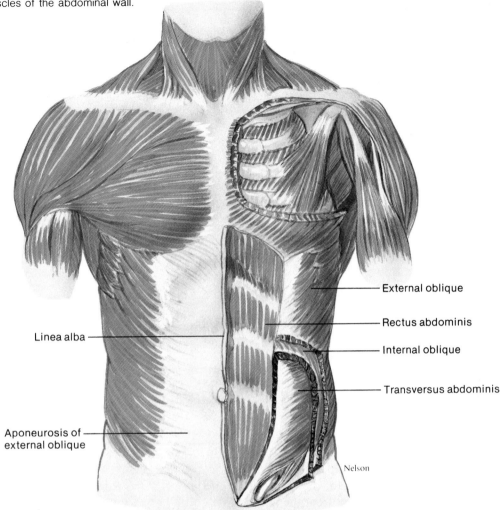

Linea alba

Aponeurosis of
external oblique

External oblique

Rectus abdominis

Internal oblique

Transversus abdominis

Nelson

Extensors. The **extensor carpi radialis longus** runs along the lateral side of the forearm, connecting the humerus to the hand. It functions to extend the wrist and assists in abducting the hand.

The **extensor carpi radialis brevis** is a companion of the extensor carpi radialis longus and is located medially to it. This muscle runs from the humerus to metacarpal bones and functions to extend the wrist. It also assists in abducting the hand.

The **extensor carpi ulnaris** is located along the posterior surface of the ulna and connects the humerus to the hand. It acts to extend the wrist and assists in adducting it.

The **extensor digitorum** runs medially along the back of the forearm. It connects the humerus to the posterior surface of the phalanges and functions to extend the fingers.

Muscles of the Abdominal Wall

Although the walls of the chest and pelvic regions are supported directly by bone, those of the abdomen are not. Instead, the anterior and lateral walls of the abdomen are composed of broad, flattened muscles arranged in layers. These muscles connect the rib cage and vertebral column to the pelvic girdle. A band of tough connective tissue, called the **linea alba,** extends from the xiphoid process of the sternum to the symphysis pubis. It serves as an attachment for some of the abdominal wall muscles.

Contraction of these muscles decreases the size of the abdominal cavity and increases the pressure inside. This action helps to press air out of the lungs during forceful exhalation, and also aids in defecation, urination, vomiting, and childbirth.

The abdominal wall muscles (fig. 9.38 and chart 9.10) include the following:

> *External oblique*
> *Internal oblique*
> *Transversus abdominis*
> *Rectus abdominis*

Chart 9.10 Muscles of the abdominal wall

Muscle	Origin	Insertion	Action
External oblique	Outer surfaces of lower ribs	Outer lip of iliac crest and linea alba	Tenses abdominal wall and compresses abdominal contents
Internal oblique	Crest of ilium and inguinal ligament	Cartilages of lower ribs, linea alba, and crest of pubis	Same as above
Transversus abdominis	Costal cartilages of lower ribs, processes of lumbar vertebrae, lip of iliac crest, and inguinal ligament	Linea alba and crest of pubis	Same as above
Rectus abdominis	Crest of pubis and symphysis pubis	Xiphoid process of sternum and costal cartilages	Same as above, also flexes vertebral column

Chart 9.11 Muscles of the pelvic outlet

Muscle	Origin	Insertion	Action
Levator ani	Pubic bone and ischial spine	Coccyx	Supports pelvic viscera, and provides sphincterlike action in anal canal and vagina
Coccygeus	Ischial spine	Sacrum and coccyx	Same as above
Superficial transversus perinei	Ischial tuberosity	Central tendon	Supports pelvic viscera
Bulbospongiosus	Central tendon	Males: urogenital diaphragm and fascia of penis	Males: assists emptying urethra
		Females: pubic arch and root of clitoris	Females: constricts vagina
Ischiocavernosus	Ischial tuberosity	Pubic arch	Assists the function of the bulbospongiosus

The **external oblique** is a broad, thin sheet of muscle whose fibers slant downward from the lower ribs to the pelvic girdle and the linea alba. When this muscle contracts, it tenses the abdominal wall and compresses the contents of the abdominal cavity.

Similarly, the **internal oblique** is a broad, thin sheet of muscle located beneath the external oblique. Its fibers run up and forward from the pelvic girdle to the lower ribs. Its function is similar to that of the external oblique.

The **transversus abdominis** forms a third layer of muscle beneath the external and internal obliques. Its fibers run horizontally from the lower ribs, lumbar vertebrae, and ilium to the linea alba and pubic bones. It functions in the same manner as the external and internal obliques.

The **rectus abdominis** is a long, straplike muscle that connects the pubic bones to the ribs and sternum. It functions with the other abdominal wall muscles to compress the contents of the abdominal cavity, and it also helps to flex the vertebral column.

Muscles of the Pelvic Outlet

The outlet of the pelvis is spanned by two muscular sheets—a deeper **pelvic diaphragm** and a more superficial **urogenital diaphragm**. The pelvic diaphragm forms the floor of the pelvic cavity, and the urogenital diaphragm fills the space within the pubic arch. The muscles of the male and female pelvic outlets include the following. (See figs. 9.39, 9.40, and chart 9.11.)

Pelvic diaphragm
Levator ani
Coccygeus
Urogenital diaphragm
Superficial transversus perinei
Bulbospongiosus
Ischiocavernosus

Pelvic Diaphragm. The **levator ani** muscles form a thin sheet across the pelvic outlet. They are connected at the midline posteriorly by a ligament that extends from the tip of the coccyx to the anal canal. Anteriorly, they are separated in the male by the urethra

Fig. 9.39 Muscles of the male pelvic outlet.

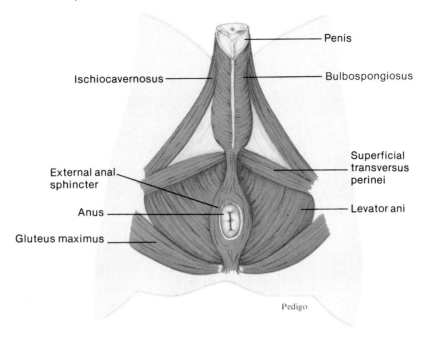

Penis

Ischiocavernosus

Bulbospongiosus

External anal sphincter

Superficial transversus perinei

Anus

Levator ani

Gluteus maximus

Pedigo

Fig. 9.40 Muscles of the female pelvic outlet.

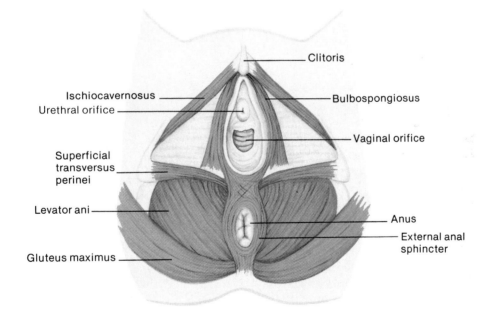

Clitoris

Ischiocavernosus
Urethral orifice

Bulbospongiosus

Vaginal orifice

Superficial transversus perinei

Levator ani

Anus

External anal sphincter

Gluteus maximus

Fig. 9.41 Muscles of the anterior right thigh.

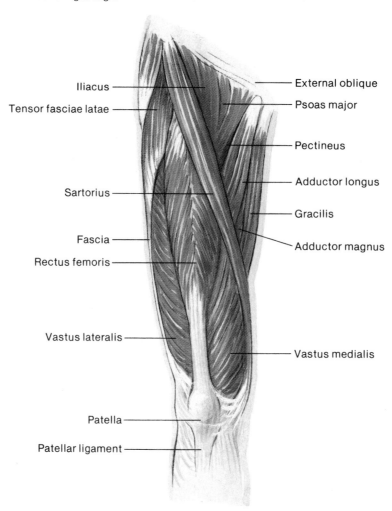

Iliacus

Tensor fasciae latae

Sartorius

Fascia

Rectus femoris

Vastus lateralis

Patella

Patellar ligament

External oblique

Psoas major

Pectineus

Adductor longus

Gracilis

Adductor magnus

Vastus medialis

and the anal canal; and in the female by the urethra, vagina, and anal canal. These muscles help to support the pelvic viscera and provide sphincterlike action in the anal canal and vagina.

The **coccygeus** is a fan-shaped muscle that extends from the ischial spine to the coccyx and sacrum. It aids the levator ani in its functions.

Urogenital Diaphragm. The **superficial transversus perinei** consists of a small bundle of muscle fibers that passes medially from the ischial tuberosity along the posterior border of the urogenital diaphragm. It assists other muscles in supporting the pelvic viscera.

In males, the **bulbospongiosus** muscles are united and surround the base of the penis. They assist in emptying the urethra. In females, these muscles are separated medially by the vagina and act to constrict the vaginal opening.

The **ischiocavernosus muscle** is a tendinous structure that extends from the ischial tuberosity to the margin of the pubic arch. It assists the function of the bulbospongiosus muscle.

Muscles That Move the Thigh

The muscles that move the thigh are attached to the femur and to some part of the pelvic girdle. They can be separated into anterior and posterior groups. The muscles of the anterior group act primarily to flex the thigh; those of the posterior group extend, abduct, or rotate it. The muscles in these groups (see figs. 9.41, 9.42, 9.43, and chart 9.12,) include the following:

Anterior group
Psoas major
Iliacus
Posterior group
Gluteus maximus
Gluteus medius
Gluteus minimus
Tensor fasciae latae

Fig. 9.42 Muscles of the lateral right thigh.

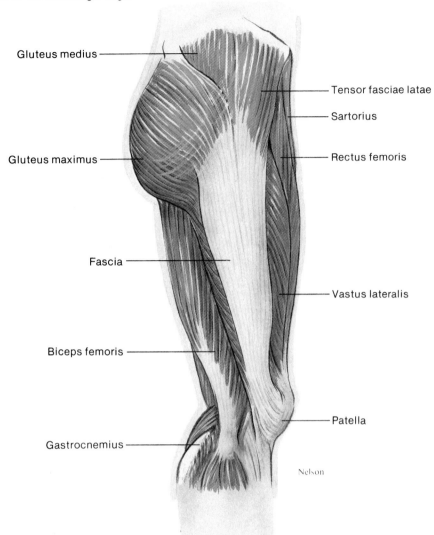

Gluteus medius

Tensor fasciae latae

Sartorius

Gluteus maximus

Rectus femoris

Fascia

Vastus lateralis

Biceps femoris

Patella

Gastrocnemius

Nelson

Chart 9.12   Muscles that move the thigh

Muscle	Origin	Insertion	Action
Psoas major	Lumbar intervertebral disks, bodies and transverse processes of lumbar vertebrae	Lesser trochanter of femur	Flexes thigh
Iliacus	Iliac fossa of ilium	Lesser trochanter of femur	Flexes thigh
Gluteus maximus	Sacrum, coccyx, and posterior surface of ilium	Posterior surface of femur and iliotibial tract	Extends leg at hip
Gluteus medius	Lateral surface of ilium	Greater trochanter of femur	Abducts and rotates thigh medially
Gluteus minimus	Lateral surface of ilium	Greater trochanter of femur	Same as gluteus medius
Tensor fasciae latae	Anterior iliac crest	Lateral condyle of tibia near fascia	Abducts, flexes, and rotates thigh medially
Adductor longus	Pubic bone near symphysis pubis	Posterior surface of femur	Adducts, flexes, and rotates thigh laterally
Adductor magnus	Ischial tuberosity	Posterior surface of femur	Adducts, extends, and rotates thigh laterally
Gracilis	Lower edge of symphysis pubis	Medial surface of tibia	Adducts thigh, flexes, and rotates leg medially at knee

Fig. 9.43 Muscles of the posterior right thigh.

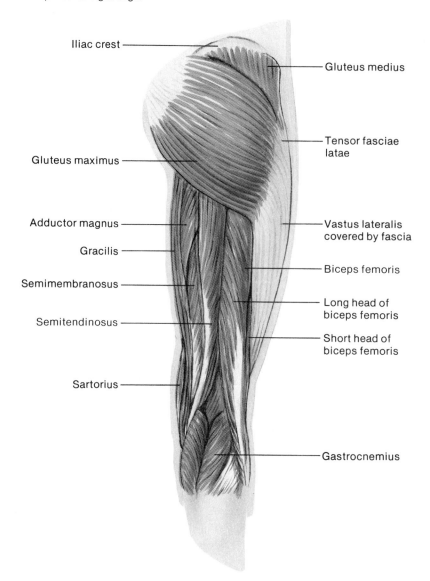

Iliac crest

Gluteus maximus

Adductor magnus

Gracilis

Semimembranosus

Semitendinosus

Sartorius

Gluteus medius

Tensor fasciae latae

Vastus lateralis covered by fascia

Biceps femoris

Long head of biceps femoris

Short head of biceps femoris

Gastrocnemius

Still another group of muscles attached to the femur and pelvic girdle functions to adduct the thigh. This group includes:

Adductor longus

Adductor magnus

Gracilis

Anterior Group. The **psoas major** is a long, thick muscle that connects the lumbar vertebrae to the femur. It functions to flex the thigh.

The **iliacus,** a large, fan-shaped muscle, lies along the lateral side of the psoas major. The iliacus and the psoas major are the primary flexors of the thigh, and they serve to advance the leg in walking movements.

Posterior Group. The **gluteus maximus** is the heaviest muscle in the body and covers a large part of the buttock. It connects the ilium, sacrum, and coccyx to the femur and acts to extend the thigh. The gluteus maximus causes the leg to straighten at the hip when a person walks, runs, or climbs. It is also used to raise the body from a sitting position.

The **gluteus medius** is partly covered by the gluteus maximus. Its fibers extend from the ilium to the femur, and they function to abduct the thigh and rotate it medially.

The **gluteus minimus** lies beneath the gluteus medius and is its companion in attachments and functions.

The **tensor fasciae latae** connects the ilium to the fascia of the thigh, which continues downward to the tibia. This muscle functions to abduct and flex the thigh and to rotate it medially.

Chart 9.13 Muscles that move the lower leg

Muscle	Origin	Insertion	Action
Hamstring group			
Biceps femoris	Ischial tuberosity and posterior surface of femur	Head of fibula and lateral condyle of tibia	Flexes and rotates leg laterally and extends thigh
Semitendinosus	Ischial tuberosity	Medial surface of tibia	Flexes and rotates leg medially and extends thigh
Semimembranosus	Ischial tuberosity	Medial condyle of tibia	Flexes and rotates leg medially and extends thigh
Sartorius	Spine of ilium	Medial surface of tibia	Flexes leg and thigh, abducts thigh, rotates thigh laterally, and rotates leg medially
Quadriceps femoris group		Patella by common tendon, which continues as patellar ligament to tibial tuberosity	Extends leg at knee
Rectus femoris	Spine of ilium and margin of acetabulum		
Vastus lateralis	Greater trochanter and posterior surface of femur		
Vastus medialis	Medial surface of femur		
Vastus intermedius	Anterior and lateral surfaces of femur		

Thigh Adductors. The **adductor longus** is a long, triangular muscle that runs from the pubic bone to the femur. It functions to adduct the thigh and assists in flexing and rotating it laterally.

The **adductor magnus** is the largest adductor of the thigh. It is a triangular muscle that connects the ischium to the femur. It adducts the thigh and assists in extending and rotating it laterally.

The **gracilis** is a long, straplike muscle that passes from the pubic bone to the tibia. It functions to adduct the thigh and to flex and rotate the leg medially at the knee.

Muscles That Move the Lower Leg

The muscles that move the lower leg connect the tibia or fibula to the femur or the pelvic girdle. (See figs. 9.41, 9.42, and 9.43, and chart 9.13.) They can be separated into two major groups—those that cause flexion at the knee and those that cause extension at the knee. The muscles of these groups include the following:

> Flexors
> *Biceps femoris*
> *Semitendinosus*
> *Semimembranosus*
> *Sartorius*
> Extensor
> *Quadriceps femoris*

Flexors. As its name implies, the **biceps femoris** has two heads, one attached to the ischium and the other attached to the femur. This muscle passes along the back of the thigh on the lateral side and connects to the proximal ends of the fibula and tibia. The biceps femoris is one of the hamstring muscles, and its tendon (hamstring) can be felt as a lateral ridge behind the knee. This muscle functions to flex and rotate the leg laterally and to extend the thigh.

The **semitendinosus** is another of the hamstring muscles. It is a long, bandlike muscle on the back of the thigh toward the medial side, connecting the ischium to the proximal end of the tibia. The semitendinosus is so named because it becomes tendinous in the middle of the thigh, continuing to its insertion as a long, cordlike tendon. It functions to flex and rotate the leg medially and to extend the thigh.

The **semimembranosus** is the third hamstring muscle and is the most medially located muscle in the back of the thigh. It connects the ischium to the tibia and functions to flex and rotate the leg medially and to extend the thigh.

The **sartorius** is an elongated straplike muscle that passes obliquely across the front of the thigh and then descends over the medial side of the knee. It connects the ilium to the tibia and functions to flex the leg and the thigh. It can also abduct the thigh, rotate it laterally, and assist in rotating the leg medially.

Fig. 9.44 Muscles of the anterior right lower leg.

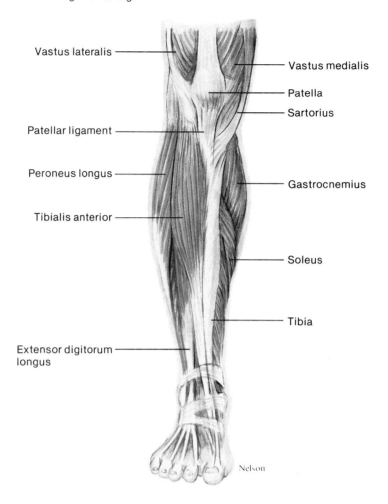

Vastus lateralis

Vastus medialis

Patella

Sartorius

Patellar ligament

Peroneus longus

Gastrocnemius

Tibialis anterior

Soleus

Tibia

Extensor digitorum longus

Nelson

Extensor. The large, fleshy muscle called **quadriceps femoris** occupies the front and sides of the thigh and is the primary extensor of the knee. It is composed of four parts—*rectus femoris, vastus lateralis, vastus medialis,* and *vastus intermedius.* These parts connect the ilium and femur to a common *patellar tendon,* which passes over the front of the knee and attaches to the patella. This tendon then continues as the *patellar ligament* to the tibia.

The muscles that move the lower leg are summarized in chart 9.13.

Muscles That Move the Ankle, Foot, and Toes

A number of muscles that function to move the ankle, foot, and toes are located in the lower leg. (See figs. 9.44, 9.45, 9.46, and chart 9.14.) They attach the femur, tibia, and fibula to various bones of the foot and are responsible for a variety of movements—moving the foot upward (dorsal flexion) or downward (plantar flexion), and turning the sole of the foot inward (inversion) or outward (eversion). These muscles include the following:

Dorsal flexors
Tibialis anterior
Peroneus tertius
Extensor digitorum longus
Plantar flexors
Gastrocnemius
Soleus
Flexor digitorum longus
Invertor
Tibialis posterior
Evertor
Peroneus longus

Fig. 9.45 Muscles of the lateral right lower leg.

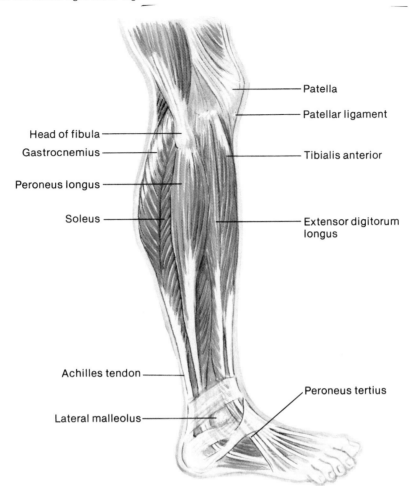

Patella

Patellar ligament

Head of fibula

Gastrocnemius

Peroneus longus

Soleus

Tibialis anterior

Extensor digitorum longus

Achilles tendon

Lateral malleolus

Peroneus tertius

Chart 9.14   Muscles that move the ankle, foot, and toes

Muscle	Origin	Insertion	Action
Tibialis anterior	Lateral condyle and lateral surface of tibia	Tarsal bone (cuneiform) and first metatarsal	Dorsal flexion and inversion of foot
Peroneus tertius	Anterior surface of fibula	Dorsal surface of fifth metatarsal	Dorsal flexion and eversion of foot
Extensor digitorum longus	Lateral condyle of tibia and anterior surface of fibula	Dorsal surfaces of second and third phalanges of four lateral toes	Dorsal flexion and eversion of foot and extension of toes
Gastrocnemius	Lateral and medial condyles of femur	Posterior surface of calcaneus	Plantar flexion of foot and flexion of leg at knee
Soleus	Head and shaft of fibula and posterior surface of tibia	Posterior surface of calcaneus	Plantar flexion of foot
Flexor digitorum longus	Posterior surface of tibia	Distal phalanges of four lateral toes	Plantar flexion and inversion of foot and flexion of four lateral toes
Tibialis posterior	Lateral condyle and posterior surface of tibia and posterior surface of fibula	Tarsal and metatarsal bones	Plantar flexion and inversion of foot
Peroneus longus	Lateral condyle of tibia and head and shaft of fibula	Tarsal and metatarsal bones	Plantar flexion and eversion of foot; also supports arch

Fig. 9.46 Muscles of the medial right lower leg.

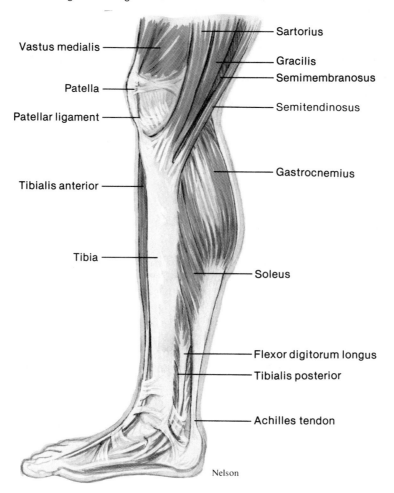

- Vastus medialis
- Patella
- Patellar ligament
- Tibialis anterior
- Tibia
- Sartorius
- Gracilis
- Semimembranosus
- Semitendinosus
- Gastrocnemius
- Soleus
- Flexor digitorum longus
- Tibialis posterior
- Achilles tendon

Nelson

Dorsal Flexors. The **tibialis anterior** is an elongated, spindle-shaped muscle located on the front of the lower leg. It arises from the surface of the tibia, passes medially over the distal end of the tibia, and attaches to bones of the ankle and foot. Contraction of the tibialis anterior causes dorsal flexion and inversion of the foot.

The **peroneus tertius** is a muscle of variable size that connects the fibula to the lateral side of the foot. It functions in dorsal flexion and eversion of the foot.

The **extensor digitorum longus** is situated along the lateral side of the lower leg just behind the tibialis anterior. It arises from the proximal end of the tibia and the shaft of the fibula. Its tendon divides into four parts as it passes over the front of the ankle. These parts continue over the surface of the foot and attach to the four lateral toes. The actions of the extensor digitorum longus include dorsal flexion of the foot, eversion of the foot, and extension of the toes.

Plantar Flexors. The **gastrocnemius,** on the back of the lower leg, forms part of the calf. It arises by two heads from the femur. The distal end of this muscle joins the strong *Achilles tendon,* which descends to the heel and attaches to the calcaneus. The gastrocnemius is a powerful plantar flexor of the foot, that aids in pushing the body forward when a person walks or runs. It also functions to flex the leg at the knee.

The **soleus** is a thick, flat muscle located beneath the gastrocnemius. These two muscles make up the calf of the leg. The soleus arises from the tibia and fibula, and it extends to the heel by way of the Achilles tendon. It acts with the gastrocnemius to cause plantar flexion of the foot.

The **flexor digitorum longus** extends from the posterior surface of the tibia to the foot. Its tendon passes along the plantar surface of the foot. There it divides into four parts that attach to the terminal bones of the four lateral toes. This muscle assists in plantar flexion of the foot, flexion of the four lateral toes, and inversion of the foot.

Invertor. The **tibialis posterior** is the deepest of the muscles on the back of the lower leg. It connects the fibula and tibia to ankle bones by means of a tendon that curves under the medial malleolus. This muscle assists in inversion and plantar flexion of the foot.

Evertor. The **peroneus longus** is a long, straplike muscle located on the lateral side of the lower leg. It connects the tibia and the fibula to the foot by means of a stout tendon that passes behind the lateral malleolus. It functions in eversion of the foot, assists in plantar flexion, and helps support the arch of the foot.

Clinical Terms Related to the Muscular System

convulsion (kun-vul'shun)—involuntary contraction of muscles.

electromyography (e-lek''tro-mi-og'rah-fe)—technique for recording the electrical changes that occur in muscle tissues.

fibrillation (fi''brĭ-la'shun)—spontaneous contractions of individual muscle fibers, producing rapid and uncoordinated activity within a muscle.

fibrosis (fi-bro'sis)—degenerative disease in which skeletal muscle tissue is replaced by fibrous connective tissue.

fibrositis (fi''bro-si'tis)—inflammatory condition of fibrous connective tissues, especially in the muscle fascia. This disease is also called muscular rheumatism.

muscular dystrophy (mus'ku-lar dis'tro-fe)—a progressively crippling disease of unknown cause in which the muscles gradually weaken and atrophy.

myalgia (mi-al'je-ah)—pain resulting from any muscular disease or disorder.

myasthenia gravis (mi''as-the'ne-ah gra'vis)—chronic disease characterized by muscles that are weak and easily fatigued. It results from a disorder at some of the neuromuscular junctions so that a stimulus is not transmitted from the motor neuron to the muscle fiber.

myokymia (mi''o-ki'me-ah)—persistent quivering of a muscle.

myology (mi-ol'o-je)—the study of muscles.

myoma (mi-o'mah)—a tumor composed of muscle tissue.

myopathy (mi-op'ah-the)—any muscular disease.

myositis (mi''o-si'tis)—inflammation of skeletal muscle tissue.

myotomy (mi-ot'o-me)—the cutting of muscle tissue.

myotonia (mi''o-to'ne-ah)—prolonged muscular spasm.

paralysis (pah-ral'ĭ-sis)—loss of ability to move a body part.

shin splints (shin splints)—a soreness on the front of the lower leg due to straining the flexor digitorum longus, often as a result of walking up and down hills.

torticollis (tor''ti-kol'is)—condition in which the neck muscles, such as the sternocleidomastoids, contract involuntarily. It is more commonly called wryneck.

Chapter Summary

Introduction

Individual muscles are the organs of the muscular system.

When they are stimulated, muscle cells contract and pull at their attachments.

Muscles are responsible for body movements, posture maintenance, and heat production.

Structure of a Skeletal Muscle

Skeletal muscles are composed of nerve, vascular, and various connective tissues as well as skeletal muscle tissue.

1. Connective tissue coverings
   a. Skeletal muscles are covered with fascia.
   b. Other connective tissues invest cells and groups of cells within the muscle's structure.
2. Skeletal muscle fibers
   a. Each fiber represents a single muscle cell which is the unit of contraction.
   b. Muscle fibers are cylindrical cells with numerous nuclei.
   c. The cytoplasm contains mitochondria, a sarcoplasmic reticulum, and myofibrils composed of actin and myosin.
   d. Striations are produced by the arrangement of the actin and myosin filaments.
   e. Transverse tubules extend from the cell membrane into the cytoplasm and are associated with the sarcoplasmic reticulum.
3. The neuromuscular junction
   a. Muscle fibers are stimulated to contract by motor neurons.
   b. The motor end plate of a muscle fiber lies on one side of a neuromuscular junction.
   c. In response to a nerve impulse, the end of the motor nerve fiber secretes a neurotransmitter, which diffuses across the junction and stimulates the muscle fiber.
4. Motor units
   a. One motor neuron and the muscle fibers associated with it constitute a motor unit.
   b. Finer movements can be produced in muscles whose motor units contain small numbers of muscle fibers.

Skeletal Muscle Contraction

Muscle fiber contraction results from a sliding movement within the myofibrils in which actin and myosin filaments merge.

1. Role of actin and myosin
   a. Globular portions of myosin filaments can form cross-bridges with actin filaments.
   b. The reaction between actin and myosin filaments generates the force of contraction.
   c. When a fiber is at rest, troponin and tropomyosin molecules interfere with cross-bridge formation.
   d. When calcium ions are present, the inhibition is removed.
2. Stimulus for contraction
   a. Muscle fiber is usually stimulated by acetylcholine released from the end of a motor nerve fiber.
   b. Acetylcholine causes muscle fiber to conduct an impulse that reaches the deep parts of the fiber by means of the transverse tubules.
   c. An impulse signals the sarcoplasmic reticulum to release calcium ions.
   d. Cross-bridges form between actin and myosin, and the actin filaments move inward.
   e. Muscle fiber relaxes when calcium ions are transported back into the sarcoplasmic reticulum.
   f. Cholinesterase decomposes acetylcholine.
3. Energy sources for contraction
   a. ATP supplies the energy for muscle fiber contraction.
   b. Creatine phosphate stores energy that can be used to synthesize ATP as it is decomposed.
   c. Active muscles depend upon cellular respiration.
4. Oxygen supply and cellular respiration
   a. Anaerobic respiration results in a gain of only a few ATP molecules, while aerobic respiration results in the gain of many more ATP molecules.
   b. Oxygen is carried from the lungs to the body cells by hemoglobin in red blood cells.
   c. Myoglobin in muscle cells stores some oxygen.
5. Oxygen debt
   a. During rest or moderate exercise, oxygen is supplied to muscles in sufficient concentration to support aerobic respiration.
   b. During strenuous exercise, an oxygen deficiency may develop, and lactic acid may accumulate as a result of anaerobic respiration.
   c. The amount of oxygen needed to convert accumulated lactic acid to glucose and to restore the supplies of ATP and creatine phosphate is called oxygen debt.

6. Muscle fatigue
   a. A fatigued muscle loses its ability to contract.
   b. Muscle fatigue is usually due to the effects of an accumulation of lactic acid.
   c. Athletes usually produce less lactic acid than nonathletes because of their increased ability to supply oxygen and nutrients to muscles.
7. Heat production
   a. Muscles represent an important source of body heat.
   b. Most of the energy released by cellular respiration is lost as heat.

Muscular Responses

1. Threshold stimulus is the minimal stimulus needed to elicit a muscular contraction.
2. All-or-none-response
   a. If a muscle fiber contracts at all, it will contract completely.
   b. Motor units respond in an all-or-none manner.
   c. When a muscle is stimulated, some motor units respond while others do not.
3. Twitch contraction
   a. A myogram is a recording of a muscular contraction.
   b. The myogram of a twitch includes a latent period, a period of contraction, and a period of relaxation.
   c. The refractory period is the time following a contraction during which a muscle remains unresponsive to stimulation.
4. Sustained contractions
   a. A rapid series of stimuli may produce a summation of twitches and a sustained contraction.
   b. When contractions fuse, the strength of contraction may increase.
   c. Even when a muscle is at rest, its fibers usually remain partially contracted.
5. Isometric and isotonic contractions
   a. When a muscle contracts but its attachments do not move, the contraction is called isometric.
   b. When a muscle contracts and its ends are pulled closer together, the contraction is called isotonic.
   c. Most body movements involve both isometric and isotonic contractions.
6. Use and disuse of skeletal muscle
   a. Muscles that are forcefully exercised tend to enlarge and become stronger.
   b. Muscles that are not used tend to decrease in size and become weaker.

Smooth Muscles

The contractile mechanisms of smooth and cardiac muscles are similar to that of skeletal muscle.

1. Smooth muscle fibers
   a. Cells contain filaments of actin and myosin.
   b. They lack transverse tubules and the sarcoplasmic reticula are not well developed.
   c. Types include multiunit smooth muscle and visceral smooth muscle.
   d. Visceral smooth muscle displays rhythmicity and is self-exciting.
   e. Movement of material through visceral organs is often aided by peristalsis.
2. Smooth muscle contraction
   a. Smooth muscle can maintain a contraction for a longer time with a given amount of energy than can skeletal muscle.
   b. Smooth muscles can change their lengths without changing their tautness.

Cardiac Muscle

1. The ends of adjacent cells are separated by intercalated disks, which hold the cells together.
2. Cardiac muscle remains contracted for a longer time than skeletal muscle.
3. A network of fibers contracts as a unit and responds to stimulation in an all-or-none manner.
4. Cardiac muscle is self-exciting, rhythmic, and remains refractory until a contraction is completed.

Body Movements

The type of movement produced by a muscle depends on the way it is attached on either side of a joint.

1. Origin and insertion
   a. The movable end of a muscle is its insertion, and the immovable end is its origin.
   b. Some muscles have more than one origin or insertion.
   c. Sometimes the end of a muscle changes its function in different body movements.
2. Interaction of skeletal muscles
   a. Skeletal muscles function in groups.
   b. A prime mover is responsible for most of a movement. Synergists aid prime movers. Antagonists can resist the movement of a prime mover.
   c. Smooth movements depend upon antagonists' giving way to the actions of prime movers.
3. Types of movements
   a. Muscles acting at freely movable joints produce movements in different directions and in different planes.
   b. Movements include flexion, extension, hyperextension, abduction, adduction, rotation, circumduction, supination, pronation, eversion, inversion, elevation, depression, protraction, and retraction.

Major Skeletal Muscles

Muscle names often describe sizes, shapes, locations, actions, number of attachments, or direction of fibers.

1. Muscles of facial expression
   a. These muscles lie beneath the skin of the face and scalp and are used to communicate feelings.
   b. They include the epicranius, orbicularis oculi, orbicularis oris, buccinator, zygomaticus, and platysma.
2. Muscles of mastication
   a. These muscles are attached to the mandible and are used in chewing.
   b. They include the masseter, temporalis, medial pterygoid, and lateral pterygoid.
3. Muscles that move the head
   a. Head movements are produced by muscles in the neck and upper back.
   b. They include the sternocleidomastoid, splenius capitis, semispinalis capitis, and longissimus capitus.
4. Muscles that move the pectoral girdle
   a. Most of these muscles connect the scapula to nearby bones and are closely associated with muscles that move the upper arm.
   b. They include the trapezius, rhomboideus major, levator scapulae, serratus anterior, and pectoralis minor.
5. Muscles that move the upper arm
   a. These muscles connect the humerus to various regions of the pectoral girdle, ribs, and vertebral column.
   b. They include the coracobrachialis, pectoralis major, teres major, latissimus dorsi, supraspinatus, deltoid, subscapularis, infraspinatus, and teres minor.
6. Muscles that move the forearm
   a. These muscles connect the radius and ulna to the humerus or pectoral girdle.
   b. They include the biceps brachii, brachialis, brachioradialis, triceps brachii, supinator, pronator teres, and pronator quadratus.
7. Muscles that move the wrist, hand, and fingers
   a. These muscles arise from the distal end of the humerus and from the radius and ulna.
   b. They include the flexor carpi radialis, flexor carpi ulnaris, palmaris longus, flexor digitorum profundus, extensor carpi radialis longus, extensor carpi radialis brevis, extensor carpi ulnaris, and extensor digitorum.
8. Muscles of the abdominal wall
   a. These muscles connect the rib cage and vertebral column to the pelvic girdle.
   b. They include the external oblique, internal oblique, transversus abdominus, and rectus abdominis.

9. Muscles of the pelvic outlet
   a. These muscles form the floor of the pelvic cavity and fill the space of the pubic arch.
   b. They include the levator ani, coccygeus, superficial transversus perinei, bulbospongiosus, and ischiocavernosus.
10. Muscles that move the thigh
   a. These muscles are attached to the femur and to some part of the pelvic girdle.
   b. They include the psoas major, iliacus, gluteus maximus, gluteus medius, gluteus minimus, tensor fasciae latae, adductor longus, adductor magnus, and gracilis.
11. Muscles that move the lower leg
   a. These muscles connect the tibia or fibula to the femur or pelvic girdle.
   b. They include the biceps femoris, semitendinosus, semimembranosus, sartorius, and quadriceps femoris.
12. Muscles that move the ankle, foot, and toes
   a. These muscles attach the femur, tibia, and fibula to various bones of the foot.
   b. They include the tibialis anterior, peroneus tertius, extensor digitorum longus, gastrocnemius, soleus, flexor digitorum longus, tibialis posterior, and peroneus longus.

Application of Knowledge

1. Why do you think athletes generally perform better if they warm up by exercising lightly before a competitive event?

2. Following childbirth, a woman may experience urinary incontinence when sneezing or coughing. What muscles of the pelvic floor should be strengthened by exercise to help control this problem?

3. What steps might be taken to minimize atrophy of skeletal muscles in patients who are confined to bed for prolonged times?

Review Activities

Part A

1. List the general functions of muscles.

2. Identify the general locations of skeletal, smooth, and cardiac muscle tissues.

3. Distinguish between fascia and aponeurosis.

4. List the major parts of a skeletal muscle fiber and describe the function of each part.

5. Describe a neuromuscular junction and a neurotransmitter substance.

6. Define *motor unit* and explain how the number of fibers within a unit affects muscular contractions.

7. Describe the major events that occur when a muscle fiber contracts.

8. Explain how ATP and creatine phosphate are related and how these substances function in the muscle fiber contraction mechanism.

9. Describe how oxygen is supplied to muscles.

10. Describe how oxygen debt may develop.

11. Explain how muscles may become fatigued and how a person's physical condition may affect tolerance to fatigue.

12. Define *threshold stimulus*.

13. Explain what is meant by an all-or-none response.

14. Sketch a myogram of a single muscular twitch and identify the latent period, period of contraction, and period of relaxation.

15. Define *refractory period*.

16. Explain how a skeletal muscle can be stimulated to produce a sustained contraction.

17. Distinguish between tetanic contraction and muscle tone.

18. Distinguish between isometric and isotonic contractions and explain how each is used in body movements.

19. Distinguish between hypertrophy and atrophy and explain how each may be caused.

20. Compare the structures of a smooth and a skeletal muscle fiber.

21. Distinguish between multiunit and visceral smooth muscles.

22. Define *peristalsis* and explain its function.

23. Compare the characteristics of smooth and skeletal muscle contractions.

24. Compare the structures of cardiac and skeletal muscles.

25. Distinguish between a muscle's origin and its insertion.

26. Define *prime mover, synergist,* and *antagonist.*

Part B

Match the muscles in column I with the descriptions and functions in column II.

I | II

1. Buccinator
2. Epicranius
3. Medial pterygoid
4. Platysma
5. Rhomboideus major
6. Splenius capitus
7. Temporalis
8. Zygomaticus

A. Inserted on the coronoid process of the mandible.
B. Draws the corner of the mouth upward.
C. Can raise and adduct the scapula.
D. Can pull the head into an upright position.
E. Consists of two parts— the frontalis and the occipitalis.
F. Compresses the cheeks.
G. Extends over the neck from the chest to the face.
H. Pulls the jaw to the side in grinding movements.

9. Biceps brachii
10. Brachialis
11. Deltoid
12. Latissimus dorsi
13. Pectoralis major
14. Pronator teres
15. Teres minor
16. Triceps brachii

I. Primary extensor of the elbow.
J. Pulls the shoulder back and downward.
K. Abducts the arm.
L. Rotates the arm laterally.
M. Pulls the arm forward and across the chest.
N. Rotates the arm medially.
O. Strongest flexor of the elbow.
P. Strongest supinator of the forearm.

17. Biceps femoris
18. External oblique
19. Gastrocnemius
20. Gluteus maximus
21. Gluteus medius
22. Gracilis
23. Rectus femoris
24. Tibialis anterior

Q. Inverts the foot.
R. A member of the quadriceps femoris group.
S. A plantar flexor of the foot.
T. Compresses the contents of the abdominal cavity.
U. Heaviest muscle in the body.
V. A hamstring muscle.
W. Adducts the thigh.
X. Abducts the thigh.

Part C

What muscles can you identify in the bodies of these models whose muscles are enlarged by exercise?

Suggestions for Additional Reading

Carlson, F. D., and Wilkie, D. R. 1974. *Muscle physiology*. Englewood Cliffs, N.J.: Prentice-Hall.

Close, R. I. 1972. Dynamic properties of mammalian skeletal muscles. *Physiol. Rev.* 52:129.

Cohen, C. November 1975. The protein switch of muscle contraction. *Scientific American*.

Endo, M. 1977. Calcium release from the sarcoplasmic reticulum. *Physiol. Rev.* 57:71.

Hoyle, G. April 1970. How a muscle is turned on and off. *Scientific American*.

Huddart, H., and Hunt, S. 1975. *Visceral muscle*. New York: Halstead Press.

Lester, H. A. February 1977. The response to acetylcholine. *Scientific American*.

Margaria, R. March 1972. The sources of muscular energy. *Scientific American*.

Merton, P. A. May 1972. How we control the contraction of our muscles. *Scientific American*.

Murray, J. M., and Weber, A. February 1974. The cooperative action of muscle proteins. *Scientific American*.

Peachey, L. D. 1974. *Muscle and motility*. New York: McGraw-Hill.

Sandow, A. 1970. Skeletal muscle. *Ann. Rev. Physio.* 32:87.

Uehara, Y. 1976. *Muscle and its innervation*. Baltimore: The Williams and Wilkins Co.

Weber, A., and Murray, J. M. 1973. Molecular control mechanisms in muscle contraction. *Physiol. Rev.* 53:612.

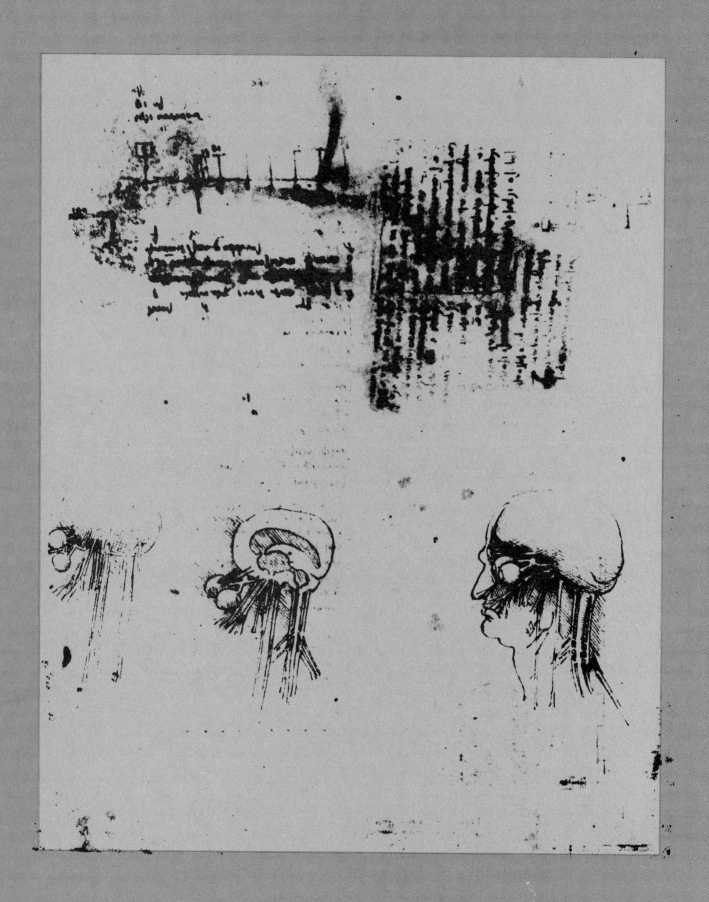

Unit 3
Integration and Coordination

The chapters of unit 3 are concerned with the structures and functions of the nervous and endocrine systems. They describe how the organs of these systems serve to keep the parts of the human body functioning together as a whole and how these organs help to maintain a stable internal environment, which is vital to the survival of the organism.

This unit includes

Chapter 10: The Nervous System
Chapter 11: Somatic and Special Senses
Chapter 12: The Endocrine System

The Nervous System

10 If a human organism is to survive, the actions of its cells, tissues, organs, and systems must be directed toward a single goal—the maintenance of homeostasis. To accomplish this, the functions of all body parts must be controlled and coordinated so that they work as a unit and respond to changes in ways that help maintain a stable internal environment.

The general task of controlling and coordinating body activities is handled by the nervous and endocrine systems. Of these, the *nervous system* provides the more rapid and precise mode of action. It controls muscular contractions and glandular secretions by means of nerve fibers that extend from the brain and spinal cord to reach all body parts. Some of these nerve fibers bring information to the brain and spinal cord from sensory receptors; this information concerns changes occurring inside and outside the body. Other fibers carry impulses away from the brain or spinal cord and stimulate muscles or glands to respond.

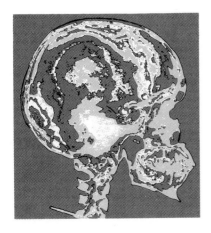

Chapter Objectives

After you have studied this chapter, you should
be able to

1. Explain the general functions of the nervous
   system.

2. Describe the general structure of a neuron.

3. Explain how an injured nerve fiber may
   regenerate.

4. Name four types of neuroglial cells and
   describe the functions of each.

5. Describe the events that lead to the
   conduction of a nerve impulse.

6. Explain how a nerve impulse is transmitted
   from one neuron to another.

7. Explain how differences in structure and
   function are used to classify neurons.

8. Name the parts of a reflex arc and describe
   the function of each.

9. Describe the coverings of the brain and
   spinal cord.

10. Describe the structure of the spinal cord
    and its major functions.

11. Name the major parts of the brain and
    describe the functions of each.

12. Distinguish between motor, sensory, and
    association areas of the cerebral cortex.

13. Describe the formation and the function
    of cerebrospinal fluid.

14. Distinguish between the limbic system
    and the reticular formation.

15. List the major parts of the peripheral
    nervous system.

16. Name the cranial nerves and list their major
    functions.

17. Describe the structure of a spinal nerve.

18. Describe the functions of the autonomic
    nervous system.

19. Distinguish between the sympathetic and the
    parasympathetic divisions of the autonomic
    nervous system.

20. Complete the review activities at the end
    of this chapter.

Fig. 10.17 Structural types of neurons: (a) multipolar neuron; (b) bipolar neuron; (c) unipolar neuron. Where can examples of each of these be found in the body?

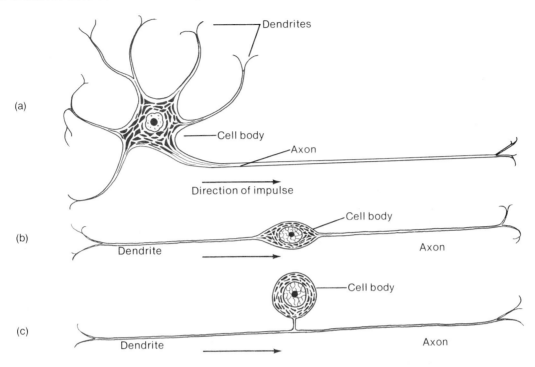

Types of Nerves

While a nerve fiber is an extension of a neuron, a **nerve** is a cordlike bundle (or group of bundles) of nerve fibers held together by layers of connective tissues, as shown in figure 10.18.

Like nerve fibers, nerves that conduct impulses into the brain or spinal cord are called **sensory nerves**, while those that carry impulses out to muscles or glands are termed **motor nerves**. Most nerves, however, include both sensory and motor fibers, and they are called **mixed nerves**.

1. Explain how neurons are classified according to structure; according to function.
2. How is a neuron related to a nerve?
3. What is a mixed nerve?

Fig. 10.18 A nerve is composed of bundles of nerve fibers held together by connective tissues.

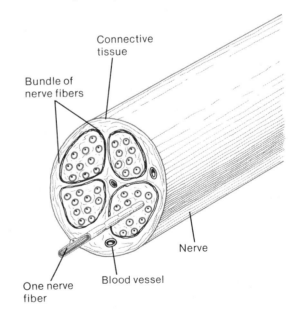

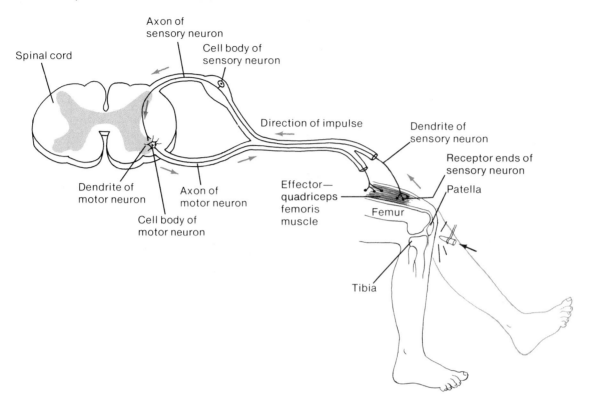

Reflex Arcs and Reflexes

The routes followed by nerve impulses as they travel through the nervous system are called *nerve pathways*. As a rule, activity on a nerve pathway begins with sensory receptors associated with sensory neurons, involves several interneurons, and may include some motor neurons.

Reflex Behavior

The simplest nerve pathways lead directly from sensory to motor neurons and include only two or three cells. Such pathways, known as **reflex arcs**, form the behavioral units of the nervous system. That is, they constitute the structural and functional bases for the simplest acts—the **reflexes**.

Reflexes are automatic, unconscious responses to changes occurring inside or outside the body. They help to control many of the body's involuntary processes such as heart rate, breathing rate, blood pressure, and digestive activities. Reflexes are also involved in swallowing, sneezing, coughing, and vomiting.

The *knee-jerk reflex* (patellar reflex) is an example of a simple reflex that employs only two neurons (fig. 10.19). This reflex is initiated by striking the patellar ligament just below the patella. As a result, the quadriceps femoris muscle, which is attached to the patella by a tendon, is pulled slightly, and stretch receptors located within the muscle are stimulated. These receptors, in turn, trigger impulses that pass along the fibers of a sensory neuron into the spinal cord. Within the spinal cord, the sensory axon forms a synapse with a dendrite of a motor neuron. The impulse then continues along the axon of the motor neuron and travels back to the quadriceps femoris. The muscle responds by contracting, and the reflex is completed as the lower leg extends.

This reflex is helpful in maintaining an upright posture. For example, when a person is standing still, if the knee begins to bend as a result of gravity pulling downward, the quadriceps femoris is stretched, the reflex is triggered, and the leg straightens again.

Another type of reflex, called a *withdrawal reflex* (fig. 10.20), occurs when a person unexpectedly touches a finger to something hot or sharp. As this happens, some of the skin receptors are activated, and sensory impulses travel to the spinal cord. There the impulses pass on to interneurons and are directed to motor neurons. The motor neurons transmit the signals to flexor muscles in the arm, and they contract

Fig. 10.20 A withdrawal reflex involves a sensory neuron, an interneuron, and a motor neuron.

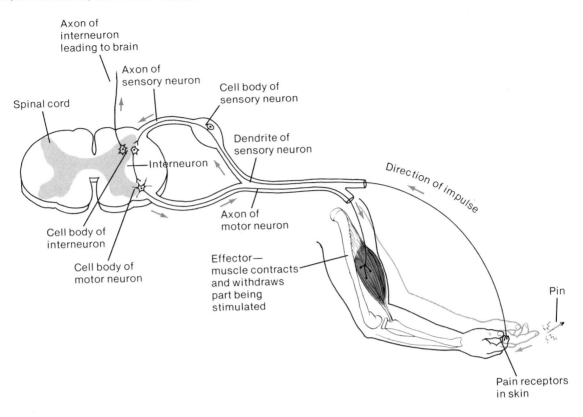

Chart 10.2   Parts of a reflex arc

Part	Description	Function
Receptor	The receptor end of a dendrite or a specialized receptor cell in a sensory organ	Sensitive to a specific type of internal or external change
Sensory neuron	Dendrite, cell body, and axon of a sensory neuron	Transmits nerve impulse from the receptor into the brain or spinal cord
Interneuron	Dendrite, cell body, and axon of a neuron within the brain or spinal cord	Conducts nerve impulse from the sensory neuron to a motor neuron
Motor neuron	Dendrite, cell body, and axon of a motor neuron	Transmits nerve impulse from the brain or spinal cord out to an effector
Effector	A muscle or gland outside the nervous system	Responds to stimulation by the motor neuron and produces the reflex or behavioral action

in response. At the same time, the antagonistic extensor muscles are inhibited, and the hand is rapidly and unconsciously withdrawn from the harmful source of stimulation.

While the flexor muscles of the stimulated side (ipsilateral side) are caused to contract, the flexor muscles of the other arm (contralateral side) are being inhibited. Furthermore, the extensor muscles on this other side are caused to contract. This phenomenon, which is called a *crossed extensor reflex*, is due to interneuron pathways within the spinal cord that allow sensory impulses arriving on one side of the cord to pass across to the other side and produce an opposite effect. (See fig. 10.21.)

Concurrent with the withdrawal reflex, other interneurons in the spinal cord carry sensory impulses to the brain, and the person becomes aware of the painful experience.

A withdrawal reflex is, of course, protective because it serves to prevent excessive tissue damage when a body part touches something that is potentially harmful.

Chart 10.2 summarizes the parts of a reflex arc.

Fig. 10.21 When the flexor muscle on one side is stimulated to contract in a withdrawal reflex, the extensor on the opposite side also contracts.

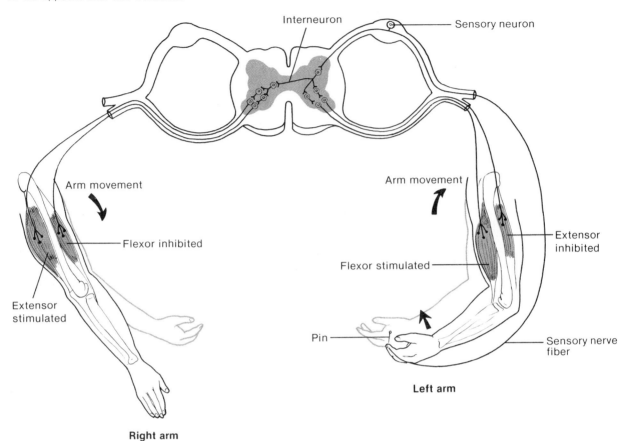

Some Clinical Uses of Reflexes

Since normal reflexes depend on normal neuron functions, reflexes are commonly used to obtain information concerning the condition of the nervous system. An anesthesiologist, for instance, may try to initiate a reflex in a patient who is being anesthetized in order to determine how the anesthetic drug is affecting nerve functions. Also, in the case of injury to some part of the nervous system, various reflexes may be tested to discover the location and extent of the damage.

If any portion of a reflex arc is injured, the normal characteristics of that arc are likely to be altered. For example, a *plantar reflex* is normally initiated by stroking the sole of the foot, and the usual response includes a flexion of the foot and toes. However, in persons who have suffered damage to certain nerve pathways (corticospinal tract) there may be an abnormal response called the *Babinski reflex*, shown in figure 10.22. In this case the reflex response is

plantar extension, in which the great toe extends upward and the smaller toes fan apart. If the injury is minor, the response may consist of plantar flexion with failure of the great toe to flex, or plantar flexion followed by plantar extension.

The Babinski reflex is, however, present normally in infants up to the age of 12 months and is thought to reflect a degree of immaturity in their corticospinal tracts.

Other reflexes that may be tested during a neurological examination include the following:

1. **Biceps-jerk reflex**. This reflex can be elicited by bending a person's arm at the elbow. The examiner's finger is placed on the inside of the bent elbow over the tendon of the biceps muscle, and the finger is tapped. The biceps contracts in response, and the forearm is flexed at the elbow.

Fig. 10.22 (a) Plantar reflex; (b) Babinski reflex. What is the significance of the Babinski reflex?

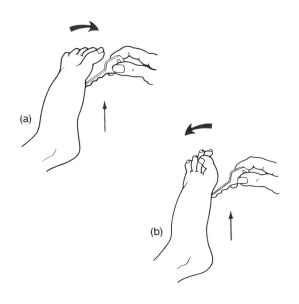

2. **Triceps-jerk reflex.** This reflex can be caused by flexing a person's arm at the elbow and tapping the short tendon of the triceps muscle close to its insertion near the tip of the elbow. The muscle contracts in response, and the forearm is extended slightly.

3. **Abdominal reflexes.** These reflexes occur when the examiner strokes the skin of the abdomen. For example, a dull pin may be drawn from the sides of the abdomen upward toward the midline and above the umbilicus. Normally, the abdominal muscles underlying the skin contract in response, and the umbilicus is moved toward the region that was stimulated.

4. **Ankle-jerk reflex.** This reflex is elicited by tapping the Achilles tendon just above its insertion on the calcaneus. The response is plantar flexion, produced by contraction of the gastrocnemius and soleus muscles.

5. **Cremasteric reflex.** This reflex is obtained in males by stroking the upper inside of the thigh. As a result, the testis on the same side is elevated by contracting muscles.

6. **Anal reflex.** This reflex is elicited by stroking the skin surrounding the anus. The anal sphincter muscles contract in response.

---

1. Explain what is meant by reflex.
2. List the parts of a reflex arc.
3. Explain how reflexes can be used to test the condition of the nervous system.

Coverings of the Central Nervous System

The central nervous system (CNS), which consists of the brain and spinal cord, is surrounded by bones and membranes. More specifically, the brain lies within the cranial cavity of the skull, and the spinal cord, which is continuous with the brain, occupies the vertebral canal within the vertebral column. Beneath these bony coverings, the brain and spinal cord are protected by membranes called *meninges* that are located between the bone and the soft tissues of the nervous system (fig. 10.23).

The Meninges

The **meninges** have three layers—a dura mater, an arachnoid mater, and a pia mater (fig. 10.23).

The **dura mater** is the outermost layer and is composed primarily of tough, white fibrous connective tissue. It is attached to the inside of the cranial cavity and forms the internal periosteum of the surrounding skull bones.

In some regions, the dura mater extends inward between lobes of the brain and forms partitions that support and protect these parts. (See chart 10.3.) In other areas, the dura mater splits into two layers forming channels called *dural sinuses*, shown in figure 10.24. Venous blood flows through these channels as it returns from the brain to vessels leading to the heart.

The dura mater continues into the vertebral canal as a strong, tubular sheath that surrounds the spinal cord. It terminates as a blind sac below the end of the cord. The membrane around the spinal cord is not attached directly to the bones of the vertebrae, but is separated by the *epidural space*, which lies between the dural sheath and the bony walls (fig. 10.25). This space is occupied by blood vessels, loose connective tissue, and fat tissue that provide a protective pad around the delicate tissues of the spinal cord.

Chart 10.3 Partitions of the dura mater

Partition	Location
Falx cerebelli	Separates the right and left cerebellar hemispheres
Falx cerebri	Extends downward into the longitudinal fissure and separates the right and left cerebral hemispheres
Tentorium cerebelli	Separates the occipital lobes of the cerebrum from the cerebellum

Fig. 10.23 The brain and spinal cord are enclosed by bone and by membranes called meninges. What are the functions of these membranes?

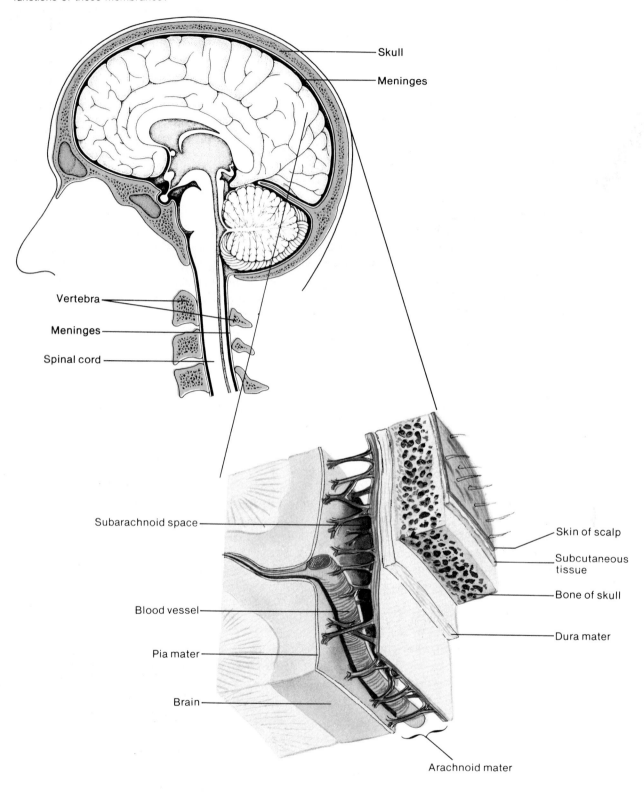

Skull

Meninges

Vertebra

Meninges

Spinal cord

Subarachnoid space

Skin of scalp

Subcutaneous tissue

Bone of skull

Blood vessel

Dura mater

Pia mater

Brain

Arachnoid mater

Fig. 10.24 Dural sinuses, which occur between layers of dura mater, form channels through which blood flows as it leaves the brain.

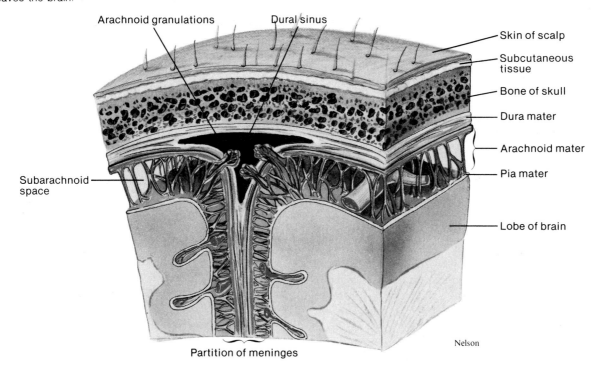

Arachnoid granulations

Dural sinus

Skin of scalp

Subcutaneous tissue

Bone of skull

Dura mater

Arachnoid mater

Pia mater

Lobe of brain

Subarachnoid space

Partition of meninges

Nelson

Fig. 10.25 The epidural space between the dural sheath and the bone of the vertebra is filled with tissues that provide a protective pad around the spinal cord.

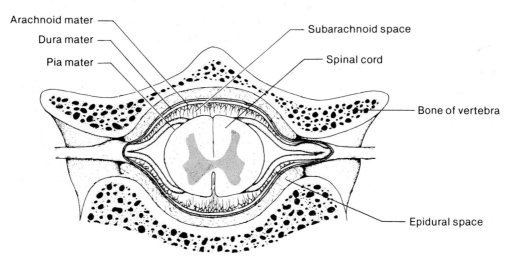

Arachnoid mater

Dura mater

Pia mater

Subarachnoid space

Spinal cord

Bone of vertebra

Epidural space

The **arachnoid mater** is a thin, netlike membrane located between the dura and pia maters. It spreads over the brain and spinal cord, but generally does not dip into the grooves and depressions on their surfaces.

Between the arachnoid and pia maters is a *subarachnoid space* that contains watery **cerebrospinal fluid.**

The **pia mater** is very thin and contains many blood vessels that aid in nourishing the underlying cells of the brain and spinal cord. This layer is attached to the surfaces of these organs and follows their irregular contours, passing over the high areas and dipping into the depressions.

An inflammation of the meninges is called *meningitis*. This condition is usually caused by the presence of certain bacteria or viruses that sometimes invade the cerebrospinal fluid. Although meningitis may involve the dura mater, it is more commonly limited to the arachnoid and pia maters.

Meningitis occurs most often in infants and children and is considered one of the more serious of the childhood infections. Possible complications of this disease include damage to the nervous system with loss of vision or hearing, paralysis, and mental retardation.

---

1. Describe the meninges.
2. Name the layers of the meninges.
3. Explain where cerebrospinal fluid occurs.

The Spinal Cord

The **spinal cord** is a slender nerve column with an oval cross section that passes downward from the brain into the vertebral canal. Although it is continuous with the brain, the spinal cord is said to begin where nerve tissue leaves the cranial cavity at the level of the foramen magnum. The cord usually tapers to a point and terminates near the intervertebral disk that separates the first and second lumbar vertebrae. (fig. 10.26.)

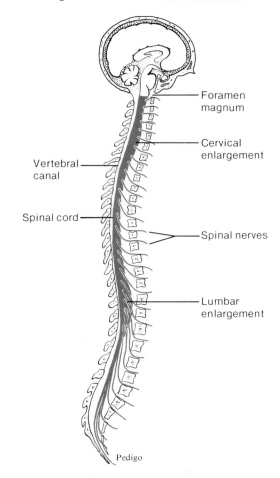

Fig. 10.26 The spinal cord begins at the level of the foramen magnum. At what level does it terminate?

Structure of the Spinal Cord

The spinal cord consists of thirty-one segments, each of which gives rise to a pair of **spinal nerves.** These nerves branch out to various body parts and connect them with the central nervous system.

In the neck region, there is a bulge in the spinal cord, called the *cervical enlargement,* that gives off nerves to the arms. A similar thickening in the lower back, the *lumbar enlargement,* gives off nerves to the legs.

Two grooves, a deep *anterior median fissure* and a shallow *posterior median sulcus,* extend the length of the spinal cord, dividing it into right and left halves. A cross section of the cord (fig. 10.27 and color plate 20) reveals that it consists of a core of gray matter surrounded by white matter. The pattern produced by the gray matter roughly resembles a butterfly with its wings outspread. The upper wings are called the *posterior horns* of the gray matter, and the lower wings are the *lateral* and *anterior horns.*

Fig. 10.27 (a) A cross section of a spinal cord.
(b) Identify the parts of the spinal cord in this micrograph.

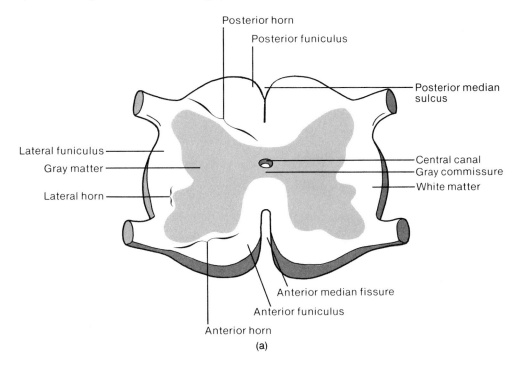

Posterior horn

Posterior funiculus

Posterior median sulcus

Lateral funiculus

Gray matter

Lateral horn

Central canal

Gray commissure

White matter

Anterior median fissure

Anterior funiculus

Anterior horn

(a)

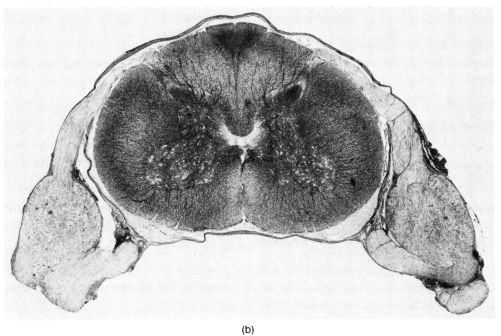

(b)

Fig. 10.28 Major ascending and descending tracts within a cross section of the spinal cord.

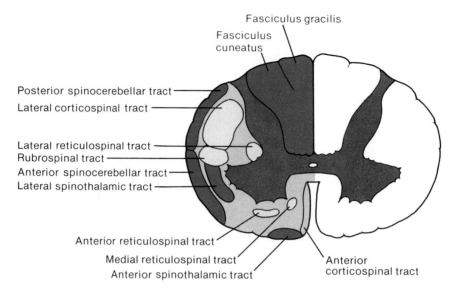

Fasciculus gracilis

Fasciculus cuneatus

Posterior spinocerebellar tract

Lateral corticospinal tract

Lateral reticulospinal tract

Rubrospinal tract

Anterior spinocerebellar tract

Lateral spinothalamic tract

Anterior reticulospinal tract

Medial reticulospinal tract

Anterior spinothalamic tract

Anterior corticospinal tract

A horizontal bar of gray matter in the middle of the spinal cord, the *gray commissure,* connects the wings of the gray matter on the right and left sides. This bar surrounds the **central canal,** which is continuous with the ventricles of the brain and contains cerebrospinal fluid.

The white matter of the spinal cord is divided into three sections on each side by the gray matter. They are known as the *anterior, lateral,* and *posterior funiculi,* and each consists of longitudinal bundles of myelinated nerve fibers, which comprise major nerve pathways called **nerve tracts.**

Functions of the Spinal Cord

The spinal cord has two major functions—to conduct nerve impulses and to serve as a center for spinal reflexes.

The tracts of the spinal cord provide a two-way system of communication between the brain and parts outside the nervous system. Those tracts that conduct impulses from body parts and carry sensory information to the brain are called **ascending tracts;** those that conduct motor impulses from the brain to muscles and glands are **descending tracts.**

The nerve fibers within these tracts are axons, and usually, all the axons within a given tract begin from cell bodies located in the same part of the nervous system and end together in some other part. The names that identify nerve tracts often reflect these common origins and terminations. For example, a *spinothalamic* tract begins in the spinal cord and carries sensory impulses to the thalamus of the brain; a *corticospinal* tract originates in the cortex of the brain and carries motor impulses downward through the spinal cord and spinal nerves to various effectors.

Ascending Tracts. Among the major ascending tracts of the spinal cord are the following. (See fig. 10.28.)

1. **Fasciculus gracilis** and **fasciculus cuneatus.** These tracts are located in the posterior funiculi of the spinal cord. Their fibers conduct sensory impulses from the skin, muscles, tendons, and joints to the brain, where they are interpreted as sensations of touch, pressure, and body movement.

2. **Spinothalamic tracts.** The *lateral* and *anterior* spinothalamic tracts are located in the lateral and anterior funiculi respectively. The lateral tracts conduct impulses to the brain and give rise to sensations of pain and temperature. Impulses carried on fibers of the anterior tracts are interpreted as touch and pressure.

Persons suffering from severe, unremitting pain are sometimes treated with a surgical operation called a *cordotomy.* This is done only as a last resort. In this procedure, fibers in the lateral spinothalamic tracts are severed above the level of pain receptors being stimulated.

3. **Spinocerebellar tracts.** The *posterior* and *anterior* spinocerebellar tracts are found near the surface in the lateral funiculi of the spinal cord. Impulses conducted on their fibers originate in the muscles of the legs and trunk and travel to the cerebellum of the brain. These impulses are necessary for the coordination of muscular movements.

Descending Tracts. The major descending tracts of the spinal cord include the following. (See fig. 10.28.)

1. **Corticospinal tracts.** The *lateral* and *anterior* corticospinal tracts occupy the lateral and anterior funiculi respectively. These tracts conduct motor impulses from the brain to spinal nerves and outward to various skeletal muscles. They function in the control of voluntary movements.

2. **Reticulospinal tracts.** The *lateral* reticulospinal tracts are located in the lateral funiculi, while the *anterior* and *medial* reticulospinal tracts are in the anterior funiculi. Motor impulses transmitted on these tracts originate in the brain and function in the control of muscular tone and in the activity of the sweat glands.

3. **Rubrospinal tracts.** The fibers of the rubrospinal tracts pass through the lateral funiculi. They carry motor impulses from the brain to skeletal muscles and are involved with muscular coordination and the control of posture.

The disease called *poliomyelitis* is caused by viruses that sometimes affect motor neurons, including those in the anterior horns of the spinal cord. When this happens, the person may develop paralysis in the muscles innervated by the affected neurons, but may not suffer any sensory losses in these parts.

In addition to serving as a pathway for various nerve tracts, the spinal cord functions in many reflexes like those described on pages 284–85. Such reflexes are called **spinal reflexes,** because the reflex arcs pass through the cord.

Spinal Cord Injuries

Injuries to the spinal cord may be caused indirectly, as by a blow to the head or by a fall, or they may be due to forces applied directly to the cord. The consequences will depend on the amount of damage sustained by the cord. The spinal cord may, for example, be compressed or distorted by a minor injury, and its functions may be disturbed only temporarily. If nerve fibers are severed, however, some of the cord's functions are likely to be permanently lost.

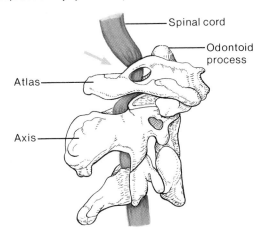

Fig. 10.29 A dislocation of the atlas may cause a compression injury to the spinal cord.

Spinal cord
Odontoid process
Atlas
Axis

Among the more common causes of direct injury to the spinal cord are gunshot wounds, stabbings, and fractures or dislocations of vertebrae. (See fig. 10.29.) Regardless of the cause, if nerve fibers in ascending tracts are cut, sensations arising from receptors below the level of the injury will be lost. If descending tracts are damaged, the result will be a loss of motor functions.

Normal spinal reflexes seem to depend on two-way communication between the spinal cord and the brain. When nerve pathways are interrupted, the cord's reflex activities in sites below the injury are usually depressed. At the same time, there is a lessening of sensations and muscular tone in the parts innervated by the affected fibers. This condition is called *spinal shock,* and it may last for days or weeks; normal reflex activity may return eventually.

When a spinal injury results in a complete loss of muscular control, the condition is called **paralysis.** Two types of paralysis result from damage to motor pathways—flaccid and spastic.

In *flaccid paralysis,* there is a total loss of tone in the muscles innervated by the damaged nerve fibers, and these muscles tend to atrophy and waste away. This type of paralysis is likely to result from damage to the corticospinal tracts.

Spastic paralysis is characterized by an increase in muscular tone and no atrophy of the muscles involved. There may be uncoordinated reflex activity during which the flexor and extensor muscles of the affected limbs alternately undergo spasms. This type of paralysis sometimes occurs when motor pathways in tracts other than the corticospinals are severed.

Less severe injuries to the spinal cord, as may be sustained from a blow to the head, whiplash in an automobile accident, or rupture of an intervertebral disk, are often accompanied by pain, weakness, and muscular atrophy in the regions supplied by the damaged nerve fibers.

---

1. Describe the structure of the spinal cord.
2. List the major ascending and descending tracts of the spinal cord and describe the function of each.
3. What type of paralysis results from damage to the corticospinal tract?
4. What signs characterize a person who suffers from spastic paralysis?

The Brain

The **brain** is the largest and most complex part of the nervous system. It occupies the cranial cavity and is composed of about one hundred billion (10^{11}) neurons and innumerable nerve fibers, by which the neurons communicate with one another and with neurons in other parts of the system.

As figure 10.30 shows, the brain can be divided into three major portions—a cerebrum, a cerebellum, and a brain stem. The **cerebrum,** which is the largest part, contains nerve centers associated with sensory and motor functions. It is also concerned with the

higher mental functions, including memory and reasoning. The **cerebellum** includes centers associated with the coordination of voluntary muscular movements. The **brain stem** contains nerve pathways by which various parts of the nervous system are interconnected, as well as nerve pathways and centers involved in the regulation of various visceral activities. (See color plate 21.)

Structure of the Cerebrum

The cerebrum consists of two large masses called **cerebral hemispheres,** which are mirror images of each other. These hemispheres are connected by a deep bridge of nerve fibers called the **corpus callosum** and are separated by a layer of dura mater called the *falx cerebri.*

The surface of the cerebrum is marked by numerous ridges or **convolutions** (gyri), which are separated by grooves. A shallow groove is called a **sulcus,** while a very deep one is a **fissure.** Although the arrangement of these elevations and depressions is complex, they form fairly distinct patterns in all brains. For example, a *longitudinal fissure* separates the right and left cerebral hemispheres, a *transverse fissure* separates the cerebrum from the cerebellum, and various sulci divide each hemisphere into lobes.

The lobes of the cerebral hemispheres (shown in fig. 10.31) are named after the skull bones that they underlie. They include the following:

Fig. 10.30 The major portions of the brain include the cerebrum, cerebellum, and brain stem.

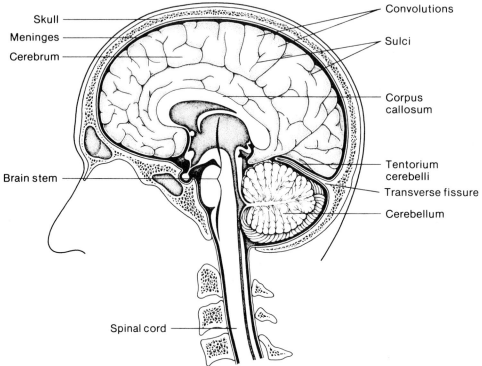

1. **Frontal lobe.** The frontal lobe forms the anterior portion of each cerebral hemisphere. It is bounded posteriorly by a *central sulcus* (fissure of Rolando) that passes out from the longitudinal fissure at a right angle, and inferiorly by a *lateral sulcus* (fissure of Sylvius) that passes out from the under surface of the brain along its sides.

2. **Parietal lobe.** The parietal lobe is posterior to the frontal lobe and is separated from it by the central sulcus.

3. **Temporal lobe.** The temporal lobe lies below the frontal lobe and is separated from it by the lateral sulcus.

4. **Occipital lobe.** The occipital lobe forms the posterior portion of each cerebral hemisphere and is separated from the cerebellum by a shelflike extension of dura mater called the *tentorium cerebelli.* There is no distinct boundary between the occipital lobe and the parietal and temporal lobes.

5. **The insula.** The insula (island of Reil) is located deep within the lateral sulcus and is covered by parts of the frontal, parietal, and temporal lobes. It is separated from them by a circular sulcus. (See fig. 10.32.)

A thin layer of gray matter called the **cerebral cortex** constitutes the outermost portion of the cerebrum. This layer, which covers the convolutions and dips into the sulci and fissures, contains countless unmyelinated neuron cell bodies.

Just beneath the cerebral cortex are masses of white matter, making up the bulk of the cerebrum. These masses contain bundles of myelinated nerve fibers that connect the neuron cell bodies of the cortex with other parts of the nervous system. Some of these fibers pass from one cerebral hemisphere to the other by way of the corpus callosum, while others carry sensory or motor impulses from portions of the cortex to nerve centers in the brain or spinal cord.

Deep within each hemisphere of the cerebrum, there are several masses of gray matter called *basal ganglia* (basal nuclei). Although the precise function of these parts is poorly understood, it is known that they contain a group of neuron cell bodies. This group serves as a relay station for motor impulses passing between the cerebral cortex, the brain stem, and the spinal cord. Impulses from basal ganglia normally inhibit motor functions and thus, aid in the control of muscular activities. (See fig. 10.32.)

Since the basal ganglia function to inhibit muscular activity, disorders in this portion of the brain may be accompanied by one of the following: a reduction in mobility (hypokinesia) due to excessive discharge of impulses from the ganglia; or involuntary movements (hyperkinesia) due to lack of inhibiting impulses from them.

Fig. 10.31 Lobes of the right cerebral hemisphere are separated by dotted lines.

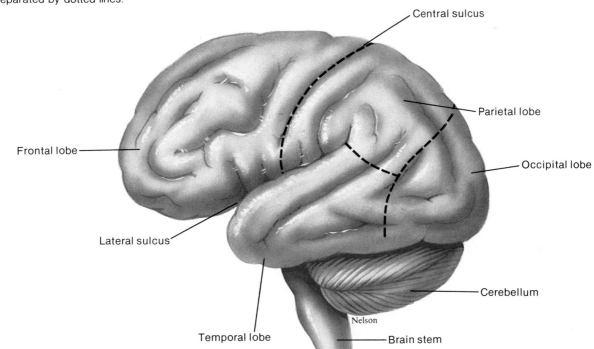

Central sulcus

Parietal lobe

Frontal lobe

Occipital lobe

Lateral sulcus

Cerebellum

Nelson

Temporal lobe

Brain stem

Fig. 10.32 A frontal section of the cerebral hemispheres.

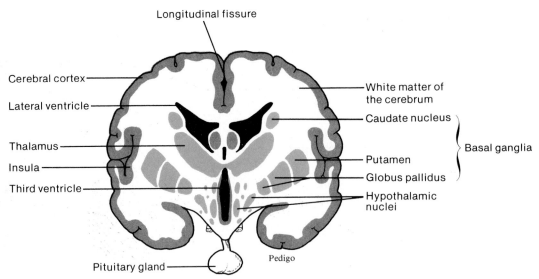

Longitudinal fissure

Cerebral cortex

Lateral ventricle

Thalamus

Insula

Third ventricle

White matter of the cerebrum

Caudate nucleus

Putamen

Globus pallidus

Basal ganglia

Hypothalamic nuclei

Pituitary gland

Pedigo

Functions of the Cerebrum

The cerebrum is concerned with the higher brain functions, in that it contains centers for interpreting sensory impulses arriving from various sense organs as well as centers for initiating voluntary muscular movements. It stores the information of memory and utilizes this information in the processes associated with reasoning. It also functions in determining a person's intelligence and personality.

Functional Regions of the Cortex. The regions of the cerebral cortex that perform specific functions have been located by using a variety of techniques. For example, persons who have suffered brain injuries or who have had portions of their brains removed surgically have been studied. Impaired abilities of these persons provide clues as to the functions of the specific parts.

In other studies, areas of cortices have been exposed surgically and stimulated mechanically or electrically. Stimulation of particular regions of the cortex is followed by responses in certain muscles or by specific sensations.

As a result of such investigations it is possible to prepare maps that indicate the functional regions of the cerebral cortex. Although there is considerable overlap in these areas, the cortex can be divided into sections known as *motor, sensory,* and *association areas.* (See fig. 10.33.)

Motor Areas. The primary motor areas of the cerebral cortex lie in the frontal lobes just anterior to the central sulcus. The nerve tissue in these regions contains numerous, large *pyramidal cells* named because of their pyramid-shaped cell bodies.

Impulses from pyramidal cells travel downward through the brain stem and into the spinal cord on the *corticospinal tracts* (pyramidal tracts). Most of the nerve fibers in these tracts cross over (decussate) from one side of the brain to the other within the brain stem and descend as the lateral corticospinal tracts. As a result of this crossing over, the motor area of the right cerebral hemisphere generally controls muscles on the left side of the body and vice versa.

Within the spinal cord, the corticospinal fibers synapse with motor neurons in the gray matter. Axons of the motor neurons lead outward through peripheral nerves to various voluntary muscles. Impulses transmitted on these pathways are primarily responsible for fine movements in skeletal muslces. (See fig. 10.34.) More specifically, as figure 10.35 shows, cells in the upper portions of the motor areas send impulses to muscles in the legs and thighs; those in the middle portions control muscles in the shoulders and arms; and those in the lower portions activate muscles of the head, face, and tongue.

Nerve tracts other than the corticospinal tracts that transmit signals from the cerebral cortex into the spinal cord are called *extrapyramidal tracts.* These include the reticulospinal and rubrospinal tracts, and they function to coordinate and control motor functions involved with the maintenance of balance and posture. Many of these fibers pass into basal ganglia on the way to the spinal cord, and as was mentioned, some of the impulses conducted on these pathways normally inhibit muscular actions.

Fig. 10.33 Some motor and sensory areas of the cerebral cortex.

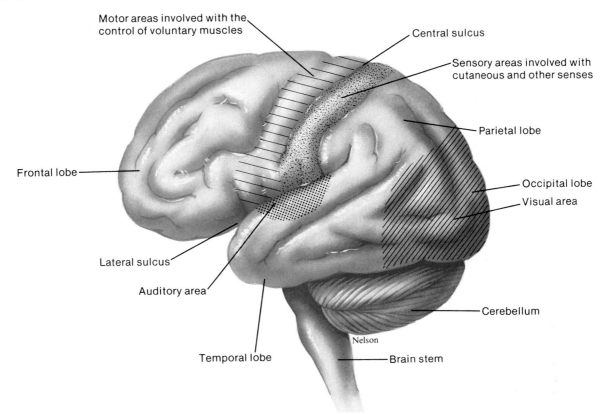

Motor areas involved with the control of voluntary muscles

Central sulcus

Sensory areas involved with cutaneous and other senses

Parietal lobe

Frontal lobe

Occipital lobe

Visual area

Lateral sulcus

Auditory area

Cerebellum

Nelson

Temporal lobe

Brain stem

Sensory Areas. Sensory areas, which occur in several lobes of the cerebrum, function in interpreting impulses that arrive from various sensory receptors. These interpretations give rise to sensations. For example, the sensations of temperature, touch, pressure, and pain from all parts of the skin arise in the anterior portions of the parietal lobes along the central sulcus (fig. 10.36). The posterior parts of the occipital lobes are concerned with vision, while the temporal lobes contain the centers for hearing. The sensory areas for taste seem to be located near the bases of the central sulci along the lateral sulci, and the sense of smell arises from centers deep within the cerebrum.

Like motor fibers, the sensory fibers cross over so that centers in the right cerebral hemisphere interpret impulses originating from the left side of the body and vice versa. The sensory areas concerned with vision, however, receive impulses from both eyes, and those for hearing receive impulses from both ears. Figure 10.37 illustrates a sensory pathway by which impulses originating in the skin enter the spinal cord, ascend to the medulla, cross over, and continue to the sensory cortex.

Association Areas. Association areas are related to memory, reasoning, verbalizing, judgment, and emotional feelings. They also function with sensory areas in analyzing sensory experiences. These areas occupy the anterior portions of the frontal lobes and are widespread in the lateral portions of the parietal, temporal, and occipital lobes. (See fig. 10.38.)

The association areas of the frontal lobes are concerned with a number of higher intellectual processes, including those necessary for concentration, planning, complex problem-solving, and judging the consequences of behavior.

The association areas of the parietal lobes aid in understanding speech and choosing words needed to express thoughts and feelings.

The association areas of the temporal lobes and the regions at the posterior ends of the lateral fissures are concerned with the interpretation of complex sensory experiences, such as those needed to read printed words. These regions also are involved with the memory of visual scenes, music, and other complex sensory patterns.

The association areas of the occipital lobes that are adjacent to the visual centers are important in combining visual images with other sensory experiences—necessary, for instance, when one recognizes another person or an object.

Fig. 10.34 Motor fibers of the corticospinal tract begin in the cerebral cortex, cross over in the medulla, and descend in the spinal cord. There they synapse with neurons whose fibers lead to spinal nerves supplying skeletal muscles.

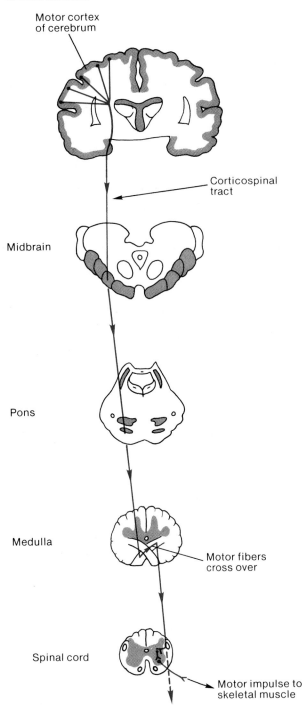

Fig. 10.35 Motor areas involved with the control of voluntary muscles.

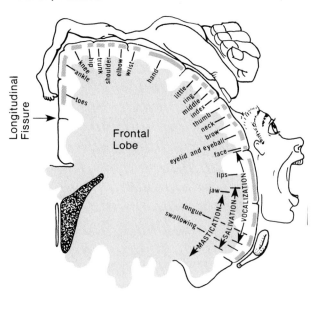

Fig. 10.36 Sensory areas involved with cutaneous and certain other senses.

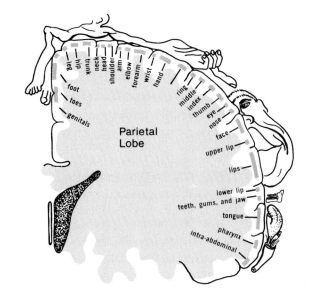

Fig. 10.37 Sensory impulses originating in skin receptors ascend in the fasciculus cuneatus tract, cross over in the medulla, and are interpreted in the sensory cortex of the cerebrum.

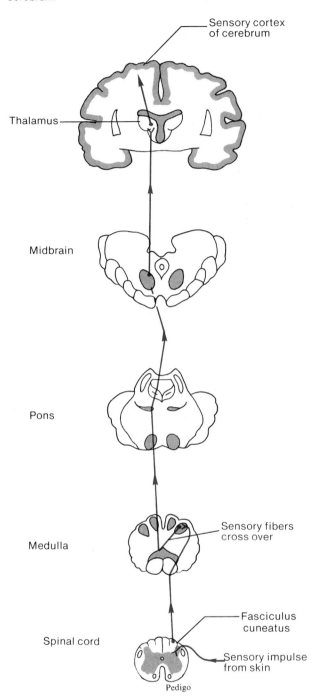

Sensory cortex of cerebrum

Thalamus

Midbrain

Pons

Medulla — Sensory fibers cross over

Spinal cord

Fasciculus cuneatus

Sensory impulse from skin

Pedigo

Of particular importance is the region where the parietal, temporal, and occipital association areas come together, near the posterior end of the lateral sulcus. This region is called the *general interpretative area,* and it plays the primary role in complex thought processing. If it is injured the person may be able to recognize words but be unable to arrange them to express a thought, or may be able to read but be unable to understand the idea presented in writing. (See fig. 10.38.)

The interpretative function of the cerebral cortex is usually more highly developed in one hemisphere than the other. Thus, one hemisphere is *dominant* in intellectual functions. In 90% of the cases the left hemisphere is dominant. In other persons the hemispheres are equally dominant, although occasionally the right one dominates.

Chart 10.4 summarizes the functions of the cerebral lobes.

If the general interpretative area of the dominant hemisphere of a child under six years of age is destroyed, the corresponding region on the opposite side may develop so that the child eventually functions nearly normally. If this happens to an adult, however, the corresponding region may develop only slight interpretative functions, and the person is likely to have a severe intellectual disability.

Chart 10.4	Functions of the cerebral lobes
Lobe	**Functions**
Frontal lobes	Motor areas control movements of voluntary skeletal muscles.
	Association areas carry on higher intellectual processes such as those required for concentration, planning, complex problem-solving, and judging the consequences of behavior.
Parietal lobes	Sensory areas are responsible for the sensations of temperature, touch, pressure, and pain from the skin.
	Association areas function in the understanding of speech and in using words to express thoughts and feelings.
Temporal lobes	Sensory areas are responsible for hearing.
	Association areas are used in the interpretation of sensory experiences and in the memory of visual scenes, music, and other complex sensory patterns.
Occipital lobes	Sensory areas are responsible for vision.
	Association areas function in combining visual images with other sensory experiences.

Fig. 10.38 Some association areas of the cerebral cortex.

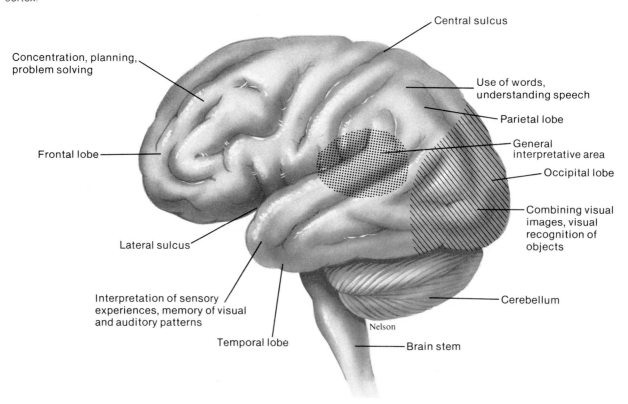

Central sulcus

Concentration, planning, problem solving

Use of words, understanding speech

Parietal lobe

Frontal lobe

General interpretative area

Occipital lobe

Combining visual images, visual recognition of objects

Lateral sulcus

Interpretation of sensory experiences, memory of visual and auditory patterns

Cerebellum

Nelson

Temporal lobe

Brain stem

Cerebral Injuries

The effects of injuries to the cerebral cortex depend on which areas are damaged and to what extent. When particular portions of the cortex are injured, the special functions of these portions are likely to be lost or at least depressed.

It is often possible to deduce the location and extent of a brain injury by determining what abilities the patient is missing. For example, if the motor areas of one frontal lobe have been damaged, the person is likely to be partially or completely paralyzed on the opposite side of the body. Similarly, if the visual cortex of one occipital lobe is injured, the person may suffer partial blindness in both eyes; if both visual cortices are damaged, the blindness may be total.

A person with damage to the frontal lobes may have difficulty in concentrating on complex mental tasks. Such an individual usually appears disorganized and is easily distracted. A person who suffers damage to association areas of the temporal lobes may have difficulty recognizing printed words or arranging words into meaningful thoughts.

Since neurons of the central nervous system seem to have little, if any, ability to regenerate, injuries to the cerebrum are likely to produce permanent loss of functions. This is particularly true when motor or sensory areas are involved. When association areas are injured, however, uninjured regions of the cerebral cortex may gradually take over some of the lost functions.

Normal brain functions are accompanied by continuous electrical activity that can be detected and recorded by placing electrodes on the surface of the head and connecting these to a recording device. The resulting patterns are called *brain waves,* and the record of these waves is an *electroencephalogram* (EEG). Such recordings are sometimes used to help diagnose abnormal brain conditions such as in epilepsy, brain injuries, or tumors.

1. List the major divisions of the brain.
2. Describe the location of the cerebral cortex.
3. What are the major functions of the cerebrum?
4. Why is an injury on one side of the brain likely to produce effects on the other side of the body?

Fig. 10.39 Anterior view of the ventricles within the cerebral hemispheres and brain stem.

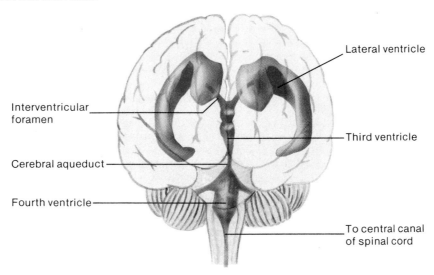

Interventricular foramen

Cerebral aqueduct

Fourth ventricle

Lateral ventricle

Third ventricle

To central canal of spinal cord

Fig. 10.40 Lateral view of the ventricles of the brain.

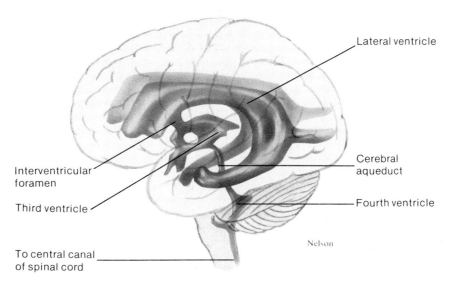

Interventricular foramen

Third ventricle

To central canal of spinal cord

Lateral ventricle

Cerebral aqueduct

Fourth ventricle

Nelson

The Ventricles and Cerebrospinal Fluid

Within the cerebral hemispheres and brain stem is a series of interconnected cavities called **ventricles,** shown in figures 10.39 and 10.40. These spaces are continuous with the central canal of the spinal cord and, like it, they are filled with cerebrospinal fluid.

The largest of the ventricles are the *lateral ventricles* (first and second ventricles), which extend into the cerebral hemispheres and occupy portions of the frontal, temporal, and occipital lobes.

A narrow space that constitutes the *third ventricle* is located in the midline of the brain, beneath the corpus callosum. This ventricle communicates with the lateral ventricles through openings (*interventricular foramina*) in its anterior end.

The *fourth ventricle* is located in the brain stem just in front of the cerebellum. It is connected to the third ventricle by a narrow canal, the *cerebral aqueduct* (aqueduct of Sylvius), which passes lengthwise through the brain stem. This ventricle is continuous with the central canal of the spinal cord and has openings in its roof that lead into the subarachnoid space of the meninges.

Cerebrospinal Fluid. The cerebrospinal fluid is secreted by tiny cauliflowerlike masses of specialized capillaries called **choroid plexuses.** These structures project outward from the inner walls of the ventricles. They are covered by epitheliallike *ependymal cells* of the neuroglia. (See fig. 10.41.)

Fig. 10.41 Cerebrospinal fluid is secreted by choroid plexuses in the walls of the ventricles. The fluid circulates through the ventricles and central canal, enters the subarachnoid space, and is reabsorbed into the blood of the dural sinuses through arachnoid granulations.

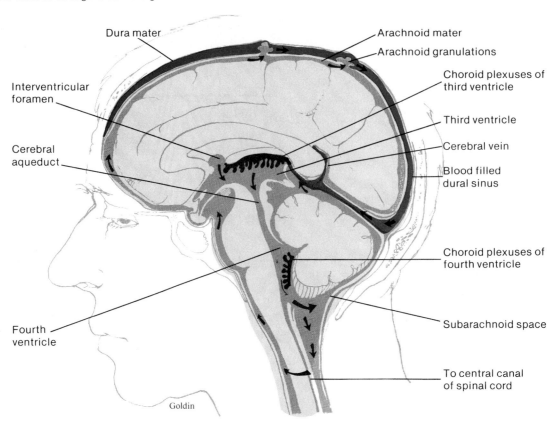

Dura mater
Interventricular foramen
Cerebral aqueduct
Fourth ventricle
Goldin
Arachnoid mater
Arachnoid granulations
Choroid plexuses of third ventricle
Third ventricle
Cerebral vein
Blood filled dural sinus
Choroid plexuses of fourth ventricle
Subarachnoid space
To central canal of spinal cord

Although choroid plexuses occur in the medial walls of the lateral ventricles and the roofs of the third and fourth ventricles, most of the cerebrospinal fluid seems to arise in the lateral ventricles. From there it circulates slowly into the third and fourth ventricles and into the central canal of the spinal cord. It also enters the subarachnoid space of the meninges by passing through the wall of the fourth ventricle near the cerebellum.

Nearly 800 milliliters of cerebrospinal fluid are secreted daily, but only about 140 milliliters are present normally. Thus, almost all the cerebrospinal fluid that is formed completes its circuit by being reabsorbed into the blood. This reabsorption occurs gradually through fingerlike structures called *arachnoid granulations* that project from the subarachnoid space into the blood-filled dural sinuses. (See fig. 10.41.)

Cerebrospinal fluid is a clear liquid that differs slightly in composition from the fluid that leaves the capillaries in other parts of the body. Specifically, it contains a greater concentration of sodium and lesser concentrations of glucose and potassium than do other extracellular fluids. Its primary function, however, seems to be protective.

As it occupies the subarachnoid space of the meninges, cerebrospinal fluid completely surrounds the brain and spinal cord. In effect, these organs float in the fluid, which protects them by absorbing shocks and other forces that might otherwise jar and damage their delicate tissues. Cerebrospinal fluid also aids in maintaining a stable ionic concentration in the central nervous system and provides a pathway to the blood for waste substances.

Because cerebrospinal fluid is secreted and reabsorbed continuously, the fluid pressure in the ventricles remains relatively constant. Sometimes, however, an infection, a tumor, or a blood clot interferes with the circulation of this fluid, and the pressure within the ventricles increases. When this happens, there is danger that the brain tissues may be injured by being forced against the skull.

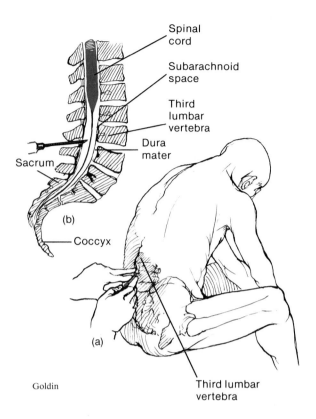

Fig. 10.42 (a) A lumbar puncture is performed by inserting a fine needle between the third and fourth lumbar vertebrae and (b) withdrawing a sample of cerebrospinal fluid from the subarachnoid space.

Spinal cord

Subarachnoid space

Third lumbar vertebra

Dura mater

Sacrum

(b)

Coccyx

(a)

Goldin

Third lumbar vertebra

In an infant whose cranial sutures have not yet united, increasing pressure within the ventricles may cause an enlargement of the cranium called *hydrocephalus* or "water on the brain." As the ventricles become dilated and the brain tissue is forced outward, the blood supply to these tissues is likely to be reduced, and the tissues may degenerate.

Hydrocephalus is often treated surgically. In this procedure, a shunt is employed to drain fluid away from the cranial cavity and into some other body region where it can be reabsorbed into the blood or excreted.

The pressure of cerebrospinal fluid can be determined by performing a *lumbar puncture*. This procedure, shown in figure 10.42, involves inserting a fine needle into the subarachnoid space between the third and fourth or the fourth and fifth lumbar vertebrae— a level below the end of the spinal cord. The pressure of the fluid, which is usually about 10 millimeters of mercury, is measured by means of a device called a *manometer*. At the same time, samples of cerebrospinal fluid may be withdrawn and tested for the presence of abnormal constituents. The presence of red blood cells, for example, is abnormal and may indicate the existence of a hemorrhage somewhere in the central nervous system. A similar procedure is sometimes used to drain excess fluid, which has accumulated as a result of injury or disease, from the subarachnoid space.

1. Where are the ventricles of the brain located?
2. Explain the pattern of cerebrospinal fluid circulation.
3. Explain how the pressure of cerebrospinal fluid can be determined.

The Brain Stem

The **brain stem**, shown in figure 10.43, is a bundle of nerve tissue that extends downward from the base of the cerebrum to the level of the foramen magnum. It consists of numerous tracts of myelinated nerve fibers and several masses of gray matter called *nuclei*. These nuclei are associated with various special senses and visceral functions. The parts of the brain stem include the diencephalon, midbrain, pons, and medulla oblongata.

The Diencephalon. The **diencephalon** is located between the cerebral hemispheres and above the midbrain; it generally surrounds the third ventricle. It is composed largely of gray matter organized into nuclei. Among these, a dense nucleus, called the *thalamus,* bulges into the third ventricle from each side. Another region of the diencephalon that includes many nuclei is called the *hypothalamus.* It lies below the thalamic nuclei and forms the lower walls and the floor of the third ventricle.

Parts associated with the hypothalamus include (a) the **optic chiasma,** where the optic nerve fibers from the eyes form a crossing; (b) the *infundibulum,* a conical process behind the optic chiasma to which the pituitary gland is attached; (c) the **pituitary gland,** which hangs from the floor of the hypothalamus; and (d) the *mamillary bodies,* which appear as two rounded structures behind the infundibulum.

The **thalamus** serves as a central relay station for sensory impulses traveling upward from other parts of the nervous system to the cerebral cortex. It receives all sensory impulses (except those associated with the sense of smell) and channels them to appropriate regions of the cortex for interpretation.

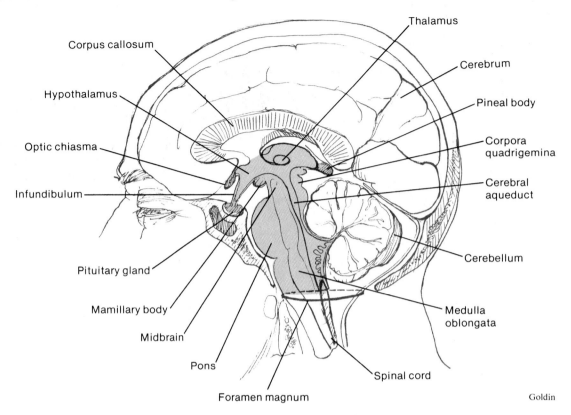

Although the cerebral cortex pinpoints the origin of sensory stimulation, the thalamus seems to produce a general awareness of certain sensations such as pain, touch, and temperature.

The **hypothalamus** is interconnected by nerve fibers to the cerebral cortex, thalamus, and other parts of the brain stem so that it can receive impulses from them and send impulses to them. The hypothalamus plays key roles in maintaining homeostasis by regulating a variety of visceral activities and by serving as a link between the nervous and endocrine systems.

Among the many important functions of the hypothalamus are the following:

1. Regulation of heart rate and arterial blood pressure.

2. Regulation of body temperature.

3. Regulation of water and electrolyte balance.

4. Control of hunger and regulation of body weight.

5. Control of movements and glandular secretions of the stomach and intestines.

6. Production of neurosecretory substances, that stimulate the pituitary gland to release various hormones.

7. Regulation of sleep and wakefulness.

In an infant, the hypothalamus is less effective in regulating body temperature than it is in older children and adults. Consequently, changes in environmental temperatures may be reflected in corresponding changes in an infant's body temperature. In a cold environment, for example, an infant's body temperature may drop to a level that threatens its life.

Structures in the general region of the diencephalon also play important roles in the control of emotional responses. For example, portions of the cerebral cortex in the medial parts of the frontal and temporal lobes are interconnected with the hypothalamus, thalamus, basal ganglia, and other deep nuclei. Together these structures comprise a complex called the limbic system.

The **limbic system** can modify the way a person acts because it functions to produce such emotional feelings as fear, anger, pleasure, and sorrow. More specifically, the limbic system seems to recognize upsets in a person's physical or psychological condition that might threaten survival. By causing pleasant or unpleasant feelings about experiences, the limbic system guides the person into behavior that is likely to increase the chance of survival.

The diencephalon also includes the **pineal gland** (body), which is a small, cone-shaped structure attached to the upper portion of the thalamus. Its possible function as an endocrine gland is discussed in chapter 12.

The Midbrain. The **midbrain** (mesencephalon) is a short section of the brain stem located between the diencephalon and the pons. It contains bundles of myelinated nerve fibers that join lower parts of the brain stem and spinal cord with higher parts of the brain. The midbrain includes several masses of gray matter that serve as reflex centers, and the *cerebral aqueduct* that connects the third and fourth ventricles. (See fig. 10.43.)

Two prominent bundles of nerve fibers on the underside of the midbrain comprise the *cerebral peduncles.* These fibers include the corticospinal tracts and are the main motor pathways between the cerebrum and lower parts of the nervous system. Beneath the cerebral peduncles are some large bundles of sensory fibers that carry impulses upward to the thalamus.

Two pairs of rounded knobs on the upper surface of the midbrain mark the location of four nuclei known collectively as *corpora quadrigemina.* These masses contain the centers for certain visual reflexes, such as those responsible for moving the eyes to view something as the head is turned. They also contain the auditory reflex centers that operate when it is necessary to move the head so that sounds can be heard more distinctly.

Near the center of the midbrain is a mass of gray matter called the *red nucleus.* This nucleus communicates with the cerebellum and with centers of the spinal cord, and it functions in reflexes concerned with the maintenance of posture.

The Pons. The **pons** appears as a rounded bulge on the underside of the brain stem where it separates the midbrain from the medulla oblongata. The dorsal portion of the pons consists largely of longitudinal nerve fibers, which relay impulses to and from the medulla oblongata and the cerebrum. Its ventral portion contains large bundles of transverse nerve fibers, which transmit impulses from the cerebrum to centers within the cerebellum.

Several nuclei of the pons relay sensory impulses from peripheral nerves to higher brain centers. Other nuclei function with centers of the medulla oblongata in regulating the rate and depth of breathing.

Medulla Oblongata. The **medulla oblongata** is an enlarged continuation of the spinal cord extending from the level of the foramen magnum to the pons. Its dorsal surface is flattened to form the floor of the fourth ventricle, and its ventral surface is marked by the corticospinal tracts, most of whose fibers cross over at this level. On each side of the medulla oblongata is an oval swelling called the *olive,* from which a large bundle of nerve fibers arises and passes to the cerebellum.

Because of its location, all ascending and descending nerve fibers connecting the brain and spinal cord must pass through the medulla oblongata. As in the spinal cord, the white matter of the medulla surrounds a central mass of gray matter. Here, however, the gray matter is broken up into nuclei that are separated by nerve fibers. Some of these nuclei relay ascending impulses to the other side of the brain stem and then onto higher brain centers. The *nucleus gracilis* and the *nucleus cuneatus,* for example, receive sensory impulses from fibers of the fasiculus gracilis and the fasiculus cuneatus and pass them on to the thalamus or the cerebellum.

Other nuclei within the medulla oblongata function as control centers for vital visceral activities. These centers include the following:

1. **Cardiac center.** Impulses originating in the cardiac center are transmitted to the heart on a pair of peripheral nerves. Impulses on these nerves can cause the heart to beat more slowly or more rapidly than it would otherwise.

2. **Vasomotor center.** Certain cells of the vasomotor center initiate impulses that travel to smooth muscles in the walls of arterioles and stimulate them to contract. This action causes vasoconstriction and a consequent rise in arteriole blood pressure. Other cells of the vasomotor center produce the opposite effect—a decrease in vasoconstriction and a drop in the arteriole blood pressure.

3. **Respiratory center.** The respiratory center functions with centers in the pons to regulate the rate and depth of breathing.

Still other nuclei within the medulla oblongata function as centers for certain nonvital reflexes, including those associated with coughing, sneezing, swallowing, and vomiting. Since the medulla also contains vital reflex centers, injuries to this part of the brain stem often are fatal.

Scattered throughout the medulla oblongata, pons, and midbrain is a complex network of nerve fibers associated with tiny islands of gray matter. This network, the **reticular formation** (reticular activating system), extends from the upper portion of the spinal cord into the diencephalon (fig. 10.44). Its intricate

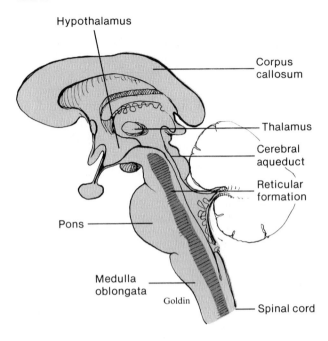

Fig. 10.44 The reticular formation extends from the upper portion of the spinal cord into the diencephalon. What is the function of this complex network of nerve fibers?

Hypothalamus

Corpus callosum

Thalamus

Cerebral aqueduct

Reticular formation

Pons

Medulla oblongata

Goldin

Spinal cord

1. List the structures of the brain stem.
2. What are the major functions of the thalamus? The hypothalamus?
3. How may the limbic system influence a person's behavior?
4. What vital reflex centers are located in the brain stem?
5. What is the function of the reticular formation?

The Cerebellum

The **cerebellum** (fig. 10.45 and color plate 22) is a large mass of tissue located below the occipital lobes of the cerebrum and posterior to the pons and medulla oblongata. It consists of two lateral hemispheres partially separated by a layer of dura mater called the *falx cerebelli.* These hemispheres are connected in the midline by a structure called the *vermis.*

Like the cerebrum, the cerebellum is composed primarily of white matter with a thin layer of gray matter, the **cerebellar cortex,** on its surface. This cortex is doubled over on itself in a series of complex folds with myelinated nerve fibers branching into them. As a result, a cut into the cerebellum reveals a treelike pattern of white matter, called the *arbor vitae,* that is surrounded by gray matter. A number of nuclei lie deep within each cerebellar hemisphere, the largest and most important of which is the *dentate nucleus.*

The cerebellum communicates with other parts of the central nervous system by means of three pairs of nerve tracts called *cerebellar peduncles* (fig. 10.46). One pair, the *inferior peduncles,* brings impulses to the cerebellum from the medulla oblongata and spinal cord. The *middle peduncles* transmit impulses from the pons to the cerebellum, and the *superior peduncles* carry impulses from the dentate nucleus to the midbrain. Some of these tracts travel upward from the midbrain to the motor areas of the cerebral cortex, while others pass downward through the pons, medulla oblongata, and spinal cord.

The cerebellum functions mainly as a reflex center in the coordination of skeletal muscle movements. The sensory impulses involved in these reflexes come from receptors, called **proprioceptors,** that are found in muscles, tendons, and joints, and from special sense organs such as the eyes and ears.

As a result of the sensory information it receives, the cerebellum becomes aware of the conditions of muscles, the attitudes of joints, and the positions of various body parts. When this information is analyzed, the cerebellum acts on it to stimulate or inhibit muscles at appropriate times, making complex muscular activities possible. The cerebellum also helps to maintain muscular tone.

system of nerve fibers interconnects centers of the hypothalamus, basal ganglia, cerebellum, and cerebrum with fibers in all the major ascending and descending tracts.

When sensory impulses reach the reticular formation, it responds by signaling the cerebral cortex, activating it into a state of wakefulness. Without this arousal, the cortex remains unaware of stimulation and cannot interpret sensory information or carry on thought processes. Thus, if the reticular formation ceases to function, as in certain injuries, the person remains unconscious.

The reticular formation also seems to act as a filter for incoming sensory impulses. Those impulses that are judged to be important, such as those originating in pain receptors, are passed onto the cerebral cortex, while others are disregarded. This selective action of the reticular formation frees the cortex from what would otherwise be a continual bombardment of sensory stimulation, and allows the cortex to concentrate upon more significant information.

The reticular formation also plays a role in regulating various motor activities so that coordinated muscular movements are smooth. It exerts an influence on spinal reflexes so that some are inhibited and others are enhanced.

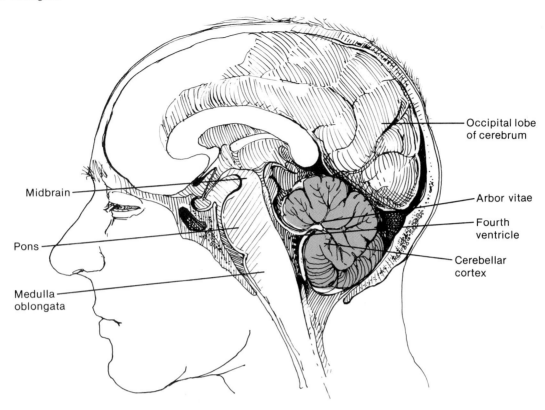

Fig. 10.46 The cerebellum communicates with other parts of the central nervous system by means of the cerebellar peduncles.

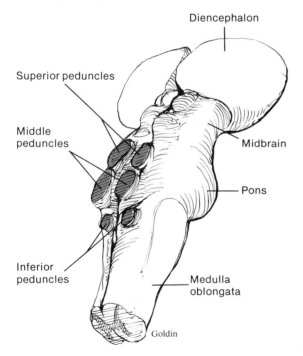

Goldin

Damage to the cerebellum is likely to result in tremors, inaccurate movements of voluntary muscles, loss of tone, reeling walk, and loss of equilibrium.

1. *Where is the cerebellum located?*
2. *What are the major functions of the cerebellum?*

The Peripheral Nervous System

The **peripheral nervous system** consists of the nerves that branch out from the central nervous system and connect it to other body parts. It includes the *cranial nerves* that arise from the brain and the *spinal nerves* that arise from the spinal cord.

This portion of the nervous system can also be subdivided into somatic and autonomic nervous systems. Generally, the **somatic system** consists of the cranial and spinal nerve fibers that connect the CNS to the skin and skeletal muscles. The **autonomic system** includes those fibers that connect the CNS to visceral organs such as the heart, stomach, intestines, and various glands. Chart 10.5 outlines the subdivisions of the nervous system.

Chart 10.5 Subdivisions of the nervous system

1. Central nervous system
   a. Brain
   b. Spinal cord
2. Peripheral nervous system
   a. Cranial nerves arising from the brain
      (1) Somatic fibers connecting to the skin and skeletal muscles
      (2) Autonomic fibers connecting to visceral organs
   b. Spinal nerves arising from the spinal cord
      (1) Somatic fibers connecting to the skin and skeletal muscles
      (2) Autonomic fibers connecting to visceral organs

The Cranial Nerves

Twelve pairs of **cranial nerves** arise from various locations on the underside of the brain. With the exception of the first pair, which springs from the cerebrum, these nerves originate from the brain stem. They pass from their sites of origin through foramina of the skull and lead to parts of the head, neck, and trunk.

Although most cranial nerves are mixed, some of those associated with special senses, such as smell and vision, contain only sensory fibers. Others that are closely involved with the activities of muscles and glands are composed primarily of motor fibers and have limited sensory functions.

When sensory fibers are present in cranial nerves, the neuron cell bodies to which the fibers are attached are located outside the brain and are usually in groups called *ganglia*. On the other hand, motor neuron cell bodies are typically located within the gray matter of the brain.

Cranial nerves are designated either by a number or a name. The numbers indicate the order in which the nerves arise from the front to the back of the brain, and the names describe their primary functions or the general distribution of their fibers. (See fig. 10.47.)

The first pair of cranial nerves, the **olfactory nerves** (I), are associated with the sense of smell and contain only sensory neurons. The cell bodies of these neurons are located in the lining of the upper nasal cavity where they serve as *olfactory receptors*. Axons from these receptors pass upward through the cribriform plates of the ethmoid bone and into *olfactory bulbs* that lie just beneath the frontal lobes of the cerebrum. Sensory impulses travel from these bulbs along *olfactory tracts* to cerebral centers where they are interpreted. The result is the sensation of smell.

Fig. 10.47 Except for the first pair, the cranial nerves arise from the brain stem. They are identified either by numbers indicating their order or by names describing their function or the general distribution of their fibers.

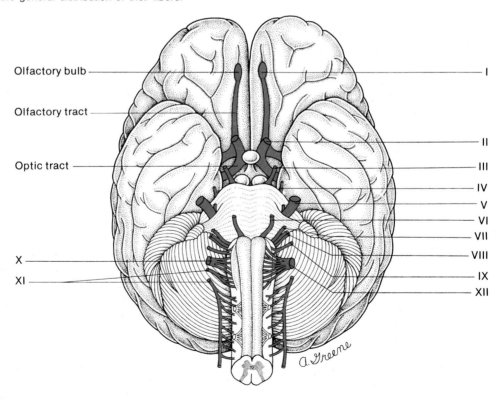

The second pair of cranial nerves, the **optic nerves** (II), lead from the eyes to the brain and are associated with the sense of vision. The sensory cell bodies of the nerve fibers occur in ganglia within the eyes, and their axons pass through the *optic foramina* of the orbits and continue into the brain.

Some of the fibers in each optic nerve cross over to the other side of the brain. Specifically, the fibers arising from the medial half of each eye cross over, while those arising from the lateral sides do not. This crossing produces the X-shaped *optic chiasma* in the hypothalamus.

Sensory impulses of the optic nerves continue along *optic tracts* to centers in the thalamus and travel from there to the visual cortices of the occipital lobes.

The third pair of cranial nerves, the **oculomotor nerves** (III), arise from the midbrain and pass into the orbits of the eyes. One component of each nerve connects to a number of voluntary muscles. These include the *levator palpebrae superioris* muscles, which function to raise the eyelids, and most of the muscles attached to the eye surfaces that cause them to move—the *superior rectus, medial rectus, inferior rectus,* and *inferior oblique* muscles.

A second portion of each oculomotor nerve is part of the autonomic nervous system, supplying involuntary muscles inside the eyes. These muscles act in adjusting the amount of light that enters the eyes and in focusing the lenses of the eyes.

Although the fibers of the oculomotor nerves are primarily motor, some sensory fibers are present. These transmit sensory information to the brain concerning the condition of various muscles.

The fourth pair, the **trochlear nerves** (IV), are the smallest cranial nerves. They arise from the midbrain and carry motor impulses to a pair of external eye muscles, the *superior obliques,* which are not supplied by the oculomotor nerves. The trochlear nerves also contain some sensory fibers that transmit information about the condition of certain muscles.

The fifth pair, the **trigeminal nerves** (V), are the largest of the cranial nerves and arise from the pons. They are mixed nerves, but their sensory portions are more extensive than their motor portions. Each sensory component includes three large branches, the ophthalmic, maxillary, and mandibular divisions (fig. 10.48).

The *ophthalmic division* consists of sensory fibers that bring impulses to the brain from the surface of the eye, the tear gland, and the skin of the anterior scalp, forehead, and upper eyelid. The fibers of the *maxillary division* carry sensory impulses from the upper teeth, upper gum, and upper lip, as well as from the mucous lining of the palate and the skin of the face. The *mandibular division* includes both motor and sensory fibers. The sensory branches transmit impulses from the scalp behind the ear, the skin of the jaw, the lower teeth, the lower gum, and the lower lip. The motor branches supply the muscles of mastication and certain muscles in the floor of the mouth.

A disorder of the trigeminal nerve called *tic douloureux* (or trigeminal neuralgia) is characterized by severe recurring pain in the face and forehead on the affected side. If the pain cannot be controlled with drugs, the sensory portion of the nerve is sometimes severed surgically. Although this procedure may relieve the pain, the patient may also lose other sensations in the parts supplied by the sensory branch. Consequently, persons who have had such surgery must be cautious when eating or drinking hot foods or liquids, because they may not sense burning. These persons must also inspect their mouths daily for the presence of food particles or damage to their cheeks from biting, which they may not feel.

The sixth pair of cranial nerves, the **abducens nerves** (VI), are quite small and originate from the pons near the medulla oblongata. They enter the orbits of the eyes and supply motor impulses to a pair of external eye muscles, the *lateral rectus* muscles. The sensory fibers of these nerves provide the brain with information concerning the condition of muscles.

The seventh pair of cranial nerves, the **facial nerves** (VII), arise from the lower part of the pons and emerge on the sides of the face. Their sensory branches are associated with taste receptors on the anterior two-thirds of the tongue, and some of their motor fibers transmit impulses to muscles of facial expression. Still other motor fibers of these nerves function in the autonomic nervous system by stimulating secretions from tear glands and certain salivary glands (submaxillary and sublingual).

The eighth pair of cranial nerves, the **vestibulocochlear nerves** (VIII, acoustic nerves), are sensory nerves and arise from the medulla oblongata. Each of these nerves has two distinct parts—a vestibular branch and a cochlear branch.

The neuron cell bodies of the *vestibular branch* fibers are located in ganglia near the vestibule and semicircular canals of the inner ear. These structures contain receptors that are sensitive to changes in the position of the head. The impulses they initiate pass into the cerebellum, where they are used in reflexes associated with the maintenance of equilibrium.

The neuron cell bodies of the *cochlear branch* fibers are located in a ganglion of the cochlea, a part of the inner ear that houses the hearing receptors.

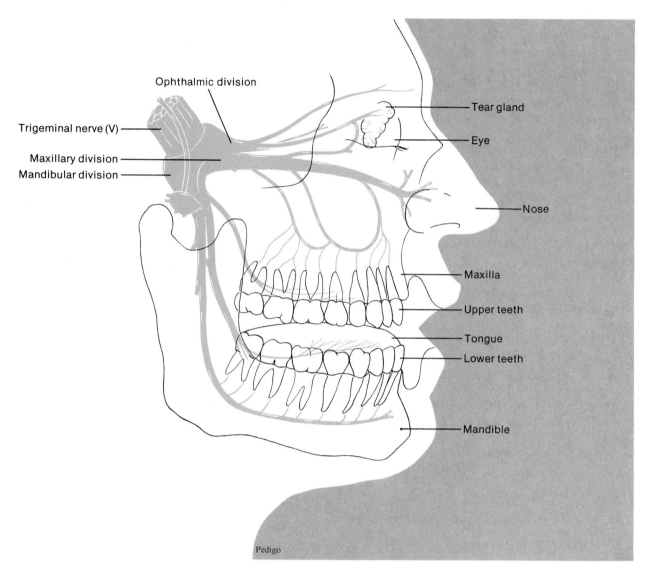

Fig. 10.48 The trigeminal nerve has three large branches that supply various parts of the head and face.

Ophthalmic division

Trigeminal nerve (V)

Maxillary division
Mandibular division

Tear gland

Eye

Nose

Maxilla

Upper teeth

Tongue
Lower teeth

Mandible

Pedigo

Impulses from this branch pass through the pons and medulla oblongata on their way to the temporal lobe for interpretation.

The ninth pair of cranial nerves, the **glosso-pharyngeal nerves** (IX), are associated with the tongue and pharynx, as the name implies. These nerves arise from the medulla oblongata, and although they are mixed nerves, their predominant fibers are sensory. These fibers carry impulses from the lining of the pharynx, tonsils, and posterior third of the tongue to the brain. Fibers in the motor component of the glossopharyngeal nerves innervate constrictor muscles in the wall of the pharynx that function in swallowing.

Other branches of the glossopharyngeal nerves function in the autonomic nervous system. Some sensory fibers, for example, conduct impulses from spe-

cial receptors in the wall of an artery in the neck (carotid artery) to a center in the medulla oblongata. This information is used in the regulation of blood pressure. Certain autonomic motor fibers lead to the parotid salivary gland in front of the ear, and can stimulate it to secrete saliva.

The tenth pair of cranial nerves, the **vagus nerves** (X), originate in the medulla oblongata and extend downward through the neck into the chest and abdomen. These nerves are mixed, and although they contain both somatic and autonomic branches, the autonomic fibers are the predominant ones.

Among the somatic components are motor fibers that carry impulses to muscles of the larynx. These fibers are associated with speech and with swallowing reflexes that employ muscles in the soft palate and pharynx. Vagal sensory fibers carry impulses from the lining of the pharynx, larynx, esophagus, and the viscera of the thorax and abdomen to the brain.

Autonomic motor fibers of the vagus nerves supply the heart and a variety of smooth muscles and glands in the visceral organs of the thorax and abdomen.

The eleventh pair of cranial nerves, the **accessory nerves** (XI), have origins in the medulla oblongata and the spinal cord, and thus have both cranial and spinal branches.

Each *cranial branch* of an accessory nerve joins a vagus nerve and carries impulses to muscles of the soft palate, pharynx, and larynx. The *spinal branch* descends into the neck and supplies motor fibers to the trapezius and sternocleidomastoid muscles.

The twelfth pair of cranial nerves, the **hypoglossal nerves** (XII), arise from the medulla oblongata and pass into the tongue. They consist primarily of motor fibers that carry impulses to muscles that move the tongue in speaking, chewing, and swallowing.

The functions of the cranial nerves are summarized in chart 10.6.

Chart 10.6 Functions of cranial nerves

Nerve		Type	Function
I	Olfactory	Sensory	Sensory fibers transmit impulses associated with the sense of smell.
II	Optic	Sensory	Sensory fibers transmit impulses associated with the sense of vision.
III	Oculomotor	Primarily motor	Motor fibers transmit impulses to muscles that raise eyelids, move eyes, adjust amount of light entering eyes, and focus lenses.
			Some sensory fibers transmit impulses associated with the condition of muscles.
IV	Trochlear	Primarily motor	Motor fibers transmit impulses to muscles that move the eyes.
			Some sensory fibers transmit impulses associated with the condition of muscles.
V	Trigeminal Ophthalmic division	Mixed	Sensory fibers transmit impulses from the surface of the eyes, the tear glands, scalp, forehead, and upper eyelids.
	Maxillary division		Sensory fibers transmit impulses from the upper teeth, upper gum, upper lip, lining of the palate, and skin of the face.
	Mandibular division		Sensory fibers transmit impulses from the scalp, skin of the jaw, lower teeth, lower gum, and lower lip.
			Motor fibers transmit impulses to muscles of mastication and muscles in floor of the mouth.
VI	Abducens	Primarily motor	Motor fibers transmit impulses to muscles that move the eyes.
			Some sensory fibers transmit impulses associated with the condition of muscles.
VII	Facial	Mixed	Sensory fibers transmit impulses associated with taste receptors of the anterior tongue.
			Motor fibers transmit impulses to muscles of facial expression, tear glands, and salivary glands.
VIII	Vestibulo-cochlear	Sensory	
	Vestibular branch		Sensory fibers transmit impulses associated with sense of equilibrium.
	Cochlear branch		Sensory fibers transmit impulses associated with the sense of hearing.
IX	Glosso-pharyngeal	Mixed	Sensory fibers transmit impulses from the pharynx, tonsils, posterior tongue, and carotid arteries.
			Motor fibers transmit impulses to muscles of pharynx used in swallowing and to salivary glands.
X	Vagus	Mixed	Motor fibers transmit impulses to muscles associated with speech and swallowing, the heart, and smooth muscles of visceral organs in the thorax and abdomen.
			Sensory fibers transmit impulses from the pharynx, larynx, esophagus, and visceral organs of the thorax and abdomen.
XI	Accessory	Motor	
	Cranial branch		Motor fibers transmit impulses to muscles of soft palate, pharynx, and larynx.
	Spinal branch		Motor fibers transmit impulses to muscles of neck and back.
XII	Hypoglossal	Motor	Motor fibers transmit impulses to muscles that move the tongue.

Cranial Nerve Injuries

The consequences of cranial nerve injuries depend on the location and extent of the injuries. For example, if only one member of a nerve pair is damaged, loss of function is limited to the affected side, but if both nerves are injured losses occur on both sides. Also, if a nerve is severed completely, the functional loss is total; if the cut is incomplete the loss may be partial.

---

1. Define the peripheral nervous system.
2. Describe the general locations of the structures innervated by the cranial nerves.
3. Name the cranial nerves and list the major functions of each.

The Spinal Nerves

Thirty-one pairs of **spinal nerves** originate from the spinal cord. They are all mixed nerves, and they provide a two-way communication system between the spinal cord and parts in the arms, legs, neck, and trunk.

Although spinal nerves are not named individually, they are grouped according to the level from which they arise, and each nerve is numbered in sequence. (See fig. 10.49.) Thus, there are eight pairs of *cervical nerves* (numbered C1 to C8), twelve pairs of *thoracic nerves* (numbered T1 to T12), five pairs of *lumbar nerves* (numbered L1 to L5), five pairs of *sacral nerves* (numbered S1 to S5), and one pair of *coccygeal nerves*.

The nerves arising from the upper part of the spinal cord pass outward nearly horizontally, while those from the lower portions of the spinal cord descend at sharp angles. This arrangement is a consequence of growth. In early life, the spinal cord extends the entire length of the vertebral column, but with age the column grows more rapidly than the cord. As a result, the adult spinal cord ends at the level between the first and second lumbar vertebrae, so the lumbar, sacral, and coccygeal nerves descend to their exits beyond the end of the cord. These descending nerves form a structure called *cauda equina,* which is shaped somewhat like a horse's tail, as figure 10.50 shows.

Each spinal nerve emerges from the cord by two short branches, or *roots,* which lie within the vertebral column. The **dorsal root** (sensory root) can be identified by the presence of an enlargement called the *dorsal root ganglion.* This ganglion contains the cell bodies of the sensory neurons whose dendrites conduct impulses inward from peripheral body parts. The axons of these neurons extend through the dorsal root and into the spinal cord, where they form synapses with dendrites of other neurons.

The **ventral root** (motor root) of each spinal nerve consists of axons from motor neurons whose cell bodies are located within the gray matter of the cord. (See fig. 10.51.)

A ventral root and a dorsal root unite to form a spinal nerve, which passes outward from the vertebral canal though an *intervertebral foramen.* Just beyond its foramen, each spinal nerve divides into several parts. One of these parts, the small *meningeal branch,* reenters the vertebral canal through the intervertebral foramen and supplies the meninges and blood vessels of the cord, as well as the intervertebral ligaments and the vertebrae.

As figure 10.52 shows, a *posterior branch* of each spinal nerve turns posteriorly and innervates muscles and skin of the back. The main portion of the nerve, the *anterior branch,* continues forward to supply the muscles and skin on the front and sides of the trunk and limbs.

The spinal nerves in the thoracic and lumbar regions have a fourth or *visceral branch,* which is part of the autonomic nervous system.

Except in the thoracic region, the anterior branches of the spinal nerves combine to form complex networks, called **plexuses,** instead of continuing directly to the peripheral body parts. In a plexus, the fibers of various spinal nerves are sorted and recombined, so that fibers associated with a particular peripheral part reach it in the same nerve, even though the fibers originate from different spinal nerves. (See figure 10.53.)

Cervical Plexuses. The **cervical plexuses** lie deep in the neck on either side. They are formed by the anterior branches of the first four cervical nerves, and fibers from these plexuses supply the muscles and skin of the neck. In addition, fibers from the third, fourth, and fifth cervical nerves pass into the right and left **phrenic nerves,** which conduct motor impulses to the muscle fibers of the diaphragm.

Brachial Plexuses. The anterior branches of the lower four cervical nerves and the first thoracic nerve give rise to the **brachial plexuses.** These nets of nerve fibers are located deep within the shoulders between the neck and the axillae (armpits). The major branches emerging from the brachial plexuses include the following:

1. *Musculocutaneous nerves,* which supply muscles of the arms on the anterior sides and the skin of the forearms.

2. *Ulnar nerves,* which supply muscles of the forearms and hands, and the skin of the hands.

3. *Median nerves,* which supply muscles of the forearms and the muscles and skin of the hands.

Fig. 10.49 There are thirty-one pairs of spinal nerves.

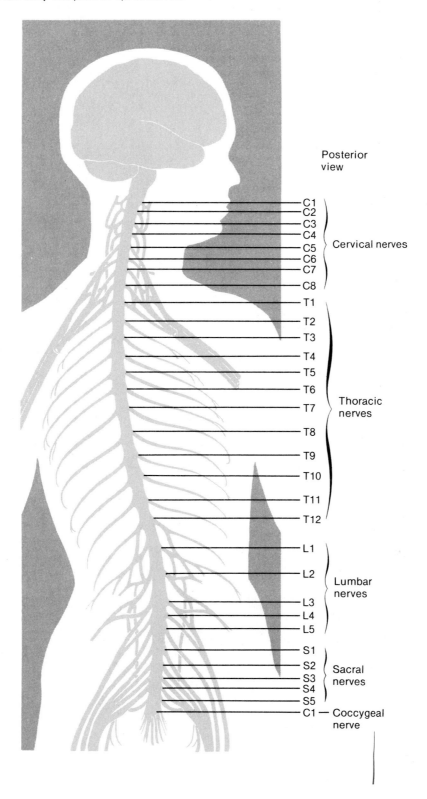

Posterior
view

C1
C2
C3
C4
C5 } Cervical nerves
C6
C7

C8

T1

T2

T3

T4

T5

T6

T7 } Thoracic
nerves

T8

T9

T10

T11

T12

L1

L2 } Lumbar
nerves

L3
L4
L5

S1

S2 } Sacral
S3   nerves
S4

S5

C1 — Coccygeal
nerve

Fig. 10.50 The spinal nerves that descend below the end of the spinal cord comprise the cauda equina.

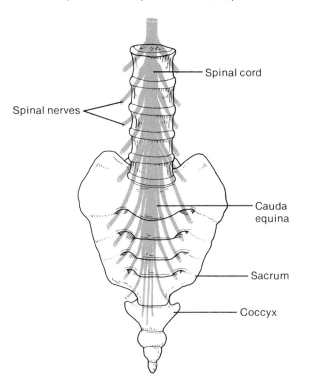

- Spinal cord
- Spinal nerves
- Cauda equina
- Sacrum
- Coccyx

Fig. 10.51 Each spinal nerve emerges from the spinal cord by a dorsal and a ventral root. What kinds of nerve fibers comprise each of these roots?

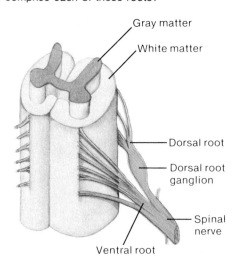

- Gray matter
- White matter
- Dorsal root
- Dorsal root ganglion
- Spinal nerve
- Ventral root

Fig. 10.52 Each spinal nerve has a posterior and an anterior branch; the thoracic and lumbar spinal nerves also have a visceral branch.

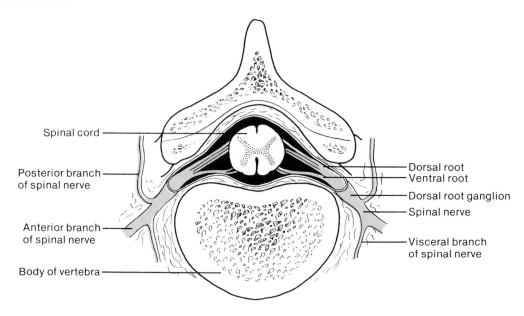

- Spinal cord
- Posterior branch of spinal nerve
- Anterior branch of spinal nerve
- Body of vertebra
- Dorsal root
- Ventral root
- Dorsal root ganglion
- Spinal nerve
- Visceral branch of spinal nerve

Fig. 10.53 The anterior branches of the spinal nerves in the thoracic region give rise to intercostal nerves. Those in other regions combine to form complex networks called plexuses.

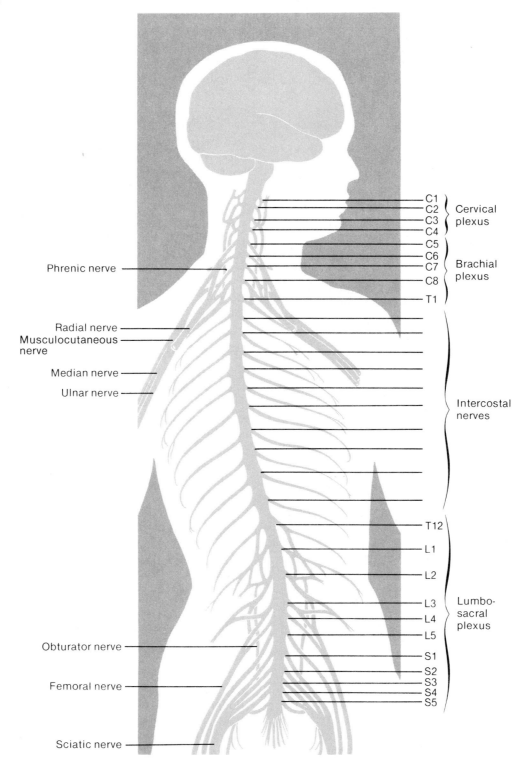

Phrenic nerve

Radial nerve
Musculocutaneous nerve

Median nerve

Ulnar nerve

Obturator nerve

Femoral nerve

Sciatic nerve

C1
C2 — Cervical
C3 — plexus
C4

C5
C6 — Brachial
C7 — plexus
C8

T1

Intercostal nerves

T12
L1
L2
L3 — Lumbo-
L4 — sacral
L5 — plexus
S1
S2
S3
S4
S5

4. *Radial nerves,* which supply muscles of the arms on the posterior sides, and the skin of the forearms and hands.

Sometimes persons whose occupations require prolonged abduction of the arm, as in painting or typing, develop intermittent or constant pain in the neck, shoulder, or some region of the arm due to excessive pressure on the brachial plexus. This condition, called *thoracic outlet syndrome,* may also be caused by congenital malformations of skeletal parts that compress the plexus during arm and shoulder movements.

Lumbosacral Plexuses. The **lumbosacral plexuses** are formed on either side by the last thoracic and the lumbar, sacral, and coccygeal nerves. These networks of nerve fibers extend from the lumbar region of the back into the pelvic cavity, giving rise to a number of motor and sensory fibers associated with the lower abdominal wall, external genitalia, buttock, thighs, legs, and feet. The major branches of these plexuses include the following:

1. *Obturator nerves,* which supply the adductor muscles of the thighs.

2. *Femoral nerves,* which divide into many branches, supplying motor impulses to muscles of the thighs and legs, and receiving sensory impulses from the skin of the thighs and lower legs.

3. *Sciatic nerves,* which are the largest and longest nerves in the body. They pass downward into the buttock and descend into the thighs, where they divide into *tibial* and *common peroneal nerves.* The many branches of these nerves supply muscles and skin in the thighs, legs, and feet.

The anterior branches of the thoracic spinal nerves do not enter a plexus. Instead, they travel into spaces between the ribs and become **intercostal nerves.** These nerves supply motor impulses to the intercostal muscles and the upper abdominal wall muscles. They also receive sensory impulses from the skin of the thorax and abdomen.

Spinal Nerve Injuries

Spinal nerves may be injured in a variety of ways including stabs, gunshot wounds, birth injuries, dislocations and fractures of vertebrae, and pressure from tumors in surrounding tissues. The nerves of the cervical plexuses are sometimes compressed by the sudden bending of the neck, called whiplash, that may occur during rear end automobile collisions. A victim of such an injury may suffer continuing headache and pain in the neck and skin, which are supplied by the cervical nerves.

If the phrenic nerves associated with the cervical plexuses are severed or damaged by a broken or dislocated vertebra, the result may be partial or complete paralysis of the diaphragm.

Nerves of a newborn's brachial plexuses are sometimes stretched or injured during childbirth when the shoulders are in a fixed position and excessive pull is applied to the head. In such cases, the newborn may suffer losses of sensory and motor functions in the arms on the injured sides, and the arms may wither as muscles atrophy.

A condition called *sciatica,* which is characterized by pain in the lower gluteal region and in the back of the thigh, occurs when the nerve roots of the lumbosacral region are irritated. This problem is usually caused by pressure from a ruptured intervertebral disk.

---

1. How are spinal nerves grouped?
2. Describe the way a spinal nerve joins the spinal cord.
3. Name and locate the major nerve plexuses.

The Autonomic Nervous System

The *autonomic nervous system* is the portion of the nervous system that functions independently (autonomously) without conscious effort on the part of the individual. This system controls the actions of smooth muscles, cardiac muscle, and most glands. It is concerned with regulating heart rate, blood pressure, breathing rate, body temperature, and other visceral functions that aid in the maintenance of homeostasis. Portions of the autonomic system are also responsive during times of emotional stress, and they serve to prepare the body to meet the demands of strenuous physical activity.

General Characteristics

Autonomic functions operate largely by reflexes in which sensory signals from visceral organs are received by nerve centers within the hypothalamus, brain stem, or spinal cord. Motor impulses travel out from these centers on peripheral nerve fibers within cranial and spinal nerves to visceral muscles or glands, which respond by contracting or secreting.

The autonomic nervous system is composed of two sections called the **sympathetic** and **parasympathetic subdivisions,** which act together. As a rule, every visceral organ is supplied with nerve fibers from each of the subdivisions. Impulses on one set of fibers tend to activate the organ, while impulses on the other set inhibit it. The actions of the organ are regulated by alternately being activated or inhibited.

Fig. 10.54 Visceral organs are usually innervated by fibers from both the parasympathetic and sympathetic divisions.

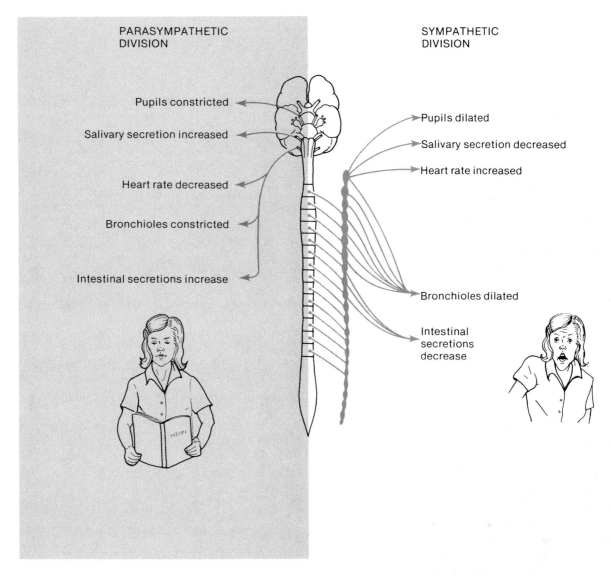

PARASYMPATHETIC DIVISION

Pupils constricted

Salivary secretion increased

Heart rate decreased

Bronchioles constricted

Intestinal secretions increase

SYMPATHETIC DIVISION

Pupils dilated

Salivary secretion decreased

Heart rate increased

Bronchioles dilated

Intestinal secretions decrease

Although the functions of the subdivisions are mixed—each activates some organs and inhibits others—they have general functional differences. The sympathetic division is concerned primarily with preparing the body for energy-expending, stressful, or emergency situations. The parasympathetic division, however, is most active under ordinary, restful conditions. It also counterbalances the effects of the sympathetic division, and restores the body to a resting state following a stressful experience. For example, during an emergency the sympathetic division will cause the heart and breathing rates to increase, and following the emergency, the parasympathetic division will slow these activities. (See fig. 10.54.)

Autonomic Nerve Fibers

The nerve fibers of the autonomic nervous system are primarily motor fibers. Unlike the motor pathways of the somatic nervous system, which usually include a single neuron between the brain or spinal cord and an effector, those of the autonomic system involve two neurons, as shown in figure 10.55. The cell body of one neuron is located in the brain or spinal cord. Its axon, the **preganglionic fiber,** leaves the CNS and forms a synapse with one or more nerve fibers whose cell bodies are housed within an autonomic ganglion outside the brain or spinal cord. The axon of such a second neuron is called a **postganglionic fiber,** and it extends to a visceral effector.

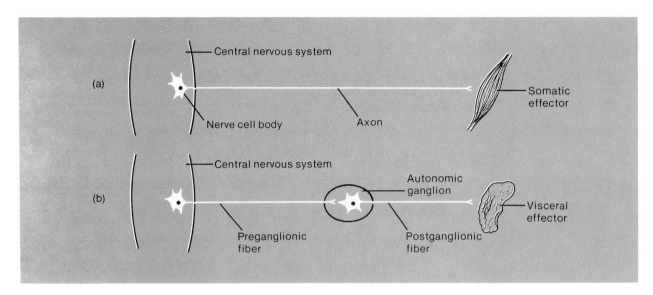

In the sympathetic division (thoracolumbar division), the preganglionic fibers begin from neurons in the gray matter of the spinal cord. (See fig. 10.56.) They leave the cord with the ventral roots of spinal nerves in the first thoracic through the second lumbar segments. After traveling a short distance, these fibers leave the spinal nerves, and each enters a member of a chain of *sympathetic ganglia.* One of these chains extends longitudinally along each side of the vertebral column.

Within a sympathetic ganglion, a preganglionic fiber forms a synapse with a second neuron. The axon of this neuron, the postganglionic fiber, typically returns to a spinal nerve and extends with it to a visceral effector. (See fig. 10.57.) An important exception to this occurs in a set of preganglionic fibers that pass through the chain ganglia and into the medulla of each adrenal gland. There the fibers end on special hormone-secreting cells that release norepinephrine and epinephrine. The function of the adrenal medulla is discussed in chapter 12.

The preganglionic fibers of the parasympathetic division (craniosacral division) arise from the *brain stem* and the *sacral region* of the spinal cord. From there, they lead outward on cranial or sacral nerves to ganglia located near or within various visceral organs. The relatively short postganglionic fibers continue from the ganglia to specific muscles or glands within these visceral organs. (See fig. 10.58.)

The parasympathetic preganglionic fibers associated with parts of the head are included in the oculomotor, facial, and glossopharyngeal nerves. Those that innervate organs of the thorax and upper abdomen are parts of the vagus nerves. (The vagus nerves carry about 75% of all parasympathetic fibers.) Preganglionic fibers arising from the sacrum are found within the branches of the second through the fourth sacral spinal nerves, and they carry impulses to visceral organs within the pelvis.

Autonomic Transmitter Substances

The preganglionic fibers of the sympathetic and parasympathetic divisions all secrete *acetylcholine;* their postganglionic fibers, however, use different transmitter substances. Most sympathetic postganglionic fibers secrete *norepinephrine* (noradrenalin), and for this reason they are called **adrenergic fibers.** The parasympathetic postganglionic fibers secrete *acetylcholine* and are called **cholinergic fibers.** These different postganglionic transmitter substances (mediators) are responsible for the different effects that the sympathetic and parasympathetic divisions have on visceral organs. (See fig. 10.59.)

Although each division can activate some effectors and inhibit others, most visceral organs are controlled primarily by one division. In other words, the divisions usually are not actively antagonistic. For example, the diameter of most blood vessels is regulated by the sympathetic division. Smooth muscles in the walls of these vessels are continuously stimulated

Fig. 10.56 The preganglionic fibers of the sympathetic division arise from the thoracic and lumbar regions of the spinal cord.

SYMPATHETIC DIVISION

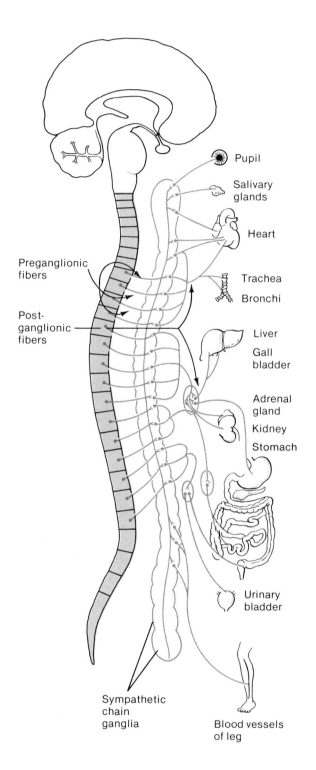

Pupil

Salivary glands

Heart

Preganglionic fibers

Trachea

Bronchi

Post-ganglionic fibers

Liver

Gall bladder

Adrenal gland

Kidney

Stomach

Urinary bladder

Sympathetic chain ganglia

Blood vessels of leg

and thus maintained in a state of partial contraction (tone). The diameter of a vessel can be increased (dilated) by decreasing the degree of sympathetic stimulation, which allows the muscular wall to relax. Conversely, the vessel can be constricted by increasing the amount of sympathetic stimulation.

The parasympathetic division is dominant in digestive gland activity. Parasympathetic impulses stimulate glandular secretions, and sympathetic impulses have little direct effect on the glands. The effects of adrenergic and cholinergic fibers on some visceral effectors are summarized in chart 10.7.

Chart 10.7 Effects of transmitter substances upon visceral effectors or actions

Visceral Effector or Action	Response to Adrenergic Fibers (sympathetic)	Response to Cholinergic Fibers (para-sympathetic)
Pupil of the eye	Dilation	Constriction
Heart rate	Increases	Decreases
Bronchioles of lungs	Dilation	Constriction
Muscles of intestinal wall	Slows peristaltic action	Speeds peristaltic action
Intestinal glands	Secretion decreases	Secretion increases
Coronary arteries	Dilation	Constriction
Blood distribution	More blood to skeletal muscles; less blood to digestive organs	More blood to digestive organs; less blood to skeletal muscles
Blood glucose concentration	Increases	Decreases
Salivary glands	Secretion decreases	Secretion increases
Tear glands	No action	Secretion
Muscles of gallbladder	Relaxation	Contraction
Muscles of urinary bladder	Relaxation	Contraction

Fig. 10.57 Sympathetic fibers leave the spinal cord in the ventral roots of spinal nerves, enter sympathetic ganglia, and synapse with other neurons that extend to visceral effectors.

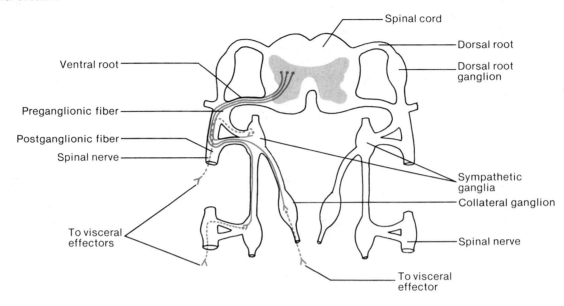

The acetylcholine released by cholinergic fibers is rapidly decomposed by the action of *cholinesterase.* (This decomposition also occurs in the neuromuscular junctions of skeletal muscle.) Thus, acetylcholine usually produces an effect for only a fraction of a second. Most of the norepinephrine released from adrenergic fibers, however, is taken back into the nerve endings by an active transport mechanism. This may take a few seconds, and during that time some molecules may diffuse into nearby tissues where they are decomposed by an enzyme. For this reason, norepinephrine is likely to produce a more prolonged effect than acetylcholine. In fact, when the adrenal medulla releases norepinephrine and epinephrine into the blood in response to sympathetic stimulation, these substances may trigger sympathetic responses in organs throughout the body that last up to 30 seconds.

Many drugs influence autonomic functions in a variety of ways. Some, like epinephrine, have effects similar to norepinephrine. Others, like ephedrine, stimulate the release of norepinephrine from sympathetic nerve endings; still others, like reserpine, inhibit sympathetic activity by preventing the synthesis of norepinephrine. Another group of drugs, which includes pilocarpine, produce parasympathetic effects, and some, like atropine, block the action of acetylcholine in visceral effectors.

Conscious Control of Autonomic Actions

Until recently it was believed that activities of the autonomic nervous system were beyond conscious control. It is now known that some people can learn to alter their visceral responses. In order to accomplish this, a person needs to receive information concerning the state of the visceral action to be controlled. Normally a person is unaware of such activities, but electronic equipment can measure blood pressure, for example, and continuously feed back this information to the person. Then the person can consciously try to produce a visceral response that will alter the blood pressure in a desired way. (See fig. 10.60.)

This procedure, called **biofeedback,** has been used successfully to control heart rates and blood pressures, and to alter the diameters of blood vessels, thus altering the amount of blood flowing into certain regions of the body. It also has been used to control migraine headaches, which are associated with changes in the muscular tone of certain blood vessels, and to control epileptic seizures.

PARASYMPATHETIC DIVISION

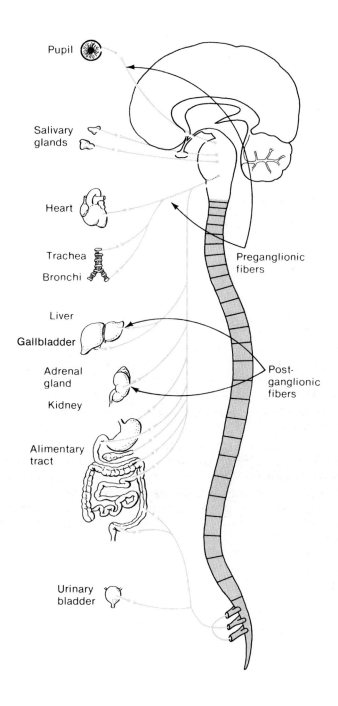

Pupil

Salivary glands

Heart

Trachea

Bronchi

Preganglionic fibers

Liver

Gallbladder

Adrenal gland

Kidney

Post-ganglionic fibers

Alimentary tract

Urinary bladder

Fig. 10.59 (a) Sympathetic fibers are adrenergic and secrete norepinephrine at the ends of the postganglionic fibers; (b) parasympathetic fibers are cholinergic and secrete acetylcholine at the ends of the postganglionic fibers.

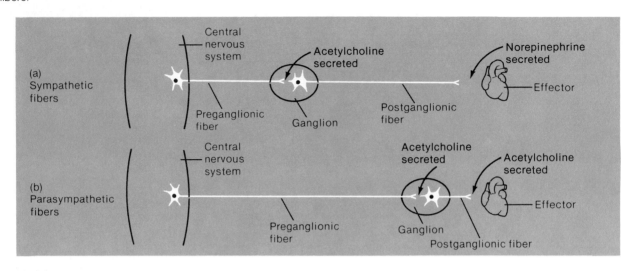

Fig. 10.60 Some persons can learn to control their visceral responses by making use of electronic equipment that monitors visceral activities.

Integration and Coordination

1. What parts of the nervous system are included in the autonomic nervous system?
2. What functions are generally controlled by the sympathetic division? By the parasympathetic division?
3. Describe a sympathetic nerve pathway. A parasympathetic nerve pathway.
4. What neurotransmitters are used in the autonomic nervous system?
5. How do the divisions of the autonomic system function to regulate visceral activities?

Clinical Terms Related to the Nervous System

analgesia (an″al-je′ze-ah—loss or reduction in the ability to sense pain, but without loss of consciousness.

analgesic (an″al-je′sik)—a pain-relieving drug.

anesthesia (an″es-the′ze-ah)—a loss of feeling.

aphasia (ah-fa′ze-ah)—a disturbance or loss in the ability to use words or to understand them, usually due to damage to cerebral association areas.

apraxia (ah-prak′se-ah)—an impairment in a person's ability to make correct use of objects.

ataxia (ah-tak′se-ah)—a partial or complete inability to coordinate voluntary movements.

cerebral palsy (ser′ĕ-bral pawl′ze)—a condition characterized by partial paralysis and lack of muscular coordination.

coma (ko′mah)—an unconscious condition in which there is an absence of responses to stimulation.

cordotomy (kor-dot′o-me)—a surgical procedure in which a nerve tract within the spinal cord is severed, usually to relieve intractable pain.

craniotomy (kra″ne-ot′o-me)—a surgical procedure in which part of the skull is opened.

electroencephalogram (EEG) (e-lek″tro-en-sef′ah-lo-gram″)—a recording of the electrical activity of the brain.

encephalitis (en″sef-ah-li′tis)—an inflammation of the brain and meninges characterized by drowsiness and apathy.

epilepsy (ep′ĭ-lep″se)—a disorder of the central nervous system that is characterized by temporary disturbances in normal brain impulses; it may be accompanied by convulsive seizures and loss of consciousness.

hemiplegia (hem″ĭ-ple′je-ah)—paralysis on one side of the body and the limbs on that side.

Huntington's chorea (hunt′ing-tunz ko-re′ah)—a rare hereditary disorder of the brain characterized by involuntary convulsive movements and mental deterioration.

laminectomy (lam″ĭ-nek′to-me)—surgical removal of the posterior arch of a vertebra, usually to relieve the symptoms of a ruptured intervertebral disk.

monoplegia (mon″o-ple′je-ah)—paralysis of a single limb.

multiple sclerosis (mul′tĭ-pl skle-ro′sis)—a disease of the central nervous system characterized by loss of myelin and the appearance of scarlike patches throughout the brain and spinal cord or both.

neuralgia (nu-ral′je-ah)—a sharp, recurring pain associated with a nerve, usually caused by inflammation or injury.

neuritis (nu-ri′tis)—an inflammation of a nerve.

paraplegia (par″ah-ple′je-ah)—paralysis of both legs.

quadriplegia (kwod″rĭ-ple′je-ah)—paralysis of all four limbs.

vagotomy (va-got′o-me)—severing of a vagus nerve.

Chapter Summary

Introduction

The actions of all body parts are directed toward the maintenance of homeostasis.

The general tasks of controlling and coordinating body activities are handled by the nervous and endocrine systems.

The nervous system provides rapid and precise control of muscles and glands.

Nerve Tissue

The organs of the nervous system contain a variety of tissues.

The brain and spinal cord constitute the central nervous system; the nerves make up the peripheral nervous system.

Nerve tissue includes neurons, which are the structural and functional units of the nervous system, and neuroglial cells.

1. Neuron structure
   a. A neuron includes a cell body, nerve fibers, and other organelles usually found in cells.
   b. Dendrites and the cell body provide receptive surfaces.
   c. A single axon arises from the cell body and may be enclosed in a myelin sheath and a neurolemma.
2. Regeneration of nerve fibers
   a. If a cell body is injured, the neuron is likely to die.
   b. If a nerve fiber is severed, its distal portion will die but the proximal portion may regenerate and reestablish its former connections provided it has a neurolemma to guide it.

Neuroglial Cells

1. Neuroglial cells occur only in the brain and spinal cord.
2. They fill spaces, support neurons, and hold nerve tissue together.
3. There are four types: astrocytes, oligodendrocytes, microglia, and ependyma.

Cell Membrane Potential

1. A cell membrane is usually polarized as a result of unequal distribution of ions.
2. Nerve cell at rest
   a. There is a high concentration of sodium ions on the outside of the membrane and a high concentration of potassium ions on the inside.
   b. There are large numbers of negatively charged ions on the inside of the cell.
   c. In a resting cell, more positive ions leave the cell than enter it, so the outside develops a positive charge with respect to the inside.

The Nerve Impulse

1. Action potential
   a. If the membrane is disturbed by a stimulus of threshold intensity, a nerve impulse is triggered.
   b. An action potential involves changes in membrane permeability and the movement of sodium and potassium ions through the membrane.
   c. Following an action potential, an active transport mechanism in the membrane reestablishes the resting potential.
   d. A wave of action potentials is a nerve impulse.
2. Refractory period
   a. The refractory period is a brief time following the passage of a nerve impulse when the membrane is unresponsive to an ordinary stimulus.
   b. During the absolute refractory period the membrane cannot be stimulated; during the relative refractory period the membrane can be stimulated with a stimulus of high intensity.
3. Impulse conduction
   a. Unmyelinated fibers conduct impulses that travel over their entire surfaces.
   b. Myelinated fibers conduct impulses that travel from node to node.
   c. Impulse conduction is more rapid on myelinated fibers with large diameters.
4. All-or-none response
   a. A nerve impulse is conducted in an all-or-none manner whenever a stimulus of threshold intensity is applied to a fiber.
   b. All the impulses conducted on a fiber are the same.

5. Some factors affecting nerve impulse conduction
   a. When calcium ions are in low concentration, spontaneous nerve impulses may travel to skeletal muscles and cause tetany.
   b. The presence of drugs like procaine decreases membrane permeability to sodium.

The Synapse

A synapse is a junction between two neurons.
A synaptic cleft is the gap between parts of two neurons at a synapse.

1. Synaptic transmission
   a. Impulses usually travel from dendrite to cell body, then along the axon to a synapse.
   b. Axons have synaptic knobs at their ends that secrete neurotransmitters.
   c. The neurotransmitter is released when a nerve impulse reaches the end of an axon, and the neurotransmitter diffuses across the synaptic cleft.
   d. When the neurotransmitter reaches the nerve fiber on the distal side of the cleft, a nerve impulse is triggered.
   e. Neurotransmitters are usually decomposed by enzymes in synaptic clefts or are otherwise removed.
2. Excitatory and inhibitory actions
   a. Neurotransmitters that trigger nerve impulses are excitatory; those that inhibit impulses are inhibitory.
   b. The effect of the synaptic knobs communicating with a neuron will depend upon which knobs are activated from moment to moment.
3. Some factors affecting synaptic transmission
   a. If impulses reach synaptic knobs at rapid rates, their supplies of neurotransmitters may be exhausted.
   b. Various drugs stimulate or inhibit the action of synaptic knobs.
4. Neuropeptides
   a. These substances are composed of amino acids in chains.
   b. Some of them serve as neurotransmitters or neuromodulators.
   c. They include enkephalins and endorphins that relieve pain sensations.

Neurons and Nerves

Neurons differ in structure and function.

1. Classification of neurons
   a. On the basis of structure, neurons can be classified as multipolar, bipolar, and unipolar.
   b. On the basis of function, neurons can be classified as sensory neurons, interneurons, or motor neurons.
2. Types of nerves
   a. Nerves are cordlike bundles of nerve fibers.
   b. Nerves can be classified as sensory, motor, or mixed, depending on which type of fibers they contain.

Reflex Arcs and Reflexes

The routes followed by nerve impulses are called nerve pathways.

1. Reflex behavior
   a. The simplest nerve pathways are reflex arcs, which consist of two or three neurons.
   b. Reflex arcs are the structural and functional units of behavior.
   c. A typical reflex arc includes a receptor, a sensory neuron, one or more interneurons, a motor neuron, and an effector.
2. Some clinical uses of reflexes
   a. Normal reflexes depend upon normal nerve functions; if portions of a reflex arc are injured, the reflex is likely to be altered.
   b. A variety of reflexes are tested during neurological examinations.

Coverings of the Central Nervous System

1. The central nervous system (CNS) consists of the brain and spinal cord.
2. The brain and spinal cord are surrounded by bone and protective membranes called meninges.
3. The meninges
   a. The meninges consist of a dura mater, arachnoid mater, and pia mater.
   b. Cerebrospinal fluid occupies the space between the arachnoid and pia maters.
   c. Meningitis is an inflammation of the meninges.

The Spinal Cord

The spinal cord is a nerve column that extends from the brain into the vertebral canal.

It terminates at the level between the first and second lumbar vertebrae.

1. Structure of the spinal cord
   a. The spinal is cord composed of thirty-one segments, each of which gives rise to a pair of spinal nerves.
   b. It is characterized by a cervical enlargement, a lumbar enlargement, and two deep longitudinal grooves that divide it into right and left halves.
   c. It has a central core of gray matter that is surrounded by white matter.
   d. The white matter is composed of bundles of myelinated nerve fibers.
2. Functions of the spinal cord
   a. The cord provides a two-way communication system between the brain and body parts outside the nervous system.
   b. Ascending tracts carry sensory impulses to the brain; descending tracts carry motor impulses to muscles and glands.
3. Spinal cord injuries
   a. The consequences of an injury depend on the location and extent of the damage sustained.
   b. Injuries may result in temporary spinal shock, sensory losses, or motor losses.
   c. Paralysis may be flaccid and accompanied by atrophy of muscles, or spastic and accompanied by uncontrolled muscular spasms.

The Brain

The brain is the largest and most complex part of the nervous system.

It is divided into a cerebrum, cerebellum, and brain stem.

1. Structure of the cerebrum
   a. The cerebrum consists of two cerebral hemispheres connected by the corpus callosum.
   b. Its surface is marked by ridges and grooves; sulci divide each hemisphere into lobes.
   c. The cerebral cortex is a thin layer of gray matter near the surface.
   d. White matter consists of myelinated nerve fibers that interconnect neurons within the nervous system and with other body parts.
2. Functions of the cerebrum
   a. The cerebrum is concerned with higher brain functions, such as interpretation of sensory impulses, control of voluntary muscles, storage of memory, thought, and reasoning.
   b. The cerebral cortex can be divided into sensory, motor, and association areas.
   c. In each individual, one hemisphere is dominant in intellectual functions.
3. Cerebral injuries
   a. The effects of cerebral injuries depend on the location and extent of the damage sustained.
   b. It is often possible to deduce the nature of an injury by observing what abilities are missing.
   c. Since neurons of the CNS are unable to regenerate, injuries are likely to produce permanent functional losses.
4. The ventricles and cerebrospinal fluid
   a. Ventricles are interconnected cavities within the cerebral hemispheres and brain stem.
   b. These spaces are filled with cerebrospinal fluid.
   c. Cerebrospinal fluid is secreted by choroid plexuses in the walls of the ventricles; it circulates through the ventricles and is reabsorbed into the blood of the dural sinuses.
   d. The pressure and composition of cerebrospinal fluid are sometimes tested for abnormalities.
5. The brain stem
   a. The brain stem extends from the base of the cerebrum to the foramen magnum.
   b. It consists of the diencephalon, midbrain, pons, and medulla oblongata.
   c. The diencephalon contains the thalamus, which serves as a central relay station for incoming sensory impulses, and the hypothalamus, which plays important roles in maintaining homeostasis.
   d. The limbic system functions to produce emotional feelings and to modify behavior.
   e. The midbrain contains reflex centers associated with eye and head movements.

f. The pons transmits impulses between the cerebrum and other parts of the nervous system and contains centers that help to regulate the rate and depth of breathing.

g. The medulla oblongata transmits all ascending and descending impulses and contains several vital and nonvital reflex centers.

h. The reticular formation acts to filter incoming sensory impulses, arousing the cerebral cortex into wakefulness whenever significant impulses are received.

6. The cerebellum
   a. The cerebellum consists of two hemispheres connected by the vermis.
   b. It is composed of white matter surrounded by a thin cortex of gray matter.
   c. The cerebellum functions primarily as a reflex center in the coordination of skeletal muscle movements and the maintenance of muscular tone.

The Peripheral Nervous System

The peripheral nervous system consists of cranial and spinal nerves that branch out from the brain and spinal cord to all body parts.

It can be divided into somatic and autonomic portions.

1. The cranial nerves
   a. Twelve pairs of cranial nerves connect the brain to parts in the head, neck, and trunk.
   b. Although most are mixed, some are pure sensory, and others are primarily motor.
   c. The names of cranial nerves indicate their primary functions or the general distributions of their fibers.
   d. Some cranial nerve fibers are somatic and others are autonomic.

2. Cranial nerve injuries
   a. The consequences of cranial nerve injuries depend on the location and the extent of the damage sustained.
   b. If one member of a nerve pair is damaged, the effect is usually limited to one side of the body; if both members are injured, functional losses occur on both sides.

3. The spinal nerves
   a. Thirty-one pairs of spinal nerves originate from the spinal cord.
   b. These mixed nerves provide a two-way communication system between the spinal cord and parts in the arms, legs, neck and trunk.
   c. Spinal nerves are grouped according to the levels from which they arise, and they are numbered in sequence.
   d. Each nerve emerges by a dorsal and a ventral root.
      (1) A dorsal root contains sensory fibers and is characterized by the presence of a dorsal root ganglion.
      (2) A ventral root contains motor fibers.

   e. Just beyond its foramen, each spinal nerve divides into several branches.
   f. Most spinal nerves combine to form plexuses in which nerve fibers are sorted and recombined so that those fibers associated with a particular part reach it together.

4. Spinal nerve injuries
   The consequences of such injuries depend on the location and the extent of the damage sustained.

The Autonomic Nervous System

The autonomic nervous system consists of the portions of the nervous system that function without conscious effort.

It is concerned primarily with the regulation of visceral activities that aid in maintaining homeostasis.

1. General characteristics
   a. Autonomic functions operate as reflex actions controlled from centers in the hypothalamus, brain stem, and spinal cord.
   b. The autonomic nervous system consists of two divisions—sympathetic and parasympathetic.
   c. The sympathetic division functions in meeting stressful and emergency conditions.
   d. The parasympathetic division is most active under ordinary conditions.

2. Autonomic nerve fibers
   a. The autonomic fibers are largely motor.
   b. Sympathetic fibers leave the spinal cord and synapse in chain ganglia.
   c. Parasympathetic fibers begin in the brain stem and sacral region of the spinal cord and synapse in ganglia near visceral organs.

3. Autonomic transmitter substances
   a. Preganglionic sympathetic and parasympathetic fibers secrete acetylcholine.
   b. Postganglionic sympathetic fibers secrete norepinephrine; postganglionic parasympathetic fibers secrete acetylcholine.
   c. The different effects of the autonomic divisions are due to different transmitter substances released by the postganglionic fibers.
   d. Most visceral organs are controlled mainly by one division.
   e. Acetylcholine acts very briefly; norepinephrine may have a more prolonged effect.

4. Conscious control of autonomic actions
   a. Some people can learn to control their visceral responses.
   b. This requires knowledge of the state of visceral activities.
   c. Electronic equipment can be used to monitor visceral activities and provide the basis for biofeedback.

Application of Knowledge

1. If a physician plans to obtain a sample of spinal fluid from a patient, what anatomical site can be safely used, and how should the patient be positioned to facilitate this procedure?

2. What functional losses would you expect to observe in a patient who has suffered injury to the right occipital lobe of the cerebral cortex?

3. Based on your knowledge of the cranial nerves, devise a set of tests to assess the normal functions of each of these nerves.

Review Activities

1. Explain the relationship between the central nervous system and the peripheral nervous system.

2. Distinguish between neurons and neuroglial cells.

3. Describe the generalized structure of a neuron and explain the functions of its parts.

4. Distinguish between myelinated and unmyelinated nerve fibers.

5. Describe how an injured nerve fiber may regenerate.

6. Discuss the functions of each type of neuroglial cell.

7. Explain how a membrane may become polarized.

8. List the changes that occur during an action potential.

9. Explain how action potentials are related to nerve impulses.

10. Define *refractory period*.

11. Define *saltatory conduction*.

12. Explain how a deficiency of calcium affects nerve impulse conduction.

13. Define *synapse*.

14. Explain how a nerve impulse is transmitted from one neuron to another.

15. Distinguish between excitatory and inhibitory actions of neurotransmitters.

16. Explain how neurons can be classified on the basis of their structure.

17. Explain how neurons can be classified on the basis of their function.

18. Distinguish between sensory, motor, and mixed nerves.

19. Define *reflex*.

20. Describe a reflex arc that consists of two neurons.

21. Explain how tests of reflex actions can be used to evaluate the condition of the nervous system.

22. Name the layers of the meninges and explain their functions.

23. Describe the structure of the spinal cord.

24. Name the major ascending and descending tracts of the spinal cord and list the functions of each.

25. Name the three major portions of the brain and describe the general functions of each.

26. Describe the general structure of the cerebrum.

27. Define *cerebral cortex* and *basal ganglia*.

28. Describe the location of the motor, sensory, and association areas of the cerebral cortex and describe the general functions of each.

29. Describe the location of the ventricles of the brain.

30. Explain how cerebrospinal fluid is produced and how it functions.

31. Name the parts of the diencephalon and describe the general functions of each.

32. Define the limbic system and explain its functions.

33. Name the parts of the midbrain and describe the general functions of each.

34. Describe the pons and its functions.

35. Describe the medulla oblongata and its functions.

36. Describe the functions of the cerebellum.

37. Distinguish between the somatic and autonomic nervous systems.

38. Name, locate, and describe the major functions of each pair of cranial nerves.

39. Explain how the spinal nerves are grouped and numbered.

40. Define *cauda equina*.

41. Describe the structure of a spinal nerve.

42. Define *plexus* and locate the major plexuses of the spinal nerves.

43. Distinguish between the sympathetic and the parasympathetic divisions of the autonomic nervous system.

44. Distinguish between a preganglionic and a postganglionic nerve fiber.

45. Trace a sympathetic nerve pathway.

46. Explain why the effects of the sympathetic and parasympathetic divisions differ.

47. Define *biofeedback*.

Suggestions for Additional Reading

Axelrod, J. June 1974. Neurotransmitters. *Scientific American.*

Barr, M. L. 1974. *The human nervous system.* 2nd ed. New York: Harper and Row.

Decara, L. V. January 1970. Learning in the autonomic nervous system. *Scientific American.*

Eccles, J. C. 1973. *The understanding of the brain.* New York: McGraw-Hill.

Evarts, E. V. September 1979. Brain mechanisms of movement. *Scientific American.*

Ferstrom, J. D., and Wurtman, R. J. February 1974. Nutrition and the brain. *Scientific American.*

Geschwind, N. September 1979. Specializations of the human brain. *Scientific American.*

Hubel, D. H. September 1979. The brain. *Scientific American.*

Hubel, D. H., and Wiesel, T. N. September 1979. Brain mechanisms of vision. *Scientific American.*

Iversen, L. L. September 1979. The chemistry of the brain. *Scientific American.*

Jacobs, B. L., and Trulson, M. E. 1979. Mechanism of action of LSD. *American Scientist* 67:396.

Kandel, E. R. September 1979. Small systems of neurons. *Scientific American.*

Kety, S. S. September 1979. Disorders of the human brain. *Scientific American.*

Keynes, R. D. September 1979. Ion channels in the nerve-cell membrane. *Scientific American.*

Kimura, D. March 1973. The asymmetry of the human brain. *Scientific American.*

Lester, H. A. February 1977. The response to acetylcholine. *Scientific American.*

Llinas, R. R. January 1975. The cortex of the cerebellum. *Scientific American.*

Nathanson, J. A., and Greengard, P. August 1977. Second messenger in the brain. *Scientific American.*

Nauta, W. J. H., and Feirtag, M. September 1979. The organization of the brain. *Scientific American.*

Noback, C. R. 1975. *The human nervous system.* 2nd ed. New York: McGraw-Hill.

Norton, W. T., and Morell, P. May 1980. Myelin. *Scientific American.*

Schwartz, J. H. April 1980. The transport of substances in nerve cells. *Scientific American.*

Shepherd, G. M. February 1978. Microcircuits in the nervous system. *Scientific American.*

Siegal, R. K. October 1977. Hallucinations. *Scientific American.*

Stevens, C. F. September 1979. The neuron. *Scientific American.*

Somatic and Special Senses

11 Before parts of the nervous system can act to control body functions, they must know what is happening inside and outside the body. This information is gathered by sensory receptors that are sensitive to environmental changes. Although receptors vary greatly in their individual characteristics, they can be grouped into two major categories. The members of one group are widely distributed throughout the skin and deeper tissues and generally have simple forms. These receptors are associated with the *somatic senses* of touch, pressure, temperature, and pain. Members of the second group function as parts of complex, specialized sensory organs that are responsible for the *special senses* of smell, taste, hearing, equilibrium, and vision.

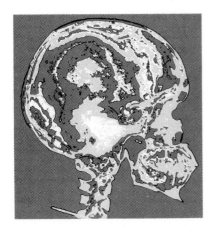

After you have studied this chapter, you should be able to

1. Name five kinds of receptors and explain the function of each.
2. Explain how a sensation is produced.
3. Distinguish between somatic and special senses.
4. Describe how pain is produced and how it may be controlled.
5. Explain the relationship between the senses of smell and taste.
6. Name the parts of the ear and explain the function of each.
7. Explain how equilibrium is maintained.
8. Name the parts of the eye and explain the function of each.
9. Describe how aging may affect sensory functions.
10. Complete the review activities at the end of this chapter.

accommodation (ah-kom″o-da′shun)

auditory (aw′di-to″re)

chemoreceptor (ke″mo-re-sep′tor)

cochlea (kok′le-ah)

dynamic equilibrium (di-nam′ik e″kwĭ-lib′re-um)

labyrinth (lab′i-rinth)

mechanoreceptor (mek″ah-no-re-sep′tor)

olfactory (ol-fak′to-re)

optic (op′tik)

photoreceptor (fo″to-re-sep′tor)

projection (pro-jek′shun)

proprioceptor (pro″pre-o-sep′tor)

referred pain (re-furd′ pān)

refraction (re-frak′shun)

sensory adaptation (sen′so-re ad″ap-ta′shun)

static equilibrium (stat′ik e″kwĭ-lib′re-um)

thermoreceptor (ther″mo-re-sep′tor)

aud-, to hear: *aud*itory—pertaining to hearing.

choroid, skinlike: *choroid* coat—middle, vascular layer of the eye.

cochlea, snail: *cochlea*—coiled tube within the inner ear.

corn-, horn: *corn*ea—transparent outer layer in the anterior portion of the eye.

iris, rainbow: *iris*—colored, muscular part of the eye.

labyrinth, maze: *labyrinth*—complex system of interconnecting chambers and tubes of the inner ear.

lacri-, tears: *lacri*mal gland—tear gland.

lut-, yellow: macula *lut*ea—yellowish spot on the retina.

macula, spot: *macula* lutea—yellowish spot on the retina.

oculi-, eye: orbicularis *oculi*—muscle associated with the eyelid.

olfact-, to smell: *olfact*ory—pertaining to the sense of smell.

palpebra, eyelid: levator *palpebra*e superioris—muscle associated with the eyelid.

scler-, hard: *scler*a—tough, outer protective layer of the eye.

therm-, heat: *therm*oreceptor—receptor sensitive to changes in temperature.

tympano-, drum: *tympan*ic membrane—the eardrum.

vitre-, glass: *vitre*ous humor—clear, jellylike substance within the eye.

As changes occur within the body and in its surroundings, *sensory receptors* are stimulated, and they trigger nerve impulses. These impulses travel on sensory pathways into the central nervous system to be processed and interpreted. As a result the person often experiences a particular type of feeling or sensation.

Receptors and Sensations

Although there are many kinds of sensory receptors, they have features in common. For example, each type of receptor is particularly sensitive to a distinct kind of environmental change and much less sensitive to other forms of stimulation.

Types of Receptors

On the basis of their sensitivities, it is possible to identify five groups of sensory receptors. The members of each group are sensitive to changes in one of the following factors:

1. *Chemical concentration.* Receptors that are stimulated by changes in concentration of chemical substances are called **chemoreceptors.** Those associated with the senses of smell and taste are of this type.

2. *Tissue damage.* Whenever tissues are damaged in any way, **pain receptors** are likely to be stimulated.

3. *Temperature change.* Receptors sensitive to temperature change are called **thermoreceptors,** and there are two different sets. One set, the *heat receptors,* responds to temperatures above body temperature; the other set, the *cold receptors,* is sensitive to temperatures below that of the body.

4. *Mechanical forces.* A number of sensory receptors are sensitive to mechanical forces, such as changes in pressure or movement of fluids. As a rule, these **mechanoreceptors** detect changes that cause them to become deformed. For example, those called *proprioceptors* (fig. 11.1) are sensitive to changes in the tensions of muscles and tendons.

5. *Light intensity.* Light or **photoreceptors** occur only in the eyes, and they respond whenever they are exposed to sufficient intensity of light energy.

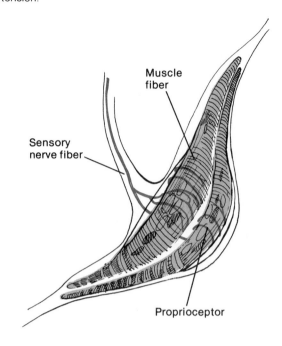

Fig. 11.1 Proprioceptors, which are closely associated with muscle fibers, are stimulated by changes in muscular tension.

Sensations

A **sensation** is a feeling that occurs when sensory receptors are stimulated. Because all the nerve impulses that travel from sensory receptors to the central nervous system are alike, different kinds of sensations must be due to the way the brain interprets the impulses rather than to differences in the receptors. In other words, when a receptor is stimulated, the resulting sensation depends on what region of the cerebral cortex receives the impulse. For example, impulses reaching one region are always interpreted as sounds, and those reaching another portion are always sensed as touch. Impulses reaching still other regions are always interpreted as pain.

It makes little difference how the impulses are stimulated. Pain receptors, for example, can be stimulated by excessive heat, cold, or pressure, but the sensation is always the same since the impulses are interpreted by the same part of the brain. Similarly, nerve impulses are sometimes triggered in visual receptors by factors other than light. When this happens, the person may "see" lights, even though no light is entering the eye, since any impulses reaching the visual cortex are interpreted as light.

At the same time a sensation is created, the brain causes it to seem to come from the receptors being stimulated. This process is called **projection,** because the brain projects the sensation back to its apparent source. Projection allows the person to pinpoint the region of stimulation. Thus, the eyes appear to see, the ears appear to hear, and so forth.

Sensory Adaptation

When they are subjected to continuous stimulation, receptors generally undergo an adjustment called **sensory adaptation.** As the receptors adapt, impulses leave them at slower rates, until finally they may completely fail to send signals. Once receptors have adapted, impulses can be triggered only if the strength of the stimulus is changed.

Sensory adaptation is experienced when a person enters a room where there is a strong odor. At first the scent seems intense, but it becomes less and less apparent as the smell receptors adapt. If the person remains in the room for a minute or more, he or she may become totally adapted to the odor and quite unaware of its presence. If the person leaves the room, however, and then reenters it, the receptors will probably be stimulated once again.

---

1. *List five general types of sensory receptors.*
2. *What do these receptors have in common?*
3. *Explain how a sensation occurs.*
4. *What is meant by sensory adaptation?*

Somatic Senses

The somatic senses include the cutaneous or dermal senses (described briefly in chapter 6), as well as those associated with muscles, joints, and visceral organs.

Touch and Pressure Receptors

Several kinds of touch and pressure (tactile) receptors have been identified (fig. 11.2). As a group, these receptors are sensitive to mechanical forces that produce distortion or displacement of tissues, and they include the following:

1. **Free ends of sensory nerve fibers.** These receptors occur commonly in epithelial tissues where they end as knobs between epithelial cells. They are associated with the sensations of touch and pressure.

2. **Meissner's corpuscles.** These structures consist of small, oval masses of flattened connective tissue cells surrounded by connective tissue sheaths. Two or more nerve fibers branch into each corpuscle and end within it as tiny knobs.

Meissner's corpuscles are especially numerous in the hairless portions of the skin, such as the lips, fingertips, palms, soles, nipples, and external genital organs. They are sensitive to the motion of objects that barely contact the skin, and impulses from them are interpreted as the sensation of light touch. They are also used when a person touches something to judge its texture.

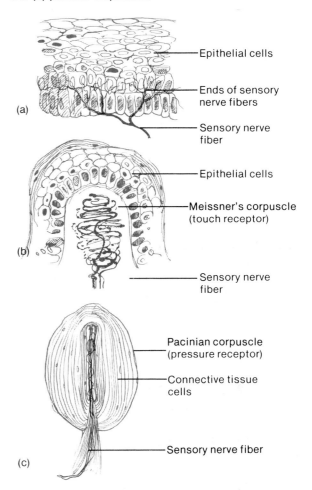

Fig. 11.2 Touch and pressure receptors include (a) free ends of sensory nerve fibers; (b) Meissner's corpuscles, and (c) pacinian corpuscles.

Epithelial cells
Ends of sensory nerve fibers
(a)
Sensory nerve fiber
Epithelial cells
Meissner's corpuscle (touch receptor)
(b)
Sensory nerve fiber
Pacinian corpuscle (pressure receptor)
Connective tissue cells
Sensory nerve fiber
(c)

3. **Pacinian corpuscles.** These sensory bodies are relatively large, elliptical structures composed of connective tissue fibers and cells. They are commonly found in the deeper subcutaneous tissues of such organs as the hands, feet, penis, clitoris, urethra, and breasts, and also occur in tendons and the ligaments of joints.

Pacinian corpuscles are stimulated by heavy pressure and are thought to be associated with the sensation of deep pressure. They may also serve to detect vibrations in tissues.

Thermoreceptors

Although there is some question about the identity of the thermoreceptors, they seem to include two types of *free nerve endings* called *heat receptors* and *cold receptors.* The heat receptors are most sensitive to temperatures above 25°C (77°F) and become unresponsive at temperatures above 45°C (113°F). As 45°C is approached, pain receptors are triggered, producing a burning sensation.

Cold receptors are most sensitive to temperatures between 10°C (50°F) and 20°C (68°F). If the temperature drops below 10°C, pain receptors are stimulated, and the person feels a freezing sensation.

Both heat and cold receptors demonstrate rapid adaptation, so that within about a minute following stimulation, the sensation of hot or cold begins to fade. A person experiences this form of sensory adaptation upon entering a tub of hot water or wading into a cold lake or stream. At first the temperature sensation may be unpleasant, but within a few moments the temperature receptors adapt, at least partially, and the person feels more comfortable.

Pain Receptors

Pain receptors also consist of free nerve fiber endings. They are widely distributed throughout the skin and internal tissues, except for the nerve tissue in the brain, which lacks pain receptors.

Pain receptors have a protective function in that they are stimulated whenever tissues are being damaged. The resulting sensation is usually perceived as unpleasant, and it serves as a signal that something should be done to remove the source of the irritation. Pain receptors along with the other cutaneous receptors are summarized in chart 11.1.

Pain receptors adapt very little, if at all, and once such a receptor has been activated, even by a single stimulus, it may continue to send impulses into the central nervous system for some time.

Although pain receptors can be stimulated by a variety of factors, including temperature, mechanical, and electrical changes, they are particularly sensitive to chemical stimulation.

The exact way that tissue damage excites pain receptors is unknown. It is thought that injuries promote the release of certain chemicals, such as hydrogen ions, potassium ions, or polypeptides (kinins), from the breakdown of proteins, histamine, and acetylcholine. Sufficient quantities of these chemicals may stimulate pain receptors. A deficiency of oxygen-rich blood (ischemia) in a tissue also stimulates pain receptors. Pain elicited during a muscle cramp, for example, seems to be related to an interruption of blood flow that occurs as the sustained contraction squeezes capillaries. At the same time, pain-stimulating chemicals accumulate excessively. (See fig. 11.3.) Increasing the blood flow through the sore tissue relieves the resulting pain, and this is why heat is sometimes applied to reduce muscle soreness. Heat causes blood vessels to dilate and thus promotes blood flow.

A similar painful condition may occur in the heart of a person suffering from the artery disease called *atherosclerosis*. In this disease, solid materials accumulate abnormally inside arteries, interfering with the flow of blood. If this happens in the arteries that supply blood to the heart muscle, the victim may feel chest pain, called *angina pectoris*, during times of physical exertion. This pain seems to be related to a decreased oxygen supply in the heart muscle and a build-up of pain-stimulating chemicals. The pain is relieved by rest, which provides time for blood to supply oxygen and to remove the offending substances.

Fig. 11.3 A sustained muscular contraction may cause an interruption of blood flow through capillaries in the muscle tissue.

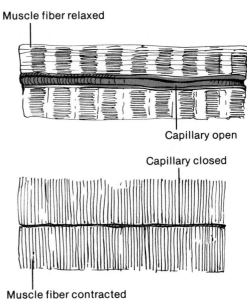

Muscle fiber relaxed

Capillary open

Capillary closed

Muscle fiber contracted

Chart 11.1 Cutaneous receptors

Type	Function	Sensation
Free nerve endings (mechanoreceptors)	Detect changes in pressure	Touch, pressure
Meissner's corpuscles (mechanoreceptors)	Detect objects moving over the skin	Touch, texture
Pacinian corpuscles (mechanoreceptors)	Detect changes in pressure	Deep pressure, vibrations
Free nerve endings (thermoreceptors for heat)	Detect changes in temperature	Heat
Free nerve endings (thermoreceptors for cold)	Detect changes in temperature	Cold
Free nerve endings (pain receptors)	Detect tissue damage	Pain

As a rule, pain receptors are the only receptors in visceral organs that produce sensations that persons are consciously aware of. These receptors seem to respond differently to stimulation than those associated with surface tissues. For example, localized damage to intestinal tissue, as may occur during surgical procedures, may not elicit any pain sensations even in a conscious person. When the same tissues are stimulated more generally by stretching, or when the smooth muscles in the intestinal walls undergo spasms, a strong pain sensation may follow. It is believed that stretching the tissues or a sustained contraction in the smooth muscle of the wall may interfere with blood flow. Once again, the resulting pain seems to be related to a decreased blood flow and an accumulation of pain-stimulating chemicals.

Another characteristic of visceral pain is that it may feel as if it is coming from some part of the body other than the part being stimulated—a phenomenon called **referred pain.** Pain originating from the heart, for example, may be referred to the left shoulder or the inside of the left arm. Pain from the lower esophagus, stomach, or small intestine may seem to be coming from the upper central (epigastric) region of the abdomen; pain from the urogenital tract may be referred to the lower central (hypogastric) region of the abdomen or to the sides between the ribs and the hip. (See fig. 11.4.)

The occurrence of referred pain seems to be related to *common nerve pathways* that are used by sensory impulses coming both from skin areas and from visceral organs. In other words, pain impulses from the heart seem to be conducted over the same nerve paths as those coming from the skin of the shoulder and the inside of the left arm, as shown in figure 11.5. Consequently, the cerebral cortex may incorrectly interpret the source of the impulses as the shoulder or arm rather than the heart.

Fig. 11.4 Surface regions to which visceral pain may be referred.

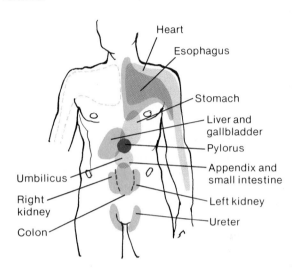

Fig. 11.5 Pain originating in the heart may feel as if it is coming from the skin because sensory impulses from these two regions follow common nerve pathways to the brain.

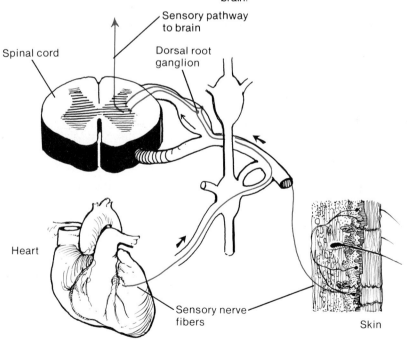

Probably the most common pain is associated with *headache.* Although brain tissue lacks pain receptors, nearly all the other tissues of the head are well supplied with them.

Most headaches seem to be related to stressful life situations that result in fatigue, emotional tension, anxiety, or frustration. These conditions are reflected in physiological changes. The changes that appear most likely to result in headache are the following:

1. Dilations of cranial blood vessels, accompanied by edema in surrounding tissues.

2. Spasms in skeletal muscles of the face, scalp, and neck.

Even though the pain of some headaches originates inside the cranium, the sensation is often referred to the outside.

Control of Pain. Most persons suffer from pains that arise from injuries and diseases, and few are able to remain indifferent to these pain sensations.

Interestingly, the intensity of stimulation needed to trigger pain receptors seems to be about the same for all individuals, yet persons vary greatly in the ways they perceive pain and respond to it. These differences seem to be related to past experiences, cultural conditioning, and the circumstances under which the painful stimuli are encountered.

When pain becomes persistent and begins to interfere with our abilities to function effectively, methods to control it are usually sought; for example, an aspirin tablet or some other pain-reducing drug may be taken. Occasionally a person develops an *intractable pain,* one that is chronic and cannot be relieved even by using narcotic drugs. In such cases, various sensory nerve pathways may be sectioned surgically. This procedure will prevent pain impulses from reaching the pain-interpreting centers of the brain, but it will also result in a permanent loss of sensation from body parts innervated by the sectioned nerve fibers.

Neuropeptides that serve as neurotransmitter substances were discussed in chapter 10. These substances include a group, called **enkephalins,** that are released by axons in the parts of the spinal cord and the brain that are within the limbic system. The limbic system, which includes the thalamus, serves as the pathway for pain impulses. Enkephalins act to inhibit such impulses, and thus they relieve pain sensations much as morphine and certain other narcotic drugs do. In fact, enkephalins seem to function by binding to the same receptors sites on neuron membranes as does morphine.

Another group of neuropeptides, called **endorphins,** with pain-suppressing, morphinelike actions occur in the pituitary gland.

It is believed that enkephalins and endorphins may be released at times when pain impulses are triggered excessively, and that in this way they provide natural pain control. These substances may also be released in response to the following conditions: treatment by *acupuncture,* in which needles are inserted into various tissues to control pain; use of heat or skin irritants (liniments) to relieve muscular soreness; or electrical stimulation of certain regions of the brain to alleviate pain.

1. Describe three types of tactile receptors.
2. What types of stimuli excite pain receptors?
3. What is referred pain?
4. Explain how neuropeptides may help to control pain.

Sense of Smell

The sense of smell, like the other special senses, is associated with complex sensory structures and makes use of cranial nerve pathways.

The smell or olfactory receptors are similar to those for taste in that they are chemoreceptors, stimulated by chemicals dissolved in liquids. These two senses function closely together and aid in food selection, since food is often smelled at the same time that it is tasted. It is often difficult to tell what part of a food sensation is due to smell and what part is really taste. For this reason, an onion tastes quite different when sampled with the nostrils closed, because much of the usual onion sensation is due to odor. Similarly, during a head cold, excessive mucus secretions may cover the smell receptors, and food may seem to lose its taste.

Olfactory Organs

The **olfactory organs,** which contain the olfactory receptors, also include various epithelial supporting cells. These organs appear as yellowish masses that cover the upper parts of the nasal cavity, the superior nasal conchae, and a portion of the nasal septum.

The **olfactory receptor cells** are bipolar neurons surrounded by columnar epithelial cells. The neurons have knobs at their distal ends that are covered by hairlike cilia. The cilia project into the nasal cavity and are thought to be the sensitive portions of the receptors (fig. 11.6).

Fig. 11.6 The olfactory receptor cells, which have cilia at their distal ends, are supported by columnar epithelial cells.

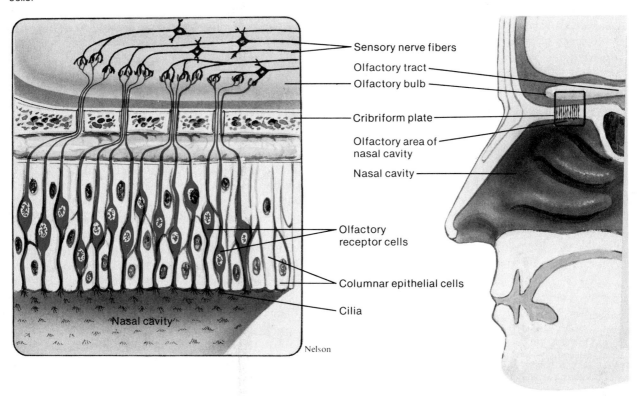

Sensory nerve fibers
Olfactory tract
Olfactory bulb
Cribriform plate
Olfactory area of nasal cavity
Nasal cavity
Olfactory receptor cells
Columnar epithelial cells
Cilia
Nasal cavity
Nelson

Chemicals that stimulate olfactory receptors enter the nasal cavity as gases; but before they can be detected, they must dissolve at least partially in the watery fluids that surround the cilia. It is believed that such chemicals must also be soluble in lipids before they can stimulate the receptors, because cilia are composed primarily of lipid materials.

Once the receptors have been stimulated, nerve impulses are triggered, and they travel along fibers of the olfactory nerves that pass through tiny openings in the cribriform plates of the ethmoid bone. These fibers lead to neurons located in enlargements called **olfactory bulbs,** which lie on either side of the crista galli of the ethmoid bone. From the olfactory bulbs, the impulses travel along the **olfactory tracts** to interpreting centers within the cerebrum.

Olfactory Stimulation

How various substances stimulate the olfactory receptors is poorly understood. One hypothesis suggests that the shapes of gaseous molecules may fit receptor sites on the cilia that have complementary shapes. A nerve impulse, according to this idea, is triggered when a molecule binds to its particular receptor site.

Attempts to classify the sensations associated with odors have met with only limited success. Many researchers, however, agree that there are at least seven groups of primary odors.

1. *Camphoraceous,* like the scent of camphor.
2. *Musky,* like the scent of musk.
3. *Floral,* like the scent of flowers.
4. *Pepperminty,* like the scent of oil of peppermint.
5. *Ethereal,* like the scent of ether.
6. *Pungent,* like the scent of spices.
7. *Putrid,* like the scent of decaying meat.

Any particular odor might be described as one of the primary odors or some combination of two or more of them.

Since the olfactory organs are located high in the nasal cavity above the usual pathway of inhaled air, a person may have to sniff and force air over the receptor areas to smell something that has a faint odor. Also, olfactory receptors undergo sensory adaptation rather rapidly, so the intensity of an olfactory sensation drops about 50% within a second following stimulation. Within a minute, the receptors may become almost insensitive to a given odor, but even though they have adapted to one scent, their sensitivity to other odors remains unchanged.

The neurons that function as olfactory receptors are the only parts of the nervous system that are in direct contact with the outside environment. Because of their exposed positions, these neurons are subject to damage, and because damaged neurons usually are not replaced, persons are likely to experience a progressive diminishing of olfactory sense with age. In fact, some investigators have estimated that a person loses about 1% of the olfactory receptors every year.

Partial or complete loss of smell is called *anosmia*. This condition may be caused by a variety of factors including inflammation of the nasal cavity lining, as occurs during a head cold or from excessive tobacco smoking, and the use of certain drugs such as adrenalin or cocaine. Persons who have injured their olfactory bulbs, olfactory tracts, or cerebral olfactory interpreting centers will experience some degree of anosmia.

Elderly persons may require special stimulation to encourage eating because their olfactory sensitivities have diminished. For this reason, it is important that their foods be visibly attractive and that mealtimes be pleasant experiences. Otherwise, disinterest in eating may contribute to malnutrition.

---

1. Where are the olfactory receptors located?
2. Trace the pathway of an olfactory impulse from a receptor to the cerebrum.
3. Why is the sense of smell likely to decrease with age?

Sense of Taste

Taste buds are the special organs of taste. They occur primarily on the surface of the tongue, but are also found in smaller numbers in the roof of the mouth and the walls of the pharynx (fig. 11.7).

Taste Receptors

Each taste bud consists of a group of modified epithelial cells, the **taste cells,** which function as receptors, along with a number of epithelial supporting cells. The entire taste bud is somewhat spherical, with an opening, the **taste pore,** on its free surface. Tiny projections (microvilli), called **taste hairs,** protrude from the outer ends of the taste cells and jut out through the taste pore. It is believed that these taste hairs are the sensitive parts of the receptor cells. (See fig. 11.8.)

Interwoven among the taste cells, and wrapped around them, is a network of nerve fibers. The ends of these fibers are in close contact with the receptor cell membranes. When a receptor cell is stimulated, an impulse is triggered on a nearby nerve fiber and is carried into the brain.

Before the taste of a particular chemical can be detected, it must be dissolved in the watery fluid surrounding the taste buds. This fluid is supplied by the salivary glands. To demonstrate its importance, blot the tongue and try to taste some dry food; then repeat the test after the tongue is moistened with saliva.

The mechanism by which various substances stimulate taste cells is not well understood. It is known, however, that taste cell membranes are normally polarized and that application of certain chemicals causes changes in their polarization. The amount of change is directly proportional to the concentration of the stimulating substance. One possible explanation holds that various substances combine with specific receptor sites on the taste hair surfaces. Such a combination is thought to be responsible for the change in membrane polarization and the generation of sensory impulses on nearby nerve fibers.

Although the taste cells in all taste buds appear very much alike microscopically, there are at least four types. Each type is sensitive to a particular kind of chemical stimulus, and consequently, there are at least four primary taste sensations. Each type of taste cell has a special pattern of distribution on the surface of the tongue. (See fig. 11.9.)

Taste Sensations

The four *primary taste sensations* are the following:
1. *Sweet,* as produced by table sugar.
2. *Sour,* as produced by vinegar.
3. *Salty,* as produced by table salt.
4. *Bitter,* as produced by caffeine or quinine.

In addition to these four, some investigators recognize two other taste sensations, which they call *alkaline* and *metallic.*

Each of the many flavors we experience daily is believed to result from one of the primary sensations or from some combination of two or more of them. The way we experience flavors also may involve the concentration of chemicals as well as the sensations of odor, texture (touch), and temperature. Furthermore, chemicals in some foods—chili peppers and ginger, for instance—may stimulate pain receptors that cause the tongue to burn.

Sweet receptors are most plentiful near the tip of the tongue. Their location is related to a child's preference to lick a candy sucker rather than chew it.

Fig. 11.7 (*a*) Taste buds on the surface of the tongue are associated with nipplelike elevations called papillae; (*b*) a taste bud contains taste cells and has an opening, the taste pore, at its free surface.

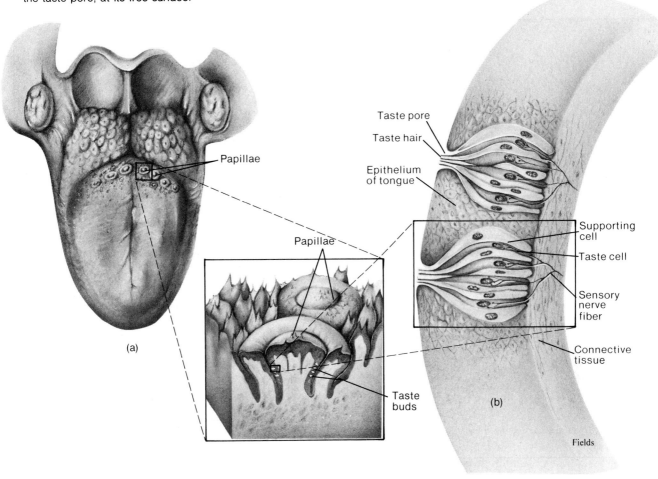

(a)

Papillae

Papillae

Taste buds

Taste pore
Taste hair
Epithelium of tongue
Supporting cell
Taste cell
Sensory nerve fiber
Connective tissue

(b)

Fields

The chemicals that stimulate these receptors are usually *organic compounds,* such as sugars and polysaccharides. A few inorganic substances, including some salts of lead and beryllium, also elicit sweet sensations.

Sour receptors occur primarily along the margins of the tongue. They are stimulated mainly by acids. The intensity of a sour sensation is roughly proportional to the concentration of the *hydrogen ions* in the substance being tasted.

Salt receptors are most common in the tip and the upper front portion of the tongue. They are stimulated by *inorganic salts* of various kinds, including certain chlorides, sulfates, iodides, and nitrates. It seems that all salty substances are electrolytes, and the quality of the sensation each produces is related to the kind of positively charged ion it releases into solution.

Fig. 11.8 A micrograph of some taste buds.

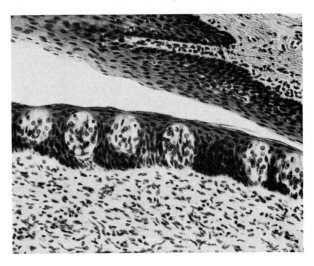

Somatic and Special Senses   339

Fig. 11.9 Patterns of taste receptor distribution: (a) sweet receptors; (b) sour receptors; (c) salt receptors; (d) bitter receptors. What kinds of chemical substances are likely to stimulate each type of taste receptor?

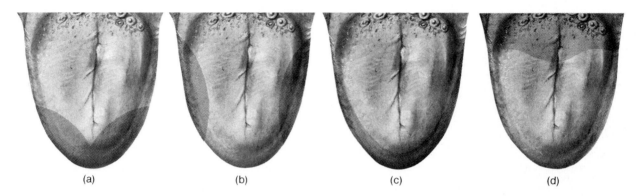

(a)   (b)   (c)   (d)

Bitter receptors are located toward the back of the tongue. They are stimulated by a variety of chemical substances, most of which are organic compounds, although some inorganic salts of magnesium and calcium produce bitter sensations, too.

One group of bitter compounds of particular interest are the *alkaloids,* which include a number of poisons such as strychnine, nicotine, and morphine. The fact that persons commonly reject bitter substances may be related to a kind of protective mechanism that causes them to avoid alkaloids in foods.

Taste receptors, like olfactory receptors, undergo sensory adaptation relatively rapidly. The resulting loss of taste can be circumvented by moving bits of food over the surface of the tongue to stimulate different receptors at different times.

Although the taste cells are located very close to the surface of the tongue and are somewhat exposed to damage by environmental factors, the sense of taste is not as likely to diminish with age as the sense of smell. This is because taste cells are reproduced continually, so that any one of these cells functions for only about a week before it is replaced.

---

1. *Why is saliva necessary for a sense of taste?*
2. *Name the four primary taste sensations.*
3. *What characteristic of taste receptors helps to maintain a sense of taste with age?*

Sense of Hearing

The organ of hearing is the **ear,** which has external, middle, and inner parts. In addition to making hearing possible, the ear functions in the sense of equilibrium, which will be discussed in a later section of this chapter.

External Ear

The external ear consists of an outer, funnellike structure called the **auricle;** an S-shaped tube, called the **external auditory meatus,** that leads inward for about 2.5 cm (1 inch), and a thin, cone-shaped eardrum, or **tympanic membrane,** that covers the deep end of the meatus. (See fig. 11.10.)

The external auditory meatus passes through the temporal bone. Near its opening the tube is guarded by hairs. It is lined with skin that contains numerous modified sweat glands called *ceruminous glands,* which secrete wax (cerumen). The hairs and wax help to keep relatively large foreign objects, such as insects, from entering the ear.

Sounds are created by vibrations and are transmitted through matter in the form of sound waves. For example, the sounds of some musical instruments are produced by vibrating strings or reeds, and the sounds of the voice are created by vibrating vocal folds in the larynx. In any case, the auricle of the ear helps to collect sound waves traveling through air and directs them into the auditory meatus.

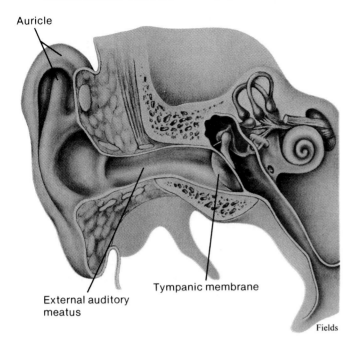

Fig. 11.10 The external ear consists of the auricle, the external auditory meatus, and the tympanic membrane.

Auricle

External auditory meatus

Tympanic membrane

Fields

After entering the meatus, the sound waves pass to the end of the tube and cause pressure changes on the eardrum. The eardrum moves back and forth in response and thus reproduces the vibrations of the sound wave source.

Middle Ear

The middle ear consists of an air-filled space in the temporal bone, called the **tympanic cavity,** that separates the external and inner ears. Three small bones or **auditory ossicles**—the *malleus* (hammer), *incus* (anvil), and *stapes* (stirrup)—are attached to the wall of the tympanic cavity by tiny ligaments. These bones form a bridge connecting the eardrum to the inner ear and function to transmit vibrations between these parts. Specifically, the malleus is attached to the eardrum, and when the eardrum vibrates, the malleus vibrates in unison with it. The malleus causes the incus to vibrate, and it passes the movement onto the stapes. The stapes is held by ligaments to a part of the inner ear called the **oval window,** and vibration of this oval window causes motion in a fluid within the inner ear. These vibrations are responsible for stimulating the receptors of hearing. (See fig. 11.11.)

Fig. 11.11 The auditory ossicles of the middle ear form a bridge that connects the tympanic membrane with the inner ear. How do these bones function to increase the force of the vibrations that are transmitted to the inner ear?

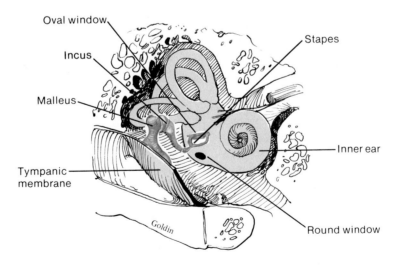

Oval window

Incus

Malleus

Tympanic membrane

Goldin

Stapes

Inner ear

Round window

Fig. 11.12 Medial view of the middle ear. Two small muscles that are attached to the malleus and stapes respectively serve as effectors in the tympanic reflex.

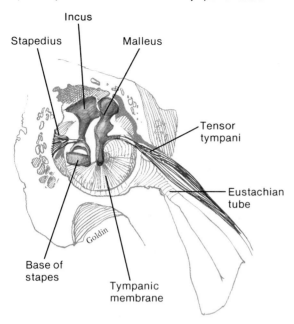

Incus

Stapedius   Malleus

Tensor tympani

Eustachian tube

Base of stapes

Tympanic membrane

Goldin

In addition to transmitting vibrations, the auditory ossicles form a lever system that increases the force of the vibrations as they are passed from the eardrum to the oval window. Also, because the auditory ossicles transmit vibrations from the relatively large surface of the eardrum to a much smaller area at the oval window, the vibration force becomes concentrated as it travels from the external to the inner ear. As a result of these two factors, the pressure applied by the stapes at the oval window is about twenty-two times greater than that exerted on the eardrum by sound waves.

The middle ear also contains two small skeletal muscles that are attached to the auditory ossicles. One of them, the *tensor tympani,* is inserted on the medial surface of the malleus, and when it contracts it pulls the bone inward. The other muscle, the *stapedius,* is attached to the posterior side of the stapes and serves to pull it outward (fig. 11.12). These muscles are the effectors in the **tympanic reflex,** which is elicited by loud sounds (fig. 11.12). When the reflex occurs, the muscles contract and the malleus and stapes are moved. As a result, the bridge of ossicles in the middle ear becomes more rigid, and its effectiveness in transmitting vibrations to the inner ear is reduced.

The tympanic reflex is a protective mechanism that reduces pressure from loud sounds that might otherwise damage the hearing receptors. The tensor tympani muscle also functions to maintain a steady pull on the eardrum. This is important because a loose tympanic membrane would not be able to transmit vibrations effectively to the auditory ossicles.

Eustachian Tube

An **eustachian tube** (auditory tube) connects each middle ear to the throat. This tube allows air to pass between the middle ear chamber and the outside of the body by way of the throat and mouth, and it is important in maintaining equal air pressure on both sides of the eardrum, which is necessary for normal hearing. (See fig. 11.13.)

The function of the eustachian tube can be experienced during rapid altitude change. A person driving down a mountain road, for example, is passing from a region of relatively low air pressure into one of higher pressure. As a person moves from a high altitude to a lower one, however, the air pressure on the outside of the membrane becomes greater and greater. The eardrum may be pushed inward, out of its normal position, and hearing may be impaired.

When the air pressure difference is great enough, some air may force its way up through the eustachian tube into the middle ear. At the same time, the pressure on both sides of the eardrum is equalized, and the drum moves back into its regular position. The person usually hears a popping sound at this moment, and normal hearing is restored. A reverse movement of air ordinarily occurs when a person moves from low altitude into a higher one.

The eustachian tube is usually closed by valve-like flaps in the throat, which may inhibit air movements into the middle ear. Swallowing, yawning, or chewing may aid in opening the valves, and these actions can hasten the equalization of air pressure if discomfort is experienced during altitude changes.

The mucous membranes that line the eustachian tubes are continuous with the lining of the middle ears. Consequently, the tubes provide a route by which mucous membrane infections may pass from the throat to the ear. For this reason, it is poor practice to pinch one nostril when blowing the nose, because pressure in the nasal cavity may then force material from the throat up the eustachian tube and into the middle ear.

Fig. 11.13 The eustachian tube extends between the middle ear and the pharynx. What is the function of this tube?

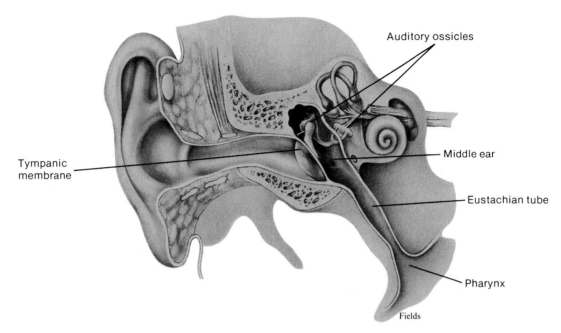

Auditory ossicles

Tympanic membrane

Middle ear

Eustachian tube

Pharynx

Fields

Inner Ear

The inner ear consists of a complex system of inter-communicating chambers and tubes called a **labyrinth;** in fact, there are two such structures in each ear—the osseous and membranous labyrinths.

The *osseous labyrinth* is a bony canal in the temporal bone; the *membranous labyrinth* lies within the osseous one and has a similar shape (fig. 11.14). Between the osseous and membranous labyrinths is a fluid, called *perilymph,* that is secreted by cells in the wall of the bony canal. The membranous labyrinth contains another fluid, called *endolymph,* whose composition is slightly different.

The parts of the labyrinths include a **cochlea** that functions in hearing, and three **semicircular canals** that function in providing a sense of equilibrium (fig. 11.15). A bony chamber, called the **vestibule,** which connects the cochlea and the semicircular canals, contains membranous structures that serve both hearing and equilibrium.

The **cochlea,** as its name suggests, is shaped like the coiled shell of a snail. Inside it contains a bony core (modiolus) and a thin bony shelf (spiral lamina) that winds around the core like the threads of a screw. The shelf divides the bony canal of the cochlea into upper and lower compartments. The upper compartment, called the *scala vestibuli,* leads from the oval window to the apex of the spiral. The lower compartment, the *scala tympani,* extends from the apex

of the cochlea to a membrane-covered opening in the wall of the inner ear called the **round window.** These compartments constitute the bony labyrinth of the cochlea, and they are filled with perilymph. At the apex of the cochlea, the fluids in the chambers can flow together through a small opening (helicotrema).

The membranous labyrinth of the cochlea is represented by the *cochlear duct* (scala media), which is filled with endolymph. It lies between the two bony compartments and ends as a closed sac at the apex of the cochlea. The cochlear duct is separated from the scala vestibuli by a *vestibular membrane* and from the scala tympani by a *basilar membrane.* (See fig. 11.16.)

The basilar membrane extends from the bony shelf of the cochlea and forms the floor of the cochlear duct. It contains many thousands of stiff, elastic fibers, whose lengths vary and become progressively longer from the base of the cochlea to its apex. It is believed that vibrations entering the perilymph at the oval window travel along the scala vestibuli and pass through the vestibular membrane to enter the cochlear duct, where they cause movements in the basilar membrane.

The shorter fibers of the basilar membrane, near the base of the cochlea, are caused to move by rapid (high frequency) vibrations. The longer fibers near the cochlear apex are set into motion by slower (low frequency) vibrations, and the fibers in between are moved by intermediate vibrations.

Somatic and Special Senses   343

Fig. 11.14 The osseous labyrinth of the inner ear is separated from the membranous labyrinth by perilymph.

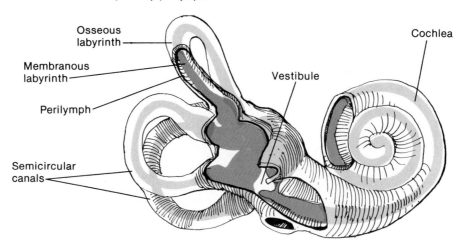

Fig. 11.15 The membranous labyrinth of the inner ear.

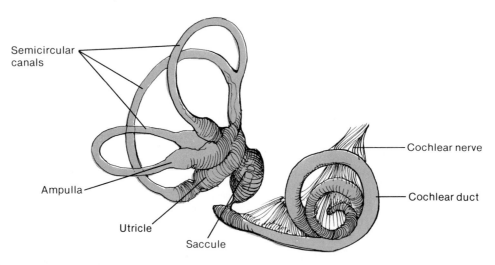

After passing through the basilar membrane, the sound vibrations enter the perilymph of the scala tympani, and their forces are dissipated to the middle ear space by movements of the membrane covering the round window. Figure 11.17 summarizes the pathway of vibrations through the parts of the middle and inner ears.

The **organ of Corti,** which contains the hearing receptors, is located on the upper surface of the basilar membrane and stretches from the apex to the base of the cochlea. Its receptor cells are arranged in rows, and they possess numerous hairlike processes that project into the endolymph of the cochlear duct. Above these cells is a *tectorial membrane* that is attached to the bony shelf of the cochlea and passes like a roof over the receptor cells, making contact with the tips of their hairs (fig. 11.18).

Because of the way these parts are arranged, different frequencies of vibrations cause motions in particular fibers of the basilar membrane. These fibers, in turn, cause specific receptor cells to move, and as the hairs of the receptors sheer back and forth against the tectorial membrane, sensory impulses are initiated in nerve fibers that are clustered at the bases of the cells. (See fig. 11.19.)

Although the human ear is able to detect sound waves with frequencies varying from about 20 to 20,000 vibrations per second, the range of greatest sensitivity is between 2,000 and 3,000 vibrations per second.

The nerve fibers associated with the hearing receptors make up parts of the cochlear branches of the vestibulocochlear nerves. They enter the auditory nerve pathways that extend into the medulla oblongata and proceed through the midbrain to the thalamic regions. From there they pass into the auditory

Fig. 11.16 (a) The cochlea consists of a coiled, bony canal with a membranous tube inside. (b) If the cochlea is unwound, the membranous tube is seen ending as a closed sac at the apex of the bony canal.

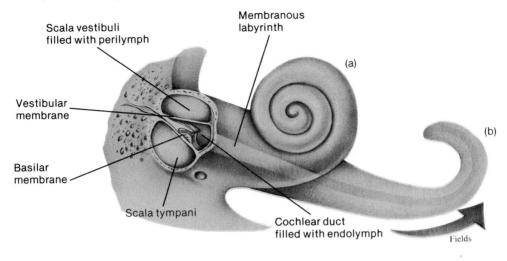

Fig. 11.17 Vibration conduction pathway within the ear.

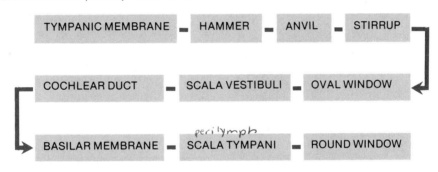

Fig. 11.18 (a) Cross section of the cochlea; (b) organ of Corti.

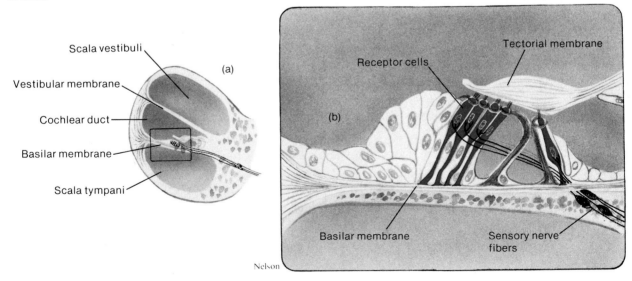

Fig. 11.19 (a) A micrograph of the organ of Corti; (b) a scanning electron micrograph of hair cells of the organ of Corti.

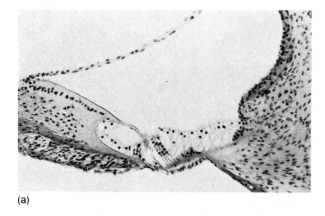

(a)

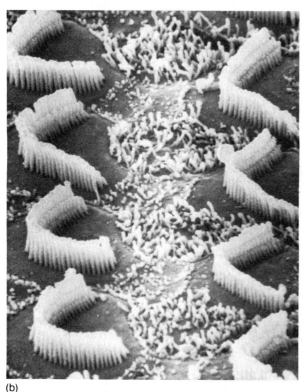

(b)

cortices of the temporal lobes of the cerebrum. On the way, some of these fibers cross over, so that impulses arising from each ear are interpreted on both sides of the brain. Consequently, damage to a temporal lobe on one side of the brain is not necessarily accompanied by complete hearing loss in the ear on that side.

Deafness

Partial or complete hearing loss can be caused by a variety of factors, including interference with the transmission of vibrations to the inner ear (conductive deafness), or damage to the cochlea, auditory nerve, or auditory nerve pathways (sensorineural deafness).

Conductive deafness may be due to a simple plugging of the external auditory meatus by an accumulation of dry wax or the presence of some foreign object. In other instances, the impairment is related to changes in the eardrum or the auditory ossicles. The eardrum, for example, may harden as a result of disease and thus be less responsive to sound waves, or it may be torn or perforated by disease or injury. The auditory ossicles, too, may be damaged or destroyed by disease or may lose their mobility.

One of the more common diseases involving the auditory ossicles is *otosclerosis*. In this condition, new bone is deposited abnormally around the base of the stapes, interfering with the motion of the ossicle needed to transmit vibrations effectively to the inner ear. Although the cause of otosclerosis is unknown, a victim's hearing often can be restored by surgical procedures. For instance, the base of the stapes may be freed by chipping away the bone that holds it fixed in position, or the stapes may be removed and replaced with a wire or plastic substitute (prosthesis).

Two tests used to help diagnose conductive deafness are the Weber and Rinne tests. In the Weber test, the handle of a vibrating tuning fork is pressed against the forehead. A person with normal hearing perceives the sound coming from directly in front, while a person with sound conduction blockage in one middle ear hears the sound coming from the impaired side.

In the Rinne test, a vibrating tuning fork is held against the bone behind the ear. After the sound is no longer heard by conduction through the bones of the skull, the fork is moved to just in front of the external auditory meatus. In middle ear conductive deafness, the vibrating fork can no longer be heard, but a normal ear will continue to hear its tone.

Sensorineural deafness can be caused by excessively loud sounds. If the exposure is of short duration, the hearing loss may be temporary, but when there is repeated and prolonged exposure, such as occurs in some foundries, near jackhammers, or on a firing range, the impairment may be permanent.

Other causes of sensorineural deafness include tumors in the central nervous system, brain damage as a result of vascular accidents, and the use of certain drugs. Some of the antibiotic drugs in the streptomycin group, for example, are known to be able to cause damage to inner ear parts. Also, some people experience hearing losses as a result of aging, a condition called *presbycusis.*

Persons over 65 years of age commonly have lost their abilities to hear high frequency sounds and they also have decreased abilities to discriminate speech sounds. Thus it is important to use lower voice tones in speaking to an elderly person. Also, the speaker should face the listener so that some lip reading is possible.

The degree of a person's hearing impairment can be measured with an *audiometer* (fig. 11.20). This device produces vibrations of known frequencies that are transmitted to the test subject through earphones. During the test, it is possible to determine the percentage of hearing loss a person has suffered at each vibration frequency.

1. How are sound waves transmitted through the external, middle, and inner ears?
2. What is the tympanic reflex?
3. Which part of the labyrinth functions in hearing?
4. Distinguish between conductive deafness and sensorineural deafness.

Fig. 11.20 An audiometer can be used to determine the percentage of hearing loss at each frequency of vibration.

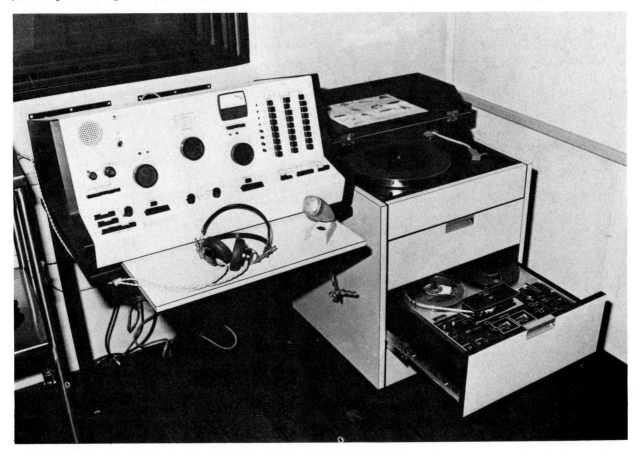

Sense of Equilibrium

The sense of equilibrium actually involves two senses—a sense of *static equilibrium* and a sense of *dynamic equilibrium*—that result from the actions of different sensory organs. The organs associated with static equilibrium function in maintaining the stability of the head and body when these parts are motionless, and the organs concerned with dynamic equilibrium aid in balancing the head and body when they are suddenly moved or rotated.

Static Equilibrium

The organs of **static equilibrium** are located within the vestibule, the bony chamber between the semicircular canals and the cochlea. More specifically, the membranous labyrinth inside the vestibule consists of two expanded chambers—an **utricle** and a **saccule.** The larger utricle communicates with the saccule and the membranous portions of the semicircular canals; the saccule in turn communicates with the cochlear duct (fig. 11.21).

On the anterior wall of the utricle is a tiny structure, called a **macula,** that contains numerous hair cells and supporting cells. The hairs of the hair cells project upward into a mass of gelatinous material that has grains of calcium carbonate (otoliths) embedded in it. These particles increase the weight of the gelatinous structure, making it more responsive to changes in position. The hair cells, which serve as sensory receptors, have nerve fibers wrapped around their bases. These fibers are associated with the vestibular portion of the vestibulocochlear nerve.

The usual stimulus to the hair cells occurs when the head is bent forward, backward, or to one side. Such movements cause the gelatinous mass of the macula to be tilted, and as it sags in response to gravity, the hairs projecting into it are bent. This action stimulates the hair cells, and they signal the nerve fibers associated with them. The resulting nerve impulses travel into the central nervous system, informing the brain as to the position of the head. The brain may act on this information by sending motor impulses to skeletal muscles, and they may contract or relax appropriately so that balance is maintained (fig. 11.22).

The macula also functions in the sense of dynamic equilibrium. For example, if the head or body is thrust forward or backward abruptly, the gelatinous mass of the macula lags slightly behind, and the hair cells are stimulated. In this way, the macula aids the brain in detecting movements such as falling.

Although there is some question as to whether the human saccule is functional, it does contain a maculalike structure on its medial wall, which may act in much the same manner as the macula of the utricle.

Dynamic Equilibrium

The three bony **semicircular canals** lie at right angles to each other, and each occupies a different plane in space. Two of them, the *superior* and the *posterior canals,* stand vertically, while the third or *lateral canal* is horizontal.

Suspended in the perilymph of each bony canal is a membranous semicircular canal that ends in a

---

Fig. 11.21 The saccule and utricle, which are expanded portions of the membranous labyrinth, are located within the bony chamber of the vestibule.

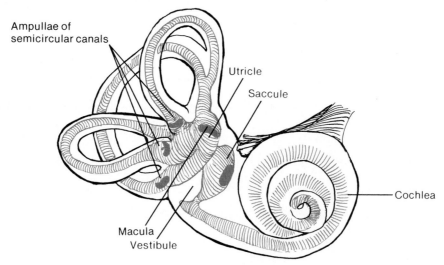

Ampullae of semicircular canals

Utricle

Saccule

Cochlea

Macula

Vestibule

Fig. 11.22 The macula is responsive to changes in the position of the head, as from (a) the head in an upright position to (b) the head bent forward.

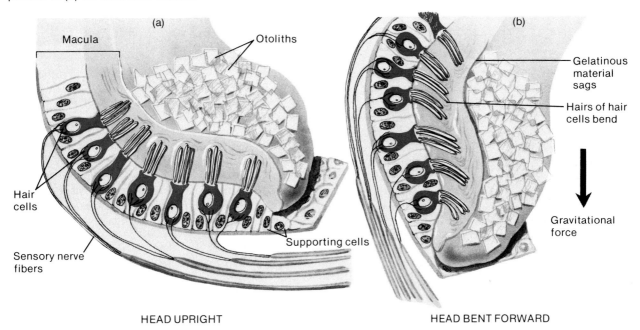

Macula

Otoliths

Hair cells

Sensory nerve fibers

Supporting cells

Gelatinous material sags

Hairs of hair cells bend

Gravitational force

HEAD UPRIGHT

HEAD BENT FORWARD

swelling, called an **ampulla.** The ampullae communicate with the utricle of the vestibule.

The sensory organs of the semicircular canals, which function in the sense of dynamic equilibrium, are located within the ampullae. Each of these organs, called a **crista ampullaris,** contains a number of sensory hair cells and supporting cells. Like the macula, the hair cells have hairs that extend upward into a dome-shaped gelatinous mass, called the *cupula.* Also, the hair cells are connected at their bases to nerve fibers that make up part of the vestibular branch of the vestibulocochlear nerve (fig. 11.23).

The hair cells of the crista are ordinarily stimulated by rapid turns of the head or body. At these times, the semicircular canals move with the head or body, but the fluid inside the membranous canals remains stationary because of inertia. This causes the cupula in one or more of the canals to be bent in a direction opposite to that of the head or body movement, and the hairs embedded in it are also bent. This bending of the sensory hairs stimulates the hair cells to signal their associated nerve fibers, and as a result, impulses travel to the brain. (See fig. 11.24.)

Parts of the cerebellum are particularly important in interpreting impulses from the semicircular canals. Analysis of such information allows the nervous system to predict the consequences of rapid body movements. The brain can then send motor impulses to stimulate appropriate skeletal muscles so that loss of balance can be prevented.

Fig. 11.23 A crista ampullaris is located within the ampulla of each semicircular canal.

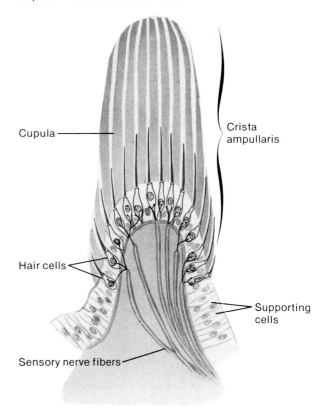

Cupula

Crista ampullaris

Hair cells

Supporting cells

Sensory nerve fibers

Fig. 11.24 (a) When the head is stationary, the cupula of the crista ampullaris remains upright. (b) When the head is moving rapidly, the cupula is bent and sensory receptors are stimulated.

(a)

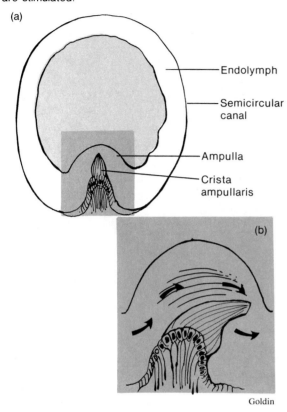

—Endolymph

—Semicircular canal

—Ampulla

—Crista ampullaris

(b)

Goldin

Some persons experience a discomfort called *motion sickness* when they are moving in a boat, airplane, or automobile. This discomfort seems to be caused by abnormal and irregular body motions that disturb the organs of equilibrium. Symptoms of motion sickness include nausea, vomiting, dizziness, headache, and prostration.

Other sensory structures also aid in the maintenance of equilibrium. For example, proprioceptors, particularly those associated with the joints of the neck, supply the brain with information concerning the position of body parts. In addition, the eyes detect changes in posture that result from body movements. Visual information is so important that a person who has suffered damage to the organs of equilibrium may be able to maintain normal balance as long as the eyes remain open and body movements are performed slowly.

1. Distinguish between the senses of static and dynamic equilibrium.
2. What structures function to provide the sense of static equilibrium? Dynamic equilibrium?
3. How does sensory information from proprioceptors help maintain equilibrium?

Sense of Sight

Although the eye contains the visual receptors and is the primary organ of sight, its functions are assisted by a number of *accessory organs*. These include the eyelids and the lacrimal apparatus, which aid in protecting the eye, and a set of extrinsic muscles, which move it.

Visual Accessory Organs

The eye, lacrimal gland, and the extrinsic muscles of the eye are housed within the pear-shaped orbital cavity of the skull. The orbit, which is lined with the periosteums of various bones, also contains fat, blood vessels, nerves, and a variety of connective tissues.

Each **eyelid** is composed of four basic layers—skin, muscle, connective tissue, and conjunctiva. The skin of the eyelid, which is the thinnest skin of the body, covers the lid's outer surface and fuses with its inner lining, the *conjunctiva,* near the margin of the lid. (See fig. 11.25.)

The muscles of the eyelids include the *orbicularis oculi,* whose fibers encircle the opening between the lids and spread out onto the cheek and forehead. This muscle acts as a sphincter and closes the lids when it contracts.

Fibers of the *levator palpebrae superioris* muscle arise from the roof of the orbit and are inserted in the connective tissue of the upper lid. When these fibers contract, the upper lid is raised and the eye opens.

The connective tissue layer of the eyelid, which helps give it form, contains numerous modified sebaceous glands (tarsal glands). The oily secretions of these glands are carried by ducts to openings along the borders of the lids. This secretion aids in preventing the lids from sticking together.

The **conjunctiva** is a mucous membrane that lines the inner surfaces of the eyelids and folds back to cover the anterior surface of the eyeball, except for its central portion (cornea). Although the conjunctiva that lines the eyelids is relatively thick, the part on the eyeball is very thin. It is also freely movable and quite transparent, so that blood vessels are clearly visible beneath it.

Fig. 11.25 Sagittal section of the eyelids and the anterior portion of the eye.

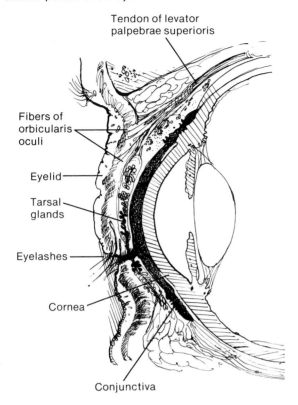

Tendon of levator palpebrae superioris

Fibers of orbicularis oculi

Eyelid

Tarsal glands

Eyelashes

Cornea

Conjunctiva

Fig. 11.26 The lacrimal apparatus consists of a tear-secreting gland and a series of ducts.

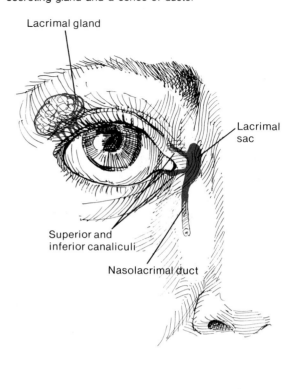

Lacrimal gland

Lacrimal sac

Superior and inferior canaliculi

Nasolacrimal duct

The *lacrimal apparatus* (fig. 11.26) consists of the **lacrimal gland,** which secretes tears, and a series of *ducts,* which carry the tears into the nasal cavity. The gland is located in the orbit, above and to the lateral side of the eye. It secretes tears continuously, and they pass out through tiny tubules, and flow downward and medially across the eye.

Tears are collected by two small ducts (superior and inferior canaliculi) whose openings can be seen on the medial borders of the eyelids. From these ducts, the fluid moves into the *lacrimal sac,* which lies in a deep groove of the lacrimal bone, and then into the *nasolacrimal duct* that empties into the nasal cavity.

Glandular cells of the conjunctiva also secrete a tearlike liquid that, together with the secretion of the lacrimal gland, keeps the surface of the eye and the lining of the lids moist and lubricated. It is known, too, that tears contain an enzyme, called *lysozyme,* that functions as an antibacterial agent, reducing the chance of eye infections.

When a person is emotionally upset by grief or disappointment, or when the conjunctiva is irritated, the tear glands are likely to secrete excessive fluids. Tears may spill over the edges of the eyelids, and the nose may fill with fluid. This response involves motor impulses carried to the lacrimal glands on parasympathetic nerve fibers.

The **extrinsic muscles** of the eye arise from the bones of the orbit and are inserted by broad tendons of the eye's tough outer surface. There are six such muscles that function to move the eye in various directions (fig. 11.27). Although any given eye movement may involve more than one of them, each muscle is associated with one primary action, as follows:

1. **Superior rectus**—rotates the eye upward and toward the midline.

2. **Inferior rectus**—rotates the eye downward and toward the midline.

3. **Medial rectus**—rotates the eye toward the midline.

4. **Lateral rectus**—rotates the eye away from the midline.

5. **Superior oblique**—rotates the eye downward and away from the midline.

6. **Inferior oblique**—rotates the eye upward and away from the midline.

The motor units of the extrinsic eye muscles contain the smallest number of muscle fibers (five to ten) of any muscles in the body. Because of this, the eyes can be moved with great precision. Also, the eyes move together so that they are aligned when looking at something. Such alignment involves complex motor adjustments that result in the contraction of certain eye muscles while their antagonists are relaxed. For

Fig. 11.27 The extrinsic muscles of the eye.

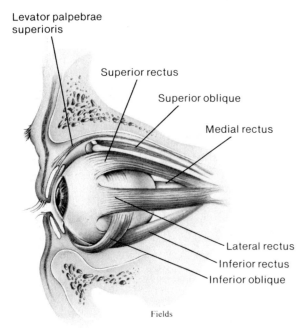

Levator palpebrae
superioris

Superior rectus

Superior oblique

Medial rectus

Lateral rectus

Inferior rectus

Inferior oblique

Fields

Chart 11.2 Muscles associated with the eye

Skeletal Muscles

Name	Innervation	Function
Orbicularis oculi	Facial nerve (VII)	Closes eye
Levator palpebrae superioris	Oculomotor nerve (III)	Opens eye
Superior rectus	Oculomotor nerve (III)	Rotates eye upward and toward midline
Inferior rectus	Oculomotor nerve (III)	Rotates eye downward and toward midline
Medial rectus	Oculomotor nerve (III)	Rotates eye toward midline
Lateral rectus	Abducens nerve (VI)	Rotates eye away from midline
Superior oblique	Trochlear nerve (IV)	Rotates eye downward and away from midline
Inferior oblique	Oculomotor nerve (III)	Rotates eye upward and away from midline

Smooth Muscles

Ciliary muscles	Oculomotor nerve (III) parasympathetic fibers	Causes suspensory ligaments to relax
Iris, circular muscles	Oculomotor nerve (III) parasympathetic fibers	Causes size of pupil to decrease
Iris, radial muscles	Sympathetic fibers	Causes size of pupil to increase

example, when the eyes move to the right, the lateral rectus of the right eye and the medial rectus of the left eye must contract. At the same time, the medial rectus of the right eye and the lateral rectus of the left eye must relax. A person whose eyes are not co-ordinated well enough to produce alignment is said to have *strabismus,* or squint.

Chart 11.2 summarizes the muscles associated with the eye.

When one eye deviates from the line of vision, the person has double vision (diplopia). If this condition persists, there is danger that changes will occur in the brain to suppress the image from the deviated eye. As a result, the turning eye may become blind (suppression amblyopia). Such monocular blindness can often be prevented if the eye deviation is treated early in life with exercises, glasses, and surgery. For this reason, vision screening programs for preschool children are very important.

1. Describe the conjunctiva.
2. What is the function of the lacrimal apparatus?
3. Describe the function of each extrinsic eye muscle.

Structure of the Eye

The eye itself is a hollow, spherical structure about 2.5 cm (one inch) in diameter. Its wall has three distinct layers—an outer *fibrous tunic,* a middle *vascular tunic,* and an inner *nervous tunic.* The spaces within the eye are filled with fluids that provide support for the wall and internal parts and help maintain its shape. Figure 11.28 shows the parts of the eye. (Also see color plate 23.)

The Outer Tunic.   The anterior 1/6 of the outer tunic bulges forward as the transparent **cornea,** which serves as the window of the eye and helps focus entering light rays. The transparency of the cornea is largely due to the fact that it contains relatively few cells and no blood vessels. Also, the cells that are present are arranged in unusually regular patterns.

Fig. 11.28 Sagittal section of the eye.

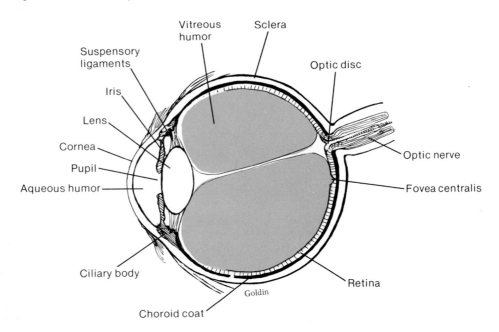

On the other hand, the cornea is well supplied with nerve fibers that enter its margin and radiate toward its center. These fibers are associated with numerous pain receptors that have very low thresholds. Cold receptors are also abundant in the cornea, but heat and touch receptors seem to be lacking.

Along its circumference, the cornea is continuous with the **sclera,** the white portion of the eye. This part makes up the posterior 5/6 of the outer tunic and is opaque due to the presence of many large, haphazardly arranged, collagenous and elastic fibers. The sclera provides protection.

In the back of the eye, the sclera is pierced by the **optic nerve** and certain blood vessels and is attached to the dura mater that encloses these structures and other parts of the central nervous system.

The Middle Tunic. The middle or vascular tunic of the eyeball (uveal layer) includes the choroid coat, the ciliary body, and the iris.

The **choroid coat,** in the posterior 5/6 of the globe, is loosely joined to the sclera and is honeycombed with blood vessels that provide nourishment to surrounding tissues. The choroid also contains numerous pigment-producing melanocytes that give it a brownish black appearance. The melanin of these cells functions to absorb excess light and thus helps to keep the eye dark inside.

The **ciliary body,** which is the thickest part of the middle tunic, extends forward from the choroid and forms an internal ring around the front of the eye. Within the ciliary body there are many radiating folds called *ciliary processes* and two distinct groups of muscle fibers that constitute the *ciliary muscles.* These structures are shown in fig. 11.29.

The transparent **lens** is held in position by a large number of strong but delicate fibers, called *suspensory ligaments,* that extend inward from the ciliary processes. The distal ends of these fibers are attached along the margin of a thin capsule that surrounds the lens. The body of the lens lies directly behind the iris and is composed of "fibers" that arise from epithelial cells. In fact, the cytoplasm of these cells makes up the transparent substance of the lens. (See color plate 24.)

The lens capsule is a clear, membranelike structure composed largely of intercellular material. It is quite elastic, and this quality keeps it under constant tension. As a result, the lens assumes a globular shape. The suspensory ligaments attached to the margin of the capsule are also under tension, and as they pull outward the capsule and the lens inside are kept somewhat flattened (fig. 11.30).

If the tension on the suspensory ligaments is relaxed, the elastic capsule rebounds, and the lens surface becomes more convex. Such a change occurs in the lens of each eye when it is focused to view a close object. This adjustment is called **accommodation.**

Fig. 11.29 Anterior portion of the eye.

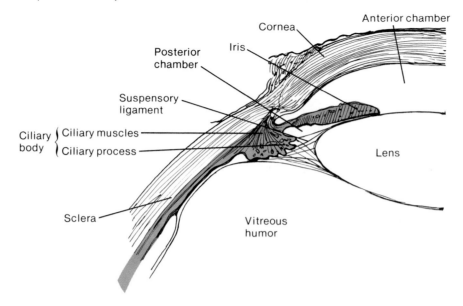

Cornea

Anterior chamber

Iris

Posterior chamber

Suspensory ligament

Ciliary body { Ciliary muscles
Ciliary process

Sclera

Lens

Vitreous humor

Fig. 11.30 The lens and ciliary body viewed from behind.

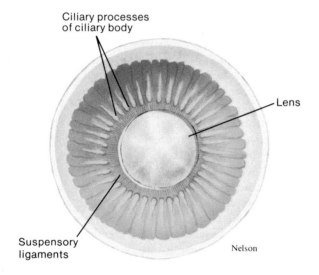

Ciliary processes of ciliary body

Lens

Suspensory ligaments

Nelson

Fig. 11.31 (*a*) How is the focus of the eye affected when the lens becomes thin as the ciliary muscle fibers relax? (*b*) How is it affected when the lens thickens as the ciliary muscle fibers contract?

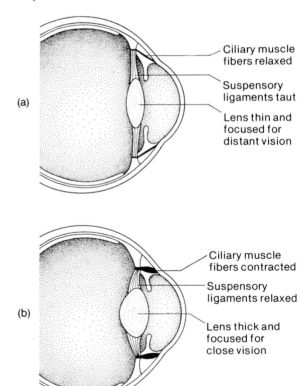

(a)

Ciliary muscle fibers relaxed

Suspensory ligaments taut

Lens thin and focused for distant vision

(b)

Ciliary muscle fibers contracted

Suspensory ligaments relaxed

Lens thick and focused for close vision

The relaxation of the suspensory ligaments during accommodation is a function of the ciliary muscles (fig. 11.31). One set of these muscle fibers is arranged in a circular pattern forming a sphincterlike structure around the ciliary processes. The fibers of the other set extend back from fixed points in the sclera to the choroid. When the circular muscle fibers contract, the diameter of the ring formed by the ciliary processes is lessened; when the other fibers contract, the choroid is pulled forward and the ciliary body is shortened. Each of these actions causes the suspensory ligaments to become relaxed, and the lens thickens in response.

Integration and Coordination

Fig. 11.32 Aqueous humor, which is secreted into the posterior chamber, circulates into the anterior chamber and leaves it through the canal of Schlemm.

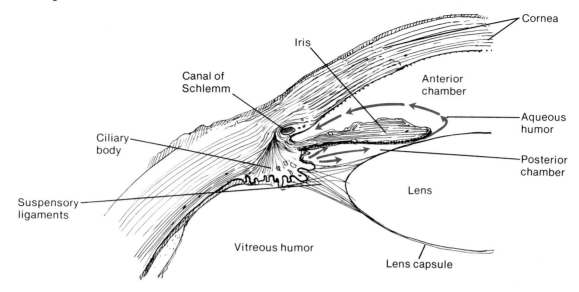

1. Describe the three layers of the eye.
2. What factors contribute to the transparency of the cornea?
3. How is the shape of the lens changed during accommodation?

The **iris** is a thin, muscular diaphragm that is seen from the outside as the colored portion of the eye. It extends forward from the periphery of the ciliary body and lies between the cornea and the lens. The iris divides the space separating these parts into an *anterior chamber* (between the cornea and the iris) and a *posterior chamber* (between the iris and the lens).

The epithelium of the ciliary body continuously secretes a watery fluid called **aqueous humor** into the posterior chamber. The fluid circulates from this chamber through the **pupil**, a circular opening in the center of the iris, and into the anterior chamber (fig. 11.32). It fills the space between the cornea and the lens and aids in maintaining the shape of the front of the eye. The aqueous humor subsequently leaves the anterior chamber through a special drainage canal (canal of Schlemm) located in its wall.

A disorder called *glaucoma* sometimes develops in the eyes as a person ages. This condition occurs when the rate of aqueous humor formation exceeds the rate of its removal. As a result, fluid accumulates in the anterior chamber of the eye, and the fluid pressure rises.

Since liquids cannot be compressed, the increasing pressure from the anterior chamber is transmitted to all parts of the eye, and in time the blood vessels that supply the receptor cells of the retina may be squeezed closed. If this happens, cells that fail to receive needed nutrients and oxygen may die, and a form of permanent blindness may result.

Glaucoma often can be controlled, because it is relatively easy to measure the pressure of the fluids within the eyes. If the pressure is known to be reaching a dangerous level, steps can be taken to reduce it. For example, drugs that promote the drainage of aqueous humor from the anterior chamber may be used, or a portion of the iris may be removed surgically to improve drainage.

The smooth muscle fibers of the iris are arranged into two groups, a circular set and a radial set. These muscles function to control the size of the pupil, the opening that light passes through as it enters the eye. The circular set acts as a sphincter, and when it contracts, the pupil gets smaller. When the radial muscle fibers contract, the diameter of the pupil increases.

Fig. 11.33 In dim light, the radial muscles of the iris are stimulated to contract, and the pupil becomes dilated. In bright light, the circular muscles of the iris are stimulated to contract, and the pupil becomes smaller.

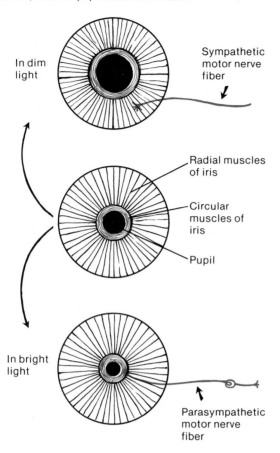

In dim light

Sympathetic motor nerve fiber

Radial muscles of iris

Circular muscles of iris

Pupil

In bright light

Parasympathetic motor nerve fiber

The size of the pupils changes constantly in response to pupillary reflexes that are triggered by such factors as light intensity, gaze, and variations in emotional state. For example, bright light elicits a reflex, and impulses travel along parasympathetic nerve fibers to the *circular muscles* of the irises. The pupils become constricted in response. Conversely, in dim light, impulses travel on sympathetic nerve fibers to the *radial muscles* of the irises, and the pupils become dilated. (See fig. 11.33.)

Since parasympathetic fibers that release acetylcholine cause the pupils to constrict, drugs that mimic the action of acetylcholine, such as pilocarpine, also cause pupillary constriction. Conversely, drugs that block parasympathetic actions, such as atropine, cause pupillary dilation.

The color of the eyes is determined largely by the amount and distribution of melanin in the irises. If melanin is present only in the epithelial cells that cover an iris's posterior surface, the iris appears blue. When this condition exists together with denser than usual tissue within the iris, it looks gray, and when melanin is present within the body of the iris as well as in the epithelial covering, it is brown.

The Inner Tunic. The inner tunic of the eye consists of the **retina,** which contains the visual receptor cells. This nearly transparent sheet of tissue is continuous with the optic nerve in the back of the eye and extends forward as the inner lining of the eyeball. It ends just behind the margin of the ciliary body.

Although the retina is thin and delicate, its structure is quite complex. It has a number of distinct layers, including pigmented epithelium, receptor cells, nerve fibers, ganglion cells, and limiting membranes. (See fig. 11.34.)

In the central region of the retina there is a yellowish spot, the **macula lutea,** that has a depression in its center called the **fovea centralis.** This depression is in the region of the retina that produces the sharpest vision.

Just medial to the fovea centralis is an area called the **optic disk** (fig. 11.35 and color plate 25). Here the nerve fibers from the retina leave the eye and become parts of the optic nerve. A central artery and vein also pass through at the optic disk. These vessels are continuous with capillary networks of the retina and, together with vessels in the underlying choroid coat, they supply blood to the cells of the inner tunic. (See color plate 26.)

Because there are no receptor cells in the region of the optic disk, it is commonly referred to as the blind spot of the eye.

The space bounded by the lens, ciliary body, and retina represents the largest compartment of the eye, and it is filled with a clear, jellylike fluid called **vitreous humor.** This fluid functions to support the internal parts of the eye and helps maintain its shape.

In summary, light waves entering the eye must pass through the cornea, aqueous humor, lens, vitreous humor, and several layers of the retina before they reach the visual receptor cells. The layers of the eye are summarized in chart 11.3.

1. How is the size of the pupil regulated?
2. Explain the origin of aqueous humor and trace its path through the eye.
3. Describe the structure of the retina.

Fig. 11.34 The retina consists of several cell layers.

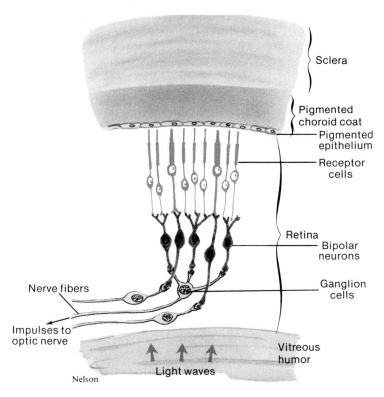

Sclera

Pigmented
choroid coat

Pigmented
epithelium

Receptor
cells

Retina

Bipolar
neurons

Ganglion
cells

Nerve fibers

Impulses to
optic nerve

Vitreous
humor

Light waves

Nelson

Fig. 11.35 Nerve fibers leave the eye in the area of the optic disk to form the optic nerve.

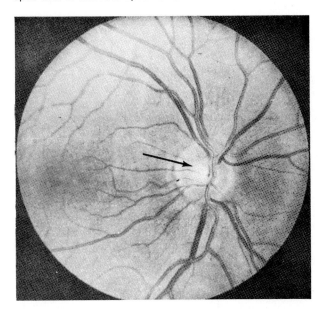

Refraction of Light

When a person sees something, it is either giving off light or light waves are being reflected from it. These waves enter the eye, and an image of what is seen focuses upon the retina. This focusing process involves a bending of the light waves—a phenomenon called **refraction.**

Refraction occurs when light waves pass at an oblique angle from a medium of one optical density into a medium of a different optical density. For example, as figure 11.36 shows, when light passes obliquely from a less dense medium such as air into a denser medium such as glass, or from air into the cornea of the eye, the light is bent toward a line perpendicular to the surface between these substances. When the surface between such refracting media is curved, a lens is formed. A lens with a *convex* surface causes light waves to converge, and a lens with a *concave* surface causes light waves to diverge. (See fig. 11.37.)

When light arrives from objects outside the eye, the light waves are refracted primarily by the convex surface of the cornea. Then the waves are refracted again by the convex surface of the lens and to a lesser extent by the surfaces of the fluids within the chambers of the eye.

Chart 11.3 Layers of the eye

Tunic	Posterior Portion	Function	Anterior Portion	Function
Outer layer	Sclera	Protection	Cornea	Light transmission
Middle layer	Choroid coat	Blood supply, pigment prevents reflection	Ciliary body, iris	Accommodation, controls light intensity
Inner layer	Retina	Photoreception, impulse transmission	None	

Fig. 11.36 When light passes at an oblique angle from air into glass, the light waves are bent toward a line perpendicular to the surface of the glass.

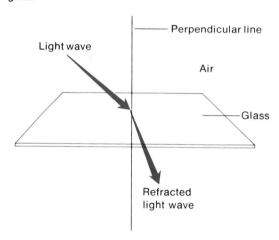

Fig. 11.37 (*a*) A lens with a convex surface causes light waves to converge; (*b*) a lens with a concave surface causes them to diverge.

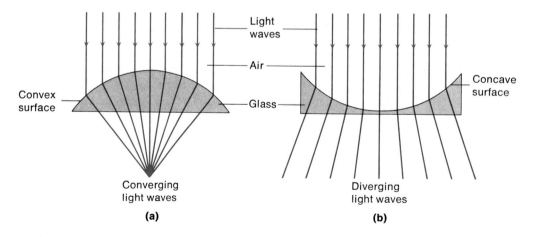

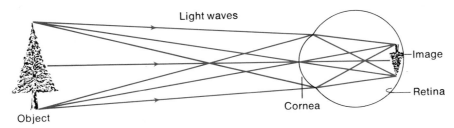

Fig. 11.38 The image of an object formed on the retina is upside-down.

If the shape of the eye is normal, light waves are focused sharply upon the retina, much as a motion picture image is focused on a screen for viewing. Unlike the motion picture image, the one formed on the retina is upside down and reversed from left to right (fig. 11.38). When the visual cortex interprets such an image, it somehow corrects this, and things are seen in their proper positions.

Light waves coming from objects more than 20 feet away are traveling in nearly parallel lines, and they are focused on the retina by the cornea and by the lens in its flattened or "at rest" condition. Light waves arriving from objects less than 20 feet away, however, reach the eye along more divergent lines—in fact, the closer the object, the more divergent the lines.

Divergent light waves tend to come into focus behind the retina unless something is done to increase the refracting power of the eye. This increase is accomplished by accommodation, which results in a thickening of the lens. As the lens thickens, light waves are converged more strongly, so that diverging light waves coming from close objects are focused on the retina.

Refraction Problems. Unfortunately, the elastic quality of the lens capsule tends to lessen with age, and persons over forty-five years of age often are unable to accommodate sufficiently to read the fine print in books and newspapers. Their eyes remain focused for distant vision. This condition is termed *presbyopia,* or farsightedness of age, and it can usually be corrected by using eyeglasses or contact lenses that make up for the eye's loss in refracting power.

Other persons suffer from problems that result when their eyeballs are too short or too long for sharp focusing. For example, if an eye is too short, light waves are not focused sharply on the retina since their point of focus lies some distance behind it. A person with this condition may be able to bring the image of distant objects into focus by accommodation, but this requires contraction of the ciliary muscles at times when these muscles are at rest in a normal eye.

Still more accommodation is needed to view closer objects, and the victim may suffer from ciliary muscle fatigue, pain, and headache whenever it is necessary to do close work.

Since people with short eyeballs commonly are unable to accommodate enough to focus on very close objects, they are said to be *farsighted* (hyperopia). Treatment for this condition involves the use of glasses or contact lenses with *convex* surfaces that cause images to be focused closer to the front of the eye.

If an eyeball is too long, light waves tend to be focused in front of the retina, and the image produced on it is blurred. In other words, the refracting power of the eye, even when the lens is flattened, is too great. Although a person with this problem may be able to focus on close objects by accommodation, distance vision is invariably poor. For this reason, the person is said to be *nearsighted* (myopia). Treatment for nearsightedness makes use of glasses or contact lenses with *concave* surfaces that cause images to be focused further from the front of the eye. (See figs. 11.39 and 11.40.)

Still another refraction problem is termed *astigmatism.* In this condition, there is a defect in the curvature of the cornea or, sometimes, in the curvature of the lens. The normal cornea has a spherical curvature, like the inside of a ball; an astigmatic cornea usually has an elliptical curvature, like the bowl of a spoon. As a result, some portions of an image are in focus on the retina, but other portions are blurred, and vision is distorted.

Without corrective lenses, astigmatic eyes tend to accommodate back and forth reflexly in an attempt to sharpen focus. The consequence of this continual action is likely to be ciliary muscle fatigue and headache.

Fig. 11.39 (*a*) If an eye is too long, the focus point of images lies in front of the retina; (*b*) in a normal eye, images focus on the retina; (*c*) if the eye is too short, the focus point of images lies behind the retina.

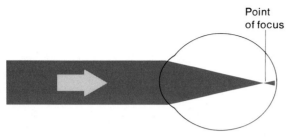

(a) Eye too long (myopia)

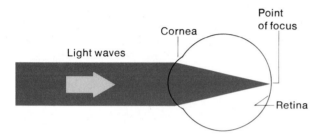

(b) Normal eye

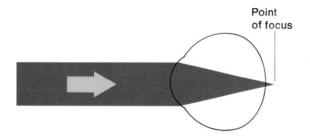

(c) Eye too short (hyperopia)

A relatively common eye disorder, particularly in older people, is called *cataract*. In this condition, the lens or its capsule slowly loses its transparency and becomes cloudy and opaque. As a result, clear images cannot be focused on the retina, and in time the person may become blind.

Cataract is usually treated by surgically removing the lens. The lens may then be replaced by an artificial one, or the loss of refractive power of the eye may be corrected with eyeglasses or contact lenses.

Fig. 11.40 (*a*) Nearsightedness is corrected by using a concave lens; (*b*) farsightedness is corrected with a convex lens.

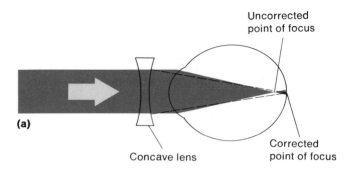

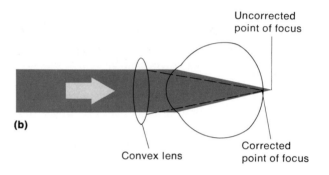

1. What is meant by refraction?
2. What parts of the eye provide refracting surfaces?
3. What common eye disorders involve errors in refraction?

Visual Receptors

The receptor cells of the eye are actually modified neurons, and there are two distinct kinds, as illustrated in figure 11.41. One group of receptors have long, thin projections at their terminal ends and are called **rods.** The cells of the other group have short, blunt projections and are called **cones.**

Instead of being located in the surface layer of the retina, the rods and cones are found in a deep portion, closely associated with a layer of pigmented epithelium. The projections from the receptors extend into the pigmented layer. (See fig. 11.42 and color plate 27.)

The pigment of the retina functions to absorb light waves that are not absorbed by the receptor cells, and with the pigment of the choroid coat, it serves to keep light from reflecting about inside the eye. Albinos, who lack melanin in all parts of their bodies including their eyes, suffer a considerable loss

Fig. 11.41 Rods and cones are the two forms of visual receptor cells.

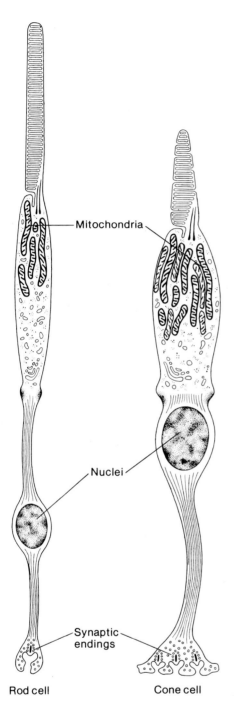

Mitochondria

Nuclei

Synaptic endings

Rod cell   Cone cell

Fig. 11.42 A micrograph of the retina.

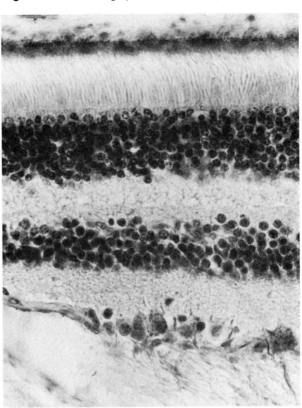

in visual sharpness (acuity). This is because light reflects from the inside of their eyes and tends to stimulate their visual receptors excessively.

The visual receptors are only stimulated when light reaches them. Thus, when a light image is focused on an area of the retina some receptors are stimulated, and impulses travel away from them to the brain.

However, the impulse leaving each receptor that is activated provides only a fragment of the information needed for the brain to interpret a total scene.

Rods and cones function differently. For example, rods are hundreds of times more sensitive to light than cones, and as a result, they enable persons to see in relatively dim light. In addition, the rods produce colorless vision, while cones can detect colors.

Still another difference involves visual acuity or the sharpness of the images perceived. Cones allow a person to see sharp images, while rods enable one to see more general outlines of objects. This characteristic is related to the fact that impulses from several rods may be transmitted to the brain on the same nerve fiber. Thus, if a point of light stimulates a rod,

Fig. 11.43 (*a*) Impulses from several rods may be transmitted to the brain on a single sensory nerve fiber; (*b*) impulses from cones are often transmitted to the brain on separate sensory nerve fibers.

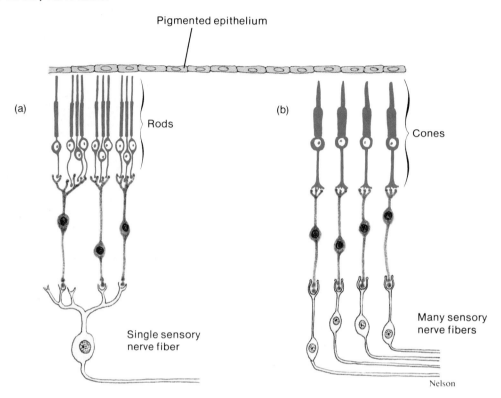

Pigmented epithelium

(a)

Rods

Single sensory nerve fiber

(b)

Cones

Many sensory nerve fibers

Nelson

the brain cannot tell which one of many receptors has actually been stimulated. Such a pooling of impulses occurs to a much lesser degree among cones, so when a cone is stimulated, the brain is able to pinpoint the stimulation more accurately. (See fig. 11.43.)

As was mentioned, the area of sharpest vision is the fovea centralis in the macula lutea. Here only cones are present and the overlying layers of the retina, as well as the retinal blood vessels, are displaced to the sides. This displacement more fully exposes the receptors to incoming light. Consequently, to view something in detail, one moves the eyes so the important part of an image falls upon the fovea.

The concentration of cones decreases in areas further away from the macula, while the concentration of rods increases in these areas. Also, the amount of impulse pooling among the rods increases toward the periphery of the retina. As a result of these factors, the visual sensations from images focused on the sides of the retina tend to be blurred compared with those focused on the central portion of the retina.

Visual Pigments. Both rods and cones contain light-sensitive pigments that decompose when they absorb light energy. The light-sensitive substance in rods is called **rhodopsin,** or visual purple. In the presence of light, rhodopsin molecules break down into molecules of a colorless protein called *scotopsin* and a yellowish substance called *retinene* (retinal), which is synthesized from vitamin A.

At the same time that a rhodopsin molecule decomposes, some energy is released that triggers a nerve impulse to travel away from the rod, through the optic nerve, and into the brain.

In bright light, nearly all of the rhodopsin in the rods of the retina is decomposed, and the sensitivity of these receptors is greatly reduced. In dim light, however, rhodopsin can be regenerated from scotopsin and retinene faster than it is broken down. This regeneration process requires cellular energy, which is provided by energy-carrying molecules of ATP. (See fig. 11.44.)

Fig. 11.44 (*a*) The decomposition of rhodopsin; (*b*) the synthesis of rhodopsin.

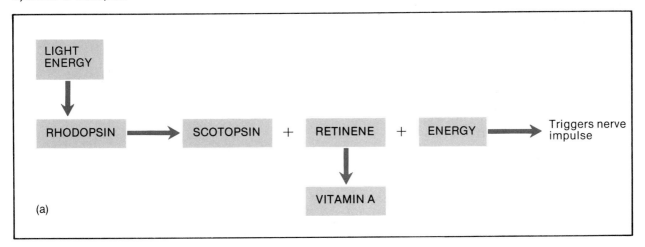

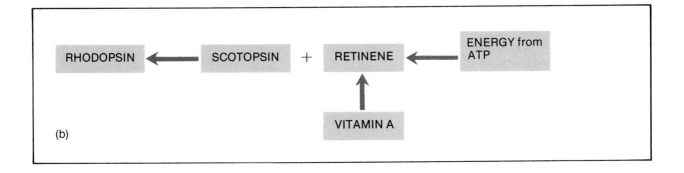

The light sensitivity of an eye whose rods have converted the available scotopsin and retinene to rhodopsin increases about 100,000 times, and the eye is said to be *dark adapted.*

A person needs a dark-adapted eye when moving from daylight into a darkened theater. At first, it is difficult to see well enough to locate a seat, but eventually the eyes become adapted to the dim light, and the vision improves.

Later, when leaving the theater and entering the sunlight, the person may feel some discomfort or even pain. This occurs at the moment that most of the rhodopsin decomposes in response to the bright light. At the same time, the light sensitivity of the eyes decreases greatly, and they become *light adapted.*

Many people, particularly children, suffer from vitamin A deficiency due to improper diet. In such cases, the quantity of retinene available for the manufacture of rhodopsin may be reduced and consequently the sensitivity of their rods may be low. This condition, called night blindness, is characterized by poor vision in dim light. Fortunately, the problem is usually easy to correct by adding vitamin A to the diet or by providing the victim with injections of vitamin A. Vitamin A deficiency remains one of the principal nutritional problems throughout the world, even though most countries have ample supplies of the plant foods that provide this needed substance.

The light-sensitive pigments of cones are similar to rhodopsin in that they are composed of retinene combined with a protein; the protein, however, differs from that in the rods. In fact, there seem to be three different kinds of visual pigments associated with cones, each containing a different protein. Also, there are apparently three different sets of cones in the retina, each containing an abundance of one of the three pigments.

The wavelength of a particular kind of light determines the color perceived from it. For example, the shortest wavelengths of visible light are perceived as violet, while the longest visible wavelengths are sensed as red. As far as cone pigments are concerned, one type (erythrolabe) is thought to be most sensitive to red light waves, another (chlorolabe) to green light waves, and a third (cyanolabe) to blue light waves. The color a person senses depends upon which set of cones or combination of sets of cones is stimulated by the light in a given image. If all three sets of cones are stimulated, the person senses the light as white, and if none are stimulated, the person senses black.

Some persons have defective color vision, apparently due to a decreased sensitivity in one or more cone sets. Such people perceive colors differently from those with normal vision and are said to be *color blind*.

Elderly persons often have diminished abilities to distinguish colors precisely. Also, their retinas may undergo dark adaptation more slowly, and they may require additional time to make visual adjustments when moving into dimly lighted areas.

Stereoscopic Vision

Stereoscopic vision (stereopsis) is vision that involves the perception of distance and depth as well as the height and width of objects. Such vision is due largely to the fact that the pupils of the eyes are 6–7 cm (about 2.5 inches) apart. Consequently, objects that are relatively close (less than 6 meters or 20 feet away) produce slightly different retinal images. That is, the right eye sees a little more of one side of an object, while the left eye sees a little more of the other side. These two images are somehow superimposed and interpreted by the visual cortex of the brain, and the result is the perception of a single object in three dimensions. (See fig. 11.45.)

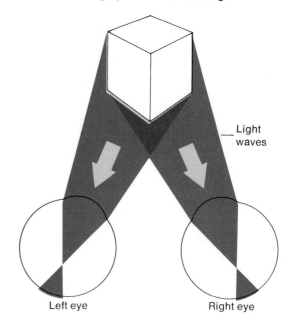

Fig. 11.45 Stereoscopic vision results from the formation of two slightly different retinal images.

Since this type of depth perception depends on vision with two eyes (binocular vision), it follows that a one-eyed person has a lessened ability to judge distance and depth accurately. To compensate when making such judgments, a person with one eye can use clues provided by the relative sizes and relative positions of familiar objects.

Visual Nerve Pathways

As is mentioned in chapter 10, the axons of the ganglion cells in the retina leave the eyes to form the *optic nerves*. Just anterior to the pituitary gland, these nerves give rise to the X-shaped *optic chiasma,* and within the chiasma some of the fibers cross over. More specifically, the fibers from the medial half of each retina cross over, while those from the lateral sides do not. Thus, fibers from the medial side of the left eye and the lateral side of the right eye form the right *optic tract;* and fibers from the medial side of the right eye and the lateral side of the left form the left optic tract.

The nerve fibers continue in the optic tracts, and just before they reach the thalamus, a few of them leave to enter nuclei that function in various visual reflexes. Most of the fibers, however, enter the thalamus and synapse in its posterior portion (lateral geniculate body). From this region the visual impulses enter nerve pathways called *optic radiations,* and they lead to the visual cortex of the occipital lobes. (See fig. 11.46.)

Fig. 11.46 The visual pathway includes the optic nerve, optic chiasma, optic tract, and optic radiations.

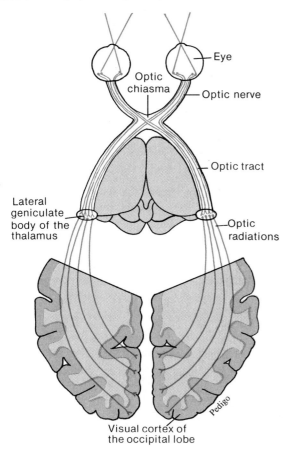

Eye

Optic chiasma

Optic nerve

Optic tract

Lateral geniculate body of the thalamus

Optic radiations

Visual cortex of the occipital lobe

Pedigo

Since each visual cortex receives impulses from each eye, a person may develop partial blindness in both eyes if either visual cortex is injured. For example, if the right visual cortex (or the right optic tract) is injured, sight may be lost in the lateral side of the right eye and the medial side of the left eye. Similarly, if the central portion of the optic chiasma, where fibers from the medial sides of the eyes cross over, is damaged, the medial sides of both eyes are blinded.

1. Distinguish between the rods and the cones of the retina.
2. Explain the roles of visual pigments.
3. What factors make stereoscopic vision possible?

Clinical Terms Related to the Senses

amblyopia (am''ble-o'pe-ah)—dimness of vision due to a cause other than a refractive disorder or lesion.

ametropia (am''ĕ-tro'pe-ah)—an eye condition characterized by inability to focus images sharply on the retina.

audiometry (aw''de-om'ĕ-tre)—the measurement of auditory acuity for various frequencies of sound waves.

blepharitis (blef''ah-ri'tis)—an inflammation of the margins of the eyelids.

causalgia (kaw-zal'je-ah)—a persistent, burning pain usually associated with injury to a limb.

conjunctivitis (kon-junk''tĭ-vi'tis)—an inflammation of the conjunctiva.

diplopia (dĭ-plo'pe-ah)—double vision, or the sensation of seeing two objects when only one is viewed.

emmetropia (em''ĕ-tro'pe-ah)—normal condition of the eyes; eyes with no refractive defects.

enucleation (e-nu''kle-a'shun)—removal of the eyeball.

exophthalmos (ek''sof-thal'mos)—condition in which the eyes protrude abnormally.

hyperalgesia (hi''per-al-je'ze-ah)—an abnormally increased sensitivity to pain.

iridectomy (ir''ĭ-dek'to-me)—the surgical removal of part of the iris.

iritis (i-ri'tis)—an inflammation of the iris.

keratitis (ker''ah-ti'tis)—an inflammation of the cornea.

labyrinthectomy (lab''ĭ-rin-thek'to-me)—the surgical removal of the labyrinth.

labyrinthitis (lab''ĭ-rin-thi'tis)—an inflammation of the labyrinth.

Ménière's disease (men''e-ārz' dĭ-zez)—an inner ear disorder characterized by ringing in the ears, increased sensitivity to sounds, dizziness, and loss of hearing.

neuralgia (nu-ral'je-ah)—pain resulting from inflammation of a nerve or a group of nerves.

neuritis (nu-ri'tis)—an inflammation of a nerve.

nystagmus (nis-tag'mus)—an involuntary oscillation of the eyes.

otitis media (o-ti'tis me'de-ah)—an inflammation of the middle ear.

otosclerosis (o''to-skle-ro'sis)—a formation of spongy bone in the inner ear, which often causes deafness by fixing the stapes to the oval window.

pterygium (tĕ-rij'e-um)—an abnormally thickened patch of conjunctiva that extends over part of the cornea.

retinitis pigmentosa (ret''ĭ-ni'tis pig''men-to'sa)—a progressive retinal sclerosis characterized by deposits of pigment in the retina and atrophy of the retina.

retinoblastoma (ret″ĭ-no-blas-to′mah)—an inherited, highly malignant tumor arising from immature retinal cells.

tinnitus (tĭ-ni′tus)—a ringing or buzzing noise in the ears.

tonometry (to-nom′ĕ-tre)—the measurement of fluid pressure within the eyeball.

trachoma (trah-ko′mah)—a virus-caused disease of the eye, characterized by conjunctivitis, that may lead to blindness.

tympanoplasty (tim″pah-no-plas′te)—the surgical reconstruction of the middle ear bones and establishment of continuity from the tympanic membrane to the oval window.

uveitis (u″ve-i′tis)—an inflammation of the uvea, which includes the iris, ciliary body, and the choroid coat.

vertigo (ver′tĭ-go)—a sensation of dizziness.

Chapter Summary

Introduction

Sensory receptors are sensitive to environmental changes and initiate impulses to the brain and spinal cord.

Somatic senses include touch, pressure, temperature, and pain.

Special senses include smell, taste, hearing, equilibrium, and vision.

Receptors and Sensations

1. Types of receptors
   a. Each type of receptor is sensitive to a distinct type of stimulus.
   b. The major types of receptors include the following:
      (1) Chemoreceptors, sensitive to changes in chemical concentration.
      (2) Pain receptors, sensitive to tissue damage.
      (3) Thermoreceptors, sensitive to temperature changes.
      (4) Mechanoreceptors, sensitive to mechanical forces.
      (5) Photoreceptors, sensitive to light.
2. Sensations
   a. Sensations are feelings resulting from sensory stimulation.
   b. A particular part of the sensory cortex always interprets impulses reaching it in the same way.
   c. The brain projects a sensation back to the region of stimulation.
3. Sensory adaptations are adjustments made by sensory receptors to continuous stimulation in which impulses are triggered at slower and slower rates.

Somatic Senses

1. Touch and pressure receptors
   a. Free ends of sensory nerve fibers are responsible for the sensations of touch and pressure.
   b. Meissner's corpuscles are responsible for the sensations of light touch.
   c. Pacinian corpuscles are responsible for the sensations of heavy pressure and vibrations.
2. Thermoreceptors include two sets of free nerve endings that serve as heat and cold receptors.
3. Pain receptors
   a. Pain receptors are free nerve endings stimulated by tissue damage. They provide a protective function, do not adapt rapidly, and can be stimulated by changes in temperature, mechanical force, and chemical concentration.
   b. The only receptors in visceral organs that provide sensations are pain receptors. These receptors are most sensitive to lack of blood flow and the presence of certain chemicals. The sensations they produce are likely to feel as if they were coming from some other part (referred pain).
4. Control of pain
   a. People vary in their perceptions and responses to pain.
   b. Intractable pain may be relieved by drugs or by sectioning nerve fibers.
   c. Enkephalins and endorphins, produced in the central nervous system, act as natural pain-suppressing neurotransmitters.

Sense of Smell

1. Olfactory organs
   a. The olfactory organs consist of receptors and supporting cells in the nasal cavity.
   b. Olfactory receptors are neurons with cilia that are sensitive to lipid-soluble chemicals.
   c. Nerve impulses travel from the olfactory receptors through the olfactory nerves, olfactory bulbs, and olfactory tracts to interpreting centers in the cerebrum.
2. Olfactory stimulation
   a. Olfactory impulses may result when various gaseous molecules combine with specific sites on the cilia of the receptor cells.
   b. Primary odors include camphoraceous, musky, floral, pepperminty, ethereal, pungent, and putrid.
   c. Olfactory receptors adapt rapidly.
   d. Olfactory receptors are often damaged by environmental factors but are not replaced.

Sense of Taste

1. Taste receptors
   a. Taste buds consist of receptor cells and supporting cells.
   b. Taste cells have taste hairs that are sensitive to particular chemicals dissolved in water.

c. Taste hair surfaces seem to have receptor sites to which chemicals combine and trigger impulses to the brain.

d. There are four primary kinds of taste cells, each particularly sensitive to a certain group of chemicals.

2. Taste sensations

a. The four primary taste sensations are sweet, sour, salty, and bitter.

b. Various taste sensations result from the stimulation of one or more sets of taste receptors.

c. Sweet receptors are most plentiful near the tip of the tongue, sour receptors along the margins, salt receptors in the tip and upper front, and bitter receptors toward the back.

Sense of Hearing

1. The external ear collects sound waves created by vibrating objects.

2. Middle ear

a. Auditory ossicles of the middle ear conduct sound waves from the tympanic membrane to the oval window of the inner ear. They also increase the force of these waves.

b. Skeletal muscles attached to the auditory ossicles act in the tympanic reflex to protect the inner ear from the effects of loud sounds.

3. Eustachian tubes connect the middle ears to the throat and function to help maintain equal air pressure on both sides of the eardrums.

4. Inner ear

a. The inner ear consists of a complex system of interconnected tubes and chambers—the osseous and membranous labyrinths. It includes the cochlea, which in turn houses the organ of Corti.

b. The organ of Corti contains the hearing receptors that are stimulated by vibrations in the fluids of the inner ear.

c. Different frequencies of vibrations are thought to stimulate different receptor cells; the human ear can detect sound frequencies from about 20 to 20,000 vibrations per second.

5. Deafness

a. Conductive deafness is caused by disorders in the external or middle ear. The most common type is due to growth of bone around the base of the stapes.

b. Sensorineural deafness is caused by damage to the cochlea, auditory nerve, auditory pathway, or temporal lobe of the brain. It can result from exposure to loud sounds, diseases, injuries, or various drugs.

Sense of Equilibrium

1. Static equilibrium is concerned with maintaining the stability of the head and body when these parts are motionless. The organs of static equilibrium are located in the vestibule.

2. Dynamic equilibrium is concerned with balancing the head and body when they are moved or rotated suddenly. The organs of this sense are located in the ampullae of the semicircular canals.

3. Other parts that help with the maintenance of equilibrium include the eyes and the proprioceptors associated with certain joints.

Sense of Sight

1. Visual accessory organs include the eyelids and lacrimal apparatus that function to protect the eye, and the extrinsic muscles that move the eye.

2. Structure of the eye

a. The wall of the eye has an outer, a middle, and an inner layer that function as follows:

(1) The outer layer (sclera) is protective, and its transparent anterior portion (cornea) refracts light entering the eye.

(2) The middle layer (choroid) is vascular and contains pigments that help to keep the inside of the eye dark.

(3) The inner layer (retina) contains the visual receptor cells.

b. The lens is a transparent, elastic structure whose shape is controlled by the action of ciliary muscles.

c. The iris is a muscular diaphragm that controls the amount of light entering the eye; the pupil is an opening in the iris.

d. Spaces within the eye are filled with fluids (aqueous and vitreous humors) that help to maintain its shape.

3. Refraction of light

a. Light waves are refracted primarily by the cornea and lens to focus an image on the retina.

b. The lens must be thickened to focus on close objects.

4. Refraction problems

a. Farsightedness of age (presbyopia) is caused by a diminishing of the elasticity of the lens.

b. Another type of farsightedness is caused by an eyeball that is too short; nearsightedness is caused by an eyeball that is too long; and astigmatism is caused by disorders in the curvatures of the cornea or lens.

5. Visual receptors

a. The visual receptors are called rods and cones.

b. Rods are responsible for colorless vision in relatively dim light, and cones are responsible for color vision.

6. Visual pigments
   a. A light-sensitive pigment in rods (rhodopsin) decomposes in the presence of light and triggers nerve impulses.
   b. Color vision seems to be related to the presence of three sets of cones containing different light-sensitive pigments, and each is sensitive to a different wavelength of light; the color perceived depends on which set or sets of cones are stimulated.
7. Stereoscopic vision
   a. Stereoscopic vision involves the perception of distance and depth.
   b. Stereoscopic vision occurs because of the formation of two slightly different retinal images that the brain superimposes and interprets as one image in three dimensions.
   c. A one-eyed person uses relative sizes and positions of familiar objects to judge distance and depth.
8. Visual nerve pathways
   a. Nerve fibers from the retina form the optic nerves.
   b. Some fibers cross over in the optic chiasma.
   c. Most of the fibers enter the thalamus and synapse with others that continue to the visual cortex.

Application of Knowledge

1. How would you explain the following observation? A person enters a tub of water and reports that it is uncomfortably warm, yet a few moments later says the water feels comfortable, even though the water temperature remains unchanged.

2. How would you explain the fact that some serious injuries, such as those produced by a bullet entering the abdomen, may be relatively painless, while others such as those involving a crushing of the skin may produce considerable discomfort?

3. Labyrinthitis is a condition in which the tissues of the inner ear are inflamed. What symptoms would you expect to observe in a patient with this disorder?

Review Activities

1. Distinguish between somatic and special senses.
2. List five groups of sensory receptors and name the environmental change to which each is sensitive.
3. Explain what is meant by the projection of a sensation.
4. Define *sensory adaptation* and provide an example of this phenomenon.
5. Describe the functions of free nerve endings, Meissner's corpuscles, and pacinian corpuscles.
6. Compare pain receptors with other types of dermal receptors in terms of structure and function.
7. Define *referred pain* and provide an example of this phenomenon.
8. Explain why people vary in their perceptions and responses to pain.
9. Explain how pain receptors may be stimulated and how the resulting sensations can be relieved.
10. Distinguish between enkephalins and endorphins.
11. Explain how the senses of smell and taste function together to create the flavor of foods.
12. Describe the olfactory organ and its function.
13. Trace a nerve impulse from the olfactory receptor to the interpreting center of the cerebrum.
14. List the seven primary olfactory sensations.
15. Explain why the olfactory sense diminishes with age.
16. Explain how the salivary glands aid the function of the taste receptors.
17. Name the four primary taste sensations and describe the patterns in which the taste receptors are distributed on the tongue.
18. Explain why taste sensation is less likely to diminish with age than olfactory sensation.
19. Distinguish between the external, middle, and inner ears.
20. Trace the path of a sound wave from the tympanic membrane to the hearing receptors.
21. Describe the functions of the auditory ossicles.
22. Describe the tympanic reflex, and explain its importance.

Plate 20   Photomicrograph of a cross section of the spinal cord.

Plate 21   What structures can you identify in this sagittal section of the brain?

Plate 22   Photomicrograph of a section of the human cerebellum. Note the layer of gray matter on the surface.

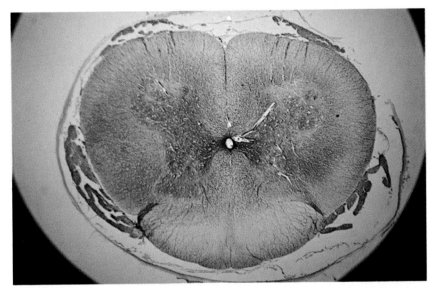

20

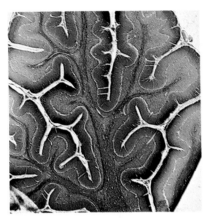

22

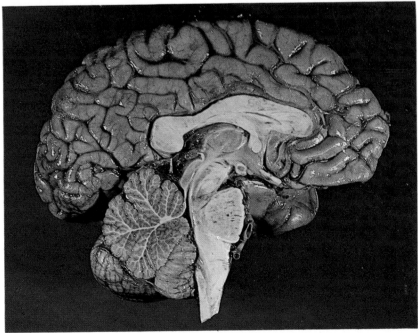

21

Plate 23 What structures can you identify in this sagittal section of the eye?

Plate 24 Anterior portion of the eye.

Plate 25 Locate the optic disk in this frontal view of the retina.

Plate 26 Posterior portion of the eye.

Plate 27 Note the layers of cells and nerve fibers in this photomicrograph of the retina.

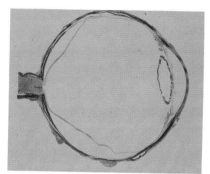

23

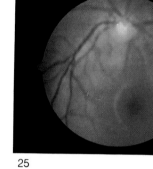

25

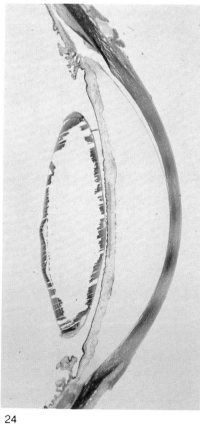

24

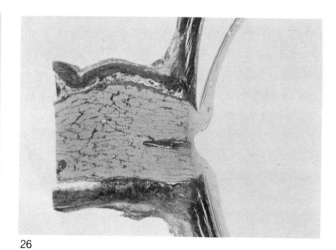

26

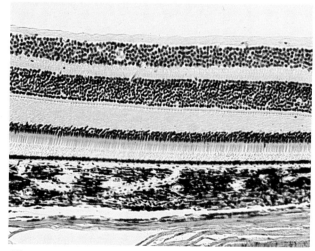

27

23. Explain the function of the eustachian tube.

24. Distinguish between the osseous and the membranous labyrinths.

25. Describe the cochlea and its function.

26. Trace a nerve impulse from the organ of Corti to the interpreting centers of the cerebrum.

27. Compare the causes of conductive and sensorineural deafness and their treatments.

28. Describe the organs of static and dynamic equilibrium and their functions.

29. Explain how the sense of vision helps to maintain equilibrium.

30. List the visual accessory organs and describe the functions of each.

31. Name the three layers of the eye wall and describe the functions of each.

32. Explain how the iris functions.

33. Distinguish between aqueous and vitreous humor.

34. Distinguish between the macula lutea and the optic disk.

35. Explain how light waves are focused on the retina.

36. Describe how the lens functions in accommodation.

37. Define *presbyopia* and identify its cause.

38. Explain the causes of near- and farsightedness and the treatments for each.

39. Distinguish between rods and cones.

40. Explain why cone vision is generally more acute than rod vision.

41. Describe the function of rhodopsin.

42. Explain how the eye becomes light and dark adapted.

43. Describe the relationship between light wavelengths and color vision.

44. Define *stereoscopic vision.*

45. Explain why a person with binocular vision is able to judge distance and depth of close objects more accurately than a one-eyed person.

46. Trace a nerve impulse from the retina to the visual cortex.

Suggestions for Additional Reading

Brindley, G. S. 1970. Central pathways of vision. *Ann. Rev. Physio.* 32:259.

Daw, N. W. 1973. Neurophysiology of color vision. *Physiol. Rev.* 53:571.

Durrant, J. D., and Lovrinic, J. H. 1977. *Bases of hearing science.* Baltimore: The Williams and Wilkins Co.

Freese, A. S. 1977. *The miracle of vision.* New York: Harper and Row.

Gombrich, E. H. September 1972. The visual image. *Scientific American.*

Green, D. M. 1976. *An introduction to hearing.* New York: Halsted Press.

Harris, J. D. 1972. Audition. *Ann. Rev. Psych.* 23:313.

Kaufman, H. E. July 1973. Corneal transplantation, a progress report. *Hosp. Prac.*

Lim, R. K. S. 1970. Pain. *Ann. Rev. Physio.* 32:269.

Pettigrew, J. D. August 1972. The neurophysiology of binocular vision. *Scientific American.*

Ross, J. March 1976. The resources of binocular perception. *Scientific American.*

Rubenstein, E. March 1980. Diseases caused by impaired communication among cells. *Scientific American.*

Rushton, W. A. H. March 1975. Visual pigments and color blindness. *Scientific American.*

Snyder, S. H. March 1977. Opiate receptors and internal opiates. *Scientific American.*

Toates, F. M. 1972. Accommodation function of the human eye. *Physio. Rev.* 52:828.

van Heyninger, R. December 1975. What happens to the human lens in cataract? *Scientific American.*

Werblin, F. S. January 1973. The control of sensitivity in the retina. *Scientific American.*

Young, R. W. October 1970. Visual cells. *Scientific American.*

The Endocrine System

12 The *endocrine system,* which consists of a variety of glandular organs, acts together with the nervous system in controlling body activities and in maintaining homeostasis. Although the general functions of these systems are closely related, the ways they accomplish their tasks differ. The nervous system, for example, transmits nerve impulses to muscles and glands that cause them to respond. The glands of the endocrine system secrete *hormones,* that are transmitted in body fluids to the tissues they affect. In contrast to the rapid and relatively brief responses stimulated by nerve impulses, the effects of endocrine secretions often require minutes, hours, or days to get started, and then their actions may continue for a considerable time.

Although some endocrine glands are responsible for specific, localized effects, those described in this chapter have more general actions. They include the pituitary gland, thyroid gland, parathyroid glands, adrenal glands, and pancreas. Other hormone-secreting glands, including those involved in digestion and reproduction, will be discussed in later chapters.

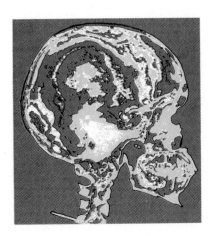

Chapter Objectives

After you have studied this chapter, you should
be able to

1. Distinguish between endocrine and exocrine
   glands.

2. Describe three major kinds of substances
   that function as hormones.

3. Explain how hormones are thought to exert
   influences on target tissues.

4. Discuss how hormone secretions are
   regulated by negative feedback
   mechanisms.

5. Explain how hormone secretions may be
   controlled by the nervous system.

6. Name and describe the location of the major
   endocrine glands of the body, and list the
   hormones they secrete.

7. Describe the general functions of these
   hormones.

8. Discuss the consequences of oversecretion
   and undersecretion of various hormones.

9. Distinguish between physical and
   psychological stress.

10. Describe the general stress response.

11. Complete the review activities at the end
    of this chapter.

adrenal cortex   (ah-dre′nal kor′teks)

adrenal medulla   (ah-dre′nal me-dul′ah)

anterior pituitary   (an-ter′e-or pĭ-tu′ĭ-tar″e)

cyclic AMP   (sik′lik)

metabolic rate   (met″ah-bol′ik rāt)

negative feedback   (neg′ah-tiv fēd′bak)

pancreas   (pan′kre-as)

parathyroid gland   (par″ah-thi′roid gland)

pineal gland   (pin′e-al gland)

posterior pituitary   (pos-ter′e-or pĭ-tu′ĭ-tar″e)

prostaglandin   (pros″tah-glan′din)

releasing factor   (re-le′sing fak′tor)

steroid   (ste′roid)

target tissue   (tar′get tish′u)

thymus gland   (thi′mus gland)

thyroid gland   (thi′roid gland)

tropic hormone   (trop′ik hōr′mon)

-crin, to secrete: endo*crine*—pertaining to internal secretions.

diuret-, to pass urine: *diuret*ic—a substance that promotes the production of urine.

endo-, within: *endo*crine gland—a gland that releases its secretion internally into a body fluid.

exo-, outside: *exo*crine gland—a gland that releases its secretion to the outside through a duct.

hyper-, above: *hyper*thyroidism—condition resulting from above normal secretion of thyroid hormone.

hypo-, below: *hypo*thyroidism—condition resulting from below normal secretion of thyroid hormone.

lact-, milk: pro*lact*in—a hormone that promotes milk production.

para-, beside: *para*thyroid glands—a set of glands located on the surface of the thyroid gland.

toc-, birth: oxy*toc*in—a hormone that stimulates uterine muscles to contract during childbirth.

tropic-, influencing: adrenocortico*tropic* hormone—a hormone secreted by the anterior pituitary gland that stimulates the adrenal cortex.

vas-, vessel: *vas*opressin—a substance that causes blood vessel walls to contract.

The term *endocrine* is used to describe glands that secrete their products internally rather than to the outside of the body. Thus, the thyroid and parathyroid glands, which release their secretions into the blood, are examples of *endocrine* (ductless) *glands*. On the other hand, sebaceous glands, sweat glands, and other glands that release their products into secretory ducts leading to the outside of the body are called *exocrine glands*. (See fig. 12.1.)

As a group, the endocrine glands help to regulate metabolic processes. They control the rates of certain chemical reactions, aid in the transport of substances through cell membranes, play vital roles in cell growth, help regulate water and electrolyte balance, and control reproduction. The functions of the endocrine system are compared with the functions of the nervous system in chart 12.1.

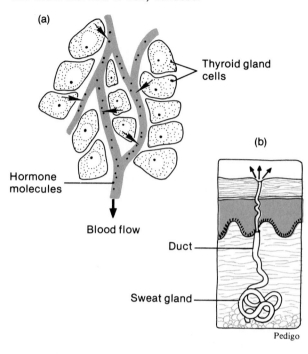

Fig. 12.1 Endocrine glands, such as (*a*) the thyroid gland, release hormones into body fluids, while exocrine glands, such as (*b*) sweat glands, release their secretions into ducts that lead to body surfaces.

Chart 12.1   A comparison of the endocrine and the nervous systems

Endocrine System	Nervous System
Secretes hormones that are carried in body fluids	Transmits nerve impulses on nerve fibers
Causes changes in the metabolic activities of target tissues	Causes muscles to contract or glands to secrete
Effects exerted relatively slowly	Effects exerted relatively rapidly
Effects generally prolonged	Effects generally brief

1. *What is the difference between an endocrine gland and an exocrine gland?*
2. *What is the general function of the endocrine system?*

Hormones and Their Actions

The chemicals secreted by endocrine glands are called *hormones*. They are released into the extracellular spaces surrounding the gland cells and are usually absorbed into the blood and carried to all parts of the body. The physiological effect of a particular hormone is often restricted to a certain tissue, called its *target tissue*, whose cells possess specific receptor sites for hormone molecules. Some hormones have a general action and affect nearly all body cells. Figure 12.2 illustrates the location of the major endocrine glands.

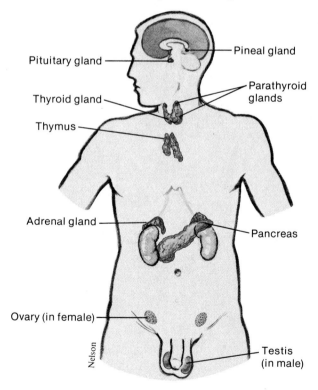

Fig. 12.2 Locations of major endocrine glands.

Fig. 12.3 Structural formulas of (a) a protein hormone, (b) a steroid hormone, and (c) an amine hormone.

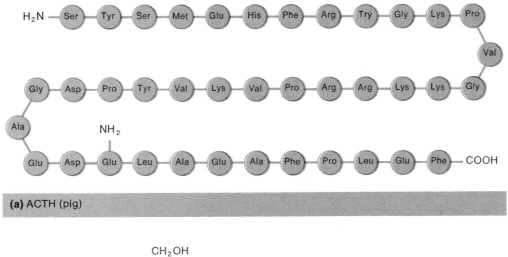

(a) ACTH (pig)

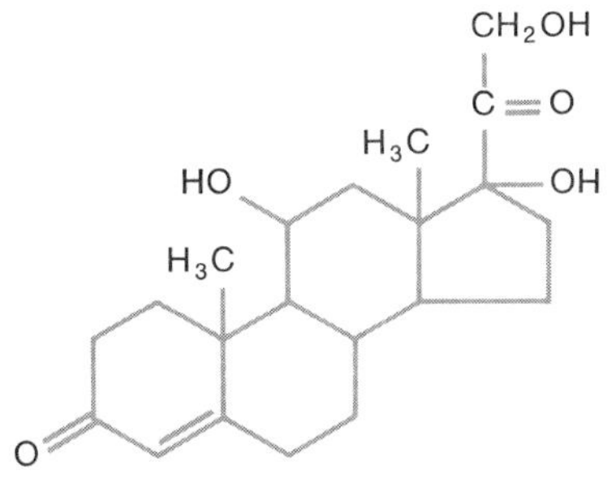

(b) Cortisol

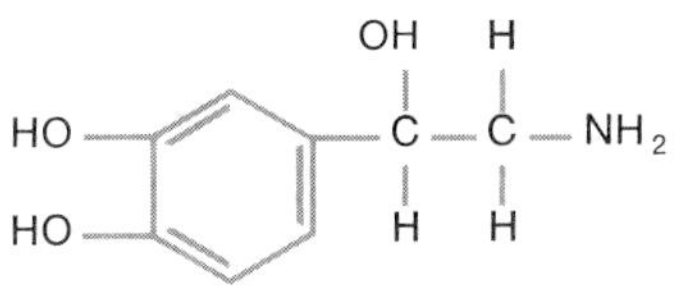

(c) Norepinephrine

Chemistry of Hormones

Each kind of hormone has a unique molecular structure and a unique action as well. All hormones are organic substances, usually steroids, proteins, or amines. (See fig. 12.3.)

Steroids are compounds whose molecules include complex rings of carbon and hydrogen atoms. The difference between one steroid and another is due to the types of atoms that are attached to these rings. Steroids are synthesized from *cholesterol,* and they include several sex hormones and the hormones secreted by the adrenal cortex (outer region of the adrenal gland).

Proteins, as was mentioned in chapter 2, are composed of amino acids bonded together in intricate chains. Glands that secrete protein hormones include the pituitary and parathyroid glands.

Amines are synthesized from certain amino acids, and each contains an amino group ($-NH_2$).

Chart 12.2 Chemistry of hormones

Type of Organic Compound	Precursor	Examples
Steroid	Cholesterol	Sex hormones, some hormones of the adrenal cortex
Protein	Amino acids	Pituitary hormones, parathyroid hormone
Amine	Amino acids	Hormones of the adrenal medulla

Amine molecules contain atoms of carbon, hydrogen, and nitrogen. The hormones secreted by the adrenal medulla (inner portion of the adrenal gland) are amines.

Chart 12.2 reviews the chemical components of hormones and gives an example of each.

Usually, hormones do not initiate new processes, rather they act by causing ordinary cellular processes to speed up or slow down. Different hormones exert their effects in different ways. Some, for example, alter the activities of enzymes that in turn cause substances to be synthesized; others alter the rates at which molecules are transported through cell membranes. The ability of a hormone to influence a particular kind of cell depends on the presence of receptor molecules in the cell or on its membranes. In other words, a hormone's target cells possess certain receptors that other cells do not.

Steroid hormones, unlike proteins and amines, are soluble in lipids such as those that make up the bulk of cell membranes. For this reason, steroid molecules can enter a cell relatively easily, and once they are inside they combine with specific protein molecules—the receptors. The steroid-protein complex may then enter the nucleus and somehow activate genes, bringing about the synthesis of particular kinds of messenger RNA molecules.

As is discussed in chapter 4, messenger RNA can leave the nucleus and enter the cytoplasm where it functions in manufacturing specific proteins. (See fig. 12.4.) Thus, steroid hormones influence cells by causing special proteins to be synthesized—proteins that may act as enzymes and alter the rates of cellular processes. (See chart 12.3.)

Many of the protein and amine hormones seem to act by combining with specific receptors on the cell membranes of their target cells. The formation of such a hormone-receptor combination activates an enzyme in the membrane called **adenyl cyclase.** This enzyme then causes ATP molecules on the inside of the cell to become molecules of **cyclic AMP** (adenosine monophosphate), and the cyclic AMP in turn causes changes in cellular processes. (See fig. 12.5.) These changes may include altering the permeability of membranes, activating other enzymes, or promoting the synthesis of proteins—changes recognized as a hormone's effects on its target cells. Thus, such a hormone acts as a messenger that stimulates its target cells. The adenyl cyclase in the cell membrane serves as a second messenger by delivering the information to the inside, thus causing changes in cellular processes. (See fig. 12.6.)

A substance called *cyclic GMP* (guanosine monophosphate) also is thought to function as a second messenger in much the same way as cyclic AMP.

Fig. 12.4 (*a*) A steroid hormone passes through a cell membrane and (*b*) combines with a receptor protein in the cytoplasm. (*c*) The steroid-protein complex enters the nucleus and (*d*) activates the synthesis of messenger RNA. (*e*) The messenger RNA leaves the nucleus and (*f*) functions in the manufacture of protein molecules.

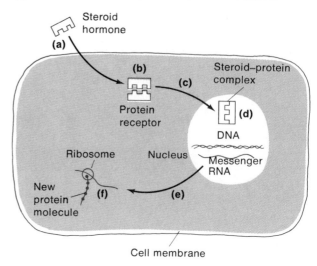

Chart 12.3 Sequence of steroid hormones' influence

1. The endocrine gland secretes steroid hormone.
2. Hormone enters the cell by passing through the cell membrane.
3. Hormone combines with a receptor protein molecule.
4. Hormone-protein complex enters the nucleus and activates the synthesis of messenger RNA.
5. Messenger RNA enters the cytoplasm and functions in manufacturing protein molecules.
6. Proteins function as enzymes and cause reactions that are recognized as the hormone's action.

Although it is not known how they accomplish this, many protein hormones enter cells. The significance of this movement is not clear; some investigators believe that the act of binding to receptor sites produces an immediate, short-term effect on a cell. Long lasting effects may be produced by actions taking place after the hormone, often with its receptor, enters the cell.

Chart 12.4 summarizes the actions of some nonsteroid hormones.

Fig. 12.5 ATP molecules are converted to cyclic AMP by the action of the enzyme, adenyl cyclase. What cellular changes may take place as a result of this reaction?

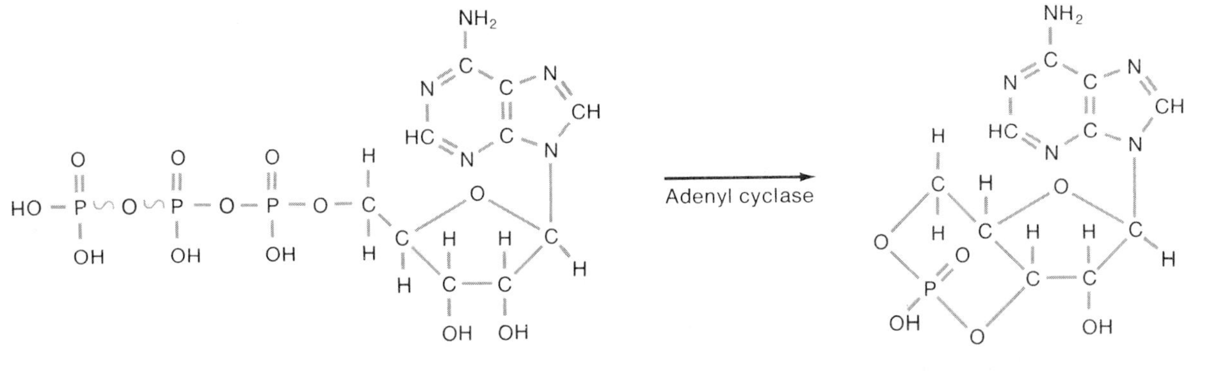

Adenyl cyclase →

ATP

Cyclic AMP

Fig. 12.6 (*a*) Hormone molecules first reach target cells by means of body fluids and (*b*) combine with receptor sites on the cell membrane. (*c*) As a result, molecules of adenyl cyclase diffuse into the cytoplasm and (*d*) cause the change of ATP into cyclic AMP. (*e*) Cyclic AMP brings about various cellular changes.

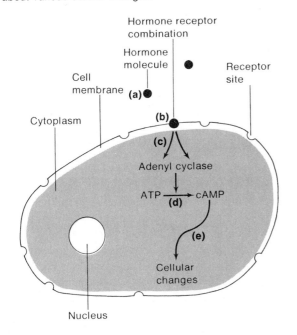

Chart 12.4 Sequence of some nonsteroid hormones' influence

1. The endocrine gland secretes hormone.
2. Hormone is carried to its target cell by body fluid.
3. Hormone combines with receptor site on the membrane of its target cell.
4. Adenyl cyclase molecules are activated within the target cell's membrane, and they diffuse into the cytoplasm.
5. Acting as an enzyme, adenyl cyclase causes ATP molecules to be converted into cyclic AMP molecules.
6. Cyclic AMP brings about cellular changes that are recognized as the hormone's actions.

Prostaglandins

Still another group of compounds that have hormonelike effects on cells is called **prostaglandins.** These substances are lipids (20-carbon fatty acids that include 5-carbon rings), and they are closely associated with the membranes of a wide variety of cells including those of the liver, kidneys, heart, lungs, thymus gland, pancreas, brain, and various reproductive organs. (See fig. 12.7.)

Although prostaglandins are present in very small quantities and are rapidly inactivated when they are released into body fluids, they have powerful effects. For example, one type causes a rapid drop in blood pressure, while a closely related type causes a rise in blood pressure. Other prostaglandins stimulate uterine wall muscles to contract, and still others inhibit glandular secretions in the stomach and stimulate urine production by the kidneys.

Fig. 12.7 A prostaglandin molecule contains a 20-carbon fatty acid with a 5-carbon ring.

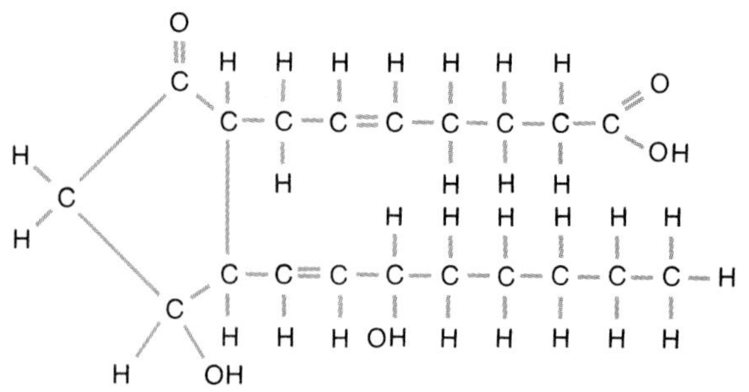

Prostaglandin E₂ (PGE₂)

It is not known how prostaglandins produce their effects, but they seem to influence the activation of adenyl cyclase in cell membranes and thus regulate the formation of cyclic AMP. In some cases, the effect is to increase the formation of cyclic AMP and in others to decrease it.

1. What is a hormone?
2. How do steroid hormones exert their effects? Nonsteroid hormones?
3. What are prostaglandins?

Control of Hormone Secretions

The concentration of each hormone in the body fluids remains relatively stable, suggesting that the rate at which a hormone is secreted is closely balanced with the rate of its usage or destruction.

Negative Feedback Systems

Typically, the concentration of a hormone in body fluid is regulated by a **negative feedback** system. As is discussed in chapter 1, such a system is activated by an imbalance of some sort, and when one occurs, information is fed back to some part that acts to correct the imbalance. For example, a negative feedback system may be used to regulate the air temperatures in a room. Such a heat-regulating system usually includes a thermostat that can be set at some desirable temperature and can sense temperature changes. The system also includes a heater that turns on and off in response to signals from the thermostat.

In practice, the thermostat is set at about 20°C (68°F). If the heater is operating it tends to overheat, but as the temperature rises above this setpoint, the thermostat senses the change and signals the heater to turn off (a negative effect). Then, when the temperature drops below the thermostat setting, this information is fed back to the heater, and it is allowed to turn on again. Thus, the system is activated by a temperature imbalance, as judged by the thermostat's setting, and the system functions to correct the imbalance.

Similarly endocrine glands have a tendency to oversecrete hormones. When the concentration of a hormone reaches a certain level, however, information is fed back to the gland often by signals from its target tissue, and further secretion is inhibited. (See fig. 12.8). Then, if the hormone concentration drops below a certain level, its effect upon the target tissue decreases, and the gland is no longer inhibited. Consequently, it begins to secrete the hormone once again.

Because of the negative feedback system, the concentration of the hormone remains relatively stable, although it will fluctuate slightly within a normal range. (See fig. 12.9.)

Excess hormones that are not used in the interactions with target cells are usually inactivated by the liver and kidneys. For this reason, a person with a liver or kidney disease may experience effects resulting from increasing concentrations of certain hormones.

Nerve Control

Other types of hormone control mechanisms involve the nervous system. For example, some endocrine glands, such as the adrenal medulla, secrete their hormones in response to nerve impulses, and no other stimulus seems able to cause a secretion.

Still another kind of control system involves interaction between an endocrine gland and the hypothalamus of the brain. In this system, neurosecretory cells in the hypothalamus secrete substances, called **releasing** (or inhibiting) **factors,** whose target tissues are in the pituitary gland. The gland responds to the releasing factor by secreting its own hormone. Then, as the gland's hormone reaches a given level in the body fluids, the hypothalamus is signaled, and its secretion of the releasing factor is inhibited. (See fig. 12.10.)

Fig. 12.8 An example of a negative feedback system: (1) gland A secretes a hormone that stimulates gland B to release another hormone; (2) the hormone from gland B inhibits the action of gland A.

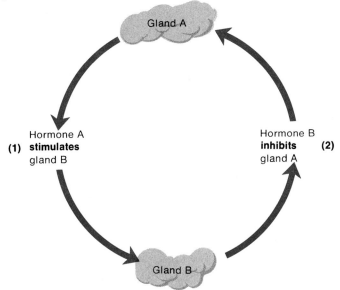

Fig. 12.9 As a result of negative feedback systems, hormone concentrations remain relatively stable although they may fluctuate slightly above and below the optimal concentrations.

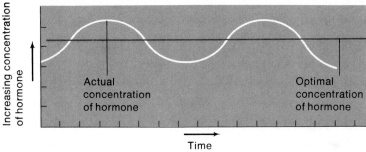

Fig. 12.10 (a) In some cases, an endocrine gland secretes its hormones in response to nerve impulses; (b) another type of gland secretes hormones in response to releasing factors secreted by the hypothalamus.

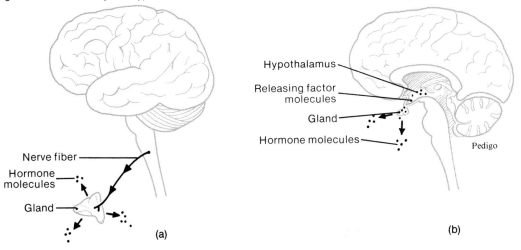

The Pituitary Gland

The **pituitary gland** (hypophysis) is about 1 cm (0.4 inch) in diameter and is located at the base of the brain. It is attached to the hypothalamus by the pituitary stalk, or *infundibulum,* and lies in the sella turcica of the sphenoid bone, as shown in figure 12.11.

This gland consists of two distinct portions—an anterior lobe (adenohypophysis) and a posterior lobe (neurohypophysis). The *anterior lobe* secretes a number of hormones, including growth hormone (GH), thyroid-stimulating hormone (TSH), adrenocorticotropic hormone (ACTH), follicle-stimulating hormone (FSH), luteinizing hormone (LH), and prolactin. Although the *posterior lobe* does not synthesize any hormones, two important ones, antidiuretic hormone (ADH) and oxytocin, are secreted by nerve fibers in its tissues.

Most of the pituitary activities are controlled by the brain. The release of hormones from the posterior lobe, for example, occurs when nerve impulses originating in the *hypothalamus* signal the axon ends of neurosecretory cells in this lobe. Secretions from the anterior lobe are typically controlled by releasing factors produced by specialized neurons of the hypothalamus (fig. 12.12). These releasing factors are

transmitted by blood, in the vessels of a capillary net in the region of the hypothalamus. These vessels merge to form the **hypophyseal portal veins** that pass downward and give rise to a capillary net in the anterior lobe of the pituitary gland. Thus, substances released into the blood from the hypothalamus are carried directly to the anterior lobe.

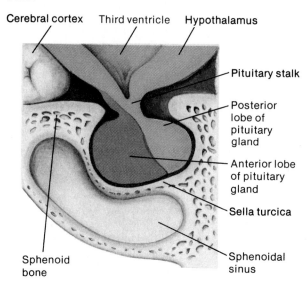

Fig. 12.11 The pituitary gland is attached to the hypothalamus and lies in the sella turcia of the sphenoid bone.

Fig. 12.12 Cells of the posterior lobe of the pituitary gland are stimulated to release hormones by nerve impulses originating in the hypothalamus; cells of the anterior lobe are stimulated by releasing factors secreted by hypothalamic neurons.

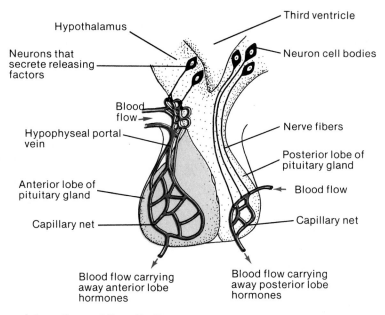

The hypothalamus (as was described in chapter 10) receives information from nearly all parts of the nervous system. This information includes data concerning a person's emotional state, body temperature, blood nutrient concentrations, and so forth. The hypothalamus sometimes acts upon such information by signaling the pituitary gland to release hormones.

---

1. Where is the pituitary gland located?
2. List the hormones secreted by the anterior lobe of the pituitary gland. Released from the posterior lobe.
3. Explain how the hypothalamus is related to the actions of the pituitary gland.

Anterior Lobe Hormones

The anterior lobe of the pituitary gland consists largely of epithelial cells arranged in blocks around many thin-walled blood vessels. Through a special staining technique, it is possible to distinguish two groups of secretory cells within this tissue. The cells of one group (acidophils) secrete the hormones GH and prolactin, while the members of the second group (basophils) produce the hormones TSH, ACTH, FSH, and LH. (See fig. 12.13.)

Growth hormone (GH or *somatotropin*) is a protein that generally stimulates body cells to increase in size and undergo more rapid cell division than usual. How GH accomplishes this action is not completely understood. It is known, however, that the hormone enhances the movement of amino acids through cell membranes and causes an increase in the rate at which cells convert these molecules into proteins. GH also causes cells to decrease the rate at which they utilize carbohydrates and to increase the rate at which they use fats. The hormone's effect on amino acids seems to be the more important one.

Although the exact mechanism for controlling growth hormone secretion is unknown, it appears to involve two substances from the hypothalamus called *growth hormone-releasing factor* (GRF) and *growth hormone release-inhibiting factor* (GIH or *somatostatin*). A person's nutritional state also seems to play a role in the control of GH, for more of it is released during periods of protein deficiency and of abnormally low blood glucose concentration. Conversely, when blood protein and glucose levels are increased, there is a resulting decrease in growth hormone secretion. Apparently the hypothalamus is able to sense changes in the concentrations of certain blood nutrients, and it releases GRF in response to some of them.

As is mentioned in chapter 8, if growth hormone is not secreted in sufficient amounts during childhood, body growth is limited, and a type of *dwarfism* (hypopituitary dwarfism) results. In this condition, body parts are usually correctly proportioned and mental development is normal. However, an abnormally low secretion of growth hormone is usually accompanied by lessened secretions from other anterior lobe hormones, leading to additional hormone deficiency symptoms. For example, a hypopituitary dwarf often fails to develop adult sexual features unless hormone therapy is provided.

Fig. 12.13 Hormones released by the anterior lobe of the pituitary gland and their target tissues.

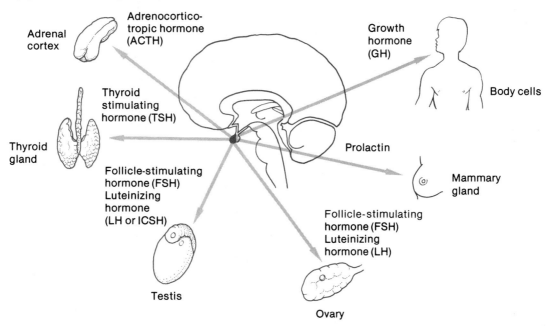

Hypopituitary dwarfism is sometimes treated by administering growth hormone, and this treatment may stimulate a rapid increase in height. The procedure, however, must be started before the epiphyseal disks of the person's long bones have become ossified. Otherwise growth in height is not possible.

An oversecretion of growth hormone during childhood may result in *gigantism*—a condition in which the person's height may exceed 8 feet. Gigantism, which is relatively rare, is usually accompanied by a tumor of the pituitary gland. In such cases, various pituitary hormones in addition to GH are likely to be secreted excessively, so that a giant often suffers from a variety of metabolic disturbances and has a shortened life expectancy. (See fig. 8.9 in chapter 8.)

If growth hormone is secreted excessively in an adult, after the epiphyses of the long bones have ossified, the person does not grow taller. The soft tissues, however, may continue to enlarge and the bones may become thicker. As a consequence, an affected individual may develop greatly enlarged hands and feet, a protruding jaw, and a large tongue and nose. This condition is called *acromegaly*, and like gigantism, it is often associated with a pituitary tumor. (See fig. 12.14.)

Prolactin is a protein, and as its name suggests, it functions in milk production in females. (It has no established function in males.) More specifically, prolactin stimulates and sustains milk production following the birth of an infant. This action is discussed in chapter 21, which deals with the reproductive system.

Although the regulation of prolactin secretion is poorly understood, it is known that at least one substance from the hypothalamus is involved. This substance, called *prolactin release-inhibiting factor* (PRIF), acts to restrain the secretion of prolactin. When PRIF is not secreted, prolactin is released.

Thyroid-stimulating hormone (TSH), which is also called *thyrotropin*, is a glycoprotein—a protein bonded to a carbohydrate. Its major function is to control the secretion of hormones from the thyroid gland.

TSH secretion is partially regulated by the hypothalamus, which secretes *thyrotropin-releasing factor* (TRF). TSH secretion is also regulated by circulating thyroid hormones that exert an inhibiting effect on the release of TRF and TSH; therefore, as the blood concentration of thyroid hormones increases, the secretions of TRF and TSH are reduced. (See fig. 12.15.)

Certain external factors influence the release of these hormones. These include exposure to extreme

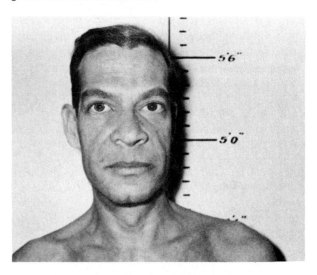

Fig. 12.14 Acromegaly is caused by an oversecretion of growth hormone in adulthood.

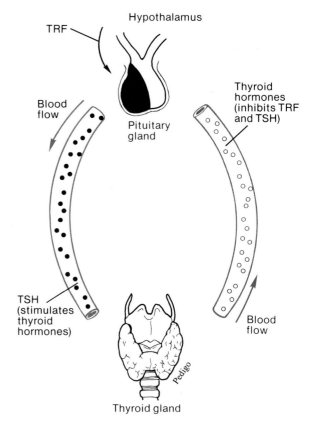

Fig. 12.15 TRF from the hypothalamus stimulates the anterior pituitary gland to release TSH. TSH stimulates the thyroid gland to release hormones, which in turn causes the hypothalamus to reduce its secretion of TRF.

cold, which is accompanied by increased hormonal secretions, and emotional stress, which sometimes triggers increased hormonal secretions and other times causes decreased secretions.

Fig. 12.16 The posterior lobe of the pituitary gland releases two hormones—antidiuretic hormone and oxytocin.

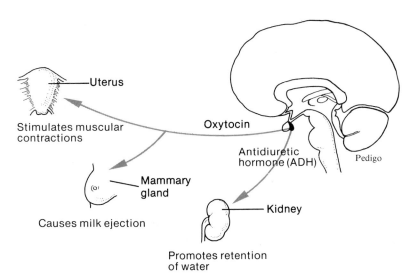

Uterus

Stimulates muscular contractions

Oxytocin

Antidiuretic hormone (ADH)

Pedigo

Mammary gland

Causes milk ejection

Kidney

Promotes retention of water

1. How does growth hormone affect the utilization of carbohydrates, fats, and proteins?
2. What is the function of prolactin?
3. How is TSH secretion regulated?

Adrenocorticotropic hormone (ACTH) is a protein that controls the manufacture and secretion of certain hormones from the outer layer, or *cortex*, of an adrenal gland. These adrenal hormones are discussed in a later section of this chapter.

The secretion of ACTH may be regulated in part by a *corticotropin-releasing factor* (CRF), which some investigators believe is released from the hypothalamus in response to decreased concentrations of adrenal cortical hormones. Also, as in the case of TSH, various forms of stress serve to stimulate the secretion of ACTH.

Both **follicle-stimulating hormone** (FSH) and **luteinizing hormone** (LH) are glycoproteins and are called *gonadotropins,* which means they exert their actions on the gonads or reproductive organs. FSH, for example, is responsible for the growth and development of egg-containing follicles in the female ovaries. It also stimulates the follicle cells to secrete a group of female sex hormones that collectively are called *estrogen.*

In males, FSH stimulates the production of sperm cells in the testes. LH (which in males is called *interstitial cell-stimulating hormone* or ICSH since it acts upon the interstitial cells of the testes) promotes the secretion of sex hormones in both males and females and is also essential for the release of egg cells from the female ovaries. Other functions of these gonadotropins and the ways they interact with each other are discussed in chapter 21.

The mechanism that regulates the secretion of gonadotropins is not well understood. It is known, however, that the hypothalamus secretes at least one *gonadotropin-releasing factor,* called *luteinizing hormone-releasing factor* (LH-RF), and perhaps another, called *follicle-stimulating hormone-releasing factor* (FSH-RF). The hypothalamus apparently fails to secrete these factors until the age of puberty, because gonadotropins are virtually absent in the body fluids of infants and children.

1. What is the function of ACTH?
2. Describe the functions of FSH and LH in a male. In a female.

Posterior Lobe Hormones

Unlike the anterior lobe of the pituitary gland, which is composed primarily of glandular epithelial cells, the posterior lobe consists largely of neuroglial cells (*pituicytes*). Instead of producing hormones, these cells seem to function as supporting structures for large numbers of nerve fibers that originate in the hypothalamus.

As was mentioned, the two hormones associated with the posterior lobe—antidiuretic hormone (ADH) and oxytocin—are actually produced by specialized neurons in the hypothalamus (fig. 12.16). These substances travel down axons through the pituitary stalk to the posterior lobe. Then the hormones are released from the axon endings in response to nerve impulses coming from the hypothalamus.

Fig. 12.17 What differences exist in the amino acid composition of these hormones?

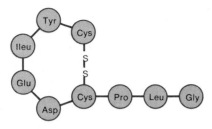

Oxytocin

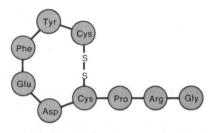

Antidiuretic hormone

Antidiuretic hormone (ADH) is a polypeptide consisting of a relatively short chain of amino acids. This is true of oxytocin also, and as figure 12.17 shows, the molecules of these substances are very similar except for amino acids in two locations.

A *diuretic* is a substance that acts to increase urine production. An *antidiuretic*, then, is a chemical that inhibits urine formation. ADH produces its antidiuretic effect by acting on the kidneys and causing them to reduce the amount of water they excrete. In this way, ADH is important in regulating the water concentration of body fluids.

Drinking alcohol (ethyl alcohol) is thought to inhibit the normal secretion of ADH. Consequently, an abnormally large volume of urine may be produced after a person drinks alcoholic beverages. The lost body fluid must later be replaced if water balance is to be maintained.

When it is present in high concentrations, ADH causes contractions of certain smooth muscles, including those in the walls of some blood vessels. As a result, the blood pressure in these vessels increases. For this reason, ADH is also called *vasopressin.* Although ADH is seldom present in sufficient quantities to elevate blood pressure, it may be released following a severe loss of blood. Blood pressure is likely to drop as a consequence of excessive bleeding, and in this situation ADH's vasopressor effect may help to restore normal pressure.

The secretion of ADH is regulated by the hypothalamus. Apparently certain neurons in this part of the brain called *osmoreceptors* are sensitive to changes in the water concentration of body fluids. If, for example, a person is sweating and losing large amounts of water, the blood becomes more and more concentrated. The osmoreceptors can sense this change and then signal the posterior lobe to release ADH. The ADH is transmitted by the blood to the kidneys, and as a result of its effects, less urine is produced. This action conserves water.

If, on the other hand, a person drinks an excess of water, the body fluids become more dilute and the release of ADH is inhibited. Consequently, the kidneys excrete more dilute urine until the water concentration of the body fluids returns to normal.

If any parts of the ADH regulating mechanism are damaged due to an injury or a tumor, the hormone may not be synthesized or released normally. The resulting ADH deficiency produces a condition called *diabetes insipidus.* This disease is characterized by an output of as much as 25 to 30 liters of very dilute urine per day and a rise in the concentration of body fluids. It is also accompanied by a sensation of great thirst.

Oxytocin also has an antidiuretic action, but it is weaker in this respect than ADH. In addition, it can cause contractions of the smooth muscles in the uterine wall. This hormone is released near the end of pregnancy and may play a role in childbirth by stimulating uterine contractions. The mechanism that triggers the release of oxytocin is not clearly understood. It is known, however, that the uterus becomes more and more sensitive to oxytocin's effects during pregnancy. Also, it is believed that stretching of uterine and vaginal tissues late in pregnancy, caused by the growing fetus, may initiate nerve impulses to the hypothalamus. The hypothalamus may then signal the posterior lobe to release oxytocin, which in turn, enhances uterine wall contractions during labor.

Chart 12.5 Hormones of the pituitary gland

Anterior Lobe

Hormone	Action	Source of Control
Growth hormone (GH) or somatotropin	Stimulates increase in size and rate of reproduction of body cells; enhances movement of amino acids through membranes	Growth hormone-releasing factor (GRF) and growth hormone release-inhibiting factor (GIH) from the hypothalamus
Prolactin	Sustains milk production after birth	Secretion restrained by prolactin release-inhibiting factor (PRIF) from the hypothalamus
Thyroid-stimulating hormone (TSH) or thyrotropin	Controls secretion of hormones from the thyroid gland	Thyrotropin-releasing factor (TRF) from the hypothalamus
Adrenocorticotropic hormone (ACTH)	Controls secretion of certain hormones from the adrenal cortex	Corticotropin-releasing factor (CRF) from the hypothalamus
Follicle-stimulating hormone (FSH)	Responsible for the development of egg-containing follicles in ovaries; stimulates follicle cells to secrete estrogen; in male stimulates the production of sperm cells	Gonadotropin-releasing factor from the hypothalamus (FSH-RF)
Luteinizing hormone (LH or ICSH in males)	Promotes secretion of sex hormones; plays role in release of egg cell in females	Gonadotropin-releasing factor from the hypothalamus (LH-RF)

Posterior Lobe

Hormone	Action	Source of Control
Antidiuretic hormone (ADH)	Causes kidneys to reduce water excretion; in high concentration, causes blood pressure to rise	Hypothalamus in response to changes in blood water concentration
Oxytocin	Causes contractions of muscles in uterine wall; causes muscles in milk ducts to contract	Hypothalamus in response to stretch in uterine and vaginal walls and stimulation of breasts

Oxytocin also has an effect upon the breasts, causing contractions in certain cells associated with milk-producing glands and their ducts. In lactating breasts, this action forces liquid from the milk glands into the milk ducts and causes the milk to be ejected from the breasts—an effect that is necessary before milk can be removed by sucking.

Sucking the nipple of a breast initiates nerve impulses that travel to the mother's hypothalamus. The hypothalamus responds by signaling the posterior lobe to release oxytocin, which in turn stimulates the release of milk. Thus, milk is normally not ejected from the milk glands and ducts until it is needed. Oxytocin has no established function in males.

Chart 12.5 reviews the hormones of the pituitary gland.

If the uterus is not contracting sufficiently to expel a fully developed fetus, commercial preparations of oxytocin are sometimes used to stimulate uterine contractions. Such preparations are often administered to the mother following childbirth to ensure that the uterine muscles contract enough to squeeze broken blood vessels closed, minimizing the danger of hemorrhage.

1. What are the functions of ADH?
2. What effects does oxytocin produce?

The Thyroid Gland

The **thyroid gland,** shown in figure 12.18, is a very vascular structure that consists of two large lobes connected by a broad isthmus. It is located just below the larynx on either side and in front of the trachea. It has a special ability to remove iodine from the blood.

Structure of the Gland. The thyroid gland is covered by a capsule of connective tissue and is made up of many secretory parts called *follicles.* The cavities of the follicles are lined with a single layer of cuboidal epithelial cells and are filled with a clear, viscous glycoprotein called *colloid.* The follicle cells produce and secrete hormones that may be stored in the colloid or released into the blood of nearby capillaries (fig. 12.18).

Fig. 12.18 The thyroid gland consists of two lobes connected anteriorly by an isthmus.

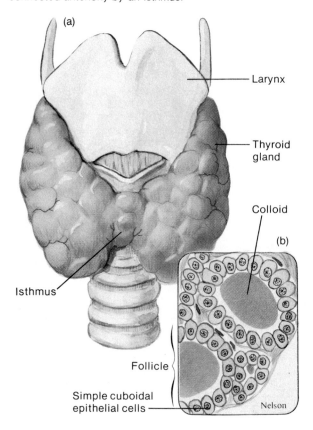

Fig. 12.19 The hormones thyroxine and triiodothyronine have very similar molecular structures.

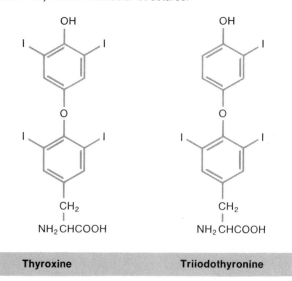

Thyroxine   Triiodothyronine

Thyroid Hormones and Their Functions

The thyroid gland produces several hormones that have marked effects on the metabolic rates of most body cells and one hormone that influences the level of blood calcium. Of the hormones that affect metabolic rates, the most important are **thyroxine** and **triiodothyronine** (fig. 12.19). They act to increase the rate of energy release from carbohydrates and the rate of protein synthesis. They also accelerate growth in young persons and stimulate activities of the nervous system. As was explained earlier, the release of these hormones is controlled by the hypothalamus and pituitary gland. (See fig. 12.15.)

Before follicle cells can produce thyroxine and triiodothyronine, they must be supplied with iodine salts (iodides). Such salts are normally obtained from foods, and after they have been absorbed from the intestine, they are carried by the blood to the thyroid gland. An efficient active transport mechanism called the *iodine pump* moves the iodides into the follicle cells, where they are used together with an amino acid (tyrosine) in the synthesis of the hormones.

Follicle cells also secrete the substance called *thyroglobulin*, which is the main ingredient of thyroid colloid. Thyroglobulin is used to store thyroid hormones whenever they are produced in excess. The stored hormones are bonded to the thyroglobulin until the hormone concentration of the body fluids drops below a certain level; then enzymes cause the hormones to be released from the colloid, and they diffuse into the blood. Once they are in the blood, thyroid hormones combine with blood proteins (alpha globulins) and are transported to body cells in this form. Although triiodothyronine is nearly five times more potent, thyroxine accounts for at least 95% of the circulating thyroid hormone.

The thyroid hormone that influences blood calcium levels is a polypeptide called **calcitonin** (fig. 12.20). This substance helps regulate the calcium level by inhibiting the rate at which calcium leaves the bones and enters the extracellular fluids.

This is accomplished by decreasing the bone resorbing activity of osteoclasts, as is described in chapter 8. At the same time, calcitonin causes an increase in the rate of calcium deposit in bone matrix by stimulating the activity of osteoblasts.

Thus, calcitonin acts to lower the concentration of blood calcium—an effect exactly opposite that promoted by parathyroid hormone. The interaction between these two substances is discussed in a later section of this chapter.

The secretion of calcitonin is thought to be controlled directly by the blood calcium level. As this level increases, so does the secretion of calcitonin.

Chart 12.6 reviews the actions and controls of the thyroid hormones.

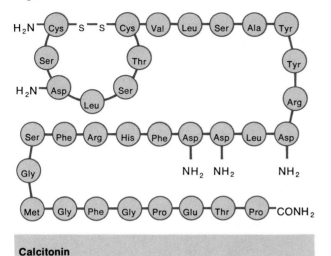

Fig. 12.20 Calcitonin is a polypeptide that helps to regulate the level of blood calcium.

Calcitonin

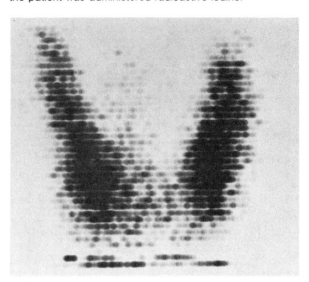

Fig. 12.21 A scan of the thyroid gland 24 hours after the patient was administered radioactive iodine.

Chart 12.6   Hormones of the thyroid gland

Hormone	Action	Source of Control
Thyroxine	Increases rate of energy release from carbohydrates; increases rate of protein synthesis; accelerates growth; stimulates activity in the nervous system	TSH from the anterior pituitary gland
Triiodothyronine	Same as above	Same as above
Calcitonin	Lowers blood calcium level by inhibiting the release of calcium from bones	Blood calcium level

1. Where is the thyroid gland located?
2. What hormones of the thyroid gland affect carbohydrate metabolism and protein synthesis?
3. What substance is essential for the production of these hormones?
4. How does the thyroid gland influence the concentration of blood calcium?

Disorders of the Thyroid Gland

Most functional disorders of the thyroid gland are characterized by *overactivity* (hyperthyroidism) or *underactivity* (hypothyroidism) of the gland cells.

Various laboratory tests are used to judge how a person's thyroid gland is functioning. One of these involves measuring the concentration of *protein-bound iodine* (PBI) in a blood sample. Nearly all the iodine associated with blood proteins is contained in thyroid hormone molecules, so the results of a PBI test furnish an index of the gland's activity.

Another test called *triiodothyronine uptake* (T_3 test) determines how much thyroid hormone is bound to the protein in a blood sample.

In the thyroxine iodine test (T_4 test) for thyroid function, the amount of iodine associated with the thyroxine in the sample is measured using one of several laboratory procedures. The value obtained is then compared with the normal range of thyroxine iodine concentrations.

Still another method for determining thyroid activity makes use of *radioactive iodine* (I–131). In this case the person drinks some distilled water containing a small amount of this radioactive substance. In 24 hours the radioactivity of the thyroid gland is measured using a scintillation counter (fig. 12.21). The results indicate how effectively the gland was able to remove iodine from the blood and concentrate it, thus giving an indication of the gland's condition.

A thyroid disorder may develop at any time during a person's life as a result of a developmental problem, an injury, a disease, or a dietary deficiency. One form of **hypothyroidism** appears in infants when their thyroid glands fail to function normally. An affected child may appear normal at birth because it has received an adequate supply of thyroid hormones

Fig. 12.22 Cretinism is due to an underactive thyroid gland during infancy and childhood. How can this condition be prevented?

Fig. 12.23 Hyperthyroidism may produce protrusion of the eyes.

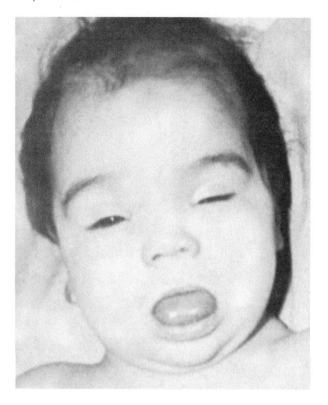

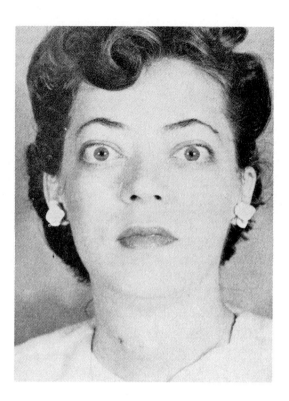

from its mother during pregnancy. When its own thyroid gland fails to produce sufficient quantities of hormones, the child soon develops a condition called *cretinism*. Cretinism is characterized by severe symptoms including stunted growth, abnormal bone formation, retarded mental development, low body temperature, and sluggishness. Without treatment with thyroid hormones within a month or so following birth, the child is likely to suffer from permanent mental retardation (fig. 12.22).

If hypothyroidism develops after growth and development have been completed, a disease called *myxedema* results. A person with this condition has an abnormally low metabolic rate, is mentally slow, usually gains a considerable amount of weight, and has swollen tissues due to an accumulation of excess body fluid (edema).

Hyperthyroidism (Graves' disease) is characterized by an elevated metabolic rate, abnormal weight loss, excessive perspiration, muscular weakness, and emotional instability. Also, the eyes are likely to protrude (exophthalmos) because of edematous swelling in the tissues behind them. At the same time, the thyroid gland is likely to enlarge, producing a *goiter* (fig. 12.23).

It is believed that hyperthyroidism may be caused by an excessive release of TSH from the anterior lobe of the pituitary gland. Overstimulation by TSH causes the thyroid cells to enlarge and secrete extra amounts of hormones. The affected person is said to have *toxic goiter*. Treatment for this condition may involve surgical removal of part or all of the enlarged gland, or the administration of a drug that blocks the production of thyroid hormones. Another method for treating toxic goiter makes use of radioactive iodine. In this procedure, a sufficient dose of radioactive iodine is administered to destroy some of the thyroid secretory cells.

A different type of goiter, called *simple* or *endemic goiter*, sometimes affects persons who live in regions where iodine is lacking in the soil and drinking water. Such a person is likely to develop an iodine deficiency, reflected in an inability to produce thyroid hormones. Since these hormones normally exert an inhibiting effect on the secretion of TSH, without such inhibition, the anterior lobe of the pituitary gland releases TSH excessively. The resulting overstimulation of the thyroid gland causes it to enlarge, but since the gland is unable to manufacture hormones, the condition is accompanied by the symptoms of hypothyroidism. (See fig. 12.24.)

Fig. 12.24 Endemic goiter is caused by an iodine deficiency.

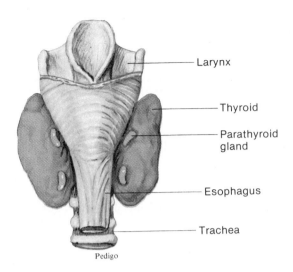

Treatment for simple goiter involves adding iodides to the diet. The condition can usually be prevented if people who live in regions where iodine deficiencies are common use table salt that contains iodides—*iodized salt.*

1. Describe four methods used for measuring the level of thyroid gland activity?
2. What effects does low thyroid gland activity have on infants? On adults?
3. What is the difference between toxic goiter and simple goiter?

The Parathyroid Glands

The **parathyroid glands** are located on the posterior surface of the thyroid gland, as shown in figure 12.25. Usually there are four of them—two associated with each of the thyroid's lateral lobes. These glands secrete a hormone that functions in the regulation of blood calcium and phosphate levels.

Structure of the Glands. Each parathyroid gland is a small, yellowish brown structure covered by a thin capsule of connective tissue. The body of the gland consists of numerous tightly packed secretory cells that are closely associated with capillary networks.

Parathyroid Hormone

The only hormone known to be secreted by the parathyroid glands is a protein called **parathyroid hormone (PTH)** or *parathormone* (fig. 12.26). This substance causes an increase in the blood calcium concentration and a decrease in the blood phosphate level. It apparently does this by influencing three types of organs—the bones, the intestine, and the kidneys.

As is described in chapter 8, the intercellular matrix of bone tissue contains a considerable amount of calcium phosphate and calcium carbonate. PTH seems to activate the bone resorbing osteoclasts and stimulate the formation of new osteoblasts. As a result of increased osteoclastic activity, calcium and phosphates are released from the bones, and the blood concentrations of these substances increase. At the same time, PTH enhances the absorption of calcium and phosphates from foods in the intestine, and this action also produces a rise in the blood levels of calcium and phosphates. PTH causes the kidneys to conserve blood calcium and to increase the excretion of phosphates in the urine.

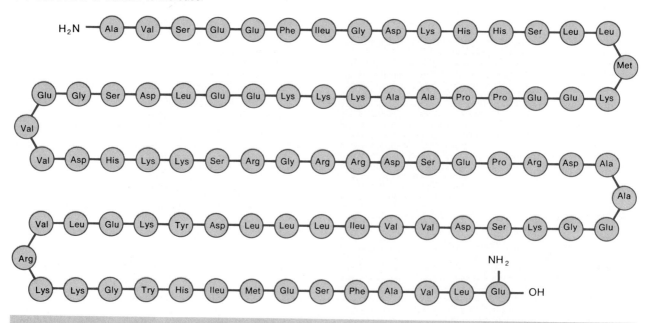

Fig. 12.26 Parathyroid hormone is a protein that causes the blood level of calcium to increase.

Parathyroid hormone

The parathyroid glands seem to be free of control by the hypothalamus. Instead, the secretion of PTH is regulated by a *negative feedback mechanism* operating between the glands and the blood calcium level. (See fig. 12.27.) As the level of blood calcium rises, less PTH is secreted; as the level of blood calcium drops, more PTH is released.

As was mentioned, *calcitonin* from the thyroid gland also acts in the regulation of blood calcium level; however, its effect is exactly opposite that of PTH. Both of these substances are released in response to changes in calcium concentrations, and together they function to ensure that a relatively constant level of blood calcium is maintained.

The homeostasis of calcium is important in a number of physiological processes. For example, as the blood calcium level drops (hypocalcemia) the nervous system becomes abnormally excitable, and impulses may be triggered spontaneously. As a result, muscles may undergo tetanic contractions, and the person may die due to a failure of respiratory movements. With an abnormally high level of blood calcium (hypercalcemia), the nervous system becomes depressed. Consequently, muscle contractions are weak and reflexes are sluggish.

Fig. 12.27 Parathyroid hormone stimulates the release of calcium from bone, the conservation of calcium by the kidneys, and the absorption of calcium by the intestine. The resulting increase in blood calcium level inhibits the secretion of this hormone.

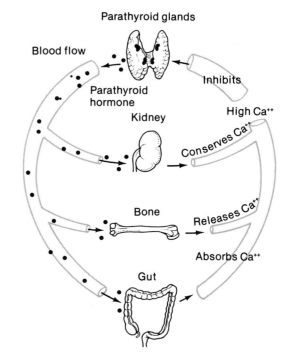

Disorders of the Parathyroid Glands

As with the thyroid gland, disorders of the parathyroids usually involve excessive hormone secretions (hyperparathyroidism) or inadequate secretions (hypoparathyroidism).

Hyperparathyroidism is most often caused by a tumor associated with a parathyroid gland. The resulting increase in PTH secretion stimulates excessive osteoclastic activity, and as bone tissue is resorbed, the bones become soft, deformed, and subject to spontaneous fractures. In extreme cases, portions of bones may disappear altogether, leaving large holes (osteoporosis) in place of the former solid structure. Later these cavities may become filled with fibrous tissue, producing a condition called *osteitis fibrosa cystica*.

The excessive calcium and phosphate released into the body fluids as a result of extra PTH secretion may be deposited in abnormal places, causing new problems such as kidney stones.

Hypoparathyroidism can result from an injury to the parathyroids or from surgical removal of these glands. In either case, decreased PTH secretion is reflected in reduced osteoclastic activity, and although the bones remain strong, the blood calcium level drops, producing the symptoms of hypocalcemia described earlier.

Fig. 12.28 An adrenal gland consists of an outer cortex and an inner medulla that represent distinctly different glands.

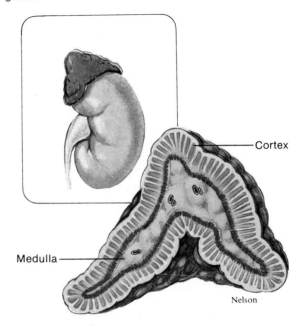

Cortex

Medulla

Nelson

The Adrenal Glands

The **adrenal glands** (suprarenal glands) are located in close association with the kidneys. A gland sits atop each kidney like a cap and is embedded in the mass of fat that encloses the kidney.

Structure of the Glands. Although the adrenal glands usually differ somewhat in size and shape, they are generally pyramidal. Each gland is very vascular and consists of two parts, as shown in figure 12.28. The central portion is termed the adrenal medulla, and the outer part is the adrenal cortex. Although these regions are not sharply divided, they represent distinct glands that secrete different hormones.

The **adrenal medulla** consists of irregularly shaped cells that are arranged in groups around blood vessels. These cells are intimately connected with the sympathetic division of the autonomic nervous system, in that they are modified postganglionic cells, and preganglionic autonomic nerve fibers lead directly to them.

The **adrenal cortex,** which makes up the bulk of the gland, is composed of closely packed masses of epithelial cells that are arranged in layers. These layers form an outer, a middle, and an inner zone of the cortex. As in the case of the medulla, the cells of the adrenal cortex are well supplied with blood vessels.

1. Where are the adrenal glands located?
2. Describe the two portions of an adrenal gland.

Hormones of the Adrenal Medulla

Cells of the adrenal medulla secrete two closely related hormones, **epinephrine** and **norepinephrine.** Both of these substances are amines (catecholamines), and they have similar structures and functions. (See fig. 12.29.) In fact, epinephrine, which makes up 80% of the medullary secretion, is produced from norepinephrine by the action of an enzyme. Chart 12.7 compares the effects of these hormones.

The effects of the medullary hormones resemble those of the sympathetic nervous system, but they last about ten times longer. These effects include an increased heart rate, elevated blood pressure, a rise in blood sugar concentration, an increased breathing rate, dilation of the airways, and decreased activity in the digestive tract.

Impulses arriving by way of sympathetic nerve fibers stimulate the adrenal medulla to release its hormones. As a rule, these impulses originate in the hypothalamus in response to various types of stress. Thus the medullary secretions function together with the sympathetic division of the autonomic nervous system in preparing the body for energy-expending action—fight or flight.

Disorders of the Adrenal Medulla

Although a lack of medullary hormones produces no significant effects, tumors in the adrenal medulla are sometimes accompanied by excessive hormone secretions. Affected persons show signs of prolonged sympathetic responses—high blood pressure, increased heart rate, elevated blood sugar, and so forth. Treatment for this condition generally involves surgical removal of the tumorous growth causing the problem.

1. Name the hormones secreted by the adrenal medulla.
2. What effects are produced by these hormones?
3. What is the usual stimulus for their release?

Hormones of the Adrenal Cortex

The cells of the adrenal cortex produce a number of different hormones; about thirty of these have been identified. These hormones are all steroids (corticosteroids), and unlike the adrenal medullary hormones—which a person can live without—those

Fig. 12.29 Epinephrine and norepinephrine have similar molecular structures and similar functions.

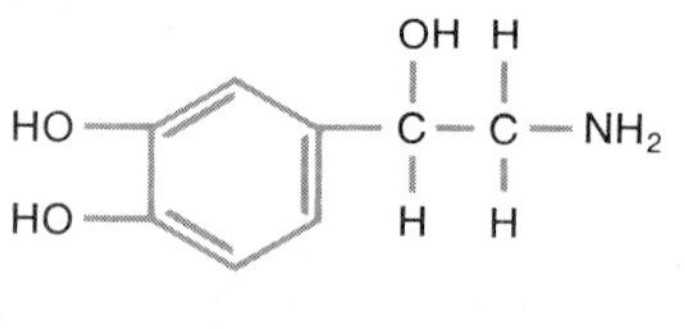

Norepinephrine

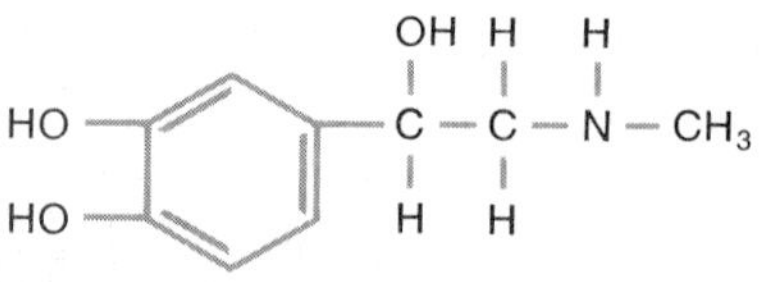

Epinephrine

Chart 12.7 Comparative effects of adrenal medulla hormones

Epinephrine	Norepinephrine
Blood pressure increases due to increased cardiac output and vasoconstriction in certain regions	Blood pressure increases to a greater degree due to generalized vasoconstriction
Rate of glycogen breakdown into glucose increases, so level of blood glucose rises	Same effect but to a lesser degree
Rate of fatty acid release from fat increases, so level of blood fatty acids rises	Same effect but to a lesser degree
Release of ACTH and TSH from the anterior pituitary gland increases	No effect

released by the cortex are vital. In fact, in the absence of cortical secretions, a person usually dies within a week unless extensive electrolyte therapy is provided.

These hormones can be grouped into three categories, each secreted by cells in a different layer of the cortex. (See fig. 12.30.)

The three groups of adrenal cortical (adrenocortical) hormones are as follows:

1. Mineralocorticoids, which help to regulate the concentrations of extracellular electrolytes.

2. Glucocorticoids, which influence the metabolism of carbohydrates, proteins, and fats.

3. Sex hormones, which affect sexual characteristics.

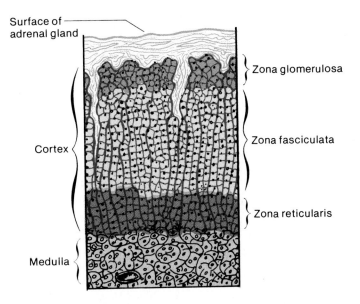

Fig. 12.30 The adrenal cortex consists of three layers or zones of cells. How do the functions of these layers differ?

Surface of adrenal gland

Zona glomerulosa

Cortex

Zona fasciculata

Zona reticularis

Medulla

Mineralocorticoids. Of the **mineralocorticoids,** which are manufactured by cells in the outer zone (zona glomerulosa) of the cortex, the most important is **aldosterone.** This hormone acts mainly through the kidneys to maintain the homeostasis of sodium and potassium ions. More specifically, aldosterone causes the kidneys to conserve sodium ions and to excrete potassium ions. At the same time, it promotes water conservation and reduces urine output.

The mechanism that controls the secretion of aldosterone is primarily responsive to the potassium ion concentration in body fluids. As the potassium concentration rises, more aldosterone is released. (See fig. 12.31.) Aldosterone secretion is also stimulated in response to decreased sodium concentration or decreased blood volume.

Glucocorticoids. For the most part **glucocorticoids** are produced in the middle zone (zona fasciculata) of the adrenal cortex. These hormones affect body cells generally, and although they affect carbohydrate metabolism (as their name suggests), they also influence protein and fat metabolism.

Of the various glucocorticoids, the hormone responsible for the greatest amount of activity is **cortisol** (hydrocortisone), a substance structurally similar to aldosterone (fig. 12.32). Its actions include the following:

1. Causes a breakdown of stored protein molecules and a consequent increase in the concentration of circulating amino acids.

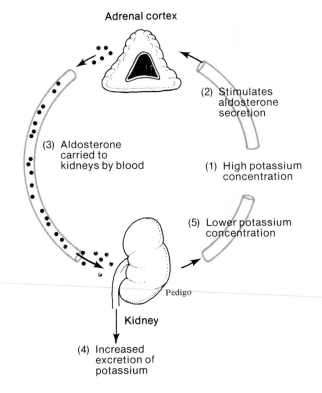

Fig. 12.31 Aldosterone secreted by the adrenal cortex causes the kidney to excrete potassium.

Adrenal cortex

(2) Stimulates aldosterone secretion

(3) Aldosterone carried to kidneys by blood

(1) High potassium concentration

(5) Lower potassium concentration

Pedigo

Kidney

(4) Increased excretion of potassium

Fig. 12.32 Cortisol and aldosterone are steroids with similar molecular structures.

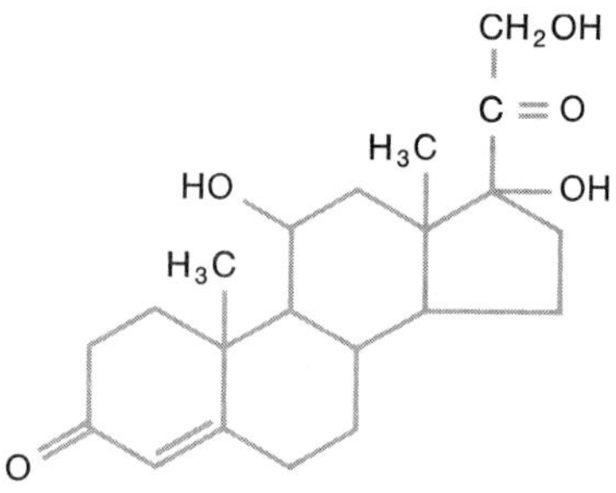

(a) Cortisol

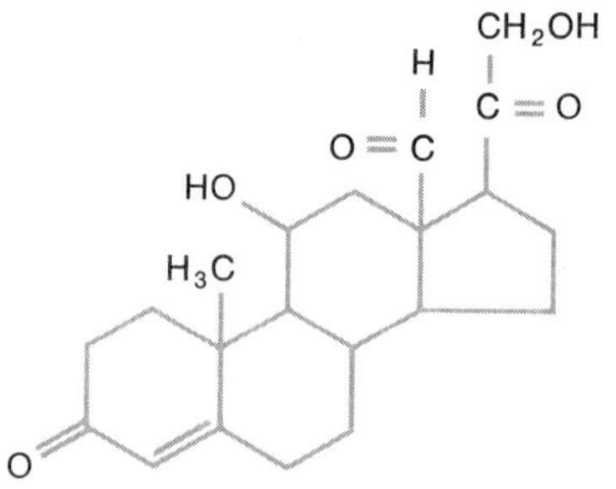

(b) Aldosterone

2. Promotes the release of fatty acids from adipose tissue, causes an increase in the use of fatty acids and a decrease in the use of glucose.

3. Stimulates liver cells to form sugar from noncarbohydrates (gluconeogenesis) such as amino acids and glycerol, and promotes an increase in blood glucose concentration.

As was discussed in an earlier section, the hypothalamus controls the anterior pituitary's secretion of ACTH, which in turn stimulates the release of cortical hormones. This action of the hypothalamus is triggered by almost all types of stress—injury, disease, exposure to temperature extremes, and so forth. One of the more significant effects of the glucocorticoids at such times seems to be the mobilization of amino acids from stored proteins. As these amino acids enter the blood, they become available for use in producing new proteins that cells may need to survive stressful conditions.

Another important action of cortisol is related to its effect upon blood vessels. Tissues subjected to stress sometimes become inflamed—red, swollen, and painful. This reaction is due in part to an increase in the permeability of the capillaries in the affected area,

which allows fluids to leak from the blood vessels and cause swelling in the surrounding tissues. Also, lysosomes of injured cells may release enzymes that produce irritation and add to the inflammation. Cortisol, however, acts to block such inflammation reactions, and although it is not clear how it accomplishes this, the cortisol may decrease the permeability of capillaries and thus inhibit the leakage of fluids. Cortisol may also stabilize lysosomal membranes so that their enzymes are not released.

Cortisol and related compounds are commonly used to treat patients with inflammatory diseases such as arthritis and allergies. They are also used to depress the inflammatory responses that accompany organ transplants and tissue grafts, and in this way help to prevent the rejection of donor tissues by the hosts.

Sex Hormones. Adrenal sex hormones are produced by cells in the inner zone (zona reticularis) of the cortex. Although the hormones of this group are primarily male types (adrenal androgens), small quantities of female hormones (adrenal estrogens and progesterone) are also present. The normal functions of these hormones are not clear, but they may supplement the supply of sex hormones from the gonads and stimulate early development of the reproductive organs. Also, there is some evidence that the adrenal androgens play a role in controlling the female sex drive.

1. How are the hormones of the adrenal cortex categorized?
2. What is the function of the mineralocorticoids?
3. What actions do glucocorticoids produce?
4. How is the secretion of cortisol controlled?

Disorders of the Adrenal Cortex

Hyposecretion of cortical hormones leads to *Addison's disease*, a condition characterized by decreased blood sodium, increased blood potassium, low blood glucose level (hypoglycemia), dehydration, low blood pressure, and increased skin pigmentation. Treatment for Addison's disease involves the use of mineralocorticoids and glucocorticoids. An untreated person is likely to live only a few days because of severe disturbances in electrolyte balance.

Hypersecretion of cortical hormones, which may be associated with an adrenal tumor or with an oversecretion of ACTH by the anterior pituitary gland, results in *Cushing's disease.* A person with this

Chart 12.8 Hormones of the adrenal cortical glands

Hormone	Action	Regulation
Mineralocorticoids (e.g., aldosterone)	Help regulate the concentration of extracellular electrolytes	Electrolyte concentrations in body fluids
Glucocorticoids (e.g., cortisol)	Influence the metabolism of carbohydrates, proteins, and fats; decrease the permeability of capillary walls	ACTH from the anterior pituitary gland in response to stress
Adrenal sex hormones	Supplement the sex hormones from the gonads	

condition suffers from changes in carbohydrate and protein metabolism and in electrolyte balance. For example, when mineralocorticoids and glucocorticoids are excessive, the blood glucose level remains high, and there is a great decrease in tissue protein. Also, sodium is retained abnormally, and as a result, tissue fluids tend to increase and the skin becomes puffy. At the same time, an increase in adrenal sex hormones may produce masculinizing effects in a female such as growth of a beard or development of a deeper voice.

Chart 12.8 summarizes the characteristics of the hormones produced by the adrenal cortex.

Treatment of Cushing's disease is usually directed toward reducing an oversecretion of ACTH. For example, a tumor associated with the pituitary gland may be removed surgically or destroyed by radiation treatment. In other cases, the adrenal glands may be partially or completely removed. After such treatment, if the patient develops symptoms of cortical insufficiency, mineralocorticoids and glucocorticoids may be administered.

---

1. *What symptoms result from the hyposecretion of cortical hormones?*
2. *What symptoms characterize the hypersecretion of these hormones?*

The Pancreas

The **pancreas** contains two major types of secretory tissues, reflecting its dual function as an exocrine gland that secretes digestive juice through a duct, and an endocrine gland that releases hormones.

Fig. 12.33 The hormone-secreting cells of the pancreas are arranged in clusters or islets surrounded by cells that secrete digestive enzymes.

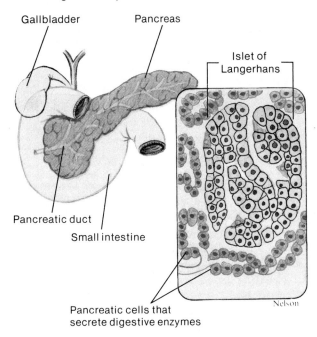

Structure of the Gland. The pancreas is an elongated, somewhat flattened organ that lies back of the stomach and behind the parietal peritoneum. (See fig. 12.33.) It is attached to the first section of the small intestine (duodenum) by a duct that transports its digestive juice to the intestine.

The endocrine portion of the pancreas consists of cells arranged in groups closely associated with blood vessels. These groups, called *islets of Langerhans*, include two distinct types of cells—the alpha cells that secrete the hormone glucagon and the beta cells that secrete the hormone insulin. (See fig. 12.34.)

The digestive functions of the pancreas are discussed in chapter 13.

Fig. 12.34 Micrograph of an islet of Langerhans and surrounding cells of the pancreas.

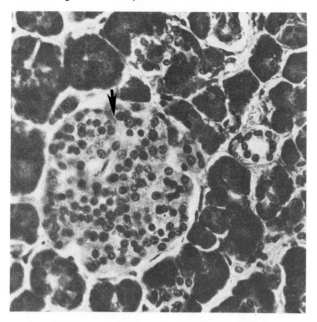

Fig. 12.35 Glucagon is a protein that promotes the conversion of glycogen into glucose.

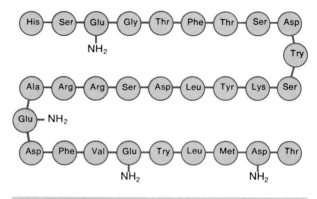

Glucagon

Hormones of the Islets of Langerhans

Glucagon is a protein that stimulates the liver to convert glycogen into glucose (glycogenolysis). This action causes the blood glucose level to rise, and for this reason glucagon is sometimes called *hyperglycemic factor*. Although its effect is similar to the action of epinephrine, glucagon is many times more effective in elevating blood sugar than epinephrine. Figure 12.35 shows its structure.

The secretion of glucagon is regulated by a negative feedback system in which a low level of blood sugar stimulates the release of the hormone. When the blood sugar level rises, glucagon is no longer secreted. This mechanism prevents hypoglycemia from occurring at times when glucose is being utilized rapidly—during periods of exercise, for example.

The hormone **insulin** is also a protein, and its main effect is exactly opposite to that of glucagon. In other words, insulin causes a decrease in the level of blood sugar. It apparently does this by promoting the facilitated diffusion (see chapter 3) of glucose through cell membranes, thus influencing sugar molecules to leave the blood and enter cells. The amino acid sequence of an insulin molecule is shown in figure 12.36.

Insulin is particularly effective in facilitating the transport of glucose into the cells of skeletal, heart, and smooth muscles and into the cells of adipose tissue. It does not seem to facilitate movement of sugar into brain cells.

Chart 12.9 Hormones of the pancreas

Hormone	Action	Source of Control
Glucagon (hyper-glycemic factor)	Stimulates the liver to convert glycogen into glucose, causing the blood glucose level to rise	Blood glucose level
Insulin	Promotes the movement of glucose through cell membranes, causing the blood glucose level to fall; stimulates the liver to convert glucose into glycogen; promotes the transport of amino acids into cells; enhances the synthesis of proteins and fats	Blood glucose level

Other effects of insulin include stimulating liver and muscle cells to convert glucose into glycogen (glycogenesis), promoting the transport of amino acids into cells, and enhancing the synthesis of proteins and fats.

As in the case of glucagon, insulin secretion is regulated by a negative feedback mechanism. When the blood glucose level rises, insulin is released; when the glucose level falls, insulin is no longer secreted (fig. 12.37). Glucagon and insulin thus function together to maintain a relatively constant blood sugar level despite great variations in the amounts of carbohydrates that are eaten. Chart 12.9 reviews these hormones.

Fig. 12.36 Amino acid sequence in a human insulin molecule.

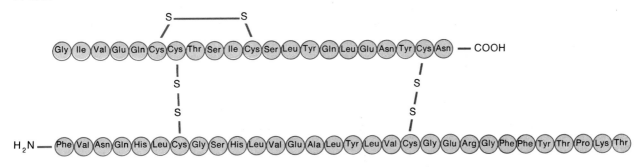

Fig. 12.37 Insulin causes a drop in the level of blood glucose, thus inhibiting the further secretion of insulin.

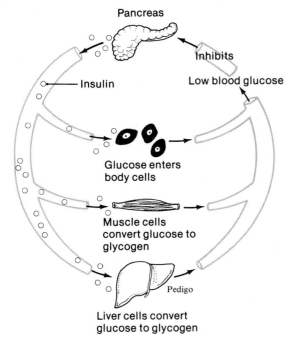

Pancreas

Inhibits

Insulin

Low blood glucose

Glucose enters body cells

Muscle cells convert glucose to glycogen

Pedigo

Liver cells convert glucose to glycogen

1. *What is the name of the endocrine portion of the pancreas?*
2. *What is the function of glucagon?*
3. *What is the function of insulin?*
4. *How are the secretions of these hormones controlled?*

Disorders of the Islets of Langerhans

A deficiency of insulin (hypoinsulinism) results in a disease called *diabetes mellitus*. This condition is characterized by severe upsets in carbohydrate metabolism as well as by disturbances in protein and fat metabolism. More specifically, there is a decrease in the utilization of glucose by cells and a consequent rise in the concentration of blood sugar (hyperglycemia). When the blood sugar reaches a certain high level, the kidneys begin to excrete the excess, and glucose appears in the urine (glycosuria). The sugar raises the osmotic pressure of the urine, normal kidney functions are upset, and more water and electrolytes (particularly sodium ions) are excreted than usual. As a result of excessive urine output (diuresis or polyuria), the affected person becomes dehydrated and experiences great thirst (polydipsia).

Meanwhile, because the body cells are unable to obtain adequate amounts of glucose, they begin to utilize abnormal quantities of fat, and this increase in fat usage is accompanied by a rise in the level of circulating fatty acids. The presence of excessive fatty acids in the blood is thought to be related to the development of certain circulatory problems. For example, diabetics tend to develop *atherosclerosis,* an artery disease in which fatty deposits build up in blood vessels and interfere with the movement of blood.

Another consequence of abnormal fat metabolism is an increase in the level of ketone bodies in the blood. As is explained in chapter 4, ketone bodies are a byproduct of fat metabolism, and although they may be excreted in the urine (ketonuria), if they accumulate excessively they can cause a condition called *acidosis.* In this instance the pH of the body fluids drops, and the affected person may lose consciousness (*diabetic coma*) and die.

Treatment of diabetes mellitus involves administering enough insulin to maintain the normal level of carbohydrate metabolism. Dosages vary with individuals, depending on such factors as diet, physical activities, and health. A patient in a diabetic coma usually requires additional therapy to correct water, electrolyte, and pH imbalances.

Since insulin is a protein and is broken down by digestive enzymes, it cannot be administered orally and is given by injection instead. Mild cases of diabetes mellitus sometimes are treated with drugs that are taken by mouth. These drugs function to stimulate islet cells of the pancreas, which may be inactive but are still able to secrete insulin.

Occasionally, a person suffers from an oversecretion of insulin (hyperinsulinism), which is followed by a drop in the blood sugar level (hypoglycemia). An affected person experiences extreme nervousness and tremor, which may be accompanied by convulsions and loss of consciousness. Hyperinsulinism may be due to a tumor involving the islet cells, but similar symptoms occur if a diabetic receives too large a dosage of insulin. In either case, the condition is commonly referred to as *insulin shock.* Treatment includes administering glucose intravenously, which usually brings the patient out of shock within a few minutes.

A diabetic is often supplied with glucagon that can be injected into the body by a family member in case the diabetic develops insulin shock and is found unconscious.

---

1. What problems are likely to be experienced by a person with hypoinsulinism?
2. What problems are likely to be experienced as a result of hyperinsulinism?

Other Endocrine Glands

Other organs that produce hormones and thus are parts of the endocrine system include the pineal gland, thymus gland, reproductive glands, and certain glands of the digestive tract.

The Pineal Gland

The **pineal gland** is a small, oval structure located deep between the cerebral hemispheres, where it is attached to the upper portion of the thalamus near the roof of the third ventricle. Its body consists largely of specialized *pineal cells* and *neuroglial cells* that provide support. (See fig. 10.43.)

Although the pineal gland does not seem to have any direct nerve connections to other brain parts, its functional cells are innervated by postganglionic sympathetic nerve fibers. Nerve impulses on these fibers apparently stimulate the pineal cells into action.

While the function of the pineal gland remains in doubt, there is evidence that it secretes at least one hormone, **melatonin.** This substance is thought to stimulate the secretion of certain releasing factors from the hypothalamus, which in turn affect the pituitary gland's secretion of gonadotropins and ACTH.

In some animals, the pineal gland seems to be controlled by varying light conditions outside their bodies. Apparently this information reaches the gland from the retinas of the eyes by means of nerve fibers in the optic tracts. As the amount of light exposure decreases, the secretion of melatonin increases. This mechanism may be involved in the regulation of some **circadian rhythms**—patterns of repeated behavior associated with environmental cycles of day and night. It is also thought that the pineal gland may help to regulate the female reproductive (menstrual) cycle by influencing the secretion of gonadotropic releasing factors by the hypothalamus.

The Thymus Gland

The **thymus gland,** which lies in the mediastinum behind the sternum and between the lungs, is relatively large in young children but diminishes in size with age. It is believed that this gland secretes a hormone, called **thymosin,** that affects the production of certain white blood cells (lymphocytes). This gland is discussed in chapter 18, which deals with the lymphatic system.

The Reproductive Glands

The reproductive organs that secrete important hormones include the **ovaries,** which produce estrogens and progesterone; the **placenta,** which produces estrogens, and progesterone, and gonadotropin; and the **testes,** which produce testosterone. These glands and their secretions are discussed in chapter 21, which concerns the reproductive systems.

The Digestive Glands

The digestive glands that secrete hormones are generally associated with the linings of the stomach and small intestine. These structures and their secretions are described in chapter 13, which deals with the digestive system.

---

1. Where is the pineal gland located?
2. What seems to be the function of the pineal gland?
3. Where is the thymus gland located?
4. Which reproductive organs secrete hormones?

Stress and Its Effects

Since survival depends upon the maintenance of homeostasis, factors that cause changes in the body's internal environment are potentially life threatening. When such dangers are sensed, nerve impulses are directed to the hypothalamus, and physiological responses are triggered that tend to resist a loss of homeostasis. These responses often include increased activity in the sympathetic division of the autonomic nervous system and increased secretion of adrenal hormones. A factor capable of stimulating such a response is called a **stressor,** and the state it produces in the body is called **stress.**

Types of Stress

The factors that produce stress are usually extreme forms of stimuli and may by physical, psychological, or a combination of these.

Physical stress may be caused by exposure to stressors that are harmful or potentially harmful to tissues. These include extreme heat or cold, decreased oxygen concentration, infections, injuries, prolonged heavy exercise, and loud sounds. Usually such physical stress is accompanied by unpleasant or painful sensations.

Psychological stress results from thoughts about real or imagined dangers, personal losses, changes in life situations, social interactions (or lack of social interactions), or factors that in any way seem to threaten a person. Feelings of anger, fear, grief, anxiety, depression, and guilt are involved with this type of stress. The factors that produce psychological stress vary greatly from person to person. Thus, a situation that is stressful to one may not affect another, and what is stressful at one time may not be at another time.

Responses to Stress

Regardless of its cause, the physiological responses to stress are directed toward maintaining homeostasis, and they usually involve a set of reactions called the *general stress syndrome,* which is largely controlled by the hypothalamus.

As is discussed in chapter 10, the hypothalamus receives information from nearly all body parts, including visceral receptors and various regions of the brain such as the cerebral cortex. At times of stress, the hypothalamus may respond to incoming impulses by activating mechanisms that generally prepare the body for fight or flight. More specifically, sympathetic impulses from the hypothalamus cause a rise in the blood glucose level, an increase in the level of blood glycerol and fatty acids, an increase in the heart rate, a rise in blood pressure, an increase in the breathing rate, dilation of the air passages, a shunting of blood from the skin and digestive organs into the skeletal muscles, and an increase in the secretion of *epinephrine* from the adrenal medulla. The epinephrine, in turn, intensifies these sympathetic responses and prolongs their effects. (See fig. 12.38.)

At the same time, the hypothalamus is likely to release *corticotropin-releasing factor* (CRF), which in turn stimulates the anterior lobe of the pituitary gland to secrete ACTH. The ACTH causes the adrenal cortex to increase its secretion of *cortisol.*

Fig. 12.38 At times of stress, the hypothalamus helps prepare the body for fight or flight by triggering sympathetic impulses to various organs. It also stimulates the release of epinephrine, which intensifies the sympathetic responses.

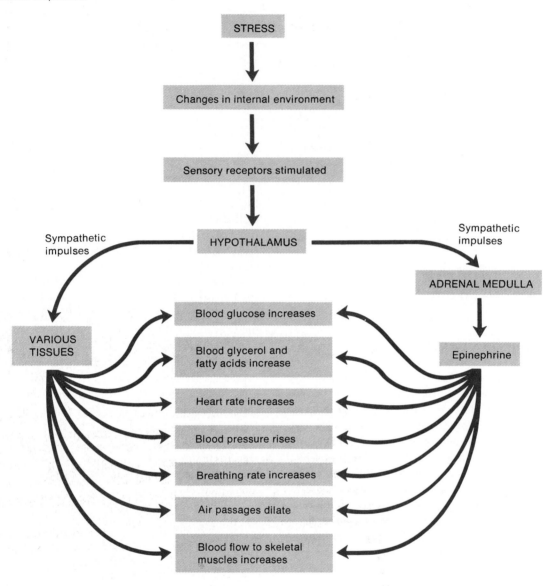

Fig. 12.39 At times of stress, the hypothalamus causes the adrenal cortex to release cortisol, which promotes reactions that resist the effects of stress.

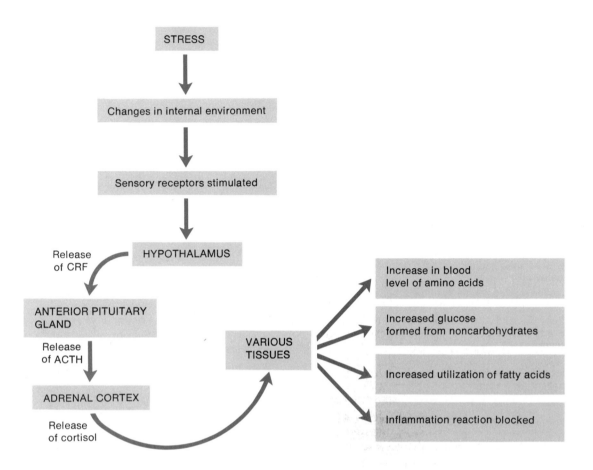

As was mentioned earlier, cortisol promotes increases in the blood level of amino acids, the utilization of fatty acids, and the formation of glucose from noncarbohydrates. Thus, while the sympathetic responses are preparing the body for physical activities that may be necessary for survival, the actions of cortisol serve to supply cells with amino acids and extra energy sources that may be needed to repair injured tissues. Cortisol also blocks inflammation reactions, apparently by decreasing the permeability of capillaries and by stabilizing lysosomal membranes. (See fig. 12.39.)

Other hormones whose secretions increase at times of stress include *glucagon* from the pancreas, *growth hormone* (GH) from the anterior pituitary, *aldosterone* from the adrenal cortex, and antidiuretic hormone (ADH) from the posterior pituitary gland.

The general effect of glucagon and growth hormone is to aid in mobilizing energy sources such as glucose, glycerol, and fatty acids and to stimulate the uptake of amino acids by cells, thus facilitating the repair of injured tissues. Aldosterone and ADH promote the retention of sodium ions and water by the kidneys. This action decreases urine output and increases blood volume—an action of particular importance if a person is bleeding excessively or sweating heavily.

Chart 12.10 summarizes these reactions to stress.

Chart 12.10 Major events in the stress response

1. As a result of stress, nerve impulses are transmitted to the hypothalamus.
2. Sympathetic impulses arising from the hypothalamus cause increases in blood glucose level, blood glycerol level, blood fatty acid level, heart rate, blood pressure, and breathing rate, as well as dilation of air passages, shunting of blood into skeletal muscles, and increased secretion of epinephrine from the adrenal medulla.
3. Epinephrine intensifies and prolongs sympathetic actions.
4. Hypothalamus secretes CRF, which stimulates the secretion of ACTH by the anterior lobe of the pituitary gland.
5. ACTH stimulates the release of cortisol by the adrenal cortex.
6. Cortisol promotes increases in the level of blood amino acids, utilization of fatty acids, and formation of glucose from noncarbohydrate sources.
7. Secretions of glucagon from the pancreas and growth hormone from the anterior pituitary increase.
8. Glucagon and growth hormone aid the mobilization of energy sources and stimulate the uptake of amino acids by cells.
9. Secretion of aldosterone from the adrenal cortex and of ADH from the posterior pituitary increases.
10. Aldosterone and ADH promote the retention of sodium ions and water by the kidneys, which increases blood volume.
11. As a result of these actions, the body is better able to resist the effects of stress.

Clinical Implications of Stress

Although stress responses are often triggered by extreme environmental changes or intense social interactions, they may also result from subtle social contacts or from a lack of social contact. Thus, a person who is socially isolated may experience stress as well as one who is anxious, fearful, or angry. A person experiencing stress may appear abnormally pale (pallor), have cool skin, and a dry mouth.

The intensity of a stress response can be judged by observing changes in a person's heart rate, breathing rate, blood pressure, and perspiration on the palms of the hands. It can also be measured by determining the concentration of epinephrine or cortisol in the blood or urine.

For reasons that are not clear, increased secretion of cortisol may be accompanied by a decrease in the activities of the *lymphatic organs,* which include the thymus, lymph nodes, and spleen. At the same time, the number of *lymphocytes* in the blood tends to decrease. Since these white blood cells, function in defending the body against infections, a person who is under stress may have a lowered resistance to infectious diseases.

In addition, the exposure to excessive amounts of cortisol may promote the development of high blood pressure, atherosclerosis, and gastrointestinal ulcers.

Most patients experience some degree of stress as a result of disease processes, pain, anxiety, or even the treatments they are receiving. Although it usually is not possible to avoid stress completely, it is often possible to reduce psychological stress by discussing problems and sources of anxiety with patients and by explaining clinical procedures so that they understand what is taking place and why.

Physical stress can be reduced by providing pleasant, comfortable external surroundings and by managing each patient's internal environment by controlling pain, administering necessary fluids and electrolytes, providing drug treatments for infections, and so forth.

1. What is meant by stress?
2. Distinguish between physical stress and psychological stress.
3. Describe the general stress syndrome.
4. In what ways can stress promote the development of disease conditions?

Some Clinical Terms Related to the Endocrine System

adrenalectomy (ah-dre″nah-lek′to-me)—surgical removal of the adrenal glands.

adrenogenital syndrome (ah-dre″no-jen′ĭ-tal sin′drōm)—a group of symptoms associated with changes in sexual characteristics as a result of increased secretion of adrenal androgens.

diabetes insipidus (di″ah-be′tēz in-sip″ĭ-dus)—a metabolic disorder characterized by a large output of dilute urine containing no sugar, and caused by a decreased secretion of ADH by the posterior pituitary gland.

exophthalmos (ek″sof-thal′mos)—an abnormal protrusion of the eyes.

hirsutism (her′sūt-izm)—excessive growth of hair.

hyperglycemia (hi″per-gli-se′me-ah)—an excess of blood glucose.

hypocalcemia (hi″po-kal-se′me-ah)—a deficiency of blood calcium.

hypoglycemia (hi″po-gli-se′me-ah)—a deficiency of blood glucose.

hypophysectomy (hi-pof″ĭ-sek′to-me)—surgical removal of the pituitary gland.

parathyroidectomy (par″ah-thi″roi-dek′to-me)—surgical removal of the parathyroid glands.

pheochromocytoma (fe-o-kro″mo-si-to′mah)—a type of tumor found in the adrenal medulla and usually accompanied by high blood pressure.

polyphagia (pol″e-fa′je-ah)—excessive eating.

Simmond's disease (sim′ondz di-zēz′)—a disorder characterized by excessive weight loss and caused by destruction of the pituitary gland.

thymectomy (thi-mek′to-me)—surgical removal of the thymus gland.

thyroidectomy (thi″roi-dek′to-me)—surgical removal of the thyroid gland.

thyroiditis (thi″roi-di′tis)—inflammation of the thyroid gland.

virilism (vir′i-lizm)—masculinization of a female.

Chapter Summary

Introduction

The endocrine and nervous systems act together in controlling body activities and maintaining homeostasis.

The actions of the nervous system are relatively rapid and brief compared with those of the endocrine system.

Endocrine glands secrete their products into body fluids, and exocrine glands secrete into ducts that lead to the outside of the body.

As a group, endocrine glands are concerned with the regulation of metabolic processes.

Hormones and Their Actions

Endocrine glands secrete hormones that have effects on specific target tissues.

1. Chemistry of hormones
   a. Each kind of hormone has a unique molecular structure and a unique action.
   b. Chemically, hormones are steroids, proteins, or amines.
2. Actions of hormones
   a. Steroid hormones seem to exert their effects by activating or suppressing the actions of specific genes.
   b. Some nonsteroid hormones act by stimulating the formation of cyclic AMP, which in turn brings about cellular changes.
3. Prostaglandins
   a. Prostaglandins are substances present in small quantities that have powerful hormonelike effects.
   b. They seem to influence the formation of cyclic AMP.

Control of Hormone Secretions

The concentration of each hormone in the body fluids remains relatively constant.

1. Negative feedback systems
   a. When an imbalance in hormone concentration occurs, information is fed back to some part that acts to correct this imbalance.
   b. Negative feedback systems are used to control the air temperature in buildings and the hormone concentrations in body fluids.
2. Nerve Control
   a. Some endocrine glands secrete their hormones in response to nerve impulses.
   b. Other glands secrete hormones in response to releasing factors secreted by the hypothalamus.

The Pituitary Gland

The pituitary gland, which is attached to the base of the brain, has an anterior lobe and a posterior lobe.

Most pituitary secretions are controlled by the hypothalamus.

1. Anterior lobe hormones
   a. The anterior lobe consists largely of epithelial cells and secretes GH, prolactin, TSH, ACTH, FSH, and LH.
   b. Growth hormone (GH)
      (1) Growth hormone stimulates body cells to increase in size and undergo an increased rate of reproduction.
      (2) The secretion of GH is controlled by growth hormone-releasing factor and growth hormone release-inhibiting factor from the hypothalamus.
      (3) In early childhood, a lack of GH results in dwarfism, and an oversecretion results in gigantism.
   c. Prolactin
      (1) Promotes breast development and stimulates milk production.
      (2) Prolactin release-inhibiting factor from the hypothalamus seems to regulate the secretion of prolactin.
   d. Thyroid-stimulating hormone (TSH)
      (1) TSH controls the secretion of hormones from the thyroid gland.
      (2) TSH secretion is regulated by the hypothalamus.
   e. Adrenocorticotropic hormone (ACTH)
      (1) ACTH controls the secretion of certain hormones from the adrenal cortex.
      (2) The secretion of ACTH is regulated by the hypothalamus.
   f. Follicle-stimulating hormone (FSH) and luteinizing hormone (LH) are gonadotropins that influence the reproductive organs.

2. Posterior lobe hormones
   a. This lobe consists largely of neuroglial cells that support nerve fibers originating in the hypothalamus.
   b. The two hormones associated with the posterior lobe are produced in the hypothalamus.
   c. Antidiuretic hormone (ADH)
      (1) ADH causes the kidneys to reduce the amount of water they excrete.
      (2) In high concentration ADH causes blood vessel walls to constrict, thus raising the blood pressure.
      (3) The secretion of ADH is regulated by the hypothalamus.
   d. Oxytocin
      (1) Oxytocin can cause muscles in the uterine wall to contract and is thought to play a role in childbirth.
      (2) It also causes contraction of certain cells associated with the production and ejection of milk.

The Thyroid Gland

The thyroid gland is located in the neck and has a special ability to remove iodine from the blood.
1. Structure of the gland
   a. The thyroid gland consists of many hollow secretory parts called follicles.
   b. The follicles are fluid-filled and store the hormones secreted by the follicle cells.
2. Thyroid hormones and their functions
   a. The thyroid hormones thyroxine and triiodothyronine cause the metabolic rate of cells to increase.
   b. The thyroid hormone calcitonin helps to regulate the level of blood calcium.
3. Disorders of the thyroid gland
   a. Most thyroid disorders involve overactivity or underactivity of the gland.
   b. Tests of protein-bound iodine, triiodothyronine uptake, thyroxine iodine concentration, and radioactive iodine uptake are used to judge the level of gland activity.
   c. Thyroid disorders include cretinism, myxedema, toxic goiter, and endemic goiter.

The Parathyroid Glands

The parathyroid glands are located on the posterior surface of the thyroid.
The parathyroid hormone helps to regulate the levels of blood calcium and phosphate.
1. Structure of the glands
   Each gland consists of secretory cells that are well supplied with capillaries.

2. Parathyroid hormone (PTH)
   a. PTH causes an increase in blood calcium and a decrease in blood phosphate levels.
   b. PTH stimulates the activity of osteoclasts in resorbing bone tissue and causes the kidneys to conserve calcium and to excrete phosphate.
   c. The parathyroid glands seem to be regulated by a negative feedback mechanism that operates between the glands and the blood.
3. Disorders of the parathyroid glands
   a. Hyperparathyroidism results in excessive loss of bone tissue.
   b. Hypoparathyroidism results in a drop in the blood calcium level, followed by muscular spasms and tetany.

The Adrenal Glands

The adrenal glands are located atop the kidneys.
1. Structure of the glands
   a. Each gland consists of a medulla and a cortex.
   b. These parts represent distinct glands that secrete different hormones.
2. Hormones of the adrenal medulla
   a. The adrenal medulla secretes epinephrine and norepinephrine.
   b. These hormones produce effects similar to those of the sympathetic nervous system.
   c. The secretion of these hormones is stimulated by sympathetic impulses originating from the hypothalamus.
3. Disorders of the adrenal medulla
   a. Tumors of the adrenal medulla sometimes cause excessive secretions.
   b. Affected persons show signs of prolonged sympathetic responses.
4. Hormones of the adrenal cortex
   a. The adrenal cortex produces a variety of vital steroid hormones that can be grouped into three major categories: mineralocorticoids, glucocorticoids, and sex hormones.
   b. The mineralocorticoids include aldosterone, which causes the kidneys to conserve sodium and water and to excrete potassium.
   c. The glucocorticoids affect carbohydrate, protein, and fat metabolism.
      (1) Cortisol is the glucocorticoid of greatest importance.
      (2) Cortisol causes amino acids to be mobilized and the permeability of the capillary walls to decrease.
   d. Adrenal sex hormones
      (1) These hormones are primarily of the male type.
      (2) They are thought to supplement the sex hormones produced by the gonads.

5. Disorders of the adrenal cortex
   a. Hyposecretion leads to Addison's disease and severe disturbances in electrolyte balance.
   b. Hypersecretion leads to Cushing's disease, which is accompanied by changes in carbohydrate and protein metabolism and in electrolyte balance.
   c. An increase in adrenal sex hormone secretion may produce masculinizing effects.

The Pancreas

The pancreas secretes digestive juices as well as hormones.
1. Structure of the gland
   a. The pancreas is located in back of the stomach and is attached to the small intestine.
   b. The endocrine portion, which is called the islets of Langerhans, secretes glucagon and insulin.
2. Hormones of the islets of Langerhans
   a. Glucagon stimulates the liver to convert glycogen into glucose, causing a rise in the level of blood glucose.
   b. Insulin promotes the movement of glucose through cell membranes and thus causes a decrease in the level of blood glucose.
3. Disorders of the islets of Langerhans
   a. Hyposecretion of insulin results in diabetes mellitus, which is characterized by upsets in metabolism of carbohydrates, proteins, and fats.
   b. Hypersecretion of insulin produces a drop in blood glucose accompanied by nervousness, convulsions, or loss of consciousness.

Other Endocrine Glands

1. The pineal gland
   a. It is attached to the thalamus near the roof of the third ventricle.
   b. It is innervated by postganglionic sympathetic nerve fibers.
   c. It secretes melatonin, which seems to stimulate the secretion of releasing factors from the hypothalamus.
   d. It may help to regulate the female reproductive cycle.
2. The thymus gland
   a. The thymus lies behind the sternum and between the lungs.
   b. Its size diminishes with age.
   c. It secretes thymosin that affects the production of lymphocytes.
3. The reproductive glands
   a. The ovaries secrete estrogens and progesterone.
   b. The placenta secretes estrogens, progesterone, and a gonadotropin.
   c. The testes secrete testosterone.
4. The digestive glands
   Certain glands of the stomach and small intestine secrete hormones.

Stress and Its Effects

Stress is the condition produced when the body responds to stressors that threaten the maintenance of homeostasis.
Stress responses involve increased activity of the sympathetic nervous system and increased secretion of adrenal hormones.
1. Types of stress
   a. Physical stress results from environmental factors that are harmful or potentially harmful to tissues.
   b. Psychological stress results from thoughts about real or imagined dangers.
   c. Factors that produce psychological stress vary from person to person and situation to situation.
2. Responses to stress
   a. Responses are directed toward maintaining homeostasis.
   b. Usually responses involve a general stress syndrome that is controlled by the hypothalamus.
      (1) Sympathetic impulses stimulate increases in blood glucose, blood glycerol, blood fatty acids, blood pressure, and breathing rate; they also cause dilation of air passages, shunting of blood into skeletal muscles, and increased secretion of epinephrine.
      (2) Hypothalamus releases corticotropin-releasing factor that stimulates the anterior pituitary gland to secrete ACTH.
      (3) ACTH stimulates the adrenal cortex to release cortisol.
      (4) Cortisol causes increases in the blood level of amino acids, utilization of fatty acids, and formation of glucose from noncarbohydrates.
      (5) Increased secretion of glucagon and growth hormone helps mobilize energy sources and stimulate the uptake of amino acids by cells.
      (6) Increased secretion of aldosterone and ADH promotes retention of sodium ions and water.
3. Clinical implications of stress
   a. Stress responses may be triggered by extreme or subtle factors or by lack of social contact.
   b. Intensity of stress responses can be judged by heart rate, breathing rate, blood pressure, and by the concentration of epinephrine or cortisol in the blood and urine.
   c. Increased levels of cortisol are accompanied by decreased activity of lymphatic organs.
      (1) The number of lymphocytes in the blood may decrease.
      (2) A person may become more susceptible to various diseases.

Application of Knowledge

1. Based on your understanding of the actions of glucagon and insulin, would a person with diabetes mellitus be likely to require more insulin or more sugar following strenuous exercise? Why?

2. What problems might result from the prolonged administration of cortisol to a person with a severe inflammatory reaction?

3. How might the environment of a patient with hyperthyroidism be modified to minimize the drain on body energy resources?

Review Activities

1. Compare the general functions of the endocrine system with those of the nervous system.

2. Explain what is meant by a ductless gland.

3. Define hormone and target tissue.

4. Explain how hormones can be grouped on the basis of their chemical structures.

5. Explain how steroid hormones exert their influence.

6. Explain how nonsteroid hormones may function through the formation of cyclic AMP.

7. Define prostaglandins.

8. Describe a negative feedback system.

9. Define releasing factor and provide an example of such a substance.

10. Describe the location and structure of the pituitary gland.

11. List the hormones secreted by the anterior lobe of the pituitary gland.

12. List the hormones released from nerve fibers in the posterior lobe of the pituitary gland.

13. Explain how pituitary gland activity is controlled by the brain.

14. Explain how growth hormone produces its effects.

15. List the major factors that affect the secretion of growth hormone.

16. Explain the causes of pituitary dwarfism and gigantism.

17. Summarize the functions of prolactin.

18. Describe the mechanism that regulates the levels of circulating thyroid hormones.

19. Explain how the secretion of ACTH is controlled.

20. List the major gonadotropins and explain the general functions of each.

21. Compare the cellular structures of the anterior and posterior lobes of the pituitary gland.

22. Describe the functions of the posterior pituitary hormones.

23. Explain how the release of ADH is regulated.

24. Describe the location and structure of the thyroid gland.

25. Name the hormones secreted by the thyroid gland and list the general functions of each.

26. Define iodine pump.

27. Distinguish between hyperthyroidism and hypothyroidism and describe the consequences of each.

28. Review four laboratory methods used to measure the level of thyroid gland activity.

29. Distinguish between cretinism and myxedema.

30. Distinguish between toxic goiter and endemic goiter.

31. Describe the location and structure of the parathyroid glands.

32. Explain the general functions of parathyroid hormone.

33. Describe the mechanism that regulates the secretion of parathyroid hormone.

34. Compare the effects of hyperparathyroidism with those of hypoparathyroidism.

35. Distinguish between the adrenal medulla and the adrenal cortex.

36. List the hormones produced by the adrenal medulla and describe their general functions.

37. Explain how the hormones of the adrenal cortex can be grouped and describe the general functions of each group.

38. Describe how the pituitary gland controls the secretion of adrenal cortical hormones.

39. Distinguish between Addison's disease and Cushing's disease.

40. Describe the location and structure of the pancreas.

41. List the hormones secreted by the pancreas and describe the general functions of each.

42. Explain how the secretion of hormones from the pancreas is regulated.

43. Explain why diabetics tend to develop atherosclerosis.

44. Define insulin shock.

45. Describe the location and general function of the pineal gland.

46. Describe the location and general function of the thymus gland.

47. Distinguish between a stressor and stress.

48. List several factors that cause physical and psychological stress.

49. Describe the general stress syndrome.

50. Review the clinical implications of stress.

Suggestions for Additional Reading

Brownstein, M. J. et al. 1980. Synthesis, transport, and release of posterior pituitary hormones. *Science* 207:373.

Catt, K. J., and Dufau, M. L. 1977. Peptide hormone receptors. *Ann. Rev. Physio.* 39:529.

Cuthbert, M. F., ed. 1973. *The prostaglandins.* Philadelphia: Lippincott.

Frohman, L. A. 1975. Neurotransmitters as regulators of endocrine functions. *Hosp. Prac.* 10:54.

Gillie, R. B. June 1971. Endemic goiter. *Scientific American.*

Goldberg, N. D. 1974. Cyclic nucleotides and cell function. *Hosp. Prac.* 9:127.

Guillemin, R., and Burgus, R. November 1972. The hormones of the hypothalamus. *Scientific American.*

Maurer, A. C. 1979. The therapy of diabetes. *American Scientist* 67:422.

McEwen, B. S. July 1976. Interactions between hormones and nerve tissue. *Scientific American.*

O'Malley, B. W., and Schrader, W. T. February 1976. The receptors of steroid hormones. *Scientific American.*

Pastan, I. August 1972. Cyclic AMP. *Scientific American.*

Pike, J. W. November 1971. Prostaglandins. *Scientific American.*

Rasmussen, H. 1974. Ions as "second messengers." *Hosp. Prac.* 9:99.

Rasmussen, H., and Pechet, M. M. October 1970. Calcitonin. *Scientific American.*

Reichlin S., et.al. 1976. Hypothalamic hormones. *Ann. Rev. Physio.* 39:389.

Schally, A. V., et.al. 1973. Hypothalamic regulatory hormones. *Science.* 179:341.

Stent, G. S. September 1972. Cellular communication. *Scientific American.*

Sutherland, E. W. 1972. Studies on the mechanism of hormone action. *Science.* 177:401.

Turner, C. D. 1976. *General endocrinology,* 6th ed. Philadelphia: W. B. Saunders Co.

Williams, R. 1974. *Textbook of endocrinology,* 5th ed. Philadelphia: W. B. Saunders Co.

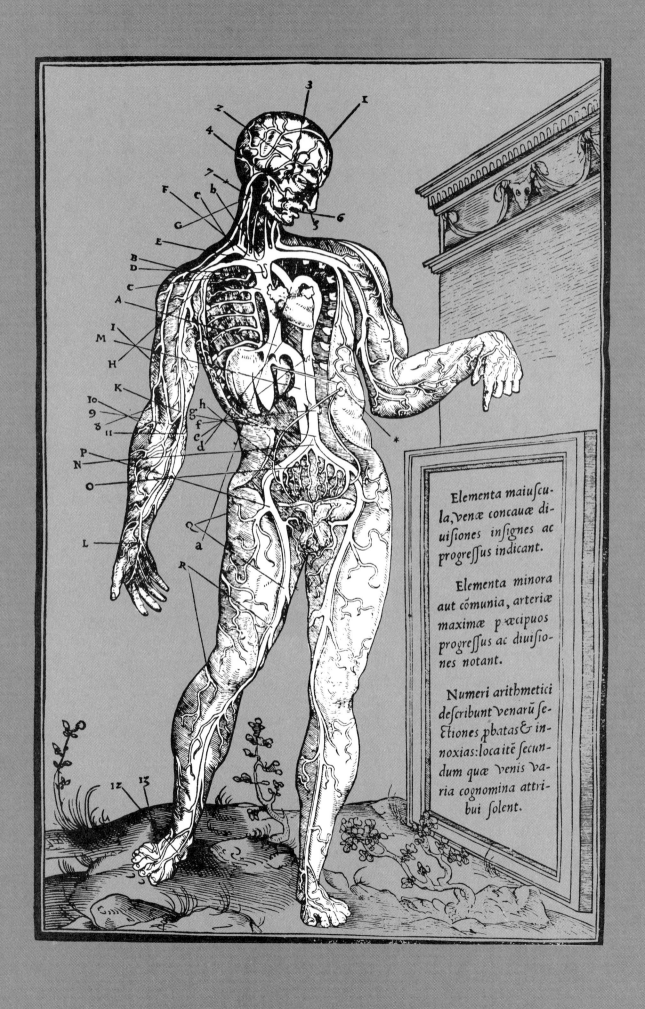

Elementa maiuscula, venæ concauæ diuisiones insignes ac progressus indicant.

Elementa minora aut cómunia, arteriæ maximæ præcipuos progressus ac diuisiones notant.

Numeri arithmetici describunt venarū sectiones pbatas & innoxias: loca itē secundum quæ venis varia cognomina attribui solent.

Unit 4
Processing and Transporting

The chapters of unit 4 are concerned with the digestive, respiratory, circulatory, lymphatic, and urinary systems. They describe how organs of these systems obtain nutrients and oxygen from outside the body, how the nutrients are altered chemically and absorbed into body fluids, and how the nutrients and oxygen are transported to body cells. They also discuss processes by which these substances are utilized by cells, how resulting wastes are transported and excreted, and how stable concentrations of various substances in body fluids are maintained.

This unit includes

The Digestive System

13 Most food substances are composed of chemicals whose molecules cannot pass through cell membranes and so cannot be absorbed by cells. The *digestive system* functions to solve this problem. Its parts are adapted to ingest foods, to secrete enzymes that modify the sizes of food molecules, to absorb the products of this digestive action, and to eliminate the unused residues.

Foods are moved through the digestive tract by muscular contractions in the wall of the tubular alimentary canal, and digestive juices, which contain digestive enzymes, are secreted into the canal by various glands and accessory organs. These functions are controlled largely by interactions between the digestive, nervous, and endocrine systems.

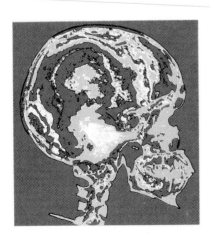

Chapter Objectives

After you have studied this chapter, you should
be able to

1. Name and describe the location of the
   organs of the digestive system and their
   major parts.

2. Describe the general functions of each
   digestive organ.

3. Describe the structure of the wall of
   the alimentary canal.

4. Explain how the contents of the alimentary
   canal are mixed and moved.

5. List the enzymes secreted by the various
   digestive organs and describe the function
   of each.

6. Describe how digestive secretions
   are regulated.

7. Explain how digestive reflexes function to
   control the movement of material through
   the alimentary canal.

8. Describe the mechanisms of swallowing,
   vomiting, and defecating.

9. Explain how the products of digestion
   are absorbed.

10. Describe some of the common disorders
    involving the digestive system.

11. Complete the review activities at the end
    of this chapter.

absorption (ab-sorp'shun)

accessory organ (ak-ses'o-re or'gan)

alimentary canal (al''ĭ-men'tar-e kah-nal')

bile (bil)

chyme (kim)

circular muscle (ser'ku-lar mus'el)

deciduous (de-sid'u-us)

gastric juice (gas'trik jōōs)

intestinal juice (in-tes'tĭ-nal jōōs)

intrinsic (in-trin'sik)

longitudinal muscle (lon''jĭ-tu'dĭ-nal mus'el)

mucous membrane (mu'kus mem'brān)

pancreatic juice (pan''kre-at'ik jōōs)

peristalsis (per''ĭ-stal'sis)

serous layer (se'rus la'er)

sphincter muscle (sfingk'ter mus'el)

villi; singular, villus (vil'i, vil'us)

aliment-, food: *aliment*ary canal—the tubelike portion of the digestive system.

cari-, decay: dental *cari*es—tooth decay.

cec-, blindness: *cec*um—blind-ended sac at the beginning of the large intestine.

chym-, juice: *chym*e—semifluid paste of food particles and gastric juice formed in the stomach.

decidu-, falling off: *decidu*ous teeth—those that are shed during childhood.

entero-, intestine: *entero*gastrone—a hormone secreted by cells in the wall of the small intestine.

frenul-, a restraint: *frenul*um—membranous fold that anchors the tongue to the floor of the mouth.

gastr-, stomach: *gastr*ic gland—portion of the stomach that secretes gastric juice.

hepat-, liver: *hepat*ic duct—duct that carries bile from the liver to the common bile duct.

hiat-, an opening: esophageal *hiat*us—opening through which the esophagus penetrates the diaphragm.

lingu-, the tongue: *lingu*al tonsil—mass of lymphatic tissue at the root of the tongue.

peri-, around: *peri*stalsis—wavelike ring of contraction that moves material along the alimentary canal.

pylor-, gatekeeper: *pylor*ic sphincter—muscle that serves as a valve between the stomach and small intestine.

vill-, hairy: *vill*i—tiny projections of mucous membrane in the small intestine.

Digestion is the process by which food substances are chemically changed into forms that can be absorbed through cell membranes. The **digestive system** includes the organs that promote this process, and it consists of an *alimentary canal* that extends from the mouth to the anus, and several *accessory organs* that release secretions into the canal. More specifically, the alimentary canal includes the mouth, pharynx, esophagus, stomach, small intestine, and large intestine; the accessory organs include the salivary glands, liver, gallbladder, and pancreas. Major organs of this system are shown in figure 13.1.

General Characteristics of the Alimentary Canal

The **alimentary canal** is a muscular tube about 9 meters (30 feet) long, which passes through the body's ventral cavity. Although it is specialized in various regions to carry on particular functions, the structure of its wall, the method by which it moves food, and its type of innervation are similar throughout its length.

Fig. 13.1 Major organs of the digestive system. What are the general functions of this system?

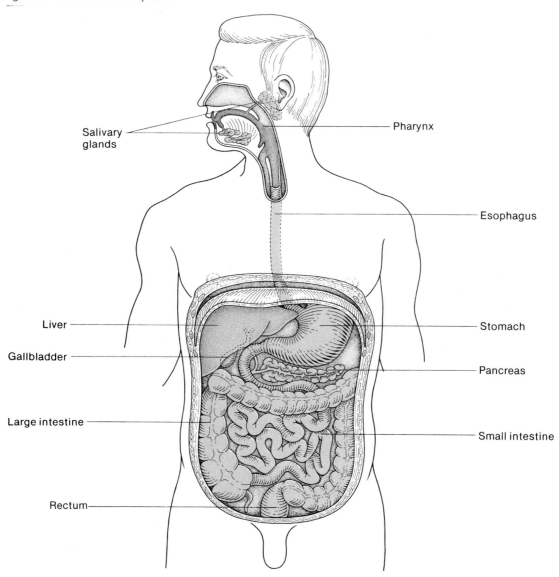

Structure of the Wall

The wall of the alimentary canal consists of four distinct layers, although the degree to which they are developed varies from part to part. Beginning with the innermost tissues, these layers, shown in figure 13.2, include the following:

1. **Mucous membrane** (*mucosa*). This layer is formed of surface epithelium, underlying connective tissue (lamina propria), and a small amount of smooth muscle. It functions to protect the tissues beneath it and to carry on absorption and secretion.

2. **Submucosa.** The submucosa contains considerable loose connective tissue as well as blood vessels, lymphatic vessels, and nerves. Its vessels serve to nourish surrounding tissues and to carry away absorbed materials.

Fig. 13.2 The wall of the alimentary canal includes four layers: mucous membrane, submucosa, muscular layer, and serous layer.

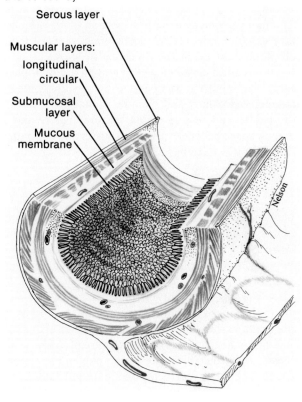

3. **Muscular layer.** This layer, which is responsible for the movements of the tube, consists of two coats of smooth muscle tissue. The fibers of the inner coat are arranged so that they encircle the tube, and when these *circular fibers* contract, the diameter of the tube is decreased. The fibers of the outer muscular coat run lengthwise, and when these *longitudinal fibers* contract, the tube is shortened.

4. **Serous layer** (*serosa*). The serous or outer covering of the tube is composed of the *visceral peritoneum*, which is formed of epithelium on the outside and connective tissue beneath. The cells of the serosa secrete serous fluid that keeps the tube's outer surface moist.

The characteristics of these layers are summarized in chart 13.1.

Movements of the Tube

The motor functions of the alimentary canal are of two basic types—mixing movements and propelling movements. (See fig. 13.3.) Mixing occurs when smooth muscles in relatively small segments of the tube undergo rhythmic contractions. When the stomach is full, for example, waves of muscular contractions move through its wall from one end to the other. These waves occur every 20 seconds or so, and their action tends to mix food substances with digestive juices secreted by the mucosa.

Propelling movements include a wavelike motion called **peristalsis.** When peristalsis occurs, a ring of contraction appears in the wall of the tube, and as it moves along, it pushes the contents ahead of it. (See fig. 13.3.) The usual stimulus for peristalsis is an expansion of the tube due to accumulation of food inside it. Such movements create sounds that can be heard through a stethoscope applied to the abdominal wall.

Chart 13.1   Layers in the wall of the alimentary canal

Layer	Composition	Function
Mucous membrane	Epithelium, connective tissue, smooth muscle	Protection, absorption, secretion
Submucosa	Loose connective tissue, blood vessels, lymphatic vessels, nerves	Nourishes surrounding tissues, transports absorbed materials
Muscular layer	Smooth muscle fibers arranged in circular and longitudinal groups	Movements of the tube and its contents
Serous layer	Epithelium, connective tissue	Protection

Fig. 13.3 (a) Mixing movements occur when small segments of the muscular wall of the alimentary canal undergo rhythmic contractions. (b) Peristaltic waves cause the contents to be moved along the canal.

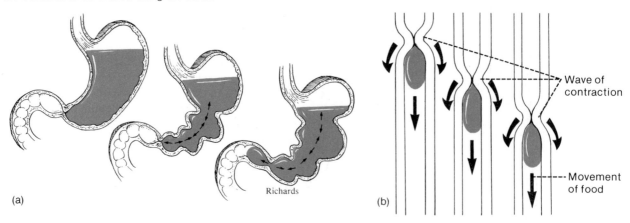

(a) Richards

(b)
Wave of contraction

Movement of food

Innervation of the Tube

The alimentary canal is innervated extensively by branches of the sympathetic and parasympathetic divisions of the autonomic nervous system. These nerve fibers are associated mainly with the tube's muscular layer, and they are responsible for maintaining muscle tone and for regulating the strength, rate, and velocity of muscular contractions.

Parasympathetic impulses generally cause an increase in the activities of the digestive system. Some of these impulses originate in the brain and are conducted on branches of the vagus nerves to the esophagus, stomach, pancreas, gallbladder, small intestine, and proximal half of the large intestine. Other parasympathetic impulses arise in the sacral region of the spinal cord and supply the distal half of the large intestine.

The effects produced by sympathetic nerve impulses usually are opposite those of the parasympathetic division. In other words, they inhibit various digestive actions. Sympathetic impulses are responsible for the contraction of certain sphincter muscles in the wall of the alimentary canal, and when they are contracted, these muscles effectively block the movement of materials through the tube.

---

1. *What organs constitute the digestive system?*
2. *Describe the wall of the alimentary canal.*
3. *Name the two types of movements that occur in the alimentary canal.*
4. *What effect do parasympathetic nerve impulses have on digestive actions? What effect do sympathetic nerve impulses have?*

The Mouth

The mouth, which is the first portion of the alimentary canal, is adapted to receive food, prepare it for digestion, and begin the digestion of starch. It also functions as an organ of speech and an organ of pleasure. The mouth is surrounded by the lips, cheeks, tongue, and palate, and includes a chamber between the palate and tongue called the *oral cavity* and a narrow space between the teeth, cheeks and lips called the *vestibule*. (See fig. 13.4.)

The Cheeks and Lips

The cheeks form the lateral walls of the mouth and consist of outer layers of skin, pads of subcutaneous fat, certain muscles associated with expression and chewing, and inner linings of stratified squamous epithelium.

The lips are highly mobile structures that surround the mouth opening. They contain skeletal muscles and a variety of sensory receptors that are useful in judging the temperature and texture of foods. Their normal reddish color is due to an abundance of blood vessels near their surfaces. The external borders of the lips mark the boundaries between the skin of the face and the mucous membrane that lines the alimentary canal.

The newborn infant's mouth is adapted for sucking. The lips are equipped with blisterlike milk pads, and the lower jaw recedes.

Fig. 13.4 The mouth is adapted for ingesting food and preparing it for digestion.

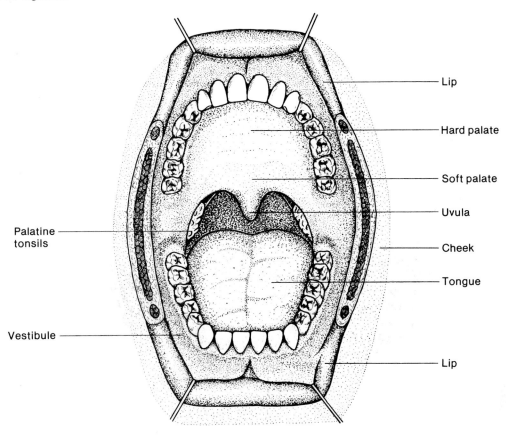

- Lip
- Hard palate
- Soft palate
- Uvula
- Cheek
- Tongue
- Lip

Palatine tonsils

Vestibule

The Tongue

The **tongue** is a thick, muscular organ that occupies the floor of the mouth and nearly fills the oral cavity when the mouth is closed. It is covered by mucous membrane and is anchored in the midline to the floor of the mouth by a membranous fold called the **frenulum**.

A person whose frenulum is too short is said to be tongue-tied. Infants with this condition may have difficulty sucking, and older children may be unable to make the tongue movements needed for normal speech. A short frenulum is usually corrected in early infancy by cutting it.

The body of the tongue is composed largely of skeletal muscle whose fibers run in several directions. These muscles aid in mixing food particles with saliva during chewing and in moving food toward the pharynx during swallowing. Rough projections, called **papillae,** on the surface of the tongue provide friction that is useful in handling food. These papillae also contain taste buds. (See fig. 13.5.)

Fig. 13.5 The surface of the tongue as viewed from above.

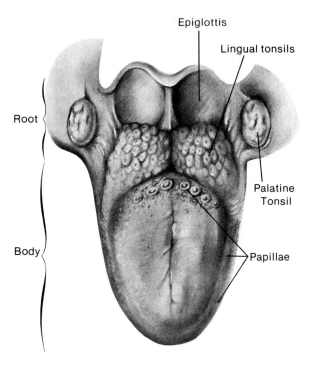

- Epiglottis
- Lingual tonsils
- Palatine Tonsil
- Papillae

Root

Body

Fig. 13.6 A sagittal section of the mouth and nasal cavity.

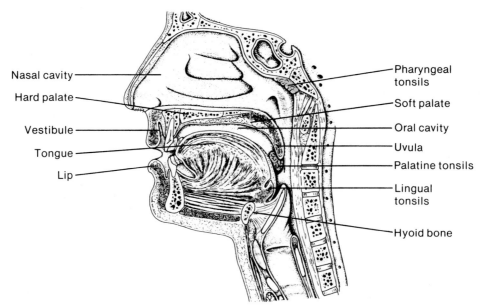

The posterior region, or *root*, of the tongue is anchored to the hyoid bone and is covered with rounded masses of lymphatic tissue called **lingual tonsils.**

The Palate

The **palate** forms the roof of the oral cavity and consists of a hard anterior part and a soft posterior part. The *hard palate* is formed by the palatine processes of the maxillary bones in front and the horizontal portions of the palatine bones in back. The *soft palate* forms a muscular arch that extends posteriorly and downward as a cone-shaped projection called the **uvula.**

During swallowing, muscles draw the soft palate and the uvula upward. This action closes the opening between the nasal cavity and the pharynx, preventing food from entering the nasal cavity.

In the back of the mouth, on either side of the tongue and closely associated with the palate, are masses of lymphatic tissue called **palatine tonsils.** These structures lie beneath the epithelial lining of the mouth and, like other lymphatic tissues, they help protect the body against infections. (See chapter 18 for more detail concerning lymphatic functions.) However, the palatine tonsils themselves are common sites of infections, and if they become inflamed, the condition is termed *tonsillitis.* Infected tonsils may become so swollen that they block the passageways of the pharynx and interfere with breathing and swallowing. Since the mucous membranes of the pharynx, eustachian tubes, and middle ears are continuous, there is always danger that such an infection may travel from the throat into the middle ears.

When tonsillitis occurs repeatedly and the condition remains unresponsive to antibiotic treatment, the tonsils are often removed. This surgical procedure is called *tonsillectomy.*

Still other masses of lymphatic tissue, called **pharyngeal tonsils** or *adenoids,* occur on the posterior wall of the pharynx, above the border of the soft palate. If these parts become enlarged and block the passage between the pharynx and the nasal cavity, they also may be removed surgically. (See fig. 13.6.)

1. What is the function of the mouth?
2. How does the tongue contribute to the function of the digestive system?
3. What is the role of the soft palate in swallowing?
4. Where are the tonsils located?

The Teeth

The teeth develop in sockets of the mandibular and maxillary bones. They are unique structures in that two sets of teeth form at different times. The members of the first set, the *primary* or **deciduous teeth,** usually erupt through the gums (gingiva) at regular intervals between the ages of six months and 2½ years. There are twenty of these decidous teeth—ten in each jaw—and they occur from the midline toward the sides in the following sequence: central incisor, lateral incisor, cuspid (canine), first molar, and second molar.

Fig. 13.7 (a) The permanent teeth of the upper jaw and (b) the lower jaw.

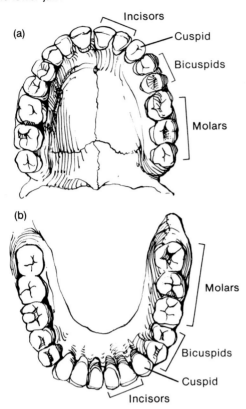

Fig. 13.8 A section of a typical tooth.

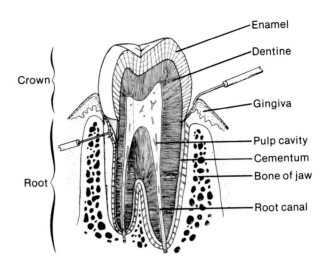

Chart 13.2 Primary and secondary teeth

Primary (deciduous) Teeth		Secondary (permanent) Teeth	
Type	Number	Type	Number
Incisor		Incisor	
central	4	central	4
lateral	4	lateral	4
Cuspid	4	Cuspid	4
		Bicuspid	
		first	4
		second	4
Molar		Molar	
first	4	first	4
second	4	second	4
		third	4
Total	20	Total	32

The deciduous teeth are usually shed in the same order they appeared. Before this happens, though, their roots are resorbed. Then the teeth are pushed out of their sockets by pressure from the developing *secondary* or *permanent* teeth. (See fig. 13.7.) This second set consists of thirty-two teeth—sixteen in each jaw—and they are arranged from the midline as follows: central incisor, lateral incisor, cuspid, first bicuspid (premolar), second bicuspid, first molar, second molar, and third molar. (See chart 13.2.)

The permanent teeth usually begin to appear about age six, but the set may not be completed until the third molars appear between seventeen and twenty-five years of age. Sometimes these molars, which are also called wisdom teeth, become wedged in abnormal positions within the jaws and fail to erupt. Such teeth are said to be *impacted*.

Teeth function to break pieces of food into smaller pieces. This action increases the surface area of the food particles and thus makes it possible for digestive enzymes to react more effectively with food molecules.

Different teeth are adapted to handle food in different ways. *Incisors* are chisel-shaped, and their sharp edges bite off relatively large pieces of food.

The *cuspids* (canines) are cone-shaped, and they are useful in grasping or tearing food. The *bicuspids* and *molars* have somewhat flattened surfaces and are specialized for grinding particles.

As fig. 13.8 shows, each tooth consists of two main portions—an exposed *crown* and a *root* that is anchored to the bone of the jaw. The crown of a tooth is covered by glossy, white *enamel* that is highly mineralized and is the hardest material in the body. Unfortunately, if enamel is damaged by abrasive action or injuries, it is not replaced. It also tends to wear away with age.

The root of a tooth is surrounded by a bonelike material called *cementum,* which fastens the tooth into its bony socket. The bulk of a tooth beneath the enamel is composed of *dentine,* a substance much like bone but somewhat harder. The dentine, in turn, surrounds the tooth's central cavity (pulp cavity), which contains blood vessels, nerves, and connective tissue (pulp). The blood vessels and nerves reach this cavity through tubular *root canals* that extend upward into the root.

Loss of teeth is most commonly associated with diseases of the gums and the dental pulp (endodontitis). Such diseases can usually be avoided by practicing good oral hygiene and obtaining regular dental treatment.

Dental caries (decay) involves decalcification of tooth enamel and is usually followed by destruction of the enamel and its underlying dentine. The result is a cavity, which must be cleaned and filled to prevent further erosion of the tooth.

Although the cause or causes of dental caries are not well understood, lack of dental cleanliness and a diet high in sugar and starch seem to promote the problem. Accumulations of food particles on the surfaces and between the teeth are thought to aid the growth of certain kinds of bacteria. These microorganisms apparently utilize carbohydrates in food particles and produce acid by-products. The acids then begin the process of destroying tooth enamel. (See fig. 13.9.)

Preventing dental caries requires brushing the teeth at least once a day, using dental floss or tape regularly to remove debris from between the teeth, and limiting the intake of sugar and starch, especially between meals. The use of fluoridated drinking water or the application of fluoride solution to children's teeth also helps prevent dental decay.

The mouth parts and their functions are summarized in chart 13.3.

Chart 13.3 Mouth parts and their digestive functions

Part	Location	Function
Cheeks	Form lateral walls of mouth	Hold food in mouth, muscles function in chewing
Lips	Surround mouth opening	Contain sensory receptors used to judge characteristics of foods
Tongue	Occupies floor of mouth	Aids in mixing food with saliva, moves food toward pharynx, contains taste receptors
Palate	Forms roof of mouth	Holds food in mouth, directs food to pharynx
Teeth	In sockets of mandibular and maxillary bones	Break food particles into smaller pieces, help mix food with saliva during chewing

Fig. 13.9 (*a*) A result of dental caries in the lower jaw and (*b*) the upper jaw. What factors seem to promote the development of caries?

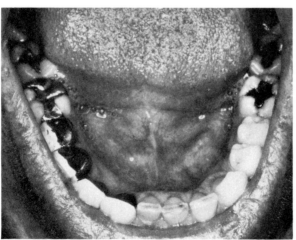

(a)

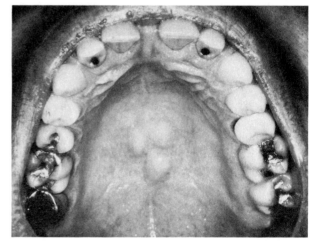

(b)

1. How do deciduous teeth differ from permanent teeth?
2. How are various types of teeth adapted to provide specialized functions?
3. Describe the structure of a tooth.
4. How can dental caries be prevented?

The Salivary Glands

The **salivary glands** function to secrete saliva. This fluid moistens food particles, helps bind them together, and begins the digestion of carbohydrates. Saliva also acts as a solvent, dissolving various food chemicals so they can be tasted, and it helps to cleanse the mouth and teeth. Bicarbonate ions (HCO_3^-) in saliva aid in regulating (buffering) its acid concentration so that its pH usually remains between 6.5 and 7.5. This is a favorable range for the action of the salivary enzyme and is important in protecting the teeth against dissolving in an excessively acid environment.

There are three major pairs of salivary glands—the parotid, submaxillary, and sublingual glands—and many minor ones that are associated with the mucous membrane of the tongue, palate, and cheeks. These glands are shown in figures 13.10 and 13.11.

Salivary Secretions

Within a salivary gland there are two types of secretory cells, called *serous* and *mucous* cells. They occur in varying proportions within different glands. The serous cells produce a watery fluid that contains a digestive enzyme, called **amylase,** that functions to split starch and glycogen molecules into disaccharides—the first step in the digestion of carbohydrates. Mucous cells secrete a thick, stringy liquid, called **mucus,** that binds food particles together and acts as a lubricant during swallowing.

Like other digestive structures, the salivary glands are innervated by branches of both sympathetic and parasympathetic nerves. Impulses arriving

Fig. 13.10 Locations of the major salivary glands.

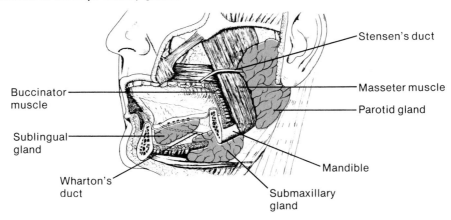

Fig. 13.11 Locations of some minor salivary glands.

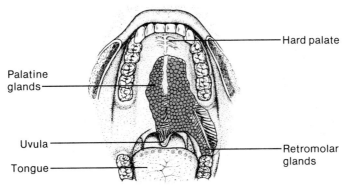

Chart 13.4 The major salivary glands

Gland	Location	Duct	Type of Secretion
Parotid glands	In front and somewhat below the ears, between the skin of the cheeks and the masseter muscles	Parotid ducts pass through the buccinator muscles and enter the mouth opposite the upper second molars	Clear, watery serous fluid rich in amylase
Submaxillary glands	In the floor of the mouth on the inside surface of the mandible	Ducts open beneath the tongue near the frenulum	Primarily serous fluid, but with some mucus; more viscous than parotid secretion
Sublingual glands	In the floor of the mouth beneath the tongue	Many separate ducts	Primarily thick, stringy mucus

on sympathetic fibers stimulate the gland cells to secrete a small quantity of viscous saliva. Parasympathetic impulses, on the other hand, elicit the secretion of a large volume of watery saliva. Such parasympathetic impulses are activated reflexly when a person sees, smells, tastes, or even thinks about pleasant foods. Conversely, if food looks, smells, or tastes unpleasant, parasympathetic activity is inhibited so that less saliva is produced, and swallowing may become difficult.

The Parotid Glands

The **parotid glands** are the largest of the salivary glands. One lies in front of and somewhat below each ear, between the skin of the cheek and the masseter muscle. A *parotid duct* (Stensen's duct) passes from the gland inward through the buccinator muscle, entering the mouth just opposite the upper second molar on either side of the jaw. These glands secrete a clear, watery fluid that is rich in **amylase.**

A person who has *mumps,* which is a virus infection, typically develops swellings in one or both of the parotid glands. Occasionally this disease also affects the submaxillary glands as well as the pancreas and the gonads.

The Submaxillary Glands

The **submaxillary** (submandibular) **glands** are located in the floor of the mouth on the inside surface of the jaw (mandible). The secretory cells of these glands are predominantly serous, although some mucous cells are present. Consequently, the submaxillary glands secrete a more viscous fluid than the parotid glands. The ducts of the submaxillary glands (Wharton's ducts) open under the tongue, near the frenulum.

The Sublingual Glands

The **sublingual glands** are the smallest of the major salivary glands. They are found on the floor of the mouth under the tongue. Their cells are primarily the mucous type, and as a result their secretions, which enter the mouth through many separate ducts, tend to be thick and stringy.

Chart 13.4 summarizes the characteristics of these glands.

1. Where are the salivary glands located?
2. What stimulates these glands to secrete saliva?
3. What is the function of saliva?

The Pharynx and Esophagus

The pharynx is a cavity behind the mouth from which the tubular esophagus leads to the stomach. Although neither of these organs contributes to the digestive process, they are important passageways, and their muscular walls function in swallowing.

Structure of the Pharynx

The **pharynx** connects the nasal and oral cavities with the larynx and esophagus (fig. 13.12). It can be divided into the following parts:

1. The **nasopharynx,** or upper portion, is located above the soft palate. It communicates with the nasal cavity and provides a passageway for air during breathing. The eustachian tubes, which connect the pharynx with the middle ears, open through the walls of the nasopharynx.

2. The **oropharynx,** or middle portion, is behind the mouth. It functions as a passageway for food moving downward from the mouth and for air moving to and from the nasal cavity.

Fig. 13.12 The pharynx connects the oral and nasal cavities with the larynx and esophagus.

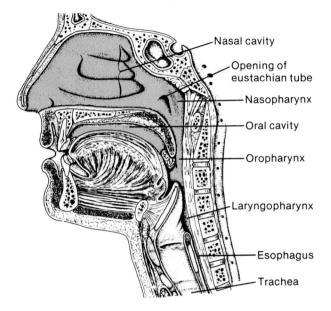

- Nasal cavity
- Opening of eustachian tube
- Nasopharynx
- Oral cavity
- Oropharynx
- Laryngopharynx
- Esophagus
- Trachea

Fig. 13.13 Muscles of the pharyngeal wall as viewed from behind.

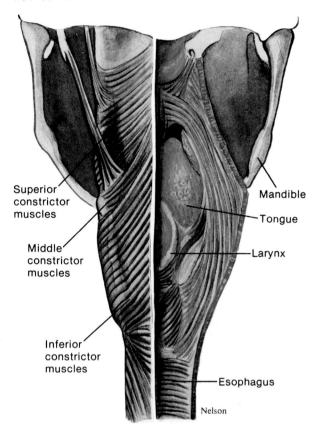

- Superior constrictor muscles
- Middle constrictor muscles
- Inferior constrictor muscles
- Mandible
- Tongue
- Larynx
- Esophagus

Nelson

3. The **laryngopharynx,** or lower portion, is just below the oropharynx. At its lower end it opens into the larynx and esophagus.

The muscles in the walls of the pharynx are arranged in inner circular and outer longitudinal groups. (See fig. 13.13.) The circular muscles, called *constrictors,* serve to pull the walls inward during swallowing. The *superior* constrictor muscles, which are attached to bony processes of the skull and mandible, curve around the upper part of the pharynx. The *middle* constrictor muscles arise from projections on the hyoid bone and fan around the middle of the pharynx. The *inferior* constrictor muscles originate from cartilage of the larynx and pass around the lower portion of the cavity. Some of the fibers in the lower part of the inferior constrictor muscles remain contracted most of the time, and this action prevents air from entering the esophagus during breathing.

Although the pharyngeal muscles are skeletal muscles, they generally are not under voluntary control. Instead, they function involuntarily in the swallowing reflex.

The Swallowing Mechanism

The act of swallowing can be divided into three stages. In the first, which is voluntary, food is chewed and mixed with saliva. Then, it is rolled into a mass (bolus) and forced into the pharynx by the tongue.

The second stage begins as the food reaches the pharynx and stimulates sensory receptors located around the pharyngeal opening. This triggers the swallowing reflex, illustrated in fig. 13.14, which includes the following actions:

1. The soft palate is raised, preventing food from entering the nasal cavity.

2. The hyoid bone and the larynx are elevated, so that food is less likely to enter the trachea.

3. The tongue is pressed against the soft palate, sealing off the oral cavity from the pharynx.

4. The longitudinal muscles in the pharyngeal wall contract, pulling the pharynx upward toward the food.

5. The lower portion of the inferior constrictor muscles relaxes, opening the esophagus.

6. The superior constrictor muscles contract, stimulating a peristaltic wave to begin in other pharyngeal muscles, and this wave forces the food into the esophagus.

During the third stage of swallowing, the food enters the esophagus and is transported to the stomach by peristalsis.

The Digestive System   423

Fig. 13.14 Steps in the swallowing reflex: (*a*) the tongue forces food into the pharynx; (*b*) the soft palate, hyoid bone, and larynx are raised; the tongue is pressed against the palate; and inferior constrictor muscles relax so that the esophagus opens; (*c*) superior constrictor muscles contract and force food into the esophagus; (*d*) peristaltic waves move food through the esophagus to the stomach.

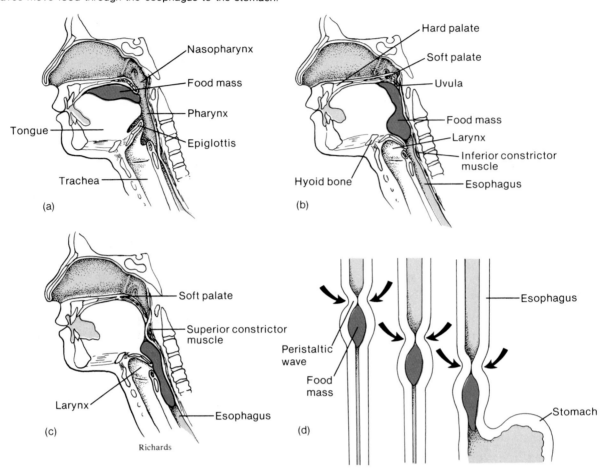

Richards

The Esophagus

The **esophagus** is a straight, collapsible tube about 25 cm (10 inches) long that functions as a passageway through the thorax. It begins at the base of the pharynx and descends behind the trachea, passing through the mediastinum. The esophagus penetrates the diaphragm through an opening, the *esophageal hiatus,* and is continuous with the stomach on the abdominal side of the diaphragm. (See figs. 13.15 and 13.16.)

Occasionally there is a weak place in the diaphragm due to a congenital defect or an injury. As a result of such weakness, a portion of the stomach, large intestine, or some other abdominal organ may protrude upward through the esophageal hiatus and into the thorax. This condition is called *hiatal hernia.*

Just above the point where the esophagus joins the stomach, some of the circular muscle fibers in its wall are thickened. These fibers are usually contracted and function to close the entrance to the stomach. In this way, they help prevent regurgitation of the stomach contents into the esophagus. If regurgitation does occur, gastric secretions from the stomach may irritate the esophageal wall, producing discomfort or pain. This condition is commonly called heartburn, because it seems to originate from behind the sternum in the region of the heart.

When peristaltic waves reach the stomach, the muscle fibers that guard its entrance relax and allow the food to enter.

1. *Describe the regions of the pharynx.*
2. *List the major events that occur during swallowing.*
3. *What is the function of the esophagus?*

Fig. 13.15 The esophagus functions as a passageway between the pharynx and the stomach.

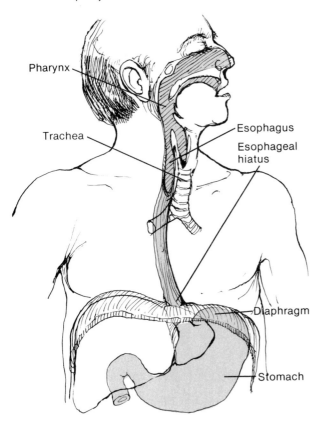

The Stomach

The stomach is a J-shaped, pouchlike organ that lies under the diaphragm in the upper left portion of the abdominal cavity. It functions to receive food from the esophagus, mix it with gastric juice, initiate the digestion of proteins, carry on a limited amount of absorption, and move food into the small intestine.

In addition to the two layers of smooth muscle—an inner circular and an outer longitudinal layer—found in other regions of the alimentary canal, some parts of the stomach have another inner layer of oblique fibers. This third muscular layer is most highly developed near the opening of the esophagus and in the body of the stomach.

Parts of the Stomach

The stomach, shown in figures 13.17 and 13.18, can be divided into cardiac, fundic, body, and pyloric regions. The *cardiac region* is a small area near the esophageal opening. The *fundic region,* which balloons above the cardiac portion, acts as a temporary storage area and sometimes becomes filled with swallowed air. The dilated *body region* is the main part of the stomach and is located between the fundic and pyloric portions, while the *pyloric region* narrows and becomes the *pyloric canal* as it approaches the junction with the small intestine.

At the end of the pyloric canal, the muscular wall is thickened, forming a powerful circular muscle called the **pyloric sphincter** (pylorus). This muscle serves as a valve that prevents regurgitation of food from the intestine back into the stomach.

Gastric Secretions

The mucous membrane that forms the inner lining of the stomach is relatively thick, and its surface is studded with many small openings. These openings, called *gastric pits,* are located at the ends of tubular **gastric glands** (fig. 13.19). Although their structure and the composition of their secretions vary in different parts of the stomach, gastric glands generally contain three types of secretory cells. One type, the *mucous cell,* occurs in the necks of the glands near the openings of the gastric pits. The other types, *chief cells* and *parietal cells,* are found in the deeper parts of the glands. (See figs. 13.19 and 13.20.) The chief cells secrete *digestive enzymes,* and the parietal cells release *hydrochloric acid.* The products of the mucous cells, chief cells, and parietal cells together form **gastric juice.**

Fig. 13.16 This cross section of the esophagus shows its muscular wall, indicated by the arrow.

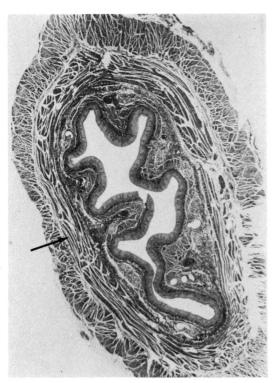

Fig. 13.17 Major regions of the stomach.

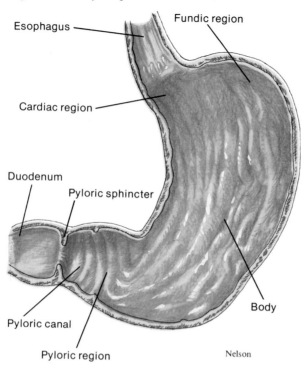

Esophagus

Fundic region

Cardiac region

Duodenum

Pyloric sphincter

Pyloric canal

Pyloric region

Body

Nelson

Fig. 13.18 X ray of a normal stomach.

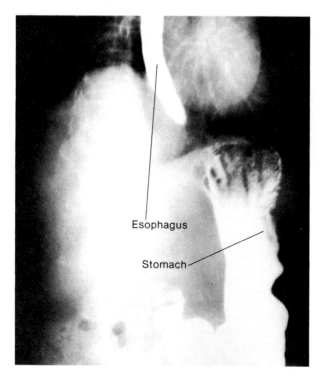

Esophagus

Stomach

Fig. 13.19 (a) The mucosa of the stomach is studded with gastric pits that are the openings of the gastric glands; (b) gastric glands include mucous cells, parietal cells, and chief cells, each type producing a different secretion.

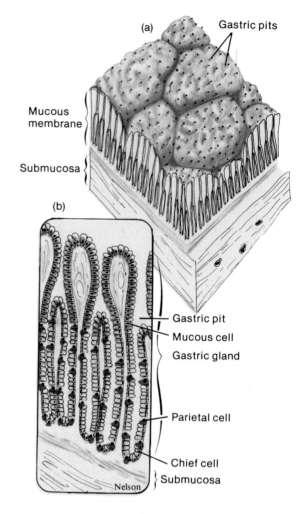

(a)

Gastric pits

Mucous membrane

Submucosa

(b)

Gastric pit

Mucous cell

Gastric gland

Parietal cell

Chief cell

Submucosa

Nelson

Fig. 13.20 A photomicrograph of the gastric mucosa.

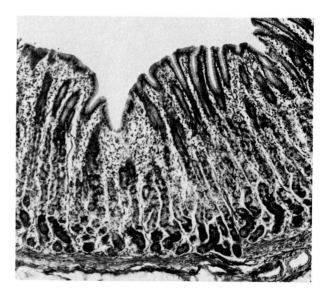

Chart 13.5 Composition of gastric juice

Component	Source	Function
Pepsinogen	Chief cells of the gastric glands	An inactive form of pepsin
Pepsin	Formed from pepsinogen in the presence of hydrochloric acid	A protein-splitting enzyme capable of digesting nearly all types of protein
Hydrochloric acid	Parietal cells of the gastric glands	Provides acid environment, needed for the conversion of pepsinogen into pepsin
Mucus	Goblet cells and mucous glands	Provides viscous, alkaline protective layer on the stomach wall
Intrinsic factor	Parietal cells of the gastric glands	Aids the absorption of vitamin B_{12}

Although gastric juice contains several digestive enzymes, **pepsin** is by far the most important. It is secreted by the chief cells as an inactive, nonerosive substance called **pepsinogen.** When pepsinogen contacts gastric juice, however, it is changed rapidly into pepsin.

Pepsin is a protein-splitting enzyme capable of beginning the digestion of nearly all types of protein in human diets. This enzyme is most active in an acid environment; the hydrochloric acid in gastric juice seems to function mainly to provide such an environment.

The mucous cells of the gastric glands secrete large quantities of mucus. In addition, the cells of the mucous membrane between these glands release a more viscous and alkaline secretion that is thought to form a protective coating on the inside of the stomach wall. This coating is especially important because pepsin is capable of digesting the proteins of the stomach tissues as well as those in foods. Thus, the coating normally prevents the stomach from digesting itself.

Still another component of gastric juice is **intrinsic factor.** This substance, which is secreted by the parietal cells of the gastric glands, aids in absorption of vitamin B_{12} from the small intestine, as is explained in chapter 14.

The substances in gastric juice are summarized in chart 13.5.

1. Where is the stomach located?
2. What is secreted by the chief cells of the gastric glands? By the parietal cells?
3. What is the most important digestive action of gastric juice?
4. How is the stomach prevented from digesting itself?

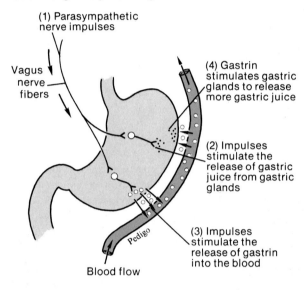

Fig. 13.21 The secretion of gastric juice is regulated in part by parasympathetic nerve impulses that stimulate the release of gastric juice and gastrin.

(1) Parasympathetic nerve impulses

Vagus nerve fibers

(4) Gastrin stimulates gastric glands to release more gastric juice

(2) Impulses stimulate the release of gastric juice from gastric glands

(3) Impulses stimulate the release of gastrin into the blood

Blood flow

Regulation of Gastric Secretions. Although gastric juice is produced continuously, the rate of production varies considerably from time to time and is under the control of neural and hormonal mechanisms. More specifically, parasympathetic impulses arriving on vagus nerve fibers stimulate gastric glands to secrete large amounts of gastric juice that is rich in hydrochloric acid and pepsin. These impulses also stimulate certain stomach cells to release a hormone, called **gastrin,** that causes the gastric glands to increase their secretory activity. (See fig. 13.21.) On the other hand, sympathetic impulses bring about a decrease in gastric gland activity.

Chart 13.6 Phases of gastric secretion

Phase	Action
Cephalic phase	Parasympathetic reflexes are triggered by sight, taste, smell, or thought of food; gastric juice is secreted in response
Gastric phase	Food in stomach chemically and mechanically stimulates the release of gastrin, which in turn stimulates the secretion of gastric juice; reflex responses also stimulate gastric juice secretion
Intestinal phase	As food enters the small intestine, it stimulates intestinal cells to release intestinal gastrin, which in turn promotes the secretion of gastric juice from the stomach wall

As a result of the ways these factors interact, three stages of gastric secretion can be recognized. They are called the cephalic, gastric, and intestinal phases. (See chart 13.6.)

The *cephalic phase* of gastric secretion begins before any food reaches the stomach and may begin before any food is eaten. In this stage, gastric secretions are stimulated by parasympathetic reflexes whenever a person tastes, smells, sees, or even thinks about food. Furthermore, the hungrier the person is, the greater the amount of gastric secretion.

The *gastric phase* starts when food enters the stomach. The presence of food triggers the release of gastrin from the stomach wall, and the gastrin in turn stimulates the production of still more gastric juice. Actually, the release of gastrin is promoted by several factors, including parasympathetic impulses, distension of the stomach wall, and the presence of certain food substances (secretagogues) in the stomach. Such chemicals as meat extracts, partially digested proteins, various spices, caffeine, and alcohol, for example, are especially stimulating to gastrin release. Food in the stomach also elicits additional parasympathetic reflexes that promote the secretion of still more gastric juice.

The *intestinal phase* begins when food leaves the stomach and enters the small intestine. When the food first contacts the intestinal wall, it apparently stimulates intestinal cells to release a hormone that again enhances gastric gland secretions. Although the actual nature of this hormone is unknown, some investigators believe it is identical to gastrin, and they call it *intestinal gastrin*.

As more food moves into the small intestine, however, the secretion of gastric juice from the stomach wall is inhibited. Apparently this inhibition involves a sympathetic reflex triggered by acid substances in the upper part of the small intestine. Also the presence of fats in this region causes the release of a hormone, **enterogastrone,** from the intestinal wall. These two actions bring about a decrease in gastric secretion and gastric motility as the small intestine fills with food.

---

1. How is the secretion of gastric juice controlled?
2. Distinguish between the cephalic, gastric, and intestinal phases of gastric secretion.
3. What is the function of enterogastrone?

Disorders in Gastric Secretions. Normally when the stomach is empty, only small quantities of gastric juice are released. However, occasionally a person's stomach secretes gastric juice abnormally, and an ulcer may develop.

An *ulcer* is an open sore in the mucous membrane resulting from a localized breakdown of the tissues. Although ulcers may occur in various parts of the alimentary canal, they often occur in the stomach. Such *gastric ulcers* are most likely to develop in the lesser curvature, which is the concave margin of the stomach near the liver. Ulcers are also common in the first portion of the small intestine, the duodenum. These *duodenal ulcers* only occur in regions exposed to gastric juice as the contents of the stomach enter the intestine. Thus, both types of ulcers are caused by the digestive action of pepsin, and are, therefore, known as *peptic ulcers.* (See fig. 13.22.)

The cause of peptic ulcers is poorly understood. It is believed that they may result from failure of mechanisms that normally protect the stomach lining from being digested or from excessive secretion of gastric juice that overwhelms the lining's mucus coating.

Persons who are emotionally stressed for prolonged periods seem susceptible to developing peptic ulcers. Apparently, their nervous systems initiate sympathetic impulses that inhibit the secretion of protective mucus or somehow stimulate excessive gastric secretions at inappropriate times.

Fortunately for ulcer victims, the cells of the mucous membrane are able to reproduce rapidly. In fact the entire lining of the stomach is thought to be replaced every few days. Peptic ulcer treatment attempts to eliminate the factors that aggravate the condition and gives the damaged tissue a chance to heal.

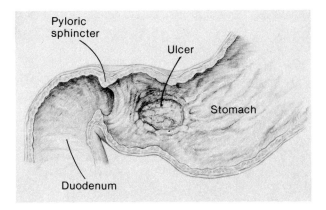

Pyloric sphincter

Ulcer

Stomach

Duodenum

Typically such treatment includes (1) removing sources of emotional stress; (2) avoiding alcohol, caffeine, and highly spiced foods that are likely to stimulate gastrin secretion, as well as avoiding specific foods that the person finds distressing; (3) providing frequent feedings so that food is usually present in the stomach; (4) using drugs such as tranquilizers that may calm a stressed nervous system and antacids that can neutralize the acid of gastric juice.

If such measures are unsuccessful, an ulcer patient may be treated surgically. For example, a portion of the stomach may be removed so that gastric secretions are greatly reduced, or branches of the vagus nerves may be cut, preventing parasympathetic impulses from reaching the gastric glands.

Since the stomach is the source of the intrinsic factor necessary for the absorption of vitamin B_{12}, persons who have their stomachs removed often develop a vitamin B_{12} deficiency disease called *pernicious anemia*. This condition is treated with injections of vitamin B_{12}.

Gastric Absorption

Although gastric enzymes begin the breakdown of proteins, the stomach wall is not well adapted to carry on the absorption of digestive products. However, small quantities of water, glucose, certain salts, alcohol, and various lipid-soluble drugs may be absorbed by the stomach.

1. *Describe an ulcer.*
2. *What seems to be the cause of peptic ulcers?*
3. *How are peptic ulcers treated?*
4. *What substances may be absorbed in the stomach?*

Filling and Emptying Actions

As more and more food enters the stomach, the smooth muscles in its wall become stretched. Although the stomach may enlarge, its muscles maintain their tone, and internal pressure normally remains unchanged. A person can eat more than the stomach can comfortably hold, and when this happens the internal pressure may rise enough so that pain receptors are stimulated. The result is a stomachache.

Following a meal, the mixing movements of the stomach wall aid in producing a semifluid paste of food particles and gastric juice called **chyme.** Peristaltic waves push the chyme toward the pyloric region of the stomach, and as it accumulates near the pyloric sphincter, this muscle begins to relax, allowing the chyme to move a little at a time into the small intestine. This process is illustrated in figure 13.23.

The rate at which the stomach empties depends on several factors, including the fluidity of the chyme and the type of food present. Liquids usually pass through the stomach quite rapidly, while solids remain until they are well mixed with gastric juice. Fatty foods may remain in the stomach from three to six hours; foods high in proteins tend to be moved through more quickly; and carbohydrates usually pass through more rapidly than either fats or proteins.

The rate at which the stomach empties is also related to age. An infant's stomach, which has a small capacity, empties very rapidly. As the person grows, the stomach enlarges, and the emptying time increases.

When the duodenum becomes filled with chyme, its internal pressure increases, and the intestinal wall is stretched. This action stimulates sensory receptors in the wall, and an **enterogastric reflex** is triggered. The name of this reflex, like those of other digestive reflexes, describes the origin and termination of reflex impulses. Thus, the enterogastric reflex begins in the small intestine (*entero*) and ends in the stomach (*gaster*).

Fig. 13.23 (a) As the stomach fills, its muscular wall becomes stretched, but the pyloric sphincter remains closed; (b) mixing movements serve to mix food and gastric juice, creating chyme; (c) peristaltic waves move the chyme toward the pyloric sphincter, which relaxes and allows some chyme to enter the duodenum.

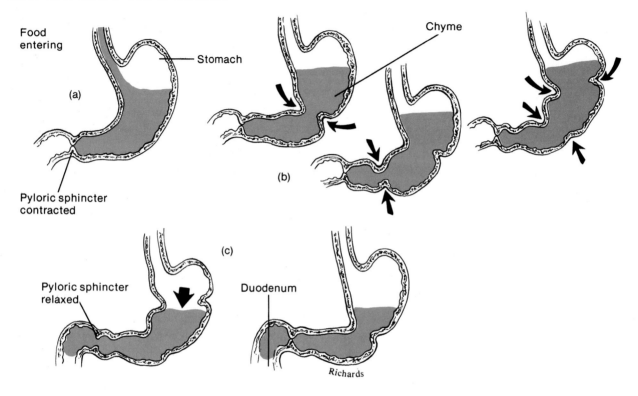

As a result of the enterogastric reflex, sympathetic impulses travel to the stomach, and peristaltic waves are inhibited. Consequently, the intestine is filled less rapidly. Also, if the chyme entering the intestine has a high fat content, the hormone enterogastrone is released from the intestinal wall, and peristalsis is inhibited even more. (See fig. 13.24.)

As the stomach contents enter the duodenum, accessory organs add their secretions to the chyme. These organs include the pancreas, liver, and gallbladder.

Vomiting results from a complex reflex that empties the stomach another way. This action is usually triggered by irritation or distension in some part of the alimentary canal, such as the stomach or intestines. Sensory impulses travel from the site of stimulation to the *vomiting center* in the medulla oblongata, and a number of motor responses follow. These include taking a deep breath, raising the soft palate and thus closing the nasal cavity, closing the opening to the trachea (glottis), relaxing the sphincterlike muscles at the base of the esophagus, contracting the diaphragm so it moves downward over the stomach, and contracting the abdominal wall

Fig. 13.24 The rate at which chyme leaves the stomach is regulated in part by the enterogastric reflex.

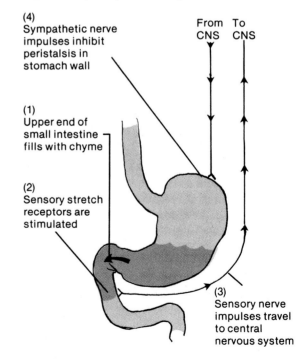

(4) Sympathetic nerve impulses inhibit peristalsis in stomach wall

From CNS   To CNS

(1) Upper end of small intestine fills with chyme

(2) Sensory stretch receptors are stimulated

(3) Sensory nerve impulses travel to central nervous system

muscles so the pressure inside the abdominal cavity is increased. As a result, the stomach is squeezed from all sides, and its contents are forced upward and out through the esophagus, pharynx, and mouth.

Activity in the vomiting center can be stimulated by various drugs (emetics), by toxins in contaminated foods, and sometimes by rapid changes in body motion. In the last situation, sensory impulses from the labyrinths of the inner ears apparently reach the vomiting center, and in some people this produces motion sickness.

The vomiting center also can be activated by stimulation of higher brain centers through sights, sounds, odors, tastes, emotional feelings, or by mechanical stimulation of the back of the pharynx.

The feeling called *nausea* seems to be due to activity in the vomiting center or in nerve centers closely associated with it.

---

1. *How is chyme produced?*
2. *What factors influence the speed with which chyme leaves the stomach?*
3. *Describe the enterogastric reflex.*
4. *Describe the vomiting reflex.*
5. *What factors may stimulate the vomiting reflex?*

The Pancreas

The shape of the **pancreas** and its general location are described in chapter 12, as are its endocrine functions. The pancreas also has an exocrine function—the secretion of digestive juice—that will be discussed here.

Structure of the Pancreas

The pancreas is closely asssociated with the small intestine and is located behind the parietal peritoneum. It extends horizontally across the posterior abdominal wall with its head in the C-shaped curve of the duodenum and its tail against the spleen. (See fig. 13.25.)

The cells that produce pancreatic juice make up the bulk of the pancreas. These cells are clustered around tiny tubes, into which they release their secretions. The smaller tubes unite to form larger ones, which in turn give rise to a *pancreatic duct* extending the length of the pancreas. This duct usually connects with the duodenum at the same place where the bile duct from the liver and gallbladder joins the duodenum. (See fig. 13.26.)

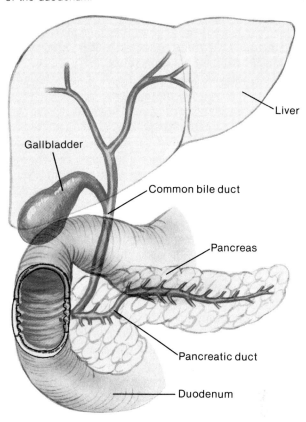

Fig. 13.25 The pancreas is located in a C-shaped curve of the duodenum.

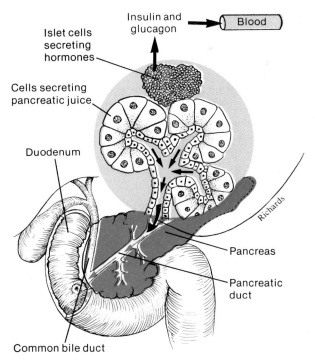

Fig. 13.26 The pancreas is composed largely of cells that secrete pancreatic juice into ducts, but it also contains islet cells that secrete hormones into the blood.

The Digestive System   431

Pancreatic Juice

Pancreatic juice contains enzymes capable of digesting carbohydrates, fats, proteins, and nucleic acids.

The carbohydrate-digesting enzyme is called *pancreatic amylase.* It splits molecules of starch or glycogen into double sugars (disaccharides); the fat-digesting enzyme, *pancreatic lipase,* breaks fat molecules into fatty acids and glycerol.

The protein-splitting enzymes are **trypsin, chymotrypsin,** and **carboxypeptidase.** Each of these acts to split the bonds between particular combinations of amino acids in proteins. Since no single enzyme can split all possible combinations, the presence of several enzymes is necessary for the complete digestion of protein molecules.

These protein-splitting enzymes are stored in inactive forms within tiny cellular structures called *zymogen granules.* They are also secreted in the inactive form and must be activated by other enzymes after they reach the small intestine. For example, the pancreatic cells release inactive *trypsinogen.* This substance becomes active trypsin when it contacts an enzyme called *enterokinase,* which is secreted by the mucosa of the small intestine. Chymotrypsin and carboxypeptidase are activated, in turn, by the presence of trypsin. This mechanism prevents the enzymatic digestion of proteins within the secreting cells and their ducts.

If something happens to block the release of pancreatic juice, it may accumulate in the duct system, and the trypsinogen may become activated. As a result, portions of the pancreas may become digested, causing a condition called *acute pancreatitis.*

In addition to the above enzymes, pancreatic juice contains two **nucleases** that break down nucleic acid molecules into nucleotides. Pancreatic juice also contains a high concentration of bicarbonate ions that makes it alkaline. This provides a favorable environment for the actions of the digestive enzymes and helps to neutralize the acidic chyme as it arrives from the stomach. At the same time, the alkaline condition in the small intestine blocks the action of pepsin, which might otherwise damage the duodenal wall.

Regulation of Pancreatic Secretion. As in the case of gastric and small intestinal secretions, the release of pancreatic juice is regulated by nerve actions as well as hormones. For example, during the cephalic and gastric phases of gastric secretion, parasympathetic impulses reach the pancreas and stimulate it

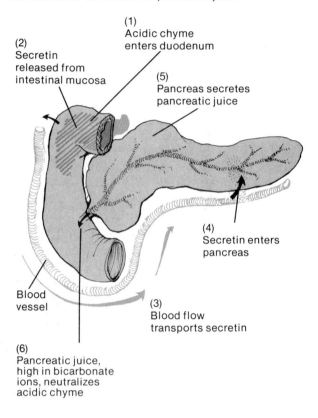

Fig. 13.27 Acidic chyme entering the duodenum from the stomach stimulates the release of secretin, which in turn stimulates the release of pancreatic juice.

(1) Acidic chyme enters duodenum

(2) Secretin released from intestinal mucosa

(5) Pancreas secretes pancreatic juice

(4) Secretin enters pancreas

(3) Blood flow transports secretin

Blood vessel

(6) Pancreatic juice, high in bicarbonate ions, neutralizes acidic chyme

to begin releasing digestive enzymes. Also, when acidic chyme enters the duodenum, the pancreas is stimulated to secrete a large quantity of fluid. This time, however, the stimulus comes from a hormone, **secretin,** which is released into the blood from the duodenal mucous membrane in response to the acid in chyme. The pancreatic juice secreted at this time contains few if any digestive enzymes. Instead, it has a high concentration of bicarbonate ions that function to neutralize the acid. (See fig. 13.27.)

The presence of chyme in the duodenum also stimulates the release of a hormone called **cholecystokinin** (cholecystokinin-pancreozymin) from the intestinal wall. As before, this hormone reaches the pancreas by way of the blood. However, it causes the secretion of pancreatic juice with a high concentration of digestive enzymes. The substances in chyme that seem most effective in eliciting the release of cholecystokinin are partially digested proteins.

---

1. Where is the pancreas located?
2. List the enzymes found in pancreatic juice.
3. What are the functions of these enzymes?
4. How is the secretion of pancreatic juice regulated?

Fig. 13.28 The liver is located in the upper right portion of the abdominal cavity and is partially surrounded by ribs.

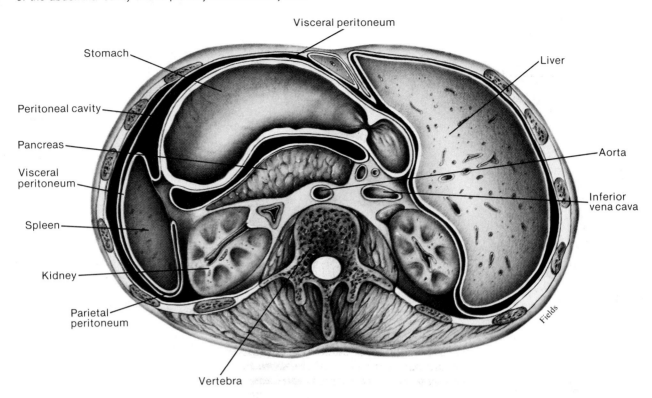

The Liver

The liver is located in the upper right portion of the abdominal cavity, just below the diaphragm. It is partially surrounded by the ribs and extends from the level of the fifth intercostal space to the lower margin of the ribs. It has a reddish brown color and is well supplied with blood vessels. (See figs. 13.28 and 13.29.)

Functions of the Liver

The liver is the largest gland in the body and has many vital functions. These include a number of metabolic activities, such as the conversion of glucose to glycogen; the formation of glucose from noncarbohydrates; the oxidation of fatty acids; the synthesis of lipoproteins, phospholipids, and cholesterol; the conversion of carbohydrates and proteins into fats; the deamination of amino acids; the formation of urea; the synthesis of many blood proteins; and the conversion of amino acids of one type into another type. The liver stores vitamins A, D, and B_{12}, as well as iron. It helps destroy red blood cells and foreign substances in the blood by phagocytosis. It alters the composition of toxic substances (detoxification) in body fluids, and secretes bile.

Since many of these functions are not directly related to the digestive system, they are discussed in other chapters. Bile secretion, however, is important to digestion and is explained in the following section.

Structure of the Liver

The liver is enclosed in a fibrous capsule and is divided by connective tissue into *lobes*—a large right lobe and a smaller left lobe. As figure 13.28 shows, the right lobe is further subdivided into the *right lobe proper,* the *quadrate lobe,* and the *caudate lobe.* Each lobe is separated into numerous tiny **hepatic lobules,** which are the functional units of the gland. (See fig. 13.30.) A lobule consists of numerous hepatic cells that radiate outward from a *central vein.* Platelike groups of these cells are separated from each other by vascular channels called **hepatic sinusoids.** Blood from the digestive tract that is carried in portal veins (see chapter 17) brings newly absorbed nutrients into the sinusoids and nourishes the hepatic cells. The blood then passes into the central vein and moves out of the liver.

Within the liver lobules, there are many fine *bile canals* that receive secretions from the hepatic cells. The canals of neighboring lobules unite to form larger ducts, and these converge to become the **hepatic ducts.**

Fig. 13.29 (a) Lobes of the liver as viewed from the front and (b) from below.

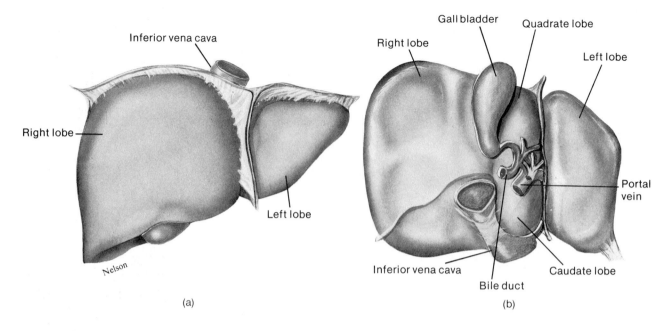

Inferior vena cava

Right lobe

Left lobe

Nelson

(a)

Gall bladder   Quadrate lobe

Right lobe   Left lobe

Portal vein

Inferior vena cava   Caudate lobe

Bile duct

(b)

Fig. 13.30 A cross section of a hepatic lobule, which is the functional unit of the liver.

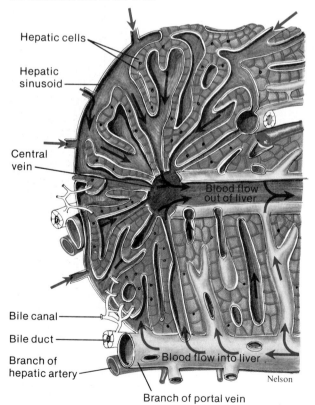

Hepatic cells

Hepatic sinusoid

Central vein

Blood flow out of liver

Bile canal

Bile duct

Branch of hepatic artery

Blood flow into liver

Nelson

Branch of portal vein

1. Describe the location of the liver.
2. Review the functions of the liver.
3. What liver function is directly related to digestion?
4. Describe an hepatic lobule.

Composition of Bile

Bile is a yellowish green liquid that is secreted continuously by hepatic cells. In addition to water, it contains *bile salts, bile pigments,* cholesterol, and various electrolytes. Of these, the bile salts are the most abundant, and they are the only substances in bile that have a digestive function.

Hepatic cells use cholesterol in the production of bile salts and release some cholesterol into the bile in the process of secreting these salts. The cholesterol apparently has no special function in bile or in the alimentary canal.

The bile pigments, *bilirubin* and *biliverdin,* are products of red blood cell breakdown and are normally excreted in the bile. (See chapter 16.) If their excretion is prevented due to obstruction of ducts, they tend to accumulate in the blood and tissues, causing a yellowish tinge in the skin and other body parts. This condition is called *obstructive jaundice.*

Fig. 13.31 The gallbladder and the ducts associated with it function to store bile and transport bile to the duodenum.

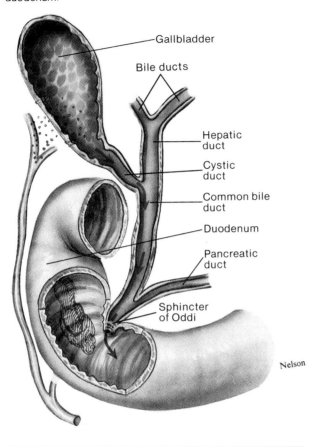

Gallbladder

Bile ducts

Hepatic duct

Cystic duct

Common bile duct

Duodenum

Pancreatic duct

Sphincter of Oddi

Nelson

The skin begins to appear yellow when the concentration of bile pigments in body fluids reaches a level about three times normal. In addition to being caused by obstructions in ducts, jaundice may occur with liver diseases in which cells are unable to secrete ordinary amounts of pigment, or with excessive destruction of red blood cells accompanied by a rapid release of pigments.

The Gallbladder and Its Functions

The **gallbladder** is a pear-shaped sac attached to the ventral surface of the liver by the **cystic duct,** which in turn joins the hepatic duct. It is lined with columnar epithelial cells and has a strong muscular layer in its wall. The gallbladder stores bile between meals, concentrates bile by reabsorbing water, and releases bile when stimulated by a hormone from the small intestine. (See fig. 13.31.)

The **common bile duct** is formed by the union of the hepatic and cystic ducts. It leads to the duodenum, where its exit is guarded by a sphincter muscle (sphincter of Oddi). This sphincter normally remains contracted, so that bile collects in the duct

Fig. 13.32 X ray of a gallbladder that contains gallstones.

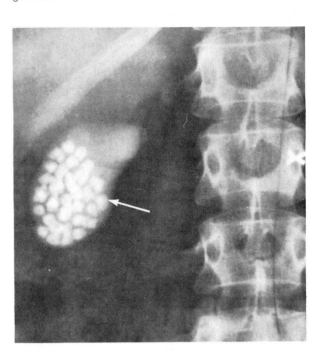

and backs up to the cystic duct. When this happens, the bile flows into the gallbladder and is stored there.

While the bile is in the gallbladder, its composition is altered because the lining reabsorbs some of the water and electrolytes. As these substances are removed, the bile salts, bile pigments, and cholesterol become increasingly concentrated. Although the cholesterol normally remains in solution, under certain conditions it may precipitate and form solid crystals. If cholesterol continues to come out of solution, these crystals become larger and larger. This may happen, for instance, if the bile is concentrated excessively, if the hepatic cells secrete too much cholesterol, or if there is an inflammation in the gallbladder. The resulting solids are called *gallstones,* and if they get into the bile duct they may block the flow of bile into the small intestine and also cause considerable pain. The X ray in figure 13.32 shows such gallstones.

Generally gallstones that cause obstructions are surgically removed. At the same time, the gallbladder is removed by the surgical procedure called *cholecystectomy.* Consequently, the person is unable to produce gallstones or store bile. Following surgery, bile continues to reach the intestine by means of the hepatic and common bile ducts.

Fig. 13.33 The gallbladder is stimulated to release bile when fat-containing chyme enters the duodenum.

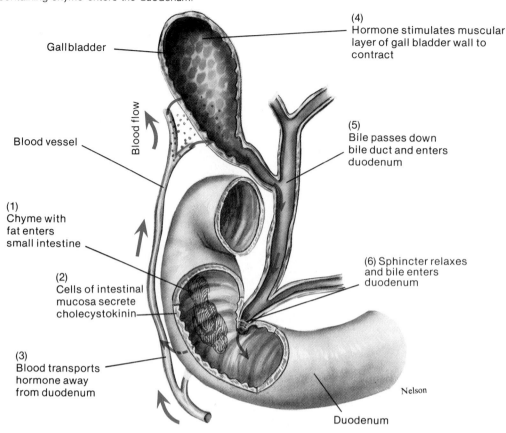

Gallbladder

Blood vessel

Blood flow

(1)
Chyme with
fat enters
small intestine

(2)
Cells of intestinal
mucosa secrete
cholecystokinin

(3)
Blood transports
hormone away
from duodenum

(4)
Hormone stimulates muscular
layer of gall bladder wall to
contract

(5)
Bile passes down
bile duct and enters
duodenum

(6) Sphincter relaxes
and bile enters
duodenum

Nelson

Duodenum

Chart 13.7   Hormones of the digestive tract

Hormone	Source	Function
Gastrin	Gastric cells, in response to the presence of food	Causes gastric glands to increase their secretory activity
Intestinal gastrin	Cells of small intestine, in response to the presence of chyme	Causes gastric glands to increase their secretory activity
Enterogastrone	Intestinal wall cells, in response to the presence of fats in the small intestine	Causes gastric glands to decrease their secretory activity and inhibits gastric motility
Secretin	Cells in the duodenal wall, in response to chyme entering the small intestine	Stimulates pancreas to secrete fluid with a high bicarbonate ion concentration
Cholecystokinin	Cells in the duodenal wall, in response to chyme entering the small intestine	Stimulates pancreas to secrete fluid with a high digestive enzyme concentration; stimulates gallbladder to contract and release bile

Regulation of Bile Release

Normally bile does not enter the duodenum until the gallbladder is stimulated to contract. The usual stimulus for such contraction involves the presence of fat in the small intestine, which triggers the release of the hormone, cholecystokinin, from the mucosa of the small intestine. The sphincter at the base of the common bile duct usually remains contracted until a peristaltic wave in the duodenal wall passes by. Then the sphincter relaxes slightly, and a squirt of bile enters the small intestine. (See fig. 13.33.)

The hormones that help control digestive functions are summarized in chart 13.7.

Digestive Functions of Bile Salts

The bile salts are the only components of bile with a digestive function. Although they do not act as digestive enzymes, these salts aid the actions of enzymes and enhance the absorption of fatty acids and certain fat-soluble vitamins.

Molecules of fats tend to clump together, forming masses called *fat globules*. Bile salts affect fat

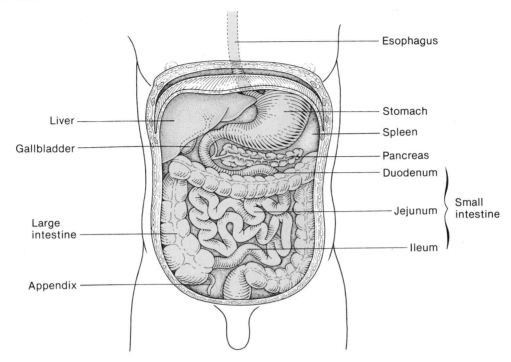

globules much as a soap or detergent does. They cause fat globules to break up into smaller droplets, an action called **emulsification.** As a result, the total surface area of the fatty substance is greatly increased, and the tiny droplets mix with water. The fat-splitting enzymes can then act on the fat molecules more effectively.

Bile salts aid in the absorption of fatty acids and cholesterol by forming complexes (micelles) that are very soluble in chyme and more easily absorbed by epithelial cells. Along with these lipids, various fat-soluble vitamins, such as vitamins A, D, E, and K, also are absorbed. Thus, if bile salts are lacking, lipids may be poorly absorbed and the person is likely to develop vitamin deficiencies.

Nearly all the bile salts are reabsorbed by the mucous membrane of the small intestine along with the fatty acids. They enter the blood and are carried to the liver where they are resecreted into the bile ducts by the hepatic cells. The small quantities that are lost in the feces are replaced by synthesis in liver cells.

---

1. *Explain how bile originates.*
2. *Describe the function of the gallbladder.*
3. *How is the secretion of bile regulated?*
4. *How does bile function in digestion?*

The Small Intestine

The small intestine is a tubular organ that extends from the pyloric sphincter to the beginning of the large intestine. With its many loops and coils, it fills much of the abdominal cavity. Although it is 5.5–6.0 meters (18–20 feet) long in a cadaver where the muscular wall is relaxed, the small intestine may be only half this long in a living person.

As was mentioned, this portion of the alimentary canal receives secretions from the pancreas and liver. It also completes the digestion of the nutrients in chyme, absorbs the various products of digestion, and transports the remaining residues to the large intestine.

Parts of the Small Intestine

The small intestine, shown in figures 13.34 and 13.35, consists of three portions, the duodenum, jejunum, and ileum.

The **duodenum,** which is about 25 centimeters (10 inches) long and 5 centimeters (2 inches) in diameter, lies behind the parietal peritoneum and is the most fixed portion of the small intestine. It follows a C-shaped path as it passes in front of the right kidney and the upper three lumbar vertebrae.

Fig. 13.35 X ray of a normal small intestine.

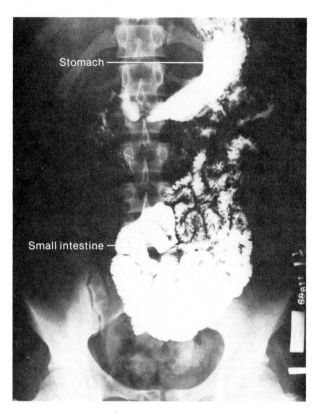

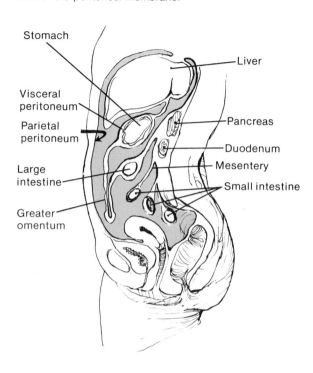

The next portion of the small intestine is the **jejunum,** and the remainder is the **ileum.** These portions are suspended from the posterior abdominal wall by a double-layered fold of peritoneum called **mesentery** (fig. 13.36). This supporting tissue contains the blood vessels, nerves, and lymphatic vessels that supply the intestinal wall.

A filmy fold of peritoneal membrane called the *greater omentum* drapes like an apron from the stomach over the transverse colon and the folds of the small intestine. If infections occur in the wall of the alimentary canal, cells from the omentum may adhere to the inflamed region and help to wall it off so that the infection is less likely to enter the peritoneal cavity.

Although there is no distinct separation between the jejunum and ileum, the diameter of the jejunum tends to be greater, and its wall is thicker, more vascular, and more active than that of the ileum.

Structure of the Small Intestinal Wall

Throughout its length, the inner wall of the small intestine has a velvety appearance. This is due to the presence of innumerable tiny projections of mucous membrane called **intestinal villi** (figs. 13.37 and 13.38). These structures project into the passageway or **lumen** of the alimentary canal, where they come in contact with the intestinal contents. They play important roles in mixing chyme with intestinal juice and in the absorption of digestive products.

Each villus consists of a layer of simple columnar epithelium and a core of connective tissue containing blood capillaries, a lymphatic capillary or **lacteal,** and nerve fibers. At their free surfaces, the epithelial cells possess many fine extensions, called *microvilli,* that create a brushlike border and greatly increase the surface area of the cells (fig. 13.39). The presence of microvilli enhances the process of absorption.

The blood and lymph capillaries function to carry away substances absorbed by the villus, while the nerve fibers act to stimulate or inhibit its activities.

Between the bases of the villi are tubular **intestinal glands** (crypts of Lieberkühn) that extend downward into the mucous membrane. The deeper layers

Fig. 13.37 Tiny projections, called villi, extend from the mucosa into the lumen of the small intestine. What are the functions of these projections?

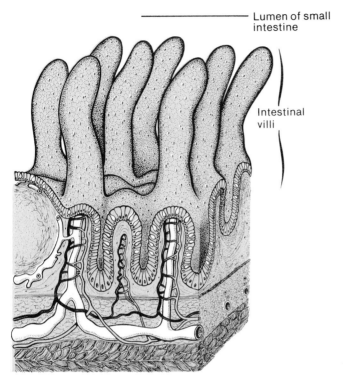

Lumen of small intestine

Intestinal villi

Simple columnar epithelium

Lacteal

Villus

Capillary network

Intestinal gland

Goblet cells

Arteriole

Venule

Lymph vessel

Fig. 13.38 A photomicrograph of intestinal villi.

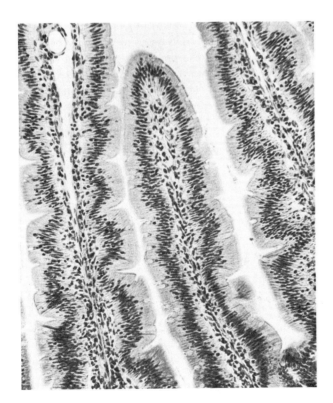

Fig. 13.39 Microvilli at the free surface of the columnar epithelial cells that line the small intestine greatly increase the surface area of these cells.

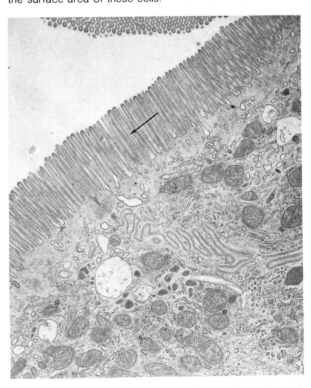

Fig. 13.40 The inner lining of the small intestine contains many circular folds, called plicae circulares.

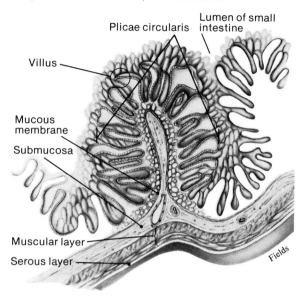

Chart 13.8 summarizes the sources and actions of the major digestive enzymes.

The amount of lactase produced in the small intestine usually reaches a maximum shortly after birth and thereafter tends to decrease. As a result, some adults produce insufficient quantities of lactase to break down the lactose or milk sugar in their diets. When this happens, the lactose from milk or certain milk products remains undigested and causes an increase in the osmotic pressure of the intestinal contents. Consequently, water is drawn from the tissues into the intestine. At the same time, intestinal bacteria may act upon the undigested sugar and produce organic acids and gases. As a result, the person may feel bloated and suffer from intestinal cramps and diarrhea.

of the small intestinal wall are much like those of other parts of the alimentary canal in that they include a submucosa, a muscular layer, and a serous layer.

The lining of the small intestine is characterized by the presence of numerous circular folds, called *plicae circulares,* that are especially well developed in the lower duodenum and upper jejunum. (See fig. 13.40.)

Secretions of the Small Intestine

In addition to mucus-secreting goblet cells that occur extensively throughout the mucosa of the small intestine, there are many specialized *mucus-secreting glands* (Brunner's glands) in the first part of the duodenum. These glands secrete large quantities of mucus in response to various stimuli.

The intestinal glands at the bases of the villi secrete great amounts of a watery fluid. The villi rapidly reabsorb this fluid, and it provides a vehicle for moving digestive products into the villi. The intestinal glands also secrete digestive enzymes, including **peptidases,** which split proteins into amino acids; **sucrase, maltase,** and **lactase,** which split the double sugars (disaccharides) sucrose, maltose, and lactose into the simple sugars (monosaccharides) glucose, fructose, and galactose; *intestinal lipase,* which splits fats into fatty acids and glycerol; *intestinal amylase,* which splits carbohydrates into double sugars; **nucleases,** which split nucleic acid molecules into nucleotides; and **enterokinase,** which activates *trypsinogen,* an enzyme secreted by the pancreas.

Regulation of Small Intestinal Secretions. Since mucus functions to protect the intestinal wall in the same way it protects the stomach lining, it is not surprising that mucus secretions are enhanced by the presence of irritants such as gastric juice. Consequently, as the stomach contents enter the small intestine, the duodenal mucous glands are stimulated to release large quantities of mucus.

Secretions from goblet cells and intestinal glands are similarly stimulated by direct contact with chyme, which provides both chemical and mechanical stimuli, and by reflexes triggered by distension of the intestinal wall. The reflex actions involve parasympathetic motor impulses that cause secretory cells to increase their activities. Some investigators believe that a hormone (enterocrinin) also may aid in the regulation of intestinal secretions. This hormone apparently is released from cells in the intestinal wall and stimulates the intestinal glands to increase their secretions.

1. Describe the parts of the small intestine.
2. Distinguish between intestinal villi and microvilli.
3. What is the function of the intestinal glands?
4. How is the secretion of these glands controlled?

Absorption in the Small Intestine

Because the large number of villi greatly increases the surface area of the intestinal mucosa, the small intestine is the most important absorbing organ of the alimentary canal. In fact, the small intestine is so effective at absorbing digestive products, water, and electrolytes, that very little absorbable material reaches its distal end.

Chart 13.8 A summary of the digestive enzymes

Enzyme	Source	Digestive Action
Salivary enzyme		
Amylase	Salivary glands	Begins carbohydrate digestion by converting starch and glycogen to disaccharides
Gastric juice enzyme		
Pepsin	Gastric glands	Begins the digestion of nearly all types of proteins
Intestinal juice enzymes		
Peptidase	Intestinal glands	Converts proteins into amino acids
Sucrase, maltase, lactase	Intestinal glands	Converts disaccharides into monosaccharides
Lipase	Intestinal glands	Converts fats into fatty acids and glycerol
Amylase	Intestinal glands	Converts starch and glycogen into disaccharides
Nuclease	Intestinal glands	Converts nucleic acids into nucleotides
Enterokinase	Intestinal glands	Activates trypsin
Pancreatic juice enzymes		
Amylase	Pancreas	Converts starch and glycogen into disaccharides
Lipase	Pancreas	Converts fats into fatty acids and glycerol
Peptidases a. Trypsin b. Chymotrypsin c. Carboxypeptidase	Pancreas	Converts proteins or partially digested proteins into amino acids
Nuclease	Pancreas	Converts nucleic acids into nucleotides

Fig. 13.41 Digestion changes complex carbohydrates into disaccharides and disaccharides into monosaccharides that are absorbed by intestinal villi and enter the blood.

Maltose + Water ⟶ Glucose + Glucose

Disaccharide   **Monosaccharides**

Carbohydrate digestion begins in the mouth as a result of the activity of salivary amylase, and it is completed in the small intestine by enzymes from the intestinal glands and pancreas. (See fig. 13.41.) The resulting monosaccharides are absorbed by the villi and enter blood capillaries. Even though some of these simple sugars may pass into the villi by diffusion, most of them are absorbed by active transport. The exact mechanism of this process, however, is poorly understood.

Protein digestion begins in the stomach as a result of pepsin activity and is completed in the small intestine by enzymes from the intestinal glands and the pancreas. During this process, large protein molecules are converted into amino acids, as shown in figure 13.42. These smaller particles are then absorbed into the villi by active transport and are carried away by the blood.

Fat molecules are digested almost entirely by enzymes from the intestinal glands and pancreas. (See fig. 13.43.) The absorption mechanism for the resulting fatty acid and glycerol molecules involves several steps, including the following: (1) the fatty acid molecules are dissolved in the epithelial cell membranes of the villi and diffuse into these cells;

Fig. 13.42 The amino acids that result from protein digestion are absorbed by intestinal villi and enter the blood.

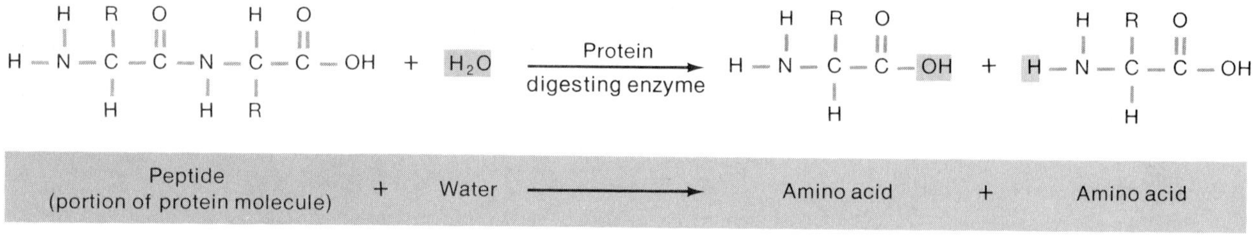

| Peptide (portion of protein molecule) | + | Water | → | Amino acid | + | Amino acid |

Fig. 13.43 Fatty acids and glycerol result from fat digestion. They are absorbed by intestinal villi, and most are resynthesized into fat molecules before they enter the blood or lymph.

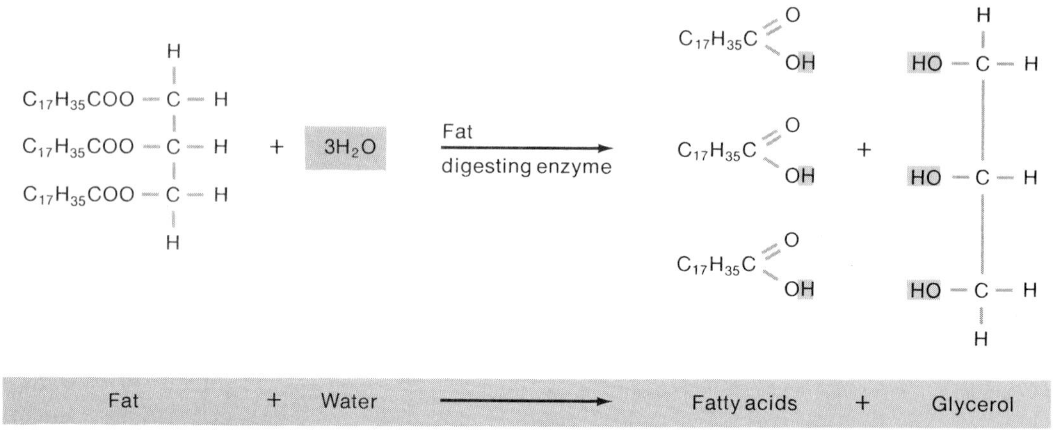

| Fat | + | Water | → | Fatty acids | + | Glycerol |

(2) the endoplasmic reticula of the cells use the fatty acids to resynthesize fat molecules similar to those previously digested; (3) these fats collect in tiny globules that become encased in protein. The resulting droplets are called *chylomicrons,* and they make their ways to the central lymphatic capillary or *lacteal* of the villus. The lymph carries the chylomicrons to the blood, and most of them are stored as fat when they leave the blood as it flows through adipose tissue. (See fig. 13.44.)

Some fatty acids with relatively short carbon chains may be absorbed directly into the blood capillary of the villus without being converted back into fat.

In addition to absorbing the products of carbohydrate, protein, and fat digestion, the intestinal villi function in the absorption of various electrolytes and water. Certain ions, such as those of sodium, potassium, chloride, nitrate, and bicarbonate, are readily absorbed; but others, including calcium, magnesium, and sulfate, are poorly absorbed.

The absorption of electrolytes usually involves active transport, and water is absorbed by osmosis. Thus, even though the intestinal contents may be hypertonic to the epithelial cells at first, as nutrients and electrolytes are absorbed, the intestinal contents tend to become hypotonic to the cells. Then, water follows the nutrients and electrolytes into the villi by osmosis.

The absorption process is summarized in Chart 13.9.

1. *What substances resulting from the digestion of carbohydrate, protein, and fat molecules are absorbed by the small intestine?*
2. *What ions are absorbed by the small intestine?*
3. *What transport mechanisms are utilized by intestinal villi?*

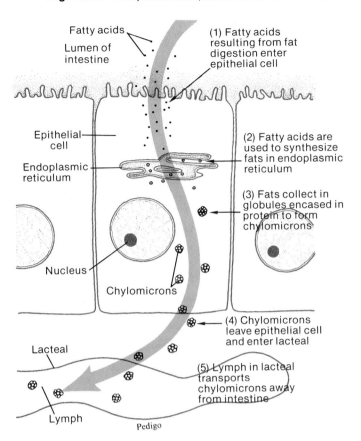

Fig. 13.44 Fatty acid absorption involves several steps.

Fatty acids
Lumen of intestine

(1) Fatty acids resulting from fat digestion enter epithelial cell

Epithelial cell

Endoplasmic reticulum

(2) Fatty acids are used to synthesize fats in endoplasmic reticulum

(3) Fats collect in globules encased in protein to form chylomicrons

Nucleus

Chylomicrons

(4) Chylomicrons leave epithelial cell and enter lacteal

Lacteal

(5) Lymph in lacteal transports chylomicrons away from intestine

Lymph

Pedigo

Chart 13.9   Intestinal absorption of nutrients

Nutrient	Absorption Mechanism	Means of Transport
Monosaccharides	Diffusion and active transport	Blood in capillaries
Amino acids	Active transport	Blood in capillaries
Fatty acids and glycerol	Diffusion into cells (a) Most fatty acids are converted back into fats and incorporated in chylomicrons for transport	Lymph in lacteals
	(b) Some fatty acids with relatively short carbon chains are transported without being converted back into fats	Blood in capillaries
Electrolytes	Diffusion and active transport	Blood in capillaries
Water	Osmosis	Blood in capillaries

Movements of the Small Intestine

Like the stomach, the small intestine carries on *mixing movements* and *peristalsis*. The mixing movements include small, ringlike contractions that occur periodically, cutting the chyme into segments over and over again. While this is happening, the villi undergo vigorous motions that aid to combine the chyme with the intestinal juice.

Another type of small intestinal action is called *pendular movement*. In this motion, a constrictive wave moves along the tube for a few segments and then reverses to move back the other way. As a result, the intestinal contents are moved to and fro.

Chyme is propelled through the small intestine by peristaltic waves. These waves are usually weak, and they stop after pushing the chyme a short distance. Consequently, food materials move relatively slowly through the small intestine, taking from 3 to 10 hours to travel its length.

As might be expected, parasympathetic impulses enhance both mixing and peristaltic movements, and sympathetic impulses inhibit them. Reflexes involving parasympathetic impulses to the small intestine sometimes originate in the stomach. For example, as the stomach fills with food and its wall becomes distended, a reflex (gastroenteric reflex) is triggered, and peristaltic activity in the small intestine is greatly increased. Another reflex is initiated when the duodenum is filled with chyme and its wall is stretched. This reflex causes the chyme to be moved through the small intestine more rapidly.

Stimulation of the small intestinal wall by overdistension or by severe irritation may elicit a strong *peristaltic rush*, that passes along its entire length. This type of movement serves to sweep the contents of the small intestine into the large intestine relatively rapidly and helps relieve the small intestine of its problem. This rapid movement of chyme prevents the normal absorption of water, nutrients, and electrolytes from the intestinal contents. The result is *diarrhea*, a condition in which defecation becomes more frequent and the stools are watery. If the diarrhea continues for a prolonged time, problems in water and electrolyte balance are likely to develop.

At the distal end of the small intestine, where the ileum joins the cecum of the large intestine, there is a sphincter muscle called the **ileocecal valve.** Normally this sphincter remains constricted, preventing the contents of the small intestine from entering. At the same time, it prevents the contents of the large intestine from backing up into the ileum. However, after a meal a reflex (gastroileal reflex) is elicited, and peristalsis in the ileum is increased. This action forces some of the contents of the small intestine into the cecum.

Fig. 13.45 Major parts of the large intestine.

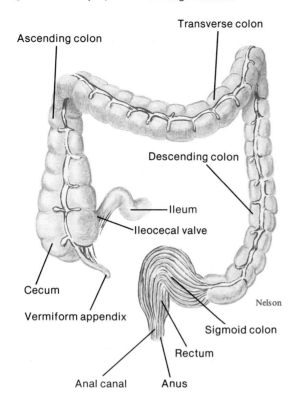

Ascending colon

Transverse colon

Descending colon

Ileum

Ileocecal valve

Cecum

Vermiform appendix

Sigmoid colon

Rectum

Anal canal   Anus

Nelson

Fig. 13.46 X ray of the large intestine. What portions can you identify?

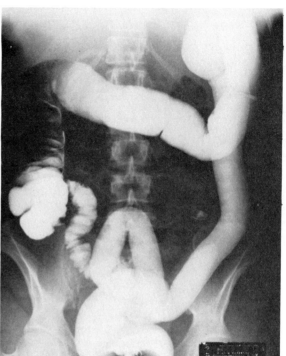

---

1. Describe the movements of the small intestine.
2. How are these movements initiated?
3. What is meant by a peristaltic rush?
4. What stimulus causes the ileocecal valve to relax?

The Large Intestine

The large intestine is so named because its diameter is greater than that of the small intestine. This portion of the alimentary canal is about 1.5 meters (5 feet) long and begins in the lower right side of the abdominal cavity, where the ileum joins the cecum. From there the large intestine (colon) travels upward on the right side, crosses obliquely to the left, and descends into the pelvis. At its distal end it opens to the outside of the body as the anus.

Although the large intestine has little or no digestive function, it serves to reabsorb water and electrolytes from the remaining chyme. It also forms and stores the feces until defecation occurs.

Parts of the Large Intestine

The large intestine, shown in figures 13.45 and 13.46, consists of the cecum, colon, rectum, and anal canal.

The **cecum**, which represents the beginning of the large intestine, is a dilated, pouchlike structure that hangs slightly below the ileocecal opening (fig. 13.47). Projecting downward from it is a narrow tube with a closed end called the **vermiform appendix.** Although the human appendix has no known digestive function, it does contain lymphatic tissue that may serve to resist infections.

Occasionally the appendix itself may become infected and inflamed, causing the condition called *appendicitis.* When this happens, the appendix is often removed surgically to prevent its rupture. If it does break open, the contents of the large intestine may enter the abdominal cavity and cause a serious infection of the peritoneum called *peritonitis.*

Fig. 13.47 The cecum represents the first portion of the large intestine.

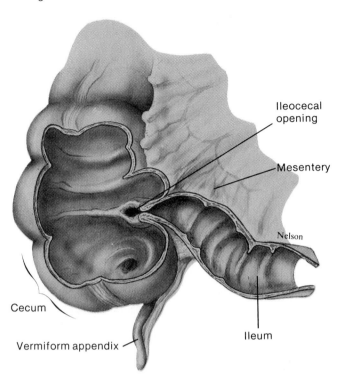

- Ileocecal opening
- Mesentery
- Nelson
- Cecum
- Vermiform appendix
- Ileum

Fig. 13.48 The rectum and anal canal are located at the distal end of the alimentary canal. Muscle fibers within the teniae coli exert tension on the wall of the large intestine and create a series of pouches or haustra.

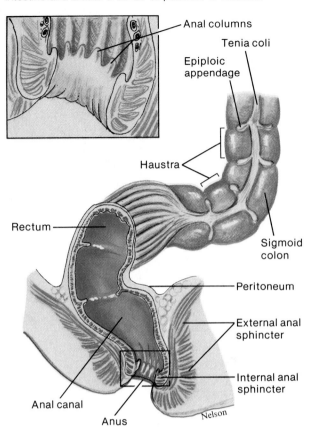

- Anal columns
- Tenia coli
- Epiploic appendage
- Haustra
- Rectum
- Sigmoid colon
- Peritoneum
- External anal sphincter
- Internal anal sphincter
- Anal canal
- Anus
- Nelson

The **colon** can be divided into four portions—the ascending, transverse, descending, and sigmoid colons. The **ascending colon** begins at the cecum and travels upward against the posterior abdominal wall to a point just below the liver. There it turns sharply to the left and becomes the **transverse colon.** The transverse colon is the longest and the most mobile part of the large intestine. It is suspended by a fold of peritoneum and tends to sag in the middle below the stomach. As the transverse colon approaches the spleen, it turns abruptly downward and becomes the **descending colon.** At the brim of the pelvis, the descending colon makes an S-shaped curve, called the **sigmoid colon,** and then becomes the rectum.

The **rectum** lies next to the sacrum and generally follows its curvature. It is firmly attached to the sacrum by the peritoneum, and it ends about two inches below the tip of the coccyx, where it becomes the anal canal (fig. 13.48).

The **anal canal** is formed by the last 2 or 3 inches of the large intestine. The mucous membrane in the canal is folded into a series of six to eight longitudinal *anal columns.* At its distal end, the canal opens to the outside as the **anus.** This opening is guarded by two sphincter muscles, an *internal anal sphincter* composed of smooth muscle, and an *external anal sphincter* of skeletal muscle.

Each anal column contains a branch of the rectal vein, and if something interferes with the blood flow in these vessels, the anal columns may become enlarged and inflamed. This condition, called *hemorrhoids,* may be aggravated by bowel movements and may be accompanied by discomfort and bleeding.

1. What is the general function of the large intestine?
2. Describe the parts of the large intestine.
3. Distinguish between the internal sphincter and the external sphincter of the anus.

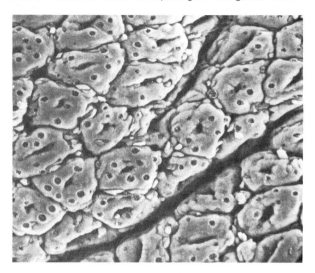

Fig. 13.49 A scanning electron micrograph of the large intestine mucosa. Notice the openings of the goblet cells.

Structure of the Large Intestinal Wall

Although the wall of the large intestine includes the same types of tissues found in other parts of the alimentary canal, it has some unique features. For example, it lacks the villi that are characteristic of the small intestine. Also, the layer of longitudinal muscle fibers does not cover its wall uniformly. Instead, the fibers are arranged in three distinct bands (teniae coli) that extend the entire length of the colon. These bands exert tension on the wall, creating a series of pouches (haustra) (fig. 13.48). The large intestinal wall also is characterized by small collections of fat (epiploic appendages) in the serous layer on its outer surface.

Functions of the Large Intestine

Although the inner lining of the large intestine is relatively smooth, its mucous membrane contains many tubular glands. Structurally these glands are similar to those of the small intestine, but they are composed almost entirely of goblet cells. Consequently, mucus is the only significant secretion of this portion of the alimentary canal. (See fig. 13.49.)

The rate of mucus secretion is controlled largely by mechanical stimulation from chyme and by parasympathetic impulses. In both cases, the goblet cells respond by increasing their production of mucus, which in turn protects the intestinal wall against the abrasive action of materials passing through it. Mucus also aids in holding particles of fecal matter together, and because it is alkaline, mucus helps to control the pH of the large intestinal contents. This is important because acids are sometimes released from the feces as a result of bacterial activity.

The chyme entering the large intestine usually has few nutrients remaining in it. It contains materials that could not be digested or absorbed by the small intestine, in addition to water, various electrolytes, mucus, and bacteria.

Absorption in the large intestine is normally limited to water and electrolytes, which usually are absorbed in the proximal half of the tube. Electrolytes, such as sodium ions, can be absorbed by active transport, while water enters the mucosa by osmosis. As a result of these processes, little sodium or water is lost in the feces.

Many bacteria (colon bacilli) normally inhabit the large intestine, and these organisms can break down some of the substances that escape the actions of enzymes. For instance, cellulose passes through the alimentary canal almost unchanged, but colon bacteria can break down this carbohydrate and utilize it as an energy source. At the same time, these bacteria synthesize certain vitamins that can be absorbed by the intestinal mucosa and used to supplement the supply of dietary vitamins. Vitamins formed in this manner include vitamins K, B_{12}, thiamine, and riboflavin. Bacterial actions in the large intestine may give rise to gases (flatus), which sometimes cause discomfort and pain.

---

1. How does the structure of the large intestine differ from that of the small intestine?
2. What substances can be absorbed by the large intestine?
3. What useful substances are produced as a result of bacteria inhabiting the large intestine?

Movements of the Large Intestine

The movements of the large intestine—mixing and peristalsis—are similar to those of the small intestine, although they are usually more sluggish. The mixing movements break the fecal matter into segments and turn it so that all portions are exposed to the mucosa. This helps in water and electrolyte absorption.

The peristaltic waves of the large intestine are different from those in the small intestine. Instead of occurring frequently, they come only two or three times each day. These waves produce *mass movements* in which a relatively large section of the colon constricts vigorously, forcing its contents to move toward the rectum. Typically, mass movements occur following a meal, as a result of a reflex that is initiated in the small intestine (duodenocolic reflex). Abnormal irritations of the mucosa can also trigger such movements. For instance, a person suffering from an inflamed colon (colitis) may experience frequent mass movements.

When it is appropriate to defecate, a person usually can voluntarily initiate a *defecation reflex* by holding a deep breath and contracting abdominal muscles. This action increases the internal abdominal pressure and forces feces into the rectum. As the rectum fills, its wall is distended and the defecation reflex is triggered. As a result, peristaltic waves in the descending colon are stimulated, and the internal anal sphincter relaxes. At the same time, other reflexes involving the sacral region of the spinal cord, cause the peristaltic waves to strengthen, the diaphragm to lower, the glottis to close, and the abdominal wall muscles to contract. These actions cause an additional increase in the internal abdominal pressure and assist in squeezing the rectum. The external anal sphincter is signaled to relax, and the feces are forced to the outside.

The Feces

As was mentioned, **feces** are composed largely of materials that could not be digested, together with water, electrolytes, mucus, and bacteria. Usually they are about 75% water, and the color is normally due to the presence of bile pigments that have been altered somewhat by bacterial actions.

The pungent odor of feces results from a variety of compounds produced by bacteria acting upon the residues. These include phenol, hydrogen, sulfide, indole, skatole, and ammonia.

---

1. How does peristalsis in the large intestine differ from peristalsis in the small intestine?
2. List the major events that occur during defecation.
3. Describe the composition of feces.

Disorders of the Large Intestine

If feces are allowed to remain in the colon too long, the material tends to become hard and dry as more and more water is removed from it. This action may produce a condition called *constipation*, in which the feces become so dry that defecation is difficult.

The causes of constipation include prolonged inhibition of the defecation reflex, lack of tone in the muscular layer of the large intestine, and muscular spasms in the intestinal wall that block the movement of feces. Constipation is sometimes treated with laxatives (cathartics), which are substances that can stimulate fecal movements.

Some laxatives act by causing an irritating effect on the intestinal mucosa, which stimulates peristaltic waves. Others prevent reabsorption of water by coating the feces with oil, or keeping the feces moist by increasing the osmotic pressure within the large intestine.

Constipation may also be treated by enemas, by increasing the fluid intake, by adding bulk foods to the diet, or by resensitizing the person to the defecation reflex.

Diarrhea creates an effect opposite to that of constipation. It occurs when the feces move more rapidly than usual through the colon and consequently remain very liquid. Among the more common causes of diarrhea are abnormally frequent peristaltic waves and oversecretions of mucus. Such actions often accompany emotional stress or intestinal irritations produced by bacterial infections, viral infections, or various chemicals.

Sometimes the mucosa of the large intestine becomes extensively ulcerated, and the person is said to have *ulcerative colitis*. The cause of this condition is unknown, but it is often associated with prolonged periods of emotional stress. In any case, the colon becomes very active, mass movements occur frequently, and the person has diarrhea most of the time.

Treatment for ulcerative colitis involves reducing nervous tension, changing the diet, and providing medication. If such measures are unsuccessful, a surgeon may perform an *ileostomy*, in which the colon is removed and a portion of the ileum is connected to an opening in the abdominal wall. As a result, the small intestinal contents can move to the outside of the body without passing through the colon.

A related procedure, called *colostomy*, is sometimes performed in cases of abscesses of the colon or intestinal obstructions, or following removal of the rectum. In this procedure, the colon is connected to an opening on the surface of the abdomen so that the rectum and anus are bypassed.

---

1. What conditions may lead to constipation?
2. What conditions may lead to diarrhea?

Some Clinical Terms Related to the Digestive System

achalasia (ak″ah-la′ze-ah)—failure of the smooth muscle to relax at some junction in the digestive tube, such as that between the esophagus and stomach.

achlorhydria (ah″klōr-hi′dre-ah)—a lack of hydrochloric acid in gastric secretions.

aphagia (ah-fa′je-ah)—an inability to swallow.

cholecystitis (ko″le-sis-ti′tis)—inflammation of the gallbladder.

cholecystolithiasis (ko″le-sis″to-li-thi′ah-sis)—the presence of stones in the gallbladder.

cirrhosis (si-ro′sis)—a liver condition in which the hepatic cells degenerate and the surrounding connective tissues thicken.

diverticulitis (di″ver-tik″u-li′tis)—inflammation of small pouches (diverticula) that sometimes form in the lining and wall of the colon.

dumping syndrome (dum′ping sin′drōm)—a set of symptoms, including diarrhea, that often occur following a gastrectomy.

dysentery (dis′en-ter″e)—an intestinal infection, caused by viruses, bacteria, or protozoans, that is accompanied by diarrhea and cramps.

dyspepsia (dis-pep′se-ah)—indigestion; difficulty in digesting a meal.

dysphagia (dis-fa′ze-ah)—difficulty in swallowing.

enteritis (en″tĕ-ri′tis)—inflammation of the intestine.

esophagitis (e-sof″ah-ji′tis)—inflammation of the esophagus.

gastrectomy (gas-trek′to-me)—partial or complete removal of the stomach.

gastritis (gas-tri′tis)—inflammation of the stomach lining.

gastrostomy (gas-tros′to-me)—the creation of an opening in the stomach wall to allow the administration of food and liquids when swallowing is not possible.

gingivitis (jin″ji-vi′tis)—inflammation of the gums.

glossitis (glŏ-si′tis)—inflammation of the tongue.

hemorrhoidectomy (hem″ŏ-roi-dek′to-me)—removal of hemorrhoids.

hemorrhoids (hem′ŏ-roids)—a condition in which veins associated with the lining of the anal canal become enlarged.

hepatitis (hep″ah-ti′tis)—inflammation of the liver.

ileitis (il″e-i′tis)—inflammation of the ileum.

pharyngitis (far″in-ji′tis)—inflammation of the pharynx.

proctitis (prok-ti′tis)—inflammation of the rectum.

pyloric stenosis (pi-lor′ik stĕ-no′sis)—a congenital obstruction at the pyloric sphincter due to an enlargement of the muscle.

pylorospasm (pi-lor′o-spazm)—a spasm of the pyloric portion of the stomach or of the pyloric sphincter.

pyorrhea (pi″o-re′ah)—an inflammation of the dental periosteum accompanied by the formation of pus.

stomatitis (sto″mah-ti′tis)—inflammation of the lining of the mouth.

vagotomy (va-got′o-me)—sectioning of the vagus nerve fibers.

Chapter Summary

Introduction
The digestive system receives food, modifies molecules, carries on absorption, and eliminates unused residues.

It consists of an alimentary canal and several accessory organs.

General Characteristics of the Alimentary Canal
Various regions of the canal are specialized to perform specific functions.
1. Structure of the wall
   a. The wall consists of four layers.
   b. These layers include mucous membrane, submucosa, muscular layer, and serous layer.
2. Movements of the tube
   a. Motor functions include mixing and propelling movements.
   b. Peristalsis is responsible for propelling movements.
3. Innervation of the tube
   a. The tube is innervated by branches of the sympathetic and parasympathetic divisions of the autonomic nervous system.
   b. Parasympathetic impulses generally cause an increase in digestive activities; sympathetic impulses generally inhibit digestive activities.
   c. Sympathetic impulses are responsible for the contraction of certain sphincter muscles that control movement through the alimentary canal.

The Mouth
The mouth is adapted to receive food and begin preparing it for digestion. It also serves as an organ of speech and pleasure.
1. The cheeks and lips
   a. Cheeks form the lateral walls of the mouth.
   b. Lips are highly mobile and possess a variety of sensory receptors useful in judging the characteristics of food.
2. The tongue
   a. The tongue is a thick, muscular organ that aids in mixing food with saliva and moving it toward the pharynx.
   b. Its rough surface aids in handling food and contains taste buds.
   c. Lingual tonsils are located on the root of the tongue.

3. The palate
   a. The palate comprises the roof of the mouth and includes hard and soft portions.
   b. The soft palate closes the opening to the nasal cavity during swallowing.
   c. Palatine tonsils are located on either side of the tongue in the back of the mouth.
   d. Tonsils consist of lymphatic tissues, but are common sites of infections and may become enlarged so that they interfere with swallowing and breathing.
4. The teeth
   a. Two sets develop in sockets of the mandibular and maxillary bones.
   b. There are twenty deciduous and thirty-two permanent teeth.
   c. They function to break food into smaller pieces, increasing the surface area of food that is exposed to digestive actions.
   d. Different kinds are adapted to handle foods in different ways, such as biting, grasping, or grinding.
   e. Each tooth consists of a crown and root and is composed of enamel, dentine, pulp, nerves, and blood vessels.
   f. Dental caries involve decalcification of enamel and erosion of the teeth.
   g. Caries can be prevented by good oral hygiene and the use of fluorides.

The Salivary Glands

Salivary glands secrete saliva, which moistens food, helps bind food particles together, begins digestion of carbohydrates, makes taste possible, helps cleanse the mouth, and regulates pH in the mouth.

1. Salivary secretions
   a. Salivary gland includes serous cells that secrete digestive enzymes and mucous cells that secrete mucus.
   b. Parasympathetic impulses stimulate the secretion of serous fluid.
2. The parotid glands
   a. These are the largest of the salivary glands.
   b. They secrete saliva rich in amylase that begins the digestion of carbohydrates.
3. The submaxillary glands
   a. These are located in the floor of the mouth.
   b. They produce saliva that is more viscous than that of the parotid glands.
4. The sublingual glands
   a. These are located in the floor of the mouth.
   b. They primarily secrete mucus.

The Pharynx and Esophagus

The pharynx and esophagus serve only as passageways.

1. Structure of the pharynx
   a. The pharynx is divided into a nasopharynx, oropharynx, and laryngopharynx.
   b. Its muscular walls contain fibers arranged in circular and longitudinal groups.

2. The swallowing mechanism
   a. The act of swallowing occurs in three stages.
      (1) Food is mixed with saliva and forced into the pharynx.
      (2) Involuntary reflexes move the food into the esophagus.
      (3) Food is transported to the stomach.
3. The esophagus
   a. The esophagus passes through the mediastinum and penetrates the diaphragm.
   b. Some circular muscle fibers at the end of the esophagus help to prevent the regurgitation of food from the stomach.

The Stomach

The stomach receives food, mixes it with gastric juice, carries on a limited amount of absorption, and moves food into the small intestine.

1. Parts of the stomach
   a. The stomach is divided into cardiac, fundic, body, and pyloric regions.
   b. The pyloric sphincter serves as a valve between the stomach and the small intestine.
2. Gastric secretions
   a. Gastric glands secrete gastric juice.
   b. Gastric juice contains pepsin, hydrochloric acid, and intrinsic factor.
3. Regulation of gastric secretions
   a. Gastric secretions are enhanced by parasympathetic impulses and the hormone, gastrin.
   b. Three stages of gastric secretion include cephalic, gastric, and intestinal phases.
   c. Presence of food in the small intestine reflexly inhibits gastric secretions.
4. Disorders of gastric secretions
   a. A breakdown of protective mechanisms or excessive secretions may lead to ulcers in the stomach or duodenum.
   b. The cause of ulcers seems to involve emotional stress.
   c. Treatment for ulcers eliminates aggravating factors and gives damaged tissues time to heal.
5. Gastric absorption
   a. The stomach is not well adapted for absorption.
   b. A few substances such as water and other small molecules may be absorbed through its wall.
6. Filling and emptying actions
   a. As the stomach fills, its wall stretches, but its internal pressure remains unchanged.
   b. Mixing movements aid in producing chyme; peristaltic waves move the chyme into the small intestine.
   c. The rate of emptying depends on the fluidity of the chyme and the type of food present.
   d. The upper part of the small intestine fills, and an enterogastric reflex causes the peristaltic waves in the stomach to be inhibited.

e. Vomiting results from a complex reflex that can be stimulated by a variety of factors.

The Pancreas
1. Structure of the pancreas
   a. The pancreas is closely associated with the duodenum.
   b. It produces pancreatic juice that is secreted into a pancreatic duct.
   c. The pancreatic duct leads to the duodenum.
2. Pancreatic juice
   a. Pancreatic juice contains enzymes that can split carbohydrates, proteins, fats, and nucleic acids.
   b. It has a high bicarbonate ion concentration that helps to neutralize chyme and causes the intestinal contents to be alkaline.
3. Regulation of pancreatic secretion
   a. Secretin from the duodenum stimulates the release of pancreatic juice that contains few digestive enzymes but has a high bicarbonate ion concentration.
   b. Cholecystokinin from the intestinal wall stimulates the release of pancreatic juice that has a high concentration of digestive enzymes.

The Liver
1. Functions of the liver
   a. It is the largest gland in the body.
   b. Liver carries on a variety of vital functions.
   c. Bile is the only secretion that directly affects digestion.
2. Structure of the liver
   a. The liver is a highly vascular organ, enclosed in a fibrous capsule, and divided into lobes.
   b. Each lobe contains hepatic lobules, the functional units of the liver.
   c. Bile from the lobules is carried by bile canals to hepatic ducts that unite to form the common bile duct.
3. Composition of bile
   a. Bile contains bile salts, bile pigments, cholesterol, and various electrolytes.
   b. Only the bile salts have digestive functions.
   c. Bile pigments are products of red blood cell breakdown.
4. The gallbladder and its functions
   a. The gallbladder stores bile between meals.
   b. Release of bile from the common bile duct is controlled by a sphincter muscle.
   c. Gallstones may sometimes form within the gallbladder.
5. Regulation of bile release
   a. Release is stimulated by cholecystokinin from the small intestine.
   b. Sphincter muscle at the base of the common bile duct relaxes as a peristaltic wave in the duodenal wall passes by.

6. Digestive functions of bile salts
   a. Bile salts emulsify fats and aid in the absorption of fatty acids, cholesterol, and certain vitamins.
   b. Bile salts are reabsorbed in the small intestine.

The Small Intestine
The small intestine extends from the pyloric sphincter to the large intestine.
It receives secretions from the pancreas and liver, completes the digestion of nutrients, absorbs the products of digestion, and transports the residues to the large intestine.
1. Parts of the small intestine
   a. The small intestine consists of the duodenum, jejunum, and ileum.
   b. It is suspended from the posterior abdominal wall by mesentery.
2. Structure of the small intestinal wall
   a. The wall is lined with villi that aid in mixing and absorption.
   b. Intestinal glands are located between the villi.
3. Secretions of the small intestine
   a. Secretions include mucus and digestive enzymes.
   b. Digestive enzymes can split molecules of sugars, proteins, fats, and nucleic acids.
4. Regulation of small intestinal secretions
   a. Secretions are enhanced by the presence of gastric juice and chyme and by the mechanical stimulation of distension.
   b. A hormone from the intestinal wall may also stimulate secretions.
5. Absorption in the small intestine
   a. Villi increase the surface area of the intestinal wall.
   b. Monosaccharides, amino acids, fatty acids, and glycerol are absorbed by the villi.
   c. Villi also absorb water and electrolytes.
   d. Fat molecules with longer chains of carbon atoms enter the lacteals of the villi; other products of digestion enter the blood capillaries of the villi.
6. Movements of the small intestine
   a. Movements include mixing, peristalsis, and pendular movements.
   b. Overdistension or irritation may stimulate a peristaltic rush and result in diarrhea.
   c. The ileoceal valve controls movement from the small intestine into the large intestine.

The Large Intestine
The large intestine functions to reabsorb water and electrolytes and to form and store feces.
1. Parts of the large intestine
   a. The large intestine consists of the cecum, colon, rectum, and anal canal.
   b. The colon is divided into ascending, transverse, descending, and sigmoid portions.

2. Structure of the large intestinal wall
   a. Basically this wall is like the wall in other parts of the alimentary canal.
   b. Unique features include a layer of longitudinal muscle fibers that do not cover the wall uniformly, and fatty appendages in the serous layer.
3. Functions of the large intestine
   a. The only significant secretion is mucus.
   b. The rate of mucus secretion is controlled by mechanical stimulation and parasympathetic impulses.
   c. Absorption is generally limited to water and electrolytes.
   d. Many bacteria inhabit the large intestine and may aid the body by synthesizing certain vitamins.
4. Movements of the large intestine
   a. Movements are similar to those in the small intestine.
   b. Mass movements occur two to three times each day.
   c. Defecation is stimulated by a defecation reflex.
5. The feces
   a. Feces consist largely of water, undigested material, mucus, and bacteria.
   b. The color is due to bile salts that have been altered by bacterial actions.
6. Disorders of the large intestine
   a. Constipation occurs when feces become excessively dry and defecation becomes difficult.
   b. Diarrhea is due to rapid movement of feces through the intestines; this interferes with water absorption.

Application of Knowledge

1. If a patient has 95% of the stomach removed (subtotal gastrectomy) as treatment for severe ulcers, how would the digestion and absorption of foods be affected? How would the patient's eating habits have to be altered? Why?

2. Why is it that a person with inflammation of the gallbladder (cholecystitis) may also develop an inflammation of the pancreas (pancreatitis)?

3. What effect is a before-dinner cocktail likely to have on digestion? Why are such beverages inadvisable for persons with ulcers?

Review Activities

1. List and describe the location of the major parts of the alimentary canal.

2. List and describe the location of the accessory organs of the digestive system.

3. Name the four layers of the wall of the alimentary canal.

4. Distinguish between mixing movements and propelling movements.

5. Define peristalsis.

6. Describe the general effects of parasympathetic and sympathetic impulses on the alimentary canal.

7. Discuss the functions of the mouth and its parts.

8. Distinguish between lingual, palatine, and pharyngeal tonsils.

9. Compare the deciduous and permanent teeth.

10. Explain how the various types of teeth are adapted to perform specialized functions.

11. Describe the structure of a tooth.

12. Explain how dental caries form and how decay may be prevented.

13. List and describe the location of the major salivary glands.

14. Explain how the secretions of the salivary glands differ.

15. Discuss the digestive functions of saliva.

16. Describe the mechanism of swallowing.

17. Explain the function of the esophagus.

18. Describe the structure of the stomach.

19. List the enzymes in gastric juice and explain the function of each.

20. Explain how gastric secretions are regulated.

21. Discuss the causes of peptic ulcers and their treatment.

22. Describe the mechanism that controls the emptying of the stomach.

23. Describe the enterogastric reflex.

24. Explain the mechanism of vomiting.

25. Describe the location of the pancreas and the pancreatic duct.

26. List the enzymes found in pancreatic juice and explain the function of each.

27. Explain how pancreatic secretions are regulated.

28. List the functions of the liver.

29. Describe the structure of the liver.

30. Describe the composition of bile.

31. Explain how gallstones may form.

32. Define cholecystokinin.

33. Explain the functions of bile salts.

34. List and describe the location of the parts of the small intestine.

35. Name the enzymes in intestinal juice and explain the function of each.

36. Explain how the secretions of the small intestine are regulated.

37. Describe the functions of the intestinal villi.

38. Explain how the movement of the intestinal contents is controlled.

39. List and describe the location of the parts of the large intestine.

40. Explain the general functions of the large intestine.

41. Describe the defecation reflex.

42. Distinguish between constipation and diarrhea.

Suggestions for Additional Reading

Borland, J. L. May 1974. Rational management of peptic ulcer disease. *Hosp. Prac.*

Bortoff, A. 1972. Digestion. *Ann. Rev. Physio.* 34:261.

———. 1976. Myogenic control of intestinal motility. *Physiol. Rev.* 56:418.

Brooks, F. P., ed. 1974. *Gastrointestinal pathophysiology.* New York: Oxford Univ. Press.

Davenport, H. W. 1977. *Physiology of the digestive tract.* 4th ed. Yearbook Medical Publishers.

———. January 1972. Why the stomach does not digest itself. *Scientific American.*

Hendrix, T. R., and Bayless, T. M. 1970. Digestion: intestinal secretion. *Ann. Rev. Physio.* 32:139.

Javitt, N. B., and McSherry, C. K. July 1973. Pathogenesis of cholesterol gallstones. *Hosp. Prac.*

Kappas, A., and Alvares, A. P. June 1975. How the liver metabolizes foreign substances. *Scientific American.*

Kretchmer, N. October 1972. Lactose and lactase. *Scientific American.*

Lieber, C. S. March 1976. The metabolism of alcohol. *Scientific American.*

Neurath, R. December 1974. Protein-digesting enzymes. *Scientific American.*

Stroud, R. M. July 1974. A family of protein-cutting proteins. *Scientific American.*

Nutrition and Metabolism

14 To maintain health, the human body must obtain a variety of chemicals in adequate amounts from its environment. These substances, which include carbohydrates, lipids, proteins, vitamins, and minerals, are called *nutrients*. They are needed to supply energy for cellular processes, to play vital roles in metabolic reactions, and to provide building materials for growth, maintenance, and repair of tissues.

The study of nutrients, their dietary sources, the ways they are absorbed and transported within the body, and their actions within cells constitutes the science of *nutrition*.

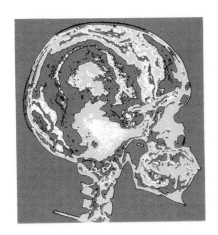

After you have studied this chapter, you should be able to

1. Define *nutrition, nutrients,* and *essential nutrients.*
2. List the major sources of carbohydrates, lipids, and proteins.
3. Describe how carbohydrates are utilized by cells.
4. Describe how lipids are utilized by cells.
5. Describe how amino acids are utilized by cells.
6. Define *nitrogen balance.*
7. Explain how the energy values of foods are determined.
8. Explain how various factors affect an individual's energy needs.
9. Define *energy balance.*
10. Explain what is meant by desirable weight.
11. List the fat-soluble and water-soluble vitamins.
12. Describe the general functions of each vitamin.
13. Distinguish between a vitamin and a mineral.
14. List the major minerals and trace elements.
15. Describe the general functions of each mineral and trace element.
16. Describe an adequate diet.
17. Explain how foods can be selected to achieve an adequate diet.
18. Distinguish between primary and secondary malnutrition.
19. Complete the review activities at the end of this chapter.

acetyl coenzyme A   (as'ĕ-til ko-en'zim)

antioxidant   (an''te-ok'sĭ-dant)

basal metabolic rate   (ba'sal met''ah-bol'ik rāt)

calorie   (kal'o-re)

calorimeter   (kal''o-rim'ĕ-ter)

citric acid cycle   (sit'rik as'id si'kl)

dynamic equilibrium   (di-nam'ik e''kwĭ-lib're-um)

energy balance   (en'er-je bal'ans)

malnutrition   (mal''nu-trish'un)

mineral   (min'er-al)

nitrogen balance   (ni'tro-jen bal'ans)

nutrient   (nu'tre-ent)

oxidation   (ok''sĭ-da'shun)

triglyceride   (tri-glis'er-īd)

vitamin   (vi'tah-min)

bas-, base: *bas*al metabolic rate—metabolic rate of body under resting (basal) conditions.

calor-, heat: *calor*ie—unit used in measurement of heat or energy content of foods.

carot-, carrot: *carot*ene—yellowish plant pigment responsible for the color of carrots and other yellowish plant parts.

mal-, bad: *mal*nutrition—poor nutrition resulting from lack of food or failure to use available foods to best advantage.

meter-, a measure: calori*meter*—instrument used to measure the caloric contents of foods.

nutri-, nourishing: *nutri*ent—chemical substance needed to nourish body cells.

obes-, fat: *obes*ity—condition in which body contains excessive fat.

pell-, skin: *pell*agra—vitamin deficiency condition that is characterized by dermatitis and other symptoms.

The **nutrients** needed by the body include the carbohydrates, lipids, proteins, vitamins, and minerals that are discussed in chapter 2.

As a group, these substances are obtained from foods, and some of them are altered by digestive processes after being ingested. Once they are processed by the digestive system, they are absorbed and transported to body cells by the blood. Within cells, nutrients enter various metabolic pathways and are used to support life processes.

Some nutrients, such as certain amino acids and fatty acids, are particularly important because they are necessary for health and cannot be synthesized by human cells. These are called **essential nutrients.**

Carbohydrates

As is discussed in chapters 2 and 4, carbohydrates are organic compounds such as sugars and starch that are used primarily to supply energy for cellular processes.

Sources of Carbohydrates

Carbohydrates are ingested in a variety of forms. These include starch from grains and certain vegetables, glycogen from meats and seafoods, disaccharides from cane sugar, beet sugar, and molasses, and simple sugars from honey and various fruits. During digestion the more complex carbohydrates are changed to monosaccharides—a form in which they can be absorbed and transported to body cells.

Utilization of Carbohydrates

The monosaccharides that are absorbed from the digestive tract include *fructose, galactose,* and *glucose.* Fructose and galactose are normally converted into glucose by the action of the liver (fig. 14.1), and glucose is the form of carbohydrate that is most commonly oxidized by cells as fuel.

If glucose is present in excessive amounts, some of it is converted to *glycogen* (glycogenesis) and stored as a glucose reserve in the liver and muscles. Glucose can be mobilized rapidly from glycogen (glycogenolysis) when it is needed to supply energy. Only a certain amount of glycogen can be stored, however, and excess glucose is usually converted into fat and stored in adipose tissue. (See fig. 14.2.)

Fig. 14.1 The monosaccharides, fructose and galactose, are converted into glucose by the liver.

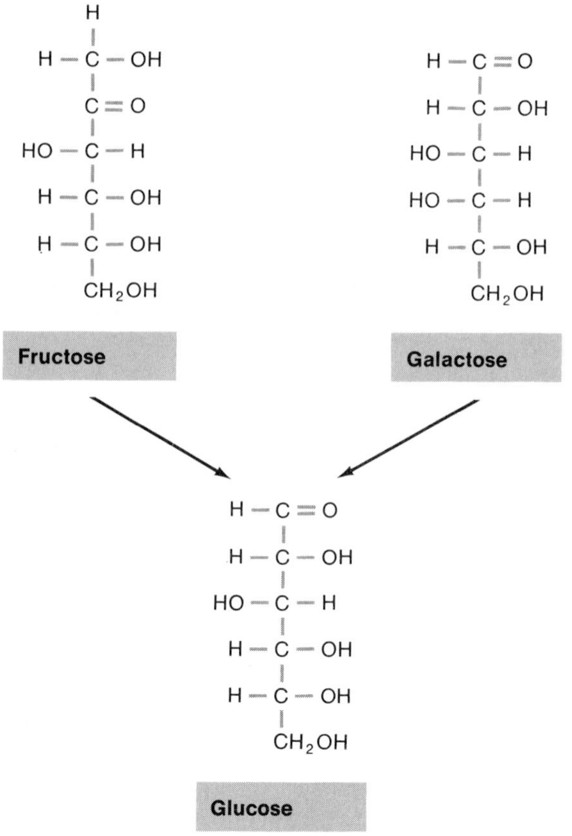

It is estimated that about 100 grams of glycogen can be stored in an adult liver and that another 200 grams can be stored in various muscular tissues. This amount of glycogen is sufficient to meet the energy demands of body cells for about 12 hours.

1. List several common sources of carbohydrates.
2. In what form are carbohydrates utilized as a cellular fuel?
3. Explain what happens to excessive amounts of glucose in the body.

Carbohydrate Requirements

Although most carbohydrates are used to supply energy, some are used to produce vital cellular substances. These include the 5-carbon sugars *ribose* and *deoxyribose* that are needed for the synthesis of the nucleic acids RNA and DNA, and the disaccharide *lactose* (milk sugar) that is produced when the breasts are actively secreting milk.

Fig. 14.2 Monosaccharides from foods are used to supply energy or are stored as glycogen and fat.

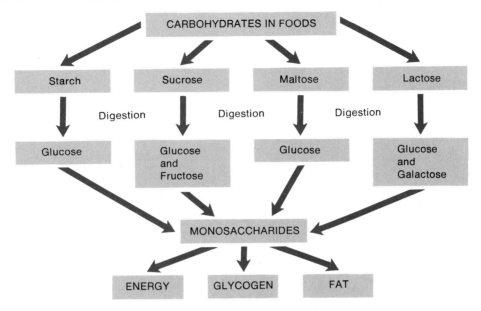

Cells generally can also obtain energy by oxidizing fatty acids. Some cells, however, such as the neurons of the central nervous system, seem to be dependent on a continuous supply of glucose for survival. Even a temporary drop in the glucose supply may produce a serious functional disorder of the nervous system or the death of nerve cells. Consequently, the presence of some carbohydrates in the body is essential; if an adequate supply is not received from foods, the liver may convert noncarbohydrates, such as amino acids from proteins, into glucose (gluconeogenesis). Thus, the need for glucose has priority over the need to manufacture proteins from available amino acids. (See fig. 14.3.)

Humans survive on widely varying amounts of carbohydrates, as a result of differences in the carbohydrate content of available or preferred foods. In the United States, for example, it is estimated that a typical adult diet supplies about 50% of total body energy in the form of carbohydrates. Since carbohydrate foods are generally inexpensive, this percentage is likely to be higher among members of lower economic groups. It is possible to maintain adequate nutrition, however, with relatively high or relatively low carbohydrate intakes provided the needs for all *essential nutrients* are met. Furthermore, poor nutrition among people with limited economic resources is more likely to be related to low intake of essential amino acids, vitamins, and minerals than to an excessive use of carbohydrate foods.

Fig. 14.3 Whenever the supply of glucose is inadequate, liver cells may convert noncarbohydrates such as amino acids into glucose.

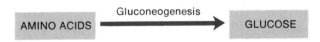

Since carbohydrates provide the primary source of fuel for cellular processes, the need for carbohydrates varies with individual energy requirements. Persons who are physically active require more fuel than those who are sedentary. The minimal requirement for carbohydrates in the human diet is unknown. It is estimated, however, that an intake of at least 100 grams daily is necessary to avoid excessive breakdown of protein and to avoid metabolic disorders that sometimes accompany excessive utilization of fats. Persons in the United States typically include 200–300 grams of carbohydrates in their daily diets.

A nutritionist might recommend that an adult diet contain enough carbohydrate to maintain a desirable body weight, enough proteins to supply the cellular needs for essential amino acids, enough fats to supply essential fatty acids and fat-soluble vitamins, and sufficient amounts of other essential vitamins and minerals.

Fig. 14.4 A triglyceride molecule consists of a glycerol portion and three fatty acid portions.

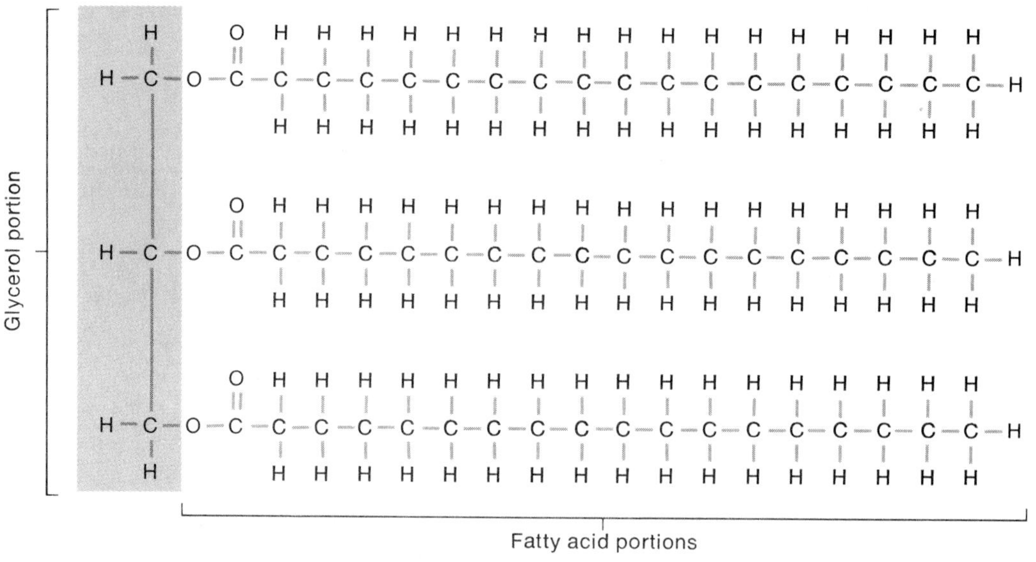

Glycerol portion

Fatty acid portions

Fig. 14.5 Cholesterol can be obtained from foods of animal origin or synthesized by body cells.

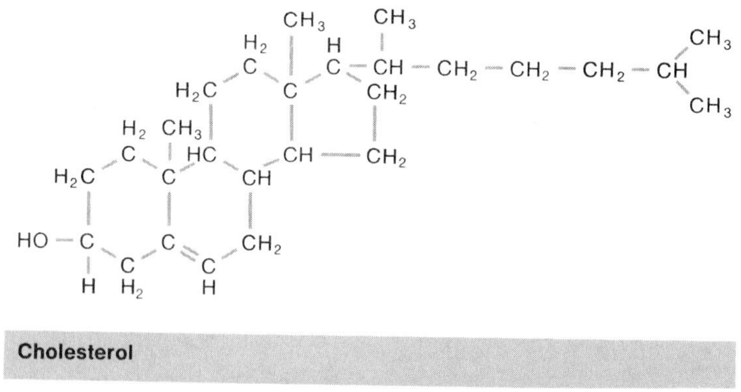

Cholesterol

1. Name two uses of carbohydrates other than supplying energy.
2. What cells are particularly dependent on a continuous supply of glucose?
3. How does the body obtain glucose when insufficient amounts of carbohydrates are ingested?
4. What is the daily minimum requirement of carbohydrates?

Lipids

Lipids comprise a group of organic compounds including fats, oils, and fatlike substances that are used to supply energy for cellular processes and to build structures such as cell membranes. Although lipids include fats, phospholipids, and cholesterol, the most common dietary lipids are fats called *triglycerides,* such as those described in chapter 4. (See fig. 14.4.)

Sources of Lipids

Triglycerides are contained in foods of both plant and animal origin. They occur, for example, in meats, eggs, milk, and lard, as well as in various nuts and plant oils such as corn oil, peanut oil, and olive oil.

Cholesterol (fig. 14.5) is obtained in relatively high concentrations from such foods as liver, egg yolk, and brain, and is present in lesser amounts in whole milk, butter, cheese, and meats. It does not occur in foods of plant origin.

Fig. 14.6 Linoleic acid is an essential fatty acid. Is this fatty acid saturated or unsaturated?

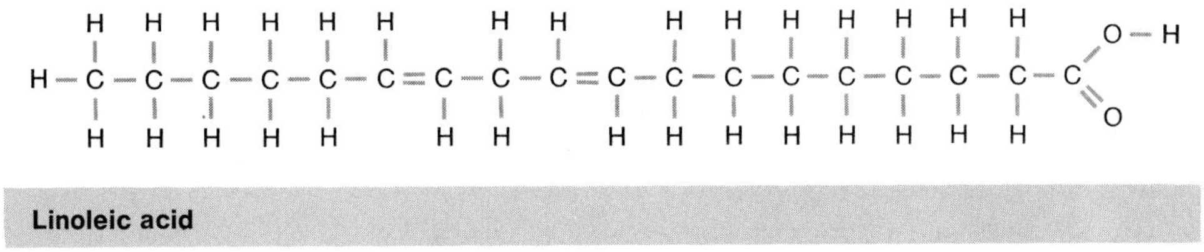

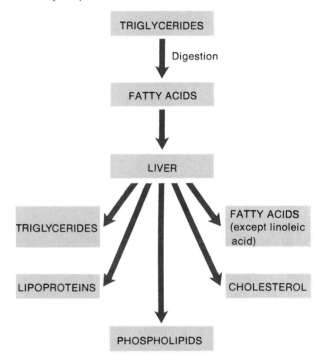

Linoleic acid

Utilization of Lipids

During digestion, triglycerides are broken down into fatty acids and glycerol, and after being absorbed, these products are transported by the lymph and blood to various tissues. The metabolism of these substances is controlled mainly by the liver and the adipose tissues.

As explained in chapter 4, the liver can convert fatty acids from one form to another, but it cannot synthesize one type of fatty acid called *linoleic acid* (fig. 14.6). This substance is an **essential fatty acid** needed for the production of certain phospholipids, which in turn are necessary for the formation of cell membranes and the transport of circulating lipids. Good sources of linoleic acid include corn oil, cottonseed oil, and soy oil.

The liver uses free fatty acids in the synthesis of triglycerides, phospholipids, and lipoproteins that may then be released into the blood. Thus, the liver is largely responsible for the control of circulating lipids. (See fig. 14.7.) In addition, it is thought to regulate the total amount of cholesterol in the body by synthesizing cholesterol and releasing it into the blood, or by removing cholesterol from the blood and excreting it into the bile. The liver also uses cholesterol in the production of bile salts.

Cholesterol is not used as an energy source, but it does provide structural material for a variety of cell parts and furnishes molecular components for the synthesis of various sex hormones and certain hormones produced by the adrenal cortex.

Excessive triglycerides are stored in adipose tissue, and if the blood lipid level drops (in response to fasting, for example), some of these triglycerides are hydrolyzed into free fatty acids and glycerol, and released into the blood.

In addition to storing energy materials, adipose tissues beneath the skin function as an insulating layer. Such insulation helps reduce the rate of body heat loss during cold weather. On the other hand, it may also interfere with heat loss during warm weather and cause considerable discomfort.

Fig. 14.7 Fatty acids are used by the liver to synthesize a variety of lipids.

Lipid Requirements

As in the case of carbohydrates, humans survive with a wide range of lipid intakes. Thus, one person's diet may provide 10% of the body's energy in the form of fats, while another's provides 40%.

Although the amounts and types of fats needed for health are unknown, linoleic acid is an essential fatty acid. Consequently, to prevent deficiency conditions from developing, nutritionists recommend that infants receive formulas in which 3% of the energy intake is in the form of linoleic acid. Such deficiencies have not been observed in adults, so it has been concluded that a typical adult diet consisting of a variety of foods provides an adequate supply of this essential nutrient.

Since fats contain fat-soluble vitamins, the intake of fats also must be sufficient to supply needed amounts of these nutrients.

The average American's diet is estimated to supply about 40% of the total body energy in the form of fats. This percentage has been increasing steadily since the early part of the present century, because people generally have decreased their intake of cereal foods and vegetables and have increased their intake of meats and dairy products. There is evidence that a high intake of saturated fats (chapter 2) and cholesterol promotes the development of cardiovascular diseases, such as atherosclerosis. Many nutritionists, therefore, are recommending that polyunsaturated fats be substituted for some of the saturated fats in the diet.

1. What foods commonly supply lipids?
2. Which fatty acid is an essential nutrient?
3. What is the role of the liver in the utilization of lipids?
4. What is the function of cholesterol?
5. How much lipid should be included in the diet?

Proteins

Proteins are organic compounds that serve as structural materials in cells, act as enzymes that regulate metabolic reactions, and are often used to supply energy.

Sources of Proteins

Foods that are rich in proteins include meats, fish, poultry, cheese, and nuts; milk, eggs, cereals, and various legumes such as beans and peas contain lesser amounts of protein.

During digestion, proteins are broken down into their component amino acids, and these smaller molecules are absorbed and transported to the body cells in the blood.

As is mentioned in chapter 4, the human body can synthesize many amino acids (chart 14.1). However, eight amino acids needed by the adult body (ten needed by growing children) cannot be synthesized, at least in adequate amounts. For this reason these are called **essential amino acids.**

Chart 14.1 Amino acids found in proteins

Alanine	Leucine (e)
Arginine (ch)	Lysine (e)
Aspartic acid	Methionine (e)
Asparagine	Phenylalanine (e)
Cysteine	Proline
Glutamic acid	Serine
Glutamine	Threonine (e)
Glycine	Tryptophan (e)
Histidine (ch)	Tyrosine
Isoleucine (e)	Valine (e)
Hydroxyproline	

Eight essential amino acids (e) cannot be synthesized by human cells and must be provided in the diet. Two additional amino acids (ch) are essential in growing children.

All the essential amino acids must be present in the body at the same time if growth and repair of tissues are to occur. If one essential molecule is missing, the process of protein synthesis cannot take place. Since essential amino acids are not stored, those that are present and are not utilized in protein synthesis are oxidized as energy sources or are converted into carbohydrates or fats.

On the basis of the kinds of amino acids that they provide, proteins can be classified as complete or incomplete. The **complete proteins,** which include those available in milk, meats, fish, poultry, and eggs, contain adequate amounts of the essential amino acids to maintain body tissues and promote normal growth and development. **Incomplete proteins,** such as *zein* in corn, which lacks the essential amino acids tryptophan and lysine, and *gelatin,* which lacks tryptophan, are unable to support tissue maintenance or normal growth and development.

A protein called *gliadin,* which occurs in wheat, is an example of a *partially complete protein.* It is deficient in the essential amino acid lysine, and although it does not contain enough lysine to promote growth, it does contain enough to maintain life.

Plant proteins typically contain less than adequate amounts of one or more of the essential amino acids. However, protein-containing plant foods can be combined in a meal so that one supplies the amino acids another lacks. For example, beans are deficient in the essential amino acid *methionine,* but they contain adequate amounts of *lysine.* Rice is deficient in lysine but contains adequate amounts of *methionine.* Thus, if beans and rice are eaten together, they complement one another, and the meal provides a suitable combination of essential amino acids.

Fig. 14.8 Amino acids resulting from the digestion of proteins are used to synthesize nonessential amino acids and proteins and to supply energy.

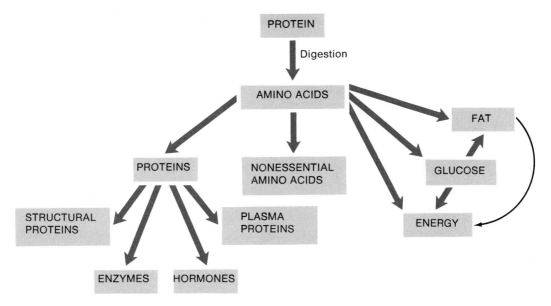

---

1. *What foods provide rich sources of proteins?*
2. *Why are some amino acids called essential?*
3. *Distinguish between a complete protein and an incomplete protein.*

Utilization of Amino Acids

As is discussed in chapter 4, amino acids are used by body cells in a variety of ways. For example, some amino acids are incorporated into protein molecules that provide cellular structure for the *actin* and *myosin* of muscle fibers, the *collagen* of connective tissue fibers, and the *keratin* of skin. Other amino acids are used to synthesize proteins that function in various body processes: hemoglobin, which transports oxygen; plasma proteins, which help to regulate water balance and control pH; enzymes, which catalyze metabolic reactions; and hormones, which help to coordinate body activities.

Amino acids also represent potential sources of energy. When they are present in excess or when the supply of carbohydrates and fats is insufficient to provide needed energy, amino acids are oxidized using various metabolic pathways. (See fig. 14.8.)

The body cells use carbohydrates preferentially as an energy source, and so a diet that supplies an adequate amount of this nutrient is said to have a *protein-sparing effect.* In other words, when sufficient carbohydrates are present, proteins remain available for tissue building and repair rather than being converted into carbohydrates for use as an energy source.

A relatively high proportion of tissue proteins may be utilized as energy sources during times of starvation. These molecules are drawn from structural parts of cells, and consequently, such utilization of structural proteins is accompanied by a wasting of the tissues.

Nitrogen Balance

In a healthy adult, proteins are continuously being built up and broken down. Although these processes occur at different rates in different tissues, the overall gain of body proteins equals the loss, so that a state of *dynamic equilibrium* exists. Since proteins contain a relatively high percentage of nitrogen, dynamic equilibrium also is characterized by a **nitrogen balance**—a condition in which the amount of nitrogen taken in is equal to the amount excreted.

A person who is starving, however, will have a *negative nitrogen balance,* because the amount of nitrogen excreted as a result of amino acid oxidation will exceed the amount replaced by the diet. A growing child, a pregnant woman, or an athlete in training is likely to have a *positive nitrogen balance,* since the amount of protein being built into new tissue probably exceeds the amount being used for energy.

Protein Requirements

In addition to supplying essential amino acids, proteins and amino acids are needed to provide molecular parts and nitrogen for the synthesis of nonessential amino acids and various nonprotein nitrogenous substances. Consequently, the amount of protein and amino acids required by individuals varies according to body size, metabolic rate, and nitrogen balance condition.

For an average adult, nutritionists recommend a daily protein intake of about 0.8 grams per kilogram (0.4 grams per pound) of body weight. For a pregnant woman, who needs to maintain a positive nitrogen balance, the recommendation is increased by an additional 30 grams of protein per day. Similarly, a nursing mother requires an additional 20 grams of protein per day to maintain a high level of milk production.

The consequences of protein deficiencies are severe, particularly among growing children. As was mentioned, a decreased protein intake is likely to result in a negative nitrogen balance and tissue wasting. It may also be accompanied by a decrease in the level of plasma proteins, and if this occurs, the osmotic pressure of the blood decreases. As a result of this osmotic change, fluids collect excessively in the tissues, producing a condition called *nutritional edema.*

Chart 14.2 summarizes the sources, uses, and requirements of protein, lipid, and carbohydrate nutrients.

Protein deficiencies in pregnant women are likely to result in anemia, miscarriages, or premature births. Children whose diets are inadequate in proteins and energy sources tend to develop a condition called *kwashiorkor,* which is characterized by nutritional edema and failure to grow (fig. 14.9). The nervous systems of such children may also fail to develop, so they may be mentally retarded.

Chart 14.2 Protein, lipid, and carbohydrate nutrients

Nutrient	Sources	Utilization	Daily Requirement
Protein	Meats, fish, poultry, cheese, nuts, milk, eggs, cereals, legumes	Production of protein molecules used to build cell structure and to function as enzymes or hormones; used in the transport of oxygen, regulation of water balance, control of pH; amino acids may be broken down and oxidized for energy or converted into carbohydrates or fats for storage	0.8 grams per kilogram of body weight; diet must include the essential amino acids
Lipid	Meats, eggs, milk, lard, plant oils	Oxidized for energy; production of triglycerides, phospholipids, lipoproteins, and cholesterol; stored in adipose tissue; glycerol portions of fat molecules may be used to synthesize glucose	Amount and type needed for health unknown; diet must supply an adequate amount of essential fatty acids and fat-soluble vitamins
Carbohydrates	Primarily from starch and sugars in foods of plant origin, and from glycogen in meats and seafoods	Oxidized for energy; used in production of ribose, deoxyribose, and lactose; stored in liver and muscles as glycogen; converted to fats and stored in adipose tissue	Varies with individual energy needs; at least 100 grams per day to prevent excessive breakdown of proteins

Fig. 14.9 How can the condition, known as kwashiorkor, be prevented?

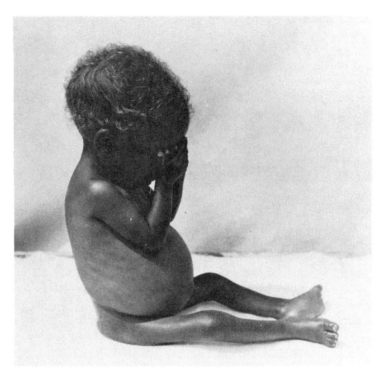

1. *What are the physiological functions of proteins?*
2. *What is meant by a negative nitrogen balance? A positive nitrogen balance?*
3. *How can edema result from inadequate nutrition?*

Energy Expenditures

Energy can be obtained from carbohydrates, fats, and proteins. Since it is required for all metabolic processes, the supply of energy is of prime importance to cell survival. If the diet is deficient in energy-supplying materials, structural molecules may gradually be consumed, and if this continues, death may result.

Energy Values of Foods

The amount of potential energy contained in a food can be expressed as **calories,** which are units of heat.

Although a calorie is commonly defined by a chemist as the amount of heat needed to raise the temperature of a gram of water by 1 degree Celsius (°C), the calorie used in the measurements of food energy is 1,000 times greater. This *large calorie* is equal to the amount of heat needed to raise the temperature of a kilogram (1000 gm) of water by 1°C (actually from 15°C to 16°C). This unit is properly called a *kilocalorie,* but it is customary in nutritional studies to refer to it simply as a calorie.

Fig. 14.10 A bomb calorimeter can be used to measure the caloric content of a food sample.

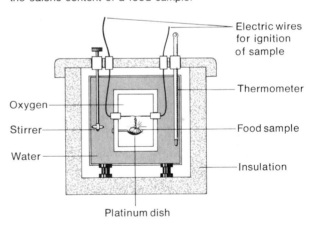

Relatively recently, the American Institute of Nutrition and the International Congress of Nutrition have recommended that a unit of energy called the *joule* (J.) be used in place of the calorie. Therefore, an increasing use of this unit to express the energy values of foods is expected to occur. One kilocalorie is equal to 4.184 kilojoules (kJ.).

The caloric contents of various foods can be determined by using an instrument, called a *bomb calorimeter,* shown in figure 14.10, that consists of

Nutrition and Metabolism   463

a metal chamber submerged in a known volume of water. The food sample being studied is dried, weighed, and placed inside the chamber. The chamber is filled with oxygen gas and submerged in the water. Then, the food is ignited and allowed to oxidize completely. As heat is released from the food, it causes the temperature of the surrounding water to rise, and the change in temperature is measured. Since the volume of the water is known, the amount of heat released from the food can be calculated in calories.

The caloric values determined this way are usually somewhat higher than the amount of energy actually released by metabolic oxidation, because nutrients generally are not completely absorbed from the digestive tract. Also, molecules of amino acids in proteins are not completely oxidized in the body. Portions of these molecules, which still contain some energy, are excreted in urea or are transformed into other nitrogenous substances. When such losses are taken into account, cellular oxidation yields about 4.18 calories from 1 gram of carbohydrates, 4.32 calories from 1 gram of protein, and 9.46 calories from 1 gram of fat.

---

1. What term is used to designate the potential energy in a food substance?
2. How can the energy value of a food be determined?
3. What is the energy value of a gram of carbohydrate? A gram of protein? A gram of fat?

Energy Requirements

The amount of energy required to support metabolic activities for 24 hours varies from person to person. The factors that influence energy needs include the individual's basal metabolic rate, degree of muscular activity, body temperature, and phase of growth.

The **basal metabolic rate** (BMR) is a measurement of the rate at which the body expends energy under *basal conditions*—the conditions that exist when the body is at rest. The apparatus used to determine this rate measures the amount of oxygen taken in by a resting person as well as the amount of carbon dioxide given off during a certain period of time. (See fig. 14.11.)

The amount of oxygen consumed by the body is directly proportional to the amount of energy released by cellular respiration. The BMR, therefore, reveals the total amount of energy expended in a given time period to support the maintenance activities of such organs as the brain, heart, lungs, liver, and kidneys.

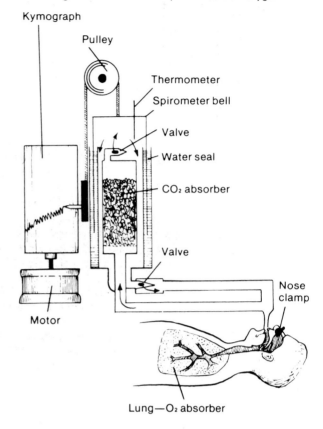

Fig. 14.11 The basal metabolic rate is determined by measuring the rate at which a person utilizes oxygen.

Although the average adult basal metabolic rate indicates a need for approximately one calorie of energy per hour for each kilogram of body weight, this need varies with such factors as sex, body size, body temperature, and level of endocrine gland activity. For example, since heat loss is directly proportional to the body surface area, and a smaller person has a relatively larger surface area, a smaller person will have a higher BMR. Males tend to have higher rates than females; as body temperature increases, the BMR increases; and as the blood level of thyroxine or epinephrine increases, so does the BMR.

Maintaining the basal metabolic rate usually requires the body's greatest expenditure of energy. The energy required to support voluntary muscular activity comes next, though this amount varies greatly with the type of activity. (See chart 14.3.) For example, energy needed to maintain posture while sitting at a desk might amount to 100 calories per hour above the basal need, while running or swimming might require 500–600 calories per hour.

Chart 14.3 Number of calories utilized by actions

Kinds of Activity	Calories (per hour)
Walking up stairs	1100
Running (a jog)	570
Swimming	500
Vigorous exercise	450
Slow walking	200
Dressing and undressing	118
Sitting at rest	100

The maintenance of body temperature may require additional energy expenditure, particularly in cold weather. In this case, extra energy may be released by involuntary muscular contractions, such as shivering, or through voluntary muscular actions.

As was mentioned in the discussion of nitrogen balance, growing children and pregnant women have special needs for nutrients because their bodies are actively engaged in the production of new tissues. They also require increased caloric intake, since tissue-building uses energy.

Energy Balance

A state of **energy balance** exists when the caloric intake in the form of foods is equal to the caloric output resulting from the basal metabolic rate, muscular activities, and so forth. Under these conditions, the body weight remains constant, except perhaps for slight variations due to changes in water content.

If, on the other hand, the caloric intake exceeds the output, a *positive* energy balance occurs, and the excessive nutrients are likely to be stored in the tissues. At the same time, the body weight increases, since an excess of 3500 calories can be stored as a pound of fat. Conversely, if the caloric output exceeds the input, the energy balance is *negative,* and stored materials are mobilized from the tissues for oxidation, causing body weight loss. Chart 14.4 gives the number of calories recommended (by height, weight, and age) for maintaining energy balance.

1. *How is the basal metabolic rate determined?*
2. *What factors influence the BMR?*
3. *What is meant by* energy balance?

Desirable Weight

The most obvious and common nutritional disorders involve caloric imbalances. Persons in underdeveloped countries, for example, are often undernourished, and obesity is rare. Conversely, in highly developed countries such as the United States, people are likely to be overnourished and overweight.

Chart 14.4 Recommended daily dietary allowances of calories for maintenance of good nutrition

	Years (from–to)	Weight (kg)	(lb)	Height (cm)	(in)	Energy (kcal)
Infants	0–6 mos	6	13	60	24	kg × 115
	6 mos–1	9	20	71	28	kg × 105
Children	1–3	13	29	90	35	1300
	4–6	20	44	112	44	1700
	7–10	28	62	132	52	2400
Males	11–14	45	99	157	62	2700
	15–18	66	145	176	69	2800
	19–22	70	154	177	70	2900
	23–50	70	154	178	70	2700
	51–75	70	154	178	70	2400
	76+	70	154	178	70	2050
Females	11–14	46	101	157	62	2200
	15–18	55	120	163	64	2100
	19–22	55	120	163	64	2100
	23–50	55	120	163	64	2000
	51–75	55	120	163	64	1800
	76+	55	120	163	64	1600
Pregnant						+300
Lactating						+500

Source: Food and Nutrition Board, *Recommended Dietary Allowances,* 9th ed., National Academy of Sciences—National Research Council, Washington, D.C., 1980, p. 23.

It is difficult to determine what is a desirable body weight. In the past, weight standards have been based on *average* weights and heights within a certain population, and the degrees of underweight and overweight have been expressed as percentage deviations from these averages. These standards reflected the gradual gain in weight that usually occurs in members of the United States population as they grow older. More recently, it has been recognized that such an increase in weight after the ages of 25 to 30 years is not necessary and may not be conducive to health. Consequently, standards of *desirable weights* have been prepared, as shown in charts 14.5 and 14.6.

These figures are based on the assumption that the average body weights for persons 25 to 30 years of age are desirable weights for all older persons, and it is recommended that each individual maintain this desirable weight throughout life.

Chart 14.5 Desirable weights for women 25 years of age and over*†
Weight in pounds according to frame (in indoor clothing)

Height
(with shoes on)
2-inch heels

Feet	Inches	Small Frame	Medium Frame	Large Frame
4	10	92–98	96–107	104–119
4	11	94–101	98–110	106–122
5	0	96–104	101–113	109–125
5	1	99–107	104–116	112–128
5	2	102–110	107–119	115–131
5	3	105–113	110–122	118–134
5	4	108–116	113–126	121–138
5	5	111–119	116–130	125–142
5	6	114–123	120–135	129–146
5	7	118–127	124–139	133–150
5	8	122–131	128–143	137–154
5	9	126–135	132–147	141–158
5	10	130–140	136–151	145–163
5	11	134–144	140–155	149–168
6	0	138–148	144–159	153–173

*Metropolitan Life Insurance Company, New York.
†For women between 18 and 25, subtract 1 pound for each year under 25.

Chart 14.6 Desirable weights for men 25 years of age and over*
Weight in pounds according to frame (in indoor clothing)

Height
(with shoes on)
1-inch heels

Feet	Inches	Small Frame	Medium Frame	Large Frame
5	2	112–120	118–129	126–141
5	3	115–123	121–133	129–144
5	4	118–126	124–136	132–148
5	5	121–129	127–139	135–152
5	6	124–133	130–143	138–156
5	7	128–137	134–147	142–161
5	8	132–141	138–152	147–166
5	9	136–145	142–156	151–170
5	10	140–150	146–160	155–174
5	11	144–154	150–165	159–179
6	0	148–158	154–170	164–184
6	1	152–162	158–175	168–189
6	2	156–167	162–180	173–194
6	3	160–171	167–185	178–199
6	4	164–175	172–190	182–204

*Metropolitan Life Insurance Company, New York.

Fig. 14.12 (*a*) An obese person is overweight and has an excess of adipose tissue. (*b*) An athlete may be overweight due to muscular hypertrophy but would not be considered obese.

(a)

(b)

Using desirable weight as the standard, *overweight* can be defined as the condition in which a person exceeds the desirable weight by 10 to 20%. A person who exceeds this standard by more than 20% is usually *obese*, though **obesity** is more correctly defined as the presence of excessive adipose tissue. In other words, being overweight and being obese are not the same. For example, as figure 14.12 shows, an athlete or a person whose work involves heavy muscular activity may be overweight, but not obese. Conversely, a sedentary person may be excessively fat, and thus obese, but not overweight.

Thus, determining what body weight a particular person should maintain for optimal health is a very complex problem, and in fact, the average weights and heights for clinically healthy persons in the United States are not known.

One of the more serious nutritional problems in the United States is overeating. It is estimated that the average daily diet in the United States provides about 3300 calories, and that about 2500 calories is probably sufficient to meet the energy needs of a reasonably active person.

1. *What is meant by* desirable weight*?*
2. *Distinguish between overweight and obesity.*

Fig. 14.13 A molecule of beta-carotene can be converted to two molecules of retinal, which in turn can be changed to retinol.

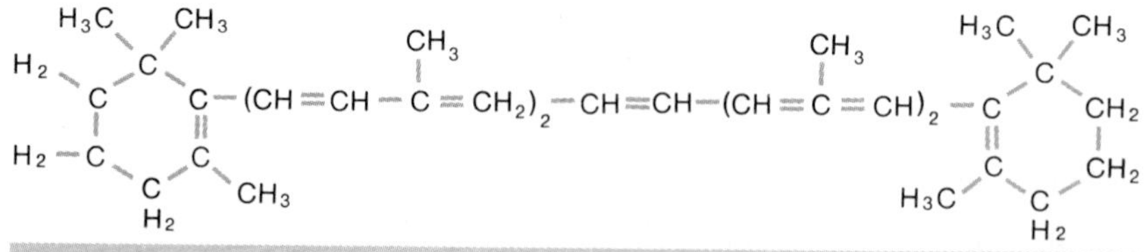

Beta—carotene

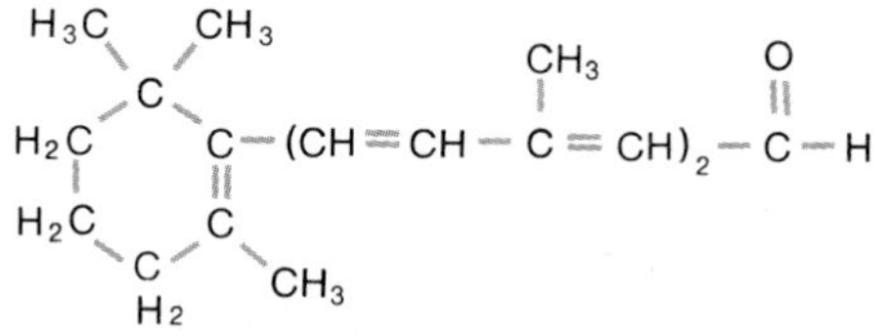

Retinal (Retinene)

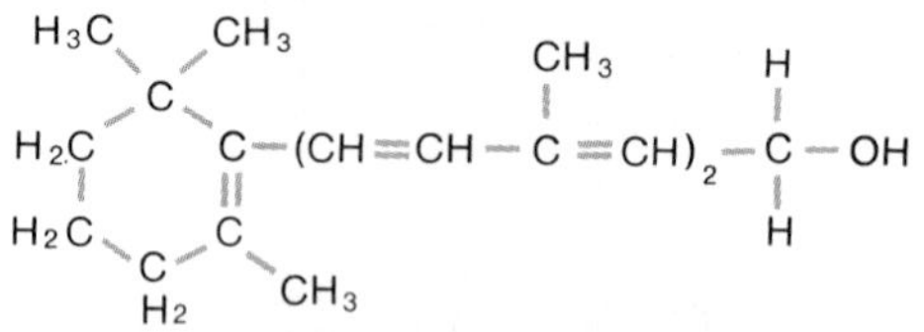

Retinol

Vitamins

Vitamins include a group of organic compounds (other than carbohydrates, lipids, and proteins) that must be present in small amounts for normal metabolic processes, but that cannot be synthesized in adequate amounts by body cells. They are, then, essential nutrients that must be supplied in foods.

For convenience, vitamins are often grouped on the basis of their solubilities, since some are soluble in fats (or fat solvents) and others are soluble in water. Those that are *fat-soluble* include vitamins A, D, E, and K; the *water-soluble* group includes the B vitamins and vitamin C.

Fat-Soluble Vitamins

Since the fat-soluble vitamins dissolve in fats, they occur in association with lipids and are influenced by the same factors that affect lipid absorption. For example, the presence of bile salts in the intestine promotes the absorption of these compounds. As a group, the fat-soluble vitamins are stored in moderate quantities within various tissues, and because they are fairly resistant to the effects of heat, they usually are not destroyed by cooking or food processing.

1. *What are vitamins?*
2. *Distinguish between fat-soluble vitamins and water-soluble vitamins.*

Vitamin A occurs in several forms, including retinol and retinal (retinene). This vitamin can be synthesized by body cells from a group of yellowish plant pigments called *carotenes*. Whenever vitamin A or its precursors are present in excess, they are stored mainly in the liver, which functions to regulate the concentrations of these substances within the body. (See fig. 14.13.)

It is estimated that the liver of a healthy adult stores enough vitamin A to supply the body needs for a year. Infants and children, on the other hand, usually lack such reserves, and consequently they are more likely to develop vitamin A deficiencies if their diets are inadequate.

Vitamin A is relatively stable to the effects of heat, acids, and alkalis, but it is readily destroyed by oxidation and is unstable in the presence of light.

Although functions of vitamin A are not fully understood, it is known that retinal is used in the synthesis of *rhodopsin* (visual purple) in the rods of the retina. Vitamin A is thought to be required for the production of *iodopsin* in the cones as well. Thus, this vitamin is needed to maintain normal vision. It is also believed to function in the synthesis of mucoproteins and mucopolysaccharides of mucus, in the development of normal bones and teeth, and in the maintenance of epithelial cells.

Vitamin A as such is obtained only from foods of animal origin like liver, whole milk, butter, and eggs. However, its precursor, carotene, is widespread in leafy green vegetables and in yellow or orange vegetables and fruits.

A deficiency of vitamin A is characterized by a condition called *night blindness,* in which a person is unable to see normally in dim light. Also, various epithelial tissues tend to undergo degenerative changes, and as a result, the body becomes more susceptible to infections by microorganisms.

An overdose of vitamin A may produce a serious condition, called *hypervitaminosis A*, that is characterized by peeling of the skin, loss of hair, nausea, headache, and dizziness. In chronic cases, growth may be inhibited, and the bones and joints may undergo degenerative changes.

1. What substance is used by body cells to synthesize vitamin A?
2. What conditions will destroy vitamin A?
3. What foods are good sources of vitamin A?

Vitamin D is a group of sterol compounds that have similar properties. One of these substances, vitamin D_3, is produced in the skin from a naturally occurring sterol (7-dehydrocholesterol) when it is exposed to ultraviolet light. Another member of the group, vitamin D_2 (cholecalciferol), is found in foods such as milk, egg yolk, and fish liver oils. This substance is also produced commercially by exposing a sterol obtained from yeasts (ergosterol) to ultraviolet light. (See fig. 14.14.)

Like other fat-soluble vitamins, vitamin D is resistant to the effects of heat and is also resistant to oxidation, acids, and alkalis. It is stored primarily in the liver and occurs to a lesser extent in the skin, brain, spleen, and bones.

Vitamin D promotes the absorption of calcium and phosphorus from the intestine and thus, helps to ensure that adequate amounts of these minerals are available in the blood for bone formation and various metabolic processes.

Although natural foods are generally poor sources of vitamin D, it is often added to food during processing. Homogenized milk, nonfat milk, and evaporated milk, for example, are typically fortified with vitamin D.

The presence of excessive amounts of vitamin D produces a toxic condition, called *hypervitaminosis D*, that is characterized by diarrhea, nausea, and weight loss. Over prolonged periods, overdoses of vitamin D can cause calcification of various soft tissues and irreversible damage to the kidneys.

In children, a deficiency of vitamin D results in the condition called *rickets,* in which the bones and teeth fail to develop normally. (See fig. 14.15.) In adults or in elderly persons who have a minimal exposure to sunlight, such a deficiency may be accompanied by *osteomalacia,* a condition in which the

Fig. 14.14 Vitamin D_3 can be synthesized in the skin. Notice the similarity between this molecule and the cholesterol molecule in figure 14.5.

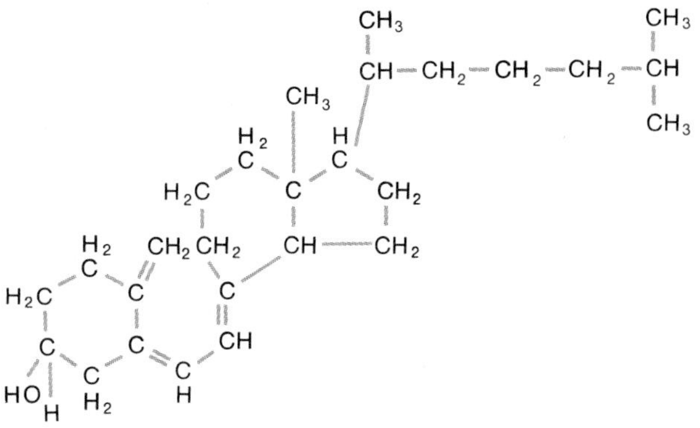

Vitamin D₃

Fig. 14.15 Rickets, which is characterized by failure of the bones and teeth to develop normally, is caused by a deficiency of vitamin D.

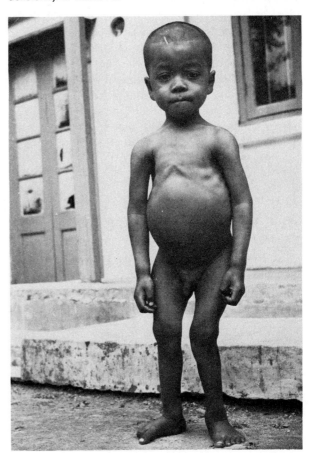

bones undergo decalcification and become weakened due to disturbances in calcium and phosphorus metabolism.

1. Where is vitamin D stored?
2. What are the functions of vitamin D?
3. What foods are good sources of vitamin D?

Vitamin E includes a group of compounds. The most active of these, metabolically, is *alpha-tocopherol.* This substance is resistant to the effects of heat, acids, and visible light, but is unstable in alkalis and in the presence of ultraviolet light or oxygen. In fact, it combines so readily with oxygen that it may prevent the oxidation of other compounds, and for this reason it is said to be an **antioxidant.**

Although vitamin E is present in all tissues, it is stored primarily in the muscles and adipose tissues. It also occurs in relatively high concentrations in the pituitary and adrenal glands.

The precise functions of vitamin E are unknown, but it apparently plays a role as an antioxidant by inhibiting the oxidation of vitamin A in the digestive tract and of polyunsaturated fatty acids in the tissues. It is also thought to help maintain the stability of cell membranes.

While vitamin E is widely distributed among foods, its richest sources are oils from cereal seeds such as wheat germ oil. Other good sources are salad oils, margarine, shortenings, fruits, and vegetables. Since this vitamin is so easily obtained, deficiency conditions are rare.

Vitamin E apparently does not pass readily through the placental membranes that separate the blood of a fetus from the blood of its mother. Consequently, a newborn child usually has a relatively low reserve of this nutrient.

1. Where is vitamin E stored?
2. What are the functions of vitamin E?
3. What foods are good sources of vitamin E?

Vitamin K, like the other fat-soluble vitamins, occurs in several chemical forms. One of these, called vitamin K_1 (phylloquinone), is found in foods, while another, called vitamin K_2, is produced by bacteria (*Escherichia coli*) that normally inhabit the human intestinal tract.

These vitamins are resistant to the effects of heat, but are destroyed by oxidation or by exposure to acids, alkalis, or light. They are stored to a limited degree in the liver.

Vitamin K functions primarily in the liver, where it is necessary for the formation of *prothrombin,* a protein required for blood clotting. Consequently, a deficiency of vitamin K is characterized by prolonged clotting time and a tendency to hemorrhage.

The richest sources of vitamin K are leafy green vegetables. Other good sources are egg yolk, pork liver, soy oil, tomatoes, and cauliflower.

The fat-soluble vitamins and their properties are summarized in chart 14.7.

Chart 14.7 Fat-soluble vitamins

Vitamin	Characteristics	Functions	Sources
Vitamin A	Occurs in several forms; synthesized from carotenes; stored in liver; stable in heat, acids, and alkalis; unstable in light	Necessary for synthesis of visual pigments, mucoproteins, and mucopolysaccharides; for normal development of bones and teeth; and for maintenance of epithelial cells	Liver, whole milk, butter, eggs, leafy green vegetables, and yellow and orange vegetables and fruits
Vitamin D	A group of sterols; resistant to heat, oxidation, acids, and alkalis; stored in liver, skin, brain, spleen, and bones	Promotes absorption of calcium and phosphorus	Produced in skin exposed to ultraviolet light; in milk, egg yolk, fish liver oils, fortified foods
Vitamin E	A group of compounds; resistant to heat and visible light; unstable in presence of oxygen and ultraviolet light; stored in muscles and adipose tissue	An antioxidant; prevents oxidation of vitamin A and polyunsaturated fatty acids; may help maintain stability of cell membranes	Oils from cereal seeds, salad oils, margarine, shortenings, fruits, and vegetables
Vitamin K	Occurs in several forms; resistant to heat but destroyed by acids, alkalis, and light; stored in liver	Needed for synthesis of prothrombin	Leafy green vegetables, egg yolk, pork liver, soy oil, tomatoes, cauliflower

Newborns sometimes may develop vitamin K deficiencies and experience abnormal bleeding because their intestines lack bacteria, and they may not be fed adequately at first. Similarly, adults who have been treated with antibiotic drugs may develop a vitamin K deficiency since such drugs often interfere with the activities of intestinal bacteria that synthesize this vitamin.

---

1. Where in the body is vitamin K synthesized?
2. What is the function of vitamin K?
3. What foods are good sources of vitamin K?

Water-Soluble Vitamins

The water-soluble vitamins include the B vitamins and vitamin C. The **B vitamins** consist of several compounds that are essential for normal cellular metabolism and are particularly involved with the oxidation of carbohydrates, lipids, and proteins. Since the B vitamins often occur together in foods, they are usually referred to as a group called the *B complex*. Members of this group differ chemically, and they have unique functions.

The B-complex vitamins include the following substances:

1. **Thiamine** or **vitamin B₁.** In its pure form, thiamine is a crystalline compound called thiamine hydrochloride. It is destroyed by exposure to heat and oxygen, especially in alkaline environments. (Its molecular structure is included in fig. 14.22.)

Thiamine functions as part of a coenzyme called *cocarboxylase,* which acts in the oxidation of carbohydrates. More specifically, thiamine is required for pyruvic acid to enter the citric acid cycle; in the absence of this vitamin, pyruvic acid accumulates in the blood. Thiamine also functions as a coenzyme in the synthesis of sugars such as ribose, which in turn are used in the production of the nucleotides of nucleic acids.

Thiamine is absorbed primarily through the wall of the duodenum and is transported by the blood to body cells. Only small amounts are stored in the tissues, and excesses are excreted in the urine.

Since this vitamin functions in the oxidation of carbohydrates, the quantity required by the cells varies with the caloric intake. It is recommended that an adult diet contain 0.5 milligram (mg) of thiamine for every 1000 calories ingested each day.

Good sources of thiamine are lean pork, other lean meats, liver, eggs, whole grain cereals, leafy green vegetables, and legumes.

A mild deficiency of thiamine is characterized by loss by appetite, fatigue, and nausea, while a prolonged deficiency leads to a disease called *beriberi.*

In the United States, beriberi occurs mainly in chronic alcoholics who substitute alcohol for needed foods. Moreover, since thiamine is required for the metabolic oxidation of alcohol, alcoholics are particularly likely to develop a thiamine deficiency.

---

2. **Riboflavin** or **vitamin B₂.** Riboflavin is a yellowish-brown crystalline substance that is relatively stable to the effects of heat, acids, and oxidation, but is destroyed by exposure to alkalis and ultraviolet light.

This vitamin functions as a part of several enzymes and coenzymes known as *flavoproteins,* which are essential for the oxidation of glucose and fatty acids, and for cellular growth. Its absorption seems to be regulated by an active transport system that controls the amount of riboflavin taken in by the intestinal mucosa. It is carried in the blood combined with blood proteins called *albumins.* Any excessive amounts of riboflavin in the blood are excreted in the urine, and any that remain unabsorbed in the intestine are lost in the feces.

The amount of riboflavin required by the body varies with the caloric intake, and it is estimated that 0.6 mg of riboflavin per 1000 calories is sufficient to meet daily cellular needs.

Riboflavin is widely distributed in foods, and rich sources include meats and dairy products. Leafy green vegetables, whole grain cereals, and enriched cereals provide lesser amounts.

Although this vitamin is essential for growth, a deficiency does not cause a clearly defined set of symptoms.

3. **Niacin.** Niacin, which is also known as *nicotinic acid,* occurs in plant tissues and is stable in the presence of heat, acids, and alkalis. After ingestion it is converted to a physiologically active form called *niacinamide* (fig. 14.16). It is in this form that the vitamin is present in foods of animal origin.

Niacin functions as part of two coenzymes (coenzyme I, also called NAD, and coenzyme II, called NADP) that play essential roles in the oxidation of glucose and in the synthesis of proteins and fats. (See fig. 14.17.) These coenzymes are also needed for the synthesis of the sugars used in the production of nucleic acids.

Niacin can be readily absorbed from foods, and can be synthesized by human cells from the essential amino acid *tryptophan.* Consequently, the daily requirement for niacin varies with the intake of tryptophan.

Fig. 14.16 Niacin from foods is converted to physiologically active niacinamide by body cells.

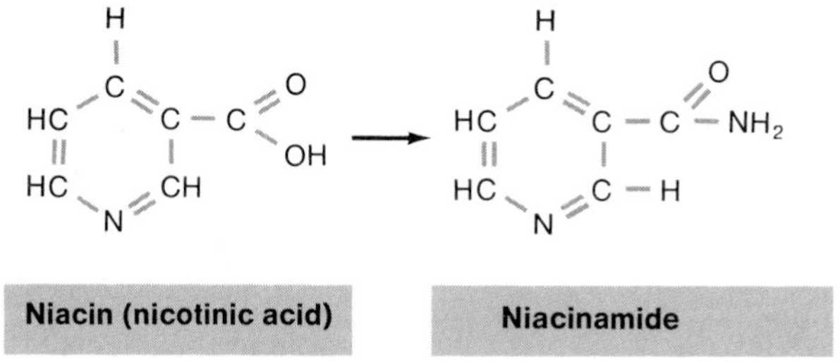

Niacin (nicotinic acid) → Niacinamide

Fig. 14.17 Niacinamide is incorporated into molecules of coenzyme I.

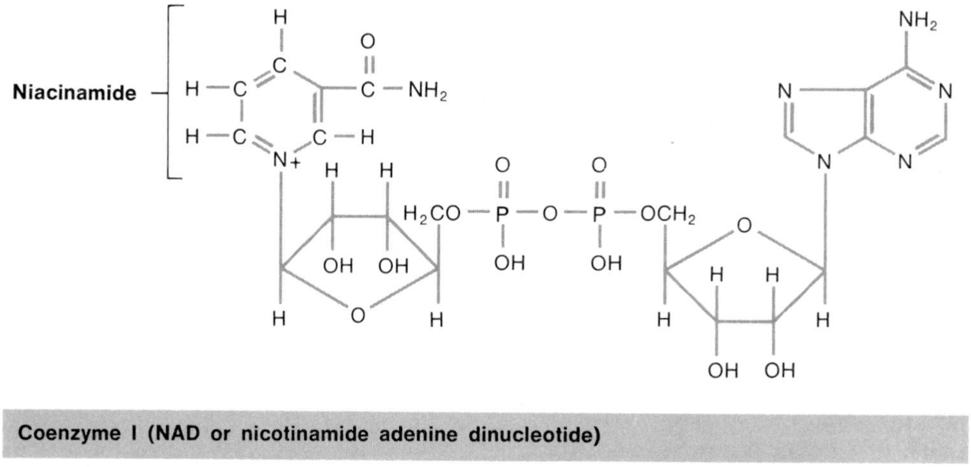

Coenzyme I (NAD or nicotinamide adenine dinucleotide)

It is believed that 60 mg of tryptophan is equivalent to 1 mg of niacin, but this equivalence seems to vary from one individual to another. With these factors in mind, nutritionists recommend a daily niacin (or niacin equivalent) intake of 6.6 mg per 1000 calories.

Rich sources of niacin (and tryptophan) include liver, lean meats, poultry, peanut butter, and legumes. Although milk is a poor source of niacin, it is a good source of tryptophan.

Historically, niacin deficiencies have been associated with diets consisting largely of corn and corn products, which are very low in niacin and lack tryptophan. Such a deficiency causes a disease, called *pellagra*, that is characterized by dermatitis, inflammation of the digestive tract, diarrhea, and various mental disorders.

Although pellagra is relatively rare in the United States today, it was a serious problem in the rural South in the early 1900s. It sometimes occurs in chronic alcoholics who have substituted alcohol for needed foods over prolonged periods of time.

4. **Vitamin B₆.** Vitamin B_6 is a group of three compounds that are closely related chemically, as figure 14.18 shows. They are called *pyridoxine, pyridoxal,* and *pyridoxamine.* These substances have similar actions and are fairly stable to heat and acids; they are destroyed by oxidation or exposure to alkalis or ultraviolet light.

They function as coenzymes that play essential roles in a wide variety of metabolic pathways, including those associated with the synthesis of proteins and various amino acids, the conversion of tryptophan to niacin, the production of antibodies, and the synthesis of nucleic acids.

Since the functions of vitamin B_6 are generally related to the metabolism of nitrogen-containing substances, the requirement for this vitamin varies with the protein content of the diet rather than with the caloric intake.

The recommended daily allowance of vitamin B_6 is 2.0 mg, but because this substance is so widespread in foods, deficiency conditions are quite rare.

Good sources of vitamin B_6 include liver, meats, fish, poultry, bananas, avocados, beans, peanuts, whole grain cereals, and egg yolk.

Fig. 14.18 Vitamin B$_6$ consists of three closely related chemical compounds.

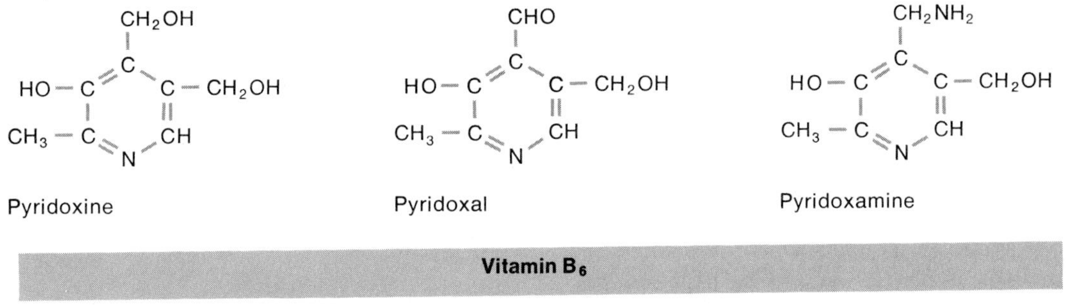

Pyridoxine

Pyridoxal

Pyridoxamine

Vitamin B$_6$

5. **Pantothenic acid.** Pantothenic acid is a yellowish oil that is destroyed by heat, acids, and alkalis. It functions as part of a complex molecule called *coenzyme A,* which in turn reacts with intermediate products of carbohydrate and fat metabolism to become *acetyl coenzyme A.* Pantothenic acid is thus essential to the energy-releasing mechanisms of cells.

A daily adult intake of 4–7 mg seems to be adequate. Most diets apparently provide sufficient amounts, since deficiencies are rare, and no clearly defined set of deficiency symptoms is known.

Good sources of pantothenic acid include meats, fish, whole grain cereals, legumes, milk, fruits, and vegetables.

6. **Cyanocobalamin** or **vitamin B$_{12}$.** Cyanocobalamin has a complex molecular structure that contains a single atom of the element *cobalt.* (See fig. 14.19.) In its pure form, this vitamin is red. It is stable to the effects of heat but is inactivated by exposure to light or by the presence of strong acids or alkalis.

The absorption of cyanocobalamin is regulated by the secretion of *intrinsic factor* from the parietal cells of the gastric glands, as explained in chapter 13. Intrinsic factor is thought to combine with cyanocobalamin and to facilitate its transfer through the epithelial lining of the small intestine and into the blood. Although the details of this transport mechanism are not well understood, it is known that calcium ions must be present for the process to take place.

Cyanocobalamin is stored in various tissues, particularly those of the liver, and an average adult body is thought to contain a reserve sufficient to supply cellular needs for 3 to 5 years.

This vitamin is essential for the functions of all cells. It serves as a part of coenzymes needed for the synthesis of nucleic acids and the metabolism of carbohydrates and fats. It also seems to play a role in the formation of myelin within the central nervous system.

Cyanocobalamin is found only in foods of animal origin, and good sources include liver, meats, poultry, fish, milk, cheese, and eggs. Since persons in the United States consume relatively large amounts

Fig. 14.19 Vitamin B$_{12}$, which has the most complex molecular structure of the vitamins, contains cobalt (Co). What is the function of this vitamin?

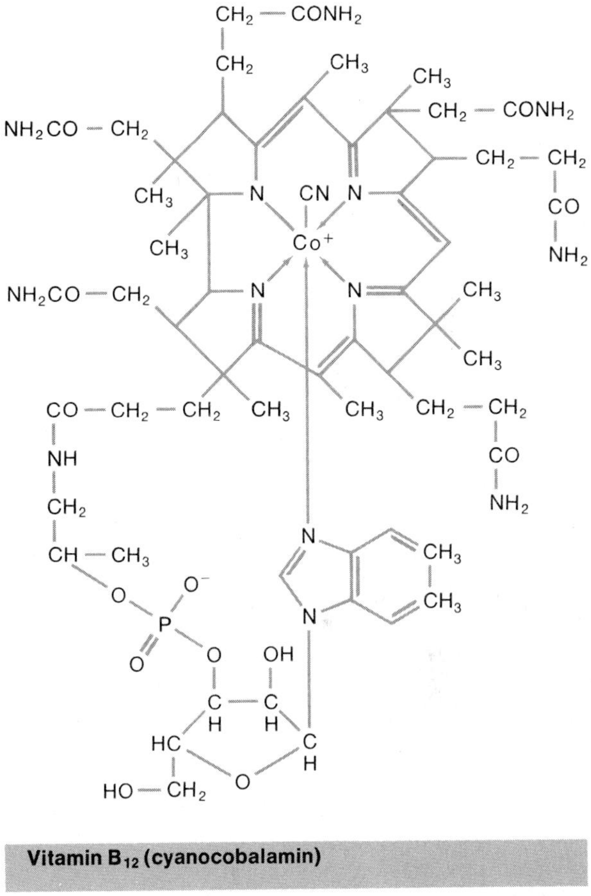

Vitamin B$_{12}$ (cyanocobalamin)

of such foods, a dietary lack of this vitamin seldom occurs, although strict vegetarians may develop a deficiency.

A cyanocobalamin deficiency may develop in association with a disorder of the absorption mechanism. Specifically, the gastric glands of some individuals fail to secrete adequate amounts of intrinsic factor. As a result, the vitamin is poorly absorbed. This leads to a condition, called *pernicious anemia,* that is characterized by the presence of abnormally

Nutrition and Metabolism   473

large red blood cells called *macrocytes.* These cells are produced when bone marrow cells fail to divide properly because of defective synthesis of DNA.

7. **Folacin** or **folic acid.** Folacin is a yellow crystalline compound that exists in several forms. It is easily oxidized in an acid environment and is destroyed by heat in alkaline solutions. Consequently, this vitamin may be lost in foods that are stored or cooked.

Folacin is readily absorbed from the digestive tract and is stored in the liver, where it is converted to a physiologically active substance called *folinic acid.*

Folinic acid functions as a coenzyme that is necessary for the metabolism of certain amino acids and for the synthesis of DNA. It also acts together with cyanocobalamin in promoting the production of normal red blood cells.

Good sources of folacin include liver, leafy green vegetables, whole grain cereals, and legumes. A deficiency of this vitamin leads to *megaloblastic anemia,* which is characterized by a reduction in the number of normal red blood cells and the presence of large, nucleated red cells.

8. **Biotin.** Biotin is a relatively simple compound that is stable to the effects of heat, acids, and light but may be destroyed by oxidation or alkalis. (The molecular structure of biotin is included in fig. 14.22.)

Biotin functions as a necessary coenzyme in a number of metabolic pathways, including those involved with the metabolism of amino acids and fatty acids. It also plays a role in the synthesis of *purines,* which are essential in the synthesis of nucleic acids.

Small quantities of biotin are stored in the tissues of metabolically active organs such as the brain, liver, and kidneys. It is believed to be manufactured by bacteria that inhabit the intestinal tract.

Biotin is widely distributed in foods, and dietary deficiencies are relatively rare. Good sources include liver, egg yolk, nuts, legumes, and mushrooms.

---

1. *What substances make up the vitamin B complex?*
2. *What foods are good sources of vitamin B complex?*
3. *Which of the B-complex vitamins can be synthesized from tryptophan?*
4. *What is the general function of each member of the B-complex?*

Fig. 14.20 Vitamin C is chemically closely related to the monosaccharides.

Vitamin C
(ascorbic acid)

Glucose
(a monosaccharide)

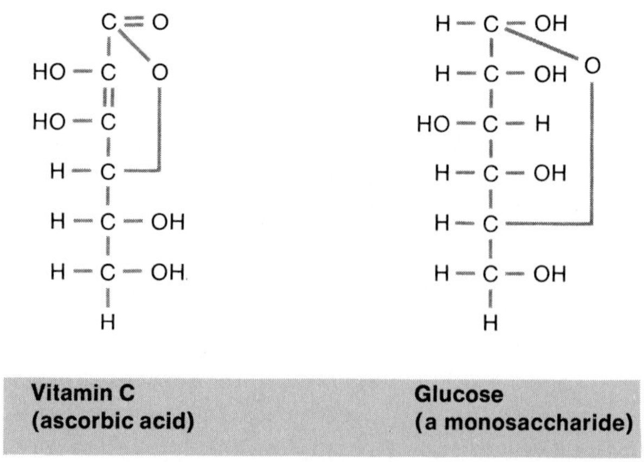

Vitamin C or **ascorbic acid.** Ascorbic acid is a crystalline compound that contains six carbon atoms. Chemically, it is closely related to the monosaccharides. (See fig. 14.20.) It is one of the least stable of the vitamins, in that it can be destroyed by oxidation, heat, light, or alkalis. It is fairly stable in acids.

The functions of ascorbic acid are poorly understood. It is known, however, to be necessary for the production of the protein, *collagen,* which is a vital part of various connective tissues such as bone, cartilage, and the fibrous connective tissue of skin. It is needed in the conversion of folacin to folinic acid and in the metabolism of certain amino acids. It also promotes the absorption of iron and the synthesis of various hormones from cholesterol.

Although this vitamin is not stored in any great amounts, tissues such as the adrenal cortex, pituitary gland, and intestinal glands contain relatively high concentrations. Excessive quantities are either excreted in the urine or oxidized.

There is considerable controversy as to how much ascorbic acid is required to maintain optimal health, and individual needs may vary. It is believed that 10 mg per day is sufficient to prevent deficiency symptoms and that 80 mg per day will saturate the tissues within a few weeks. Consequently, many nutritionists generally recommend a daily adult intake of 60 mg, which is thought to be enough to replenish normal losses and to provide a satisfactory level for cellular needs.

Ascorbic acid is fairly widespread in plant foods. Those that provide particularly high concentrations include citrus fruits, citrus juices, tomatoes, and cabbage. Potatoes, leafy green vegetables, and fresh fruits are also good sources.

Chart 14.8 Water-soluble vitamins

Vitamin	Characteristics	Functions	Sources
Thiamine (Vitamin B_1)	Destroyed by heat and oxygen, especially in alkaline environment	Part of coenzyme needed for oxidation of carbohydrates, and coenzyme needed in synthesis of ribose	Lean meats, liver, eggs, whole grain cereals, leafy green vegetables, legumes
Riboflavin (Vitamin B_2)	Stable to heat, acids, and oxidation; destroyed by alkalis and light	Parts of enzymes and coenzymes needed for oxidation of glucose and fatty acids and for cellular growth	Meats, dairy products, leafy green vegetables, whole grain cereals
Niacin (Nicotinic acid)	Stable to heat, acids, and alkalis; converted to niacinamide by cells; synthesized from tryptophan	Part of coenzymes needed for oxidation of glucose and synthesis of proteins, fats, and nucleic acids	Liver, lean meats, poultry, peanuts, legumes
Vitamin B_6	Group of three compounds; stable to heat and acids; destroyed by oxidation, alkalis, and ultraviolet light	Coenzyme needed for synthesis of proteins and various amino acids, for conversion of tryptophan to niacin, for production of antibodies, and for synthesis of nucleic acids	Liver, meats, fish, poultry, bananas, avocados, beans, peanuts, whole grain cereals, egg yolk
Pantothenic Acid	Destroyed by heat, acids, and alkalis	Part of coenzyme needed for oxidation of carbohydrates and fats	Meats, fish, whole grain cereals, legumes, milk, fruits, vegetables
Cyanocobalamin (Vitamin B_{12})	Complex, cobalt-containing compound; stable to heat; inactivated by light, strong acids, and strong alkalis; absorption regulated by intrinsic factor from gastric glands; stored in liver	Part of coenzyme needed for synthesis of nucleic acids and for metabolism of carbohydrates; plays role in synthesis of myelin	Liver, meats, poultry, fish, milk, cheese, eggs
Folacin (Folic Acid)	Occurs in several forms; destroyed by oxidation in acid environment or by heat in alkaline environment; stored in liver where it is converted into folinic acid	Coenzyme needed for metabolism of certain amino acids and for synthesis of DNA; promotes production of normal red blood cells	Liver, leafy green vegetables, whole grain cereals, legumes
Biotin	Stable to heat, acids, and light; destroyed by oxidation and alkalis	Coenzyme needed for metabolism of amino acids and fatty acids and for synthesis of nucleic acids	Liver, egg yolk, nuts, legumes, mushrooms
Ascorbic Acid (Vitamin C)	Closely related to monosaccharides; stable in acids, but destroyed by oxidation, heat, light, and alkalis	Needed for production of collagen, conversion of folacin to folinic acid, and metabolism of certain amino acids; promotes absorption of iron and synthesis of hormones from cholesterol	Citrus fruits, citrus juices, tomatoes, cabbage, potatoes, leafy green vegetables, fresh fruits

A prolonged deficiency of ascorbic acid leads to a disease, called *scurvy,* that occurs more frequently in infants and children than in adults. This condition is characterized by abnormal bone development and swollen, painful joints. There is also a tendency for the cells to pull apart, and consequently, the gums may swell and bleed easily, resistance to infection is lowered, and wounds heal slowly.

Chart 14.8 summarizes the water-soluble vitamins and their characteristics.

1. What factors cause the destruction of vitamin C?
2. What is the function of vitamin C?
3. What foods are good sources of vitamin C?

Minerals

The nutrients that have been discussed—carbohydrates, lipids, proteins, and vitamins—are all organic substances. Dietary **minerals,** however, are inorganic elements that play essential roles in human metabolism. These elements are usually extracted from the soil by plants. Humans, in turn, obtain them from plant foods or from animals that have eaten plants.

Characteristics of Minerals

Minerals are responsible for about 4% of the body weight and are most concentrated in the bones and teeth. In fact, the minerals, *calcium* and *phosphorus,* which are very abundant in these tissues, account for nearly ¾ of the body's minerals.

Minerals are usually incorporated into organic molecules. For example, phosphorus occurs in phospholipids, iron in hemoglobin, and iodine in thyroxine. However, some occur in inorganic compounds, such as the calcium phosphate of bone; others occur as free ions, such as the sodium, chlorine, and calcium ions in blood plasma.

Minerals are present in all body cells, where they comprise parts of structural materials. They function as portions of enzyme molecules, help create the osmotic pressure of body fluids, and play vital roles in the conduction of nerve impulses, the contraction of muscle fibers, the coagulation of blood, and the maintenance of pH.

The concentrations of various minerals in body fluids are regulated by homeostatic mechanisms, which ensure that the excretion of these substances will be balanced with the dietary intake. Thus, toxic excesses are avoided, while minerals that are present in limited amounts are carefully conserved.

---

1. How do minerals differ from other nutrients?
2. What are the major functions of minerals?
3. What are the most abundant minerals in the body?

Major Minerals

Calcium and phosphorus account for nearly 75% of the mineral elements in the body, and thus they are **major minerals.** Other major minerals, each of which accounts for 0.05% or more of the body weight, include potassium, sulfur, sodium, chlorine, and magnesium.

The major minerals are described as follows:

1. **Calcium.** Calcium (Ca) is widely distributed in cells and body fluids, even though 99% of the body's supply occurs in the inorganic salts of the bones and teeth. It is essential for nerve impulse conduction, muscle fiber contraction, and blood coagulation. It also serves to decrease the permeability of cell membranes and to activate certain enzymes.

The amount of calcium that is absorbed varies with a number of factors. For example, the proportion of calcium absorbed increases as the body's need for calcium increases. Vitamin D and high protein intake promote calcium absorption; increased motility of the digestive tract or an excessive intake of fats causes a decreased absorption. Consequently, the amount of calcium needed to provide an adequate cellular supply may vary from time to time. However, nutritionists believe a daily intake of 800 mg is sufficient to cover adult needs in spite of variations in absorption.

Unfortunately, only a few foods contain significant amounts of calcium. Milk and milk products are the richest sources, while leafy green vegetables such as mustard greens, turnip greens, and kale provide good sources. Since these vegetables are not very popular foods in the United States, it is difficult to maintain an adequate intake of calcium unless milk or milk products such as cheese, cottage cheese, and ice cream are regularly incorporated in the diet.

A deficiency of calcium in children may be accompanied by stunted growth, misshapen bones, and enlarged wrists and ankles. In adults such a deficiency may result in removal of calcium from the bones. As this occurs, the bones may become thinner and more fragile and may fracture easily.

2. **Phosphorus.** Phosphorus (P) is responsible for about 1% of the total body weight, and most of it is incorporated in the calcium phosphate of the bones and teeth. The remainder is distributed throughout the body cells, where it serves as a structural component and plays important roles in nearly all metabolic reactions. More specifically, phosphorus is a constituent of nucleic acids, many proteins, some enzymes, and some vitamins. It also occurs in the phospholipids of cell membranes, the energy-carrying molecules of ATP, and the phosphates of body fluids that function in the regulation of pH. (The molecular structure of ATP is shown in fig. 4.7.)

The recommended daily adult intake of phosphorus is 800 mg, and since this mineral is abundant in protein foods, diets adequate in proteins are also adequate in phosphorus. In addition to being found in protein foods such as meats, poultry, fish, cheese, and nuts, phosphorus occurs in whole grain cereals, milk, and legumes.

1. What are the functions of calcium?
2. What are the functions of phosphorus?
3. What foods are good sources of these minerals?

3. **Potassium.** Potassium (K) is widely distributed throughout the body and tends to be concentrated inside cells rather than in extracellular fluids. On the other hand, *sodium,* which has similar chemical properties, tends to be concentrated outside the cells. More specifically, the ratio of potassium to sodium within a cell is 10:1, while the ratio outside the cell is 1:28 (fig. 14.21).

Within cells, potassium functions to help maintain the intercellular osmotic pressure and to regulate pH. It promotes reactions involved in carbohydrate and protein metabolism and plays a vital role in the membrane polarization that occurs in nerve impulse conduction and muscle fiber contraction.

Nutritionists recommend a daily adult intake of 2.5 grams (2500 mg) of potassium, and since this mineral is widely distributed in foods, it is estimated that a typical adult diet provides between 2 and 6 grams each day. Although deficiencies of potassium due to an inadequate diet are relatively rare, they may occur for other reasons. For example, when a person has diarrhea, the intestinal contents may pass through the digestive tract so rapidly that potassium abosrption is greatly reduced. Vomiting or using diuretic drugs also may lead to potassium depletion. The consequences of such losses may include muscular weakness, cardiac abnormalities, and edema.

Foods rich in potassium are avocados, dried apricots, meats, milk, peanut butter, potatoes, and bananas. Citrus fruits, citrus juices, apples, carrots, and tomatoes provide lesser amounts.

1. How is potassium distributed in the body?
2. What is the function of potassium?
3. What foods are good sources of this mineral?

4. **Sulfur.** Sulfur (S) is responsible for about 0.25% of the body weight and is widely distributed through the tissues. It is particularly abundant in the skin, hair, and nails. Most of it is incorporated in molecules of the amino acids *methionine* and *cysteine.* Other sulfur-containing compounds include thiamine, insulin, and biotin. (See fig. 14.22.) In addition, sulfur is a constituent of mucopolysaccharides that are found in cartilage, tendons, and bones, and of sulfolipids that occur in the liver, kidneys, salivary glands, and brain.

Fig. 14.21 Potassium tends to be more concentrated inside cells, and sodium tends to be more concentrated in the body fluids outside cells.

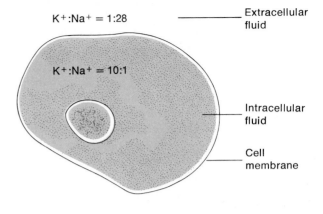

$K^+:Na^+ = 1:28$ — Extracellular fluid

$K^+:Na^+ = 10:1$

Intracellular fluid

Cell membrane

Fig. 14.22 Three examples of essential sulfur-containing nutrients.

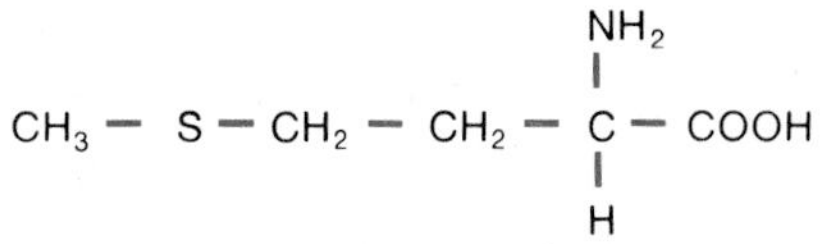

$$CH_3 - S - CH_2 - CH_2 - \underset{\underset{H}{|}}{\overset{\overset{NH_2}{|}}{C}} - COOH$$

Methionine

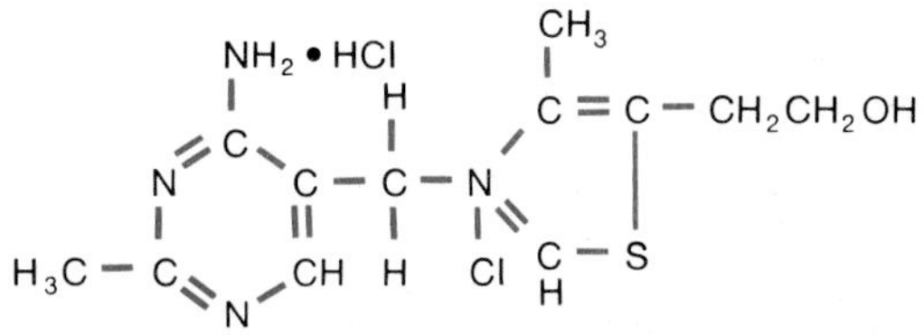

Thiamine hydrochloride (vitamin B₁)

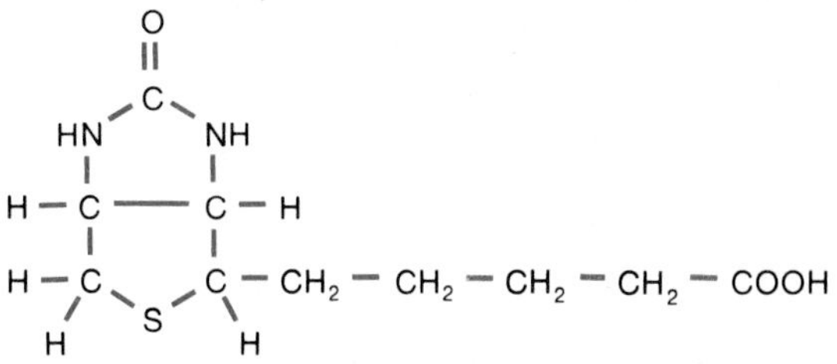

Biotin

No daily requirement for sulfur has been established. It is thought, however, that a diet providing adequate amounts of protein will also meet the body's need for sulfur. Good food sources of this mineral include meats, milk, eggs, and legumes.

5. **Sodium.** It is estimated that about 0.15% of the adult weight is due to sodium (Na), which is widely distributed throughout the body. Only about 10% of this mineral occurs inside the cells, and about 40% is found within the extracellular fluids. The remainder is bonded to the inorganic salts of bones.

Sodium is readily absorbed from foods by active transport, and the blood concentration of this element is regulated by the kidneys under the influence of the adrenal hormone *aldosterone*.

Sodium helps to maintain the osmotic concentrations of extracellular fluids and thus serves to regulate the water balance between cells and their surroundings. Its presence is necessary for nerve impulse conduction and for muscle fiber contraction, and it aids in regulating pH and transporting various substances across cell membranes.

Although a daily requirement for sodium has not been established, it is believed that the usual human diet provides more than enough to meet the body's need. On the other hand, sodium may be lost as a result of diarrhea, vomiting, kidney disorders, sweating, or using diuretics. Such losses may cause a variety of symptoms including edema, muscular cramps, and convulsions.

The amount of sodium naturally present in foods varies greatly, and it is commonly added to foods in the form of table salt (sodium chloride). In some geographic regions, there are significant concentrations of sodium in the drinking water.

Foods that have relatively high sodium contents include cured ham, sauerkraut, cheese, and graham crackers.

A limited intake of sodium is likely to be accompanied by a drop in blood pressure. Because of this, persons with cardiovascular disease are sometimes advised to restrict the amount of sodium in their diets.

---

1. In what compounds and tissues of the body is sulfur found?
2. What hormone regulates the blood concentration of sodium?
3. What are the functions of sodium?

6. **Chlorine.** Chlorine (Cl) in the form of chloride ions is closely associated with sodium. Like sodium, it is widely distributed throughout the body, although it is most highly concentrated in cerebrospinal fluid and in gastric juice.

Together with sodium, chlorine helps to maintain the osmotic concentrations of extracellular fluids, regulate pH, and maintain electrolyte balance. It is also essential for the formation of hydrochloric acid in gastric juice, and it functions in the transport of carbon dioxide by red blood cells.

Chlorine and sodium are usually obtained together in the form of table salt (sodium chloride), and as in the case of sodium, an ordinary diet usually provides considerably more chlorine than the body needs. It too can be lost excessively as a result of vomiting, diarrhea, kidney disorders, sweating, or using diuretics.

7. **Magnesium.** Magnesium (Mg) is responsible for about 0.05% of the body weight and is found in all cells. It is particularly abundant in bones in the form of phosphates and carbonates.

This mineral functions in a number of metabolic reactions including many that occur within mitochondria and are associated with the production of ATP. It also plays a role in the conversion of ATP to ADP, and thus is important in providing energy for cellular processes.

When the intake of magnesium is high, a smaller percentage is absorbed from the intestinal tract, and when the intake is low, a larger percentage is absorbed. Absorption also increases as the protein intake increases, and it decreases as the calcium and vitamin D intake increases. A reserve supply of magnesium is stored in bone tissue, and excess amounts are excreted in the urine under the influence of *aldosterone*.

The recommended daily allowance of magnesium is 300 mg for females and 350 mg for males. Because a typical diet usually provides only about 120 mg of magnesium for every 1000 calories, it may barely meet the body's need.

Good sources of magnesium include milk and dairy products (except butter), legumes, nuts, and leafy green vegetables.

The major minerals are summarized in chart 14.9.

---

1. Where is chlorine most highly concentrated in the body?
2. Where is magnesium stored?
3. What factors influence the absorption of magnesium from the intestinal tract?

Chart 14.9   Major minerals

Mineral	Distribution	Functions	Sources
Calcium (Ca)	Mostly in the inorganic salts of bones and teeth	Structure of bones and teeth; essential for nerve impulse conduction, muscle fiber contraction, and blood coagulation; increases permeability of cell membranes; activates certain enzymes	Milk, milk products, leafy green vegetables
Phosphorus (P)	Mostly in the inorganic salts of bones and teeth	Structure of bones and teeth; component in nearly all metabolic reactions; constituent of nucleic acids, many proteins, some enzymes, and some vitamins; occurs in cell membrane, ATP, and phosphates of body fluids	Meats, poultry, fish, cheese, nuts, whole grain cereals, milk, legumes
Potassium (K)	Widely distributed; tends to be concentrated inside cells	Helps maintain intracellular osmotic pressure and regulate pH; promotes metabolism; needed for nerve impulse conduction and muscle fiber contraction	Avocados, dried apricots, meats, nuts, potatoes, bananas
Sulfur (S)	Widely distributed	Essential part of various amino acids, thiamine, insulin, biotin, and mucopolysaccharides	Meats, milk, eggs, legumes
Sodium (Na)	Widely distributed; large proportion occurs in extracellular fluids and bonded to inorganic salts of bone	Helps maintain osmotic pressure of extracellular fluids and regulate water balance; needed for conduction of nerve impulses and contraction of muscle fibers; aids in regulation of pH and in transport of substances across cell membranes	Table salt, cured ham, sauerkraut, cheese, graham crackers
Chlorine (Cl)	Closely associated with sodium; most highly concentrated in cerebrospinal fluid and gastric juice	Helps maintain osmotic pressure of extracellular fluids, regulate pH, and maintain electrolyte balance; essential in formation of hydrochloric acid; aids transport of carbon dioxide by red blood cells	Same as for sodium
Magnesium (Mg)	Abundant in bones	Needed in metabolic reactions that occur in mitochondria and are associated with the production of ATP; plays role in conversion of ATP to ADP	Milk, dairy products, legumes, nuts, leafy green vegetables

Trace elements are essential minerals that occur in minute amounts, each one making up less than 0.005% of the adult body weight. They include iron, manganese, copper, iodine, cobalt, and zinc.

Iron (Fe) is found primarily in the blood, although some is present in all body cells, and a reserve supply is stored in the liver, spleen, and bone marrow. It functions as part of the *hemoglobin* molecules in red blood cells and is responsible for the ability of these molecules to carry oxygen. (See fig. 14.23.) Iron also occurs in *myoglobin* molecules, which are synthesized in muscle cells and act to store oxygen temporarily. In addition, iron functions to catalyze the formation of vitamin A and is incorporated into a number of enzymes.

It has been estimated that an adult male requires from 0.7 to 1 mg of iron daily, while females need 1.2 to 2 mg. It is also believed that although a typical diet supplies about 10–18 mg of iron each day, only 2 to 10% of this iron is absorbed. Such a diet provides a narrow margin of safety as far as iron intake is concerned, and it may fail to provide the amount needed by some individuals.

Pregnant women require additional quantities of iron in order to support the formation of a placenta, the growth and development of a fetus, and the synthesis of hemoglobin that is associated with the increase in maternal blood volume that occurs during pregnancy.

Liver is the only really rich source of dietary iron, and since liver is not a very popular food, iron is one of the more difficult nutrients to obtain from natural sources in adequate amounts. Foods that contain some iron include lean meats, dried apricots, raisins, and prunes, enriched whole grain cereals, legumes, and molasses.

Manganese (Mn) is most concentrated in the liver, kidneys, and pancreas. It is necessary for normal growth and development of skeletal structures and other connective tissues, and occurs in enzymes that are essential for the synthesis of fatty acids and cholesterol, for the formation of urea, and for the normal functions of the nervous system.

Fig. 14.23 A hemoglobin molecule contains four heme portions, each of which contains a single iron atom (Fe) that can combine with oxygen.

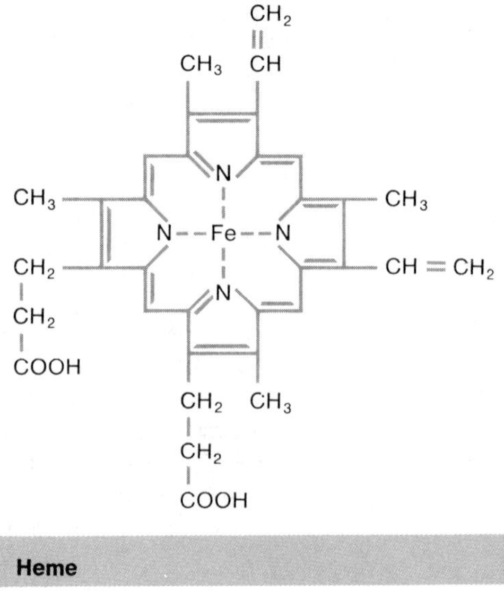

Heme

The daily requirement for manganese is unknown. The richest sources include nuts, legumes, and whole grain cereals; leafy green vegetables and fruits are good sources.

---

1. *What is the primary function of iron?*
2. *Why does the usual diet provide only a narrow margin of safety in supplying iron?*
3. *How is manganese utilized?*
4. *What foods are good sources of manganese?*

Copper (Cu) is found in all body tissues, but is most highly concentrated in the liver, heart, and brain. It is essential for the synthesis of hemoglobin, the normal development of bone, the production of melanin, and the formation of myelin within the nervous system.

It is believed that a daily intake of 2 mg of copper is sufficient to provide the needs of body cells and that a typical adult diet supplies about 2–5 mg of this mineral. Adults seldom develop copper deficiencies.

Foods rich in copper include liver, oysters, crabmeat, nuts, whole grain cereals, and legumes.

Iodine (I) occurs in minute quantities in all tissues, but is highly concentrated within the thyroid gland. Its only known function is to provide an essential component for the synthesis of *thyroid hormones*. (The molecular structures of two of these hormones are shown in fig. 12.19.)

Chart 14.10 Trace elements

Trace Element	Distribution	Functions	Sources
Iron (Fe)	Primarily in blood; stored in liver, spleen, and bone marrow	Part of hemoglobin molecule; catalyzes formation of vitamin A; incorporated into a number of enzymes	Liver, lean meats, dried apricots, raisins, enriched whole grain cereals, legumes, molasses
Manganese (Mn)	Most concentrated in liver, kidneys, and pancreas	Occurs in enzymes needed for synthesis of fatty acids and cholesterol, formation of urea, and normal functioning of the nervous system	Nuts, legumes, whole grain cereals, leafy green vegetables, fruits
Copper (Cu)	Most highly concentrated in liver, heart, and brain	Essential for synthesis of hemoglobin, development of bone, production of melanin, and formation of myelin	Liver, oysters, crabmeat, nuts, whole grain cereals, legumes
Iodine (I)	Concentrated in thyroid gland	Essential component for synthesis of thyroid hormones	Food content varies with soil content in different geographic regions; iodized table salt
Cobalt (Co)	Widely distributed	Component of cyanocobalamin; needed for synthesis of several enzymes	Liver, lean meats, poultry, fish, milk
Zinc (Zn)	Most concentrated in liver, kidneys, and brain	Constituent of several enzymes involved in digestion, respiration, bone metabolism, liver metabolism; necessary for normal wound healing and maintaining integrity of the skin	Seafoods, meats, cereals, legumes, nuts, vegetables

An intake of 1 microgram (.001 mg) of iodine per kilogram of body weight daily is thought to provide an adequate amount for most adults. Since the iodine content of foods varies with the iodine content of soils in different geographic regions, it is sometimes necessary to increase artificially the iodine in foods to prevent deficiencies from developing. The use of iodized table salt, which contains artificially added *potassium iodide* (KI), has proved to be an effective means of accomplishing this.

Cobalt (Co) is widely distributed throughout the body, since it is an essential part of *cyanocobalamin* (vitamin B_{12}) molecules. It is also thought to be necessary for the synthesis of several important enzymes.

The amount of cobalt required in the daily diet is unknown. This mineral is found in a great variety of foods, and the quantity present in the average diet is apparently sufficient to meet the body's need. Good sources of cobalt include liver, lean meats, poultry, fish, and milk.

Zinc (Zn) generally occurs in body tissues, but is most concentrated in the liver, kidneys, and brain. It is a constituent of a large number of enzymes involved in digestion, respiration, bone metabolism, and liver metabolism. It is also necessary for normal wound healing and for maintaining the integrity of the skin.

The daily requirement for zinc is believed to be about 15 mg, and most diets are thought to provide 10 to 15 mg. Since only a portion of this amount may be absorbed, zinc deficiencies may occur relatively commonly.

The richest sources of zinc are seafoods and meats; cereals, legumes, nuts, and vegetables provide lesser amounts.

Characteristics of these trace elements are summarized in chart 14.10.

---

1. How is copper utilized?
2. What is the function of iodine?
3. Why are zinc deficiencies thought to occur relatively commonly?

Chart 14.11 Recommended daily dietary allowances for various age groups in the United States

	Infants	Children	Adolescents 15–18 yrs		Adults 23–50 yrs	
	0–6 mos	*4–6 yrs*	*Males*	*Females*	*Males*	*Females*
Weight, kg (lb)	6 (13)	20 (44)	66 (145)	55 (120)	70 (154)	55 (120)
Height, cm (in)	60 (24)	112 (44)	176 (69)	163 (64)	178 (70)	163 (64)
Protein, g	kg × 2.2	30	56	46	56	44
Fat-soluble vitamins†						
Vitamin A, μg	420	500	1000	800	1000	800
Vitamin D, μg	10	10	10	10	5	5
Vitamin E activity, mg	3	6	10	8	10	8
Water soluble vitamins						
Ascorbic acid, mg	35	45	60	60	60	60
Folacin, μg	30	200	400	400	400	400
Niacin, mg	6	11	18	14	18	13
Riboflavin, mg	0.4	1.0	1.7	1.3	1.6	1.2
Thiamine, mg	0.3	0.9	1.4	1.1	1.4	1.0
Vitamin B_6, mg	0.3	1.3	2.0	2.0	2.2	2.0
Vitamin B_{12}, μg	0.5	2.5	3.0	3.0	3.0	3.0
Minerals						
Calcium, mg	360	800	1200	1200	800	800
Phosphorus, mg	240	800	1200	1200	800	800
Iodine, μg	40	90	150	150	150	150
Iron, mg	10	10	18	18	10	18
Magnesium, mg	50	200	400	300	350	300
Zinc, mg	3	10	15	15	15	15

Source: Food and Nutrition Board, *Recommended Dietary Allowances,* 9th ed., National Academy of Sciences—National Research Council, Washington, D.C., 1980.

†Microgram

Adequate Diets

An adequate diet is one that provides sufficient *calories, essential fatty acids, essential amino acids, vitamins,* and *minerals* to support optimal growth and to maintain and repair body tissues. However, since individual needs for nutrients vary greatly with age, sex, growth rate, amount of physical activity, and level of stress, as well as with genetic and environmental factors, it is not possible to design a diet that will be adequate for everyone.

On the other hand, nutrients are so widely distributed in foods that satisfactory amounts and combinations of essential substances usually can be obtained in spite of individual food preferences that are related to cultural backgrounds, lifestyles, and emotional attitudes. Chart 14.11 lists the recommended daily allowance for each of the major nutrients.

Food Selection

Although recommended daily allowances of essential nutrients are often used by nutritionists and dietitians as guides in planning adequate diets, these values are usually of limited use to the average person. A more useful aid for most persons is the idea of *basic food groups.*

A basic food group is a class of foods that will supply sufficient amounts of certain essential nutrients when a given quantity of food from the group is included in the diet. For example, one basic food plan includes four food groups, each of which provides a unique contribution toward achieving an adequate diet. In this plan, the food groups and the quantities of food required from each are as follows:

Group 1: Milk and Dairy Products. This group includes milk, cottage cheese, cream cheese, natural and processed cheese, and ice cream. It provides calcium, phosphorus, magnesium, protein, vitamin B_6, vitamin B_{12}, and riboflavin. It is recommended that an adult have two 8-ounce glasses of milk or the equivalent in other dairy foods each day.

Group 2: Fruits and Vegetables. This group includes all fruits and vegetables, and it provides vitamin C, vitamin A, iron, magnesium, and vitamin B_6. It is recommended that an adult diet contain four servings from this group each day, and that one of these be citrus fruit or some other fruit or vegetable that is particularly high in ascorbic acid content. At least every other day, one of the servings should be a dark green, yellow, or orange vegetable.

| Adults | |
| 51 yrs and over | |
Males	*Females*
70 (154)	55 (120)
178 (70)	163 (64)
56	44
1000	800
5	5
10	8
60	60
400	400
16	13
1.4	1.2
1.2	1.0
2.2	2.0
3.0	3.0
800	800
800	800
150	150
10	10
350	300
15	15

Group 3: Meat, Poultry, and Fish. This group includes all meats, poultry, and fish as well as meat substitutes such as eggs, beans, peas, and nuts. It provides protein, vitamin A, thiamine, niacin, vitamin B_6, vitamin B_{12}, riboflavin, phosphorus, magnesium, and iron. For an adult, two or more servings from this group are recommended each day.

Group 4: Breads and Cereals. This group includes all whole grain or enriched breads and cereals. It provides iron, thiamine, niacin, riboflavin, protein, phosphorus, and magnesium. Four or more servings per day from this group are recommended for adults.

Chart 14.12 provides a food guide based on these basic food groups.

Malnutrition

Malnutrition is poor nutrition that results from a lack of essential nutrients or a failure to use available foods to best advantage. It may involve *undernutrition* and include the symptoms of deficiency diseases, or it may be due to *overnutrition* arising from an excessive intake of nutrients.

The factors leading to malnutrition are varied. A deficiency condition may, for example, stem from lack of availability or poor quality of food. On the

Chart 14.12 A daily food guide

Milk Group (8-ounce cups)

2 to 3 cups for children under 9 years

3 or more cups for children 9 to 12 years

4 cups or more for teen-agers

2 cups or more for adults

3 cups or more for pregnant women

4 cups or more for nursing mothers

Meat Group

2 or more servings. Count as one serving:

2 to 3 ounces lean, cooked beef, veal, pork, lamb, poultry, fish—without bone

2 eggs

1 cup cooked dry beans, dry peas, lentils

4 tablespoons peanut butter

Vegetable and Fruit Group (½ cup serving, or 1 piece fruit, etc.)

4 or more servings per day, including:

1 serving of citrus fruit, or other fruit or vegetable as a good source of vitamin C, or 2 servings of a fair source

1 serving, at least every other day, of a dark green or deep yellow vegetable for vitamin A

2 or more servings of other vegetables and fruits, including potatoes

Bread and Cereals Group

4 or more servings daily (whole grain, enriched, or restored). Count as one serving:

1 slice bread

1 ounce ready-to-eat cereal

½ to ¾ cup cooked cereal, corn meal, grits, macaroni noodles, rice, or spaghetti

Source: "A Daily Food Guide" in *Consumers All.* Yearbook of Agriculture, 1965, U.S. Department of Agriculture, Washington, D.C., 1965, p. 394.

other hand, malnutrition may result from excessive intake of vitamin supplements, which causes hypervitaminosis, or from excessive caloric intake. Malnutrition from causes that involve the diet alone is called *primary malnutrition.*

Secondary malnutrition occurs when a person's individual characteristics make a normally adequate diet unsuitable. A person who secretes low quantities of bile salts, for example, may develop a deficiency of fat-soluble vitamins because the presence of bile salts promotes the absorption of these substances. Likewise, severe and prolonged emotional stress may lead to secondary malnutrition, because stress can cause changes in hormonal concentrations, and such changes may result in abnormal breakdown of amino acids or in excessive excretion of various nutrients.

Since older persons are generally less active, they need fewer calories to maintain desirable weights, but at the same time, they need to have favorable intakes of essential nutrients. Thus, if they eat simply to satisfy hunger, they may develop symptoms of malnutrition resulting from deficiencies of such nutrients as iron, magnesium, calcium, vitamin A, vitamin D, vitamin C, or from excesses of sodium, chlorine, or phosphorus.

1. What is meant by an adequate diet?
2. What factors influence individual needs for nutrients?
3. What are the basic food groups?
4. What is meant by primary malnutrition? By secondary malnutrition?

Some Terms Related to Nutrients and Nutrition

anorexia (an″o-rek′se-ah)—a loss of appetite.

casein (ka′se-in)—the primary protein found in milk.

celiac disease (se′le-ak dǐ-zēz′)—a digestive disorder characterized by the inability to digest or utilize fats and carbohydrates.

emaciation (e-ma″se-a′shun)—excessive leanness of the body due to tissue wasting.

extrinsic factor (ek-strin′sik fak′tor)—vitamin B_{12}.

hyperalimentation (hi″per-al″ǐ men-ta′shun)—prolonged intravenous nutrition provided for patients with severe digestive disorders.

hypercalcemia (hi″per-kal-se′me-ah)—an excessive level of calcium in the blood.

hypercalciuria (hi″per-kal″se-u′re-ah)—an excessive excretion of calcium in the urine.

hyperglycemia (hi″per-gli-se′me-ah)—an excessive level of glucose in the blood.

hyperkalemia (hi″per-kah-le′me-ah)—an excessive level of potassium in the blood.

hypoalbuminemia (hi″po-al-bu″mǐ-ne′me-ah)—a low level of albumin in the blood.

hypoglycemia (hi″po-gli-se′me-ah)—a low level of glucose in the blood.

hypokalemia (hi″po-kah-le′me-ah)—a low level of potassium in the blood.

isocaloric (i″so-kah-lo′rik)—containing equal amounts of heat energy.

lipogenesis (lip″o-jen′ě-sis)—the formation of fat.

marasmus (mah-raz′mus)—an extreme form of protein and calorie malnutrition.

nyctalopia (nik″tah-lo′pe-ah)—night blindness.

pica (pi′kah)—a hunger for substances that are not suitable foods.

polyphagia (pol″e-fa′je-ah)—excessive intake of food.

proteinuria (pro″te-ǐ-nu′re-ah)—presence of protein in the urine.

provitamin (pro-vi′tah-min)—a substance used as a precursor in vitamin synthesis.

Chapter Summary

Introduction
Nutrients include carbohydrates, lipids, proteins, vitamins, and minerals.
Essential nutrients are needed for health and cannot be synthesized by body cells.

Carbohydrates
Carbohydrates are organic compounds that are used primarily to supply cellular energy.
1. Sources of carbohydrates
   a. Carbohydrates are ingested in a variety of forms.
   b. Starch, glycogen, disaccharides, and monosaccharides are carbohydrates.
2. Utilization of carbohydrates
   a. They are absorbed as monosaccharides.
   b. Fructose and galactose are converted to glucose by the liver.
   c. Energy is released from glucose by oxidation.
   d. Excessive glucose is stored as glycogen or converted to fat.
3. Carbohydrate requirements
   a. Most are utilized to supply energy, although some are used to produce important sugars.
   b. Some cells depend on a continuous supply of glucose to survive.
   c. If inadequate amounts of glucose are available, amino acids may be converted into glucose.
   d. Humans survive with a wide range of carbohydrate intakes.
   e. Poor nutritional status is usually related to low intake of nutrients other than carbohydrates.

Lipids
Lipids are organic compounds that supply energy and are used to build cell structures. They include fats, phospholipids, and cholesterol.
1. Sources of lipids
   a. Triglycerides are obtained from foods of plant and animal origins.
   b. Cholesterol is obtained in foods of animal origin only.
2. Utilization of lipids
   a. Metabolism of triglycerides is controlled mainly by the liver and adipose tissues.
   b. The liver can alter the molecular structures of fatty acids.
   c. Linoleic acid is an essential fatty acid.
   d. Liver also regulates the amount of cholesterol by synthesizing or excreting it.
3. Lipid requirements
   a. Humans survive with a wide range of lipid intakes.

b. The amounts and types of lipids needed for health are unknown.

c. Some fats contain fat-soluble vitamins, and the intake of fats must be sufficient to supply these essential nutrients.

Proteins

Proteins are organic compounds that serve as structural materials, act as enzymes, and provide energy.

1. Sources of proteins
   a. Proteins are obtained mainly from meats, dairy products, cereals, and legumes.
   b. During digestion they are broken down into amino acids.
   c. Eight amino acids are essential for adults, while ten are essential in growing children.
   d. Complete proteins contain adequate amounts of all the essential amino acids to maintain the tissues and to promote growth.
   e. Incomplete proteins lack one or more essential amino acids.

2. Utilization of amino acids
   a. Amino acids are incorporated into various structural and functional proteins, including enzymes.
   b. During starvation, tissue proteins may be used as energy sources, and thus the tissues waste away.

3. Nitrogen balance
   a. In healthy adults, the gain of protein equals the loss of protein, and a nitrogen balance exists.
   b. A starving person has a negative nitrogen balance, while a growing child, a pregnant woman, or an athlete in training usually has a positive nitrogen balance.

4. Protein requirements
   a. Proteins and amino acids are needed to supply essential amino acids and nitrogen for the synthesis of various nitrogen-containing molecules.
   b. The consequences of protein deficiencies are particularly severe among growing children.

Energy Expenditures

Energy is of prime importance to survival and may be obtained from carbohydrates, fats, or proteins.

1. Energy values of foods
   a. The potential energy values of foods are expressed in calories.
   b. When energy losses due to incomplete absorption and incomplete oxidation are taken into account, one gram of carbohydrate or one gram of protein yields about 4 calories, while one gram of fat yields about 9 calories.

2. Energy requirements
   a. The amount of energy required varies from person to person.
   b. Factors that influence energy requirements include basal metabolic rate, muscular activity, body temperature, and nitrogen balance.

3. Energy balance
   a. Energy balance exists when caloric intake equals caloric output.
   b. If balance is positive, body weight increases; if balance is negative, body weight decreases.

4. Desirable weight
   a. The most common nutritional disorders involve caloric imbalances.
   b. Average weights of persons 25–30 years of age are thought to be desirable for all older persons.
      (1) A person who exceeds the desirable weight by 10–20% is called overweight.
      (2) A person whose body contains an excess of fatty tissue is said to be obese.

Vitamins

Vitamins are organic compounds (other than carbohydrates, lipids, and proteins) that are essential for normal metabolic processes and cannot be synthesized by body cells in adequate amounts.

1. Fat-soluble vitamins
   a. General characteristics
      (1) Occur in association with lipids and are influenced by the same factors that affect lipid absorption.
      (2) Fairly resistant to the effects of heat, and thus are not destroyed by cooking or food processing.
   b. Vitamin A
      (1) Occurs in several forms, is synthesized from carotenes, and is stored in the liver.
      (2) Functions in the production of pigments necessary for vision.
   c. Vitamin D
      (1) Represented by a group of related sterols.
      (2) Vitamin D_3 is produced in the skin when it is exposed to ultraviolet light; vitamin D_2 occurs in certain foods.
      (3) Vitamin D promotes the absorption of calcium and phosphorus.
   d. Vitamin E
      (1) Represented by a group of compounds that are antioxidants.
      (2) Stored in muscles and adipose tissue.
      (3) Precise functions are unknown, but seems to prevent oxidation of vitamin A and polyunsaturated fatty acids, and to stabilize cell membranes.
   e. Vitamin K
      (1) Vitamin K_1 occurs in foods; K_2 is produced by certain bacteria that normally inhabit the intestinal tract.
      (2) Stored to a limited degree in the liver.
      (3) Utilized in the production of prothrombin that is needed for normal blood clotting.

2. Water-soluble vitamins
   a. General characteristics
      (1) Group includes the B vitamins and vitamin C.

(2) B vitamins make up a group called the B complex and are generally involved with the oxidation of carbohydrates, lipids, and proteins.

 b. Vitamin B complex
  (1) Thiamine
   (a) Functions as parts of coenzymes that act in the oxidation of carbohydrates and in the synthesis of essential sugars.
   (b) Small amounts are stored in the tissues; excess is excreted in the urine.
   (c) Quantities needed vary with caloric intake.
  (2) Riboflavin
   (a) Functions as parts of several enzymes and coenzymes that are essential to the oxidation of glucose and fatty acids.
   (b) Absorption is regulated by an active transport system; excess is excreted in the urine.
   (c) Quantities needed vary with caloric intake.
  (3) Niacin
   (a) Functions as parts of coenzymes needed for the oxidation of glucose and for the synthesis of proteins and fats.
   (b) Can be synthesized from tryptophan; daily requirement varies with the tryptophan intake.
  (4) Vitamin B_6
   (a) A group of compounds that function as coenzymes needed by a variety of metabolic pathways involved in the synthesis of proteins, various amino acids, antibodies, and nucleic acids.
   (b) Requirement varies with protein intake.
  (5) Pantothenic acid
   (a) Functions as part of coenzyme A and is thus essential for energy-releasing mechanisms.
   (b) Daily requirement is not known.
  (6) Cyanocobalamin
   (a) Molecule contains cobalt.
   (b) Absorption is regulated by the secretion of intrinsic factor from the gastric glands.
   (c) Functions as part of coenzymes needed for the synthesis of nucleic acids and for the metabolism of carbohydrates and fats.
  (7) Folacin
   (a) Converted by the liver to physiologically active folinic acid.
   (b) Functions as a coenzyme needed for the metabolism of certain amino acids, the synthesis of DNA, and the normal production of red blood cells.

  (8) Biotin
   (a) Functions as a coenzyme needed for the metabolism of amino acids and fatty acids, and for the synthesis of purines.
   (b) Stored in metabolically active organs.
 c. Ascorbic acid (vitamin C)
  (1) Closely related chemically to monosaccharides.
  (2) Functions are poorly understood but is thought to be needed for the production of collagen, the metabolism of certain amino acids, and the absorption of iron.
  (3) Not stored in any great amounts; excess is excreted in the urine.

Minerals

1. Characteristics of minerals
 a. Responsible for about 4% of body weight.
 b. About 75% of the minerals are found in bones and teeth as calcium and phosphorus.
 c. Minerals are usually incorporated into organic molecules, although some occur in inorganic compounds or as free ions.
 d. They comprise structural materials, function in enzymes, and play vital roles in various metabolic processes.
 e. Mineral concentrations are generally regulated by homeostatic mechanisms.
2. Major minerals
 a. Calcium
  (1) Essential for the formation of bones and teeth, the conduction of nerve impulses, the contraction of muscle fibers, the coagulation of blood, and the activation of various enzymes.
  (2) Absorption is affected by existing calcium concentration, vitamin D, protein intake, and motility of the digestive tract.
 b. Phosphorus
  (1) Incorporated for the most part into the salts of bones and teeth.
  (2) Plays roles in nearly all metabolic reactions as a constituent of nucleic acids, proteins, enzymes, and some vitamins.
  (3) Also occurs in the phospholipids of cell membranes, in ATP, and in phosphates of body fluids.
 c. Potassium
  (1) Tends to be concentrated inside cells.
  (2) Functions in maintenance of osmotic pressure, regulation of pH, metabolism of carbohydrates and proteins, the conduction of nerve impulses, and the contraction of muscle fibers.
 d. Sulfur
  (1) Incorporated for the most part into the molecular structures of certain amino acids.
  (2) Also included in thiamine, insulin, biotin, and mucopolysaccharides.

e. Sodium
   (1) Most occurs in extracellular fluids or is bonded to the inorganic salts of bone.
   (2) Blood concentration is regulated by the kidneys under the influence of aldosterone.
   (3) Helps maintain osmotic concentrations and regulate water balance and pH.
   (4) Essential for the conduction of nerve impulses, the contraction of muscle fibers, and the movement of substances across cell membranes.
f. Chlorine
   (1) Closely associated with sodium in the form of chloride ions.
   (2) Acts with sodium to help maintain osmotic pressure, regulate pH, and maintain electrolyte balance.
   (3) Essential for the formation of hydrochloric acid and for the transport of carbon dioxide by red blood cells.
g. Magnesium
   (1) Particularly abundant in the bones as phosphates and carbonates.
   (2) Functions in the production of ATP and in the conversion of ATP to ADP.
   (3) Reserve supply stored in the bones; excesses excreted in the urine.
3. Trace elements
   a. Iron
      (1) Occurs primarily in hemoglobin of red blood cells and in myoglobin of muscles.
      (2) Reserve supply stored in the liver, spleen, and bone marrow.
      (3) Needed to catalyze the formation of vitamin A; incorporated into various enzymes.
   b. Manganese
      (1) Most concentrated in the liver, kidneys, and pancreas.
      (2) Necessary for normal growth and development of skeletal structures and other connective tissues; essential for the synthesis of fatty acids, cholesterol, and urea.
   c. Copper
      (1) Most concentrated in the liver, heart, and brain.
      (2) Needed for synthesis of hemoglobin, development of bones, production of melanin, and formation of myelin.
   d. Iodine
      (1) Most highly concentrated in the thyroid gland.
      (2) Provides an essential component for the synthesis of thyroid hormones.
      (3) Often added to foods in the form of iodized table salt.
   e. Cobalt
      (1) Widely distributed throughout the body.
      (2) Essential part of cyanocobalamin and probably needed for the synthesis of several important enzymes.

f. Zinc
   (1) Most concentrated in the liver, kidneys, and brain.
   (2) A constituent of several enzymes involved with digestion, respiration, bone metabolism, and liver metabolism.

Adequate Diets

An adequate diet provides sufficient calories and essential nutrients to support optimal growth and maintenance and repair of tissues.

Individual needs vary so greatly that it is not possible to design a diet that is adequate for everyone.

1. Food selection
   a. Basic food groups are a useful aid for planning an adequate diet.
   b. A basic food group is a class of foods that will supply certain essential nutrients in sufficient amounts if a given quantity of food from the group is included in the daily diet.
   c. One plan includes four basic food groups: milk and dairy products; fruits and vegetables; meat, poultry, and fish; breads and cereals.
2. Malnutrition
   a. Poor nutrition is due to lack of foods or failure to make best use of available foods.
   b. Primary malnutrition is due to poor diet.
   c. Secondary malnutrition is due to an individual characteristic that makes a normal diet unsuitable.

Application of Knowledge

1. How would you explain the fact that the blood sugar level of a person whose diet is relatively low in carbohydrates remains stable?

2. Calculate the carbohydrate, lipid, protein, and caloric content of your diet for a 24-hour period. Assuming that this 24-hour sample is representative of your normal eating habits, what improvements could be made in its composition?

3. Examine the label information on the packages of a variety of dry breakfast cereals. Which types of cereals provide the best sources of vitamins and minerals?

Review Activities

1. Define *essential nutrient*.

2. List some common sources of carbohydrates.

3. Explain what happens to excessive amounts of glucose in the body.

4. Explain why a temporary drop in the glucose level may produce functional disorders of the nervous system.

5. List some of the factors that affect an individual's need for carbohydrates.

6. Define *triglyceride*.

7. List some common sources of lipids.

8. Describe the role of the liver in fat metabolism.

9. Explain the function of adipose tissue in regulating the level of blood lipids.

10. Discuss the functions of cholesterol.

11. List some common sources of proteins.

12. Distinguish between essential and nonessential amino acids.

13. Distinguish between complete and incomplete proteins.

14. Review the major functions of amino acids.

15. Define *nitrogen balance*.

16. Explain why a protein deficiency may be accompanied by edema.

17. Define *calorie*.

18. Explain how the caloric values of foods are determined.

19. Define *basal metabolic rate*.

20. Explain how a BMR is determined.

21. List some of the factors that affect the BMR.

22. Define *energy balance*.

23. Explain how a person's desirable weight is determined.

24. Distinguish between overweight and obesity.

25. Discuss the general characteristics of fat-soluble vitamins.

26. List the fat-soluble vitamins and describe the major functions of each.

27. List some good sources for each of the fat-soluble vitamins.

28. Explain what is meant by the vitamin B complex.

29. List the water-soluble vitamins and describe the major functions of each.

30. List some good sources for each of the water-soluble vitamins.

31. Discuss the general characteristics of the mineral nutrients.

32. List the major minerals and describe the major functions of each.

33. List some good sources for each of the major minerals.

34. Distinguish between a major mineral and a trace element.

35. List the trace elements and describe the major functions of each.

36. List some good sources for each of the trace elements.

37. Define *adequate diet*.

38. Explain how a basic food group plan can be used to achieve an adequate diet.

39. Define *malnutrition*.

40. Distinguish between primary and secondary malnutrition.

Suggestions for Additional Reading

Connell, A. M. March 1976. Dietary fiber and diverticular disease. *Hosp. Prac.*

Danowski, T. S. April 1976. The management of obesity. *Hosp. Prac.*

Deevey, E. S. September 1970. Mineral cycles. *Scientific American.*

Delwiche, C. C. September 1970. The nitrogen cycle. *Scientific American.*

Fernstrom, J. D., and Wurtman, R. J. February 1974. Nutrition and the brain. *Scientific American.*

Fleck, H. 1976. *Introduction to nutrition.* 3rd ed. New York: Macmillan.

Freedland, R. A., and Briggs, S. 1977. *A biochemical approach to nutrition.* New York: Halsted Press.

Gillie, R. B. June 1971. Endemic goiter. *Scientific American.*

Guthrie, H. A. 1975. *Introductory nutrition.* St. Louis: C. V. Mosby.

Harpstead, D. D. August 1971. High-lysine corn. *Scientific American.*

Lappe, M. 1975. *Diet for a small planet.* Rev. ed. New York: Ballantine Books.

Maugh, T. H. 1973. Trace elements: a growing appreciation of the effects on man. *Science* 181:253.

Mayer, J. 1974. *Human nutrition.* Springfield, Ill.: Charles C. Thomas, Publisher.

Robinson, C. H. 1973. *Fundamentals of normal nutrition.* 2nd ed. New York: Macmillan.

Scrimshaw, N. S., and Young, V. R. September 1976. The requirements of human nutrition. *Scientific American.*

Williams, E. R. December 1975. Making vegetarian diets nutritious. *Amer. J. Nurs.*

Winick, M. June 1974. Childhood obesity. *Nutrition Today.*

Young, V. R., and Scrimshaw, N. S. October 1971. The physiology of starvation. *Scientific American.*

The Respiratory System

15 Before body cells can oxidize nutrients as sources of energy, they must be supplied with oxygen. Also, the carbon dioxide that results from the oxidation must be excreted. These two general processes—obtaining oxygen and removing carbon dioxide—are the primary functions of the *respiratory system.*

In addition, the respiratory organs filter particles from incoming air, help to control the temperature and water content of the air, aid in producing the sounds used in speech, and play important roles in the sense of smell and the regulation of pH.

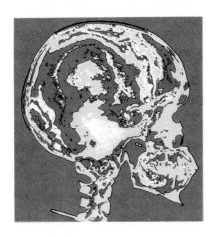

After you have studied this chapter, you should be able to

1. List the general functions of the respiratory system.

2. Name and describe the location of the organs of the respiratory system.

3. Describe the functions of each organ of the respiratory system.

4. Explain how inspiration and expiration are accomplished.

5. Name and define each of the respiratory air volumes.

6. List several nonrespiratory air movements and explain how each occurs.

7. Explain how various respiratory disorders may interfere with breathing.

8. Locate the respiratory center and explain how it controls normal breathing.

9. Discuss how various factors affect the respiratory center.

10. Describe the structure and function of the respiratory membrane.

11. Discuss how various respiratory disorders may affect gas exchanges.

12. Explain how oxygen and carbon dioxide are transported in the blood.

13. Review the major events that occur during cellular respiration.

14. Explain how oxygen is utilized by cells.

15. Complete the review activities at the end of this chapter.

alveolus (al-ve′o-lus)

bronchial tree (brong′ke-al tre)

carbonic anhydrase (kar-bon′ik an-hi′drās)

cellular respiration
(sel′u-lar res′′pĭ-ra′ shun)

citric acid cycle (sit′rik as′id si′kl)

expiration (ek′′spĭ-ra′shun)

glottis (glot′is)

hemoglobin (he′′mo-glo′bin)

hyperventilation (hi′′per-ven′′tĭ-la′shun)

inspiration (in′′spĭ-ra′shun)

partial pressure (par′shıl presh′ur)

pleural cavity (ploo′ral kav′ĭ-te)

respiratory center (re-spi′rah-to′′re sen′ter)

respiratory membrane
(re-spi′rah-to′′re mem′brān)

respiratory volume (re-spi′rah-to′′re vol′ūm)

surface tension (ser′fas ten′shun)

surfactant (ser-fak′tant)

alveol-, a small cavity: *alveol*us—a microscopic air sac within a lung.

bronch-, the windpipe: *bronch*us—a primary branch of the trachea.

carcin-, a spreading sore: *carcin*oma—a type of cancer.

carin-, keellike: *carin*a—a ridge of cartilage between the right and left bronchi.

cric-, a ring: *cric*oid cartilage—a ring-shaped mass of cartilage at the base of the larynx.

hem-, blood: *hem*oglobin—pigment in red blood cells that serves to transport oxygen and carbon dioxide.

tuber-, a swelling: *tuber*culosis—a disease characterized by the formation of fibrous masses within the lungs.

The **respiratory system** consists of a group of passages that filter incoming air and transport it from the outside of the body into the lungs, and numerous microscopic air sacs in which gas exchanges take place.

The parts of the respiratory system, shown in figure 15.1, can be divided into two sets or tracts. Those organs outside the thorax constitute the *upper respiratory tract,* and those within the thorax comprise the *lower respiratory tract.*

Organs of the Respiratory System

The organs of the respiratory system include the nose, nasal cavity, sinuses, pharynx, larynx, trachea, bronchial tree, and lungs.

The Nose

The nose is covered with skin and is supported internally by bone and cartilage. Its two *nostrils* (external nares) provide openings where air can enter and leave the nasal cavity. These openings are guarded by numerous internal hairs that help prevent the entrance of relatively coarse particles sometimes carried by the air.

The Nasal Cavity

The **nasal cavity,** which is a hollow space behind the nose, is divided medially into right and left portions by the **nasal septum.** This cavity is separated from the cranial cavity by the cribriform plate of the ethmoid bone and from the mouth by the hard palate.

The nasal septum is usually straight at birth, although it is sometimes bent as a result of a birth injury. It remains straight throughout early childhood, but with age it tends to bend toward one side or the other. Such a *deviated septum* may create an obstruction in the nasal cavity that makes breathing difficult. In a newborn this problem can be life threatening because a newborn cannot breathe through its mouth.

As figure 15.2 shows, **nasal conchae** (turbinate bones) curl out from the lateral walls of the nasal cavity on each side, dividing the cavity into passageways called the *superior, middle,* and *inferior meatuses.*

Fig. 15.1   Major organs of the respiratory system.

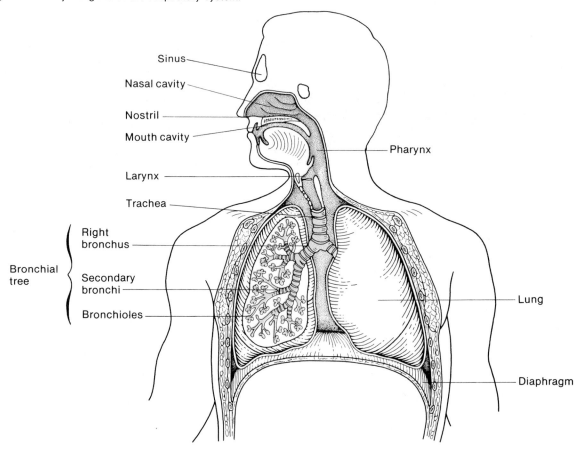

The upper posterior portion of the nasal cavity, below the cribriform plate, is slitlike, and its lining contains the olfactory receptors that function in the sense of smell. The remainder of the cavity serves to conduct air to and from the nasopharynx.

The *mucous membrane* that lines the nasal cavity includes an extensive network of blood vessels and normally appears pinkish. As air passes over the membrane, heat radiates from the blood and warms the air. In this way, the temperature of the incoming air quickly adjusts to that of the body. In addition, the incoming air tends to become moistened by evaporation of water from the mucous lining. The sticky mucus secreted by the mucous membrane entraps dust and other small particles entering with the air.

Many of the cells within the mucous membrane are *ciliated* (fig. 15.3), and as these cilia move, a thin layer of mucus and any entrapped particles are pushed toward the pharynx. When the mucus reaches the pharynx, it is swallowed. In the stomach, any microorganisms in the mucus, including disease-causing forms, are likely to be destroyed by the action of gastric juice. Thus, the filtering mechanism provided by the mucous membrane not only prevents particles from reaching the lower air passages, but also helps to prevent respiratory infections. (See fig. 15.4.)

---

1. *What organs constitute the respiratory system?*
2. *What is the function of the mucous membrane that lines the nasal cavity?*
3. *What is the function of the cilia in the cells of this lining?*

Fig. 15.3 Ciliated cells line the air passages of the respiratory tract. The arrow indicates the cilia, which are highly magnified in this scanning electron micrograph.

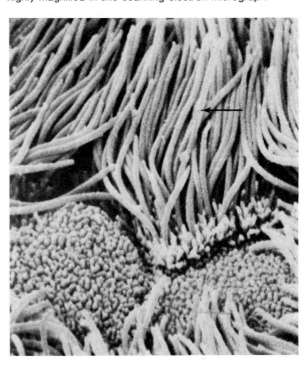

Fig. 15.2 Nasal conchae divide the nasal cavity into passageways: the superior, middle, and inferior meatuses.

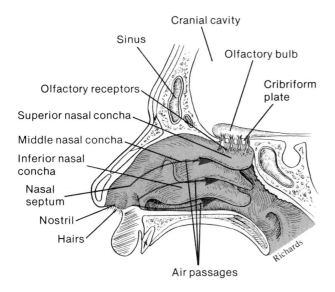

Fig. 15.4 Cilia move mucus and trapped particles from the nasal cavity to the pharynx, where they are swallowed and passed into the stomach.

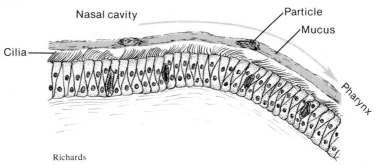

The Sinuses

As discussed in chapter 8, the **sinuses** (paranasal sinuses) are air-filled spaces located within (and named from) the *maxillary, frontal, ethmoid,* and *sphenoid bones* (fig. 15.5). These spaces open into the nasal cavity and are lined with mucous membranes that are continuous with the lining of the nasal cavity. Consequently, mucus secretions can drain from the sinuses into the nasal cavity. If this drainage is blocked by swollen membranes, which sometimes accompany nasal infections or allergic reactions, the accumulating fluids may cause increasing pressure within a sinus and a painful sinus headache.

Although the sinuses function mainly to reduce the weight of the skull, they also serve as resonant chambers that affect the quality of the voice.

It is possible to illuminate a person's frontal sinus in a darkened room by holding a small flashlight just beneath the eyebrow. Similarly, the maxillary sinuses can be illuminated by holding the flashlight in the mouth.

1. Where are the sinuses located?
2. What are the functions of the sinuses?

The Pharynx

The **pharynx,** or throat, is located behind the mouth cavity and between the nasal cavity and larynx. It functions as a passageway for food traveling from the oral cavity to the esophagus and for air passing between the nasal cavity and larynx (fig. 15.6). It also aids in creating the sounds of speech.

The Larynx

The **larynx** is an enlargement in the airway at the top of the trachea and below the pharynx. It serves as a passageway for air moving in and out of the trachea and functions to prevent foreign objects from entering the trachea. In addition, it houses the *vocal cords* and, because of this, is commonly called the voice box.

The larynx is composed primarily of muscles and cartilages that are bound together by elastic tissue. The largest of the cartilages (shown in fig. 15.7) are the thyroid, cricoid, and epiglottic cartilages. These structures occur singly, and the other laryngeal cartilages—the arytenoid, corniculate, and cuneiform cartilages—are paired.

Fig. 15.5 (*a*) X ray of a skull from the front and (*b*) from the side showing air-filled sinuses within the bones.

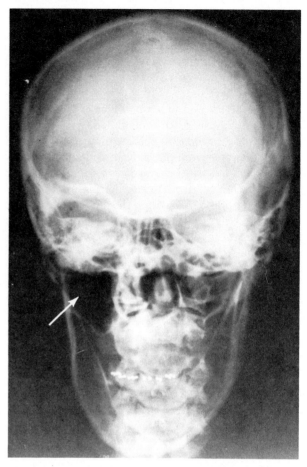

(a)

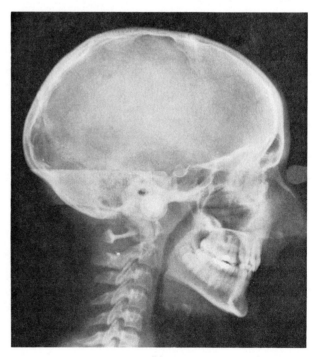

(b)

Fig. 15.6 The pharynx functions as a common passageway for food and air.

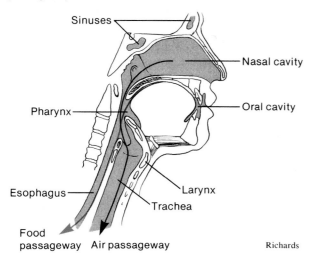

Sinuses

Nasal cavity

Pharynx

Oral cavity

Esophagus

Larynx

Trachea

Food passageway   Air passageway

Richards

Fig. 15.7 (a) Anterior view and (b) posterior view of the larynx.

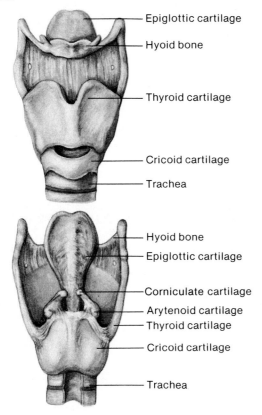

Epiglottic cartilage

Hyoid bone

Thyroid cartilage

Cricoid cartilage

Trachea

Hyoid bone

Epiglottic cartilage

Corniculate cartilage

Arytenoid cartilage

Thyroid cartilage

Cricoid cartilage

Trachea

The **thyroid cartilage** was named for the thyroid gland that covers its lower part. This cartilage is the shieldlike structure that protrudes in the front of the neck and is sometimes called the Adam's apple. This protrusion typically is more prominent in males than in females because of an effect of male sex hormones on the development of the larynx.

The **cricoid cartilage** lies below the thyroid cartilage and marks the lowermost portion of the larynx.

The **epiglottic cartilage** is attached to the upper border of the thyroid cartilage and supports a flaplike structure called the **epiglottis.** The epiglottis usually stands upright and allows air to enter the larynx. During swallowing, however, the larynx is raised by muscular contractions, and the epiglottis is pressed downward by the base of the tongue. As a result, the epiglottis partially covers the opening into the larynx, helping to prevent foods and liquids from entering the air passages.

The pyramid-shaped **arytenoid cartilages** are located above and on either side of the cricoid cartilage, and attached to their tips are the tiny, conelike **corniculate cartilages.** These cartilages serve as attachments for muscles that help to regulate tension on the vocal cords during speech and aid in closing the larynx during swallowing.

The **cuneiform cartilages** are small cylindrical parts found in the mucous membrane between the epiglottic and the arytenoid cartilages. They function to stiffen the soft tissues in this region.

Inside the larynx, two pairs of horizontal folds in the mucous membrane extend inward from the lateral walls. The upper folds (vestibular folds) are called *false vocal cords* because they do not function in the production of sounds. Muscle fibers within these folds help to close the larynx during swallowing.

The lower folds are the *true vocal cords*. (Both pairs of cords are shown in fig. 15.8.) They contain elastic fibers and are responsible for vocal sounds, which are created by forcing air between the vocal cords. This action generates sound waves that can be formed into words by changing the shapes of the pharynx and oral cavity and by using the tongue and lips.

The *pitch* of the vocal sound is controlled by changing the tension on the cords. This is accomplished by contracting or relaxing various laryngeal muscles. Increasing the tension produces a higher pitched sound, and decreasing the tension creates a lower pitch.

The *intensity* (loudness) of a vocal sound is related to the force of the air passing over the vocal cords. Louder sounds are produced by using stronger blasts of air to cause the vocal cords to vibrate.

Fig. 15.8 (*a*) Frontal section and (*b*) sagittal section of the larynx.

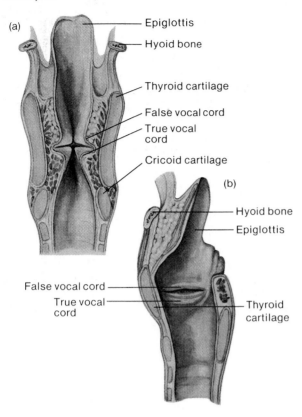

During normal breathing, the vocal cords remain relaxed, and the opening between them, called the **glottis,** appears as a triangular slit. When food or liquid is swallowed, however, the glottis is closed by muscles within the larynx, and this prevents foreign substances from entering the trachea. (See fig. 15.9.)

The mucous membrane that lines the larynx continues to filter the incoming air by entrapping particles and moving them toward the pharynx by ciliary action.

Occasionally the mucous membrane of the larynx becomes inflamed and swollen as a result of an infection or an irritation from inhaled vapors. When this happens, the vocal cords may not vibrate as freely as before, and the voice may become hoarse. This condition is called *laryngitis,* and although it is usually mild, laryngitis is potentially dangerous because the swollen tissues may obstruct the airway and interfere with breathing. In such cases it may be necessary to provide a passageway by inserting a tube into the trachea through the nose, mouth, or a surgical opening in the trachea.

1. *What part of the respiratory tract is shared with the alimentary canal?*
2. *How do the vocal cords function to produce sounds?*
3. *What is the condition of the glottis during breathing? During swallowing?*

The Trachea

The **trachea,** or windpipe, is a flexible cylindrical tube about 2.5 cm (1 in) in diameter and 12.5 cm (5 in) in length. It extends downward in front of the esophagus and into the thoracic cavity, where it splits into right and left bronchi, as shown in figure 15.10.

The inner wall of the trachea is lined with a ciliated mucous membrane that contains many goblet cells. As before, this membrane continues to filter the incoming air and to move entrapped particles upward to the pharynx.

Within the tracheal wall there are about twenty C-shaped pieces of hyaline cartilage arranged one above the other. The open ends of these incomplete rings are directed posteriorly, and the gaps between their ends are filled with smooth muscle and connective tissues. (See fig. 15.11.) These cartilaginous rings prevent the trachea from collapsing and blocking the airway. At the same time, the soft tissues that complete the rings in the back allow the nearby esophagus to expand as it carries food to the stomach.

Fig. 15.9 (*a*) Vocal cords as viewed from above with the glottis closed and (*b*) with the glottis opened.

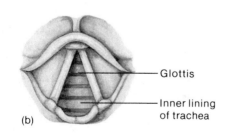

Fig. 15.10 The trachea transports air between the larynx and the bronchi.

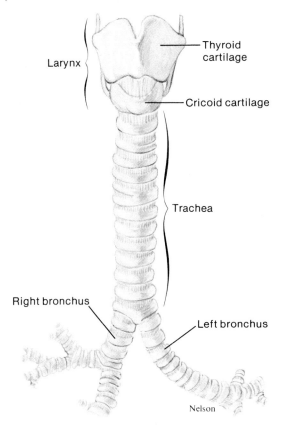

Fig. 15.11 Cross section of the trachea. What is the significance of the C-shaped rings of hyaline cartilage in the wall of the trachea?

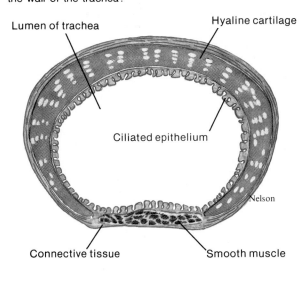

If the trachea becomes obstructed as a result of swollen tissues, excessive secretions, or the presence of foreign objects, it may be necessary to create an external opening in this tube so that air can bypass the obstruction. (See fig. 15.12). This is called a *tracheostomy*. It may be a life-saving procedure, because if the trachea is blocked, asphyxiation can occur within a few minutes.

The Bronchial Tree

The **bronchial tree** consists of the branched airways leading from the trachea to the microscopic air sacs (figs. 15.13 and 15.14). It begins with the right and left **bronchi** that arise from the trachea at the level of the fifth thoracic vertebrae. The openings of these tubes are separated by a ridge of cartilage called the *carina*. Each bronchus, accompanied by large blood vessels, enters its respective lung.

The right bronchus is shorter and straighter than the left one. Consequently, the right bronchus is more likely to receive any foreign objects accidentally aspirated by a young child.

Fig. 15.12 A tracheostomy may be performed to allow air to bypass an obstruction within the larynx.

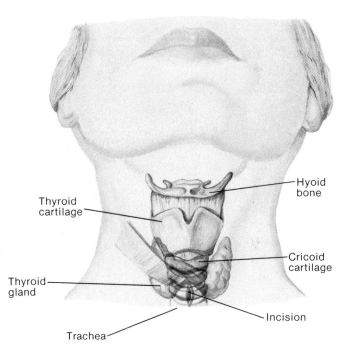

A short distance from its origin, each bronchus divides into secondary bronchi, which in turn branch into finer and finer tubes. These smaller tubes, termed **bronchioles,** continue to divide, giving rise to very thin tubes called **alveolar ducts.** These ducts terminate in masses of microscopic air sacs called **alveoli,** as shown in figures 15.15 and 15.16.

Fig. 15.13 The bronchial tree consists of the passageways that connect the trachea and the alveoli.

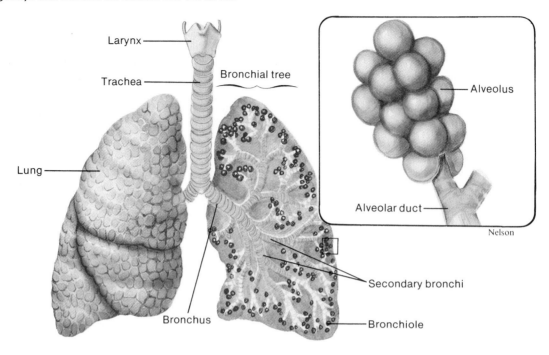

Larynx

Trachea

Bronchial tree

Lung

Alveolus

Alveolar duct

Nelson

Secondary bronchi

Bronchus

Bronchiole

Fig. 15.14 This bronchogram reveals many branches of the bronchial tree.

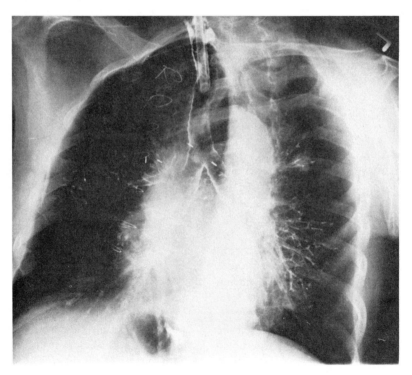

Fig. 15.15 The tubes of the alveolar ducts end in tiny alveoli; each of these is surrounded by a capillary net.

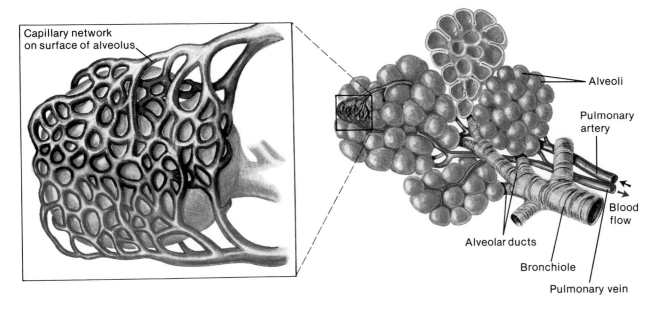

Capillary network on surface of alveolus

Alveoli

Pulmonary artery

Blood flow

Alveolar ducts

Bronchiole

Pulmonary vein

Fig. 15.16 Alveoli appear as open spaces in this photomicrograph of human lung tissue.

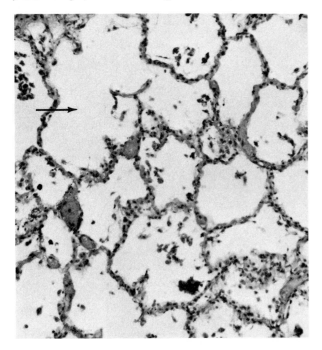

The larger branches of the bronchial tree are supported by cartilaginous rings similar to those of the trachea, but as the tubes become smaller and smaller, the C-shaped rings are replaced by cartilaginous plates. In the finest tubes, the cartilage disappears altogether.

As the quantity of cartilage within the respiratory passages decreases, there is a change in the type of cells that line the tubes. The lining of the larger tubes consists of ciliated columnar epithelium; in the finer tubes, the lining is cuboidal epithelium; and in the alveoli, it is simple squamous epithelium.

The branches of the bronchial tree serve as air passages, distributing incoming air to the alveoli in all parts of the lungs. The alveoli, in turn, provide a large surface area through which gas exchanges can occur. During these exchanges, oxygen diffuses through the alveolar walls and enters the blood in nearby capillaries, while carbon dioxide diffuses from the blood through these walls and enters the alveoli. (See fig. 15.17.)

It is estimated that there are about 300 million alveoli in an adult lung and that these spaces have a total surface area of between 70 and 80 square meters. (This is roughly equivalent to the area of the floor in a room measuring 25 feet by 30 feet.)

1. What is the function of the cartilaginous rings in the tracheal wall?
2. How do the right and left bronchi differ in structure?
3. Describe the changes in structure that occur in the respiratory tubes as they become smaller and smaller.

Fig. 15.17 Oxygen diffuses from the alveolus into the capillary, while carbon dioxide diffuses from the capillary into the alveolus.

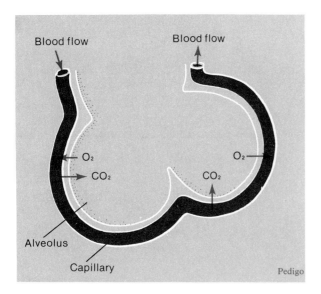

Blood flow

Blood flow

O_2

O_2

CO_2

CO_2

Alveolus

Capillary

Pedigo

The Lungs

The lungs are soft, spongy, cone-shaped organs located in the thoracic cavity. The right and left lungs are separated medially by the heart and the mediastinum, and they are enclosed by the diaphragm and the thoracic cage.

Each lung occupies most of the thoracic space on its side and is suspended in this cavity by its attachments, which include a bronchus and some large blood vessels. These parts enter the lung on its medial surface through a region called the **hilus.** A layer of serous membrane, the **visceral pleura,** is firmly attached to the surface of each lung, and this membrane folds back at the hilus to become the **parietal pleura.** The parietal pleura, in turn, forms part of the mediastinum and lines the inner wall of the thoracic cavity.

The potential space between the visceral and parietal pleurae is called the **pleural cavity,** and it contains a thin film of serous fluid. This fluid lubricates the adjacent pleural surfaces, reducing friction as they move against one another during breathing. (See fig. 15.18.)

Fig. 15.18 Transverse section through the thorax. What force acts to hold the pleural membranes together so that the pleural cavity is only a potential space?

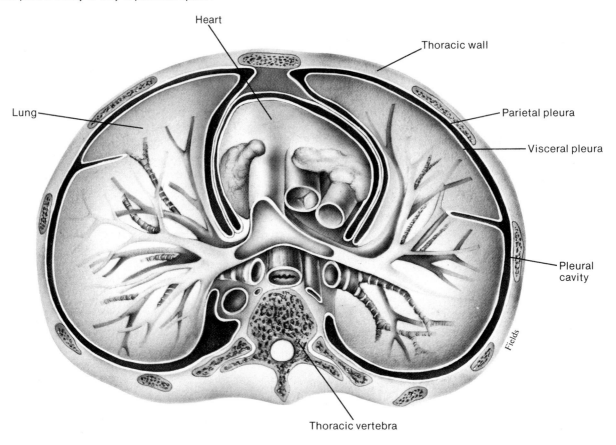

Heart

Thoracic wall

Lung

Parietal pleura

Visceral pleura

Pleural cavity

Fields

Thoracic vertebra

Chart 15.1 Parts of the respiratory system

Part	Description	Function
Nose	Part of face centered above the mouth and below the space between the eyes	Nostrils provide entrance to nasal cavity; internal hairs begin to filter incoming air
Nasal cavity	Hollow space behind nose	Conducts air to pharynx; mucous lining filters, warms, and moistens air
Sinuses	Hollow spaces in various bones of the skull	Reduce weight of the skull; serve as resonant chambers
Pharynx	Chamber behind mouth cavity and between nasal cavity and larynx	Passageway for air moving from nasal cavity to larynx and for food moving from mouth cavity to esophagus
Larynx	Enlargement at the top of the trachea	Passageway for air; prevents foreign objects from entering trachea; houses vocal cords
Trachea	Flexible tube that connects larynx with bronchial tree	Passageway for air; mucous lining continues to filter air
Bronchial tree	Branched tubes that lead from the trachea to the alveoli	Conducts air from the trachea to the alveoli; mucous lining continues to filter air
Lungs	Soft, cone-shaped organs that occupy a large portion of the thoracic cavity	Contain the air passages, alveoli, blood vessels, connective tissues, lymphatic vessels, and nerves of the lower respiratory tract

As figure 15.19 shows, the right lung is larger than the left one, and it is divided into three parts, called the superior, middle, and inferior *lobes*. The left lung consists of two parts, a superior and an inferior lobe.

Each lobe is supplied by a major branch of the bronchial tree. It also has connections to blood and lymphatic vessels and is enclosed by connective tissues. A lobe is further subdivided by connective tissue into **lobules,** and each of these units contains a terminal bronchiole together with its alveolar ducts, alveoli, and associated blood and lymphatic vessels. Thus, the substance of a lung includes air passages, air sacs, blood vessels, connective tissues, lymphatic vessels, and nerves.

Chart 15.1 summarizes the characteristics of the major parts of the respiratory system.

---

1. Where are the lungs located?
2. What is the function of the serous fluid within the pleural cavity?
3. How does the structure of the right lung differ from that of the left lung?

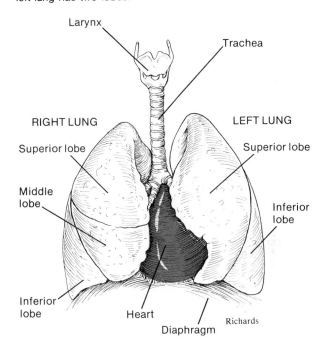

Fig. 15.19 The right lung consists of three lobes; the left lung has two lobes.

Mechanism of Breathing

Breathing, which is also called *pulmonary ventilation,* entails the movement of air from outside the body into the bronchial tree and alveoli, followed by a reversal of this air movement. These actions are termed **inspiration** (inhalation) and **expiration** (exhalation), and they are accompanied by changes in the size of the thoracic cavity.

Inspiration

Atmospheric pressure, due to the weight of air, is the force that causes air to move into the lungs. At sea level, this pressure is sufficient to support a column of mercury about 760 mm (30 in) high in a tube. (See fig. 15.20.) Thus, normal air pressure is said to equal 760 mm of mercury (Hg).

Fig. 15.20 Atmospheric pressure is sufficient to support a column of mercury 760 mm high.

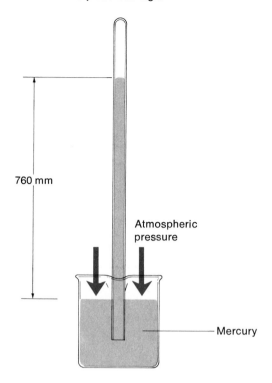

760 mm

Atmospheric pressure

Mercury

Fig. 15.21 When the lungs are at rest, the pressure on the inside of the lungs is equal to the pressure on the outside of the thorax.

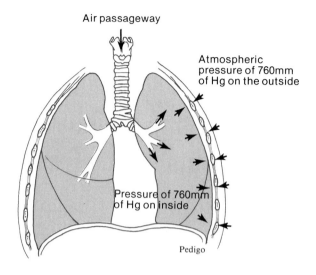

Air passageway

Atmospheric pressure of 760mm of Hg on the outside

Pressure of 760mm of Hg on inside

Pedigo

Air pressure is exerted on all surfaces in contact with air, and since persons breathe air, the inside surfaces of their lungs are also subjected to pressure. In other words, the pressures on the inside of the lungs and on the outside of the thoracic wall are about the same. (See fig. 15.21.)

If the pressure on the inside of the lungs decreases, air from the outside will then be pushed into the airways by atmospheric pressure. This is what happens during inspiration. Muscle fibers in the dome-shaped *diaphragm* are stimulated to contract by impulses carried on the phrenic nerves. As this happens, the diaphragm moves downward, the size of the thoracic cavity is enlarged, and the pressure within the cavity (intrathoracic pressure) is reduced. At the same time, air is forced into the airways by atmospheric pressure, and the lungs expand.

While the diaphragm is contracting and moving downward, the *external intercostal muscles* and the anterior *internal intercostal muscles* between the ribs may also be stimulated to contract. This action raises the ribs and elevates the sternum, so that the size of the thoracic cavity increases even more, and the pressure inside is further reduced. (See fig. 15.22.)

The expansion of the lungs is aided by the fact that the parietal pleura, on the inner wall of the thoracic cavity, and the visceral pleura, attached to the surface of the lungs, are separated only by a thin film of serous fluid. The *water molecules* in this fluid have a great attraction for one another, and the resulting **surface tension** holds the moist surfaces of the pleural membranes tightly together. Consequently, when the thoracic wall is moved upward and outward by the action of the intercostal muscles, the parietal pleura is moved too, and the visceral pleura follows it. This helps to expand the lungs in all directions. The steps in inspiration are summarized in chart 15.2.

The attraction between adjacent moist membranes is sufficient to cause the collapse of the tiny air sacs, which also have moist surfaces. Normally the cells of the lungs secrete a substance, called **surfactant,** that acts to reduce surface tension and prevents such collapse.

Chart 15.2   Major events in inspiration

1. Nerve impulses are carried on phrenic nerves to muscle fibers in the diaphragm causing them to contract.
2. As the dome-shaped diaphragm moves downward, the size of the thoracic cavity is increased.
3. At the same time, the external intercostal muscles may contract, raising the ribs and causing the size of the thoracic cavity to increase still more.
4. As the size of the thoracic cavity is increased, its internal pressure is decreased.
5. Atmospheric pressure, which is relatively greater on the outside, forces air into the respiratory tract through the air passages.
6. The lungs become inflated.

Fig. 15.22 The size of the thoracic cavity enlarges as the diaphragm and various intercostal muscles contract.

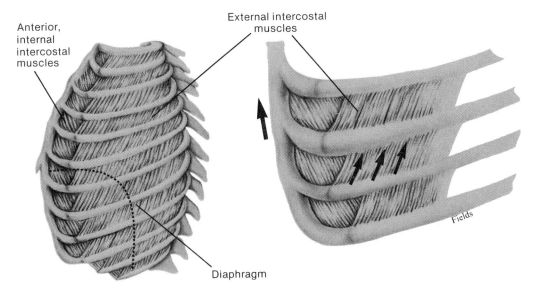

Anterior, internal intercostal muscles

External intercostal muscles

Fields

Diaphragm

Sometimes the lungs of a newborn fail to produce enough surfactant, and the newborn's breathing mechanism may be unable to overcome the force of surface tension. Consequently, the lungs cannot be ventilated and the newborn is likely to die of suffocation. This condition is called *hyaline membrane disease* (respiratory distress syndrome), and it is the primary cause of respiratory difficulty in immature newborns and those born to mothers with certain forms of diabetes mellitus.

If a person needs to take a deeper than normal breath, the diaphragm and external intercostal muscles can usually be contracted to a greater degree. Additional muscles, such as the pectoralis minors and sternocleidomastoids, can be used to pull the thoracic cage further upward and outward. (See fig. 15.23.)

The ease with which the lungs can be expanded as a result of pressure changes occurring during breathing is called *compliance*. Conditions that tend to obstruct air passages, destroy lung tissue, or impede lung expansion in other ways are said to decrease compliance.

Expiration

The force responsible for normal expiration comes from the *elastic recoil* of tissues. The lungs and thoracic wall, for example, contain a considerable amount of elastic tissue, and as the lungs expand during inspiration, these tissues are stretched. As the diaphragm lowers, the abdominal organs beneath it are compressed. As the diaphragm and the external intercostal muscles relax following inspiration, these elastic tissues cause the lungs and thoracic cage to recoil, and they return to their original positions. Similarly, the abdominal organs spring back into their previous locations, pushing the diaphragm upward. Each of these actions tends to increase the pressure on the lungs, so that the air inside is forced out through the respiratory passages.

The recoil of the elastic fibers within the lung tissues tends to reduce the pressure in the pleural cavity. Thus the pressure between the pleural membranes is usually less than atmospheric pressure.

Since the visceral and parietal pleural membranes are held together by surface tension, there is normally no actual space in the pleural cavity between them. However, if the thoracic wall is punctured, atmospheric air may enter the pleural cavity and create a real space between the membranes. This condition is called *pneumothorax*, and when it occurs the lung on the affected side may collapse because of its elasticity.

Fig. 15.23 (*a*) Shape of the thorax at the end of normal inspiration. (*b*) Shape of the thorax at the end of maximal inspiration, aided by contraction of the sternocleidomastoid and pectoralis minor muscles.

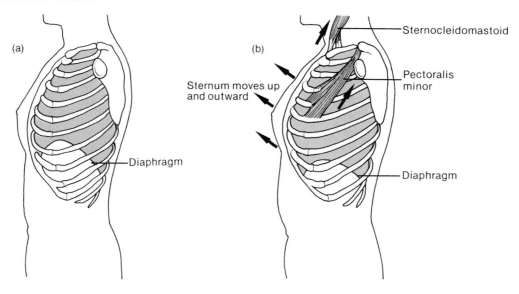

Fig. 15.24 (*a*) Normal expiration is due to elastic recoil of the thoracic wall and abdominal organs; (*b*) maximal expiration is aided by contraction of the abdominal wall and posterior internal intercostal muscles.

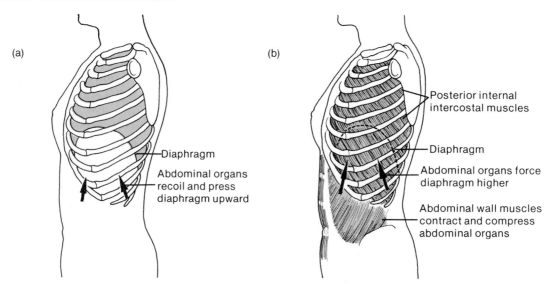

If a person needs to exhale more air than normal, the posterior *internal intercostal muscles* can be contracted. These muscles pull the ribs and the sternum down and inward increasing the pressure in the lungs. Also, the *abdominal wall muscles,* including the external and internal obliques, transversus abdominis, and rectus abdominis, can be used to squeeze the abdominal organs inward. Thus, the abdominal wall muscles cause the pressure in the abdominal cavity to increase and force the diaphragm still higher against the lungs (fig. 15.24). As a result of these actions, additional air can be squeezed out of the lungs.

Chart 15.3 summarizes these steps in expiration.

Fig. 15.25 Respiratory air volumes.

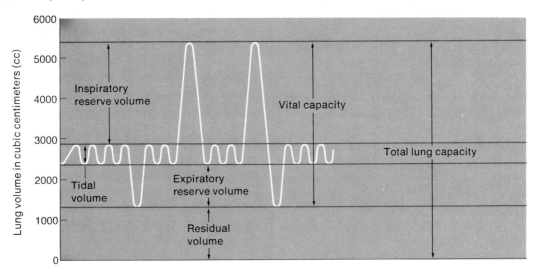

Chart 15.3 Major events in expiration

1. The diaphragm and external respiratory muscles relax.

2. Elastic tissues of the lungs, thoracic cage, and abdominal organs, which were stretched during inspiration, suddenly recoil.

3. Tissues recoiling around the lungs cause the air pressure within the lungs to increase.

4. Air is squeezed out of the lungs into the air passages.

---

1. Describe the events in inspiration.
2. How does surface tension aid expansion of the lungs during inspiration?
3. What force is responsible for normal expiration?

Respiratory Air Volumes

The amount of air that enters the lungs during a normal, quiet inspiration is about 500 cc. Approximately the same volume leaves during a normal expiration. This volume is termed the **tidal volume.**

During forced inspiration, a quantity of air in addition to the tidal volume enters the lungs. This additional volume is called the *inspiratory reserve volume* (complemental air), and it equals about 3000 cc.

During forced expiration, about 1000 cc of air in addition to the tidal volume can be expelled from the lungs. This quantity is called the *expiratory reserve volume* (supplemental air). However, even after the most forceful expiration, about 1500 cc of air remains in the lungs. This volume is called *residual volume.*

Residual air remains in the lungs at all times, and consequently, newly inhaled air is always mixed with air that is already in the lungs. This prevents the oxygen and carbon dioxide concentrations in the lungs from fluctuating excessively with each breath.

If the *inspiratory reserve volume* (3000 cc) is combined with the *tidal volume* (500 cc) and the *expiratory reserve volume* (1000 cc), the total is termed the **vital capacity** (4500 cc). This volume is the maximum amount of air a person can exhale after taking the deepest breath possible.

The *vital capacity* plus the *residual volume* equals the *total lung capacity,* which is about 6000 cc. This total varies with age, sex, and body size. (See fig. 15.25.)

Some of the air that enters the respiratory tract during breathing fails to reach the alveoli. This volume (about 150 ml) remains in the passageways of the trachea, bronchi, and bronchioles. Since gas exchanges do not occur through the walls of these passages, this air is said to occupy *dead space.*

The respiratory air volumes are summarized in chart 15.4.

With the exception of the residual volume, which is measured using special techniques, respiratory air volumes can be determined by using an instrument, called a *spirometer,* shown in figure 15.26. The results of such measurements may be useful in evaluating the courses of various diseases like emphysema, pneumonia, or lung cancer in which there are losses in functional lung tissue. The results also may be valuable in studying the progress of bronchial asthma and other diseases accompanied by obstructions of the air passages.

Chart 15.4 Respiratory air volumes

Volume	Quantity of Air	Description
Tidal volume (TV)	500 cc	Volume moved in or out of the lungs during quiet breathing
Inspiratory reserve volume (IRV)	3000 cc	Volume that can be inhaled during forced breathing in addition to tidal volume
Expiratory reserve volume (ERV)	1000 cc	Volume that can be exhaled during forced breathing in addition to tidal volume
Vital capacity (VC)	4500 cc	Maximum volume of air that can be exhaled after taking the deepest breath possible: VC = TV + IRV + ERV
Residual volume (RV)	1500 cc	Volume that remains in the lungs at all times
Total lung capacity (TLC)	6000 cc	Total volume of air that the lungs can hold: TLC = VC + RV

1. What is meant by tidal volume?
2. Distinguish between inspiratory and expiratory reserve volumes.
3. How is vital capacity measured?
4. How is the total lung capacity calculated?

Nonrespiratory Air Movements

Air movements that occur in addition to breathing are called *nonrespiratory movements.* They are used to clear air passages, as in coughing and sneezing, or to express emotional feelings, as in laughing and crying.

Nonrespiratory movements usually result from *reflexes,* although sometimes they are initiated voluntarily. A cough, for example, can be produced through conscious effort or may be triggered by the presence of a foreign object in an air passage.

The act of *coughing* involves taking a deep breath, closing the glottis, and forcing air upward from the lungs against the closure. Then the glottis is suddenly opened, and a blast of air is forced upward from the lower respiratory tract. Usually this rapid rush of air will remove the substance that triggered the reflex.

Fig. 15.26 A spirometer is used to measure respiratory air volumes.

The most sensitive areas of the air passages are in the larynx and regions near the branches of the major bronchi. The distal portions of the bronchioles (respiratory bronchioles), alveolar ducts, and alveoli lack a nerve supply. Consequently, before any material in these parts can trigger a cough reflex, it must be moved into the larger passages of the respiratory tract.

A *sneeze* is much like a cough, but it clears the upper respiratory passages rather than lower ones. This reflex is usually initiated by a mild irritation in the lining of the nasal cavity and, in response, a blast of air is forced up through the glottis. This time, the air is directed into the nasal passages by depressing the uvula and thus closing the opening between the pharynx and the oral cavity.

Laughing involves taking a breath and releasing it in a series of short expirations. *Crying* consists of very similar movements, and sometimes it is necessary to note a person's facial expressions in order to distinguish laughing from crying.

A *hiccup* is caused by sudden inspiration due to a spasmodic contraction of the diaphragm while the glottis is closed. The sound of the hiccup is created by air striking the vocal folds. The reason for hiccups is not well understood, and this movement apparently serves no useful function.

Yawning is thought to aid respiration by providing an occasional deep breath. During normal, quiet breathing, not all of the alveoli are ventilated, and some blood may pass through the lungs without

Chart 15.5 Nonrespiratory air movements

Air Movement	Mechanism	Function
Coughing	Deep breath is taken, glottis is closed, and air is forced against the closure; suddenly the glottis is opened and a blast of air passes upward	Clears lower respiratory passages
Sneezing	Same as coughing, except air moving upward is directed into the nasal cavity by depressing the uvula	Clears upper respiratory passages
Laughing	Deep breath is released in a series of short expirations	Expresses emotional happiness
Crying	Same as laughing	Expresses emotional sadness
Hiccuping	Diaphragm contracts spasmodically while the glottis is closed	No useful function
Yawning	Deep breath taken	Ventilates a large proportion of the alveoli and aids oxygenation of the blood

becoming well oxygenated. It is believed that this low blood oxygen concentration somehow triggers the yawn reflex, and in response, a very deep breath is taken. The deep breath ventilates a large proportion of the alveoli, and the blood oxygen level rises.

Chart 15.5 summarizes nonrespiratory air movements.

1. What nonrespiratory air movements help to clear the air passages?
2. What nonrespiratory air movements are used to express emotions?
3. What seems to be the function of a yawn?

Some Breathing Disorders

Although respiratory disorders are caused by a variety of factors, some are characterized by inadequate ventilation. This group includes paralysis of various breathing muscles, bronchial asthma, emphysema, and lung cancer.

Paralysis of breathing muscles is sometimes caused by injuries to the respiratory center or to spinal nerve tracts that transmit motor impulses. In other instances, paralysis may be due to a disease, such as *poliomyelitis,* that affects parts of the central nervous system and injures motor neurons. The consequences of such paralysis depend on which muscles are affected. Sometimes, by increasing their responses, other muscles are able to compensate for functional losses of a paralyzed muscle. If unaffected muscles are unable to ventilate the lungs adequately, a person must be provided with some type of mechanical breathing device in order to survive.

Bronchial asthma is a condition commonly caused by an *allergic reaction* to foreign substances in the respiratory tract. Typically the foreign substance is a plant pollen that enters with inhaled air.

As a result of this reaction, the walls of the small bronchioles become edematous, the cells lining the respiratory tubes secrete abnormally large amounts of thick mucus, and the smooth muscles in these tubes contract. These muscles cause the bronchioles to constrict, reducing the diameters of the air passages. As these changes occur, the person finds it increasingly difficult to breathe and produces a characteristic wheezing sound, as air moves through narrowed passages.

An asthmatic person usually finds it harder to force air out of the lungs than to bring it in. This is because inspiration involves powerful breathing muscles, and as they contract, the lungs expand. This expansion of the tissues helps to open the air passages. Expiration, on the other hand, is usually a passive process due to elastic recoil of stretched tissues. Also, it causes compression of the tissues and decreases the diameters of the bronchioles, and this adds to the problem of moving air through the narrowed air passages.

In addition to plant pollens, factors that trigger asthmatic reactions (allergens) include various foods, dust, animal dander, and mold spores. Asthma may also be aggravated by respiratory infections, air pollutants, tobacco smoke, changes in weather, or emotional disturbances. The reactions can be prevented by avoiding the offending factors, but this is often difficult.

The symptoms of asthma are often treated with drugs, such as epinephrine, that cause the smooth muscles in the air passages to relax, thus reducing the narrowing of these passages.

Emphysema is a progressive, degenerative disease characterized by the destruction of many alveolar walls. As a result, clusters of small air sacs merge to form larger chambers, so that the total surface area of the respiratory membrane decreases. At the same time, the alveolar walls tend to lose their elasticity, and the capillary networks associated with the alveoli become less abundant. (See fig. 15.27.)

Because of the loss of tissue elasticity, the person with emphysema finds it increasingly difficult to force air out of the lungs. As was mentioned, normal expiration involves an elastic recoil of inflated tissues; in emphysema, abnormal muscular efforts are required to produce this movement.

The cause of emphysema is not clear, but some investigators believe it develops in response to prolonged exposure to respiratory irritants, such as those in tobacco smoke and polluted air. Emphysema is becoming one of the more common respiratory disorders among older persons, although it is not limited to this group.

Lung cancer, like other cancers, involves an uncontrolled growth of abnormal cells. These cells develop in and around the normal tissues and deprive them of nutrients. In effect, the cancer cells cause the death of normal cells by crowding them out.

Some cancerous growths in the lungs result from cancer cells that spread (metastasis) from other parts of the body, such as the breasts, alimentary tract, liver, or kidneys.

Primary pulmonary cancer, which begins in the lungs, is the most common form of cancer in males today, and is becoming increasingly common among females. Primary pulmonary cancer may arise from epithelial cells, connective tissue cells, or various blood cells. The most common form arises from epithelium, and is called *bronchogenic carcinoma*, which means it originates from the cells that line the tubes of the bronchial tree. This type of cancer seems to occur in response to excessive irritation such as that produced by prolonged exposure to tobacco smoke. (See fig. 15.28.)

Once lung cancer cells have appeared, they are likely to produce masses that obstruct air passages and reduce the amount of alveolar surface available for gas exchanges. Furthermore, bronchogenic carcinoma is likely to spread to other tissues relatively quickly and establish secondary cancers. The sites of such secondary cancer commonly includes lymph nodes, liver, bones, brain, and kidneys.

Fig. 15.27 (*a*) As emphysema develops, the alveoli tend to merge, (*b, c*) forming larger and larger chambers.

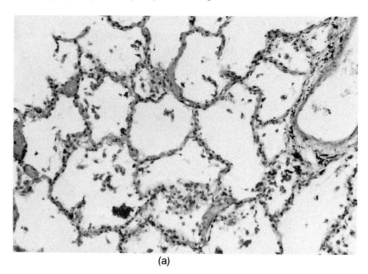

(a)

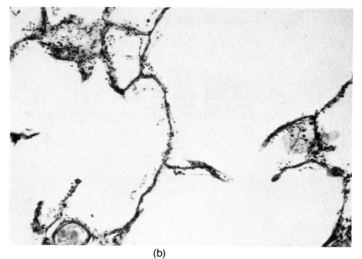

(b)

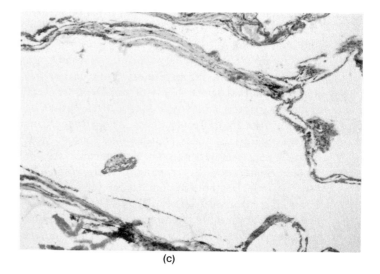

(c)

Fig. 15.28 Lung cancer usually starts in the lining (epithelium) of the bronchus. (*a*) The normal lining shows (*4*) columnar cells with (*2*) hairlike cilia, (*3*) goblet cells that secrete (*1*) mucus, and (*5*) basal cells from which new columnar cells arise. (*6*) A basement membrane separates the epithelial cells from (*7*) the underlying connective tissue. (*b*) In the first stage of lung cancer, the basal cells divide repeatedly. The goblet cells secrete excessive mucus, and the cilia function less efficiently in moving the heavy mucus secretion. (*c*) With the continued multiplication of basal cells, the columnar and goblet cells are displaced. The basal cells penetrate the basement membrane and invade the deeper connective tissue.

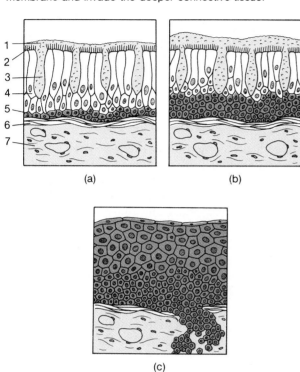

(a)   (b)

(c)

The progress of lung cancer is often difficult to control. Usually it is treated by surgical removal of the diseased portions of the lungs, by exposure to ionizing radiation, and by the use of drugs (chemotherapy). Despite these treatments, the death rate among lung cancer victims remains extremely high.

1. *What factors make breathing difficult for a person with bronchial asthma?*
2. *Why does a person with emphysema have difficulty with expiration?*
3. *What is meant by primary pulmonary cancer?*

Fig. 15.29 The respiratory center is located in the pons and the medulla oblongata.

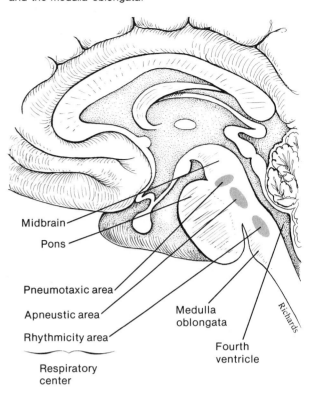

Midbrain
Pons
Pneumotaxic area
Apneustic area
Rhythmicity area
Respiratory center
Medulla oblongata
Fourth ventricle

Control of Breathing

Although respiratory muscles can be controlled voluntarily, normal breathing is a rhythmic, involuntary act that continues even when a person is unconscious.

The Respiratory Center

Breathing is controlled by a poorly defined region in the brain stem called the **respiratory center**. This center initiates impulses that travel on cranial and spinal nerves to various breathing muscles, and it is able to adjust the rate and depth of breathing. As a result, cellular needs for a supply of oxygen and the removal of carbon dioxide are met even during periods of strenuous physical exercise.

The neurons that make up the respiratory center are widely scattered throughout the pons and the upper ⅔ of the medulla oblongata. However, three areas of the center are of special interest. They are the rhythmicity area of the medulla, and the apneustic and pneumotaxic areas of the pons, shown in figure 15.29.

As its name implies, the **medullary rhythmicity area** is responsible for establishing the basic rhythm of breathing; but the neurons of this area cannot produce the usual smooth pattern of breathing movements by themselves. Nerve impulses from other parts of the nervous system, including the apneustic and pneumotaxic areas, contribute to this pattern.

For example, if impulses from the apneustic and pneumotaxic areas are prevented from reaching the medullary rhythmicity area, the breathing movements consist of short inspirations followed by prolonged expirations. On the other hand, stimulation of the **apneustic area** produces forceful, prolonged inspirations and weak expirations. Stimulation of the **pneumotaxic area** inhibits this apneustic type of breathing. Also, activity in the pneumotaxic area can change the breathing rate. Thus, although the apneustic and pneumotaxic areas are not necessary for the basic rhythm of breathing, they can modify that rhythm considerably.

The mechanism responsible for maintaining the basic breathing pattern is not well understood. It seems to involve two groups of intermingled neurons in the *medullary rhythmicity area*. One group is concerned primarily with inspiration, and the other functions in expiration.

It is believed that when one of the inspiratory neurons becomes excited, it stimulates other neurons, which in turn stimulate still others. This action seems to establish a self-perpetuating circuit, and impulses travel around this circuit until the neurons become fatigued in about 2 seconds. Meanwhile, impulses also travel out from the inspiratory circuit to inspiratory muscles that contract and cause inspiration. At the same time, other impulses from the inspiratory circuit are directed to an expiratory circuit, inhibiting it.

When the inspiratory circuit becomes fatigued, the expiratory neurons, which are no longer inhibited, become excited, and impulses begin to flow through the expiratory circuit. Impulses from this circuit cause expiration. At the same time, inhibiting impulses are sent to the neurons of the inspiratory circuit until the expiratory circuit becomes fatigued.

Thus, normal breathing seems to result from alternating activity in the neurons of the inspiratory and the expiratory circuits. The action of the medullary rhythmicity center is summarized in chart 15.6.

---

1. Where is the respiratory center located?
2. Describe how the respiratory center functions to maintain a normal breathing pattern.

Chart 15.6 Activities of the medullary rhythmicity center

1. Inspiratory neurons excite one another so that a self-perpetuating circuit is established.
2. Impulses from the inspiratory circuit stimulate inspiration, while other impulses from this circuit inhibit activity of expiratory neurons.
3. The neurons of the inspiratory circuit become fatigued in a short time, then stop inhibiting expiratory neurons.
4. Expiratory neurons excite one another and a self-perpetuating circuit is established.
5. Impulses from the expiratory circuit stimulate expiration, while other impulses from this circuit inhibit activity of inspiratory neurons.
6. The neurons of the expiratory circuit become fatigued in a short time, then stop inhibiting inspiratory neurons.
7. The cycle is repeated.

Factors Affecting Breathing

In addition to the controls exerted by the respiratory center, breathing rate and depth are influenced by a variety of other factors. These include the degree to which the lung tissues are stretched, the presence of certain chemicals in body fluids, and the person's emotional state.

An inflation reflex and a deflation reflex (Hering-Breuer reflexes) help to regulate the depth of breathing and to maintain the respiratory rhythm. The *inflation reflex* occurs when *stretch receptors* in parts such as the visceral pleura, bronchioles, and alveoli are stimulated during inspiration. The resulting sensory impulses travel via the vagus nerves to the respiratory center, and further inspiration is inhibited (fig. 15.30). This action prevents overinflation of the lungs.

The *deflation reflex* is activated during expiration, when the stretch receptors are no longer being stimulated. It inhibits the expiration circuit in the respiratory center and thus reduces the amount of deflation that might otherwise occur.

Chemical factors in the blood that affect breathing include carbon dioxide, hydrogen ions, and oxygen. More specifically, breathing rate increases as the blood concentrations of carbon dioxide or hydrogen ions increase, or when the concentration of oxygen decreases.

Low blood oxygen seems to have little direct effect upon the respiratory center. Instead, changes in blood oxygen levels are sensed by *chemoreceptors* in specialized structures called the *carotid* and *aortic bodies*, which are located in the walls of certain large arteries (carotid arteries and aorta) in the neck and thorax (fig. 15.31). When these receptors are stimulated, impulses are transmitted to the respiratory center, and the breathing rate is increased. This

Fig. 15.30 In the process of inspiration, motor impulses travel from the respiratory center to the diaphragm and external intercostal muscles, which contract and cause the lungs to expand. This stimulates stretch receptors to send inhibiting impulses to the respiratory center.

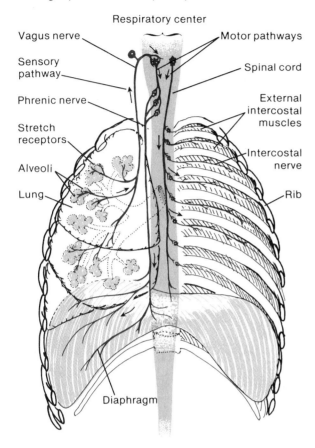

Fig. 15.31 Chemoreceptors in the carotid and aortic bodies are sensitive to changes in blood levels of carbon dioxide and oxygen. What role do these receptors play in the control of respiration?

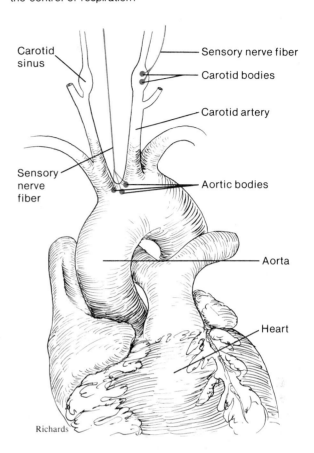

mechanism usually is not triggered until the blood oxygen concentration reaches a very low level. Thus oxygen seems to play only a minor role in the control of normal respiration.

Although the chemoreceptors also seem to be stimulated by changes in the blood levels of carbon dioxide and hydrogen ions, these substances have a much more powerful effect when they act directly on the respiratory center. In fact, these substances seem to be the most important chemical factors in the control of normal breathing.

An exception to the normal pattern of chemical control may occur in patients who have chronic obstructive pulmonary diseases such as asthma and bronchitis. Over a period of time, these patients seem to become adapted to the presence of high levels of carbon dioxide, and in such persons low oxygen levels may serve as effective respiratory stimuli.

Increasing blood levels of carbon dioxide and of hydrogen ions have the same effects on the respiratory center, and in each instance the response is an increase in the breathing rate. The similarity of these effects is related to the fact that carbon dioxide combines with water in blood or cerebrospinal fluid to form *carbonic acid* (H_2CO_3):

$$CO_2 + H_2O \rightarrow H_2CO_3$$

The carbonic acid thus formed soon becomes ionized, releasing *hydrogen ions* (H^+) and *bicarbonate ions* (HCO_3^-).

$$H_2CO_3 \rightarrow H^+ + HCO_3^-$$

Apparently, it is the presence of hydrogen ions that influences the respiratory center rather than the presence of carbon dioxide molecules. In any event, a person's breathing rate increases when air rich in carbon dioxide is inhaled.

Physical factors affecting breathing are summarized in chart 15.7.

Chart 15.7 Factors affecting breathing

Factor	Receptors Stimulated	Response	Effect
Stretch of tissues	Stretch receptors in visceral pleura, bronchioles, and alveoli	Inhibits inspiration	Prevents overinflation of lungs
Lack of stretch	Same as above	Inhibits expiration	Reduces amount of deflation
Low blood oxygen	Chemoreceptors in carotid and aortic bodies	Increases breathing rate	Increases blood oxygen level
High blood carbon dioxide	Same as above and direct effect on the respiratory center	Increases breathing rate	Decreases blood carbon dioxide level
High blood hydrogen ion	Same as above	Increases breathing rate	Decreases blood hydrogen ion level

Air to which additional carbon dioxide has been added is sometimes used to stimulate the rate and depth of breathing.

Ordinary air is about 0.04% carbon dioxide. If a patient inhales air containing 4% carbon dioxide, the breathing rate usually doubles.

The normal breathing pattern may also be altered if a person is emotionally upset. Fear, for example, typically causes an increased breathing rate, as does pain. Thus a person may gasp in response to a sudden fright or the chill of a cold shower.

Then, too, because the respiratory muscles are voluntary, the breathing pattern can be altered consciously. In fact, breathing can be stopped altogether, at least for a time.

If a person decides to stop breathing, the blood concentrations of carbon dioxide and hydrogen ions begin to rise, and the concentration of oxygen falls. These changes stimulate the respiratory center, and soon the need to inhale overpowers the desire to hold the breath. On the other hand, a person can increase the breath-holding time by breathing rapidly and deeply in advance. This action, termed **hyperventilation**, causes a lowering of the blood carbon dioxide level. Following hyperventilation, it takes longer than usual for the carbon dioxide concentration to reach the level needed to produce an overwhelming effect upon the respiratory center.

Sometimes a person who is emotionally upset may hyperventilate, become dizzy, and lose consciousness. This effect seems to be due to a lowered carbon dioxide level accompanied by a pH change (respiratory alkalosis). Since a normal level of carbon dioxide is needed to maintain the tone of smooth muscles in blood vessels, a drop in the carbon dioxide level may be followed by dilation of blood vessels and a decrease in blood pressure. As a result, the oxygen supply to the brain cells may be inadequate and the person may faint.

Hyperventilation should never be used to help in holding the breath while swimming, because the person who has hyperventilated may lose consciousness under water and drown.

1. Describe the inflation and deflation reflexes.
2. What chemical factors affect breathing?
3. How does hyperventilation result in a decreased respiratory rate?

Exercise and Breathing Rate

When a person engages in moderate to heavy physical exercise, the amount of oxygen utilized by the skeletal muscles increases greatly, and so does the amount of carbon dioxide that is produced. Since decreased blood oxygen and increased blood carbon dioxide are stimulating to the respiratory center, it is not surprising that exercise is accompanied by an increase in the breathing rate. However, the breathing rate during physical exercise often increases to a greater degree than would be expected from the changes in blood oxygen and carbon dioxide levels alone.

This observation has led investigators to search for other factors that might act to increase the breathing rate during exercise. The mechanism that seems to be responsible for most of the increase involves the *cerebral cortex* and *proprioceptors* associated with joints. Specifically, the cortex seems to transmit stimulating impulses to the respiratory center whenever it signals skeletal muscles to contract. At the same time, the muscular movements stimulate proprioceptors, and a *joint reflex* is triggered. In this reflex, sensory impulses are transmitted from the proprioceptors to the respiratory center, and the breathing rate increases.

In caring for an immobilized patient, a nurse may employ range-of-motion exercises and frequent changes in position that stimulate proprioceptors and trigger joint reflexes. Such actions cause an increase in the breathing rate and thus help the patient to ventilate.

Whenever an increase in the breathing rate occurs during exercise, there must also be an increase in the blood flow if the needs of body cells are to be met. Physical exercise thus places an increased demand on both the respiratory and the circulatory systems. If either of these systems fails to keep up with cellular demands, the person will begin to feel out of breath. This feeling is usually due to an inability of the circulatory system to move enough blood between the lungs and the body cells, rather than to an inability of the respiratory system to move enough air.

1. Explain how the joint reflex affects breathing during exercise.
2. Why is a person likely to feel out of breath following vigorous exercise?

Alveolar Gas Exchanges

While other parts of the respiratory system function in moving air into and out of the air passages, the alveoli carry on the vital process of exchanging gases between the air and the blood.

The Alveoli

The **alveoli** are microscopic air sacs clustered at the distal ends of the finest respiratory tubes. Each alveolus consists of a tiny space surrounded by a thin wall that separates it from adjacent alveoli. There are

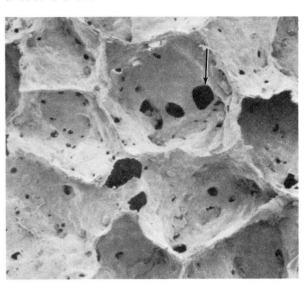

Fig. 15.32 Alveolar pores allow air to pass from one alveolus to another.

minute openings, called **alveolar pores**, in the walls of some alveoli, and although the function of these pores is not clear, they may permit air to pass from one alveolus to another. (See fig. 15.32.) This arrangement may provide alternate air pathways if the passages in some portions of the lung become obstructed.

The Respiratory Membrane

The wall of an alveolus consists of an inner lining of simple squamous epithelium and a dense network of capillaries that are also lined with simple squamous cells. Thin basement membranes separate the layers of these flattened cells, and in the spaces between them there are elastic and collagenous fibers that help to support the wall. As figure 15.33 shows, there are at least two thicknesses of epithelial cells and basement membranes between the air in an alveolus and the blood in a capillary. These layers make up the **respiratory membrane** through which gas exchanges occur.

Diffusion through the Respiratory Membrane. As is described in chapter 3, gas molecules diffuse from regions where they are in higher concentration toward regions where they are in lower concentration. Similarly, gases move from regions of higher pressure toward regions of lower pressure, and it is the *pressure* of a gas that determines the rate at which it will diffuse from one region to another.

Measured by volume, ordinary air is about 78% nitrogen, 21% oxygen, and 0.04% carbon dioxide. It also contains small amounts of certain other gases that have little or no physiological importance.

In a mixture of gases, such as air, each gas is responsible for a portion of the total weight or pressure produced by the mixture. The amount of pressure each gas creates is directly related to its concentration in the mixture. For example, since air is 21% oxygen, this gas is responsible for 21% of the atmospheric pressure. Since 21% of 760 mm of Hg is equal to 160 mm of Hg, it is said that the **partial pressure** of oxygen, symbolized P_{O_2}, in atmospheric air is 160 mm of Hg. Similarly, it can be calculated that the partial pressure of carbon dioxide (P_{CO_2}) in air is 0.3 mm of Hg.

When a mixture of gases dissolves in blood, each gas exerts its own partial pressure in proportion to its dissolved concentration. Furthermore, each gas will diffuse between the liquid and its surroundings, and as figure 15.34 shows, this movement will tend to equalize its partial pressures in the two regions.

For example, the P_{CO_2} in capillary blood is 46 mm of Hg, while the P_{CO_2} in alveolar air is 39 mm. As a consequence of the difference between these partial pressures, carbon dioxide diffuses from the blood, where its pressure is higher, through the respiratory membrane and into the alveolar air. When blood leaves the lungs, its P_{CO_2} is 40 mm of Hg, which is about the same as the P_{CO_2} of the alveolar air.

Similarly, the P_{O_2} of capillary blood is 40 mm of Hg, while that of alveolar air is 101 mm. Thus, oxygen diffuses from the alveolar air into the blood, and when the blood leaves the lungs it has a P_{O_2} of 100 mm of Hg.

---

1. Describe the structure of the respiratory membrane.
2. What is meant by partial pressure?
3. What causes oxygen and carbon dioxide to move across the respiratory membrane?

Some Disorders Involving Gas Exchange

As was mentioned, the disorder called *emphysema* is accompanied by a progressive decrease in the surface area available for gas exchanges, because the alveoli tend to merge and form larger chambers. The functional surface of an emphysematous lung may be reduced to ¼ that of a normal lung. This greatly decreases the lung's ability to carry on gas exchanges by diffusion, and its ability to oxygenate blood is further decreased by a loss of capillaries. Similar problems develop in pneumonia, tuberculosis, and a condition called atelectasis.

Fig. 15.33 The respiratory membrane consists of the walls of the alveolus and the capillary.

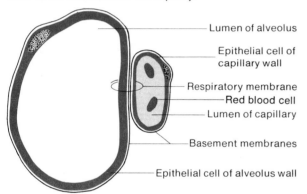

- Lumen of alveolus
- Epithelial cell of capillary wall
- Respiratory membrane
- Red blood cell
- Lumen of capillary
- Basement membranes
- Epithelial cell of alveolus wall

Fig. 15.34 Gas exchanges occur between the air of the alveolus and the blood of the capillary as a result of differences in partial pressures.

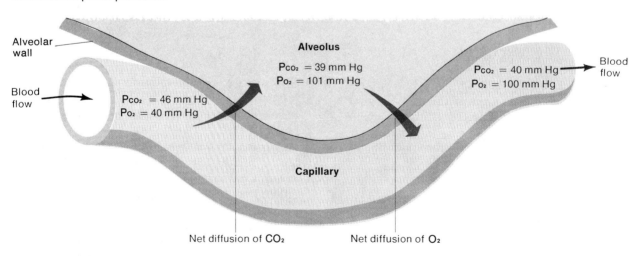

Alveolar wall

Blood flow

$P_{CO_2} = 46$ mm Hg
$P_{O_2} = 40$ mm Hg

Alveolus

$P_{CO_2} = 39$ mm Hg
$P_{O_2} = 101$ mm Hg

$P_{CO_2} = 40$ mm Hg
$P_{O_2} = 100$ mm Hg

Blood flow

Capillary

Net diffusion of CO_2 Net diffusion of O_2

When a person has *pneumonia,* the alveoli become filled with fluids and blood cells. Pneumonia is usually due to an acute infection by bacterial organisms called *pneumococci,* or to the presence of certain viruses.

In response to such infections, the linings of the alveoli become edematous and abnormally permeable. As a result, fluids and white blood cells enter the air sacs in abnormal quantities. As the alveoli fill up, the surface area available for gas exchanges is greatly reduced.

This type of infection is likely to spread to other regions of the respiratory tract and may involve lobes (lobar pneumonia) or an entire lung. The death rate in untreated cases of pneumonia is relatively high.

Tuberculosis is a bacterial disease caused by the *tubercle bacillus.* In this condition, fibrous connective tissue develops around the sites of infection, creating structures called *tubercles.* By walling off the bacteria, the tubercles help to inhibit their spread. Sometimes the protective mechanism fails, and the bacteria become widely distributed throughout the lungs.

In the later stages of tuberculosis, other kinds of bacteria are likely to appear and cause secondary infections in the lungs. There may be extensive destruction of lung tissues, which greatly reduces the surface area of the respiratory membrane. In addition, the widespread development of fibrous tissue within a tubercular lung causes an increase in the thickness of the respiratory membrane, which further restricts gas exchange.

Atelectasis refers to the collapse of a lung or some part of it, together with the collapse of the blood vessels that supply the affected region.

Although atelectasis can be caused by a variety of factors, it is commonly due to an obstruction of a respiratory tube. Such a blockage may result from an abnormal secretion of mucus or the presence of some foreign object. In either case, the air in the alveoli beyond the obstruction tends to be absorbed by the lung tissues. As the air pressure in the alveoli decreases, their elastic walls cause them to collapse, and they become nonfunctional. (See fig. 15.35.)

Fortunately for persons affected by this condition, when a portion of a lung collapses, the functional regions that remain often are able to carry on enough gas exchange to sustain the body cells.

Hyaline membrane disease, which was discussed earlier, is a special form of atelectasis that is characterized by the collapse of alveoli in an infant's lungs due to a lack of surfactant.

Fig. 15.35 (*a*) If an obstruction occurs in a respiratory passageway, air in the alveoli beyond the obstruction is absorbed. (*b*) Why does this cause the affected portion of the lung to collapse?

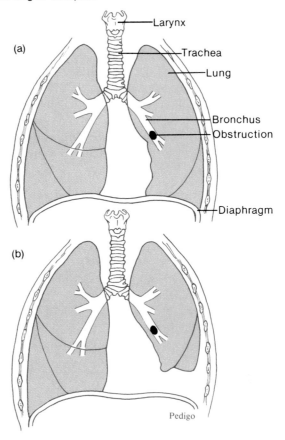

1. How is alveolar gas exchange affected by emphysema?
2. How does pneumonia affect gas exchange?
3. In what way is the effect of tuberculosis similar to that of emphysema?
4. What conditions may lead to atelectasis?

Transport of Gases

The transport of oxygen and carbon dioxide between the lungs and the body cells is a function of the blood. As these gases enter the blood, they dissolve in the liquid portion (plasma). They soon combine chemically with various blood components, and most are carried in combination with other atoms or molecules.

Oxygen Transport

Almost all the oxygen carried in the blood is combined with the compound, **hemoglobin,** that occurs within the red blood cells. It is responsible for the color of these cells.

Hemoglobin has a complex molecular structure that consists of two subunits, called *heme* and *globin*. Globin is a protein that contains 574 amino acids arranged in four polypeptide chains. Each chain is associated with a heme group, and each heme group contains an atom of iron. Each iron atom can combine loosely with an oxygen molecule. Thus, as oxygen dissolves in blood, it combines rapidly with hemoglobin, and the result of this chemical reaction is a new substance called **oxyhemoglobin**.

The amount of oxygen that combines with hemoglobin is determined by the P_{O_2}. The greater the P_{O_2}, the more oxygen will combine with hemoglobin, until the hemoglobin molecules are saturated with oxygen.

Although the proportion of oxygen in air remains the same (about 21%) at high altitudes, the P_{O_2} decreases. Consequently, if a person breathes high altitude air, oxygen diffuses less rapidly from the alveoli into the blood, and the hemoglobin is less likely to become saturated with oxygen. Such a person may develop the symptoms of an oxygen deficiency (hypoxia). For example, at 8000 feet a person may experience anxiety, restlessness, an increased breathing rate, and a rapid pulse. At 12,500 feet, where the P_{O_2} is only 100 mm of Hg rather than 160 mm of Hg at sea level, the symptoms may include drowsiness, mental fatigue, headache, and nausea. As altitude increases these conditions are likely to increase in intensity until the person loses consciousness at about 23,000 feet, after which death may occur.

The chemical bonds that form between oxygen and hemoglobin molecules are relatively unstable, and as the P_{O_2} decreases, oxygen is released from oxyhemoglobin molecules (fig. 15.36). This happens in tissues where cells have used oxygen in their respiratory processes, and the free oxygen diffuses from the blood into nearby cells, as shown in figure 15.37.

Fig. 15.36 The oxyhemoglobin dissociation curve indicates that increased quantities of oxygen are released as the P_{O_2} decreases.

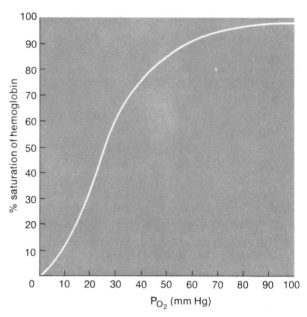

Oxyhemoglobin dissociation at 38°C

Fig. 15.37 (a) Oxygen molecules entering the blood from the alveolus bond to hemoglobin, forming oxyhemoglobin. (b) In the regions of the body cells, oxyhemoglobin releases oxygen.

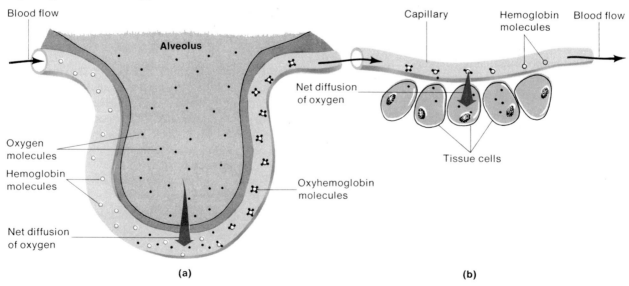

(a)

(b)

The amount of oxygen that is released from oxyhemoglobin is affected by several factors, including the blood concentration of carbon dioxide, the pH, and the temperature. Thus, as the concentration of carbon dioxide (P_{CO_2}) increases, oxyhemoglobin tends to release more oxygen. (See fig. 15.38.) Also, as the blood becomes more acidic or as the temperature increases, more oxygen is released. (See figs. 15.39 and 15.40.)

Because of these factors, more oxygen is released from the blood to skeletal muscle during periods of exercise, since increased muscular activity causes an increase in the P_{CO_2}, a decrease in the pH value, and a rise in the local temperature. At the same time, less active cells receive relatively smaller amounts of oxygen.

Fig. 15.38 The amount of oxygen released from oxyhemoglobin increases as the P_{CO_2} increases.

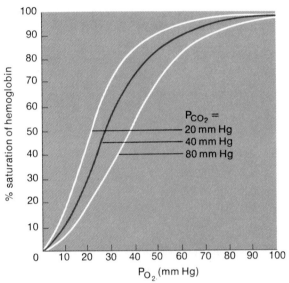

Oxyhemoglobin dissociation at 38°C

Fig. 15.39 The amount of oxygen released from oxyhemoglobin increases as the pH decreases.

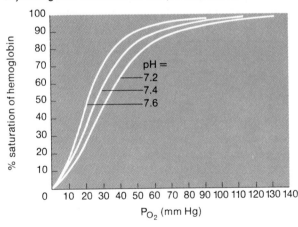

Oxyhemoglobin dissociation at 38°C

Carbon Monoxide

Carbon monoxide (CO) is a toxic gas produced in gasoline engines as a result of incomplete combustion of fuel. It is also a component of tobacco smoke. Its toxic effect occurs because it combines with hemoglobin more effectively than oxygen. Furthermore, carbon monoxide does not dissociate readily from the hemoglobin. Thus, when a person breathes carbon monoxide, increasing quantities of hemoglobin become unavailable for oxygen transport, and the body cells soon begin to suffer.

A victim of carbon monoxide poisoning may be treated by administering oxygen in high concentration, to replace some of the carbon monoxide bonded to hemoglobin molecules. Carbon dioxide is usually administered simultaneously to stimulate the respiratory center, which in turn causes an increase in the breathing rate. Rapid breathing is desirable because it helps to reduce the concentration of carbon monoxide in the alveoli.

Fig. 15.40 The amount of oxygen released from oxyhemoglobin increases as the temperature increases.

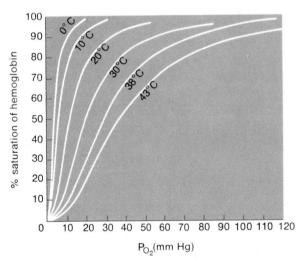

Oxyhemoglobin dissociation at various temperatures

Carbon Dioxide Transport

Blood flowing through the capillaries of body tissues gains carbon dioxide because the tissues have a relatively high P_{CO_2}. This carbon dioxide is transported to the lungs in one of three forms: it may be carried as carbon dioxide dissolved in blood, as part of a compound formed by bonding to hemoglobin, or as part of a bicarbonate ion. (See fig. 15.41.)

The amount of carbon dioxide that dissolves in blood is determined by its partial pressure. The higher the P_{CO_2} of the tissues, the more carbon dioxide will go into solution. However, only about 7% of the carbon dioxide is transported in this form.

Unlike oxygen, which combines with the iron atoms of hemoglobin molecules, carbon dioxide bonds with the amino groups ($-NH_2$) of these molecules. Consequently, oxygen and carbon dioxide do not compete for bonding sites, and both gases can be transported by a hemoglobin molecule at the same time.

When carbon dioxide combines with hemoglobin, a loosely bonded compound called **carbamino-hemoglobin** is formed. This substance decomposes readily in regions where the P_{CO_2} is low, and thus its carbon dioxide is released. Although this method of

Fig. 15.41 Carbon dioxide produced by tissue cells is transported in the blood either in a dissolved state, combined with hemoglobin, or in the form of bicarbonate ions (HCO_3^-).

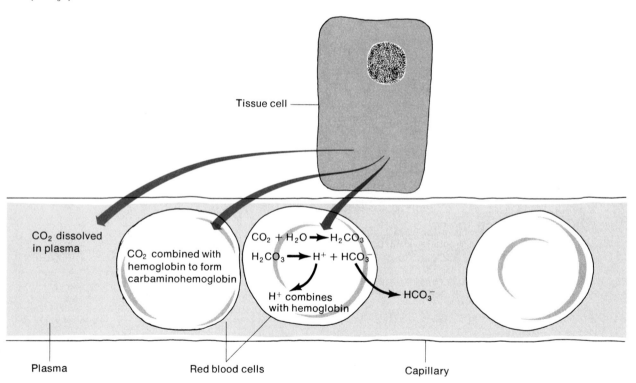

transporting carbon dioxide is theoretically quite effective, carbaminohemoglobin forms relatively slowly, and it is believed that only a small percentage of the total carbon dioxide is carried this way.

The most important carbon dioxide transport mechanism involves the formation of **bicarbonate ions** (HCO_3^-). As was mentioned, carbon dioxide reacts with water to form carbonic acid (H_2CO_3). Although this reaction occurs slowly in blood plasma, much of the carbon dioxide diffuses into red blood cells, and these cells contain an enzyme called **carbonic anhydrase** that speeds the reaction between carbon dioxide and water.

The resulting carbonic acid dissociates almost immediately, releasing hydrogen ions (H^+) and bicarbonate ions (HCO_3^-).

$$H_2CO_3 \rightarrow H^+ + HCO_3^-$$

Most of the hydrogen ions combine quickly with hemoglobin molecules, and are thus prevented from accumulating and causing a change in the blood pH. The bicarbonate ions tend to diffuse out of the red blood cells and enter the blood plasma. It is estimated that nearly 90% of the carbon dioxide transported in blood is carried in this form.

As the bicarbonate ions (HCO_3^-) leave the red blood cells and enter the plasma, *chloride ions* (Cl^-) are electrically repelled, and they move from the plasma into the red cells. This exchange in position of the two negatively charged ions, shown in figure 15.42, maintains the ionic balance between the red cells and the plasma. It is termed the **chloride shift.**

Fig. 15.42 As bicarbonate ions (HCO_3^-) diffuse out of the red blood cell, chloride ions (Cl^-) from the plasma diffuse into the cell, thus maintaining the electrical balance between ions. This exchange of ions is called the chloride shift.

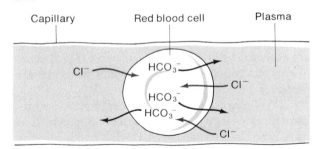

When blood passes through the capillaries of the lungs, it loses its dissolved carbon dioxide by diffusion into the alveoli. This occurs in response to the relatively low P_{CO_2} of the alveolar air. At the same time, hydrogen ions and bicarbonate ions in the red blood cells recombine to form carbonic acid molecules, and under the influence of carbonic anhydrase, the carbonic acid gives rise to carbon dioxide and water.

$$H^+ + HCO_3^- \rightarrow H_2CO_3 \rightarrow CO_2 + H_2O$$

Carbaminohemoglobin also releases its carbon dioxide, and as carbon dioxide continues to diffuse out of the blood, an equilibrium is established between the P_{CO_2} of the blood and the P_{CO_2} of the alveolar air. This process is summarized in figure 15.43.

Chart 15.8 summarizes the transport of blood gases.

---

1. Describe three ways carbon dioxide can be carried from body cells to the lungs.
2. How is it possible for hemoglobin to carry oxygen and carbon dioxide at the same time?
3. What is meant by the chloride shift?
4. How is carbon dioxide released from the blood into the lungs?

Chart 15.8   Gases transported in blood

Gas	Reaction Involved	Substance Transported
Oxygen	Combines with iron atoms of hemoglobin molecules	Oxyhemoglobin
Carbon dioxide	Small amount dissolves in plasma	Carbon dioxide
	Small amount combines with the amino groups of hemoglobin molecules	Carbamino-hemoglobin
	Greatest amount reacts with water to form carbonic acid; the carbonic acid then dissociates to release hydrogen ions and bicarbonate ions	Bicarbonate ions

Fig. 15.43 In the lungs, carbon dioxide diffuses from the blood into the alveoli.

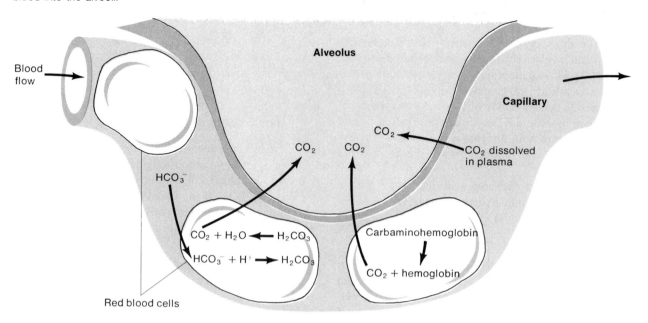

Utilization of Oxygen

The most important use of oxygen by body cells occurs during cellular respiration, even though certain cells may utilize oxygen for other purposes.

Cellular Respiration

As is discussed in chapter 4, **cellular respiration** involves the molecular breakdown of substances, such as glucose, and the release of energy from them.

To review, this process involves two phases. The first phase, which takes place in the cytoplasm of cells, occurs in the absence of oxygen and is therefore *anaerobic*. For glucose, this anaerobic phase includes the conversion of glucose to pyruvic acid and is accompanied by the transfer of a relatively small amount of energy of ATP molecules.

The second phase takes place within the mitochondria and is *aerobic*, since it requires the presence of oxygen. During this phase, the pyruvic acid is converted into acetyl coenzyme A, and these molecules enter a complex series of chemical reactions that constitute the *citric acid cycle*.

Citric Acid Cycle

A molecule of acetyl coenzyme A contains two carbon atoms, and as it enters the citric acid cycle, it combines with a 4-carbon atom molecule (oxaloacetic acid) to form a 6-carbon molecule of *citric acid*.

The citric acid in turn is oxidized, first to a 5-carbon molecule (α ketoglutaric acid) and then to a 4-carbon molecule (oxaloacetic acid). This 4-carbon molecule can then combine with another molecule of acetyl coenzyme A, and the cycle is completed, as shown in figure 15.44.

During the change of the 6-carbon molecule to a 5-carbon structure, and again during the change from a 5-carbon to a 4-carbon structure, molecules of carbon dioxide and hydrogen atoms are released. For each molecule of acetyl coenzyme A that enters the cycle, two carbon dioxide molecules and eight hydrogen atoms emerge. Thus, the two carbon atoms of the acetyl coenzyme A that entered the citric acid cycle are converted to carbon dioxide.

The carbon dioxide diffuses out of the cell and is transported away by the blood. The hydrogen atoms are accepted by available oxygen under the influence of enzymes called *oxidases*, and water molecules are formed.

While molecules of acetyl coenzyme A are broken down, energy is being released. Some of it is transferred to molecules of ATP that can be used to provide energy for various metabolic processes. The rest of the energy is lost as heat.

---

1. *What is meant by cellular respiration?*
2. *Distinguish between the anaerobic and the aerobic phases of cellular respiration.*
3. *Describe the major events that occur during the citric acid cycle.*

Fig. 15.44 Oxygen functions to accept hydrogen atoms that are released from the citric acid cycle, and this action results in the formation of water molecules.

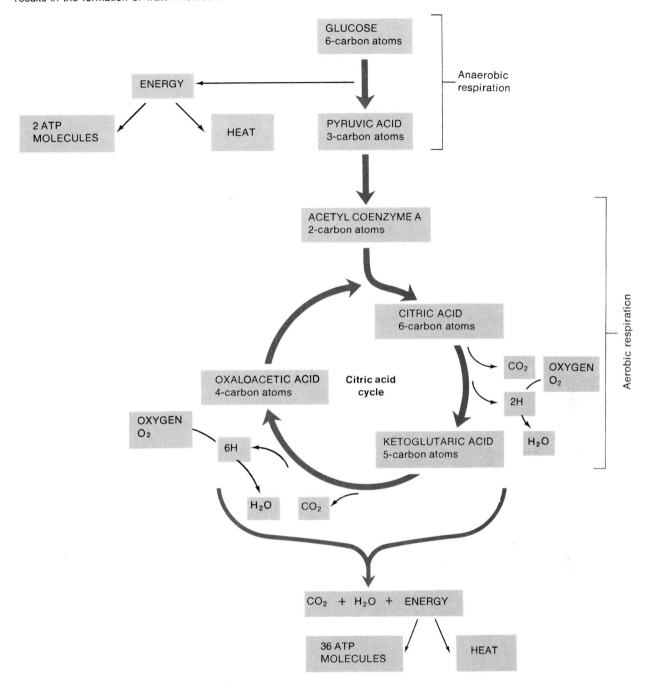

Some Clinical Terms Related to the Respiratory System

anoxia (ah-nok'se-ah)—an absence or deficiency of oxygen within tissues.

apnea (ap-ne'ah)—temporary absence of breathing.

asphyxia (as-fik'se-ah)—condition in which there is a deficiency of oxygen and an excess of carbon dioxide in the blood and tissues.

bradypnea (brad''e-ne'ah)—abnormally slow breathing.

bronchiolectasis (brong''ke-o-lek'tah-sis)—chronic dilation of the bronchioles.

bronchitis (brong-ki'tis)—inflammation of the bronchial lining.

Cheyne-Stokes respiration (chān stōks res''pī-ra'shun)—irregular breathing characterized by a series of shallow breaths that increase in depth and rate, followed by breaths that decrease in depth and rate.

eupnea (ūp-ne'ah)—normal breathing.

hypercapnia (hi''per-kap'ne-ah)—excessive carbon dioxide in the blood.

hyperpnea (hi''perp-ne'ah)—increased depth and rate of breathing.

hyperventilation (hi''per-ven''tĭ-la'-shun)—prolonged, rapid, and deep breathing.

hypoxemia (hi''pok-se'me-ah)—a deficiency in the oxygenation of the blood.

hypoxia (hi-pok'se-ah)—a diminished availability of oxygen in the tissues.

lobar pneumonia (lo'ber nu-mo'ne-ah)—pneumonia that affects an entire lobe of a lung.

pleurisy (ploo'rĭ-se)—inflammation of the pleural membranes.

pneumothorax (nu''mo-tho'raks)—entrance of air into the space between the pleural membranes, followed by collapse of the lung.

rhinitis (ri-ni'tis)—inflammation of the nasal cavity lining.

sinusitis (si''nŭ-si'tis)—inflammation of the sinus cavity lining.

tachypnea (tak''ip-ne'ah)—rapid, shallow breathing.

tracheotomy (tra''ke-ot'o-me)—an incision in the trachea for exploration or the removal of a foreign object.

Chapter Summary

Introduction

The respiratory system functions to obtain oxygen, release carbon dioxide, filter incoming air, control the temperature and water content of incoming air, produce sounds, and play roles in the sense of smell and the regulation of pH.

The respiratory system includes the nose, nasal cavity, sinuses, pharynx, larynx, trachea, bronchial tree, and lungs.

Organs of the Respiratory System

1. The nose
   a. The nose is a part of the face supported by bone and cartilage.
   b. Nostrils provide entrances for air.
2. The nasal cavity
   a. The nasal cavity is a space behind the nose.
   b. It is divided medially by the nasal septum.
   c. Nasal conchae divide the cavity into passageways.
   d. Mucous membrane lining filters, warms, and moistens incoming air.
   e. Particles trapped in the mucus are carried to the pharynx and swallowed.
3. The sinuses
   a. Sinuses are spaces in the bones of the skull that open into the nasal cavity.
   b. They are lined with mucous membrane that is continuous with the lining of the nasal cavity.
4. The pharynx
   a. The pharynx is located behind the mouth and between the nasal cavity and the larynx.
   b. It functions as a common passage for air and food.
   c. It aids in creating vocal sounds.
5. The larynx
   a. The larynx is an enlargement at the top of the trachea.
   b. It serves as a passageway for air and helps prevent foreign objects from entering the trachea.
   c. It is composed of muscles and cartilages, some of which are single while others are paired.
   d. It contains the vocal cords, which produce sounds by vibrating as air passes over them.
      (1) The pitch of a sound is related to the tension on the cords.
      (2) The intensity of a sound is related to the force of the air passing over the cords.
6. The trachea
   a. The trachea extends into the thoracic cavity in front of the esophagus.
   b. It divides into right and left bronchi.
   c. The mucous lining continues to filter incoming air.
   d. The wall is supported by cartilaginous rings.

7. The bronchial tree
    a. The bronchial tree consists of branched air passages that lead from the trachea to the air sacs.
    b. Alveoli are located at the distal ends of the finest tubes.
    c. Branches distribute air to all parts of the lungs.
8. The lungs
    a. The left and right lungs are separated by the mediastinum and enclosed by the diaphragm and the thoracic cage.
    b. The visceral pleura is attached to the surface of the lungs; parietal pleura lines the thoracic cavity.
    c. The right lung has three lobes, and the left lung has two.
    d. Each lobe is composed of lobules, which contain alveoli, blood vessels, and supporting tissues.

Mechanism of Breathing

Inspiration and expiration movements are accompanied by changes in the size of the thoracic cavity.
1. Inspiration
    a. Air is forced into the lungs by atmospheric pressure.
    b. Inspiration occurs when the pressure inside the thorax is reduced.
    c. Pressure within the thorax is reduced when the diaphragm moves downward and the thoracic cage moves upward and outward.
    d. Expansion of the lungs is aided by surface tension that holds the pleural membranes tightly together.
    e. The alveoli do not collapse because of the presence of surfactant that reduces surface tension.
2. Expiration
    a. The force of expiration comes from the elastic recoil of tissues.
    b. Expiration can be aided by thoracic and abdominal wall muscles that pull the thoracic cage downward and inward and compress the abdominal organs.
3. Respiratory air volumes
    a. The amount of air that normally moves in and out during quiet breathing is the tidal volume.
    b. Additional air that can be inhaled is the inspiratory reserve volume; additional air that can be exhaled is the expiratory reserve volume.
    c. Residual air remains in the lungs and is mixed with newly inhaled air.
    d. The vital capacity is the maximum amount of air a person can exhale after taking the deepest breath possible.
    e. The total lung capacity is equal to the vital capacity plus the residual air volume.

4. Nonrespiratory air movements
    a. Nonrespiratory air movements occur in addition to breathing.
    b. They include coughing, sneezing, laughing, crying, hiccuping, and yawning.
5. Some breathing disorders
    a. Breathing disorders are those that result in inadequate ventilation; they include paralysis of breathing muscles, bronchial asthma, emphysema, and lung cancer.
    b. Paralysis may be due to injuries to the nervous system or to diseases such as poliomyelitis.
    c. Bronchial asthma may result from an allergic reaction accompanied by a reduction in the diameters of the bronchioles.
    d. Emphysema involves destruction of alveolar walls and a loss of elasticity in the respiratory tissues.
    e. Lung cancer may produce overgrowths that obstruct air passages and reduce alveolar surfaces.

Control of Breathing

Normal breathing is rhythmic and involuntary, although the respiratory muscles can be controlled voluntarily.
1. The respiratory center
    a. This center is located in the brain stem, and includes parts of the medulla oblongata and pons.
    b. It initiates impulses that travel to breathing muscles.
    c. It can adjust the rate and depth of breathing.
    d. Maintaining a normal breathing pattern involves alternating actions of the inspiratory and expiratory circuits, which can inhibit each other.
2. Factors affecting breathing
    a. Tissue stretch, presence of certain chemicals, and emotional state affect breathing.
    b. Inflation and deflation reflexes help to regulate breathing and maintain rhythm.
    c. Increasing levels of carbon dioxide and hydrogen ions cause the rate to increase; a decreasing level of oxygen causes the rate to increase.
    d. Hyperventilation causes the carbon dioxide level to decrease.
3. Exercise and breathing rate
    a. During exercise, carbon dioxide increases and oxygen decreases, so that the respiratory center is stimulated.
    b. The cerebral cortex sends impulses to the respiratory center at the same time it stimulates skeletal muscles to contract.
    c. As joints are moved, proprioceptors are stimulated and joint reflexes are triggered, which in turn stimulate the respiratory center.
    d. When a person feels out of breath, it is usually due to an inability of the circulatory system to supply cells with their needs.

Alveolar Gas Exchanges

Alveoli carry on gas exchanges between the air and the blood.

1. The alveoli
   a. The alveoli are tiny air sacs clustered at the distal ends of the finest respiratory tubes.
   b. Some alveoli have openings into adjacent air sacs that provide alternate pathways for air when passages are obstructed.
2. The respiratory membrane
   a. This membrane consists of the alveolar and capillary walls.
   b. Gas exchanges take place through these walls.
3. Diffusion through the respiratory membrane
   a. The partial pressure of a gas is determined by the amount of that gas in a mixture or the amount dissolved in a liquid.
   b. Gases diffuse from regions of higher partial pressure toward regions of lower partial pressure.
   c. Oxygen diffuses from the alveolar air into the blood; carbon dioxide diffuses from the blood into the alveolar air.
4. Some disorders involving gas exchange
   a. Emphysema, pneumonia, tuberculosis, and atelectasis are conditions in which the surface areas available for gas exchange are reduced.
   b. In pneumonia, the alveoli become filled with fluid and cells.
   c. In tuberculosis, masses of fibrous connective tissue form in the sites of infection.
   d. Atelectasis involves partial or complete lung collapse.

Transport of Gases

Blood transports gases between the lungs and the body cells.

1. Oxygen transport
   a. Oxygen is transported in combination with hemoglobin molecules.
   b. The resulting oxyhemoglobin is relatively unstable and releases its oxygen in regions where the P_{O_2} is low.
2. Carbon monoxide
   a. Carbon monoxide forms as a result of incomplete combustion of fuels.
   b. It combines with hemoglobin more readily than oxygen and forms a stable compound.
   c. Carbon monoxide is toxic because the hemoglobin with which it combines is no longer available for oxygen transport.
3. Carbon dioxide transport
   a. Carbon dioxide may be carried in solution, bonded to hemoglobin, or as a bicarbonate ion.
   b. Most carbon dioxide is transported in the form of bicarbonate ions.
   c. The enzyme, carbonic anhydrase, speeds the reaction between carbon dioxide and water to form carbonic acid.
   d. Carbonic acid dissociates to release hydrogen ions and bicarbonate ions.

Utilization of Oxygen

1. Cellular respiration
   a. Cellular respiration occurs in two phases.
      (1) The anaerobic phase takes place in the cytoplasm.
      (2) The aerobic phase takes place in the mitochondria.
   b. During aerobic respiration, a complex series of chemical reactions results in the breakdown of acetyl coenzyme A and the formation of carbon dioxide and water.
2. Citric acid cycle
   a. Carbon atoms of acetyl coenzyme A are converted into carbon dioxide.
   b. Hydrogen atoms that are released are combined with oxygen to form water molecules.
   c. Some of the resulting energy is transferred to molecules of ATP, and the rest escapes as heat.
   d. Oxaloacetic acid combines with another molecule of acetyl coenzyme A to complete the cycle.

Application of Knowledge

1. If the upper respiratory passages are bypassed with a tracheostomy, how might the air entering the trachea be different from air normally passing through this canal? What problems may this create for the patient?

2. In certain respiratory disorders, such as emphysema, the capacity of the lungs to recoil elastically is reduced. Which respiratory air volumes will be affected by a condition of this type? Why?

3. What changes would you expect to occur in the relative concentrations of blood oxygen and carbon dioxide in a patient who breathes rapidly and deeply for a prolonged time? Why?

4. If a person has stopped breathing and is receiving pulmonary resuscitation, would it be better to administer pure oxygen or a mixture of oxygen and carbon dioxide? Why?

Review Activities

1. Describe the general functions of the respiratory system.

2. Explain how the nose and nasal cavity function in filtering incoming air.

3. Name and describe the locations of the major sinuses, and explain how a sinus headache may occur.

4. Distinguish between the pharynx and the larynx.

5. Name and describe the locations and functions of the cartilages of the larynx.

6. Distinguish between the false vocal cords and the true vocal cords.

7. Compare the structure of the trachea and the branches of the bronchial tree.

8. Distinguish between visceral pleura and parietal pleura.

9. Name and describe the locations of the lobes of the lungs.

10. Explain how normal inspiration and forced inspiration are accomplished.

11. Define *surface tension* and explain how it aids the breathing mechanism.

12. Define *surfactant* and explain its function.

13. Explain how normal expiration and forced expiration are accomplished.

14. Distinguish between vital capacity and total lung capacity.

15. Compare the mechanisms of coughing and sneezing and explain the function of each.

16. Explain the function of yawning.

17. Discuss how poliomyelitis may result in paralysis of breathing muscles.

18. Describe the cause of bronchial asthma and explain its effect upon breathing.

19. Define *emphysema* and explain how it affects breathing.

20. Describe the location of the respiratory center and explain how it functions in the control of breathing.

21. Describe the inflation and deflation reflexes.

22. Explain why increasing blood levels of carbon dioxide and hydrogen ions have similar effects upon the respiratory center.

23. Discuss the effects of emotions on breathing.

24. Define *hyperventilation* and explain how it affects the respiratory center.

25. Discuss the mechanism that affects the breathing rate during physical exercise.

26. Define *respiratory membrane* and explain its function.

27. Explain the relationship between the partial pressure of a gas and its rate of diffusion.

28. Summarize the gas exchanges that occur through the respiratory membrane.

29. Compare the respiratory effects of pneumonia and tuberculosis.

30. Define *atelectasis* and explain its possible causes.

31. Describe how oxygen is transported in blood.

32. Explain why carbon monoxide is toxic.

33. List three ways that carbon dioxide can be transported in blood.

34. Explain the function of carbonic anhydrase.

35. Define *chloride shift*.

36. Distinguish between anaerobic and aerobic respiration.

37. Explain the major steps in the citric acid cycle.

38. Explain how oxygen is utilized during cellular respiration.

Suggestions for Additional Reading

Avery, M. E., et al. April 1973. The lung of the newborn infant. *Scientific American.*

Cohen, A. B., and Gold, W. M. 1975. Defense mechanisms of the lungs. *Ann. Rev. Physio.* 37:325.

Comroe, J. H. 1974. *Physiology of respiration: an introduction,* 2nd ed. Chicago: Year Book Medical Publishers.

Crandall, E. D. 1976. Pulmonary gas exchange. *Ann. Rev. Physio.* 38:69.

Cumming, G., and Semple, S. 1973. *Disorders of the respiratory system.* Philadelphia: J.B. Lippincott Co.

Fraser, R. G., and Pare, J. A. 1971. *Structure and function of the lung.* Philadelphia: W. B. Saunders.

Guz, A. 1975. Regulation of respiration in man. *Ann. Rev. Physio.* 37:303.

Hock, R. J. February 1970. The physiology of high altitude. *Scientific American.*

Lehninger, A. L. September 1961. How cells transform energy. *Scientific American.*

Motley, H. L. 1973. Pulmonary emphysema. *Hosp. Med.* 9:11.

Naeye, R. L. 1974. Hypoxemia and the sudden infant death syndrome. *Science* 186:837.

Naeye, R. L. April 1980. Sudden infant death. *Scientific American.*

Slonim, N. B., and Chapin, J. L. 1971. *Respiratory physiology,* 2nd ed. St. Louis: C. V. Mosby.

West, J. B. 1972. Respiration. *Ann. Rev. Physio.* 34:91.

Winter, P. M., and Lowenstein, E. November 1969. Acute respiratory failure. *Scientific American.*

Wyman, R. J. 1977. Neural generation of the breathing rhythm. *Ann. Rev. Physio.* 39:417.

The Blood

16 The blood, heart, and blood vessels constitute the *circulatory system*, and provide a link between the body's internal parts and its external environment. More specifically, the blood transports nutrients from the digestive tract, oxygen from respiratory organs to the body cells, and wastes from these cells to the respiratory and excretory organs. It transports hormones from the endocrine glands to target tissues and bathes the body cells in a liquid of relatively stable composition. It also aids in temperature control by distributing heat from skeletal muscles and other active organs to all body parts. Thus, the blood provides vital support for cellular activities and aids in maintaining a favorable cellular environment.

The heart and the closed system of blood vessels provide the apparatus for moving blood throughout the body. These organs constitute the *cardiovascular system*, which will be discussed in chapter 17.

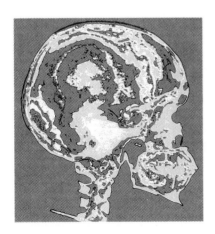

Chapter Outline

Chapter Objectives

After you have studied this chapter, you should be able to

1. Describe the general characteristics of blood, and discuss its major functions.

2. Distinguish between the various types of cells found in blood.

3. Explain how blood cell counts are made and how they are used.

4. Discuss the life cycle of a red blood cell.

5. Explain how red cell production is controlled.

6. Describe some of the more common disorders involving blood cells.

7. List the major components of blood plasma, and describe the functions of each.

8. Define *hemostasis*, and explain the mechanisms that help to achieve it.

9. Review the major steps in blood coagulation, and explain how coagulation can be prevented.

10. Explain the basis for blood typing and the methods used to avoid adverse reactions following blood transfusions.

11. Describe how blood reactions may occur between fetal and maternal tissues and how such reactions can be prevented.

12. Complete the review activities at the end of this chapter.

agglutinin (ah-gloo'tĭ-nin)

agglutinogen (ag''loo-tin'o-jen)

albumin (al-bu'min)

antibody (an'tĭ-bod''e)

coagulation (ko-ag''u-la'shun)

embolus (em'bo-lus)

erythrocyte (ĕ-rith'ro-sīt)

erythropoietin (ĕ-rith''ro-poi'ĕ-tin)

fibrinogen (fi-brin'o-jen)

globulin (glob'u-lin)

hemostasis (he''mo-sta'sis)

leukocyte (lu'ko-sīt)

macrophage (mak'ro-fāj)

plasma (plaz'mah)

platelet (plāt'let)

thrombus (throm'bus)

viscosity (vis-kos'ĭ-te)

agglutin-, to glue together: *agglutin*ation—the clumping together of red blood cells.

bil-, bile: *bil*irubin—pigment excreted in bile.

-crit, to separate: hemato*crit*—percentage of cells in a blood sample, determined by separating the cells from the plasma.

embol-, a stopper: *embol*us—an obstruction in a blood vessel.

erythr-, red: *erythr*ocyte—a red blood cell.

hemo-, blood: *hemo*globin—red pigment responsible for the color of blood.

hepar-, liver: *hepar*in—anticoagulant secreted by liver cells.

leuko-, white: *leuko*cyte—a white blood cell.

-lys, to break up: fibrino*lys*in—protein-splitting enzyme that can digest fibrin.

macro-, large: *macro*phage—a large phagocytic cell.

-osis, increased production: leukocyt*osis*—condition in which white blood cells are produced in abnormally large numbers.

-poiet, making: erythro*poiet*in—a hormone that stimulates the production of red blood cells.

poly-, many: *poly*cythemia—condition in which red blood cells appear in abnormally large numbers.

-stas, a halting: hemo*stas*is—arrest of bleeding from damaged blood vessels.

thromb-, a clot: *thromb*ocyte—blood platelet that plays a role in the formation of a blood clot.

As is mentioned in chapter 5, blood is a type of connective tissue whose cells are suspended in a liquid intercellular material.

Blood and Blood Cells

Whole blood is slightly heavier and three to four times more viscous than water. Its cells, which are formed mostly in red bone marrow, include *red blood cells, white blood cells,* and some cellular fragments called *platelets.* (See fig. 16.1.)

Volume and Composition of Blood

The volume of blood varies with body size. It also varies with changes in fluid and electrolyte concentrations, the amount of fat tissue present, and certain other factors. An average-size male (70 kg or 154 lbs), however, will have a blood volume of about 5 liters (5.2 quarts).

If a blood sample is allowed to stand in a tube for a while, the cells, which are over 99% red cells, will become separated from the liquid portion of the blood and settle to the bottom. This separation can be speeded by centrifuging the sample so that the cells quickly become packed into the lower part of the centrifuge tube, as shown in figure 16.2. When the amount of cells and liquid is measured, the percentage of each in the blood sample can be calculated.

A blood sample is usually close to 45% red cells. This percentage is called **hematocrit**. The remaining 55% of a blood sample consists of clear, straw-colored **plasma**.

In addition to red cells, the solids of the blood include *white blood cells* and *blood platelets*. The plasma is composed of a complex mixture that includes water, amino acids, proteins, carbohydrates, lipids, vitamins, hormones, electrolytes, and cellular wastes.

---

1. What are the major components of blood?
2. What factors affect blood volume?
3. How is hematocrit determined?

Characteristics of Red Blood Cells

Red blood cells, or **erythrocytes**, are tiny, biconcave disks that are thin near their centers and thicker around their rims. This special shape is related to the red cell's function of transporting gases, in that it provides an increased surface area for gas diffusion. (See fig. 16.3 and color plate 28.) The shape also places the cell membrane closer to various interior parts, where oxygen-carrying *hemoglobin* is found.

Fig. 16.1 Blood consists of a liquid portion, called plasma, and a solid portion that includes red blood cells, white blood cells, and platelets.

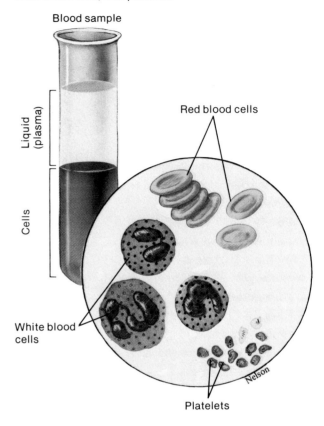

Fig. 16.2 If a blood-filled capillary tube is centrifuged, the red cells become packed in the lower portion, and the percentage of red cells (hematocrit) can be determined.

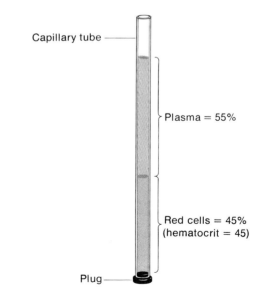

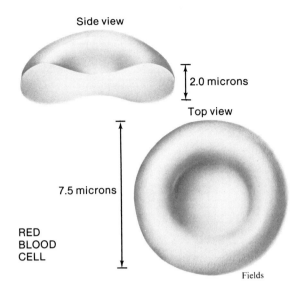

Side view

2.0 microns

Top view

7.5 microns

RED
BLOOD
CELL

Fields

Each red blood cell is about ⅓ hemoglobin by volume, and this substance is responsible for the color of the blood. Specifically, when the hemoglobin is combined with oxygen, the resulting *oxyhemoglobin* is bright red, and when the oxygen is released, the remaining *deoxyhemoglobin* is darker. In fact, blood rich in deoxyhemoglobin may appear bluish when it is viewed through blood vessel walls.

A person experiencing a prolonged oxygen deficiency (hypoxia) may develop a symptom called *cyanosis*. In this condition, the skin and mucous membranes appear bluish due to an abnormally high concentration of deoxyhemoglobin in the blood. Cyanosis may also occur as a result of exposure to low temperature. In this case, superficial blood vessels become constricted, and more oxygen than usual is removed from the blood flowing through them.

Although red blood cells have nuclei during the early stages of their development in the red bone marrow, these nuclei are lost as the cells mature. This characteristic, like the shape of a red blood cell, seems to be related to the function of transporting oxygen, since the space previously occupied by the nucleus is available to hold hemoglobin.

Even though the nucleus is missing, the cytoplasmic enzymes of a red blood cell remain functional, and the cell is able to carry on vital energy-releasing processes. With age, however, the cell becomes less and less active. Red blood cells have limited life spans that average about 120 days.

1. Describe a red blood cell.
2. What is the function of hemoglobin?
3. What is the average life expectancy of a red blood cell?

Red Blood Cell Counts

The number of red blood cells in a cubic millimeter (mm^3) of blood is called the *red cell count*. Although this number varies from time to time even in health, the normal range for adult males is 4,600,000–6,200,000 cells per mm^3, while that for adult females is 4,200,000–5,400,000 cells per mm^3. For children, the normal range is 4,500,000–5,100,000 cells per mm^3. The number of red cells generally increases following exercise, a large meal, a rise in temperature, or an increase in altitude.

Since the number of circulating red blood cells is closely related to the blood's *oxygen-carrying capacity*, any changes in this number may be significant. For this reason, red cell counts are routinely made to help in diagnosing and evaluating the courses of various diseases.

In the past, blood cell counts were made using a special glass slide called a *hemocytometer*. This device is designed so the cells in a small, but known, volume of blood can be observed and counted with the aid of a microscope. Today clinical laboratories are often equipped with electronic counters that detect and record the number of blood cells forced through a tiny opening in the instrument. (See fig. 16.4.)

Destruction of Red Blood Cells

Red blood cells are quite elastic and flexible, and they readily change shape as they pass through small blood vessels. With age, however, these cells become more fragile, and they are often damaged or ruptured when they squeeze through capillaries, particularly those in active muscles.

Fig. 16.4 An electronic blood cell counter.

Fig. 16.5 When a hemoglobin molecule is decomposed, the heme portions are broken down into iron (Fe) and biliverdin. Most of the biliverdin is then converted to bilirubin.

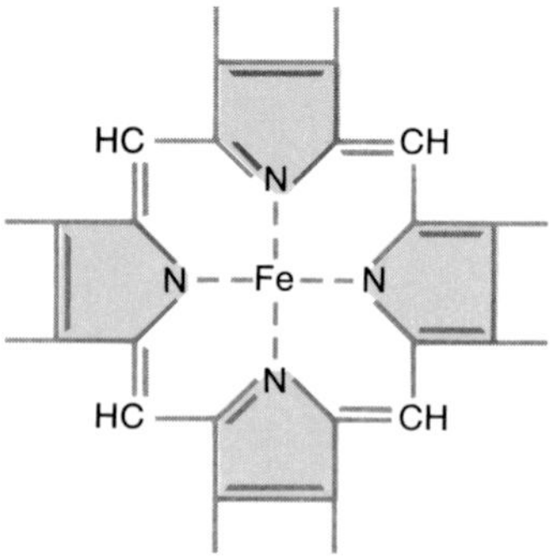

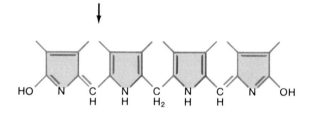

Heme

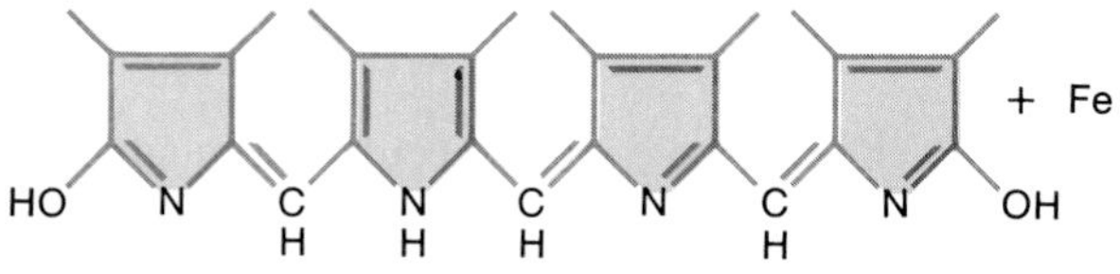

Biliverdin ($C_{33}H_{34}O_6N_4$)

Bilirubin ($C_{33}H_{36}O_6N_4$)

Damaged red cells are phagocytized and destroyed, primarily in the liver and spleen, by large reticuloendothelial cells called **macrophages**, which are described in chapter 5. The fragments and contents of ruptured cells are also phagocytized and digested by macrophages.

The hemoglobin molecules from red cells undergoing destruction are broken down into molecules of *heme* and *globin*. The heme is further decomposed into iron and a greenish pigment called **biliverdin**. The iron may be carried by the blood to the blood-cell-forming tissue in the red bone marrow, where it is reused in the synthesis of a new hemoglobin. Otherwise it may be stored in liver cells in the form of an iron-protein complex called *ferritin*. In time, the biliverdin is converted to an orange pigment called **bilirubin**. These substances are excreted in bile as bile pigments. (See fig. 16.5.) Chart 16.1 summarizes the process of red cell destruction.

It is estimated that 1/15 of the red blood cells are removed from circulation each day. Yet the number of cells in the circulating blood remains relatively stable. This observation suggests that a *homeostatic mechanism* regulates the rate of red cell production.

Chart 16.1 Major events in red blood cell destruction

1. Cells are damaged while squeezing through the capillaries of active tissues.
2. Damaged red cells are phagocytized by macrophages in the liver and spleen.
3. Hemoglobin from the red cells is decomposed into heme and globin.
4. Heme is decomposed into iron and biliverdin.
5. Iron is made available for reuse in the synthesis of new hemoglobin or is stored in the liver as ferritin.
6. Some biliverdin is converted into bilirubin.
7. Biliverdin and bilirubin are excreted in *bile* as bile pigments.

Fig. 16.6 Hemocytoblasts in red bone marrow give rise to erythroblasts, which in turn produce erythrocytes.

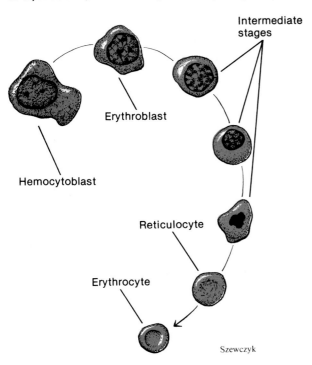

Intermediate stages

Erythroblast

Hemocytoblast

Reticulocyte

Erythrocyte

Szewczyk

Fig. 16.7 Low P_{O_2} causes the release of erythropoietin from the kidneys and liver. This substance stimulates the production of red blood cells, and these cells carry additional oxygen to the tissues.

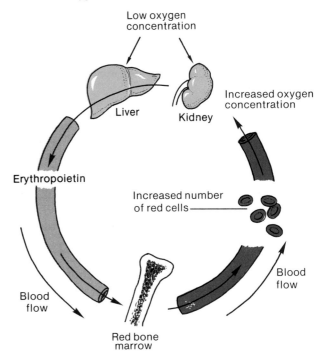

Low oxygen concentration

Liver

Kidney

Increased oxygen concentration

Erythropoietin

Increased number of red cells

Blood flow

Blood flow

Red bone marrow

About ⅓ of all newborns develop a mild disorder called *physiologic jaundice* within a few days following birth. In this condition, as in other forms of jaundice, the skin and eyes become yellowish due to an accumulation of bilirubin in the tissues.

Physiologic jaundice is thought to be the result of immature liver cells that are somewhat ineffective in excreting bilirubin into the bile. The condition is treated by feedings that promote bowel movements and by exposure to fluorescent light that reduces the concentration of bilirubin in the tissues.

---

1. *What is the normal red blood cell count for a male? For a female?*
2. *What is the function of macrophages?*
3. *What are the end products of hemoglobin breakdown?*

Red Blood Cell Production and Its Control

As is mentioned in chapter 8, red blood cells are formed in the yolk sac, liver, and spleen of an embryo; but after an infant is born, they are produced almost exclusively by cells that line the vascular sinuses of the red bone marrow.

Within the red marrow, cells, called **hemocytoblasts**, give rise to **erythroblasts** that can synthesize hemoglobin. (See fig. 16.6.) The erythroblasts also reproduce and give rise to many daughter cells whose nuclei soon shrink and disappear. The resulting cells are **erythrocytes**, and they contain no nuclear materials when they are finally released into circulation. Some of these young red cells, however, may contain a netlike structure (reticulum) that represents the endoplasmic reticulum. Such cells are called **reticulocytes**.

The rate of red blood cell formation seems to be controlled by a *negative feedback mechanism* involving a hormone called **erythropoietin**. In response to prolonged oxygen deficiency, this hormone is released, primarily from the kidneys and to a lesser extent from the liver.

At high altitudes, for example, where the P_{O_2} is reduced, the amount of oxygen delivered to the tissues decreases. As figure 16.7 shows, this drop in oxygen concentration triggers the release of erythropoietin, which travels via the blood to the red bone marrow and stimulates increased red cell production.

Chart 16.2 Major events in the hormonal control of red blood cell production

1. The kidney and liver tissues experience an oxygen deficiency.
2. These tissues release the hormone erythropoietin.
3. Erythropoietin travels to the red bone marrow and stimulates an increase in the production of red cells.
4. As increasing numbers of red blood cells are released into the circulation, the oxygen-carrying capacity of the blood rises.
5. The oxygen concentration in the kidney and liver tissues increases, and the release of erythropoietin decreases.

Chart 16.3 Dietary factors affecting red blood cell production

Substance	Source	Function
Vitamin B_{12}	Absorbed from small intestine in the presence of intrinsic factor	Necessary for the synthesis of DNA
Folic Acid	Absorbed from small intestine	Necessary for the synthesis of DNA
Iron	Absorbed from small intestine, conserved during red cell destruction and made available for reuse	Necessary for the synthesis of hemoglobin

After a few days, large numbers of newly formed red cells begin to appear in the circulating blood, and the increased rate of production continues as long as the kidney and liver tissues experience an oxygen deficiency. Eventually, the number of red cells in circulation may be sufficient to supply these tissues with their oxygen needs. When this happens or when the Po_2 returns to normal, the release of erythropoietin ceases, and the rate of red cell production is reduced.

Other conditions that produce a low Po_2 and thus stimulate erythropoietin release include the following: loss of blood, which results in a decreased oxygen-carrying capacity of the circulatory system; and chronic lung diseases, which cause a decrease in the respiratory surface area available for gas exchanges.

Chart 16.2 summarizes the steps in the regulation of red blood cell production.

1. *Where are red blood cells produced?*
2. *How is the production of red blood cells controlled?*

Dietary Factors Affecting Red Blood Cell Production. Red blood cell production is significantly influenced by the availability of two of the B-complex vitamins—*vitamin B_{12}* and *folic acid.* These substances are required for the synthesis of DNA molecules, and so they are needed by all cells for growth and reproduction. Since cellular reproduction occurs at a particularly high rate in red-blood-cell-forming tissue, this tissue is especially affected by a lack of either of these vitamins.

A lack of vitamin B_{12} usually is due to a disorder in the stomach lining, rather than a dietary deficiency. As is mentioned in chapters 13 and 14, the parietal cells of the gastric glands normally secrete a substance called *intrinsic factor* that promotes the absorption of vitamin B_{12}.

In the absence of vitamin B_{12}, the red bone marrow forms red blood cells that are abnormally large, irregularly shaped, and have particularly thin membranes. These cells tend to be damaged more easily than normal red cells, and consequently they have shorter lives. The result is a condition called *pernicious anemia.*

Since vitamin B_{12} is absorbed poorly by persons with pernicious anemia, this disorder is usually treated with injections of the deficient vitamin rather than with oral ingestion.

As was mentioned, *iron* is needed for hemoglobin synthesis and normal red blood cell production. It can be absorbed from food as it passes through the small intestine; also, much of the iron released during the decomposition of hemoglobin from damaged red blood cells is made available for reuse. Because of this conservation mechanism, relatively small quantities of iron are needed in the daily diet.

In the absence of an adequate supply of iron, a person may develop a condition called *hypochromic anemia,* which is characterized by the presence of small, pale red blood cells as those shown in figure 16.8.

The dietary factors that affect red blood cell production are summarized in chart 16.3.

Fig. 16.8 (a) Normal human erythrocytes; (b) erythrocytes from a person with hypochromic anemia. (The larger cells with the stained nuclei are white blood cells.)

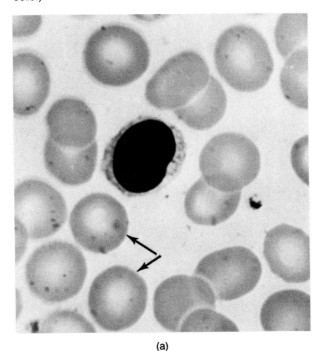

(a)

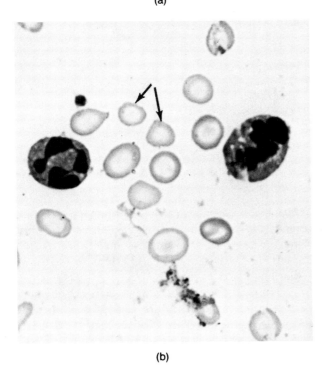

(b)

During pregnancy, the enlarging uterus and its developing vascular system create a demand for more blood. Consequently, the maternal blood volume usually increases greatly. Since this increase requires accelerated production of red blood cells, a pregnant woman must take additional iron to meet the needs of the blood-cell-forming tissues.

1. What vitamins are necessary for red blood cell production?
2. Why is iron needed for the normal development of red blood cells?

Red Blood Cell Disorders

Whenever a rapid loss or an inadequate production of red cells causes a deficiency of these cells, the resulting condition is termed **anemia**. In disorders of this type, the oxygen-carrying capacity of the blood is reduced, and the person may appear pale and lack energy. Among the more common types of anemia are hemorrhagic anemia, aplastic anemia, and hemolytic anemia.

Hemorrhagic anemia is caused by an abnormal loss of blood and may be mild or severe depending on the amount of bleeding that occurs. Increased production of red cells usually makes up for a blood loss within a few weeks. If the blood loss is prolonged, however, the person may not be able to obtain enough iron from food to replace the lost hemoglobin. As a result, even though the normal number of red cells is reestablished, they may lack a normal concentration of hemoglobin. This condition is called *hypochromic anemia*.

Aplastic anemia is due to a malfunction in the red bone marrow and is characterized by decreased production of red blood cells. This condition is sometimes caused by an overexposure to X ray or some other form of *ionizing radiation*. It can also be caused by the toxic effects of various chemicals or adverse drug reactions.

Hemolytic anemia is characterized by an abnormally high rate of red blood cell rupture (hemolysis). Such cellular breakdown may be caused by a variety of factors including hereditary defects, parasitic infections, and adverse drug reactions.

Fig. 16.9 Anemia may result in an increased work load on the heart by causing a decrease in blood viscosity.

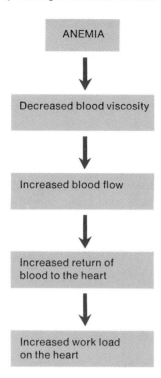

ANEMIA

↓

Decreased blood viscosity

↓

Increased blood flow

↓

Increased return of blood to the heart

↓

Increased work load on the heart

Fig. 16.10 Granulocytes include neutrophils, eosinophils, and basophils. How are these cells distinguished from each other?

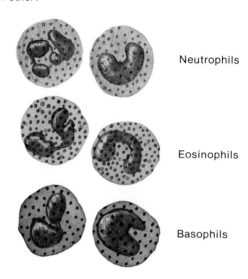

Neutrophils

Eosinophils

Basophils

Various forms of anemia are usually accompanied by a change in the **viscosity** of the blood. Viscosity is a physical property of a fluid and is related to the ease with which its molecules slide past one another. A fluid with a high viscosity tends to be sticky like syrup, and one with a low viscosity flows easily like water.

The viscosity of blood depends largely on the concentration of red blood cells. Anemic blood with decreased viscosity flows relatively easily through the blood vessels and returns to the heart in abnormally great volumes. This effect tends to increase the work load on the heart and can result in heart failure during periods of heavy physical exercise. (See fig. 16.9.)

Another type of red blood cell disorder is called **polycythemia**. In this condition, there is an abnormal increase in the number of circulating red cells. It may result from stimulation of the erythrocyte-forming tissues, as occurs when a person remains at a high altitude or engages in prolonged physical activity. It may also occur when the blood becomes concentrated from excessive loss of body fluids that accompanies heavy sweating or shock. Polycythemia also may be caused by a tumorous condition in the red-cell-forming tissues, causing the production of abnormally large numbers of cells.

A general effect of polycythemia is to increase the viscosity of the blood so that blood flow through the vessels is decreased. As a result of sluggish movement of blood, an abnormal amount of hemoglobin may be *deoxygenated*, and the skin may appear bluish (cyanotic).

1. What is anemia?
2. Describe the causes of three common types of anemia.
3. Define viscosity.
4. How does polycythemia affect the viscosity of the blood?

Types of White Blood Cells

White blood cells, also called **leukocytes,** function primarily to defend the body against microorganisms. Although these cells do most of their work outside the circulatory system, they use the blood for transportation to sites of infection.

Normally, five types of white cells can be found in circulating blood. They are distinguished by size, the nature of the cytoplasm, the shape of the nucleus, and the staining characteristics. For example, some types have granular cytoplasm and make up a group called *granulocytes*, while others lack cytoplasmic granules and are called *agranulocytes*.

A **granulocyte** is typically about twice the size of a red cell. Members of this group include three types of white cells—neutrophils, eosinophils, and basophils—that are shown in figure 16.10. These cells develop in the red bone marrow and are produced from *hemocytoblasts* in much the same manner as red cells.

Neutrophils are characterized by the presence of fine cytoplasmic granules that stain pinkish in neutral stain. The nucleus of a neutrophil is lobed and consists of two to five parts connected by thin strands of chromatin. For this reason, neutrophils are also called *polymorphonuclear leukocytes*, which means their nuclei take many forms. Neutrophils account for 54 to 62% of the white cells in a normal blood sample.

Eosinophils contain coarse, uniformly sized cytoplasmic granules that stain deep red in acid stain. The nucleus usually has only two lobes (bilobed). These cells make up 1 to 3% of the total number of circulating leukocytes.

Basophils are similar to eosinophils in size and in the shape of their nuclei. However, they have fewer, more irregularly shaped cytoplasmic granules that stain deep blue in basic stain. This type of leukocyte usually accounts for less than 1% of the white cells.

The leukocytes of the **agranulocyte** group, shown in figure 16.11, include monocytes and lymphocytes. While monocytes generally arise from red bone marrow, lymphocytes are formed in organs of the lymphatic system as well as in the red bone marrow.

Monocytes are the largest cells found in the blood, having diameters two to three times greater than red cells. Their nuclei vary in shape and are described as round, kidney-shaped, oval, or lobed. They usually make up 3 to 9% of the leukocytes in a blood sample.

Although large **lymphocytes** are sometimes found in the blood, usually they are only slightly larger than the red cells. Typically, a lymphocyte contains a relatively large, round nucleus with a thin rim of cytoplasm surrounding it. These cells account for 25 to 33% of the circulating white blood cells.

---

1. Distinguish between granulocytes and agranulocytes.
2. List five types of white blood cells, and explain how they differ from one another.

White Blood Cell Counts

The procedure used to count white blood cells is similar to that used for counting red cells. Before a *white cell count* is made, however, the red cells in the blood sample are destroyed so they will not be mistaken for white cells. Normally there are from 5000 to 10,000 white cells per mm³ of blood.

Since the number of white blood cells may change in response to abnormal conditions, white blood cell counts are of clinical interest. For example, some infectious diseases are accompanied by a rise in the number of circulating white cells, and if the total

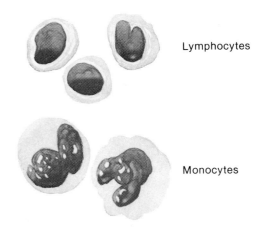

Fig. 16.11 Agranulocytes include monocytes and lymphocytes.

Lymphocytes

Monocytes

number of white cells exceeds 10,000 per mm³ of blood, the person is said to have **leukocytosis**. This occurs during acute infections, such as appendicitis. It may also appear following vigorous exercise, emotional disturbances, or excessive loss of body fluids.

If the total white cell count drops below 5000 per mm³ of blood, the condition is called **leukopenia**. Such a deficiency may accompany typhoid fever, influenza, measles, mumps, chicken pox, and poliomyelitis. It may also occur as a result of various anemias or poisoning from lead, arsenic, or mercury.

A *differential white blood cell count* is one in which the percentages of the various types of leukocytes in a blood sample are determined. This test is useful because the relative proportions of white cells may change in particular diseases. Neutrophils, for instance, usually increase during bacterial infections, while eosinophils may increase during certain parasitic infections and allergies.

---

1. What is the normal white blood cell count?
2. Distinguish between leukocytosis and leukopenia.
3. What is a differential white blood cell count?

Functions of White Blood Cells

As was mentioned, white blood cells generally function to protect the body against microorganisms. For example, some leukocytes phagocytize microorganisms that get into the body, and others produce antibodies that react with such foreign particles to destroy or disable them.

Fig. 16.12 When bacteria invade the tissues, leukocytes migrate into the region and destroy bacteria by phagocytosis.

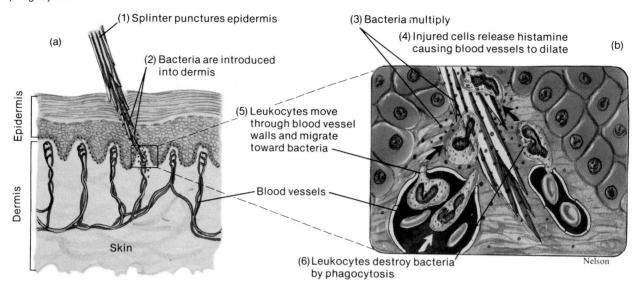

(1) Splinter punctures epidermis

(a)

(2) Bacteria are introduced into dermis

(3) Bacteria multiply

(4) Injured cells release histamine causing blood vessels to dilate

(b)

Epidermis

Dermis

(5) Leukocytes move through blood vessel walls and migrate toward bacteria

Blood vessels

Skin

(6) Leukocytes destroy bacteria by phagocytosis

Nelson

Leukocytes can squeeze between the cells that make up blood vessel walls. This movement, called **diapedesis**, allows the white cells to leave the circulation. Once outside the blood, they move through interstitial spaces with an amebalike self-propulsion, called *ameboid motion.*

The most mobile and active phagocytic leukocytes are the *neutrophils* and the *monocytes*. Although the neutrophils are unable to ingest particles much larger than bacterial cells, monocytes can engulf relatively large objects. Both types contain numerous *lysosomes* that are filled with digestive enzymes capable of breaking down various proteins and lipids like those in bacterial cell membranes. As a result of their activities, neutrophils and monocytes often become so engorged with digestive products and bacterial toxins that they also die.

When tissues are invaded by microorganisms, the body cells often respond by releasing substances that cause local blood vessels to dilate. One such substance, **histamine**, also causes an increase in the permeability of nearby capillaries and the walls of small veins (venules). As these changes occur, the tissues become reddened and large quantities of fluids leak into the interstitial spaces. The swelling produced by this *inflammatory reaction* tends to delay the spread of the invading microorganisms into other regions. At the same time, leukocytes are somehow attracted by chemicals released from damaged cells, and they move toward these substances. This phenomenon is called **positive chemotaxis**, and it results in large numbers of white blood cells moving quickly into inflamed areas. (See fig. 16.12.)

As bacteria, leukocytes, and damaged cells accumulate in an inflamed area, a thick fluid, called *pus*, often forms, and continues to do so as long as the invading microorganisms are active. If the pus cannot escape to the outside of the body or into a body cavity, it may remain walled in the tissues for some time. Eventually it may be absorbed by the surrounding cells.

The function of *eosinophils* is uncertain, although there is evidence that they act to detoxify foreign proteins that enter the body through the lungs or intestinal tract. They may also release enzymes that break down blood clots and help remove certain products of immune responses (antigen-antibody complexes) described in chapter 18.

Some of the cytoplasmic granules of *basophils* contain a blood-clot-inhibiting substance, called *heparin*, while other granules contain *histamine*. It is thought that basophils may aid in preventing intravascular blood clot formation by releasing heparin, and that they may cause an increase in blood flow to injured tissues by releasing histamine. Following a fatty meal, heparin seems to speed the removal of fat particles from the blood.

Lymphocytes play an important role in the process of *immunity*. Some, for example, can form **antibodies** that act against various foreign substances that enter the body. This function is discussed in chapter 18.

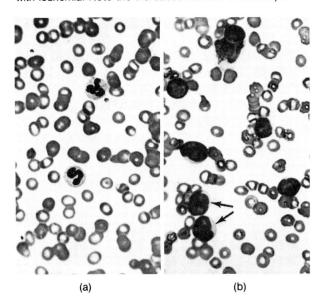

(a)   (b)

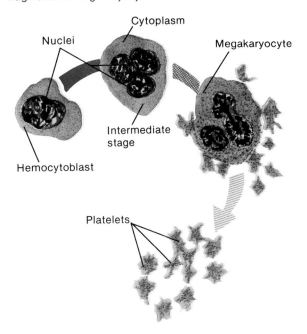

Nuclei

Cytoplasm

Megakaryocyte

Intermediate stage

Hemocytoblast

Platelets

Other lymphocytes seem to be multipotential cells that can change into other types of blood cells, such as monocytes or erythrocytes, as the body needs demand.

1. What are the primary functions of white blood cells?
2. How do white blood cells reach microorganisms that are outside blood vessels?
3. Which white blood cells are the most active phagocytes?
4. What are the functions of eosinophils and basophils?

Leukemia

Leukemia is a form of cancer characterized by an uncontrolled production of leukocytes that fail to mature. (See fig. 16.13.) There are two major types of leukemia. *Myeloid leukemia* results from an abnormal production of granulocytes by the red bone marrow, while *lymphoid leukemia* is accompanied by increased formation of lymphocytes from lymph nodes. In both types, the cells produced usually fail to differentiate and remain nonfunctional. Thus, even though large numbers of neutrophils may be formed in myeloid leukemia, these cells have little ability to phagocytize, and the victim has a lowered resistance to infections.

Eventually, the cells responsible for the overproduction of leukocytes tend to spread from the bone marrow or lymph nodes to other parts of the body. As with other forms of cancer, the leukemic cells finally appear in such great numbers that they crowd out the normal, functioning cells.

Leukemias are classified as *acute* or *chronic*. An acute condition appears suddenly, the symptoms progress rapidly, and death occurs in a few months when the condition is untreated. Chronic forms begin more slowly and may remain undetected for many months. Without treatment, life expectancy is about 3 years.

Acute lymphoid leukemia is primarily a disease of children; the chronic form usually occurs after 50 years of age. Acute myeloid leukemia occurs at any age, but is more frequent in adults; chronic myeloid leukemia is primarily a disease of adults between 20 and 50 years of age.

Blood Platelets

Platelets are also called **thrombocytes**, and although they are not complete cells, they have important functions. Platelets arise from giant, multinucleated cells in the red bone marrow, called **megakaryocytes**. These cells give off tiny cytoplasmic fragments, and as they detach and enter the circulation, they become platelets. (See figs. 16.14 and 16.15.)

Fig. 16.15 (*a*) An electron micrograph of human blood cells: *E*, erythrocyte; *L*, lymphocyte; *M*, monocyte; *N*, neutrophil; *P*, platelet. You can judge the actual sizes of these cells by the line in the *lower right* corner, which represents 1 micron (*μ*). (*b*) Platelets more highly magnified.

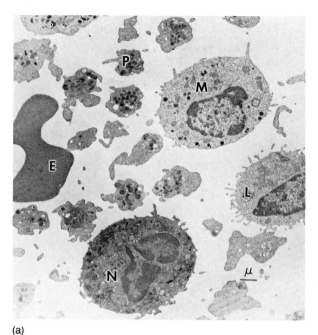

(a)

(b)

Each platelet is a round disk that lacks a nucleus and is less than half the size of a red blood cell. It is capable of ameboid movement. In normal blood, the platelet count will vary from 130,000 to 360,000 platelets per mm^3.

Platelets help close breaks in damaged blood vessels, by sticking to the broken surfaces. In the presence of such breaks, platelets release a substance, called **serotonin**, that causes smooth muscles in the vessel walls to contract—an action that reduces blood flow. In addition, platelets function to initiate the formation of blood clots, as is explained in a later section of this chapter.

Chart 16.4 summarizes the characteristics of blood cells and platelets.

1. *What is leukemia?*
2. *Distinguish between myeloid and lymphoid leukemia.*
3. *What is the function of blood platelets?*
4. *What is the normal blood platelet count?*

Blood Plasma

Plasma is the straw-colored, liquid portion of the blood in which the cells and platelets are suspended. It is approximately 92% water and contains a complex mixture of organic and inorganic substances that function in a variety of ways. These functions include transporting nutrients, gases, and vitamins, regulating fluid and electrolyte balances, and maintaining a favorable pH. Figure 16.16 shows the chemical makeup of plasma.

Plasma Proteins

The most abundant dissolved substances (solutes) in plasma are **plasma proteins**. These proteins remain in the blood and interstitial fluids and ordinarily are not used as energy sources. By employing laboratory techniques, it is possible to separate the plasma proteins into three main groups—albumins, globulins, and fibrinogen. The members of each group differ in their chemical structures and in their physiological functions.

Albumins account for about 60% of the plasma proteins and have the smallest molecules of these proteins. They are synthesized in the liver, and because they are so plentiful, they are of particular significance in maintaining the *osmotic pressure* of the blood.

Chart 16.4 Cellular components of blood

Component	Description	Number Present	Function
Red blood cell (erythrocyte)	Biconcave disk without nucleus, about ⅓ hemoglobin	4,000,000 to 6,000,000 per mm³	Transports oxygen and carbon dioxide
White blood cells (leukocytes)		5000 to 10,000 per mm³	Aids in defense against infections by microorganisms
Granulocytes	About twice the size of red cells, cytoplasmic granules present		
1. Neutrophil	Nucleus with two to five lobes, cytoplasmic granules stain pink in neutral stain	54 to 62% of white cells present	Destroys relatively small particles by phagocytosis
2. Eosinophil	Nucleus bilobed, cytoplasmic granules stain red in acid stain	1 to 3% of white cells present	Helps to detoxify foreign substances, secretes enzymes that break down clots and removes products of immune reactions
3. Basophil	Nucleus lobed, cytoplasmic granules stain blue in basic stain	Less than 1% of white cells present	Releases anticoagulant, heparin, and histamine
Agranulocytes	Cytoplasmic granules absent		
1. Monocyte	Two to three times larger than red cell, nuclear shape varies from round to lobed	3 to 9% of white cells present	Destroys relatively large particles by phagocytosis
2. Lymphocyte	Only slightly larger than red cell, nucleus nearly fills cell	25 to 33% of white cells present	Forms antibodies of immunity
Platelet (thrombocyte)	Cytoplasmic fragment	130,000 to 360,000 per mm³	Helps control blood loss from broken vessels

Fig. 16.16 Plasma, which is primarily water, contains about 7% protein and 1.5% of other substances.

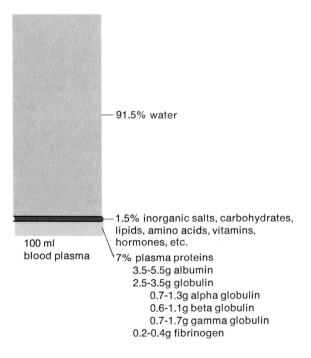

— 91.5% water

100 ml blood plasma

— 1.5% inorganic salts, carbohydrates, lipids, amino acids, vitamins, hormones, etc.

7% plasma proteins
  3.5-5.5g albumin
  2.5-3.5g globulin
    0.7-1.3g alpha globulin
    0.6-1.1g beta globulin
    0.7-1.7g gamma globulin
  0.2-0.4g fibrinogen

As is explained in chapter 3, whenever the concentration of dissolved substances changes on either side of a cell membrane, water is likely to move through the membrane toward the region where the dissolved molecules are in higher concentration. For this reason, it is important that the concentration of dissolved substances in plasma remain relatively stable. Otherwise, water will tend to leave the blood and enter the tissues, or leave the tissues and enter the blood by *osmosis*. The presence of albumins (and other plasma proteins) adds to the osmotic pressure of the plasma and aids in regulating the water balance between the blood and the tissues. It also helps to control the blood volume, which in turn is directly related to the blood pressure.

The concentration of plasma proteins may decrease significantly if a person is starving or has a protein-deficient diet. Similarly, the plasma protein level may drop if liver disease interferes with the synthesis of these proteins. In each case, as the blood protein concentration decreases the osmotic pressure of the blood decreases, and water tends to accumulate in the intercellular spaces, causing edema.

Chart 16.5 Plasma proteins

Protein	Percentage of Total	Origin	Function
Albumin	60%	Liver	Helps in the maintenance of blood osmotic pressure
Globulin	36%		
Alpha globulins		Liver	Transport of lipids and fat-soluble vitamins
Beta globulins		Liver	Same as above
Gamma globulins		Lymphatic tissues	Constitute antibodies of immunity
Fibrinogen	4%	Liver	Plays key role in blood clot formation

The **globulins**, which make up about 36% of the plasma proteins, can be further separated into fractions called *alpha globulins*, *beta globulins*, and *gamma globulins*. The alpha and beta globulins are synthesized in the liver, and they have a variety of functions including the transport of lipids and fat-soluble vitamins. The gamma globulins are produced in lymphatic tissues, and they include the proteins that function as *antibodies of immunity*.

Fibrinogen plays a primary role in the blood clotting mechanism. It is synthesized in the liver and has the largest molecules of the plasma proteins.

Chart 16.5 summarizes the characteristics of the plasma proteins.

1. *List three types of plasma proteins.*
2. *How does albumin help to maintain a water balance between the blood and the tissues?*
3. *Which of the globulins functions in immunity?*
4. *What is the role of fibrinogen?*

Nutrients and Gases

The *plasma nutrients* include amino acids, simple sugars, and various lipids that have been absorbed from the digestive tract. Glucose, for example, is transported by the plasma from the intestine to the liver, where it may be stored as glycogen or changed into fat. If the blood glucose level drops below the normal range, glycogen may be converted back into glucose, as is described in chapter 12.

Recently absorbed amino acids also are carried to the liver, where they may be used in the manufacture of proteins, such as those found in the plasma, or deaminated and used as an energy source.

The lipids of plasma include fats, phospholipids, and cholesterol. Generally these lipids are combined with proteins in complexes called **lipoproteins**, such as the *chylomicrons* described in chapter 13.

In addition to chylomicrons, which are present only temporarily following the digestion of a fatty meal, plasma contains smaller particles of lipoproteins composed of triglycerides, phospholipids, cholesterol, and protein in varying proportions.

Since fats are less dense than proteins, as the proportion of lipid in the lipoprotein increases, the density of the particle decreases. Conversely, as the proportion of lipid decreases, the density increases.

On the basis of their densities, which reflect their composition, these particles can be classified as follows: *very low density lipoproteins* (VLDL), which have a relatively high concentration of triglycerides; *low density lipoproteins* (LDL), which have a relatively high concentration of cholesterol; and *high density lipoproteins* (HDL), which have a relatively high concentration of protein and a lower concentration of lipid.

Most of these lipoproteins are produced in the liver, and although their function is not well understood, they seem to be involved in transporting lipids from the liver to various tissues.

Some people have an inherited condition in which their plasma concentration of low density lipoproteins is abnormally high. Since these lipoproteins contain a high degree of cholesterol, this condition is characterized by excessive blood cholesterol (hypercholesteremia). This high cholesterol level is associated with increased deposits of cholesterol in the walls of arteries and the development of atherosclerosis.

A **phospholipid** molecule is like a fat molecule in which one of the fatty acid portions has been replaced by a phosphoric acid-containing group. (See fig. 16.17.) Two important phospholipids found in plasma are called *lecithin* and *cephalin*. Among other activities, these lipids aid in the absorption of fatty acids by helping to transport them through the intestinal wall and into the lacteals of the intestinal villi.

Cholesterol, which is commonly present in foods of animal origin, is absorbed from the intestine without being digested. As is explained in chapter 14, the plasma concentration of cholesterol is largely regulated by the liver.

The most important blood gases are oxygen and carbon dioxide. While plasma also contains considerable amounts of dissolved nitrogen, this gas ordinarily has no physiological function.

Fig. 16.17 (a) A fat molecule contains a glycerol portion and three fatty acids; (b) in a phospholipid molecule, one fatty acid is replaced by a phosphoric acid-containing group.

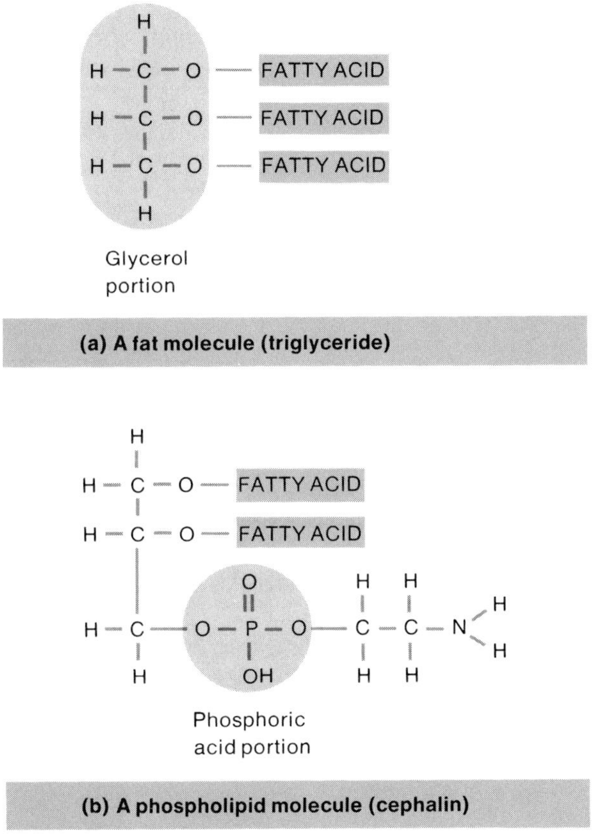

Glycerol portion

(a) A fat molecule (triglyceride)

Phosphoric acid portion

(b) A phospholipid molecule (cephalin)

1. What nutrients are found in blood plasma?
2. How are fats transported in plasma?
3. What is one function of lecithin and cephalin?
4. What gases occur in plasma?

Nonprotein Nitrogenous Substances

Molecules that contain nitrogen atoms but are not proteins comprise a group called **nonprotein nitrogenous substances.** Within the plasma this group includes amino acids, urea, uric acid, creatine, and creatinine. The *amino acids* are present as a result of protein digestion and amino acid absorption. *Urea* and *uric acid* are the products of protein and nucleic acid catabolism, respectively, and *creatinine* results from the metabolism of *creatine.*

As is discussed in chapter 9, creatine occurs as **creatine phosphate** in muscle and brain tissues, and in the blood. This substance functions to store high energy phosphate bonds, much like those of ATP molecules.

Normally, the level of nonprotein nitrogenous (NPN) substances remains relatively stable, because protein intake and utilization are balanced with the excretion of nitrogenous wastes. However, about half of the NPN is urea, which is ordinarily excreted by the kidneys. A rise in the plasma NPN level may suggest a kidney disorder, although such an increase may also occur as a result of excessive protein catabolism or the presence of an infection.

One test of kidney function is called *blood urea nitrogen* or *BUN.* In this test the concentration of blood urea is determined. Since the level of this substance is normally fairly low, any increase in its concentration usually reflects an inability of the kidneys to excrete urea.

Plasma Electrolytes

Plasma contains a variety of *electrolytes* that have been absorbed from the intestine or have been released as by-products of cellular metabolism. They include sodium, potassium, calcium, magnesium, chloride, bicarbonate, phosphate, and sulfate ions. Of these, sodium and chloride ions are the most abundant.

Such ions are important in maintaining the osmotic pressure and the pH of the plasma, and like other plasma constituents, they are regulated so that their blood concentrations remain relatively stable. These electrolytes are discussed in chapter 20 in connection with water and electrolyte balance.

1. What is meant by a nonprotein nitrogenous substance?
2. Why does kidney disease cause an increase in the blood concentration of these substances?
3. What are the sources of plasma electrolytes?

Hemostasis

The term **hemostasis** refers to the stoppage of bleeding, which is vitally important when blood vessels are damaged. Following such an injury, several things may occur that help to prevent excessive blood loss. These include blood vessel spasm, platelet plug formation, and blood coagulation.

These mechanisms are most effective in minimizing blood losses from relatively small vessels, such as arterioles, capillaries, and venules. Injuries to larger vessels, particularly arteries, may result in a severe hemorrhage and may require special treatment for its control.

The Blood

Fig. 16.18   Steps in platelet plug formation.

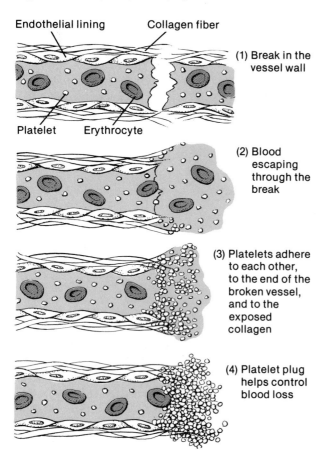

Endothelial lining   Collagen fiber

(1) Break in the vessel wall

Platelet   Erythrocyte

(2) Blood escaping through the break

(3) Platelets adhere to each other, to the end of the broken vessel, and to the exposed collagen

(4) Platelet plug helps control blood loss

Blood Vessel Spasm

When an arteriole or a venule is cut or broken, the smooth muscles in its wall are stimulated to contract, and blood loss is decreased almost immediately. In fact, the ends of a severed vessel may be closed completely by such a *spasm*. This effect seems to result from direct stimulation of the vessel wall as well as from reflexes elicited by pain receptors in the injured tissues.

Although the reflex response may last only a few minutes, the effect of the direct stimulation usually continues for about 30 minutes, and by then the platelet plug and blood coagulation mechanisms normally are operating. Also, as the platelet plug forms, the platelets release *serotonin,* and its vasoconstricting action helps to maintain a prolonged vascular spasm.

Platelet Plug Formulation

As was mentioned earlier, platelets tend to stick to the exposed ends of injured blood vessels. Actually they adhere to any rough surfaces, particularly to the *collagen* in connective tissue that underlies the endothelial lining of blood vessels.

When platelets contact this collagen, their shapes change drastically, and numerous processes begin to protrude from their membranes. At the same time, they tend to stick to each other creating a *platelet plug* in the vascular break. Such a plug may be able to control blood loss from a small opening, but a larger one may require the aid of a blood clot to halt bleeding.

The steps in platelet plug formation are shown in figure 16.18.

---

1. What is meant by hemostasis?
2. How does blood vessel spasm help to control bleeding?
3. Describe the formation of a platelet plug.

Blood Coagulation

Coagulation, which is the most effective of the hemostatic mechanisms, results in the formation of a *blood clot.* This process may be initiated by a variety of factors and may involve an extrinsic or an intrinsic clotting mechanism.

The *extrinsic clotting mechanism* is triggered by the release of chemical substances from the damaged tissues. The *intrinsic mechanism* is stimulated by contact with foreign surfaces in the absence of tissue damage.

Although the mechanism by which blood coagulates is poorly understood, it is known that many substances function in the process. Some of these substances promote coagulation, and others inhibit it. Whether or not the blood coagulates depends on the balance that exists between these two groups of factors. Normally the anticoagulants prevail, and the blood does not clot. As a result of trauma, however, substances that favor coagulation may increase in concentration, and the blood may coagulate.

The basic event in blood clot formation is the conversion of the soluble plasma protein *fibrinogen* into relatively insoluble threads of the protein **fibrin.** This change is triggered by the activation of certain plasma proteins that remain inactive until they are affected by the presence of still other protein factors.

Chart 16.6 and figure 16.19 summarize the three primary hemostatic mechanisms.

Extrinsic Clotting Mechanism.   When tissues are damaged, the extrinsic clotting mechanism initiates a series of reactions resulting in the production of a substance called *prothrombin activator.* This series of changes seems to depend upon the presence of *calcium ions* as well as certain proteins and phospholipids for its completion.

Chart 16.6 Hemostatic mechanisms

Mechanism	Stimulus	Effect
Blood vessel spasm	Direct stimulus to vessel wall or to pain receptors triggers reflex; serotonin released from platelets	Smooth muscles in vessel wall contract; vasoconstriction helps to maintain prolonged vessel spasm
Platelet plug formation	Exposure of platelets to rough surfaces or to collagen of connective tissue	Platelets adhere to rough surfaces and to each other, forming a plug
Blood coagulation	Cellular damage and blood contact with foreign surfaces results in the production of substances that favor coagulation	Blood clot forms as a result of a series of reactions terminating in the conversion of fibrinogen into fibrin

Chart 16.7 Major events in the extrinsic blood clotting mechanism

1. Tissue damage results in the release of prothrombin activator.
2. In the presence of calcium ions, prothrombin activator causes prothrombin to become thrombin.
3. Thrombin acts as an enzyme that causes small portions of fibrinogen molecules to split off.
4. Remaining portions of the fibrinogen molecules attract other molecules like themselves.
5. Fibrinogen molecules join to form long threads of fibrin.
6. Fibrin threads stick to exposed surfaces of damaged blood vessels, creating a meshwork.
7. Blood cells and platelets become entangled in the meshwork of fibrin, and the resulting mass is a blood clot.

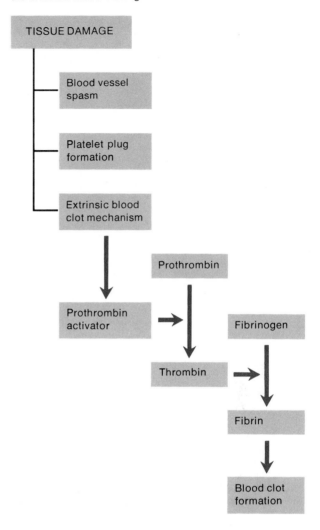

Fig. 16.19 Hemostasis following tissue damage is likely to involve blood vessel spasm, platelet plug formation, and the extrinsic blood-clotting mechanism.

Prothrombin is an alpha globulin that is continually produced by the liver and thus is normally present in plasma. In the presence of calcium ions, prothrombin is converted into **thrombin** by the action of prothrombin activator. Thrombin, in turn, acts as an enzyme and causes a reaction in molecules of fibrinogen. In this reaction, small portions of the fibrinogen molecules are split off, and the remaining pieces develop attractions for other molecules like themselves. These activated fibrinogen molecules join, end to end, forming long threads of *fibrin*. The production of fibrin threads is also enhanced by the presence of calcium ions and other protein factors.

Once threads of fibrin have formed, they tend to stick to the exposed surfaces of damaged blood vessels and create a meshwork in which various blood cells and platelets become entangled. (See fig. 16.20.) The resulting mass is the *blood clot,* which may effectively block a vascular break and prevent further loss of blood. The major steps of blood clotting are summarized in chart 16.7.

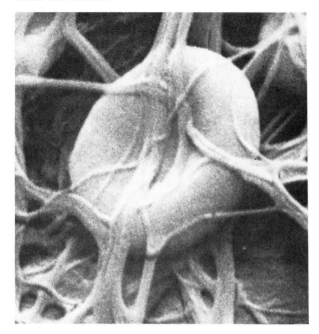

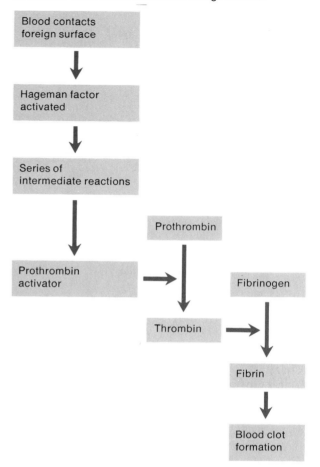

The amount of prothrombin activator that appears in the blood is directly proportional to the degree of tissue damage. Once a blood clot begins to form, it promotes still more clotting. This happens because thrombin also acts directly on blood clotting factors other than fibrinogen, and it can cause prothrombin to form still more thrombin. This type of self-initiating action is an example of a **positive feedback system** (more commonly called a vicious circle).

Normally, the formation of a massive clot throughout the blood system is prevented by blood movement, which rapidly carries excessive thrombin away and thus keeps its concentration too low to enhance further clotting. As a result, blood clot formation is usually limited to blood that is standing still, and clotting stops where a clot comes in contact with circulating blood.

Sometimes the clotting mechanism is activated in widespread regions of the circulatory system. This condition, called *disseminated intravascular clotting* (DIC), is usually associated with the presence of bacteria or bacterial toxins in the blood. As a result, many small clots may appear and obstruct blood flow into various tissues and organs. As the plasma clotting factors and platelets become depleted, the patient may develop a tendency to bleed.

Intrinsic Clotting Mechanism. A secondary clotting mechanism is initiated by the activation of a substance called the *Hageman factor.* This occurs when blood is exposed to a foreign surface such as collagen or when blood is stored in a glass container. In the presence of *calcium ions,* the activated factor triggers a complex series of changes leading to the formation of *prothrombin activator.* The subsequent steps of blood clot formation are the same as those described for the extrinsic mechanism. (See fig. 16.21.)

Regardless of the mechanism involved, after a blood clot has formed, it soon begins to retract, apparently due to platelet activity. The tiny processes extending from the platelet membranes adhere to strands of fibrin within the clot, and these processes seem to contract. The blood clot becomes smaller, and the edges of the broken vessel are pulled closer together. At the same time, a fluid, called **serum,** is

Fig. 16.22 (a) If a sample of plasma remains in a glass tube, a blood clot forms. (b) The plasma minus the clotting factors is called serum.

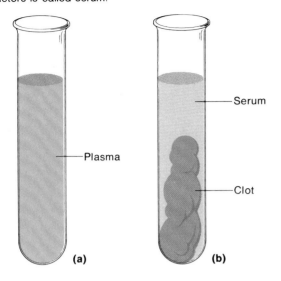

Plasma

(a)

Serum

Clot

(b)

Chart 16.8 Factors that inhibit blood clot formation

Factor	Action
Smooth lining of blood vessel	Prevents activation of intrinsic blood clotting mechanism
Negative charge of blood vessel lining	Repels platelets
Fibrin threads	Adsorb thrombin
Antithrombin in plasma	Interferes with the action of thrombin
Heparin from mast cells and basophils	Interferes with the formation of prothrombin activator

squeezed from the clot. Serum is essentially plasma minus all of its fibrinogen and most of the other factors involved in the clotting mechanism. (See fig. 16.22.)

Blood clots that form in ruptured vessels are soon invaded by *fibroblasts*. These cells produce fibrous connective tissue throughout the clots, which helps to strengthen and seal vascular breaks. Many clots, including those that form in tissues as a result of blood leakage (hematomas), disappear in time. This dissolution involves the activation of a plasma protein, called *profibrinolysin,* that is incorporated into blood clots along with other proteins. Profibrinolysin apparently is activated by substances released from the tissues surrounding the clot. When it is activated, profibrinolysin becomes *fibrinolysin* (plasmin), a protein-splitting enzyme that can digest fibrin threads and other proteins associated with blood clots. This action may cause a whole clot to dissolve.

Clots that fill large blood vessels are seldom removed by natural processes.

1. Distinguish between extrinsic and intrinsic clotting mechanisms.
2. What is the basic event in blood clot formation?
3. What prevents the formation of massive clots throughout the blood system?
4. How does blood clot formation contribute to the repair of a damaged blood vessel?

Prevention of Coagulation

In a normal vascular system, spontaneous formation of blood clots is prevented in part by the endothelium of the blood vessels. This smooth lining carries a negative electrical charge that tends to repel platelets and various clotting factors. As long as these vascular characteristics are maintained, the coagulation mechanism is not initiated.

Also, when a clot is forming, fibrin threads adsorb thrombin, thus helping to prevent the spread of the clotting reaction. Any additional thrombin is likely to be inactivated by an alpha globulin, called *antithrombin,* normally present in the plasma. This substance binds to thrombin and blocks its action on fibrinogen.

In addition, certain cells including basophils and mast cells, which are common in the connective tissue surrounding capillaries, secrete the anticoagulant *heparin.* This substance interferes with the formation of prothrombin activator, prevents the action of thrombin on fibrinogen, and promotes the removal of thrombin by antithrombin and fibrin adsorption.

Heparin-secreting cells are particularly abundant in the liver and lungs, where capillaries are likely to trap small blood clots that commonly occur in the slow-moving blood of veins. These cells are thought to secrete heparin continually, thus helping to prevent additional clotting in the circulatory system. Chart 16.8 summarizes the clot-preventing factors.

Because of its strong anticoagulant action, heparin is widely used in the treatment of conditions in which blood clots are likely to form abnormally. Administration of heparin by injection almost immediately increases the time required for blood coagulation.

Coagulation Disorders

Coagulation disorders fall into two main groups—those that result in excessive bleeding and those that produce abnormal blood clotting.

As the liver plays an important role in the synthesis of various plasma proteins, such as prothrombin, it is not surprising that liver diseases are often accompanied by a tendency to bleed. Also, bile salts from the liver are necessary for the efficient absorption of *vitamin K* from the intestine, and this vitamin is essential for the synthesis of prothrombin. If the liver fails to produce enough bile, or if the bile ducts become obstructed, a vitamin K deficiency is likely to develop, and the ability to form blood clots may be diminished. For this reason, vitamin K is often administered to patients with liver diseases or bile duct obstructions before they are treated surgically.

Newborns sometimes have a tendency to hemorrhage because they have relatively low blood levels of certain coagulation factors. These factors usually rise in concentration during the first few weeks of life. Meanwhile, such infants are usually given preventive treatment with vitamin K, which serves to reduce the chance of bleeding.

In other cases, a tendency to bleed is related to abnormally low platelet counts. This condition, called **thrombocytopenia,** is said to occur whenever the platelet count drops below 100,000 platelets per mm³ of blood.

The cause of thrombocytopenia is usually unknown, but various factors that damage red bone marrow, such as excessive exposure to ionizing radiation or adverse drug reactions, may produce this condition. Affected persons bleed easily, numerous capillary hemorrhages usually occur throughout their body tissues, and their skin typically exhibits many small, bruiselike spots.

Hemophilia is a hereditary disease that appears almost exclusively in males. Hemophiliacs are deficient in a blood factor necessary for coagulation, and consequently, they usually experience repeated episodes of serious bleeding. Often hemophiliacs die at a young age.

There are several types of hemophilia, and each is caused by the lack of a different blood factor. However, the symptoms of these conditions are so similar that they are difficult to distinguish. In addition to the tendency to hemorrhage severely following minor injuries, hemophilia is characterized by frequent nosebleeds, large intramuscular hematomas, blood in the urine (hematuria), and severe pain and disability from bleeding into joints and body cavities.

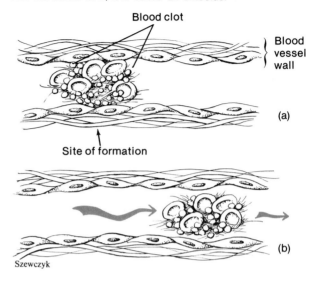

Fig. 16.23 (*a*) A blood clot that remains at its site of formation is called a thrombus; (*b*) if a thrombus moves with the blood flow, it is called an embolus.

Treatment for hemophilia may involve pressure or packing of accessible bleeding sites in an effort to control blood loss. Transfusions are used to replace missing blood factors. For example, a common type of hemophilia caused by a deficiency of a substance, called blood factor VIII, may be treated with fresh plasma, fresh-frozen plasma, or plasma concentrates (cryoprecipitates).

If a blood clot forms in a vessel abnormally, it is termed a **thrombus.** If the clot becomes dislodged or if a fragment of it breaks loose and is carried away by the blood flow, it is called an **embolus.** Generally, emboli continue to move until they reach narrow places in vessels where they become lodged and interfere with the blood flow. (See fig. 16.23.)

Such abnormal clot formations are often associated with conditions that cause changes in the endothelial linings of vessels. In *atherosclerosis,* for example, arterial linings are changed by accumulations of fatty deposits. These changes may initiate the clotting mechanism (fig. 16.24).

Coagulation also may occur in blood that is flowing too slowly. In this instance, the concentration of clot-promoting substances may increase to a critical level instead of being carried away by more rapidly moving blood, and a clot may form.

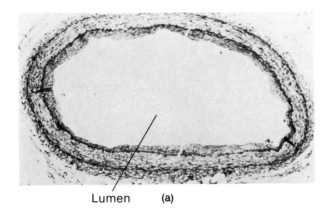

Lumen   (a)

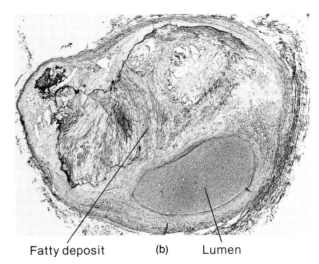

Fatty deposit   (b)   Lumen

Sometimes a thrombus or an embolus obstructs a vessel that supplies a vital organ, such as the heart or the brain, and may cause death. Clots that obstruct vessels of the heart or brain are usually thrombi that form in these organs. The most common sites of emboli are the vessels of the lungs.

1. How does the lining of a blood vessel help to prevent blood clot formation?
2. How does heparin help to prevent blood clot formation?
3. What conditions lead to excessive blood clot formation?
4. Distinguish between a thrombus and an embolus.

Blood Groups and Transfusions

Early attempts to transfer blood from one person to another produced varied results. Sometimes the person receiving the transfusion was aided by the procedure. Other times the recipient suffered a blood reaction in which the red blood cells clumped together, obstructing vessels and producing other serious consequences.

Eventually, it was discovered that human blood is not all alike. Instead, each individual was found to have a particular combination of substances in his or her blood. Some of these substances reacted with those in another person's blood. These discoveries led to the development of procedures for typing blood. It is now known that safe transfusions of whole blood depend upon properly matching the blood types of the donors and recipients.

Agglutinogens and Agglutinins

The clumping of red cells following a transfusion reaction is called **agglutination.** This phenomenon is due to the presence of substances called **agglutinogens** (antigens) in the red cell membranes and substances called **agglutinins** (antibodies) dissolved in the plasma.

Blood typing involves identifying the agglutinogens that are present in a person's red cells. Although there are many different agglutinogens associated with human erythrocytes, only a few of them are likely to produce serious transfusion reactions. These include the agglutinogens of the ABO group and those of the Rh group.

Avoiding the mixture of certain kinds of agglutinogens and agglutinins prevents adverse transfusion reactions.

The ABO Blood Group

The *ABO blood group* is based upon the presence (or absence) of two major agglutinogens in red cell membranes—*agglutinogen A* and *agglutinogen B.* The erythrocytes of each person contain one of the four following combinations of agglutinogens: only A, only B, both A and B, or neither A nor B.

Chart 16.9 Agglutinogens and agglutinins of the ABO blood group

Blood type	Agglutinogen	Agglutinin
A	A	anti-B
B	B	anti-A
AB	A and B	Neither anti-A nor anti-B
O	Neither A nor B	Both anti-A and anti-B

Chart 16.10 Preferred and permissible blood types for transfusions

Blood Type	Preferred Transfusion	Permissible Transfusion
A	A	O
B	B	O
AB	AB	A, B, O
O	O	none

A person with only agglutinogen A is said to have *type A blood;* a person with only agglutinogen B has *type B blood;* one with both agglutinogen A and B has *type AB blood;* and one with neither agglutinogen A nor B has *type O blood.* Thus, all humans have one of four possible blood types—A, B, AB, or O.

Certain agglutinins in the plasma accompany the agglutinogens in the red cell membranes of each person's blood. Specifically, whenever agglutinogen A is absent, an agglutinin called *anti-A* is present; and whenever agglutinogen B is absent, an agglutinin called *anti-B* is present. Therefore, persons with type A blood also have agglutinin anti-B in their plasma; those with B blood have agglutinin anti-A; those with type AB blood have neither agglutinin; and those with type O blood have both agglutinin anti-A and anti-B. (See fig. 16.25.)

Chart 16.9 summarizes the agglutinogens and agglutinins of the ABO blood group.

Since an agglutinin of one kind will react with an agglutinogen of the same kind and cause red blood cells to clump together, such combinations are avoided whenever possible. Actually, the major concern in blood transfusion procedures is that the cells in the *transfused blood* not be agglutinated by the agglutinins in the recipient's plasma. For this reason, a person with type A (anti-B) blood should never be given blood of type B or AB, because the red cells of both types would be agglutinated by the anti-B in the recipient's type A blood. Likewise, a person with type B (anti-A) blood should never be given type A or AB blood and a person with type O (anti-A and anti-B) blood should never be given type A, B, or AB blood. (See fig. 16.26.)

Since type AB blood lacks both anti-A and anti-B agglutinins, it would appear that an AB person could receive a transfusion of blood of any other type. For this reason, type AB persons are sometimes called *universal recipients.* It should be noted, however, that type A (anti-B) blood, type B (anti-A) blood, and type O (anti-A and anti-B) blood still contain agglutinins (either anti-A or anti-B) that could cause agglutination of type AB cells. Consequently, it is always best to use donor blood (AB) of the same type as the recipient blood (AB). If the matching type is not available and type A, B, or O is used, it should be transfused slowly so that the donor blood is well diluted by the recipient's larger blood volume. This precaution usually avoids serious reactions between the donor's agglutinins and the recipient's agglutinogens.

Similarly, because type O blood lacks agglutinogens A and B, it would seem that this type could be transfused into persons with blood of any other type. A person with type O blood, therefore, is sometimes called a *universal donor.* Type O blood, however, does contain both anti-A and anti-B agglutinins, and if it is given to a person with blood type A, B, or AB, it too should be transfused slowly to minimize the chance of an adverse reaction.

Chart 16.10 summarizes preferred blood type for normal transfusions and permissible blood type for emergency transfusions.

The agglutinogens of the ABO group are *inherited factors* that are present in the red cell membranes at the time of birth. The plasma agglutinins begin to appear spontaneously, for unknown reasons, about 2 to 8 months after birth, and they reach a maximum concentration between 8 and 10 years of age. A person's blood type cannot be changed by a transfusion or by any other procedure.

Blood substitutes, such as isotonic saline solution or isotonic glucose solution, are sometimes administered to persons during emergencies. These substitutes will increase the blood volume only temporarily. If a patient has suffered a severe hemorrhage, a transfusion of whole blood may be necessary for survival.

Fig. 16.25 Each blood type is characterized by a different combination of agglutinogens and agglutinins.

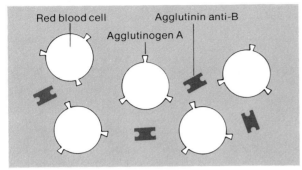

Red blood cell   Agglutinin anti-B
Agglutinogen A

Type A blood

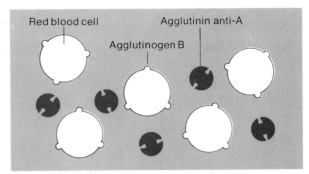

Red blood cell   Agglutinin anti-A
Agglutinogen B

Type B blood

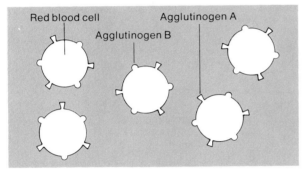

Red blood cell   Agglutinogen A
Agglutinogen B

Type AB blood

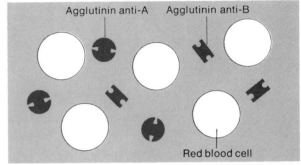

Agglutinin anti-A   Agglutinin anti-B

Red blood cell

Type O blood

Fig. 16.26 (a) If red blood cells with agglutinogen A are added to blood containing agglutinin anti-A, (b) the agglutinins will react with the agglutinogens of the red blood cells and cause them to clump together.

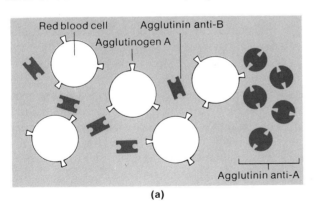

Red blood cell   Agglutinin anti-B
Agglutinogen A

Agglutinin anti-A

(a)

Agglutinated red blood cells

(b)

Fig. 16.27 (*a*) If an Rh-negative woman is pregnant with
an Rh-positive fetus, (*b*) some of the fetal red blood cells
with Rh agglutinogens may enter the maternal blood at the
time of birth. (*c*) As a result, the woman's cells may
produce anti-Rh agglutinins.

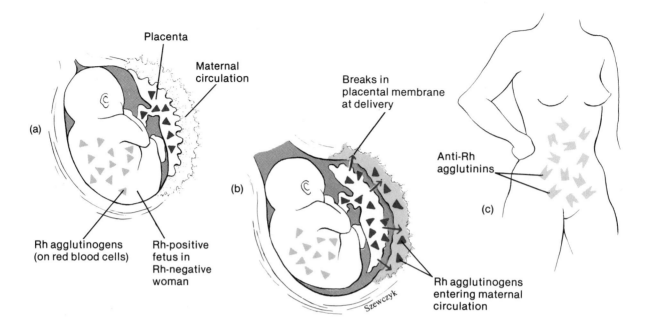

1. Distinguish between agglutinogens and agglutinins.
2. What is meant by blood type?
3. What is the main concern when blood is
   transfused from one individual to another?
4. Why is a type AB person called a universal
   recipient?

The Rh Blood Group

The *Rh blood group* was named after the *rhesus
monkey,* in which it was first observed. In humans
this group is based upon several different Rh agglutinogens (factors). The most important of these is
agglutinogen D; however, if any of the Rh factors are
present in the red cell membranes, the blood is said
to be *Rh positive.* Conversely, if the red cells lack Rh
agglutinogens, the blood is called *Rh negative.*

As in the case of agglutinogens A and B, the
presence (or absence) of an Rh agglutinogen is an
inherited trait. Unlike anti-A and anti-B, agglutinins
for Rh (*anti-Rh*) do not appear spontaneously. Instead, they form only in Rh-negative persons in response to special stimulation.

If an Rh-negative person receives a transfusion
of Rh-positive blood, the recipient's antibody-produc-
ing cells will be stimulated by the presence of the Rh
agglutinogen and will begin producing *anti-Rh agglutinin.* Generally there are no serious consequences
from this initial transfusion, but if the Rh-negative
person—who is now sensitized to Rh-positive blood—
receives a subsequent transfusion of Rh-positive
blood, the donor's red cells are likely to agglutinate.

A related condition may occur when an Rh-
negative woman is pregnant with an Rh-positive fetus
for the first time. Such a pregnancy may be uneventful; however, at the time of this infant's birth, the
placental membranes, which had separated the maternal blood from the fetal blood, may be broken, and
some of the infant's Rh-positive blood cells may get
into the maternal circulation. These Rh-positive cells
may then stimulate the maternal tissues to begin producing anti-Rh agglutinins. (See fig. 16.27.)

If the mother, who has already developed anti-
Rh agglutinin, becomes pregnant with a second Rh-
positive fetus, these anti-Rh agglutinins can pass
through the placental membrane and react with the
fetal red cells, causing them to agglutinate. The fetus
then develops a disease called **erythroblastosis fetalis.**
(See fig. 16.28.)

Chart 16.11 reviews the events that can lead to
this condition.

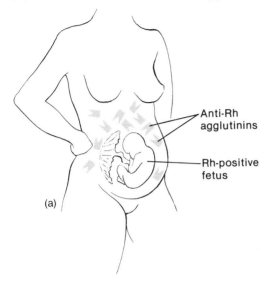

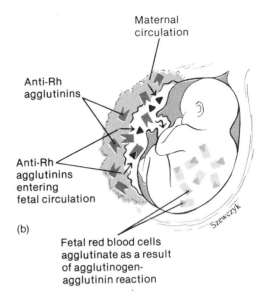

Chart 16.11 Possible events leading to erythroblastosis fetalis

1. Rh-negative woman becomes pregnant with her first Rh-positive child.
2. Pregnancy is uneventful, but at the time of birth some Rh-positive red cells enter the maternal circulation through damaged placental tissues.
3. Maternal tissues produce anti-Rh agglutinins.
4. A second Rh-positive child is conceived.
5. Anti-Rh agglutinins from the maternal circulation pass through the placental membranes and enter the fetal blood.
6. The fetus develops erythroblastosis fetalis as the maternal anti-Rh agglutinins react with the Rh agglutinogens of the fetal red blood cells and cause them to agglutinate.

In any blood reaction involving agglutinogens and agglutinins, the agglutinated red cells usually degenerate or are destroyed by reticuloendothelial cells. At the same time, hemoglobin and other red cell contents are released, and the blood concentration of *free hemoglobin* increases greatly. Some of this hemoglobin may diffuse out of the vascular system and enter the body tissues, where it is gradually converted into bilirubin. As a result, the tissues develop a yellowish stain, which is a condition called *jaundice*.

Free hemoglobin may also pass into the kidneys and interfere with the vital functions of these organs, so that a person with a blood transfusion reaction may later experience kidney (renal) failure.

The severity of a blood transfusion reaction depends on a number of factors, including the degree of incompatibility between the blood of the donor and that of the recipient, the quantity of blood that is transfused, and the rate at which the transfusion is administered.

If a reaction occurs, the patient is likely to experience anxiety, breathing difficulty, facial flushing, headache, and severe pain in the neck, chest, and lumbar area.

Infants with erythroblastosis fetalis are usually jaundiced and severely anemic. As their blood-cell-forming tissues respond to the need for more red cells, various immature erythrocytes including *erythroblasts* are released into the blood. (The presence of these immature cells is related to the name of the disease.)

An affected infant may suffer permanent brain damage as a result of bilirubin precipitating in the brain tissues and injuring neurons. This condition is called *kernicterus*, and if the infant survives, it may have motor or sensory losses and exhibit mental deficiencies.

Treatment for erythroblastosis fetalis usually involves exposing the affected infant to *fluorescent light*. Bilirubin is a light-sensitive substance, and this exposure causes a decrease in the blood bilirubin level.

In more severe cases, the infant's Rh-positive blood may be replaced slowly with Rh-negative blood. This procedure is called an *exchange transfusion*, and it reduces the concentration of bilirubin in the infant's tissues in addition to removing the agglutinating red cells, the anti-Rh agglutinins, the free hemoglobin, and the other products of erythrocyte destruction. It also provides a temporary supply of red cells that will not be agglutinated by any remaining anti-Rh agglutinins. In time, the infant's blood-cell-forming tissues will replace the donor's blood cells with Rh-positive cells, but by then the maternal agglutinins will have disappeared.

Erythroblastosis fetalis can be prevented in future offspring by treating Rh-negative mothers with a special blood serum within 72 hours following the birth of each Rh-positive child. This serum is obtained from the blood of another Rh-negative person who has formed anti-Rh agglutinin. Thus, it is able to inactivate any Rh-positive cells that may have entered the maternal blood at the time of the infant's birth. The maternal tissues, consequently, are not stimulated to manufacture anti-Rh agglutinins, and the mother's blood should not be a hazard to her next Rh-positive child.

Crossmatching Blood

Every person's blood contains a great variety of substances, and although the agglutinogens of the ABO and Rh groups are the most likely to produce serious blood reactions, other factors sometimes cause problems. For this reason, it is a good practice to determine whether samples of recipient and donor blood will produce agglutination of red cells before administering a transfusion to a patient. This procedure, called *crossmatching*, involves mixing a suspension of donor cells in some recipient serum, and then mixing a suspension of recipient cells in some donor serum. If the red cells do not agglutinate in either case, it is probably safe to give the transfusion.

---

1. Under what conditions might a person with Rh-negative blood develop Rh agglutinins?
2. What happens to red blood cells that are agglutinated?
3. How does exposure to fluorescent light help a newborn with erythroblastosis fetalis?
4. What is meant by crossmatching blood?

Some Clinical Terms Related to the Blood

anisocytosis (an-i″so-si-to′sis)—condition in which there is an abnormal variation in the size of erythrocytes.

antihemophilic plasma (an″tĭ-he″mo-fil′ik plaz′mah)—normal blood plasma that has been processed to preserve an antihemophilic factor.

Christmas disease (kris′mas dĭ-zēz′)—a hereditary bleeding disease that is due to a deficiency in a clotting factor; also called *hemophilia B*.

citrated whole blood (sit′rāt-ed hōl blud)—normal blood in a solution of acid citrate solution to prevent coagulation.

dried plasma (drīd plaz′mah)—normal blood plasma that has been vacuum dried to prevent the growth of microorganisms.

hemochromatosis (he″mo-kro″mah-to′sis)—a disorder in iron metabolism in which excessive iron is deposited in the tissues.

hemorrhagic telangiectasia (hem″o-raj′ik tel-an″je-ek-ta′ze-ah)—a hereditary disorder in which there is a tendency to bleed from localized lesions of capillaries.

heparinized whole blood (hep′er-ĭ-nizd″ hōl blud)—normal blood in a solution of heparin to prevent coagulation.

macrocytosis (mak″ro-si-to′sis)—condition characterized by the presence of abnormally large erythrocytes.

microcytosis (mi″kro-si-to′sis)—condition characterized by the presence of abnormally small erythrocytes.

neutrophilia (nu″tro-fil′e-ah)—condition in which there is an increase in the number of circulating neutrophils.

normal plasma (nor′mal plaz′mah)—plasma from which the blood cells have been removed by centrifugation or sedimentation.

packed red cells (pakd red selz)—a concentrated suspension of red blood cells from which the plasma has been removed.

pancytopenia (pan″si-to-pe′ne-ah)—condition characterized by an abnormal depression of all the cellular components of blood.

poikilocytosis (poi″kĭ-lo-si-to′sis)—condition in which the erythrocytes are irregularly shaped.

pseudoagglutination (su″do-ah-gloo″tĭ-na′shun)—clumping of erythrocytes due to some factor other than an agglutinogen-agglutinin reaction.

purpura (per′pu-rah)—a disease characterized by spontaneous bleeding into the tissues and through the mucous membrane.

spherocytosis (sfēr″o-si-to′sis)—a hereditary form of hemolytic anemia characterized by the presence of spherical erythrocytes (spherocytes).

thalassemia (thal″ah-se′me-ah)—a group of hereditary hemolytic anemias characterized by the presence of very thin, fragile erythrocytes.

von Willebrand's disease (fon vil'ĕ-brandz
dĭ-zēz')—a hereditary condition due to a deficiency
of an antihemophilic blood factor and capillary
defects, which is characterized by bleeding from
the nose, gums, and genitalia.

Chapter Summary

Introduction
The circulatory system includes the blood and
cardiovascular system.
Blood functions to transport nutrients, oxygen,
wastes, and hormones, as well as aid in
temperature control.

Blood and Blood Cells
Blood is a type of connective tissue whose cells are
suspended in liquid.
1. Volume and composition of blood
   a. Volume varies with body size, fluid and
      electrolyte balance, and fat content.
   b. It can be separated into solid and liquid
      portions.
      (1) The cellular portion is mostly red blood
          cells.
      (2) Plasma includes water, nutrients,
          hormones, electrolytes, and cellular
          wastes.
2. Characteristics of red blood cells
   a. Red blood cells are biconcave disks whose
      shapes provide increased surface area and
      place their cell membranes close to internal
      parts.
   b. They contain hemoglobin that combines
      loosely with oxygen.
   c. The mature forms lack nuclei, but contain
      enzymes needed for energy-releasing
      processes.
   d. The average red blood cell lives 120 days.
3. Red blood cell counts
   a. The red blood cell count equals the number
      of cells per mm^3 of blood.
   b. The average count is from 4,000,000 to
      6,000,000 cells per mm^3.
   c. Red cell count is related to the oxygen-
      carrying capacity of the blood and is used to
      aid in diagnosing and evaluating the courses
      of diseases.
4. Destruction of red blood cells
   a. Red cells are fragile and are damaged while
      moving through capillaries.
   b. Damaged red cells are phagocytized by cells
      in the liver and spleen.
   c. Hemoglobin molecules are decomposed and
      the iron they contain is conserved.
   d. The number of red blood cells remains
      relatively stable.

5. Red blood cell production and its control
   a. During fetal development, red cells are
      formed in the yolk sac, liver, and spleen;
      later almost all red cells are produced by the
      red bone marrow.
   b. The rate of red cell production is controlled
      by a negative feedback mechanism that
      involves a hormone from the kidneys and
      liver.
      (1) This hormone is released in response to
          low oxygen levels.
      (2) Low oxygen levels may be caused by
          high altitude, loss of blood, or chronic
          lung disease.
6. Dietary factors affecting red blood cell
   production
   a. Production is affected by the availability of
      vitamin B$_{12}$ and folic acid that are needed for
      DNA synthesis.
   b. Iron is needed for hemoglobin synthesis; lack
      of dietary iron may result in hypochromic
      anemia.
7. Red blood cell disorders
   a. Loss of red cells or inadequate production of
      these cells results in anemia.
      (1) Hemorrhagic anemia is caused by blood
          loss.
      (2) Aplastic anemia is caused by malfunction
          of red marrow cells.
      (3) Hemolytic anemia is caused by abnormal
          destruction of red cells.
   b. In anemia, blood viscosity decreases and
      blood returns to the heart in abnormally large
      volumes.
   c. In polycythemia, the red cell number is
      abnormally increased.
      (1) It can be caused by stimulation of the
          red-cell-forming tissue, a tumor, or
          dehydration.
      (2) The general effect is to increase the
          blood viscosity.
8. Types of white blood cells
   a. White blood cells function to defend the body
      against infections by microorganisms.
   b. Granulocytes include neutrophils, eosinophils,
      and basophils.
   c. Agranulocytes include monocytes and
      lymphocytes.
9. White blood cell counts
   a. Normal counts vary from 5000 to 10,000
      cells per mm^3.

b. Number of white cells may change in abnormal conditions such as infections, emotional disturbances, or excessive loss of body fluids.

c. A differential white cell count indicates the percentages of various types of leukocytes present.

10. Functions of white blood cells
   a. Some phagocytize foreign particles; others produce antibodies of immunity.
   b. Leukocytes may be stimulated by the presence of chemicals released by damaged cells and move toward these chemicals.
   c. Basophils release heparin, which inhibits blood clotting.
   d. Some lymphocytes are multipotential cells and can change into other types of blood cells.

11. Leukemia
   a. Leukemia is a form of cancer characterized by an uncontrolled production of white cells that fail to mature.
   b. Two major types exist.
      (1) Myeloid leukemia involves the bone marrow.
      (2) Lymphoid leukemia involves the lymph nodes.
   c. Leukemic cells remain nonfunctional.

12. Blood platelets
   a. Blood platelets are fragments of giant cells that become detached and enter the circulation.
   b. The normal count varies from 130,000 to 360,000 platelets per mm³.
   c. They function to help close breaks in blood vessels.

Blood Plasma

Plasma is the liquid part of the blood, composed of water and a mixture of organic and inorganic substances.

It functions to transport nutrients and gases, regulate fluid and electrolyte balance, and maintain pH.

1. Plasma proteins
   a. These remain in blood and interstitial fluids and are not normally used as energy sources.
   b. Three major groups exist.
      (1) Albumins help maintain the osmotic pressure of blood.
      (2) Globulins function to transport lipids and fat-soluble vitamins, and they include the antibodies of immunity.
      (3) Fibrinogen functions in blood clotting.

2. Nutrients and gases
   a. Nutrients include amino acids, simple sugars, and lipids.
      (1) Glucose is stored in the liver as glycogen and is released whenever the blood glucose level falls.
      (2) Amino acids are used to synthesize proteins and are deaminated for use as energy sources.
      (3) Lipids are present as chylomicrons and lipoproteins that function in the transport of these lipids.
   b. Gases in plasma include oxygen, carbon dioxide, and nitrogen.

3. Nonprotein nitrogenous substances (NPN)
   a. These are composed of molecules that contain nitrogen atoms but are not proteins.
   b. They include amino acids, urea, uric acid, creatine, and creatinine.
      (1) Urea and uric acid are products of catabolic metabolism.
      (2) Creatinine results from the metabolism of creatine.
   c. NPN usually remains stable; an increase may indicate a kidney disorder.

4. Plasma electrolytes
   a. Plasma electrolytes are obtained by absorption from the intestines and are released as by-products of cellular metabolism.
   b. They include ions of sodium, potassium, calcium, magnesium, chlorine, bicarbonate, phosphate, and sulfate.
   c. They are important in the maintenance of osmotic pressure and pH.

Hemostasis

Hemostasis refers to the stoppage of bleeding.

Hemostatic mechanisms are most effective in controlling blood loss from small vessels.

1. Blood vessel spasm
   a. Smooth muscles in walls of arterioles and arteries contract reflexly following injury.
   b. Platelets release serotonin that stimulates vasoconstriction and helps to maintain vessel spasm.

2. Platelet plug formation
   a. Platelets adhere to rough surfaces and exposed collagen.
   b. Platelets stick together at the sites of injuries and form platelet plugs in broken vessels.

3. Blood coagulation
   a. Blood clotting is the most effective means of hemostasis and may be initiated by extrinsic or intrinsic mechanisms.
   b. Clot formation depends on the balance between substances that promote clotting and those that inhibit clotting.
   c. The basic event is the conversion of soluble fibrinogen into insoluble fibrin.
   d. Factors that promote clotting include the presence of prothrombin activator, prothrombin, and calcium ions.
   e. After forming, the clot retracts and pulls the edges of a broken vessel closer together.
   f. The clot is invaded by fibroblasts that form connective tissue throughout the clot.

g. The clot may eventually be destroyed by the action of protein-splitting enzymes.
4. Prevention of coagulation
   a. The smooth lining of blood vessels repels platelets.
   b. As a clot forms, fibrin absorbs thrombin and prevents the reaction from spreading.
   c. Antithrombin interferes with the action of excessive thrombin.
   d. Some cells secrete heparin, an anticoagulant.
5. Coagulation disorders
   a. There are two main groups of coagulation disorders—those resulting in excessive bleeding and those that produce abnormal clotting.
      (1) Hemophilia is a hereditary disease characterized by the lack of a clotting factor and a tendency to bleed severely from minor injuries.
      (2) Abnormal clotting may be associated with liver diseases or a low platelet count.
   b. A thrombus is a blood clot in a vessel; an embolus is a clot or fragment of a clot that has moved in a vessel.

Blood Groups and Transfusions

Blood can be typed on the basis of the substances it contains.

Blood substances of certain types will react adversely with other types.

1. Agglutinogens and agglutinins
   a. Red blood cell membranes may contain agglutinogens and blood plasma may contain agglutinins.
   b. Blood typing involves identifying the agglutinogens present in the red cell membranes.
2. The ABO blood group
   a. Blood can be grouped according to the presence or absence of agglutinogens A and B.
   b. Whenever agglutinogen A is absent, agglutinin anti-A is present; whenever agglutinogen B is absent, agglutinin anti-B is present.
   c. Adverse transfusion reactions are avoided by preventing the mixing of red cells that contain an agglutinogen with plasma that contains the corresponding agglutinin.
   d. Adverse reactions involve agglutination (clumping) of the red blood cells.
3. The Rh blood group
   a. Rh agglutinogens are present in the red cell membranes of Rh-positive blood; they are absent in Rh-negative blood.
   b. If an Rh-negative person is exposed to Rh-positive blood, anti-Rh agglutinins are produced in response.
   c. Mixing Rh-positive red cells with plasma that contains anti-Rh agglutinins results in agglutination of the positive cells.
   d. If an Rh-negative female is pregnant with an Rh-positive fetus, some of the positive cells may enter the maternal blood at the time of birth and stimulate the maternal tissues to produce anti-Rh agglutinins.
   e. Anti-Rh agglutinins in maternal blood may pass through the placental tissues and react with the red cells of an Rh-positive fetus, causing it to develop erythroblastosis fetalis.
   f. Erythroblastosis fetalis can be prevented by treating an Rh-negative female with a serum that contains anti-Rh agglutinin following the birth of every Rh-positive child.
4. Crossmatching blood
   a. ABO and Rh blood factors are the ones most likely to cause adverse transfusion reactions; other blood factors may cause problems.
   b. Crossmatching is used to determine if donor cells will react with recipient plasma and if recipient cells will react with donor plasma.

Application of Knowledge

1. What changes would you expect to occur in the hematocrit of a person who is dehydrating? Why?

2. If a patient with an inoperable cancer is treated by using a drug that reduces the rate of cell division, what changes might occur in the patient's white blood cell count? How might the patient's environment be modified to compensate for the effects of these changes?

3. Hypochromic anemia is relatively common among aging persons who are admitted to hospitals for other conditions. What environmental and sociological factors might promote this?

Review Activities

1. List the major functions of blood.

2. Define *hematocrit* and explain how it is determined.

3. Describe a red blood cell.

4. Distinguish between oxyhemoglobin and deoxyhemoglobin.

5. Explain how a red blood cell count is made.

6. Describe the life cycle of a red blood cell.

7. Distinguish between biliverdin and bilirubin.

8. Define *erythropoietin* and explain its function.

9. Explain how vitamin B_{12} and folic acid deficiencies affect red blood cell production.

10. Compare the causes of hemorrhagic, aplastic, and hemolytic anemias.

11. Define *viscosity* and explain how this property is affected by anemia.

12. Define *polycythemia* and list several possible causes for this condition.

13. Distinguish between granulocytes and agranulocytes.

14. Name five types of leukocytes and list the major functions of each.

15. Explain the significance of white blood cell counts as aids to diagnosing diseases.

16. Distinguish between myeloid and lymphoid leukemia.

17. Describe a blood platelet and explain its functions.

18. Name three types of plasma proteins and list the major functions of each.

19. Define *lipoprotein*.

20. Distinguish between low density lipoprotein and high density lipoprotein.

21. Define *phospholipid*.

22. Define *nonprotein nitrogenous substances* and name those commonly present in plasma.

23. Name several plasma electrolytes.

24. Define *hemostasis*.

25. Explain how blood vessel spasms are stimulated following an injury.

26. Explain how a platelet plug forms.

27. List the major steps leading to the formation of a blood clot.

28. Distinguish between fibrinogen and fibrin.

29. Provide an example of a positive feedback system.

30. Define *serum*.

31. Explain how a blood clot may be removed naturally from a blood vessel.

32. Describe how blood coagulation may be prevented.

33. Review the function of vitamin K.

34. Define *hemophilia*.

35. Distinguish between thrombus and embolus.

36. Distinguish between agglutinogen and agglutinin.

37. Explain the basis of ABO blood types.

38. Explain why a person with blood type AB is sometimes called a universal recipient.

39. Explain why a person with blood type O is sometimes called a universal donor.

40. Distinguish between Rh-positive and Rh-negative blood.

41. Describe how a person may become sensitized to Rh-positive blood.

42. Define *erythroblastosis fetalis* and explain how this condition may develop.

43. Explain how erythroblastosis fetalis can be prevented.

44. Describe the procedure of crossmatching blood and explain its importance in preventing transfusion reactions.

Suggestions for Additional Reading

Adamson, J. W., and Finch, C. A. 1975. Hemoglobin function, oxygen affinity, and erythropoietin. *Ann. Rev. Physio.* 37:351.

Bank, A. et al. 1980. Disorders of human hemoglobin. *Science* 207:486.

Child, J., et al. 1972. Blood transfusions. *Amer. J. Nurs.* 72:1602.

Custer, R. P. 1975. *An atlas of the blood and bone marrow*, 2nd ed. Philadelphia: W. B. Saunders Co.

Erslev, A. J. 1975. *Pathophysiology of blood.* Philadelphia: W. B. Saunders Co.

Perutz, M. F. December 1978. Hemoglobin structure and respiratory transport. *Scientific American.*

Platt, W. R. 1975. *Color atlas and textbook of hematology.* Philadelphia: J. B. Lippincott Co.

Rapaport, S. I. 1971. *Introduction to hematology.* New York: Harper and Row.

Ratnoff, O. D., and Bennett, B. 1973. The genetics of hereditary disorders of blood coagulation. *Science* 179:1291.

Seegers, W. H. 1969. Blood clotting mechanism: three basic reactions. *Ann. Rev. Physio.* 269:31.

Sergis, E., and Hilgartner, M. W. 1972. Hemophilia. *Amer. J. Nurs.* 72:11.

Simone, J. V. July 1974. Childhood leukemia. *Hosp. Prac.*

Wood, W. B. February 1971. White blood cells vs bacteria. *Scientific American.*

Zucker, M. B. June 1980. The functioning of blood platelets. *Scientific American.*

The Cardiovascular System

17 The *cardiovascular system* is the portion of the circulatory system that includes the heart and blood vessels. It functions to move blood between the body cells and organs of the integumentary, digestive, respiratory, and urinary systems, which communicate with the external environment.

In performing this function, the heart acts as a pump that forces blood through the blood vessels. The blood vessels, in turn, form a closed system of ducts that transports blood and allows exchanges of gases, nutrients, and wastes between the blood and the body cells.

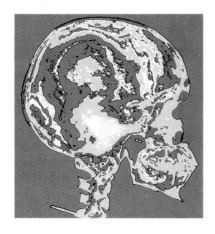

After you have studied this chapter, you should be able to

1. Name the organs of the cardiovascular system and discuss their functions.

2. Name and describe the location of the major parts of the heart and discuss the function of each part.

3. Trace the pathway of blood through the heart and the vessels of the coronary circulation.

4. Discuss the cardiac cycle and explain how it is controlled.

5. Identify the parts of a normal ECG pattern and discuss the significance of this pattern.

6. Define *cardiac arrhythmia* and describe several forms of arrhythmia.

7. Compare the structures and functions of the major types of blood vessels.

8. Describe the mechanism that aids in the return of venous blood to the heart.

9. Explain how blood pressure is created and controlled.

10. Compare the pulmonary and systemic circuits of the cardiovascular system.

11. Identify and describe the location of the major arteries and veins of the pulmonary and systemic circuits.

12. Complete the review activities at the end of this chapter.

arrhythmia (ah-rith'me-ah)

arterial pathway (ar-te're-al path'wā)

atrium (a'tre-um)

cardiac conduction system
  (kar'de-ak kon-duk'shun sis'tem)

cardiac cycle (kar'de-ak si'kl)

cardiac output (kar'de-ak owt'poot)

diastolic pressure (di''ah-stol'ik presh'ur)

electrocardiogram (e-lek''tro-kar'de-o-gram'')

functional syncytium (funk'shun-al sin-sish'e-um)

myocardium (mi''o-kar'de-um)

pacemaker (pās'māk-er)

pericardium (per''ı-kar'de-um)

peripheral resistance (pe-rif'er-al re-zis'tans)

pulmonary circuit (pul'mo-ner''e sur'kit)

sphygmomanometer (sfig''mo-mah-nom'ĕ-ter)

systemic circuit (sis-tem'ik sur'kit)

systolic pressure (sis-tol'ik presh'ur)

vasoconstriction (vas''o-kon-strik'shun)

vasodilation (vas''o-di-la'shun)

venous pathway (ve'nus path'wā)

ventricle (ven'trĭ-kl)

viscosity (vis-kos'ĭ-te)

angio-, a vessel: *angio*tensin—a substance that causes blood vessels to constrict.

brady-, slow: *brady*cardia—an abnormally slow heartbeat.

diastol-, a dilation: *diastol*ic pressure—the blood pressure that occurs when the ventricle is relaxed (thus dilated).

ectop-, out of place: *ectop*ic beat—a heartbeat that occurs before it is expected in a normal series of cardiac cycles.

edem-, a swelling: *edem*a—condition in which fluids accumulate in the tissues and cause them to swell.

-gram, something written: electrocardio*gram*—a recording of the electrical changes that occur in the heart muscle during a cardiac cycle.

myo-, muscle: *myo*cardium—the muscle tissue within the wall of the heart.

papill-, nipple: *papill*ary muscle—a small mound of muscle within a chamber of the heart.

phleb-, vein: *phleb*itis—an inflammation of a vein.

scler-, hard: *scler*osis—condition in which a blood vessel wall loses its elasticity and becomes hard.

syn-, together; *syn*cytium—a mass of merging cells that act together.

systol-, contraction: *systol*ic pressure—the blood pressure that occurs during a ventricular contraction.

tachy-, rapid: *tachy*cardia—an abnormally fast heartbeat.

A functional cardiovascular system is vital for survival, because without circulation, tissues lack a supply of oxygen and nutrients, and waste substances begin to accumulate. Under these conditions cells soon begin to undergo irreversible changes that quickly lead to the death of the organism. The general pattern of the cardiovascular system is shown in figure 17.1.

The Heart

The heart is a cone-shaped, muscular pump located within the mediastinum of the thorax and resting upon the diaphragm.

Size and Location of the Heart

Although the size of the heart varies with body size, it is generally about 14 cm (5.5 in) long and 9 cm (3.5 in) wide in an average adult.

The heart is enclosed laterally by the lungs, posteriorly by the backbone, and anteriorly by the sternum (fig. 17.2). Its *base*, which is attached to several large blood vessels, lies beneath the second rib. Its distal end extends downward and to the left, terminating as a bluntly pointed *apex* at the level of the fifth intercostal space. For this reason, it is possible to sense the *apical heartbeat* by feeling or listening to the chest wall between the fifth and sixth ribs, about 7.5 cm (3 in) to the left of the midline.

Coverings of the Heart

The heart and the proximal ends of the large vessels to which it is attached are enclosed by a double-layered **pericardium.** The inner layer of this membrane is called the *visceral pericardium* (epicardium), and it consists of a thin, serous covering closely applied to the surface of the heart. At the base of the heart, the visceral pericardium turns back upon itself and becomes the serous part of a loose-fitting parietal pericardium.

The *parietal pericardium* is a tough, protective sac composed largely of white fibrous connective tissue. It is attached to the central portion of the diaphragm, the back of the sternum, the vertebral column, and the large blood vessels emerging from the heart. Between the parietal and visceral membranes is a potential space, the *pericardial cavity,* that contains a small amount of serous fluid. (See fig. 17.3.) This fluid serves to reduce friction between the pericardial membranes as the heart moves within them.

Fig. 17.1 The cardiovascular system functions to transport blood between the body cells and organs that communicate with the external environment.

Fig. 17.2 The heart is located behind the sternum, where it rests upon the diaphragm.

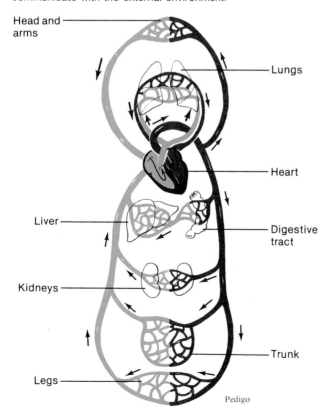

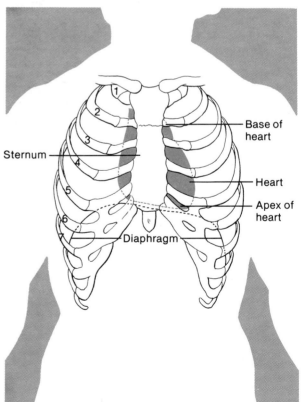

If the pericardium becomes inflamed due to a bacterial or viral infection, the condition is called *pericarditis*. As a result of this disease, the layers of the pericardium sometimes become stuck together by adhesions, and this may interfere with heart movements. If this happens, surgery may be required to separate the surfaces and make unrestricted heart actions possible again.

1. Where is the heart located?
2. Where would you listen to hear the apical heartbeat?
3. Distinguish between the visceral pericardium and the parietal pericardium.
4. What is the function of the fluid in the pericardial cavity?

Wall of the Heart

The wall of the heart is composed of three distinct layers—an outer epicardium, a middle myocardium, and an inner endocardium. (See fig. 17.4.)

The outer **epicardium** provides a protective layer and is composed of the visceral pericardium. This serous membrane consists of connective tissue covered by epithelium and includes blood capillaries, lymph capillaries, and nerve fibers. The deeper portion often contains fat, particularly along the paths of larger blood vessels.

The middle layer, or **myocardium,** is relatively thick and consists largely of the cardiac muscle tissue responsible for forcing blood out of the heart chambers. The muscle fibers are arranged in planes separated by connective tissues that are richly supplied with blood and lymph capillaries and nerve fibers.

Fig. 17.3 The pericardial cavity is a potential space between the visceral and parietal pericardial membranes.

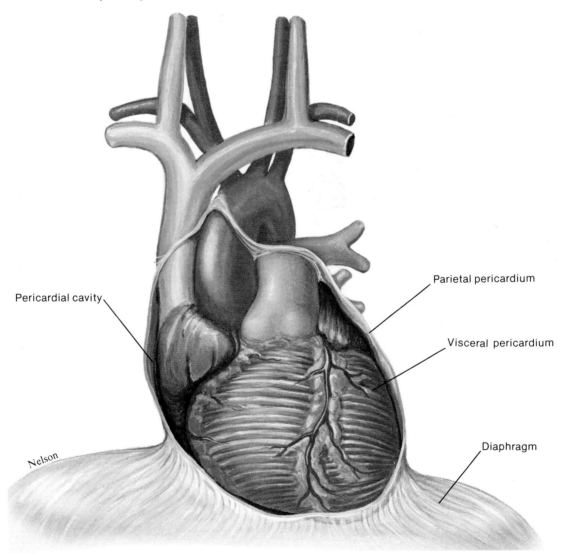

Parietal pericardium

Visceral pericardium

Pericardial cavity

Diaphragm

Nelson

Fig. 17.4 The wall of the heart consists of three layers: the endocardium, the myocardium, and the epicardium.

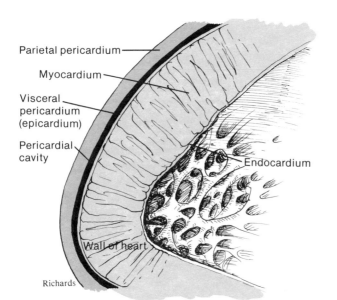

Parietal pericardium

Myocardium

Visceral pericardium (epicardium)

Pericardial cavity

Endocardium

Wall of heart

Richards

Fig. 17.5 The chambers of the heart.

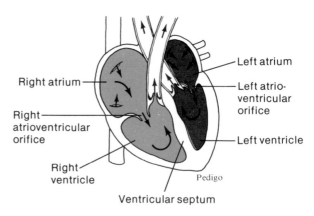

Right atrium

Right atrioventricular orifice

Right ventricle

Left atrium

Left atrioventricular orifice

Left ventricle

Pedigo

Ventricular septum

Chart 17.1   Wall of the heart

Layer	Composition	Function
Epicardium (visceral pericardium)	Serous membrane of connective tissue covered with epithelium and including blood capillaries, lymph capillaries, and nerve fibers	Protective outer covering
Myocardium	Cardiac muscle tissue separated by connective tissues and including blood capillaries, lymph capillaries, and nerve fibers	Muscular contractions that force blood from the heart chambers
Endocardium	Membrane of epithelium and connective tissues, including elastic and collagenous fibers, blood vessels, and specialized muscle fibers	Protective inner lining of the chambers and valves

The inner layer, or **endocardium,** consists of epithelium and connective tissue that contains many elastic and collagenous fibers. The connective tissue also contains blood vessels and some specialized cardiac muscle fibers called *Purkinje fibers.*

The endocardium lines all of the heart chambers and covers structures, such as the heart valves, that project into them. This inner lining is also continuous with the linings of the blood vessels (endothelium) attached to the heart.

Chart 17.1 summarizes the characteristics of these three layers of the heart.

An inflammation of the endocardium is termed *endocarditis.* This condition sometimes accompanies bacterial diseases, such as scarlet fever or syphilis, and may produce lasting effects by damaging the valves of the heart.

Heart Chambers and Valves

Internally, the heart is divided into four hollow chambers, two on the left and two on the right. The upper chambers, called **atria** (singular, *atrium*), have relatively thin walls and receive blood from veins. The lower chambers, the **ventricles,** force blood out of the heart into arteries.

The atrium and ventricle on the right side are separated from those on the left by a *septum.* The atrium on each side communicates with its corresponding ventricle through an opening called the **atrioventricular orifice,** which is guarded by an *atrioventricular valve* (*A-V valve*). These parts of the heart are shown in figure 17.5.

Grooves on the surface of the heart mark the divisions between its chambers, and they also contain major blood vessels that supply the heart tissues. The deepest of these grooves is the **atrioventricular** (coronary) **sulcus,** which encircles the heart between the atrial and ventricular portions. Two *interventricular*

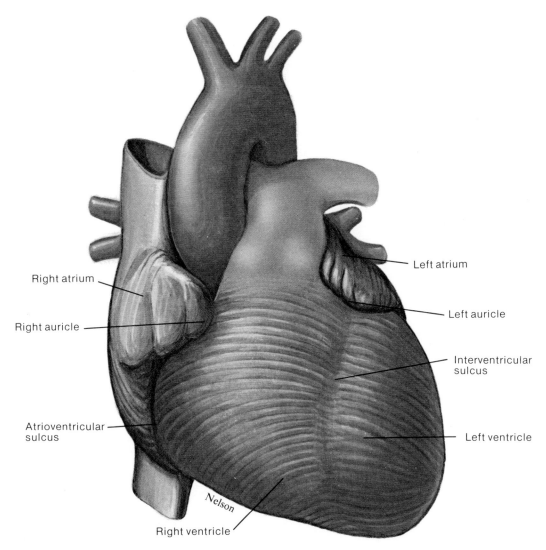

(anterior and posterior) *sulci* indicate the location of the septum that separates the right and left ventricles. Small, earlike projections, called **auricles,** extend outward from the atria. (See fig. 17.6 and color plate 29.)

---

1. *Describe the layers of the heart wall.*
2. *Name and locate the four chambers of the heart.*
3. *Name the orifices that occur between the upper and the lower chambers of the heart.*
4. *Name the structure that separates the right and left sides of the heart.*

The right atrium receives blood from two large veins—the *superior vena cava* and *inferior vena cava.* These return blood that is low in oxygen from various body parts. A smaller vein, the *coronary sinus,* also drains blood into the right atrium from the wall of the heart.

The atrioventricular orifice between the right atrium and the right ventricle is guarded by a large **tricuspid valve,** which is composed of three leaflets or cusps as its name implies. This valve permits blood to move from the right atrium into the right ventricle and prevents passage in the opposite direction. The cusps fold back out of the way when the blood pressure is greater on the atrial side, and they close the orifice when the pressure is greater on the ventricular side.

Strong, fibrous strings, called *chordae tendineae,* are attached to the cusps on the ventricular side. These strings originate from small mounds of muscle tissue, the **papillary muscles,** that project inward from the walls of the ventricle. When the cusps close, the chordae tendineae prevent them from everting back into the atrium.

The right ventricle has a much thinner muscular wall than the left ventricle. This chamber pumps blood a relatively short distance to the lungs,

Fig. 17.7 A frontal section of the heart.

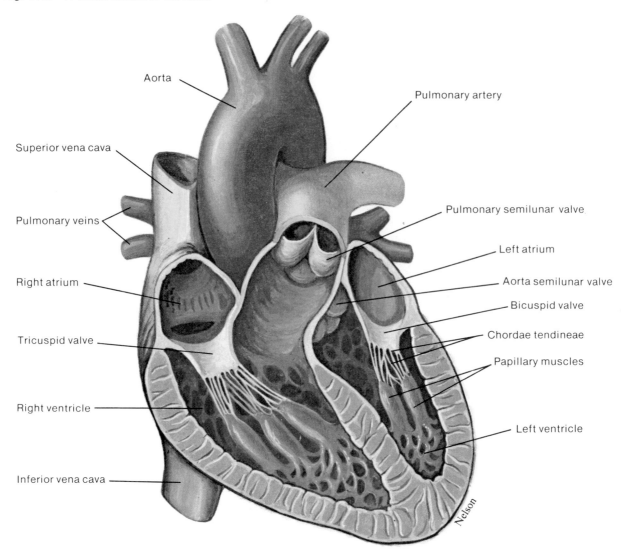

while the left ventricle must force blood to all other parts of the body.

When the muscular wall of the right ventricle contracts, the blood inside its chamber is put under increasing pressure, and the tricuspid valve closes. As a result, the only exit is through the *pulmonary artery* that leads to the lungs. At the base of this artery is a **pulmonary semilunar valve** that consists of three cusps. This valve opens as the right ventricle contracts. When the ventricular muscles relax, however, blood begins to back up in the pulmonary artery. This causes the semilunar valve to close, thus preventing a return flow into the ventricular chambers.

The left atrium receives blood from the lungs through four *pulmonary veins*—two from the right lung and two from the left lung. Blood passes from the left atrium into the left ventricle through the atrioventricular orifice, which is guarded by a valve.

This valve consists of two leaflets, and its is appropriately named the **bicuspid (mitral) valve.** It prevents blood from flowing back into the left atrium from the ventricle. Like the tricuspid valve, the bicuspid valve is aided by chordae tendineae and papillary muscles.

When the left ventricle contracts, the bicuspid valve closes, and the only exit is through a large artery called the *aorta.* Its branches distribute blood to all parts of the body.

At the base of the aorta, there is an **aortic semilunar valve** that consists of three cusps. It opens and allows blood to leave the left ventricle as it contracts. When the ventricular muscles relax, this valve closes and prevents blood from backing up into the ventricles. Chart 17.2 summarizes the heart valves that are shown also in figure 17.7.

Chart 17.2 Valves of the heart

Valve	Location	Function
Tricuspid valve	Right atrioventricular orifice	Prevents blood from moving from right ventricle into right atrium during ventricular contraction
Pulmonary semilunar valve	Entrance to pulmonary artery	Prevents blood from moving from pulmonary artery into right ventricle during ventricular relaxation
Bicuspid (mitral) valve	Left atrioventricular orifice	Prevents blood from moving from left ventricle into left atrium during ventricular contraction
Aortic semilunar valve	Entrance to aorta	Prevents blood from moving from aorta into left ventricle during ventricular relaxation

As an aid to diagnosing certain heart disorders, a procedure called *cardiac catheterization* is sometimes used. In this procedure, a long, thin tube, or catheter, is passed through a blood vessel and into a chamber of the heart or an associated vessel, as shown in figure 17.8. The location of the catheter can be observed by means of a fluoroscope. When it is in a desired position, blood samples can be removed for testing, blood pressures can be determined, or a radiopaque substance can be released into the circulatory system so that blood vessels associated with the heart will appear in X rays.

1. *Which blood vessels carry blood into the right atrium?*
2. *Where does the blood go after it leaves the right ventricle?*
3. *Which blood vessels carry blood into the left atrium?*
4. *What prevents blood from flowing back into the ventricles when they are relaxed?*

Skeleton of the Heart

At their proximal ends, the pulmonary artery and aorta are surrounded by rings of dense fibrous connective tissue. These are continuous with others that encircle the atrioventricular orifices. The rings provide firm attachments for the heart valves and for various muscle fibers. In addition, they prevent the outlets of the atria and ventricles from dilating during

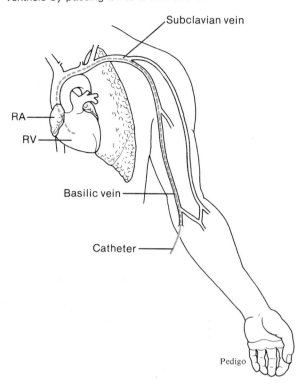

Fig. 17.8 A catheter may be introduced into the right ventricle by passing it into a vein of the arm.

Subclavian vein

RA

RV

Basilic vein

Catheter

Pedigo

myocardial contraction. The fibrous rings together with other masses of dense fibrous tissue in the interventricular septum constitute the skeleton of the heart (fig. 17.9).

Path of Blood through the Heart

Blood that is low in oxygen enters the right atrium through the venae cavae and the coronary sinus. As the right atrial wall contracts, blood passes through the right atrioventricular orifice and enters the chamber of the right ventricle. (See fig. 17.10.)

When the right ventricular wall contracts, the tricuspid valve closes the right atrioventricular orifice, and the blood moves into the pulmonary artery and its branches (pulmonary circuit). From these vessels, it enters the capillaries associated with the alveoli of the lungs. Gas exchanges occur between the blood in the capillaries and the air in the alveoli, and the freshly oxygenated blood returns to the heart through the pulmonary veins that lead to the left atrium.

The left atrial wall contracts, and the blood moves through the left atrioventricular orifice and into the chamber of the left ventricle. When the left ventricular wall contracts, the bicuspid valve closes the left atrioventricular orifice, and the blood passes into the aorta and its branches (systemic circuit).

Fig. 17.9 The skeleton of the heart consists of fibrous rings to which the heart valves are attached.

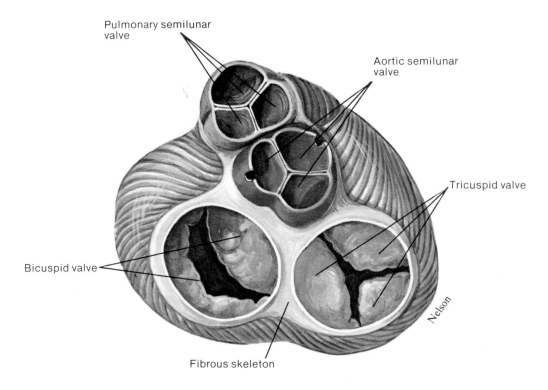

Pulmonary semilunar valve

Aortic semilunar valve

Tricuspid valve

Bicuspid valve

Fibrous skeleton

Nelson

Fig. 17.10 The right ventricle forces blood to the lungs, while the left ventricle forces blood to all other body parts. How does the composition of the blood in these two chambers differ?

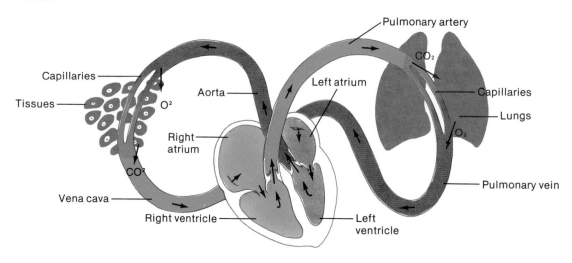

Pulmonary artery

Capillaries

Tissues

O^2

Aorta

Left atrium

Capillaries

Lungs

Right atrium

CO_2

CO^2

O_2

Vena cava

Pulmonary vein

Right ventricle

Left ventricle

Blood Supply to the Heart

Blood is supplied to the tissues of the heart by the first two branches of the aorta, called the right and left **coronary arteries.** Their openings lie just beyond the aortic semilunar valve.

One branch of the left coronary artery, the *circumflex artery*, follows the *atrioventricular sulcus* between the left atrium and the left ventricle. Another branch, the *anterior interventricular artery*, travels in the *anterior interventricular sulcus*. The right coronary artery proceeds along the atrioventricular sulcus between the right atrium and the right ventricle and branches over the posterior surface of the heart. (See figs. 17.11, 17.12, and color plate 30.)

Since the heart must beat continually to supply blood to the body cells, the myocardial cells require

Fig. 17.11 The coronary arteries provide blood to the tissues of the heart.

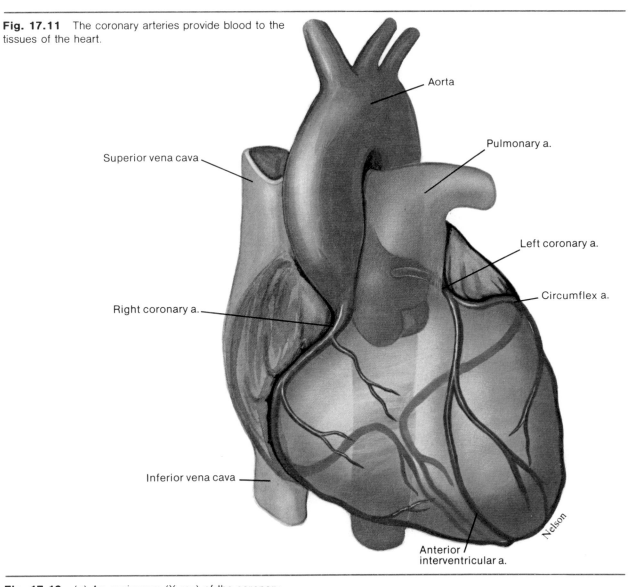

Aorta

Pulmonary a.

Superior vena cava

Left coronary a.

Circumflex a.

Right coronary a.

Inferior vena cava

Anterior interventricular a.

Nelson

Fig. 17.12 (*a*) An angiogram (X ray) of the coronary arteries; (*b*) a cast of the coronary arteries and their major branches.

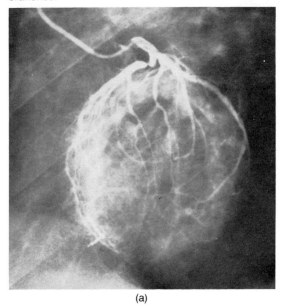

(a)

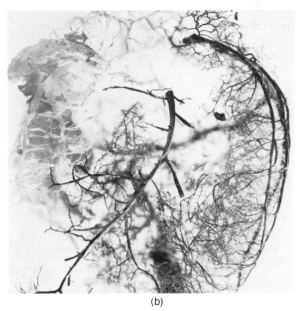

(b)

The Cardiovascular System   569

Fig. 17.13 The cardiac veins carry blood to the coronary sinus that empties into the right atrium.

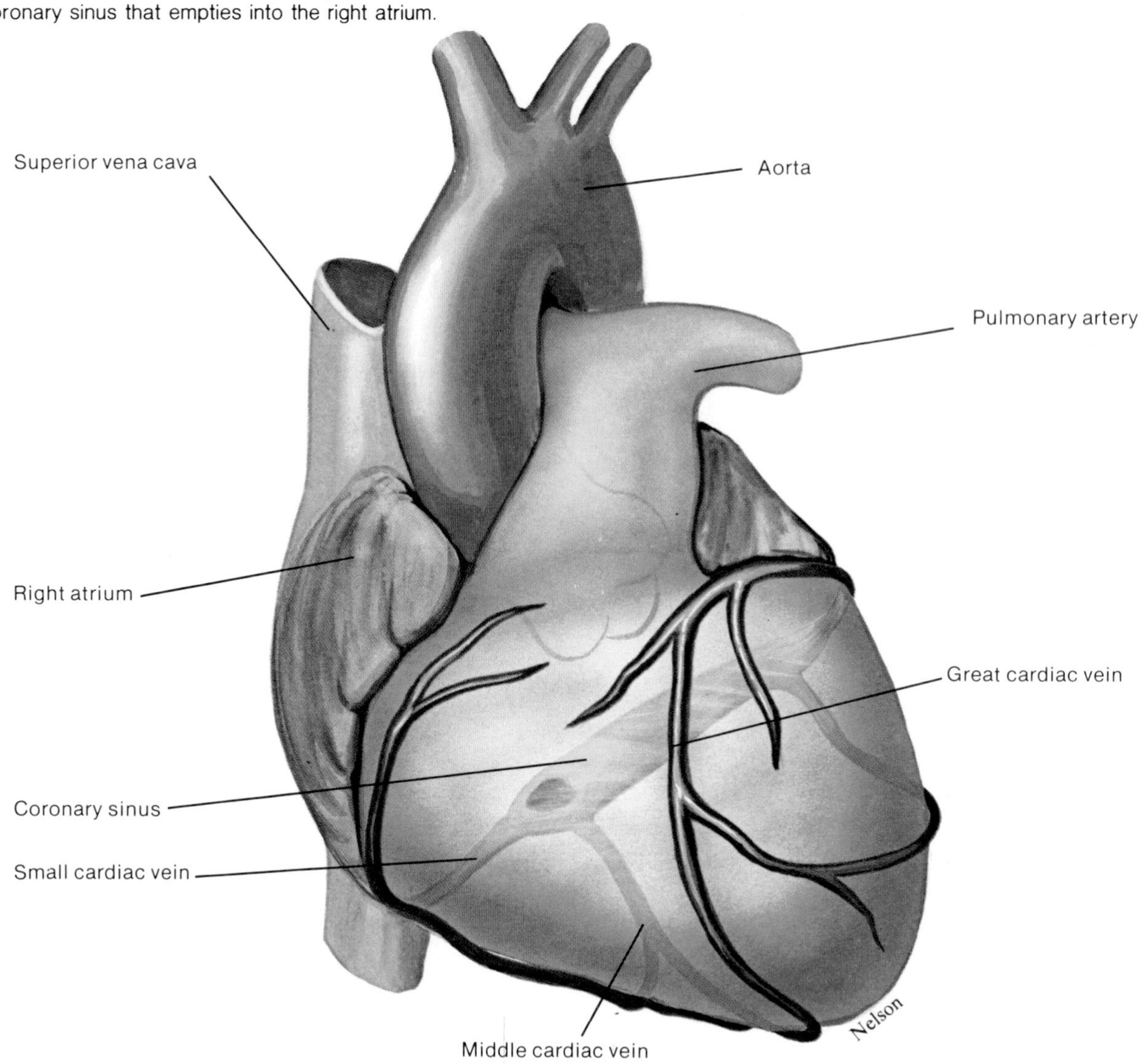

Superior vena cava

Aorta

Pulmonary artery

Right atrium

Great cardiac vein

Coronary sinus

Small cardiac vein

Nelson

Middle cardiac vein

a constant supply of freshly oxygenated blood. This tissue contains many capillaries fed by branches of the coronary arteries. The larger branches of these arteries have **anastomoses,** or interconnections, between vessels that provide alternate pathways for blood.

If a branch of a coronary artery becomes abnormally constricted or obstructed by a thrombus or embolus, the myocardial cells it supplies may experience a blood deficiency, called *ischemia.* Sometimes a portion of the heart dies because of ischemia, and this condition, a *myocardial infarction,* or more commonly a heart attack, is one of the leading causes of death.

In most body parts, blood flow in arteries reaches a peak during ventricular contraction. Blood flow in the vessels of the myocardium, however, is poorest during ventricular contraction. This is because the muscle fibers of the myocardium compress nearby vessels as they contract, and this action interferes with blood flow. Conversely, during ventricular relaxation, the myocardial vessels are no longer compressed, and blood flow increases.

Blood that has passed through the capillaries of the myocardium is drained by branches of **cardiac veins,** whose paths roughly parallel those of the coronary arteries. As figure 17.13 shows, these veins join an enlarged vessel, the **coronary sinus,** located in the atrioventricular sulcus and empty into the right atrium.

Fig. 17.14 (*a*) The ventricles fill with blood during ventricular diastole and (*b*) empty during ventricular systole.

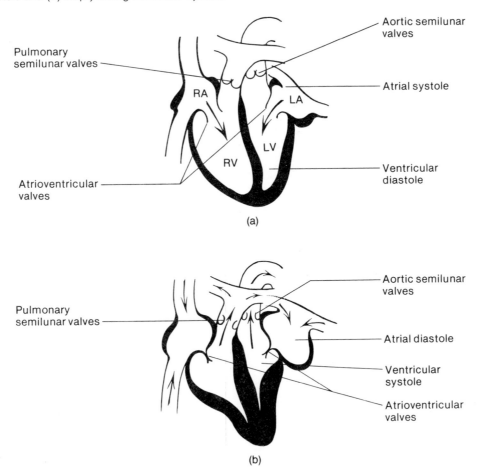

Pulmonary semilunar valves

Aortic semilunar valves

Atrial systole

RA

LA

LV

RV

Ventricular diastole

Atrioventricular valves

(a)

Pulmonary semilunar valves

Aortic semilunar valves

Atrial diastole

Ventricular systole

Atrioventricular valves

(b)

1. *What structures make up the skeleton of the heart?*
2. *Review the path of blood through the heart.*
3. *What vessels supply blood to the heart itself?*
4. *How does blood return from the cardiac tissues to the right atrium?*

The Cardiac Cycle

Although the previous discussion described the actions of the heart chambers one at a time, they do not function independently. Instead, their actions are regulated so that the atrial walls contract while the ventricular walls are relaxed, and ventricular walls contract while the atrial walls are relaxed. Such a series of contractions, shown in figure 17.14, constitutes a complete heartbeat or **cardiac cycle.** At the end of each cycle, the atria and the ventricles remain relaxed for a moment, and then a new cycle begins.

During a cardiac cycle, the pressure within the chambers rises and falls. For example, when the atria are relaxed, blood flows into them from the large, attached veins. As these chambers fill, the pressure inside gradually increases. About 70% of the entering blood flows directly into the ventricles through the atrioventricular orifices before the atrial walls contract. Then, during atrial contraction (atrial systole), the atrial pressure rises suddenly, forcing the remaining 30% of the atrial contents into the ventricles. This is followed by atrial relaxation (atrial diastole). (See fig. 17.15.)

As the ventricles contract (ventricular systole), the A-V valves guarding the atrioventricular orifices close and bulge back into the atria, causing the atrial pressure to rise sharply. The atrial pressure soon falls, however, as blood flows out of the ventricles into the arteries. During the ventricular contraction, the A-V valves remain closed and the atrial pressure gradually increases as the atria fill with blood. When the ventricles relax (ventricular diastole), the A-V valves open, blood flows through them into the ventricles, and the atrial pressure drops to a low point.

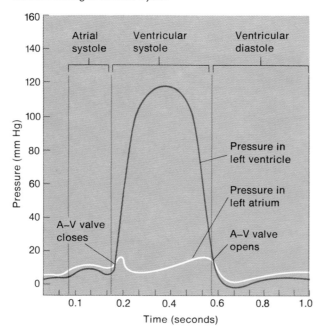

Pressure in the ventricles is low while they are filling, but when the atria contract, the ventricular pressure increases slightly. Then, as the ventricles contract, the ventricular pressure rises sharply, and as soon as the pressure exceeds that in the atria, the A-V valves close. The ventricular pressure continues to increase until it exceeds the pressure in the pulmonary artery and aorta. Then, the semilunar valves open, and blood is ejected from the ventricles into these arteries. When the ventricles are nearly empty, the ventricular pressure begins to drop, and it continues to drop as the ventricles relax. When the ventricular pressure is less than that in the arteries, the semilunar valves are closed by arterial blood flowing back toward the ventricles. As soon as the ventricular pressure falls below that of the atria, the A-V valves open, and the ventricles begin to fill once more.

Heart Sounds

The sounds associated with a heartbeat can be heard with a stethoscope and are described as *lub*-dup sounds. These sounds are due to vibrations in heart tissues that are created as blood flow is suddenly speeded or slowed with the contraction and relaxation of heart chambers and the opening and closing of valves.

The first part of a heart sound occurs during the ventricular contraction, when the A-V valves are closing. The second part occurs during ventricular relaxation, when the semilunar valves are closing.

Sometimes, during inspiration, the interval between the closure of the pulmonary and the aortic semilunar valves is long enough so that a sound related to each of these events can be heard. In this case, the second heart sound is said to be split.

Heart sounds are of particular interest because they provide information concerning the condition of the heart valves. For example, inflammation accompanying endocarditis may cause changes in the shapes of the valvular cusps (valvular stenosis). Then, when they close, the closure may be incomplete, and some blood may leak back through the valve. When this happens, an abnormal sound called a *murmur* may be heard. The seriousness of a heart murmur depends on the amount of valvular damage. Fortunately for those who have serious problems, it is possible to repair damaged valves or to replace them by open heart surgery. (See fig. 17.16.)

With the aid of a stethoscope, it is possible to hear sounds associated with the aortic and pulmonary semilunar valves by listening from the second intercostal space on either side of the sternum. The *aortic sound* is heard on the right, and the *pulmonic sound* is heard on the left.

The sound associated with the bicuspid (mitral) valve can be heard from the fifth intercostal space at the nipple line on the left. That of the tricuspid valve can be heard at the tip of the sternum.

Cardiac Muscle Fibers

As is mentioned in chapter 9, cardiac muscle fibers function much like those of skeletal muscles. In cardiac muscle, however, the fibers are interconnected in branching networks that spread in all directions through the heart. When any portion of this net is stimulated, an impulse travels to all of its parts, and the whole structure contracts as a unit.

A mass of merging cells that act together in this way is called a **functional syncytium.** There are two such structures in the heart—one in the atrial walls and another in the ventricular walls. These masses of muscle fibers are separated from each other by portions of the heart's fibrous skeleton, except for only in a small area of the right atrial floor. In this region, the *atrial syncytium* and the *ventricular syncytium* are connected by fibers of the cardiac conduction system.

Fig. 17.16 As a result of damage, the closure of a heart valve may be incomplete. An artificial prosthesis may be inserted to replace a damaged valve.

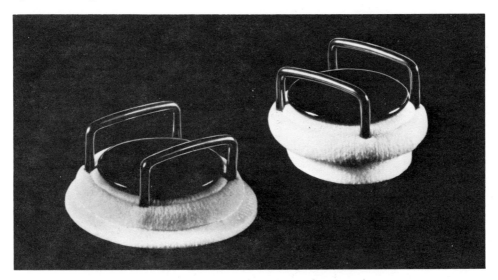

1. Describe the pressure changes that occur in the atria and ventricles during a cardiac cycle.
2. What causes heart sounds?
3. What is meant by a functional syncytium?
4. Where are the functional syncytia of the heart located?

Cardiac Conduction System

Throughout the heart there are clumps and strands of specialized cardiac muscle tissue, whose fibers contain only a few myofibrils. Instead of contracting, these parts function to initiate and distribute impulses (cardiac impulses) throughout the myocardium, and they comprise the **cardiac conduction system.**

A key portion of this conduction system is called the **sinoatrial (S-A) node.** It consists of a small mass of specialized muscle tissue just beneath the epicardium. It is located in the posterior wall of the right atrium, below the opening of the superior vena cava, and its fibers are continuous with those of the *atrial syncytium.*

The cells of the S-A node have a special ability to excite themselves. Without being stimulated by nerve fibers or any other outside agents, these cells initiate impulses that spread into the myocardium and stimulate cardiac muscle fibers to contract. Furthermore, this activity is rhythmic. The S-A node initiates one impulse after another, seventy to eighty times a minute. Thus, it is responsible for the rhythmic contractions of the heart and is often called the **pacemaker.**

As an impulse travels from the S-A node into the atrial syncytium, the right and left atria contract almost simultaneously. Because the atrial and ventricular syncytia are separated by the fibrous skeleton of the heart, the impulse is delayed in reaching the ventricles. During this time, the atria empty and the ventricles fill with blood.

Instead of passing directly into the walls of the ventricles, the cardiac impulse passes along fibers of the conduction system that are connected to atrial muscles. These fibers lead to a mass of specialized muscle tissue called the **atrioventricular (A-V) node.** This node, located in the floor of the right atrium near the interatrial septum, provides the only normal conduction pathway between the atrial and ventricular syncytia.

When the cardiac impulse reaches the A-V node, it passes into a group of fibers called the **A-V bundle** (bundle of His). This bundle enters the upper part of the interventricular septum, and divides into right and left branches that lie just beneath the endocardium. About halfway down the septum, the branches give rise to enlarged **Purkinje fibers.**

Sometimes a branch of the A-V bundle is injured, and it no longer conducts impulses normally. If this occurs, cardiac impulses may reach the ventricles at different times, and they may not contract together. This condition is called a *bundle branch block.*

Fig. 17.17 The cardiac conduction system. What is the function of this system?

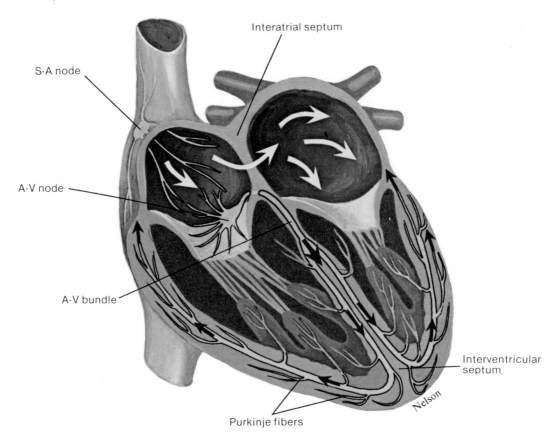

Interatrial septum

S-A node

A-V node

A-V bundle

Interventricular septum

Purkinje fibers

Nelson

The Purkinje fibers spread from the interventricular septum, into the papillary muscles that project inward from the ventricular walls, and continue downward to the apex of the heart. There they curve around the tips of the ventricles and pass upward over the lateral walls of these chambers. Along the way, the Purkinje fibers give off many small branches that terminate in cardiac muscle fibers. These parts of the conduction system are shown in figure 17.17.

The muscle fibers in the ventricular walls are arranged in irregular whorls, so that when they are stimulated by impulses on Purkinje fibers, the ventricular walls contract with a twisting motion (fig. 17.18). This action squeezes or wrings the blood out of the ventricular chambers and forces it into the arteries.

1. What kinds of tissues make up the cardiac conduction system?
2. How is a cardiac impulse initiated?
3. How is this impulse transmitted from the atrium to the ventricles?

Fig. 17.18 The muscle fibers within the ventricular walls are arranged in patterns of whorls. The fibers of groups (a) and (b) surround both ventricles.

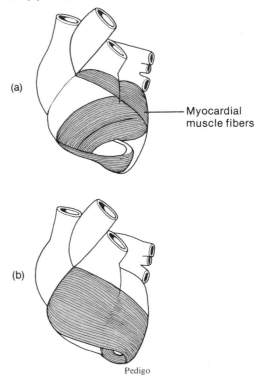

(a)

Myocardial muscle fibers

(b)

Pedigo

Fig. 17.19 This instrument is used to record an ECG.

Fig. 17.20 A normal ECG.

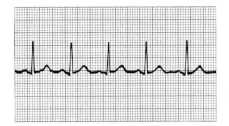

Fig. 17.21 In an ECG pattern, the P wave results from depolarization of the atria, the QRS complex results from depolarization of the ventricles, and the T wave results from repolarization of the ventricles.

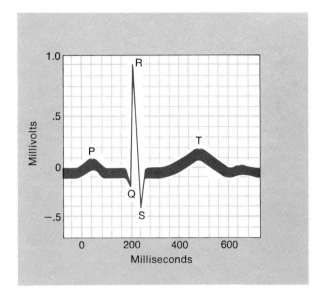

The Electrocardiogram

An **electrocardiogram** (ECG) is a recording of the electrical changes that occur in the *myocardium* during a cardiac cycle. These changes result from the depolarization and repolarization associated with the contraction of muscle fibers. Because body fluids can conduct electrical currents, such changes can be detected on the surface of the body.

To record an *ECG,* metal electrodes are placed in certain locations on the skin. These electrodes are connected by wires to an instrument (shown in fig. 17.19) that responds to very weak electrical changes by causing a pen or stylus to mark on a moving strip of paper. When the instrument is operating, up-and-down movements of the pen correspond to electrical changes occurring within the body as a result of myocardial activity.

Since the paper moves past the pen at a known rate, the distance between pen deflections can be used to measure the time elapsing between various phases of the cardiac cycle.

As figure 17.20 illustrates, the ECG pattern includes several deflections, or *waves,* during each cardiac cycle. Between cycles, the muscle fibers remain polarized, and no detectable electrical changes occur. Consequently, the pen simply marks along the base line as the paper moves through the instrument. When the S-A node triggers a cardiac impulse, however, the atrial fibers are stimulated to depolarize, and an electrical change occurs. As a result, the pen is deflected, and when this electrical change is completed, the pen returns to the base position. This pen movement produces a *P wave* that is caused by the depolarization of the atrial fibers just before they contract.

When the cardiac impulse reaches the ventricular fibers, they are stimulated to depolarize. Because the ventricular walls are much more extensive than those of the atria, the amount of electrical change is greater, and the pen is deflected to a greater degree than before. Once again, when the electrical change is completed, the pen returns to the base line, leaving a mark called the *QRS complex*. This wave appears just prior to the contraction of the ventricular walls.

Near the end of the ECG pattern, the pen is deflected for a third time, producing a *T wave*. This wave is caused by electrical changes occurring as the ventricular muscle fibers become repolarized. The record of the atrial repolarization is missing from the pattern, because the atrial fibers repolarize at the same time that the ventricular fibers depolarize. The recording of the atrial repolarization is thus obscured by the QRS complex. (See fig. 17.21.)

To summarize: The P wave is caused by atrial depolarization, the QRS complex is caused by ventricular depolarization, and the T wave is caused by ventricular repolarization.

Fig. 17.22 A prolonged QRS complex may result from damage to the Purkinje fibers.

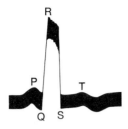

Fig. 17.23 The activities of the S-A and A-V nodes can be altered by autonomic nerve impulses.

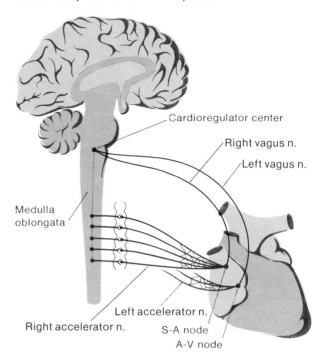

ECG patterns are especially important because they allow a physician to assess the heart's ability to conduct impulses and thus to judge its condition. For example, the time period between the beginning of a P wave and the beginning of a QRS complex (*P-Q or P-R interval*) indicates how long it takes for the cardiac impulse to travel from the S-A node through the A-V node and into the ventricular walls. If ischemia or other problems involving the fibers of the A-V conduction pathways are present, this P-Q interval sometimes increases. Similarly, if the Purkinje fibers are injured, the duration of the QRS complex may increase, because it may take longer for an impulse to spread throughout the ventricular walls. (See fig. 17.22.)

1. What is an electrocardiogram?
2. What cardiac event is represented by the P wave? By the QRS complex? By the T wave?

Regulation of the Cardiac Cycle

The primary function of the heart is to pump blood to the body cells, and when the needs of these cells change, the quantity of blood pumped must change also. For example, during strenuous exercise, the amount of blood required by skeletal muscles increases greatly, and the rate of the heartbeat increases in response to this need. Since the S-A node normally controls the heart rate, changes in this rate often involve factors that affect the pacemaker. These include motor impulses on parasympathetic and sympathetic nerve fibers.

The parasympathetic fibers that supply the heart arise from neurons in the medulla oblongata and make up parts of the *vagus nerves*. Most of these fibers branch to the S-A node and the A-V node. When nerve impulses reach their endings, these fibers secrete acetylcholine that causes a decrease in S-A and A-V nodal activity. As a result, the rate of heartbeat decreases.

The vagus nerves seem to carry impulses continually to the S-A and A-V nodes. These impulses impose a braking action on the heart. Consequently, parasympathetic activity can cause the heart rate to change in either direction. An increase in impulses causes a slowing of the heart, and a decrease in impulses releases the parasympathetic brake and allows the heartbeat to increase.

Sympathetic fibers reach the heart by means of the *accelerator nerves,* whose branches join the S-A and A-V nodes as well as other areas of the atrial and ventricular myocardium. (See fig. 17.23.) The endings of these fibers secrete norepinephrine in response to nerve impulses, and this substance causes an increase in the rate and the force of myocardial contractions.

A normal balance between the inhibitory effects of the parasympathetic fibers and the excitatory effects of the sympathetic fibers is maintained by the cardiac center of the medulla oblongata. In this region of the brain, masses of neurons function as *cardioinhibitor* and *cardioaccelerator reflex centers*. These centers receive sensory impulses from various parts of the circulatory system and relay motor impulses to the heart in response.

For example, there are *pressoreceptors* in certain regions of the aorta (aortic sinus and aortic arch) and the carotid arteries (carotid sinuses). These receptors are sensitive to changes in blood pressure, and if the pressure rises, they signal the cardioinhibitor center in the medulla. In response, this reflex center

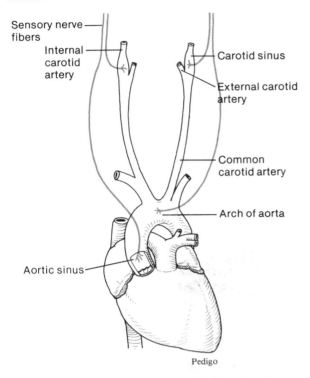

Sensory nerve fibers
Internal carotid artery
Carotid sinus
External carotid artery
Common carotid artery
Arch of aorta
Aortic sinus
Pedigo

Fig. 17.25 Tachycardia is characterized by a rapid heartbeat.

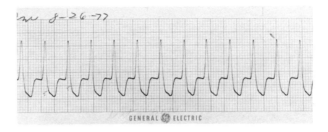

GENERAL ELECTRIC

sends *parasympathetic* motor impulses to the heart, causing the heart rate and force of contraction to decrease. This also causes blood pressure to drop toward the normal level. (See fig. 17.24.)

Another regulatory reflex involves pressoreceptors in the venae cavae near the entrances to the right atrium. If the blood pressure increases abnormally in these vessels, the receptors signal the cardioaccelerator center, and *sympathetic* impulses flow to the heart. As a result, the heart rate and force of contraction increase, and the venous blood pressure is reduced.

Two other factors that influence heart rate are temperature and various ions. Heart action is increased by a rising body temperature, which accounts for the fact that heart rate usually increases during fever. On the other hand, cardiac activity is decreased by abnormally low body temperature. Consequently, a patient's body temperature is sometimes deliberately lowered (hypothermia) to slow the heart during surgery.

Of the ions that influence heart action, the most important are potassium (K^+) and calcium (Ca^{++}). Although homeostatic mechanisms normally maintain the concentrations of these substances within narrow ranges, these mechanisms sometimes fail, and the consequences can be serious or even fatal.

An excess of *potassium ions* (hyperkalemia), for example, seems to alter the usual polarized state of the cardiac muscle fibers, and the result is a decrease in the rate and force of contractions. In fact, if the potassium ion concentration is very high, the conduction of cardiac impulses may be blocked, and heart action may suddenly stop (cardiac arrest). Conversely, if the potassium concentration falls below normal (hypokalemia), the heart may develop a serious abnormal rhythm (arrhythmia). This condition can also be life threatening.

Excessive *calcium ions* (hypercalcemia) cause increased heart actions, and there is a danger that the heart will undergo a prolonged contraction. Conversely, low calcium (hypocalcemia) depresses heart action. This effect probably occurs because calcium ions help initiate the muscle contraction mechanism, as is described in chapter 9.

1. What nerves supply parasympathetic fibers to the heart? Sympathetic fibers?
2. How do parasympathetic and sympathetic impulses help control heart rate?
3. How do changes in body temperature affect heart rate?
4. Describe the effects on the heart of abnormal amounts of potassium and calcium.

Abnormal Heart Actions

Although slight variations in heart actions sometimes occur normally, marked changes in the usual rate or rhythm may suggest cardiovascular disease. Such abnormal actions, termed cardiac **arrhythmias,** include the following:

1. **Tachycardia.** Tachycardia is characterized by an abnormally fast heartbeat, usually over one-hundred beats per minute. This condition may be caused by such factors as an increase in body temperature, nodal stimulation by sympathetic fibers, the presence of various drugs or hormones, or heart disease. Figure 17.25 shows the ECG of a tachycardic heart.

Fig. 17.26 Bradycardia is characterized by a slow heartbeat.

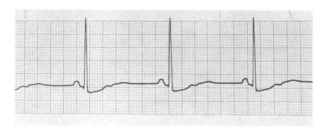

Fig. 17.28 Atrial flutter is characterized by an abnormally rapid rate of atrial contraction.

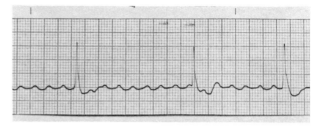

Fig. 17.27 Ectopic beats are usually caused by impulses originating from sites other than the S-A node.

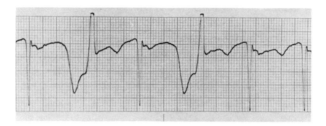

Fig. 17.29 Atrial fibrillation is characterized by contractions that are rapid and uncoordinated. Why is this condition less serious than ventricular fibrillation?

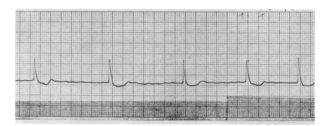

2. **Bradycardia.** Bradycardia means a slow heart rate, usually one that is less than sixty beats per minute. It may be caused by decreased body temperature, nodal stimulation by parasympathetic impulses, or the presence of certain drugs. Also, athletes sometimes have unusually slow heartbeats because their hearts have developed the ability to pump a greater than normal volume of blood with each beat. Apparently the resulting increase in cardiac output stimulates various reflexes to inhibit heart actions, and thus the heart is slowed. (See the ECG in fig. 17.26.)

3. **Premature heartbeats.** A premature or *ectopic beat* is one that occurs before it is expected in a normal series of cardiac cycles. (See ECG in fig. 17.27.) Such an occurrence is probably caused by cardiac impulses originating from unusual (ectopic) regions of the heart. That is, the impulse originates from a site other than the S-A node. They may arise from ischemic tissues or from muscle fibers that are irritated by the effects of disease or drugs.

4. **Flutter.** A heart chamber is said to flutter when it is contracting regularly, but at a very rapid rate, such as 250–350 contractions per minute. Although normal hearts may flutter occasionally, this condition is more likely to be due to damage to the myocardium. (See fig. 17.28.)

5. **Fibrillation.** Fibrillation is also characterized by rapid heart actions but, unlike flutter, the accompanying contractions are *uncoordinated* because small regions of the myocardium contract and relax

independently of all other areas. As a result, the myocardium fails to contract as a whole, and the walls of the fibrillating chambers are completely ineffective in pumping blood.

Although a person may survive atrial fibrillation, because venous blood pressure may continue to force blood into the ventricles, ventricular fibrillation is very likely to cause death. (See fig. 17.29.)

The ventricles can be stimulated to fibrillate by a variety of factors, including ischemia associated with obstruction of a coronary artery (coronary occlusion), electric shock, traumatic injury to the heart or chest wall, or toxic effects of various drugs. Once the ventricles are fibrillating, normal cardiac rhythm is not likely to be restored unless the heart can be *defibrillated.* Under such conditions, if circulation is not reestablished within a few minutes, the person may suffer irreparable damage to the cerebral cortex and may die.

Defibrillation is accomplished by exposing a fibrillating myocardium to a strong electric current for a short time (fig. 17.30). This action causes all of the muscle fibers within the myocardium to become depolarized simultaneously, so that all contractile activity ceases. It is hoped that, in a short time, the S-A node will begin to function, and a normal rhythm will be reestablished.

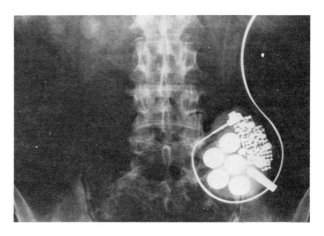

Fig. 17.31 When the A-V node is the functional pacemaker, the P wave may be inverted (or absent), and the PR interval reduced.

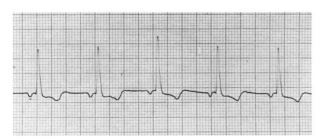

Sometimes a patient suffering from a conduction disorder can be helped by an *artificial pacemaker.* Such a device contains a small, battery- or nuclear-powered electrical stimulator (generator pack). When the artificial pacemaker is installed, electrodes may be threaded through veins into the right ventricle, and the stimulator may be implanted beneath the skin in the shoulder or abdomen (fig. 17.32). It then transmits rhythmic electrical impulses to the heart, and the myocardium responds by contracting rhythmically. Although some artificial pacemakers operate at a fixed rate, other *demand pacemakers* respond to the changing needs of body cells for increased or decreased blood supplies.

6. **Arrhythmias due to conduction disorders.** Any interference or block in cardiac impulse conduction may cause arrhythmia. The type of arrhythmia produced, however, varies with the location and extent of the block. Related to this is the fact that certain cardiac tissues other than the S-A node can function as pacemakers.

The S-A node usually initiates seventy to eighty heartbeats per minute. If the S-A node is damaged, impulses originating in the A-V node may travel upward into the atrial myocardium and downward into the ventricular walls, stimulating them to contract. Under the influence of the A-V node acting as a *secondary pacemaker,* the heart may continue to pump blood, but at a rate of forty to sixty beats per minute. (See fig. 17.31.) Similarly, the Purkinje fibers can initiate cardiac impulses, causing the heart to contract fifteen to forty times per minute.

1. *Distinguish between tachycardia and bradycardia.*
2. *Define* fibrillation.
3. *How may arrhythmia be related to a conduction disorder?*

Blood Vessels

The vessels of the cardiovascular system form a closed circuit of tubes that carry blood from the heart to the body cells and back again. These tubes include arteries, arterioles, capillaries, venules, and veins. As figure 17.33 illustrates, the arteries and arterioles conduct blood away from the ventricles of the heart and lead to the capillaries. The capillaries function to exchange substances between the blood and the body cells, and the venules and veins return blood from the capillaries to the atria.

Fig. 17.33 Arteries and arterioles conduct blood away from the heart, while venules and veins conduct it toward the heart.

Fig. 17.34 The wall of an artery.

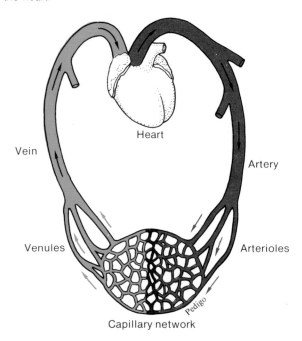

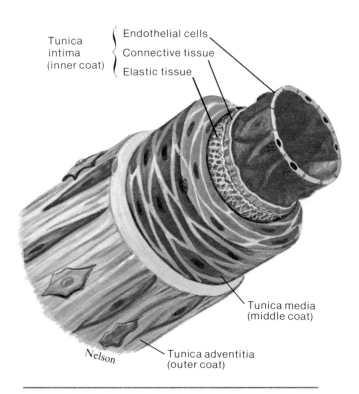

Arteries and Arterioles

Arteries are strong, elastic vessels that are adapted for carrying blood under high pressure. These vessels subdivide into progressively thinner tubes and eventually give rise to fine branches called **arterioles.**

The wall of an artery consists of three distinct layers, or *tunics,* shown in figure 17.34. (Also see color plate 31.) The innermost layer (tunica intima) is composed of a layer of epithelium, called *endothelium,* resting on a connective tissue membrane that is rich in elastic and collagenous fibers.

The middle tunic (tunica media) makes up the bulk of the arterial wall. It includes smooth muscle fibers that encircle the tube and a thick layer of elastic connective tissue.

The outer layer (tunica adventitia) is relatively thin and consists chiefly of connective tissue with irregularly arranged elastic and collagenous fibers. This layer attaches the artery to the surrounding tissues. It also contains minute vessels (vasa vasorum) that provide blood to the more external cells of the artery wall.

The endothelial lining provides a smooth surface so that blood cells and platelets can flow without being damaged. The connective tissues give the vessel a tough elasticity that allows it to withstand the force of high blood pressure and, at the same time, to stretch and accommodate the sudden increase in blood volume that accompanies each ventricular contraction.

Fig. 17.35 (*a*) Lumen diameter of a vessel with normal tone; (*b*) diameter as a result of vasoconstriction; (*c*) diameter as a result of vasodilation.

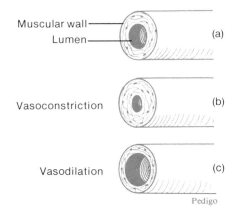

The smooth muscles in the walls of arteries and arterioles are innervated by sympathetic branches of the autonomic nervous system. Impulses on these *vasomotor* fibers cause the smooth muscles to contract, thus reducing the diameter of the vessel. This action is called **vasoconstriction.** If such vasomotor impulses are inhibited, the muscle fibers relax and the diameter of the vessel increases. In this case, the artery is said to undergo **vasodilation.** Changes in the diameters of arteries, shown in figure 17.35, greatly influence the flow and pressure of the blood.

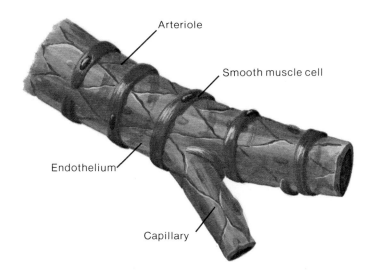

Arteriole

Smooth muscle cell

Endothelium

Capillary

Arterioles, which are microscopic continuations of arteries, give off branches called *metarterioles* that, in turn, join capillaries. Although the walls of the larger arterioles have three layers, similar to those of arteries, these walls become thinner and thinner as the arterioles approach the capillaries. The wall of a very small arteriole consists only of an endothelial lining and some smooth muscle, surrounded by a small amount of connective tissue (fig. 17.36).

The arteriole and metarteriole walls are adapted for vasoconstriction and vasodilation in that their muscle fibers respond to impulses from the autonomic nervous system by contracting or relaxing. Thus, these vessels function in helping to control the flow of blood into the capillaries.

Sometimes metarterioles are connected directly to venules, and the blood entering them can bypass the capillaries. These connections between arteriole and venous pathways, shown in figure 17.37, are called *arteriovenous shunts*.

1. Describe the wall of an artery.
2. What is the function of the smooth muscle in the arterial wall?
3. How is the structure of an arteriole different from that of an artery?

Fig. 17.37 Some metarterioles provide arteriovenous shunts by connecting arterioles directly to venules.

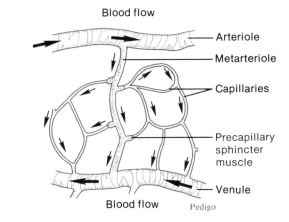

Blood flow

Arteriole

Metarteriole

Capillaries

Precapillary sphincter muscle

Venule

Blood flow   Pedigo

Capillaries

Capillaries are the smallest blood vessels. They form the connections between the smallest arterioles and the smallest venules. Capillaries are essentially extensions of the inner linings of these larger vessels, in that their walls consist of endothelium—a single layer of squamous epithelial cells. (See fig. 17.36.) These thin walls form the semipermeable membranes through which substances in the blood are exchanged for substances in the tissue fluid surrounding body cells. (See fig. 17.38.)

Fig. 17.38 Oxygen and nutrients leave the blood of capillaries at their arteriole ends, while wastes enter the blood at the venule ends.

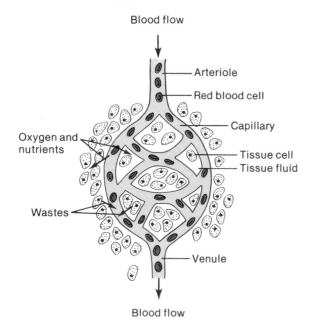

Blood flow

Arteriole

Red blood cell

Capillary

Oxygen and nutrients

Tissue cell
Tissue fluid

Wastes

Venule

Blood flow

The density of capillaries within tissues varies directly with the tissues' rates of metabolism. Thus, muscle and nerve tissues, which utilize relatively large quantities of oxygen and nutrients, are richly supplied with capillaries, while hyaline cartilage, the epidermis, and the cornea, whose metabolic rates are very slow, lack capillaries.

The patterns of capillary arrangement also differ in various body parts. For example, some capillaries pass directly from arterioles to venules, while others lead to highly branched networks. Such arrangements make it possible for blood to follow different pathways through a tissue and meet the varying demands of its cells. During periods of exercise, for example, blood can be directed into the capillary networks of the skeletal muscles, where cells are experiencing an increasing need for oxygen and nutrients. At the same time, blood can bypass some of the capillary nets in the tissues of the digestive tract, where the demand for blood is less critical. Conversely, when a person is relaxing after a meal, blood can be shunted from the inactive skeletal muscles into the capillary networks of the digestive organs, where it is needed to support the processes of digestion and absorption.

The distribution of blood in the various capillary pathways is regulated mainly by smooth muscles that encircle the capillary entrances. As figure 17.37 shows, these muscles form *precapillary sphincters* that may close a capillary by contracting or open it by relaxing. How the sphincters are controlled is not clear, but they seem to respond to the demands of the cells supplied by their individual capillaries. When the cells need oxygen and nutrients, the sphincter relaxes; and when the need is met, the sphincter may contract again.

1. Describe the wall of a capillary.
2. What is the function of a capillary?
3. How is blood flow into capillaries controlled?

Exchanges in the Capillaries. The vital function of exchanging gases, nutrients, and metabolic by-products between the blood and the tissue fluid surrounding body cells occurs in the capillaries. The substances exchanged move through the capillary walls primarily by the processes of diffusion, filtration, and osmosis, described in chapter 3. Of these processes, diffusion provides the most important means of transfer.

It is by diffusion that molecules and ions move from regions where they are more highly concentrated toward regions where they are in lower concentration. Since blood entering capillaries of the systemic circuit generally carries relatively high concentrations of oxygen and nutrients, these substances diffuse through the capillary walls and enter the tissue fluid. Conversely, the concentrations of carbon dioxide and various wastes are generally more highly concentrated in the regions of body cells, and they tend to diffuse into the capillary blood.

The paths followed by these substances depend primarily on their solubilities in lipids. Those that are soluble in lipid, such as oxygen, carbon dioxide, and fatty acids, can diffuse through most areas of the cell membranes that make up the capillary wall, because the membranes are largely lipid. Lipid insoluble substances, such as water, sodium ions, and chloride ions, diffuse through pores in the cell membranes and slit-like openings between the endothelial cells that form the capillary wall. (See fig. 17.39.)

Fig. 17.39 Electron micrograph of a capillary cross section. Note that the capillary wall consists of a single endothelial cell and that the capillary lumen is occupied by a red blood cell.

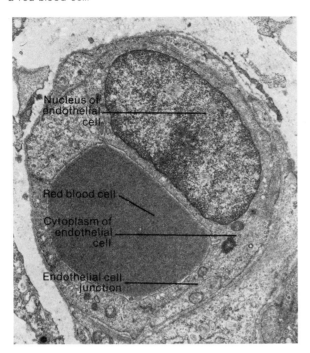

In the brain, the endothelial cells of the capillary walls are more tightly fused (tight junctions) than those in other body regions. Consequently, some substances that readily leave capillaries in other tissues enter brain tissues only slightly or not at all. This resistance to movement is called the *blood-brain barrier,* and it is of particular interest because it prevents certain drugs from entering the brain tissues or cerebrospinal fluid in sufficient concentrations to effectively treat diseases.

The capillaries of the hypothalamus are an exception, since they are permeable. This allows the hypothalamus to respond to changes in the chemical composition of plasma, as described in chapter 10.

Plasma proteins generally remain in the blood because they are not soluble in the lipid portions of the capillary membranes, and their molecular size is too great to permit diffusion through the membrane pores or openings between the endothelial cells.

Filtration involves the forcing of molecules through a membrane by *hydrostatic pressure.* In capillaries, the force is provided by blood pressure generated by contractions of the ventricular walls.

Blood pressure is also responsible for moving blood through the arteries and arterioles. Pressure tends to decrease, however, as the distance from the heart increases, because friction (peripheral resistance) between the blood and the vessel walls slows the flow. For this reason, blood pressure is greater in arteries than in arterioles, and greater in arterioles than in capillaries. It is similarly greater at the arteriole end of a capillary than at the venule end.

Blood components generally fail to pass through the walls of arteries and arterioles because these structures are too thick. When blood reaches the thin-walled capillaries, however, the hydrostatic pressure of the blood serves to increase the rate by which substances diffuse through the capillary wall. This filtration effect occurs primarily at the arteriole ends of capillaries, while diffusion takes place along their entire lengths.

The plasma proteins, which remain in the capillaries, help to make the *osmotic pressure* of the blood greater (hypertonic) than that of the tissue fluid. Although the capillary blood has a greater osmotic attraction for water than does the tissue fluid, this attraction is overcome by the greater force of the blood pressure. As a result, the net movement of water and dissolved substances is outward at the arteriole end of the capillary.

More specifically the force, including *hydrostatic pressure,* that tends to move fluid outward at the arteriole end of a capillary is 36.3 millimeters of mercury (mm Hg), while the *osmotic pressure* of the blood, which tends to move fluid inward, is only 28 mm Hg. Consequently, there is a net movement of water and dissolved substances outward. However, the blood pressure decreases as the blood moves through the capillary, and at the venule end, the outward force equals 21.3 mm Hg, while the osmotic pressure of the blood remains unchanged at 28 mm Hg. Thus there is a net movement of water and dissolved materials into the venule end of the capillary by osmosis. This process is shown in figure 17.40.

Normally, more fluid leaves the capillaries than returns to them, and the excess is collected and returned to the venous circulation by *lymphatic vessels.* This mechanism is discussed in chapter 18.

Sometimes unusual events cause an increase in the permeability of the capillaries, and an excessive amount of fluid is likely to enter the interstitial spaces. This may occur, for instance, following a traumatic injury to the tissues or in response to certain chemicals such as *histamine,* which increase membrane permeability. In any case, so much fluid may leak out of the capillaries that the lymphatic drainage is overwhelmed, and the affected tissues become swollen and painful. This condition is called *edema.*

Fig. 17.40 Substances leave capillaries because of a net outward filtration pressure; other substances enter the capillaries because of a net inward force of osmotic pressure.

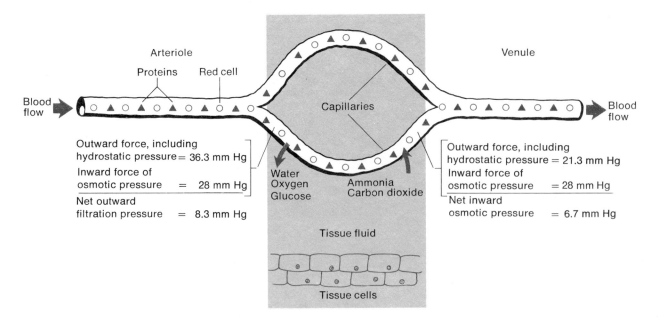

Arteriole

Proteins   Red cell

Blood flow

Capillaries

Venule

Blood flow

Outward force, including hydrostatic pressure = 36.3 mm Hg
Inward force of osmotic pressure = 28 mm Hg
Net outward filtration pressure = 8.3 mm Hg

Water
Oxygen
Glucose

Ammonia
Carbon dioxide

Outward force, including hydrostatic pressure = 21.3 mm Hg
Inward force of osmotic pressure = 28 mm Hg
Net inward osmotic pressure = 6.7 mm Hg

Tissue fluid

Tissue cells

A failing heart—one that is unable to pump blood out of the ventricles as rapidly as it enters—is often accompanied by edema in other parts of the body. This occurs because the blood backs up into the veins, venules, and capillaries, and the blood pressure in these vessels increases. As a result of this increased *back pressure,* the osmotic pressure of the blood in the venule ends of the capillaries is less effective in attracting water from tissue fluid, and the tissues tend to become edematous. This is true particularly of the tissues in the lower extremities if the person is upright, or of the back if the person is supine. In the terminal stages of heart failure, edema may become widespread, and fluid may begin to accumulate in the peritoneal cavity of the abdomen. This condition is called *ascites.*

1. What forces are responsible for the exchange of substances between the blood and tissue fluid?
2. Why is the fluid movement out of a capillary greater at its arteriole end than at its venule end?
3. Since more fluid leaves the capillary than returns to it, how is the remainder returned to the vascular system?

Venules and Veins

Venules are the microscopic vessels that continue from the capillaries and merge to form **veins.** The veins, which carry blood back to the atria, follow pathways that roughly parallel those of the arteries.

The walls of veins are similar to those of arteries in that they are composed of three distinct layers. However, the middle layer of the venous wall is poorly developed. Consequently, veins have thinner walls and contain less smooth muscle and less elastic tissue than arteries. (See fig. 17.41 and color plate 32.)

Many veins, particularly those in the arms and legs, contain flaplike *valves* that project inward from their linings. These valves, shown in figure 17.42, are usually composed of two leaflets that close if the blood begins to back up in a vein. In other words, the valves aid in returning blood to the heart, since the valves open as long as the flow is toward the heart, but close if it is in the opposite direction.

In addition to providing pathways for blood returning to the heart, the veins function as blood reservoirs that can be drawn on in times of need. For example, if a hemorrhage accompanied by a drop in arterial blood pressure occurs, the muscular walls of veins are stimulated reflexly by sympathetic nerve impulses. The resulting venous constrictions help to raise the blood pressure. This mechanism ensures a nearly normal blood flow even when as much as 25% of the blood volume has been lost. Figure 17.43 illustrates the relative volumes of blood in the veins and other blood vessels.

Fig. 17.41 The wall of a vein.

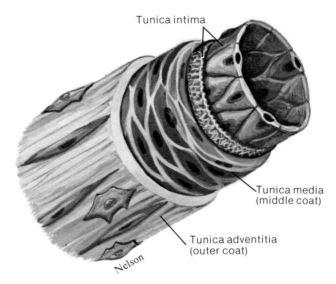

Tunica intima

Tunica media (middle coat)

Tunica adventitia (outer coat)

Nelson

Fig. 17.42 (*a*) Venous valves allow blood to move toward the heart but (*b*) prevent blood from moving away from the heart.

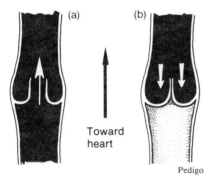

(a)

(b)

Toward heart

Pedigo

Fig. 17.43 Most of the blood volume is contained within the veins and venules.

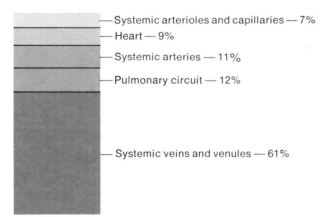

—Systemic arterioles and capillaries — 7%

—Heart — 9%

—Systemic arteries — 11%

—Pulmonary circuit — 12%

—Systemic veins and venules — 61%

Normal distribution of blood volume

The characteristics of the blood vessels are summarized in chart 17.3.

Some Blood Vessel Disorders

It is estimated that nearly half of all deaths in the United States are due to the arterial disease called **atherosclerosis**. This condition is characterized by an accumulation of soft masses of fatty materials, particularly cholesterol, on the inside of the arterial wall. Such deposits are called *plaque*, and as they develop they tend to protrude into the lumens of the vessels and interfere with the blood flow (fig. 17.44). Furthermore, plaque often creates a surface that can initiate the formation of a blood clot. As a result, persons with atherosclerosis may develop thrombi or emboli that cause blood deficiency (*ischemia*) or tissue death (*necrosis*) downstream from the obstruction.

The walls of affected arteries also tend to undergo degenerative changes during which they lose their elasticity and become hardened or *sclerotic*. This stage of the disease is called **arteriosclerosis**, and when it occurs, there is danger that a sclerotic vessel will rupture under the force of blood pressure.

Although the cause of atherosclerosis is not well understood, the disease seems to be associated with such factors as excessive use of saturated fats and refined carbohydrates in the diet, elevated blood pressure, cigarette smoking, overweight, and lack of physical exercise. Emotional and genetic factors also may increase the susceptibility to atherosclerosis.

Phlebitis, or inflammation of a vein, is a relatively common disorder, and although it may occur in association with an injury or infection or as an aftermath of surgery, it sometimes develops for no apparent reason.

The Cardiovascular System 585

Chart 17.3   Characteristics of blood vessels

Vessel	Type of Wall	Function
Artery	Thick, strong wall with three layers—endothelial lining, middle layer of smooth muscle and elastic tissue, and outer layer of connective tissue	Carries high pressure blood from heart to arterioles
Arteriole	Thinner wall than artery but with three layers; smaller arterioles have endothelial lining, some smooth muscle tissue, and small amount of connective tissue	Connects artery to capillary; helps to control blood flow into capillary by undergoing vasoconstriction or vasodilation
Capillary	Single layer of squamous epithelium	Provides semipermeable membrane through which nutrients, gases, and wastes are exchanged between blood and tissue cells; connects arteriole to venule
Venule	Thinner wall, less smooth muscle and elastic tissue than arteriole	Connects capillary to vein
Vein	Thinner wall than artery but with similar layers; middle layer more poorly developed; some with flaplike valves	Carries low pressure blood from venule to heart; valves prevent back flow of blood; serves as blood reservoir

Fig. 17.44 As atherosclerosis develops, masses of fatty materials accumulate beneath the inner linings of certain arteries and protrude into their lumens. (a) Normal arteriole; (b, c, d) accumulation of plaque on the inner wall of the arteriole.

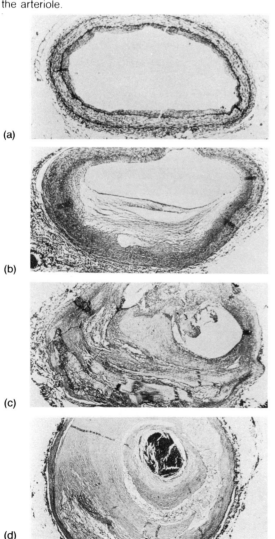

(a)

(b)

(c)

(d)

If such an inflammation is restricted to a superficial vein, the blood flow may be rechanneled through other vessels. If it occurs in a deep vein, however, the consequences can be quite serious, particularly if the blood within the affected vessel clots and blocks the normal circulation. This condition is called *thrombophlebitis.*

Varicose veins are distinguished by the presence of abnormal and irregular dilations in superficial veins, particularly those in the lower legs. This condition is usually associated with prolonged, increased back pressure within the affected vessels due to the force of gravity, as occurs when a person stands. The problem can also be aggravated when venous blood flow is obstructed by crossing the legs or by sitting in a chair so that its edge presses against the popliteal area behind the knee.

Excessive back pressure causes the veins to stretch and their diameters to increase. However, since the valves within these vessels do not change size, they soon lose their abilities to block the backward flow of blood, and blood tends to accumulate in the enlarged regions. (See fig. 17.45.)

Increased venous pressure is also accompanied by rising pressure within the venules and capillaries that supply the veins. Consequently, tissues in affected regions typically become edematous and painful.

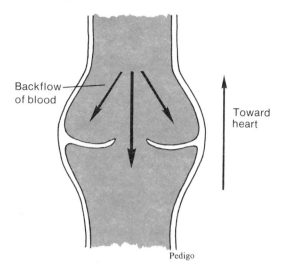

Backflow of blood

Toward heart

Pedigo

Although some people seem to inherit a weakness in their venous valves, *varicose veins* are most common in persons who stand or sit for prolonged periods. Pregnancy and obesity also seem to favor the development of this condition. The discomfort associated with varicose veins can sometimes be relieved by elevating the legs or by wearing support hosiery that are put on before a person arises in the morning thus preventing the vessel dilation that occurs upon standing. In other cases, surgical removal of the affected veins may be necessary to reduce the problem.

1. How does the structure of a vein differ from that of an artery?
2. What is the difference between atherosclerosis and arteriosclerosis?
3. How does the venous circulation help to maintain blood pressure when blood is lost by hemorrhage?
4. What factors promote the development of varicose veins?

Blood Pressure

Blood pressure is the force exerted by the blood against the inner walls of blood vessels. Although such a force occurs throughout the vascular system, the term *blood pressure* is most commonly used to refer to arterial pressure.

Arterial Blood Pressure

Arterial blood pressure rises and falls in a pattern corresponding to the phases of the cardiac cycle. That is, when the ventricles contract (ventricular systole), their walls squeeze the blood inside their chambers and force it into the pulmonary artery and aorta. As a result, the pressures in these arteries rise sharply. The maximum pressure achieved during ventricular contraction is called the **systolic pressure**. When the ventricles relax (ventricular diastole), the arterial pressure drops, and the lowest pressure that remains in the arteries before the next ventricular contraction is termed the **diastolic pressure.**

The surge of blood entering the arterial system during a ventricular contraction causes the elastic walls of the arteries to swell, but the pressure drops almost immediately as the contraction is completed, and the arterial walls recoil. This alternate expanding and recoiling of an arterial wall can be felt as a *pulse* in an artery that runs close to the surface. Figure 17.46 shows several sites where a pulse can be detected. The radial artery, for example, courses near the surface at the wrist and is commonly used to sense a person's radial pulse.

The radial pulse rate is usually equal to the rate at which the left ventricle is contracting, and for this reason it can be used to determine the heart rate. A pulse also can reveal something about blood pressure, because an elevated pressure produces a pulse that feels full, while a low pressure is accompanied by a pulse that is easily compressed.

Measurement of Arterial Blood Pressure

Arterial blood pressure can be measured by using an instrument called a *sphygmomanometer* (shown in fig. 17.47). This device consists of an inflatable rubber cuff that is connected by tubing to a compressible bulb and a glass tube containing a column of mercury. The bulb is used to pump air into the cuff, and the pressure produced is indicated by a rise in the mercury column. Thus, the pressure in the cuff can be expressed in millimeters of mercury (mm Hg). A pressure of 100 mm Hg, for example, would be enough to force the mercury column upward for a distance of 100 mm.

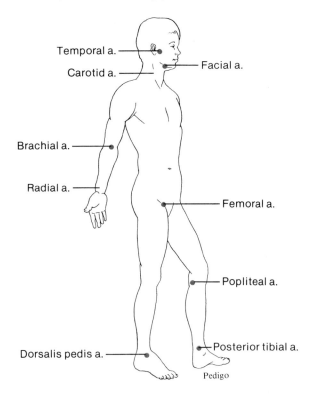

Fig. 17.47 A sphygmomanometer is used to measure arterial blood pressure.

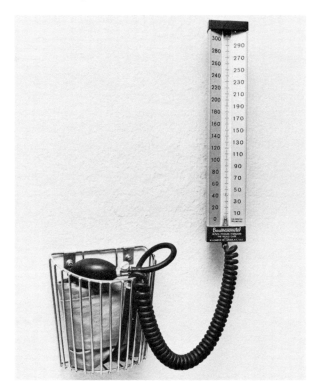

To measure blood pressure, the cuff of the sphygmomanometer is usually wrapped around the upper arm so it surrounds the brachial artery. Air is pumped into the cuff using the bulb until the cuff pressure exceeds the pressure in the brachial artery. As a result, the vessel is squeezed closed, and its blood flow is stopped. At this moment, if the bell of a stethoscope is placed over the brachial artery at the distal border of the cuff, no sounds can be heard from the vessel because the blood flow has been interrupted. As air is slowly released from the cuff, the air pressure inside it decreases. When the cuff pressure is approximately equal to the systolic blood pressure within the brachial artery, the artery opens enough for a small amount of blood to spurt through. This movement produces a sharp sound that can be heard through the stethoscope, and the height of the mercury column when this first sound is heard represents the *arterial systolic pressure* (SP).

As the cuff pressure continues to drop, a series of increasingly louder sounds can be heard. Then the sounds become abruptly muffled and finally disappear. The height of the mercury column when the sounds become abruptly muffled represents the *arterial diastolic pressure* (DP).

The results of a blood pressure measurement are reported as a fraction, such as 120/80. In this notation the upper number indicates the systolic pressure in mm Hg, and the lower number indicates the diastolic pressure in mm Hg (SP/DP). Figure 17.48 shows how these pressures vary throughout the systemic circuit.

The difference between the systolic and diastolic pressures (SP−DP), which is called the *pulse pressure* (PP), is generally about 40 mm Hg.

The average pressure in the arterial system also is of interest, since it represents the force that is effective throughout the cardiac cycle for driving blood to the tissues. This force, called the *mean arterial pressure*, is approximated by adding the diastolic pressure and one-third of the pulse pressure (DP+⅓PP).

---

1. Distinguish between systolic and diastolic blood pressure.
2. What cardiac event is responsible for the systolic pressure? The diastolic pressure?
3. Define mean arterial pressure.

Fig. 17.48 Blood pressure decreases as the distance from the left ventricle increases.

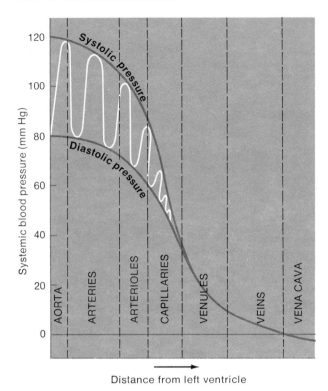

Distance from left ventricle

Fig. 17.49 Some factors that influence arterial blood pressure.

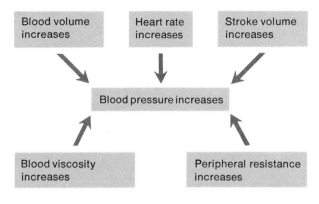

When a physically inactive person engages in mild to moderate exercise, the cardiac output usually increases as a result of an increase in heart rate. The stroke volume usually does not increase unless the exercise is strenuous.

An athlete's heart, however, is likely to increase the stroke volume much earlier and more markedly in response to exercise.

Factors That Influence Arterial Blood Pressure

The arterial pressure depends on a variety of factors, including heart action, blood volume, resistance to flow and the viscosity of the blood. (See fig. 17.49.)

Heart Action. In addition to creating blood pressure by forcing blood into the arteries, the heart action determines how much blood will enter the arterial system with each ventricular contraction, as well as the rate of this fluid output.

The volume of blood discharged from the ventricle with each contraction is called the **stroke volume** and equals about 70 ml. The volume discharged from the ventricle per minute is called the **cardiac output**. It is calculated by multiplying the stroke volume by the heart rate in beats per minute. (Cardiac output=stroke volume × heart rate.) Thus, if the stroke volume is 70 ml, and the heart rate is 72 beats per minute, the cardiac output is 5040 ml per minute.

Blood pressure varies directly with the cardiac output. If either the stroke volume or the heart rate increases, so does the cardiac output and, as a result, the blood pressure rises. Conversely, if the stroke volume or the heart rate decreases, so do the cardiac output and the blood pressure.

Blood Volume. The **blood volume** is equal to the sum of the blood cell and plasma volumes in the vascular system. Although the blood volume varies somewhat with age, body size, and sex, for adults it usually remains about 5 liters.

Blood pressure is directly proportional to the volume of blood within the vascular system. Thus, any changes in blood volume are accompanied by changes in blood pressure. For example, if blood volume is reduced by a hemorrhage, the blood pressure drops. If the normal blood volume is restored by a blood transfusion, normal pressure may be reestablished.

Arterial blood volume, and consequently arterial blood pressure, also vary directly with cardiac output. Thus, if the stroke volume or the heart rate increases, the arterial system may be forced to accept a greater volume of blood than can flow into the peripheral blood vessels. Consequently, the blood volume that the arteries must accommodate is increased even when the ventricles are relaxed. For this reason, an increased cardiac output is reflected in an elevated *diastolic pressure*. On the other hand, an increase in the force of ventricular contraction produces an elevated *systolic pressure*.

Peripheral Resistance. Friction between the blood and the walls of the blood vessels creates a force called **peripheral resistance**, which hinders blood flow. This force must be overcome by blood pressure if the blood is to continue flowing. Consequently, factors that alter peripheral resistance cause changes in blood pressure.

For example, if the smooth muscles in the walls of arterioles contract, the peripheral resistance of these constricted vessels increases. Blood tends to back up into the arteries supplying the arterioles, and the arterial pressure rises. Dilation of the arterioles has the opposite effect—peripheral resistance lessens, and the arterial blood pressure drops in response.

As was mentioned, the arterial walls are quite elastic, and when the ventricles discharge a surge of blood, the arteries swell. Almost immediately, the elastic tissues recoil, and the vessel walls press against the blood inside. This action helps to force the blood onward against the peripheral resistance created by the arterioles and capillaries. It also tends to convert the intermittent flow of blood, which is characteristic of the arterial system, into a more continuous movement through the capillaries.

Viscosity. As mentioned in chapter 16, the viscosity of a fluid is related to the ease with which its molecules flow past one another. The greater the viscosity, the greater the resistance to flowing.

The presence of blood cells and plasma proteins increases the viscosity of the blood. Since the greater the blood's resistance to flowing, the greater the force needed to move it through the vascular system, it is not surprising that blood pressure rises as blood viscosity increases and drops as viscosity decreases.

Although the viscosity of blood normally remains relatively stable, any condition that alters the concentrations of blood cells or plasma proteins may cause changes in viscosity. For example, anemia and hemorrhage may be accompanied by a decreasing viscosity and a consequent drop in blood pressure, while polycythemia, which causes an increase in viscosity, may be accompanied by a rise in blood pressure.

---

1. How is cardiac output calculated?
2. What is the relationship between cardiac output and blood pressure?
3. How does blood volume affect blood pressure?
4. What is the relationship between peripheral resistance and blood pressure? Between viscosity and blood pressure?

Fig. 17.50 Cardiac output is related to the volume of blood entering the heart.

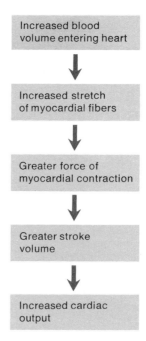

Control of Blood Pressure

Two important mechanisms for maintaining normal arterial pressure involve the regulation of cardiac output and peripheral resistance.

Cardiac output depends on the volume of blood discharged from the ventricle with each contraction (stroke volume) and the rate of heartbeat. These actions are affected by mechanical, neural, and chemical factors.

For example, the volume of blood entering the ventricle affects the stroke volume. As the blood enters, myocardial fibers in the ventricular wall are mechanically stretched. Within limits, the greater the length of these fibers, the greater the force with which they contract. This relationship between fiber length and force of contraction is called *Starling's law of the heart.* Because of it, the heart can respond to the demands placed on it by the varying quantities of blood that return from the venous system.

In other words, the more blood that enters the heart from the veins, the stronger the ventricular contraction, the greater the stroke volume, and the greater the cardiac output. (See fig. 17.50.) The less blood that returns from the veins, the weaker the ventricular contraction, the lesser the stroke volume, and the lesser the cardiac output.

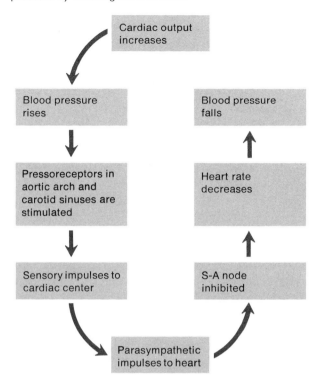

This mechanism ensures that the volume of blood discharged from the heart is equal to the volume entering its chambers. Consequently, the volume of blood that enters the right atrium from the venae cavae is normally equal to the volume that leaves the left ventricle and enters the aorta.

The neural regulation of heart rate was described previously in this chapter. Pressoreceptors (baroreceptors) located in the walls of the aortic arch and carotid sinuses are sensitive to changes in blood pressure. If the arterial pressure rises, nerve impulses travel from the receptors to the *cardiac center* of the medulla oblongata. This center relays parasympathetic impulses to the S-A node in the heart, and its rate decreases in response. As a result of this *cardioinhibitor reflex*, the cardiac output is reduced, and blood pressure falls toward the normal level. Figure 17.51 summarizes this mechanism.

Conversely, if the arterial blood pressure drops, the *cardioaccelerator reflex*, involving sympathetic impulses to the S-A node, is initiated, and the heart beats faster. This response increases the cardiac output, and the arterial pressure rises.

Epinephrine and *norepinephrine* are two chemicals that influence heart rate and consequently alter cardiac output and blood pressure. These substances cause an increase both in the heart rate and in the force with which the myocardium contracts, so that the cardiac output increases and the blood pressure rises.

Other factors that cause an increase in heart rate and a rise in blood pressure include emotional responses such as fear and anger, physical exercise, and increasing body temperature.

Peripheral resistance is regulated primarily by changes in the diameters of arterioles. Since blood vessels with smaller diameters offer greater resistance to blood flow, factors that cause arteriole vasoconstriction bring about increasing peripheral resistance, while those causing vasodilation produce a decrease in resistance.

The *vasomotor center* of the medulla oblongata continually sends sympathetic impulses to the smooth muscles in the arteriole walls. As a result, these muscles are kept in a state of tonic contraction, which helps to maintain the peripheral resistance associated with normal blood pressure. Since the vasomotor center is responsive to changes in blood pressure, it can cause an increase in peripheral resistance by increasing its outflow of sympathetic impulses, or can cause a decrease in such resistance by decreasing its sympathetic outflow. In the latter case, the vessels undergo vasodilation as the sympathetic stimulation is decreased.

For instance, as figure 17.52 illustrates, whenever the arterial blood pressure suddenly rises, the pressoreceptors in the aortic arch and carotid sinuses signal the vasomotor center, and the sympathetic outflow to the arteriole walls is inhibited. The resulting vasodilation causes a decrease in peripheral resistance, and the blood pressure falls toward the normal level.

Control of vasoconstriction and vasodilation by the vasomotor center is especially important in the arterioles of the *abdominal viscera* (splanchnic region). These vessels, if fully dilated, could accept nearly all the blood of the body and cause the arterial pressure to approach zero. Thus, control of their diameters is essential in the regulation of normal peripheral resistance.

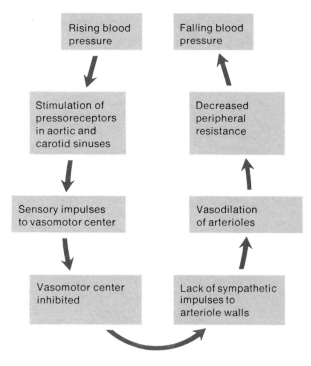

Chemical substances, including carbon dioxide, oxygen and hydrogen ions, also influence peripheral resistance by affecting the smooth muscles in the walls of arterioles and metarterioles and the actions of precapillary sphincters. An increasing P_{CO_2}, a decreasing P_{O_2}, and a decreasing pH cause relaxation of these muscles and a consequent drop in blood pressure. In addition, epinephrine and norepinephrine cause vasoconstriction of many vessels, followed by a rise in blood pressure; even though epinephrine causes vasodilation of the vessels within the skeletal muscles.

1. What factors affect cardiac output?
2. Define Starling's law of the heart.
3. What is the function of the pressoreceptors in the walls of the aortic arch and carotid sinuses?
4. How does the vasomotor center control the diameter of arterioles?

Venous Blood Flow

Blood pressure decreases as the blood moves through the arterial system and into the capillary networks. In fact, little pressure remains at the venule ends of the capillaries. (See fig. 17.48.) Therefore, blood flow through the venous system is not the direct result of heart action, but depends on other factors, such as skeletal muscle contraction, breathing movements, and vasoconstriction of veins.

For example, when skeletal muscles contract, they thicken and press on nearby vessels, squeezing the blood inside. As was mentioned, many veins, particularly those in the arms and legs, contain flaplike *semilunar valves* that project inward from their linings. These valves offer little resistance to blood flowing toward the heart, but they close if blood moves in the opposite direction. (See fig. 17.42.) Consequently, as *skeletal muscles* exert pressure on veins with valves, some blood is moved from one valve section to another. This massaging action of contracting skeletal muscles helps to push blood through the venous system toward the heart.

Respiratory movements provide another means of moving venous blood. During inspiration, the pressure within the thoracic cavity is reduced as the diaphragm contracts and the rib cage moves upward and outward. At the same time, the pressure within the abdominal cavity is increased as the diaphragm presses downward on the abdominal viscera. Consequently, blood tends to be squeezed out of the abdominal veins and into the thoracic veins. Back flow into the legs is prevented by the valves in the veins of the legs.

During exercise, these respiratory movements act together with skeletal muscle contractions to increase the return of venous blood.

Another mechanism that promotes the return of venous blood involves constriction of veins. When venous pressure is low, sympathetic reflexes stimulate smooth muscles in the walls of veins to contract. Such *venoconstriction* serves to increase the venous pressure and forces more blood toward the heart.

Also, as was mentioned, the veins provide a blood reservoir that can adapt its capacity to changes in the blood volume. If blood is lost and the blood pressure decreases, venoconstriction can help to return the blood pressure to normal by forcing blood out of this reservoir.

Central Venous Pressure

Since all the systemic veins drain into the right atrium, the pressure within this heart chamber is called the *central venous pressure*.

This pressure is of special interest because it has an effect on the pressure within the peripheral veins. For example, if the heart is beating weakly, the central venous pressure increases, and blood tends to back up in the venous network, causing its pressure to increase also. If the heart is beating forcefully, however, the central venous pressure and the pressure within the venous network decrease.

Other factors that increase the flow of blood into the right atrium, and thus cause the central venous pressure to become elevated, include an increase in blood volume and widespread venoconstriction.

---

1. What is the function of the venous valves?
2. How do skeletal muscles affect venous blood flow?
3. How do respiratory movements affect venous blood flow?
4. What factors stimulate venoconstriction?

High Blood Pressure

High blood pressure, or **hypertension**, is characterized by a persistently elevated arterial pressure, and is one of the more common diseases of the cardiovascular system.

Since the cause of high blood pressure is often unknown, it may be called *essential* (or idiopathic) *hypertension*. Sometimes the elevated pressure is related to another problem, such as arteriosclerosis or kidney disease.

Arteriosclerosis, for example, is accompanied by decreasing elasticity of the arterial walls and narrowing of the lumens of these vessels. Both of these effects promote an increase in blood pressure.

Kidney diseases often produce changes that interfere with blood flow to kidney cells. In response, the affected tissues may release an enzyme, called **renin**, that acts upon certain plasma proteins to form a substance called **angiotensin**.

Angiotensin is a powerful vasoconstrictor whose action increases the peripheral resistance in the arterial system, and this causes the arterial pressure to rise. Angiotensin also causes *aldosterone* to be released from the adrenal cortex. This hormone promotes the retention of salts and water by the kidneys, and the resulting increase in blood volume causes an additional increase in blood pressure. (See fig. 17.53.)

Normally this mechanism ensures that a decrease in blood flow to the kidneys is followed by an increase in arterial pressure and that results in an increase in blood flow to the kidneys. If the decreased blood flow is the result of a disease, such as atherosclerosis, the mechanism may cause high blood pressure and promote further deterioration of the arterial system.

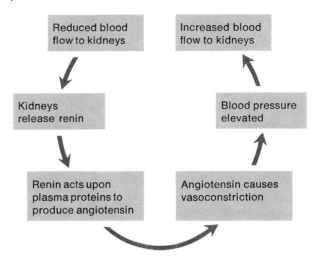

Fig. 17.53 A mechanism that acts to elevate blood pressure.

Although high blood pressure can often be reduced by the use of drugs or special low salt diets, the effects of prolonged, uncontrolled hypertension can be very serious. For example, the left ventricle must contract with greater force in order to discharge a normal volume of blood against the increased arterial pressure. As a result of the increased work load, the myocardium tends to thicken, and the heart becomes enlarged. As the muscle tissue increases, the coronary blood vessels are less and less able to supply it with enough blood, and the affected person may develop the painful condition called *angina pectoris*.

The discomfort of angina pectoris usually occurs during physical activity or an emotional disturbance, and is relieved by rest. It may take the form of pain or a sensation of heavy pressure, tightening, or squeezing in the chest. Although it is usually felt in the region behind the sternum or in the anterior portion of the upper thorax, the pain may radiate to other parts, including the neck, jaw, throat, arm, shoulder, elbow, back, or upper abdomen.

Hypertension also enhances the development of atherosclerosis. As a result, various arteries, including those of the coronary circuit, tend to accumulate plaque that may cause arterial occlusions. The victim then may suffer from a *coronary thrombosis* or a *coronary embolism*. Similar changes in the arteries of the brain increase the chances of a *cerebral vascular accident* (CVA)—a cerebral thrombosis, embolism, or hemorrhage—more commonly called a stroke.

Paths of Circulation

The blood vessels of the cardiovascular system can be divided into two major pathways—a pulmonary circuit and a systemic circuit. The **pulmonary circuit** consists of those vessels that carry blood from the heart to the lungs and back to the heart, while the **systemic circuit** is responsible for carrying blood from the heart to all other parts of the body and back again. (See fig. 17.54.)

The circulatory pathways described in the following sections are those of an adult. The fetal pathways, which are somewhat different, are described in chapter 22.

The Pulmonary Circuit

Blood enters the pulmonary circuit as it leaves the right ventricle and passes into the *pulmonary artery.* The pulmonary artery extends upward and posteriorly from the heart and, about 2 inches above its origin, divides into right and left branches. These branches penetrate the right and left lungs respectively. Within the lungs, they divide into *lobar branches* (three on the right side and two on the left) that accompany the main divisions of the bronchi into the lobes of the lungs. After repeated divisions, the lobar branches give rise to arterioles that continue into the capillary networks associated with the walls of the alveoli. (See fig. 17.55.)

The blood in the arteries and arterioles of the pulmonary circuit has a relatively low concentration of oxygen and a relatively high concentration of carbon dioxide. As is explained in chapter 15, gas exchanges occur between the blood and the air as the blood moves through the *pulmonary capillaries.*

Since the right ventricle contracts with less force than the left ventricle, the arterial pressure in the pulmonary circuit is less than that in the systemic circuit. Consequently, the pulmonary capillary pressure is relatively low.

The force tending to move fluid out of a pulmonary capillary is 23 mm Hg, while the force tending to pull fluid into it is 22 mm Hg. Thus, such a capillary has a net filtration pressure of 1 mm Hg. This pressure is responsible for a slight, continuous flow of fluid into the narrow interstitial space between the capillary and the alveolus.

Fig. 17.54 The pulmonary circuit consists of the vessels that carry blood between the heart and the lungs; all other vessels are included in the systemic circuit.

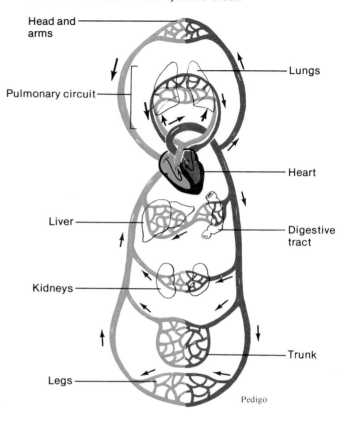

The epithelial cells of the alveolar membranes are so tightly joined that ions of sodium, chlorine, and potassium and molecules of glucose and urea that enter the interstitial space usually fail to enter the alveoli. This helps to maintain a relatively high osmotic pressure in the interstitial fluid. Consequently, any water that gets into the alveoli is rapidly moved back into the interstitial space by osmosis. This mechanism helps to keep the alveoli from filling with fluid, and they are said to remain dry. (See fig. 17.56.)

Fluid in the interstitial space may be drawn back into the pulmonary capillaries by the osmotic pressure of the blood, or it may be returned to the circulation by means of lymphatic vessels described in chapter 18.

As a result of the gas exchanges occurring between the blood and the alveolar air, blood entering the venules of the pulmonary circuit is rich in oxygen and low in carbon dioxide.

These venules merge to form small veins, and they in turn converge to form still larger ones. Four *pulmonary veins,* two from each lung, return blood to the left atrium, and this completes the vascular loop of the pulmonary circuit.

Fig. 17.55 Blood is carried to the lungs through branches of the pulmonary arteries and returns to the heart through pulmonary veins.

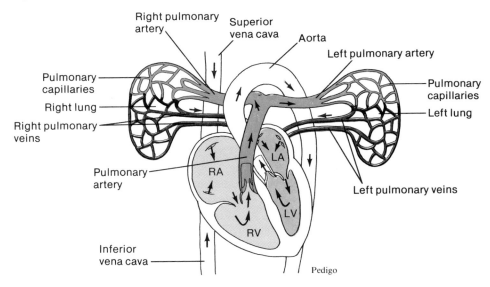

Fig. 17.56 The alveoli normally remain dry because the cells of the alveolar wall are tightly joined and water is drawn out of the alveoli by the high osmotic pressure of the interstitial fluid.

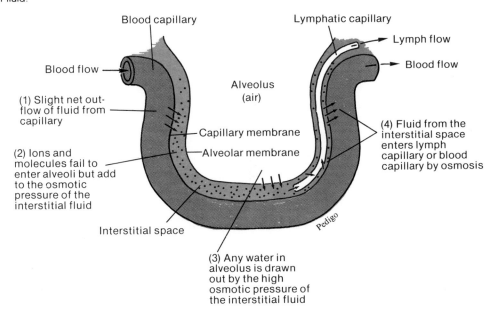

A condition called *pulmonary edema* sometimes accompanies a failing heart or a damaged bicuspid valve. If the left ventricle is weak, for example, it may be unable to move the normal volume of blood into the systemic circuit. As the blood backs up into the pulmonary circuit, the pressure in the pulmonary capillaries rises, and the interstitial spaces may become flooded with fluid. Increasing pressure in the interstitial fluid may cause the alveolar membranes to rupture, and fluid may enter the alveoli more rapidly than it can be removed. As a result, the alveolar surface available for gas exchange is reduced, and the person may suffocate.

The Systemic Circuit

The freshly oxygenated blood received by the left atrium is forced into the systemic circuit by the contraction of the left ventricle. This circuit includes the aorta and its branches that lead to all body tissues, as well as the companion system of veins that returns blood to the right atrium.

1. *Distinguish between the pulmonary and systemic circuits.*
2. *Trace a drop of blood through the pulmonary circuit from the right ventricle.*
3. *Explain why the alveoli normally remain dry.*

The Arterial System

The **aorta** is the largest artery in the body. It extends upward from the left ventricle, arches over the heart to the left, and descends just in front of the vertebral column. Figure 17.57 shows the aorta and its main branches.

Fig. 17.57 The principal branches of the aorta. (*a.* stands for *artery.*)

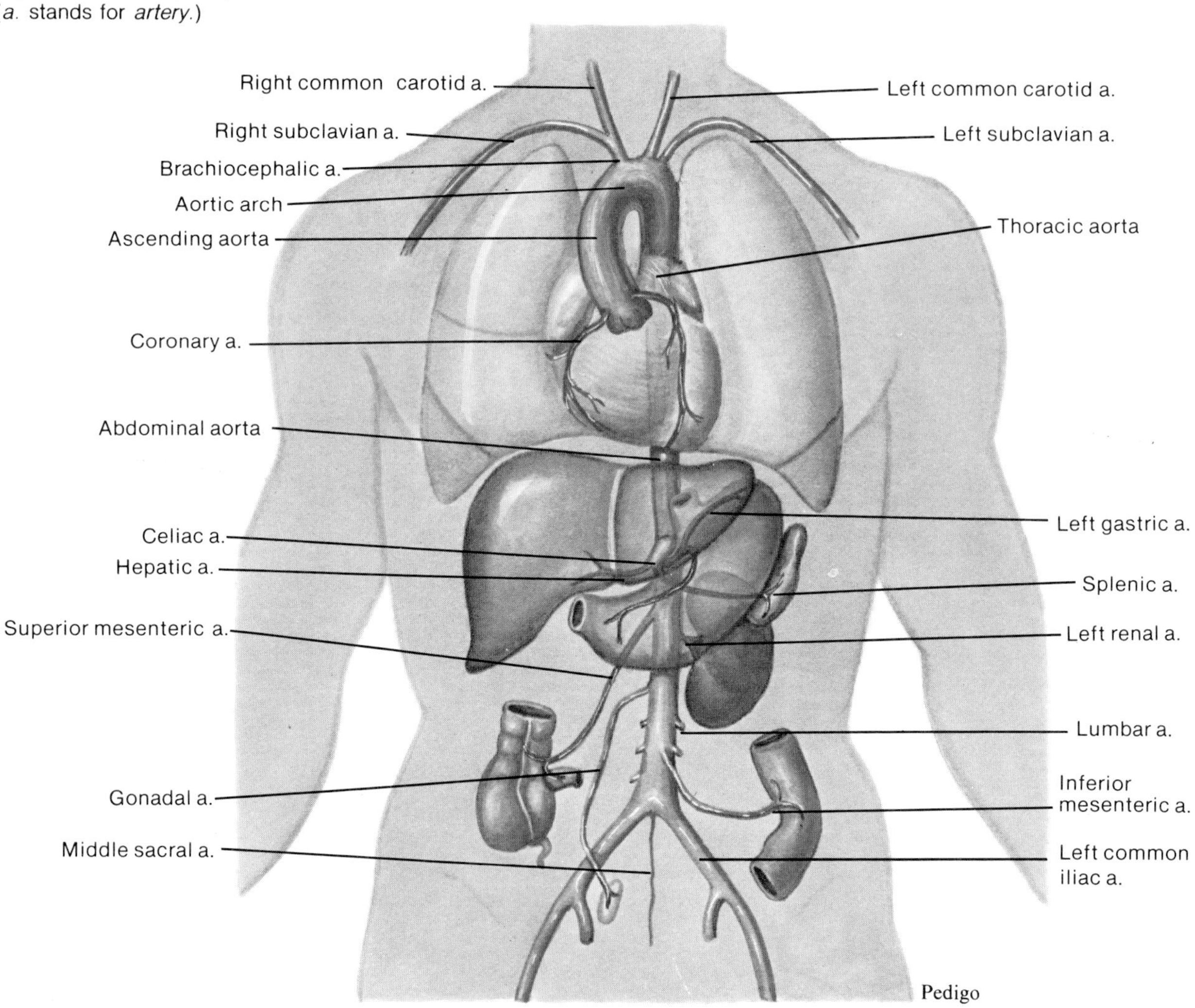

Right common carotid a.
Right subclavian a.
Brachiocephalic a.
Aortic arch
Ascending aorta
Coronary a.
Abdominal aorta
Celiac a.
Hepatic a.
Superior mesenteric a.
Gonadal a.
Middle sacral a.

Left common carotid a.
Left subclavian a.
Thoracic aorta
Left gastric a.
Splenic a.
Left renal a.
Lumbar a.
Inferior mesenteric a.
Left common iliac a.

Pedigo

Principal Branches of the Aorta

The first portion of the aorta is called the *ascending aorta.* At its base are located the three cusps of the aortic semilunar valve, and opposite each one is a swelling in the aortic wall called an **aortic sinus.** The right and left *coronary arteries* spring from two of these sinuses. Blood flow into these arteries is intermittent and is driven by the elastic recoil of the aortic wall following a contraction of the left ventricle.

As was mentioned, several small structures called **aortic bodies** occur within the epithelial lining of the aortic sinuses. These bodies contain pressoreceptors that function to control blood pressure and chemoreceptors that are sensitive to the blood concentrations of oxygen and carbon dioxide.

Three major arteries originate from the *arch of the aorta* (aortic arch). They are the brachiocephalic (innominate) artery, the left common carotid artery, and the left subclavian artery.

The **brachiocephalic artery,** as its name suggests, supplies blood to the tissues of the arm and head. It is the first branch from the aortic arch, and rises upward through the mediastinum to a point near the junction of the sternum and the right clavicle. There it divides, giving rise to the right **common carotid artery,** which carries blood to the right side of the neck and head, and the right **subclavian artery,** which leads into the right arm. Branches of the subclavian artery also supply blood to parts of the shoulder, neck, and head.

The left *common carotid artery* and the left *subclavian artery* are respectively the second and third branches of the aortic arch. They supply blood to regions on the left side of the body corresponding to those supplied by their counterparts on the right.

Although the upper part of the *descending aorta* is positioned to the left of the midline, it gradually moves medially and finally lies directly in front of the vertebral column at the level of the twelfth thoracic vertebra.

The portion of the descending aorta above the diaphragm is known as the **thoracic aorta,** and it gives off numerous small branches to the thoracic wall and the thoracic visceral organs. These branches include *bronchial, pericardial,* and *esophageal arteries,* which supply blood to the structures for which they were named. Other branches become *mediastinal arteries,* supplying various tissues within the mediastinum, and *posterior intercostal arteries* that pass into the thoracic wall.

Below the diaphragm, the descending aorta becomes the **abdominal aorta,** and it gives off branches to the abdominal wall and various abdominal visceral organs. These branches include the following:

1. **Celiac artery.** This single vessel gives rise to the left *gastric, splenic,* and *hepatic arteries,* which supply upper portions of the digestive tube, the spleen, and the liver, respectively.

2. **Phrenic arteries.** These paired arteries supply blood to the diaphragm.

3. **Superior mesenteric artery.** The superior mesenteric is a large unpaired artery that branches to many parts of the intestinal tract, including the jejunum, ileum, cecum, ascending colon, and transverse colon.

4. **Suprarenal arteries.** This pair of vessels supplies blood to the adrenal glands.

5. **Renal arteries.** The renal arteries pass laterally from the aorta into the kidneys. Each artery then divides into several lobar branches within the kidney tissues. (See color plate 33.)

6. **Gonadal arteries.** In a female, paired *ovarian arteries* arise from the aorta and pass into the pelvis to supply the ovaries. In a male, *spermatic arteries* originate in similar locations. They course downward and pass through the body wall by way of the *inguinal canal* to supply the testes.

7. **Inferior mesenteric artery.** Branches of this single artery lead to the descending colon, the sigmoid colon, and the rectum.

8. **Lumbar arteries.** Three or four pairs of lumbar arteries arise from the posterior surface of the aorta in the region of the lumbar vertebrae. These arteries supply various muscles of the skin and the posterior abdominal wall.

9. **Middle sacral artery.** This small, single vessel descends medially from the aorta along the anterior surfaces of the lower lumbar vertebrae. It carries blood to the sacrum and coccyx.

The descending aorta terminates near the brim of the pelvis, where it divides into the right and left *common iliac arteries.* These vessels supply blood to the lower regions of the abdominal wall, the pelvic organs, and the lower extremities.

Chart 17.4 summarizes these main branches of the aorta.

If the aortic wall becomes weakened as a result of injury or disease, a blood-filled dilation or sac may appear in the vessel. This condition is called an *aortic aneurysm,* and it occurs most frequently in the abdominal aorta below the level of the renal arteries. An aneurysm usually increases in size with time, and the swelling is likely to rupture unless the affected portion of the artery is removed and replaced with a vessel graft.

Chart 17.4 Aorta and its principal branches

Portion of Aorta	Major Branch	General Regions or Organs Supplied
Ascending aorta	Right and left coronary arteries	Heart
Arch of aorta	Brachiocephalic artery	Right arm, right side of head
	Left common carotid artery	Left side of head
	Left subclavian artery	Left arm
Descending aorta		
Thoracic aorta	Bronchial artery	Bronchi
	Pericardial artery	Pericardium
	Esophageal artery	Esophagus
	Mediastinal artery	Mediastinum
	Intercostal artery	Thoracic wall
Abdominal aorta	Celiac artery	Organs of upper digestive system
	Phrenic artery	Diaphragm
	Superior mesenteric artery	Portions of small and large intestines
	Suprarenal artery	Adrenal gland
	Renal artery	Kidney
	Gonadal artery	Ovary or testis
	Inferior mesenteric artery	Lower portions of large intestine
	Lumbar artery	Posterior abdominal wall
	Middle sacral artery	Sacrum and coccyx
	Common iliac artery	Lower abdominal wall, pelvic organs, and leg

Arteries to the Neck and Head

Blood is supplied to parts within the neck and head through branches of the subclavian and common carotid arteries. (See fig. 17.58.) The main divisions of the subclavian artery to these regions are the vertebral, thyrocervical, and costocervical arteries. The common carotid communicates to these parts by means of the internal and external carotid arteries.

The **vertebral arteries** arise from the subclavian arteries in the base of the neck, near the tips of the lungs. They pass upward through the foramina of the transverse processes of the cervical vertebrae and enter the skull by way of the foramen magnum. Along their paths, these vessels supply blood to the vertebrae and to the ligaments and muscles associated with them.

Within the cranial cavity, the vertebral arteries unite to form a single *basilar artery*. This vessel passes along the ventral brain stem and gives rise to branches leading to the pons and cerebellum. The basilar artery terminates by dividing into two *posterior cerebral arteries* that supply portions of the occipital and temporal lobes of the cerebrum. The posterior cerebral arteries also help to form an arterial circle at the base of the brain, called the **circle of Willis,** that creates a connection between the vertebral artery and internal carotid artery systems. (See fig. 17.59.) The union of these systems provides alternate pathways through which blood can reach brain tissues in the event of an arterial occlusion.

The **thyrocervical arteries** are short vessels that give off branches to the thyroid glands, parathyroid glands, larynx, trachea, esophagus, and pharynx, as well as to various muscles in the neck, shoulder, and back.

The **costocervical arteries**, which are the third vessels to branch from the subclavians, carry blood to muscles in the neck, back, and thoracic wall.

The left and right *common carotid arteries* ascend deeply within the neck on either side. At the level of the upper laryngeal border, they divide to form the internal and external carotid arteries.

The **external carotid artery** courses upward on the side of the head, giving off branches to various structures in the neck, face, jaw, scalp, and base of the skull. The main vessels that originate from this artery include the following:

1. *Superior thyroid artery* to the hyoid bone, larynx, and thyroid gland.

2. *Lingual artery* to the tongue, muscles of the tongue, and salivary glands beneath the tongue.

3. *Facial artery* to the pharynx, palate, chin, lips, and nose.

4. *Occipital artery* to the scalp on the back of the skull, the meninges, the mastoid process, and various muscles in the neck.

5. *Posterior auricular artery* to the ear and scalp over the ear.

The external carotid artery terminates by dividing into *maxillary* and *superficial temporal arteries*. The maxillary artery supplies blood to the teeth, gums, jaws, cheek, nasal cavity, eyelids, and meninges. The temporal artery extends to the parotid salivary gland and to various surface regions of the face and scalp.

Fig. 17.58 The main arteries of the head and neck. (*a.* stands for *artery.*)

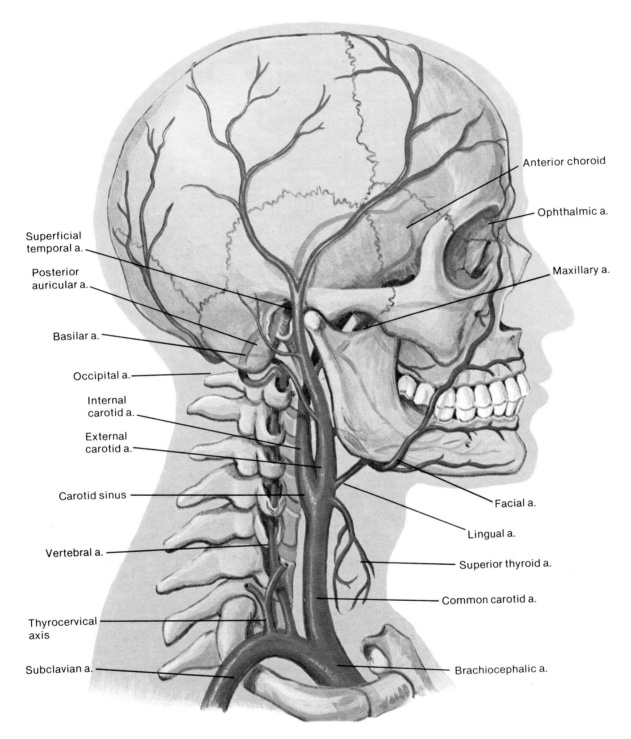

Superficial temporal a.

Posterior auricular a.

Basilar a.

Occipital a.

Internal carotid a.

External carotid a.

Carotid sinus

Vertebral a.

Thyrocervical axis

Subclavian a.

Anterior choroid

Ophthalmic a.

Maxillary a.

Facial a.

Lingual a.

Superior thyroid a.

Common carotid a.

Brachiocephalic a.

Fig. 17.59 The circle of Willis is formed by the anterior cerebral arteries, which are connected by the anterior communicating artery, and the posterior cerebral arteries, which are connected to the internal carotid arteries by the posterior communicating arteries. (*a.* stands for *artery*.)

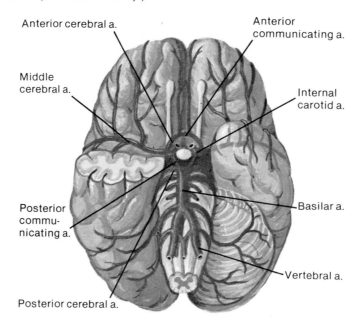

Anterior cerebral a.

Middle cerebral a.

Posterior communicating a.

Posterior cerebral a.

Anterior communicating a.

Internal carotid a.

Basilar a.

Vertebral a.

The **internal carotid artery** follows a deep course upward along the pharynx to the base of the skull. Entering the cranial cavity, it provides the major blood supply to the brain. Its major branches include the following:

1. *Ophthalmic artery* to the eyeball and various muscles and accessory organs within the orbit.

2. *Posterior communicating artery* that forms part of the circle of Willis.

3. *Anterior choroid artery* to the choroid plexus within the lateral ventricle of the brain and to a variety of nerve structures within the brain.

The internal carotid artery terminates by dividing into *anterior* and *middle cerebral arteries*. These vessels supply the interior of the cerebrum and the cerebral cortex.

Near the base of each internal carotid artery is an enlargement called a **carotid sinus.** Like the aortic sinuses, these structures contain pressoreceptors that function in the reflex control of blood pressure. A number of small epithelial masses, called **carotid bodies,** also occur in the wall of the carotid sinus. These bodies contain chemoreceptors that act with the chemoreceptors of the aortic bodies in regulating various circulatory and respiratory actions.

Arteries to the Shoulder and Arm

The subclavian artery, after giving off branches to the neck, continues into the upper arm. (See fig. 17.60.) It passes between the clavicle and the first rib and becomes the axillary artery.

The **axillary artery** supplies branches to structures in the axilla and the chest wall, including the skin of the shoulder, part of the mammary gland, the upper end of the humerus, the shoulder joint, and various muscles in the back, shoulder, and chest. As this vessel leaves the axilla, it becomes the brachial artery.

The **brachial artery** courses along the humerus to the elbow. It gives rise to a *deep brachial artery* that curves posteriorly around the humerus and supplies the triceps muscle. Shorter branches pass into the muscles on the anterior side of the upper arm, while others descend on each side to the elbow and interconnect with arteries in the forearm. The resulting arterial network allows blood to reach the lower arm even if some portion of the distal brachial artery becomes obstructed.

Within the elbow, the brachial artery divides into an ulnar and a radial artery. The **ulnar artery** leads downward on the ulnar side of the forearm to the wrist. Some of its branches join the anastomosis around the elbow joint, while others supply blood to flexor and extensor muscles in the lower arm.

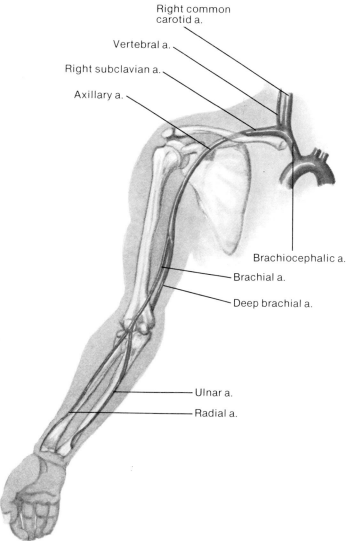

Right common
carotid a.

Vertebral a.

Right subclavian a.

Axillary a.

Brachiocephalic a.

Brachial a.

Deep brachial a.

Ulnar a.

Radial a.

The **radial artery,** a continuation of the brachial artery, travels along the radial side of the forearm to the wrist. As it nears the wrist, it comes close to the surface and provides a convenient vessel for taking the pulse (radial pulse).

Various branches of the radial artery join the anastomosis of the elbow and supply lateral muscles of the forearm.

At the wrist, branches of the ulnar and radial arteries join to form an interconnecting network of vessels. Arteries arising from this network supply blood to structures in the wrist, hand, and fingers.

Arteries to the Thoracic and Abdominal Walls

Blood reaches the thoracic wall through several vessels, including branches from the subclavian artery and the thoracic aorta. (See fig. 17.61.)

The subclavian artery contributes to this supply through a branch called the **internal thoracic artery.**

This vessel originates in the base of the neck and passes downward on the pleura and behind the cartilages of the upper six ribs. It gives off two *anterior intercostal arteries* to each of the upper six intercostal spaces, which supply the intercostal muscles, other intercostal tissues, and the mammary glands.

Posterior intercostal arteries arise from the thoracic aorta and enter the intercostal spaces between the third through the eleventh ribs. These arteries give off branches that supply the intercostal muscles, the vertebrae, the spinal cord, and various deep muscles of the back.

The blood supply to the anterior abdominal wall is provided primarily by branches of the *internal thoracic* and *external iliac arteries.* Structures in the posterior and lateral abdominal wall are supplied by paired vessels originating from the abdominal aorta, including the *phrenic* and *lumbar arteries* mentioned previously.

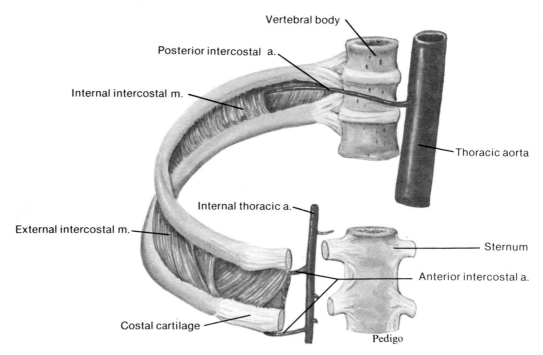

Fig. 17.61 Arteries that supply the thoracic wall. (*a.* stands for *artery*; *m.* for *muscle*.)

Vertebral body

Posterior intercostal a.

Internal intercostal m.

Thoracic aorta

External intercostal m.

Internal thoracic a.

Sternum

Anterior intercostal a.

Costal cartilage

Pedigo

Arteries to the Pelvis and Leg

The abdominal aorta divides to form the **common iliac arteries** at the level of the pelvic brim, and these vessels provide blood to the pelvic organs, gluteal region, and legs.

Each common iliac artery descends a short distance and divides into an internal (hypogastric) and an external branch. The **internal iliac artery** gives off numerous branches to various pelvic muscles and visceral structures, as well as to the gluteal muscles and the external genitalia. Important branches of this vessel include the following (see fig. 17.62):

1. *Iliolumbar artery* to the ilium and muscles of the back.

2. *Superior* and *inferior gluteal arteries* to the gluteal muscles, pelvic muscles, and skin of the buttocks.

3. *Internal pudendal artery* to muscles in the distal portion of the alimentary canal, the external genitalia, and the hip joint.

4. *Superior* and *inferior vesical arteries* to the urinary bladder. In males these vessels also supply the seminal vesicles and the prostate gland.

5. *Middle rectal artery* to the rectum.

6. *Uterine artery* to the uterus and vagina in females.

The **external iliac artery** provides the main blood supply to the legs. (See fig. 17.63.) It passes downward along the brim of the pelvis and gives off two large branches—an *inferior epigastric artery* and a *deep circumflex iliac artery*. These vessels supply muscles and skin in the lower abdominal wall.

Midway between the pubic symphysis and the anterior superior iliac spine of the ilium, the external iliac artery passes beneath the inguinal ligament and becomes the femoral artery.

The **femoral artery,** which passes fairly close to the anterior surface of the upper thigh, gives off many branches to the muscles and superficial tissues of the thigh. These branches also supply the skin of the groin and lower abdominal wall. Important subdivisions of the femoral artery include the following:

1. *Superficial circumflex iliac artery* to lymph nodes and skin of the groin.

2. *Superficial epigastric artery* to the skin of the lower abdominal wall.

3. *Superficial* and *deep external pudendal arteries* to the skin of the lower abdomen and the external genitalia.

4. *Profunda femoris artery,* which is the largest branch of the femoral artery, to the hip joint and various muscles of the thigh.

5. *Deep genicular artery* to distal ends of thigh muscles and to an anastomosis around the knee joint.

Fig. 17.62 Main branches of the internal iliac artery. (*a.* stands for *artery.*)

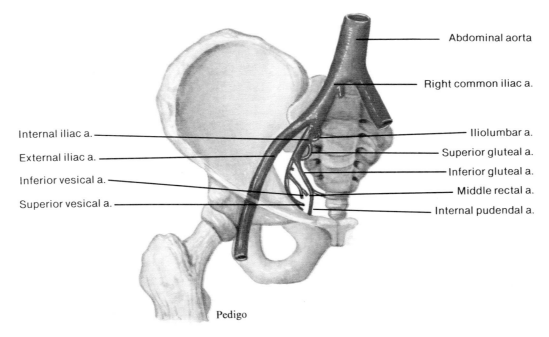

Abdominal aorta

Right common iliac a.

Internal iliac a.

External iliac a.

Inferior vesical a.

Superior vesical a.

Iliolumbar a.

Superior gluteal a.

Inferior gluteal a.

Middle rectal a.

Internal pudendal a.

Pedigo

As the femoral artery reaches the proximal border of the space behind the knee (popliteal fossa), it becomes the **popliteal artery.** Branches of this artery supply blood to the knee joint and to certain muscles in the thigh and calf. Also, many of its branches join the anastomosis of the knee and thus help to provide alternate pathways for blood in the case of arterial obstructions. At the lower border of the popliteal fossa, the popliteal artery divides into the anterior and posterior tibial arteries.

The **anterior tibial artery** passes downward between the tibia and the fibula, giving off branches to the skin and muscles in the anterior and lateral regions of the lower leg. It also communicates with the anastomosis of the knee and a network of arteries around the ankle. This vessel continues into the foot as the *dorsalis pedis artery,* which supplies blood to the foot and toes.

The **posterior tibial artery,** the larger of the two popliteal branches, descends beneath the calf muscles, giving off branches to skin, muscles, and other tissues of the lower leg along the way. Some of these vessels join the anastomoses of the knee and ankle.

The largest branch of the posterior tibial artery is the *peroneal artery,* which travels downward along the fibula and contributes to the anastomosis of the ankle. As it passes between the medial malleolus and the heel, this vessel divides into the *medial* and *lateral plantar arteries.* Branches from these arteries supply blood to tissues of the heel, foot, and toes.

The major vessels of the arterial system are shown in figure 17.64.

---

1. Describe the structure of the aorta.
2. Name the vessels that arise from the aortic arch.
3. Name the branches of the thoracic and abdominal aorta.
4. What vessels supply blood to the head? The arm? The abdominal wall? The leg?

Fig. 17.63 Main branches of the external iliac artery.
(*a.* stands for *artery.*)

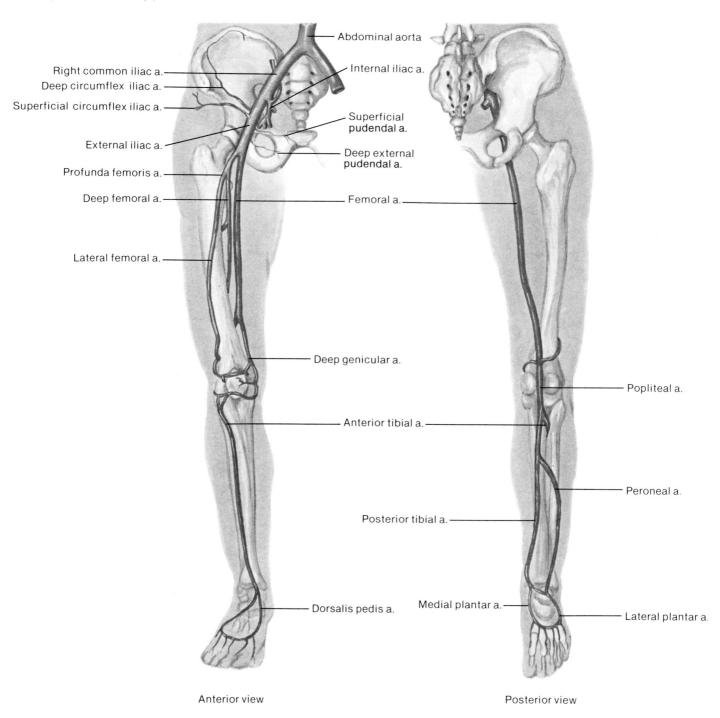

Abdominal aorta

Right common iliac a.

Deep circumflex iliac a.

Superficial circumflex iliac a.

Internal iliac a.

Superficial pudendal a.

External iliac a.

Deep external pudendal a.

Profunda femoris a.

Deep femoral a.

Femoral a.

Lateral femoral a.

Deep genicular a.

Popliteal a.

Anterior tibial a.

Peroneal a.

Posterior tibial a.

Dorsalis pedis a.

Medial plantar a.

Lateral plantar a.

Anterior view

Posterior view

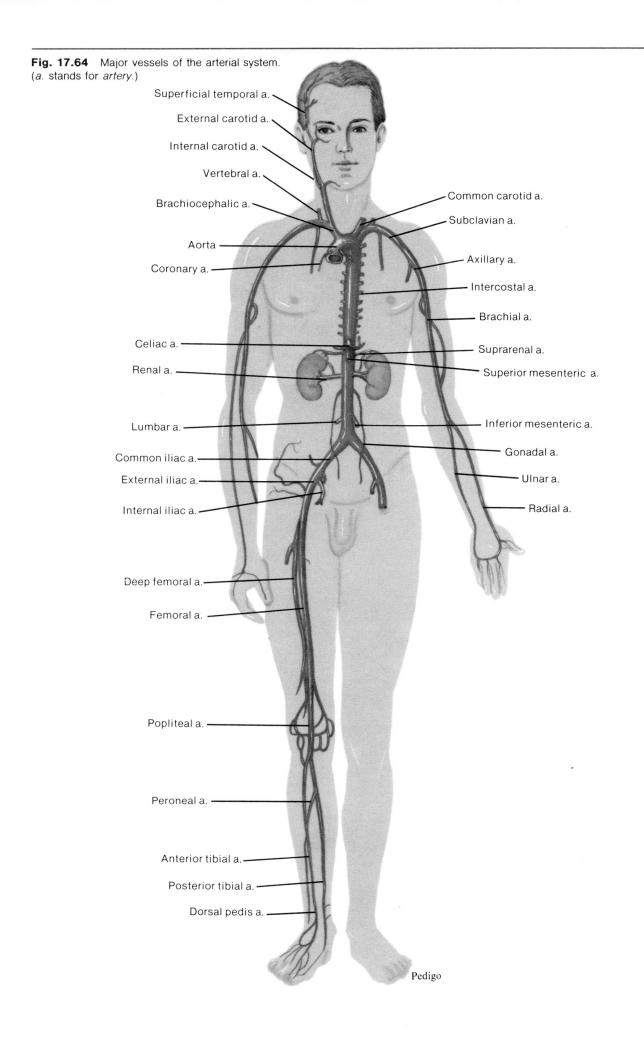

Fig. 17.64 Major vessels of the arterial system. (*a.* stands for *artery.*)

Superficial temporal a.

External carotid a.

Internal carotid a.

Vertebral a.

Brachiocephalic a.

Aorta

Coronary a.

Celiac a.

Renal a.

Lumbar a.

Common iliac a.

External iliac a.

Internal iliac a.

Deep femoral a.

Femoral a.

Popliteal a.

Peroneal a.

Anterior tibial a.

Posterior tibial a.

Dorsal pedis a.

Common carotid a.

Subclavian a.

Axillary a.

Intercostal a.

Brachial a.

Suprarenal a.

Superior mesenteric a.

Inferior mesenteric a.

Gonadal a.

Ulnar a.

Radial a.

Pedigo

The Venous System

Venous circulation is responsible for returning blood to the heart after exchanges of gases, nutrients, and wastes have been made between the blood and the body cells.

Characteristics of Venous Pathways

The vessels of the venous system begin with the merging of capillaries into venules, venules into small veins, and small veins into larger ones. Unlike arterial pathways, however, those of the venous system are difficult to follow. This is because the vessels are commonly interconnected in irregular networks, so that many unnamed tributaries may join to form a relatively large vein.

On the other hand, larger veins typically parallel the courses taken by named arteries, and these veins often have the same names as their companions in the arterial system. Thus, with some exceptions, the name of a major artery also provides the name of the vein next to it. For example, the renal vein parallels the renal artery, the common iliac vein accompanies the common iliac artery and so forth.

The veins that carry blood from the lungs and myocardium back to the heart have already been described. The veins from all other parts of the body converge into two major pathways that lead to the right atrium. They are the *superior* and *inferior venae cavae.*

Veins from the Head and Neck

Blood from the face, scalp, and superficial regions of the neck is drained by the **external jugular veins.** These vessels descend on either side of the neck, passing over the sternocleidomastoid muscles and beneath the platysma. They empty into the *right* and *left subclavian veins* in the base of the neck. (See fig. 17.65.)

The **internal jugular veins,** which are somewhat larger than the externals, arise from numerous veins and venous sinuses of the brain and from deep veins

Fig. 17.65 The main veins of the head and neck. (*v.* stands for *vein.*)

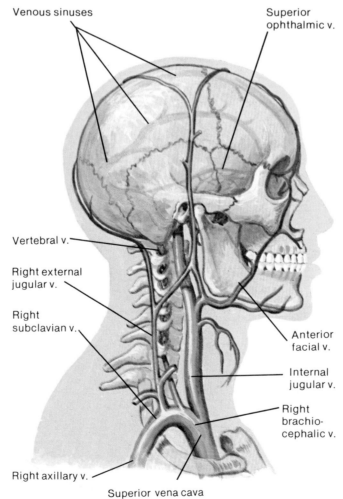

Venous sinuses

Superior ophthalmic v.

Vertebral v.

Right external jugular v.

Right subclavian v.

Anterior facial v.

Internal jugular v.

Right brachiocephalic v.

Right axillary v.

Superior vena cava

in various parts of the face and neck. They pass downward through the neck beside the common carotid arteries and also join the subclavian veins. These unions of the internal jugular and subclavian veins form large **brachiocephalic** (innominate) **veins** on each side. These vessels then merge in the mediastinum and give rise to the **superior vena cava,** which enters the right atrium.

Veins from the Arm and Shoulder

The arm is drained by a set of deep veins and a set of superficial ones. The deep veins generally parallel the arteries in each region and are given similar names, such as *radial vein, ulnar vein, brachial vein,* and *axillary vein.* The superficial veins are interconnected in complex networks just beneath the skin. They also communicate with the deep vessels of the arm, providing many alternate pathways through which blood can leave the tissues. (See fig. 17.66.)

The main vessels of the superficial network are the basilic and cephalic veins. They arise from anasto-

moses in the hand and wrist on the radial and ulnar sides, respectively.

The **basilic vein** passes along the back of the forearm on the ulnar side for a distance and then curves forward to the anterior surface below the elbow. It continues ascending on the medial side until it reaches the middle of the upper arm. There it penetrates the tissues deeply and joins the *brachial vein.* As the basilic and brachial veins merge, they form the *axillary vein.*

The **cephalic vein** courses upward on the lateral side of the arm from the hand to the shoulder. In the shoulder it pierces the tissues and empties into the axillary vein. Beyond the axilla, the axillary vein becomes the *subclavian vein.*

In the bend of the elbow, a *median cubital vein* ascends from the cephalic vein on the lateral side of the arm to the basilic vein on the medial side. This vein is often used as a site for *venipuncture,* when it is necessary to remove a sample of blood for examination or to add fluids to the blood.

Fig. 17.66 The main veins of the arm and shoulder. (*v.* stands for *vein.*)

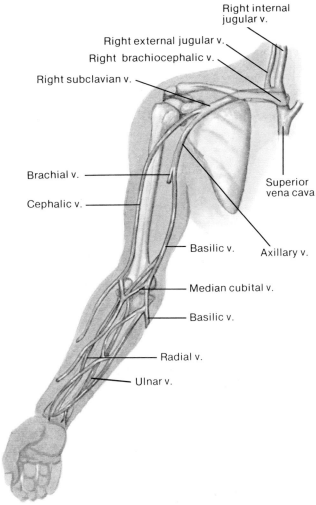

Fig. 17.67 Veins that drain the thoracic wall.
(v. stands for vein.)

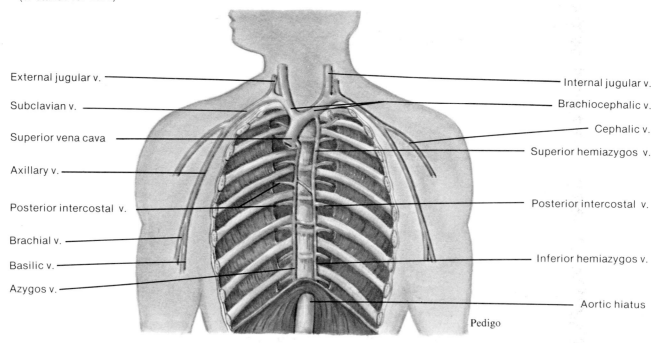

External jugular v.
Subclavian v.
Superior vena cava
Axillary v.
Posterior intercostal v.
Brachial v.
Basilic v.
Azygos v.

Internal jugular v.
Brachiocephalic v.
Cephalic v.
Superior hemiazygos v.
Posterior intercostal v.
Inferior hemiazygos v.
Aortic hiatus

Pedigo

Veins from the Abdominal and Thoracic Walls

The abdominal and thoracic walls are drained mainly by tributaries of the brachiocephalic and azygos veins. For example, the *brachiocephalic vein* receives blood from the *internal thoracic vein,* which generally drains the tissues supplied by the internal thoracic artery. Some *intercostal veins* also empty into the brachiocephalic vein. (See fig. 17.67.)

The **azygos vein** originates in the dorsal abdominal wall and ascends through the mediastinum on the right side of the vertebral column to join the superior vena cava. It drains most of the muscular tissue in the abdominal and thoracic walls.

Tributaries of the azygos vein include the *posterior intercostal veins,* which drain the intercostal spaces on the right side, and the *superior* and *inferior hemiazygos veins,* which connect to the posterior intercostal veins on the left. Right and left *ascending lumbar veins,* whose tributaries include vessels from the lumbar and sacral regions, also connect to the azygos system.

Veins from the Abdominal Viscera

Although veins usually carry blood directly to the atria of the heart, those that drain the abdominal viscera are exceptions. (See fig. 17.68.) They originate in the capillary networks of the stomach, intestines, pancreas, and spleen, and carry blood from these organs through a **portal vein** to the liver. There it enters capillarylike **hepatic sinusoids.** This unique venous pathway is called the **hepatic portal system.**

The tributaries of the portal vein include the following vessels:

1. Right and left *gastric veins* from the stomach.

2. *Superior mesenteric vein* from the small intestine, ascending colon, and transverse colon.

3. *Splenic vein* from a convergence of several veins draining the spleen, pancreas, and a portion of the stomach. Its largest tributary, the *inferior mesenteric vein,* brings blood upward from the descending colon, sigmoid colon, and rectum.

About 80% of the blood flowing to the liver in the hepatic portal system comes from capillaries in the stomach and intestines and is rich in nutrients. As is discussed in chapter 13, the liver acts on these nutrients in a variety of ways. For example, it monitors the blood glucose concentration and converts excesses into glycogen and fat for storage, or it releases glucose from glycogen if the blood glucose level drops below normal.

Plate 28   Describe the functions of each of the blood cells illustrated in this diagram.

Plate 29   Anterior view of a human heart.

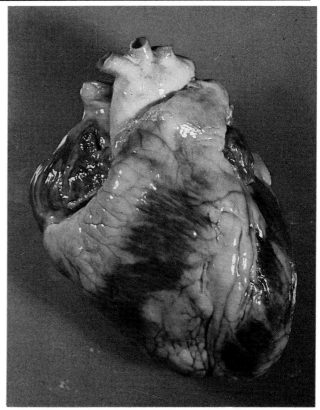

29

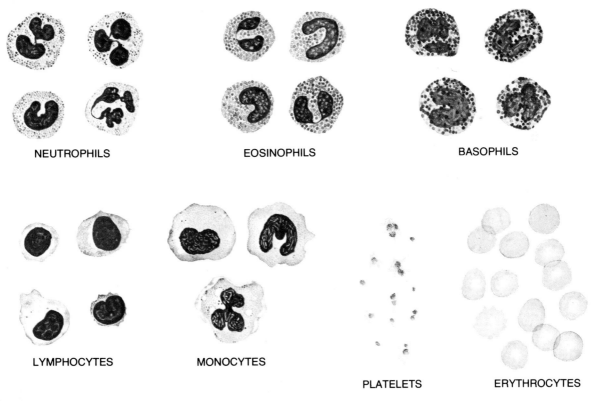

NEUTROPHILS   EOSINOPHILS   BASOPHILS

LYMPHOCYTES   MONOCYTES

PLATELETS   ERYTHROCYTES

K. Talaro

Plate 30   Cast of the blood vessels associated with the heart.

Plate 31   Cross section of an artery.

Plate 32   Cross section of a vein.

Plate 33   Cast of the renal arteries.

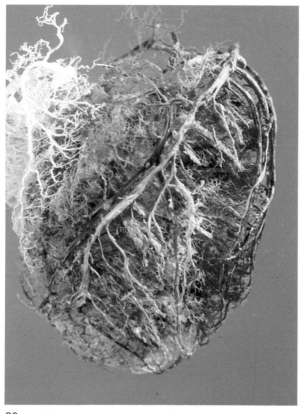

30

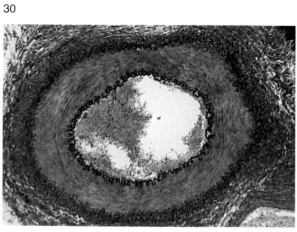

31

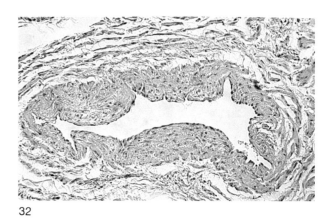

32

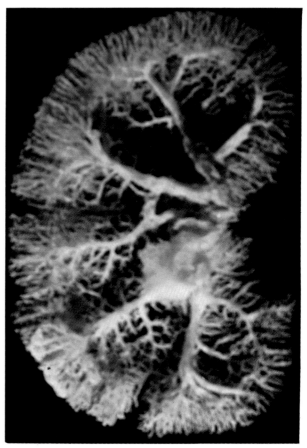

33

Fig. 17.68 Veins that drain the abdominal viscera.

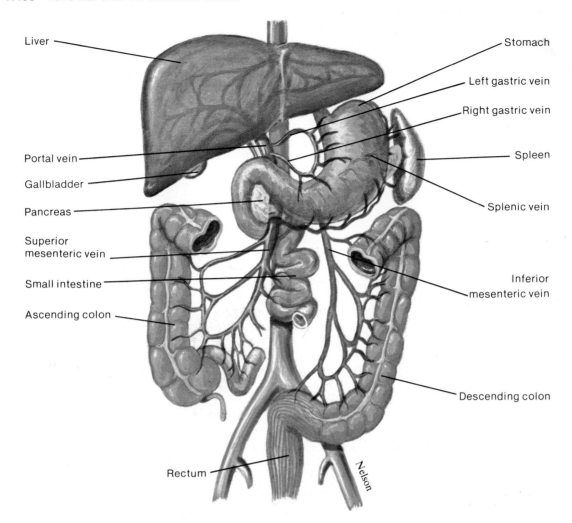

Liver

Portal vein

Gallbladder

Pancreas

Superior mesenteric vein

Small intestine

Ascending colon

Rectum

Stomach

Left gastric vein

Right gastric vein

Spleen

Splenic vein

Inferior mesenteric vein

Descending colon

Nelson

Similarly, the liver helps to regulate the blood concentrations of recently absorbed amino acids and lipids by modifying their molecules into forms usable by cells, oxidizing them, or changing them into storage forms. The liver also functions to store certain vitamins and to detoxify harmful substances.

The blood in the portal vein nearly always contains bacteria that have entered through the intestinal capillaries. These microorganisms are removed by the phagocytic action of large *Kupffer cells* lining the hepatic sinusoids. Consequently, nearly all of the bacteria have been removed from the portal blood before it leaves the liver.

After passing through the hepatic sinusoids of the liver, blood in the hepatic portal system is carried through a series of merging vessels into **hepatic veins.** These veins empty into the *inferior vena cava,* and thus the blood is returned to the general circulation.

Other veins also empty into the inferior vena cava as it ascends through the abdomen. They include the *lumbar, gonadal, renal, suprarenal,* and *phrenic veins.* Generally these vessels drain regions that are supplied by arteries with corresponding names.

Veins from the Leg and Pelvis

As in the arm, veins that drain blood from the leg can be divided into deep and superficial groups. (See fig. 17.69.)

The deep veins of the lower leg, such as the *anterior* and *posterior tibial veins,* have names that correspond with the arteries they accompany. At the level of the knee, these vessels form a single trunk, the **popliteal vein.** This vein continues upward through the thigh as the **femoral vein,** which in turn becomes the **external iliac vein** just behind the inguinal ligament.

Fig. 17.69   The main veins of the leg and pelvis.

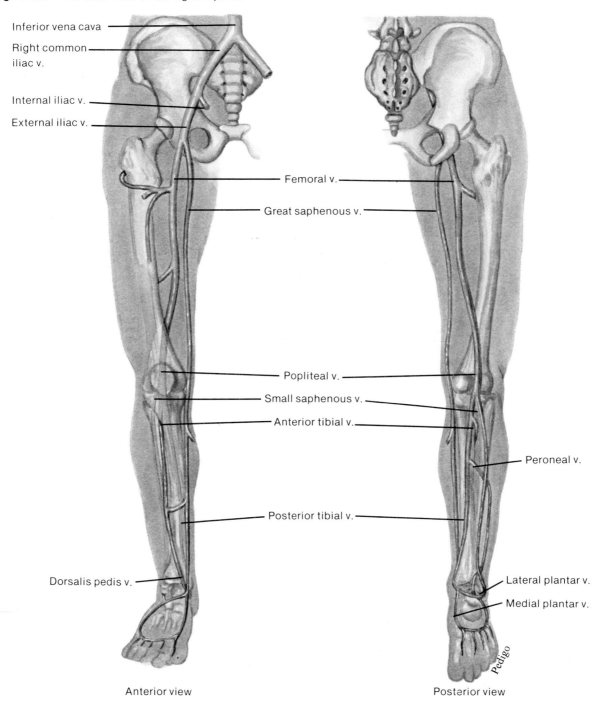

Inferior vena cava

Right common
iliac v.

Internal iliac v.

External iliac v.

Femoral v.

Great saphenous v.

Popliteal v.

Small saphenous v.

Anterior tibial v.

Peroneal v.

Posterior tibial v.

Dorsalis pedis v.

Lateral plantar v.

Medial plantar v.

Pedigo

Anterior view

Posterior view

The superficial veins of the foot and leg interconnect to form a complex network beneath the skin. These vessels drain into two major trunks—the small and great saphenous veins.

The **small saphenous vein** begins in the lateral portion of the foot and passes upward behind the lateral malleolus. It ascends along the back of the calf, enters the popliteal fossa, and joins the *popliteal vein.*

The **great saphenous vein**, which is the longest vein in the body, originates on the medial side of the foot. It ascends in front of the medial malleolus and extends upward along the medial side of the leg. In the thigh just below the inguinal ligament, it penetrates deeply and joins the femoral vein. Near its termination, the great saphenous vein receives tributaries from a number of vessels that drain the upper thigh, the groin, and the lower abdominal wall.

In addition to communicating freely with each other, the saphenous veins communicate extensively with the deep veins of the leg. As a result, there are many pathways by which blood can be returned to the heart from the lower extremities.

Whenever a person stands, the *saphenous veins* are subjected to increased blood pressure because of gravitational forces. When such pressure is prolonged, these veins are especially prone to developing abnormal dilations characteristic of varicose veins.

In the pelvic region, blood is carried away from organs of the reproductive, urinary, and digestive systems by vessels leading to the **internal iliac vein.** This vein is formed by tributaries corresponding to the branches of the internal iliac artery, such as the *gluteal, pudendal, vesical, rectal, uterine,* and *vaginal veins.* Typically, these veins have many interconnections and form complex networks (plexuses) in the regions of the rectum, urinary bladder, and prostate gland (in the male) or uterus and vagina (in the female).

The internal iliac veins originate deep within the pelvis and ascend to the pelvis brim. There they unite with the right and left *external iliac veins* to form the **common iliac veins.** These vessels in turn merge to produce the *inferior vena cava* at the level of the fifth lumbar vertebra.

Figure 17.70 shows the major vessels of the venous system.

---

1. Name the veins that return blood to the right atrium.
2. What major veins drain blood from the head? The arm? The abdominal viscera? The leg?

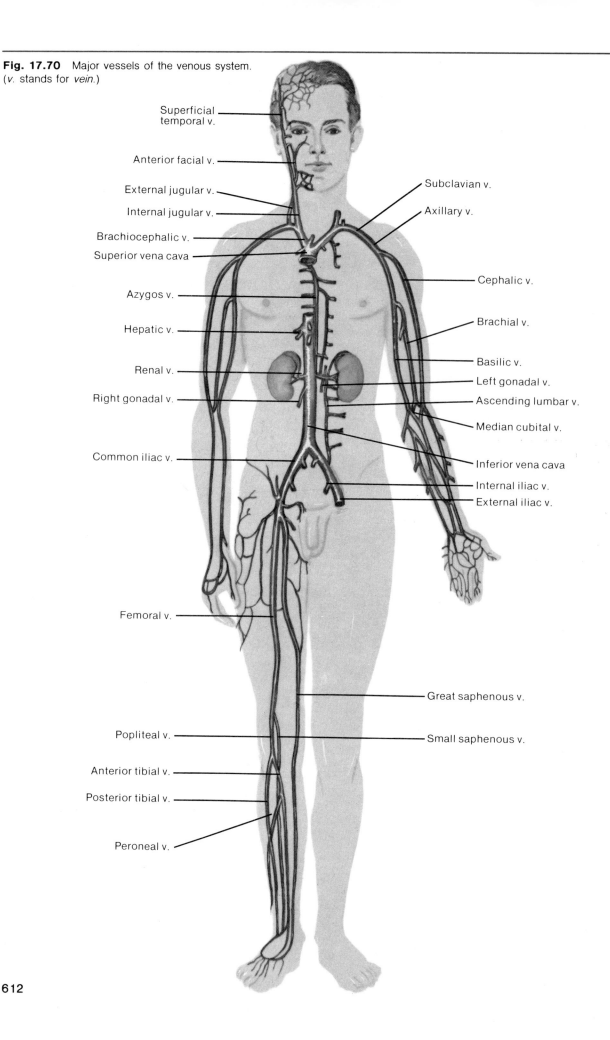

Fig. 17.70 Major vessels of the venous system. (*v.* stands for *vein.*)

Superficial temporal v.

Anterior facial v.

External jugular v.

Internal jugular v.

Brachiocephalic v.

Superior vena cava

Azygos v.

Hepatic v.

Renal v.

Right gonadal v.

Common iliac v.

Femoral v.

Popliteal v.

Anterior tibial v.

Posterior tibial v.

Peroneal v.

Subclavian v.

Axillary v.

Cephalic v.

Brachial v.

Basilic v.

Left gonadal v.

Ascending lumbar v.

Median cubital v.

Inferior vena cava

Internal iliac v.

External iliac v.

Great saphenous v.

Small saphenous v.

Some Clinical Terms Related to the Cardiovascular System

aneurysm (an′u-rizm)—a saclike swelling in the wall of a blood vessel, usually an artery.

angiocardiography (an″je-o-kar″de-og′rah-fe)—injection of radiopaque solution into the vascular system for X-ray examination of the heart and pulmonary circuit.

angiospasm (an′je-o-spazm″)—a muscular spasm in the wall of a blood vessel.

arteriography (ar″te-re-og′rah-fe)—injection of radiopaque solution into the vascular system for X-ray examination of arteries.

asystole (a-sis′to-le)—condition in which the myocardium fails to contract.

cardiac tamponade (kar′de-ak tam″po-nād′)—compression of the heart by an accumulation of fluid within the pericardial cavity.

congestive heart failure (kon-jes′tiv hart fāl′yer)—condition in which the heart is unable to pump an adequate amount of blood to the body cells.

cor pulmonale (kor pul-mo-na′le)—a heart-lung disorder characterized by pulmonary hypertension and hypertrophy of the right ventricle.

embolectomy (em″bo-lek′to-me)—removal of an embolus through an incision in a blood vessel.

endarterectomy (en″dar-ter-ek′to-me)—removal of the inner wall of an artery to reduce an arterial occlusion.

palpitation (pal″pĭ-ta′shun)—an awareness of a heartbeat that is unusually rapid, strong, or irregular.

pericardiectomy (per″ĭ-kar″de-ek′to-me)—an incision of the pericardium.

phlebitis (flĕ-bi′tis)—inflammation of a vein, usually in the legs.

phlebosclerosis (fleb″o-sklĕ-ro′sis)—abnormal thickening or hardening of the walls of veins.

phlebotomy (flĕ-bot′o-me)—incision of a vein for the purpose of withdrawing blood.

sinus rhythm (si′nus rithm)—normal cardiac rhythm regulated by the S-A node.

thrombophlebitis (throm″bo-flĕ-bi′tis)—formation of a blood clot in a vein in response to inflammation of the venous wall.

valvotomy (val-vot′o-me)—an incision of a valve.

venography (ve-nog′rah-fe)—injection of radiopaque solution into the vascular system for X-ray examination of veins.

Chapter Summary

Introduction

The cardiovascular system includes the heart, which pumps blood, and the blood vessels, which transport the blood.

The Heart

1. Size and location of the heart
   a. The heart is about 14 cm long and 9 cm wide.
   b. It is located within the mediastinum and rests on the diaphragm.
2. Coverings of the heart
   a. The heart is enclosed in a pericardium.
   b. The pericardial cavity is a potential space between the visceral and parietal membranes.
3. Wall of the heart
   a. The wall of the heart is composed of three layers.
   b. These layers include epicardium, myocardium, and endocardium.
4. Heart chambers and valves
   a. The heart is divided into four chambers—two atria and two ventricles—that communicate through atrioventricular orifices.
   b. Right chambers and valves
      (1) The right atrium receives blood from the venae cavae and coronary sinus.
      (2) The right atrioventricular orifice is guarded by the tricuspid valve.
      (3) The right ventricle pumps blood through the pulmonary artery.
      (4) The base of the pulmonary artery is guarded by a pulmonary semilunar valve.
   c. Left chambers and valves
      (1) The left atrium receives blood from the pulmonary veins.
      (2) The left atrioventricular orifice is guarded by the bicuspid valve.
      (3) The left ventricle pumps blood into the aorta.
      (4) The base of the aorta is guarded by an aortic semilunar valve.
5. Skeleton of the heart
   a. The skeleton of the heart consists of fibrous rings that enclose bases of the pulmonary artery, aorta, and atrioventricular orifices.
   b. The fibrous rings provide attachments for valves and muscle fibers and prevent the orifices from dilating excessively during ventricular contractions.

6. Path of blood through the heart
   a. Deoxygenated blood enters the right side of the heart from the venae cavae and is pumped into the pulmonary circulation.
   b. Blood is oxygenated in the lungs and returns to the left side of the heart through the pulmonary veins.
   c. From the left ventricle, it moves into the aorta.
7. Blood supply to the heart
   a. Blood is supplied through the coronary arteries.
   b. It is returned to the right atrium through the cardiac veins and coronary sinus.
8. The cardiac cycle
   a. Atria contract while ventricles relax; ventricles contract while atria relax.
   b. Pressure within the chambers rises and falls in repeated cycles.
9. Heart sounds
   a. Sounds can be described as *lub*-dup.
   b. Sounds are due to vibrations produced as a result of blood and valve movements.
   c. The first part of the sound occurs as A-V valves are closing and the second part is associated with the closing of semilunar valves.
10. Cardiac muscle fibers
    a. Fibers are interconnected to form a functional syncytium.
    b. If any part of the syncytium is stimulated, the whole structure contracts as a unit.
    c. Except for a small region in the floor of the right atrium, the atrial syncytium is separated from the ventricular syncytium by the fibrous skeleton.
11. Cardiac conduction system
    a. This system is composed of specialized muscle tissue and functions to initiate and conduct depolarization waves through the myocardium.
    b. Impulses from the S-A node pass to the A-V node, A-V bundle, and Purkinje fibers.
    c. Muscle fibers in the ventricular walls are arranged in whorls that squeeze blood out of the ventricles when they contract.
12. The electrocardiogram (ECG)
    a. The ECG is a recording of the electrical changes occurring in the myocardium during a cardiac cycle.
    b. The pattern contains several waves.
       (1) The P wave represents atrial depolarization.
       (2) The QRS complex represents ventricular depolarization.
       (3) The T wave represents ventricular repolarization.

13. Regulation of the cardiac cycle
    a. Heartbeat is affected by physical exercise, body temperature, and the concentration of various ions.
    b. S-A and A-V nodes are innervated by branches of sympathetic and parasympathetic nerve fibers.
       (1) Parasympathetic impulses cause heart action to decrease; sympathetic impulses cause heart action to increase.
       (2) Autonomic impulses are regulated by the cardiac center in the medulla oblongata.
14. Abnormal heart actions
    a. Abnormal rate or rhythm is called cardiac arrhythmia.
    b. Arrhythmias include tachycardia, bradycardia, ectopic beats, flutter, and fibrillation.
    c. If the S-A node becomes inactive, the A-V node may function as a secondary pacemaker.

Blood Vessels

The blood vessels form a closed circuit of tubes that transport blood between the heart and body cells.

Tubes include arteries, arterioles, capillaries, venules, and veins.

1. Arteries and arterioles
   a. Arteries are adapted to carry high pressure blood.
   b. Arterioles are branches of arteries.
   c. Walls of these vessels consist of layers of endothelium, smooth muscle, and connective tissue.
   d. The smooth muscles are innervated by autonomic fibers that can stimulate vasoconstriction or vasodilation.
2. Capillaries
   a. Capillaries form connections between arterioles and venules.
   b. The capillary wall consists of a single layer of cells that form a semipermeable membrane.
   c. Blood flow into the capillary is controlled by the precapillary sphincter.
3. Exchanges in capillaries
   a. Gases, nutrients, and metabolic by-products are exchanged between the capillary blood and tissue fluid.
   b. Diffusion provides the most important means of transport.
   c. Diffusion pathways depend upon lipid solubilities.
   d. Plasma proteins generally remain in the blood.
   e. Filtration, which is due to the hydrostatic pressure of blood, causes a net outward movement of fluid at the arterial end of a capillary.
   f. Osmosis causes a net inward movement of fluid at the venule end of the capillary.
   g. Some factors cause fluids to accumulate excessively in the tissues, causing edema.

4. Venules and veins
   a. Venules continue from capillaries and merge to form veins.
   b. Venous walls are similar to arterial walls, but are thinner and contain less muscle and elastic tissue.
5. Some blood vessel disorders
   a. Disorders include atherosclerosis, which causes nearly half of all deaths in the United States.
   b. Varicose veins are caused by back pressure in veins that forces them to dilate and makes their valves ineffective.
   c. Phlebitis, associated with injury or infection, may block normal circulation.

Blood Pressure

Blood pressure is the force exerted by the blood against the inside of the blood vessels.
1. Arterial blood pressure
   a. Arterial blood pressure is created primarily by heart action; it rises and falls with phases of the cardiac cycle.
   b. Systolic pressure occurs when the ventricle contracts; diastolic pressure occurs when the ventricle relaxes.
2. Measurement of arterial blood pressure
   a. Arterial blood pressure can be determined using a sphygmomanometer.
   b. Results are reported as a fraction with the systolic pressure above and the diastolic pressure below.
   c. Pulse pressure is the difference between the systolic and diastolic pressures; mean arterial pressure is approximated by adding diastolic pressure and ⅓ the pulse pressure.
3. Factors that influence arterial blood pressure
   a. Heart action, blood volume, resistance to flow, and blood viscosity influence arterial blood pressure.
   b. Arterial pressure increases as the cardiac output, blood volume, peripheral resistance, or blood viscosity increases.
4. Control of blood pressure
   a. Blood pressure is controlled in part by the mechanisms that regulate cardiac output and peripheral resistance.
   b. Cardiac output depends on the volume of blood discharged from the ventricle with each beat and the rate of heartbeat.
      (1) The more blood that enters the heart, the stronger the ventricular contraction, the greater the stroke volume, and the greater the cardiac output.
      (2) Heart rate is regulated by the cardiac center of the medulla oblongata.
   c. Regulation of peripheral resistance involves changes in the diameter of arterioles, which is controlled by the vasomotor center of the medulla oblongata.

5. Venous blood flow
   a. Venous blood flow is not a direct result of heart action; it depends on skeletal muscle contraction, breathing movements, and venoconstriction.
   b. Many veins contain flaplike valves that prevent blood from backing up.
6. Central venous pressure
   a. Central venous pressure is the pressure in the right atrium.
   b. It is influenced by factors that alter inflow of blood to the right atrium.
   c. It affects pressure within the peripheral veins.
7. High blood pressure
   a. The cause of hypertension is usually unknown, but it may be related to atherosclerosis or kidney disease.
      (1) Atherosclerosis is accompanied by a decrease in the elasticity of arteriole walls and a narrowing of the vessel lumens.
      (2) Kidney disease may stimulate the release of substances that promote a rise in the blood pressure.
   b. Hypertension enhances the development of atherosclerosis and kidney disease.

Paths of Circulation

1. The pulmonary circuit
   a. This circuit is composed of vessels that carry blood from the right ventricle to the lungs, the pulmonary capillaries, and the vessels that lead back to the left atrium.
   b. Pulmonary capillaries contain higher pressure than those of the systemic circuit.
   c. Tightly joined epithelial cells of alveoli walls prevent most substances from entering the alveoli.
   d. Water is rapidly drawn out of the alveoli into the interstitial fluid by osmotic pressure, so alveoli remain dry.
2. The systemic circuit
   a. This circuit is composed of vessels that lead from the heart to the body cells and back to the heart.
   b. It includes the aorta and its branches.

The Arterial System

1. Principal branches of the aorta
   a. Branches of the ascending aorta include the right and left coronary arteries.
   b. Branches of the aortic arch include the brachiocephalic, left common carotid, and left subclavian arteries.
   c. Branches of the descending aorta include thoracic and abdominal groups.
   d. The descending aorta terminates by dividing into right and left common iliac arteries.
2. Arteries to the neck and head
   These include branches of the subclavian and common carotid arteries.

3. Arteries to the shoulder and arm
   a. The subclavian artery passes into the upper arm and in various regions is called the axillary and the brachial artery.
   b. Branches of the brachial artery include the ulnar and radial arteries.
4. Arteries of the thoracic and abdominal walls
   a. The thoracic wall is supplied by branches of the subclavian artery and thoracic aorta.
   b. The abdominal wall is supplied by branches of the abdominal aorta and other arteries.
5. Arteries of the pelvis and leg
   The common iliac artery supplies the pelvic organs, gluteal region, and leg.

The Venous System

1. Characteristics of venous pathways
   a. Veins are responsible for returning blood to the heart.
   b. Larger veins usually parallel the courses of major arteries.
2. Veins from the head and neck
   a. These regions are drained by the jugular veins.
   b. Jugular veins unite with subclavian veins to form the brachiocephalic veins.
3. Veins from the arm and shoulder
   a. The arm is drained by sets of superficial and deep veins.
   b. Major superficial veins are the basilic and cephalic veins.
   c. The median cubital vein in the bend of the elbow is often used as a site for venipuncture.
4. Veins from the abdominal and thoracic walls
   These are drained by tributaries of the brachiocephalic and azygos veins.
5. Veins from the abdominal viscera
   a. Blood from the abdominal viscera generally enters the hepatic portal system and is carried to the liver.
   b. Blood in the portal system is rich in nutrients.
   c. Liver functions to regulate the blood levels of glucose, amino acids, and lipids.
   d. Phagocytic cells in the liver remove bacteria from the portal blood.
   e. From the liver, the blood is carried by hepatic veins to the inferior vena cava.
6. Veins from the leg and pelvis
   a. These are drained by sets of deep and superficial veins.
   b. Deep veins include the tibial veins, and the superficial veins include the saphenous veins.

Application of Knowledge

1. Based upon your understanding of the way capillary blood flow is regulated, do you think it is wiser to rest or to exercise following a heavy meal? Give a reason for your answer.
2. If a patient develops a blood clot in the femoral vein of the left leg and a portion of the clot breaks loose, where is the blood flow likely to carry the embolus? What symptoms is this condition likely to produce?
3. In the case of a patient with a heart weakened by damage to the myocardium, would it be better to keep the legs raised or lowered below the level of the heart? Why?

Review Activities

1. Describe the general structure, function, and location of the heart.
2. Distinguish between the visceral pericardium and the parietal pericardium.
3. Compare the layers of the cardiac wall.
4. Identify and describe the location of the chambers and the valves of the heart.
5. Describe the skeleton of the heart and explain its function.
6. Trace the path of blood through the heart.
7. Trace the path of blood through the coronary circulation.
8. Describe a cardiac cycle.
9. Describe the pressure changes that occur in the atria and ventricles during a cardiac cycle.
10. Explain the origin of heart sounds.
11. Describe the arrangement of the cardiac muscle fibers.
12. Distinguish between the S-A node and A-V node.
13. Explain how the cardiac conduction system functions in the control of the cardiac cycle.
14. Describe a normal ECG pattern and explain the significance of its various waves.
15. Discuss how the nervous system functions in the regulation of the cardiac cycle.
16. Explain how plasma levels of potassium and calcium affect the cardiac cycle.
17. Define *cardiac arrhythmia* and describe several forms.

18. Distinguish between an artery and an arteriole.

19. Explain how vasoconstriction and vasodilation are controlled.

20. Describe the structure and function of a capillary.

21. Explain how the blood flow through a capillary is controlled.

22. Explain how diffusion functions in the exchange of substances between plasma and tissue fluid.

23. Explain why water and dissolved substances leave the arteriole end of a capillary and enter the venule end.

24. Describe the effect of histamine on a capillary.

25. Define *edema* and identify some possible causes of this condition.

26. Distinguish between a venule and a vein.

27. Explain how veins function as blood reservoirs.

28. Distinguish between atherosclerosis and arteriosclerosis.

29. Discuss some of the factors that increase susceptibility to atherosclerosis.

30. Describe varicose veins and explain why they occur.

31. Distinguish between systolic and diastolic blood pressures.

32. Name several factors that influence blood pressure and explain how each produces its effect.

33. Describe how blood pressure is controlled.

34. List the factors that promote the flow of venous blood.

35. Define *central venous pressure*.

36. Define *hypertension* and explain some possible causes of this condition.

37. Distinguish between the pulmonary and systemic circuits of the cardiovascular system.

38. Trace the path of blood through the pulmonary circuit.

39. Explain why the alveoli normally remain dry.

40. Describe the aorta and name its principal branches.

41. On a diagram, locate and identify the major arteries that supply the abdominal visceral organs.

42. On a diagram, locate and identify the major arteries that supply parts in the head and neck.

43. On a diagram, locate and identify the major arteries that supply parts in the shoulder and arm.

44. On a diagram, locate and identify the major arteries that supply parts in the thoracic and abdominal walls.

45. On a diagram, locate and identify the major arteries that supply parts in the pelvis and leg.

46. Describe the relationship between the major venous pathways and the major arterial pathways.

47. On a diagram, locate and identify the major veins that drain parts in the head and neck.

48. On a diagram, locate and identify the major veins that drain parts in the arm and shoulder.

49. On a diagram, locate and identify the major veins that drain parts in the abdominal and thoracic walls.

50. On a diagram, locate and identify the major veins that drain parts of the abdominal viscera.

51. Review the actions of the liver on nutrients carried in the portal veins.

52. On a diagram, locate and identify the major veins that drain parts of the leg and pelvis.

Suggestions for Additional Reading

Baez, S. 1977. Microcirculation. *Ann. Rev. Physio.* 39:391.

Benditt, E. P. February 1977. The origin of atherosclerosis. *Scientific American.*

Berne, R. M., and Levy, M. N. 1977. *Cardiovascular physiology,* 3rd ed. St. Louis: C. V. Mosby.

Braunwald, E. 1974. *The myocardium: failure and infarction.* New York: H. P. Publishing.

Burton, A. C. 1972. *Physiology and biophysics of the circulation.* Chicago: Year Book Medical Publishers.

DeBakey, M., and Gotton, A. 1977. *The living heart.* New York: David McKay Co.

Fozzard, H. A. 1977. Heart: excitation-contraction coupling. *Ann. Rev. Physio.* 39:201.

Genest, J. August 1974. Basic mechanisms in benign essential hypertension. *Hosp. Prac.*

Hurst, J. W., ed. 1974. *The heart,* 3rd ed. New York: McGraw-Hill.

Ross, R., and Glomset, J. A. 1973. Atherosclerosis and the arterial smooth muscle cell. *Science.* 180:1332.

Rushmer, R. F. 1976. *Structure and function of the cardiovascular system,* 2nd ed. Philadelphia: W. B. Saunders.

Scheuer, J., and Tipton, C. M. 1977. Cardiovascular adaptation to physical training. *Ann. Rev. Physio.* 39:221.

Westfall, V. A. February 1976. Electrical and mechanical events in the cardiac cycle. *Amer. J. Nurs.*

The Lymphatic System

18 During the exchange of substances between the blood and tissue fluid, more fluid leaves the capillaries than returns to them. If the fluid remaining in the interstitial spaces were allowed to accumulate, the hydrostatic pressure in tissues would increase. The *lymphatic system* helps to prevent such an imbalance by providing pathways through which tissue fluid can be transported as lymph from the interstitial spaces to veins, where it becomes part of the blood.

The lymphatic system also helps to defend the tissues against infections by filtering particles from the lymph and by supporting the activities of lymphocytes that furnish immunity against specific disease-causing agents.

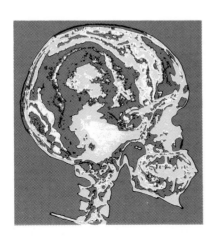

Chapter Outline

Chapter Objectives

After you have studied this chapter, you should be able to

1. Describe the general functions of the lymphatic system.

2. Describe the location of the major lymphatic pathways.

3. Describe how tissue fluid and lymph are formed and explain the function of lymph.

4. Explain how lymphatic circulation is maintained and the consequence of lymphatic obstruction.

5. Describe a lymph node and its major functions.

6. Describe the location of the major chains of lymph nodes.

7. Discuss the functions of the thymus and spleen.

8. Distinguish between specific and nonspecific body defenses and provide examples of each.

9. Explain how lymphocytes are formed and how they function in immune mechanisms.

10. Name the major types of immunoglobulins and discuss their origins and functions.

11. Distinguish between primary and secondary immune responses.

12. Distinguish between active and passive immunity.

13. Explain how allergic reactions and tissue rejection reactions are related to immune mechanisms.

14. Complete the review activities at the end of this chapter.

allergen (al'er-jen)

antibody (an'tĭ-bod''e)

antigen (an'tĭ-jen)

autoimmune (aw''to-ı-mūn')

complement (kom'plĕ-ment)

immunity (ĭ-mu'nĭ-te)

immunoglobulin (im''u-no-glob'u-lin)

interferon (in''ter-fēr'on)

lymph (limf)

lymphatic pathway (lim-fat'ik path'wa)

lymph node (limf nōd)

lymphocyte (lim'fo-sīt)

macrophage (mak'ro-fāj)

pathogen (path'o-jen)

reticuloendothelial tissue
  (rĕ-tik''u-lo-en''do-the'le-al tish'u)

spleen (splēn)

thymus (thi'mus)

vaccine (vak'sēn)

auto-, self: *auto*immune disease—condition in which the immune system attacks the body's own tissues.

gen-, to be produced: aller*gen*—substance that stimulates an allergic response.

humor-, fluid: *humor*al immunity—immunity resulting from soluble antibodies in body fluids.

immun-, free: *immun*ity—resistance to (freedom from) a specific disease.

inflamm-, setting on fire: *inflamm*ation—a condition characterized by localized redness, heat, swelling, and pain in the tissues.

nod-, knot: *nod*ule—a small mass of lymphocytes surrounded by connective tissue.

patho-, disease: *patho*gen—a disease-causing agent.

The lymphatic system is closely associated with the *cardiovascular system,* since it includes a network of vessels that assist in circulating body fluids. These vessels transport excess fluid away from the interstitial spaces and return it to the bloodstream. (See fig. 18.1.)

Lymphatic Pathways

The lymphatic pathways begin as lymphatic capillaries. These tiny tubes merge to form larger lymphatic vessels, and they, in turn, lead to collecting ducts that unite with veins in the thorax.

Lymphatic Capillaries

Lymphatic capillaries are microscopic, closed-ended tubes that extend into the interstitial spaces of most tissues, forming complex networks. (See fig. 18.2.) The walls of these vessels, like those of blood capillaries, consist of a single layer of squamous epithelial cells. This thin wall makes it possible for tissue fluid from the interstitial space to enter the lymphatic capillary. Once the fluid is inside the capillary, it is called **lymph.**

The villi of the small intestine contain specialized lymphatic capillaries called *lacteals,* described in chapter 13. These vessels are responsible for transporting recently absorbed fats away from the digestive tract.

Lymphatic Vessels

Lymphatic vessels, which are formed by merging lymphatic capillaries, have walls similar to those of veins. That is, their walls are composed of three layers—an endothelial lining, a middle layer of smooth muscle and elastic fibers, and an outer layer of connective tissue. Also like veins, the lymphatic vessels have flaplike *valves* that help to prevent the backflow of lymph. Figure 18.3 shows one of these valves.

Typically, lymphatic vessels lead to specialized organs called **lymph nodes,** and after leaving these structures, the vessels merge to form still larger lymphatic trunks.

Fig. 18.2 Lymph capillaries are microscopic closed-ended tubes that begin in the interstitial spaces of most tissues.

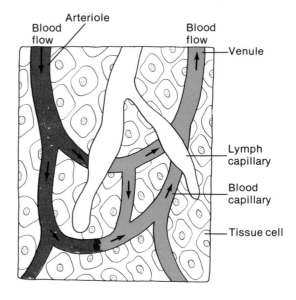

Fig. 18.3 A photomicrograph of the flaplike valve within a lymphatic vessel. What is the function of this valve?

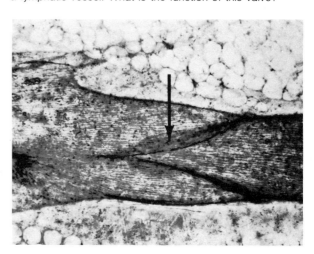

Fig. 18.1 Lymphatic vessels transport fluid from interstitial spaces to the bloodstream.

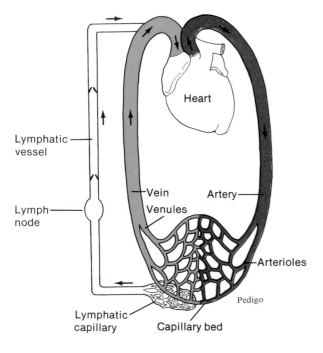

Lymphatic Trunks and Collecting Ducts

The **lymphatic trunks** drain lymph from relatively large regions of the body, and they are named for the regions they serve. For example, the *lumbar trunk* drains lymph from the legs, lower abdominal wall, and the pelvic organs; the *intestinal trunk* drains organs of the abdominal viscera; the *intercostal* and *bronchomediastinal trunks* receive lymph from portions of the thorax; the *subclavian trunk* drains the arm; and the *jugular trunk* drains portions of the neck and head. These lymphatic trunks then join one of two **collecting ducts**—the thoracic duct or the right lymphatic duct. Figure 18.4 shows the location of the major lymphatic trunks and collecting ducts, and figure 18.5 shows a lymphangiogram or X ray of the lymphatic pathways.

The **thoracic duct** is the larger and longer of the collecting ducts. It begins in the abdomen, passes upward through the diaphragm beside the aorta, ascends in front of the vertebral column through the mediastinum, and empties into the left subclavian vein near the junction of the left jugular vein. This duct drains lymph from the intestinal, lumbar, and intercostal trunks, as well as from the left subclavian, left jugular, and left bronchomediastinal trunks.

The **right lymphatic duct** originates in the right thorax by the union of the right jugular, right subclavian, and right bronchomediastinal trunks. It empties into the right subclavian vein near the junction of the right jugular vein.

After leaving the two collecting ducts, lymph enters the venous system and becomes part of the plasma just before the blood returns to the right atrium.

Fig. 18.4 Lymphatic vessels merge into larger lymphatic trunks, which in turn drain into collecting ducts.

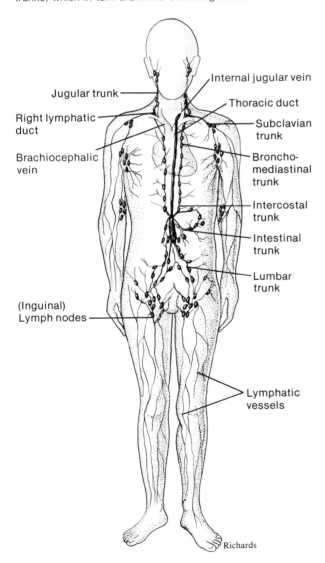

Jugular trunk

Right lymphatic duct

Brachiocephalic vein

Internal jugular vein

Thoracic duct

Subclavian trunk

Broncho-mediastinal trunk

Intercostal trunk

Intestinal trunk

Lumbar trunk

(Inguinal) Lymph nodes

Lymphatic vessels

Richards

Fig. 18.5 Locate the lymphatic vessels and lymph nodes in this lymphangiogram of the pelvic region.

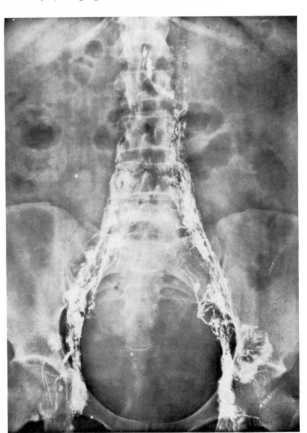

Chart 18.1 Typical lymphatic pathway

Tissue fluid
  leaves the interstitial space and becomes
Lymph
  as it enters the
Lymphatic capillary,
  which merges with other capillaries to form the
Afferent lymphatic vessel,
  which enters the
Lymph node,
  where lymph is filtered and leaves via the
Efferent lymphatic vessel,
  which merges with other vessels to form the
Lymphatic trunk,
  which merges with other trunks and joins the
Collecting duct,
  which empties into the
Subclavian vein,
  where lymph is added to the blood.

To summarize: Lymph from the lower body regions, left arm, and left side of the head and neck enters the thoracic duct; lymph from the right side of the head and neck, right arm, and right thorax enters the right lymphatic duct. (See fig. 18.6.) Chart 18.1 traces a typical lymphatic pathway.

The skin is richly supplied with lymphatic capillaries. Consequently, if the skin is broken or something is injected into the skin, such as venom from a stinging insect, foreign particles are likely to enter the lymphatic system relatively rapidly.

1. What is the general function of the lymphatic system?
2. Through what lymphatic vessels would some lymph pass in traveling from a leg to the bloodstream?

Tissue Fluid and Lymph

Lymph is essentially tissue fluid that has entered a lymphatic capillary. The formation of lymph is, then, closely associated with the formation of tissue fluid.

Tissue Fluid Formation

As is explained in chapter 16, *tissue fluid* originates from blood plasma. This fluid is composed of water and dissolved substances that leave the capillaries as a result of diffusion and filtration.

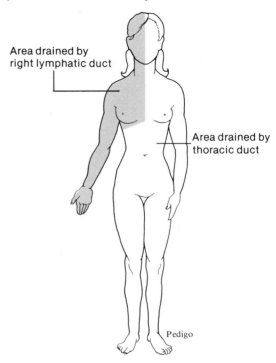

Fig. 18.6 The right lymphatic duct drains lymph from the upper right side of the body, while the thoracic duct drains lymph from the rest of the body.

Area drained by right lymphatic duct

Area drained by thoracic duct

Pedigo

Although tissue fluid contains various nutrients and gases found in plasma, it generally lacks proteins of large molecular size. Some proteins with smaller molecules do leak out of the capillaries and enter the interstitial space. Usually, these proteins are not reabsorbed when water and other dissolved substances move back into the venule ends of capillaries by diffusion and osmosis. As a result, the protein concentration of the tissue fluid tends to rise, causing the *osmotic pressure* of the fluid to rise also.

Lymph Formation

As the osmotic pressure of the tissue fluid rises, it interferes with the osmotic reabsorption of water by the capillaries. The volume of fluid in the interstitial spaces then tends to increase, as does the pressure within the spaces.

It is thought that this increasing interstitial pressure is responsible for forcing some of the tissue fluid into the lymphatic capillaries, where it becomes lymph. (See fig. 18.7.)

Fig. 18.7 As tissue fluid enters a lymphatic capillary, it becomes lymph. What factors are responsible for this movement of tissue fluid?

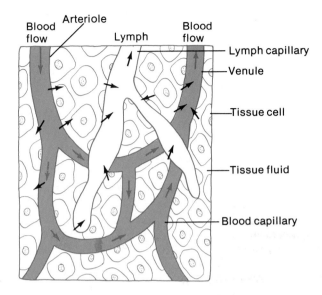

Fig. 18.8 Tissue fluid enters lymphatic capillaries through flaplike valves between adjacent epithelial cells.

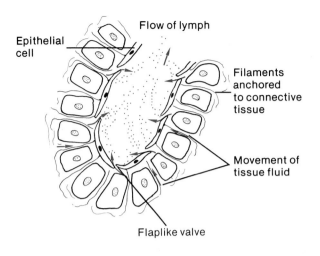

Function of Lymph

Most of the protein molecules that leak out of the blood capillaries are carried away by lymph and are returned to the bloodstream. At the same time, lymph transports various foreign particles, such as bacterial cells or viruses that may have entered the tissue fluids, to lymph nodes.

Although these proteins and foreign particles cannot easily enter blood capillaries, the lymphatic capillaries are especially adapted to receive them. Specifically, the endothelial cells that form the walls of these vessels are arranged so that the edge of one cell overlaps the edge of an adjacent cell, but is not attached to it. This arrangement, shown in figure 18.8, creates flaplike valves in the lymphatic capillary wall, which are pushed inward when the pressure is greater on the outside of the capillary, but close when the pressure is greater on the inside.

The epithelial cells of the lymphatic capillary wall are also attached to surrounding connective tissue cells by thin filaments, so that the lumen of a lymphatic capillary remains open even when the pressure outside is increased.

---

1. *How would you explain the relationship between tissue fluid and lymph?*
2. *How does the presence of protein in tissue fluid affect the formation of lymph?*
3. *What are the major functions of lymph?*

Movement of Lymph

Although the entrance of lymph into the lymphatic capillaries is influenced by the *osmotic pressure* of tissue fluid, the movement of lymph through the lymphatic vessels is controlled largely by *muscular activity.*

Flow of Lymph

Lymph, like venous blood, is under relatively low pressure and may not flow readily through the lymphatic vessels without the aid of outside forces. These forces include contraction of skeletal muscles, pressure changes due to the action of breathing muscles, and contraction of smooth muscles in the walls of larger lymphatic vessels.

As *skeletal muscles* contract, they compress lymphatic vessels. This squeezing action causes the lymph inside a vessel to move, but since lymphatic vessels contain valves that prevent backflow, the lymph can only move toward a collecting duct. Similarly, the smooth muscles in the walls of larger lymphatic vessels may contract and compress the lymph inside. This action also helps to force the fluid onward.

Respiratory muscles aid the circulation of lymph (as they do that of venous blood) by creating relatively low pressure in the thorax during inhalation. At the same time, the pressure in the abdominal cavity is increased by the contracting diaphragm. Consequently, lymph (and venous blood as well) is squeezed out of the abdominal vessels and into the thoracic vessels. Once again, backflow of lymph (and blood) is prevented by valves within the vessels.

Since the actions of skeletal and breathing muscles promote the circulation of lymph, it is not surprising that the flow of lymph is greatest during periods of physical exercise.

Obstruction of Lymph Movement

Because of the continuous movement of fluid from interstitial spaces into blood capillaries and lymphatic capillaries, the volume of fluid in these spaces remains stable. Conditions sometimes occur, however, that interfere with lymph movement, and tissue fluids accumulate in the spaces, causing a form of *edema*.

Lymphatic vessels may be obstructed as a result of surgical procedures in which portions of the lymphatic system are removed. Since the affected pathways can no longer drain lymph from the tissues, proteins tend to accumulate in the interstitial spaces. This causes an increase in the osmotic pressure of the tissue fluid, which in turn promotes the accumulation of water within the tissues.

Since lymphatic vessels tend to carry particles away from tissues, these vessels may also transport cancer cells and promote their spread to other sites (metastasis). For this reason, the lymphatic tissues (lymph nodes) in the axillary regions are commonly excised during the surgical removal of cancerous breast tissue (mastectomy). As a result, lymphatic drainage from the arm and other nearby tissues may be obstructed, and these parts may become edematous following surgery.

1. What factors promote the flow of lymph?
2. What is the consequence of lymphatic obstruction?

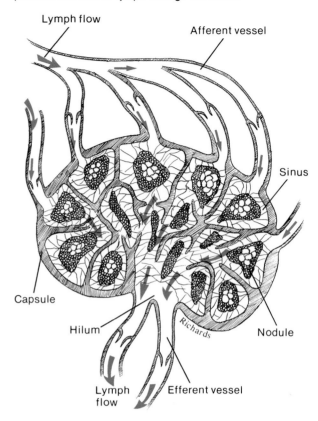

Fig. 18.9 A section of a lymph node. What factors promote the flow of lymph through the node?

Lymph flow

Afferent vessel

Sinus

Capsule

Hilum

Richards

Nodule

Lymph flow

Efferent vessel

Lymph Nodes

Lymph nodes (lymph glands) are structures located along the lymphatic pathways. They contain large numbers of *lymphocytes* that are vital in the defense against invasion by microorganisms.

Structure of a Lymph Node

Lymph nodes vary in size and shape; however, they are usually less than 2.5 cm (1 in) in length and are somewhat bean shaped. A section of a typical lymph node is illustrated in figure 18.9.

The indented region of a bean-shaped node is called the **hilum,** and it is the portion through which blood vessels enter the structure. The lymphatic vessels leading to a node (afferent vessels) enter separately at various points on its convex surface, while the lymphatic vessels leaving the node (efferent vessels) exit from the hilum along with the blood vessels.

Each lymph node is enclosed by a capsule of white fibrous connective tissue. This tissue also extends into the node and partially divides it into compartments that contain dense masses of lymphocytes. These masses, called **nodules,** represent the structural units of the node.

Spaces within the nodules, called **lymph sinuses,** provide channels for lymph circulation as it passes through the node.

Lymph then enters a lymph node through an *afferent* lymphatic vessel, moves slowly through the lymph sinuses, and leaves through an *efferent* lymphatic vessel.

Sometimes lymphatic vessels become inflamed due to a bacterial infection. When this happens in superficial vessels, painful reddish streaks may appear beneath the skin—for example, in an arm or leg. This condition is called *lymphangitis,* and it is usually followed by *lymphadenitis,* an inflammation of the lymph nodes. Affected nodes may become greatly enlarged and quite painful.

Nodules also occur singly or in groups associated with the mucous membranes of the respiratory and digestive tracts. The *tonsils,* described in chapter 13, are composed of partially encapsulated lymph nodules. Also, aggregations of nodules, called *Peyer's patches,* are scattered throughout the mucosal lining of the ileum of the small intestine.

Locations of Lymph Nodes

Lymph nodes generally occur in groups or chains along the larger lymphatic vessels. Although they are widely distributed throughout the body, they are lacking in the tissues of the central nervous system.

The major locations of lymph nodes, shown in figure 18.10, are as follows:

1. **Cervical region.** Nodes in the cervical region occur along the lower border of the mandible, in front of and behind the ears, and deep within the neck along the paths of the larger blood vessels. These nodes are associated with the lymphatic vessels that drain the skin of the scalp and face, as well as the tissues of the nasal cavity and pharynx.

2. **Axillary region.** In the underarm region, nodes receive lymph from vessels that drain the arm, the wall of the thorax, the mammary gland (breast), and the upper wall of the abdomen.

3. **Inguinal region.** Nodes in the inguinal region receive lymph from the legs, the external genitalia, and the lower abdominal wall.

4. **Pelvic cavity.** Within the pelvic cavity, nodes occur primarily along the paths of the iliac blood vessels. They receive lymph from the lymphatic vessels of the pelvic viscera.

5. **Abdominal cavity.** Within this cavity, nodes occur in chains along the main branches of the mesenteric arteries and the abdominal aorta. These nodes receive lymph from the abdominal viscera.

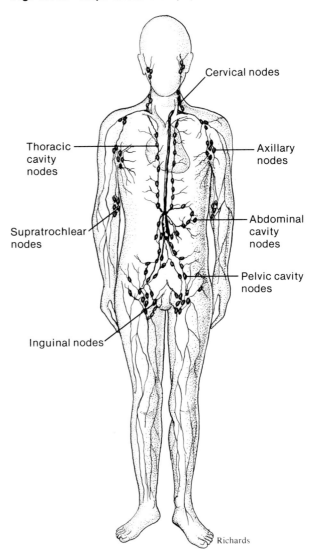

Fig. 18.10 Major locations of lymph nodes.

Cervical nodes

Thoracic cavity nodes

Axillary nodes

Supratrochlear nodes

Abdominal cavity nodes

Pelvic cavity nodes

Inguinal nodes

Richards

6. **Thoracic cavity.** Nodes of the thoracic cavity occur within the mediastinum and along the trachea and bronchi. They receive lymph from the thoracic viscera and from the internal wall of the thorax.

The supratrochlear lymph nodes (cubital lymph nodes), which are located superficially on the medial side of the elbow, often become enlarged in children as a result of infections associated with many cuts and scrapes on the hands.

The Lymphatic System 627

As was mentioned, lymph nodes contain large numbers of *lymphocytes.* These nodes, in fact, are centers for lymphocyte production, although such cells are produced in other tissues as well. Lymphocytes act against foreign particles, such as bacterial cells and viruses, that are carried to the lymph nodes by the lymphatic vessels.

The nodes also contain phagocytic cells that engulf and destroy foreign substances, damaged cells, and cellular debris.

Thus, the primary functions of lymph nodes are the production of lymphocytes, which help to defend the body against microorganisms, and the filtration of foreign particles and cellular debris from lymph before it is returned to the bloodstream.

---

1. *How would you distinguish between a lymph node and a lymph nodule?*
2. *In what body regions are lymph nodes most abundant?*
3. *What are the major functions of lymph nodes?*

Thymus and Spleen

Two other lymphatic organs whose functions are closely related to those of the lymph nodes are the thymus and the spleen.

The Thymus

The **thymus,** shown in figure 18.11, is a soft, bilobed structure whose lobes are surrounded by connective tissue. It is located within the mediastinum, in front of the aorta and behind the upper part of the sternum, extending from the root of the neck to the pericardium. Although the thymus varies in size from person to person, it is usually relatively large during infancy and early childhood. After puberty it tends to decrease in size, and in an adult it may be quite small. In an elderly person, the thymus is often largely replaced by fat and connective tissue.

This organ is composed of lymphatic tissue that is subdivided into *lobules* by connective tissues extending inward from its surface (fig. 18.12). The lobules contain large numbers of lymphocytes. The majority of these cells (thymocytes) remain inactive, however, some of them develop into a group (T-lymphocytes) that leaves the thymus and functions in immunity.

The thymus may also secrete a hormone called *thymosin,* which is thought to stimulate the activity of these lymphocytes after they leave the thymus and migrate to other lymphatic tissues.

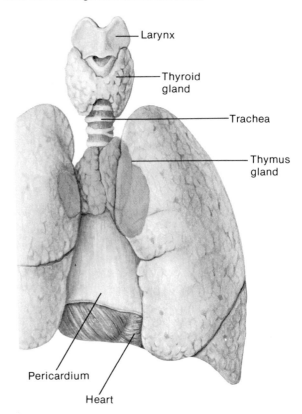

Fig. 18.11 The thymus gland is a bilobed organ located between the lungs and above the heart.

— Larynx

— Thyroid gland

— Trachea

— Thymus gland

Pericardium

Heart

The Spleen

The *spleen* is the largest of the lymphatic organs. As figure 18.13 shows, it is located in the upper left portion of the abdominal cavity, just beneath the diaphragm and behind the stomach.

The spleen resembles a large lymph node in some respects. It is, for example, enclosed in connective tissue that extends inward from the surface and partially divides the organ into chambers or *lobules.* It also has a *hilum* on one surface through which blood vessels enter. However, unlike a lymph node, the spaces (venous sinuses) within the chambers of the spleen are filled with *blood* instead of lymph. (See fig. 18.14.)

In addition to being very vascular, the spleen is soft and elastic, and can be distended by blood filling its venous sinuses. These characteristics are related to the spleen's function as a *blood reservoir.* During times of rest when circulation of blood is decreased, some blood can be stored within the spleen by vasodilation of its blood vessels. Then, during periods of exercise or in response to anoxia or hemorrhage, these blood vessels are stimulated to undergo vasoconstriction, and some of the stored blood is expelled into the circulation.

Fig. 18.12 A cross section of the thymus gland.

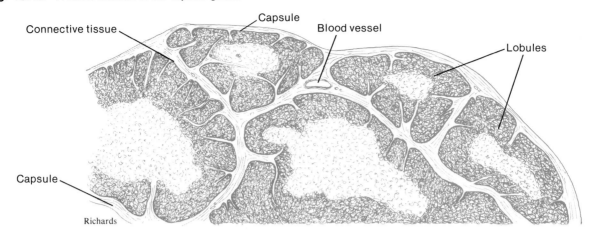

Connective tissue

Capsule

Blood vessel

Lobules

Capsule

Richards

Fig. 18.13 A spleen is located beneath the diaphragm in the upper left portion of the abdominal cavity.

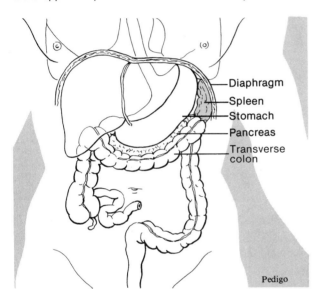

Diaphragm

Spleen

Stomach

Pancreas

Transverse colon

Pedigo

Fig. 18.14 The spleen resembles a large lymph node.

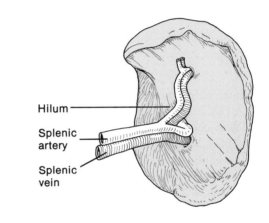

Hilum

Splenic artery

Splenic vein

Fig. 18.15 A cross section of the spleen.

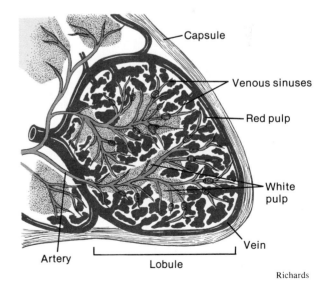

Capsule

Venous sinuses

Red pulp

White pulp

Vein

Artery

Lobule

Richards

Within the lobules of the spleen, the tissues are called *pulp* and are of two types—white pulp and red pulp. The *white pulp* is distributed throughout the spleen in tiny islands. This tissue is composed of lymphatic nodules (splenic nodules) similar to those found in lymph nodes. The *red pulp,* which fills the remaining spaces of the lobules, surrounds the venous sinuses. This pulp contains relatively large numbers of red blood cells, which are responsible for its color, along with lymphocytes and other kinds of cells normally found in circulating blood. (See fig. 18.15.)

Chart 18.2 Major organs of the lymphatic system

Organ	Location	Function
Lymph nodes	In groups or chains along the paths of larger lymphatic vessels	Center for lymphocyte production; house T-lymphocytes and B-lymphocytes that are responsible for immunity; phagocytes filter foreign particles and cellular debris from lymph
Thymus	Within the mediastinum behind the upper portion of the sternum	Houses lymphocytes, changes undifferentiated lymphocytes into T-lymphocytes
Spleen	In upper left portion of abdominal cavity beneath the diaphragm and behind the stomach	Serves as blood reservoir; phagocytes filter foreign particles, damaged red blood cells, and cellular debris from the blood; houses lymphocytes

Splenic pulp also contains numerous large *phagocytes* (macrophages) in the linings of its venous sinuses. These cells engulf and destroy foreign particles that may be carried in the blood as it flows through the sinuses. Thus, the spleen filters blood much as the lymph nodes filter lymph. In addition, the phagocytes help to destroy damaged red blood cells and the remains of ruptured cells carried in the blood.

The lymphocytes of the spleen, like those of the thymus, lymph nodes, and nodules, help to defend the body against infections.

Chart 18.2 summarizes the characteristics of the major organs of the lymphatic system.

During fetal development, pulp cells of the spleen function to produce blood cells, much as red bone marrow cells do in later life. As the time of birth approaches, this splenic function ceases. However, in certain diseases, such as *erythroblastosis fetalis,* in which large numbers of red blood cells are destroyed, the splenic pulp cells may resume their blood-cell-forming activity.

---

1. Why are the thymus and spleen considered to be organs of the lymphatic system?
2. What are the major functions of the thymus and the spleen?

Body Defenses against Infection

Infection suggests the presence of some kind of disease-causing agent within the body. Such agents are termed **pathogens,** and they include viruses and microorganisms such as bacteria, fungi, and protozoans, as well as various parasitic worms and other forms of life.

The human body is equipped with a variety of *defense mechanisms* that help to prevent the entrance of pathogens or act to destroy them if they do enter the tissues. Some of these mechanisms are quite general, or *nonspecific,* in that they protect against many types of pathogens. These mechanisms include species resistance, mechanical barriers, actions of enzymes, interferon, inflammation, and phagocytosis.

Other defense mechanisms are very *specific* in their actions, protecting against particular disease-causing agents and are responsible for the type of resistance called *immunity.*

Nonspecific Resistance

Species Resistance

Species resistance refers to the fact that a given kind of organism or *species* (such as humans) develops diseases that are unique to it. At the same time, a species is somehow resistant to diseases that affect other species. For example, humans are subject to infections by the microorganisms that cause gonorrhea and syphilis, but other animal species are generally resistant to these diseases. Similarly, humans are resistant to certain forms of malaria and tuberculosis that affect various birds.

Mechanical Barriers

The *skin* and the *mucous membranes* lining the tubes of the respiratory, digestive, urinary, and reproductive systems create **mechanical barriers** against the entrance of infectious agents. As long as these barriers remain unbroken, many pathogens are unable to penetrate them. In addition, the mucus-coated ciliated epithelium, described in chapter 15, that lines the respiratory passages acts to entrap and sweep particles out of the airways and into the pharynx, where they are swallowed.

Enzymatic Actions

Enzymatic actions against disease-causing agents are due to the presence of enzymes in various body fluids. Gastric juice, for example, contains the protein-splitting enzyme *pepsin*. It also has a low pH due to the presence of hydrochloric acid, and the combined effect of these substances is lethal to many pathogens that reach the stomach. Similarly, tears contain an enzyme called *lysozyme* that has an antibacterial action against certain pathogens that get into the eyes.

Interferon

Interferon is the name given to a group of proteins produced by cells in response to the presence of *viruses*. Although the effect of interferon is nonspecific, it somehow interferes with the reproduction (replication) of viruses and thus helps to control the development of the diseases they cause. Furthermore, interferon released from infected cells can be taken in by other, noninfected cells. Consequently, the noninfected cells become protected against the viral infection, and the spread of the pathogenic viruses is inhibited.

Inflammation

Inflammation is a tissue response to injury from mechanical forces, chemical irritants, exposure to extreme temperatures, or exposure to excessive radiation. It may also accompany infections.

The major symptoms of inflammation include localized redness, swelling, heat, and pain. The *redness* is a result of blood vessel dilation and the consequent increase of blood volume within the affected tissues (hyperemia). This effect, coupled with an increase in the permeability of the capillaries, is responsible for tissue *swelling* (edema). The *heat* is due to the presence of blood from deeper body parts, which is generally warmer than that near the surface, while the *pain* results from the stimulation of pain receptors in the affected tissues.

White blood cells tend to accumulate at the sites of inflammation. Some of these cells help to control pathogens by *phagocytosis*. In the case of bacterial infections, the resulting mass of white blood cells, bacterial cells, and damaged tissue cells may create a thick fluid called *pus*.

Body fluids (exudate) also tend to collect in inflamed tissues. These fluids contain *fibrinogen* and other clotting elements. As a result of clotting, a network of fibrin threads may develop within a region of

damaged tissue. Later, *fibroblasts* may appear and form fibers around the affected area until it is enclosed in a sac of fibrous connective tissue. This action serves to confine an infection and to prevent the spread of pathogens to adjacent tissues.

A *pimple* or *boil* is an example of such a confined infection. An abscess of this type should never be squeezed, because the pressure acting upon the inflamed tissues may force pathogenic microorganisms into adjacent tissues and cause the infection to spread.

Once an infection has been controlled, phagocytic cells tend to remove dead cells and other debris from the site of inflammation, and the damaged tissues are replaced by cellular reproduction.

The process of inflammation and healing is outlined in chart 18.3.

Phagocytosis

As is mentioned in chapter 15, the most active phagocytic cells in the blood are *neutrophils* and *monocytes*. These wandering cells can leave the bloodstream by squeezing between the cells of blood vessel walls (diapedesis). They are attracted toward sites of inflammation by chemicals released from injured tissues. As is explained in chapter 16, neutrophils are able to engulf and digest smaller particles, while monocytes can phagocytize somewhat larger objects.

Chart 18.3 Major actions that may occur during an inflammation response

Blood vessels dilate.
Capillary permeability increases.
   Tissues become red, swollen, warm, and painful.
White blood cells invade the region.
   Pus may form as white blood cells, bacterial cells, and cellular debris accumulate.
Body fluids seep into the area.
   A clot containing threads of fibrin may form.
Fibroblasts appear
   A connective tissue sac may be formed around the injured tissues.
Phagocytes are active.
   Dead cells and other debris are removed.
Cells reproduce.
   Newly formed cells replace injured ones.

Another important group of phagocytes are the *macrophages* (histiocytes) that remain fixed in various tissues. Macrophages, described in chapter 5, are widely distributed throughout the body and, with certain other phagocytes that are associated with the linings of blood vessels in the bone marrow, liver, spleen, and lymph nodes, they form the **reticuloendothelial tissue** (system).

The name *reticuloendothelium* describes two characteristics of cells that make up this tissue. They are capable of forming fibrous networks or *reticula*, and they are often associated with the linings of blood vessels, the *endothelium*. For example, the phagocytic cells in the linings of the lymph sinuses of the lymph nodes, the venous sinuses of the spleen, and the sinusoids of the liver are reticuloendothelial cells.

The cells that line the vascular sinuses of the red bone marrow are also reticuloendothelial cells. Although these cells are especially concerned with the formation of red and white blood cells, they are also capable of phagocytosis. In fact, they are adapted to remove very small particles, such as foreign protein molecules, from the blood.

As a result of reticuloendothelial activities, foreign particles are removed from lymph as it moves from the interstitial spaces to the bloodstream. Any such particles that reach the blood are likely to be removed by phagocytes located in the vessels and tissues of the spleen, liver, or bone marrow.

---

1. What is meant by an infection?
2. Explain six nonspecific defense mechanisms.
3. Define reticuloendothelial tissue.

Immunity

Immunity is resistance to specific foreign agents, such as pathogens, or the toxins they release. It involves a number of *immune mechanisms,* in which certain cells recognize the presence of particular foreign substances and act to destroy them. The cells that function in immune mechanisms include lymphocytes and macrophages.

Origin of Lymphocytes

Lymphocytes originate from bone marrow cells (stem cells), although later they are produced in various lymphatic tissues.

Before they become *differentiated,* developing lymphocytes are released from the marrow and are carried away by the blood. About half of them reach the thymus gland, where they remain for a time.

Within the thymus, these undifferentiated cells undergo changes, and thereafter they are called *T-lymphocytes* (thymus-derived lymphocytes). Later, the T-lymphocytes are transported away from the thymus by the blood and tend to reside in various organs of the lymphatic system.

Lymphocytes that have been released from the bone marrow and do not reach the thymus gland are called *B-lymphocytes* (bone-marrow-derived lymphocytes). These cells are also distributed by the blood and tend to settle in lymphatic organs together with the T-lymphocytes. (See figs. 18.16 and 18.17.)

Functions of Lymphocytes

Prior to birth, certain cells apparently note the kinds of proteins that are present within the organism's tissues and body fluids. Subsequently, the proteins that are present, called self-proteins, can be distinguished from nonself or foreign proteins, and when foreign proteins are recognized, lymphocytes can produce immune reactions against them.

The foreign proteins to which lymphocytes respond are called **antigens,** and the responses to them are very specific. Before a lymphocyte can recognize a particular antigen, however, it must be *sensitized* or *programmed* to react to that antigen and no others.

The process by which lymphocytes become sensitized to particular antigens is poorly understood. It seems to involve activities of *macrophages,* which may serve as processing cells. For example, a macrophage may engulf an antigen or antigen-bearing particle that has entered the body, somehow combine this antigen with RNA, and pass the resulting antigen-RNA complexes to lymphocytes. In this way, the lymphocytes may acquire the information needed to recognize specific antigens.

Ultimately, the ability to recognize specific antigens seems to involve the presence of *receptor molecules* that protrude from the surfaces of sensitized lymphocytes. Apparently these receptor molecules and the specific antigen molecules they recognize have complementary shapes—as the shape of a specific key fits a particular lock.

T-lymphocytes and B-lymphocytes respond to their particular antigens in different ways. T-lymphocytes attach themselves to antigen-bearing cells, such as bacterial cells, and interact with them directly—that is, with cell to cell contact. This type of response, called **cellular immunity,** may involve the T-lymphocyte's secretion of a specific toxic substance (lymphotoxin) that is lethal to the cell being attacked.

Fig. 18.16 Bone marrow releases undifferentiated lymphocytes, which become B-lymphocytes within the blood or develop into T-lymphocytes within the thymus gland.

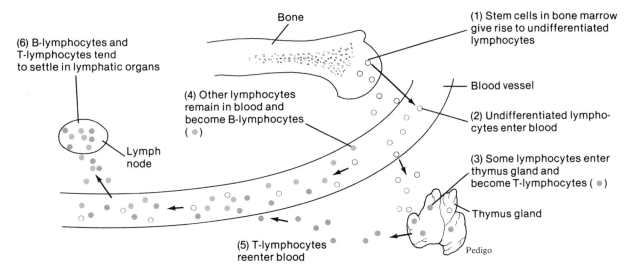

Bone

(1) Stem cells in bone marrow give rise to undifferentiated lymphocytes

Blood vessel

(2) Undifferentiated lympho-cytes enter blood

(3) Some lymphocytes enter thymus gland and become T-lymphocytes (•)

Thymus gland

Pedigo

(6) B-lymphocytes and T-lymphocytes tend to settle in lymphatic organs

Lymph node

(4) Other lymphocytes remain in blood and become B-lymphocytes (•)

(5) T-lymphocytes reenter blood

T-lymphocytes are also thought to provide an important defense against viral infections. Viruses grow inside cells, where they are somewhat protected. Since most viruses cause antigens to be produced on the membranes of infected cells, T-lymphocytes can detect such cells and destroy them.

T-lymphocytes seem to be able to recognize and destroy certain kinds of cancer cells. This has led some investigators to believe that interference with normal T-lymphocyte function may be a factor in the development of some cancers. Furthermore, since T-lymphocyte function seems to decline with age, this may help to explain why some types of cancer appear more commonly in elderly persons.

Fig. 18.17 Scanning electron micrograph of a human circulating lymphocyte.

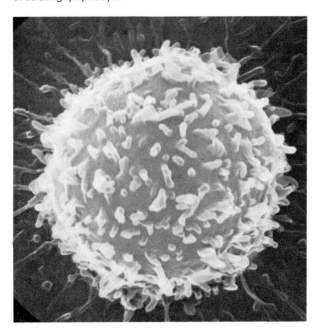

Chart 18.4   A comparison of T-lymphocytes and B-lymphocytes

	T-lymphocyte	B-lymphocyte
Origin of undifferentiated cell	Bone marrow	Bone marrow
Site of differentation	Thymus gland	Outside the thymus gland
Primary location	Various lymphatic organs	Various lymphatic organs
Programmed to respond to specific antigens by	Macrophage	Macrophage
Primary function	Responsible for cellular immunity; secretes specific toxins; triggers action of B-lymphocytes	Responsible for humoral immunity; secretes specific antibodies into body fluids.

B-lymphocytes act indirectly against the antigens they recognize by producing and secreting globular proteins called **antibodies.** Antibodies are carried by body fluids and react in various ways with specific antigens or antigen-bearing particles to destroy them. This type of response is called **humoral immunity.**

Figure 18.18 summarizes the development of cellular and humoral immunity.

T-lymphocytes and B-lymphocytes interact with each other in complex ways. Although these interactions are not well understood, it is thought that T-lymphocytes must somehow trigger B-lymphocytes before they can produce antibodies efficiently. Consequently, both types of lymphocytes are required for normal immune responses to occur. Chart 18.4 compares T- and B-lymphocytes.

---

1. *Define* immunity.
2. *How would you explain the difference between a T-lymphocyte and a B-lymphocyte?*
3. *How would you explain the difference between an antigen and an antibody?*

Types of Antibodies

The antibodies produced and secreted by B-lymphocytes are all soluble globular proteins. These proteins are called **immunoglobulins,** and they constitute a portion of the *gamma globulin* fraction of plasma proteins.

Some persons partially or completely lack the ability to form immunoglobulins. As a consequence, they cannot produce antibodies and are very susceptible to developing infectious diseases. This condition is termed *agammaglobulinemia.*

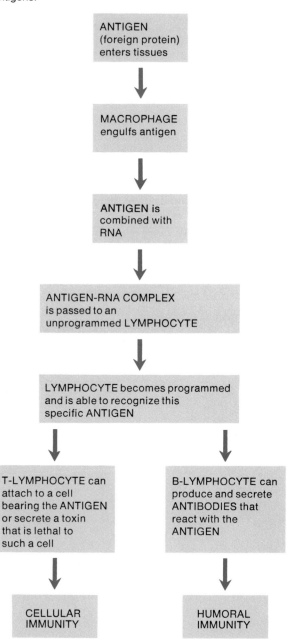

Fig. 18.18 Possible mechanisms by which lymphocytes become programmed to recognize and respond to specific antigens.

ANTIGEN (foreign protein) enters tissues

MACROPHAGE engulfs antigen

ANTIGEN is combined with RNA

ANTIGEN-RNA COMPLEX is passed to an unprogrammed LYMPHOCYTE

LYMPHOCYTE becomes programmed and is able to recognize this specific ANTIGEN

T-LYMPHOCYTE can attach to a cell bearing the ANTIGEN or secrete a toxin that is lethal to such a cell

B-LYMPHOCYTE can produce and secrete ANTIBODIES that react with the ANTIGEN

CELLULAR IMMUNITY

HUMORAL IMMUNITY

Chart 18.5 Characteristics of three major immunoglobulins

Type	Occurrence	Functions against	Mode of Action
IgG	Tissue fluid and plasma	Bacterial cells, viruses, and toxins	Forms insoluble precipitates; acts with complement to activate enzymes that attack foreign particles in various ways
IgA	Secretions of exocrine glands	Viruses that affect the respiratory and digestive systems	Forms insoluble precipitates; causes agglutination
IgM	Plasma	Reacts with antigens that occur naturally on certain red blood cells	Causes agglutination

There are five major types of immunoglobulins, three of which constitute the bulk of the circulating antibodies. They are immunoglobulin G, which accounts for about 80% of the antibodies; immunoglobulin A, which makes up about 13%; and immunoglobulin M, which is responsible for about 6%. Immunoglobulins D and E account for the remainder.

Immunoglobulin G (IgG) occurs in plasma and tissue fluids and is particularly effective against bacterial cells, viruses, and various toxins. It functions by causing antigen-bearing particles to form insoluble precipitates or by acting together with a group of proteins collectively called **complement** (complement system). These proteins, which occur in plasma and body fluids, are inactive forms of enzymes. When IgG antibodies form complexes with antigens, large numbers of the complement enzymes are activated in response, and they attack the antigen-bearing particles in a variety of ways. Some, for example, digest the membranes of foreign cells (lysis). Others alter cell membranes so the cells become more susceptible to phagocytosis by neutrophils and macrophages (opsonization). Still others cause cells to clump together (agglutination), attract phagocytes into the area (chemotaxis), or promote inflammation. (See fig. 18.19.)

Immunoglobulin A (IgA) is commonly found in the secretions of various *exocrine glands*. It occurs in milk, tears, nasal fluid, gastric juice, intestinal juice, bile, and urine. IgA helps to control certain respiratory viruses and various pathogens responsible for digestive disturbances. It seems to act by causing antigens to form insoluble precipitates and by causing antigen-bearing cells to clump together (agglutination).

Immunoglobulin A can pass from a nursing mother to her baby via milk and other breast secretions. These antibodies are thought to provide the infant with some protection against digestive disturbances that might otherwise cause serious problems.

Fig. 18.19 IgG antibodies may activate complement enzymes that attack antigen-bearing particles in a variety of ways.

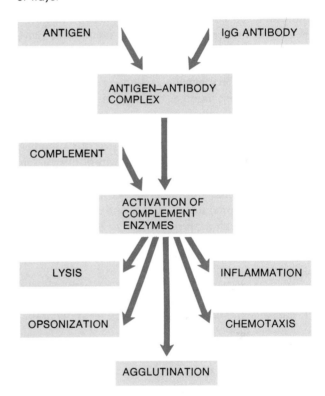

Immunoglobulin M (IgM) is the type of antibody that occurs naturally in blood plasma (agglutinins anti-A and anti-B) described in chapter 16. In transfusion reactions, these substances cause the agglutination of red blood cells, by reacting with the antigens (agglutinogens A and B) on their membranes.

Chart 18.5 summarizes the characteristics of the major immunoglobulins.

Primary and Secondary Immune Responses

When the body cells first encounter a particular antigen, the immune reaction that follows is called a **primary immune response.** During this response, lymphocytes are programmed to act against the specific antigen that has entered the tissues.

If the B-lymphocytes are triggered to produce antibodies, they typically undergo a period of rapid reproduction, and the newly formed offspring cells develop into **plasma cells.** The plasma cells, in turn, continue to produce and secrete the appropriate antibodies for a time. (See fig. 18.20.)

If the antigen that stimulates the primary immune response is present on certain viruses or bacterial cells, T-lymphocytes may be activated and function as the *effector cells* that attack the invaders.

Since the lymphocytes involved in a primary immune response become programmed to recognize the specific antigen that stimulated them, these lymphocytes can respond more rapidly if the same antigen is encountered in the future. Such a reaction is called a **secondary immune response.**

A primary immune response might be stimulated in a person who is exposed to a particular pathogen, such as a live virus, for the first time. When this happens, detectable concentrations of antibodies usually appear in the body fluids within five to ten days following the exposure. If the same pathogen is encountered some time later, a secondary immune response may produce additional antibodies within a day or two. Such a person is said to have developed an *active immunity* to the pathogen involved.

---

1. How do the functions of various immunoglobulins differ?
2. Distinguish between a primary and a secondary immune response.

Types of Immunity

One type of immunity is called *naturally acquired active immunity.* It occurs when a person who has been exposed to a live pathogen develops a disease, and becomes resistant to that pathogen as a result of a primary immune response.

Another type of active immunity can be produced in response to a **vaccine.** Such a substance contains an antigen that can stimulate a primary immune response against a particular disease-causing agent but is unable to produce the symptoms of that disease.

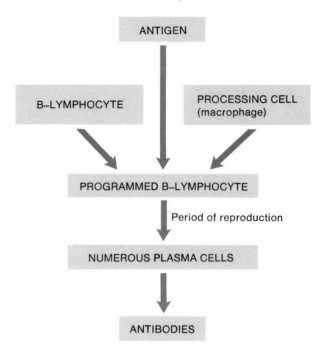

Fig. 18.20 A primary immune response occurs when body cells first encounter an antigen.

A vaccine, for example, might contain a virus that has been killed or weakened sufficiently so it cannot cause a serious infection or produce the severe symptoms of an infection. Its antigens, however, still retain the characteristics needed to stimulate a primary immune response. Thus, a person who has been vaccinated is said to develop *artificially acquired active immunity.*

Vaccines are available to stimulate the development of active immunity against a great variety of diseases. These include such diseases as typhoid fever, cholera, whooping cough, diphtheria, tetanus, polio, measles, mumps, and smallpox.

Sometimes a person needs protection against a disease-causing microorganism but lacks the time needed to develop active immunity. In such a case, it may be possible to provide the person with an injection of ready-made antibodies. These antibodies may be obtained from *gamma globulin* separated from the blood of persons who have already developed immunity against the disease in question.

A person who receives an injection of gamma globulin is said to have *artificially acquired passive immunity.* This type of immunity is called passive

Chart 18.6 Types of immunity

Type	Stimulus	Result
Naturally acquired active immunity	Exposure to live pathogens	Symptoms of a disease and stimulation of an immune response
Artificially acquired active immunity	Exposure to a vaccine containing weakened or dead pathogens	Stimulation of an immune response without the severe symptoms of a disease
Artificially acquired passive immunity	Injection of gamma globulin containing antibodies	Immunity for a short time without stimulating an immune response
Naturally acquired passive immunity	Antibodies passed to fetus from mother with active immunity	Short-term immunity for infant, without stimulating an immune response

because the antibodies involved are not produced by the recipient's cells. Such immunity is relatively short-term, seldom lasting more than a few months. Furthermore, the recipient's lymphocytes remain unprogrammed to respond against the pathogens for which the protection was needed. Consequently, the person continues to be susceptible to those pathogens in the future.

The types of immunity are summarized in Chart 18.6.

During pregnancy, certain antibodies (immunoglobin G) are able to pass from the maternal blood through the placental membrane and into the fetal bloodstream. As a result, the fetus acquires some immunity against the pathogens for which the mother has developed active immunities. In this case, the fetus is said to have *naturally acquired passive immunity,* which may remain effective for six months to a year after birth.

Allergic Reactions

Allergic reactions are closely related to immune responses in that both involve the combining of antigens with antibodies. Allergic reactions, however, are likely to be excessive or violent, and may produce tissue damage.

Persons who develop allergies have inherited the ability to produce abnormal immune reactions (hypersensitivities) to certain common antigens, such as plant pollens or house dust.

An antigen that stimulates an allergic reaction is called an **allergen.** In one type of allergic reaction, the first encounter with an allergen stimulates the cells to produce antibodies against it. As a result, the person is said to become sensitized to the allergen.

The antibodies formed in response to an allergen belong to the immunoglobulin E group. They become attached to widely distributed cells called **mast cells,** which are described in chapter 5.

When antibodies are present and they encounter allergens, an allergen-antibody reaction occurs. As a result of this reaction, the mast cells are somehow altered, and they release various substances, including *histamine,* that have various physiological effects. For example, blood vessels may dilate, tissues may become edematous, or smooth muscles may contract. Such actions are responsible for various symptoms of allergic reactions including hives, hayfever, asthma, eczema, and gastric disturbances.

Allergic reactions are usually short lived, because they are terminated by a special type of T-lymphocyte called **suppressor cells.** These cells produce a substance called *suppressor factor* that interferes with the production of the antibodies needed for the reaction.

The events leading to an allergic reaction are summarized in figure 18.21.

Transplantation and Tissue Rejection

It is occasionally desirable to transplant some tissue or an organ from one person to another to replace a nonfunctional, damaged, or lost body part. In such cases, there is a danger that the recipient's cells may recognize the donor's tissues as being foreign. This triggers the recipient's immune mechanisms, which may act to destroy the donor tissue. Such a response is called a **tissue rejection reaction.**

Tissue rejection involves the activities of lymphocytes and the production of antibodies. This response is similar to those that occur when any foreign substances are present. (See fig. 18.22.) However, the greater the antigenic difference between the proteins of the recipient and the donor tissues, the more rapid and severe the rejection reaction will be. Thus, the reaction can sometimes be minimized by matching recipient and donor tissues. This means locating a donor whose tissues are antigenically similar to those of the person needing a transplant—a procedure much like matching the blood of a donor with that of a recipient before giving a blood transfusion.

Fig. 18.21 (a) An initial contact with an allergen may stimulate the production of antibodies. (b) Subsequent contacts with this allergen may trigger allergic reactions.

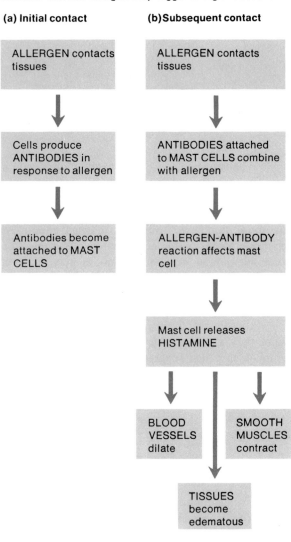

(a) Initial contact

ALLERGEN contacts tissues

↓

Cells produce ANTIBODIES in response to allergen

↓

Antibodies become attached to MAST CELLS

(b) Subsequent contact

ALLERGEN contacts tissues

↓

ANTIBODIES attached to MAST CELLS combine with allergen

↓

ALLERGEN-ANTIBODY reaction affects mast cell

↓

Mast cell releases HISTAMINE

↓

BLOOD VESSELS dilate   SMOOTH MUSCLES contract

↓

TISSUES become edematous

Fig. 18.22 An organ transplant may be followed by a tissue rejection reaction.

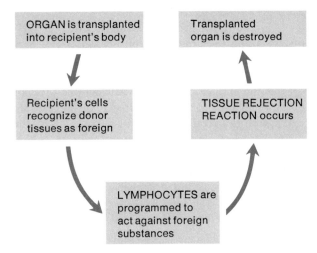

ORGAN is transplanted into recipient's body

↓

Recipient's cells recognize donor tissues as foreign

↓

LYMPHOCYTES are programmed to act against foreign substances

↑

TISSUE REJECTION REACTION occurs

↑

Transplanted organ is destroyed

One matching procedure makes use of lymphocytes from circulating blood. These cells contain both antigens and receptor sites for recognizing foreign antigens. In the test, lymphocytes from a potential donor and a potential recipient are mixed and cultured. The antigenic difference between these cells can then be judged by observing certain cellular responses, including enlargement, amount of DNA synthesized, and the rate of cell division. Minimal antigenic difference results in the best match.

Another approach to reducing the rejection of transplanted tissue involves the use of **immunosuppressive drugs.** These substances act by interfering with the recipient's immune mechanisms. A drug may, for example, suppress the formation of antibodies, or it may cause the destruction of lymphocytes and so prevent them from producing antibodies.

Unfortunately, the use of immunosuppressive drugs leaves the recipient relatively unprotected against infections. Although the drug may prevent a tissue rejection reaction, the recipient may develop a serious infectious disease that is difficult to control.

Autoimmune Diseases

As was mentioned, immune responses are usually directed toward nonself or foreign proteins, while self-proteins are tolerated by the immune mechanism. Occasionally, something happens to change this, and the tolerance to self-proteins is lost. As a result, immune responses may be directed toward a person's own tissues, and antibodies are produced that attack them. Such a condition is called an **autoimmune disease** (autoallergy).

Although the mechanism responsible for these diseases is not well understood, they are more common in older persons, and they may involve viral or bacterial infections. Some investigators believe that the release of abnormally large quantities of antigens may occur when infectious agents cause tissue damage. These agents may also cause body proteins to change into forms that are reacted against. At the same time, the activity of suppressor cells, which normally limits this type of reaction, seems to be repressed.

Examples of autoimmune diseases include *rheumatic fever* in which tissues of the heart and joints are attacked after exposure to certain bacterial (streptococcus) toxins; *myasthenia gravis* in which muscle cells are damaged and their acetylcholine receptor sites blocked; and *systemic lupus erythematosus* in which various connective tissues are destroyed.

1. Explain the difference between active and passive immunities.
2. In what ways is an allergic reaction related to an immune reaction?
3. In what ways is a tissue rejection reaction related to an immune response?
4. Define autoimmune disease.

Some Clinical Terms Related to the Lymphatic System and Immunity

anaphylaxis (an''ah-fi-lak'sis)—hypersensitivity to the presence of a foreign substance.

asplenia (ah-sple'ne-ah)—the absence of a spleen.

autograft (aw'to-graft)—transplantation of tissue from one part of a body to another part of the same body.

histocompatibility (his''to-kom-pat'' ĭ-bil' ĭ-te)—compatibility between the tissues of a donor and a recipient based on antigenic similarities.

homograft (ho'mo-graft)—transplantation of tissue from one person to another person.

lymphadenectomy (lim-fad'' ĕ-nek'to-me)—surgical removal of lymph nodes.

lymphadenopathy (lim-fad'' ĕ-nop'ah-the)—enlargement of the lymph nodes.

lymphadenotomy (lim-fad'' ĕ-not'o-me)—an incision into a lymph node.

lymphocytopenia (lim''fo-si''to-pe'ne-ah)—an abnormally low concentration of lymphocytes in the blood.

lymphocytosis (lim''fo-si''to'sis)—an abnormally high concentration of lymphocytes in the blood.

lymphoma (lim-fo'mah)—a tumor composed of lymphatic tissue.

lymphosarcoma (lim''fo-sar-ko'mah)—a cancer within the lymphatic tissue.

splenectomy (sple-nek'to-me)—surgical removal of the spleen.

splenitis (sple-ni'tis)—an inflammation of the spleen.

splenomegaly (sple''no-meg'ah-le)—an abnormal enlargement of the spleen.

splenotomy (sple-not'o-me)—incision of the spleen.

thymectomy (thi-mek'to-me)—surgical removal of the thymus gland.

thymitis (thi-mi'tis)—an inflammation of the thymus gland.

Chapter Summary

Introduction

Lymphatic system functions to transport excess tissue fluid to the bloodstream and to help defend the body against invasion by microorganisms.

Lymphatic Pathways

1. Lymphatic capillaries
   a. Lymphatic capillaries are microscopic closed-ended tubes that extend into interstitial spaces.
   b. They receive lymph through their thin walls.
   c. Lacteals are lymphatic capillaries in the villi of the small intestine.
2. Lymphatic vessels
   a. Lymphatic vessels are formed by the merging of lymphatic capillaries.
   b. They have walls like veins with valves that prevent backflow of lymph.
   c. They lead to lymph nodes and then merge into lymphatic trunks.
3. Lymphatic trunks and collecting ducts
   a. These drain lymph from relatively large body regions.
   b. Trunks lead to two collecting ducts within the thorax.
   c. Collecting ducts join the subclavian veins.

Tissue Fluid and Lymph

1. Tissue fluid formation
   a. Tissue fluid originates from blood plasma and includes water and dissolved substances that passed through the capillary wall.
   b. It generally lacks proteins, but some smaller protein molecules leak into interstitial spaces.
   c. As protein concentration of tissue fluid increases, osmotic pressure increases also.
2. Lymph formation
   a. Rising osmotic pressure in tissue fluid interferes with the return of water to the blood capillaries.
   b. Increasing pressure within interstitial spaces forces some tissue fluid into lymphatic capillaries, and this fluid becomes lymph.
3. Function of lymph
   a. Lymph returns protein molecules to the bloodstream.
   b. It transports foreign particles to the lymph nodes.

Movement of Lymph

1. Flow of lymph
   a. Lymph is under low pressure and may not flow readily without aid from external forces.
   b. Forces that aid movement of lymph include squeezing action of skeletal muscles and low pressure in the thorax created by breathing movements.

Lymph Nodes

1. Structure of a lymph node
   a. Lymph nodes are usually bean shaped with blood vessels and efferent lymphatic vessels attached to the indented region; afferent lymphatic vessels enter at points on the convex surface.
   b. Lymph nodes are enclosed in connective tissue that extends into the nodes and divides them into nodules.
   c. Nodules contain masses of lymphocytes and spaces through which lymph flows.
2. Locations of lymph nodes
   a. Lymph nodes generally occur in groups or chains along the paths of larger lymphatic vessels.
   b. They occur primarily in cervical, axillary, and inguinal regions and within the pelvic, abdominal, and thoracic cavities.
3. Functions of lymph nodes
   a. Lymph nodes are centers for production of lymphocytes that act against foreign particles.
   b. They contain phagocytic cells that remove foreign particles from lymph.

Thymus and Spleen

1. The thymus
   a. The thymus is a soft, bilobed organ located within the mediastinum.
   b. It tends to decrease in size after puberty.
   c. It is composed of lymphatic tissue that is subdivided into lobules.
   d. Lobules contain lymphocytes, most of which are inactive.
   e. Some lymphocytes leave the thymus and function in immunity.
   f. The thymus may secrete a hormone called thymosin, which stimulates lymphocytes that have migrated to other lymphatic tissues.
2. The spleen
   a. The spleen is located in the upper left portion of the abdominal cavity.
   b. It resembles a large lymph node that is encapsulated and subdivided into lobules by connective tissue.
   c. Spaces within lobules are filled with blood.
   d. It acts as a blood reservoir, while phagocytes and lymphocytes filter foreign particles from the blood.

Body Defenses against Infection

Infection is caused by the presence of pathogens. The body is equipped with specific and nonspecific defenses against infection.

Nonspecific Resistance

1. Species resistance
   Each species of organism is resistant to certain diseases that may affect other species, but is susceptible to diseases that other species may be able to resist.
2. Mechanical barriers
   Mechanical barriers include skin and mucous membranes, which prevent entrance of many pathogens as long as they remain unbroken.
3. Enzymatic actions
   Enzymes of gastric juice are lethal to many pathogens, and enzymes in tears have antibacterial actions.
4. Interferon
   Interferon is a group of proteins produced by cells in response to the presence of viruses; it can interfere with the reproduction and spread of viruses.
5. Inflammation
   a. Inflammation is a tissue response to injury or infection.
   b. The response includes localized redness, swelling, heat, and pain.
   c. Chemicals released by damaged tissues attract various white blood cells to the site of inflammation.
   d. Clotting may occur in body fluids that accumulate in affected tissues.
   e. Fibrous connective tissue may form a sac around the injured tissue and thus prevent the spread of infection.
6. Phagocytosis
   a. Most active phagocytes in blood are neutrophils and monocytes; macrophages remain fixed in tissues.
   b. Phagocytes associated with the linings of blood vessels in bone marrow, liver, spleen, and lymph nodes constitute the reticuloendothelial tissue.
   c. Phagocytes function to remove foreign particles from tissues and body fluids.

Immunity

1. Origin of lymphocytes
   a. Lymphocytes originate in bone marrow and are released into the blood before they become differentiated.
   b. Some reach the thymus where they become T-lymphocytes.
   c. Others become B-lymphocytes in the blood.
   d. Both T- and B-lymphocytes tend to reside in organs of the lymphatic system.
2. Functions of lymphocytes
   a. Lymphocytes are programmed by macrophages to respond to the presence of specific antigens.
   b. T-lymphocytes attack antigens or antigen-bearing cells directly and provide a defense against viruses.

c. B-lymphocytes produce antibodies and secrete them into body fluids; specific antibodies act against specific antigens.

d. Normal immune responses require the interaction of T- and B-lymphocytes.

3. Types of antibodies

a. Antibodies are composed of soluble proteins called immunoglobulins.

b. There are five major types of immunoglobulins; IgG, IgA, and IgM make up most of the circulating antibodies.

c. Immunoglobulins act in various ways, causing antigens or antigen-bearing particles to form insoluble precipitates, or causing cells to undergo lysis, opsonization, or agglutination. Others attract phagocytes or produce inflammation.

4. Primary and secondary immune responses

a. A primary response is stimulated by the first encounter with an antigen.

b. In primary response, lymphocytes are programmed to act against the antigen.

c. After being programmed, lymphocytes can carry out a rapid secondary response if the antigen is encountered again.

5. Types of immunity

a. A person who encounters a pathogen and has a primary immune response develops naturally acquired active immunity.

b. A person who receives vaccine containing a dead or weakened pathogen develops artificially acquired active immunity.

c. A person who receives an injection of gamma globulin that contains ready-made antibodies has artificially acquired passive immunity.

d. Active immunity lasts much longer than passive immunity.

6. Allergic reactions

a. Allergic reactions involve combining of antigens with antibodies; reactions are likely to be excessive or violent and may cause tissue damage.

b. A person with an allergy has inherited an ability to carry on an abnormal immune reaction.

c. Allergic reaction may damage mast cells, which in turn release histamine and other chemicals.

d. Released chemicals are responsible for the symptoms of the allergic reaction: hives, hayfever, asthma, eczema, or gastric disturbances.

e. Reaction is terminated by suppressor cells releasing suppressor factor.

7. Transplantation and tissue rejection

a. If tissue is transplanted from one person to another, the recipient's cells may recognize the donor's tissue as foreign and act against it.

b. Tissue rejection reaction may be reduced by matching the donor and recipient tissues or by using immunosuppressive drugs.

c. Immunosuppressive drugs interfere with the recipient's immune mechanisms and cause the recipient to be very susceptible to infection.

8. Autoimmune diseases

a. Autoimmune diseases are due to immune reactions directed toward a person's own tissues.

b. They are more common in older persons and following viral or bacterial infections.

c. They may be due to antigens released from damaged tissues or to changes in body proteins.

Application of Knowledge

1. Based on your understanding of the functions of lymph nodes, how would you explain the fact that enlarged nodes are often removed for microscopic examination as an aid to diagnosing certain disease conditions?

2. Why is it true that an injection into the skin is, to a large extent, an injection into the lymphatic system.

3. Explain the fact that vaccination provides long-lasting protection against a disease, while gamma globulin provides only short-term protection.

Review Activities

1. Explain how the lymphatic system is related to the cardiovascular system.

2. Trace the general pathway of lymph from the interstitial spaces to the bloodstream.

3. Identify and describe the location of the major lymphatic trunks and collecting ducts.

4. Distinguish between tissue fluid and lymph.

5. Describe the primary functions of lymph.

6. Explain how physical exercise promotes lymphatic circulation.

7. Explain how a lymphatic obstruction leads to edema.

8. Describe the structure of a lymph node and list its major functions.

9. Locate the major body regions occupied by lymph nodes.

10. Describe the structure and functions of the thymus gland.

11. Describe the structure and functions of the spleen.

12. Distinguish between specific and nonspecific body defenses against infection.

13. Explain what is meant by species resistance.

14. Name two mechanical barriers to infection.

15. Describe how enzymatic actions function as defense mechanisms.

16. Define *interferon* and explain its action.

17. List the major symptoms of inflammation and explain why each occurs.

18. Identify the major phagocytic cells in the blood and other tissues.

19. Define the *reticuloendothelial tissue* and explain its importance.

20. Discuss the origin and function of T-lymphocytes and B-lymphocytes.

21. Distinguish between an antigen and an antibody.

22. Explain how macrophages are thought to function in programming lymphocytes.

23. Describe how lymphocytes are thought to recognize specific antigens.

24. Distinguish between cellular immunity and humoral immunity.

25. List the major types of immunoglobulin and describe their similarities and differences.

26. Explain the function of complement.

27. Distinguish between a primary and a secondary immune response.

28. Define *plasma cell*.

29. Distinguish between active and passive immunity.

30. Define a *vaccine* and explain its action.

31. Explain the relationship between an allergic reaction and an immune response.

32. Distinguish between an antigen and an allergen.

33. List the major events leading to an allergic reaction.

34. Explain the relationship between a tissue rejection and an immune response.

35. Describe two methods used to reduce the severity of a tissue rejection reaction.

36. Define *autoimmune disease*.

37. Explain how an autoimmune disease might develop.

Suggestions for Additional Reading

Bellanti, J. A. 1971. *Immunology.* Philadelphia: W. B. Saunders.

Burke, D. C. April 1977. The status of interferon. *Scientific American.*

Capra, J. D., and Edmundson, A. B. January 1977. The antibody combining site. *Scientific American.*

Claman, H. N. 1973. The new cellular immunology. *BioScience* 23:10.

Cunningham, B. A. October 1977. The structure and functions of histocompatibility antigens. *Scientific American.*

Edelman, G. M. August 1970. The structure and function of antibodies. *Scientific American.*

Hilleman, M. R., and Tytell, A. A. July 1971. The induction of interferon. *Scientific American.*

Jerne, N. K. July 1973. The immune system. *Scientific American.*

Mayer, M. M. November 1973. The complement system. *Scientific American.*

Old, L. J. May 1977. Cancer immunology. *Scientific American.*

Sell, S. 1972. *Immunology, immunopathology, and immunity.* New York: Harper and Row.

Talmage, D. W. 1979. Recognition and memory in the cells of the immune system. *American Scientist* 67:174.

Weiss, L. 1972. *The cells and tissues of the immune system.* Englewood Cliffs, N.J.: Prentice-Hall.

The Urinary System

19 Cells form a variety of wastes as by-products of metabolic processes, and if these substances are allowed to accumulate, their effects are likely to be toxic.

Body fluids, such as blood and lymph, serve to carry wastes away from the tissues that produce them. Other parts remove these wastes from the blood and transport them to the outside. The respiratory system, for example, removes carbon dioxide and water from the blood, and the *urinary system* removes water, salts, and nitrogenous wastes. In both systems, the wastes are carried to the outside through tubular organs.

The urinary system also helps to maintain the normal concentrations of water and electrolytes within body fluids. It helps to regulate the volume of body fluids and aids in the control of red blood cell production and blood pressure. Thus, the urinary system plays a key role in maintaining homeostasis.

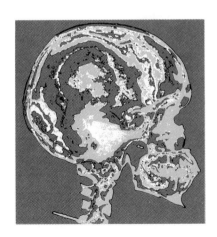

Chapter Outline

Chapter Objectives

After you have studied this chapter, you should be able to

1. Name the organs of the urinary system and list their general functions.

2. Describe the locations of the kidneys and the structure of a kidney.

3. Trace the pathway of blood through the major vessels within a kidney.

4. Describe a nephron and explain the functions of its parts.

5. Explain how glomerular filtrate is produced and describe its composition.

6. Discuss the role of tubular reabsorption in the formation of urine.

7. Explain why the osmotic concentration of the glomerular filtrate changes as it travels through a renal tubule.

8. Describe a countercurrent mechanism and explain how it helps concentrate urine.

9. Define *tubular secretion* and explain its role in urine formation.

10. Describe the structure of the ureters, urinary bladder, and urethra.

11. Discuss the process of micturition and explain how it is controlled.

12. Complete the review activities at the end of this chapter.

afferent arteriole   (af′er-ent ar-te′re-ōl)

Bowman's capsule   (bo′manz kap′sūl)

countercurrent mechanism
  (kown′ter kur′ent mek′ah-nizm)

detrusor muscle   (de-truz′or mus′l)

efferent arteriole   (ef′er-ent ar-te′re-ōl)

glomerulus   (glo-mer′u-lus)

micturition   (mik″tu-rish′un)

nephron   (nef′ron)

peritubular capillary   (per″i-tū′bu-lar kap′ĭ-ler″e)

renal corpuscle   (re′nal kor′pusl)

renal cortex   (re′nal kor′teks)

renal medulla   (re′nal mĕ-dul′ah)

renal plasma threshold
  (re′nal plaz′mah thresh′old)

renal tubule   (re′nal tu′būl)

calyc-, a small cup: major *calyc*es—cuplike divisions of the renal pelvis.

detrus-, to force away: *detrus*or muscle—muscle within bladder wall that causes urine to be expelled.

glom-, little ball: *glom*erulus—cluster of capillaries within a renal capsule.

juxta-, near to: *juxta*medullary nephron—a nephron located near the renal medulla.

nephr-, pertaining to the kidney: *nephr*on—functional unit of a kidney.

mict-, to pass urine: *mict*urition—process of expelling urine from the bladder.

papill-, nipple: renal *papill*ae—small elevations that project into the renal sinus.

ren-, kidney: *ren*al cortex—outer region of the kidney.

trigon-, a triangular shape: *trigon*e—triangular area on the internal floor of the bladder.

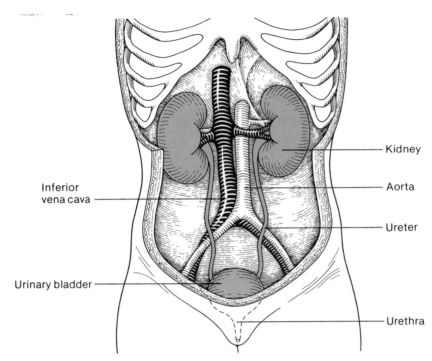

Kidney

Aorta

Ureter

Inferior vena cava

Urinary bladder

Urethra

The urinary system consists of a pair of glandular *kidneys*, which remove substances from the blood and form urine; a pair of tubular *ureters*, which transport urine away from the kidneys; a saclike *urinary bladder*, which serves as a urine reservoir; and a tubular *urethra*, which conveys urine to the outside of the body. These organs are shown in figure 19.1.

The Kidney

A **kidney** is a reddish brown, bean-shaped organ with a smooth surface. It is about 12 cm long, 6 cm wide, and 3 cm thick (4.7 × 2.3 × 1.2 in) in an adult, and is enclosed in a tough, fibrous capsule (tunic fibrosa).

Location of the Kidneys

The kidneys lie on either side of the vertebral column in a depression high on the posterior wall of the abdominal cavity.

Although the positions of the kidneys may vary slightly with changes in posture and with breathing movements, their upper and lower borders are generally at the levels of the twelfth thoracic and third lumbar vertebrae, respectively. The left kidney usually remains about 1.5 to 2 cm higher than the right one.

The kidneys are positioned *retroperitoneally*, which means they are behind the parietal peritoneum and against the deep muscles of the back. As figure 19.2 shows, they are held in position by connective tissue (renal fascia) and masses of adipose tissue (renal fat) that surround them.

Structure of a Kidney

The lateral surface of each kidney is convex, while its medial side is deeply concave. The resulting medial depression leads into a hollow chamber called the **renal sinus**, which is completely surrounded by kidney tissue. The entrance to this sinus is termed the *hilum*, and through it pass various blood vessels, nerves, lymphatic vessels, and the ureter. (See fig. 19.3.)

The superior end of the ureter is expanded to form a funnel-shaped part called the **renal pelvis**, which is located inside the renal sinus. The pelvis is divided into two or three tubes called *major calyces* (singular, *calyx*), and they in turn are divided into several (eight to fourteen) *minor calyces*.

A series of small elevations project into the renal sinus from the kidney tissue forming the sinus wall. These projections are called *renal papillae*, and each of them is pierced by tiny openings into a minor calyx.

Fig. 19.2 The kidneys are located behind the parietal peritoneum and are surrounded and supported by adipose tissue.

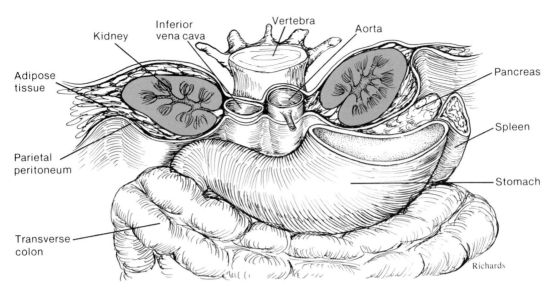

The substance of the kidney is divided into two distinct regions—an inner medulla and an outer cortex. The **renal medulla** is composed of conical masses of tissue called *renal pyramids*, whose bases are directed toward the convex surface of the kidney, and whose apexes form the renal papillae. The tissue of the medulla appears striated due to the presence of microscopic tubules leading from the cortex to the renal papillae.

The **renal cortex**, which appears somewhat granular, forms a shell around the medulla. Its tissue dips into the medulla between adjacent renal pyramids to form *renal columns*. The granular appearance of the cortex is due to the random arrangement of tiny tubules associated with **nephrons**, the functional units of the kidney.

During fetal development of a human kidney, several pyramidal masses or lobes fuse to form the final structure. Consequently, lobulation is evident on the surface of a fetal kidney, and this condition sometimes persists into adulthood.

1. *Where are the kidneys located?*
2. *Describe the structure of a kidney.*
3. *Name the functional unit of the kidney.*

Renal Blood Vessels

Blood is supplied to a kidney by means of a **renal artery** that arises from the abdominal aorta. This artery enters a kidney through the hilum and gives off several branches, the *interlobar arteries*, that pass between the medullary pyramids.

At the junction between the medulla and cortex, the interlobar arteries branch to form a series of incomplete arches, the *arciform arteries* (arcuate arteries), which in turn give rise to *interlobular arteries*. Lateral branches of the interlobular arteries, called **afferent arterioles**, lead to the nephrons in the renal cortex.

Venous blood is returned through a series of vessels that correspond generally to the arterial pathways. For example, the venous blood passes through interlobular, arciform, interlobar, and renal veins. The **renal vein** then joins the inferior vena cava as it courses through the abdominal cavity. (The renal arteries and veins are shown in fig. 19.4.)

The Nephrons

Structure of a Nephron. A kidney contains about one million nephrons, each consisting of a **renal corpuscle** and a **renal tubule**. (See fig. 19.3.)

A renal corpuscle (malpighian corpuscle) is composed of a tangled cluster of blood capillaries, called a **glomerulus**, and a thin-walled, saclike structure, called **Bowman's capsule**, that surrounds the glomerulus.

Fig. 19.3 (*a*) Longitudinal section of a kidney; (*b*) a renal pyramid containing nephrons; (*c*) a single nephron.

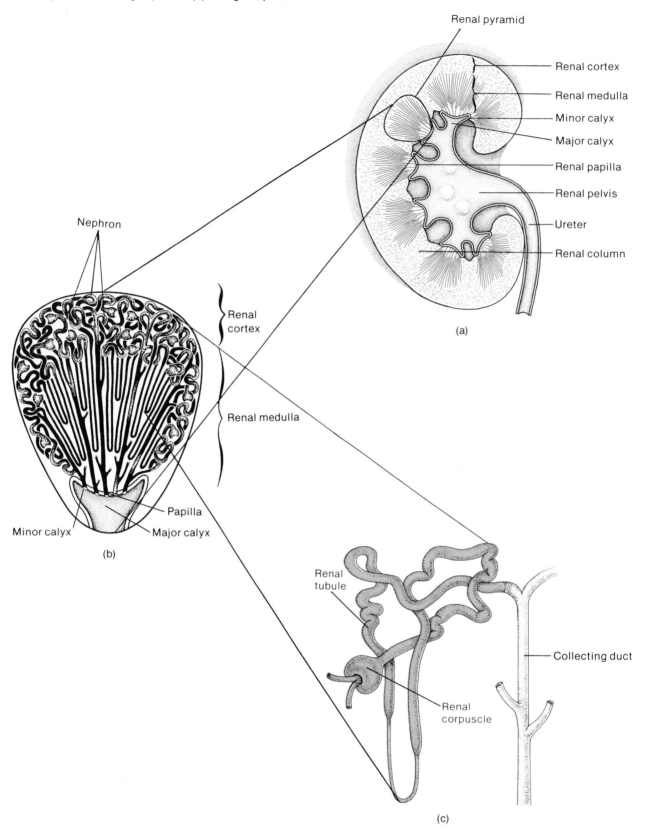

Renal pyramid

Renal cortex

Renal medulla

Minor calyx

Major calyx

Renal papilla

Renal pelvis

Ureter

Renal column

(a)

Nephron

Renal cortex

Renal medulla

Minor calyx

Papilla

Major calyx

(b)

Renal tubule

Collecting duct

Renal corpuscle

(c)

Fig. 19.4 Main branches of the renal artery and vein.

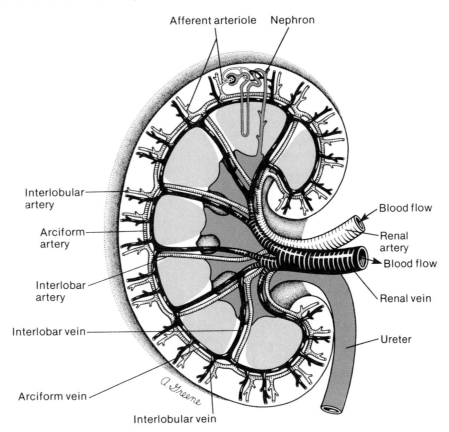

The Bowman's capsule is an expansion at the closed end of a renal tubule. It is composed of two layers of squamous epithelial cells: a visceral layer that closely covers the glomerulus, and an outer parietal layer that is continuous with the visceral layer and with the wall of the renal tubule.

The renal tubule leads away from the Bowman's capsule and becomes highly coiled. This portion is appropriately named the *proximal convoluted tubule*.

The proximal tubule then dips toward the renal pelvis. As it follows a straight path into the deeper layers of the cortex, it is called the *descending limb of the loop of Henle*. The tubule soon curves back toward its renal corpuscle and becomes the *ascending limb of the loop of Henle*.

The ascending limb returns in a straight line to the region of the renal corpuscle, where it becomes highly coiled again, and is called the *distal convoluted tubule*. This distal portion is shorter than the proximal tubule, and its convolutions are less complex.

Near its beginning, the distal tubule contacts the afferent arteriole close to its entrance into the glomerulus. At this point, cells of the tubule and arteriole form a specialized structure (juxtaglomerular apparatus) from which *renin* is released. This substance, as is described in chapter 17, causes vasoconstriction and an increase in arterial blood pressure in response to decreased blood flow through the kidney.

Several distal convoluted tubules merge in the renal cortex to form a *collecting duct*, which in turn passes into the renal medulla, becoming larger and larger as it is joined by other collecting ducts. The resulting tube (papillary duct) empties into a minor calyx through an opening in a renal papilla. The parts of a nephron are shown in figure 19.5.

Cortical and Juxtamedullary Nephrons. Most nephrons have corpuscles located in the renal cortex near the surface of the kidney. They are called *cortical nephrons*, and they have relatively short loops of Henle that usually do not reach the renal medulla.

Another group, called *juxtamedullary nephrons*, have corpuscles close to the renal medulla, and their loops of Henle extend deep into the medulla. Although they represent only about 20% of the total, these nephrons play an important role in regulating the process of concentrating urine. (See fig. 19.6.)

Fig. 19.5 The parts of the renal tubule connect the Bowman's capsule with a collecting duct.

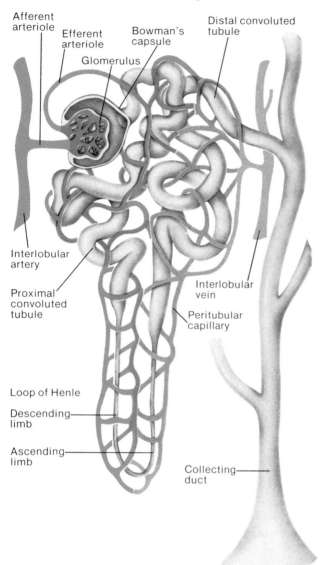

Fig. 19.6 Cortical nephrons are close to the surface of a kidney; juxtamedullary nephrons are near the renal medulla.

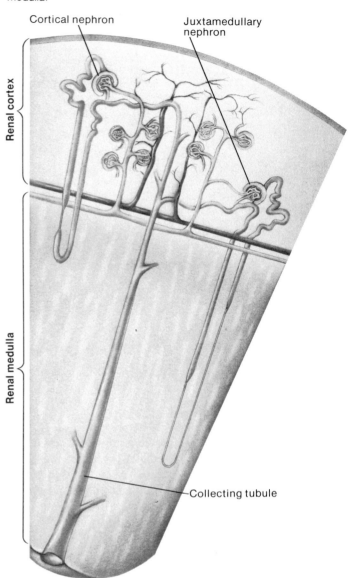

Blood Supply of a Nephron. The capillary cluster that forms a glomerulus arises from an **afferent arteriole**. After passing through the capillary of the glomerulus, blood enters an **efferent arteriole** (rather than a venule) whose diameter is somewhat less than that of the afferent vessel.

The efferent arteriole branches into a complex, freely anastomosing network of capillaries that surrounds the various portions of the renal tubule. This network is called the **peritubular capillary system**. (See fig. 19.5.)

Special branches of this system, which receive blood primarily from the efferent arterioles of the juxtamedullary nephrons, form capillary loops called *vasa recta.* These loops dip into the renal medulla and are closely associated with the loops of the juxtamedullary nephrons.

After flowing through the vasa recta, blood is returned to the renal cortex, where it joins blood from other branches of the peritubular capillary system and enters the venous system of the kidney. (See fig. 19.7.)

The term *nephritis* refers to an inflammation of the kidney. *Glomerulonephritis*, for example, involves inflammation of the glomeruli, while *pyelonephritis* is an inflammation of the renal pelvis.

Fig. 19.7 The capillary loop of the vasa recta is closely associated with the loop of Henle of a juxtamedullary nephron.

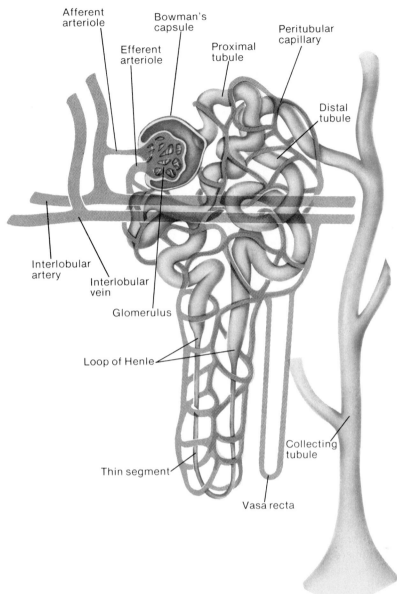

1. Describe the characteristics of the blood supply to the kidney.
2. Name the parts of a nephron.
3. Distinguish between a cortical and a juxtamedullary nephron.
4. Describe the characteristics of the blood supply to a nephron.

Urine Formation

The primary functions of the nephrons include the removal of waste substances from the blood and the regulation of water and electrolyte concentrations within the body fluids. The end product of these func-

tions is **urine**, which is excreted to the outside of the body carrying with it wastes, excess water, and excess electrolytes.

Urine formation involves these processes: *filtration,* into renal tubules, of various substances from the plasma within glomerular capillaries; *reabsorption,* into the plasma, of some of these substances; *secretion,* into the renal tubules, of other substances from the plasma within the peritubular capillaries.

Glomerular Filtration

Urine formation begins when water and various dissolved substances are forced out of the glomerular capillaries by blood pressure (hydrostatic pressure).

Fig. 19.8 The first step in urine formation is the filtration of substances through the glomerular membrane into Bowman's capsule.

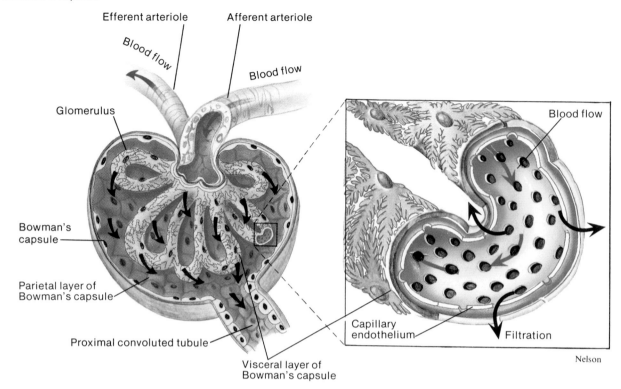

Nelson

The filtration of these materials through the capillary walls is much like the filtration that occurs at the arteriole ends of other capillaries throughout the body. However, the glomerular capillaries are many times more permeable than the capillaries in other tissues. (See fig. 19.8.)

The resulting **glomerular filtrate** has about the same composition as the filtrate that becomes tissue fluid elsewhere. That is, glomerular filtrate is largely water and contains essentially the same substances as blood plasma, except for the larger protein molecules, which the filtrate lacks. More specifically, glomerular filtrate contains water, glucose, amino acids, urea, uric acid, creatine, creatinine, and ions of sodium, chlorine, potassium, calcium, bicarbonate, phosphate, and sulfate. The relative concentrations of some of these substances in plasma, glomerular filtrate, and urine are shown in chart 19.1.

As the filtrate leaves the glomerular capillaries, it is received by the enlarged end of the renal tubule—the Bowman's capsule. The rate of filtration is directly proportional to the *hydrostatic pressure* of the blood. In other words, factors that tend to cause an increase in blood pressure within the glomerulus will promote filtration, while factors causing a decrease in pressure will inhibit it.

Since the glomerular capillary is located between two arterioles—the *afferent* and *efferent arter-*

Chart 19.1 Relative concentrations of certain substances in plasma, glomerular filtrate, and urine

Substance	Concentrations (mEq/l)		
	Plasma	Glomerular Filtrate	Urine
Sodium (Na^+)	142	142	128
Potassium (K^+)	5	5	60
Calcium (Ca^{+2})	4	4	5
Magnesium (Mg^{+2})	3	3	15
Chlorine (Cl^-)	103	103	134
Bicarbonate (HCO_3^-)	27	27	14
Sulfate (SO_4^{-2})	1	1	33
Phosphate (PO_4^{-3})	2	2	40

Substance	Concentrations (mg/100 ml)		
	Plasma	Glomerular Filtrate	Urine
Glucose	100	100	0
Urea	26	26	1820
Uric acid	4	4	53
Creatinine	1	1	196

ioles—any change in the diameters of these vessels is likely to cause a hydrostatic pressure change within the glomerulus, accompanied by a change in the glomerular filtration rate. For example, if the afferent arteriole, through which blood enters the glomerulus, becomes constricted, blood flow is diminished, the hydrostatic pressure is decreased, and the filtration rate drops. If the efferent arteriole, through which blood leaves the glomerulus, becomes constricted, blood backs up into the glomerulus, the glomerular hydrostatic pressure is increased, and the filtration rate rises. Converse effects are produced by vasodilation of these vessels.

In *glomerulonephritis*, the glomerular capillaries are inflamed and become more permeable to proteins. Consequently, proteins appear in the glomerular filtrate and are excreted in the urine (proteinuria). At the same time, the protein concentration in the blood plasma decreases (hypoproteinemia), and this causes a drop in the osmotic pressure of the blood. As a result, the movement of tissue fluid into the capillaries is decreased, and edema develops.

Normally, *sympathetic nerve impulses* cause proportional amounts of vasoconstriction in the afferent and efferent arterioles, and the glomerular filtration rate remains relatively stable. But when there is excessive sympathetic stimulation, the vessels may constrict so much that blood is largely shunted to other body parts, and the glomerular filtration rate may approach zero.

It might be expected that any rise in arterial pressure would cause a corresponding rise in the glomerular filtration rate, but this usually does not occur. Instead, rises in arterial pressure are accompanied by vasoconstriction of the afferent arterioles, and the glomerular filtration rate usually increases only slightly. This phenomenon is called *autoregulation*. On the other hand, if the arterial blood pressure drops significantly, as may occur in *shock*, the glomerular hydrostatic pressure falls and the filtration rate may decrease considerably. In fact a reduction of kidney function is a factor that sometimes causes the effects of shock to be irreversible.

If urine excretion is prevented by obstructions in renal tubules or excretory ducts leading to the outside, fluid tends to back up in the tubules. As a consequence, the hydrostatic pressure within Bowman's capsules increases, interfering with glomerular filtration. A stone that blocks a ureter, for example, or an enlarged prostate gland that obstructs the urethra may cause a significant decrease in glomerular filtration.

The *osmotic pressure* of the plasma is another important factor that influences glomerular filtration. In capillaries, the hydrostatic pressure acting to force

Fig. 19.9 Some factors that affect the rate of glomerular filtration.

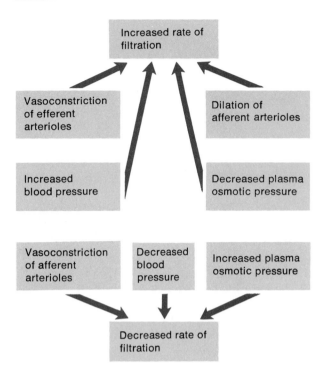

water and dissolved substances outward is opposed by the effect of the plasma osmotic pressure that attracts water inward. Actually, as filtration occurs, proteins remaining in the plasma cause the osmotic pressure within the glomerular capillary to rise. When this pressure reaches a certain high level, filtration ceases. Conversely, conditions that tend to decrease plasma osmotic pressure, such as a decrease in plasma protein concentration, cause an increase in the filtration rate. (See fig. 19.9.)

In an average adult, the glomerular filtration rate for the nephrons of both kidneys is about 125 ml per minute, or 180,000 ml (180 liters) in 24 hours. Assuming that the plasma volume is about 3 liters (3.2 quarts), the production of 180 liters of filtrate in 24 hours means that all of the plasma must be filtered through the glomeruli about sixty times each day. (See fig. 19.10.)

Since this 24-hour volume is nearly 45 gallons, it is obvious that all of it is not excreted as urine. Instead, most of the fluid that passes through the renal tubules is reabsorbed and reenters the plasma.

The volume of plasma filtered by the kidneys is also related to the amount of *surface area* within the glomerular capillaries. This surface area is estimated to be about 2 square meters—approximately equal to the surface area of an adult's skin.

Fig. 19.10 (a) Relative amounts of glomerular filtrate and (b) urine formed in 24 hours.

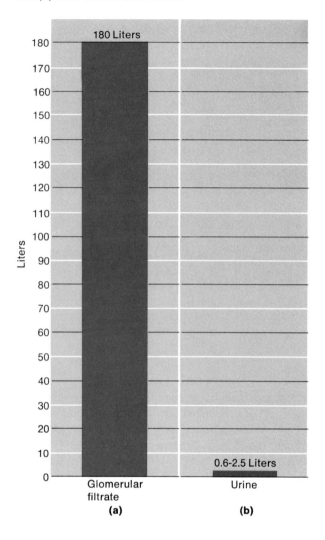

(a)

(b)

---

1. What general processes are involved with urine formation?
2. What force is responsible for glomerular filtration?
3. What factors influence the rate of glomerular filtration?

Tubular Reabsorption

If the composition of the glomerular filtrate entering the renal tubule is compared with that of the urine leaving the tubule, it is clear that changes have occurred as the fluid passed through the tubule. (See chart 19.1.) For example, glucose is present in the filtrate, but is absent in the urine. Also, urea and uric acid are considerably more concentrated in urine than they are in the glomerular filtrate. Such changes in fluid composition are largely the result of **tubular reabsorption**, a process by which substances are trans-

Fig. 19.11 Reabsorption is the process by which substances are transported from the glomerular filtrate into the blood of the peritubular capillary. What substances are reabsorbed in this manner?

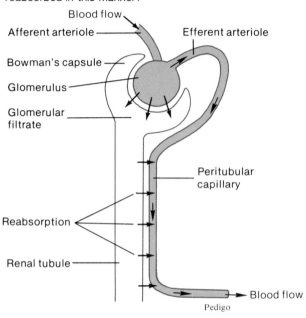

ported out of the glomerular filtrate, through the epithelium of the renal tubule, and into the blood of the peritubular capillary. (See fig. 19.11.)

Since the efferent arteriole is narrower than the peritubular capillary, blood flowing from the former into the latter is under relatively low pressure. Also, the wall of this capillary is more permeable than that of other capillaries. Both of these factors enhance the rate of fluid reabsorption from the renal tubule.

Although tubular reabsorption occurs throughout the renal tubule, most of it occurs in the proximal convoluted portion. Various segments of the tubule, however, are adapted to reabsorb specific substances using particular modes of transport. Glucose reabsorption, for example, occurs primarily through the walls of the proximal tubule by active transport, while water is reabsorbed throughout the length of the renal tubule by osmosis, a passive process.

As is described in chapter 3, an active transport mechanism depends on the presence of carrier molecules in a cell membrane. These carriers transport passenger molecules through the membrane, release them, and return to the other side to transport more passenger molecules. Such a mechanism has a *limited transport capacity*; that is, it can only transport a certain number of molecules in a given amount of time, because the number of carriers is limited. (See fig. 19.12.)

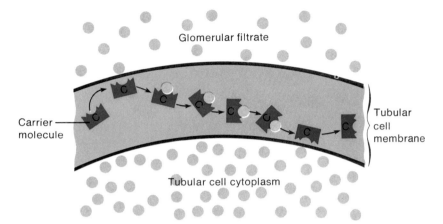

Usually all of the glucose in the glomerular filtrate is reabsorbed, since there are enough carrier molecules to transport them. Sometimes the plasma glucose concentration increases, and if it reaches a critical level, called the *renal plasma threshold*, there will be more glucose molecules in the filtrate than the active transport mechanism can handle. As a result, some glucose will remain in the filtrate and be excreted in the urine.

The appearance of glucose in the urine is called *glucosuria* (or *glycosuria*). This condition may occur following the administration of glucose intravenously or in a patient with diabetes mellitus. If the cause is diabetes mellitus, the blood glucose concentration rises because of insufficient insulin from the pancreas.

Amino acids also enter the glomerular filtrate and are reabsorbed in the proximal convoluted tubule, apparently by three different active transport mechanisms. Each mechanism is thought to reabsorb a different group of amino acids, whose members have molecular similarities. As a result of their actions, only a trace of amino acids normally remains in urine.

Although the glomerular filtrate is nearly free of protein, a minute quantity of *albumin* may be present. These proteins have relatively small molecules, and they are thought to be reabsorbed by *pinocytosis*. Once they are inside an epithelial cell, the proteins are probably converted to amino acids.

Excessive protein in the urine, called *proteinuria*, is one of the more common symptoms of kidney disease, and it usually reflects an increase in the permeability of the glomerular capillaries. Although albumins are most likely to appear in the urine, other plasma proteins also may be present.

Other substances reabsorbed by the epithelium of the proximal convoluted tubule include creatine, lactic acid, citric acid, uric acid, ascorbic acid (vitamin C), phosphate ions, sulfate ions, calcium ions, potassium ions, and sodium ions. As a group, these substances are reabsorbed by active transport mechanisms with *limited transport capacities* (like that of glucose). Such a substance usually does not appear in the urine until its concentration in the glomerular filtrate exceeds its particular threshold.

Sodium and Water Reabsorption. Substances that remain in the renal tubule tend to become more and more concentrated as water is reabsorbed from the filtrate. Most of this water reabsorption occurs *passively* by *osmosis* in the proximal convoluted tubule and is closely associated with the active reabsorption of sodium ions. In fact, if sodium reabsorption increases, water reabsorption increases; if sodium reabsorption decreases, water reabsorption decreases also.

About 70% of the *sodium ion reabsorption* occurs in the proximal segment of the renal tubule by active transport (sodium pump mechanism). As these positively charged ions (Na^+) are moved through the tubular wall, negatively charged ions including chloride (Cl^-), phosphate (PO_4^{-3}), and bicarbonate

(HCO$_3$$^-$) accompany them. This movement of negatively charged ions is due to the electrochemical attraction between particles of opposite charge. It is termed **passive transport** because it does not require a direct expenditure of cellular energy.

As more and more sodium ions are actively transported into the peritubular capillary, along with various negatively charged ions, the concentration of solutes within the peritubular blood is increased. Furthermore, since water moves through cell membranes from regions of lesser solute concentration (hypotonic) toward regions of greater solute concentration (hypertonic), water is transported by osmosis from the renal tubule into the peritubular capillary. Because of the movement of solutes and water into the peritubular capillary, the volume of fluid within the renal tubule is greatly reduced. (See fig. 19.13.)

1. How is the peritubular capillary adapted for reabsorption?
2. What substances present in glomerular filtrate are not normally present in urine?
3. What mechanisms are responsible for reabsorption of solutes from the glomerular filtrate?
4. Define renal plasma threshold.
5. Describe the role of passive transport in urine formation.

Regulation of Urine Concentration and Volume

Sodium ions continue to be reabsorbed by active transport as the tubular fluid moves through the loop of Henle, the distal convoluted segment, and the collecting duct. Consequently, almost all the sodium that enters the renal tubule as glomerular filtrate may be reabsorbed before the urine is excreted.

Water also continues to be reabsorbed passively by osmosis in various segments of the renal tubule, and as this occurs, the osmotic concentration of the tubular fluid changes. These changes result from a *countercurrent mechanism* and the actions of certain *hormones*.

The **countercurrent mechanism** involves the loops of Henle of the juxtamedullary nephrons. The descending and ascending limbs of these U-shaped structures lie parallel and very close to one another. The mechanism is named for the fact that fluid moving down the descending limb creates a current that is counter to that of the fluid moving up in the ascending limb.

The tubular fluid in the proximal segment is *isotonic* to the plasma of the peritubular capillary blood; however, it becomes *hypertonic* when it moves through the descending limb. Then, as it moves through the ascending limb, the tubular fluid becomes *hypotonic*.

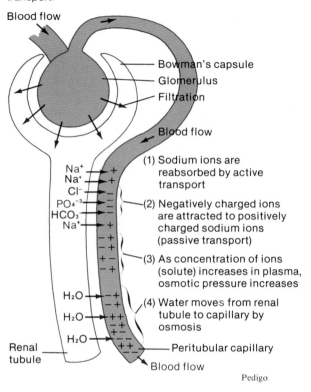

Fig. 19.13 Water reabsorption by osmosis occurs in response to the reabsorption of sodium by active transport.

Blood flow

Bowman's capsule
Glomerulus
Filtration

Blood flow

Na$^+$
Na$^+$
Cl$^-$
PO$_4$$^{-3}$
HCO$_3$
Na$^+$

(1) Sodium ions are reabsorbed by active transport

(2) Negatively charged ions are attracted to positively charged sodium ions (passive transport)

(3) As concentration of ions (solute) increases in plasma, osmotic pressure increases

H$_2$O
H$_2$O
H$_2$O

(4) Water moves from renal tubule to capillary by osmosis

Renal tubule

Peritubular capillary

Blood flow

Pedigo

These changes in the osmotic concentration are due primarily to activity in the epithelium of the ascending limb. Specifically, the lining in the upper portion of this limb (thick segment) is impermeable to water. However, the epithelium does carry on the active reabsorption of chloride ions, and sodium ions follow them passively. As NaCl accumulates, the interstitial fluid outside the ascending limb becomes *hypertonic,* while the tubular fluid inside becomes *hypotonic* since it is losing its solute.

The epithelium of the descending limb (thin segment) is quite permeable. Because this segment is surrounded by the hypertonic fluid created by the ascending limb, water tends to leave the descending limb by osmosis. Thus, the contents of the descending limb become more and more concentrated, or *hypertonic* to the plasma of the peritubular capillary.

Also, since the concentration of NaCl in the medullary interstitial fluid is high, NaCl reenters the descending limb by diffusion. This makes the fluid in the descending limb still more concentrated (hypertonic). (See fig. 19.14.)

Thus, NaCl moves from the descending limb into the ascending limb, is reabsorbed into the medullary interstitial fluid, and diffuses back into the descending limb again. Each time this circuit is

Fig. 19.14 (a) Fluid in the ascending limb becomes hypotonic as solute is reabsorbed; (b) fluid in the descending limb becomes hypertonic as it loses water by osmosis and gains solute by diffusion.

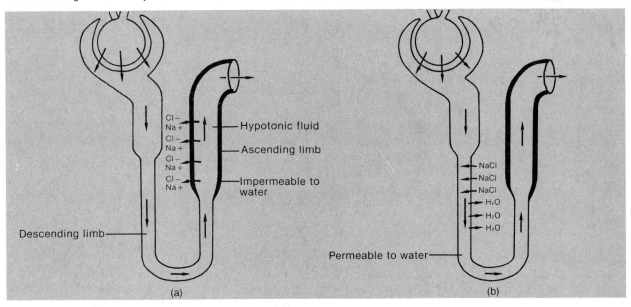

(a)

(b)

Fig. 19.15 (a) As NaCl completes the countercurrent circuit again and again, it becomes more concentrated; (b) as a result, an NaCl concentration gradient is established in the medullary interstitial fluid.

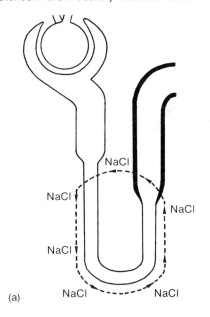

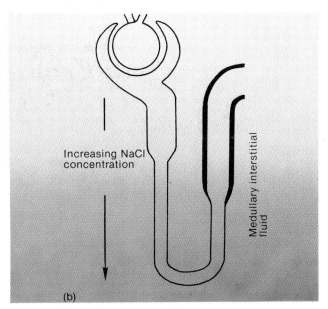

completed, the concentration of NaCl increases or multiplies. For this reason, the mechanism is called a *countercurrent multiplier.*

As a result of this mechanism, the NaCl concentration of the medullary interstitial fluid is greatest near the tip of the loops of Henle and decreases progressively toward the renal cortex. As is explained

later, this concentration gradient is important to the process of concentrating urine. (See fig. 19.15.)

The maintenance of the NaCl concentration gradient is aided by another countercurrent mechanism operating in the *vasa recta.* In this case, blood flows relatively slowly down the descending portion of the vessel loop, and NaCl enters it by diffusion.

Fig. 19.16 A countercurrent mechanism in the vasa recta helps to maintain the NaCl concentration gradient in the medullary interstitial fluid.

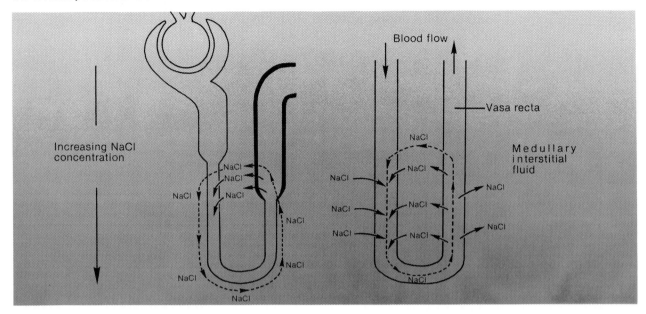

Then, as the blood moves back up toward the renal cortex, most of the NaCl diffuses from the blood and reenters the medullary interstitial fluid. Consequently, little NaCl is carried away from the renal medulla. (See fig. 19.16.)

The tubular fluid reaching the distal convoluted tubule is hypotonic to its surroundings. The cells lining this segment and the collecting duct that follows, continue to reabsorb sodium and chloride ions, but they are quite impermeable to water. Thus, water tends to accumulate inside the tubule and may be excreted as dilute urine.

The action of antidiuretic hormone (ADH) greatly increases the water permeability of these tubular cells. As is discussed in chapter 12, ADH is produced by specialized neurons in the *hypothalamus* and is released in the posterior lobe of the pituitary gland in response to a decreasing concentration of water in the blood.

When it reaches the kidney, ADH causes an increase in the permeability of the epithelial linings of the distal convoluted tubule and the collecting duct, and water moves rapidly out of these segments by osmosis. Consequently, the urine volume is reduced, and it becomes more concentrated (hypertonic).

This concentrating of the urine continues as it passes down the *collecting duct*; this occurs because the hypertonic interstitial fluid of the renal medulla surrounds the collecting duct. (See fig. 19.17.)

To summarize, ADH stimulates the production of concentrated urine, which contains soluble wastes and other substances in a minimum of water; it also inhibits the loss of body fluids whenever there is a danger of dehydration. If the water concentration of the body fluids is excessive, ADH secretion is decreased. In the absence of ADH, the epithelial linings of the distal segment and the collecting duct become impermeable to water, less water is reabsorbed, and the urine tends to be more dilute.

Urea and Uric Acid Excretion

Urea is a by-product of amino acid metabolism. Consequently, its plasma concentration is directly related to the amount of protein in the diet. Urea enters the renal tubule by filtration, and about 50% of it is reabsorbed (passively) by diffusion while the remainder is excreted in the urine.

Urea is also involved in more complex movements within the kidney, including a countercurrent multiplier mechanism that causes it to be concentrated in the medullary interstitial fluid. As a result of this, urea becomes more highly concentrated in the urine than would otherwise be possible.

Uric acid, which results from the metabolism of certain organic bases (purines) in nucleic acids, is reabsorbed by active transport. Although this mechanism seems able to reabsorb all the uric acid normally present in glomerular filtrate, about 10% of the amount filtered is excreted in the urine. This amount is apparently *secreted* into the renal tubule.

Fig. 19.17 (a) The distal convoluted tubule and collecting duct are impermeable to water, so water may be excreted as dilute urine; (b) if ADH is present, however, these segments become permeable, and water is reabsorbed by osmosis into the hypertonic medullary interstitial fluid.

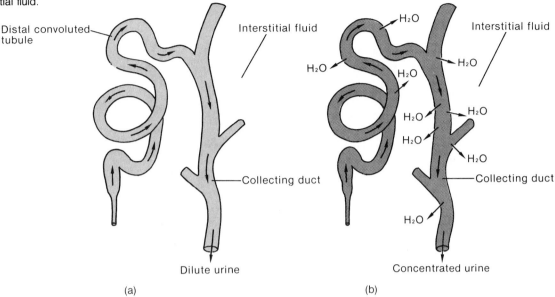

(a) (b)

Gout is a disorder in which the plasma concentration of uric acid becomes abnormally high. Since uric acid is a relatively insoluble substance, it tends to precipitate when it is present in excess. As a result, crystals of uric acid may be deposited in joints and other tissues, where they produce inflammation and extreme pain. The joints of the great toes are most commonly affected, but other joints in the hands and feet are often involved.

This condition, which often seems to be inherited, is sometimes treated by administering drugs that inhibit the reabsorption of uric acid and, thus, increase its excretion.

1. Describe the countercurrent mechanism.
2. What role does the hypothalamus play in the concentrating of urine?
3. Explain how urea and uric acid are excreted.

Tubular Secretion

Tubular secretion (tubular excretion) is the process by which certain substances are transported from the plasma of the peritubular capillary into the fluid of the renal tubule. As a result, the amount of a particular substance excreted in the urine may be greater than the amount filtered from the plasma in the glomerulus. (See fig. 19.18.)

Fig. 19.18 Secretory mechanisms act to move substances from the plasma of the peritubular capillary into the fluid of the renal tubule.

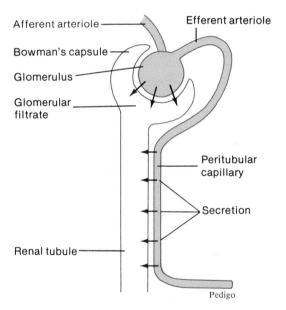

Some substances are secreted by active transport mechanisms similar to those that function in reabsorption. However, the *secretory mechanisms* transport substances in the opposite direction. For example, certain organic compounds, including penicillin and histamine, are actively secreted into the tubular fluid by the epithelium of the proximal convoluted segment.

Hydrogen ions are also actively secreted. In this case, the proximal segment of the renal tubule is specialized to secrete large quantities of hydrogen ions between plasma and tubular fluid, where hydrogen ion concentrations are similar. The distal segment and collecting duct are specialized to transport small quantities of hydrogen ions between plasma and urine, where hydrogen ion concentrations may vary.

Although most of the *potassium ions* in the glomerular filtrate are actively reabsorbed in the proximal convoluted tubule, some may be secreted passively in the distal segment and collecting duct. During this process, the active reabsorption of sodium ions out of the tubular fluid creates a negative electrical charge within the tube. Since positively charged potassium ions (K^+) are attracted to regions that are negatively charged, these ions move through the tubular epithelium and enter the tubular fluid.

To summarize, urine is formed as a result of the following: glomerular filtration of materials from blood plasma; reabsorption of substances, such as glucose, amino acids, proteins, creatine, lactic acid, citric acid, uric acid, ascorbic acid, phosphate ions, sulfate ions, calcium ions, potassium ions, sodium ions, water, and urea; and secretion of substances, such as penicillin, histamine, phenobarbital, hydrogen ions, ammonia, and potassium ions. (See chart 19.2.)

---

1. *Define* tubular secretion.
2. *What substances are actively secreted? Passively secreted?*
3. *How does the reabsorption of sodium affect the secretion of potassium?*

Since the kidneys serve to remove substances from the blood, the rate at which a particular substance is removed or cleared is indicative of kidney function. A test called *inulin clearance,* for example, is often used to judge the rate of glomerular filtration. In this test, the polysaccharide, inulin, is infused into the blood at a constant rate, and the levels of inulin in samples of plasma and urine are determined. The volume of plasma that is cleared of inulin per minute is calculated. The results of this test are useful in detecting damage to the glomeruli and for observing the progress of a glomerular disease.

Composition of Urine

The composition of urine varies considerably from time to time because of differences in dietary intake

Chart 19.2 Functions of nephron parts

Part	Function
Renal Corpuscle	
Glomerulus	Filtration of water and dissolved substances from plasma
Bowman's capsule	Receives glomerular filtrate
Renal Tubule	
Proximal convoluted tubule	Reabsorption of glucose, amino acids, creatine, lactic acid, citric acid, uric acid, ascorbic acid, phosphate ions, sulfate ions, calcium ions, potassium ions, and sodium ions by active transport
	Reabsorption of proteins by pinocytosis
	Reabsorption of water by osmosis
	Reabsorption of chloride ions, and other negatively charged ions by electrochemical attraction
	Active secretion of substances such as penicillin, histamine, and hydrogen ions
Descending limb of loop of Henle	Reabsorption of water by osmosis
Ascending limb of loop of Henle	Reabsorption of chloride ions by active transport and passive reabsorption of sodium ions
Distal convoluted tubule	Reabsorption of sodium ions by active transport
	Reabsorption of water by osmosis
	Active secretion of hydrogen ions
	Passive secretion of potassium ions by electrochemical attraction

and physical activity. In addition to containing about 95% water, it usually contains *urea* from the catabolic metabolism of amino acids, *uric acid* from the metabolism of nucleic acids, and *creatinine* from the metabolism of creatine. It may also contain a trace of *amino acids,* as well as a variety of *electrolytes* whose concentrations tend to vary directly with the amounts included in the diet. (See chart 19.1.)

Abnormal constituents of urine include *glucose, proteins, hemoglobin, ketones,* and various *blood cells.* The significance of such substances in urine, however, may depend on the amounts present and on other factors. For example, glucose may appear in urine following a large intake of carbohydrates, proteins may appear following vigorous physical exercise, and ketones may appear following a prolonged fast. Also, some pregnant women have glucose in their urine toward the end of pregnancy.

The volume of urine produced by the kidneys usually varies between 0.6 and 2.5 liters per day. The exact volume is influenced by such factors as the fluid intake, the environmental temperature, the relative humidity of the surrounding air, and the person's emotional condition, respiratory rate, and body temperature. An output of 50–60 cc of urine per hour is considered normal, and an output of less than 30 cc per hour may be an indication of kidney failure.

The kidneys of infants and young children are unable to concentrate urine and conserve water as effectively as those of adults. Consequently, such young persons produce relatively large volumes of urine and tend to lose water rapidly.

1. List the normal constituents of urine.
2. What is the normal hourly output of urine? The minimal hourly output?

Elimination of Urine

After being formed, urine passes from the collecting ducts through openings in the renal papillae and enters the major and minor calyces of the kidney. From there it passes through the renal pelvis and is conveyed by a ureter to the urinary bladder. It is excreted to the outside of the body by means of the urethra.

The Ureter

The **ureter**, which is a tubular organ about 25 cm (10 in) long, begins as the funnel-shaped renal pelvis. It extends downward behind the parietal peritoneum and parallel to the vertebral column. Within the pelvic cavity, it courses forward and medially to join the urinary bladder from underneath.

The wall of the ureter is composed of three layers of tissues. The inner layer, or *mucous coat,* includes several thicknesses of epithelial cells and is continuous with the linings of the renal tubules and the urinary bladder. The middle layer, or *muscular coat,* consists largely of smooth muscle fibers arranged in circular and longitudinal bundles. The outer layer, or *fibrous coat,* is composed of connective tissue. (See fig. 19.19.)

Fig. 19.19 Cross section of a ureter.

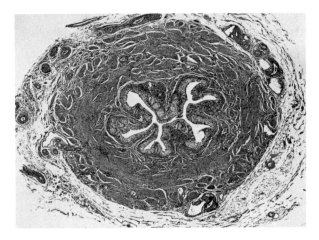

Since the linings of the ureters and the urinary bladder are continuous, infectious agents such as bacteria may ascend from the bladder into the ureters. An inflammation of the bladder, which is called *cystitis,* occurs more commonly in women than men because the female urethral pathway is shorter. An inflammation of the ureter is called *ureteritis.*

Although the ureter is simply a tube leading from the kidney to the urinary bladder, its muscular wall helps to move urine. Muscular peristaltic waves that originate in the renal pelvis force urine along the length of the ureter. These waves are initiated by the presence of urine in the renal pelvis, and their frequency is related to the rate of urine formation. If the rate of urine formation is high, a peristaltic wave may occur every few seconds; if the rate is low, a wave may occur every few minutes.

When such a peristaltic wave reaches the urinary bladder, it causes a jet of urine to spurt into the bladder. The opening through which the urine enters is covered by a flaplike fold of mucous membrane. This fold acts as a valve, allowing urine to move inward from the ureter but preventing it from backing up.

If a ureter becomes obstructed, as when a small kidney stone (renal calculus) is present, strong peristaltic waves are initiated in the proximal portion of the tube. Such waves may help to move the stone into the bladder. At the same time, the presence of a stone usually stimulates a sympathetic reflex (ureterorenal reflex) that results in constriction of the renal arterioles and reduces the production of urine in the kidney on the affected side.

Kidney stones, which are usually composed of uric acid, calcium oxalate, calcium phosphate, or magnesium phosphate, sometimes form in the renal pelvis. If such a stone passes into a ureter it may stimulate severe pain. This pain usually begins in the region of the kidney and tends to radiate into the abdomen, pelvis, and legs. It may also be accompanied by nausea and vomiting.

1. Describe the structure of a ureter.
2. How is urine moved from the renal pelvis?
3. What prevents urine from backing up from the urinary bladder into the ureters?
4. How does an obstruction in a ureter affect urine production?

The Urinary Bladder

The **urinary bladder** is a hollow, distensible, muscular organ. It is located within the pelvic cavity, behind the pubic symphysis and below the parietal peritoneum. (See fig. 19.20.) In a male, it lies against the rectum posteriorly, and in a female it contacts the anterior walls of the uterus and vagina.

Although the bladder is somewhat spherical, its shape is altered by the pressures of surrounding organs. When it is empty, the inner wall of the bladder is thrown into many folds, but as it fills with urine, the wall becomes smoother. At the same time, the superior surface of the bladder expands upward into a dome.

When it is greatly distended, the bladder pushes above the pubic crest and into the region between the abdominal wall and the parietal peritoneum. The dome can reach the level of the umbilicus and press against the coils of the small intestine.

The internal floor of the bladder consists of a triangular area called the *trigone*, which has an opening at each of its three angles. (See fig. 19.21.) Posteriorly, at the base of the trigone, the openings are those of the ureters. The *internal urethral orifice*, which opens into the urethra, is located anteriorly at the apex of the trigone. The trigone generally remains in a fixed position even though the rest of the bladder changes shape during distension and contraction, thus preventing a reflux of urine into the ureters.

The wall of the urinary bladder consists of four layers. The inner layer, or *mucous coat*, includes several thicknesses of epithelial cells, similar to those lining the ureters and the upper portion of the urethra. This tissue, called *transitional epithelium*, is adapted to changes in tension. More precisely, its thickness

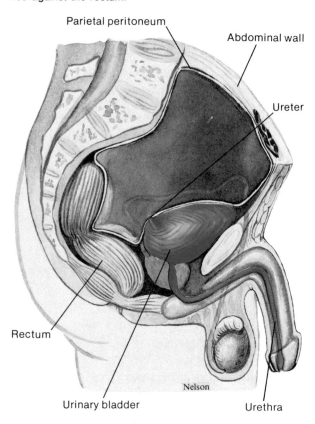

Fig. 19.20 The urinary bladder is located within the pelvic cavity and behind the pubic symphysis. In a male, it lies against the rectum.

Parietal peritoneum

Abdominal wall

Ureter

Rectum

Nelson

Urinary bladder

Urethra

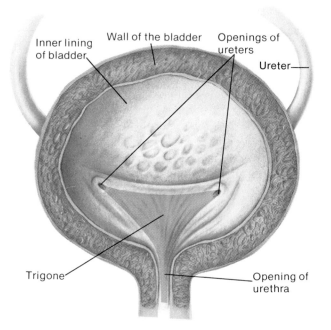

Fig. 19.21 The trigone of the urinary bladder is marked by the openings of the ureters and urethra.

Inner lining of bladder

Wall of the bladder

Openings of ureters

Ureter

Trigone

Opening of urethra

Fig. 19.22 As the urinary bladder expands, its wall becomes thinner.

Wall of bladder in distended condition

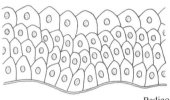

Wall of bladder in collapsed condition

Pedigo

changes as the bladder expands and contracts, so that during distension it may be only two or three cells thick, while during contraction it may be five or six cells thick. (See fig. 19.22.)

The second layer of the wall is the *submucous coat.* It consists of connective tissue and contains many elastic fibers.

The third layer, or *muscular coat,* is composed primarily of coarse bundles of smooth muscle fibers. These bundles are interlaced in all directions and depths, and together they comprise the **detrusor muscle.** This muscle is supplied with parasympathetic nerve fibers that function in the micturition reflex.

The outer layer, or *serous coat,* consists of the parietal peritoneum. This layer occurs only on the upper surface of the bladder. Elsewhere, the outer coat is composed of fibrous connective tissue.

---

1. Describe the trigone of the urinary bladder.
2. Describe the structure of the bladder wall.
3. What kind of nerve fibers supply the detrusor muscle?

Micturition

Micturition, or urination, is the process by which urine is expelled from the urinary bladder. It involves the contraction of the detrusor muscle and may be aided by contractions of muscles in the abdominal wall and pelvic floor, and fixation of the thoracic wall and diaphragm. Micturition also involves the relaxation of the *external urethral sphincter.* This muscle surrounds the urethra about 2 cm (0.8 in) from the bladder and is composed of voluntary muscle tissue.

The need to urinate is usually stimulated by distension of the bladder wall as it fills with urine. As the wall expands, stretch receptors are stimulated, and the micturition reflex is triggered.

Although an infant is unable to control the voluntary muscles associated with micturition, voluntary control of micturition becomes possible as various portions of the brain and spinal cord mature.

The *micturition reflex center* is located in the sacral segments of the spinal cord. When it is signalled by sensory impulses from the stretch receptors, *parasympathetic* motor impulses travel out to the detrusor muscle, and it undergoes rhythmic contractions in response. This action is accompanied by a sensation of urgency.

Although the urinary bladder may hold as much as 600 ml of urine, the desire to urinate is usually experienced when it contains about 150 ml. Then, as the volume of urine increases to 300 ml or more, the sensation of fullness becomes increasingly uncomfortable.

Since the external urethral sphincter can be consciously controlled, it ordinarily remains contracted until a decision is made to urinate. This control is aided by nerve centers in the *midbrain* and *cerebral cortex* that are able to inhibit the micturition reflex. When a person decides to urinate, the external urethral sphincter is allowed to relax, and the micturition reflex is no longer inhibited. Nerve centers within the *pons* and *hypothalamus* may function to make the micturition reflex more effective. Consequently, the detrusor muscle contracts, and urine is excreted to the outside through the urethra. Within a few moments, the neurons of the micturition reflex seem to fatigue, the detrusor muscle relaxes, and the bladder begins to fill with urine again.

This process is outlined in chart 19.3.

Damage to the spinal cord above the sacral region may result in loss of voluntary control of urination. However, if the micturition reflex center and its sensory and motor fibers are uninjured, micturition may continue to occur reflexly. In this case, the bladder collects urine until its walls are stretched enough to trigger a micturition reflex, and the detrusor muscle contracts in response. This condition is called an *automatic bladder.*

Chart 19.3 Major events of micturition

1. Urinary bladder becomes distended as it fills with urine.

2. Stretch receptors in the bladder wall are stimulated, and they signal the micturition center in the spinal cord.

3. Parasympathetic nerve impulses travel to the detrusor muscle, which responds by contracting rhythmically.

4. The need to urinate is sensed as urgent.

5. Urination is prevented by voluntary contraction of the external urethral sphincter and by inhibition of the micturition reflex by impulses from the midbrain and cerebral cortex.

6. Following the decision to urinate, the external urethral sphincter is relaxed, and the micturition reflex is facilitated by impulses from the pons and hypothalamus.

7. The detrusor muscle contracts, and urine is expelled through the urethra.

8. Neurons of the micturition reflex center fatigue, the detrusor muscle relaxes, and the bladder begins to fill with urine again.

The Urethra

The **urethra** is a tube that conveys urine from the urinary bladder to the outside of the body. Its wall is lined with mucous membrane and contains a relatively thick layer of smooth muscle tissue, whose fibers are generally directed longitudinally. It also contains numerous mucous glands, called *urethral glands,* that secrete mucus into the urethral canal.

In a female, the urethra is about 4 cm long (1.6 in). It passes forward from the bladder, courses below the pubic symphysis, and empties between the labia minora. Its opening, the *external urethral orifice* (urinary meatus), is located anterior to the vaginal opening and about 2.5 cm (1 in) posterior to the clitoris.

In a male, the urethra, which functions both as a urinary canal and a passageway for cells and secretions from various reproductive organs, can be divided into three sections—the prostatic urethra, the membranous urethra, and the penile urethra. (See fig. 19.23.)

The **prostatic urethra** is about 2.5 cm long (1 in) and passes from the urinary bladder through the *prostate gland,* which is located just below the bladder. Ducts from various reproductive structures join the urethra in this region.

Fig. 19.23   Section of the male urethra.

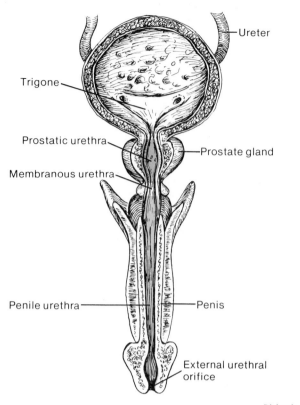

Trigone

Prostatic urethra

Membranous urethra

Penile urethra

Ureter

Prostate gland

Penis

External urethral orifice

Richards

The **membranous urethra** is about 0.5 cm long (0.2 in). It begins just distal to the prostate gland, passes through the urogenital diaphragm, and is surrounded by the fibers of the external urethral sphincter muscle.

The **penile urethra** is about 15 cm long (6 in) and passes through the corpus spongiosum of the penis, where it is surrounded by erectile tissue. This portion of the urethra terminates with the *external urethral orifice* at the tip of the penis.

1. Describe the events of micturition.
2. How is it possible to inhibit the micturition reflex?
3. Describe the structure of the urethra.
4. How does the urethra of a male differ from that of a female?

Some Clinical Terms Related to the Urinary System

anuria (ah-nu're-ah)—an absence of urine due to failure of kidney function or to an obstruction in a urinary pathway.

bacteriuria (bak-te''re-u're-ah)—bacteria in the urine.

cystectomy (sis-tek'to-me)—surgical removal of the urinary bladder.

cystitis (sis-ti'tis)—inflammation of the urinary bladder.

cystoscope (sis'to-skōp)—instrument used for visual examination of the interior of the urinary bladder.

cystotomy (sis-tot'o-me)—an incision into the wall of the urinary bladder.

diuresis (di''u-re'sis)—an increased production of urine.

dysuria (dis-u're-ah)—painful or difficult urination.

enuresis (en''u-re'sis)—uncontrolled urination.

hematuria (hem''ah-tu're-ah)—blood in the urine.

nephrectomy (nĕ-frek'to-me)—surgical removal of a kidney.

nephrolithiasis (nef''ro-lĭ-thi'ah-sis)—presence of a stone(s) in the kidney.

nephroptosis (nef''rop-to'sis)—a movable or displaced kidney.

oliguria (ol''ĭ-gu're-ah)—a scanty output of urine.

polyuria (pol''e-u're-ah)—an excessive output of urine.

pyelolithotomy (pi''ĕ-lo-lĭ-thot'o-me)—removal of a stone from the renal pelvis.

pyelotomy (pi''ĕ-lot'o-me)—incision into the renal pelvis.

pyuria (pi-u're-ah)—pus in the urine.

uremia (u-re'me-ah)—condition in which substances ordinarily excreted in the urine accumulate in the blood.

ureteritis (u-re''ter-i'tis)—inflammation of the ureter.

urethritis (u''re-thri'tis)—inflammation of the urethra.

Chapter Summary

Introduction

The urinary system functions to remove wastes from the blood, regulate the concentrations of fluids and electrolytes, help control red blood cell production and blood pressure, and excrete urine.

The system consists of kidneys, ureters, urinary bladder, and urethra.

The Kidney

1. Location of the kidneys
   a. The kidneys are on either side of the vertebral column, high on the posterior wall of the abdominal cavity.
   b. They are positioned behind the parietal peritoneum and held in place by adipose and connective tissue.
2. Structure of a kidney
   a. A kidney contains a hollow renal sinus surrounded by kidney tissue.
   b. The ureter expands into the renal pelvis, which in turn is divided into major and minor calyces.
   c. Renal papillae project into renal sinus.
   d. Kidney tissue is divided into medulla and cortex.
3. Renal blood vessels
   a. Arterial blood flows through the renal artery, interlobar arteries, arciform arteries, interlobular arteries, and afferent arterioles.
   b. Venous blood returns through a series of vessels that correspond to those of the arterial pathways.
4. The nephrons
   a. Structure of the nephron
      (1) The nephron is the functional unit of the kidney.
      (2) It consists of a renal corpuscle and a renal tubule.
         (a) The corpuscle consists of glomerulus and Bowman's capsule.
         (b) Portions of the renal tubule include the proximal convoluted tubule, the loop of Henle (ascending and descending limbs), the distal convoluted tubule, and the collecting duct.
      (3) The collecting duct empties into the minor calyx of the renal pelvis.
   b. Cortical and juxtamedullary nephrons
      (1) Cortical nephrons are most numerous and have corpuscles near the surface of the kidney.
      (2) Juxtamedullary nephrons have corpuscles near the medulla.

c. Blood supply of a nephron
  (1) The glomerular capillary receives blood from the afferent arteriole and passes it to the efferent arteriole.
  (2) The efferent arteriole gives rise to the peritubular capillary system that surrounds the renal tubule.
  (3) Capillary loops, called vasa recta, dip down into the medulla.

Urine Formation

Nephrons function to remove wastes from blood and to regulate water and electrolyte concentrations.

Urine is the end product of these functions, which involve filtration, reabsorption, and secretion of substances from renal tubules.

1. Glomerular filtration
  a. Water and dissolved materials are forced out of the glomerular capillary by blood pressure.
  b. Composition of filtrate is similar to tissue fluid.
  c. Glomerular filtrate is received by Bowman's capsule.
  d. Rate of filtration
    (1) This varies with the hydrostatic pressure of blood.
    (2) It varies with diameters of afferent and efferent arterioles.
    (3) It remains relatively stable in the normal blood pressure range, but decreases if blood pressure drops significantly.
    (4) It varies with the osmotic pressure of plasma.
  e. The kidneys produce about 125 ml of glomerular fluid per minute, most of which is reabsorbed.
  f. Volume of filtrate varies with the surface area of the glomerular capillary.
2. Tubular reabsorption
  a. Substances are selectively reabsorbed from the glomerular filtrate.
  b. The peritubular capillary is adapted for reabsorption.
    (1) It carries low pressure blood.
    (2) It is very permeable.
  c. Most reabsorption occurs in the proximal tubule.
  d. Various substances are reabsorbed in particular segments of the renal tubule by different modes of transport.
    (1) Glucose and amino acids are reabsorbed by active transport.
    (2) Water is reabsorbed by osmosis.
    (3) Proteins are reabsorbed by pinocytosis.
  e. Active transport mechanisms have limited transport capacities.
  f. If the concentration of a substance in the filtrate exceeds its renal plasma threshold, the excess is excreted in the urine.
  g. Substances that remain in the filtrate are concentrated as water is reabsorbed.

h. Sodium ions are reabsorbed by active transport.
  (1) As positively charged sodium ions are transported out of the filtrate, negatively charged ions accompany them.
  (2) Water is passively reabsorbed by osmosis as sodium ions are actively reabsorbed.
3. Regulation of urine concentration and volume
  a. Most sodium is reabsorbed before urine is excreted.
  b. Sodium is concentrated in the renal medulla by the countercurrent mechanism.
    (1) Chloride ions are actively reabsorbed in the ascending limb, and sodium ions follow them passively.
    (2) Tubular fluid in the ascending limb becomes hypotonic as it loses solutes.
    (3) Water leaves the descending limb by osmosis, and NaCl enters this limb by diffusion.
    (4) Tubular fluid in the descending limb becomes hypertonic as it loses water and gains NaCl.
    (5) As NaCl repeats this circuit, its concentration in the medulla increases.
  c. The vasa recta countercurrent mechanism helps to maintain the NaCl concentration in the medulla.
  d. The distal tubule and collecting duct are impermeable to water so it tends to be excreted in urine.
  e. ADH from the posterior pituitary gland causes the permeability of the distal tubule and collecting duct to increase, and thus promotes the reabsorption of water.
4. Urea and uric acid excretion
  a. Urea is a by-product of amino acid metabolism.
    (1) It is reabsorbed passively by diffusion.
    (2) About 50% is excreted in urine.
    (3) A countercurrent mechanism helps in the excretion of urea.
  b. Uric acid results from the metabolism of nucleic acids.
    (1) Most is reabsorbed by active transport.
    (2) Some is secreted into the renal tubule.
5. Tubular secretion
  a. This is the process by which certain substances are transported from the plasma to the tubular fluid.
  b. Some substances are secreted actively.
    (1) These include various organic compounds and hydrogen ions.
    (2) Hydrogen ions are secreted by the proximal and distal segments of the renal tubule.
  c. Potassium ions are secreted passively in the distal segment and collecting duct where they are attracted by the negative charge that develops in the lumen of the tubule.

6. Composition of urine
   a. Urine is about 95% water, and it usually contains urea, uric acid, and creatinine.
   b. It may contain a trace of amino acids and varying amounts of electrolytes, depending upon the dietary intake.
   c. The volume of urine varies with the fluid intake and with certain environmental factors.

Excretion of Urine

1. The ureter
   a. The ureter is a tubular organ that extends from the kidney to the urinary bladder.
   b. Its wall has mucous, muscular, and fibrous layers.
   c. Peristaltic waves in the ureter force urine to the bladder.
   d. Obstruction in the ureter stimulates strong peristaltic waves and a reflex that causes the kidney to decrease urine production.
2. The urinary bladder
   a. The urinary bladder is a distensible organ that stores urine and forces it into the urethra.
   b. The openings for the ureters and urethra are located at the three angles of the trigone in the floor of the urinary bladder.
   c. Muscle fibers in the wall form the detrusor muscle.
3. Micturition
   a. Micturition is the process by which urine is expelled.
   b. It involves contraction of the detrusor muscle and relaxation of the external urethral sphincter.
   c. Micturition reflex
      (1) Stretch receptors in the bladder wall are stimulated by distension.
      (2) The micturition reflex center in the sacral spinal cord sends parasympathetic motor impulses to the detrusor muscle.
      (3) Urination can be controlled by means of the voluntary external urethral sphincter and nerve centers in the brain that can inhibit micturition reflex.
      (4) When the decision to urinate is made, the external urethral sphincter is allowed to relax and nerve centers in the brain act to facilitate the micturition reflex.
4. The urethra
   a. The urethra conveys urine from the bladder to the outside.
   b. In females, it empties between the labia minora.
   c. In males, it conveys products of reproductive organs as well as urine.
      (1) There are three portions of the male urethra: prostatic, membranous, and penile.
      (2) It empties at the tip of the penis.

Application of Knowledge

1. If an infant is born with a narrowing of the renal arteries, what effect would this condition have on the volume of urine produced? Explain your answer.
2. If a patient who has had major abdominal surgery receives intravenous fluids equal to the volume of blood lost during surgery, would you expect the volume of urine produced to be greater or less than normal? Why?
3. If a physician prescribed oral penicillin therapy for a patient with an infection of the urinary bladder, how would you describe for the patient the route by which the drug will reach the bladder?

Review Activities

1. Name the organs of the urinary system and list their general functions.
2. Describe the external and internal structure of a kidney.
3. Name the vessels through which blood passes as it travels from the renal artery to the renal vein.
4. Distinguish between a renal corpuscle and a renal tubule.
5. Name the parts through which fluid passes as it travels from the glomerulus to the collecting duct.
6. Describe the location of the source of renin.
7. Distinguish between cortical and juxtamedullary nephrons.
8. Distinguish between filtration, reabsorption, and secretion as they relate to urine formation.
9. Compare the composition of the glomerular filtrate with that of blood plasma.
10. Explain how the diameters of the afferent and efferent arterioles affect the rate of glomerular filtration.
11. Explain how changes in the osmotic pressure of the blood plasma may affect the rate of glomerular filtration.
12. Discuss the reason tubular reabsorption is said to be a selective process.
13. Explain how the peritubular capillary is adapted for reabsorption.
14. Explain why active transport mechanisms have limited transport capacities.
15. Define *renal plasma threshold* and explain its significance in tubular reabsorption.

16. Explain how amino acids and proteins are reabsorbed.

17. Describe the effect of sodium reabsorption on the reabsorption of negatively charged ions.

18. Explain how sodium reabsorption affects water reabsorption.

19. Explain how hypotonic tubular fluid is produced in the ascending limb of the loop of Henle.

20. Explain why fluid in the descending loop of Henle is hypertonic.

21. Describe the function of ADH.

22. Explain how urine may become concentrated as it moves through the collecting duct.

23. Compare the processes by which urea and uric acid are reabsorbed.

24. Explain how the renal tubule is adapted to carry on the secretion of hydrogen ions.

25. Explain how potassium ions may be secreted passively.

26. List the more common wastes found in urine and their sources.

27. List some factors that affect the volume of urine produced each day.

28. Describe the structure and function of a ureter.

29. Explain how the muscular wall of the ureter aids in moving urine.

30. Discuss what happens if a ureter becomes obstructed.

31. Describe the structure and location of the urinary bladder.

32. Define *detrusor muscle*.

33. Describe the micturition reflex.

34. Explain how the micturition reflex can be voluntarily controlled.

35. Compare the urethra of a female with that of a male.

Suggestions for Additional Reading

Anderson, B. 1977. Regulation of body fluids. *Ann. Rev. Physio.* 39:185.

Bauman, J. W., and Chinard, F. P. 1975. *Renal function: physiological and medical aspects.* St Louis: C.V. Mosby.

Gilmore, J. P. 1972. *Renal physiology.* Baltimore: The Williams and Wilkins Co.

Harvey, R. J. 1976. *The kidneys and the internal environment.* New York: Halsted Press.

Keitzer, W. A., and Huffman, G. C. 1971. *Urodynamics.* Springfield, Ill.: C. C. Thomas.

Lassiter, W. E. 1975. Kidney. *Ann. Rev. Physio.* 37:371.

Orloff, J., and Burg, M. 1971. Kidney. *Ann. Rev. Physio.* 33:83.

Pitts, R. F. 1974. *Physiology of the kidney and body fluids.* 3rd ed. Chicago: Year Book Medical Publishers.

Share, L., et al. 1972. Regulation of body fluids. *Ann. Rev. Physio.* 34:235.

Sullivan, L. P. 1974. *Physiology of the kidney.* Philadelphia: Lea and Febiger.

Vander, A. J. 1975. *Renal physiology.* New York: McGraw-Hill.

Water and Electrolyte Balance

20 Cell functions and, indeed, cell survival depend on homeostasis—the existence of a stable cellular environment. In such an environment, body cells are continually supplied with oxygen and nutrients, and the waste products resulting from their metabolic activities are continually carried away. At the same time, the concentrations of water and dissolved electrolytes in the cellular fluids and their surroundings remain constant. This condition requires the maintenance of a *water* and *electrolyte balance*.

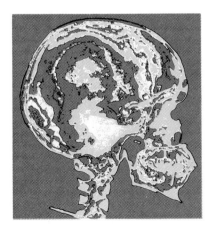

After you have studied this chapter, you should be able to

1. Explain what is meant by water and electrolyte balance and discuss the importance of this balance.

2. Describe how body fluids are distributed within compartments, how the fluid composition differs between compartments, and how fluids move from one compartment to another.

3. List the routes by which water enters and leaves the body and explain how water input and output are regulated.

4. Explain how electrolytes enter and leave the body and how the input and output of electrolytes are regulated.

5. Discuss the factors leading to dehydration and water intoxication and describe the consequences of each.

6. List several causes of edema and explain how each produces its effect.

7. Describe the consequences of sodium and potassium imbalance.

8. Explain what is meant by acid-base balance.

9. Describe how hydrogen ion concentrations are expressed.

10. List the major sources of hydrogen ions in the body.

11. Distinguish between strong and weak acids and bases.

12. Explain how changing pH values of body fluids are minimized by chemical buffer systems, the respiratory center, and the kidneys.

13. Distinguish between acidosis and alkalosis, explain how these conditions may arise, and describe how they are controlled.

14. Complete the review activities at the end of this chapter.

acid   (as′id)

acidosis   (as″i-do′sis)

alkaline   (al′kah-līn)

alkalosis   (al″kah-lo′sis)

base   (bās)

buffer system   (buf′er sis′tem)

dehydration   (de″hi-dra′shun)

edema   (ĕ-de′mah)

electrolyte balance   (e-lek′tro-līt bal′ans)

extracellular   (ek″strah-sel′u-lar)

intracellular   (in″trah-sel′u-lar)

osmoreceptor   (oz″mo-re-sep′tor)

transcellular   (trans-sel′u-lar)

water balance   (wot′er bal′ans)

water intoxication   (wot′er in-tok″sĭ-ka′shun)

de-, separation from: *de*hydration—removal of water from cells or body fluids.

edem-, swelling: *edem*a—swelling due to an abnormal accumulation of extracellular fluid.

-emia, a blood condition: hypoprotein*emia*—an abnormally low concentration of blood plasma protein.

extra-, outside: *extra*cellular fluid—fluid outside of body cells.

im- (or in-), not: *im*balance—condition in which factors are not in equilibrium.

intra-, within: *intra*cellular fluid—fluid within body cells.

neutr-, neither one nor the other: *neutr*al—solution that is neither acidic nor basic.

-osis, a state of: acid*osis*—condition in which hydrogen ion concentration is abnormally high.

-uria, a urine condition: keto*uria*—presence of ketone bodies in the urine.

The term *balance* suggests a state of equilibrium, and in the case of water and electrolytes, it means the quantities entering the body are equal to the quantities leaving it. Maintaining such a balance requires mechanisms that ensure that lost water and electrolytes will be replaced and any excesses will be expelled. As a result, the quantities within the body are relatively stable at all times.

It is important to remember that water balance and electrolyte balance are interdependent, since the electrolytes are dissolved in the water of body fluids. Consequently, anything that alters the concentrations of electrolytes will necessarily alter the concentration of water by adding solutes to it or by removing solutes from it. Likewise, anything that changes the concentration of water will change the concentrations of electrolytes by making them more concentrated or more diluted.

Distribution of Body Fluids

Water and electrolytes are not evenly distributed throughout the tissues. Instead, they occur in regions or *compartments* that contain fluids of varying compositions. Movement of water and electrolytes between these compartments is regulated, so that their distribution remains stable.

Fluid Compartments

The body of an average adult male is about 63% water by weight. This water (about 40 liters), together with its dissolved electrolytes, is distributed into two major compartments—an intracellular compartment and an extracellular compartment. (See fig. 20.1.)

While the average adult male body is about 63% water by weight, the average adult female body is only about 52% water. This difference is related to the fact that adipose tissue has a relatively low water content, and generally there is a greater development of subcutaneous adipose tissue in females. Also, obese people of either sex contain proportionally less body water than do lean people.

The *intracellular fluid compartment* includes all the water and electrolytes enclosed by cell membranes. In other words, intracellular fluid is the fluid within cells, and in an adult it represents about 63% of the total body water.

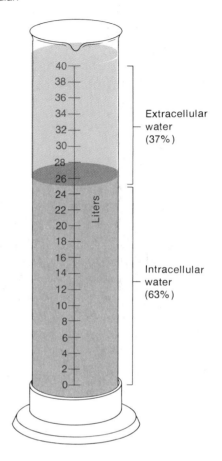

Fig. 20.1 Of the 40 liters of water in the body of an average adult male, about 63% is intracellular and 37% is extracellular.

The *extracellular fluid compartment* includes all the fluid outside cells—within the tissue spaces (interstitial fluid), the blood vessels (plasma), and the lymphatic vessels (lymph). A specialized fraction of the extracellular fluid, called **transcellular fluid,** includes *cerebrospinal fluid* of the central nervous system, *aqueous and vitreous humors* of the eyes, *synovial fluid* of the joints, *serous fluid* within the various body cavities, and fluid *secretions* of glands. All together, the fluids of the extracellular compartment constitute about 37% of the total body water (fig. 20.2).

Composition of Body Fluids

Extracellular fluids generally have similar compositions. They are characterized by relatively high concentrations of sodium, chloride, and bicarbonate ions, and lesser concentrations of potassium, calcium, magnesium, phosphate, and sulfate ions. The plasma fraction of extracellular fluid contains considerably more protein than either interstitial fluid or lymph.

Fig. 20.2 Fluid within the intracellular compartment is separated from fluid in the extracellular compartment by cell membranes. What membranes separate the various components of the extracellular fluid compartment?

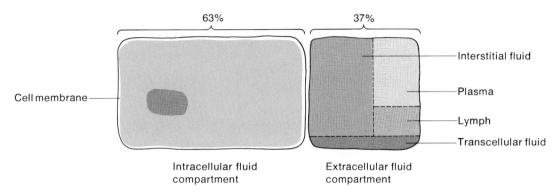

Total body water

Cell membrane

Intracellular fluid compartment

Extracellular fluid compartment

Interstitial fluid
Plasma
Lymph
Transcellular fluid

Intracellular fluid contains relatively high concentrations of potassium, phosphate, and magnesium ions. It includes somewhat greater concentrations of calcium and sulfate ions, and lesser concentrations of sodium, chloride, and bicarbonate ions than extracellular fluid. Intracellular fluid also has a greater concentration of protein than plasma. These relative concentrations are shown in figure 20.3.

---

1. How are fluid balance and electrolyte balance interdependent?
2. Describe the normal distribution of water within the body.
3. What electrolytes are in higher concentrations in extracellular fluid? In intracellular fluid?
4. How does the concentration of protein vary in various body fluids?

Movement of Fluid between Compartments

The movement of water and electrolytes from one compartment to another is largely regulated by two factors—hydrostatic pressure and osmotic concentration. For example, as is explained in chapter 17, fluid leaves the plasma at the arteriole ends of capillaries and enters the interstitial spaces because of the filtration effect of *hydrostatic pressure* (blood pressure). It returns to the plasma from the interstitial spaces at the venule ends of capillaries because of *osmotic pressure*. Likewise, as is mentioned in chapter 18, fluid leaves the interstitial spaces and enters the lymph capillaries because of osmotic pressure that develops within these spaces. As a result of the circulation of lymph, interstitial fluid is returned to the plasma.

The movement of fluid between the intracellular and extracellular compartments is similarly controlled by such pressures. However, since the hydrostatic pressure within the cells and the surrounding interstitial fluid ordinarily is equal and remains stable, any fluid movement that occurs is likely to be the result of changes in osmotic pressure. (See fig. 20.4.)

For example, because the sodium concentration in the interstitial fluid is especially high, a decrease in this concentration will cause a net movement of water from the extracellular compartment into the intracellular compartment by osmosis. As a consequence, cells will tend to swell. Conversely, if the concentration of sodium in the interstitial fluid increases, the net movement of water will be outward from the intracellular compartment, and the cells will shrink as they lose water.

---

1. What factors control movement of water and electrolytes from one fluid compartment to another?
2. How does the sodium concentration within body fluids affect the net movement of water between compartments?

Fig. 20.3 Extracellular fluid is relatively high in concentrations of sodium, chloride, and bicarbonate ions; intracellular fluid is relatively high in concentrations of potassium, magnesium, and phosphate ions.

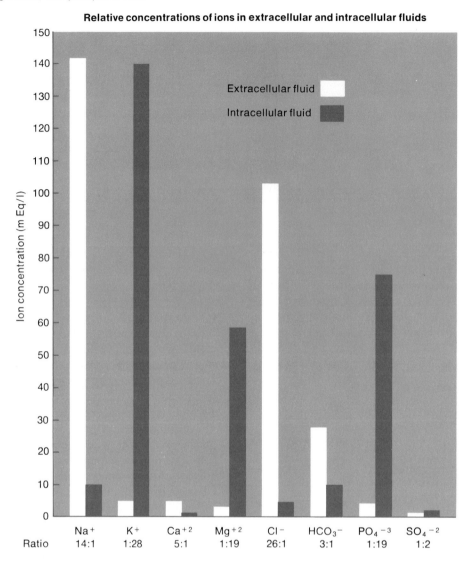

Relative concentrations of ions in extracellular and intracellular fluids

Fig. 20.4 Net movements of fluids between compartments result from differences in hydrostatic and osmotic pressures.

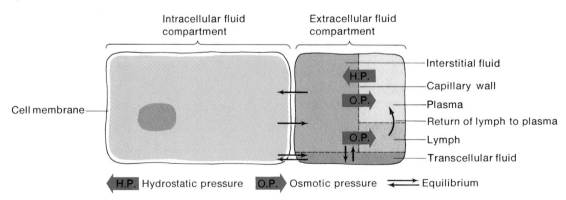

Processing and Transporting

Water Balance

Water balance exists when the total intake of water is equal to the total loss of water.

Water Intake

Although the volume of water gained each day varies from individual to individual, an average adult living in a moderate environment takes in about 2500 ml.

Of this amount, probably 60% (1500 ml) will be drinking water or beverages, while another 30% (750 ml) will be in moist food. The remaining 10% (250 ml) will be a by-product of oxidative metabolism of various nutrients, which is called **water of metabolism.** (See fig. 20.5.)

Regulation of Water Intake

The primary regulator of water intake is thirst, and although the thirst mechanism is poorly understood, it seems to involve the osmotic pressure of extracellular fluid and a *thirst center* in the hypothalamus of the brain.

As water is lost from the body, the osmotic pressure of the extracellular fluid increases. *Osmoreceptors* in the thirst center are thought to be stimulated by such an osmotic change, and as a result, the hypothalamus causes the person to feel thirsty and be motivated to seek water. The feeling of thirst is usually accompanied by dryness of the mouth, which is related to the loss of extracellular water and a consequent decreased flow of saliva.

The thirst mechanism is usually triggered whenever the total body water is decreased by 1 or 2%. As a person drinks water in response to thirst, the act of drinking and the resulting distension of the stomach wall seems to trigger nerve impulses that inhibit the mechanism. Thus, the person usually stops drinking long before the swallowed water has been absorbed. This inhibition helps to prevent the person from drinking more than is required to replace the quantity lost. Thus, it avoids development of an imbalance of the opposite type. Chart 20.1 summarizes the steps in this mechanism.

The managers of bars and taverns often provide salty snacks without charge for their customers. As the patrons enjoy these snacks, their thirst mechanisms are triggered, and they are likely to order more liquid refreshments. Thus, the patrons satisfy their thirsts for fluid as well as the tavernkeeper's thirst for revenue.

Fig. 20.5 Major sources of body water.

Average daily intake of water

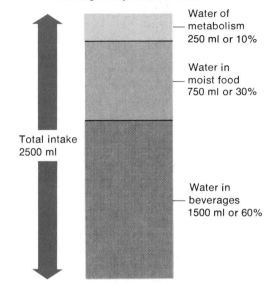

Total intake 2500 ml

Water of metabolism 250 ml or 10%

Water in moist food 750 ml or 30%

Water in beverages 1500 ml or 60%

Chart 20.1   Regulation of water intake

1. The body loses 1 to 2% of its water.
2. Osmoreceptors in the thirst center are stimulated by an increase in the osmotic pressure of extracellular fluid due to water loss.
3. Activity in the hypothalamus causes the person to feel thirsty, and seek water in response.
4. The act of drinking and distension of the stomach by water stimulate nerve impulses that inhibit the thirst center.
5. Water is absorbed through the walls of the stomach and small intestine.
6. Osmotic pressure of extracellular fluid decreases.

---

1. What is meant by water balance?
2. Where is the thirst center located?
3. What mechanism stimulates fluid intake? What mechanism inhibits it?

Water Output

Water normally enters the body only through the mouth, but it can be lost by a variety of routes. These include obvious losses in urine, feces, and sweat, as well as insensible losses that occur by diffusion of water through the skin (insensible perspiration) and evaporation of water from the lungs during breathing.

If an average adult takes in 2500 ml of water each day, then 2500 ml must be eliminated if water balance is to be maintained. Of this volume, perhaps 60% (1500 ml) will be lost in urine, 6% (150 ml) in feces, and 6% (150 ml) in sweat. About 28% (700 ml) will be lost by diffusion through the skin and evaporation from the lungs. (See fig. 20.6.) These percentages will, of course, vary with such environmental factors as temperature and relative humidity and with the amount of physical exercise.

Water lost by sweating is a necessary part of the body's temperature control mechanism; water lost in feces accompanies the elimination of undigested food materials; and water lost by diffusion and evaporation is largely unavoidable. Therefore, the primary regulator of water output involves urine production.

Regulation of Water Balance

As is discussed in chapter 19, the volume of water excreted in the urine is regulated mainly by activity in the distal convoluted tubule and collecting duct of the nephron. The epithelial linings of these segments of the renal tubule remain relatively impermeable to water unless antidiuretic hormone (ADH) is present.

Diuretics are substances that promote the production of urine. A number of common substances, including coffee and tea, have diuretic effects, as do a variety of drugs used to reduce the volume of body fluids.

Although diuretics produce their effects in different ways, some, such as alcohol and various narcotic drugs, promote urine formation by inhibiting the release of ADH. Others, including a group called *mercurial diuretics,* inhibit the reabsorption of sodium and other solutes in the renal tubules. As a consequence, the osmotic pressure of the tubular fluid increases, and the reabsorption of water by osmosis is reduced.

As is explained in chapter 12, the release of ADH involves osmoreceptors in the hypothalamus. If the plasma becomes more concentrated because of excessive water loss, these osmoreceptors tend to become dehydrated and shrink. This change triggers impulses that signal the posterior pituitary gland to release ADH. The ADH is carried by the blood to the kidneys where it causes an increase in the permeability of the distal tubule and collecting duct. Consequently, water reabsorption increases, and water is conserved. This action resists further osmotic change in the plasma. In fact, the *osmoreceptor-ADH mechanism* can reduce a normal urine production of 1500 ml per day to about 500 ml per day at times when the body is tending to become dehydrated.

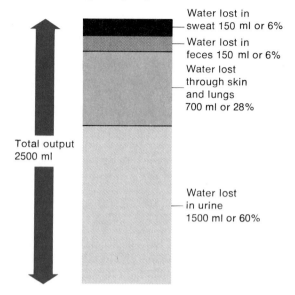

Fig. 20.6 Routes by which body water is lost. What factors influence water loss by each of these routes?

Average daily output of water

Water lost in sweat 150 ml or 6%

Water lost in feces 150 ml or 6%

Water lost through skin and lungs 700 ml or 28%

Total output 2500 ml

Water lost in urine 1500 ml or 60%

If, on the other hand, excessive amounts of water are taken into the body, the plasma becomes less concentrated, and the osmoreceptors swell as they receive extra water by osmosis. In this instance, the release of ADH is inhibited, and the distal segment and collecting duct remain impermeable to water. Consequently, less water is reabsorbed and a greater volume of urine is produced. Chart 20.2 summarizes the steps in this mechanism.

Chart 20.2 Events in regulation of water balance

In Case of Dehydration

1. Extracellular fluid becomes more concentrated.
2. Osmoreceptors in the hypothalamus are stimulated by change in osmotic pressure of body fluid.
3. The hypothalamus signals the posterior pituitary gland to release ADH into the blood.
4. Blood carries ADH to the kidneys.
5. ADH causes distal convoluted tubules and collecting ducts to increase water reabsorption.
6. The output of urine decreases, and water is conserved.

In Case of Excessive Water Intake

1. Extracellular fluid becomes osmotically less concentrated.
2. Osmoreceptors in the hypothalamus are stimulated by this change.
3. The posterior pituitary gland decreases the release of ADH.
4. Renal tubules decrease water reabsorption.
5. The output of urine increases, and excess water is excreted.

Fig. 20.7 Electrolyte balance exists when the intake of electrolytes from all sources equals the output of electrolytes.

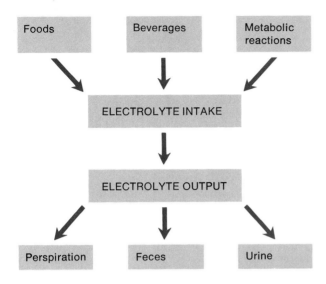

Fig. 20.7 Electrolyte balance exists when the intake of electrolytes from all sources equals the output of electrolytes.

In the condition called *diabetes insipidus,* the ability to produce ADH is greatly reduced or lost altogether because of an injury affecting the hypothalamus or the posterior pituitary gland. Consequently, the person produces a very large volume of dilute urine, and the daily output may reach 10 to 15 liters. At the same time, the person feels very thirsty and may be able to maintain a water balance by drinking large quantities of water.

---

1. By what routes is water lost from the body?
2. What is the primary regulator of water loss?
3. What types of water loss are unavoidable?
4. What role does the hypothalamus play in the regulation of water balance?

Electrolyte Balance

An **electrolyte balance** exists when the quantities of the various electrolytes gained by the body are equal to those lost (fig. 20.7).

Electrolyte Intake

The electrolytes of greatest importance to cellular functions are those that release ions of sodium, potassium, calcium, magnesium, chloride, sulfate, phosphate, and bicarbonate. These substances are obtained primarily from foods, but may also occur in drinking water and other beverages. In addition, some electrolytes occur as by-products of various metabolic reactions.

Regulation of Electrolyte Intake

Ordinarily, a person obtains sufficient electrolytes by responding to hunger and thirst. When there is a severe electrolyte deficiency, however, a person may experience a *salt craving,* which is a strong desire to eat salty foods.

Electrolyte Output

Some electrolytes are lost from the body by perspiration. The quantities leaving by this route will vary with the amount of perspiration excreted. Greater quantities are lost on warmer days and during times of strenuous exercise. Also, varying amounts of electrolytes are lost with the feces. The greatest electrolyte output occurs in the urine as a result of kidney functions.

---

1. What electrolytes are most important to cellular functions?
2. What mechanisms ordinarily regulate electrolyte intake?
3. By what routes are electrolytes lost from the body?

Regulation of Electrolyte Balance

The concentrations of positively charged ions, such as sodium (Na^+), potassium (K^+), and calcium (Ca^{+2}), are particularly important. Certain levels of these ions, for example, are necessary for the conduction of nerve impulses, the contraction of muscle fibers, and for maintaining the normal permeability of cell membranes. Thus, it is vital that their concentrations be regulated.

Sodium ions account for nearly 90% of the positively charged ions in extracellular fluid. The primary mechanism regulating these ions involves the kidneys and the hormone, *aldosterone.* As is explained in chapter 12, this hormone is secreted by the adrenal cortex. Its presence causes an increase in sodium reabsorption in the ascending limb of the loop of Henle, the distal convoluted tubule, and the collecting duct of the renal tubule.

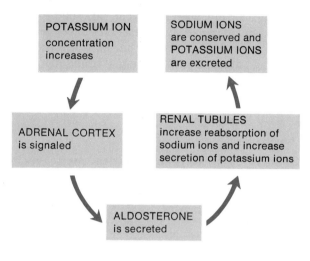

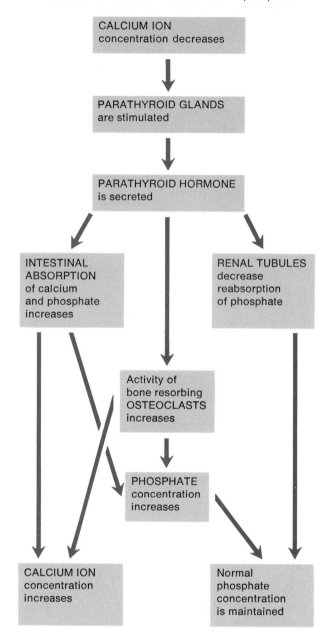

Addison's disease is characterized by a severe salt deficiency that develops because of a lack of aldosterone. In the absence of this hormone, the normal concentration of extracellular sodium is depleted within a few days.

Aldosterone also functions in regulating potassium. In fact, the most important stimulus for its secretion is a rising potassium concentration, which seems to stimulate the cells of the adrenal cortex directly. This hormone enhances the reabsorption of sodium and causes the secretion of potassium at the same time.

As was discussed in chapter 19, this action occurs because of the negative electrical charge that develops within the renal tubule as sodium ions are actively transported outward. The positively charged potassium ions (K^+) in the extracellular fluid of the kidney are attracted to this negative charge and are passively secreted into the urine. Consequently, as sodium ions are conserved, potassium ions are excreted. (See fig. 20.8.)

As was mentioned in chapter 12, the concentration of *calcium ions* in extracellular fluid is regulated mainly by the parathyroid glands. Whenever the calcium concentration drops below the normal level, these glands are directly stimulated, and they secrete *parathyroid hormone* in response.

Parathyroid hormone causes increased activity in bone-reabsorbing cells (osteoclasts). As a result of their actions, the levels of calcium and phosphate in the extracellular fluid increase.

Parathyroid hormone also stimulates the absorption of calcium and phosphate from the intestine.

Concurrently it causes a decrease in phosphate reabsorption by the renal tubule. The net effect of the hormone, then, is to increase the *calcium* concentration of the extracellular fluid, but to maintain a normal *phosphate* concentration by promoting the excretion of the excess phosphate. (See fig. 20.9.)

The renal tubule also helps regulate the calcium concentration by conserving calcium through reabsorption when the extracellular concentration is low, and by allowing excess calcium to be excreted when the level is high.

Fig. 20.10 If excessive amounts of extracellular fluids are lost, cells become dehydrated by osmosis. What happens to the concentration of electrolytes within the intracellular compartment at the same time?

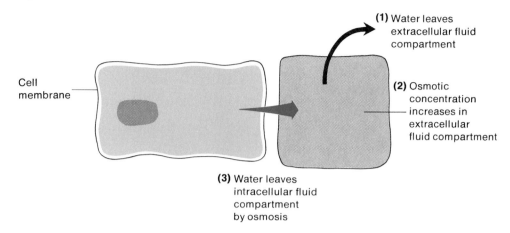

Cell membrane

(1) Water leaves extracellular fluid compartment

(2) Osmotic concentration increases in extracellular fluid compartment

(3) Water leaves intracellular fluid compartment by osmosis

Generally, the concentrations of negatively charged ions are controlled secondarily by the regulatory mechanisms that control positively charged ions. Chloride ions (Cl^-), for example (the most abundant negatively charged ions in extracellular fluid), are passively reabsorbed from the renal tubule as a result of the active reabsorption of sodium ions. That is, the chloride ions are electrically attracted to the positively charged sodium ions and accompany them as they are reabsorbed.

Some negatively charged ions, such as phosphate (PO_4^{-3}) and sulfate (SO_4^{-2}), are also partially regulated by active transport mechanisms that have limited transport capacities. Thus, if the extracellular phosphate concentration is low, the phosphate ions in the renal tubules are conserved. On the other hand, if the renal plasma threshold is exceeded, the excess phosphate will be excreted in the urine.

---

1. What is the action of aldosterone?
2. How is the level of sodium and potassium ions controlled?
3. How is calcium regulated?
4. What mechanism functions to regulate the concentrations of most negatively charged ions?

Some Disorders in Water and Electrolyte Balance

Disorders involving water and electrolyte balance can be grouped into three broad categories: deficiencies, excesses, and unfavorable proportions of substances.

Dehydration

Dehydration is a deficiency condition that occurs when the output of water exceeds the intake. This condition may develop following excessive sweating, or as a result of prolonged water deprivation accompanied by continued water output. In either case, as water is lost, the extracellular fluid becomes increasingly more concentrated, and water tends to leave cells by osmosis, causing them to dehydrate. (See fig. 20.10.) Dehydration may also accompany illnesses in which excessive fluids are lost as a result of prolonged vomiting or diarrhea. During dehydration, the skin and mucous membranes of the mouth feel dry, body weight is lost, and a high fever may develop as the body temperature-regulating mechanism becomes less effective due to a lack of water. In severe cases, as waste products accumulate in the extracellular fluid, symptoms of cerebral disturbances including mental confusion, delirium, and coma may develop.

The kidneys of infants are less able to conserve water than those of adults. Consequently, infants are more likely to become dehydrated than adults.

Elderly persons also are more susceptible to developing water imbalances, because the sensitivity of their thirst mechanisms tend to decrease with age, and physical disabilities may make it difficult to obtain adequate fluids.

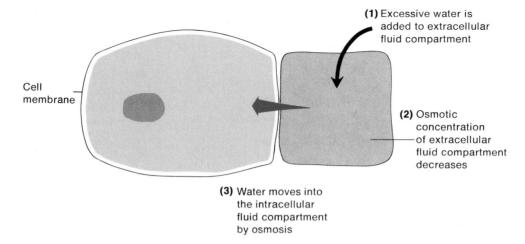

Cell membrane

(1) Excessive water is added to extracellular fluid compartment

(2) Osmotic concentration of extracellular fluid compartment decreases

(3) Water moves into the intracellular fluid compartment by osmosis

In *diabetes mellitus,* there may be an excess of glucose in the urine as a result of insulin deficiency. The presence of glucose increases the osmotic pressure of the urine in the renal tubule, and this interferes with the osmotic reabsorption of water. Consequently, the diabetic tends to excrete an abnormally large volume of urine, develop more concentrated extracellular fluid, and experience cellular dehydration.

The treatment for dehydration involves the replacement of lost water and electrolytes. It is important to note that if the water alone is replaced, the extracellular fluid will become more dilute than normal, and this may produce a condition called *water intoxication.*

Water Intoxication

Water intoxication is characterized by the presence of hypotonic extracellular fluid. As was mentioned, this condition may occur as a result of administering pure water to a dehydrated person, or it may develop in a person who drinks water for a prolonged period faster than the kidneys can excrete the excess.

In either instance, as the water is absorbed, the extracellular fluid becomes hypotonic to the cells. Then the cells tend to swell as water enters them in abnormal quantities by osmosis (fig. 20.11).

The symptoms of water intoxication are related mainly to a decrease in the extracellular sodium concentration, and they include edema, painful muscular contractions (heat cramps), convulsions, confusion, and coma. Treatment of this condition usually involves restricting water intake and administering hypertonic salt solutions.

1. *What changes does dehydration cause in the composition of body fluids?*
2. *What are some causes of dehydration?*
3. *Define* water intoxication.
4. *What conditions may lead to water intoxication?*

Edema

Edema is characterized by an abnormal accumulation of extracellular fluid within the interstitial spaces. It can be caused by a variety of factors, including a decrease in the plasma protein concentration (hypoproteinemia), obstructions in lymphatic vessels, increased venous pressure, and increased capillary permeability.

Hypoproteinemia may result from such conditions as liver disease, in which there is a failure to synthesize plasma proteins; kidney disease (glomerulonephritis), in which the glomerular capillary is damaged allowing protein to escape into the urine; or starvation, in which the intake of amino acids is insufficient to support the synthesis of plasma proteins.

In each of these instances, the plasma protein concentration is decreased, and this is reflected in decreased plasma osmotic pressure. Consequently, the normal return of tissue fluid to the venule ends of capillaries is reduced, and tissue fluid tends to accumulate in the interstitial spaces.

As is discussed in chapter 18, *lymphatic obstructions* may result from various surgical procedures. In such cases, back pressure develops in the lymphatic vessels, interfering with the normal movement of tissue fluid into them. At the same time, proteins that are ordinarily removed by lymphatic circulation tend to accumulate in the interstitial spaces. The presence of these proteins causes the osmotic pressure of the interstitial fluid to rise, and this effect causes still more fluid to be attracted into the interstitial spaces.

A similar problem involving back pressure occurs in veins as a result of venous obstructions or faulty valves, as in *varicose veins*. Increased venous back pressure may also accompany a failing heart that is unable to move blood out of the veins at a normal rate. As a consequence, the capillary hydrostatic pressure is increased, and this interferes with the return of tissue fluid from the interstitial spaces into the venule ends of capillaries. (See fig. 20.12.)

If the outflow of blood from the liver into the vena cava is blocked, the blood pressure within the liver and portal blood vessels increases greatly. As a result, fluid with a high concentration of protein tends to be exuded from the surfaces of the liver and intestine into the peritoneal cavity. This causes a rise in the osmotic pressure of the abdominal fluid, which in turn attracts more water into the peritoneal cavity by osmosis. This condition, called *ascites,* is characterized by an uncomfortable distension of the abdomen.

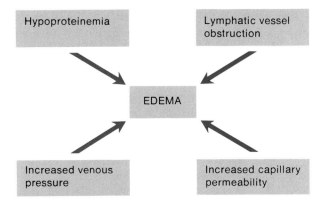

Fig. 20.12 Some factors that promote edema.

Edema may also result from increased capillary permeability accompanying an *inflammation reaction.* As is described in chapter 18, this reaction occurs in response to tissue damage and usually involves the release of chemicals such as histamine from damaged cells. Histamine causes an increase in capillary permeability, so that excessive amounts of fluid tend to leak out of the capillary and enter the interstitial spaces.

Chart 20.3 summarizes the factors that result in edema.

---

1. *Define* edema.
2. *Why does edema accompany hypoproteinemia?*
3. *How does obstruction of lymphatic circulation lead to edema?*
4. *What characteristic of the inflammatory response causes edema?*

Chart 20.3 Factors associated with edema

Factor	Cause	Effect
Low plasma protein concentration	Liver disease and failure to synthesize proteins; kidney disease and loss of proteins in urine; lack of protein in diet due to starvation	Plasma osmotic pressure decreases; less fluid enters venule ends of capillaries by osmosis
Obstruction of lymph vessels	Surgical removal of portions of lymphatic pathways	Back pressure in lymph vessels interferes with movement of fluid from interstitial spaces into lymph capillaries
Increased venous pressure	Venous obstructions or faulty venous valves	Back pressure in veins interferes with return of fluid from interstitial spaces into venule ends of capillaries
Inflammation	Tissue damage	Capillaries become abnormally permeable; fluid leaks from plasma into interstitial spaces

Imbalance in Sodium and Potassium Concentrations

Extracellular fluids are characterized by higher sodium ion concentrations, and intracellular fluid by higher potassium ion concentrations. The renal regulation of sodium and of potassium are closely related in that reabsorption of sodium is accompanied by passive secretion (and excretion) of potassium. Thus, it is not surprising that conditions involving sodium imbalance are closely related to those involving potassium imbalance.

Such disorders can be summarized as follows:

1. **Low sodium concentration** (hyponatremia). Possible causes of sodium deficiencies include prolonged sweating, vomiting, or diarrhea; renal disease in which sodium is inadequately reabsorbed; adrenal cortex disorders in which aldosterone secretion is insufficient (Addison's disease) to promote the reabsorption of sodium; and excessive intake of drinking water.

Possible effects of *hyponatremia* include the development of extracellular fluid that is hypotonic to cells, which promotes the movement of water into the cells by osmosis. This is accompanied by the symptoms of *water intoxication*.

2. **High sodium concentration** (hypernatremia). Possible causes of elevated sodium concentration include excessive water losses by diffusion and evaporation, as may occur during high fever; or increased water loss accompanying diabetes insipidus, in which ADH secretion is insufficient to maintain water conservation by the renal tubules.

Possible effects of *hypernatremia* include disturbances of the central nervous system, such as confusion, stupor, and coma.

3. **Low potassium concentration** (hypokalemia). Possible causes of potassium deficiency include excessive release of aldosterone by the adrenal cortex (Cushing's syndrome), causing increased renal secretion of potassium; use of diuretic drugs that promote the excretion of potassium; kidney disease; and prolonged vomiting or diarrhea.

Possible effects of *hypokalemia* include muscular weakness or paralysis, respiratory difficulty, and cardiac disturbances such as atrial or ventricular arrhythmias.

4. **High potassium concentration** (hyperkalemia). Possible causes of elevated potassium concentration include renal disease in which potassium excretion is decreased; use of drugs that promote renal conservation of potassium; insufficient secretion of aldosterone by the adrenal cortex (Addison's disease); or a shift of potassium from the intracellular fluid to the extracellular fluid accompanying an increase in hydrogen ion concentration (acidosis).

Possible effects of *hyperkalemia* include paralysis of skeletal muscles and severe cardiac disturbances such as cardiac arrest.

---

1. Why are the symptoms of hyponatremia similar to those of water intoxication?
2. What signs might appear in a person with hypernatremia?
3. What conditions lead to hypokalemia?
4. What are the possible effects of hyperkalemia?

Acid-Base Balance

As is discussed in chapter 2, electrolytes that ionize in water and release hydrogen ions are called *acids*. Substances that combine with hydrogen ions are called *bases*. The concentrations of acids and bases within body fluids must be controlled if homeostasis is to be maintained. The regulation of hydrogen ions is most important, since slight changes in hydrogen ion concentrations can alter the rates of enzyme-controlled metabolic reactions, cause shifts in the distribution of other ions, or modify the actions of various hormones. Thus, acid-base balance is primarily concerned with the regulation of hydrogen ion concentrations, which are measured in pH units. (See the Appendix for a detailed explanation of pH.)

Sources of Hydrogen Ions

Most of the hydrogen ions in the body fluids originate as by-products of metabolic processes, although small quantities may enter in foods.

The major metabolic sources include the following:

1. **Aerobic respiration of glucose.** This process results in the production of carbon dioxide and water. Carbon dioxide diffuses out of cells and reacts with water in extracellular fluid to form *carbonic acid*.

$$CO_2 + H_2O \longrightarrow H_2CO_3$$

The resulting carbonic acid then ionizes to release hydrogen ions and bicarbonate ions.

$$H_2CO_3 \longrightarrow H^+ + HCO_3^-$$

2. **Anaerobic respiration of glucose.** When glucose is utilized anaerobically, *lactic acid* is produced and hydrogen ions are added to body fluids.

3. **Incomplete oxidation of fatty acids.** The incomplete oxidation of fatty acids results in the production of *acidic ketone bodies* that cause the hydrogen ion concentration to increase.

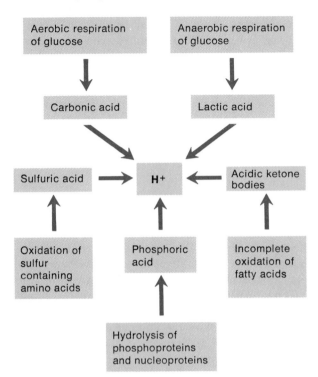

Fig. 20.13 Some metabolic processes that serve as sources of hydrogen ions.

4. Oxidation of amino acids containing sulfur. The oxidation of sulfur-containing amino acids results in the production of *sulfuric acid* (H_2SO_4) that ionizes to release hydrogen ions and sulfate ions.

5. Breakdown (hydrolysis) of phosphoproteins and nucleoproteins. Phosphoproteins and nucleoproteins contain phosphorus, and their oxidation results in the production of *phosphoric acid* (H_3PO_4).

The acids produced by various metabolic processes vary in strength, and so their effects on the hydrogen ion concentration of body fluids vary also. (See fig. 20.13.)

---

1. Explain why the regulation of hydrogen ion concentration is so important.
2. What are the major sources of hydrogen ions in the body?

Strengths of Acids and Bases

Acids that ionize more completely are termed *strong acids*, while those that ionize less completely are termed *weak acids*. For example, the hydrochloric acid (HCl) of gastric juice is a strong acid, while the carbonic acid (H_2CO_3) produced when carbon dioxide reacts with water is weak.

The degrees to which such acids ionize can be indicated by the lengths of the arrows in chemical equations. That is, for hydrochloric acid, a long arrow indicates a considerable amount of ionization; for carbonic acid, a short arrow indicates a smaller amount of ionization.

$$HCl \rightleftharpoons H^+ + Cl^-$$

$$H_2CO_3 \rightleftharpoons H^+ + HCO_3^-$$

The arrows pointing to the left indicate that these chemical reactions are *reversible*, and once again the lengths of the arrows represent the degrees to which the reverse reactions tend to proceed.

Bases are substances that, like hydroxyl ions (OH^-), will combine with hydrogen ions. *Chloride ions* (Cl^-) and *bicarbonate ions* (HCO_3^-) are also bases. Furthermore, since chloride ions combine less readily with hydrogen ions (as indicated by the short arrow pointing to the left in the above equation), they are *weak* bases, while the bicarbonate ions, which combine more readily with hydrogen ions, are *strong bases*.

Once again, the degrees to which these substances react with hydrogen ions can be indicated by the lengths of the arrows in chemical equations, as follows:

$$Cl^- + H^+ \rightleftharpoons HCl$$

$$HCO_3^- + H^+ \rightleftharpoons H_2CO_3$$

Regulation of Hydrogen Ion Concentration

The concentrations of hydrogen ions in body fluids are regulated primarily by acid-base buffer systems, the activity of the respiratory center in the brain stem, and the functions of nephrons in the kidneys.

Acid-Base Buffer Systems. **Acid-base buffer systems** occur in all body fluids and are usually composed of sets of two or more chemical substances. Such chemicals can combine with acids or bases when they occur in excess, thus helping to reduce the pH change to a minimum. More specifically, the substances of a buffer system can convert strong acids, which tend to release large quantities of hydrogen ions, into weak acids, which release fewer hydrogen ions. Likewise, they can combine with strong bases and change them into weak bases.

The three most important acid-base buffer systems in body fluids are these:

1. **Bicarbonate buffer system.** The bicarbonate buffer system, which is present in intracellular and extracellular body fluids, consists of carbonic acid (H_2CO_3) and sodium bicarbonate ($NaHCO_3$). If a strong acid, like hydrochloric acid, is present in body fluid, it reacts with the sodium bicarbonate. The products are carbonic acid, which is a weak acid, and sodium chloride. Consequently, the increase in the hydrogen ion concentration in the body fluid is minimized.

$$HCl + NaHCO_3 \longrightarrow H_2CO_3 + NaCl$$
(strong acid)   (weak acid)

If, on the other hand, a strong base like sodium hydroxide ($NaOH$) is present, it reacts with the carbonic acid. The products are sodium bicarbonate ($NaHCO_3$), which is a weak base, and water. Thus, a shift toward a more alkaline state is minimized.

$$NaOH + H_2CO_3 \longrightarrow NaHCO_3 + H_2O$$
(strong base)   (weak base)

2. **Phosphate buffer system.** The phosphate acid-base buffer system is also present in intracellular and extracellular body fluids. It is particularly important as a regulator of the hydrogen ion concentration in the tubular fluid of the nephrons and the urine. This buffer system consists of two phosphate compounds—sodium monohydrogen phosphate (Na_2HPO_4) and sodium dihydrogen phosphate (NaH_2PO_4).

If a strong acid is present, it reacts with the sodium monohydrogen phosphate to produce a weaker acid (sodium dihydrogen phosphate) and sodium chloride.

$$HCl + Na_2HPO_4 \longrightarrow NaH_2PO_4 + NaCl$$
(strong acid)   (weak acid)

If a strong base is present, it reacts with the sodium dihydrogen phosphate, and again the products are a weak base (sodium monohydrogen phosphate) and water.

$$NaOH + NaH_2PO_4 \longrightarrow Na_2HPO_4 + H_2O$$
(strong base)   (weak base)

3. **Protein buffer system.** The protein acid-base buffer system is the most important buffer system in plasma and intracellular fluid. It consists of the plasma proteins and various proteins within cells, including the hemoglobin of red blood cells.

Proteins are composed of amino acids bonded together in complex chains. Some of these amino acids have groups of atoms, called *carboxyl groups,* that are freely exposed. Under certain conditions, a carboxyl group ($-COOH$) can become ionized, and a hydrogen ion is released.

$$-COOH \longrightarrow -COO^- + H^+$$

Some of the amino acids within a protein molecule also contain freely exposed *amino groups* ($-NH_2$). Under certain conditions, these amino groups can accept hydrogen ions.

$$-NH_2 + H^+ \longrightarrow -NH_3^+$$

Thus, protein molecules can function as acids by releasing hydrogen ions from their carboxyl groups, or as bases by accepting hydrogen ions into their amino groups. This special property allows protein molecules to operate as an acid-base buffer system. In the presence of excess hydrogen ions, the $-COO^-$ portions of the protein molecules accept hydrogen ions and become $-COOH$ groups again. This action decreases the number of free hydrogen ions in body fluid and minimizes the amount of pH change that occurs.

In the presence of excess hydroxyl ions (OH^-), the $-NH_3^+$ groups of protein molecules give up hydrogen ions and become $-NH_2$ groups again. These hydrogen ions then combine with the hydroxyl ions to form water molecules. Once again, the pH change is reduced to a minimum, since water is a neutral substance.

$$H^+ + OH^- \longrightarrow H_2O$$

The hemoglobin in red blood cells provides an example of a specific protein that functions to buffer hydrogen ions. As is explained in chapter 15, carbon dioxide, which is produced by the cellular oxidation of glucose, diffuses through the capillary wall and enters the plasma and red blood cells. The red cells contain an enzyme, *carbonic anhydrase,* that speeds the reaction between carbon dioxide and water and results in carbonic acid.

$$CO_2 + H_2O \longrightarrow H_2CO_3$$

The carbonic acid quickly dissociates, releasing hydrogen ions and bicarbonate ions (HCO_3^-).

$$H_2CO_3 \longrightarrow H^+ + HCO_3^-$$

The hydrogen ions are accepted by the hemoglobin molecules, and thus, the pH change that would otherwise occur is reduced to a minimum.

Individual amino acids in body fluids can also function as acid-base buffers by accepting or giving up hydrogen ions, because every amino acid has an amino group ($-NH_2$) and a carboxyl group ($-COOH$) as part of its molecule.

Chart 20.4 Chemical acid-base buffer systems

Buffer System	Constituents	Actions
Bicarbonate system	Sodium bicarbonate ($NaHCO_3$)	Converts strong acid into weak acid
	Carbonic acid (H_2CO_3)	Converts strong base into weak base
Phosphate system	Sodium monohydrogen phosphate (Na_2HPO_4)	Converts strong acid into weak acid
	Sodium dihydrogen phosphate (NaH_2PO_4)	Converts strong base into weak base
Protein system (and amino acids)	$-COO^-$ group of a molecule	Accepts hydrogen ions in presence of excess acid
	$-NH_3{}^+$ group of a molecule	Releases hydrogen ions in presence of excess base

To summarize, acid-base buffer systems act to take in hydrogen ions when body fluids are becoming more acidic, and give up hydrogen ions when the fluids are becoming more alkaline. The buffer systems accomplish this action by converting stronger acids into weaker ones or by converting stronger bases into weaker ones, as summarized in chart 20.4.

Although buffer systems cannot prevent a pH change, they are able to reduce it to a minimum. Furthermore, the various acid-base buffer systems in body fluids act together to buffer each other. Consequently, whenever the hydrogen ion concentration begins to drift, the chemical balances within all of the buffer systems change too, and the drift in pH is resisted.

---

1. What is the difference between a strong acid or base and a weak acid or base?
2. How does a chemical buffer system work?
3. List the major buffer systems of the body.

The Respiratory Center. The region in the brain stem called the **respiratory center** helps to regulate hydrogen ion concentrations in body fluids by controlling the rate and depth of breathing. Specifically, if the cells increase their production of carbon dioxide, as occurs during periods of physical exercise, the production of carbonic acid increases. As the carbonic acid dissociates, the concentration of hydrogen ions increases, and the pH of the fluids tends to drop.

Fig. 20.14 An increase in carbon dioxide production is followed by an increase in carbon dioxide elimination.

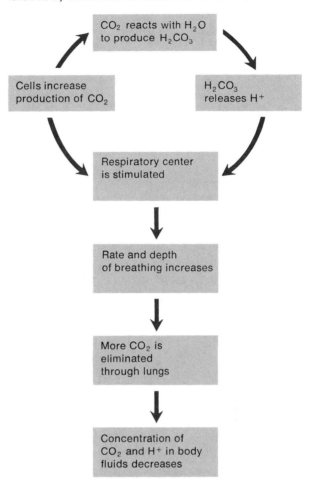

Such an increasing concentration of carbon dioxide and the consequent increase in hydrogen ion concentration in the plasma directly stimulate the cells of the respiratory center. These substances also stimulate the *carotid* and *aortic bodies*, mentioned in chapter 15, and as a result, impulses are transmitted to the respiratory center.

In response, the respiratory center causes the depth and rate of breathing to increase, so that a greater amount of carbon dioxide is excreted through the lungs. This loss of carbon dioxide is accompanied by a drop in the hydrogen ion concentration in the body fluids, since the released carbon dioxide comes from carbonic acid. (See fig. 20.14.)

$$H_2CO_3 \longrightarrow CO_2 + H_2O$$

Conversely, if the body cells are less active, the concentrations of carbon dioxide and hydrogen ions in the plasma are relatively low. As a consequence, the breathing rate and depth are decreased. In time this may result in an accumulation of carbon dioxide in the body fluids, and as the pH drops, the respiratory center is stimulated to increase the rate and depth of breathing once again.

Thus, the activity of the respiratory center is altered in response to shifts in the plasma pH, reducing these shifts to a minimum. Since most of the hydrogen ions in the body fluids originate from carbonic acid produced when carbon dioxide reacts with water, the respiratory regulation of hydrogen ion concentration is of considerable importance.

The Kidneys.   The nephrons of the kidneys help to regulate hydrogen ion concentration of the body fluids by secreting hydrogen ions. As is discussed in chapter 19, these ions are secreted into the urine of the renal tubules primarily by the epithelial cells that line the proximal and distal convoluted tubules and the collecting ducts.

This mechanism is important in balancing the quantities of sulfuric acid, phosphoric acid, and various organic acids that appear in body fluids as by-products of metabolic processes.

The metabolism of certain amino acids, for example, results in the formation of sulfuric and phosphoric acids. Consequently, a diet high in proteins (which give rise to amino acids when they are digested) will be accompanied by an excessive production of acids. The kidneys compensate for such gains in acids by altering the rate of hydrogen ion secretion into the urine. Thus, a shift in the pH of body fluids is resisted. (See fig. 20.15.)

Once hydrogen ions reach the urine, they are buffered by phosphates that were filtered into the fluid of the renal tubule. Ammonia also aids in this buffering action.

By deamination of certain amino acids, the cells of the renal tubules produce ammonia (NH_3). It diffuses readily through the cell membranes and enters the urine. Since ammonia is a weak base, it can accept hydrogen ions, and the result is *ammonium ions* (NH_4^+).

$$H^+ + NH_3 \longrightarrow NH_4^+$$

Cell membranes are quite impermeable to ammonium ions. Consequently, these ions are trapped in the urine as they form, and they are excreted from the body with the urine.

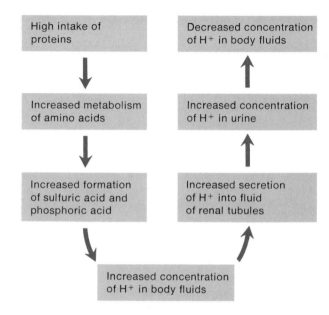

Fig. 20.15   If the concentration of hydrogen ions in body fluids increases, the renal tubules increase their secretion of hydrogen ions into the urine.

As a result of a poorly understood mechanism, a prolonged increase in the hydrogen ion concentration of body fluids is accompanied by an increase in the production of ammonia by cells of the renal tubules. This mechanism helps to transport excessive hydrogen ions to the outside, and to control the pH of the urine.

The various regulators of hydrogen ion concentration operate at different rates. Acid-base buffers, for example, function rapidly and can convert strong acids or bases into weak acids or bases almost immediately. For this reason, these chemical buffer systems are sometimes called the body's *first line of defense* against shifts in pH.

Physiological buffer systems, such as the respiratory and renal mechanisms, function more slowly, and constitute *secondary defenses*. The respiratory mechanism may require several minutes to begin resisting a change in pH, while the renal mechanism may require several hours or even days to control a changing hydrogen ion concentration. (See fig. 20.16.)

1. *How does the respiratory system help to regulate acid-base balance?*
2. *How do the kidneys handle excessive concentrations of hydrogen ions?*
3. *What are the differences between the chemical and physiological buffer systems?*

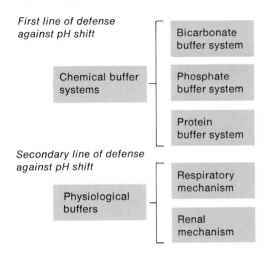

Fig. 20.16 Chemical buffers act rapidly, while physiological buffers may require several minutes to begin resisting a change in pH.

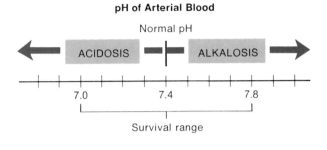

Fig. 20.17 If the pH of the arterial blood drops to 7.0 or rises to 7.8 for more than a few minutes, the person usually cannot survive.

Fig. 20.18 Acidosis results from an accumulation of acids or a loss of bases; alkalosis results from a loss of acids or an accumulation of bases.

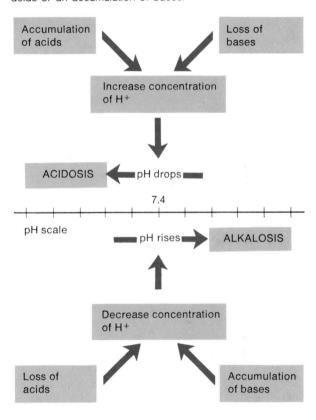

Some Disorders in Acid-Base Balance

Ordinarily the hydrogen ion concentration of body fluids is maintained within very narrow pH ranges by the actions of chemical and physiological buffer systems. However, abnormal conditions may cause disturbances in the acid-base balance, and the pH may drop producing acidosis, or it may rise producing alkalosis.

Acidosis and Alkalosis

The pH of arterial blood is normally 7.4, and if this value drops below 7.4, the person is said to have *acidosis*. If the pH rises above 7.4, the condition is called *alkalosis*. Such shifts in the pH of body fluids may be life-threatening, and in fact, a person usually cannot survive if the pH drops to 7.0 or rises to 7.8 for more than a few minutes (fig. 20.17).

Acidosis results from an accumulation of acids or a loss of bases, both of which are accompanied by abnormal increases in the hydrogen ion concentrations of body fluids. Conversely, alkalosis results from a loss of acids or an accumulation of bases accompanied by a decrease in hydrogen ion concentrations. (See fig. 20.18.)

There are two major types of acidosis—*respiratory acidosis* and *metabolic acidosis*. Respiratory acidosis is caused by factors that produce an increase in carbon dioxide, accompanied by an increase in the concentration of the respiratory acid, carbonic acid.

Metabolic acidosis is due to an abnormal accumulation of any other acids in the body fluids, or to a loss of bases.

Similarly, there are two major types of alkalosis—*respiratory alkalosis* and *metabolic alkalosis*. Respiratory alkalosis is caused by an excessive loss of carbon dioxide and a consequent loss of carbonic acid. Metabolic alkalosis is due to an excessive loss of hydrogen ions or to the gain of bases.

Fig. 20.19 Some factors that lead to respiratory acidosis.

Fig. 20.20 Some factors that lead to metabolic acidosis.

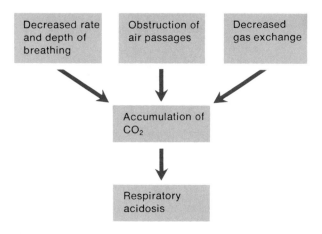

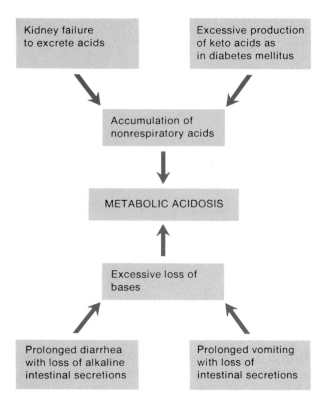

Respiratory Acidosis

Since respiratory acidosis involves an accumulation of carbon dioxide, it can be caused by factors that hinder pulmonary ventilation (fig. 20.19). These include the following:

1. Injury to the respiratory center of the brain stem, followed by decreased rate and depth of breathing.

2. Obstructions in air passages that interfere with the movement of air into the alveoli.

3. Diseases that decrease gas exchanges, such as pneumonia; or those that cause a reduction in surface area of the respiratory membrane, such as emphysema.

As a result of such conditions, the amount of carbonic acid occurring in the body fluids increases, the concentration of hydrogen ions increases, and the pH value drops. This shift in pH may be resisted by the action of chemical buffers, such as hemoglobin. At the same time, the respiratory center may be stimulated by the increasing concentrations of carbon dioxide and hydrogen ions, so that the breathing rate and depth are increased, and the carbon dioxide concentration is thereby lowered. Also, the kidneys may begin to secrete increasing quantities of hydrogen ions.

Eventually, the pH of the body fluids may return to normal as a result of the actions of these chemical and physiological buffers. When this happens, the acidosis is said to be *compensated*.

The symptoms of respiratory acidosis include depression of activities in the central nervous system, characterized by drowsiness, disorientation, and stupor. These signs are usually accompanied by evidence of respiratory insufficiency such as labored breathing and cyanosis. In *uncompensated acidosis,* the person may become comatose and die.

Metabolic Acidosis

Metabolic acidosis involves an accumulation of nonrespiratory acids or the loss of bases. (See fig. 20.20.) Factors that may lead to this condition include the following:

1. Kidney disease in which there is a failure to excrete acids produced by metabolic processes (uremic acidosis).

2. Prolonged vomiting in which the alkaline contents of the upper intestine are lost in addition to the stomach contents. (If only the stomach contents are lost, the result is metabolic alkalosis.)

3. Prolonged diarrhea in which there is an excessive loss of alkaline intestinal secretions.

4. Diabetes mellitus, in which some fatty acids are converted into acidic ketone bodies. These ketone bodies include *acetoacetic acid, beta-hydroxybutyric acid,* and *acetone.* Normally these substances are produced in relatively small quantities, and they are oxidized by cells as energy sources. However, if fats are being utilized at an abnormally high rate, as may occur in diabetes mellitus, ketone bodies may accumulate faster than they can be oxidized. At such times, these compounds may be excreted in the urine (ketouria); in addition, acetone, which is volatile, may be excreted by the lungs and impart a fruity odor to

the breath. More seriously, the accumulation of ace-toacetic acid and beta-hydroxybutyric acid may cause a drop in pH (ketonemic acidosis).

These acids also tend to combine with bicarbonate ions in the urine. As a result, these ions are excreted excessively, and this loss interferes with the function of the bicarbonate acid-base buffer system.

In each case, the pH tends to shift toward lower values. However, this shift is resisted by the following: the actions of chemical buffer systems, which accept excessive hydrogen ions; the respiratory center, which causes the breathing rate and depth to increase; and the kidneys, which secrete increasing quantities of hydrogen ions.

Respiratory Alkalosis

Respiratory alkalosis develops as a result of *hyperventilation,* described in chapter 15. This action is accompanied by excessive loss of carbon dioxide and consequent decreases in carbonic acid and hydrogen ion concentrations. (See fig. 20.21).

Commonly, hyperventilation occurs during periods of anxiety, although it may also accompany fever or poisoning from salicylates such as aspirin. At high altitudes, hyperventilation may occur in response to low oxygen pressure. In any event, carbon dioxide is lost excessively by rapid, deep breathing, and the pH value of body fluids increases.

This shift in pH is resisted by chemical buffers, such as hemoglobin, that release hydrogen ions. Also, since the concentrations of carbon dioxide and hydrogen ions are lower, the respiratory center is stimulated to a lesser degree. As a result, the tendency to hyperventilate is inhibited, thus reducing further loss of carbon dioxide. At the same time, the kidneys decrease their secretion of hydrogen ions, and the urine becomes alkaline as bases are excreted.

The symptoms of respiratory alkalosis include lightheadedness, agitation, dizziness, and tingling sensations. In severe cases, impulses may be triggered spontaneously on peripheral nerves, and muscles may respond by developing tetany.

Metabolic Alkalosis

Metabolic alkalosis results from excessive loss of hydrogen ions or from a gain in bases, both of which are accompanied by a rise in pH of the body fluids. (See fig. 20.22.)

This condition may occur following gastric drainage (lavage) or prolonged vomiting in which only the stomach contents are lost. Since gastric juice is very acidic, its loss leaves the body fluids with a net increase of basic substances and a pH shift toward alkaline values.

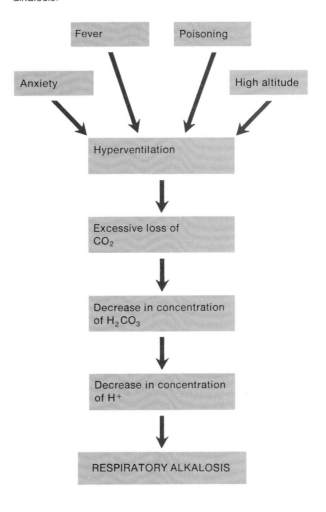

Fig. 20.21 Some factors that lead to respiratory alkalosis.

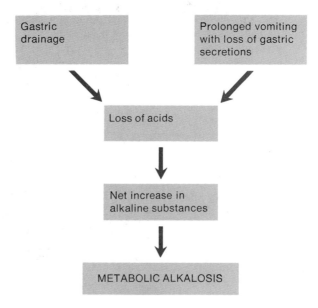

Fig. 20.22 Some factors that lead to metabolic alkalosis.

Sometimes metabolic alkalosis develops as a result of ingesting excessive amounts of antacids, such as sodium bicarbonate. These substances commonly are taken to relieve the symptoms of indigestion, heartburn, or other digestive disorders.

The resulting pH shift is resisted by chemical buffer systems, such as proteins and phosphates, that release hydrogen ions, by a decrease in the rate and depth of breathing, and by a decrease in the renal secretion of hydrogen ions.

1. What factors lead to respiratory acidosis? To metabolic acidosis?
2. Define compensated respiratory acidosis.
3. What problems may lead to respiratory alkalosis? To metabolic alkalosis?
4. What are the symptoms of respiratory acidosis? Of respiratory alkalosis?

Some Clinical Terms Related to Water and Electrolyte Balance

acetonemia (as″ĕ-to-ne′me-ah)—the presence of abnormal amounts of acetone in the blood.

acetonuria (as″ĕ-to-nu′re-ah)—the presence of abnormal amounts of acetone in the urine.

albuminuria (al-bu″mĭ-nu′re-ah)—the presence of albumin in the urine.

antacid (ant-as′id)—a substance that neutralizes an acid.

anuria (ah-nu′re-ah)—the absence of urine excretion.

azotemia (az″o-te′me-ah)—an accumulation of nitrogenous wastes in the blood.

diuresis (di″u-re′sis)—increased production of urine.

glycosuria (gli″ko-su′re-ah)—the presence of excessive sugar in the urine.

hyperkalemia (hi″per-kah-le′me-ah)—the presence of excessive potassium in the blood.

hypernatremia (hi″per-na-tre′me-ah)—the presence of excessive sodium in the blood.

hyperuricemia (hi″per-u″rĭ-se′me-ah)—the presence of excessive uric acid in the blood.

hypoglycemia (hi″po-gli-se′me-ah)—an abnormally low level of blood sugar.

ketonuria (ke″to-nu′re-ah)—the presence of ketone bodies in the urine.

ketosis (ke″to′sis)—acidosis due to the presence of excessive ketone bodies in the body fluids.

proteinuria (pro″te-ĭ-nu′re-ah)—the presence of protein in the urine.

uremia (u-re′me-ah)—toxic condition resulting from the presence of excessive amounts of nitrogenous wastes in the blood.

Chapter Summary

Introduction

The maintenance of water and electrolyte balance requires that the quantities of these substances entering the body equal the quantities leaving. Altering the water balance necessarily affects the electrolyte balance.

Distribution of Body Fluids

1. Fluid compartments
   a. Intracellular fluid compartments include the fluids and electrolytes enclosed by cell membranes.
   b. Extracellular fluid compartments include all fluids and electrolytes outside cell membranes.
      (1) Interstitial fluid
      (2) Plasma
      (3) Lymph
      (4) Transcellular fluid within body cavities
2. Composition of body fluids
   a. Extracellular fluid
      (1) Extracellular fluid is characterized by high concentrations of sodium, chloride, and bicarbonate ions with lesser amounts of potassium, calcium, magnesium, phosphate, and sulfate ions.
      (2) Plasma contains more protein than either interstitial fluid or lymph.
   b. Intracellular fluid contains relatively high concentrations of potassium, phosphate, and magnesium ions; it also contains greater concentrations of calcium and sulfate ions and lesser concentrations of sodium, chloride, and bicarbonate ions than extracellular fluid.
3. Movement of fluid between compartments
   a. Movements are regulated by hydrostatic pressure and osmotic pressure.
      (1) Fluid leaves plasma because of hydrostatic pressure and returns to plasma because of osmotic pressure.
      (2) Fluid enters lymph vessels because of osmotic pressure.
      (3) Movement in and out of cells is regulated primarily by osmotic pressure.
   b. Sodium concentrations are especially important in the regulation of fluid movements.

Water Balance

1. Water intake
   a. Volume varies from person to person.
   b. Most water enters as liquid or in moist foods.
   c. Some water is produced by oxidative metabolism.
2. Regulation of water intake
   a. The thirst mechanism is the primary regulator.
   b. The act of drinking and distension of the stomach inhibit the thirst mechanism.

3. Water output
   a. Water is lost in a variety of ways.
      (1) It is excreted in urine, feces, and sweat.
      (2) Insensible loss occurs through evaporation from skin and lungs.
   b. The primary regulator involves urine production.
4. Regulation of water balance
   a. Water balance is regulated mainly by the distal convoluted tubule and collecting duct of the nephron.
      (1) ADH from the posterior pituitary gland stimulates water reabsorption in these segments.
      (2) The mechanism involving ADH can reduce normal output of 1500 ml to 500 ml per day.
   b. If excess water is taken in, the ADH mechanism is inhibited.

Electrolyte Balance

1. Electrolyte intake
   a. The most important electrolytes in body fluids are ions of sodium, potassium, calcium, magnesium, chloride, sulfate, phosphate, and bicarbonate.
   b. These are obtained in foods and beverages or as by-products of metabolic processes.
2. Regulation of electrolyte intake
   a. Electrolytes are usually obtained in sufficient quantities in response to hunger and thirst mechanisms.
   b. In severe deficiency, a person may experience salt craving.
3. Electrolyte output
   a. Electrolytes are lost through perspiration, feces, and urine.
   b. Quantities lost vary with temperature and physical exercise.
   c. The greatest loss is usually in urine.
4. Regulation of electrolyte balance
   a. Concentrations of sodium, potassium, and calcium ions in body fluids are particularly important.
   b. Regulation of sodium ions involves the secretion of aldosterone from the adrenal glands.
   c. Regulation of potassium ions also involves aldosterone.
   d. Regulation of calcium ions involves parathyroid hormone.
   e. In general, negatively charged ions are regulated secondarily by the mechanisms that control positively charged ions.
      (1) Chloride ions are passively reabsorbed in renal tubules as sodium ions are actively reabsorbed.
      (2) Some negatively charged ions, such as phosphate ions, are partially reabsorbed by active transport mechanisms with limited capacities.

Some Disorders in Water and Electrolyte Balance

1. Dehydration
   a. Dehydration occurs when water output exceeds input.
   b. As extracellular water is lost, water leaves cells by osmosis.
   c. Dehydration may occur as a result of excessive sweating, prolonged water deprivation, vomiting, or diarrhea.
   d. As water is lost, electrolytes are lost too; when water is replaced, electrolytes must be replaced.
2. Water intoxication
   a. Water intoxication results when water is replaced following dehydration without replacing electrolytes.
   b. Extracellular fluid becomes hypotonic to cells, and cells receive water by osmosis.
3. Edema
   a. Edema is caused by decreased plasma protein concentration, obstruction of lymph vessels, increased venous pressure, and increased capillary permeability.
   b. Decreased plasma protein concentration
      (1) Decreased plasma protein concentration may accompany liver disease, renal failure, or starvation.
      (2) As plasma protein decreases, return of fluid from the interstitial spaces to the plasma is inhibited.
   c. Lymphatic obstruction
      (1) Lymphatic obstruction may be caused by some surgical procedures.
      (2) As back pressure increases, less fluid enters the lymph capillaries.
   d. Increased venous pressure
      (1) Increased venous pressure may be due to venous obstructions or faulty venous valves.
      (2) As venous back pressure increases, less fluid enters the venule ends of capillaries.
   e. Inflammation is accompanied by an increase in capillary permeability and the leakage of fluid into interstitial spaces.
4. Imbalance of sodium and potassium concentrations
   a. Low sodium concentration
      (1) Low sodium concentration is caused by sweating, vomiting, diarrhea, renal disease, or adrenal cortex disorders.
      (2) This results in extracellular fluid that is hypotonic to cells.
   b. High sodium concentration
      (1) High sodium concentration is caused by excessive water loss, by sweating or evaporation, or by lack of ADH.
      (2) This results in disturbances in the central nervous system.

c. Low potassium concentration
   (1) Low potassium concentration is caused by excessive release of aldosterone, use of certain diuretic drugs, or prolonged vomiting or diarrhea.
   (2) It results in muscular weakness, respiratory difficulty, and cardiac disturbances.
d. High potassium concentration
   (1) High potassium concentration is caused by renal disease, use of drugs that promote renal conservation of potassium, insufficient secretion of aldosterone, or acidosis.
   (2) It results in paralysis and severe cardiac disturbances.

Acid-Base Balance

Acids are electrolytes that release hydrogen ions. Bases are substances that combine with hydrogen ions.

1. Sources of hydrogen ions
   a. Aerobic respiration of glucose
      (1) Aerobic respiration of glucose produces carbon dioxide, which reacts with water to form carbonic acid.
      (2) Carbonic acid dissociates to release hydrogen ions and bicarbonate ions.
   b. Anaerobic respiration of glucose gives rise to lactic acid.
   c. Incomplete oxidation of fatty acids gives rise to acidic ketone bodies.
   d. Oxidation of certain amino acids gives rise to sulfuric acid.
   e. Hydrolysis of phosphoproteins and nucleoproteins gives rise to phosphoric acid.
2. Strengths of acids and bases
   a. Acids vary in the degrees to which they ionize.
      (1) Strong acids, such as hydrochloric acid, ionize more completely.
      (2) Weak acids, such as carbonic acid, ionize less completely.
   b. Bases vary in strength also.
      (1) Strong bases, such as hydroxyl ions, combine readily with hydrogen ions.
      (2) Weak bases, such as chloride ions, combine with hydrogen ions less readily.
3. Regulation of hydrogen ion concentration
   a. Acid-base buffer systems
      (1) These buffer systems are composed of sets of two or more substances.
      (2) They function to convert strong acids into weaker ones, or strong bases into weaker ones.
      (3) They include the bicarbonate buffer system, phosphate buffer system, and protein buffer system.
      (4) Buffer systems cannot prevent pH changes, but they can reduce these changes to a minimum.

b. Respiratory center
   (1) The respiratory center is located in the brain stem.
   (2) It helps regulate pH by controlling the rate and depth of breathing.
   (3) As carbon dioxide and hydrogen ion concentrations increase, the respiratory center is stimulated; the breathing rate and depth increase, and the carbon dioxide concentration decreases.
   (4) If the carbon dioxide and hydrogen ion concentrations are low, the respiratory center inhibits breathing.
c. Kidneys
   (1) Nephrons regulate pH by secreting hydrogen ions.
   (2) Hydrogen ions in urine are buffered by phosphates.
   (3) Ammonia produced by renal cells helps transport hydrogen ions to the outside of the body.
d. Chemical buffers act rapidly; physiological buffers act less rapidly.

Some Disorders in Acid-Base Balance

1. Acidosis and alkalosis
   a. Acidosis results from an accumulation of acids or a loss of bases.
   b. Alkalosis results from a loss of acids or an accumulation of bases.
   c. Two major types of acidosis are respiratory and metabolic acidosis.
   d. Two major types of alkalosis are respiratory and metabolic alkalosis.
2. Respiratory acidosis
   a. This type is caused by factors that promote an accumulation of carbon dioxide, such as injury to the respiratory center, obstructed air passages, and respiratory diseases.
   b. It results in an excessive concentration of hydrogen ions.
   c. A shift in pH is resisted by chemical buffers and physiological buffers.
   d. Symptoms include depression of the central nervous system and signs of respiratory insufficiency.
3. Metabolic acidosis
   a. This involves an accumulation of nonrespiratory acids or a loss of bases.
   b. It is caused by kidney disease, prolonged vomiting, prolonged diarrhea, and diabetes mellitus.
   c. A shift in pH is resisted by buffer systems.
   d. Symptoms are similar to those of respiratory acidosis.
4. Respiratory alkalosis
   a. Develops as a result of hyperventilation, which is accompanied by a decrease in carbon dioxide concentration.
   b. It occurs during periods of anxiety, fever, or salicylate poisoning.

c. It may occur in high altitudes when the respiratory center is stimulated by low oxygen pressure.

d. A shift in pH is resisted by buffer systems.

e. Symptoms include lightheadedness, dizziness, tingling sensations, and tetany.

5. Metabolic alkalosis

a. This results from an excessive loss of hydrogen ions or a gain of bases.

b. It may occur following a loss of gastric secretions by drainage or vomiting.

c. A shift in pH is resisted by buffer systems.

Application of Knowledge

1. An elderly, semiconscious patient is tentatively diagnosed as having acidosis. What component of the arterial blood will be most valuable in determining whether the acidosis is of metabolic or respiratory origin?

2. Some time ago a news story reported the death of several newborn infants due to an error in which sodium chloride was substituted for sugar in their formula. What symptoms would this produce? Why are infants more prone to the hazard of excess salt intake than adults?

3. Explain the threat to fluid and electrolyte balance in the following situation: a patient is being nutritionally maintained on concentrated solutions of hydrolyzed protein that are administered through a gastrostomy tube.

Review Activities

1. Explain how water balance and electrolyte balance are interdependent.

2. Name the body fluid compartments and describe their locations.

3. Explain how the fluids within these compartments differ in composition.

4. Describe how fluid movements between compartments are controlled.

5. Prepare a water budget to illustrate how the input of water equals the output of water.

6. Define water of metabolism.

7. Explain how water intake is regulated.

8. Explain how the hypothalamus functions in the regulation of water output.

9. List the electrolytes of greatest importance in body fluids.

10. Explain how electrolyte intake is regulated.

11. List the routes by which electrolytes leave the body.

12. Explain how the adrenal cortex functions in the regulation of electrolyte output.

13. Describe the role of the parathyroid glands in regulating electrolyte balance.

14. Describe the role of the renal tubule in regulating electrolyte balance.

15. Define dehydration and list several factors that promote it.

16. Define water intoxication and explain how it can be avoided.

17. List several causes of edema and explain how each produces this effect.

18. Explain why sodium and potassium concentrations are closely related.

19. Describe the effects of sodium imbalance.

20. Describe the effects of potassium imbalance.

21. Distinguish between an acid and a base.

22. List five sources of hydrogen ions in body fluids and name an acid that originates from each source.

23. Distinguish between a strong acid and a weak acid, and name an example of each.

24. Distinguish between a strong base and a weak base, and name an example of each.

25. Explain, in general terms, how an acid-base buffer system functions.

26. Describe how the bicarbonate buffer system resists shifts in pH.

27. Explain why a protein has acidic as well as basic properties.

28. Describe how a protein functions as a buffer system.

29. Describe the function of hemoglobin as a buffer of carbonic acid.

30. Explain how the respiratory center functions in the regulation of acid-base balance.

31. Explain how the kidneys function in the regulation of acid-base balance.

32. Describe the role of ammonia in the transport of hydrogen ions to the outside of the body.

33. Distinguish between a chemical buffer system and a physiological buffer system.

34. Distinguish between acidosis and alkalosis.

35. Distinguish between respiratory acidosis and metabolic acidosis.

36. List several factors that promote respiratory acidosis.

37. List several factors that promote metabolic acidosis.

38. Explain why hyperventilation leads to respiratory alkalosis.

39. Explain why loss of gastric secretions leads to metabolic alkalosis.

Suggestions for Additional Reading

Anderson, B. 1977. Regulation of body fluids. *Ann. Rev. Physio.* 39:185.

Burk, S. R. 1972. *The composition and function of body fluids.* St. Louis: C.V. Mosby.

Deetjen, P. et al. 1974. *Physiology of the kidney and water balance.* New York: Springer-Verlag.

Hills, A. G. 1973. *Acid-base balance.* Baltimore: The Williams and Wilkins Co.

Pitts, R. F. 1974. *Physiology of the kidney and body fluids,* 3rd ed. Chicago: Year Book Medical Publishers.

Share, L., et al. 1972. Regulation of body fluids. *Ann. Rev. Physio.* 34:235.

Valtin, H. 1973. *Renal function: mechanism for preserving fluid and solute balance in health.* Boston: Little, Brown.

Vander, A. J. 1975. *Renal Physiology.* New York: McGraw-Hill.

Weldy, N. J. 1972. *Body fluids and electrolytes.* St. Louis: C.V. Mosby.

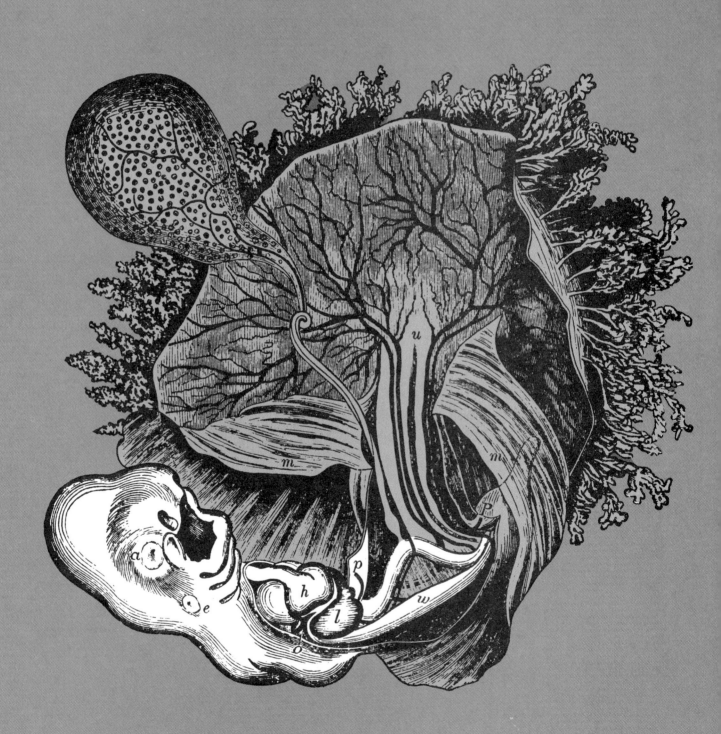

Unit 5
The Human Life Cycle

The chapters of unit 5 are concerned with the reproduction, growth, and development of the human organism. They describe how the organs of the male and female reproductive systems function to produce an embryo and how this offspring grows and develops, before and after birth, as it passes through the phases of its life cycle. The final chapter explains the genetic determination of individual traits.

This unit includes

Chapter 21: The Reproductive Systems
Chapter 22: Human Growth and Development
Chapter 23: Human Genetics

The Reproductive Systems

21 The male and female *reproductive systems* are specialized to produce offspring. These systems are unique in that their functions are not necessary for the survival of each individual. Instead, their functions are vital to the continuation of the human species. Without reproduction by at least some of its members, the species would soon become extinct.

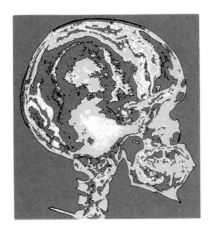

After you have studied this chapter, you should
be able to

1. State the general functions of the
   reproductive systems.

2. Name the parts of the male reproductive
   system and describe the general functions
   of each.

3. Describe the structure of a testis and
   explain how sperm cells are formed.

4. Trace the path followed by sperm cells from
   their site of formation to the outside and
   explain the environmental changes that
   occur along this path.

5. Describe the structure of the penis and
   explain how its parts function to produce
   an erection.

6. Explain how hormones control the activities
   of the male reproductive organs and how
   they are related to the development of male
   secondary characteristics.

7. Name the parts of the female reproductive
   system and describe the general functions
   of each.

8. Describe the structure of an ovary and
   explain how egg cells and follicles are
   formed.

9. Trace the path followed by an egg cell after
   ovulation.

10. Describe how hormones control the
    activities of the female reproductive system
    and how they are related to the development
    of female secondary sexual characteristics.

11. Describe the major events that occur during
    a menstrual cycle.

12. Define *pregnancy* and describe the process
    of fertilization.

13. Describe the hormonal changes that occur
    during pregnancy.

14. Describe the birth process and explain the
    role of hormones in this process.

15. Review the structure and function of the
    mammary glands.

16. List the common methods of contraception
    and explain how each interferes with normal
    reproductive processes.

17. Complete the review activities at the end
    of this chapter.

androgen (an'dro-jen)

cleavage (klēv'ij)

contraception (kon''trah-sep'shun)

ejaculation (e-jak''u-la'shun)

emission (e-mish'un)

estrogen (es'tro-jen)

fertilization (fer''tǐ-lǐ-za'shun)

follicle (fol'ǐ-kl)

gonadotropin (go-nad''o-trōp'in)

implantation (im''plan-ta'shun)

menopause (men'o-pawz)

menstrual cycle (men'stroo-al si'kl)

oogenesis (o''o-jen'ē-sis)

ovulation (o''vu-la'shun)

placenta (plah-sen'tah)

pregnancy (preg'nan-se)

progesterone (pro-jes'tě-rōn)

puberty (pu'ber-te)

seminal fluid (se'men-al floo'id)

spermatogenesis (sper''mah-to-jen'ē-sis)

andr-, man: _andr_ogens—male sex hormones.

contra-, opposed to: _contra_ception—the prevention of fertilization or of implantation.

crur-, lower part: _crur_a—diverging parts at the base of the penis by which it is attached to the pelvic arch.

ejacul-, to shoot forth: _ejacul_ation—process by which seminal fluid is expelled from the male reproductive tract.

fimb-, a fringe: _fimb_riae—irregular extensions on the margin of the infundibulum of the uterine tube.

follic-, small bag: _follic_le—ovarian structure that contains an egg cell.

genesis, origin: spermato_genesis_—process by which sperm cells are formed.

gubern-, to guide: _gubern_aculum—fibromuscular cord that guides the descent of a testis.

labi-, lip: _labi_a minora—flattened, longitudinal folds that extend along the margins of the vestibule.

mens-, month: _mens_trual cycle—monthly female reproductive cycle.

mons, mountain: _mons_ pubis—rounded elevation of fatty tissue overlying the pubic symphysis in a female.

puber-, adult: _puber_ty—time when a person becomes able to function reproductively.

The organs of the reproductive systems are concerned with the general process of reproduction, and each is adapted for specialized tasks. Some, for example, are concerned with the production of sex cells, and others function to sustain the lives of these cells or to transport them from one location to another. Still other parts produce and secrete hormones. These hormones regulate the formation of sex cells, and play roles in the development and maintenance of male and female sexual characteristics.

Organs of the Male Reproductive System

The primary function of the male reproductive system is to produce male sex cells (germ cells), called **sperm cells** (spermatozoa). The *primary organs* of the male system are the two *testes* in which these sperm cells are formed. The other structures of the male reproductive system are termed *accessory organs*, and they include two groups—the internal and external reproductive organs. (See fig. 21.1).

The Testes

The **testes** are ovoid structures, about 5 cm (2 in) in length and 3 cm (1.2 in) in diameter. Each is suspended by a spermatic cord within the cavity of the saclike *scrotum.*

Descent of the Testes

In a male fetus, the testes originate from masses of tissue located behind the parietal peritoneum, near the developing kidneys. Usually about a month before birth, these organs descend to regions in the lower abdominal cavity and pass through the abdominal wall into the scrotum.

The process by which a testis descends is poorly understood, but it seems to be aided by a fibro-muscular cord called the **gubernaculum.** This cord is attached to the developing testis and extends into the inguinal region of the abdominal cavity. It passes through the abdominal wall and is fastened to the skin on the outside. As the testis descends, apparently guided by the gubernaculum, it passes through the

---

Fig. 21.1 Organs of the male reproductive system.

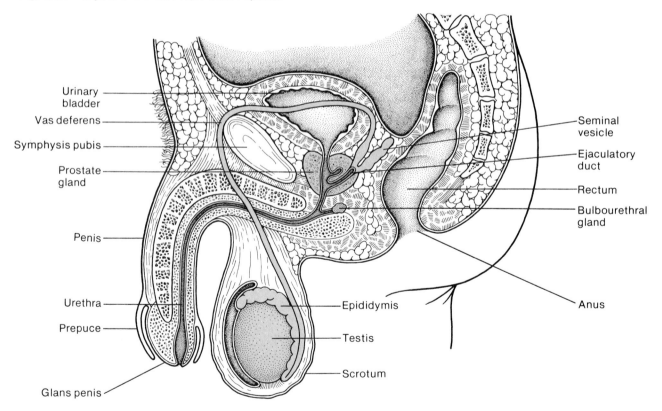

The Human Life Cycle

inguinal canal of the abdominal wall and enters the scrotum, where it remains anchored by the gubernaculum. The testis carries with it a developing *vas deferens* and various blood vessels and nerves. These structures later form parts of the **spermatic cord** by which the testis is suspended in the scrotum. (See fig. 21.2.)

If the testes fail to descend into the scrotum, they will not produce sperm cells. In fact, in this condition, called *cryptorchidism,* the cells that produce sperm cells degenerate, and the male is sterile.

This failure to produce sperm cells is apparently caused by the unfavorable temperature of the abdominal cavity, which is a few degrees higher than the scrotal temperature.

Fig. 21.2 During fetal development, each testis descends through an inguinal canal and enters the scrotum. What is the function of the gubernaculum when this occurs?

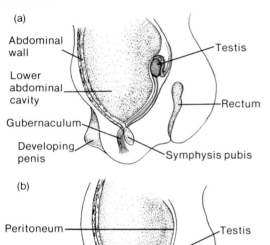

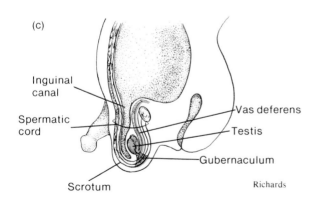

Nurses and physicians routinely palpate the scrotal sacs of newborns in order to determine if the testes have descended. If they have not, it is sometimes possible to induce their descent by administering certain hormones. This condition may also be treated by a surgical procedure in which the testes are relocated in the scrotum.

1. What are the primary organs of the male reproductive system?
2. Describe the descent of the testes.
3. What happens if the testes fail to descend into the scrotum?

Structure of the Testes

Each testis is enclosed by a tough, white fibrous capsule called the *tunica albuginea.* Along its posterior border, the connective tissue thickens and extends into the organ forming a mass called the *mediastinum testis.* From this structure, thin dividers of connective tissue, called *septa,* pass into the testis and divide it into about 250 *lobules.*

The lobules contain one to four highly coiled, convoluted **seminiferous tubules,** each of which is up to 70 cm (28 in) long when uncoiled. These tubules course posteriorly and become united into a complex network of channels called the *rete testis.* The rete testis is located within the mediastinum testis and gives rise to several ducts that join the **epididymis.** The epididymis, in turn, is coiled on the outer surface of the testis.

The seminiferous tubules are lined with a specialized tissue called **germinal epithelium.** The cells of this tissue are stratified, and they function to produce the male sex cells. Other specialized cells, called **interstitial cells** (cells of Leydig), are located in the spaces between the seminiferous tubules. They function in the production and secretion of male sex hormones. (See figs. 21.3 and 21.4.)

1. Describe the structure of a testis.
2. Where are the sperm cells produced within the testes?
3. What cells produce male sex hormones?

Fig. 21.3 Sagittal section of a testis.

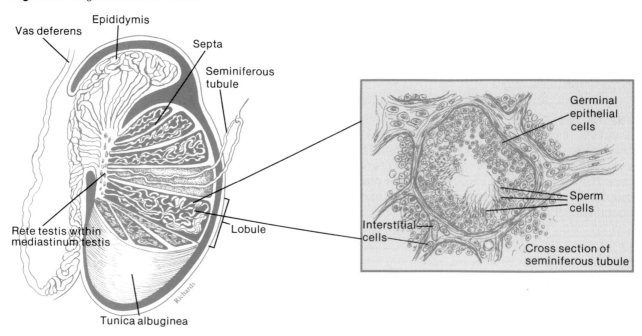

Fig. 21.4 Scanning electron micrograph of cross sections of human seminiferous tubules.

Formation of Sperm Cells

The germinal epithelium consists of two types of cells—supporting cells (Sertoli's cells) and spermatogenic cells. The *supporting cells* are tall, columnar cells that extend the full thickness of the epithelium from its base to the lumen of the seminiferous tubule. Numerous, thin processes project from these cells, filling the spaces between nearby spermatogenic cells. They function to support and nourish the *spermatogenic cells,* which give rise to sperm cells.

In a young male, all the spermatogenic cells are undifferentiated and are called *spermatogonia.* During early adolescence, hormones stimulate these cells to change. The spermatogonia begin to undergo mitosis, and some of them enlarge to become *primary spermatocytes.*

Primary spermatocytes each contain 46 chromosomes in their nuclei, which is the usual number of chromosomes for human cells. However, when primary spermatocytes divide, they do so by a special type of cell division called *meiosis,* which is described in chapter 22. During this process the chromosome number of the offspring cells is reduced by one-half. Consequently, the offspring cells, called *secondary spermatocytes,* contain only 23 chromosomes each.

Soon after they are formed, the secondary spermatocytes each divide again, and the result is four *spermatids* with 23 chromosomes each. Finally, the spermatids become transformed into mature *sperm cells.* Thus, each primary spermatocyte gives rise to four sperm cells by a process of differentiation, which is called **spermatogenesis.** (See fig. 21.5.)

Fig. 21.5 Spermatogonia give rise to primary spermatocytes by mitosis; the spermatocytes in turn give rise to sperm cells by meiosis.

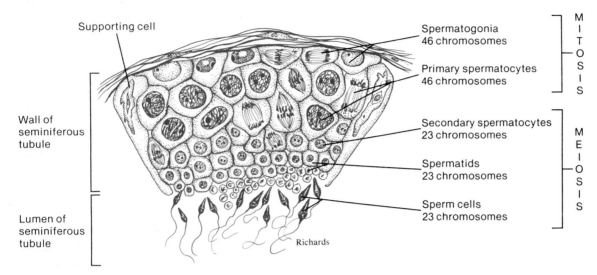

Supporting cell

Spermatogonia
46 chromosomes

Primary spermatocytes
46 chromosomes

MITOSIS

Wall of
seminiferous
tubule

Secondary spermatocytes
23 chromosomes

Spermatids
23 chromosomes

Sperm cells
23 chromosomes

MEIOSIS

Lumen of
seminiferous
tubule

Richards

The spermatogonia are located near the base of the germinal epithelium. As spermatogenesis occurs, cells in more advanced stages are pushed along the sides of supporting cells toward the lumen of the seminiferous tubule.

Near the base of the epithelium, membranous processes from adjacent supporting cells are fused by specialized junctions (occluding junctions) into complexes that divide the tissue into two layers. The spermatogonia are located on one side of this barrier, and the cells in more advanced stages are on the other side. This membranous complex seems to help maintain a favorable environment for the development of sperm cells by preventing the movement of certain large molecules from the interstitial fluid of the basal epithelium into the region of the differentiating cells.

Spermatogenesis occurs continually throughout the reproductive life of a male. The resulting sperm cells collect in the lumens of the seminiferous tubules. Then they pass through the rete testis to the epididymis, where they remain for a time and mature.

A mature sperm cell, shown in figure 21.6, is a tiny, tadpole-shaped structure about 0.06 mm long. It consists of a flattened head, a cylindrical body, and an elongated tail.

The *head* of a sperm cell, which is oval in outline, is composed primarily of a nucleus and contains the chromatin of 23 chromosomes. It has a small part at its anterior end called the *acrosome,* which is thought to aid the sperm cell in penetrating an egg cell at the time of fertilization.

Fig. 21.6 Parts of a mature sperm cell.

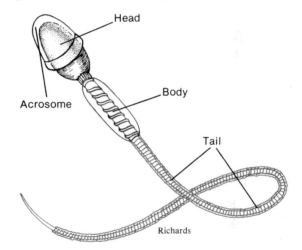

Head

Body

Acrosome

Tail

Richards

The *body* of the sperm cell contains a central, filamentous core and a large number of mitochondria arranged in a spiral. The *tail* consists of several longitudinal fibrils enclosed in an extension of the cell membrane. The tail also contains a relatively large quantity of ATP, which is thought to provide the energy for the lashing movement that propels the sperm cell through fluid. The photomicrograph in figure 21.7 shows some mature sperm cells.

1. *Explain the function of supporting cells in the germinal epithelium.*
2. *Describe the process of spermatogenesis.*
3. *Describe the structure of a sperm cell.*

Fig. 21.7 Scanning electron micrograph of a human sperm cell.

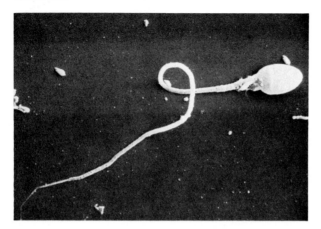

Fig. 21.8 Cross section of a human epididymis.

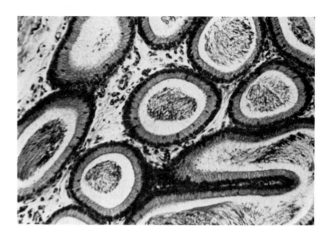

Male Internal Accessory Organs

The *internal accessory organs* of the male reproductive system include the epididymides, vasa deferentia, ejaculatory ducts, and urethra, as well as the seminal vesicles, prostate gland, and bulbourethral glands.

The Epididymis

The **epididymis** (plural, epididymides) is a tightly coiled, threadlike tube that is about 6 meters (20 ft) long. (See figs. 21.1 and 21.8.) This tube emerges from the top of the testis, descends along its posterior surface, and then courses upward to become the *vas deferens.*

As mentioned, immature sperm cells move from the tubules of the testis into the epididymis, where they are stored and undergo *maturation.* The inner lining of the epididymis is composed of pseudostratified columnar cells that bear nonmotile cilia. These cells are thought to secrete glycogen, which helps sustain the lives of the stored sperm cells.

In response to sexual stimulation, smooth muscles in the walls of the epididymis and ducts of the testis undergo peristaltic contractions that propel the tubular contents into the vas deferens. Similarly, peristaltic contractions in the wall of the vas deferens force the sperm cells into the ejaculatory duct and urethra. This movement of sperm cells is called *emission.*

The Vas Deferens

The **vas deferens** (plural vasa deferentia; also called ductus deferens) is a muscular tube about 45 cm (18 in) long. It begins at the lower end of the epididymis and passes upward along the medial side of the testis to become part of the spermatic cord. It passes through the inguinal canal, enters the abdominal cavity outside of the parietal peritoneum, and courses over the pelvic brim. From there, it extends backward and medially into the pelvic cavity, where it ends behind the urinary bladder.

Near its termination, the vas deferens becomes dilated into a portion called the *ampulla.* Just outside the prostate gland, the tube becomes slender again and unites with the duct of a seminal vesicle. The fusion of these two ducts forms an **ejaculatory duct,** which passes through the substance of the prostate gland and empties into the urethra through a slitlike opening. (See fig. 21.9.)

The Seminal Vesicle

A **seminal vesicle** is a convoluted, saclike structure about 5 cm (2 in) long that is attached to the vas deferens near the base of the bladder.

The glandular tissue lining the inner wall of the seminal vesicle secretes a slightly alkaline fluid. This fluid is thought to help regulate the pH of the tubular contents as sperm cells are conveyed to the outside. The secretion of the seminal vesicle contains a variety of nutrients and is rich in *fructose,* a monosaccharide that is thought to provide sperm cells with an energy source. It also contains *prostaglandins,* which are thought to stimulate muscular contractions within the female reproductive organs and thus aid the movement of sperm cells toward the female egg cell.

Fig. 21.9 The vas deferens extends from the epididymis to the ejaculatory duct.

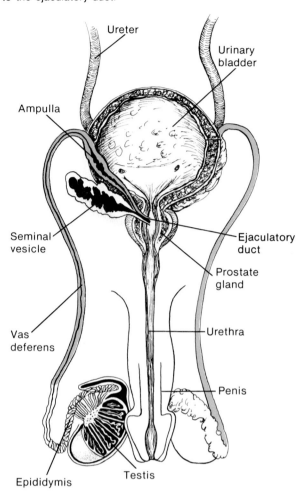

Fig. 21.10 The prostate gland surrounds the urethra.

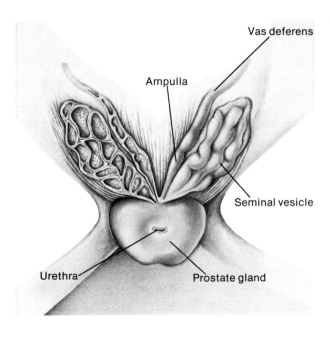

At the time of ejaculation, the contents of the seminal vesicles are emptied into the ejaculatory ducts, greatly increasing the volume of the fluid that is discharged from the vas deferens. This fluid, which contains sperm cells and the secretion of the seminal vesicle, is called *seminal fluid* or *semen*. (See figs. 21.1 and 21.9.)

---

1. *Describe the structure of the epididymis.*
2. *Trace the path of the vas deferens.*
3. *What is the function of the seminal vesicles?*

The Prostate Gland

The **prostate gland** is a chestnut-shaped structure about 4 cm (1.6 in) across and 3 cm (1.2 in) thick that surrounds the beginning of the urethra, just below the urinary bladder. It is enclosed by connective tissue and is composed of many branched tubular glands. These glands are separated by septa of connective tissue and smooth muscle that extend inward from the capsule. Their ducts open into the urethra. (See fig. 21.10.)

The prostate gland secretes a thin, milky fluid with an alkaline pH. It functions to neutralize the seminal fluid, which is acidic due to an accumulation of metabolic wastes produced by stored sperm cells. It also enhances the motility of the sperm cells, which remain relatively immobile in the acidic contents of the epididymis. In addition, the prostatic fluid aids to neutralize the acidic secretions of the vagina, thus helping to sustain sperm cells that enter the female reproductive tract.

The prostate gland releases its secretions into the urethra as a result of smooth muscle contractions in its capsular wall. This release occurs during emission, as the contents of the vas deferens and the seminal vesicles are entering from the ejaculatory ducts, and thus the volume of the seminal fluid is increased still more.

Although the prostate gland is relatively small in male children, it begins to grow in early adolescence and reaches its adult size a few years later. As a rule, its size remains unchanged between the ages of 20 and 50 years of age. In older males the prostate gland usually enlarges. When this happens it tends to squeeze the urethra and interfere with urine excretion.

The treatment for an enlarged prostate gland is usually surgical. If the obstruction is slight, the procedure may be performed through the urethral canal and is called a *transurethral prostatic resection.*

The prostate gland is a common site of cancer in older males. Such cancers are usually stimulated to grow more rapidly by the male sex hormone, testosterone, and are inhibited by the female sex hormone, estrogen. Consequently, treatment for this type of cancer may involve removal of the testes, which are the main source of testosterone, and the administration of estrogen.

The Bulbourethral Glands

The **bulbourethral glands** (Cowper's glands) are two small structures about the size of peas, which are located below the prostate gland among the fibers of the external urethral sphincter muscle. (See figs. 21.1 and 21.11).

These glands are composed of numerous tubes whose epithelial linings secrete a mucuslike fluid. This fluid is released in response to sexual stimulation and lubricates the end of the penis in preparation for sexual intercourse (coitus).

Seminal Fluid

The **seminal fluid** (semen) conveyed by the urethra to the outside consists of sperm cells and secretions of the seminal vesicles, prostate gland, and bulbourethral glands. It has a slightly alkaline pH (about 7.5) and a milky appearance. It contains a variety of nutrients, as well as prostaglandins, that function to enhance sperm cell survival and movement through the female reproductive tract.

The volume of seminal fluid released by emission varies from 2 to 6 ml, while the average number of sperm cells present in the fluid is about 120 million per ml.

Fig. 21.11 Section of a human bulbourethral gland.

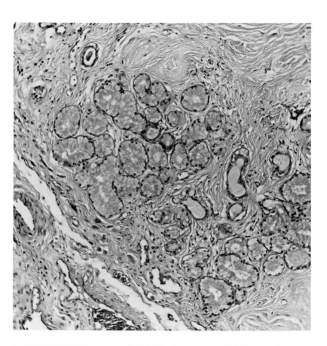

Some males seem to be infertile because of a deficiency of sperm cells in their seminal fluid. Although estimates of the number of sperm cells necessary for fertility vary, 20 million cells per ml of seminal fluid in a release of 3 to 5 ml is often cited as the minimum for fertility.

Sperm cells remain immobile while they are in the ducts of the testis and epididymis, but become activated as they are mixed with the secretions of the accessory glands during emission.

Although sperm cells are able to live for many weeks in the ducts of the male reproductive tract, they tend to survive for only a day or two after being expelled to the outside even when they are maintained at body temperature.

---

1. *Where is the prostate gland located?*
2. *What is the function of its secretion?*
3. *What is the function of the bulbourethral glands?*
4. *What are the characteristics of seminal fluid?*

Male External Reproductive Organs

The male external reproductive organs are the scrotum, which encloses the testes, and the penis, through which the urethra passes.

Fig. 21.12 (*a*) Sagittal section of the penis; (*b*) cross section of the penis.

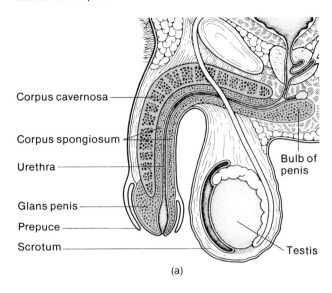

(a)

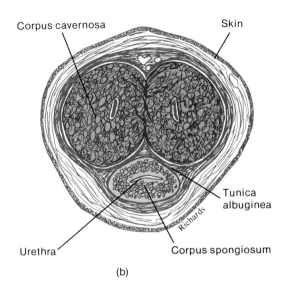

(b)

The Scrotum

The **scrotum** is a pouch of skin and subcutaneous tissue that hangs from the lower abdominal region behind the penis.

Although its subcutaneous tissue lacks fat, the scrotal wall contains a layer of smooth muscle fibers that constitute the *dartos muscle.* When these muscle fibers are contracted, the scrotal skin becomes wrinkled and is held close to the testes; when the fibers are relaxed, the scrotum hangs more loosely.

The scrotum is divided into chambers by a medial septum, and each chamber is occupied by a testis. Each chamber also contains a serous membrane that provides a covering for the front and sides of the testis and epididymis. This covering helps ensure that the testis will move smoothly within the scrotum. (See fig. 21.1.)

The Penis

The **penis** is a cylindrical organ that functions to convey urine and seminal fluid through the urethra to the outside. It is also specialized to become enlarged and stiffened by a process called *erection,* so that it can be inserted into the female vagina during sexual intercourse.

The *body,* or shaft, of the penis is composed of three columns of erectile tissue, including a pair of dorsally located *corpora cavernosa* and a single *corpus spongiosum* below. These columns are surrounded by skin, a thin layer of subcutaneous tissue,

and a layer of elastic tissue. In addition, each column is enclosed by a tough capsule of white fibrous connective tissue called a *tunica albuginea.*

The corpus spongiosum, through which the urethra extends, is enlarged at its distal end to form a sensitive, cone-shaped **glans penis**. This glans covers the ends of the corpora cavernosa and bears the *external urethral meatus.* The skin of the glans is very thin and hairless. Also a loose fold of skin called the *prepuce* (foreskin) begins just behind the glans and extends forward to cover it as a sheath. The prepuce commonly is removed during infancy by a surgical procedure called *circumcision.*

At the *root* of the penis, the columns of erectile tissue become separated. The corpora cavernosa diverge laterally in the perineum and are firmly attached to the medial surfaces of the pubic arch by connective tissue. These diverging parts form the *crura* of the penis. The single corpus spongiosum is enlarged between the crura as the *bulb* of the penis, which is attached to membranes of the perineum. (See fig. 21.12.)

---

1. *Describe the structure of the penis.*
2. *What is circumcision?*
3. *How is the penis attached to the perineum?*

The masses of erectile tissue within the body of the penis contain networks of vascular spaces. These spaces are lined with endothelium and are separated from each other by cross bars of smooth muscle and connective tissue.

Ordinarily, the vascular spaces are kept small as a result of partial contractions in the smooth muscle fibers. However, during sexual stimulation, the smooth muscles become relaxed. At the same time, *parasympathetic* nerve impulses pass from the sacral portion of the spinal cord to arteries leading into the penis and cause them to dilate. As a result, arterial blood under relatively high pressure enters the vascular spaces of the erectile tissue, and the flow of venous blood away from the penis is partially reduced. Consequently, blood accumulates in the erectile tissues, and the penis swells, elongates, and becomes erect. (See fig. 21.13.)

The culmination of sexual stimulation is called *orgasm* and involves a pleasurable feeling of physiological and psychological release. Orgasm is accompanied by emission and ejaculation.

As was mentioned, *emission* is the movement of sperm cells together with glandular secretions into the urethra. *Ejaculation* is the expulsion of seminal fluid from the urethra of the erect penis. These movements occur in response to reflexes employing centers in the lumbar and sacral portions of the spinal cord. As a result of these reflexes, *sympathetic* impulses travel to smooth muscles in the walls of the testicular ducts, epididymides, vasa deferentia, and ejaculatory ducts, causing peristaltic contractions. At the same time, impulses stimulate rhythmic contractions of the seminal vesicles and prostate gland. Other impulses stimulate certain skeletal muscles at the base of the erectile columns of the penis to contract rhythmically. This causes increased pressure within the erectile tissue and aids in forcing the seminal fluid through the urethra to the outside.

The sequence of events during emission and ejaculation is regulated so that the fluid from the bulbourethral glands is expelled first, followed by the release of fluid from the prostate gland, the passage of sperm cells, and finally the ejection of fluid from the seminal vesicles. (See fig. 21.14.)

Immediately after ejaculation, sympathetic impulses cause vasoconstriction of the arteries that supply the erectile tissue with blood, reducing the inflow of blood. The smooth muscles within the walls of the vascular spaces partially contract again, and the veins of the penis carry the excess blood out of these spaces. The penis gradually returns to its former flaccid condition, and usually another erection and ejaculation cannot be triggered for a period of 10 to 30 minutes or longer.

Fig. 21.13 Mechanism that causes erection of the penis.

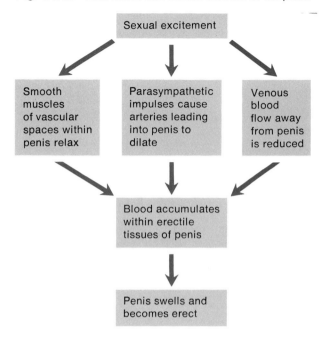

The functions of the male reproductive organs are summarized in chart 21.1.

Chart 21.1 Functions of male reproductive organs

Organ	Function
Testes	
Seminiferous tubules	Production of sperm cells
Interstitial cells	Production and secretion of male sex hormones
Epididymis	Storage and maturation of sperm cells; conveys sperm cells to vas deferens
Vas deferens	Conveys sperm cells to ejaculatory duct
Seminal vesicle	Secretes alkaline fluid containing nutrients and prostaglandins; fluid helps neutralize acidic seminal fluid
Prostate gland	Secretes alkaline fluid that helps neutralize acidic seminal fluid and enhances motility of sperm cells
Bulbourethral gland	Secretes fluid that lubricates end of penis
Scrotum	Encloses and protects testes
Penis	Conveys urine and seminal fluid to outside of body; inserted into vagina during sexual intercourse; glans penis is richly supplied with sensory nerve endings associated with feelings of pleasure during sexual stimulation

The Human Life Cycle

Fig. 21.14 Mechanism that results in emission and ejaculation.

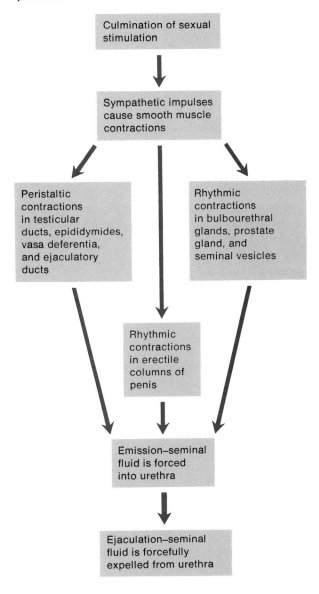

Culmination of sexual stimulation

Sympathetic impulses cause smooth muscle contractions

Peristaltic contractions in testicular ducts, epididymides, vasa deferentia, and ejaculatory ducts

Rhythmic contractions in bulbourethral glands, prostate gland, and seminal vesicles

Rhythmic contractions in erectile columns of penis

Emission—seminal fluid is forced into urethra

Ejaculation—seminal fluid is forcefully expelled from urethra

Spontaneous emissions and ejaculations commonly occur in adolescent males during sleep. These *nocturnal emissions* are apparently caused by changes in hormonal concentrations that accompany adolescent development.

1. How is blood flow into the erectile tissues of the penis controlled?
2. Distinguish between orgasm, emission, and ejaculation.
3. Review the events associated with emission and ejaculation.

Hormonal Control of Male Reproductive Functions

Male reproductive functions are largely controlled by hormones secreted by the *anterior pituitary gland* and the *testes.* These hormones are responsible for the initiation and maintenance of sperm cell production and for the development and maintenance of male secondary sexual characteristics.

Pituitary Hormones

Prior to about 10 years of age, the male body is reproductively immature. It remains childlike, and the spermatogenic cells of the testes remain undifferentiated. Then a series of changes are triggered that lead to the development of a reproductively functional adult. Although the mechanism that initiates such changes is not well understood, it involves the action of the hypothalamus of the brain.

As is explained in chapter 12, the hypothalamus secretes one, and perhaps two, *gonadotropin-releasing factors* (GRF). These substances, luteinizing hormone-releasing factor (LH-RF) and follicle-stimulating hormone-releasing factor (FSH-RF), enter blood vessels leading to the anterior lobe of the pituitary gland and stimulate it to secrete the **gonadotropins,** called *luteinizing hormone* (LH) and *follicle-stimulating hormone* (FSH). LH, which is also called *interstitial cell-stimulating hormone* (ICSH), promotes the development of the interstitial cells of the testes, and they in turn secrete male sex hormones. FSH stimulates the spermatogenic cells of the testes to undergo spermatogenesis.

Male Sex Hormones

As a group, the male sex hormones are termed **androgens,** and they include substances produced by the adrenal cortices and the testes. Among the androgens, the hormone called **testosterone** is the most important.

Testosterone is produced and secreted by interstitial cells of the testes under the influence of LH (ICSH). It is responsible for the development and maintenance of male secondary sexual characteristics and is also needed for the final steps of spermatogenesis. Although FSH stimulates primary spermatocytes to undergo meiosis, sperm cells do not mature completely unless testosterone is present. (See fig. 21.15.)

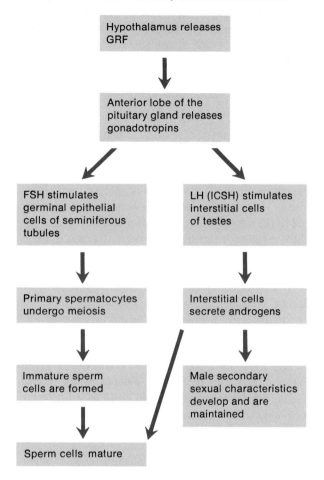

Fig. 21.15 Mechanism by which the hypothalamus controls the maturation of sperm cells and the development of male secondary sexual characteristics.

Hypothalamus releases GRF

Anterior lobe of the pituitary gland releases gonadotropins

FSH stimulates germinal epithelial cells of seminiferous tubules

LH (ICSH) stimulates interstitial cells of testes

Primary spermatocytes undergo meiosis

Interstitial cells secrete androgens

Immature sperm cells are formed

Male secondary sexual characteristics develop and are maintained

Sperm cells mature

The *primary sexual characteristic* of a male is the presence of sperm-producing testes. The *secondary sexual characteristics* are other features that are usually associated with the adult male body. They include the following:

1. Enlargement of the penis, scrotum, prostate gland, seminal vesicles, and bulbourethral glands.

2. Increased growth of body hair, particularly on the face, chest, axillary region, and pubic region, but sometimes accompanied by decreased growth of hair on the scalp.

3. Enlargement of the larynx and thickening of the vocal folds, accompanied by the development of a lower-pitched voice.

4. Thickening of the skin.

5. Increased muscular growth accompanied by the development of broader shoulders and a relatively narrow waist.

6. Thickening and strengthening of the bones.

Testosterone causes bones to thicken and strengthen by promoting the deposition of calcium salts in osseous tissues. Because of this, testosterone is sometimes used to treat elderly persons suffering from *osteoporosis*, a condition in which the bones thin due to excessive bone resorption.

To summarize, under the influence of FSH, the testes of a male child first begin to produce sperm cells, and under the influence of LH (ICSH), the interstitial cells produce and secrete testosterone. The testosterone, in turn, stimulates the development of male secondary sexual characteristics and promotes the maturation of developing sperm cells. Sometime between 13 and 15 years of age, a male usually becomes reproductively functional. This stage in development is called *puberty.*

If the testes are removed prior to puberty, the person lacks a source of testosterone and fails to develop male secondary sexual characteristics. Such a castrated male is called a *eunuch.* The external reproductive organs remain small, the voice fails to deepen, and the male musculature and hair pattern fail to develop. However, a eunuch may grow taller than expected, since testosterone normally enhances the ossification of the epiphyses.

Regulation of Male Sex Hormones

The degree to which male secondary sexual characteristics develop is directly related to the amount of testosterone secreted by the interstitial cells. This quantity is regulated by a *negative feedback system* involving the hypothalamus. (See fig. 21.16).

As the concentration of testosterone in the blood increases, the hypothalamus becomes inhibited, and its stimulation of the anterior pituitary gland by GRF is decreased. As the pituitary's secretion of LH (ICSH) is reduced, the amount of testosterone released by the interstitial cells is reduced also.

But as the blood level of testosterone drops, the hypothalamus becomes less inhibited, and it once again stimulates the pituitary gland to release LH. The increasing secretion of LH causes the interstitial cells to release more testosterone, and its blood level rises.

Thus, the concentration of testosterone in the male body is regulated so that it remains relatively constant.

Fig. 21.16 A negative feedback mechanism operating between the anterior lobe of the pituitary gland and the testes controls the concentration of testosterone.

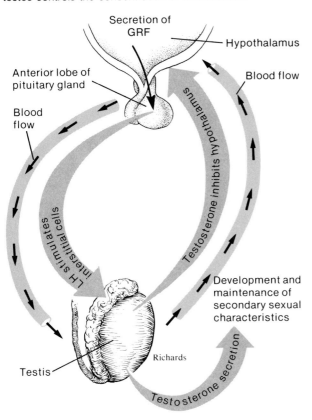

The amount of testosterone secreted by the testes usually declines gradually after about 25 years of age. Consequently, even though sexual activity may be continued into old age, males typically experience a decrease in sexual functions as they grow older. This decrease is sometimes called the *male climacteric*.

1. What initiates the changes associated with male sexual maturity?
2. Describe several male secondary sexual characteristics.
3. List the functions of testosterone.
4. Explain how the secretion of male sex hormones is regulated.

Organs of the Female Reproductive System

The organs of the female reproductive system can be grouped into primary, internal accessory, and external parts. The primary organs are the *ovaries,* which produce the female sex cells or **egg cells.**

The Ovaries

The **ovaries** are solid, ovoid structures measuring about 3.5 cm (1.4 in) in length, 2 cm (0.8 in) in width, and 1 cm (0.4 in) in thickness. They are located, one on each side, in a shallow depression (ovarian fossa) of the lateral wall of the pelvic cavity. (See fig. 21.17 and color plate 34.)

Attachments of the Ovaries

Each ovary is attached to several ligaments that help to hold it in position. The largest of these, formed by a fold of peritoneum, is called the *broad ligament.* It is also attached to the uterine tubes and uterus.

At its upper end, the ovary is held by a small fold of peritoneum, called the *suspensory ligament,* that contains the ovarian blood vessels and nerves. At its lower end, it is attached to the uterus by a rounded, cord-like thickening of the broad ligament, called the *ovarian ligament.* (See fig. 21.18.)

Descent of the Ovaries

Like the testes in a male fetus, the ovaries in a female fetus originate from masses of tissue behind the parietal peritoneum, near the developing kidneys. During development, these structures descend to locations just below the pelvic brim, where they remain attached to the lateral pelvic wall.

Structure of the Ovaries

The tissues of an ovary can be divided into two regions, an inner *medulla* and an outer *cortex.* However, these regions are not distinctly separated.

The ovarian medulla is composed largely of loose connective tissue and contains numerous blood vessels, lymphatic vessels, and nerve fibers. The ovarian cortex is composed of more compact tissue and has a somewhat granular appearance due to the presence of tiny masses of cells called *ovarian follicles.*

Fig. 21.17 Organs of the female reproductive system.

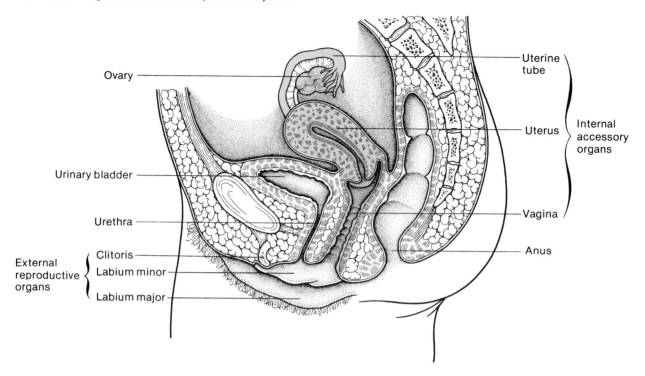

Labels in figure:
- Ovary
- Urinary bladder
- Urethra
- Clitoris
- Labium minor
- Labium major
- External reproductive organs
- Uterine tube
- Uterus
- Vagina
- Anus
- Internal accessory organs

Fig. 21.18 The ovaries are located on each side against the lateral walls of the pelvic cavity.

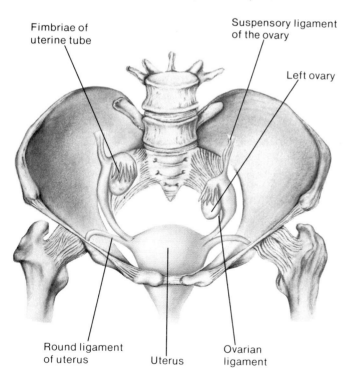

Labels in figure:
- Fimbriae of uterine tube
- Suspensory ligament of the ovary
- Left ovary
- Round ligament of uterus
- Uterus
- Ovarian ligament

The free surface of the ovary is covered by a layer of cuboidal cells, called the **germinal epithelium.** This germinal epithelium gives rise to sex cells, as does its counterpart in the testes. (See fig. 21.19.)

1. *What are the primary organs of the female reproductive system?*
2. *Describe the descent of the ovary.*
3. *Describe the structure of an ovary.*

Formation of Egg Cells

Unlike sex cell production in males, which begins at puberty and continues throughout adulthood, egg cell production in females begins before birth and ceases relatively early in life.

During fetal development, small groups of cells grow inward from the germinal epithelium of the ovary and form numerous **ovarian follicles.** Each follicle consists of a single *primary oocyte* surrounded by a group of epithelial cells (follicular cells).

A primary oocyte undergoes meiosis (shortly before ovulation), and the resulting offspring cells, have one-half as many chromosomes in their nuclei. Thus the chromosome number is reduced from 46 in the primary oocyte to 23 in the offspring cells.

When a primary oocyte divides, the division of the cellular cytoplasm is grossly unequal. One of the

Fig. 21.19 Section of a human ovary: (*a*) germinal epithelium; (*b*) primary follicles.

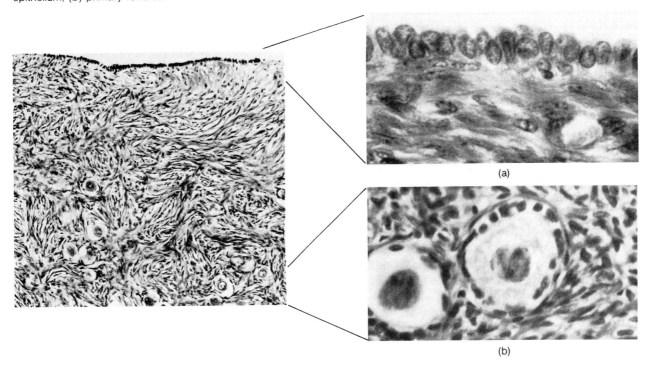

(a)

(b)

resulting cells, called a *secondary oocyte,* is quite large, and the other, called the *first polar body,* is very small.

The large secondary oocyte functions as an *egg cell* (ovum) in that it can be fertilized by a sperm cell. If this happens, the egg cell divides unequally to produce a tiny *second polar body* and a relatively large fertilized egg cell or **zygote.** Meanwhile, the first polar body may also divide to form two more polar bodies. (See fig. 21.20.)

Thus, the result of this process, which is called **oogenesis,** is one secondary oocyte or egg cell and a polar body. After being fertilized, the egg cell divides to produce a second polar body and a zygote, which can give rise to an embryo. Although the first polar body may divide again, the polar bodies have no further function and they soon degenerate.

At the time of birth, a female may have as many as 500,000 primary oocytes in each ovary. This number may decline to 250,000 by puberty. These cells have not completed their meiotic divisions and are in a resting state. All but a few hundred of the primary oocytes eventually degenerate without giving rise to egg cells.

---

1. *How does the production of egg cells differ from that of sperm cells in terms of time span?*
2. *Describe the major events of oogenesis.*

Fig. 21.20 During the process of oogenesis, a single egg cell (secondary oocyte) results from the meiosis of a primary oocyte. If the egg is fertilized it forms a second polar body and becomes a zygote.

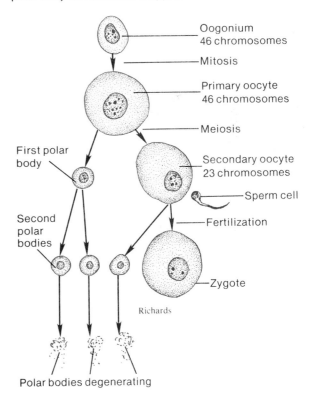

Oogonium
46 chromosomes

Mitosis

Primary oocyte
46 chromosomes

Meiosis

First polar body

Secondary oocyte
23 chromosomes

Sperm cell

Second polar bodies

Fertilization

Zygote

Richards

Polar bodies degenerating

Maturation of a Follicle

The ovarian follicles containing primary oocytes are called *primary follicles,* and they remain relatively unchanged throughout childhood. At puberty, however, the hormone, FSH, is released by the anterior pituitary gland. This hormone stimulates the ovaries to enlarge and some of the primary follicles to undergo *maturation.* (See color plate 35.)

In the process of maturation, the oocyte of a primary follicle grows larger, and the follicular cells surrounding the oocyte begin to divide actively. These follicular cells soon organize themselves into layers, and a cavity appears in the cellular mass. As the cavity forms, it becomes filled with a clear *follicular fluid* that bathes the oocyte.

Cells located outside the maturing follicle differentiate into two layers around it. One is an *inner vascular layer* (theca interna), composed largely of loose connective tissue and blood vessels; the other is an *outer fibrous layer* (theca externa), composed of tightly packed fibers.

The follicular cells continue to divide, the fluid-filled follicular cavity continues to enlarge, and the oocyte is pressed to one side within the follicle. Consequently, the follicle greatly increases in size, and eventually it bulges outward on the surface of the ovary like a blister.

The oocyte within such a mature follicle is a large, spherical cell surrounded by a thick, tough membrane (zona pellucida). It is enclosed by a mantle of follicular cells (corona radiata). Processes from these follicular cells extend through the zona pellucida and are thought to supply the oocyte with nutrients.

Although as many as twenty primary follicles may begin the process of maturation at any one time, as a rule only one follicle will reach full development, and the others will degenerate.

Ovulation

As a follicle matures, its primary oocyte undergoes oogenesis, giving rise to an egg cell and a first polar body. These cells are released from the follicle by the process called ovulation.

Ovulation is stimulated by hormones from the anterior pituitary gland, which apparently cause the mature follicle to swell rapidly and eventually rupture. When this happens, the follicular fluid, accompanied by the egg cell, oozes outward from the surface of the ovary and enters the peritoneal cavity. The maturation and eventual expulsion of the egg cell are shown in figure 21.21.

After it is expelled from the ovary, the egg cell and one or two layers of follicular cells surrounding it are usually propelled to the opening of a nearby *uterine tube.* If the egg cell is not fertilized by union with a sperm cell within a relatively short time, it will degenerate.

---

1. What causes a primary follicle to mature?
2. What changes occur in a follicle and its oocyte during maturation?
3. What causes ovulation?
4. What happens to an egg cell following ovulation?

Fig. 21.21 As a follicle matures, the egg cell enlarges and becomes surrounded by a mantle of follicular cells and fluid. Eventually the mature follicle ruptures to release the egg cell.

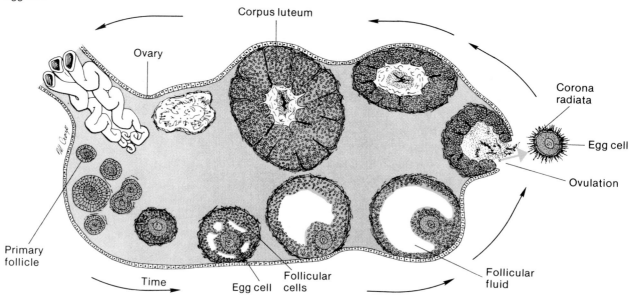

　　　The Human Life Cycle

Female Internal Accessory Organs

The *internal accessory organs* of the female reproductive system include a pair of uterine tubes, a uterus, and a vagina.

The Uterine Tubes

The **uterine tubes** (fallopian tubes or oviducts), which convey egg cells toward the uterus, are suspended by portions of the broad ligament, and have openings into the peritoneal cavity near the ovaries. Each tube, which is about 10 cm (4 in) long and 0.7 cm (0.3 in) in diameter, passes medially to the uterus, penetrates its wall, and opens into the uterine cavity.

Near the ovary the uterine tube expands to form a funnel-shaped *infundibulum,* which partially encircles the ovary, medially. On its margin the infundibulum bears a number of irregular, branched extensions, called *fimbriae.* (See fig. 21.22.) Although the infundibulum generally does not touch the ovary, one of the larger extensions (ovarian fimbria) is connected directly to it.

The wall of a uterine tube consists of an inner mucosal layer, a middle muscular layer, and an outer covering of peritoneum. The mucosal layer is drawn into numerous longitudinal folds and is lined with simple columnar epithelial cells, some of which are *ciliated.* (See fig. 21.23.) The epithelium secretes mucus, and the cilia beat toward the uterus. These actions apparently help to draw the egg cell and expelled follicular fluid into the infundibulum following ovulation.

Ciliary action also aids the transport of the egg cell down the uterine tube, and peristaltic contractions in the tube's muscular layer help force the egg along.

The Uterus

The **uterus,** whose function is to receive the embryo resulting from a fertilized egg cell and to sustain its life during development, is a hollow, muscular organ shaped somewhat like an inverted pear.

The *broad ligament,* which is also attached to the ovaries and uterine tubes, extends from the lateral walls of the uterus to the pelvic walls and floor, creating a septum across the pelvic cavity. (See fig. 21.22.) A fibrous sheet along the sides of the lower uterus and vagina form the *cardinal ligament,* which provides a deep continuation of this septum. Also, a flattened band of tissue within the broad ligament, called the *round ligament,* connects the upper end of the uterus to the pelvic wall.

Although the size of the uterus changes greatly during pregnancy, in its usual state it is about 7 cm (2.8 in) long, 5 cm (2 in) wide (at its broadest point), and 2.5 cm (1 in) in diameter. The uterus is located medially within the anterior portion of the pelvic cavity, above the vagina, and is usually bent forward over the urinary bladder.

The upper two-thirds of the uterus, or *body,* has a dome-shaped top and is joined by the uterine tubes that enter its wall at its broadest part.

The lower one-third of the uterus is the tubular **cervix,** which extends downward into the upper portion of the vagina. The cervix surrounds the opening called the *cervical orifice* (ostium uteri), through which the uterus communicates with the vagina.

Fig. 21.22 The funnel-shaped infundibulum of the uterine tube partially encircles the ovary. What factors aid the movement of an egg cell into the infundibulum following ovulation?

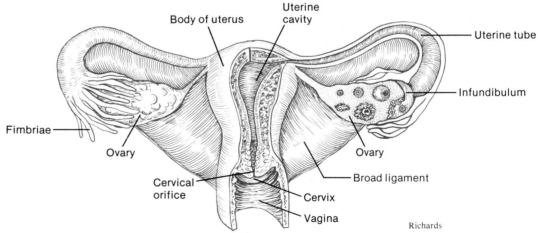

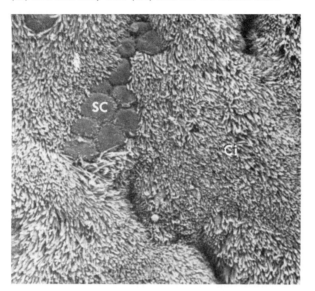

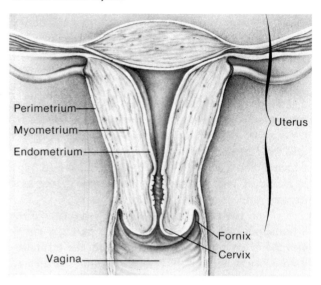

Cancer developing within the tissues of the uterus or cervix can usually be detected by means of a relatively simple and painless procedure called the *Pap* (Papanicolaou) *smear test.* This technique involves removing a tiny sample of tissue by scraping, smearing the sample on a glass slide, staining it, and examining it for the presence of abnormal cells.

Since this test can reveal a uterine or cervical cancer in the early stages of development, when it may be completely cured, the American Cancer Society recommends that all women over 20 years of age have a Pap test at regular intervals.

The wall of the uterus is relatively thick and is composed of three layers, as figure 21.24 shows. The **endometrium,** the inner mucosal layer lining the uterine cavity, is covered with columnar epithelium and contains numerous tubular glands. The **myometrium,** a very thick, muscular layer, consists largely of bundles of smooth muscle fibers arranged in longitudinal, circular, and spiral patterns, and interlaced with connective tissues. During the monthly female reproductive cycles and during pregnancy, these layers undergo extensive changes. (These changes are described in a later section of this chapter.) The **perimetrium** of the uterus is the outer serosal layer. It is composed of the peritoneal layer of the broad ligament that covers the body of the uterus and part of the cervix.

The Vagina

The **vagina** is a fibromuscular tube, about 9 cm (3.6 in) in length, extending from the uterus to the vestibule. It conveys uterine secretions to the outside, receives the erect penis during sexual intercourse, and transports the fetus during the birth process.

The vagina extends upward and back from the vestibule into the pelvic cavity. It is located posterior to the urinary bladder and urethra and anterior to the rectum, and is attached to these parts by connective tissues. The upper one-fourth of the vagina is separated from the rectum by a pouch (rectouterine pouch). The tubular vagina also surrounds the end of the cervix, and the recesses that occur between the vaginal wall and the cervix are termed *fornices.* (See fig. 21.24.)

The fornices are clinically important since they are relatively thin-walled and allow the internal abdominal organs to be palpated during a physical examination. Also, the posterior fornix, which is somewhat longer than the others, provides a surgical access to the peritoneal cavity through the vagina.

The *vaginal orifice* opens into the vestibule as a slit and is partially closed by a thin membrane of connective tissue and stratified squamous epithelium called the **hymen.** A central opening of varying size allows uterine and vaginal secretions to pass to the outside.

Fig. 21.25 Female external reproductive organs and vestibular bulbs.

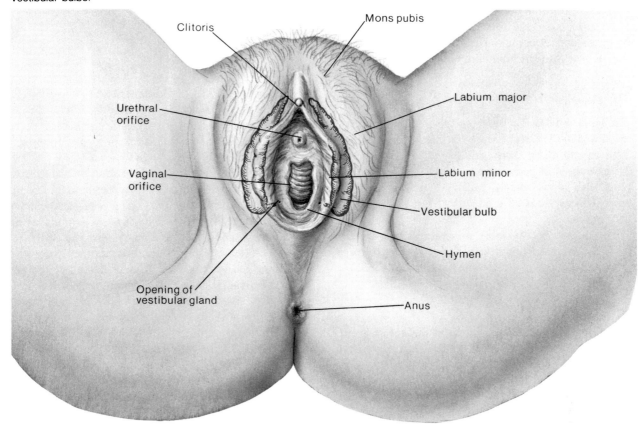

Clitoris

Mons pubis

Urethral orifice

Labium major

Vaginal orifice

Labium minor

Vestibular bulb

Hymen

Opening of vestibular gland

Anus

Ordinarily the hymen is ruptured during the first sexual intercourse, and it becomes permanently divided into two or three portions. Sometimes, the hymen may not be torn during intercourse, or it may be so resistant to penetration that a surgical incision is needed before intercourse can be performed.

The vaginal wall consists of three layers—an inner mucosa, a middle muscularis, and an outer fibrous layer. The *mucosal layer* is lined with stratified squamous epithelium and is drawn into numerous longitudinal and transverse ridges (vaginal rugae). This layer contains only a few glands. The mucus found in the lumen of the vagina comes primarily from the glands of the uterus.

The *muscular layer* consists mainly of smooth muscle fibers arranged in longitudinal and circular patterns. At the lower end of the vagina there is a thin band of striated muscle. This band helps to close the vaginal opening, but the *levator ani muscle* is primarily responsible for closing this orifice.

The *fibrous layer* consists of dense fibrous connective tissue interlaced with elastic fibers, and it serves to attach the vagina to surrounding organs.

1. How is an egg cell moved along a uterine tube?
2. Describe the structure of the uterus.
3. What is the function of the uterus?
4. Describe the structure of the vagina.

Female External Reproductive Organs

The *external organs* of the female reproductive system include the labia majora, labia minora, clitoris, vestibule, and vestibular glands. As a group, these structures that surround the openings of the urethra and vagina compose the **vulva.** They are shown in figure 21.25.

The Labia Majora

The **labia majora** enclose and protect the other external reproductive organs. They correspond to the scrotum of the male and are composed primarily of rounded folds of adipose tissue covered by skin. On the outside this skin includes numerous hairs, sweat glands, and sebaceous glands, while on the inside it is thinner and hairless.

The labia majora lie closely together and are separated longitudinally by a cleft (pudendal cleft) that includes the urethral and vaginal openings. At their anterior ends, the labia merge to form a medial, rounded elevation of fatty tissue called the *mons pubis,* which overlies the pubic symphysis. At their posterior ends, the labia are somewhat tapered, and they merge into the perineum near the anus.

The Labia Minora

The **labia minora** are flattened longitudinal folds located within the cleft between the labia majora. These folds extend along either side of the vestibule. They are composed of connective tissue that is richly supplied with blood vessels, causing a pinkish appearance. This tissue is covered with stratified squamous epithelium.

Posteriorly, the labia minora merge with the labia majora, while anteriorly they converge to form a hoodlike covering around the clitoris.

The Clitoris

The **clitoris** is a small projection at the anterior end of the vulva between the labia minora. Although most of it is embedded in tissues, it is usually about 2 cm (0.8 in) long and 0.5 cm (0.2 in) in diameter. The clitoris is somewhat similar in structure to the penis of the male. More specifically, it is composed of two columns of erectile tissue called *corpora cavernosa.* These columns are separated by a septum and are surrounded by a covering of dense fibrous connective tissue.

At the root of the clitoris, the corpora cavernosa diverge to form *crura,* which in turn are attached to the sides of the pubic arch. At its anterior end, a small mass of erectile tissue forms a **glans,** which is richly supplied with sensory nerve fibers.

The Vestibule

The **vestibule** of the vulva is the space enclosed by the labia minora. The vagina opens into the posterior portion of the vestibule, while the urethra opens in the midline just anterior to the vagina and about 2.5 cm (1 in) behind the glans of the clitoris.

A pair of **vestibular glands** (Bartholin's glands), which correspond to the bulbourethral glands of the male, lie one on either side of the vaginal opening. Their ducts open into the vestibule near the lateral margins of the vaginal orifice.

Beneath the mucosa of the vestibule on either side is a mass of vascular erectile tissue. These structures, shown in figure 21.25, are called *vestibular bulbs.* They are separated from each other by the vagina and urethra, and they extend forward from the level of the vaginal opening to the clitoris.

1. *What is the male counterpart of the labia majora? Of the clitoris?*
2. *What structures are located within the vestibule?*

Erection, Lubrication, and Orgasm

The erectile tissues of the female, like those of the male, respond to sexual stimulation. Thus, during periods of sexual excitement, *parasympathetic* nerve impulses pass out from the sacral portion of the spinal cord and cause dilation of the arteries leading to the clitoris and vestibular bulbs. The inflow of blood to these parts is increased, and the erectile tissues swell.

At the same time, the vagina expands and elongates, and parasympathetic impulses stimulate the vestibular glands to secrete mucus into the vestibule. This secretion moistens and lubricates the tissues surrounding the vestibule and the lower end of the vagina, thus facilitating the insertion of the penis into the vagina. Also, since mucus usually continues to be secreted from these glands during sexual intercourse, it helps to prevent irritation of tissues that might occur if the vagina remained dry.

The clitoris is abundantly supplied with sensory nerve fibers that are especially sensitive to local stimulation, and the culmination of such stimulation is the pleasurable sense of physiological and psychological release called *orgasm.*

Just prior to orgasm, the tissues of the outer third of the vagina become engorged with blood and swell. This action serves to increase the friction on the penis during intercourse. As orgasm is triggered, a series of reflexes involving the sacral and lumbar portions of the spinal cord are initiated.

In response to these reflexes, the muscles of the perineum contract rhythmically, and the muscular walls of the uterus and uterine tubes become active. These muscular contractions are thought to aid the transport of sperm cells through the female reproductive tract toward the upper ends of the uterine tubes, where there may be an egg cell.

Following the orgasm, the inflow of blood into the erectile tissues is reduced, and the muscles of the perineum and reproductive tract tend to relax. Consequently the organs return to a state similar to that prior to sexual stimulation.

The various functions of the female reproductive organs are summarized in chart 21.2.

1. *What events result from parasympathetic stimulation of the female reproductive organs?*
2. *What changes take place in the vagina just prior to and during female orgasm?*
3. *What is the response of the uterus and the uterine tubes to orgasm?*

Chart 21.2 Functions of the female reproductive organs

Organ	Function
Ovary	Production of egg cells and female sex hormones
Uterine tube	Conveys egg cell toward uterus; site of fertilization; conveys developing embryo to uterus
Uterus	Protects and sustains life of embryo during pregnancy
Vagina	Conveys uterine secretions to outside of body; receives erect penis during coitus; transports fetus during birth process
Labia majora	Encloses and protects other external reproductive organs
Labia minora	Forms margins of vestibule; protects openings of vagina and urethra
Clitoris	Glans is richly supplied with sensory nerve endings associated with feeling of pleasure during sexual stimulation
Vestibule	Space between labia minora that includes vaginal and urethral openings
Vestibular glands	Secrete fluid that moistens and lubricates vestibule

Hormonal Control of Female Reproductive Functions

Female reproductive functions are largely controlled by hormones secreted by the *anterior pituitary gland* and the *ovaries*. These hormones are responsible for the development and maintenance of female secondary sexual characteristics, the maturation of egg cells, and changes that occur during the monthly reproductive cycles.

Pituitary Hormones

A female child's body remains reproductively immature until gonadotropin-releasing factors (GRF) are secreted from the hypothalamus. These substances enter blood vessels leading to the anterior pituitary gland, and the pituitary gland releases *gonadotropins* in response. These gonadotropins include the hormones FSH and LH. They play primary roles in the control of sex cell maturation and in the development and maintenance of secondary sexual characteristics in the female, as they do in the male. These occurrences in the female, however, are much more complex because hormonal concentrations in the female body undergo cyclic changes rather than remaining relatively stable as they do in the adult male.

Fig. 21.26 Mechanism by which female secondary sexual characteristics are stimulated to develop.

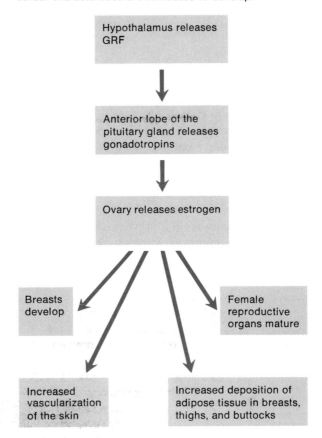

Female Sex Hormones

Several different female sex hormones are secreted by various tissues, including the ovaries, adrenal cortices, and the placenta (during pregnancy). These hormones belong to two major groups that are referred to as **estrogen** and **progesterone.**

The primary source of *estrogen* (in a nonpregnant female) is the ovaries, and it corresponds in its effects to testosterone in the male. That is, estrogen is mainly responsible for the development of female secondary sexual characteristics (fig. 21.26) and for the maintenance of these traits, which include the following:

1. Enlargement of the vagina, uterus, uterine tubes, ovaries, and external reproductive organs.

2. Development of the breasts and the ductile system of the mammary glands within the breasts.

3. Increased deposition of adipose tissue in the subcutaneous layer generally, and particularly in the breasts, thighs, and buttocks.

4. Increased vascularization of the skin.

Excess estrogens are metabolized primarily in the liver, and the products are excreted in the bile and urine. For this reason, females with liver disorders sometimes experience the symptoms of increased estrogen activity.

Certain other changes that occur in females at puberty seem to be related to *androgen* concentrations. For example, increased growth of hair in the pubic and axillary regions seems to be due to the presence of androgen secreted by the adrenal cortices. Conversely, the development of the female skeletal configuration, which includes narrow shoulders and broad hips, seems to be related to a lack of androgen.

The ovaries are the primary source of *progesterone* (in a nonpregnant female). The effects of this hormone are related mainly to changes that occur in the uterus during the rhythmic reproductive cycles.

1. What factors initiate sexual maturity in a female?
2. Name the two major female sex hormones.
3. What is the function of estrogen?
4. What is the function of androgen in a female?

Female Reproductive Cycles

The female reproductive cycles or **menstrual cycles** are characterized by regular, recurring changes in the uterine lining that culminate in menstrual bleeding. Such cycles usually begin near the thirteenth year of life and continue into middle age, at which time the cycles cease.

A female's first menstrual cycle (menarche) is initiated when the hypothalamus secretes gonadotropin-releasing factors (GRF), which in turn stimulate the anterior pituitary gland to secrete FSH (follicle-stimulating hormone) and LH (luteinizing hormone). As its name implies, FSH acts upon the ovary to stimulate the maturation of a *follicle*, and during this development the follicular cells secrete increasing amounts of estrogen (Fig. 21.27 shows a mature follicle.)

In a young female, this estrogen is responsible for the development of various secondary sexual characteristics. The estrogen secreted during future menstrual cycles is responsible for continuing the development of these traits and for their maintenance. Chart 21.3 summarizes the hormonal control of female secondary sexual characteristics.

The increasing concentration of estrogen during the early part of a menstrual cycle causes changes in the uterine lining, including thickening of the glandular endometrium. This activity continues for about 14 days. Meanwhile, the developing follicle has com-

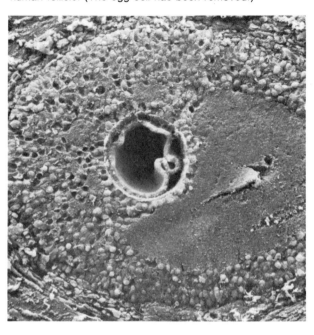

Fig. 21.27 Scanning electron micrograph of a mature human follicle. (The egg cell has been removed.)

Chart 21.3 Hormonal control of female secondary sexual characteristics

1. The hypothalamus releases GRF, which stimulates the anterior pituitary gland.
2. The anterior pituitary gland secretes FSH.
3. FSH stimulates the maturation of a follicle.
4. Follicular cells produce and secrete estrogen.
5. Estrogen is responsible for the development and maintenance of most female secondary sexual characteristics.
6. Levels of androgen affect other secondary sexual characteristics, including skeletal growth and growth of hair.
7. Progesterone, secreted by the ovaries, affects cyclical changes in the uterus.

pleted its maturation, and by the fourteenth day of the cycle it appears on the surface of the ovary as a blisterlike bulge.

For reasons that are not clear, about this time the anterior pituitary gland releases a relatively large quantity of LH and an increased amount of FSH. The resulting surge in concentrations of these hormones seems to cause the mature follicle to swell rapidly and rupture. Ovulation occurs as the follicular fluid, accompanied by the egg cell, leaves the follicle and enters the peritoneal cavity. The egg cell is then drawn into the uterine tube.

Following ovulation, the remnants of the follicle left in the ovary undergo rapid changes. The space occupied by the follicular fluid fills with blood that soon clots, and the follicular cells enlarge greatly to

Fig. 21.28 Major steps in the female reproductive cycle: (*1*) the hypothalamus releases GRF; (*2*) the anterior lobe of the pituitary gland secretes FSH, which stimulates the maturation of an ovarian follicle; (*3*) follicular cells secrete estrogen, which causes the uterine wall to thicken; (*4*) the anterior pituitary gland secretes LH, which stimulates ovulation and causes follicular cells to form the corpus luteum; (*5*) the corpus luteum secretes estrogen and progesterone, which stimulate further development of the uterine wall; (*6*) the corpus luteum degenerates, and tissues from the uterine wall slough away as concentrations of estrogen and progesterone decrease.

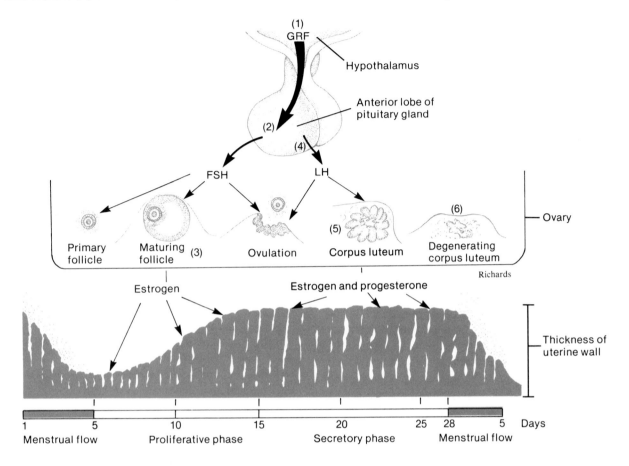

form a new glandular structure within the ovary, called **corpus luteum**. (See fig. 21.21.)

Although the follicular cells secrete minute quantities of progesterone during the first part of the menstrual cycle, corpus luteum cells secrete very large quantities of progesterone and estrogen during the last half of the cycle. Consequently, as a corpus luteum becomes established the blood concentration of progesterone increases sharply.

Progesterone acts on the endometrium of the uterus, causing it to become more vascular and glandular. It also stimulates the uterine glands to secrete increasing quantities of glycogen and lipids. As a result, the endometrial tissues of the uterus become filled with fluids containing nutrients and electrolytes that provide a favorable environment for the development of an embryo.

Since high blood concentrations of estrogen and progesterone inhibit the release of GRF from the hypothalamus and gonadotropins from the pituitary gland, no other follicles are stimulated to develop during the time the corpus luteum is active. However, if the egg cell that was released at ovulation is not fertilized by a sperm cell, the corpus luteum begins to degenerate about the twenty-fourth day of the cycle.

When the corpus luteum ceases to function, the concentrations of estrogen and progesterone decline rapidly, and in response, blood vessels in the endometrium become constricted. This action reduces the supply of oxygen and nutrients to the thickened uterine lining, and these tissues (desidua) soon disintegrate and slough away. At the same time, blood escapes from damaged capillaries, creating a flow of blood and cellular debris that passes through the vagina as the *menstrual flow*. This flow usually begins about the twenty-eighth day of the cycle and continues for 3 to 5 days, while the estrogen concentration is relatively low.

The beginning of the menstrual flow marks the end of a menstrual cycle and the beginning of a new cycle. This cycle is diagrammed in figure 21.28, and summarized in chart 21.4.

The Reproductive Systems 723

Chart 21.4 Major events in a menstrual cycle

1. The anterior pituitary gland secretes FSH.
2. FSH stimulates maturation of a follicle.
3. Follicular cells produce and secrete estrogen.
   a. Estrogen maintains secondary sexual traits.
   b. Estrogen causes uterine lining to thicken.
4. The anterior pituitary gland secretes a relatively large amount of LH, which stimulates ovulation.
5. Follicular cells become corpus luteum cells, which secrete estrogen and progesterone.
   a. Estrogen continues to stimulate uterine wall development.
   b. Progesterone stimulates uterine lining to become more glandular and vascular.
   c. Estrogen and progesterone inhibit the secretion of FSH and LH from the anterior pituitary gland.
6. If egg cell is not fertilized, corpus luteum degenerates, and no longer secretes estrogen and progesterone.
7. As concentrations of estrogen and progesterone decline, blood vessels in the uterine lining constrict.
8. The uterine lining disintegrates and sloughs away, producing menstrual flow.
9. The anterior pituitary gland, which is no longer inhibited, again secretes FSH.
10. The cycle is repeated.

Since the blood concentration of estrogen is low at the beginning of the cycle, the hypothalamus and pituitary gland are no longer inhibited. Consequently, the concentration of FSH and LH soon increases, and a new follicle is stimulated to mature. As this follicle secretes estrogen, the lining of the uterus undergoes repair and the endometrium begins to thicken again.

Figure 21.29 summarizes the changing hormone concentrations during the menstrual cycle.

Menopause

After puberty, menstrual cycles continue to occur at more or less regular intervals into the late forties, at which time they usually become increasingly irregular and after a few months or years cease altogether. This period in life is called **menopause** (female climacteric).

The cause of menopause seems to be aging of the ovaries. After about 35 years of cycling, apparently few primary follicles remain to be stimulated by pituitary gonadotropins. Consequently, follicles no longer mature, ovulation does not occur, and the blood level of estrogen decreases greatly.

As a result of low estrogen concentrations, the female secondary sexual characteristics usually undergo changes. The vagina, uterus, and uterine tubes may decrease in size, as may the external reproductive organs. The pubic and axillary hair may

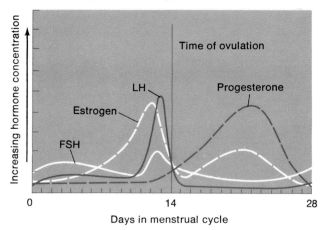

Fig. 21.29 Changes in relative concentrations of hormones during a menstrual cycle.

become thinner, and the breasts may regress. Furthermore, since estrogen no longer inhibits the pituitary secretion of FSH and LH, these hormones are released continuously. This increase in gonadotropin concentration, coupled with a decrease in estrogen concentration, seems to be responsible for some unpleasant symptoms occasionally experienced by women in menopause. For example, they may feel a sensation of heat in the face and upper body, called a hot flash. In addition, menopausal women may feel irritable, worried, or fatigued. Figure 21.30 lists some of the hormonal changes that may occur during menopause.

Since a decreasing level of estrogen may be accompanied by a reduction in the size of the vagina and a tendency for the vaginal tissues to become drier, after menopause a woman may experience some discomfort during sexual intercourse.

Although women usually do not require medical treatment for the symptoms of menopause, those with severe discomforts may be provided with daily doses of estrogen to relieve these symptoms.

1. Trace the events of the female menstrual cycle.
2. What effect does progesterone have on the endometrium of the uterus?
3. What causes the menstrual flow?
4. What are some changes that may occur at menopause?

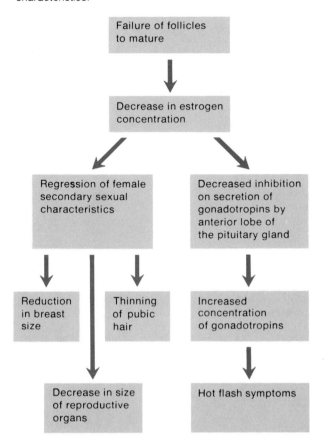

Fig. 21.30 Failure of follicles to mature results in a decreasing concentration of estrogen and may be accompanied by a regression of female secondary sexual characteristics.

Failure of follicles to mature

↓

Decrease in estrogen concentration

Regression of female secondary sexual characteristics

Decreased inhibition on secretion of gonadotropins by anterior lobe of the pituitary gland

Reduction in breast size

Thinning of pubic hair

Increased concentration of gonadotropins

Decrease in size of reproductive organs

Hot flash symptoms

Pregnancy

Pregnancy is the condition of having an offspring developing within the uterus. It results from the *fertilization* of an egg cell by a sperm cell.

Transport of Sex Cells

Ordinarily, before fertilization can occur, an egg cell (secondary oocyte) must be released by ovulation and be carried into a uterine tube by the action of the ciliated epithelium lining the tube.

During sexual intercourse, seminal fluid containing sperm cells is usually deposited in the vagina near the cervix. To reach the egg cell, sperm cells must then be transported upward through the uterus and uterine tube. This transport is aided by the lashing movements of the sperm's tail and by muscular contractions within the walls of the uterus and uterine tube, which are thought to be stimulated by prostaglandins in the seminal fluid. Also, under the influence of the high estrogen concentrations during the

first part of the menstrual cycle, the uterus and cervix contain a thin, watery secretion that promotes sperm transport and survival. (See fig. 21.31.)

The sperm transport mechanism is relatively ineffective, however, because even though as many as 300 million to 500 million sperm cells may be deposited in the vagina by a single ejaculation, only a few hundred sperm cells ever reach an egg cell.

Studies indicate that an egg cell may survive for only 12 to 24 hours following ovulation, while sperm cells may live up to 72 hours within the female reproductive tract. Consequently, sexual intercourse probably must occur no more than 72 hours before ovulation or 24 hours following ovulation if fertilization is to take place.

The first sperm cells are thought to reach the upper portions of the uterine tube within an hour following sexual intercourse. Although many sperm cells may reach an egg cell, only one will participate in fertilization. (See fig. 21.32.)

Fertilization

When a sperm cell reaches an egg cell, it moves through the follicular cells that adhere to the egg's surface (corona radiata) and penetrates the *zona pellucida* that surrounds the egg cell membrane. This penetration seems to be aided by an enzyme (hyaluronidase), released by the acrosome in the sperm head. This enzyme apparently allows the sperm cell to digest its way through the corona radiata and the zona pellucida. (See fig. 21.33.)

As the sperm cell penetrates the zona pellucida, this covering becomes impenetrable by any other sperm cells. Even though others may begin to enter, they are inactivated by an unknown mechanism and fail to get through.

Once a sperm cell reaches the egg cell membrane, it passes through the membrane and enters the cytoplasm. During this process, the sperm cell loses its tail, and its head swells to form a nucleus. As explained earlier, the egg cell divides unequally to form a relatively large cell, and a tiny second polar body, which is expelled. The nuclei of the egg cell and that of the sperm cell then come together in the center of the larger cell. Their nuclear membranes disappear, and their chromosomes combine, thus completing the process of **fertilization.** This process is diagrammed in figure 21.34.

Fig. 21.31 The paths of the egg and sperm cells through the female reproductive tract. What factors aid the movements of these cells?

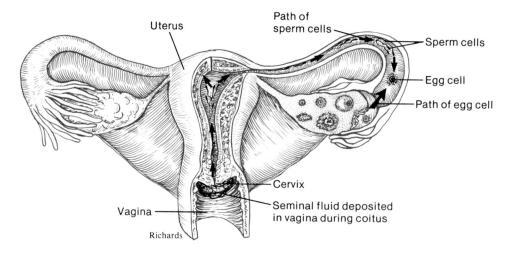

Fig. 21.32 Although many sperm cells may reach an egg cell, only one will fertilize it.

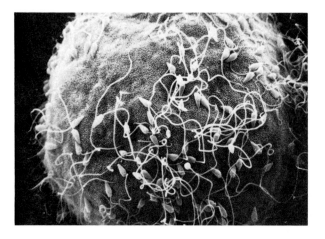

Fig. 21.33 A sperm cell penetrates the zona pellucida surrounding an egg cell with the aid of enzymes released from the sperm head.

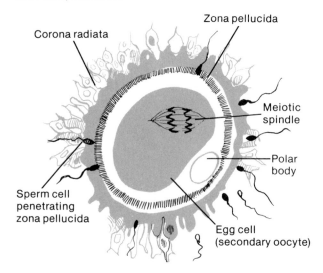

Since the sperm cell and the egg cell each provide 23 chromosomes, the result of fertilization is a cell with 46 chromosomes—the usual number of a human cell. This cell, called a **zygote,** is the first cell of the future offspring.

1. What factors enhance motility of sperm cells following sexual intercourse?
2. Where does fertilization normally take place?
3. List the events that occur during fertilization.

Early Embryonic Development

Shortly after it is formed, the zygote undergoes *mitosis,* giving rise to two daughter cells. These cells in turn divide into four cells, which divide into eight cells, and so forth. With each subsequent division, the resulting daughter cells are smaller and smaller. Consequently, this phase in development is termed **cleavage.**

Meanwhile, the tiny mass of cells is moved through the uterine tube to the cavity of the uterus. This movement is aided by the action of cilia of the

Fig. 21.34 The process of fertilization: (a) after a sperm cell enters it, a secondary oocyte divides unequally to form a polar body and a zygote; (b) the nucleus of the sperm cell enlarges and approaches the nucleus of the egg cell; (c) the nuclear membranes disappear and the chromosomes of the two nuclei combine; (d) the result is a cell with 46 chromosomes.

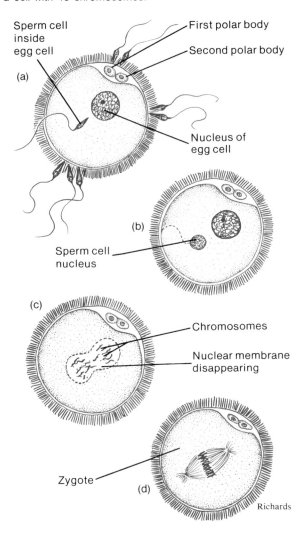

Sperm cell inside egg cell

First polar body

Second polar body

(a)

Nucleus of egg cell

(b)

Sperm cell nucleus

(c)

Chromosomes

Nuclear membrane disappearing

Zygote

(d)

Richards

tubular epithelium and by weak peristaltic contractions of smooth muscles in the tubular wall. Secretions from the epithelial lining are thought to provide the developing organism with nutrients.

The trip to the uterine cavity takes about 3 days, and at the end of this time the structure consists of a solid ball (morula) of about sixteen cells.

Once inside the uterus, it remains free within the uterine cavity for about 3 days, during which the zona pellucida of the original egg cell degenerates, and the structure, which now consists of a hollow ball of cells (blastocyst), begins to attach itself to the uterine lining. By the end of the first week of development, it is superficially *implanted* in the endometrium. (See fig. 21.35.)

About this time, certain cells within the blastocyst organize themselves into a group that will give rise to the body of the offspring. This marks the beginning of the embryonic period of development. The offspring is termed an **embryo** until the end of the seventh week, when it is called a **fetus**.

Eventually, the outer cells of the embryo together with cells of the maternal endometrium form a complex vascular structure called the **placenta.** This organ serves to attach the embryo to the uterine wall, to exchange nutrients, gases, and wastes between the maternal blood and the embryonic blood, and to secrete hormones.

Occasionally developing offspring may become implanted in tissues outside the uterus, including those of a uterine tube, ovary, cervix, or an organ in the abdominal cavity. The result is called an *ectopic pregnancy.* Most commonly, this condition occurs within a uterine tube and is termed a tubal pregnancy.

In a tubal pregnancy, the tube usually ruptures as the embryo enlarges. This is accompanied by severe pain and heavy bleeding through the vagina. The treatment involves prompt surgical removal of the embryo and repair or removal of the damaged uterine tube.

1. What is a zygote? A morula?
2. How does an embryo become implanted in the uterine wall?
3. How does a developing offspring obtain nutrients and oxygen?

Hormonal Changes during Pregnancy

During the usual menstrual cycle, the corpus luteum degenerates about 2 weeks after ovulation. Consequently, the estrogen and progesterone levels decline rapidly, the uterine lining is no longer maintained, and the endometrium sloughs away as menstrual flow. If this occurs following implantation, the embryo will be lost (spontaneously aborted).

The mechanism that normally prevents such a termination of pregnancy involves a hormone, called HCG (human chorionic gonadotropin), that is secreted by cells of the *implanted embryo.* This hormone has properties similar to LH, and it causes the corpus luteum to be maintained and to continue secreting high levels of estrogen and progesterone. Thus, the uterine wall continues to grow and develop.

Fig. 21.35 Stages in early development.

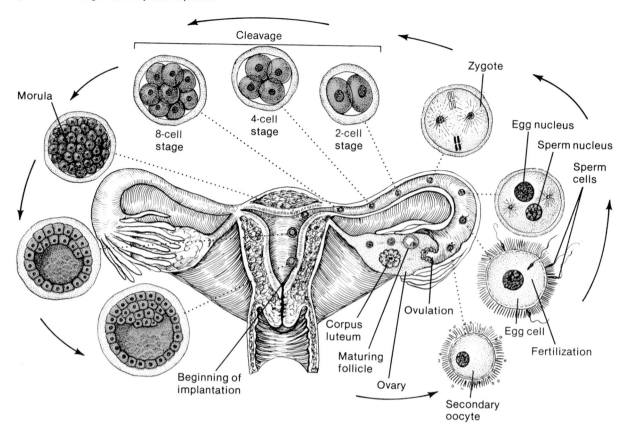

At the same time, the estrogen and progesterone suppress the release of FSH and LH from the pituitary gland, so that normal menstrual cycles are inhibited. (See fig. 21.36.)

Females are sometimes infertile because the anterior pituitary glands produce insufficient quantities of gonadotropic hormones, and ovulation does not occur. Treatment may include administration of HCG, which, like LH, can stimulate ovulation. HMG (human menopausal gonadotropin), a substance rich in LH and FSH, that can be extracted from the urine of postmenopausal women may also be used for treatment of this deficiency. Either treatment, however, may overstimulate the ovaries and cause many follicles to release egg cells simultaneously, resulting in multiple births.

The secretion of HCG continues at a high level for about two months, then declines to a relatively low level by the end of four months. Although the corpus luteum is maintained throughout the pregnancy, its function as a source of hormones becomes

Fig. 21.36 Mechanism that prevents the loss of the uterine wall during the early stages of pregnancy.

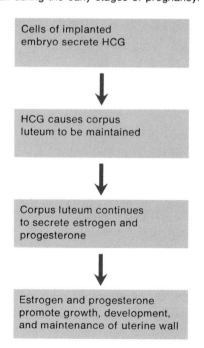

Fig. 21.37 Relative concentrations of three hormones in the blood of a pregnant woman.

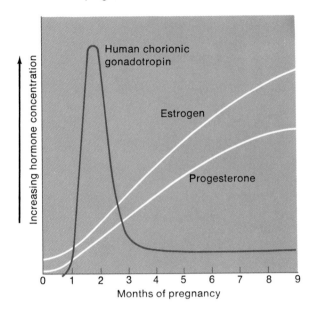

Chart 21.5 Hormonal changes during pregnancy

1. Following implantation, embryonic cells secrete HCG.
2. HCG causes the corpus luteum to be maintained and to continue secreting estrogen and progesterone.
3. As the placenta develops, it secretes large quantities of estrogen and progesterone.
4. Placental estrogen and progesterone
   a. Stimulate the uterine lining to continue development.
   b. Maintain the uterine lining.
   c. Inhibit the secretion of FSH and LH from the anterior pituitary gland.
   d. Stimulate development of the mammary glands.
   e. Progesterone inhibits uterine contractions.
   f. Estrogen causes enlargement of reproductive organs and relaxation of ligaments of pelvic joints.
5. The placenta also secretes placental lactogen that stimulates breast development.

less important after the first three months of pregnancy (first trimester). This is due to the fact that the placenta is usually well developed by this time, and the placental tissues secrete high levels of estrogen and progesterone. (See fig. 21.37.)

For the remainder of the pregnancy, *placental estrogen* and *progesterone* maintain the uterine wall. The placenta also secretes a hormone called **placental lactogen.** This hormone is thought to stimulate breast development and preparation for milk secretion, a function that is aided by placental estrogen and progesterone. Placental progesterone also inhibits the smooth muscles in the myometrium so that uterine contractions are suppressed until it is time for the birth process to begin.

The high concentration of placental estrogen during pregnancy causes enlargement of the vagina and external reproductive organs, as well as relaxation of the ligaments holding the pubic symphysis and sacroiliac joints together. This latter action allows for greater movement at these joints and thus aids the passage of the fetus through the birth canal. Chart 21.5 summarizes the hormonal changes of pregnancy.

---

1. What mechanism is responsible for maintaining the uterine wall during pregnancy?
2. What is the source of HCG during the first few months of pregnancy?
3. What is the source of hormones that sustain the uterine wall during the latter part of pregnancy?

Other Changes during Pregnancy

A number of other changes occur in a woman's body as a result of the increased demands of a growing fetus. For example, as the fetus increases in size, the uterus enlarges greatly, and instead of being confined to its normal location in the pelvic cavity, it extends upward and may eventually reach the level of the ribs. At the same time, the abdominal organs are displaced upward and compressed against the diaphragm. Also, as the uterus enlarges, it tends to press on the urinary bladder and cause the woman to experience a need to urinate frequently.

As the placenta grows and develops, it requires more blood, and as the fetus enlarges, it needs more oxygen and produces greater amounts of wastes that must be excreted. Consequently, the mother's blood volume, cardiac output, breathing rate, and urine production all tend to increase in response to the fetal demands.

The fetal need for increasing amounts of nutrients is reflected in an increased dietary intake by the mother. Her intake must supply adequate vitamins, minerals, and proteins for both herself and the fetus. The fetal tissues have a greater capacity to capture available nutrients than do the maternal tissues. Consequently, if the mother's diet is inadequate, her body will usually show symptoms of a deficiency condition before fetal growth is affected.

Pregnancy usually continues for 40 weeks (280 days) or about 9 calendar months (10 lunar months), if it is measured from the beginning of the last menstrual cycle. The pregnancy terminates with the *birth process* (parturition).

Although the mechanism causing birth is not well understood, a variety of factors seem to be involved. For example, progesterone suppresses uterine contractions during pregnancy. Estrogen, however, tends to excite such contractions. After the seventh month, the placental secretion of estrogen increases to a greater degree than the secretion of progesterone. As a result of the rising concentration of estrogen, the contractility of the uterine wall is enhanced.

Also, as is discussed in chapter 12, the stretching of the uterine and vaginal tissues late in pregnancy is thought to initiate nerve impulses to the hypothalamus. The hypothalamus, in turn, signals the posterior pituitary gland, which responds by releasing the hormone, **oxytocin.**

Oxytocin is a powerful stimulator of uterine contractions, and its effect, combined with the greater excitability of the myometrium due to the decline in progesterone secretion, may be involved in initiating labor.

Labor is the term for the process whereby muscular contractions force the fetus through the birth canal. Once labor starts, rhythmic contractions that begin at the top of the uterus and travel down its length force the contents of the uterus toward the cervix.

Since the fetus is usually positioned with its head downward, labor contractions force the head against the cervix. This action causes stretching of the cervix, which is thought to elicit a reflex that stimulates still stronger labor contractions. Thus, a *positive feedback system* operates in which uterine contractions result in more intense uterine contractions until a maximum effort is achieved (fig. 21.38). At the same time, dilation of the cervix reflexly stimulates an increased release of oxytocin from the pituitary gland.

During childbirth the tissues of the perineum are sometimes torn by the stretching that occurs as the infant passes through the birth canal. For this reason, an incision may be made along the midline of the perineum from the vestibule to within 1.5 cm (0.6 in) of the anus before the birth is completed. This procedure, called an *episiotomy*, ensures that the perineal tissues are cut cleanly rather than torn.

Fig. 21.38 The birth process involves this positive feedback mechanism.

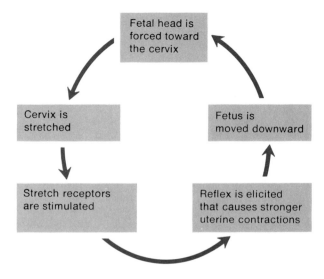

Chart 21.6   Some factors involved in the labor process

1. As the time of birth approaches, secretion of progesterone declines and its inhibiting effect on uterine contractions is lessened.
2. Stretching of uterine tissues causes reflex release of oxytocin from the posterior pituitary gland.
3. Oxytocin may stimulate uterine contractions, and may be involved in initiating labor.
4. As the fetal head causes the cervix to stretch, a positive feedback mechanism results in stronger and stronger uterine contractions and a greater release of oxytocin.
5. Abdominal wall muscles are stimulated by positive feedback to contract with greater and greater force.
6. The fetus is forced through the birth canal to the outside of the body.

As labor continues, abdominal wall muscles are stimulated to contract by a positive feedback mechanism, and they also aid in forcing the fetus through the cervix and vagina to the outside. (See chart 21.6.)

Following the birth of the fetus (usually within 10 to 15 minutes), the placenta, which remains inside the uterus, becomes separated from the uterine wall and is expelled by uterine contractions through the birth canal. This expulsion, which is termed the afterbirth, is accompanied by bleeding because vascular tissues are damaged in the process. However, the loss of blood is usually minimized by continued contraction of the uterus that constricts the bleeding vessels.

Figure 21.39 illustrates the steps of the birth process.

Fig. 21.39 Major steps in the birth process.

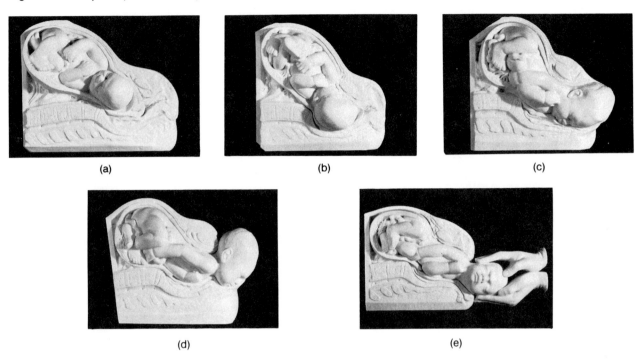

(a) (b) (c)

(d) (e)

For several weeks following childbirth, the uterus becomes smaller by a process called *involution.* Also, its endometrium sloughs off and is discharged through the vagina. This is followed by a return of an epithelial lining characteristic of a nonpregnant female.

---

1. List some of the physiological changes that occur in a woman's body during pregnancy.
2. Describe the events thought to initiate labor.
3. Explain how dilation of the cervix affects labor.
4. How is bleeding controlled naturally after the placenta is expelled?

The Mammary Glands

The **mammary glands** are accessory organs of the female reproductive system that are specialized to secrete milk following pregnancy.

Location of the Glands

The mammary glands are located in the subcutaneous tissue of the anterior thorax within hemispherical elevations, called *breasts.* The breasts overlie the *pec-toralis major* muscles and extend from the second to the sixth ribs, and from the sternum to the axillae.

A *nipple* is located near the tip of each breast at about the level of the fourth intercostal space, and it is surrounded by a circular area of pigmented skin called the *areola.* (See fig. 21.40.)

Structure of the Glands

The mammary glands originate from parts of the skin and, in fact, are modified sweat glands.

A mammary gland is composed of fifteen to twenty lobes, each of which includes tubular glands (alveolar glands), and a duct that leads to the nipple and opens to the outside. The lobes are separated from each other by dense connective and adipose tissues. These tissues also support the glands and attach them to the fascia of the underlying pectoral muscles. Other connective tissue, which forms dense strands called *suspensory ligaments,* extends inward from the dermis of the breast to the fascia, helping to support the weight of the breast.

Fig. 21.40 Sagittal section of a breast.

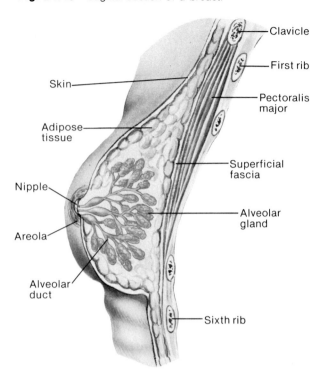

Clavicle

First rib

Skin

Pectoralis major

Adipose tissue

Superficial fascia

Nipple

Alveolar gland

Areola

Alveolar duct

Sixth rib

Breast cancer, which is one of the more common types of cancer in women, usually begins as a small, painless lump.

Since an early diagnosis of such a tumor is of prime importance in successful treatment, the American Cancer Society recommends that women examine their breasts each month, paying particular attention to the upper, outer portions. The examination should be made just after menstruation when the breasts are usually soft, and any lump that is discovered should be immediately checked by a physician.

Development of the Breasts

The mammary glands of male and female children are similar. As children reach *puberty*, the male glands fail to develop, and the female glands are stimulated to develop by ovarian hormones. As a result, the alveolar glands and ducts enlarge, and fat is deposited so that the breasts become surrounded by adipose tissue, except for the region of the areola.

During pregnancy, placental estrogen and progesterone stimulate further development of the mammary glands. Estrogen causes the ductile systems to grow and become branched and to have large quantities of fat deposited around them. Progesterone, on the other hand, stimulates the development of the alveolar glands. These changes are also promoted by the presence of placental lactogen.

As a consequence of hormonal activity, the breasts typically double in size during pregnancy, and the mammary glands become capable of secreting milk. However, no milk is produced because the high levels of estrogen and progesterone that occur during pregnancy inhibit the hypothalamus. The hypothalamus, in turn, suppresses the secretion of the hormone **prolactin** by the anterior pituitary gland.

Production and Secretion of Milk

Following childbirth and the expulsion of the placenta, the maternal blood levels of estrogen and progesterone decline rapidly. Consequently, the hypothalamus is no longer inhibited, and it signals the pituitary gland to release prolactin.

Prolactin stimulates the mammary glands to secrete large quantities of milk. This hormonal effect does not occur for 2 or 3 days following birth, and in the meantime, the glands secrete a few milliliters of a fluid called *colostrum* each day. Although colostrum contains some of the nutrients found in milk, it lacks fat.

The milk produced under the influence of prolactin does not flow readily through the ductile system of the mammary gland, but must be actively ejected by contraction of specialized *myoepithelial cells* surrounding the ducts. The contraction of these cells and the consequent ejection of milk through the nipple result from a reflex action.

This reflex is elicited when the breast is sucked or the nipple or areola is otherwise mechanically stimulated. Then sensory impulses travel to the hypothalamus, which signals the posterior pituitary gland to release oxytocin. The oxytocin reaches the breast by means of the blood, and it stimulates the myoepithelial cells of the ductile system to contract. Consequently, milk is ejected into a suckling infant's mouth in about 30 seconds. (See fig. 21.41.)

As long as milk is removed from the breasts, prolactin and oxytocin continue to be released, and milk continues to be produced. If milk is not removed regularly, the hypothalamus causes the secretion of prolactin to be inhibited, and within about 1 week, the mammary glands lose their capacity to produce milk.

Although it is possible for a woman to become pregnant during the period that she breast-feeds her child, menstrual cycles seem to be inhibited, at least for a time, while the mammary glands are active. Although the mechanism responsible for this effect is not well understood, it is thought that prolactin may suppress the release of gonadotropins from the anterior pituitary gland. In any event, menstrual cycles may not begin for some time following the birth

Fig. 21.41 Mechanism that causes the ejection of milk from the breasts. What happens to milk production if the milk is not regularly removed from the breasts?

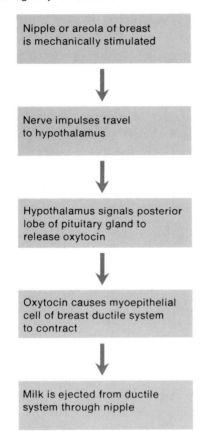

Nipple or areola of breast is mechanically stimulated

↓

Nerve impulses travel to hypothalamus

↓

Hypothalamus signals posterior lobe of pituitary gland to release oxytocin

↓

Oxytocin causes myoepithelial cell of breast ductile system to contract

↓

Milk is ejected from ductile system through nipple

of an infant that is breast-fed. On the other hand, after several months of breast-feeding, FSH is usually released, and the monthly reproductive cycles are re-established.

Chart 21.7 summarizes the hormonal effects involved in producing milk.

1. Describe the structure of a mammary gland.
2. How does pregnancy affect the mammary glands?
3. What stimulates the mammary glands to produce milk?
4. How is milk stimulated to flow into the ductile system of a mammary gland?

Chart 21.7 Hormonal control of the mammary glands

Before Pregnancy (beginning of puberty)

1. Ovarian hormones secreted during menstrual cycles stimulate alveolar glands and ducts of mammary glands to develop.

During Pregnancy

1. Estrogen causes the ductile system to grow and branch.
2. Progesterone stimulates development of the alveolar glands.
3. Placental lactogen promotes development of the breasts.
4. Secretion of the prolactin is inhibited by estrogen and progesterone, so no milk is produced.

Following Childbirth

1. Estrogen and progesterone concentrations decline, so prolactin secretion is no longer inhibited.
2. The anterior pituitary gland secretes prolactin, which stimulates milk production.
3. Mechanical stimulation of the breasts causes reflex release of oxytocin from the posterior pituitary gland.
4. Oxytocin stimulates ejection of milk from the ducts.
5. As long as milk is removed, more prolactin is released; if milk is not removed, milk production ceases.

Birth Control

Birth control is the voluntary regulation of the number of offspring a woman will produce and of the time they will be conceived. This control usually involves some method of **contraception**, designed to avoid the fertilization of an egg cell following sexual intercourse or to prevent the implantation of an embryo.

A variety of contraceptive methods are commonly practiced, including coitus interruptus, rhythm method, mechanical barriers, chemical barriers, oral contraceptives, intrauterine devices, and surgical methods. The relative effectiveness of these methods is summarized in chart 21.8.

Coitus Interruptus

Coitus interruptus involves withdrawing the penis from the vagina before ejaculation, thus preventing the entrance of sperm cells into the female reproductive tract. This method of contraception often proves unsatisfactory and may result in pregnancy, since some males find it emotionally difficult to withdraw just prior to ejaculation. Also, small quantities of seminal fluid containing sperm cells may be expelled from the penis before ejaculation occurs.

Chart 21.8 Effectiveness of contraceptive methods

Method	Pregnancies per 100 Women per Year[a]	
	High[b]	Low
No contraceptive[c]	80	40
Coitus interruptus	23	15
Condom[d]	17	8
Chemicals (spermicides)[e]	40	9
Diaphragm and jelly	28	11
Rhythm[f]	58	14
Oral contraceptives	2	0.03
IUD	8	3
Sterilization	0.003	0

[a] Data describe the number of women per hundred who will become pregnant in a one-year period while using a given method.
[b] High and low values represent best and worst estimates from various demographic and clinical studies.
[c] In the complete absence of contraceptive practice, eight out of ten women can expect to become pregnant within one year.
[d] Effectiveness increases if spermicidal jelly or cream is used in addition.
[e] Aerosol foam is considered to be the best of the chemical barriers.
[f] Use of a clinical thermometer to record daily temperatures increases effectiveness.

From E. Peter Volpe, *Man, Nature, and Society* (Dubuque, Ia.: Wm. C. Brown Co. Publishers, 1979). Used by permission of the publisher.

Rhythm Method

The *rhythm method,* like coitus interruptus, requires no artificial devices or chemicals. Instead, it requires abstinence from sexual intercourse a few days before and a few days after ovulation.

Since ovulation theoretically occurs on the fourteenth day of a 28-day menstrual cycle, it might seem easy to avoid intercourse near ovulation. Few women, however, have absolutely regular menstrual cycles, and the lengths of the cycles vary from time to time. Further, the variable part of the cycle occurs before ovulation, for regardless of the length of a cycle, the menstrual flow almost always begins 13 to 15 days following ovulation. Inasmuch as the length of a cycle cannot be predicted ahead of time, it is almost impossible to predict the time of ovulation accurately. Thus, it is not surprising that the rhythm method results in a relatively high rate of pregnancy.

Another disadvantage of this method is that it requires adherence to a particular pattern of behavior and thus restricts spontaneity in sexual activity.

The effectiveness of the rhythm method can sometimes be increased by measuring and recording the woman's body temperature when she awakes each morning for several months.

Since the body temperature typically rises about 0.6 degrees Fahrenheit immediately following ovulation, this procedure may allow a woman to more accurately predict the "unsafe times" in her reproductive cycle. On the other hand, many women apparently do not show such a change in body temperature at ovulation, and furthermore, the body temperature may vary in response to other factors such as illnesses or emotional upsets.

1. Why is coitus interruptus an unreliable method of contraception?
2. Describe the idea behind the rhythm method of contraception.
3. What factors make the rhythm method less reliable than some other methods of contraception?

Mechanical Barriers

Mechanical barriers are sometimes employed to prevent sperm cells from entering the female reproductive tract during sexual intercourse. One such device, for use by males, is called a *condom.* It consists of a thin rubber sheath that is placed over the erect penis before intercourse to prevent seminal fluid from entering the vagina upon ejaculation.

The condom provides a relatively inexpensive method of contraception that can be purchased without a prescription, and it also protects the user against contracting venereal diseases. However, men often feel a condom decreases the sensitivity of the penis during intercourse. Also, its use has the disadvantage of interrupting the sex act.

Another rubber device, which is used by females, is the *diaphragm.* It is a cup-shaped structure with a flexible ring forming the rim. The diaphragm is inserted into the vagina so that it covers the cervix and thus prevents the entrance of sperm cells into the uterus.

To be effective, a diaphragm must be fitted for size by a physician, be inserted properly, and be used in conjunction with a chemical spermicide that is applied to the surface adjacent to the cervix and to the rim of the diaphragm. Also, the diaphragm must be left in position for several hours following sexual intercourse.

Although the diaphragm can be inserted into the vagina sometime before sexual contact is to occur, so that its use does not interrupt the sex act, some women object to it because they find the insertion of the diaphragm distasteful or uncomfortable.

Chemical Barriers

Chemical barriers for the purpose of contraception include a variety of creams, foams, and jellies with spermicidal properties. Within the vagina, such chemicals create an environment that is very unfavorable for sperm cells.

Chemical barriers are fairly easy to use, but have a relatively high failure rate when used alone. They are best used along with a rubber diaphragm.

Oral Contraceptives

An *oral contraceptive* is commonly called the pill, because this method employs a series of drug-containing tablets taken by the woman.

The most common forms of oral contraceptives contain estrogen-like and progesterone-like substances. The mechanism by which these drugs produce their contraceptive effects is not well understood, but they are thought to inhibit the release of gonadotropins from the anterior pituitary gland and thus prevent ovulation.

Proper use of oral contraceptives requires motivation and planning. However, if they are used correctly, they prevent pregnancy almost 100% of the time. Although these hormone-like drugs usually do not produce serious side effects, some women, particularly those over 35 years of age who smoke, may experience an undesirable tendency to form intravascular clots if they use the pill.

As was mentioned in chapter 12, *prostaglandins* are compounds that have a variety of powerful physiological effects. One of them (PGE_2) is known to be able to prevent implantation of an embryo in the uterine wall. It is being investigated for use as a possible "morning after pill"—a contraceptive that might be effective when used after sexual intercourse.

Fig. 21.42 Intrauterine devices (IUD).

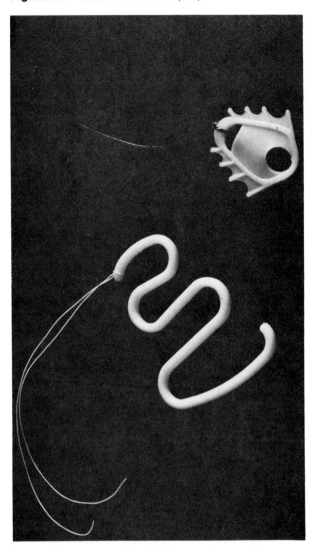

1. Describe two methods of contraception that make use of mechanical barriers.
2. How can the effectiveness of chemical contraceptives be increased?
3. What substances are contained in oral contraceptives?
4. How does an oral contraceptive prevent pregnancy?

Intrauterine Devices

An *intrauterine device*, or *IUD*, is a small, solid object in the form of a ring, coil, spiral, or loop that can be placed within the uterine cavity by a physician (fig. 21.42). Although such a device is relatively effective in preventing pregnancy, the mechanism by which it produces its effect is unknown. However, it is believed

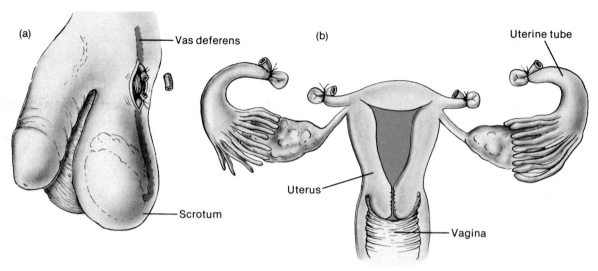

that the IUD somehow interferes with the implantation of an embryo within the uterine wall, perhaps by causing an inflammatory reaction in the uterine tissues.

Unfortunately, an IUD may be expelled from the uterus spontaneously, or it may produce unpleasant side effects, such as pain or excessive bleeding. On occasion, it may injure the uterus or produce other serious health problems.

Surgical Methods

Surgical methods of contraception involve sterilization of either the male or female.

In the male, a small section of each vas deferens is removed near the epididymis, and the cut ends of the ducts are tied. This procedure is called *vasectomy*, and it is a relatively simple operation that produces no side effects. After a vasectomy, sperm cells cannot leave the epididymis, and thus they are not included in the seminal fluid.

However, sperm cells may already be present in portions of the ducts distal to the cuts. Consequently the sperm count of the seminal fluid released by ejaculation may not reach zero for several weeks.

The corresponding procedure in the female is called *tubal ligation*. In this instance the uterine tubes are cut and tied so that sperm cells cannot reach the egg cell.

Neither a vasectomy nor a tubal ligation produces any changes in the hormonal concentrations or the sexual drives of the individuals involved. These sterilization procedures, shown in figure 21.43, provide the most reliable forms of contraception.

Although it is theoretically possible to reverse the effects of vasectomy or tubal ligation by surgically reconnecting the severed tubes, such procedures are quite difficult and have a relatively low rate of success.

1. How is an IUD thought to prevent pregnancy?
2. Describe the surgical methods of contraception for a female and for a male.

Some Clinical Terms Related to the Reproductive Systems

abortion (ah-bor'shun)—the spontaneous or deliberate termination of pregnancy; a spontaneous abortion is commonly termed a miscarriage.

amenorrhea (a-men"o-re'ah)—the absence of menstrual flow, usually due to a disturbance in hormonal levels.

cesarean section (sĕ-sa're-an sek'shun)—the delivery of a fetus through an abdominal incision.

conization (ko"nĭ-za'shun)—the surgical removal of a cone of tissue from the cervix for examination.

curettage (ku"rĕ-tahzh')—surgical procedure in which the cervix is dilated and the endometrium of the uterus is scraped (commonly called D and C).

dysmenorrhea (dis"men-ŏ-re'ah)—painful menstruation.

endometritis (en"do-me-tri'tis)—inflammation of the uterine lining.

epididymitis (ep"ĭ-did"ĭ-mi'tis)—inflammation of the epididymis.

gestation (jes-ta'shun)—the entire period of pregnancy.

hematometra (hem"ah-to-me'trah)—an accumulation of menstrual blood within the uterine cavity.

hysterectomy (his"tĕ-rek'to-me)—the surgical removal of the uterus.

mastitis (mas"ti'tis)—inflammation of a mammary gland.

oophorectomy (o"of-o-rek'to-me)—surgical removal of an ovary.

oophoritis (o"of-o-ri'tis)—inflammation of an ovary.

orchiectomy (or"ke-ek'to-me)—surgical removal of a testis.

orchitis (or-ki'tis)—inflammation of a testis.

prostatectomy (pros"tah-tek'to-me)—surgical removal of a portion or all of the prostate gland.

prostatitis (pros"tah-ti'tis)—inflammation of the prostate gland.

salpingectomy (sal"pin-jek'to-me)—surgical removal of a uterine tube.

vaginitis (vaj"ĭ-ni'tis)—inflammation of the vaginal lining.

varicocele (var'ĭ-ko-sēl")—distension of the veins within the spermatic cord.

Chapter Summary

Introduction
Various reproductive organs produce sex cells, help sustain the lives of these cells, or transport them from place to place; other parts produce sex hormones.

Organs of the Male Reproductive System
Primary organs are the testes, which produce sperm cells.

Accessory organs include internal and external reproductive organs.

The Testes
1. Descent of the testes
   a. Testes originate behind the parietal peritoneum near the level of the developing kidneys.
   b. The gubernaculum guides the descent of testes into the lower abdominal cavity and through the inguinal canal.
   c. Undescended testes fail to produce sperm cells because of the relatively high abdominal temperature.
2. Structure of the testes
   a. The testes are composed of lobules separated by connective tissue and filled with seminiferous tubules.
   b. Seminiferous tubules unite to form the rete testis that joins the epididymis.
   c. Seminiferous tubules are lined with germinal epithelium that produces sperm cells.
   d. Interstitial cells that produce male sex hormones occur between the seminiferous tubules.
3. Formation of sperm cells
   a. Germinal epithelium consists of supporting cells and spermatogenic cells.
      (1) Supporting cells support and nourish the spermatogenic cells.
      (2) Spermatogenic cells give rise to spermatogonia.
   b. Sperm cells are produced from spermatogonia by the process of spermatogenesis.
      (1) The number of chromosomes in offspring cells is reduced by one-half (46 to 23).
      (2) The product of spermatogenesis is four sperm cells from each primary spermatocyte.
   c. Membranous processes of adjacent supporting cells form a barrier within the germinal epithelium.
      (1) The barrier separates early and advanced stages of spermatogenesis.
      (2) It helps provide favorable environment for differentiating cells.
   d. The sperm cell has a head, body, and tail.
   e. Immature sperm cells mature as they are stored in the epididymis.

Male Internal Accessory Organs

1. The epididymis
   a. The epididymis is a tightly coiled tube on the outside of the testis that leads into the vas deferens.
   b. It stores immature sperm cells as they mature.
   c. Peristaltic waves in its wall aid in ejaculation.
2. The vas deferens
   a. The vas deferens is a muscular tube that forms part of the spermatic cord.
   b. It passes through the inguinal canal, enters the abdominal cavity, courses medially into the pelvic cavity, and ends behind the urinary bladder.
   c. It fuses with the duct from the seminal vesicle to form the ejaculatory duct.
3. The seminal vesicle
   a. The seminal vesicle is a saclike structure attached to the vas deferens.
   b. It secretes alkaline fluid that contains nutrients and prostaglandins.
   c. This secretion is added to sperm cells during ejaculation to form seminal fluid.
4. The prostate gland
   a. This gland surrounds the urethra just below the bladder.
   b. It secretes thin, milky fluid that neutralizes seminal fluid and vaginal secretions, and enhances the motility of sperm cells.
   c. It may enlarge in older males and interfere with urination.
5. The bulbourethral glands
   a. These are two small structures beneath the prostate gland.
   b. They secrete fluid that serves as a lubricant for the penis in preparation for sexual intercourse.
6. Seminal fluid
   a. Seminal fluid is composed of sperm cells and secretions of seminal vesicles, the prostate gland, and bulbourethral glands.
   b. This fluid is slightly alkaline and contains nutrients and prostaglandins.
   c. It activates sperm cells.

Male External Reproductive Organs

1. The scrotum
   a. Pouch of skin and subcutaneous tissue that encloses the testes.
   b. Dartos muscle in the scrotal wall causes the skin of the scrotum to be held close to the testes or to hang loosely.
2. The penis
   a. The penis functions to convey urine and seminal fluid.
   b. It is specialized to become erect for insertion into the vagina during sexual intercourse.
   c. The body of the penis is composed of three columns of erectile tissue surrounded by connective tissue.
   d. The root of the penis is attached to the pelvic arch and membranes of the perineum.

3. Erection, orgasm, and ejaculation
   a. During erection, vascular spaces within erectile tissue become engorged with blood as arteries dilate and venous outflow is reduced.
   b. Orgasm is the culmination of sexual stimulation and is accompanied by emission and ejaculation.
   c. Movement of seminal fluid occurs as result of sympathetic reflexes.
   d. Following ejaculation, the penis becomes flaccid.

Hormonal Control of Male Reproductive Functions

1. Pituitary hormones
   The male body remains reproductively immature until the hypothalamus releases GRF, which stimulates the anterior pituitary gland to release gonadotropins.
   a. FSH stimulates the germinal epithelium to undergo spermatogenesis.
   b. LH (ICSH) stimulates the interstitial cells to produce male sex hormones.
2. Male sex hormones
   a. Male sex hormones are called androgens.
   b. Testosterone is the most important androgen.
      (1) It is necessary for maturation of sperm cells.
      (2) It is responsible for the development and maintenance of male secondary sexual characteristics.
3. Regulation of male sex hormones
   a. Testosterone concentration is regulated by a negative feedback mechanism.
      (1) As its concentration rises, the hypothalamus is inhibited and pituitary secretion of gonadotropins is reduced.
      (2) As the concentration falls, the hypothalamus signals the pituitary to secrete gonadotropins.
   b. The concentration of testosterone remains relatively stable from day to day.

Organs of the Female Reproductive System

The primary organs of the female reproductive system are the ovaries, which produce egg cells.
Accessory organs include internal and external reproductive organs.

The Ovaries

1. Attachments of the ovaries
   a. The ovaries are held in position by several ligaments.
   b. Ligaments include broad, suspensory, and ovarian ligaments.
2. Descent of the ovaries
   a. The ovaries descend from behind the parietal peritoneum near the developing kidneys.
   b. They are attached to the pelvic wall just below the pelvic brim.

3. Structure of the ovaries
   a. The ovaries are divided into medulla and cortex.
   b. The medulla is composed of connective tissue, blood vessels, lymphatic vessels, and nerves.
   c. The cortex contains ovarian follicles and is covered by germinal epithelium.
4. Formation of egg cells
   a. Egg cell production begins before birth and ceases relatively early in life.
   b. Germinal epithelium gives rise to follicles that contain primary oocytes.
   c. A primary oocyte undergoes oogenesis and gives rise to a single egg cell (secondary oocyte) in which the chromosome number is reduced by one-half (46 to 23).
   d. A relatively small number of primary oocytes give rise to egg cells.
5. Maturation of a follicle
   a. Primary follicles contain primary oocytes.
   b. Such a follicle may mature under the influence of FSH.
   c. During maturation the oocyte enlarges, the follicular cells multiply, and a fluid-filled cavity appears within the follicle.
6. Ovulation
   a. Oogenesis is completed as the follicle matures.
   b. The resulting egg cell is released when the follicle ruptures.
   c. After ovulation, the egg cell is drawn into the opening of the uterine tube.

Female Internal Accessory Organs
1. The uterine tubes
   a. These tubes convey egg cells toward the uterus.
   b. The end of each tube is expanded and its margin bears irregular extensions.
   c. Movement of egg cell into opening is aided by ciliated cells that line the tube and by peristaltic contractions in the wall of the tube.
2. The uterus
   a. The uterus receives the embryo and sustains its life during development.
   b. The cervix of the uterus is partially enclosed by the vagina.
   c. The uterine wall includes endometrium, myometrium, and perimetrium.
3. The vagina
   a. The vagina connects the uterus to the vestibule.
   b. It serves to receive the erect penis, to convey uterine secretions to the outside, and to transport the fetus during birth.
   c. The vaginal orifice is partially closed by a thin membrane.
   d. Its wall consists of a mucosa, muscularis, and outer fibrous coat.

Female External Reproductive Organs
1. The labia majora
   a. The labia majora are rounded folds of fatty tissue and skin that enclose and protect the other external reproductive parts.
   b. The upper ends form a rounded, fatty elevation over the pubic symphysis.
2. The labia minora
   a. These are flattened, longitudinal folds between the labia majora.
   b. They form the sides of the vestibule and anteriorly form the hoodlike covering of the clitoris.
3. The clitoris
   a. The clitoris is a small projection at the anterior end of the vulva.
   b. It is composed of two columns of erectile tissue.
   c. Its root is attached to the sides of the pubic arch.
4. The vestibule
   a. The vestibule is the space between the labia majora that encloses the vaginal and urethral openings.
   b. Vestibular glands secrete mucus into the vestibule during sexual excitement.
5. Erection, lubrication, and orgasm.
   a. During periods of sexual stimulation, erectile tissues of the clitoris and vestibular bulbs become engorged with blood and swell.
   b. Vestibular glands secrete mucus into the vestibule and vagina, which lubricates these parts during sexual intercourse.
   c. During orgasm, muscles of the perineum, uterine wall, and uterine tubes contract rhythmically.

Hormonal Control of Female Reproductive Functions
1. Pituitary hormones
   a. The female body begins to mature when the hypothalamus releases GRF, which signals the anterior pituitary gland to secrete gonadotropins.
   b. Gonadotropins play primary roles in control of sex cell maturation and development and maintenance of female secondary sexual characteristics.
2. Female sex hormones
   a. The most important female sex hormones are estrogen and progesterone.
      (1) Estrogen from the ovaries is responsible for development and maintenance of most female secondary sexual characteristics.
      (2) Progesterone functions to cause changes in the uterus.

3. Female reproductive cycles
   a. These are called menstrual cycles; they are characterized by regularly recurring changes in the uterine lining culminating in menstrual flow.
   b. The cycle is initiated by FSH, which stimulates the maturation of a follicle.
   c. The maturing follicle secretes estrogen, which is responsible for maintaining the secondary sexual traits and causing the uterine lining to thicken.
   d. Ovulation is triggered when the anterior pituitary gland secretes a relatively large amount of LH and an increased amount of FSH.
   e. Following ovulation, follicular cells give rise to corpus luteum.
      (1) The corpus luteum secretes progesterone, which causes the uterine lining to become more vascular and glandular.
      (2) If an egg cell is not fertilized, the corpus luteum begins to degenerate.
      (3) As levels of estrogen and progesterone decline, the uterine lining disintegrates, causing menstrual flow.
   f. During this cycle, estrogen and progesterone inhibit the hypothalamus and the pituitary gland; as the concentrations of estrogen and progesterone fall, the pituitary secretes FSH and LH again, stimulating a new cycle.
4. Menopause
   a. Eventually the ovaries cease responding to FSH, and cycling ceases.
   b. This results in a low estrogen concentration and a continually high concentration of FSH and LH.

Pregnancy
1. Transport of sex cells
   a. The egg cell is carried to the uterine tube by ciliary action.
   b. To move, a sperm cell lashes its tail; the motion is aided by muscular contractions in the uterus and uterine tube.
2. Fertilization
   a. A sperm cell penetrates an egg cell with the aid of an enzyme.
   b. When a sperm cell penetrates an egg cell, the entrance of any other sperm cells is prevented.
   c. When the nuclei of sperm and egg cells fuse, the process of fertilization is complete.
   d. The product of fertilization is a zygote with 46 chromosomes.
3. Early embryonic development
   a. Cells undergo mitosis, giving rise to smaller and smaller cells.
   b. The developing offspring is moved down the uterine tube to the uterus, where it becomes implanted in the endometrium.
   c. Offspring is called an embryo from the second through the seventh week of development; thereafter it is a fetus.
   d. Eventually the embryonic and maternal cells together form a placenta.
4. Hormonal changes during pregnancy
   a. Embryonic cells produce HCG that causes the corpus luteum to be maintained.
   b. Placental tissue produces high concentrations of estrogen and progesterone.
      (1) Estrogen and progesterone maintain the uterine wall and inhibit the secretion of FSH and LH.
      (2) Progesterone causes uterine contractions to be suppressed.
      (3) Estrogen causes enlargement of the vagina and relaxation of the ligaments that hold the pelvic joints together.
   c. The placenta also secretes placental lactogen that stimulates development of the breasts.
5. Other changes during pregnancy
   a. The uterus enlarges greatly.
   b. The woman's blood volume, cardiac output, breathing rate, and urine production increase.
   c. The woman's dietary intake increases, but if intake is inadequate, fetal tissues have priority for use of available nutrients.
6. The birth process
   a. Pregnancy usually lasts 40 weeks.
   b. During pregnancy, estrogen excites uterine contractions and progesterone inhibits uterine contractions.
   c. A variety of factors are involved with the birth process.
      (1) Secretion of progesterone decreases.
      (2) The posterior pituitary gland releases oxytocin.
      (3) Uterine muscles are stimulated to contract, and labor begins.
      (4) A positive feedback mechanism causes stronger contractions and greater release of oxytocin.
   d. Following the birth of infant, placental tissues are expelled.

The Mammary Glands
1. Location of the glands
   a. The mammary glands are located in subcutaneous tissue of the anterior thorax within the breasts.
   b. Breasts extend between the second and sixth ribs and from sternum to axillae.
2. Structure of the glands
   a. The mammary glands are composed of lobes that contain tubular glands.
   b. Lobes are separated by dense connective and adipose tissues.
   c. The mammary glands are connected to the nipple by ducts.

3. Development of the breasts
   a. Male breasts remain nonfunctional.
   b. Estrogen stimulates female breast development.
      (1) Alveolar glands and ducts enlarge.
      (2) Fat is deposited around and within breasts.
   c. During pregnancy the breasts change.
      (1) Estrogen causes ductile system to grow.
      (2) Progesterone causes development of alveolar glands.
      (3) No milk is produced because estrogen and progesterone inhibit the secretion of prolactin.
4. Production and secretion of milk
   a. Following childbirth, estrogen and progesterone concentrations decline.
      (1) The anterior pituitary secretes prolactin.
      (2) The mammary glands begin to secrete milk.
   b. Reflex response to mechanical stimulation of the nipple causes the posterior pituitary to release oxytocin, which causes milk to be ejected from ducts.
   c. As long as milk is removed from glands, more milk is produced; if milk is not removed, production ceases.
   d. During the period of milk production, menstrual cycles are partially inhibited.

Birth Control

Voluntary regulation of the number of children a woman will produce and the time they are conceived is called birth control.
This usually involves some method of contraception.
1. Coitus interruptus
   a. Coitus interruptus is withdrawal of the penis from the vagina before ejaculation.
   b. Some seminal fluid may be expelled from the penis before ejaculation.
2. Rhythm method
   a. Abstinence from sexual intercourse a few days before and after ovulation is the rhythm method.
   b. It is almost impossible to predict the time of ovulation accurately.
3. Mechanical barriers
   a. The condom is used by males.
   b. The diaphragm is used by females.
4. Chemical barriers
   a. Spermicidal creams, foams, and jellies are chemical barriers to conception.
   b. These provide an unfavorable environment in the vagina for sperm survival.
5. Oral contraceptives
   a. Tablets that contain estrogen- and progesterone-like substances are taken by the woman.
   b. They inhibit release of gonadotropins from the anterior pituitary and thus prevent ovulation.

c. When used correctly, the method is almost 100% effective.
   d. Some women have undesirable side effects.
6. Intrauterine devices
   a. An IUD is a solid object inserted in the uterine cavity.
   b. It is thought to prevent pregnancy by interfering with implantation.
   c. It may be expelled spontaneously or produce undesirable side effects.
7. Surgical methods
   a. These involve sterilization procedures.
      (1) Vasectomy is performed in males.
      (2) Tubal ligation is performed in females.
   b. This is the most reliable form of contraception.

Application of Knowledge

1. What changes, if any, might be expected to occur in the secondary sexual characteristics of an adult male following removal of one testis? Following removal of both testes? Following removal of the prostate gland?

2. How would you explain the fact that new mothers sometimes experience cramps in their lower abdomen when they begin to nurse their babies?

3. If a woman who is considering having a tubal ligation asks, "Will the operation cause me to go through my change of life?" how would you answer?

Review Activities

1. List the general functions of the reproductive systems.

2. Distinguish between the primary and accessory male reproductive organs.

3. Describe the descent of the testes.

4. Define *cryptorchidism*.

5. Describe the structure of a testis.

6. List the major steps in spermatogenesis.

7. Explain the function of the supporting cells in the testis.

8. Describe a sperm cell.

9. Describe the epididymis and explain its function.

10. Trace the path of the vas deferens from the epididymis to the ejaculatory duct.

11. On a diagram, locate the seminal vesicles and describe the composition of their secretion.

12. On a diagram, locate the prostate gland and describe the composition of its secretion.

13. On a diagram, locate the bulbourethral glands and explain the function of their secretion.

14. Define *seminal fluid*.

15. Describe the structure of the scrotum.

16. Describe the structure of the penis.

17. Explain the mechanism that produces an erection of the penis.

18. Distinguish between emission and ejaculation.

19. Explain the mechanism of ejaculation.

20. Define *GRF* and explain its role in the control of male reproductive functions.

21. Distinguish between androgen and testosterone.

22. List several male secondary sexual characteristics.

23. Define *puberty*.

24. Explain how the concentration of testosterone is regulated.

25. Describe how the ovaries are held in position.

26. Describe the descent of the ovaries.

27. Describe the structure of an ovary.

28. List the major steps in oogenesis.

29. Describe how a follicle matures.

30. Define *ovulation*.

31. On a diagram, locate the uterine tubes and explain their function.

32. Describe the structure of the uterus.

33. Describe the structure of the vagina.

34. Distinguish between the labia majora and the labia minora.

35. On a diagram, locate the clitoris and describe its structure.

36. Define *vestibule*.

37. Explain the role of GRF in regulating female reproductive functions.

38. List several female secondary sexual characteristics.

39. Define *menstrual cycle*.

40. Explain the roles of estrogen and progesterone in the menstrual cycle.

41. Summarize the major events in a menstrual cycle.

42. Define *menopause*.

43. Describe how male and female sex cells are transported within the female reproductive tract.

44. Describe the process of fertilization.

45. Define *cleavage*.

46. Define *implantation*.

47. List the major functions of the placenta.

48. Explain the major hormonal changes that occur during pregnancy.

49. Describe the major nonhormonal changes that occur during pregnancy.

50. Discuss the events that occur during the birth process.

51. Describe the structure of a mammary gland.

52. Explain the roles of prolactin and oxytocin in milk production and secretion.

53. Define *contraception*.

54. List several methods of contraception and explain how each interferes with normal reproductive functions.

Suggestions for Additional Reading

Epel, D. November 1977. The program of fertilization. *Scientific American*.

Goldstein, B. 1976. *Human sexuality*. New York: McGraw-Hill.

Hardin, G. 1970. *Birth control*. Indianapolis: Bobbs-Merrill.

Hatcher, R. R. 1977. *Contraceptive technology, 1976–1977*. New York: Wiley.

Jones, K. L., et al. 1973. *Sex*, 2nd ed. New York: Harper and Row.

Katchadourian, H. A., and Lunde, D. T. 1972. *Fundamentals of human sexuality*. New York: Holt, Rinehart and Winston.

Marx, J. L. 1973. Birth control: current technology, future prospects. *Science* 179:1222.

Odell, W. D., and Moyer, D. L. 1971. *Physiology of reproduction*. St. Louis: C. V. Mosby.

O'Malley, B. W., and Means, A. R. 1974. Female steroid hormones and target cell nuclei. *Science* 183:610.

Parkes, A. 1976. *Patterns of sexuality and reproduction*. New York: Oxford Univ. Press.

Parr, E. L. June 1973. Contraception with intrauterine devices. *Bioscience*.

Pengelley, E. T. 1974. *Sex and human life*. Reading, Mass.: Addison-Wesley.

Segal, S. J. September 1974. The physiology of human reproduction. *Scientific American*.

Swanson, H. D. 1974. *Human reproduction: biology and social change*. New York: Oxford Univ. Press.

Human Growth and Development

22 The products of the female and male reproductive systems are egg cells and sperm cells respectively. When an egg cell and a sperm cell unite, a zygote is formed by the process of fertilization. Such a single-celled zygote is the first cell of an offspring, and it is capable of giving rise to an adult of the subsequent generation. The processes by which this is accomplished are called *growth* and *development*.

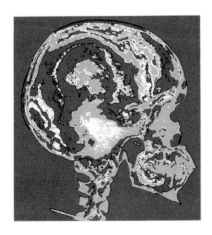

Chapter Outline

Chapter Objectives

After you have studied this chapter, you should
be able to

1. Distinguish between growth and development.

2. Describe the major events that occur during
   the period of cleavage.

3. Explain how the primary germ layers originate
   and list the structures produced by each
   layer.

4. Describe the formation and function of the
   placenta.

5. Define *fetus* and describe the major events
   that occur during the fetal stage of
   development.

6. Trace the general path of blood through the
   fetal circulatory system.

7. Describe the major physiological adjustments
   that occur in the newborn.

8. Name the stages of development that occur
   between the neonatal period and death, and
   list the general characteristics of each stage.

9. Complete the review activities at the end of
   this chapter.

amnion   (am'ne-on)

chorion   (ko're-on)

cleavage   (klēv'ij)

embryo   (em'bre-o)

fetus   (fe'tus)

germ layer   (jerm la'er)

neonatal   (ne''o-na'tal)

placenta   (plah-sen'tah)

postnatal   (pōst-na'tal)

prenatal   (pre-na'tal)

senescence   (sĕ-nes'ens)

umbilical cord   (um-bil'ĭ-kal kord)

zygote   (zi'gōt)

allant-, sausage-shaped: *allant*ois—tubelike structure that extends from the yolk sac into the connecting stalk of the embryo.

chorio-, skin: *chorio*n—outermost membrane that surrounds the fetus and other fetal membranes.

cleav-, to divide: *cleav*age—period of development characterized by a division of the zygote into smaller and smaller cells.

lacun-, a pool: *lacun*a—space between the chorionic villi that is filled with maternal blood.

morul-, mulberry: *morul*a—embryonic structure consisting of a solid ball of about 16 cells, thus appearing somewhat like a mulberry.

nat-, to be born: pre*nat*al—period of development before birth.

sen-, old: *sen*escence—process of growing old.

troph-, nourishment: *troph*oblast—cellular layer that surrounds the inner cell mass and helps nourish it.

umbil-, the navel: *umbil*ical cord—structure attached to the fetal navel (umbilicus) that connects the fetus to the placenta.

Fig. 22.1 (*a*) Growth involves an increase in size; (*b*) development refers to the process of changing from one phase of life to another.

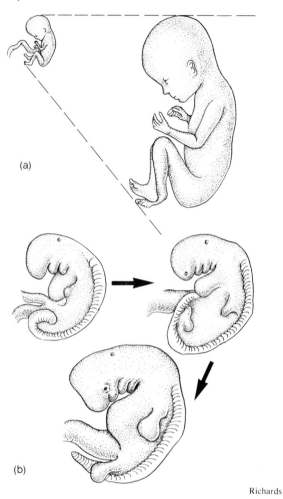

(a)

(b)

Richards

Fig. 22.2 During the period of cleavage, the cells divide by mitosis and become smaller and smaller.

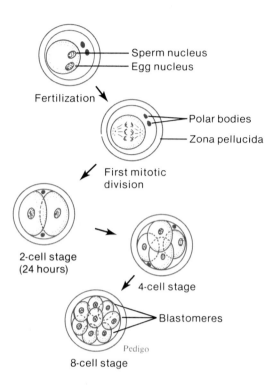

Sperm nucleus
Egg nucleus

Fertilization

Polar bodies
Zona pellucida

First mitotic division

2-cell stage (24 hours)

4-cell stage

Blastomeres

Pedigo

8-cell stage

Growth refers to an increase in size. In a human, growth usually reflects an increase in cell numbers as a result of *mitosis,* followed by enlargement of the newly formed cells and enlargement of the body.

Development, on the other hand, is the continuous process by which an individual changes from one life phase to another. These life phases include a **prenatal period,** which begins with the fertilization of an egg cell and ends at birth, and a **postnatal period,** which begins at birth and ends with death in old age. (See fig. 22.1.)

Prenatal Period

The prenatal period of development usually lasts for 40 weeks (ten lunar months) and can be divided into a period of cleavage, an embryonic stage, and a fetal stage.

Period of Cleavage

As is described in chapter 21, fertilization occurs within a uterine tube. About 24 hours after the zygote is formed, it undergoes mitosis, giving rise to two daughter cells. These cells in turn divide to form four cells, they divide into eight cells, and so forth. With each subsequent division, the resulting cells are smaller and smaller. This distribution of the zygote's contents into smaller and smaller cells is called *cleavage,* and the cells produced in this way are called *blastomeres.* (See fig. 22.2.)

It is usually not possible to determine the actual time of fertilization because reliable records concerning sexual activities are seldom available. However, the approximate time can be calculated by adding fourteen days to the date of the onset of the last menstruation. This time can then be used to calculate the fertilization age of an embryo. The expected time of birth can be estimated by adding 266 days to the fertilization date. Most fetuses are born within 10 to 15 days of this calculated time.

The mass of cells that is formed by cleavage is still enclosed in the zona pellucida of the original egg cell, and in about three days it consists of a solid ball of about sixteen cells, called a *morula.*

Fig. 22.3 (*a*) A morula consists of a solid ball of cells; (*b*) a blastocyst has a fluid-filled cavity.

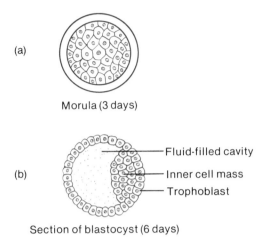

(a)

Morula (3 days)

(b)

Fluid-filled cavity
Inner cell mass
Trophoblast

Section of blastocyst (6 days)

Fig. 22.4 (*a*) About the sixth day of development, the blastocyst contacts the uterine wall and (*b*) begins to become implanted within the wall.

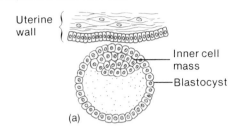

Uterine wall

Inner cell mass
Blastocyst

(a)

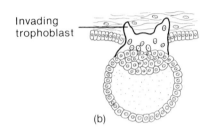

Invading trophoblast

(b)

The morula enters the uterine cavity and remains unattached for about 3 days. During this time, the zona pellucida degenerates, and the morula develops a fluid-filled central cavity. Once this cavity appears, the morula becomes a hollow ball of cells called a **blastocyst.** (See fig. 22.3.)

Within the blastocyst, cells in one region group together to form an *inner cell mass* that eventually gives rise to the **embryo** proper—the body of the developing offspring. The cells forming the wall of the blastocyst make up the *trophoblast,* which will create structures that assist the embryo in its development.

Sometimes two ovarian follicles release egg cells simultaneously, and if both are fertilized the resulting zygotes can develop into fraternal (dizygotic) twins. Such twins are no more genetically alike than any brothers or sisters from the same parents. In other instances, twins develop from a single fertilized egg (monozygotic twins). This can happen if two inner cell masses form within a blastocyst and each produces an embryo. Twins of this type usually share a single placenta, and they are genetically identical. Thus, they are always the same sex and are very similar in appearance.

About the sixth day, the blastocyst begins to attach itself to the uterine lining. This attachment is apparently aided by its secretion of proteolytic enzymes that digest a portion of the endometrium. The blastocyst sinks into the resulting depression, becoming completely buried in the uterine lining. At the same time, the uterine lining is stimulated to thicken

Fig. 22.5 Photomicrograph of a human embryo undergoing implantation in the uterine wall.

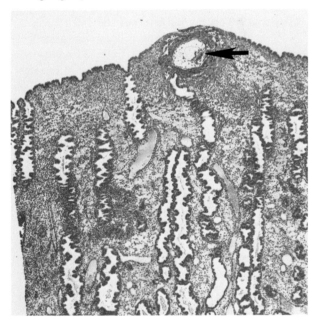

below the implanting blastocyst, and cells of the trophoblast begin to produce tiny, fingerlike processes that grow into the endometrium. (See fig. 22.4.)

This process of **implantation** occurs near the end of the first week of development, and completes the period of cleavage. (See fig. 22.5.)

Usually the blastocyst becomes implanted in the upper posterior wall of the uterus. Sometimes, however, implantation occurs in the lower portion near the cervix, and as the placenta develops, it may partially or totally cover the opening of the cervix. This condition is called *placenta previa,* and it is likely to produce complications since any expansion of the cervix may damage placental tissues and cause bleeding.

1. Distinguish between growth and development.
2. What changes characterize the period of cleavage?
3. How does a blastocyst become attached to the endometrium of the uterus?
4. In what ways does the endometrium respond to the activities of the blastocyst?

Embryonic Stage

The **embryonic stage** of development extends from the second week through the eighth week and is characterized by the formation of the placenta, the development of the main internal organs, and the appearance of the major external body structures.

Early in this stage, the cells of the inner cell mass become organized into a flattened **embryonic disk** that consists of two distinct layers—an outer *ectoderm* and an inner *endoderm.* (See fig. 22.6.) A short time later, a third layer of cells, the *mesoderm,* forms between the ectoderm and endoderm. These three layers of cells are called the **primary germ layers,** and they are responsible for forming all of the body organs.

More specifically, *ectodermal cells* give rise to the nervous system, portions of special sensory organs, the epidermis, hair, nails, glands of the skin, and the linings of the mouth and anal canal. *Mesodermal cells* form all types of muscle tissue, bone tissue, bone marrow, blood, blood vessels, lymphatic vessels, various connective tissues, internal reproductive organs, kidneys, and the epithelial linings of the body cavities. *Endodermal cells* produce the epithelial linings of the digestive tract, respiratory tract, urinary bladder, and urethra.

During the fourth week of development (fig. 22.7), the flat embryonic disk is transformed into a cylindrical structure, which is attached to the developing placenta by a *connecting stalk* (body stalk). By this time, the head and jaws are appearing, the heart is beating and forcing blood through blood vessels, and tiny buds that will give rise to the arms and legs are forming.

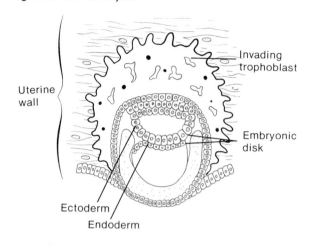

Fig. 22.6 The cells of the embryonic disk first become organized into two layers.

Uterine wall

Invading trophoblast

Embryonic disk

Ectoderm

Endoderm

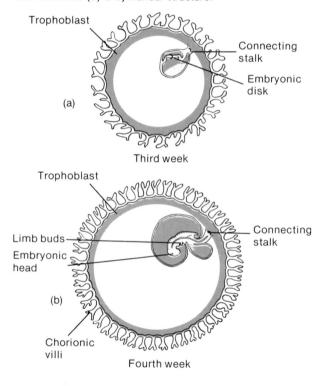

Fig. 22.7 (*a*) During the fourth week, the flat embryonic disk becomes (*b*) a cylindrical structure.

Trophoblast

Connecting stalk

Embryonic disk

(a)

Third week

Trophoblast

Limb buds

Embryonic head

Connecting stalk

(b)

Chorionic villi

Fourth week

During the fifth through the seventh weeks, as shown in figure 22.8, the head grows rapidly and becomes rounded and erect. The face, which is developing eyes, nose, and mouth, becomes more humanlike. The arms and legs elongate, and fingers and toes appear.

By the end of the seventh week, all the main internal organs have become established, and as these

Fig. 22.8 In the fifth through the seventh weeks of development, the embryonic body and face develop a humanlike appearance.

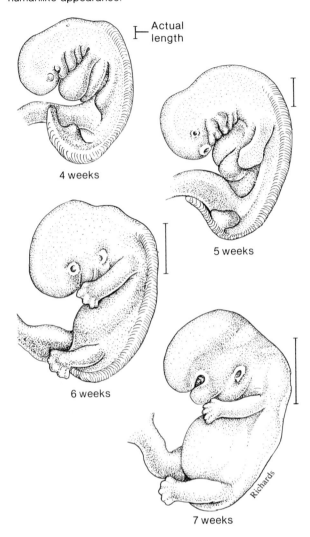

⊢ Actual length

4 weeks

5 weeks

6 weeks

7 weeks

Richards

Fig. 22.9 By the end of the seventh week of development, all the major internal organs have been established.

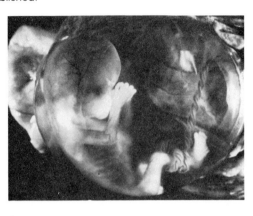

Fig. 22.10 The placental membrane consists of the epithelial wall of an embryonic capillary and the epithelial wall of a chorionic villus. What is the significance of this membrane?

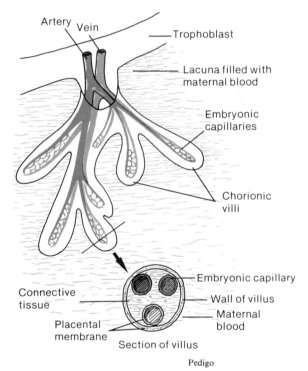

Artery Vein
Trophoblast
Lacuna filled with maternal blood
Embryonic capillaries
Chorionic villi
Connective tissue
Placental membrane
Embryonic capillary
Wall of villus
Maternal blood
Section of villus

Pedigo

structures enlarge, they affect the shape of the body (fig. 22.9). Consequently, the body takes on a humanlike appearance.

Meanwhile, the embryo continues to become implanted within the uterus. As mentioned, early in this process slender projections grow out from the trophoblast into the surrounding endometrium. These extensions, which are called **chorionic villi,** become branched, and by the end of the fourth week of development they are well formed.

While the chorionic villi are developing, embryonic blood vessels appear within them, and these vessels are continuous with those passing through the connecting stalk to the body of the embryo. At the same time, irregular spaces called **lacunae** are eroded around and between the villi. These spaces become filled with maternal blood that escapes from eroded endometrial blood vessels.

A thin membrane separates embryonic blood within the capillary of a chorionic villus from maternal blood in a lacuna. This membrane, called the **placental membrane,** is composed of the epithelium of the villus and the epithelium of the capillary. (See fig. 22.10.) Through this membrane, exchanges take place between maternal and embryonic blood. Oxygen and nutrients diffuse from the maternal blood into the embryonic blood, and carbon dioxide and other wastes diffuse from the embryonic blood into the maternal blood. Various substances also move through the placental membrane by active transport and pinocytosis.

Most drugs are able to pass freely through the placental membrane, and so substances ingested by the mother may affect the fetus. Thus, fetal drug addiction may occur following the mother's use of drugs such as heroin.

Similarly, depressant drugs administered to the mother during labor can produce effects within the fetus and may, for example, depress the activity of its respiratory system.

1. What major events occur during the embryonic stage of development?
2. What tissues and structures develop from ectoderm? From mesoderm? From endoderm?
3. Describe the structure of a chorionic villus.
4. How are substances exchanged between the embryonic blood and the maternal blood?

Until about the end of the eighth week, the chorionic villi cover the entire surface of the former trophoblast, which is now called the **chorion** (fig. 22.11). However, as the embryo and the chorion surrounding it continue to enlarge, only those villi that remain in contact with the endometrium endure. The others degenerate, and the portions of the chorion to which they were attached become smooth. Thus, the region of the chorion still in contact with the uterine wall is restricted to a disk-shaped area that becomes the **placenta.**

The embryonic portion of the placenta is composed of the chorion and its villi; the maternal portion is composed of the area of the uterine wall (decidua basalis) to which the villi are attached. When it is fully formed, the placenta appears as a reddish brown disk, about 20 cm (8 in) long and 2.5 cm (1 in) thick. It usually weighs about 0.5 kg (1 lb). Figure 22.12 shows the structure of the placenta.

While the placenta is forming from the chorion, another membrane, called the **amnion,** develops around the embryo. This second membrane begins to appear during the second week. Its margin is attached around the edge of the embryonic disk, and fluid, called **amniotic fluid,** fills the space between them.

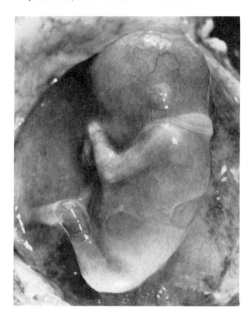

Fig. 22.11 By the end of the eighth week of development, chorionic villi cover the entire surface of the former trophoblast, now called the chorion.

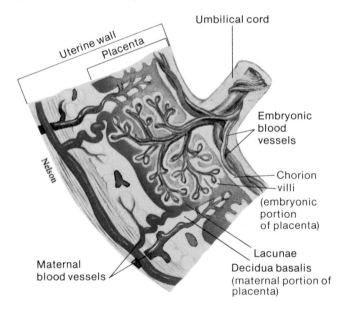

Fig. 22.12 The placenta consists of an embryonic portion and a maternal portion.

As the embryo is transformed into a cylindrical structure, the margins of the amnion are folded around it so that the embryo is enclosed by the amnion and surrounded by amniotic fluid. As this process continues, the amnion envelops the tissues on the underside of the embryo, by which it is attached to the chorion and the developing placenta. In this manner, as figure 22.13 illustrates, the **umbilical cord** is formed.

Fig. 22.13 (a, b, c) As the amnion develops, it surrounds the embryo, and (d) the umbilical cord is formed. What structures comprise this cord?

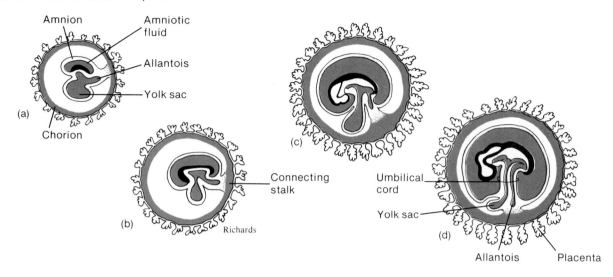

Fig. 22.14 The umbilical cord contains two arteries and one vein.

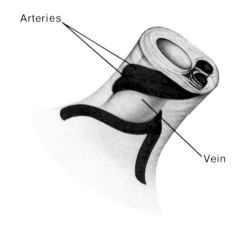

Fig. 22.15 As the amniotic cavity enlarges, the amnion contacts the chorion, and the two membranes fuse to form the amniochorionic membrane.

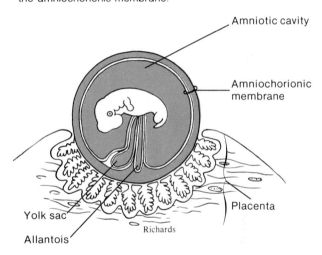

When the umbilical cord is fully developed, it is about 1 cm (0.5 in) in diameter and about 55 cm (22 in) in length. It begins at the umbilicus of the embryo and is inserted into the central region of the placenta. As figure 22.14 shows, the cord contains three blood vessels—two *umbilical arteries* and an *umbilical vein*—through which blood passes between the embryo and the placenta.

The umbilical cord also functions to suspend the embryo in the amniotic cavity, where the amniotic fluid provides a watery environment in which the embryo can grow freely without being compressed by surrounding tissues. The fluid also protects the embryo against being jarred by movements of the mother's body.

Eventually the amniotic cavity becomes so enlarged that the membrane of the amnion contacts the thicker chorion around it, and the two membranes become fused into an *amniochorionic membrane*. (See fig. 22.15.)

In addition to the amnion and chorion, two other embryonic membranes appear during development. They are the yolk sac and the allantois.

The **yolk sac** appears during the second week, and it is attached to the underside of the embryonic disc. It functions to form blood cells in the early stages of development and gives rise to the cells that later become sex cells. Portions of the yolk sac also enter into the formation of the embryonic digestive tube.

Part of this membrane becomes incorporated into the umbilical cord, while the remainder lies in the cavity between the chorion and the amnion near the placenta.

The **allantois** forms during the third week as a tube extending from the early yolk sac into the connecting stalk of the embryo. It functions in the formation of blood cells and gives rise to the umbilical arteries and vein.

The *embryonic stage* is completed at the end of the eighth week. It is the most critical period of development, for during this time the embryo becomes implanted within the uterine wall and all the essential external and internal body parts are formed. Any disturbances in the developmental processes occurring during the embryonic stage are likely to result in major malformations or malfunctions.

Agents that cause congenital malformations by affecting an embryo during its period of rapid growth and development are called teratogens. Such factors include exposure to radiation, the presence of various drugs, and infections caused by certain microorganisms. Exposure to radiation, for example, tends to produce abnormalities within the central nervous system. The drug, thalidomide, interferes with the normal development of the limbs, and the presence of the virus causing rubella (German measles) may result in malformations of the heart as well as cataract and congenital deafness.

By the beginning of the eighth week, the embryo is usually 30 mm (1.2 in) in length and weighs less than 5 grams (0.2 oz). Although its body is quite unfinished, it is clearly recognizable as a human being. (See fig. 22.16 and color plate 36.) From this time on, it is called a **fetus.**

The eighth week of development occurs two weeks after a pregnant woman has missed her second menstrual period. Since these first few weeks are critical periods in development, it is important that a woman seek health care as soon as she thinks she may be pregnant.

Pregnancy can be diagnosed with 95% accuracy after about 10 days following the first missed menstrual period by using a test that detects the presence of human chorionic gonadotropins (HCG) in a woman's urine.

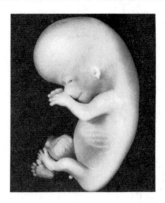

Fig. 22.16 By the beginning of the eighth week of development, the embryonic body is clearly recognizable as a human, and thereafter it is called a fetus.

1. Describe the development of the amnion.
2. What blood vessels are found in the umbilical cord?
3. What is the function of amniotic fluid?
4. What is the significance of the yolk sac?

Fetal Stage

The **fetal stage** of development begins at the eighth week and continues to the time of birth. During this period, the existing body structures continue to grow and mature, and only a few new parts appear. However, the rate of growth is great, and the body proportions change considerably. For example, at the beginning of the fetal stage, the head is disproportionately large and the legs are relatively short. (See fig. 22.17)

During the third lunar month, growth in body length is accelerated, while the growth of the head slows. The arms achieve the relative length they will maintain throughout development, and ossification centers appear in most of the bones. By the twelfth week, the external reproductive organs are distinguishable as male or female. (See color plate 37.)

In the fourth lunar month, the body grows very rapidly and reaches a length of 13 to 17 cm (5 to 6.8 in). The legs lengthen considerably, and the skeleton continues to ossify.

In the fifth lunar month, the rate of growth decreases somewhat. The legs achieve their final relative proportions, and the skeletal muscles become active so that the mother may feel fetal movements (quickening). Some hair appears on the head, and the skin becomes covered with fine, downy hair (lanugo).

Plate 34 Photomicrograph of a mammalian ovary. Note the clusters of primary follicles near the surface.

Plate 35 Photomicrograph of a maturing follicle.

Plate 36 What structures can you identify in this photograph of a human embryo?

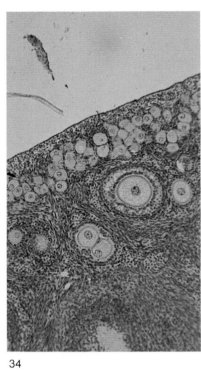

34

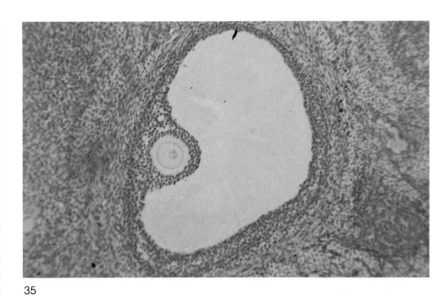

35

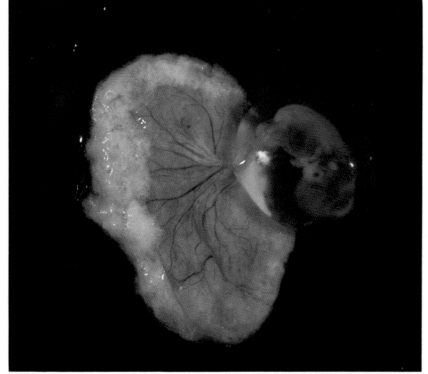

36

Plate 37 Human embryos and fetuses at various stages
of development:
(*a*) embryo in uterine wall; (*b*) embryo, ca. 6 weeks;
(*c*) fetus, ca. 9 weeks; (*d*) fetus, ca. 10 weeks; (*e*) fetus,
ca. 12 weeks.

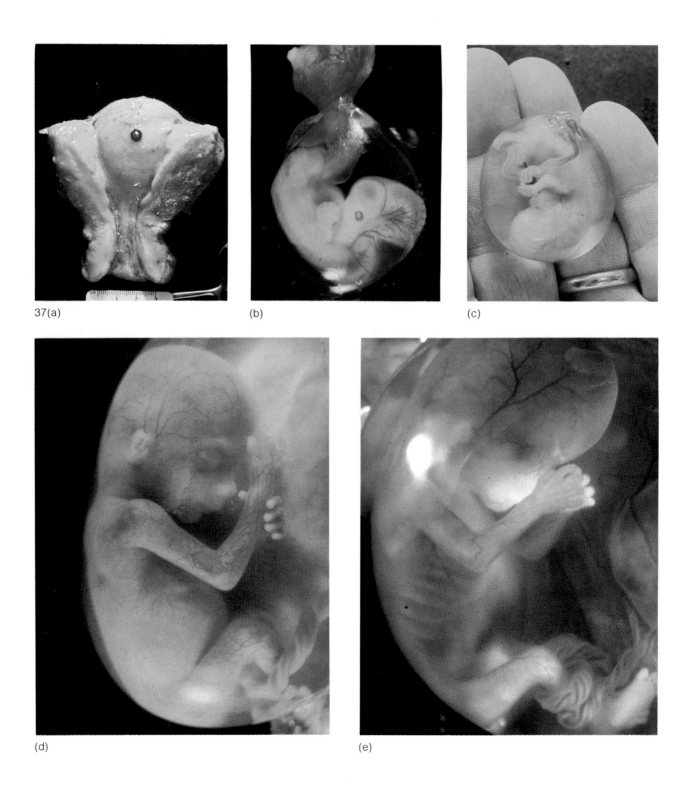

37(a)

(b)

(c)

(d)

(e)

Fig. 22.17 During fetal development, the body proportions change considerably.

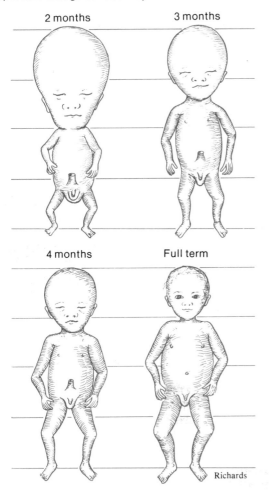

2 months 3 months

4 months Full term

Richards

Fig. 22.18 During the fifth month of development, the fetal limbs achieve their final relative proportions.

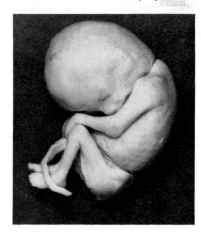

The skin is also coated with a cheesy mixture of sebum from the sebaceous glands and dead epidermal cells. (See fig. 22.18.)

During the sixth lunar month, the body gains a substantial amount of weight. The eyebrows and eyelashes appear. The skin is quite wrinkled and translucent. It is also reddish, due to the presence of dermal blood vessels.

In the seventh month, the skin becomes smoother as fat is deposited in the subcutaneous tissues. The eyelids, which fused together during the third month, reopen. At the end of the seventh month, a fetus is about 37 cm (14.8 in) in length. A fetus of this age occasionally will survive if born, but most such premature fetuses succumb to respiratory problems.

Premature fetuses that are born near the end of the seventh month have a greater chance for survival if their lungs are sufficiently developed. Such fetuses have respiratory membranes sufficiently thin to permit exchanges of oxygen and carbon dioxide by diffusion, and they produce enough surfactant to reduce the surface tension within the alveoli.

In the eighth lunar month, the fetal skin is still reddish and somewhat wrinkled. The testes of males descend into the scrotum. A fetus born at the end of the eighth lunar month may survive if given proper care.

During the ninth lunar month, the fetus reaches a length of about 47 cm (18.8 in). The skin is smooth, and the body appears chubby due to an accumulation of subcutaneous fat. The reddishness of the skin fades to pinkish or bluish pink, even in fetuses of dark-skinned parents, because melanin is not produced until the skin is exposed to light. Thus, it may be difficult to determine a newborn's race by its skin color during the first few hours or days of life. Infants born at the end of the ninth lunar month have an excellent chance of surviving.

At the end of the tenth lunar month, the fetus is said to be full term. It is about 50 cm (19.5 in) long and weighs 6 to 8 lbs. The skin has lost its downy hair, but is still covered with sebum and dead epidermal cells. The scalp is usually covered with hair, the fingers and toes have well developed nails, and the bones of the head are ossified. As figure 22.19 shows, the fetus is positioned upside down with its head toward the cervix (*vertex position*) and is ready to be born.

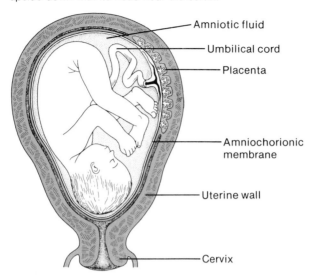

Amniotic fluid

Umbilical cord

Placenta

Amniochorionic membrane

Uterine wall

Cervix

Chart 22.1 summarizes the stages of prenatal development.

A set of abnormalities sometimes observed in infants is thought to result from the mother drinking as little as 3 oz of alcohol per day during pregnancy. This set of symptoms, which is called the *fetal alcohol syndrome,* includes incomplete growth, reduced head size, mental retardation, defective body organs, and various brain dysfunctions.

1. What major changes characterize the fetal stage of development?
2. When can the sex of a fetus be determined?
3. What factors favor the survival of a fetus born at the end of the seventh month?
4. How is a fetus usually positioned within the uterus at the end of the tenth lunar month?

Fetal Circulation

Throughout the fetal stage, the maternal blood supplies the fetus with oxygen and nutrients and carries away its wastes. These substances diffuse between the maternal and fetal blood through the placental membrane, and they are carried to and from the fetal body by means of the umbilical vessels (Figure 22.20). Consequently, the fetal vascular system must be adapted to intrauterine life in special ways, and its pattern of blood flow must differ from that of an adult.

Chart 22.1	Stages in prenatal development	
Stage	**Time Period**	**Some Major Events**
Period of cleavage	First week	Cells undergo mitosis, blastocyst forms; inner cell mass appears; blastocyst becomes implanted in uterine wall
Embryonic stage	Second through eighth week	Inner cell mass becomes embryonic disk; primary germ layers form; embryo proper becomes cylindrical; main internal organs and external body structures appear; placenta and umbilical cord form; embryo proper is suspended in amniotic fluid
Fetal stage	Ninth week to birth	Existing structures continue to grow; ossification centers appear in bones; reproductive organs develop; arms and legs achieve final relative proportions; muscles become active; skin is covered with sebum and dead epidermal cells; head becomes positioned toward cervix

In fetal circulation, the *umbilical vein* transports blood rich in oxygen and nutrients from the placenta to the fetal body. This vein enters the body through the umbilical ring and travels along the anterior abdominal wall to the liver. About half the blood passes into the liver, and the rest enters a vessel called the **ductus venosus** that bypasses the liver.

The ductus venosus travels a short distance and joins the inferior vena cava. There the oxygenated blood from the placenta is mixed with deoxygenated blood from the lower parts of the fetal body. This blood continues through the vena cava to the right atrium.

In an adult heart, blood from the right atrium enters the right ventricle and is pumped through the pulmonary artery to the lungs. However, in the fetus the lungs are nonfunctional, and the blood largely bypasses them. More specifically, as blood relatively rich in oxygen enters the right atrium of the fetal heart, a large proportion of it is shunted directly into the left atrium through an opening in the atrial septum. This opening is called the **foramen ovale,** and blood passes through it because the blood pressure in the right atrium is somewhat greater than that in the

Fig. 22.20 Oxygen and nutrients enter the fetal blood from the maternal blood by diffusion; waste substances enter the maternal blood from the fetal blood by diffusion.

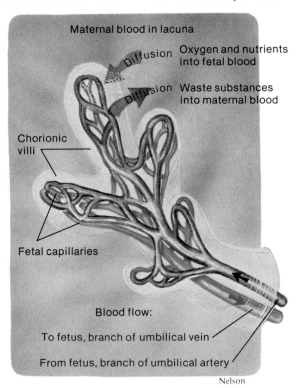

Maternal blood in lacuna

Diffusion — Oxygen and nutrients into fetal blood

Diffusion — Waste substances into maternal blood

Chorionic villi

Fetal capillaries

Blood flow:

To fetus, branch of umbilical vein

From fetus, branch of umbilical artery

Nelson

Chart 22.2 Fetal circulatory adaptations

Adaptation	Function
Umbilical vein	Carries oxygenated blood from placenta to fetus
Ductus venosus	Conducts about half the blood from the umbilical vein directly to the inferior vena cava, thus bypassing the liver
Foramen ovale	Conveys large proportion of blood entering the right atrium from the inferior vena cava, through the atrial septum, and into the left atrium, thus bypassing the lungs
Ductus arteriosus	Conducts some blood from the pulmonary artery to the aorta, thus bypassing the lungs
Umbilical arteries	Carry blood from the internal iliac arteries to the placenta for reoxygenation

The more highly oxygenated blood that enters the left atrium through the foramen ovale is mixed with a small amount of deoxygenated blood returning from the pulmonary veins. This mixture moves into the left ventricle and is pumped into the aorta. Some of it reaches the myocardium by means of the coronary arteries, and some reaches the tissues of the brain through the carotid arteries.

The blood carried by the descending aorta is partially oxygenated and partially deoxygenated. Some of it is carried into the branches of the aorta that lead to various parts in the lower regions of the body. The rest passes into the *umbilical arteries* which branch from the internal iliac arteries and lead to the placenta. There the blood is reoxygenated.

The umbilical cord usually contains two arteries and one vein. In a small percentage of newborns, there is only one umbilical artery. Since this condition is often associated with various other cardiovascular disorders, the number of vessels within the severed cord is routinely counted following a birth.

This pattern of fetal circulation, shown in figure 22.21, becomes fully developed near the middle of the prenatal period. (Chart 22.2 summarizes its major features.) At the time of birth, important adjustments must occur when the placenta ceases to function and breathing begins.

1. *Which umbilical vessel carries oxygen rich blood to the fetus?*
2. *What is the function of the ductus venosus?*
3. *How does fetal circulation allow blood to bypass the lungs?*
4. *What characteristic of the fetal lungs tends to shunt blood away from them?*

left atrium. Furthermore, a small valve (septum primum) located on the left side of the atrial septum overlies the foramen ovale and helps to prevent blood from moving from left to right.

The rest of the blood entering the right atrium, as well as a large proportion of the deoxygenated blood entering from the superior vena cava, passes into the right ventricle and out through the pulmonary artery.

The vessels of the pulmonary circuit have a high resistance to blood flow since the lungs are collapsed, and the vessels are somewhat compressed. Enough blood reaches the lung tissues, however, to sustain them.

Most of the blood in the pulmonary artery bypasses the lungs by entering a fetal vessel called **ductus arteriosus,** which connects the pulmonary artery to the descending portion of the aortic arch. As a result, blood with a relatively low oxygen concentration returning to the heart in the superior vena cava bypasses the lungs. At the same time it is prevented from entering the portion of the aorta that provides branches leading to the heart and brain.

Fig. 22.21 The general pattern of fetal circulation. How does this pattern of circulation differ from that of an adult?

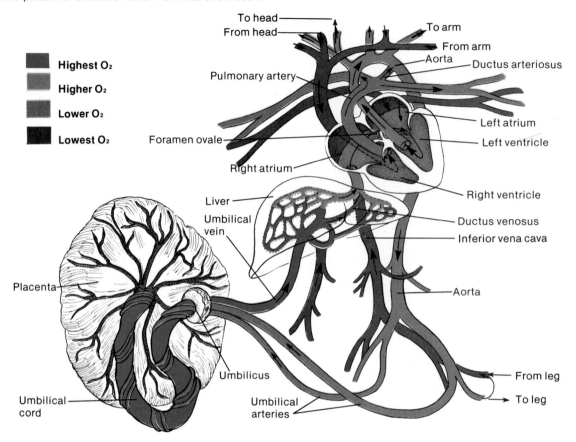

Highest O₂

Higher O₂

Lower O₂

Lowest O₂

To head
From head
To arm
From arm
Aorta
Pulmonary artery
Ductus arteriosus
Left atrium
Foramen ovale
Left ventricle
Right atrium
Right ventricle
Liver
Ductus venosus
Umbilical vein
Inferior vena cava
Placenta
Aorta
Umbilicus
From leg
Umbilical cord
Umbilical arteries
To leg

Postnatal Period

The postnatal period of development lasts from birth until death and can be divided into a neonatal period, infancy, childhood, adolescence, adulthood, and senescence.

Neonatal Period

The **neonatal period**, which extends from birth to the end of the first four weeks, begins very abruptly at birth. (See fig. 22.22.) Physiological adjustments must be made quickly, because the newborn must suddenly do for itself those things that the mother's body has been doing for it. Thus, the newborn must carry on respiration, obtain nutrients, digest nutrients, excrete wastes, regulate body temperature, and so forth. However, its most immediate need is to obtain oxygen and excrete carbon dioxide, so its first breath is critical.

The first breath must be particularly forceful, because the newborn's lungs are collapsed, and the airways are small and offer considerable resistance to air movement. Also, surface tension tends to hold the moist membranes of the lungs together.

Fig. 22.22 The neonatal period extends from birth to the end of the first 4 weeks after birth.

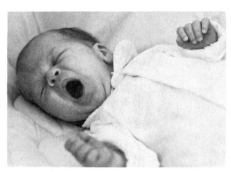

Fortunately, the lungs of a full term fetus secrete *surfactant* (see chapter 15), which reduces surface tension, and after the first powerful breath begins to expand the lungs, breathing becomes easier.

It is not clear whether the first breath is stimulated by one or several factors. Those that may be involved include an increasing level of carbon dioxide, a decreasing pH, low oxygen concentration, a drop in body temperature, and mechanical stimulation that occurs during and after the birth process. (See fig. 22.23.)

Fig. 22.23 Some factors that help to stimulate the newborn's first breath.

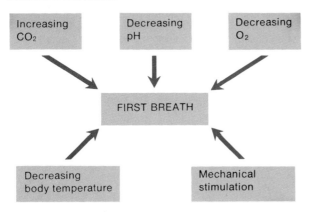

Fig. 22.24 Major changes that occur in the newborn's circulatory system.

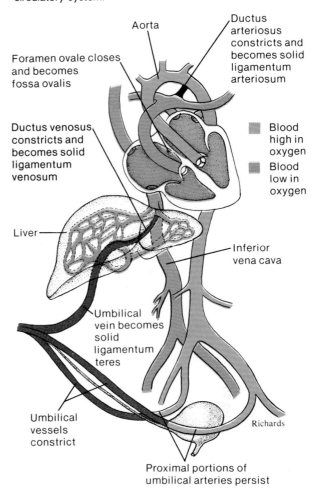

Prior to birth, the fetus depends primarily on glucose and fatty acids obtained from the mother's blood as energy sources. The newborn, on the other hand, is suddenly without an external source of nutrients, and its mother's milk will not be produced for 2 to 3 days. Furthermore, the newborn has a relatively high rate of metabolism, and its liver, which is not fully mature, may be unable to supply enough glucose to support the metabolic needs. Consequently, the newborn typically utilizes stored fat as an energy source.

---

1. Define the neonatal period of development.
2. Why must the first breath of an infant be particularly forceful?
3. What factors seem to stimulate the first breath?
4. What does a newborn use for an energy supply during the first few days after birth?

As a rule, the newborn's kidneys are unable to produce concentrated urine, so the newborn excretes a relatively dilute fluid. For this reason, the newborn may become dehydrated and develop a water and electrolyte imbalance. Also, certain of the newborn's homeostatic control mechanisms may function imperfectly. The temperature regulating system, for example, may be unable to maintain a constant body temperature. As a consequence, the body temperature may be unstable during the first few days of life and respond to slight stimuli by fluctuating above or below the normal level.

As was mentioned, when the circulation of blood through the placenta ceases and the lungs begin to function at birth, adjustments are made in the circulatory system that was formerly adapted to the needs of intrauterine life. (See fig. 22.24.)

For example, following birth, the umbilical vessels constrict. The arteries close first, and if the umbilical cord is not clamped or severed for a minute or so, blood continues to flow from the placenta to the newborn through the umbilical vein, adding to the newborn's blood volume.

The proximal portions of the umbilical arteries persist in the adult as the *superior vesical arteries* that supply blood to the urinary bladder. The more distal portions become solid cords (lateral umbilical ligaments). The umbilical vein becomes the cordlike *ligamentum teres* that extends from the umbilicus to the liver in an adult.

Similarly, the ductus venosus constricts shortly after birth and is represented in the adult as a fibrous cord (ligamentum venosum), which is superficially embedded in the wall of the liver.

The foramen ovale closes as a result of blood pressure changes occurring in the right and left atria as fetal vessels constrict. More precisely, as blood ceases to flow from the umbilical vein into the inferior vena cava, the blood pressure in the right atrium drops. Also, as the lungs expand with the first breathing movements, the resistance to blood flow through the pulmonary circuit decreases, more blood enters the left atrium through the pulmonary veins, and the blood pressure in the left atrium increases.

As pressure in the left atrium rises and that in the right atrium falls, the valve (septum primum) on the left side of the atrial septum closes the foramen ovale. This valve gradually fuses with the tissues along the margin of the foramen. In an adult, the side of the previous opening is marked by a depression called the *fossa ovalis.*

In some newborns, the foramen ovale remains open (patent), and since the blood pressure in the left atrium is greater than that in the right atrium, some blood may flow from the left chamber to the right chamber.

Although this condition usually produces no symptoms, it may cause a heart murmur.

The ductus arteriosus, like the other fetal vessels, constricts after birth. This constriction seems to be stimulated by a substance called *bradykinin,* which is released from the lungs during their initial expansions. Bradykinin is thought to act when the oxygen concentration of the aortic blood rises as a result of breathing. After the ductus arteriosus has closed, blood can no longer bypass the lungs by moving from the pulmonary artery directly into the aorta. In an adult, the ductus arteriosus is represented by a cord called the *ligamentum arteriosum.*

The ductus arteriosus sometimes fails to close. Although the cause of this condition is usually unknown, it often occurs in newborns whose mothers were infected with rubella virus (German measles) during the first three months of embryonic development.

After birth, the blood pressure in the aorta is normally greater than that in the pulmonary artery. Therefore, a patent ductus arteriosus usually results in the flow of some blood from the aorta into the pulmonary artery.

These changes in the newborn's circulatory system do not occur abruptly. Although the constriction of the ductus arteriosus may be functionally complete within 15 minutes, the permanent closure of the foramen ovale may take up to a year. Chart 22.3 summarizes these circulatory changes.

---

1. *How do the kidneys of a newborn differ from those of an adult?*
2. *What portions of the umbilical arteries continue to function into adulthood?*
3. *What is the fate of the foramen ovale? The ductus arteriosus?*
4. *When is the closure of the ductus arteriosus functionally complete?*

Infancy

The period of development extending from the end of the first four weeks to 1 year is called **infancy**. During this time, the infant grows rapidly and may triple its birth weight. Its teeth begin to erupt through the gums, and its muscular and nervous systems mature so that coordinated muscular activities become possible. Consequently, the infant is soon able to sit,

---

Chart 22.3 Circulatory adjustments in the newborn

Structure	Adjustment	In the Adult
Umbilical vein	Becomes constricted	Becomes ligamentum teres that extends from the umbilicus to the liver
Ductus venosus	Becomes constricted	Becomes ligamentum venosum that is superficially embedded in the wall of the liver
Foramen ovale	Is closed by valvelike septum primum as blood pressure in right atrium decreases and pressure in left atrium increases	Valve fuses along margin of foramen ovale and is marked by a depression called fossa ovalis
Ductus arteriosus	Constricts in response to bradykinin released from the expanding lungs	Becomes ligamentum arteriosum that extends from the pulmonary artery to the aorta
Umbilical arteries	Distal portions become constricted	Distal portions become lateral umbilical ligaments; proximal portions function as superior vesical arteries

Fig. 22.25 Infancy extends from the end of 4 weeks after birth until 1 year of age.

Fig. 22.26 Childhood begins at the end of the first year and is completed at puberty.

creep, and stand, to reach for objects and hold them, and to follow objects visually. (See fig. 22.25.)

Infancy also is characterized by the beginning of the ability to communicate with others: the infant learns to smile, laugh, and respond to some sounds. By the end of the first year, the infant may be able to say two or three words.

Since infancy (as well as childhood) is a period of rapid growth, the infant has particular nutritional needs. For example, in addition to an energy source, the body requires proteins to provide the amino acids necessary to form new tissues, calcium and vitamin D to promote the development and ossification of skeletal structures, iron to support blood cell formation, and vitamin C for the production of structural tissues such as cartilage and bone.

Childhood

The period called **childhood** begins at the end of the first year and is completed at puberty. (See fig. 22.26.) During this period, growth continues at a high rate. The deciduous teeth appear and are subsequently replaced by permanent ones. The child develops a high degree of voluntary muscular control and learns to walk, run, and climb. Bladder and bowel controls are established. The child learns to communicate effectively by speaking, and later, usually learns to read, write, and reason objectively. At the same time the child is maturing emotionally.

1. Define infancy.
2. What developmental changes characterize this phase of life?
3. Define childhood.
4. What developmental changes characterize this phase of life?

Fig. 22.27 Adolescence is the period between puberty and adulthood.

Adolescence

Adolescence is the period of development between puberty and adulthood. As is discussed in chapter 21, puberty is a period of anatomical and physiological changes resulting in reproductively functional individuals. These changes are, for the most part, hormonally controlled, and they include the appearance of secondary sexual characteristics and growth spurts in the muscular and skeletal systems.

Females usually experience these changes somewhat earlier than males, so that early in adolescence, females may be taller and stronger than their male peers. (See fig. 22.27.) On the other hand, females achieve their full growth at earlier ages, and in late adolescence the average male is taller and stronger than the average female.

Fig. 22.28 Adulthood extends from adolescence to old age.

Fig. 22.29 The changes that characterize old age begin during adulthood and culminate with death.

The periods of rapid growth in adolescence, which usually begin between the ages of 11 and 13 in females and between 13 and 15 in males, cause increased demands for certain nutrients. In addition to energy sources, foods must provide ample amounts of proteins, vitamins, and minerals to support the growth of new tissues.

Adolescence is also characterized by increasing levels of motor skill, intellectual ability, and emotional maturity.

Adulthood

Adulthood (maturity) extends from adolescence to old age. Since the anatomy and physiology of the adult has been the basis for chapters 1–21, little more needs to be said about this period of development.

Ordinarily the adult remains relatively unchanged anatomically or physiologically for many years. (See fig. 22.28.) However, sometime after the age of 30, depending on hereditary factors and the environmental factors to which the person has been exposed, degenerative changes usually begin to occur. For example, skeletal muscles tend to lose strength as more and more connective tissue appears within the muscles; the circulatory system becomes less efficient as the lumens of arterioles and arteries become narrowed due to accumulations of fatty deposits; the skin tends to become loose and wrinkled as elastic fibers in the dermis undergo changes; and the hair on the head turns gray as melanin-producing enzyme systems cease to function.

Also, the capacity of the sex cell-producing tissues decline. Females experience menopause in later adulthood, and males experience losses in sexual functions that accompany a decrease in the secretion of the male sex hormones.

Senescence

Senescence is the process of growing old, and it culminates in the death of the individual. This process involves a continuation of the degenerative changes that begin during adulthood. (See fig. 22.29.) As a result, the body becomes less and less able to cope with the demands placed on it by the individual and by the environment.

The cause of senescence is not well understood, and there is no generally accepted explanation for it. However, the degenerative changes that occur seem to involve a variety of factors. Some are due to the prolonged use of body parts. For example, the cartilages covering the ends of bones at joints may wear away, leaving the joints stiff and painful.

Other degenerative changes are caused by disease processes that interfere with vital functions, such as gas exchanges or blood circulation. Still others seem to be due to cellular changes, including alterations in metabolic rates and in distribution of body fluids. The rate of cell division may decline, and the ability of cells to carry on immune responses may decrease. As a result, the person becomes less able to repair damaged tissue and is more susceptible to disease.

Senescence is typically accompanied by decreasing efficiency of the central nervous system. As a result, the person may experience a loss in intellectual functions. Also, the physiological coordinating capacity of the nervous system may decrease, and homeostatic mechanisms may fail to operate effectively.

Death usually results from mechanical disturbances in the cardiovascular system or from disease processes that affect vital organs.

Chart 22.4 summarizes the major phases of life and their characteristics.

Chart 22.4 Stages in postnatal development

Stage	Time Period	Some Major Events
Neonatal period	Birth to end of fourth week	Newborn begins to carry on respiration, obtain nutrients, digest nutrients, excrete wastes, regulate body temperature, and make circulatory adjustments
Infancy	End of fourth week to one year	Growth rate is high, teeth begin to erupt, muscular and nervous systems mature so coordinated activities are possible, communication begins
Childhood	One year to puberty	Growth rate is high, deciduous teeth erupt and are replaced by permanent teeth, high degree of muscular control is achieved, bladder and bowel controls are established, intellectual abilities mature
Adolescence	Puberty to adulthood	Person becomes reproductively functional and emotionally more mature, growth spurts occur in skeletal and muscular systems, high levels of motor skills are developed, intellectual abilities increase
Adulthood	Adolescence to old age	Person remains relatively unchanged anatomically and physiologically, degenerative changes begin to occur
Senescence	Old age to death	Degenerative changes continue, body becomes less and less able to cope with the demands placed upon it, death usually results from mechanical disturbances in the cardiovascular system or from disease processes that affect vital organs

1. What changes characterize adolescence?
2. Define adulthood.
3. What changes characterize adulthood?
4. What changes characterize senescence?

Some Clinical Terms Related to Growth and Development

ablatio placentae (ab-la′she-o plah-cen′tā)—a premature separation of the placenta from the uterine wall.

amniocentesis (am″ne-o-sen-te′sis)—a technique in which a sample of amniotic fluid is withdrawn from the amniotic cavity by inserting a hollow needle through the mother's abdominal wall.

dizygotic twins (di″zi-got′ik twinz)—twins resulting from the fertilization of two ova by two sperm cells.

hydatid mole (hi′dah-tid mōl)—a type of uterine tumor that originates from placental tissue.

hydramnios (hi-dram′ne-os)—the presence of excessive amniotic fluid.

intrauterine transfusion (in″trah-u′ter-in trans-fu′zhun)—transfusion administered by injecting blood into the fetal peritoneal cavity before birth.

lochia (lo′ke-ah)—vaginal discharge following childbirth.

meconium (mĕ-ko′ne-um)—the first fecal discharge of a newborn.

monozygotic twins (mon″o-zi-got′ik twinz)—twins resulting from the fertilization of one ovum by one sperm cell.

perinatology (per″ĭ-na-tol′o-je)—branch of medicine concerned with the fetus after 25 weeks of development and with the newborn for the first four weeks after birth.

postpartum (pōst-par′tum)—occurring after birth.

teratology (ter″ah-tol′o-je)—the study of abnormal development and congenital malformations.

trimester (tri-mes′ter)—each third of the total period of pregnancy.

ultrasonography (ul″trah-son-og′rah-fe)—technique used to visualize the size and position of fetal structures by means of ultrasonic sound waves.

Chapter Summary

Introduction

Growth refers to an increase in size; development is the process of changing from one phase of life to another.

Prenatal Period

1. Period of cleavage
   a. Fertilization occurs in a uterine tube and results in a zygote.
   b. The zygote undergoes mitosis, and the daughter cells divide.
   c. Each subsequent division produces smaller and smaller cells.
   d. A solid ball of cells (morula) is formed, and it becomes a hollow ball, called a blastocyst.
   e. The blastocyst becomes implanted in the uterine wall.
      (1) Enzymes digest the endometrium around the blastocyst.
      (2) Fingerlike processes from the blastocyst penetrate into the endometrium.
   f. The inner cell mass that gives rise to the embryo proper forms within the blastocyst.
   g. The period of cleavage lasts through the first week of development.
2. Embryonic stage
   a. This stage extends from the second to the eighth week.

b. It is characterized by development of the placenta and the main internal and external body structures.
c. The cells of the inner cell mass become arranged into primary germ layers.
   (1) Ectoderm gives rise to the nervous system, portions of the skin, the lining of the mouth, and the lining of the anal canal.
   (2) Mesoderm gives rise to muscles, bones, blood vessels, lymphatic vessels, reproductive organs, kidneys, and linings of body cavities.
   (3) Endoderm gives rise to linings of the digestive tract, respiratory tract, urinary bladder, and urethra.
d. The embryonic disk becomes cylindrical and is attached to the developing placenta by the connecting stalk.
e. The embryo develops head, face, arms, legs, and mouth, and appears more human.
f. Chorionic villi develop and are surrounded by spaces filled with maternal blood.
g. The placental membrane consists of the epithelium of villi and the epithelium of capillaries inside villi.
   (1) Oxygen and nutrients diffuse from the maternal blood through the membrane and into the fetal blood.
   (2) Carbon dioxide and other wastes diffuse from the fetal blood through the membrane and into the maternal blood.
h. The placenta develops in the disk-shaped area where the chorion remains in contact with the uterine wall.
   (1) The embryonic portion consists of chorion and its villi.
   (2) The maternal portion consists of the uterine wall to which villi are attached.
i. Fluid-filled amnion develops around the embryo.
j. The umbilical cord is formed as amnion envelops the tissues attached to the underside of the embryo.
   (1) The umbilical cord includes two arteries and a vein.
   (2) It suspends the embryo in the amniotic cavity.
k. The chorion and amnion become fused.
l. The yolk sac forms on the underside of the embryonic disk.
   (1) It gives rise to blood cells and cells that later form sex cells.
   (2) It helps form the digestive tube.
m. The allantois extends from the yolk sac into the connecting stalk.
   (1) It forms blood cells.
   (2) It gives rise to the umbilical vessels.
n. By the beginning of the eighth week, the embryo is recognizable as a human and is called a fetus.

3. Fetal stage
   a. This stage extends from the beginning of the eighth week and continues until birth.
   b. Existing structures grow and mature; only a few new parts appear.
   c. The body enlarges, arms and legs achieve final relative proportions, the skin is covered with sebum and dead epidermal cells, the skeleton continues to ossify, the muscles become active, and fat is deposited in subcutaneous tissue.
   d. If a fetus is born prematurely:
      (1) It has a poor chance of surviving if born at the end of the seventh lunar month.
      (2) It may survive at the end of the eighth lunar month.
      (3) It has an excellent chance for survival at the end of the ninth lunar month.
   e. Fetus is full term at the end of the tenth lunar month.
      (1) It is about 50 cm long and weighs 6–8 lbs.
      (2) It is positioned with its head toward the cervix.
4. Fetal circulation
   a. Blood is carried between the placenta and fetus by umbilical vessels.
   b. Blood enters the fetus through the umbilical vein and partially bypasses the liver by means of the ductus venosus.
   c. Blood enters the right atrium and partially bypasses the lungs by means of the foramen ovale.
   d. Blood entering the pulmonary artery partially bypasses the lungs by means of the ductus arteriosus.
   e. Blood enters the umbilical arteries by means of the internal iliac arteries.

Postnatal Period
1. Neonatal period
   a. This period extends from birth to the end of the fourth week.
   b. The newborn must begin to carry on respiration, obtain nutrients, excrete wastes, and regulate its body temperature.
   c. The first breath must be powerful to expand the lungs.
      (1) Surfactant reduces surface tension.
      (2) The first breath is stimulated by a variety of factors.
   d. The liver is immature and unable to supply sufficient glucose, so the newborn depends primarily on stored fat as an energy source.
   e. Immature kidneys cannot concentrate urine very well.
      (1) The newborn may become dehydrated.
      (2) Water and electrolyte imbalance may develop.
   f. Homeostatic mechanisms may function imperfectly and body temperature may be unstable.

g. The circulatory system undergoes changes when placental circulation ceases.
   (1) Umbilical vessels constrict.
   (2) The ductus venosus constricts.
   (3) The foramen ovale is closed by a valve as blood pressure in the right atrium falls and pressure in the left atrium rises.
   (4) Bradykinin released from the expanding lungs stimulates constriction of the ductus arteriosus.
2. Infancy
   a. Infancy extends from the end of the fourth week to 1 year of age.
   b. Infancy is a period of rapid growth.
      (1) The muscular and nervous systems mature, and coordinated activities become possible.
      (2) Communication begins.
   c. Rapid growth depends on an adequate intake of proteins, vitamins, and minerals in addition to energy sources.
3. Childhood
   a. Childhood extends from the end of the first year to puberty.
   b. It is characterized by rapid growth, development of muscular control, and establishment of bladder and bowel control.
4. Adolescence
   a. Adolescence extends from puberty to adulthood.
   b. It is characterized by physiological and anatomical changes that result in a reproductively functional individual.
   c. Females may be taller and stronger than males in early adolescence, but the situation reverses in late adolescence.
   d. Adolescents develop high levels of motor skills, their intellectual abilities increase, and they continue to mature emotionally.
5. Adulthood
   a. Adulthood extends from adolescence to old age.
   b. The adult remains relatively unchanged physiologically and anatomically for many years.
   c. After age 30, degenerative changes usually begin to occur.
      (1) Skeletal muscles lose strength.
      (2) The circulatory system becomes less efficient.
      (3) The skin loses its elasticity.
      (4) The capacity to produce sex cells declines.
6. Senescence
   a. Senescence is the process of growing old.
   b. Degenerative changes continue, and the body becomes less able to cope with demands placed upon it.
   c. Changes occur because of prolonged use, effects of disease, and cellular alterations.

d. Aging person usually experiences losses in intellectual functions and physiological coordinating capacities.
e. Death usually results from mechanical disturbances in the cardiovascular system or from disease processes that affect vital organs.

Application of Knowledge

1. If an aged relative came to live with you, what special provisions could you make in your household environment and routines that would demonstrate your understanding of the changes brought on by aging processes?
2. Why is it important for a middle-aged adult who has neglected physical activity for many years to see a physician for a physical examination before beginning an exercise program?

Review Activities

1. Define *growth* and *development*.
2. Describe the process of cleavage.
3. Distinguish between a blastomere and a blastocyst.
4. Describe the formation of the inner cell mass and explain its significance.
5. Describe the process of implantation.
6. Explain how the primary germ layers form.
7. List the major body parts derived from ectoderm.
8. List the major body parts derived from mesoderm.
9. List the major body parts derived from endoderm.
10. Describe the formation of the placenta and list its functions.
11. Define *placental membrane*.
12. Distinguish between the chorion and the amnion.
13. Explain the function of amniotic fluid.
14. Describe the formation of the umbilical cord.
15. Explain how the yolk sac and the allantois are related and list the functions of each.
16. Explain why the embryonic period of development is so critical.
17. Define *fetus*.
18. List the major developmental changes that occur during the fetal stage.
19. Describe a full term fetus.

20. Explain how the fetal circulatory system is adapted for intrauterine life.

21. Trace the pathway of blood from the placenta to the fetus and back to the placenta.

22. Distinguish between a newborn and an infant.

23. Explain why the first breath must be particularly forceful.

24. List some factors that are thought to act as stimuli for the first breath.

25. Explain why newborns tend to develop water and electrolyte imbalances.

26. Describe the circulatory changes that occur in the newborn.

27. Describe the characteristics of an infant.

28. Distinguish between a child and an adolescent.

29. Define *adulthood*.

30. List some of the degenerative changes that begin during adulthood.

31. Define *senescence*.

32. List some of the factors that seem to promote senescence.

Suggestions for Additional Reading

Balinsky, B. I. 1970. *An introduction to embryology.* Philadelphia: W. B. Saunders.

Beaconsfield, P. et al. August 1980. The placenta. *Scientific American.*

Ebert, J. D., and Sussex, I. M. 1970. *Interacting systems in development.* New York: Holt, Rinehart and Winston.

Gardner, L. I. July 1972. Deprivation dwarfism. *Scientific American.*

Gellis, S. S., ed. 1973. *Year book of pediatrics.* Chicago: Year Book Medical Publishers.

Marx, J. L. 1974. Aging research I: cellular theories of senescence. *Science* 186:1105.

—————. 1974. Aging research II: pacemakers for aging. *Science* 186:1196.

Moore, K. L. 1974. *Before we are born: basic embryology and birth defects.* Philadelphia: W. B. Saunders.

—————. 1974. *The developing human: clinically oriented embryology.* Philadelphia: W. B. Saunders.

Morison, R. S. 1971. Death: process or event. *Science* 173:695.

Patten, B. M., and Carlson, B. M. 1974. *Foundations of embryology.* 3rd ed. New York: McGraw-Hill.

Rugh, R., and Shettles, L. B. 1971. *From conception to birth.* New York: Harper and Row.

Streissguth, A. P. et al. 1980. Teratogenic effects of alcohol in humans and laboratory animals. *Science* 209:353.

Timiras, P. S. 1972. *Developmental physiology and aging.* New York: Macmillan.

—————. September 1978. Biological perspectives on aging. *American Scientist.*

Human Genetics

23 The complex changes an embryo undergoes as it grows and develops is controlled by mechanisms that ensure the resulting offspring will resemble its parents. At the same time, these mechanisms guarantee the offspring will differ from its parents and will become an individual with unique characteristics.

The study of the similarities and differences between parents and their offspring and the mechanisms responsible for them is called *genetics*.

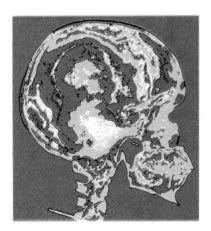

After you have studied this chapter, you should be able to

1. Explain how heredity and environment influence the development of individual characteristics.

2. Distinguish between genes and chromosomes and explain why genes and chromosomes are paired.

3. Explain why gene expression may differ with different allelic combinations.

4. Describe how environmental factors may influence gene expression.

5. Outline the process of meiosis and explain why cells resulting from this process contain different combinations of genetic information.

6. Explain the patterns by which single traits are transmitted from parents to offspring.

7. Describe how chromosomes control the inheritance of sex.

8. Explain the pattern by which a trait controlled by a sex-linked gene is inherited.

9. Define *nondisjunction* and explain its significance in the appearance of chromosome disorders.

10. Explain what is meant by *genetic counseling*.

11. Complete the review activities at the end of this chapter.

allele   (ah-lēl′)

autosome   (aw′to-sōm)

dominant   (dom′ĭ-nant)

gene   (jēn)

genetics   (jĕ-net′iks)

genotype   (je′no-tɪp)

heredity   (hĕ-red′ĭ-te)

heterozygous   (het″er-o-zi′gus)

homozygous   (ho″mo-zi′gus)

meiosis   (mi-o′sis)

multiple allele   (mul′tĭ-pl ah-lēl′)

mutation   (mu-ta′shun)

nondisjunction   (non″dis-jungk′shun)

phenotype   (fe′no-tɪp)

recessive   (re-ses′iv)

sex chromosome   (seks kro′mo-sōm)

sex-linked gene   (seks-linkt′ jēn)

alb-, white: *alb*ino—condition in which normal skin pigments are lacking.

centr-, a point: *centr*omere—region at which paired chromatids are held together.

hem-, blood: *hem*ophilia—condition in which bleeding is prolonged due to a defect in the blood clotting mechanism.

hetero-, different: *hetero*zygous—condition in which the alleles of a gene pair are different.

homo-, same: *homo*logous chromosomes—a pair of chromosomes that contain similar genetic information.

pheno-, visible: *pheno*type—appearance of an individual due to the ways genes are expressed.

syn-, together: *syn*apsis—process by which homologous chromosomes become tightly intertwined.

tri-, three: *tri*somy—condition in which one kind of chromosome is triply represented.

Although genetics is a relatively new field of study, knowledge of genetic mechanisms has expanded rapidly in recent times. It is known that individual characteristics result from the interaction of two major factors—heredity and environment.

Heredity is the transmission of genetic information from parents to offspring. This information is passed in the form of DNA molecules contained in the nuclei of egg and sperm cells, and it provides chemical instructions for human development. **Environment,** on the other hand, includes the chemical, physical, and biological factors in the surroundings of an individual that act to influence his or her characteristics.

Genes and Chromosomes

As is explained in chapter 4, genetic information within DNA molecules tells a cell how to construct specific kinds of protein molecules, which in turn function as structural materials, enzymes, and other vital substances.

The Gene

The portion of a DNA molecule containing the information for producing one kind of protein molecule is called a **gene.** The genes within the egg and sperm cells that unite to form a zygote instruct the zygote how to synthesize particular proteins. Some of these proteins act as enzymes, which in turn promote the metabolic reactions necessary for the growth and development of a unique individual. (See fig. 23.1.)

It takes one gene to produce a particular kind of enzyme. Therefore, many sets of genes are required for the synthesis of enzymes responsible for an offspring's specific sex, eye color, hair color, skin color, blood type, and so forth.

The total number of genes in a human set is unknown. However, many kinds of enzymes, structural proteins, and proteins with special functions occur in cells. Since at least one gene is needed for the synthesis of each kind of protein, some investigators estimate that the human set includes as many as 50,000 genes.

Fig. 23.1 A gene is a portion of a DNA molecule.

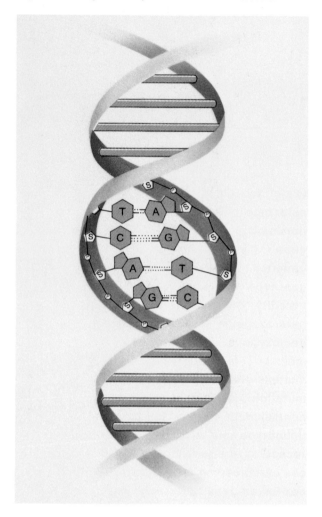

Chromosome Numbers

Within a cell, the DNA molecules, and thus the genes, are housed within *chromosomes*. These structures, which are described in chapter 3, consist of proteins (histones) as well as DNA. They appear as rod-shaped bodies in the nucleus when a cell undergoes cell division. (See fig. 23.2.)

The number of chromosomes within the cells of organisms varies with species. For the human species (*Homo sapiens*), the number is 46. Thus, a human zygote contains 46 chromosomes, 23 of which were received from its female parent by means of an egg cell and 23 from its male parent by means of a sperm cell. (See fig. 23.3.)

Fig. 23.2 Chromosomes appear in a cell as it undergoes cell division.

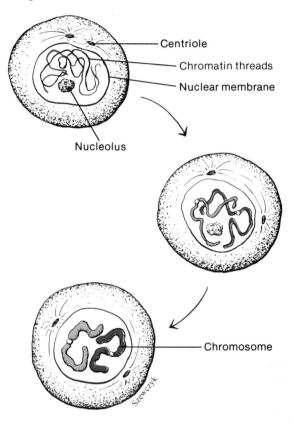

- Centriole
- Chromatin threads
- Nuclear membrane

Nucleolus

Chromosome

Szewczyk

Fig. 23.3 A typical human cell contains 46 chromosomes.

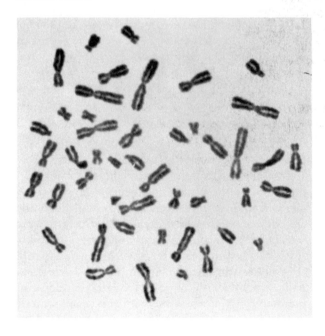

Fig. 23.4 A set of human chromosomes arranged in pairs of homologous chromosomes. One chromosome of each pair has come from each parent.

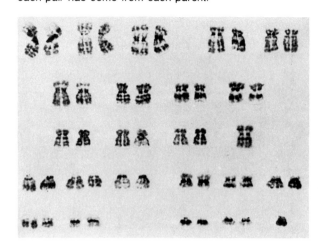

Chromosome Pairs

The 23 chromosomes a zygote receives from its female parent contain a complete set of genetic information for the growth and development of an offspring, as do the 23 chromosomes it receives from its male parent. The chromosomes of these sets that contain similar genetic material are said to comprise **homologous pairs.** That is, the chromosome of the maternal set containing the information for a particular set of traits and the chromosome of the paternal set containing information for these same traits make up a homologous pair. The chromosome set shown in figure 23.4 is arranged in homologous pairs.

During interphase, the stage in a cell's life cycle between mitotic divisions described in chapter 3, the chromosomes and the genes they contain become duplicated. One member of each duplicated chromosome is transmitted to each daughter cell, so that the daughter cells receive identical sets of homologous chromosomes and thus receive identical sets of genes.

Gene Pairs

Since the chromosomes within a cell are paired, the genes within the chromosomes are paired also. Such a pair of genes is said to consist of two **alleles,** and because the members of a gene pair are parts of homologous chromosomes, one allele of each pair originates from the female parent and the other orginates from the male parent.

Sometimes the alleles of a gene pair are alike. That is, they contain exactly the same genetic information. They may, for example, both contain information for the synthesis of a particular enzyme needed for the production of normal skin pigment (melanin). A zygote receiving such a pair of identical alleles is called **homozygous** for the genes involved, and in this case would develop into an individual with normal skin pigmentation.

In other instances, the members of a gene pair are different. One allele may contain information for the synthesis of the skin pigment enzyme, while its allele is a *mutant gene* that lacks information for enzyme production. When a zygote contains a gene pair whose alleles differ, it is said to be **heterozygous** for the genes involved.

---

1. *Define a gene.*
2. *What is the relationship between genes and chromosomes?*
3. *What is an* allele?
4. *What is meant by* homozygous? *By* heterozygous?

Gene Expression

The particular combination of genes present in a zygote and its subsequent daughter cells is said to constitute a **genotype.** The appearance of the individual that develops as a result of the ways the genes are expressed is termed the **phenotype.**

Dominant and Recessive Genes

A zygote that is *homozygous* for normal skin pigmentation will develop into an offspring with normally colored skin. Similarly, a zygote that is *heterozygous* for skin pigmentation—containing one normal gene and one mutant gene—will also develop into an offspring with normal skin pigment.

In this heterozygous condition, one gene (normal pigmentation) is expressed but its allele (lack of pigmentation) is not. In such a case, the expressed gene is called **dominant** and the unexpressed gene is called **recessive.**

For convenience, dominant genes are symbolized with capital letters (A), and recessive genes are symbolized with small letters (a). Thus, the *genotype* of an individual who is homozygous for normal pigmentation due to a dominant gene is symbolized AA; a heterozygous pair is symbolized Aa. In both of these instances, the *phenotypes* of the individuals would be normal pigmentation, since a dominant gene is present in each genotype.

Chart 23.1 Some traits determined by single pairs of dominant and recessive alleles

Phenotype Due to Expression of Dominant Gene	Phenotype Due to Expression of Recessive Gene
Full lips	Thin lips
Dark hair	Light hair
Free ear lobes	Attached ear lobes
Bridge of nose convex	Bridge of nose concave
Dark eyes (brown iris)	Light eyes (blue iris)
Farsightedness	Normal vision
Astigmatism	Normal vision
Extra fingers or toes (polydactyly)	Normal numbers of fingers and toes
Freckles	Lack of freckles
Ability to roll tongue into U-shape	Lack of this ability
Dimples in cheeks	Lack of dimples
Feet with normal arches	Flatfeet
Otosclerosis with hearing loss	Normal hearing

A zygote that receives two recessive genes for skin pigmentation has the genotype *aa* and will develop into an individual who lacks skin pigment. The phenotype of such an individual is called *albino.* (See fig. 23.5.)

Thus, three genotypes are possible in the instance of a gene pair involving dominant and recessive alleles. They are *homozygous dominant* (*AA*), *heterozygous* (*Aa*), and *homozygous recessive* (*aa*). However, only two phenotypes are possible, since the dominant gene is expressed whenever it is present—in both the homozygous dominant (*AA*) and the heterozygous (*Aa*) individuals. The recessive gene is expressed only in the homozygous recessive (*aa*) condition.

Chart 23.1 lists some traits due to dominant and recessive alleles.

Incomplete Dominance

Sometimes the alleles of a gene pair occur in two forms that are expressed differently, and neither is dominant to the other. In other words, when such genes are paired in a heterozygous individual, both are partially expressed. Genes of this type are said to blend or to be *incompletely dominant.*

For example, two forms of a gene needed for the synthesis of hemoglobin are *incompletely dominant.* A person with the genotype H^1H^1 develops normal hemoglobin, while one with the genotype H^2H^2 develops hemoglobin with abnormal molecules. These molecules tend to form long chains when they are exposed to low oxygen concentrations, and they cause

Fig. 23.5 Albinism is due to the inheritance of two recessive genes for skin pigment.

the red blood cells containing them to become distorted or *sickle-shaped*. Consequently, a person with the genotype *H²H²* is said to inherit *sickle-cell anemia*. (See fig. 23.6.)

A heterozygous individual, with the genotype *H¹H²*, develops some normal and some abnormal hemoglobin. In this case, the red blood cells may exhibit sickling if oxygen concentrations are relatively low, but usually the red cells remain normal. This condition is called *sickle-cell trait.*

The gene for sickle-cell anemia occurs most commonly in persons with Central African ancestry. About 8% of the United States population with such ancestry possess this gene. As a result, about 1 child in 170 produced by this group develops sickle-cell anemia. This condition is characterized by symptoms associated with lack of tissue oxygen, resulting from the blockage of capillaries by sickled cells. These symptoms include severe joint and abdominal pain, skin ulcers, and chronic kidney disease.

Fig. 23.6 Red blood cells from a person with sickle-cell anemia become elongated or sickled when the oxygen concentration is low.

In the previous examples, only two forms of allelic genes were involved. In other instances, alleles occur in several forms, and such a group is called **multiple alleles.**

In multiple alleles, each person carries a single pair of genes, but since the different forms of the gene can occur in various combinations, several genotypes are possible. The inheritance of ABO *blood type,* for example, involves multiple alleles.

As is described in chapter 16, a person's ABO blood type is related to the *agglutinogens* present on the red blood cell membranes. Specifically, there are two agglutinogens involved—*agglutinogen A* and *agglutinogen B.* If the red cells have only agglutinogen A, the blood type is A; if only agglutinogen B is present, the blood type is B; if both agglutinogens A and B are present, the blood type is AB; and if neither agglutinogen A nor B is present, the blood type is O.

These blood types are determined by various combinations of three allelic genes that can be symbolized I^A, I^B, and i^O. In this instance, genes I^A and I^B are equally dominant and are said to be **codominant genes.** They are both dominant to gene i^O. Since each person inherits only two genes for ABO blood type, and there are three forms of the gene, several genotypes are possible. They are: $I^A I^A$, $I^A i^O$, $I^B I^B$, $I^B i^O$, $I^A I^B$, and $i^O i^O$.

If gene I^A is present, agglutinogen A develops on the red blood cell membranes, and if gene I^B is present, agglutinogen B is produced. The presence of gene i^O causes either no agglutinogen to be produced or such a weak agglutinogen that it is usually insignificant. Thus, persons with genotypes $I^A I^A$ or $I^A i^O$ have blood type A. Those with genotypes $I^B I^B$ or $I^B i^O$ have blood type B. Genotype $I^A I^B$ produces type AB blood, while genotype $i^O i^O$ results in type O blood.

Thus, as far as the ABO blood group is concerned, every person belongs to one of six genotypes and has one of four phenotypes.

Genotypes	Phenotypes
$I^A I^A$, $I^A i^O$	blood type A
$I^B I^B$, $I^B i^O$	blood type B
$I^A I^B$	blood type AB
$i^O i^O$	blood type O

Blood types are sometimes used as genetic evidence to prove a particular male is *not* the father of a particular child. In such cases, when the child's genotype includes a gene not included in the genotype of either the male or the mother, the male is clearly not the father.

Although evidence of this kind can be used to demonstrate a male is *not* the father of a particular child, it cannot be used to prove he *is* the father, because other males would have the same genotype for blood type as the male in question.

1. Distinguish between genotype and phenotype.
2. What is meant by dominant gene? By recessive gene?
3. Define incomplete dominance.
4. Explain why there are six genotypes but only four phenotypes in the ABO blood group.

Polygenic Inheritance

Although some traits—such as the presence or absence of skin pigment, the production of normal or abnormal hemoglobin, and the development of a certain blood type—seem to be determined by single pairs of genes, other traits are controlled by several pairs. Inheritance of this type is called *polygenic,* and examples of such inheritance include body stature and the concentration of skin pigment.

Height seems to be determined by at least four pairs of genes located in different positions within the chromosome set. Each pair causes a small effect in the development of height.

Using the symbol, T, for a gene causing tallness and, t, for a gene causing shortness, the genotype of a very tall person might be $T^1 T^1 \; T^2 T^2 \; T^3 T^3 \; T^4 T^4$, and a very short person might be $t^1 t^1 \; t^2 t^2 \; t^3 t^3 \; t^4 t^4$. Persons with various combinations of these genes for tallness and shortness, such as $T^1 T^1 \; T^2 t^2 \; T^3 T^3 \; t^4 t^4$ or $T^1 t^1 \; T^2 t^2 \; t^3 t^3 \; t^4 t^4$, would grow to intermediate heights. In fact, because many different combinations of these genes are possible, people exhibit considerable variation in height. (See fig. 23.7.)

Similarly, the amount of melanin produced in the skin seems to be determined by several pairs of genes, and as before, at least four pairs seem to be involved. Consequently, the genotype of a person with very dark skin might be symbolized $M^1 M^1 \; M^2 M^2 \; M^3 M^3 \; M^4 M^4$, and one with very light skin might be $m^1 m^1 \; m^2 m^2 \; m^3 m^3 \; m^4 m^4$. Persons with various combinations of genes for dark and light skin would develop skin of intermediate colors and would exhibit a broad range of variation. (See fig. 23.8.)

Fig. 23.7 Variation in height is due to different combinations of several pairs of genes.

Fig. 23.8 Variation in skin color is due in part of different combinations of several pairs of genes. What other factors influence skin color?

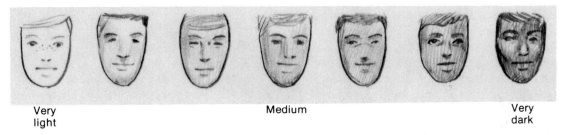

Very
light

Medium

Very
dark

Fig. 23.9 The gene responsible for freckles is more fully expressed when the skin is exposed to sunlight.

No sunlight   Exposed to sun

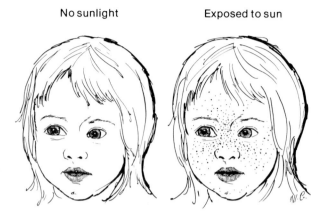

Environmental Factors

Although the traits of height and concentration of skin pigment are determined by several pairs of genes, the *development* of these and other hereditary traits is influenced by environmental factors as well. Whether or not an individual will achieve the stature that is hereditarily possible, for example, is affected by such environmental factors as diet and disease. Even if a person's genotype is for extreme tallness, the person may only grow to an intermediate height if the diet is inadequate in certain essential nutrients or if the body is affected by a disease that interferes with bone growth and development.

Similarly, the amount of pigment deposited in the skin is influenced by environmental factors. If the skin is exposed to ultraviolet light from a sunlamp or from sunlight, the skin becomes more heavily pigmented. Conversely, if such exposure is avoided, the concentration of pigment is decreased. (See fig. 23.9.)

To summarize, the growth and development of a zygote is controlled by the paired genes it receives from an egg cell and a sperm cell. These genes encode genetic information for the synthesis of specific enzyme molecules, and the enzymes in turn promote the metabolic reactions that cause the development of such characteristics as sex, blood type, eye color, hair color, skin color, height, concentration of skin pigment, and so forth.

The way in which a particular gene is expressed depends on the nature of the gene and on the way in which it is combined with other genes. Gene expression is also influenced by environmental factors that may enhance the development of a trait or inhibit it.

Some genes, called *lethal genes,* cause abnormal development that ultimately results in the death of the individual who inherits them.

If, for example, a set of lethal genes causes abnormal heart development, an embryo might die about the fourth week, when its heart normally begins to function. If the lethal genes cause abnormal development of the kidneys, however, death might not occur until after birth, when the newborn must function without the aid of its mother's organs.

1. How does polygenic inheritance make many variations of a given trait possible?
2. How may environmental factors influence gene expression?

Chromosomes and Sex Cell Production

Although the offspring of a set of parents have many characteristics in common, they also differ because they originate from different egg and sperm cells and develop in different environments. Furthermore, different egg and sperm cells (from the same parents) produce zygotes with different genotypes, because the sex cells of each individual carry unique combinations of genes. This uniqueness results from the process of meiosis.

Fig. 23.10 Spermatogenesis is a meiotic process that involves two successive divisions.

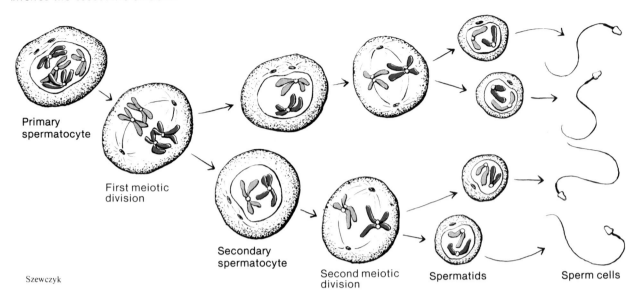

Primary spermatocyte

First meiotic division

Secondary spermatocyte

Second meiotic division

Spermatids

Sperm cells

Szewczyk

Meiosis

Meiosis occurs during *spermatogenesis* and *oogenesis,* and as a result of it, the chromosome numbers of the resulting sex cells are reduced by one-half. This process involves two successive divisions—a first and second *meiotic division.* In the first meiotic division, the homologous chromosomes of the parent cell are separated, and the chromosome number is reduced in the daughter cells. During the second meiotic division, the chromosomes act much as they do in *mitosis,* and the result is the formation of sex cells. (See fig. 23.10.)

Like mitosis, described in chapter 3, meiosis is a continuous process without marked interruptions between steps. However, for convenience, the process can be divided into stages as follows:

1. **First meiotic prophase.** This stage, shown in figure 23.11, resembles prophase in mitosis, although it lasts somewhat longer and is more complex. During it, the individual chromosomes appear as thin threads within the nucleus. The threads become shorter and thicker, the nucleoli disappear, the nuclear membrane fades away, and the spindle fibers become organized.

Throughout this stage, the individual chromosomes, which became duplicated during the interphase prior to the beginning of meiosis, are composed of two identical chromatids. Each pair of chromatids is held together at a central region called the *centromere.*

Fig. 23.11 Stages in the first meiotic prophase.

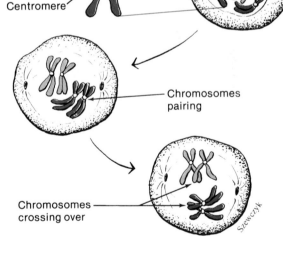

Chromatin threads

Identical chromatids

Centromere

Chromosomes pairing

Chromosomes crossing over

Szewczyk

Fig. 23.12 (*a*) Pairing of homologous chromosomes; (*b*) chromatids crossing over; (*c*) results of crossing over.

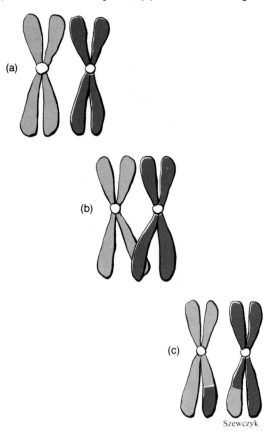

Szewczyk

Fig. 23.13 The first meiotic metaphase.

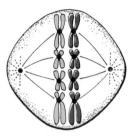

Fig. 23.14 The first meiotic anaphase.

Fig. 23.15 The first meiotic telophase.

As the prophase continues, homologous chromosomes approach each other, lie side by side, and become tightly intertwined. This process of pairing, shown in figure 23.12, is called *synapsis,* and during it the chromatids of the homologous chromosomes contact one another at various points. In fact, the chromatids may break in one or more places and exchange parts, forming chromatids with new combinations of genetic information. For example, since one chromosome of a homologous pair is of maternal origin and the other is of paternal origin, an exchange of parts between homologous chromatids results in chromatids that contain genetic information from both maternal and paternal sources. This process of exchanging genetic information is called **crossing over.**

2. **First meiotic metaphase.** During the first metaphase, shown in figure 23.13, the synaptic pairs of chromosomes become lined up about midway between the poles of the developing spindle. Each pair, consisting of two chromosomes and four chromatids, becomes associated with a spindle fiber.

3. **First meiotic anaphase.** In mitosis, the centromeres that hold the chromatids of each chromosome pair together divide during anaphase, and the chromatids are free to move toward opposite ends of the spindle. However, during meiosis the centromeres *do not* divide (fig. 23.14), and as a result, the homologous chromosomes become separated, and the chromatids of each chromosome move together to one end of the spindle. Thus, each daughter cell receives only one member of an homologous pair of chromosomes, and the chromosome number is thereby reduced by one-half.

4. **First meiotic telophase.** Meiotic telophase is similar to mitotic telophase in that the parent cell divides into two daughter cells. (See fig. 23.15.) Also during this phase, nuclear membranes appear around the chromosome sets, the nucleoli reappear, and the spindle fibers fade away.

The first meiotic telophase is followed by a short *interphase* and the beginning of the second meiotic division. This division is essentially the same as mitosis, and it can also be divided into *prophase, metaphase, anaphase,* and *telophase.* These stages are shown in figure 23.16.

Fig. 23.16 The second meiotic division is similar to mitosis.

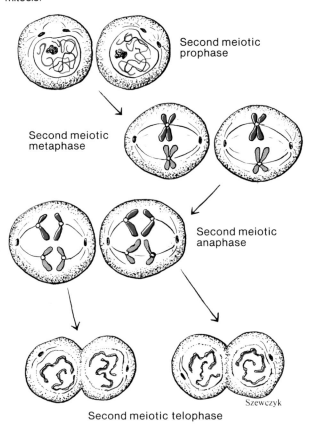

Second meiotic prophase

Second meiotic metaphase

Second meiotic anaphase

Second meiotic telophase

Szewczyk

Fig. 23.17 As a result of crossing over, the genetic information held within the sperm cells varies from cell to cell.

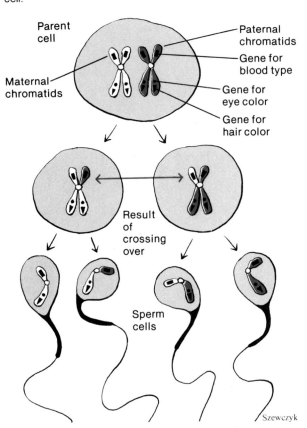

Parent cell

Maternal chromatids

Paternal chromatids

Gene for blood type

Gene for eye color

Gene for hair color

Result of crossing over

Sperm cells

Szewczyk

During the *second meiotic prophase,* the chromosomes reappear. They are still composed of pairs of chromatids, and the chromatids are held together by centromeres. Near the end of this phase, the chromosomes move into positions midway between the poles of the developing spindle.

In *second meiotic metaphase,* the double-stranded chromosomes become attached to spindle fibers, and during *second meiotic anaphase,* the centromeres divide so that the chromatids are free to move to opposite poles of the spindle. As a result of *second meiotic telophase,* the cell divides into two daughter cells, and new nuclei are organized around the sets of single-stranded chromosomes.

Results of Meiosis

In spermatogenesis, meiosis results in the formation of *four sperm cells,* each of which contains 23 single-stranded chromosomes. These chromosomes contain a complete set of genetic information, but only one gene of each type, whereas the original parent cell (spermatogonium) contained a pair of each type of

gene. Furthermore, the genetic information varies from sperm cell to sperm cell because of the crossing over that occurred during the first meiotic prophase. For example, as figure 23.17 shows, one sperm cell may contain genes for the development of eye color and hair color from an original maternal chromosome and a gene for blood type from an original paternal chromosome. Another sperm cell may contain a gene for maternal eye color and paternal hair color and blood type, while a third type contains genes for maternal hair color and blood type and a paternal gene for eye color.

In oogenesis, meiosis results in *one egg cell* (secondary oocyte) and a nonfunctional polar body, which may divide. If the egg cell is fertilized, a second polar body is produced. (See fig. 23.18.) As is true in sperm cells, egg cells contain 23 single-stranded chromosomes whose combinations of genetic information differ from egg cell to egg cell because of crossing over.

Oögenesis

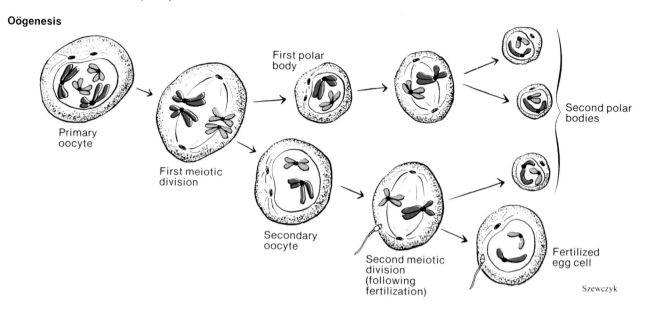

1. *Distinguish between meiosis and mitosis.*
2. *Describe the major events that occur during meiosis.*
3. *What is the final result of meiosis?*

Inheritance of Single Traits

Genes occur in pairs, but as a result of meiosis, a sex cell contains only one member of each gene pair. Consequently, some sex cells contain the original *maternal gene* of a gene pair, and the other sex cells contain the original *paternal gene* of such a pair.

Dominant and Recessive Inheritance

As was mentioned earlier, some genes occur in two forms—*dominant* and *recessive*. In pigmentation, for example, the dominant gene for the development of normal pigment can be symbolized as *A*, while its recessive allele can be symbolized as *a*. Thus, a person who is homozygous dominant for this trait has the genotype *AA* and will appear normal. A person who is heterozygous has the genotype *Aa* and will also appear normal, since a dominant gene for pigmentation is present. A person who is homozygous recessive, however, has the genotype *aa* and will be *albino*.

Since a person with genotype *AA* has two identical genes, all the sex cells produced by this individual will receive a dominant gene (*A*). Likewise all of the sex cells produced by a person with genotype *aa* will receive a recessive gene (*a*). However, a person with the genotype *Aa* will produce equal numbers of two kinds of sex cells. One kind will have a dominant gene (*A*), and the other kind will have a recessive gene (*a*). (See fig. 23.19.)

When information concerning the genotypes of parents is available, it is often possible to predict what types of offspring can be produced as well as the proportions in which they are likely to occur. For example, since a male parent with genotype *AA* produces only *A*-bearing sperm cells, and a female parent with genotype *aa* produces only *a*-bearing egg cells, all the zygotes resulting from the fusion of such sex cells will have the genotype *Aa*:

	Male Parent	Female Parent
Genotype:	*AA*	*aa*
Phenotype:	Normal pigment	Albino
Sex cells:	All *A*	All *a*

Egg cells

a

	a
A	*Aa*

Possible zygotes: (Sperm cells)

Offspring genotypes: All *Aa*
Offspring phenotypes: All normal pigment

Fig. 23.19 In the case of alleles *A* and *a* for skin pigmentation, there are two possible phenotypes and three possible genotypes. The sex cells each person produces reflect the composition of his or her genotype.

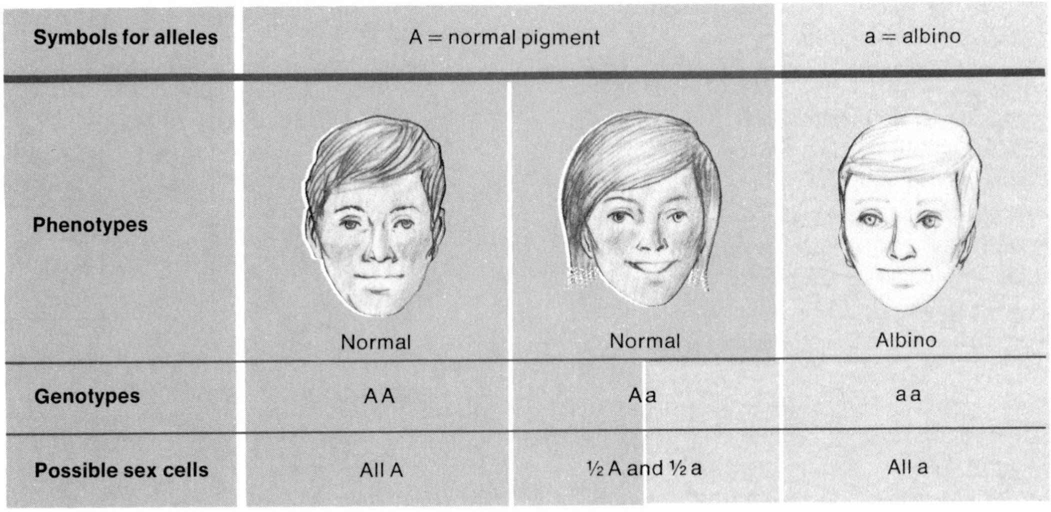

Symbols for alleles	A = normal pigment		a = albino
Phenotypes	Normal	Normal	Albino
Genotypes	A A	A a	a a
Possible sex cells	All A	½ A and ½ a	All a

If, on the other hand, the male parent has the genotype *Aa* and the female parent is *aa*, offspring of two possible genotypes can be produced—*Aa* and *aa*:

	Male Parent	Female Parent
Genotype:	*Aa*	*aa*
Phenotype:	Normal pigment	Albino
Sex cells:	½ *A* and ½ *a*	All *a*

Egg cells

		a
Sperm cells	½ *A*	½ *Aa*
	½ *a*	½ *aa*

Possible zygotes:

Offspring genotypes: ½ *Aa* and ½ *aa*
Offspring phenotypes: ½ normal pigment and ½ albino

In this case, ½ of the sperm cells carry the gene *A* and ½ carry the gene *a*. Since sperm cells have equal chances of fertilizing an egg cell, it is expected that ½ of the offspring from these parents will have the genotype *Aa* and ½ will have the genotype *aa*. Because the presence of a dominant gene (*A*) produces normal pigment, those offspring with genotype *Aa* will have normal pigmentation, and those with genotype *aa* will be albino.

A more complex situation arises if both the parents have the genotype *Aa*. In this instance, each parent produces two kinds of sex cells in equal numbers ½ *A* and ½ *a*—and so the offspring resulting from such parents can have three possible genotypes—*AA, Aa,* or *aa*:

	Male Parent	Female Parent
Genotype:	*Aa*	*Aa*
Phenotype:	Normal pigment	Normal pigment
Sex cells:	½ *A* and ½ *a*	½ *A* and ½ *a*

Egg cells

		½ *A*	½ *a*
Sperm cells	½ *A*	¼ *AA*	¼ *Aa*
	½ *a*	¼ *Aa*	¼ *aa*

Possible zygotes:

Offspring genotypes: ¼ *AA*, ½ *Aa*, ¼ *aa*
Offspring phenotypes: ¾ normal pigment and ¼ albino

In this case, ¼ (25%) of the offspring are expected to have the genotype *AA*, ½ (50%) are expected to be *Aa*, and ¼ (25%) are expected to be *aa*. However, since those with genotypes *AA* or *Aa* will appear normal, only ¼ of the offspring from such parents are expected to be albino (*aa*).

Since chance events, such as the fertilization of a particular kind of egg cell by a particular kind of sperm cell, occur independently, the genotype of one child has no effect upon the genotypes of other children produced by a set of parents. Thus each child conceived by parents with genotypes *Aa* and *Aa* has a 25% chance of being *AA* (normal), a 50% chance of being *Aa* (normal, but carrying the albino gene), and a 25% chance of being *aa* (albino).

Incompletely Dominant Inheritance

As was mentioned, there are two forms of a particular gene that function in the production of hemoglobin. One of the alleles (H^1) is responsible for the formation of normal hemoglobin, and the other (H^2) results in abnormal molecules that cause *sickling* of red blood cells. Since genes H^1 and H^2 are *incompletely dominant*, a person with genotype H^1H^1 has normal hemoglobin, one with genotype H^1H^2 develops sickle-cell *trait,* and one with genotype H^2H^2 suffers from sickle-cell *anemia.*

The kinds of offspring expected from parents with various combinations of codominant genes can be determined as follows:

1. Parent genotypes H^1H^1 and H^2H^2:

	Male Parent	Female Parent
Genotype:	H^1H^1	H^2H^2
Phenotype:	Normal	Sickle-cell anemia
Sex cells:	All H^1	All H^2

Egg cells
H^2

Possible zygotes: Sperm cells H^1 | H^1H^2

Offspring genotypes: All H^1H^2
Offspring phenotypes: All sickle-cell trait

2. Parent genotypes H^1H^2 and H^2H^2:

	Male Parent	Female Parent
Genotype:	H^1H^2	H^2H^2
Phenotype:	Sickle-cell trait	Sickle-cell anemia
Sex cells:	½ H^1 and ½ H^2	All H^2

Egg cells
H^2

Possible zygotes: Sperm cells ½ H^1 | ½ H^1H^2 ; ½ H^2 | ½ H^2H^2

Offspring genotypes: ½ H^1H^2 and ½ H^2H^2
Offspring phenotypes: ½ sickle-cell trait and ½ sickle-cell anemia

3. Parent genotypes H^1H^2 and H^1H^2:

	Male Parent	Female Parent
Genotype:	H^1H^2	H^1H^2
Phenotype:	Sickle-cell trait	Sickle-cell trait
Sex cells:	½ H^1 and ½ H^2	½ H^1 and ½ H^2

Egg cells
½ H^1 ½ H^2

Possible zygotes: Sperm cells
½ H^1 | ¼ H^1H^1 | ¼ H^1H^2
½ H^2 | ¼ H^1H^2 | ¼ H^2H^2

Offspring genotypes: ¼ H^1H^1, ½ H^1H^2, ¼ H^2H^2
Offspring phenotypes: ¼ normal, ½ sickle-cell trait, ¼ sickle-cell anemia

---

1. What offspring genotypes can be produced by parents with the genotypes AA and aa? AA and Aa?
2. How many offspring genotypes can be produced by parents with genotypes Aa and Aa? How many offspring phenotypes?
3. If both parents have sickle-cell trait, what will be the proportions of the genotypes and phenotypes of offspring that can be produced?

Sex Chromosomes

The sex of an individual is inherited in much the same manner as other traits. However, in this instance whole chromosomes are involved rather than a few paired genes.

Sex Inheritance

Within the human set of 46 chromosomes, there are 22 pairs called **autosomes** (which are chromosomes other than sex chromosomes) and one pair of **sex chromosomes.** These sex chromosomes are responsible for the development of the contrasting characteristics associated with *male* and *female.* In other words, the kinds of sex chromosomes present in a zygote determine whether the embryonic reproductive organs (which are essentially neutral in the early stages of development) will form into male or female structures.

Sex chromosomes are of two types. One type, which is relatively large, is called an **X chromosome,** and the other, which is relatively small, is called a **Y chromosome.** A female set of sex chromosomes consists of two X chromosomes, and consequently all egg cells contain a single *X chromosome.* The male set consists of an X and a Y chromosome, and so equal numbers of two types of sperm cells are produced. One-half of the sperm cells possess an *X chromosome,* and ½ possess a *Y chromosome.* This difference is shown in figure 23.20.

The sex of an offspring is determined by the type of sperm cell that fertilizes the egg cell. If the sperm involved carries an X chromosome, it will be paired with the X chromosome of the egg cell, and the resulting XX combination will direct the development of a female. If, on the other hand, a Y-bearing sperm cell fertilizes the egg, the XY combination causes the development of a male offspring. Since X-bearing and Y-bearing sperm cells occur in equal numbers, it is expected that equal numbers of males and females will result from the chance meeting of egg and sperm cells.

	Male Parent	Female Parent
Sex chromosomes:	XY	XX
Sex cells:	½ X and ½ Y	All X

Egg cells

X

		X
Sperm cells	½ X	½ XX
	½ Y	½ XY

Possible zygotes:

Offspring sex: ½ males (XY) and ½ females (XX)

However, for reasons that are not clear, an unusually large proportion of males appear. In fact, in the United States population, the ratio of males to females at birth is 106:100.

Even though greater numbers of males than females are born, the death rate among males is higher. Consequently, the sex ratio between males and females changes as they age. This change in sex ratio is reflected in the following data:

Age	Sex Ratio (males:females)
birth	106:100
18 years	100:100
50 years	85:100
85 years	50:100
100 years	20:100

Fig. 23.20 (a) A set of human male chromosomes and (b) a set of human female chromosomes arranged in pairs of homologous chromosomes. How are they different?

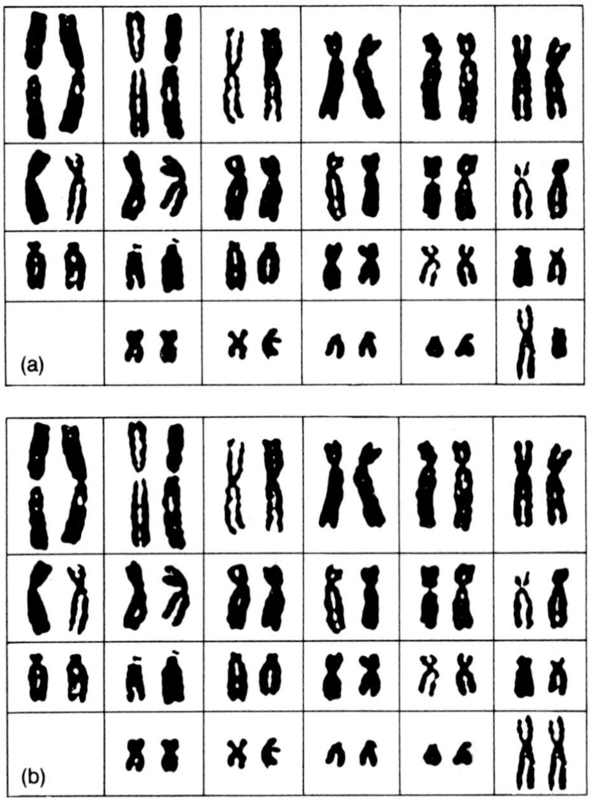

Sex-linked Inheritance

In addition to directing the differentiation of the embryonic reproductive organs, the X chromosomes contain genes responsible for the development of certain traits unrelated to sex. However, since these traits are determined by genes located on the X chromosomes, the patterns by which they are inherited differ in each sex. For example, they are more likely to be expressed in males (XY) than in females (XX).

This phenomenon is related to the fact that the X and Y sex chromosomes are quite different in size, and most of the genes on the X chromosome do not have alleles on the relatively small Y chromosome. Consequently, in males the genes of the single X chromosome are expressed even though they may be recessive, while in females a trait determined by such a recessive gene is expressed only if it is present on both X chromosomes (homozygous recessive). Traits determined by recessive genes located on X chromosomes are called **sex linked.**

Some traits are due to genes located within a portion of the Y chromosome for which there is no homologous region in an X chromosome. These traits appear only in males and are passed from fathers to their sons. This type of genetic transmission is called *holandric inheritance. Hypertrichosis,* which is characterized by excessively hairy ears, is an example of such a trait.

Color blindness is an example of a sex-linked trait. In this instance, normal color vision is due to the presence of a dominant gene (C). This gene is needed for the development of functional color receptors (cones) in the retina of the eye. The recessive allele (c) causes the production of defective receptors.

A male will have normal color vision if his X chromosome carries the dominant gene, which can be symbolized X^C. If, however, a male receives one recessive gene for color blindness (X^c), he will be color-blind, since there is no corresponding gene on the Y chromosome to overcome its effect.

A female, on the other hand, will have normal color vision if either of her X chromosomes carries the dominant gene (X^C). Thus, both the homozygous dominant ($X^C X^C$) and the heterozygous female ($X^C X^c$) will develop normal vision. Only a homozygous recessive female ($X^c X^c$) will be color-blind.

A color-blind male only occurs when the female parent carries recessive genes for color blindness, since the X chromosome of a male offspring is always received from his mother:

	Normal Male Parent	Color-blind Female Parent
Sex chromosomes:	$X^C Y$	$X^c X^c$
Sex cells:	½ X^C and ½ Y	All X^c

Egg cells

X^c

		Egg cells X^c
Sperm cells	½ X^C	½ $X^C X^c$
	½ Y	½ $X^c Y$

Possible zygotes for offspring:

Offspring genotypes:

Females—$X^C X^c$
Males—$X^c Y$

Offspring phenotypes:

Females—normal
Males—color-blind

Although all the daughters in this case will have normal vision, they will carry the gene for color blindness.

In the case of a homozygous dominant female parent ($X^C X^C$) and a male parent with color blindness ($X^c Y$), all of the children will develop normal vision. However, all the daughters of a colorblind male parent will receive the recessive gene for color blindness, since all daughters receive the X chromosome of the father. These female offspring will be heterozygous ($X^C X^c$) and will be carriers of the gene for color blindness.

	Color-blind Male Parent	Normal Female Parent
Sex chromosomes:	$X^c Y$	$X^C X^C$
Sex cells:	½ X^c and ½ Y	All X^C

Egg cells

X^C

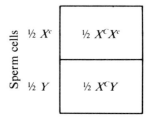

		Egg cells X^C
Sperm cells	½ X^c	½ $X^C X^c$
	½ Y	½ $X^C Y$

Possible zygotes for offspring:

Offspring genotypes:

Females—$X^C X^c$
Males—$X^C Y$

Offspring phenotypes:

Females—normal
Males—normal

If the female parent has normal vision but carries the gene for color blindness ($X^C X^c$), and the male parent has normal vision ($X^C Y$), all the female offspring will have normal vision, but ½ of them are expected to be carriers of the recessive gene. At the same time, ½ of the male offspring are expected to have normal vision ($X^C Y$) and ½ are expected to be color-blind ($X^c Y$).

	Normal Male Parent	Normal Female Parent
Genotypes:	$X^C Y$	$X^C X^c$
Sex cells:	½ X^C and ½ Y	½ X^C and ½ X^c

Egg cells

		½ X^C	½ X^c
Sperm cells	½ X^C	¼ $X^C X^C$	¼ $X^C X^c$
	½ Y	¼ $X^C Y$	¼ $X^c Y$

Possible zygotes:

Offspring genotypes:

Females—½ $X^C X^C$, ½ $X^C X^c$
Males—½ $X^C Y$, ½ $X^c Y$

Offspring phenotypes:

Females—normal
Males—½ normal, ½ color-blind

Hemophilia, sometimes called the bleeder's disease, is also due to a sex-linked recessive gene. In this instance, the presence of a single recessive gene in the male ($X^h Y$) or of two recessive genes in a female ($X^h X^h$) causes a defect in the blood clotting mechanism. As a result, prolonged bleeding may accompany an otherwise minor injury.

Like color blindness, hemophilia is more common in males. The recessive gene is received by male offspring from their mothers and is passed from affected fathers to their daughters.

Actually there are two major forms of hemophilia caused by recessive genes carried on the X chromosomes. *Hemophilia A* is characterized by the lack of plasma protein called *antihemophilic globulin* (blood clotting factor VIII). *Hemophilia B* (Christmas disease) is due to a deficiency of another blood substance called *plasma thromboplastin component* (blood clotting factor IX or Christmas factor).

Chart 23.2 lists some other traits related to sex-linked genes.

An inherited disease called *pseudohemophilia* (von Willebrand's disease) is caused by the presence of a dominant gene on an autosome. In this condition, the platelets are defective and bleeding is prolonged. However, since the responsible gene is on an autosome, this disease is not sex linked.

Chart 23.2 Some traits related to sex-linked recessive genes

Trait	Characteristics
Brown enamel	Tooth enamel appears brown rather than white
Coloboma iridis	A fissure in the iris of the eye causes the pupil to appear slitlike rather than round
Congenital night blindness	Vision is poor in dim light, but is relatively normal in bright light; however, this condition is not due to a deficiency of vitamin A
Microphthalmia	Eyes fail to develop normally and instead remain small and nonfunctional
Optic atrophy	Blindness results from degeneration of the optic nerves

1. How is the sex of an offspring determined?
2. What is meant by sex-linked inheritance?
3. Why do sex-linked traits appear most commonly in males?

Baldness is a hereditary trait that is due to the expression of *sex-influenced genes.* Such genes act differently in males than in females; that is, they act as dominant genes in the hormonal environment of an adult male and as recessive genes in the hormonal environment of an adult female. Consequently, baldness occurs more frequently in males than in females.

Chromosome Disorders

Chromosome disorders are characterized by the presence of abnormal numbers of chromosomes. Since the human set usually contains 46 chromosomes, any other number is considered to be abnormal.

Cause of Chromosome Disorders

As was described earlier, during meiosis, homologous chromosomes pair and undergo synapsis. Later these homologous pairs separate and move to opposite ends of the spindle. As a result, each daughter cell receives one chromosome of a homologous pair.

If, however, one pair of homologous chromosomes fails to separate during meiosis, both members of that pair will move into one daughter cell, while the other daughter cell will receive no chromosome

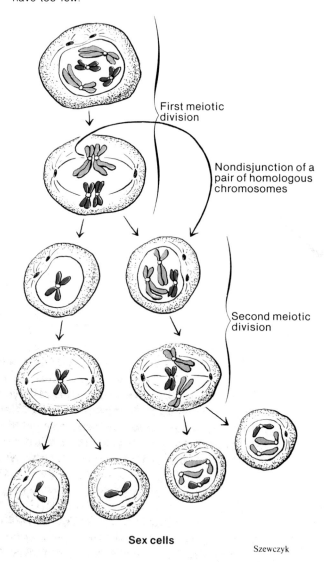

First meiotic division

Nondisjunction of a pair of homologous chromosomes

Second meiotic division

Sex cells

Szewczyk

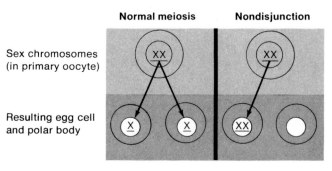

Normal meiosis   Nondisjunction

Sex chromosomes (in primary oocyte)

Resulting egg cell and polar body

Apparently only one *X* chromosome of an *XX* set is functional, and in many female cells the other, nonfunctional *X* chromosome is found lying against the nuclear membrane, where it forms a tiny structure called a *Barr body.*

In some hospitals, cells of the chorion of each newborn are stained and observed for the presence of Barr bodies. In this way, it is possible to determine if there are abnormal numbers of *X* chromosomes in the cells.

of that type. Such a failure of homologous chromosomes to separate is called **nondisjunction.** When it occurs, some of the resulting cells have too many chromosomes in their sets, while others have too few. (See fig. 23.21.)

Sex Chromosome Disorders

If nondisjunction occurs during *oogenesis,* some of the resulting egg cells may receive two X chromosomes (*XX*), while others receive none (fig. 23.22). Both of these types of egg cells can be fertilized by sperm cells, and the zygotes formed can undergo development.

If an *XX* egg cell is fertilized by a *Y*-bearing sperm cell, the resulting zygote will have 47 chromosomes and the genotype *XXY*. This abnormal combination of sex chromosomes causes the development of an individual with male sex organs. However, the testes usually fail to produce sperm cells, so that the person is sterile. This condition, which is called *Klinefelter's syndrome,* is also characterized by tall stature, a somewhat feminine musculature, and partial development of the breasts. (See fig. 23.23.)

An *XX* egg cell fertilized by an X-bearing sperm cell becomes a zygote with the genotype *XXX*. The individual who develops from such a zygote is a relatively normal female, and although some *XXX* females are sterile, others have produced offspring. This condition is sometimes called the *super female.*

Females with genotypes *XXXX* and *XXXXX* are known to exist, and although a consistent pattern of traits is lacking in such individuals, the chances of developing mental retardation seem to increase markedly as the number of *X* chromosomes in the genotype increases.

Fig. 23.23 A person with Klinefelter's syndrome has the sex chromosome genotype XXY. How might such a genotype come about?

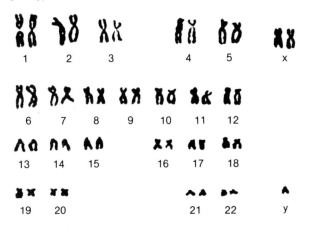

Fig. 23.24 A person with Turner's syndrome has the sex chromosome genotype XO.

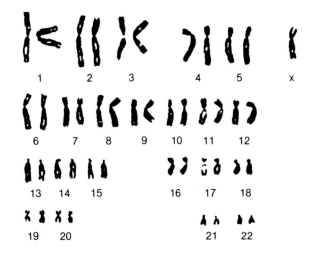

If an egg cell without an *X* chromosome is fertilized by an *X*-bearing sperm cell, the resulting *XO* zygote develops into an individual with *Turner's syndrome*. (See fig. 23.24.) In this instance, the person is female, but usually fails to mature reproductively. The sex organs remain juvenile, the breasts fail to develop, the stature is short, and loose folds of skin appear in the posterior region of the neck.

As they age, older persons seem to develop increasing numbers of body cells with 45 chromosomes. In males, this abnormal number seems to involve a lack of Y chromosomes, and in females, cells tend to miss an X chromosome. These losses apparently result from abnormal mitotic divisions, but their significance to the aging process is unknown.

Autosomal Abnormalities

Nondisjunction sometimes occurs among homologous pairs of autosomes, and as a result, a cell may receive an extra autosome. For example, if nondisjunction occurs during oogenesis, an egg cell may appear with two chromosomes of the same type. If such an egg cell is fertilized by a normal sperm cell, the zygote will have three chromosomes of one kind—a condition that is called *trisomy*.

In humans, *Down's syndrome* results from trisomy in which a zygote receives three autosomes of a particular type (chromosome 21). (See fig. 23.25.) The individual who develops from such a zygote usually has folds around the eyes resembling those of persons with Mongolian ancestry. Consequently, Down's syndrome is sometimes called *mongolism*.

Fig. 23.25 A person with Down's syndrome has three members of chromosome 21.

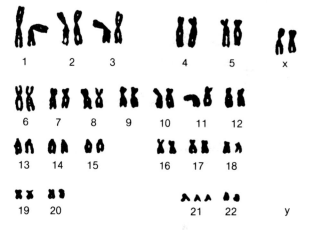

This condition is also characterized by mental retardation, short stature, and stubby hands and feet. In addition, the reproductive organs usually remain underdeveloped, and malformations of the heart may be present.

Although the cause of nondisjunction leading to *Down's syndrome* is unknown, this condition is more likely to occur in the offspring of older mothers than in those of younger mothers. In fact, Down's syndrome occurs only about once in two-thousand live births among women of early childbearing age, but it occurs about once in fifty live births among women over 40 years of age.

Other chromosome abnormalities resulting from triply represented autosomes include *Edward's syndrome* (trisomy 18) in which the fetus develops heart and kidney defects, and *Patau's syndrome* (trisomy 13) in which the infant is usually blind and has heart abnormalities. With either abnormality, the offspring seldom lives longer than a few weeks following birth.

---

1. What is meant by nondisjunction?
2. What are some sex chromosome disorders due to nondisjunction?
3. Define trisomy.
4. What are the consequences of trisomy?

Medical Genetics

Medical genetics is the branch of genetics that is concerned with the relationship between inheritance and disease.

Heredity and Disease

Hereditary diseases comprise a diverse group of disorders including abnormalities of blood cells (such as sickle-cell anemia), defects in blood clotting mechanisms (such as hemophilia), and mental retardation (such as Down's syndrome). The causes of such diseases are called *mutations*—changes in the genes or chromosomes that can be passed from parent to offspring by means of sex cells. (See chapter 4.) Although some of these diseases (such as hemophilia) are due to mutations of single genes, others are caused by chromosomal disorders, as in the trisomy leading to Down's syndrome.

Genetic Counseling

Genetic counseling is a service by which persons, who have produced children with hereditary diseases or who are members of families in which such disorders occur, can obtain information about the chances of genetic disorders being transmitted to their offspring. Sometimes a genetic counselor can make rather precise predictions. For example, a counselor could tell parents who both have sickle-cell trait that their chances of having a child with sickle-cell anemia are one in four or 25%. However, if one parent is normal for hemoglobin and the other parent has sickle-cell anemia, the chance of their producing a child with sickle-cell anemia is zero, although all of their children will inherit sickle-cell trait.

The accuracy of such a prediction depends on a correct diagnosis of the genetic disorder, understanding of the pattern by which the disorder is inherited, and knowledge of the parents' genotypes. Unfortunately, some hereditary traits behave differently in different families. A trait may, for example, act as an autosomal recessive in one family and as a sex-linked recessive in another. Consequently, a genetic counselor may have to determine how a particular trait is transmitted within a family before making a prediction.

In addition, the counselor must be concerned with the psychological effects such a prediction may have upon the individuals involved, and must help them weigh the probability of a genetic defect occurring against the probability of desirable traits that might appear in their child—traits that could compensate to some degree for the possible defect. At the same time, the counselor must try to remain objective and avoid enforcing a particular set of values upon the prospective parents, since the final decision about childbearing is theirs.

Detection of Genetic Disorders

Sometimes it is possible to determine the genotypes of particular individuals by performing blood tests. *Sickle-cell trait*, for example, can be detected by exposing red blood cells to low oxygen concentrations. In the test, a drop of blood is mixed with a substance that deoxygenates it, and if the red cells contain abnormal hemoglobin, they assume a sickle shape.

Similarly, blood tests can be used to reveal carriers (heterozygotes) of the recessive genes involved in the development of the following: *hemophilia*, in which bleeding is prolonged; *Tay-Sach's disease*, which leads to early blindness, deterioration of mental and physical abilities, and death; and *Cooley's anemia* (thalassemia major), which is accompanied by severe defects in the red blood cells and invariably leads to death.

Tay-Sachs disease involves failure to produce an enzyme needed for the normal metabolism of lipids. As a result, lipids accumulate abnormally in nerve cells of the brain and interfere with their functions.

Although this genetic disorder is most common in persons of middle and northern European Jewish descent, it may also occur in non-Jewish populations.

If prospective parents are concerned about the genetic condition of a developing fetus (prenatal diagnosis), information can sometimes be gained by testing fetal cells, obtained by means of *amniocentesis*. In this procedure, which can be performed in a physician's office using sterile techniques, a small quantity of *amniotic fluid* is withdrawn from the amniotic cavity by passing a hollow needle through the abdominal and uterine walls. When this is done between the fourteenth and sixteenth weeks of development, cells of fetal origin (fibroblasts) are usually present in the fluid, and they can be cultured in the laboratory and tested for abnormalities.

If a serious chromosomal disorder, such as the trisomy leading to Down's syndrome, is detected, a therapeutic abortion may be performed to prevent the expression of the genetic defect.

Such fetal cells can also be tested for various hemoglobin disorders, including thalassemia and sickle-cell anemia. Even though fetal fibroblasts obtained by amniocentesis do not synthesize hemoglobin, they contain a complete set of human DNA molecules. These molecules can be removed from the cells, and certain portions of their gene structures can be analyzed. This technique utilizes special enzymes (restriction enzymes) that break the DNA nucleotide sequences within the DNA molecules at particular points, producing precise patterns of fragments. The nature of these fragments reveals the presence of normal or abnormal genes for hemoglobin synthesis.

---

1. *What is medical genetics?*
2. *What is the purpose of genetic counseling?*
3. *How can amniocentesis be used in diagnosing chromosomal disorders?*

Some Hereditary Disorders of Clinical Interest

cri du chat syndrome (krē dō chat sin′ drōm)—characterized by peculiar mewing (cat cry) vocal sounds, microcephaly, and severe mental retardation; due to a partial deletion in one autosome (probably number 5).

cystic fibrosis (sis′tik fi-bro′sis)—characterized by formation of thick mucus in the pancreas and lungs, which interferes with normal digestion and breathing; an expression of autosomal recessive genes.

familial cretinism (fah-mil′e-al kre′tĭ-nizm)—characterized by lack of thyroid secretions due to a defect in the iodine transport mechanism. Untreated children are dwarfed, sterile, and usually mentally retarded; an expression of autosomal recessive genes.

galactosemia (gah-lak″to-se′me-ah)—characterized by an inability to metabolize galactose, a component of milk sugar. This results in cataract, mental retardation, and damaged liver; an expression of autosomal recessive genes.

gout (gowt)—characterized by an accumulation of uric acid in the blood and tissues due to abnormal metabolism of substances called purines; an expression of an autosomal dominant gene.

hepatic porphyria (hĕ-pat′ik por-fēr′e-ah)—characterized by abdominal pain, gastrointestinal disorders, and neurologic disturbances due to abnormal metabolism of substances called porphyrins; an expression of an autosomal dominant gene.

hepatolenticular degeneration (Wilson's disease) (hep″ah-to-len-tik′u-lar de-jen″ĕ-ra′shun)—characterized by an increase in the absorption of copper and the accumulation of copper in the brain and liver. This results in degenerative changes in the brain and cirrhosis of the liver; an expression of autosomal recessive genes.

hereditary hemochromatosis (he-red′ĭ-ter″e he″mo-kro″mah-to′sis)—characterized by an accumulation of iron in the liver, pancreas, and heart. This results in cirrhosis of the liver, diabetes, and heart failure; an expression of a sex-influenced autosomal dominant gene.

hereditary leukomelanopathy (he-red′ĭ-ter″e lu″ko-mel″ah-nop′ah-the)—characterized by decreased pigmentation in the skin, eyes, and hair, and abnormalities of the white blood cells that are accompanied by increased susceptibility to infections and early death; an expression of autosomal recessive genes.

Huntington's chorea (hunt′ing-tunz ko-re′ah)—characterized by uncontrolled twitching of voluntary muscles and deterioration of mental capacities. This condition usually does not appear until after maturity, so affected persons often transmit the mutant gene to their children before the symptoms develop; an expression of an autosomal dominant gene.

Marfan's syndrome (mar-fahnz′ sin′drōm)—characterized by extremities of abnormally great lengths, dislocation of the lenses, and congenital defects of the cardiovascular system; an expression of an autosomal dominant gene.

nephrogenic diabetes insipidus (nef″ro-jen′ik di″ah-be′tēz in-sip′ĭ-dus)—characterized by production of large volumes of very dilute urine and consequent dehydration due to failure of the distal convoluted tubules to reabsorb water in response to ADH; an expression of a sex-linked recessive gene.

phenylketonuria (PKU) (fen″il-ke″to-nu′re-ah)—characterized by an inability to normally metabolize the amino acid, phenylalanine. This results in nerve and brain damage and is accompanied by mental retardation; an expression of autosomal recessive genes.

pseudohypertrophic muscular dystrophy
(su″do-hi″per-trof′ik mus′ku-lar dis′tro-fe)—
characterized by progressive atrophy of the
muscles, which usually begins during childhood
and leads to death in adolescence; an expression
of a sex-linked recessive gene.

retinitis pigmentosa (ret″ĭ-ni′tis pig″men-to′sa)—
characterized by progressive atrophy of the retina
which causes blindness; an expression of a sex-
linked recessive gene.

Tay-Sach's disease (ta saks′ dĭ-zēz′)—
characterized by early blindness, deterioration of
mental and physical abilities, and death; an
expression of autosomal recessive genes.

thalassemia major (Cooley's anemia)
(thal″ah-se′me-ah mā′jĕr)—characterized by very
thin and fragile red blood cells resulting in anemia
and a short life expectancy; an expression of
autosomal recessive genes.

Chapter Summary

Introduction
Individual characteristics result from an interaction of
heredity and environment.
Heredity is the transmission of genetic information
from parents to offspring; environment includes
chemical, physical, and biological factors in the
surroundings.

Genes and Chromosomes
1. The gene
   a. A gene is a portion of a DNA molecule that
      contains information for the production of one
      kind of protein molecule.
   b. Genes within a zygote instruct it to synthesize
      particular proteins, which function as enzymes
      that promote metabolic reactions needed for
      growth and development.
2. Chromosome numbers
   a. Human body cells normally contain 46
      chromosomes.
   b. Twenty-three chromosomes of a zygote are
      received from the female parent and 23 are
      received from the male parent.
3. Chromosome pairs
   a. Chromosomes from male and female parents
      contain complete sets of genetic information.
   b. Within these sets, chromosomes that contain
      similar genetic information comprise
      homologous chromosomes.
   c. During mitosis, daughter cells receive identical
      sets of homologous chromosomes.
4. Gene pairs
   a. Since chromosomes are paired, the genes
      within the chromosomes are paired also.
   b. A pair of genes consists of two alleles.
   c. A zygote that contains a pair of identical
      alleles is homozygous; if the alleles of a pair
      are different, the zygote is heterozygous.

Gene Expression
The combination of genes present in an individual's
cells constitutes a genotype; the appearance of
the individual is its phenotype.
1. Dominant and recessive genes
   a. In the heterozygous condition, if one gene is
      expressed and its allele is unexpressed, the
      expressed gene is called dominant and the
      allele is called recessive.
   b. Dominant genes are symbolized by capital
      letters and recessive genes are symbolized by
      small letters.
   c. Three genotypes are possible in the instance
      of a gene pair involving dominant and
      recessive alleles, but only two phenotypes are
      possible.
2. Incomplete dominance
   a. When the alleles of a gene pair are expressed
      differently and neither is dominant to the other,
      they are called incompletely dominant.
   b. Incompletely dominant alleles result in three
      possible genotypes and three different
      phenotypes.
3. Multiple alleles
   a. When alleles occur in several forms, the group
      of genes is called multiple alleles.
   b. The inheritance of ABO blood types involves
      multiple alleles.
   c. In ABO blood type inheritance, every person
      belongs to one of six genotypes and has one
      of four phenotypes.
4. Polygenic inheritance
   a. Some traits are determined by single gene
      pairs, while others are determined by several
      pairs of genes.
   b. Body stature and concentration of skin
      pigment are examples of traits determined by
      polygenic inheritance.
   c. Polygenic inheritance results in a broad range
      of phenotypes, so that people vary greatly in
      their heights and the concentration of pigment
      in their skins.
5. Environmental factors
   a. Environmental factors influence the ways
      genes are expressed.
   b. Body stature is influenced by genes and also
      by such environmental factors as diet and
      disease.
   c. Concentration of skin pigment is influenced by
      genes and also by exposure to ultraviolet
      light.

Chromosomes and Sex Cell Production
Sex cells carry unique combinations of genes as a
result of meiosis.
1. Meiosis
   a. Meiosis involves two successive divisions
      (1) In first meiotic division, homologous
          chromosomes of the parent cell are
          separated and the chromosome number is
          reduced by ½.

(2) In second meiotic division, the chromosomes act as they do in mitosis.

(3) Meiosis results in the formation of sex cells.

b. Meiosis is a continuous process without marked interruptions between steps, but for convenience it can be divided into stages.

(1) Stages of the first meiotic division include prophase, metaphase, anaphase, and telophase.

(2) Stages of second meiotic division also include prophase, metaphase, anaphase, and telophase.

2. Results of meiosis

a. In spermatogenesis, four sperm cells are produced.

(1) Each sperm cell contains 23 chromosomes.

(2) The genetic information within the sperm cells varies from cell to cell as a result of crossing over.

b. In oogenesis, one egg cell (secondary oocyte) and three nonfunctional polar bodies are produced.

(1) Each egg cell contains 23 chromosomes.

(2) The genetic information within the egg cells varies from cell to cell as a result of crossing over.

Inheritance of Single Traits

1. Dominant and recessive inheritance

a. In pigmentation, the gene for normal pigment is symbolized as *A,* and its recessive allele is symbolized as *a.*

(1) *AA* is homozygous dominant. *Aa* is heterozygous, and *aa* is homozygous recessive.

(2) *AA* and *Aa* individuals appear normal, while *aa* individuals are albino.

b. When the genotypes of parents are known, it is possible to predict the types of offspring that they can produce.

2. Incompletely dominant inheritance

a. Three genotypes and three phenotypes can result from a pair of incompletely dominant genes.

b. When the genotypes of the parents are known, it is possible to predict the types of offspring that they can produce.

Sex Chromosomes

1. Sex inheritance

a. A set of human chromosomes consists of 22 pairs of autosomes and one pair of sex chromosomes.

b. The sex chromosomes are responsible for the development of characteristics associated with male and female.

(1) A male set of sex chromosomes consists of an *X* and a *Y* chromosome.

(2) A female set of sex chromosomes consists of two *X* chromosomes.

c. The sex of an offspring is determined by the type of sperm cell that fertilizes the egg cell.

2. Sex-linked inheritance

a. Sex chromosomes contain genes responsible for certain traits that are unrelated to sex.

b. The patterns in which these traits are inherited differ in each sex.

(1) In males, genes on the single *X* chromosome are expressed even if they are recessive.

(2) In females, such recessive genes are expressed only if they are present on both *X* chromosomes.

c. Traits determined by recessive genes located on *X* chromosomes are called sex-linked.

(1) Color blindness and hemophilia are examples of sex-linked traits.

(2) Males receive sex-linked traits from their mothers.

(3) All the daughters of a male parent expressing a sex-linked trait will receive the recessive gene for that trait.

Chromosome Disorders

Disorders are characterized by the presence of abnormal numbers of chromosomes.

1. Cause of chromosome disorders

a. If homologous chromosomes fail to separate during meiosis, one daughter cell will receive two chromosomes of one type while the other daughter cell will receive no chromosome of that type.

b. Such a failure of homologous chromosomes to separate is called nondisjunction.

2. Sex chromosome disorders

a. Nondisjunction during oogenesis can result in egg cells with two *X* chromosomes and egg cells with no *X* chromosome.

b. Such abnormal egg cells can be fertilized and the resulting zygotes may develop into offspring.

(1) If an *XX* egg cell is fertilized by a *Y*-bearing sperm cell, the resulting *XXY* combination leads to Klinefelter's syndrome.

(2) If an *XX* egg cell is fertilized by an *X*-bearing sperm cell, the resulting *XXX* combination produces a relatively normal female.

(3) If an egg cell without an *X* chromosome is fertilized by an *X*-bearing sperm cell, the resulting *XO* combination leads to Turner's syndrome.

3. Autosomal abnormalities

a. Nondisjunction may occur among autosomes and result in the presence of two chromosomes of the same type within an egg cell.

b. If such an egg cell is fertilized by a normal sperm cell, the resulting zygote will have three chromosomes of one kind, called trisomy.

c. Down's syndrome, Edward's syndrome, and Patau's syndrome are due to triply represented autosomes.

Medical Genetics
1. Heredity and disease
   a. Hereditary diseases include a diverse group of disorders.
   b. Such diseases are caused by mutations.
   c. Parents often seek advice concerning the chances of a hereditary disease affecting their future children.
2. Genetic counseling
   a. A service by which individuals can obtain information about the chances of genetic disorders being transmitted to their offspring is called genetic counseling.
   b. Genetic counselors can sometimes make precise predictions concerning the probability of a future child being affected by an inherited disease.
   c. The accuracy of such predictions depends upon the following:
      (1) Correct diagnosis of the genetic disorder;
      (2) Understanding of the pattern by which the disorder is inherited;
      (3) Knowledge of the parents' genotypes.
3. Detection of genetic disorders
   a. Genotypes can sometimes be determined by means of blood tests.
   b. Fetal cells are sometimes cultured and tested for the presence of chromosome disorders.
   c. Such cells are obtained from amniotic fluid of the pregnant female parent.
   d. If a serious disorder is detected, a therapeutic abortion may be performed to prevent expression of the genetic defect.
   e. Fetal cells can also be tested for the presence of certain genetic disorders.

Application of Knowledge

1. Using the principles of human genetics, how could you support the stand taken by many civil and religious authorities forbidding marriages between persons who are first or second cousins?

2. If a young couple with sickle-cell trait, who have a child with sickle-cell anemia, said to you, "Our first child has sickle-cell anemia, but now we can have three more children before we have another one with this disease," how would you respond?

3. If a 49-year-old woman, who thought she was menopausal and thus failed to seek medical attention when she began missing her menstrual periods, finds that she is five months pregnant, what special kinds of tests are likely to be ordered by her physician? Why? Could such a woman obtain a legal abortion in your state at this stage of pregnancy?

Review Activities

1. Identify two major factors that influence the development of individual characteristics.

2. Define a *gene*.

3. Discuss the origin of the 46 chromosomes in a human zygote.

4. Define *homologous chromosomes*.

5. Distinguish between homozygous and heterozygous.

6. Distinguish between genotype and phenotype.

7. Explain what is meant by a dominant gene and its recessive allele.

8. Define *incomplete dominance*.

9. Explain how ABO blood type inheritance is controlled by multiple alleles.

10. Describe how environmental factors may influence the expression of the genes that control pigmentation of the skin.

11. Outline the process of meiosis.

12. Explain the significance of synapsis during meiosis.

13. Describe the genotypes and phenotypes of the offspring expected from the following parents. (*A* represents the dominant gene for normal pigmentation, and *a* represents its recessive allele.)
    a. *AA* and *Aa*
    b. *Aa* and *aa*
    c. *Aa* and *Aa*

14. Describe the genotypes and phenotypes of the offspring expected from the following parents. (H^1 represents the normal gene for hemoglobin production, and H^2 represents the incompletely dominant allele that causes the formation of abnormal hemoglobin.)
    a. H^1H^1 and H^2H^2
    b. H^1H^1 and H^1H^2
    c. H^1H^2 and H^1H^2

15. Distinguish between autosomes and sex chromosomes.

16. Explain how sex is inherited.

17. Explain why ½ of the human zygotes are expected to develop into males and ½ into females.

18. Define *sex-linked genes*.

19. Explain why sex-linked genes are always expressed more frequently in males than in females.

20. Describe the genotypes and phenotypes of the offspring expected from the following parents:
    a. Color-blind male and normal (homozygous dominant) female.
    b. Normal male and color-blind female.

21. Distinguish between hemophilia A and hemophilia B.

22. Explain how nondisjunction may lead to chromosome disorders.

23. Explain how an individual with a sex chromosome combination of *XXY* might occur.

24. Explain how an individual with a sex chromosome combination of *XO* might occur.

25. Define *autosomal trisomy*.

26. Define *amniocentesis* and explain its use.

Suggestions for Additional Reading

Brady, R. O. February 1973. Hereditary fat-metabolism diseases. *Scientific American.*

Caspersson, T., and Zech, L. September 1972. Chromosome identification by fluorescence. *Hosp. Prac.*

Ferguson-Smith, M. A. April 1970. Chromosomal abnormalities II: sex chromosome defects. *Hosp. Prac.*

Fraser, F. C. January 1971. Genetic counseling, *Hosp. Prac.*

Friedmann, T. November 1971. Prenatal diagnosis of genetic disease. *Scientific American.*

Fuchs, F. June 1980. Genetic amniocentesis. *Scientific American.*

Gardner, E. J. 1972. *Principles of genetics.* 4th ed. New York: Wiley.

German, J. L. 1970. Studying human chromosomes today. *Amer. Scientist* 58:182.

Hirschhorn, K. February 1970. Chromosomal abnormalities I: autosomal defects. *Hosp. Prac.*

McKusick, V. A., and Ruddle, F. H. 1977. The status of the gene map of the human chromosome. *Science* 196:390.

Merrell, D. J. 1975. *An introduction to genetics.* New York: W. W. Norton.

Mertens, T. R. 1975. *Human genetics: readings on the implications of genetics engineering.* New York: Wiley.

Moody, P. A. 1975. *Genetics in man.* 2nd ed. New York: W. W. Norton.

Motulsky, A. G. 1974. Brave new world? Screening for phenylketonuria and related conditions. *Science* 185:656.

Nyhan, W. L. 1976. *The heredity factor: genes, chromosomes, and you.* New York: Grosset and Dunlap.

Ruddle, F. H., and Kucherlapati, R. S. July 1974. Hybrid cells and human genes. *Scientific American.*

Scheinfeld, A. 1972. *Heredity in humans.* New York: Lippincott.

Thompson, J. S., and Thompson, M. W. 1973. *Genetics in medicine.* Philadelphia: W. B. Saunders.

Winchester, A. M. 1971. *Human genetics.* 2nd ed. Englewood Cliffs, N.J.: Prentice-Hall.

Appendix A
Cellular Respiration

As is discussed in chapter 4, during cellular respiration a 6-carbon glucose molecule is decomposed by enzymatic actions into two 3-carbon pyruvic acid molecules. This phase of respiration is called *glycolysis*, which means the splitting of glucose.

Glycolysis involves a series of chemical reactions, as illustrated in figure A.1. In the first step of this series, a phosphate group from an ATP molecule is added to the glucose molecule, forming glucose-6-phosphate. The atoms within the glucose-6-phosphate molecule are then rearranged slightly to form fructose-6-phosphate, and a phosphate group from a second ATP is added to it to produce fructose-1,6-phosphate. These beginning steps require the conversion of ATP molecules to ADP, and thus they involve the use of cellular energy.

In the next step, fructose-1,6-phosphate is changed into two molecules of glyceraldehyde-3-phosphate. Each of these molecules, in turn, is converted to pyruvic acid in the following manner:

a. An inorganic phosphate group is added to glyceraldehyde-3-phosphate to form 1,3-diphosphoglyceric acid. At the same time, two hydrogen atoms are released.

b. 1,3-diphosphoglyceric acid is changed to 3-phosphoglyceric acid. As this occurs, some energy in the form of a high energy phosphate is transferred from the 1,3-diphosphoglyceric acid to an ADP molecule, converting the ADP to ATP.

c. A slight alteration of 3-phosphoglyceric acid occurs to form 2-phosphoglyceric acid.

d. A change in 2-phosphoglyceric acid converts it into phosphoenolpyruvic acid.

e. Finally, a high energy phosphate is transferred from the phosphoenolpyruvic acid to an ADP molecule, converting it to ATP. A molecule of pyruvic acid remains.

Since two molecules of glyceraldehyde-3-phosphate are involved, two molecules of pyruvic acid are produced by these reactions. Also a total of four hydrogen atoms are released (step *a*) and four ATP molecules are formed (two in step *b* and two in step *e*). However, because two molecules of ATP are utilized early in glycolysis, there is a net gain of only two ATP molecules during this phase of cellular respiration.

In the absence of oxygen, the resulting pyruvic acid molecules may be converted into lactic acid, as is discussed in chapter 9. In the presence of oxygen, however, each pyruvic acid molecule is combined with a molecule of coenzyme A (obtained from the vitamin, pantothenic acid) to form the substance called acetyl coenzyme A. As this occurs, two hydrogen atoms are released for each molecule of acetyl coenzyme A formed. The acetyl coenzyme A is then broken down by means of the citric acid cycle (Kreb's cycle), which is illustrated in figure A.2.

An acetyl coenzyme A molecule enters the citric acid cycle by combining with a molecule of oxaloacetic acid to form citric acid. The citric acid is then changed by a series of reactions back into oxaloacetic acid, and the cycle is completed.

As citric acid is produced, coenzyme A is released and thus can be used again and again in the formation of acetyl coenzyme A from pyruvic acid molecules.

During various steps in the citric acid cycle, carbon dioxide and hydrogen atoms are released. More specifically, for each glucose molecule metabolized in the presence of oxygen, two molecules of acetyl coenzyme A enter the citric acid cycle, and as a result of the cycle, four carbon dioxide molecules and sixteen hydrogen atoms are released. At the same time, two more molecules of ATP are formed.

The released carbon dioxide dissolves in the cellular fluid and is transported by the blood. Most of the hydrogen atoms released from the citric acid cycle, and also those released during glycolysis and during the formation of acetyl coenzyme A, react immediately with a substance called nicotinamide adenine dinucleotide (NAD), which is obtained from the vitamin, niacin. NAD serves as a hydrogen carrier, and in the next step of respiration the hydrogen is changed into hydrogen ions and is oxidized.

The oxidation process involves a series of enzymatically controlled reactions in which oxygen supplied from the blood is changed to hydroxyl ions (OH^-), and these ions combine with the hydrogen ions to form water. As this occurs, a large amount of energy is released, and much of it is used to synthesize ATP molecules.

As a result of oxidizing the hydrogen released during the metabolism of a glucose molecule, thirty-four ATP molecules are formed. Also, since there is a net gain of two ATP molecules from glycolysis, and two more are produced as a result of the citric acid cycle, a total of thirty-eight ATP molecules result from the breakdown of each glucose molecule.

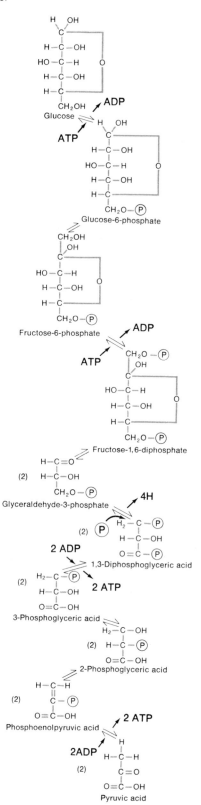

Fig. A.2 Chemical reactions of the citric acid cycle.

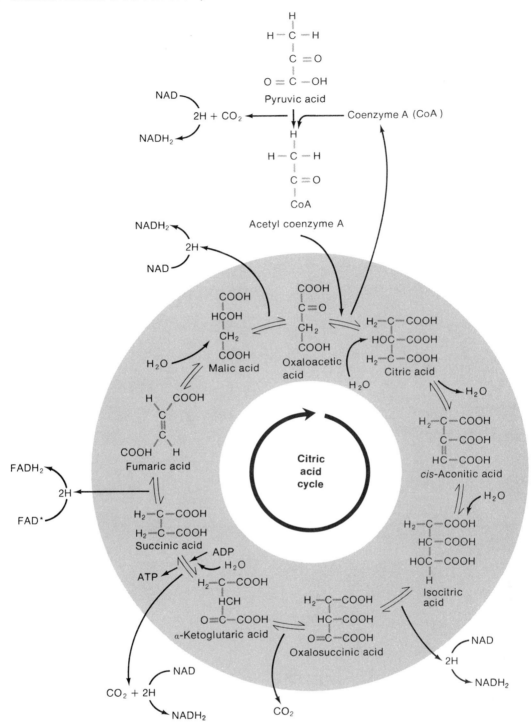

*At this point in the cycle, the hydrogen₂ carrier is FAD (flavine adenine dinucleotide).

Appendix B
Hydrogen Ion Concentration

As is discussed in chapter 2, the concentration of *hydrogen ions* in a solution can be expressed in grams of ions per liter or in pH units, in which the pH value is equal to the negative logarithm of the hydrogen ion concentration. For example, a solution with a hydrogen ion concentration of 0.1 grams per liter has a pH value of 1.0; a concentration of 0.01 g H^+/l has pH 2.0; 0.001 g H^+/l has pH 3.0, and so forth. Thus, between each whole number on the pH scale (which extends from pH 0 to pH 14.0), there is a tenfold difference in the hydrogen ion concentration, which decreases as pH rises.

In pure water, which ionizes only slightly, the hydrogen ion concentration is 0.0000001 g/l, and the pH is 7.0. Since water ionizes to release equal numbers of acidic hydrogen ions (H^+) and basic hydroxyl ions (OH^-), it is said to be *neutral*.

$$H_2O \rightarrow H^+ + OH^-$$

Therefore, solutions with more hydrogen ions than hydroxyl ions are said to be *acidic;* they have pH values of less than 7.0. Solutions with fewer hydrogen ions than hydroxyl ions are said to be *basic* (alkaline); they have pH values of more than 7.0. Chart A.1 illustrates the relationship between hydrogen ion concentration and pH.

Chart A.1 Hydrogen ion concentrations and pH

Grams of H^+ per liter	pH	
1.0	0	
0.1	1	
0.01	2	
0.001	3	Increasingly acidic
0.0001	4	
0.00001	5	
0.000001	6	
0.0000001	7	Neutral—neither acidic
0.00000001	8	nor basic
0.000000001	9	
0.0000000001	10	
0.00000000001	11	Increasingly basic
0.000000000001	12	
0.0000000000001	13	
0.00000000000001	14	

Appendix C
Recording an Action Potential

As is described in chapter 10, a sequence of membrane potential changes called an **action potential** occurs whenever a polarized nerve (or muscle cell) membrane is disturbed by a stimulus of threshold intensity or above.

Because an action potential occurs in about 1/1000 second, an instrument that responds rapidly is needed to record the sequence of changes. The instrument usually used is the **cathode ray oscilloscope.** This instrument includes a *cathode ray tube* that looks and functions somewhat like the picture tube of a television. (See fig. A.3.) It contains an *electron gun,* which fires a concentrated beam of electrons from the neck toward the face of the tube. The face is coated with a thin layer of fluorescent material that produces a spot of light where the electron beam strikes it.

Horizontal deflection plates are located on either side of the electron beam within the cathode ray tube. These plates can be electrically charged by the action of a *sweep circuit* of the oscilloscope. This circuit causes one of the horizontal deflection plates to become positively charged while the other becomes negatively charged. When this happens, the beam of electrons (which are negatively charged) is drawn toward the positive plate and repelled by the negative one. As a result, the electron beam moves across the face of the tube and produces a glowing horizontal line, called a *trace.*

The sweep circuit can be adjusted to move the electron beam across the face of the cathode ray tube from left to right at a known velocity. Each time the beam reaches the right edge of the tube, it almost instantly jumps back to the left side and begins a new horizontal trace.

Fig. A.3 A cathode ray oscilloscope can be used to record the action potential of a nerve fiber.

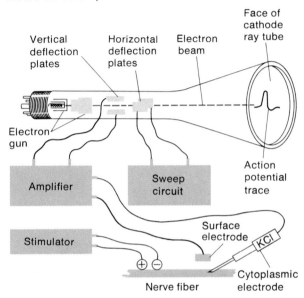

Other plates called *vertical deflection plates* are located above and below the electron beam within the cathode ray tube. If these plates become electrically charged, the electron beam moves up or down.

When the oscilloscope is being used to record an action potential of a nerve fiber, electrodes from an electronic *amplifier* are placed in contact with the fiber, and the amplifier is connected to the vertical deflection plates. Any change in nerve fiber membrane potential is detected by the electrodes, and a signal is transmitted to the amplifier. The amplifier increases the intensity of the signal in direct proportion to the amount of change occurring in the membrane potential and causes the vertical deflection

Fig. A.4 An action potential as it might be recorded on the face of a cathode ray tube.

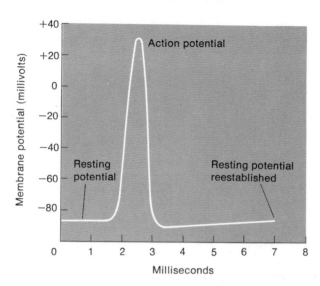

plates to become charged. If this happens when the electron beam is moving across the face of the cathode ray tube, its horizontal trace will be deflected up or down in direct proportion to the change in membrane potential detected by the electrodes.

Figure A.4 illustrates an action potential (monophasic) of the type recorded on an oscilloscope when one electrode is placed on the surface of a nerve fiber membrane and the other electrode is inserted into the cytoplasm of the fiber. Note that the fiber's resting potential is about −85 millivolts. As the action potential occurs following stimulation of the fiber, the potential momentarily becomes positive, and then the negative resting potential is reestablished.

Appendix D
Units of Measurement and Their Equivalents

Apothecaries' Weights and Their Metric Equivalents

1 grain (gr) =
0.05 scruple (s)
0.017 dram (dr)
0.002 ounce (oz)
0.0002 pound (lb)
0.065 gram (g)
65. milligrams (mg)

1 scruple (s) =
20. grains (gr)
0.33 dram (dr)
0.042 ounce (oz)
0.004 pound (lb)
1.3 grams (g)
1,300. milligrams (mg)

1 dram (dr) =
60. grains (gr)
3. scruples (s)
0.13 ounce (oz)
0.010 pound (lb)
3.9 grams (g)
3,900. milligrams (mg)

1 ounce (oz) =
480. grains (gr)
24. scruples (s)
8. drams (dr)
0.08 pound (lb)
31.1 grams (g)
31,100. milligrams (mg)

1 pound (lb) =
5,760. grains (gr)
288. scruples (s)
96. drams (dr)
12. ounces (oz)
373. grams (g)
373,000. milligrams (mg)

Apothecaries' Volumes and Their Metric Equivalents

1 minim (min) =
0.017 fluid dram (fl dr)
0.002 fluid ounce (fl oz)
0.0001 pint (pt)
0.06 milliliter (ml)
0.06 cubic centimeter (cc)

1 fluid dram (fl dr) =
60. minims (min)
0.13 fluid ounce (fl oz)
0.008 pint (pt)
3.70 milliliters (ml)
3.70 cubic centimeters (cc)

1 fluid ounce (fl oz) =
480. minims (min)
8. fluid drams (fl dr)
0.06 pint (pt)
29.6 milliliters (ml)
29.6 cubic centimeters (cc)

1 pint (pt) =
7,680. minims (min)
128. fluid drams (fl dr)
16. fluid ounces (fl oz)
473. milliliters (ml)
473. cubic centimeters (cc)

Metric Weights and Their Apothecaries' Equivalents

1 gram (g) =
0.001 kilogram (kg)
1,000. milligrams (mg)
1,000,000. micrograms (μg)
15.4 grains (gr)
0.032 ounce (oz)

1 kilogram (kg) =
1,000. grams (g)
1,000,000. milligrams (mg)
1,000,000,000. micrograms (μg)
32. ounces (oz)
2.7 pounds (lb)

1 milligram (mg) =
0.000001 kilogram (kg)
0.001 gram (g)
1,000. micrograms (μg)
0.0154 grains (gr)
0.000032 ounce (oz)

Metric Volumes and Their Apothecaries' Equivalents

1 liter (l) =
1,000. milliliters (ml)
1,000. cubic centimeters (cc)
2.1 pints (pt)
270. fluid drams (fl dr)
34. fluid ounces (fl oz)

1 milliliter (ml) =
0.001 liter (l)
1. cubic centimeter (cc)
16.2 minims (min)
0.27 fluid dram (fl dr)
0.034 fluid ounce (fl oz)

Approximate Equivalents of Household Measures

1 teaspoon (tsp) =
4. milliliters (ml)
4. cubic centimeters (cc)
1. fluid dram (fl dr)

1 tablespoon (tbsp) =
15. milliliters (ml)
15. cubic centimeters (cc)
0.5 fluid ounce (fl oz)
3.7 teaspoons (tsp)

1 cup (c) =
240. milliliters (ml)
240. cubic centimeters (cc)
8. fluid ounces (fl oz)
0.5 pint (pt)
16. tablespoons (tbsp)

1 quart (qt) =
960. milliliters (ml)
960. cubic centimeters (cc)
2. pints (pt)
4. cups (c)
32. fluid ounces (fl oz)

Conversion of Units
from One Form to Another

Refer to the preceding equivalency lists when converting one unit to another equivalent unit.

To convert a unit shown in bold type to one of the equivalent units listed immediately below it, multiply the first number (bold type unit) by the appropriate equivalent unit listed below it.

Sample problems:

1. Convert 320 grains into scruples (1 gr = 0.05 s).

$$320 \text{ gr} \times \frac{0.05 \text{ s}}{1 \text{ gr}} = 16.0 \text{ s}$$

2. Convert 320 grains into drams (1 gr = 0.017 dr).

$$320 \text{ gr} \times \frac{0.017 \text{ dr}}{1 \text{ gr}} = 5.44 \text{ dr}$$

3. Convert 320 grains into grams (1 gr = 0.065 g).

$$320 \text{ gr} \times \frac{0.065 \text{ g}}{1 \text{ gr}} = 20.8 \text{ g}$$

Body Temperatures
in °Fahrenheit and °Celsius

°F	°C	°F	°C
95.0	35.0	100.0	37.8
95.2	35.1	100.2	37.9
95.4	35.2	100.4	38.0
95.6	35.3	100.6	38.1
95.8	35.4	100.8	38.2
96.0	35.5	101.0	38.3
96.2	35.7	101.2	38.4
96.4	35.8	101.4	38.6
96.6	35.9	101.6	38.7
96.8	36.0	101.8	38.8
97.0	36.1	102.0	38.9
97.2	36.2	102.2	39.0
97.4	36.3	102.4	39.1
97.6	36.4	102.6	39.2
97.8	36.6	102.8	39.3
98.0	36.7	103.0	39.4
98.2	36.8	103.2	39.6
98.4	36.9	103.4	39.7
98.6	37.0	103.6	39.8
98.8	37.1	103.8	39.9
99.0	37.2	104.0	40.0
99.2	37.3	104.2	40.1
99.4	37.4	104.4	40.2
99.6	37.6	104.6	40.3
99.8	37.7	104.8	40.4
		105.0	40.6

To convert °F to °C
Subtract 32 from °F and multiply by 5/9.

$$\underline{\hspace{2cm}} \text{ °F} - 32 \times 5/9 =$$
$$\underline{\hspace{2cm}} \text{ °C}$$

To convert °C to °F
Multiply °C by 9/5 and add 32.

$$\underline{\hspace{2cm}} \text{ °C} \times 9/5 + 32 =$$
$$\underline{\hspace{2cm}} \text{ °F}$$

Appendix E
Some Laboratory Tests of Clinical Importance

Common Tests Performed on Blood

Test	Normal Values (adult)	Clinical Significance
Acetone and acetoacetate (serum)	0.3–2.0 mg/100 ml	Values increase in diabetic acidosis, toxemia of pregnancy, fasting, and high-fat diet.
Albumin-globulin ratio or A/G ratio (serum)	1.5:1 to 2.5:1	Ratio of albumin to globulin is lowered in kidney diseases and malnutrition.
Albumin (serum)	3.2–5.5 gm/100 ml	Values increase in multiple myeloma and decrease with proteinuria and as a result of severe burns.
Ammonia (plasma)	50–170 μg/100 ml	Values increase in severe liver disease, pneumonia, shock, and congestive heart failure.
Amylase (serum)	80–160 Somogyi units/100 ml	Values increase in acute pancreatitis, intestinal obstructions, and mumps. They decrease in chronic pancreatitis, cirrhosis of the liver, and toxemia of pregnancy.
Bilirubin, total (serum)	0.3–1.1 mg/100 ml	Values increase in conditions causing red blood cell destruction or biliary obstruction.
Blood urea nitrogen or BUN (plasma or serum)	10–20 mg/100 ml	Values increase in various kidney disorders and decrease in liver failure and during pregnancy.
Calcium (serum)	9.0–11.0 mg/100 ml	Values increase in hyperparathyroidism, hypervitaminosis D, and respiratory conditions that cause a rise in CO_2 concentration. They decrease in hypoparathyroidism, malnutrition, and severe diarrhea.
Carbon dioxide (serum)	24–30 mEq/l	Values increase in respiratory diseases, intestinal obstruction, and vomiting. They decrease in acidosis, nephritis, and diarrhea.
Chloride (serum)	96–106 mEq/l	Values increase in nephritis, Cushing's syndrome, and hyperventilation. They decrease in diabetic acidosis, Addison's disease, diarrhea, and following severe burns.
Cholesterol, total (serum)	150–250 mg/100 ml	Values increase in diabetes mellitus and hypothyroidism. They decrease in pernicious anemia, hyperthyroidism, and acute infections.

Test	Normal Values (adult)	Clinical Significance
Creatine phosphokinase or CPK (serum)	Men: 0–20 IU/l Women: 0–14 IU/l	Values increase in myocardial infarction and skeletal muscle diseases such as muscular dystrophy.
Creatine (serum)	0.2–0.8 mg/100 ml	Values increase in muscular dystrophy, nephritis, severe damage to muscle tissue, and during pregnancy.
Creatinine (serum)	0.7–1.5 mg/100 ml	Values increase in various kidney diseases.
Erythrocyte count or red cell count (whole blood)	Men: 4,600,000–6,200,000/cu mm Women: 4,200,000–5,400,000/cu mm Children: 4,500,000–5,100,000/cu mm (varies with age)	Values increase as a result of severe dehydration or diarrhea and decrease in anemia, leukemia, and following severe hemorrhage.
Fatty acids, total (serum)	190–420 mg/100 ml	Values increase in diabetes mellitus, anemia, kidney disease, and hypothyroidism. They decrease in hyperthyroidism.
Globulin (serum)	2.5–3.5 gm/100 ml	Values increase as a result of chronic infections.
Glucose (plasma)	70–115 mg/100 ml	Values increase in diabetes mellitus, liver diseases, nephritis, hyperthyroidism, and pregnancy. They decrease in hyperinsulinism, hypothyroidism, and Addison's disease.
Hematocrit (whole blood)	Men: 40–54 ml/100 ml Women: 37–47 ml/100 ml Children: 35–49 ml/100 ml (varies with age)	Values increase in polycythemia due to dehydration or shock. They decrease in anemia and following severe hemorrhage.
Hemoglobin (whole blood)	Men: 14–18 gm/100 ml Women: 12–16 gm/100 ml Children: 11.2–16.5 gm/100 ml (varies with age)	Values increase in polycythemia, obstructive pulmonary diseases, congestive heart failure, and at high altitudes. They decrease in anemia, pregnancy, and as a result of severe hemorrhage or excessive fluid intake.
Iron (serum)	75–175 µg/100 ml	Values increase in various anemias and liver disease. They decrease in iron deficiency anemia.
Iron-binding capacity (serum)	250–410 µg/100 ml	Values increase in iron deficiency anemia and pregnancy. They decrease in pernicious anemia, liver disease, and chronic infections.
Lactic acid (whole blood)	6–16 mg/100 ml	Values increase with muscular activity and in congestive heart failure, severe hemorrhage, and shock.
Lactic dehydrogenase or LDH (serum)	90–200 milliunits/ml	Values increase in pernicious anemia, myocardial infarction, liver diseases, acute leukemia, and widespread carcinoma.
Lipids, total (serum)	450–850 mg/100 ml	Values increase in hypothyroidism, diabetes mellitus, and nephritis. They decrease in hyperthyroidism.
Oxygen saturation (whole blood)	Arterial: 94–100% Venous: 60–85%	Values increase in polycythemia and decrease in anemia and obstructive pulmonary diseases.

Test	Normal Values (adult)	Clinical Significance
pH (whole blood)	7.35–7.45	Values increase due to vomiting, Cushing's syndrome, and hyperventilation. They decrease as a result of hypoventilation, severe diarrhea, Addison's disease, and diabetic acidosis.
Phosphatase, acid (serum)	1.0–5.0 King-Armstrong units/ml	Values increase in cancer of the prostate gland, hyperparathyroidism, certain liver diseases, myocardial infarction, and pulmonary embolism.
Phosphatase, alkaline (serum)	5–13 King-Armstrong units/ml	Values increase in hyperparathyroidism (and in other conditions that promote resorption of bone), liver diseases, and pregnancy.
Phospholipids (serum)	6–12 mg/100 ml as lipid phosphorus	Values increase in diabetes mellitus and nephritis.
Phosphorus (serum)	3.0–4.5 mg/100 ml	Values increase in kidney diseases, hypoparathyroidism, acromegaly, and hypervitaminosis D. They decrease in hyperparathyroidism.
Platelet count (whole blood)	150,000–350,000/cu mm	Values increase in polycythemia and certain anemias. They decrease in acute leukemia and aplastic anemia.
Potassium (serum)	3.5–5.0 mEq/l	Values increase in Addison's disease, hypoventilation, and conditions that cause severe cellular destruction. They decrease in diarrhea, vomiting, diabetic acidosis, and chronic kidney disease.
Protein, total (serum)	6.0–8.0 gm/100 ml	Values increase in severe dehydration and shock. They decrease in severe malnutrition and hemorrhage.
Protein-bound iodine or PBI (serum)	3.5–8.0 μg/100 ml	Values increase in hyperthyroidism and liver disease. They decrease in hypothyroidism.
Prothrombin time (serum)	12–14 sec (one stage)	Values increase in certain hemorrhagic diseases, liver disease, vitamin K deficiency, and following the use of various drugs.
Sedimentation rate, Westergren (whole blood)	Men: 0–15 mm/hr Women: 0–20 mm/hr	Values increase in infectious diseases, menstruation, pregnancy, and as a result of severe tissue damage.
Sodium (serum)	136–145 mEq/l	Values increase in nephritis and severe dehydration. They decrease in Addison's disease, myxedema, kidney disease, and diarrhea.
Thyroxine or T_4 (serum)	2.9–6.4 μg/100 ml	Values increase in hyperthyroidism and pregnancy. They decrease in hypothyroidism.
Thromboplastin time, partial (plasma)	35–45 sec	Values increase in deficiencies of blood factors VIII, IX, and X.
Transaminases or SGOT (serum)	5–40 units/ml	Values increase in myocardial infarction, liver disease, and diseases of skeletal muscles.
Uric acid (serum)	Men: 2.5–8.0 mg/100 ml Women: 1.5–6.0 mg/100 ml	Values increase in gout, leukemia, pneumonia, toxemia of pregnancy, and as a result of severe tissue damage.

Test	Normal Values *(adult)*	Clinical Significance
White blood cell count, differential (whole blood)	Neutrophils 54–62% Eosinophils 1–3% Basophils 0–1% Lymphocytes 25–33% Monocytes 3–7%	Neutrophils increase in bacterial diseases; lymphocytes and monocytes increase in viral diseases; eosinophils increase in collagen diseases, allergies, and in the presence of intestinal parasites.
White blood cell count, total (whole blood)	5,000–10,000/cu mm	Values increase in acute infections, acute leukemia, and following menstruation. They decrease in aplastic anemia and as a result of drug toxicity.

Common Tests Performed on Urine

Test	Normal Values	Clinical Significance
Acetone and acetoacetate	0	Values increase in diabetic acidosis.
Albumin, qualitative	0 to trace	Values increase in kidney disease, hypertension, and heart failure.
Ammonia	20–70 mEq/l	Values increase in diabetes mellitus and liver diseases.
Bacterial count	Under 10,000/ml	Values increase in urinary tract infection.
Bile and bilirubin	0	Values increase in melanoma and biliary tract obstruction.
Calcium	Under 250 mg/24 hr	Values increase in hyperparathyroidism and decrease in hypoparathyroidism.
Creatinine clearance	100–140 ml/min	Values increase in renal diseases.
Creatinine	1–2 gm/24 hr	Values increase in infections and decrease in muscular atrophy, anemia, leukemia, and kidney diseases.
Glucose	0	Values increase in diabetes mellitus and various pituitary gland disorders.
17-hydroxycorticosteroids	2–10 mg/24 hr	Values increase in Cushing's syndrome and decrease in Addison's disease.
Phenylpyruvic acid	0	Values increase in phenylketonuria.
Urea clearance	Over 40 ml blood cleared of urea/min	Values increase in renal diseases.
Urobilinogen	0–4 mg/24 hr	Values increase in liver diseases and hemolytic anemia. They decrease in complete biliary obstruction and severe diarrhea.
Urea	25–35 gm/24 hr	Values increase as a result of excessive protein breakdown. They decrease as a result of impaired renal function.
Uric acid	0.6–1.0 gm/24 hr as urate	Values increase in gout and decrease in various kidney diseases.

Glossary

The words in this Glossary are followed by a phonetic guide to pronunciation. The markings in this guide may be confusing at first, because they are unlike the diacritical marks that appear in an ordinary dictionary. This is a simplified system that is standard in medical usage and terminology.

Any unmarked vowel that ends a syllable or stands alone as a syllable is long. In this system, then, the word *play* would be spelled *pla.*

Any unmarked vowel that is followed by a consonant has the short sound. The word *tough,* for instance, would be spelled *tuf.*

If a long vowel does appear in the middle of a syllable (followed by a consonant), then it is marked with the macron (ˉ), the standard sign for a long vowel. For instance, the word *plate* would be phonetically spelled plāt.

Similarly, if a vowel stands alone or ends a syllable, but should have the short sound, it is marked with a breve (ˇ).

abdomen (ab-do′men)   Portion of the body between the diaphragm and the pelvis.

abduction (ab-duk′shun)   Movement of a body part away from the midline.

absorption (ab-sorp′shun)   The taking in of substances by cells or membranes.

accessory organs (ak-ses′o-re or′ganz)   Organs that supplement the functions of other organs; accessory organs of the digestive and reproductive systems.

accommodation (ah-kom″o-da′shun)   Adjustment of the lens for close vision.

acetone (as′e-tōn)   One of the ketone bodies produced as a result of the oxidation of fats.

acetylcholine (as″ĕ-til-ko′lēn)   Substance secreted at the axon ends of many neurons that transmits a nerve impulse across a synapse.

acetyl coenzyme A (as′ĕ-til ko-en′zīm)   An intermediate compound produced during the oxidation of carbohydrates and fats.

Achilles tendon (ah-kil′ēz ten′don)   Tendon in the back of the heel that connects muscles in the posterior lower leg to the calcaneus.

acid (as′id)   A substance that ionizes in water to release hydrogen ions.

acidosis (as″ĭ-do′sis)   Condition in which there is a relative increase in the acid content of body fluids.

ACTH   Adrenocorticotropic hormone.

actin (ak′tin)   A protein in a muscle fiber that, together with myosin, is responsible for contraction and relaxation.

action potential (ak′shun po-ten′shal)   The sequence of electrical changes occurring when a nerve cell membrane is exposed to a stimulus that exceeds its threshold.

activation energy (ak″tĭ-va′shun en′er-je)   Energy needed to initiate a chemical reaction.

active site (ak′tiv sīt)   Region of an enzyme molecule that combines temporarily with a substrate.

active transport (ak′tiv trans′port)   Process that requires an expenditure of energy to move a substance across a cell membrane; usually moved against the concentration gradient.

acupuncture (ak′u-pungk″chūr)   Procedure in which needles are inserted into various tissues to control pain sensations.

adaptation (ad″ap-ta′shun)   Adjustment to environmental conditions.

adduction (ah-duk′shun)   Movement of a body part toward the midline.

adenoids (ad′ē-noids)   The pharyngeal tonsils located in the nasopharynx.

adenosine diphosphate (ah-den′o-sēn di-fos′fāt)   ADP; molecule created when the terminal phosphate is lost from a molecule of adenosine triphosphate.

adenosine triphosphate (ah-den′o-sēn tri-fos′fāt)   An organic molecule that stores energy and releases energy for use in cellular processes.

ADH   Antidiuretic hormone.

adipose tissue (ad′ĭ-pōs tish′u)   Fat-storing tissue.

adolescence (ad″o-les′ens)   Period of life between puberty and adulthood.

ADP   Adenosine diphosphate.

adrenal cortex (ah-dre′nal kor′teks)   The outer portion of the adrenal gland.

adrenal glands (ah-dre′nal glandz)   Endocrine glands located on the tops of the kidneys.

adrenaline (ah-dren′ah-lin)   Epinephrine.

adrenal medulla (ah-dre′nal me-dul′ah)   The inner portion of the adrenal gland.

adrenergic fiber (ad″ren-er′jik fi′ber)   A nerve fiber that secretes norepinephrine at the terminal end of its axon.

adrenocorticotropic hormone (ah-dre′no-kor″te-ko-trōp′ik hōr′mōn)   ACTH; hormone secreted by the anterior lobe of the pituitary gland that stimulates activity in the adrenal cortex.

adulthood (ah-dult′hood)   Period of life that extends from adolescence to old age.

aerobic respiration (a″er-ōb′ik res″pĭ-ra′shun)   Phase of cellular respiration that requires the presence of oxygen.

afferent arteriole (af'er-ent ar-te're-ōl) Vessel that supplies blood to the glomerulus of a nephron within the kidney.

agglutination (ah-gloo''ti-na'shun) Clumping together of blood cells in response to a reaction between an agglutinin and an agglutinogen.

agglutinin (ah-gloo'ti-nin) A substance that reacts with an agglutinogen; an antibody.

agglutinogen (ag''loo-tin'o-jen) A substance that stimulates the formation of agglutinins; a foreign substance.

agranulocytes (a-gran'u-lo-sit) A nongranular leukocyte.

albumin (al-bu'min) A plasma protein that helps to regulate the osmotic concentration of the blood.

aldosterone (al-dos'ter-ōn) A hormone, secreted by the adrenal cortex, that functions in regulating sodium and potassium concentrations and water balance.

alimentary canal (al''i-men'tar-e kah-nal') The tubular portion of the digestive tract that leads from the mouth to the anus.

alkaline (al'kah-lin) Pertaining to or having the properties of a base or alkali.

alkaloid (al'kah-loid) A group of organic substances that are usually bitter in taste and have toxic effects.

alkalosis (al''kah-lo'sis) Condition in which there is a relative increase in the alkaline content of body fluids.

allantois (ah-lan'to-is) A structure that appears during embryonic development and functions in the formation of umbilical blood vessels.

alleles (ah-lēls') Genes that occupy corresponding positions on homologous chromosomes.

allergen (al'er-jen) A foreign substance capable of stimulating an allergic reaction.

all-or-none response (al'or-nun' re-spons') Phenomenon in which a muscle fiber contracts completely when it is exposed to a stimulus of threshold strength.

alpha-tocopherol (al''fah-to-kof'er-ol) Vitamin E.

alveolar ducts (al-ve'o-lar dukts') Fine tubes that carry air to the air sacs of the lungs.

alveolar pores (al-ve'o-lar pōrz) Minute openings in the walls of air sacs, which permit air to pass from one alveolus to another.

alveolus (al-ve'o-lus) An air sac of a lung; a saclike structure.

amine (am'in) A type of nitrogen-containing organic compound, including the hormones secreted by the adrenal medulla.

amino acid (ah-me'no as'id) An organic compound of relatively small molecular size that contains an amino group ($-NH_2$) and a carboxyl group ($-COOH$); the structural unit of a protein molecule.

amniocentesis (am''ne-o-sen-te'sis) A procedure in which a sample of amniotic fluid is removed through the abdominal wall of a pregnant woman.

amnion (am'ne-on) An embryonic membrane that encircles a developing fetus and contains amniotic fluid.

amniotic cavity (am''ne-ot'ik kav'ĭ-te) Fluid-filled space enclosed by the amnion.

amniotic fluid (am''ne-ot'ik floo'id) Fluid within the amniotic cavity that surrounds the developing fetus.

ampulla (am-pul'ah) An expansion at the end of each semicircular canal that contains a crista ampullaris.

amylase (am'i-lās) An enzyme that functions to hydrolyze starch.

anabolic metabolism (an''ah-bol'ik mē-tab'o-lizm) Metabolic process by which larger molecules are formed from smaller ones; anabolism.

anaerobic respiration (an-a''er-ōb'ik res''pĭ-ra'shun) Phase of cellular respiration that occurs in the absence of oxygen.

anal canal (a'nal kah-nal') The last 2 or 3 inches of the large intestine that opens to the outside as the anus.

anaphase (an'ah-fāz) Stage in mitosis during which duplicate chromosomes move to opposite poles of the cell.

anaplasia (an''ah-pla'ze-ah) A change in which mature cells become more primitive.

anastomosis (ah-nas''to-mo'sis) A union of nerve fibers or blood vessels to form an intercommunicating network.

anatomy (ah-nat'o-me) Branch of science dealing with the form and structure of body parts.

androgen (an'dro-jen) A male sex hormone such as testosterone.

anemia (ah-ne'me-ah) A condition characterized by a deficiency of red blood cells or of hemoglobin.

angiotensin (an''je-o-ten'sin) A vasoconstricting substance that is produced when blood flow to the kidneys is reduced, causing an increase in the blood pressure.

anoxia (an-ok'se-ah) Condition in which the oxygen concentration of the tissues is abnormally low.

antagonist (an-tag'o-nist) A muscle that acts in opposition to a prime mover.

antebrachium (an''te-bra'ke-um) The forearm.

antecubital (an''te-ku'bi-tal) The region in front of the elbow joint.

anterior (an-te're-or) Pertaining to the front; the opposite of posterior.

anterior pituitary (an-te're-or pi-tu''i-tār''e) The front lobe of the pituitary gland.

antibody (an'tĭ-bod''e) A specific substance produced by cells in response to the presence of an antigen; it reacts with the antigen.

anticoagulant (an''tĭ-ko-ag'u-lant) A substance that inhibits the action of the blood clotting mechanism.

antidiuretic hormone (an''tĭ-di''u-ret'ik hōr'mōn) Hormone released from the posterior lobe of the pituitary gland that enhances the conservation of water by the kidneys.

antigen (an'tĭ-jen) A substance that stimulates cells to produce antibodies.

antioxidant (an''tĭ-ok'sĭ-dant) A substance that inhibits the oxidation of another substance.

antithrombin (an''tĭ-throm'bin) A substance that inhibits the action of thrombin and thus inhibits the blood clotting mechanism.

anus (a'nus) Inferior outlet of the digestive tube.

aorta (a-or'tah) Major systemic artery that receives blood from the left ventricle.

aortic body (a-or'tik bod'e) A structure associated with the wall of the aorta that contains a group of chemoreceptors.

aortic semilunar valve (a-or'tik sem''ĭ-lu'nar valv) Flaplike structures in the wall of the aorta near its origin that prevent blood from returning to the left ventricle of the heart.

aortic sinus (a-or'tik si'nus) Swelling in the wall of the aorta that contains pressoreceptors.

apneustic area (ap-nu'stik a're-ah) A portion of the respiratory control center located in the pons.

apocrine gland (ap'o-krin gland) A type of sweat gland that responds during periods of emotional stress.

aponeurosis (ap''o-nu-ro'sis) A sheetlike tendon by which certain muscles are attached to other parts.

appendicular (ap''en-dik'u-lar) Pertaining to the arms or legs.

appendix (ah-pen'diks) A small, tubular appendage that extends outward from the cecum of the large intestine.

aqueous humor (a'kwe-us hu'mor) Watery fluid that fills the anterior and posterior chambers of the eye.

arachnoid (ah-rak'noid) Delicate, weblike middle layer of the meninges; arachnoid mater.

arachnoid granulation (ah-rak'noid gran''u-la'shun) Fingerlike structures that project from the subarachnoid space of the meninges into blood-filled dural sinuses and function in the reabsorption of cerebrospinal fluid.

arbor vitae (ar'bor vi'ta) Treelike pattern of white matter seen in a section of cerebellum.

areola (ah-re'o-lah) Pigmented region surrounding the nipple of the mammary gland or breast.

arrector pili muscle (ah-rek'tor pil'i mus'l) Smooth muscle in the skin associated with a hair follicle.

arrhythmia (ah-rith'me-ah) Abnormal heart action characterized by a loss of rhythm.

arterial pathway (ar-te're-al path'wa) Course followed by blood as it travels from the heart to the body cells.

arteriole (ar-te're-ōl) A small branch of an artery that communicates with a capillary network.

arteriosclerosis (ar-te''re-o-skle-ro'sis) Condition in which the walls of arteries thicken and lose their elasticity; hardening of the arteries.

artery (ar'ter-e) A vessel that transports blood away from the heart.

arthritis (ar-thri'tis) Condition characterized by inflammation of joints.

articular cartilage (ar-tik'u-lar kar'tĭ-lij) Hyaline cartilage that covers the ends of bones in synovial joints.

articulation (ar-tik''u-la'shun) The joining together of parts at a joint.

ascending colon (ah-send'ing ko'lon) Portion of the large intestine that passes upward on the right side of the abdomen from the cecum to the lower edge of the liver.

ascending tracts (ah-send'ing trakts) Groups of nerve fibers in the spinal cord that transmit sensory impulses upward to the brain.

ascorbic acid (as-kor'bik as'id) One of the water-soluble vitamins; vitamin C.

assimilation (ah-sim''ĭ-la'shun) The action of changing absorbed substances into forms that differ chemically from those entering.

association area (ah-so''se-a'shun a're-ah) Region of the cerebral cortex related to memory, reasoning, judgment, and emotional feelings.

astigmatism (ah-stig'mah-tizm) Visual defect due to errors in refraction caused by abnormal curvatures in the surface of the cornea or lens.

astrocyte (as'tro-sit) A type of neuroglial cell that functions to connect neurons to blood vessels.

atherosclerosis (ath''er-o-skle-ro'sis) Condition in which fatty substances accumulate abnormally on the inner linings of arteries.

atmospheric pressure (at''mos-fer'ik presh'ur) Pressure exerted by the weight of the air; about 1034 grams per square centimeter at sea level.

atom (at'om) Smallest particle of an element that has the properties of that element.

atomic number (ah-tom'ik num'ber) Number equal to the number of protons in an atom of an element.

atomic weight (ah-tom'ik wāt) Number approximately equal to the number of protons plus the number of neutrons in an atom of an element.

ATP Adenosine triphosphate.

ATPase Enzyme that causes ATP molecules to release the energy stored in the terminal phosphate bonds.

atrioventricular bundle (a''tre-o-ven-trik'u-lar bun'dl) Group of specialized fibers that conduct impulses from the atrioventricular node to the ventricular muscle of the heart; A-V bundle.

atrioventricular node (a''tre-o-ven-trik'u-lar nōd) Specialized mass of muscle fibers located in the interatrial septum of the heart; functions in the transmission of the cardiac impulses from the sinoatrial node to the ventricular walls.

atrioventricular orifice (a''tre-o-ven-trik'u-lar or'ĭ-fis) Opening between the atrium and the ventricle on one side of the heart.

atrioventricular sulcus (a''tre-o-ven-trik'u-lar sul'kus) A groove on the surface of the heart that marks the division between an atrium and a ventricle.

atrioventricular valve (a''tre-o-ven-trik'u-lar valv) Cardiac valve located between an atrium and a ventricle.

atrium (a'tre-um) A chamber of the heart that receives blood from veins.

atrophy (at'ro-fe) A wasting away or decrease in size of an organ or tissue.

audiometer (aw''de-om'ĕ-ter) An instrument used to measure the acuity of hearing.

auditory (aw'di-to''re) Pertaining to the ear or to the sense of hearing.

auditory ossicle (aw'di-to''re os'ĭ-kl) A bone of the middle ear.

auricle (aw'ri-kl) An earlike structure; the portion of the heart that forms the wall of an atrium.

autoimmune disease (aw''to-ĭ-mūn' dĭ-zēz') Disorder characterized by an immune response directed toward a person's own tissues; autoallergy.

autonomic nervous system (aw''to-nom'ik ner'vus sis'tem) Portion of the nervous system that functions to control the actions of the visceral organs and skin.

autosome (aw'to-sōm) A chromosome other than a sex chromosome.

A-V bundle (bun'dl) A group of fibers that conduct cardiac impulses from the A-V node to the Purkinje fibers; bundle of His.

A-V node (nōd) Atrioventricular node.

axial skeleton (ak'se-al skel'ĕ-ton) Portion of the skeleton that supports and protects the organs of the head, neck, and trunk.

axillary (ak'sĭ-ler''e) Pertaining to the armpit.

axon (ak'son) A nerve fiber that conducts a nerve impulse away from a neuron cell body.

basal ganglion (ba'sal gang'gle-on) Mass of gray matter located deep within a cerebral hemisphere of the brain.

basal metabolic rate (ba'sal met''ah-bol'ic rāt) Rate at which metabolic reactions occur when the body is at rest.

base (bās) A substance that ionizes in water to release hydroxyl ions (OH⁻) or other ions that combine with hydrogen ions.

basement membrane (bās'ment mem'brān) A layer of nonliving material that anchors epithelial tissue to underlying connective tissue.

basophil (ba'so-fil) White blood cell characterized by the presence of cytoplasmic granules that become stained by basophilic dye.

beta oxidation (ba'tah ok"sī-da'shun) Chemical process by which fatty acids are converted to molecules of acetyl coenzyme A.

bicarbonate ion (bi-kar'bon-āt i'on) HCO_3^-

bicuspid tooth (bi-kus'pid tooth) A premolar that is specialized for grinding hard particles of food.

bicuspid valve (bi-kus'pid valv) Heart valve located between the left atrium and the left ventricle; mitral valve.

bile (bil) Fluid secreted by the liver and stored in the gallbladder.

bilirubin (bil"ĭ-roo'bin) A bile pigment produced as a result of hemoglobin breakdown.

biliverdin (bil"ĭ-ver'din) A bile pigment produced as a result of hemoglobin breakdown.

biochemistry (bi"o-kem'is-tre) Branch of science dealing with the chemistry of living organisms.

biofeedback (bi"o-fēd'bak) Procedure in which electronic equipment is used to help a person learn to consciously control certain visceral responses.

biotin (bi'o-tin) A water-soluble vitamin; a member of the vitamin B complex.

bipolar neuron (bi-po'lar nu'ron) A nerve cell whose cell body has only two processes, one serving as an axon and the other as a dendrite.

blastocyst (blas'to-sist) An early stage of embryonic development that consists of a hollow ball of cells.

B-lymphocyte (lim'fo-sit) Lymphocyte that reacts against foreign substances in the body by producing and secreting antibodies.

BMR Basal metabolic rate.

Bowman's capsule (bo'manz kap'sūl) Proximal portion of a renal tubule that encloses the glomerulus of a nephron.

brachial (bra'ke-al) Pertaining to the arm.

bradycardia (brad"e-kar'de-ah) An abnormally slow heart rate or pulse rate.

brain stem (brān stem) Portion of the brain that includes the midbrain, pons, and medulla oblongata.

bronchial tree (brong'ke-al trē) The bronchi and their branches that function to carry air from the trachea to the alveoli of the lungs.

bronchiole (brong'ke-ōl) A small branch of a bronchus within the lung.

bronchus (brong'kus) A branch of the trachea that leads to a lung.

buccal (buk'al) Pertaining to the mouth and the inner lining of the cheeks.

buffer (buf'er) A substance that can react with a strong acid or base to form a weaker acid or base and thus resist a change in pH.

bulbourethral glands (bul"bo-u-re'thral glandz) Glands that secrete a viscous fluid into the male urethra at times of sexual excitement.

bursa (bur'sah) A saclike, fluid-filled structure, lined with synovial membrane, that occurs near a joint.

bursitis (bur-si'tis) Inflammation of a bursa.

calcification (kal"sī-fī-ka'shun) The process by which salts of calcium are deposited within a tissue.

calcitonin (kal"sī-to'nin) Hormone secreted by the thyroid gland that helps to regulate the level of blood calcium.

calorie (kal'o-re) A unit used in the measurement of heat energy and the energy values of foods.

calorimeter (kal"o-rim'ē-ter) A device used to measure the heat energy content of foods; bomb calorimeter.

canaliculus (kan"ah-lik'u-lus) Microscopic canals that interconnect the lacunae of bone tissue.

cancellous bone (kan'se-lus bōn) Bone tissue with a lattice-work structure; spongy bone.

capillary (kap'ĭ-ler"e) A small blood vessel that connects an arteriole and a venule.

carbaminohemoglobin (kar"bah-me'no-he"mo-glo'bin) Compound formed by the union of carbon dioxide and hemoglobin.

carbohydrate (kar"bo-hi'drāt) An organic compound that contains carbon, hydrogen, and oxygen, with a 2:1 ratio of hydrogen to oxygen atoms.

carbonic anhydrase (kar-bon'ik an-hi'drās) Enzyme that promotes the reaction between carbon dioxide and water to form carbonic acid.

carbon monoxide (kar'bon mon-ok'sid) A toxic gas that combines readily with hemoglobin to form a relatively stable compound; CO.

carboxypeptidase (kar-bok"se-pep'ti-dās) A protein-splitting enzyme found in pancreatic juice.

cardiac conduction system (kar'de-ak kon-duk'shun sis'tem) System of specialized muscle fibers that conducts cardiac impulses from the S-A node into the myocardium.

cardiac cycle (kar'de-ak si'kl) A series of myocardial contractions that constitute a complete heartbeat.

cardiac muscle (kar'de-ak mus'l) Specialized type of muscle tissue found only in the heart.

cardiac output (kar'de-ak owt'poot) A quantity calculated by multiplying the stroke volume by the heart rate in beats per minute.

cardiac veins (kar'de-ak vāns) Blood vessels that return blood from the venules of the myocardium to the coronary sinus.

carina (kah-ri'nah) A cartilaginous ridge located between the openings of the right and left bronchi.

carotene (kar'o-tēn) A yellow, orange, or reddish pigment that occurs in plants and from which vitamin A can be synthesized.

carotid bodies (kah-rot'id bod'ēz) Masses of chemoreceptors located in the wall of the internal carotid artery near the carotid sinus.

carpals (kar'pals) Bones of the wrist.

carpus (kar'pus) The wrist; the wrist bones as a group.

cartilage (kar'tĭ-lij) Type of connective tissue in which cells are located within lacunae and are separated by a semi-solid matrix.

cartilaginous bone (kar"tĭ-laj'i-nus bōn) Bone that appears in the form of a cartilaginous model, which is largely replaced by bony tissue during development.

catabolic metabolism (kat"ah-bol'ik mē-tab'o-lism) Metabolic process by which large molecules are broken down into smaller ones; catabolism.

catalase (kat'ah-lās) An enzyme that causes the decomposition of hydrogen peroxide.

catalyst (kat'ah-list) A substance that increases the rate of a chemical reaction but is not permanently altered by the reaction.

cataract (kat'ah-rakt) Condition characterized by loss of transparency of the lens of the eye.

catecholamine (kat"ĕ-kol-am'in) A type of organic compound that includes epinephrine and norepinephrine.

cauda equina (kaw'da ek-win'a) A group of spinal nerves that extends below the distal end of the spinal cord.

cecum (se'kum) A pouchlike portion of the large intestine to which the small intestine is attached.

celiac (se'le-ak) Pertaining to the abdomen.

cell (sel) The structural and functional unit of an organism.

cell body (sel bod'e) Portion of a nerve cell that includes a cytoplasmic mass and a nucleus and from which the nerve fibers extend.

cellular respiration (sel'u-lar res''pĭ-ra'shun) Process by which energy is released from organic compounds within cells.

cementum (se-men'tum) Bonelike material that fastens the root of a tooth into its bony socket.

central canal (sen'tral kah-nal') Tube within the spinal cord that is continuous with the ventricles of the brain and contains cerebrospinal fluid.

central nervous system (sen'tral ner'vus sis'tem) Portion of the nervous system that consists of the brain and spinal cord; CNS.

centriole (sen'tre-ōl) A cellular organelle that functions in the organization of the spindle during mitosis.

centromere (sen'tro-mēr) Portion of a chromosome to which the spindle fiber attaches during mitosis.

centrosome (sen'tro-sōm) Cellular organelle consisting of two centrioles.

cephalic (sĕ-fal'ik) Pertaining to the head.

cerebellar cortex (ser''ĕ-bel'ar kor'teks) The outer layer of the cerebellum.

cerebellum (ser''ĕ-bel'um) Portion of the brain that coordinates skeletal muscle movement.

cerebral aqueduct (ser'ĕ-bral ak'wĕ-dukt'') Tube that connects the third and fourth ventricles of the brain.

cerebral cortex (ser'ĕ-bral kor'teks) Outer layer of the cerebrum.

cerebral hemisphere (ser'ĕ-bral hem'ĭ-sfēr) One of the large, paired structures that together constitute the cerebrum of the brain.

cerebrospinal fluid (ser''ĕ-bro-spi'nal floo'id) Fluid that occupies the ventricles of the brain, the subarachnoid space of the meninges, and the central canal of the spinal cord.

cerebrum (ser'ĕ-brum) Portion of the brain that occupies the upper part of the cranial cavity.

cerumen (sĕ-roo'men) Waxlike substance produced by cells that line the canal of the external ear.

cervical (ser'vĭ-kal) Pertaining to the neck or to the cervix of the uterus.

cervix (ser'viks) Narrow, inferior end of the uterus that leads into the vagina.

chemoreceptor (ke''mo-re-sep'tor) A receptor that is stimulated by the presence of certain chemical substances.

chief cell (chēf sel) Cell of gastric gland that secretes various digestive enzymes, including pepsinogen.

childhood (child'hood) Phase of the human life cycle that begins at the end of the first year and ends at puberty.

chloride shift (klo'rĭd shift) Movement of chloride ions from the blood plasma into red blood cells as bicarbonate ions diffuse out of the red blood cells into the plasma.

cholecystokinin (ko''le-sis''to-ki'nin) Hormone secreted by the small intestine that stimulates the release of pancreatic juice from the pancreas and bile from the gallbladder.

cholesterol (ko-les'ter-ol) A lipid produced by body cells that is used in the synthesis of steroid hormones and is excreted into the bile.

cholinergic fiber (ko''lin-er'jik fi'ber) A nerve fiber that secretes acetylcholine at the terminal end of its axon.

cholinesterase (ko''lin-es'ter-ās) An enzyme that causes the decomposition of acetylcholine.

chondrin (kon'drin) A protein that occurs in the intercellular substance of cartilage tissues.

chondrocyte (kon'dro-sɪt) A cartilage cell.

chorion (ko're-on) Embryonic membrane that forms the outermost covering around a developing fetus and contributes to the formation of the placenta.

chorionic villi (ko''re-on'ik vil'i) Projections that extend from the outer surface of the chorion and help attach an embryo to the uterine wall.

choroid coat (ko'roid kōt) The vascular, pigmented middle layer of the wall of the eye.

choroid plexus (ko'roid plek'sus) Mass of specialized capillaries from which cerebrospinal fluid is secreted into a ventricle of the brain.

chromatid (kro'mah-tid) A member of a duplicate pair of chromosomes.

chromatin (kro'mah-tin) Nuclear material that gives rise to chromosomes during mitosis.

chromosome (kro'mo-sōm) Rodlike structure that appears in the nucleus of a cell during mitosis; contains the genes responsible for heredity.

chylomicron (ki''lo-mi'kron) A microscopic droplet of fat, found in the blood following the digestion of fats.

chyme (kīm) Semifluid mass of food materials that passes from the stomach to the small intestine.

chymotrypsin (ki''mo-trip'sin) A protein-splitting enzyme found in pancreatic juice.

cilia (sil'e-ah) Microscopic, hairlike processes on the exposed surfaces of certain epithelial cells.

ciliary body (sil'e-er''e bod'e) Structure associated with the choroid layer of the eye that secretes aqueous humor and contains the ciliary muscle.

circadian rhythm (ser''kah-de'an rithm) A pattern of repeated behavior associated with the cycles of night and day.

circle of Willis (sir'kl uv wil'is) An arterial ring located on the ventral surface of the brain.

circular muscles (ser'ku-lar mus'lz) Muscles whose fibers are arranged in circular patterns, usually around an opening or in the wall of a tube; sphincter muscles.

circumduction (ser''kum-duk'shun) Movement of a body part, such as a limb, so that the end follows a circular path.

citric acid cycle (sit'rik as'id si'kl) A series of chemical reactions by which various molecules are oxidized and energy is released from them; Kreb's cycle.

cleavage (klēv'ij) The early successive divisions of embryonic cells into smaller and smaller cells.

clitoris (kli'to-ris) Small erectile organ located in the anterior portion of the female vulva; corresponds to the penis of the male.

CNS Central nervous system.

coagulation (ko-ag''u-la'shun) The clotting of blood.

cocarboxylase (ko''kar-bok'sĭ-lās) A coenzyme that is synthesized from thiamine and acts in the oxidation of carbohydrates.

cochlea (kok'le-ah) Portion of the inner ear that contains the receptors of hearing.

coenzyme (ko-en'zim) A nonprotein substance that is necessary to complete the structure of an enzyme molecule.

coenzyme A (ko-en'zim) Acetyl coenzyme A.

collagen (kol'ah-jen) Protein that occurs in the white fibers of connective tissues and in the matrix of bone.

collateral (ko-lat'er-al) A branch of a nerve fiber or blood vessel.

colon (ko'lon) The large intestine.

color blindness (kul'er blind'nes) An inability to distinguish colors normally.

colostrum (ko-los'trum) The first secretion of the mammary glands following the birth of an infant.

common bile duct (kom'mon bil dukt) Tube that transports bile from the cystic duct to the duodenum.

complete protein (kom-plēt' pro'te-in) A protein that contains adequate amounts of the essential amino acids to maintain body tissues and to promote normal growth and development.

compound (kom'pownd) A substance composed of two or more elements joined by chemical bonds.

condom (kon'dum) A rubber sheath used to cover the penis during sexual intercourse; used as a contraceptive.

conduction (kon-duk'shun) Process by which body heat moves into the molecules of cooler objects in contact with the body surface.

condyle (kon'dil) A rounded process of a bone, usually at the articular end.

cones (kōns) Color receptors located in the retina of the eye.

congenital (kon-jen'ĭ-tal) Any condition that exists at the time of birth.

conjunctiva (kon''junk-ti'vah) Membranous covering on the anterior surface of the eye.

connective tissue (kŏ-nek'tiv tish'u) One of the basic types of tissue that includes bone, cartilage, and various fibrous tissues.

contraception (kon''trah-sep'shun) The prevention of fertilization of the egg cell or the development of an embryo.

convection (kon-vek'shun) The transmission of heat from one substance to another through the circulation of heated air particles.

convolution (kon''vo-lu'shun) An elevation on the surface of a structure caused by an infolding of the structure upon itself.

cornea (kor'ne-ah) Transparent anterior portion of the outer layer of the eye wall.

coronary artery (kor'o-na''re ar'ter-e) An artery that supplies blood to the wall of the heart.

coronary sinus (kor'o-na''re si'nus) A large vessel on the posterior surface of the heart into which the cardiac veins drain.

corpus callosum (kor'pus kah-lo'sum) A mass of white matter within the brain, composed of nerve fibers connecting the right and left cerebral hemispheres.

corpus luteum (kor'pus lut'e-um) Structure that forms from the tissues of a ruptured ovarian follicle and functions to secrete female hormones.

corpus striatum (kor'pus stri-a'tum) Portion of the cerebrum that includes certain basal ganglia.

cortex (kor'teks) Outer layer of an organ such as the adrenal gland, cerebrum, or kidney.

cortical nephron (kor'tĭ-kl nef'ron) A nephron with its corpuscle located in the renal cortex.

cortisol (kor'tĭ-sol) A glucocorticoid secreted by the adrenal cortex.

costal (kos'tal) Pertaining to the ribs.

covalent bond (ko'va-lent bond) Chemical bond created by the sharing of electrons between atoms.

cranial (kra'ne-al) Pertaining to the cranium.

cranial nerve (kra'ne-al nerv) Nerve that arises from the brain.

creatine phosphate (kre'ah-tin fos'fāt) A substance present in muscle that acts to store energy.

crest (krest) A ridgelike projection of a bone.

cretinism (kre'tĭ-nizm) A condition resulting from a lack of thyroid secretion in an infant.

cricoid cartilage (kri'koid kar'tĭ-lij) A ringlike cartilage that forms the lower end of the larynx.

crista ampullaris (kris'tah am-pul'ar-is) Sensory organ located within a semicircular canal that functions in the sense of dynamic equilibrium.

crossing over (kros'ing o'ver) The exchange of genetic material between homologous chromosomes during meiosis.

crossmatching (kros'mach''ing) A procedure used to determine whether donor and recipient blood samples will agglutinate.

cubital (ku'bi-tal) Pertaining to the forearm.

cuspid (kus'pid) A canine tooth.

cutaneous (ku-ta'ne-us) Pertaining to the skin.

cyanocobalamin (si''ah-no-ko-bal'ah-min) Vitamin B_{12}.

cyanosis (si''ah-no'sis) A condition characterized by a bluish coloration of the skin due to a decreased blood oxygen concentration.

cyclic AMP (sik'lik) A substance produced from ATP that causes a variety of changes in cells.

cystic duct (sis'tik dukt) Tube that connects the gallbladder to the common bile duct.

cytocrine secretion (si'to-krin se-kre'shun) Process by which melanocytes transfer granules of melanin into adjacent epithelial cells.

cytoplasm (si'to-plazm) The contents of a cell surrounding its nucleus.

deamination (de-am''ĭ-na'shun) Chemical process by which amino groups ($-NH_2$) are removed from amino acid molecules.

deciduous teeth (de-sid'u-us tēth) Teeth that are shed and replaced by permanent teeth.

decomposition (de-kom''po-zish'un) The breakdown of molecules into simpler compounds.

defecation (def''ē-ka'shun) The discharge of feces from the rectum through the anus.

dehydration (de''hi-dra'shun) Excessive loss of water.

dehydration synthesis (de''hi-dra'shun sin'thē-sis) Anabolic process by which molecules are joined together to form larger molecules.

dendrite (den'drit) Nerve fiber that transmits impulses toward a neuron cell body.

dental caries (den'tal kar'ēz) Process by which teeth become decalcified and decayed.

dentine (den'tēn) Bonelike substance that forms the bulk of a tooth.

deoxyhemoglobin (de-ok''sĭ-he''mo-glo'bin) Hemoglobin that lacks oxygen.

depolarization (de-po''lar-ĭ-za'shun) The loss of an electrical charge on the surface of a membrane.

dermis (der'mis) The thick layer of the skin beneath the epidermis.

descending colon (de-send'ing ko'lon) Portion of the large intestine that passes downward along the left side of the abdominal cavity to the brim of the pelvis.

descending tracts (de-send'ing trakts) Groups of nerve fibers that carry nerve impulses downward from the brain through the spinal cord.

desmosome (des'mo-sōm) A specialized junction between cells, which serves as a ''spot weld.''

detrusor muscle (de-trūz'or mus'l) Muscular wall of the urinary bladder.

dextrose (dek'strōs) Glucose.

diabetes insipidus (di''ah-be'tēz in-sip'ĭ-dus) Condition characterized by an abnormally great production of urine due to a deficiency of antidiuretic hormone.

diabetes mellitus (di''ah-be'tēz mel-li'tus) Condition characterized by a high blood glucose level and the appearance of glucose in the urine due to a deficiency of insulin.

dialysis (di-al'ĭ-sis) Process by which smaller molecules are separated from larger ones in a liquid.

diapedesis (di''ah-pĕ-de'sis) Process by which leukocytes squeeze between the cells that make up the walls of blood vessels.

diaphragm (di'ah-fram) A sheetlike structure composed largely of muscle and connective tissue that separates the thoracic and abdominal cavities; also a caplike rubber device inserted in the vagina to be used as a contraceptive.

diaphysis (di-af'ĭ-sis) The shaft of a long bone.

diastole (di-as'to-le) Phase of the cardiac cycle during which a heart chamber wall is relaxed.

diastolic pressure (di-a-stol'ik presh'ur) Arterial blood pressure during the diastolic phase of the cardiac cycle.

diencephalon (di''en-sef'ah-lon) Portion of the brain in the region of the third ventricle that includes the thalamus and hypothalamus.

differentiation (dif''er-en''she-a'shun) Process by which cells become structurally and functionally specialized during development.

diffusion (dĭ-fu'zhun) Random movement of molecules from a region of higher concentration toward one of lower concentration.

digestion (di-jes'chun) The process by which larger molecules of food substances are broken down into smaller molecules that can be absorbed; hydrolysis.

dipeptide (di-pep'tid) A molecule composed of two amino acids bonded together.

disaccharide (di-sak'ah-rīd) A sugar produced by the union of two monosaccharide molecules.

distal (dis'tal) Further from the midline or origin; opposite of proximal.

diuretic (di''u-ret'ik) A substance that causes an increased production of urine.

DNA Deoxyribonucleic acid.

dominant gene (dom'ĭ-nant jēn) The gene of a gene pair that is expressed while its allele is not expressed.

dorsal root (dor'sal rōōt) The sensory branch of a spinal nerve by which it joins the spinal cord.

dorsal root ganglion (dor'sal rōōt gang'gle-on) Mass of sensory neuron cell bodies located in the dorsal root of a spinal nerve.

dorsum (dors'um) Pertaining to the back surface of a body part.

ductus arteriosus (duk'tus ar-te''re-o'sus) Blood vessel that connects the pulmonary artery and the aorta in a fetus.

ductus venosus (duk'tus ven-o'sus) Blood vessel that connects the umbilical vein and the inferior vena cava in a fetus.

duodenum (du''o-de'num) The first portion of the small intestine that leads from the stomach to the jejunum.

dural sinus (du'ral si'nus) Blood-filled channel formed by the splitting of the dura mater into two layers.

dura mater (du'rah ma'ter) Tough outer layer of the meninges.

dynamic equilibrium (di-nam'ik e''kwĭ-lib're-um) The maintenance of balance when the head and body are suddenly moved or rotated.

eccrine gland (ek'rin gland) Sweat gland that functions in the maintenance of body temperature.

ECG Electrocardiogram; EKG.

ectoderm (ek'to-derm) The outermost layer of the primary germ layers, responsible for forming certain embryonic body parts.

edema (ĕ-de'mah) An excessive accumulation of fluid within the tissue spaces.

effector (ĕ-fek'tor) Organ, such as a muscle or gland, that responds to stimulation.

efferent arteriole (ef'er-ent ar-te're-ol) Arteriole that conducts blood away from the glomerulus of a nephron.

ejaculation (e-jak''u-la'shun) Discharge of sperm-containing seminal fluid from the male urethra.

elastin (e-las'tin) Protein that comprises the yellow, elastic fibers of connective tissue.

electrocardiogram (e-lek''tro-kar'de-o-gram'') A recording of the electrical activity associated with the heartbeat; ECG or EKG.

electrolyte (e-lek'tro-lit) A substance that ionizes in water solution.

electrolyte balance (e-lek'tro-lit bal'ans) Condition that exists when the quantities of electrolytes entering the body equal those leaving it.

electron (e-lek'tron) A small, negatively charged particle that revolves around the nucleus of an atom.

electrovalent bond (e-lek''tro-va'lent bond) Chemical bond formed between two ions as a result of the transfer of electrons.

element (el'ĕ-ment) A basic chemical substance.

embolus (em'bo-lus) A substance, such as a blood clot or bubble of gas, that is carried by the blood and obstructs a blood vessel.

embryo (em'bre-o) An organism in its earliest stages of development.

emission (e-mish'un) The movement of sperm cells from the vas deferens into the ejaculator duct and urethra.

emphysema (em''fĭ-se'mah) A condition characterized by abnormal enlargement of the air sacs of the lungs.

emulsification (e-mul''sĭ-fĭ'ka'shun) Process by which fat globules are caused to break up into smaller droplets by the action of bile salts.

enamel (e-nam'el) Hard covering on the exposed surface of a tooth.

endocardium (en''do-kar'de-um) Inner lining of the heart chambers.

endocrine gland (en'do-krin gland) A gland that secretes hormones directly into the blood or body fluids.

endoderm (en'do-derm) The innermost layer of the primary germ layers responsible for forming certain embryonic body parts.

endolymph (en'do-limf) Fluid contained within the membranous labyrinth of the inner ear.

endometrium (en''do-me'tre-um) The inner lining of the uterus.

endomysium (en''do-mis'e-um) The sheath of connective tissue surrounding each skeletal muscle fiber.

endoplasmic reticulum (en-do-plaz'mic rĕ-tik'u-lum) Cytoplasmic organelle composed of a system of interconnected membranous tubules and vesicles.

endorphin (en-dor'fin) A neuropeptide that occurs in the pituitary gland and has a pain-suppressing action.

endothelium (en''do-the'le-um) The layer of epithelial cells that forms the inner lining of blood vessels and heart chambers.

energy (en'er-je) An ability to cause something to move and thus to do work.

energy balance (en'er-je bal'ans) Condition that exists when the caloric intake of the body equals its caloric output.

enkephalin (en-kef'ah-lin) A neuropeptide that occurs in the brain and spinal cord and inhibits pain impulses, thus relieving pain sensations.

enterogastrone (en''ter-o-gas'trōn) A hormone secreted from the intestinal wall that inhibits gastric secretion and motility.

enzyme (en'zim) A protein that is synthesized by a cell and acts as a catalyst in a specific cellular reaction.

eosinophil (e''o-sin'o-fil) White blood cell characterized by the presence of cytoplasmic granules that become stained by acidic dye.

ependyma (ĕ-pen'dĭ-mah) Membrane composed of neuroglial cells that lines the ventricles of the brain.

epicardium (ep''ĭ-kar'de-um) The visceral portion of the pericardium located on the surface of the heart.

epicondyle (ep''ĭ-kon'dil) A projection of a bone located above a condyle.

epidermis (ep''ĭ-der'mis) Outer epithelial layer of the skin.

epididymis (ep''ĭ-did'ĭ-mis) Highly coiled tubule that leads from the seminiferous tubules of the testis to the vas deferens.

epidural space (ep''ĭ-du'ral spās) The space between the dural sheath of the spinal cord and the bone of the vertebral canal.

epigastric region (ep''ĭ-gas'trik re'jun) The upper middle portion of the abdomen.

epiglottis (ep''ĭ-glot'is) Flaplike cartilaginous structure located at the back of the tongue near the entrance to the trachea.

epimysium (ep''ĭ-mis'e-um) The outer sheath of connective tissue surrounding a skeletal muscle.

epinephrine (ep''ĭ-nef'rin) A hormone secreted by the adrenal medulla during times of stress.

epiphyseal disk (ep''ĭ-fiz'e-al disk) Cartilaginous layer within the epiphysis of a long bone that functions as a growing region.

epiphysis (ĕ-pif'ĭ-sis) The end of a long bone.

epithelium (ep''ĭ-the'le-um) The type of tissue that covers all free body surfaces.

equilibrium (e''kwĭ-lib're-um) A state of balance between two opposing forces.

erythroblast (ĕ-rith'ro-blast) An immature red blood cell.

erythroblastosis (ĕ-rith''ro-blas-to'sis) Condition characterized by the presence of erythroblasts in the circulating blood.

erythrocyte (ĕ-rith'ro-sit) A red blood cell.

erythropoiesis (ĕ-rith''ro-poi-e'sis) Red blood cell formation.

erythropoietin (ĕ-rith''ro-poi'ĕ-tin) Substance released by the kidneys and liver that promotes red blood cell formation.

esophageal hiatus (ĕ-sof''ah-je'al hi-a'tus) Opening in the diaphragm through which the esophagus passes.

esophagus (ĕ-sof'ah-gus) Tubular portion of the digestive tract that leads from the pharynx to the stomach.

essential amino acid (ĕ-sen'shal ah-me'no as'id) Amino acid required for health that cannot be synthesized in adequate amounts by body cells.

essential fatty acid (ĕ-sen'shal fat'e as'id) Fatty acid required for health that cannot be synthesized in adequate amounts by body cells.

estrogen (es'tro-jen) Hormone that stimulates the development of female secondary sexual characteristics.

eustachian tube (u-sta'ke-an tūb) Tube that connects the middle ear to the pharynx; auditory tube.

evaporation (e''vap'o-ra-shun) Process by which a liquid changes into a gas.

eversion (e-ver'zhun) Movement in which the sole of the foot is turned outward.

exchange reaction (eks-chānj' re-ak'shun) A chemical reaction in which parts of two kinds of molecules trade positions.

excretion (ek-skre'shun) Process by which metabolic wastes are eliminated.

exocrine gland (ek'so-krin gland) A gland that secretes its products into a duct or onto a body surface.

expiration (ek''spĭ-ra'shun) Process of expelling air from the lungs.

extension (ek-sten'shun) Movement by which the angle between parts at a joint is increased.

extracellular (ek''strah-sel'u-lar) Outside of cells.

extrapyramidal tract (ek''strah-pi-ram'ĭ-dal trakt) Nerve tracts, other than the corticospinal tracts, that transmit impulses from the cerebral cortex into the spinal cord.

extremity (ek-strem'ĭ-te) A limb; an arm or leg.

facet (fas'et) A small, flattened surface of a bone.

facilitated diffusion (fah-sil'ĭ-tāt''id dĭ-fu'zhun) Diffusion in which substances are moved through membranes by carrier molecules.

fallopian tube (fah-lo'pe-an tūb) Tube that transports an egg cell from the region of the ovary to the uterus; oviduct or uterine tube.

fascia (fash'e-ah) A sheet of fibrous connective tissue that encloses a muscle.

fasciculus (fah-sik'u-lus) A small bundle of muscle fibers.

fat (fat) Adipose tissue; or an organic substance whose molecules contain glycerol and fatty acids.

fatty acid (fat'e as'id) An organic substance that serves as a building block for a fat molecule.

feces (fe'sēz) Material expelled from the digestive tract during defecation.

ferritin (fer'ĭ-tin) An iron-protein complex in which iron is stored in liver cells.

fertilization (fer''tĭ-lĭ-za'shun) The union of an egg cell and a sperm cell.

fetus (fe'tus) A human embryo after seven weeks of development.

fibril (fi'bril) A tiny fiber or filament.

fibrillation (fi''brĭ-la'shun) Uncoordinated contraction of muscle fibers.

fibrin (fi'brin) Insoluble, fibrous protein formed from fibrinogen during blood coagulation.

fibrinogen (fi-brin′o-jen) Plasma protein that is converted into fibrin during blood coagulation.

fibrinolysin (fi″brĭ-nol′ĭ-sin) A protein-splitting enzyme that can digest the substance of a blood clot.

fibroblast (fi′bro-blast) Cell that functions to produce fibers and other intercellular materials in connective tissues.

filtration (fil-tra′shun) Movement of material through a membrane as a result of hydrostatic pressure.

fissure (fish′ūr) A narrow cleft separating parts, such as the lobes of the cerebrum.

flaccid paralysis (flak′sid pah-ral′ĭ-sis) A condition characterized by total loss of tone in the muscles innervated by damaged nerve fibers.

flexion (flek′shun) Bending at a joint so that the angle between bones is decreased.

follicle (fol′ĭ-kl) A pouchlike depression or cavity.

follicle-stimulating hormone (fol′ĭ-kl stim′u-la″ting hōr′mōn) A substance secreted by the anterior pituitary gland that stimulates the development of an ovarian follicle in a female or the production of sperm cells in a male; FSH.

follicular cells (fŏ-lik′u-lar selz) Ovarian cells that surround a developing egg cell and secrete female sex hormones.

fontanel (fon″tah-nel′) Membranous region located between certain cranial bones in the skull of a fetus or infant.

foramen (fo-ra′men) An opening, usually in a bone or membrane (plural, foramina).

foramen magnum (fo-ra′men mag′num) Opening in the occipital bone of the skull through which the spinal cord passes.

foramen ovale (fo-ra′men o-val′e) Opening in the interatrial septum of the fetal heart.

formula (fōr′mu-lah) A group of symbols and numbers used to express the composition of a compound.

fossa (fos′ah) A depression in a bone or other part.

fovea (fo′ve-ah) A tiny pit or depression.

fovea centralis (fo′ve-ah sen-tral′is) Region of the retina, consisting of densely packed cones, which is responsible for the greatest visual acuity.

fracture (frak′tūr) A break in a bone.

frenulum (fren′u-lum) A fold of tissue that serves to anchor and limit the movement of a body part.

frontal (frun′tal) Pertaining to the region of the forehead.

FSH Follicle-stimulating hormone.

galactose (gah-lak′tōs) A monosaccharide component of the disaccharide, lactose.

gallbladder (gawl′blad-er) Saclike organ associated with the liver that stores and concentrates bile.

gamete (gam′ēt) A sex cell; either an egg cell or a sperm cell.

ganglion (gang′gle-on) A mass of neuron cell bodies, usually outside the central nervous system.

gastric gland (gas′trik gland) Gland within the stomach wall that secretes gastric juice.

gastric juice (gas′trik jōōs) Secretion of the gastric glands within the stomach.

gastrin (gas′trin) Hormone secreted by the stomach lining that stimulates the secretion of gastric juice.

gene (jēn) Portion of a DNA molecule that contains the information needed to synthesize an enzyme.

genetic code (jĕ-net′ik kōd) System by which information for synthesizing proteins is built into the structure of DNA molecules.

genetics (jĕ-net′iks) The study of the mechanism by which characteristics are passed from parents to offspring.

genotype (je′no-tip) The combination of genes present within a zygote or within the cells of an individual.

germinal epithelium (jer′mĭ-nal ep″ĭ-the′le-um) Tissue within an ovary or testis that gives rise to sex cells.

germ layers (jerm la′ers) Layers of cells within an embryo that form the body organs during development.

globin (glo′bin) The protein portion of a hemoglobin molecule.

globulin (glob′u-lin) A type of protein that occurs in blood plasma.

glomerulus (glo-mer′u-lus) A capillary tuft located within the Bowman's capsule of a nephron.

glottis (glot′is) Slitlike opening between the true vocal folds or vocal cords.

glucagon (gloo′kah-gon) Hormone secreted by the pancreatic islets of Langerhans that causes the release of glucose from glycogen.

glucocorticoid (gloo″ko-kor′tĭ-koid) Any one of a group of hormones secreted by the adrenal cortex that influence carbohydrate, fat, and protein metabolism.

gluconeogenesis (gloo″ko-ne″o-jen′ĕ-sis) The synthesis of glucose from noncarbohydrate materials, such as amino acid molecules.

glucose (gloo′kōs) A monosaccharide found in the blood that serves as the primary source of cellular energy.

glucosuria (gloo″ko-su′re-ah) The presence of glucose in urine.

gluteal (gloo′te-al) Pertaining to the buttocks.

glycerol (glis′er-ol) An organic compound that serves as a building block for fat molecules.

glycogen (gli′ko-jen) A polysaccharide that functions to store glucose in the liver and muscles.

glycolysis (gli-kol′ĭ-sis) The conversion of glucose to pyruvic acid during cellular respiration.

goblet cell (gob′let sel) An epithelial cell that is specialized to secrete mucus.

goiter (goi′ter) A condition characterized by the enlargement of the thyroid gland.

Golgi apparatus (gol′je ap″ah-ra′tus) A cytoplasmic organelle that functions in preparing cellular products for secretion.

gonad (go′nad) A sex-cell-producing organ; an ovary or testis.

gonadotropin (go-nad″o-trōp′in) A hormone that stimulates activity in the gonads.

granulocyte (gran′u-lo-sīt) A leukocyte that contains granules in its cytoplasm.

gray matter (grā mat′er) Region of the central nervous system that generally lacks myelin and thus appears gray.

groin (groin) Region of the body between the abdomen and thighs.

growth (grōth) Process by which a structure enlarges.

growth hormone (grōth hōr′mōn) A hormone released by the anterior lobe of the pituitary gland that promotes the growth of the organism.

hair follicle (hār fol′ĭ-kl) Tubelike depression in the skin in which a hair develops.

haversian canal (ha-ver′shan kah-nal′) Tiny channel in bone tissue that contains a blood vessel.

haversian system (ha-ver′shan sis′tem)   A group of bone cells and the haversian canal that they surround; the basic unit of structure in osseous tissue.

hematocrit (he-mat′o-krit)   The volume percentage of red blood cells within a sample of whole blood.

hematoma (he″mah-to′mah)   A mass of coagulated blood within tissues or a body cavity.

hematopoiesis (hem″ah-to-poi-e′sis)   The production of blood and blood cells; hemopoiesis.

heme (hem)   The iron-containing portion of a hemoglobin molecule.

hemocytoblast (he″mo-si′to-blast)   A cell that gives rise to blood cells.

hemoglobin (he″mo-glo′bin)   Pigment of red blood cells responsible for the transport of oxygen.

hemolysis (he-mol′ĭ-sis)   The rupture of red blood cells accompanied by the release of hemoglobin.

hemopoiesis (he″mo-poi-e′sis)   The production of blood and blood cells; hematopoiesis.

hemorrhage (hem′ō-rij)   Loss of blood from the circulatory system; bleeding.

hemostasis (he″mo-sta′sis)   The stoppage of bleeding.

heparin (hep′ah-rin)   A substance that interferes with the formation of a blood clot; an anticoagulant.

hepatic (hĕ-pat′ik)   Pertaining to the liver.

hepatic lobule (hĕ-pat′ik lob′ūl)   A functional unit of the liver.

hepatic sinusoid (hĕ-pat′ik si′nŭ-soid)   Vascular channel within the liver.

heredity (hĕ-red′ĭ-te)   The transmission of genetic information from parent to offspring.

heterozygote (het″er-o-zi′gōt)   An individual who possesses different alleles in a gene pair.

heterozygous (het″er-o-zi′gus)   Pertaining to a heterozygote.

histamine (his′tah-min)   A substance released from cells subjected to stressful conditions.

histology (his-tol′o-je)   The study of the structure and function of tissues.

homeostasis (ho″me-o-sta′sis)   A state of equilibrium in which the internal environment of the body remains relatively constant.

homozygote (ho″mo-zi′gōt)   An individual possessing identical alleles in a gene pair.

homozygous (ho″mo-zi′gus)   Pertaining to a homozygote.

hormone (hōr′mōn)   A substance secreted by an endocine gland that is transmitted in the blood or body fluids.

humoral immunity (hu′mor-al ĭ-mu′nĭ-te)   Resistance to the effects of specific disease-causing agents due to the presence of circulating antibodies.

hydrolysis (hi-drol′ĭ-sis)   The splitting of a molecule into smaller portions by the addition of a water molecule.

hydrostatic pressure (hi″dro-stat′ik presh′ur)   Pressure exerted by fluids, such as blood pressure.

hydroxyapatite (hi-drok″se-ap′ah-tĭt)   A type of crystalline calcium phosphate found in bone matrix.

hydroxyl ion (hi-drok′sil i′on)   OH⁻.

hymen (hi′men)   A membranous fold of tissue that partially covers the vaginal opening.

hyperglycemia (hi″per-gli-se′me-ah)   An excessive level of blood glucose.

hyperkalemia (hi″per-kah-le′me-ah)   An excessive concentration of blood potassium.

hypernatremia (hi″per-nah-tre′me-ah)   An excessive concentration of blood sodium.

hyperparathyroidism (hi″per-par″ah-thi′roi-dizm)   An excessive secretion of parathyroid hormone.

hyperplasia (hi″per-pla′ze-ah)   An increased production and growth of new cells.

hypertension (hi″per-ten′shun)   Excessive blood pressure.

hyperthyroidism (hi″per-thi′roi-dizm)   An excessive secretion of thyroid hormones.

hypertonic (hi″per-ton′ik)   Condition in which a solution contains a greater concentration of dissolved particles than the solution with which it is compared.

hypertrophy (hi-per′tro-fe)   Enlargement of an organ or tissue.

hyperventilation (hi″per-ven″tĭ-la′shun)   Breathing that is abnormally deep and prolonged.

hypervitaminosis (hi″per-vi″tah-mĭ-no′sis)   Excessive intake of vitamins.

hypochondriac region (hi″po-kon′dre-ak re′jun)   The portion of the abdomen on either side of the middle or epigastric region.

hypogastric region (hi″po-gas′trik re′jun)   The lower middle portion of the abdomen.

hypoglycemia (hi″po-gli-se′me-ah)   Abnormally low concentration of blood glucose.

hypokalemia (hi″po-kah-le′me-ah)   A low concentration of blood potassium.

hyponatremia (hi″po-nah-tre′me-ah)   A low concentration of blood sodium.

hypoparathyroidism (hi″po-par″ah-thi′roi-dizm)   An undersecretion of parathyroid hormone.

hypophysis (hi-pof′ĭ-sis)   The pituitary gland.

hypoproteinemia (hi″po-pro″te-ĭ-ne′me-ah)   A low concentration of blood proteins.

hypothalamus (hi″po-thal′ah-mus)   A portion of the brain located below the thalamus and forming the floor of the third ventricle.

hypothyroidism (hi″po-thi′roi-dizm)   A low secretion of thyroid hormones.

hypotonic (hi″po-ton′ik)   Condition in which a solution contains a lesser concentration of dissolved particles than the solution to which it is compared.

hypoxia (hi-pok′se-ah)   A deficiency of oxygen in the tissues.

ileocecal valve (il′e-o-se′kal valv)   Sphincter valve located at the distal end of the ileum where it joins the cecum.

ileum (il′e-um)   Portion of the small intestine between the jejunum and cecum.

iliac region (il′e-ak re′jun)   Portion of the abdomen on either side of the lower middle or hypogastric region.

ilium (il′e-um)   One of the bones of an os coxa or hipbone.

immunity (ĭ-mu′nĭ-te)   Resistance to the effects of specific disease-causing agents.

immunoglobulin (im″u-no-glob′u-lin)   Globular plasma proteins that function as antibodies of immunity.

immunosuppressive drugs (im″u-no-sŭ-pres′iv drugz)   Substances that inhibit the formation of antibodies.

implantation (im″plan-ta′shun)   The embedding of an embryo in the lining of the uterus.

impulse (im′puls)   A wave of depolarization conducted along a nerve fiber or muscle fiber.

incisor (in-si′zor)   One of the front teeth that is adapted for cutting food.

inclusion (in-kloo′zhun)   A mass of lifeless chemical substance within the cytoplasm of a cell.

incomplete protein (in″kom-plēt′ pro′te-in)   A protein that lacks essential amino acids.

infancy (in′fan-se)   Period of life from the end of the first 4 weeks to 1 year of age.

inferior (in-fēr'e-or) Situated below something else; pertaining to the lower surface of a part.

inflammation (in''flah-ma'shun) A tissue response to stress that is characterized by dilation of blood vessels and an accumulation of fluid in the affected region.

infrared ray (in''frah-red' ra) A form of radiation energy, with wavelengths longer than visible light, by which heat moves from warmer surfaces to cooler surroundings.

infundibulum (in''fun-dib'u-lum) The stalk by which the pituitary gland is attached to the base of the brain.

ingestion (in-jes'chun) The taking of food or liquid into the body by way of the mouth.

inguinal (ing'gwĭ-nal) Pertaining to the groin region.

inguinal canal (ing'gwĭ-nal kah-nal') Passage through which a testis descends into the scrotum.

inorganic (in''or-gan'ik) Pertaining to chemical substances that lack carbon.

insertion (in-ser'shun) The end of a muscle that is attached to a movable part.

inspiration (in''spĭ-ra'shun) Act of breathing in; inhalation.

insula (in'su-lah) A cerebral lobe located deep within the lateral sulcus.

insulin (in'su-lin) A hormone secreted by the pancreatic islets of Langerhans that functions in the control of carbohydrate metabolism.

integumentary (in-teg-u-men'tar-e) Pertaining to the skin and its accessory organs.

intercalated disk (in-ter''kah-lāt'ed disk) Membranous boundary between adjacent cardiac muscle cells.

intercellular (in''ter-sel'u-lar) Between cells.

intercellular fluid (in''ter-sel'u-lar floo'id) Tissue fluid located between cells, other than blood cells.

intracellular junction (in''ter-sel'u-lar jungk'shun) A connection between the membranes of adjacent cells.

interferon (in''ter-fēr'on) A substance produced by cells that inhibits the multiplication of viruses.

interneuron (in''ter-nu'ron) A neuron located between a sensory neuron and a motor neuron.

interphase (in'ter-fāz) Period between two cell divisions when a cell is carrying on its normal functions.

interstitial cell (in''ter-stish'al sel) A hormone-secreting cell located between the seminiferous tubules of the testis.

interstitial fluid (in''ter-stish'al floo'id) Same as intercellular fluid.

intervertebral disk (in''ter-ver'tĕ-bral disk) A layer of fibrocartilage located between the bodies of adjacent vertebrae.

intestinal gland (in-tes'tĭ'nal gland) Tubular gland located at the base of a villus within the intestinal wall.

intestinal juice (in-tes'tĭ'nal jōōs) The secretion of the intestinal glands.

intracellular (in''trah-sel'u-lar) Within cells.

intracellular fluid (in''trah-sel'u-lar floo'id) Fluid within cells.

intrauterine device (in''trah-u'ter-in de-vīs') A solid object placed in the uterine cavity for purposes of contraception; IUD.

intrinsic factor (in-trin'sik fak'tor) A substance produced by the gastric glands that promotes the absorption of vitamin B_{12}.

inversion (in-ver'zhun) Movement in which the sole of the foot is turned inward.

involuntary (in-vol'un-tār''e) Not consciously controlled; functions automatically.

ion (i'on) An atom or a group of atoms with an electrical charge.

ionization (i''on-i-za'shun) Chemical process by which substances dissociate into ions.

iris (i'ris) Colored muscular portion of the eye that surrounds the pupil and regulates its size.

irritability (ir''ĭ-tah-bil'ĭ-te) The ability of an organism to react to changes taking place in its environment.

ischemia (is-ke'me-ah) A deficiency of blood in a body part.

isometric contraction (i''so-met'rik kon-trak'shun) Muscular contraction in which the muscle fails to shorten.

isotonic contraction (i''so-ton'ik kon-trak'shun) Muscular contraction in which the muscle shortens.

isotonic solution (i''so-ton'ik so-lu'shun) A solution that has the same concentration of dissolved particles as the solution with which it is compared.

isotope (i'so-tōp) An atom that has the same number of protons as other atoms of an element but has a different number of neutrons in its nucleus.

IUD An intrauterine device.

jejunum (je-joo'num) Portion of the small intestine located between the duodenum and the ileum.

joint (joint) The union of two or more bones; an articulation.

juxtamedullary nephron (juks''tah-med'u-lār-e nef'ron) A nephron with its corpuscle located near the renal medulla.

keratin (ker'ah-tin) Protein present in the epidermis, hair, and nails.

keratinization (ker''ah-tin''ĭ-za'shun) The process by which cells form fibrils of keratin and become hardened.

kernicterus (ker-nik'ter-us) Condition in which bilirubin precipitates in the brain tissues of an infant affected with erythroblastosis fetalis.

ketogenesis (ke''to-jen'ĕ-sis) The formation of ketone bodies.

ketone body (ke'tōn bod'e) Type of compound produced during fat catabolism, including acetone, acetoacetic acid, and betahydroxybutyric acid.

ketosis (ke''to'sis) A condition in which the concentration of ketone bodies in body fluids is abnormally increased.

kilocalorie (kil'o-kal''o-re) One thousand calories.

kilogram (kil'o-gram) A unit of weight equivalent to 1000 grams.

Kreb's cycle (krebz si'kl) The citric acid cycle.

kyphosis (ki-fo'sis) An abnormally increased convex curvature in the thoracic portion of the vertebral column.

labor (la'bor) The process of childbirth.

labyrinth (lab'ĭ-rinth) The system of interconnecting tubes within the inner ear, which includes the cochlea, vestibule, and semicircular canals.

lacrimal gland (lak'rĭ-mal gland) Tear secreting gland.

lactase (lak'tās) Enzyme that converts lactose into glucose and galactose.

lactation (lak-ta'shun) The production of milk by the mammary glands.

lacteal (lak'te-al) A lymphatic vessel associated with a villus of the small intestine.

lactic acid (lak'tik as'id) An organic substance formed from pyruvic acid during anaerobic respiration.

lactose (lak'tōs) A disaccharide that occurs in milk; milk sugar.

lacuna (lah-ku'nah) A hollow cavity.

lamella (lah-mel'ah) A layer of matrix in bone tissue.

laryngopharynx (lah-ring″go-far′ingks)   The lower portion of the pharynx near the opening to the larynx.

larynx (lar′ingks)   Structure located between the pharynx and trachea that houses the vocal cords.

latent period (la′tent pe′re-od)   Time lapse between the application of a stimulus and the beginning of a response in a muscle fiber.

lateral (lat′er-al)   Pertaining to the side.

leukocyte (lu′ko-sīt)   A white blood cell.

leukocytosis (lu″ko-si-to′sis)   An abnormally large increase in the number of white blood cells.

leukopenia (lu″ko-pe′ne-ah)   An abnormally low number of leukocytes in the blood.

lever (lev′er)   A simple mechanical device consisting of a rod, fulcrum, weight, and a source of energy that is applied to some point on the rod.

ligament (lig′ah-ment)   A cord or sheet of connective tissue by which two or more bones are bound together at a joint.

limbic system (lim′bik sis′tem)   A group of interconnected structures within the brain that function to produce various emotional feelings.

linea alba (lin′e-ah al′bah)   A narrow band of tendinous connective tissue located in the midline of the anterior abdominal wall.

lingual (ling′gwal)   Pertaining to the tongue.

lipase (li′pās)   A fat-digesting enzyme.

lipid (lip′id)   A fat, oil, or fatlike compound that usually has fatty acids in its molecular structure.

lipoprotein (lip″o-pro′te-in)   A complex of lipid and protein.

lordosis (lor-do′sis)   An abnormally increased concave curvature in the lumbar portion of the vertebral column.

lumbar (lum′bar)   Pertaining to the region of the loins.

lumen (lu′men)   Space within a tubular structure such as a blood vessel or intestine.

luteinizing hormone (lu′te-in-īz″ing hōr′mōn)   A hormone secreted by the anterior pituitary gland that controls the formation of corpus luteum in females and the secretion of testosterone in males; LH.

lymph (limf)   Fluid transported by the lymphatic vessels.

lymph node (limf nōd)   A mass of lymphoid tissue located along the course of a lymphatic vessel.

lymphocyte (lim′fo-sīt)   A type of white blood cell produced in lymphatic tissue.

lysosome (li′so-sōm)   Cytoplasmic organelle that contains digestive enzymes.

macrocyte (mak′ro-sīt)   A large red blood cell.

macrophage (mak′ro-fāj)   A large phagocytic cell.

macroscopic (mak″ro-skop′ik)   Large enough to be seen with the unaided eye.

macula lutea (mak′u-lah lu′te-ah)   A yellowish depression in the retina of the eye that is associated with acute vision.

malignant (mah-lig′nant)   The power to threaten life; cancerous.

malnutrition (mal″nu-trish′un)   A condition resulting from an improper diet.

maltase (mawl′tās)   An enzyme that converts maltōse into glucose.

maltose (mawl′tōs)   A disaccharide composed of two glucose molecules.

mammary (mam′ar-e)   Pertaining to the breast.

marrow (mar′o)   Connective tissue that occupies the spaces within bones.

mast cell (mast sel)   A cell to which antibodies, formed in response to allergens, become attached.

mastication (mas″tĭ-ka′shun)   Chewing movements.

matrix (ma′triks)   The intercellular substance of connective tissue.

matter (mat′er)   Anything that has weight and occupies space.

meatus (me-a′tus)   A passageway or channel, or the external opening of such a passageway.

mechanoreceptor (mek″ah-no-re-sep′tor)   A sensory receptor that is sensitive to mechanical stimulation such as changes in pressure or tension.

medial (me′de-al)   Toward or near the midline.

mediastinum (me″de-ah-sti′num)   Tissues and organs of the thoracic cavity that form a septum between the lungs.

medulla (mĕ-dul′ah)   The inner portion of an organ.

medulla oblongata (mĕ-dul′ah ob″long-gah′tah)   Portion of the brain stem located between the pons and the spinal cord.

medullary cavity (med′u-lār″e kav′ĭ-te)   Cavity within the diaphysis of a long bone occupied by marrow.

megakaryocyte (meg″ah-kar′e-o-sīt)   A large bone marrow cell that functions to produce blood platelets.

meiosis (mi-o′sis)   Process of cell division by which egg and sperm cells are formed.

melanin (mel′ah-nin)   Dark pigment normally found in skin and hair.

melanocyte (mel′ah-no-sīt″)   Melanin-producing cell.

melatonin (mel″ah-to′nin)   A hormone thought to be secreted by the pineal gland.

membranous bone (mem′brah-nus bōn)   Bone that develops from layers of membranous connective tissue.

menarche (mĕ-nar′ke)   The first menstrual period.

meninges (mĕ-nin′jēz)   A group of three membranes that covers the brain and spinal cord (singular, *meninx*).

menopause (men′o-pawz)   Termination of the menstrual cycles.

menstrual cycle (men′stroo-al si′kl)   The female reproductive cycle that is characterized by regularly reoccurring changes in the uterine lining.

menstruation (men″stroo-a′shun)   Loss of blood and tissue from the uterus at the end of a female reproductive cycle.

mesentery (mes′en-ter″e)   A fold of peritoneal membrane that attaches an abdominal organ to the abdominal wall.

mesoderm (mez′o-derm)   The middle layer of the primary germ layers, responsible for forming certain embryonic body parts.

messenger RNA (mes′in-jer)   Molecule of RNA that transmits information for protein synthesis from the nucleus of a cell to the cytoplasm.

metabolic rate (met″ah-bol′ic rāt)   The rate at which chemical changes occur within the body.

metabolism (mĕ-tab′o-lizm)   All of the chemical changes that occur within cells considered together.

metacarpals (met″ah-kar′pals)   Bones of the hand between the wrist and finger bones.

metaphase (met′ah-fāz)   Stage in mitosis when chromosomes become aligned in the middle of the spindle.

metastasis (mĕ-tas′tah-sis)   The spread of disease from one body region to another.

metatarsals (met″ah-tar′sals)   Bones of the foot between the ankle and toe bones.

microglia (mi-krog'le-ah) A type of neuroglial cell that helps support neurons and acts to carry on phagocytosis.

micron (mi'kron) A unit of measurement; 1/1000 of a millimeter.

microscopic (mi''kro-skop'ik) Too small to be seen with the unaided eye.

microtubule (mi''kro-tu'būl) A minute, hollow rod found in the cytoplasm of cells.

micturition (mik''tu-rish'un) Urination.

midbrain (mid'brān) A small region of the brain stem located between the diencephalon and pons.

mineralocorticoid (min''er-al-o-kor'tĭ-koid) Any one of a group of hormones secreted by the adrenal cortex that influences the concentrations of electrolytes in body fluids.

mitochondrion (mi''to-kon'dre-on) Cytoplasmic organelle that contains enzymes responsible for aerobic respiration (plural, *mitochondria*).

mitosis (mi-to'sis) Process by which body cells divide to form two identical daughter cells.

mitral valve (mi'tral valv) Heart valve located between the left atrium and the left ventricle; bicuspid valve.

mixed nerve (mikst nerv) Nerve that includes both sensory and motor nerve fibers.

molar (mo'lar) A rear tooth with a somewhat flattened surface adapted for grinding food.

molecule (mol'ĕ-kūl) A particle composed of two or more atoms bonded together.

monocyte (mon'o-sit) A type of white blood cell that functions as a phagocyte.

monosaccharide (mon''o-sak'ah-rīd) A simple sugar, such as glucose or fructose, that represents the structural unit of a carbohydrate.

morula (mor'u-lah) An early stage in embryonic development; a solid ball of cells.

motor area (mo'tor a're-ah) A region of the brain from which impulses to muscles or glands originate.

motor end plate (mo'tor end plāt) Enlargement at the terminal end of a motor neuron axon that secretes a neurotransmitter.

motor nerve (mo'tor nerv) A nerve that consists of motor nerve fibers.

motor neuron (mo'tor nu'ron) A neuron that transmits impulses from the central nervous system to an effector.

motor unit (mo'tor unit) A motor neuron and the muscle fibers associated with it.

mucosa (mu-ko'sah) The membrane that lines tubes and body cavities that open to the outside of the body; mucous membrane.

mucous cell (mu'kus sel) Glandular cell that secretes mucus.

mucous membrane (mu'kus mem'brān) Mucosa.

mucus (mu'kus) Fluid secretion of the mucous cells.

multiple alleles (mul'tĭ-pl ah-lēls') Situation in which the alleles that constitute a gene pair occur in several forms.

multiple motor unit summation (mul'tĭ-pl mo'tor u'nit sum-mā'shun) A sustained muscle contraction of increasing strength resulting from responses by many motor units.

multipolar neuron (mul'tĭ-po'lar nu'ron) Nerve cell that has many processes arising from its cell body.

mutagenic (mu'tah-jen'ik) Pertaining to a factor that can cause mutations.

mutation (mu-ta'shun) A change in the genetic information of a chromosome.

myelin (mi'ĕ-lin) Fatty material that forms a sheathlike covering around some nerve fibers.

myocardium (mi''o-kar'de-um) Muscle tissue of the heart.

myofibril (mi''o-fi'bril) Contractile fibers found within muscle cells.

myoglobin (mi''o-glo'bin) A pigmented compound found in muscle tissue that acts to store oxygen.

myogram (mi'o-gram) A recording of a muscular contraction.

myometrium (mi''o-me'tre-um) The layer of smooth muscle tissue within the uterine wall.

myoneural junction (mi''o-nu'ral jungk'shun) Site of union between a motor neuron axon and a muscle fiber.

myopia (mi-o'pe-ah) Nearsightedness.

myosin (mi'o-sin) A protein, that, together with actin, is responsible for muscular contraction and relaxation.

myxedema (mik''sĕ-de'mah) Condition resulting from a deficiency of thyroid hormones in an adult.

nasal cavity (na'zal kav'ĭ-te) Space within the nose.

nasal concha (na'zal kong'kah) Shell-like bone extending outward from the wall of the nasal cavity; a turbinate bone.

nasal septum (na'zal sep'tum) A wall of bone and cartilage that separates the nasal cavity into two portions.

nasopharynx (na''zo-far'ingks) Portion of the pharynx associated with the nasal cavity.

negative feedback (neg'ah-tiv fēd'bak) A mechanism that is activated by an imbalance and acts to correct it.

neonatal (ne''o-na'tal) Pertaining to the period of life from birth to the end of 4 weeks.

nephron (nef'ron) The functional unit of a kidney, consisting of a renal corpuscle and a renal tubule.

nerve (nerv) A bundle of nerve fibers.

neuroglia (nu-rog'le-ah) The supporting tissue within the brain and spinal cord, composed of neuroglial cells.

neurolemma (nu''ro-lem'ah) Sheath on the outside of some nerve fibers due to the presence of Schwann cells.

neuromodulator (nu''ro-mod'u-lā-tor) A substance that alters a neuron's response to a neurotransmitter.

neuromuscular junction (nu''ro-mus'ku-lar jungk'shun) Myoneural junction.

neuron (nu'ron) A nerve cell that consists of a cell body and its processes.

neuropeptide (nu''ro-pep'tĭd) A peptide that occurs in the brain and seems to function as a neurotransmitter or neuromodulator.

neurotransmitter (nu''ro-trans-mit'er) Chemical substance secreted by the terminal end of an axon that stimulates a muscle fiber contraction or an impulse in another neuron.

neutron (nu'tron) An electrically neutral particle found in an atomic nucleus.

neutrophil (nu'tro-fil) A type of phagocytic leukocyte.

niacin (ni'ah-sin) A vitamin of the B-complex group; nicotinic acid.

niacinamide (ni''ah-sin-am'id) The physiologically active form of niacin.

nicotinic acid (nik''o-tin'ik as'id) Niacin.

Nissl bodies (nis'l bod'ēz) Membranous sacs that occur within the cytoplasm of nerve cells and have ribosomes attached to their surfaces.

nitrogen balance (ni'tro-jen bal'ans) Condition in which the quantity of nitrogen ingested equals the quantity excreted.

nondisjunction (non''dis-jungk'shun) The failure of a pair of chromosomes to separate during meiosis.

nonelectrolyte (non''e-lek'tro-līt) A substance that does not dissociate into ions when it is dissolved.

nonprotein nitrogenous substance (non-pro'te-in ni-troj'ē-nus sub'stans) A substance, such as urea or uric acid, that contains nitrogen but is not a protein.

norepinephrine (nor''ep-ĭ-nef'rin) A neurotransmitter substance released from the axon ends of some nerve fibers.

nuclease (nu'kle-ās) An enzyme that causes nucleic acids to decompose.

nucleic acid (nu-kle'ik as'id) A substance composed of nucleotides bonded together; RNA or DNA.

nucleolus (nu-kle'o-lus) Small structure that occurs within the nucleus of a cell and contains RNA.

nucleoplasm (nu'kle-o-plazm'') The contents of the nucleus of a cell.

nucleosome (nu'kle-o-sōm) A beadlike particle within a chromatin fiber composed of DNA and protein.

nucleotide (nu'kle-o-tid'') A component of a nucleic acid molecule, consisting of a sugar, a nitrogenous base, and a phosphate group.

nucleus (nu'kle-us) A body that occurs within a cell and contains relatively large quantities of DNA; the dense core of an atom that is composed of protons and neutrons.

nutrient (nu'tre-ent) A chemical substance that must be supplied to the body from its environment.

nutrition (nu-trish'un) The study of the sources, actions, and interactions of nutrients.

obesity (o-bēs'ĭ-te) An excessive accumulation of adipose tissue; usually the condition of exceeding the desirable weight by more than 20%.

occipital (ok-sip'ĭ-tal) Pertaining to the lower, back portion of the head.

olfactory (ol-fak'to-re) Pertaining to the sense of smell.

olfactory nerves (ol-fak'to-re nervz) The first pair of cranial nerves, which conduct impulses associated with the sense of smell.

oligodendrocyte (ol''ĭ-go-den'dro-sīt) A type of neuroglial cell that functions to connect neurons to blood vessels and to form myelin.

oocyte (o'o-sīt) An immature egg cell.

oogenesis (o''o-jen'ē-sis) The process by which an egg cell forms from an oocyte.

ophthalmic (of-thal'mik) Pertaining to the eye.

optic (op'tik) Pertaining to the eye.

optic chiasma (op'tik ki-az'mah) X-shaped structure on the underside of the brain created by a partial crossing over of fibers in the optic nerves.

optic disk (op'tik disk) Region in the retina of the eye where nerve fibers leave to become part of the optic nerve.

oral (o'ral) Pertaining to the mouth.

organ (or'gan) A structure consisting of a group of tissues that performs a specialized function.

organelle (or''gah-nel') A living part of a cell that performs a specialized function.

organic (or-gan'ik) Pertaining to carbon-containing substances.

organism (or'gah-nizm) An individual living thing.

orifice (or'ĭ-fis) An opening.

origin (or'ĭ-jin) End of a muscle that is attached to a relatively immovable part.

oropharynx (o''ro-far'ingks) Portion of the pharynx in the posterior part of the oral cavity.

osmoreceptor (oz''mo-re-sep'tor) Receptor that is sensitive to changes in the osmotic pressure of body fluids.

osmosis (oz-mo'sis) Diffusion of water through a selectively permeable membrane due to the existence of a concentration gradient.

osmotic (oz-mot'ik) Pertaining to osmosis.

osmotic pressure (oz-mot'ik presh'ur) The amount of pressure needed to stop osmosis; the potential pressure of a solution due to the presence of nondiffusible solute particles in the solution.

osseous tissue (os'e-us tish'u) Bone tissue.

ossification (os''ĭ-fĭ-ka'shun) The formation of bone tissue.

osteoblast (os'te-o-blast'') A bone-forming cell.

osteoclast (os'te-o-klast'') A cell that causes the erosion of bone.

osteocyte (os'te-o-sit) A bone cell.

otolith (o'to-lith) A small particle of calcium carbonate associated with the receptors of equilibrium.

otosclerosis (o''to-sklĕ-ro'sis) Abnormal formation of spongy bone within the ear that may interfere with the transmission of sound vibrations to hearing receptors.

oval window (o'val win'do) Opening between the stapes and the inner ear.

ovarian (o-va're-an) Pertaining to the ovary.

ovary (o'var-e) The primary reproductive organ of a female; an egg-cell-producing organ.

oviduct (o'vĭ-dukt) A tube that leads from the ovary to the uterus; uterine tube or fallopian tube.

ovulation (o''vu-la'shun) The release of an egg cell from a mature ovarian follicle.

ovum (o'vum) A mature egg cell.

oxidase (ok'sĭ-dās) An enzyme that promotes oxidation.

oxidation (ok''sĭ-da'shun) Process by which oxygen is combined with a chemical substance.

oxygen debt (ok'sĭ-jen det) The amount of oxygen that must be supplied following physical exercise to convert accumulated lactic acid to glucose.

oxyhemoglobin (ok''sĭ-he''mo-glo'bin) Compound formed when oxygen combines with hemoglobin.

oxytocin (ok''sĭ-to'sin) Hormone released by the posterior lobe of the pituitary gland that causes contraction of smooth muscles in the uterus and mammary glands.

pacemaker (pās'māk-er) Mass of specialized muscle tissue that controls the rhythm of the heartbeat; the sinoatrial node.

pain receptor (pān re''sep'tor) Sensory nerve ending associated with the feeling of pain.

palate (pal'at) The roof of the mouth.

palatine (pal'ah-tīn) Pertaining to the palate.

palmar (pahl'mar) Pertaining to the palm of the hand.

pancreas (pan'kre-as) Glandular organ in the abdominal cavity that secretes hormones and digestive enzymes.

pancreatic (pan''kre-at'ik) Pertaining to the pancreas.

pantothenic acid (pan''to-then'ik as'id) A vitamin of the B-complex group.

papilla (pah-pil'ah) Tiny nipplelike projection.

papillary muscle (pap'ĭ-ler''e mus'l) Muscle that extends inward from the ventricular walls of the heart and to which the chordae tendineae are attached.

paralysis (pah-ral′i-sis) Loss of ability to control voluntary muscular movements, usually due to a disorder of the nervous system.

parasympathetic division (par″ah-sim″pah-thet′ik dī-vizh′un) Portion of the autonomic nervous system that arises from the brain and sacral region of the spinal cord.

parathormone (par″ah-thōr′mōn) Hormone secreted by the parathyroid glands that helps to regulate the level of blood calcium and phosphate.

parathyroid glands (par″ah-thi′roid glandz) Small endocrine glands that are embedded in the posterior portion of the thyroid gland.

parietal (pah-ri′ĕ-tal) Pertaining to the wall of an organ or cavity.

parietal cell (pah-ri′ĕ-tal sel) Cell of a gastric gland that secretes hydrochloric acid and intrinsic factor.

parietal pleura (pah-ri′ĕ-tal ploo′rah) Membrane that lines the inner wall of the thoracic cavity.

parotid glands (pah-rot′id glandz) Large salivary glands located on the sides of the face just in front and below the ears.

partial pressure (par′shal presh′ur) The pressure produced by one gas in a mixture of gases.

parturition (par″tu-rish′un) The process of childbirth.

pathogen (path′o-jen) Any disease-causing agent.

pathology (pah-thol′o-je) The study of disease.

pectoral (pek′tor-al) Pertaining to the chest.

pectoral girdle (pek′tor-al ger′dl) Portion of the skeleton that provides support and attachment for the arms.

pelvic (pel′vik) Pertaining to the pelvis.

pelvic girdle (pel′vik ger′dl) Portion of the skeleton to which the legs are attached.

pelvis (pel′vis) Bony ring formed by the sacrum and os coxae.

penis (pe′nis) External reproductive organ of the male through which the urethra passes.

pepsin (pep′sin) Protein-splitting enzyme secreted by the gastric glands of the stomach.

pepsinogen (pep-sin′o-jen) Inactive form of pepsin.

peptidase (pep′tĭ-dās) An enzyme that causes the breakdown of polypeptides.

peptide (pep′tīd) Compound composed of two or more amino acid molecules joined together.

peptide bond (pep′tīd bond) Bond that forms between the carboxyl group of one amino acid and the amino group of another.

pericardial (per″ĭ-kar′de-al) Pertaining to the pericardium.

pericardium (per″ĭ-kar′de-um) Serous membrane that surrounds the heart.

perilymph (per′ĭ-limf) Fluid contained in the space between the membranous and osseous labyrinths of the inner ear.

perimysium (per″ĭ-mis′e-um) Sheath of connective tissue that encloses a bundle of striated muscle fibers.

perineal (per″ĭ-ne′al) Pertaining to the perineum.

perineum (per″ĭ-ne′um) Body region between the scrotum or urethral opening and the anus.

periosteum (per″e-os′te-um) Covering of fibrous connective tissue on the surface of a bone.

peripheral (pĕ-rif′er-al) Pertaining to parts located near the surface or toward the outside.

peripheral nervous system (pĕ-rif′er-al ner′vus sis′tem) The portions of the nervous system outside the central nervous system.

peripheral resistance (pĕ-rif′er-al re-zis′tans) Resistance to blood flow due to friction between the blood and the walls of the blood vessels.

peristalsis (per″ĭ-stal′sis) Rhythmic waves of muscular contraction that occur in the walls of various tubular organs.

peritoneal (per″ĭ-to-ne′al) Pertaining to the peritoneum.

peritoneal cavity (per″ĭ-to-ne′al kav′ĭ-te) The potential space between the parietal and visceral peritoneal membranes.

peritoneum (per″ĭ-to-ne′um) A serous membrane that lines the abdominal cavity and encloses the abdominal viscera.

peritubular capillary (per″ĭ-tu′bu-lar kap′ĭ-ler″e) Capillary that surrounds a renal tubule and functions in reabsorption and secretion during urine formation.

permeable (per′me-ah-bl) Open to passage or penetration.

pH The negative logarithm of the hydrogen ion concentration used to indicate the acid or alkaline condition of a solution.

phagocytosis (fag″o-si-to′sis) Process by which a cell engulfs and digests solid substances.

phalanx (fa′langks) A bone of a finger or toe.

pharynx (far′ingks) Portion of the digestive tube between the mouth and esophagus.

phenotype (fe′no-tīp) The appearance of an individual due to the action of a particular set of genes.

phospholipid (fos″fo-lip′id) A lipid that contains phosphorus.

photoreceptor (fo″to-re-sep′tor) A nerve ending that is sensitive to light energy.

physiology (fiz″e-ol′o-je) The branch of science dealing with the study of body functions.

pia mater (pi′ah ma′ter) Inner layer of meninges that encloses the brain and spinal cord.

pineal gland (pin′e-al gland) A small structure located in the central part of the brain.

pinocytosis (pin″o-si-to′sis) Process by which a cell engulfs droplets of fluid from its surroundings.

pituitary gland (pĭ-tu′ĭ-tār″e gland) Endocrine gland that is attached to the base of the brain and consists of anterior and posterior lobes; the hypophysis.

placenta (plah-sen′tah) Structure by which an unborn child is attached to its mother's uterine wall and through which it is nourished.

plantar (plan′tar) Pertaining to the sole of the foot.

plasma (plaz′mah) Fluid portion of circulating blood.

plasma protein (plaz′mah pro′te-in) Any of several proteins normally found dissolved in blood plasma.

platelet (plāt′let) Cytoplasmic fragment formed in the bone marrow that functions in blood coagulation.

pleural (ploo′ral) Pertaining to the pleura or membranes investing the lungs.

pleural cavity (ploo′ral kav′ĭ-te) Potential space between the pleural membranes.

pleural membranes (ploo′ral mem′brānz) Serous membranes that enclose the lungs.

plexus (plek′sus) A network of interlaced nerves or blood vessels.

pneumotaxic area (nu″mo-tax′ik a′re-ah) A portion of the respiratory control center located in the pons of the brain.

polar body (po′lar bod′e) Small, nonfunctional cell produced as a result of meiosis during egg cell formation.

polarization (po''lar-ĭ-za'shun) The development of an electrical charge on the surface of a cell membrane due to an unequal distribution of ions on either side of the membrane.

polycythemia (pol''e-si-the'me-ah) An excessive concentration of red blood cells.

polymorphonuclear leukocyte (pol''e-mor''fo-nu'kle-ar lu'ko-sit) A leukocyte or white blood cell with an irregularly lobed nucleus.

polypeptide (pol''e-pep'tid) A compound formed by the union of many amino acid molecules.

polysaccharide (pol''e-sak'ah-rid) A carbohydrate composed of many monosaccharide molecules joined together.

pons (ponz) A portion of the brain stem above the medulla oblongata and below the midbrain.

popliteal (pop''lĭ-te'al) Pertaining to the region behind the knee.

posterior (pos-tēr'e-or) Toward the back; opposite of anterior.

postganglionic fiber (pōst''gang-gle-on'ik fi'ber) Autonomic nerve fiber located on the distal side of a ganglion.

postnatal (pōst-na'tal) After birth.

precursor (pre-ker'sor) Substances from which another substance is formed.

preganglionic fiber (pre''gang-gle-on'ik fi'ber) Autonomic nerve fiber located on the proximal side of a ganglion.

pregnancy (preg'nan-se) The condition in which a female has a developing offspring in her uterus.

prenatal (pre-na'tal) Before birth.

presbycusis (pres''bĭ-ku'sis) Loss of hearing that accompanies old age.

presbyopia (pres''be-o'pe-ah) Condition in which the eye loses its ability to accommodate due to loss of elasticity in the lens.

pressoreceptor (pres''o-re-sep'tor) A receptor that is sensitive to changes in pressure.

primary reproductive organs (pri'ma-re re''pro-duk'tiv or'ganz) Sex-cell-producing parts; testes in males and ovaries in females.

prime mover (prīm mōōv'er) Muscle that is mainly responsible for a particular body movement.

profibrinolysin (pro''fi-brĭ-no-li'sin) The inactive form of fibrinolysin.

progesterone (pro-jes'tĕ-rōn) A female hormone secreted by the corpus luteum of the ovary and by the placenta.

projection (pro-jek'shun) Process by which the brain causes a sensation to seem to come from the region of the body being stimulated.

prolactin (pro-lak'tin) Hormone secreted by the anterior pituitary gland that stimulates the production of milk from the mammary glands.

pronation (pro-na'shun) Movement in which the palm of the hand is moved downward or backward.

prophase (pro'fāz) Stage of mitosis during which chromosomes become visible.

proprioceptor (pro''pre-o-sep'tor) A sensory nerve ending that is sensitive to changes in tension of a muscle or tendon.

prostaglandins (pros''tah-glan'dins) A group of compounds that have powerful, hormonelike effects.

prostate gland (pros'tāt gland) Gland located around the male urethra below the urinary bladder that adds its secretion to seminal fluid during ejaculation.

protein (pro'te-in) Nitrogen-containing organic compound composed of amino acid molecules joined together.

prothrombin (pro-throm'bin) Plasma protein that functions in the formation of blood clots.

proton (pro'ton) A positively charged particle found in an atomic nucleus.

protraction (pro-trak'shun) A forward movement of a body part.

proximal (prok'sĭ-mal) Closer to the midline or origin; opposite of distal.

puberty (pu'ber-te) Stage of development in which the reproductive organs become functional.

pulmonary (pul'mo-ner''e) Pertaining to the lungs.

pulmonary circuit (pul'mo-ner''e ser'kit) System of blood vessels that carries blood between the heart and lungs.

pulse (puls) The surge of blood felt through the walls of arteries due to the contraction of the ventricles of the heart.

pupil (pu'pil) Opening in the iris through which light enters the eye.

Purkinje fibers (pur-kin'je fi'berz) Specialized muscle fibers that conduct the cardiac impulse from the A-V bundle into the ventricular walls.

pyloric sphincter muscle (pi-lor'ik sfingk'ter mus'l) Sphincter muscle located between the stomach and the duodenum; pylorus.

pyramidal cell (pĭ-ram'ĭ-dal sel) A large, pyramid-shaped neuron found within the cerebral cortex.

pyridoxine (pir''ĭ-dok'sēn) A vitamin of the B-complex group; vitamin B_6.

pyruvic acid (pi-roo'vik as'id) An intermediate product of carbohydrate oxidation.

radiation (ra''de-a'shun) A form of energy that includes visible light, ultraviolet light, and X rays; means by which body heat is lost in the form of infrared rays.

receptor (re''sep'tor) Part located at the distal end of a sensory dendrite that is sensitive to stimulation.

recessive gene (re-ses'iv jēn) An allele of a gene pair that is not expressed while the other allele is expressed.

rectum (rek'tum) The terminal end of the digestive tube between the sigmoid colon and the anus.

red marrow (red mar'o) Blood-cell-forming tissue located in spaces within bones.

referred pain (re-ferd' pān) Pain that feels as if it is originating from a part other than the site being stimulated.

reflex (re'fleks) A rapid, automatic response to a stimulus.

reflex arc (re'fleks ark) A nerve pathway, consisting of a sensory neuron, interneuron, and motor neuron, that forms the structural and functional bases for a reflex.

refraction (re-frak'shun) A bending of light as it passes from one medium into another medium with a different density.

refractory period (re-frak'to-re pe're-od) Time period following stimulation during which a neuron or muscle fiber will not respond to a stimulus.

renal (re'nal) Pertaining to the kidney.

renal corpuscle (re'nal kor'pusl) Part of a nephron that consists of a glomerulus and a Bowman's capsule; malpighian corpuscle.

renal cortex (re'nal kor'teks) The outer portion of a kidney.

renal medulla (re'nal mĕ-dul'ah) The inner portion of a kidney.

renal pelvis (re'nal pel'vis) The hollow cavity within a kidney.

renal tubule (re'nal tu'būl) Portion of a nephron that extends from the renal corpuscle to the collecting duct.

renin (re'nin) Substance released from the kidneys that causes a rise in blood pressure.

reproduction (re″pro-duk′shun) The process by which an offspring is formed.

resorption (re-sorp′shun) The process by which something that has been secreted is absorbed.

respiration (res″pĭ-ra′shun) Cellular process by which energy is released from nutrients.

respiratory center (re-spi′rah-to″re sen′ter) Portion of the brain stem that controls the depth and rate of breathing.

respiratory membrane (re-spi′rah-to″re mem′brān) Membrane composed of a capillary and an alveolar wall through which gases are exchanged between the blood and air.

response (re-spons′) The action resulting from a stimulus.

resting potential (res′ting po-ten′shal) The difference in electrical charge between the inside and outside of an undisturbed nerve cell membrane.

reticular formation (rĕ-tik′u-lar fōr-ma′shun) A complex network of nerve fibers within the brain stem that functions in arousing the cerebrum.

reticulocyte (rĕ-tik′u-lo-sīt) A young red blood cell that has a network of fibrils in its cytoplasm.

reticuloendothelial tissue (rĕ-tik″u-lo-en″do-the′le-al tish′u) Tissue composed of widely scattered phagocytic cells.

retina (ret′ĭ-nah) Inner layer of the eye wall that contains the visual receptors.

retinal (ret′ĭ-nal) A form of vitamin A; retinene.

retinene (ret′ĭ-nēn) Substance used in the production of rhodopsin, a light-sensitive pigment of the rods of the eye.

retraction (rĕ-trak′shun) Movement of a part toward the back.

retroperitoneal (ret″ro-per″ĭ-to-ne′al) Located behind the peritoneum.

rhodopsin (ro-dop′sin) Light-sensitive substance that occurs in the rods of the retina; visual purple.

rhythmicity area (rith-mis′ĭ-te a′re-ah) A portion of the respiratory control center located in the medulla.

riboflavin (ri″bo-fla′vin) A vitamin of the B-complex group; vitamin B_2.

ribonucleic acid (ri″bo-nu-kle′ik as′id) A nucleic acid that contains ribose sugar; RNA.

ribose (ri′bōs) A five-carbon sugar found in RNA molecules.

ribosome (ri′bo-sōm) Cytoplasmic organelle that consists largely of RNA and functions in the synthesis of proteins.

RNA Ribonucleic acid.

rod (rod) A type of light receptor that is responsible for colorless vision.

rotation (ro-ta′shun) Movement by which a body part is turned on its longitudinal axis.

round window (rownd win′do) A membrane-covered opening between the inner ear and the middle ear.

saccule (sak′ūl) A saclike cavity that makes up part of the membranous labyrinth of the inner ear.

sagittal (saj′i-tal) A plane or section that divides a structure into right and left portions.

salivary gland (sal′ĭ-ver-e gland) A gland associated with the mouth that secretes saliva.

salt (sawlt) A compound produced by a reaction between an acid and a base.

saltatory conduction (sal′tah-tor-e kon-duk′shun) A type of nerve impulse conduction in which the impulse seems to jump from one node to the next.

S-A node (nōd) Sinoatrial node.

sarcolemma (sar″ko-lem′ah) The cell membrane of a muscle fiber.

sarcomere (sar′ko-mēr) The structural and functional unit of a myofibril.

sarcoplasm (sar′ko-plazm) The cytoplasm within a muscle fiber.

sarcoplasmic reticulum (sar′ko-plaz′mik rĕ-tik′u-lum) Membranous network of channels and tubules within a muscle fiber, corresponding to the endoplasmic reticulum of other cells.

saturated fatty acid (sat′u-rāt″ed fat′e as′id) Fatty acid molecule that lacks double bonds between the atoms of its carbon chain.

Schwann cell (shwahn sel) Cell that surrounds a fiber of a peripheral nerve and forms the neurolemmal sheath.

sclera (skle′rah) White fibrous outer layer of the eyeball.

scoliosis (sko″le-o′sis) Abnormal lateral curvature of the vertebral column.

scrotum (skro′tum) A pouch of skin that encloses the testes.

sebaceous gland (se-ba′shus gland) Gland of the skin that secretes sebum.

sebum (se′bum) Oily secretion of the sebaceous glands.

secretin (se-kre′tin) Hormone secreted from the small intestine that stimulates the release of pancreatic juice from the pancreas.

semicircular canal (sem″ĭ-ser′ku-lar kah-nal′) Tubular structure within the inner ear that contains the receptors responsible for the sense of dynamic equilibrium.

seminal fluid (sem′ĭ-nal floo′id) Fluid discharged from the male reproductive tract at ejaculation that contains sperm cells and the secretions of various glands; semen.

seminiferous tubule (sem″ĭ-nif′er-us tu′būl) Tubule within the testes in which sperm cells are formed.

semipermeable (sem″ĭ-per′me-ah-bl) Condition in which a membrane is permeable to some molecules and not to others; selectively permeable.

senescence (sĕ-nes′ens) The process of growing old.

sensation (sen-sa′shun) A feeling resulting from the interpretation of sensory nerve impulses by the brain.

sensory area (sen′so-re a′re-ah) A portion of the cerebral cortex that receives and interprets sensory nerve impulses.

sensory nerve (sen′so-re nerv) A nerve composed of sensory nerve fibers.

sensory neuron (sen′so-re nu′ron) A neuron that transmits an impulse from a receptor to the central nervous system.

serotonin (se″ro-to′nin) A vasoconstricting substance that is released by blood platelets when blood vessels are broken and thus helps to control bleeding.

serous cell (se′rus sel) A glandular cell that secretes a watery fluid with a high enzyme content.

serous fluid (ser′us floo′id) The secretion of a serous membrane.

serous membrane (ser′us mem′brān) Membrane that lines a cavity without an opening to the outside of the body.

serum (se′rum) The fluid portion of coagulated blood.

sesamoid bone (ses′ah-moid bōn) A round bone that may occur in tendons adjacent to joints.

sex chromosome (seks kro′mo-sōm) A chromosome responsible for the development of characteristics associated with maleness or femaleness; an *X* or *Y* chromosome.

sex-linked trait (seks-linkt′ trāt) Trait determined by a recessive gene located on an *X* chromosome.

sigmoid colon (sig′moid ko′lōn) S-shaped portion of the large intestine between the descending colon and the rectum.

simple sugar (sim′pl shoog′ar) A monosaccharide.

sinoatrial node (si″no-a′tre-al nōd) Group of specialized tissue in the wall of the right atrium that initiates cardiac cycles; the pacemaker; S-A node.

sinus (si′nus) A cavity or hollow space in a bone or other body part.

skeletal muscle (skel′ĕ-tal mus′l) Type of muscle tissue found in muscles attached to skeletal parts.

smooth muscle (smooth mus′l) Type of muscle tissue found in the walls of hollow visceral organs; visceral muscle.

sodium pump (so′de-um pump) The active transport mechanism that functions to concentrate sodium ions on the outside of a cell membrane.

solute (sol′ūt) The substance that is dissolved in a solution.

solvent (sol′vent) The liquid portion of a solution in which a solute is dissolved.

somatic cell (so-mat′ik sel) Any cell of the body other than the sex cells.

somatotropin (so″mah-to-tro′pin) Growth hormone.

spastic paralysis (spas′tik pah-ral′ĭ-sis) A form of paralysis characterized by an increase in muscular tone without atrophy of the muscles involved.

special sense (spesh′al sens) Sense that involves receptors associated with specialized sensory organs, such as the eyes and ears.

spermatid (sper′mah-tid) An intermediate stage in the formation of sperm cells.

spermatocyte (sper-mat′o-sit) An early stage in the formation of sperm cells.

spermatogenesis (sper″mah-to-jen′ĕ-sis) The production of sperm cells.

spermatogonium (sper″mah-to-go′ne-um) Undifferentiated spermatogenic cell found in the germinal epithelium of a seminiferous tubule.

spermatozoa (sper″mah-to-zo′ah) Male reproductive cells; sperm cells.

sphincter (sfingk′ter) A circular muscle that functions to close an opening or the lumen of a tubular structure.

sphygmomanometer (sfig′mo-mah-nom′ĕ-ter) Instrument used for measuring blood pressure.

spinal (spi′nal) Pertaining to the spinal cord or to the vertebral canal.

spinal cord (spi′nal kord) Portion of the central nervous system extending downward from the brain stem through the vertebral canal.

spinal nerve (spi′nal nerv) Nerve that arises from the spinal cord.

spleen (splēn) A large, glandular organ located in the upper left region of the abdomen.

spongy bone (spunj′e bōn) Bone that consists of bars and plates separated by irregular spaces; cancellous bone.

squamous (skwa′mus) Flat or platelike.

starch (starch) A polysaccharide that is common in foods of plant origin.

static equilibrium (stat′ik e″kwĭ-lib′re-um) The maintenance of balance when the head and body are motionless.

sterility (stĕ-ril′ĭ-te) An inability to produce offspring.

steroid (ste′roid) An organic substance whose molecules include complex rings of carbon and hydrogen atoms.

stimulus (stim′u-lus) A change in the environmental conditions that is followed by a response by an organism or cell.

stomach (stum′ak) Digestive organ located between the esophagus and the small intestine.

strabismus (strah-biz′mus) A condition characterized by lack of visual coordination; crossed eyes.

stratified (strat′ĭ-fid) Arranged in layers.

stratum corneum (stra′tum kor′ne-um) Outer horny layer of the epidermis.

stratum germinativum (stra′tum jer′mĭ-na″tiv-um) The deepest layer of the epidermis in which the cells undergo mitosis.

stress (stres) Condition produced by factors causing potentially life-threatening changes in the body's internal environment.

stressor (stres′or) A factor capable of stimulating a stress response.

stroke volume (strōk vol′ūm) The amount of blood discharged from the ventricle with each heartbeat.

structural formula (struk′cher-al fōr′mu-lah) A representation of the way atoms are bonded together within a molecule, using the symbols for each element present and lines to indicate chemical bonds.

subarachnoid space (sub″ah-rak′noid spās) The space within the meninges between the arachnoid mater and the pia mater.

subcutaneous (sub″ku-ta′ne-us) Beneath the skin.

sublingual (sub-ling′gwal) Beneath the tongue.

submaxillary (sub-mak′sĭ-ler″e) Below the maxilla.

submucosa (sub″mu-ko′sah) Layer of connective tissue that underlies a mucous membrane.

substrate (sub′strāt) The substance upon which an enzyme acts.

sucrose (soo′krōs) A disaccharide; table sugar.

sulcus (sul′kus) A shallow groove, such as that between adjacent convolutions on the surface of the brain.

superficial (soo″per-fish′al) Near the surface.

superior (su-pe′re-or) Pertaining to a structure that is higher than another structure.

supination (soo″pĭ-na′shun) Rotation of the forearm so that the palm faces upward when the arm is outstretched.

suppressor cell (sŭ-pres′or sel) A special type of T-lymphocyte that functions to interfere with the production of antibodies responsible for allergic reactions.

surface tension (ser′fas ten′shun) Force that tends to hold moist membranes together due to an attraction that water molecules have for one another.

surfactant (ser-fak′tant) Substance produced by the lungs that reduces the surface tension within the alveoli.

suture (soo′cher) An immovable joint, such as that between adjacent flat bones of the skull.

sympathetic nervous system (sim″pah-thet′ik ner′vus sis′tem) Portion of the autonomic nervous system that arises from the thoracic and lumbar regions of the spinal cord.

symphysis (sim′fĭ-sis) A slightly movable joint between bones separated by a pad of fibrocartilage.

synapse (sin′aps) The junction between the axon end of one neuron and the dendrite or cell body of another neuron.

synaptic knob (sĭ-nap'tik nob) Tiny enlargement at the end of an axon that secretes a neurotransmitter substance.

syncytium (sin-sish'e-um) A mass of merging cells.

syndrome (sin'drōm) A group of symptoms that together characterize a disease condition.

synergist (sin'er-jist) A muscle that assists the action of a prime mover.

synovial fluid (sĭ-no've-al floo'id) Fluid secreted by the synovial membrane.

synovial joint (sĭ-no've-al joint) A freely movable joint.

synovial membrane (sĭ-no've-al mem'bran) Membrane that forms the inner lining of the capsule of a freely movable joint.

synthesis (sin'thĕ-sis) The process by which substances are united to form more complex substances.

system (sis'tem) A group of organs that act together to carry on a specialized function.

systemic circuit (sis-tem'ik ser'kit) The vessels that conduct blood between the heart and all body tissues except the lungs.

systole (sis'to-le) Phase of cardiac cycle during which a heart chamber wall is contracted.

systolic pressure (sis-tol'ik presh'ur) Arterial blood pressure during the systolic phase of the cardiac cycle.

tachycardia (tak"e-kar'de-ah) An abnormally rapid heartbeat.

target tissue (tar'get tish'u) Specific tissue on which a hormone acts.

tarsus (tar'sus) The bones that form the ankle.

taste bud (tāst bud) Organ containing the receptors associated with the sense of taste.

telophase (tel'o-fāz) Stage in mitosis during which daughter cells become separate structures.

tendon (ten'don) A cordlike or bandlike mass of white fibrous connective tissue that connects a muscle to a bone.

testis (tes'tis) Primary reproductive organ of a male; a sperm-cell-producing organ.

testosterone (tes-tos'tĕ-rōn) Male sex hormone secreted by the interstitial cells of the testes.

tetany (tet'ah-ne) A continuous, forceful muscular contraction.

thalamus (thal'ah-mus) A mass of gray matter located at the base of the cerebrum in the wall of the third ventricle.

thermoreceptor (ther"mo-re-sep'tor) A sensory receptor that is sensitive to changes in temperature; a heat receptor.

thiamine (thi'ah-min) Vitamin B_1.

thoracic (tho-ras'ik) Pertaining to the chest.

threshold stimulus (thresh'old stim'u-lus) The level of stimulation that must be exceeded to elicit a nerve impulse or a muscle contraction.

thrombocyte (throm'bo-sit) A blood platelet.

thrombocytopenia (throm"bo-si"to-pe'ne-ah) A low number of platelets in the circulating blood.

thrombus (throm'bus) A blood clot in a blood vessel that remains at its site of formation.

thymus (thi'mus) A two-lobed glandular organ located in the mediastinum behind the sternum and between the lungs.

thyroglobulin (thi"ro-glob'u-lin) Substance secreted by cells of the thyroid gland that serves to store thyroid hormones.

thyroid gland (thi'roid gland) Endocrine gland located just below the larynx and in front of the trachea that secretes thyroid hormones.

thyrotropin (thi"ro-trōp'in) Hormone secreted by the anterior pituitary gland that stimulates the thyroid gland to secrete hormones; TSH.

thyroxine (thi-rok'sin) A hormone secreted by the thyroid gland.

tissue (tish'u) A group of similar cells that performs a specialized function.

T-lymphocyte (lim'fo-sit) Lymphocytes that interact directly with antigen-bearing particles and are responsible for cellular immunity.

trabecula (trah-bek'u-lah) Branching bony plate that separates irregular spaces within spongy bone.

trachea (tra'ke-ah) Tubular organ that leads from the larynx to the bronchi.

transcellular fluid (trans"sel'u-lar floo'id) A portion of the extracellular fluid, including the fluid within special body cavities.

transfer RNA (trans'fer) Molecule of RNA that carries an amino acid to a ribosome in the process of protein synthesis.

transverse colon (trans-vers' ko'lon) Portion of the large intestine that extends across the abdomen from right to left below the stomach.

tricuspid valve (tri-kus'pid valv) Heart valve located between the right atrium and the right ventricle.

triglyceride (tri-glis'er-īd) A lipid composed of three fatty acids combined with a glycerol molecule.

triiodothyronine (tri"i-o"do-thi'ro-nēn) One of the thyroid hormones.

trisomy (tri'so-me) Condition in which a cell contains three chromosomes of a particular type instead of two.

trochanter (tro-kan'ter) A broad process on a bone.

trochlea (trok'le-ah) A pulley-shaped structure.

trophoblast (trof'o-blast) The outer cells of a blastocyst that help to form the placenta and other embryonic membranes.

tropic hormone (trōp'ik hor'mōn) A hormone that has an endocrine gland as its target tissue.

tropomyosin (tro"po-mi'o-sin) Protein that functions in blocking muscle contraction until calcium ions are present.

troponin (tro'po-nin) Protein that functions with tropomyosin to block muscle contraction until calcium ions are present.

trypsin (trip'sin) An enzyme in pancreatic juice that acts to break down protein molecules.

tubercle (tu'ber-kl) A small, rounded process on a bone.

tuberosity (tu"bĕ-ros'ĭ-te) An elevation or protuberance on a bone.

twitch (twich) A brief muscular contraction followed by relaxation.

tympanic membrane (tim-pan'ik mem'brān) A thin membrane that covers the auditory canal and separates the external ear from the middle ear; the eardrum.

umbilical cord (um-bil'ĭ-kal kord) Cordlike structure that connects the fetus to the placenta.

umbilical region (um-bil'ĭ-kal re'jun) The central portion of the abdomen.

umbilicus (um-bil'ĭ-kus) Region to which the umbilical cord was attached; the navel.

unipolar neuron (u"nĭ-po'lar nu'ron) A neuron that has a single nerve fiber extending from its cell body.

unsaturated fatty acid (un-sat'u-rāt"ed fat'e as'id) Fatty acid molecule that has one or more double bonds between the atoms of its carbon chain.

urea (u-re'ah) A nonprotein nitrogenous substance produced as a result of protein metabolism.

ureter (u-re'ter) A muscular tube that carries urine from the kidney to the urinary bladder.

urethra (u-re'thrah) Tube leading from the urinary bladder to the outside of the body.

uterine (u'ter-in) Pertaining to the uterus.

uterine tube (u'ter-in tūb) Tube that extends from the uterus on each side toward an ovary and functions to transport sex cells; fallopian tube or oviduct.

uterus (u'ter-us) Hollow muscular organ located within the female pelvis in which a fetus develops.

utricle (u'trĭ-kl) An enlarged portion of the membranous labyrinth of the inner ear.

uvula (u'vu-lah) A fleshy portion of the soft palate that hangs down above the root of the tongue.

vaccine (vak'sēn) A substance that contains antigens and is used to stimulate the production of antibodies.

vacuole (vak'u-ōl) A space or cavity within the cytoplasm of a cell.

vagina (vah-ji'nah) Tubular organ that leads from the uterus to the vestibule of the female reproductive tract.

varicose veins (var'ĭ-kos vānz) Abnormally swollen and enlarged veins, especially in the legs.

vasa recta (va'sah rek'tah) A branch of the peritubular capillary that receives blood from the efferent arterioles of juxtamedullary nephrons.

vascular (vas'ku-lar) Pertaining to blood vessels.

vas deferens (vas def'er-ens) Tube that leads from the epididymis to the urethra of the male reproductive tract (plural, *vasa deferentia*).

vasoconstriction (vas''o-kon-strik'shun) A decrease in the diameter of a blood vessel.

vasodilation (vas''o-di-la'shun) An increase in the diameter of a blood vessel.

vasopressin (vas''o-pres'in) Antidiuretic hormone.

vein (vān) A vessel that carries blood toward the heart.

vena cava (vēn'ah kāv'ah) One of two large veins that convey deoxygenated blood to the right atrium of the heart.

ventral root (ven'tral rōot) Motor branch of a spinal nerve by which it is attached to the spinal cord.

ventricle (ven'trĭ-kl) A cavity, such as those of the brain that are filled with cerebrospinal fluid, or those of the heart that contain blood.

venule (ven'ūl) A vessel that carries blood from capillaries to a vein.

vermiform appendix (ver'mĭ-form ah-pen'diks) Appendix.

villus (vil'us) Tiny, fingerlike projection that extends outward from the inner lining of the small intestine.

visceral (vis'er-al) Pertaining to the contents of a body cavity.

visceral peritoneum (vis'er-al per''ĭ-to-ne'um) Membrane that covers the surfaces of organs within the abdominal cavity.

visceral pleura (vis'er-al ploo'rah) Membrane that covers the surfaces of the lungs.

viscosity (vis-kos'ĭ-te) The tendency for a fluid to resist flowing due to the friction of its molecules on one another.

vitamin (vi'tah-min) An organic substance other than a carbohydrate, lipid, or protein that is needed for normal metabolism but cannot be synthesized in adequate amounts by the body.

vitreous humor (vit're-us hu'mor) The substance that occupies the space between the lens and retina of the eye.

vocal cords (vo'kal kordz) Folds of tissue within the larynx that create vocal sounds when they vibrate.

Volkmann's canal (fōlk'mahnz kah-nal') A transverse channel that interconnects haversian canals within compact bone.

voluntary (vol'un-tār''e) Capable of being consciously controlled.

vulva (vul'vah) The external reproductive parts of the female that surround the opening of the vagina.

water balance (wot'er bal'ans) Condition in which the quantity of water entering the body is equal to the quantity leaving it.

water intoxication (wot'er in-tok''sĭ-ka'shun) Condition in which the extracellular body fluids become abnormally diluted due to the presence of excessive water.

water of metabolism (wot'er uv mĕ-tab'o-lizm) Water produced as a by-product of metabolic processes.

wave summation (wāv sum-ma'shun) A sustained muscle contraction that occurs when a series of twitches fuse together; summation of twitches.

yellow marrow (yel'o mar'o) Fat storage tissue found in the cavities within bones.

zygote (zi'gōt) Cell produced by the fusion of an egg and sperm; a fertilized egg cell.

zymogen granule (zi-mo'jen gran'ūl) A cellular structure that stores inactive forms of protein-splitting enzymes in a pancreatic cell.

Visual Credits

Cover photo: Manfred Kage from Peter Arnold, Inc.

Chapter 1

The Bettmann Archive: 1.1; Diane Nelson and Associates: 1.2, 1.4, 1.5; Fine Line Illustrations, Inc.: 1.3.

Chapter 2

Fine Line Illustrations, Inc.: 2.1, 2.4–2.22; Martin M. Rotker, TAURUS PHOTOS: 2.2; © CARROLL H. WEISS, RBP, 1973: 2.3.

Chapter 3

L. B. Shuttles: 3.1; Diane Nelson and Associates: 3.2, 3.3, 3.6, 3.8c, 3.28, 3.29 (left), 3.30 (left), 3.31 (left), 3.32 (left), 3.33 (left), 3.34; Courtesy of K. R. Porter: 3.4 (left), 3.8a, 3.12a; Martin M. Rotker, TAURUS PHOTOS: 3.7; From Mader, Sylvia S., INQUIRY INTO LIFE, 2d ed., © 1976, 1979 Wm. C. Brown Company Publishers, Dubuque, Iowa. Reprinted by permission: 3.4 (right), 3.5, 3.8b, 3.9b, 3.10, 3.11b, 3.12b, 3.14b; Courtesy of the Upjohn Company: 3.9a; © Charles P. Havel, CPH Biomedical Photography: 3.11a; Dr. Joseph Gall. Reproduced from *Journal of Cell Biology*, 31 (1966). By copyrighted permission of the Rockefeller University Press: 3.13; *Cell Ultrastructure* by William A. Jensen and Roderick B. Parks, © 1967 by Wadsworth Publishing Co., Inc., Belmont, California. Reprinted by permission of the publisher: 3.14a; Fine Line Illustrations, Inc.: 3.15, 3.16, 3.17, 3.19–3.27, 3.35; John Robaton, Camera 5: 3.18; Courtesy of Carolina Biological Supply: 3.29 (right)–3.33 (right); Oscar Auerbach, M.D., Veterans Administration Hospital, East Orange, New Jersey: 3.36.

Chapter 4

Fine Line Illustrations, Inc.: 4.1–4.4, 4.6–4.24; *Man, Nature, and Society: An Introduction to Biology* by E. Peter Volpe, Copyright © 1975 by Wm. C. Brown Company Publishers, Dubuque, Iowa: 4.5.

Chapter 5

Diane Nelson and Associates: 5.1 (left), 5.2 (left), 5.3 (left), 5.4 (left), 5.5 (left), 5.6, 5.7, 5.9 (left), 5.10 (left), 5.11 (left), 5.12 (left), 5.13 (left), 5.14 (left), 5.15 (left), 5.16 (left), 5.17 (left), 5.18 (left), 5.19 (left), 5.20; Photomicrograph by Edwin Reschke: 5.1 (right), 5.4 (right), 5.14 (right); Courtesy of Carolina Biological Supply: 5.2 (right), 5.5 (right), 5.9 (right), 5.10 (right), 5.11 (right), 5.12 (right), 5.13 (right), 5.15 (right), 5.17 (right), 5.18 (right); © Charles P. Havel, CPH Biomedical Photography: 5.3 (right), 5.19 (right); Courtesy of the American Lung Association: 5.8; Armed Forces Institute of Pathology: 5.16 (right).

Chapter 6

Diane Nelson and Associates: 6.1, 6.2, 6.5, 6.6, 6.8, 6.9, 6.12; Martin M. Rotker, TAURUS PHOTOS: 6.3; *Andrew's Diseases of the Skin*, A. N. Domonkos, © W. B. Saunders Co., 1971: 6.4; Edwin A. Reschke: 6.7; Fine Line Illustrations, Inc.: 6.10; UPI: 6.11.

Chapter 7

Diane Nelson and Associates: 7.1–7.11, 7.13–7.14; Fine Line Illustrations, Inc.: 7.12.

Chapter 8

Diane Nelson and Associates: 8.1–8.4, 8.6, 8.7, 8.10, 8.11, 8.14, 8.16–8.31, 8.33–8.39, 8.41, 8.43, 8.44, 8.46–8.48, 8.50–8.52, 8.54–8.56, 8.58, 8.60–8.62, 8.64, 8.66–8.71; Ted Conde, St. Joseph's Hospital: 8.5, 8.65; Martin M. Rotker, TAURUS PHOTOS: 8.8, 8.32b, 8.42, 8.49, 8.53, 8.57, 8.59, 8.63; Ewing Galloway: 8.9; Courtesy of Kodak: 8.12, 8.32a, 8.45; Fine Line Illustrations, Inc.: 8.13, 8.15; From Sherrill, Claudine, ADAPTED PHYSICAL EDUCATION AND RECREATION: A MULTIDISCIPLINARY APPROACH © 1976 Wm. C. Brown Company Publishers, Dubuque, Iowa. Reprinted by permission: 8.40.

Chapter 9

Diane Nelson and Associates: 9.1–9.4, 9.6–9.8, 9.17, 9.19, 9.20, 9.22–9.46; Courtesy of H. E. Huxley, Cambridge University: 9.5; Fine Line Illustrations, Inc.: 9.9–9.13, 9.15, 9.16; Courtesy of Narco Scientific: 9.14; From Sherrill, Claudine, ADAPTED PHYSICAL EDUCATION AND RECREATION: A MULTIDISCIPLINARY APPROACH © 1976 Wm. C. Brown Company Publishers, Dubuque, Iowa. Reprinted by permission: 9.18; Courtesy of Dr. Paul Heidger, University of Iowa: 9.21; John Carl Mese: photos of muscular models, pp. 261–62.

Chapter 10

Diane Nelson and Associates: 10.1–10.4, 10.6, 10.7, 10.13–10.22, 10.24, 10.26–10.29, 10.31–10.46, 10.48–10.59; Courtesy of J. D. Robertson: 10.5; Fine Line Illustrations, Inc.: 10.8–10.12; From Mader, Sylvia S., INQUIRY INTO LIFE, 2d ed., © 1976, 1979 Wm. C. Brown Company Publishers, Dubuque, Iowa. Reprinted by permission: 10.23, 10.30, 10.47; *Anatomy and Physiology Laboratory Textbook* by Harold J. Benson and Stanley E. Gunstream, Complete 2d edition. Copyright © 1970, 1976 by Wm. C. Brown Company Publishers, Dubuque, Iowa: 10.25; Per H. Kjeldsen, University of Michigan: 10.27b; Jean-Claude Lejeune: 10.60.

Chapter 11

Diane Nelson and Associates: 11.1–11.7, 11.9–11.16, 11.18, 11.21–11.34, 11.41, 11.43, 11.46; Per H. Kjeldsen, University of Michigan: 11.8, 11.19b; Fine Line Illustrations, Inc.: 11.17, 11.36–11.40, 11.44, 11.45; Dr. Richard G. Kessel, University of Iowa: 11.19a; Martin M. Rotker, TAURUS PHOTOS: 11.20, 11.42; Reproduced with permission from Asbury T. Vaughn, *General Ophthalmology*, 8th ed., Lange, 1977: 11.35.

Chapter 12

Diane Nelson and Associates: 12.1, 12.2, 12.10–12.13, 12.15, 12.16, 12.18, 12.25, 12.27, 12.28, 12.30, 12.31, 12.33, 12.37; Fine Line Illustrations, Inc.: 12.3–12.9, 12.17, 12.19, 12.20, 12.26, 12.29, 12.32, 12.35, 12.36, 12.38, 12.39; Lester V. Bergmann Associates: 12.14, 12.22; Edwin A. Reschke: 12.21; Courtesy of F. A. Davis Company, Philadelphia, and Dr. R. H. Kampmeier: 12.23; © Charles P. Havel, CPH Biomedical Photography: 12.24; © Victor B. Eichler, BIO-ART: 12.34.

Chapter 13

From Mader, Sylvia S. INQUIRY INTO LIFE, 2d ed., © 1976, 1979 Wm. C. Brown Company Publishers, Dubuque, Iowa. Reprinted by permission: 13.1, 13.34, 13.37; Diane Nelson and Associates: 13.2, 13.3, 13.5, 13.7, 13.13–13.15, 13.17, 13.19, 13.21, 13.23–13.31, 13.33, 13.36, 13.40, 13.44, 13.45, 13.47, 13.48; *Anatomy and Physiology Laboratory Textbook* by Harold J. Benson and Stanley E. Gunstream, Complete 2d edition. Copyright © 1970, 1976 by Wm. C. Brown Company Publishers, Dubuque, Iowa: 13.4, 13.6, 13.8, 13.10, 13.11, 13.12; © Charles P. Havel, CPH Biomedical Photography: 13.9; W. B. Saunders Company and Dr. F. E. Templeton: 13.16; Armed Forces Institute of Pathology: 13.18, 13.35; Per H. Kjeldsen, University of Michigan: 13.20, 13.38; *Concepts in Biology* by Eldon D. Enger, Andrew H. Gibson, J. Richard Kormelink, Frederick C. Ross, Rodney J. Smith, Copyright © 1976 by Wm. C. Brown Company Publishers, Dubuque, Iowa: 13.22; © Carroll H. Weiss, RBP: 13.32; Courtesy of K. R. Porter: 13.39; Fine Line Illustrations, Inc.: 13.41–13.43; James L. Shaffer: 13.46; Dr. Richard G. Kessel, University of Iowa: 13.49.

Chapter 14

Fine Line Illustrations, Inc.:
14.1–14.8, 14.10, 14.13, 14.14,
14.16–14.23; United Nations Photo:
14.9. Reprinted with permission of
Macmillan Publishing Co., Inc. from
THE HUMAN BODY: ITS
STRUCTURE AND PHYSIOLOGY by
Sigmund Grollman. Copyright
© 1974 by Sigmund Grollman:
14.11; Bob Coyle: 14.12a; Freelance
Photo Guild: 14.12b; United Nations
Photo, WHO: 14.15.

Chapter 15

From Mader, Sylvia S., INQUIRY
INTO LIFE, 2d ed., © 1976, 1979
Wm. C. Brown Company Publishers,
Dubuque, Iowa. Reprinted by
permission: 15.1; Diane Nelson and
Associates: 15.2, 15.4, 15.6–15.13,
15.15, 15.17–15.19, 15.21–15.24,
15.29, 15.31, 15.35; Reproduced
with permission of the American
Lung Association: 15.3, 15.32;
Courtesy of Eastman Kodak: 15.5;
Eastman Kodak Company,
Radiography Markets Division,
Rochester, New York: 15.14; Oscar
Auerbach, M.D., Veterans
Administration Hospital, East Orange,
New Jersey: 15.16, 15.27; Fine Line
Illustrations, Inc.: 15.20, 15.25,
15.33, 15.34, 15.36–15.44; Richard
Merrill: 15.26; *Man, Nature, and
Society: An Introduction to Biology*
by E. Peter Volpe, Copyright © 1975
by Wm. C. Brown Company
Publishers, Dubuque, Iowa: 15.28;
*Laboratory Studies in Biology:
Observations and Their Implications*
by Chester A. Lawson, Ralph W.
Lewis, Mary Alice Burmester, and
Garret Hardin, W. H. Freeman and
Company, Copyright © 1955: 15.30.

Chapter 16

Diane Nelson and Associates: 16.1,
16.3, 16.6, 16.7, 16.10–16.12,
16.14, 16.18, 16.23, 16.27, 16.28;
Fine Line Illustrations, Inc.: 16.2,
16.5, 16.9, 16.16, 16.17, 16.19,
16.21, 16.22, 16.25, 16.26; Fox-
Naro Labs: 16.4; Armed Forces
Institute of Pathology: 16.8; Courtesy
of the American Cancer Society:
16.13; Courtesy of Joseph R.
Goodman, Veterans Administration
Hospital, San Francisco: 16.15;
Courtesy of E. Bernstein and
E. Kairinen, *Science*, Cover, 27
August 1971, Vol. 173. Copyright
1971 by the American Association
for the Advancement of Science:
16.20; © CARROLL H. WEISS, RBP,
1973: 16.24.

Chapter 17

Diane Nelson and Associates:
17.1–17.11, 17.13, 17.17, 17.18,
17.24, 17.34–17.38, 17.41, 17.42,
17.45, 17.46, 17.54–17.63,
17.65–17.69; Courtesy of Kodak:
17.12; Functional Human Anatomy,
2d ed., James E. Crouch.
Philadelphia: Lea & Febiger, 1972:
17.14; Fine Line Illustrations, Inc.:
17.15, 17.21–17.23, 17.40, 17.43,
17.48–17.53; Courtesy of Medical
Engineering Corporation: 17.16;
James L. Shaffer: 17.19,
17.25–17.29, 17.31; *Anatomy and
Physiology Laboratory Textbook* by
Harold J. Benson and Stanley E.
Gunstream, Complete 2d edition.
Copyright © 1970, 1976 by Wm. C.
Brown Company Publishers,
Dubuque, Iowa: 17.20; Martin M.
Rotker: 17.30, 17.32, 17.47;
Courtesy of Dr. Paul Heidger,
University of Iowa: 17.39; Courtesy
of the American Heart Association:
17.44; From Mader, Sylvia S.,
INQUIRY INTO LIFE, 2d ed., © 1976,
1979 Wm. C. Brown Company
Publishers, Dubuque, Iowa. Reprinted
by permission: 17.33; *Laboratory
Manual and Study Guide for
Anatomy and Physiology*, 3d ed., by
Edwin B. Steen, Copyright © 1976
by Edwin B. Steen: 17.64, 17.70.

Chapter 18

Diane Nelson and Associates: 18.1,
18.2, 18.4, 18.6–18.16; Courtesy of
Carolina Biological Supply: 18.3;
Courtesy of Kodak: 18.5; Courtesy
of Memorial Sloan-Kettering Cancer
Center: 18.17; Fine Line Illustrations,
Inc.: 18.18–18.22.

Chapter 19

From Mader, Sylvia S., INQUIRY
INTO LIFE, 2d ed., © 1976, 1979
Wm. C. Brown Company Publishers,
Dubuque, Iowa. Reprinted by
permission: 19.1, 19.3, 19.4; Diane
Nelson and Associates: 19.2,
19.5–19.8, 19.11, 19.13–19.18,
19.20–19.23; Fine Line Illustrations:
19.9, 19.10, 19.12; Courtesy of
Carolina Biological Supply: 19.19.

Chapter 20

Fine Line Illustrations, Inc.:
20.1–20.22.

Chapter 21

From Mader, Sylvia S. INQUIRY INTO
LIFE, 2d ed., © 1976, 1979 Wm. C.
Brown Company Publishers,
Dubuque, Iowa. Reprinted by
permission: 21.1, 21.12, 21.17;
Diane Nelson and Associates: 21.2,
21.3, 21.5, 21.6, 21.9, 21.10, 21.16,
21.18, 21.20–21.22, 21.24, 21.25,
21.28, 21.31, 21.33–21.35, 21.40,
21.43; Dr. Richard G. Kessel,
University of Iowa: 21.4, 21.7, 21.23,
21.27; Ward's Natural Science
Establishment: 21.8, 21.11, 21.19
(left); Fine Line Illustrations, Inc.:
21.13–21.15, 21.26, 21.29, 21.30,
21.36–21.38, 21.41; Martin M.
Rotker, TAURUS PHOTOS: 21.19a;
Per H. Kjeldsen, University of
Michigan: 21.19b; Dr. Jack Kath:
21.32; Courtesy of the Cleveland
Health Museum: 21.39; Courtesy of
the Pennsylvania State University:
21.42.

Chapter 22

Diane Nelson and Associates:
22.1–22.4, 22.6–22.8, 22.10,
22.12–22.15, 22.17, 22.20, 22.21,
22.24; Carnegie Institute of
Washington, Department of
Embryology, Davis Division: 22.5;
© CARROLL H. WEISS, RPB: 22.9,
22.16; Courtesy of the Carnegie
Institute of Washington, Baltimore,
Maryland: 22.11; Armed Forces
Institute of Pathology: 22.18; From
Mader, Sylvia S., INQUIRY INTO
LIFE, 2d ed., © 1976, 1979 Wm. C.
Brown Company Publishers,
Dubuque, Iowa. Reprinted by
permission: 22.19; Peter Gridley,
Freelance Photographers Guild:
22.22; Fine Line Illustrations, Inc.:
22.23; Florence Sharp: 22.25;
James L. Bowman: 22.26; David S.
Strickler: 22.27; Rick Smolan: 22.28;
Religious News Service: 22.29.

Chapter 23

Fine Line Illustrations, Inc.: 23.1,
23.22; Diane Nelson and Associates:
23.2, 23.10–23.12, 23.16–23.18,
23.21; Courtesy of the March of
Dimes: 23.3, 23.20; Courtesy of the
Upjohn Company, Kalamazoo,
Michigan: 23.4; Wide World Photos:
23.5, 23.6; Rick Smolan: 23.7;
Concepts in Biology by Eldon D.
Enger, Andrew H. Gibson, J. Richard
Kormelink, Frederick C. Ross,
Rodney J. Smith, Copyright © 1976
by Wm. C. Brown Company
Publishers, Dubuque, Iowa: 23.8,
23.9, 23.13–23.15, 23.19; *The
Science of Genetics* by William
Hexter and Henry T. Yost, Jr.,
© 1976 Prentice-Hall, Inc.,
Englewood Cliffs, New Jersey:
23.23, 23.24, 23.25.

Appendixes

Fine Line Illustrations, Inc.: A.1–A.4.

Unit Art

Unit 1: Guy de Chaulic
(1298?–1368) giving an anatomical
lecture. From Oxford manuscript.
The Bettmann Archive, Inc.; Unit 2:
Anatomical figures with circles
around them to indicate proportions.
Copper engraving by Crisostoman
Martinez, Spanish engraver living in
Paris (ca. 1694). The Bettmann
Archive, Inc.; Unit 3: Sketches of the
head and brain by Leonardo Da
Vinci. The Bettmann Archive, Inc.;
Unit 4: Blood vessels of the human
body by Charles Estienne. From De
dissectione partium corpons humani,
Paris 1545. The Bettmann Archive,
Inc.; Unit 5: From Leishman, after
Coste. *Gray's Anatomy* by Henry
Gray. Edited by T. Pickering Pick
and Robert Howden. New York:
Crown Publishers, Inc., 1977.

Color Plates

John W. Hole: plates 1, 2, 3, 4, 5;
Courtesy of the Ealing Corporation,
South Natick: plates 6, 7; © Manfred
Kage from Peter Arnold, Inc.: plates
8, 14, 15, 22; Edwin A. Reschke:
plates 9, 10, 11, 12, 13, 19, 20, 31,
34, 35; Courtesy of Dr. Lee C. Chiu,
University of Iowa Hospitals and
Clinics: plates 17, 18; Courtesy of
Ohio Nuclear: plate 16; Courtesy of
Igaku Shoin, Ltd.: plates 21, 29;
© CARROLL H. WEISS, RBP, 1973:
plates 23, 24, 25, 26, 30; Per
H. Kjeldsen, University of Michigan at
Ann Arbor: plate 27; *Anatomy and
Physiology Laboratory Textbook* by
Harold J. Benson and Stanley E.
Gunstream, Complete 2d edition.
Copyright © 1976 by Wm. C. Brown
Company Publishers, Dubuque, Iowa:
plate 28; © Victor B. Eichler, BIO-
ART: plate 32; Courtesy of Stuart
Fox, Los Angeles Community
College: plate 33; © Martin M.
Rotker, TAURUS PHOTOS: plate 36;
© DONALD YEAGER, 1973: plate 37
(a–e).

Index

movements of tube, 415
structure of wall, 415
alkaline (basic), 27, 682
alkaline taste, 338
alkaloids, 340
alkalosis, 687, 688, 689-90
metabolic, 689-90
respiratory, 689
allantois, 752
alleles, 769
multiple, 772
some traits determined by single pairs of dominant and recessive (chart), 770
allergen, 507, 637
allergic reactions, 637
bronchial asthma and, 507
all-or-none nerve response, 226, 279
alopecia, 128
alpha globulins, 542
alpha particle, 35
alphatocopherol, 470
alveolar ducts, 497
alveolar gas exchanges, 513-15
alveoli, 513
disorders of, 514-15
respiratory membrane, 513-14
alveolar glands, 99, 731
alveolar pores, 513
alveoli, 497, 513, 595
amblyopia, 365
ameboid motion of white blood cells, 538
amenorrhea, 737
ametropia, 365
amines, as hormones, 375
amino acids, 31, 32, 33, 460
essential, 79, 460
found in proteins (chart), 460
genetic code for certain (chart), 84
in plasma, 543
reabsorption in kidneys, 655
structural formulas, 32
in urine, 660
utilization of, 461
amino acids containing sulfur, oxidation of, 683
amino groups, 684
ammonium ions, 686
amniocentesis, 761, 787
amniochorionic membrane, 751
amnion membrane, 750, 751
amniotic fluid, 750, 751, 787
AMP (adenosine monophosphate), cyclic, 376
amphetamine, 281
amphiarthroses, 202
ampulla, 344, 349, 706
amylase, 421, 422
anabolic metabolism, 70
anaerobic, 224
cellular glucose respiration, 74, 520, 682
anal canal, 445
anal columns, 445
analgesia, 323

analgesic, 323
anal reflex, 287
anal sphincter, external and internal, 445
anaphase
first and second meiotic, 776
in mitosis, 60, 61
anaphylaxis, 639
anaplasia, 63
anastomoses, 570
anatomical position, 139
anatomy, defined, 9
androgens, 711
and development of female secondary sexual characteristics, 721
anemia, 535
aplastic, 535
hemolytic, 535
hemorrhagic, 535
hypochromic, 534, 535
megaloblastic, 474
pernicious, 473, 534
sickle-cell, 771
anesthesia, 286, 323
aneurysm, 613
angina pectoris, 334, 593
angiocardiography, 613
angiospasm, 613
angiotensin, 593
anhydrase, carbonic, 519
anisocytosis, 554
ankle, muscles that move the, 254-57
ankle bones. See tarsals
ankle-jerk reflex, 287
ankylosis, 207
anorexia, 484
anosmia, 338
anoxia, 522
antacid, 690
antagonists, 233
antebrachium, 143
antecubital, 143
anterior branch of spinal nerves, 312
anterior cerebral artery, 600
anterior chamber of eye, 355
anterior choroid artery, 600
anterior corticospinal tract, 293
anterior crest of tibia, 198
anterior funiculi, 292
anterior horns of spinal cord, 290
anterior intercostal arteries, 601
anterior interventribular artery, 568
anterior interventricular sulcus, 568
anterior lobe hormones of pituitary gland, 381-83, 711, 720
anterior median fissure, 290
anterior spinocerebellar tract, 293
anterior spinothalamic tract, 292
anterior superior iliac spine, 195
anterior tibial artery, 603
anterior tibial vein, 609

anterior (ventral) relative position, 141
anti-A agglutinin, 550
anti-B agglutinin, 550
antibodies, 538, 542, 549
types of, 634-35
antidiuretic, defined, 384
antidiuretic hormone. See ADH
antigen-antibody complexes, 538
antigens, 549, 632
antihemophilic globulin, 783
antihemophilic plasma, 554
antioxidant, 470
anti-Rh agglutinin, 552
antithrombin, 547
antrum of Highmore, 176
anulus fibrosus, 184
anuria, 665, 690
anus, 445
anvil of middle ear, 341
aorta, 566, 596
abdominal, 597
aneurysm, 597
arch of, 576, 597
ascending, 597
bodies of, 510, 597, 685
descending, 597
principal branches of, 597-98
semilunar valves of, 566
sinus of, 576, 597
sound of, 572
thoracic, 597
aorta and its principal branches (chart), 598
apex of heart, 562
aphagia, 448
aphasia, 323
apical heartbeat, 562
aplastic anemia, 535
apnea, 522
apneustic area of pons, breathing and, 509-10
apocrine glands, 100, 122
aponeuroses, 216
appendages of skin, 120-23
appendicitis, 444
appendicular portion, 134
appendicular skeleton, 167
applied sciences, terms, 15
apraxia, 323
aqueduct of Sylvius, 301
aqueous humor, 355, 672
arachnoid granulations, 302
arachnoid matter, 287, 288, 290
arbor vitae, 306
arch of aorta, 597
arches of the foot, 201
arciform arteries of kidney, 647
arcuate arteries, 647
areola of breast, 731
arginine, 460
arm, upper, muscles that move the, 243-44
arms. See upper limbs
arrector pili muscle, 121
in section of skin, 117
arrhythmias of heart, 577-79
arterial blood pressure, 587
factors influencing, 589-90
blood volume, 589

heart action, 589
peripheral resistance, 590
viscosity, 590
measurement of, 587-89
See also blood pressure
arterial system, 579-81, 596-605
aorta, principal branches of, 597-98
to neck and head, 598-600
to pelvis and leg, 602-4
to shoulder and arm, 600-601
to thoracic and abdominal walls, 601, 602
diastolic pressure (DP), 587
mean pressure, 588
systolic pressure (SP), 587
arteriography, 613
arterioles, 579-80
arteriosclerosis, 585, 593
arteriovenous shunts, 580
arthralgia, 207
arthritis, 206
arthrocentesis, 207
articular cartilage, 154, 203
articulations. See joints
artificial kidney, 53
artificially acquired active immunity, 636
artificially acquired passive immunity, 636
artificial pacemakers, 579
arytenoid cartilages of larynx, 494-95
ascending aorta, 597
ascending colon, 444, 445
ascending limb of loop of Henle, 649
ascending lumbar veins, 608
ascending tracts of spinal cord, 292, 292-93
ascites, 584, 681
ascorbic acid (vitamin C), 474
asparagine, 460
aspartic acid, 460
asphyxia, 522
asplenia, 639
assimilation, defined, 10
association areas of cerebral cortex, 297-99
association neurons, 282
asthma, bronchial, 507
astigmatism, 359
astrocytes, 275
asystole, 613
ataxia, 323
atelectasis, 515
atherosclerosis, 334, 397, 548, 585, 586
athlete's foot (tinea pedis), 128
atlas, 182, 293
atmospheric pressure, 11
atomic number, 21
atomic radiations, 22
atomic weight, 21
atoms, 20
bonding of, 23-25
level of complexity, 13
structure, atomic, 20-21